SI Base Units

length	meter	m	electric current	ampere	A
mass	kilogram	kg	thermodynamic temperature	Kelvin	K
time	second	s	amount of substance	mole	mol

SI Derived or Related Units

frequency	hertz	Hz	$= s^{-1}$
force	newton	N	$= m\ kg\ s^{-2}$
	kilogram-force	kgf	= 9.80665 N
	dyne	dyne	$= 10^{-5}$ N
energy	joule	J	$= N\ m = kg\ m^2\ s^{-2}$
	erg	erg	$= 10^{-7}$ J
	electron volt	eV	$= 1.60218 \times 10^{-19}$ J
	calorie	cal	= 4.184 J
	bar L	bar L	= 100 J
	atomic unit	au	$= 4.35974 \times 10^{-18}$ J
power	watt	W	$= J\ s^{-1} = V\ A$
volume	liter	L	$= 10^{-3}\ m^3$
pressure	pascal	Pa	$= N\ m^{-2}$
	bar	bar	$= 10^5$ Pa
	atmosphere	atm	= 1.01325 bar
	torr	Torr	= 1/760 atm
	millimeter of mercury	mm Hg	= Torr
electric charge	coulomb	C	= A s
electric potential, electromotive force	volt	V	$= J\ C^{-1}$
electric resistance	ohm	Ω	$= V\ A^{-1}$
electric conductance	siemens	S	$= \Omega^{-1}$
electric capacitance	farad	F	$= C\ V^{-1}$
magnetic flux density	tesla	T	$= V\ s\ m^{-2}$
	gauss	G	$= 10^{-4}$ T
inductance	henry	H	$= V\ A^{-1}\ s$
electric dipole moment	coulomb meter		= C m
	Debye	D	$= 3.3356 \times 10^{-30}$ C m
magnetic moment		$A\ m^2$	$= J\ T^{-1}$

Prefixes

z	a	f	p	n	μ	m	c	d
zepto	atto	femto	pico	nano	micro	milli	centi	deci
10^{-21}	10^{-18}	10^{-15}	10^{-12}	10^{-9}	10^{-6}	10^{-3}	10^{-2}	10^{-1}

da	h	k	M	G	T	P	E	Z
deka	hecto	kilo	mega	giga	tera	peta	exa	zetta
10^{1}	10^{2}	10^{3}	10^{6}	10^{9}	10^{12}	10^{15}	10^{18}	10^{21}

Solutions Manual for Principles of Physical Chemistry

Third Edition

Edited by Hans Kuhn, David H. Waldeck, and Horst-Dieter Försterling

WILEY

Published by John Wiley & Sons, Inc., Hoboken, New Jersey.
Published simultaneously in Canada.

For general information on our other products and services or for technical support, please contact our Customer Care Department within the United States at (800) 762-2974, outside the United States at (317) 572-3993 or fax (317) 572-4002.

Wiley also publishes its books in a variety of electronic formats. Some content that appears in print may not be available in electronic formats. For more information about Wiley products, visit our web site at www.wiley.com

Library of Congress Cataloging-in-Publication Data applied for:

Hardback ISBN: 9781119852902

Cover Design: Wiley
www.wiley.com

Set in 9.5/12.5pt STIXTwoText by Straive, Chennai, India

Contents

Preface Third Edition

This book poses many exercises and problems for the students to use as tools to gain mastery over the methods and concepts introduced in the textbook. The exercises are intended to be activities for you to perform that will reinforce the concepts and important equations introduced in the textbook, whereas the problems are intended to be more challenging questions that often ask you to integrate concepts or to analyze data. The problems use data from published works and in the later chapters you are asked to read research articles and answer questions on them. The goal of this approach is to enhance your appreciation for science as a community activity.

Learning physical chemistry includes both the mastering of skills and the integration of new concepts into knowledge and understanding. Like other endeavors, for example, sports or music, daily practice is important for developing your skills. By applying effort on skills and knowledge accumulation on a daily basis, you will find that you develop a mastery and deeper understanding of physical chemistry over a period of time.

August 2023 *David H. Waldeck and Horst-Dieter Försterling*

Acknowledgment

Here we acknowledge those who have been important to the preparation of this third edition of Solutions Manual for Principles of Physical Chemistry. We are greatly indebted to Michael Leventhal and his colleagues at John Wiley & Sons, Ltd for their encouragement and support during all stages of this third edition. We are indebted to Professor Jeffry Madura (who died in 2017) who collaborated on the original version of the solutions manual. David Waldeck thanks his wife, Janet, and children, Aaron and Anna, for their support and understanding during this endeavor. Horst-Dieter Försterling is thankful for the great support by his family during the time of developing this book and its precursors, first of all to his wife Inge (who died in 2019) for their encouragement.

1

Wave–Particle Duality

1.1 Exercises

E1.1 Consider a microwave source that is generating 2.0 GHz electromagnetic radiation. Compute the wavelength of the microwaves. If this microwave source was used in an oven of width 30 cm, how many wavelengths of the microwave can be included across the oven's width?

Compute the energy per photon for the 2.0 GHz frequency. If a cup containing 250 mL of water is irradiated by this source, how many photons must be absorbed to raise the temperature of the water from 25 °C to 80 °C (a nice temperature for a cup of tea). For simplicity, assume that the water density is 1.0 g/mL, that the heat capacity is 4.184 J/(g °C), and that they do not change over the temperature range.

Solution

First we calculate the wavelength of a 2.0 GHz microwave and then compare it to the oven's width. The wavelength and frequency are related by $\lambda = c_0/\nu$ with c_0 being the speed of light (2.998×10^{10} cm s^{-1}), so

$$\lambda = \frac{2.998 \times 10^{10}\ \text{cm s}^{-1}}{2.0 \times 10^{9}\ \text{s}^{-1}} = 1.5 \times 10^{1}\ \text{cm} = 15.0\ \text{cm}$$

Hence, the oven is about 2λ wide.

Here we calculate the energy in a 2.0 GHz photon and compare it to the energy needed to warm the water (assuming no extraneous losses). The energy and frequency are related by $E = h\nu$, so that the energy per photon is

$$E = h\nu = 6.626 \times 10^{-34}\ \text{J s} \cdot 2.0 \times 10^{9}\ \text{s}^{-1} = 1.3 \times 10^{-24}\ \text{J}$$

The amount of energy the water must absorb is $Q = mC \cdot \Delta T$, where m is the mass of water (250 mL or 250 g), C is the heat capacity, and ΔT is the change in temperature (80 − 25) °C. Thus

$$Q = 250\ \text{g} \cdot 4.184\ \text{J g}^{-1}\ {}^\circ\text{C}^{-1} \cdot 55\ {}^\circ\text{C} = 57.530\ \text{kJ}$$

so that the number of 2.0 GHz photons will be

$$\frac{Q}{E} = \frac{57.530\ \text{kJ}}{1.3 \times 10^{-24}\ \text{J}} = 4.4 \times 10^{28}$$

Because we have ignored any extraneous losses (e.g., heat conduction to the container and convective cooling), this value is a lower bound.

E1.2 Consider an ultraviolet light source that generates 300 nm electromagnetic radiation. Compute the frequency of the ultraviolet light. If one photon of this light is absorbed by an organic molecule, how much energy does the molecule gain? Is this energy enough to break a carbon–carbon bond in the molecule? Use a "typical" carbon–carbon bond energy of 5.8×10^{-19} J for your comparison. Perform the same calculations for a photon of wavelength 600 nm and a photon of wavelength 1200 nm. Perform your comparisons using the energy units of J and of eV.

Solution

The wavelength and frequency are related by $\nu = c_0/\lambda$ with c_0 being the speed of light (2.998×10^{8} m s^{-1}), so $\nu = (2.998 \times 10^{8}\ \text{m s}^{-1})/(300 \times 10^{-9}\ \text{m}) = 9.96 \times 10^{14}\ \text{s}^{-1}$.

Solutions Manual for Principles of Physical Chemistry, Third Edition. Edited by Hans Kuhn, David H. Waldeck, and Horst-Dieter Försterling.

The energy and frequency are related by $E = h\nu$, so the energy per photon $E = (6.626 \times 10^{-34}\ \text{J s})\ (9.96 \times 10^{14}\ \text{s}^{-1}) = 6.60 \times 10^{-19}$ J.
This amount of energy is "just" sufficient to break a bond of 5.8×10^{-19} J.
The corresponding energies for 600 and 1200 nm photons are $\frac{1}{2} \cdot 6.60 \times 10^{-19} = 3.30 \times 10^{-19}$ J and $\frac{1}{4} \cdot 6.60 \times 10^{-19} = 1.65 \times 10^{-19}$ J, neither of which is sufficient to break the typical carbon-carbon double bond.
The corresponding energies in eV (1.6×10^{-19} J = 1 eV) are 4.12 eV (300 nm light), 2.06 eV (600 nm light), and 1.03 eV (1200 nm light).

E1.3 Consider an electron with a kinetic energy of 1.0 eV (i.e., it has been accelerated across a 1 V potential difference). Compute the momentum $p = mv$ of this electron. Compare this momentum to that of a "typical" N_2 gas molecule at room temperature (consider the gas molecule to have a speed of 500 m/s).
Compute this electron's speed. At what fraction of light speed (3.00×10^8 m/s) is the electron moving?
Compute this electron's wavelength. Compare this wavelength to the diameter of a hydrogen atom (*ca.* 128 pm). Perform this same calculation for a 10 eV electron and a 100 eV electron. Comment on the trends in your values. How many electron wavelengths can fit into a hydrogen atom at these different energies?

Solution
The momentum $p = mv$ is related to the kinetic energy by $E_{\text{kin}} = \frac{1}{2}mv^2 = p^2/(2m)$, so we find the momentum by

$$p_{\text{electron}} = \sqrt{2mE_{\text{kin}}} = \sqrt{2(9.11 \times 10^{-31}\ \text{kg})(1.0\ \text{eV})(1.602 \times 10^{-19}\ \text{J eV}^{-1})}$$
$$= 5.4 \times 10^{-25}\ \text{kg m s}^{-1}$$

The momentum of a "typical" gas-phase nitrogen molecule (N_2) is

$$p_{N_2\ \text{molecule}} = mv = 4.650 \times 10^{-26}\ \text{kg} \cdot 500\text{m s}^{-1}$$
$$= 2.32 \times 10^{-23}\ \text{kg m s}^{-1}$$

which is about 43 times greater than the momentum of the electron.
The electron's speed is

$$v = \frac{p}{m} = \frac{5.403 \times 10^{-25}\ \text{kg m s}^{-1}}{9.11 \times 10^{-31}\ \text{kg}} = 5.931 \times 10^5\ \text{m/s}$$

This value is 0.002, or 0.2%, of the speed of light! While this speed is significant, it is still small enough to neglect relativistic effects.
The electron's wavelength can be calculated using the de Broglie relationship, so that

$$\Lambda = \frac{h}{mv} = \frac{h}{p} = \frac{h}{\sqrt{2mE_{kin}}} = \frac{6.626 \times 10^{-34}\ \text{J} \cdot \text{s}}{5.40 \times 10^{-25}\ \text{kg m s}^{-1}} = 1.23\ \text{nm}$$

where we have used 5.40×10^{-25} kg m s^{-1} for the momentum of the electron. This wavelength is 9 to 10 times larger than the characteristic size of an H atom.
For 10 eV electrons $\Lambda = 0.388$ nm, and for 100 eV electrons $\Lambda = 123$ pm. The electron wavelength decreases as the square root of its kinetic energy and a 100 eV electron has a wavelength that is similar to the diameter of an H-atom.

E1.4 A typical value for a particle's kinetic energy at 25 °C is 6.21×10^{-21} J. Use this value of the kinetic energy to estimate the speed of spheres with different masses; i.e.,

a) ping pong ball (2.60 g)
b) a 10.0 μ diameter polystyrene bead (0.300 g/cm^3 = 300 kg/m^3)
c) a 50.0 nm radius colloidal particle of Ag (10.5 g/cm^3 = 10,500 kg/m^3)
d) Buckminster fullerene (C_{60}) (0.720 kg/mol)
e) He atom (4.0 g/mol = 4.0×10^{-3} kg/mol).

Use these speeds and masses to estimate the de Broglie wavelength of these spheres. Comment on the trend in your wavelengths. For which, if any, of these particles would you expect their wave properties to be important? If the kinetic energy was decreased by 100 times, how would your wavelengths change? Do you think that wave properties would be important under these circumstances?

Solution

The speed and kinetic energy are related by

$$E_{\text{kin}} = \frac{1}{2}mv^2 \text{ so that } v = \sqrt{2E_{\text{kin}}/m}$$

Hence we find

a) for the ping pong ball

$$v = \sqrt{\frac{2 \cdot 6.21 \times 10^{-21}\ \text{J}}{2.6 \times 10^{-3}\ \text{kg}}} = 2.19 \times 10^{-9}\ \text{m/s}$$

b) for the polystyrene bead we first compute its mass by

$$m = \left(\frac{4}{3}\right)\pi\left(\frac{10 \times 10^{-6}}{2}\right)^3 \text{m}^3 \left(300\ \frac{\text{kg}}{\text{m}^3}\right) = 1.57 \times 10^{-13}\ \text{kg}$$

and then find its speed by

$$v = \sqrt{\frac{2 \cdot 6.21 \times 10^{-21}\ \text{J}}{1.57 \times 10^{-13}\ \text{kg}}} = 2.81 \times 10^{-4}\ \text{m/s}$$

c) for the silver colloid particle we first compute its mass by

$$m - \left(\frac{4}{3}\right)\pi\left(50 \times 10^{-9}\right)^3 \text{m}^3 \left(10500\ \frac{\text{kg}}{\text{m}^3}\right) = 5.50 \times 10^{-18}\ \text{kg}$$

and then find its speed by

$$v = \sqrt{\frac{2 \cdot 6.21 \times 10^{-21}\ \text{J}}{5.50 \times 10^{-18}\ \text{kg}}} = 5.79 \times 10^{-2}\ \text{m/s}$$

d) for the Buckminsterfullerene we first compute its mass by

$$m = \frac{0.720\ \text{kg mol}^{-1}}{6.022 \times 10^{23}\ \text{mol}^{-1}} = 1.19 \times 10^{-24}\ \text{kg}$$

and then find its speed by

$$v = \sqrt{\frac{2 \cdot 6.21 \times 10^{-21}\ \text{J}}{1.19 \times 10^{-24}\ \text{kg}}} = 102\ \text{m/s}$$

e) for the He atom we first compute its mass by

$$m = \frac{4.0 \times 10^{-3}\ \text{kg mol}^{-1}}{6.022 \times 10^{23}\ \text{mol}^{-1}} = 6.64 \times 10^{-27}\ \text{kg}$$

and then find its speed by

$$v = \sqrt{\frac{2 \cdot 6.21 \times 10^{-21}\ \text{J}}{6.64 \times 10^{-27}\ \text{kg}}} = 1370\ \text{m/s}$$

To find the de Broglie wavelengths Λ, we use the fundamental relation

$$\Lambda = \frac{h}{mv}$$

By way of example, we consider the Ag colloid particle and calculate

$$\Lambda = \frac{6.626 \times 10^{-34}\ \text{J s}}{5.50 \times 10^{-18}\ \text{kg} \cdot 5.79 \times 10^{-2}\ \text{m/s}} = 2.08 \times 10^{-15}\ \text{m}$$

Proceeding in a like manner for each of the cases above we find

particle	Λ /m
ping-pong ball	1.16×10^{-22}
polystyrene bead	1.50×10^{-17}
Ag particle	2.08×10^{-15}
fullerene, C_{60}	5.43×10^{-12}
He atom	7.28×10^{-11}

These numbers suggest that it is not necessary to consider the wave nature of these particles under these conditions; i.e., the wavelength is small compared to the size of structures from which it might collide so that diffraction is not important.

E1.5 Describe the photoelectric effect experiment.
a) Provide a sketch of the apparatus.
b) State the implications of the experiment.
c) Describe what is observed in the experiment and how it relates to the experiment's implications.

Solution
a) Fig. 1.2a of the textbook gives a schematic of the photoelectric effect apparatus.
b) The principal implication is that light can behave has a particle.
c) The two observations are that the stopping potential depends on the light frequency and not on intensity, while the number of photoelectrons depends on light intensity and not frequency. These results are exactly the opposite of the behavior that one expects for a classical wave and are exactly what would be expected if the light behaved as a particle.

E1.6 Consider the diffraction of photons, electrons, and neutrons from an aperture with diameter d. Consider the case where d is 1 cm and the case where it is 10^{-7} cm.
a) If you direct a light beam onto the aperture, how large must the wavelength be so that diffraction can be observed? What is the frequency of the light you found?
b) If you direct an electron beam onto the aperture, how large must the speed of the electrons be so that diffraction can be observed?
c) We assume that the de Broglie relationship holds not only for electrons, but also for any particle. How large must the speed of the neutrons be for the aperture to diffract a neutron beam?
Do not be disturbed if the answers to these exercises are not experimentally feasible. The goal is to clarify the content of Equations (1.7) and (1.9)

Solution
Diffraction occurs when the wavelength of the wave is approximately the same as the size of the aperture. Considering the size of the aperture as 1 cm and 10^{-7} cm,
a) For an aperture of 1 cm, the wavelength is 1 cm, and the corresponding frequency is 2.998×10^{10} cm s^{-1} / 1 cm $= 2.998 \times 10^{10}$s^{-1}. For a 10^{-7} cm aperture, the wavelength is 10^{-7} cm, and the corresponding frequency is 2.998×10^{10} cm s$^{-1}/10^{-7}$ cm $= 2.998 \times 10^{17}$s^{-1}.
b) We need to find the electron's wavelength through the de Broglie relationship, $\Lambda = h/(mv)$. For a 1 cm wavelength $v = (6.626 \times 10^{-34}$ J s$)/(9.11 \times 10^{-31}$ kg$\cdot$0.01 m$) = 7.273 \times 10^{-2}$ cm s^{-1}. For a 10^{-7} cm wavelength $v = (6.626 \times 10^{-34}$ J s$)/(9.11 \times 10^{-31}$ kg$)$ $(10^{-9}$ m$) = 7.273 \times 10^{5}$ cm s^{-1}.
c) We need to find the neutron's wavelength through the de Broglie relationship, $\Lambda = h/(mv)$. For a 1 cm wavelength $v = (6.626 \times 10^{-34}$ J s$)/(1.675 \times 10^{-27}$ kg$\cdot$0.01 m$) = 3.956 \times 10^{-5}$ cm s^{-1}. For a 10^{-7} cm wavelength $v = (6.626 \times 10^{-34}$ J s$)/(1.675 \times 10^{-27}$ kg$\cdot 10^{-9}$ m$) = 3.956 \times 10^{2}$ cm s^{-1}.

E1.7 If photons are particles they have momentum. Compute the momentum of a 590 nm photon. Compare this momentum to that of a Na atom moving at a speed of 900 m/s, which is a typical value at 1200 °C. Assume that 590 nm photons collide head on with the sodium atom so that the momentum exchange is twice the photon momentum, how many photons are needed to 'stop' the sodium atom?

Solution

Again we employ the de Broglie relationship,

$$\lambda = \frac{h}{mv} = \frac{h}{p}$$

and find that the momentum of a photon is

$$p = \frac{h}{\lambda}$$

Thus for a wavelength of $\lambda = 590$ nm we obtain

$$p_{590} = \frac{6.626 \times 10^{-34}\ \mathrm{J\ s}}{590 \times 10^{-9}\ \mathrm{m}} = 1.123 \times 10^{-27}\ \mathrm{kg\ m\ s^{-1}}$$

This photon momentum should be compared with the sodium atom's momentum, which is

$$p_{\mathrm{Na}} = \left(\frac{22.989 \times 10^{-3}}{6.022 \times 10^{23}}\ \mathrm{kg}\right)(900\ \mathrm{m/s}) = 3.436 \times 10^{-23}\ \mathrm{kg\ m\ s^{-1}}$$

Thus interactions with about 30,600 photons would be required to slow a sodium atom to a stop. Processes of this sort are used in the cooling and trapping of atoms (see *Laser Trapping of Neutral Particles*, by S. Chu, Scientific American, February 1992, p. 71).

E1.8 Imagine you build an experimental apparatus in which you can use a photon to eject an electron from an H-atom; i.e., absorb a photon and generate a proton and a free electron. This is called a photoelectron experiment. Imagine that you perform this experiment on single hydrogen atoms, in a chamber where the atoms are surrounded by a thousand electron detectors (numbered ed1 through ed1000), so that all sides (all 4π steradians) are sensed and the detected electron's position can be reported. In more technical language, imagine measuring the full angular distribution of photoejected electrons. In addition, assume that the hydrogen atoms are in their ground electronic state, and that the photons irradiate the sample isotropically with photons of energy much higher than that which is needed to eject an electron from the hydrogen atom.

a) If the experiment is performed on a single hydrogen atom, what is the probability that *ed375* detects the photoejected electron? If the experiment is performed on one-hundred hydrogen atoms in succession, what is the probability that *ed375* detects the first photoejected electron? What is the probability that *ed375* detects any photoelectron?
b) Imagine a related experiment in which hydrogen atoms that are initially excited are injected into a chamber and they emit light. Perform experiments of the same type as in part (a) but detecting photons instead of electrons. Does your analysis change? Explain!

Solution

a) The photoelectrons should be emitted with equal probability into all 4π steradians. The probability of *ed375* detecting the photoejected electron or the first electron is thus 1/1000. If done 100 times in succession, the probability is now 100 (1/1000) = 0.010 for *ed375* to detect any electron.
b) Assuming that photoelectrons and photons are both emitted isotropically (independently of direction) there is no change in the analysis.

E1.9 Using Fig. 1.6, calculate the distance x between the intensity maxima, P_1 and P_2, in terms of the wavelength of the incident light and the parameters d and a. If your apparatus has $d = 1.7$ μm and $a = 1$ m, and you illuminate the slit with monochromatic light of $\lambda = 640$ nm (red light of a laser pointer), what is the distance x in the figure?

Solutions

Using Fig. E1.9 we can calculate the distance x between the intensity maxima, P_1 and P_2. The point P_1 is the location where a light ray from each slit traverses the same distance to the screen. For this reason the phases of the light waves arriving at point P_1 are the same, and we obtain an intensity maximum at point P_1. The point P_2 is the position where the light ray from the lower slit travels an extra distance of one wavelength and constructively interferes with a light ray from the upper slit. Thus, at point P_2 we find the next intensity maximum: the distances r_1 and r_2 differ by λ; thus, the phases of the two waves arriving at point P_2 are the same.

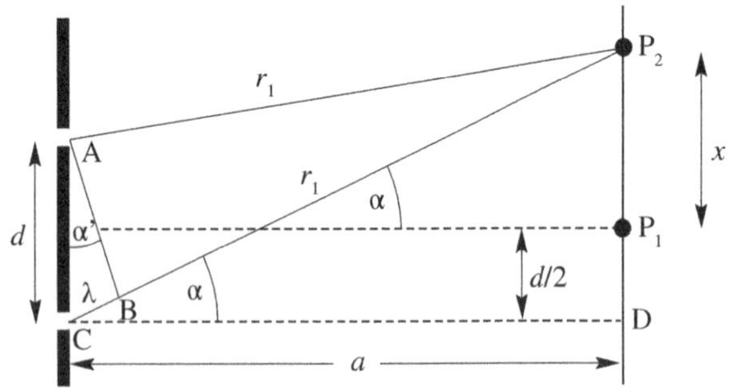

Figure E1.9 Diffraction of light on a double slit.

We restrict our consideration to the case that the distance d between the slits is much smaller than the distance between the points P_1 and P_2. In this case the angle ABC is approximately 90°, and we can set $\alpha' \approx \alpha$ (the triangle ABC is approximately a right triangle). Then we obtain

$$\frac{x + d/2}{a} = \tan\alpha \text{ (triangle } P_2 \text{ CD)}, \quad \frac{\lambda}{d} = \sin\alpha \text{ (triangle ABC)} \tag{1E9.1}$$

In addition, we can use trigonometry to relate the tangent of the angle to its sine, so that

$$\tan\alpha = \frac{\sin\alpha}{\cos\alpha} = \frac{\sin\alpha}{\sqrt{1 - \sin^2\alpha}}$$

Now we can combine these results to relate the separation x between the intensity maxima to the wavelength and the parameters d and a

$$\frac{x + d/2}{a} = \frac{\lambda/d}{\sqrt{1 - (\lambda/d)^2}} \tag{1E9.2}$$

We can rearrange this expression to solve for x and obtain

$$x = \frac{a\lambda}{d}\frac{1}{\sqrt{1 - \left(\frac{\lambda}{d}\right)^2}} - \frac{d}{2}$$

Because the distance x is much larger than $d/2$, we can simplify this expression as

$$x = \frac{a\lambda}{d}\frac{1}{\sqrt{1 - \left(\frac{\lambda}{d}\right)^2}}$$

As an example, we illuminate a double slit of width $d = 1.7\ \mu\text{m}$ with the light of a red laser pointer ($\lambda = 640$ nm). For a screen at a distance of $a = 1$ m we calculate

$$x = \frac{(1\text{ m} \cdot 640\,\text{nm})}{1.7\ \mu\text{m}}\frac{1}{\sqrt{1 - \left(\frac{640\text{ nm}}{1.7\ \mu\text{m}}\right)^2}} = 0.376\text{ m} \cdot 1.079 = 40.6\text{ cm}$$

We can solve this equation for the wavelength λ.

$$\lambda = \frac{a}{xd}\frac{1}{\sqrt{1 - \left(\frac{\lambda}{d}\right)^2}}$$

Note that the remaining term with λ/d is a correction term, and we can approximate it by setting $\tan\alpha \approx \sin\alpha$ in Equation (1E9.1) leading to

$$\frac{x}{a} = \frac{\lambda}{d} \tag{1E9.3}$$

and

$$\lambda = \frac{xd}{a}\sqrt{1 - \left(\frac{\lambda}{d}\right)^2} = \frac{xd}{a}\sqrt{1 - \left(\frac{x}{a}\right)^2}$$

With the data $a = 1$ m, $d = 1.7 \times 10^{-6}$ m, and $x = 40.6$ cm from our example for λ we obtain

$$\lambda = \frac{0.406\ \text{m} \cdot 1.7 \times 10^{-6}\ \text{m}}{1\ \text{m}} \sqrt{1 - \left(\frac{0.406\ \text{m}}{1\ \text{m}}\right)^2} = 690 \times 10^{-9}\ \text{m} \cdot 0.914 = 631\ \text{nm}$$

This is less than the expected 640 nm by 1.4%.
We can improve the calculation by using Equation (1E9.2) instead of Equation (1E9.3)

$$\frac{x}{a} = \frac{\lambda}{d} \frac{1}{\sqrt{1 - \left(\frac{\lambda}{d}\right)^2}} = \frac{\lambda}{d} \frac{1}{\sqrt{1 - \left(\frac{x}{a}\right)^2}}$$

or

$$\frac{\lambda}{d} = \frac{x}{a}\sqrt{1 - \left(\frac{x}{a}\right)^2}$$

leading to the improved equation

$$\lambda = \frac{xd}{a}\sqrt{1 - \left(\frac{\lambda}{d}\right)^2} = \frac{xd}{a}\sqrt{1 - \left[\left(\frac{x}{a}\right)^2\left(1 - \left(\frac{x}{a}\right)^2\right)\right]}$$

With the data $a = 1$ m, $d = 1.7 \times 10^{-6}$ m, and $x = 40.6$ cm for λ we obtain the correct value:

$$\lambda = \frac{0.406\ \text{m} \cdot 1.7 \times 10^{-6}\ \text{m}}{1\ \text{m}} \cdot \sqrt{1 - \left[\left(\frac{0.406\ \text{m}}{1\ \text{m}}\right)^2\left(1 - \left(\frac{0.406\ \text{m}}{1\ \text{m}}\right)^2\right)\right]} = 690 \times 10^{-9}\ \text{m} \cdot 0.928 = 640\ \text{nm}$$

Note that optical measurements are very precise and it is a good choice to use equally precise equations to describe the experiments.

1.2 Problems

P1.1 Consider the following data, taken from O. W. Richardson and K. T. Compton, Phil. Mag. 24 (1913) 575, for the photoemission of electrons from a metal substrate. E_{kin} is the kinetic energy of the photoelectrons and λ is the wavelength of the light. Analyze these data using a least squares analysis. Find the workfunction E_{work} of the metal and determine a value for Planck's constant. The workfunction is the minimum energy that is needed to remove an electron from the metal and place it at the detector some macroscopic distance away.

Sodium		Copper	
E_{kin}/eV	λ/nm	E_{kin}/eV	λ/nm
0.60	436	0.35	260
1.00	366	0.48	254
1.50	313	0.73	230
2.30	254	1.02	210
3.00	210	1.25	200

Solutions
In analogy to the discussion in Section 1.2.2.5, we plot the kinetic energy versus the photon frequency $\nu = c/\lambda$. We begin by constructing a data table for an electron's maximum kinetic energy and the frequency of the light for each metal (Fig. P1.1). By way of example, we calculate the frequency for light with a wavelength of 43.6×10^{-6} cm $= 43.6 \times 10^{-8}$ m, as

$$\nu = \frac{c}{\lambda} = \frac{2.998 \times 10^8\ \text{m s}^{-1}}{43.6 \times 10^{-8}\ \text{m}} = 6.88 \times 10^{14}\ \text{s}^{-1}$$

and for a kinetic energy $E_{kin} = 0.60$ eV we find

$$E_{kin} = 0.60\ \text{eV} \cdot \frac{1.602 \times 10^{-19}\ \text{J}}{1\ \text{eV}} = 0.96 \times 10^{-19}\ \text{J}$$

Proceeding in like manner, for sodium we find the values in the table

E_{kin} / eV	λ/nm	E_{kin} / 10^{-20} J	ν / 10^{14} s^{-1}
0.60	436	9.6	6.88
1.00	366	16.0	8.19
1.50	313	24.0	9.58
2.30	254	36.9	11.8
3.00	210	48.1	14.3

and for copper we find the values in the table below.

E_{kin} / eV	λ / nm	E_{kin} / 10^{-20} J	ν / 10^{14} s^{-1}
0.35	260	5.6	11.53
0.48	254	7.7	11.80
0.73	230	11.7	13.03
1.02	210	16.3	14.28
1.25	200	20.0	14.99

In Figure P1.1, we plot data as E_{kin} versus ν. The slopes of the lines through the data should both be equal to h, and the x-intercept gives the frequency corresponding to the work function for the metal.

For sodium, the slope is $h = 5.29 \times 10^{-34}$ J s and the x-intercept is $\nu = 5.06 \times 10^{14}$ s^{-1}. Thus the work function is $h\nu = 6.626 \times 10^{-34}$ J s$\cdot 5.06 \times 10^{14}$ s^{-1} $= 3.35 \times 10^{-19}$ J or 2.09 eV. The value reported in the *CRC Handbook of Chemistry and Physics* is 2.75 eV. The value obtained for the slope differs by about 20% from the more modern and precise value of h.

For copper, the slope is $h = 3.88 \times 10^{-34}$ J s and the x-intercept is $\nu = 9.97 \times 10^{14}$ s^{-1}. Thus the work function is $h\nu = 6.626 \times 10^{-34}$ J s$\cdot 9.97 \times 10^{14}$ s^{-1} $= 6.61 \times 10^{-19}$ J or 4.12 eV. The values reported in the *CRC Handbook of Chemistry and Physics* range from 4.48 eV to 4.94 eV. The value obtained for the slope differs substantially, about 40%, from the modern value of h.

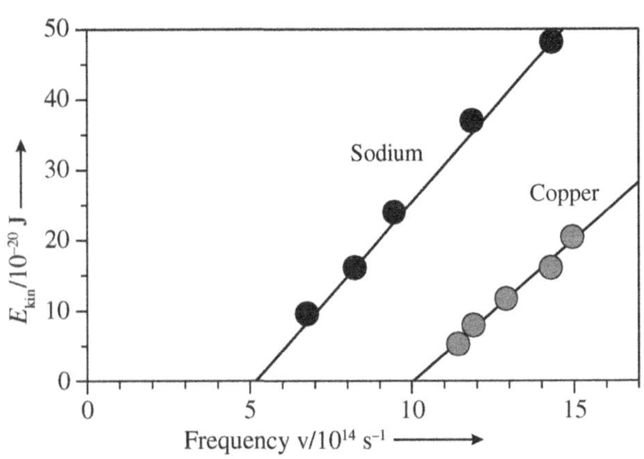

Figure P1.1 Photoelectric effect. Kinetic energy of the ejected electrons versus frequency ν of the photon for sodium (black) and copper (grey). The solid lines are regression curves.

P1.2 Find a formula for the amplitude distribution on the screen in the double-slit experiment in Fig. 1.6. For arbitrary distances r_1 and r_2 between the slits and the point P_2, the resulting wave displacement φ on the screen can be written as a superposition of the waves φ_1 and φ_2 originating from each slit, so that

$$\varphi = \varphi_1 + \varphi_2 = A\ \sin\left(\frac{2\pi\,(vt - r_1)}{\lambda}\right) + A\ \sin\left(\frac{2\pi\,(vt - r_2)}{\lambda}\right)$$

Show that the resulting waveform can be written as

$$\varphi = 2A\ \sin\left(\frac{\pi\,(2vt - r_1 - r_2)}{\lambda}\right)\ \cos\left(\frac{\pi\,(r_2 - r_1)}{\lambda}\right) = \psi(x)\ \sin\left(\frac{\pi\,(2vt - r_1 - r_2)}{\lambda}\right)$$

where the wave's amplitude is $\psi(x) = 2A\cos\left(\pi\left(r_2 - r_1\right)/\lambda\right)$. Simplify this expression for $\psi(x)$ further by realizing that $r_2 - r_1 = d\,x/a$ for $x \ll a$. Identify where the wave amplitude has extrema: minima and maxima.

Solutions

We generalize the treatment in Section 1.1.2 (shown in Fig. 1.6) for arbitrary distances r_1 and r_2 between the slits and the point P_2. From the wave displacements φ_1 and φ_2 we calculate the resulting displacement φ at point P_2.

$$\varphi = \varphi_1 + \varphi_2 = A\ \sin\left(\frac{2\pi(vt - r_1)}{\lambda}\right) + A\ \sin\left(\frac{2\pi(vt - r_2)}{\lambda}\right)$$

With the relation $\sin\alpha + \sin\beta = 2\ \sin\left(\frac{\alpha+\beta}{2}\right)\ \cos\left(\frac{\alpha-\beta}{2}\right)$, we get

$$\varphi = 2A\ \sin\left(\frac{\pi(2vt - r_1 - r_2)}{\lambda}\right)\cos\left(\frac{\pi(r_2 - r_1)}{\lambda}\right)$$

This corresponds to a wave with the maximum displacement

$$\psi = 2A\cdot\cos\left(\frac{\pi(r_2 - r_1)}{\lambda}\right)$$

Because $r_2 - r_1 = xd/a$ for $x \ll a$ and $d/2 \ll x$, we then obtain

$$\psi = 2A\cdot\cos\left(\frac{\pi d}{\lambda a}x\right)$$

The maxima of ψ are then at the positions $x = 0, \lambda a/d, 2\lambda a/d$, and so on, while the minima are at $x = \lambda a/(2d)$, $3\lambda a/(2d)$, and so on.

P1.3 The essential ideas of the photoelectric effect can be applied to atoms and molecules, as well. Consider a sample of Ne gas that is exposed to high energy electromagnetic radiation; i.e., light. If a 58.5 eV source of UV light shines on a sample of Ne gas one observes two distinct distributions of photoelectron kinetic energies at 17.5 eV and at 44.5 eV. In addition, if X ray photons of 1.28 keV shine on a sample of Ne three distinct distributions of photoelectron energies can be seen at 417 eV, 1.24 keV, and 1.27 keV. What information about the electronic structure of a neon atom can you obtain from these data? Make a hypothesis about your observations. How might you test your hypothesis? Describe a set of experiments that could be performed.

Solutions

The 58.5 eV source tells us that there are at least two energy levels, and they are located at energies of (58.5 – 17.5) = 41.0 eV and (58.5 – 44.5) = 14.0 eV below the vacuum level (the zero of energy is a free electron with zero excess kinetic energy). Similarly, the 1.28 keV source tells us that there are at least three energy levels, and these three are located at (1280 – 417) eV= 863 eV, (1280 – 1240) eV= 40 eV, and (1280 – 1270) eV= 10 eV below the vacuum level. Note that we expect the error in these measurements to be of the order of 10 eV, whereas they are significantly less in the earlier experiments.

These two experiments give comparable results for the two more weakly bound electronic energy states (within experimental error/significant digits). One way to support this hypothesis is to look at the UV absorption spectrum for lines in the appropriate region.

P1.4 This problem explores the shape of a diffraction pattern made by a wave passing through a single slit. The intensity profile $I(\theta)$ of a wave passing through a slit of width w is given by

$$I(\theta) = I_{\max}\left(\frac{\sin\left(\frac{\pi w}{\lambda}\sin(\theta)\right)}{\frac{\pi w}{\lambda}\sin(\theta)}\right)^2$$

where $I_{\max}$ is the maximum intensity, λ is the wavelength, and θ is the angular distribution of the intensity pattern. Using a spread sheet program (e.g., Excel) plot this function over the range of -90° to $+90^\circ$ for five different values of the ratio w/λ. Comment on how the diffraction pattern changes as this parameter is changed. In Exercise E1.4 you calculated de Broglie wavelengths for different particles. Based on those calculations and the results of your diffraction calculations, comment on the likelihood of being able to observe diffraction for the different particles in Exercise E1.4.

Solutions

Figure P1.4a illustrates a wave incident on a slit and defines the angle θ.

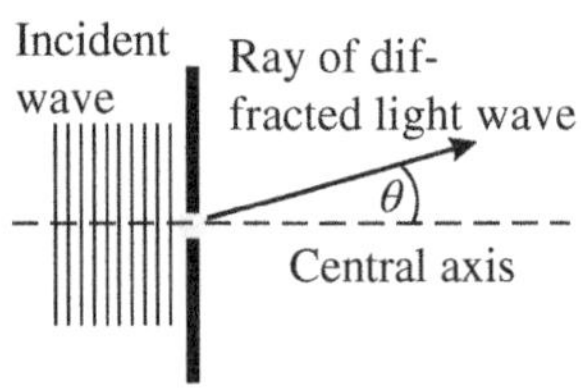

Figure P1.4a Diffraction on a slit.

Figure P1.4b plots the equation for three values of $a = w/\lambda$; namely $a = 0.4$, $a = 2.0$, and $a = 10.0$. As the plots show there is less diffraction when the wavelength of the light is small compared to the width of the aperture; i.e., the dashed black line shows a sharp transmission of the particle at a single well-defined angle. In contrast when the wavelength of the particle becomes similar to or larger than the slit, the wave is strongly diffracted and the amplitude appears at high angles.

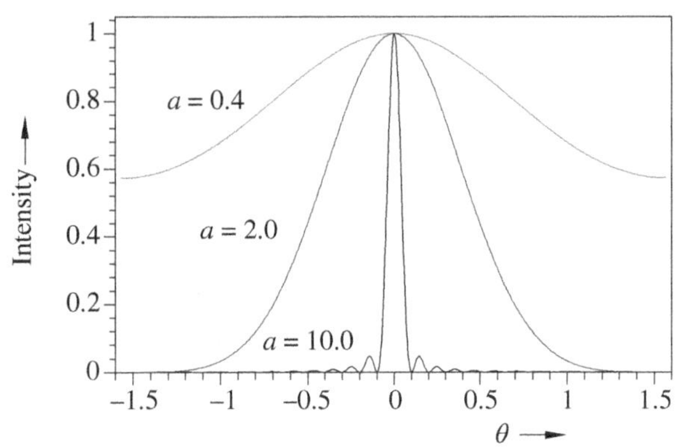

Figure P1.4b Diffraction on a slit: intensity distributions are plotted for three different values of the parameter $a = w/\lambda$.

Based on this analysis we would expect that He atoms could diffract from solid surfaces because their de Broglie wavelength is similar in magnitude to interatomic spacing in solids. In contrast, the other particles in E1.4 have de Broglie wavelengths that are quite small, and it would be much more difficult to observe diffraction with them.

P1.5 The intensity distribution behind a slit of width w is a special case of the distribution behind a grating with N grooves spaced at distance d. For a grating the intensity is

$$I(\theta) = const \cdot \left(\frac{\sin\left(\frac{\pi N d}{\lambda}\sin(\theta)\right)}{\sin\left[\frac{\pi d}{\lambda}\sin(\theta)\right]}\right)^2 \tag{1P5.1}$$

Calculate the intensity distribution for gratings with different numbers of grooves. Choose $d = 1.7\ \mu m$, $\lambda = 500$ nm and perform calculations for $N = 2, 10, 100$, and 1000. Based on your results comment on how the groove number affects the resolution of a wavelength determination.

Solutions

The equation is plotted below in two different plots. The $N = 100$ result is shown in both panels so that comparisons can be drawn. In each case the function is normalized to unity height by dividing its value by N^2. Note that the limiting value of $I(\theta)$ is N^2 at $\theta = 0$. This means that the peak intensity increases substantially with N.

From Fig. P1.5 it is clear that increasing N, decreases the linewidth and thereby increasing the precision for determining λ.

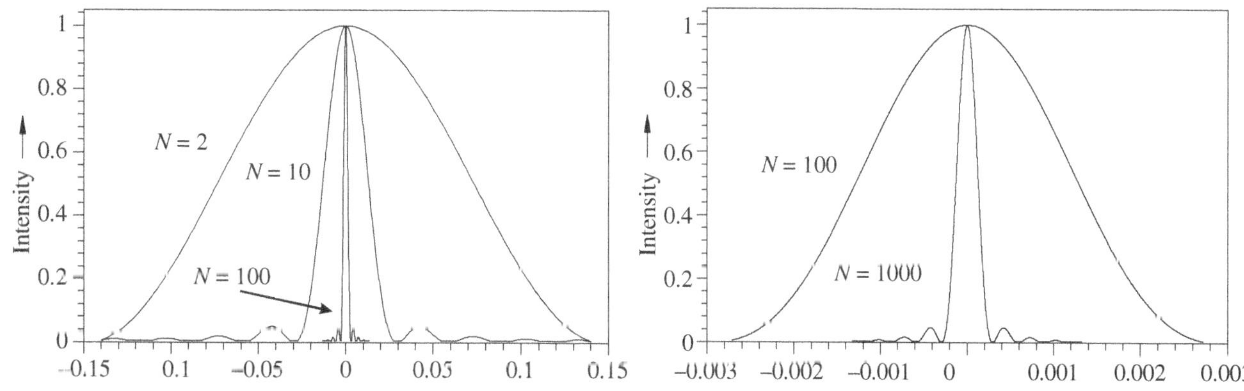

Figure P1.5 Equation (1P5.1) is plotted in two figures. The left panel shows plots for $N = 2$, $N = 10$, and $N = 100$. The right panel shows the plots for $N = 100$ and $N = 1000$. Note the change in scale for the two different abscissas and that the magnitude of the plots are normalized to 1.

P1.6 In the early part of the twentieth century, many of the experiments that laid the foundation for quantum mechanics and wave-particle duality were performed. Because of their importance, many of these findings were recognized by awarding the Nobel Prize. Select a Nobel Prize lecture for one of these "classic" studies, explain, and summarize the findings. See https://www.nobelprize.org.

Solutions

No answer is given because of the number of possible Nobel prize choices and the diversity of acceptable responses.

P1.7 In order to gain some elementary experience with probabilities, consider the following thought experiment. Imagine a thin, rectangular, metal film (ca. 100 nm thick). Now place electron detectors on each side (left and right) of the metal film to face each surface of the metal. Imagine shining photons of energy greater than the metal's work function on the film and collecting the electrons. Assume that we can make the film thick enough so that it behaves like the bulk metal, but thin enough that the light source goes completely through the film. In this case, the electrons could escape out of either face of the metal film with equal probability. Can you predict which face (left or right) emits an electron first? How does the distribution (left versus right) of collected electrons evolve as the experiment proceeds, for the first 100 electrons that are detected?

You can answer these questions by making an analogy to the case of a two-sided coin and denote one side "heads" and the other "tails."

a) If the coin is flipped once, will it give heads? How many times will a coin flip yield heads if it is flipped one-hundred times? Comment on your ability to accurately predict the result of these two sets of experiments?
b) Using your coin, perform the experiment by flipping it one-hundred times and recording the results. Before each time you flip the coin make a prediction of the outcome, record the outcome. What percentage of the time was your result correct?
c) Using your data from b), calculate the fraction of total times heads was obtained, after successive flipping of the coin. Make a plot of your fraction versus the number of times you flip the coin. How rapidly does your data converge to the limiting value? Compare your results to your predictions in part a).

d) Apply your understanding to the electron experiment.

Solutions

a) On average the probability is one-half. For 100 coin flips, we expect to see "heads" about 50 times. Because it is probabilistic, the number may not be exactly 50.
b) A variable answer here, based on the individual experiment.
c) The plot should show that the observed value for the fraction converges on one-half as the number of trials increases.
d) You cannot predict which surface will emit an electron first though after counting a large number of electrons, you will find that $\frac{1}{2}$ are emitted from each surface.

P1.8 In Section 1.4 we discussed how the outcome of many events can be predicted with certainty, but that of individual events cannot, since quantum mechanics only predicts probabilities. Let us use this feature of probability to make a certain prediction. You will need to obtain or borrow a globe of the planet earth, hopefully fairly precise and not too fragile. Can you predict the land-to-water ratio of the planet's surface? Can you measure the land-to-water ratio of the planet's surface? Rather than use a deterministic method (e.g., a ruler), we will use a probabilistic method to determine the land-to-water ratios. Perform the following experiment a number of different times. Toss the globe into the air (or to a friend) and catch it. Record whether your right index finger landed on water or land. Perform this experiment over (and over) and calculate the fraction of times your right index finger ended up on water. Plot your fraction versus the number of trials and determine the value at which it converges. How does this value relate to the known value? Comment on the agreement between your result and the known value. How can your experiment be improved?

Solutions

In the limit of a large number of observations approximately 70% will result in landing in water. This type of testing is called Monte Carlo testing, a simulation method which is useful in physical chemistry when probabilistic arguments are made.

2

Essential Aspects of Structure and Bonding

2.1 Exercises

E2.1 The potential energy $V(r)$ of interaction between a proton (charge of $+e$) and an electron (charge of $-e$) is given, according to Coulomb's law, by

$$V(r) = -\frac{e^2}{4\pi\varepsilon_0 r}$$

where $e = 1.602 \times 10^{-19}$ C, $\varepsilon_0 = 8.854 \times 10^{-12}$ C^2 m^{-1} J^{-1}, and r is the distance between the particles. Use V to compute the electrostatic force f_{electric} between these two particles at a distance of 300 pm. Compute the gravitational force f_{gravity} between these two particles at a distance of 300 pm. Note that $m_e = 9.109 \times 10^{-31}$ kg, $m_p = 1.67 \times 10^{-27}$ kg, and the gravitational constant is $G = 6.67 \times 10^{-11}$ N m^2/kg^2. Compare f_{gravity} to f_{electric}. Is it reasonable to ignore gravitational effects when we describe the interaction of an electron and a proton?

Solution
The force is the derivative of the potential energy with respect to the distance, thus

$$f_{\text{electric}} = -\frac{dV}{dr} = \frac{e^2}{4\pi\varepsilon_0}\frac{1}{r^2}$$

and the electrostatic force for the situation in the problem is given by

$$f_{\text{electric}} = \frac{(1.602 \times 10^{-19}\ \text{C})^2}{4\pi\ (8.854 \times 10^{-12}\ \text{C}^2\ \text{m}^{-1}\ \text{J}^{-1})\ (300 \times 10^{-12}\ \text{m})^2} = 2.56 \times 10^{-9}\ \text{N}$$

For comparison the gravitational force is given by

$$f_{\text{gravity}} = G \cdot \frac{m_e\ m_p}{r^2} = \left(6.67 \times 10^{-11}\ \frac{\text{N} \cdot \text{m}^2}{\text{kg}^2}\right) \cdot \frac{9.109 \times 10^{-31}\ \text{kg} \cdot 1.67 \times 10^{-27}\ \text{kg}}{(300 \times 10^{-12}\ \text{m})^2} = 1.13 \times 10^{-48}\ \text{N}$$

and the force due to gravity is much less (30 orders of magnitude) than that arising from electrostatics.

E2.2 Consider two charged particles, an electron and a proton. Compute the electrostatic energy between a proton and an electron at 100 pm, 1.0 nm, and 10.0 nm. Express your answer in J and in electron volts (1 eV = 1.602×10^{-19} J). Evaluate the distance at which the potential energy is 13.6 eV below the dissociation limit (i.e., zero). Compare this distance to the average distance $\bar{r}$ that we found in Section 2.4.2 for the H atom.

Extension: Use a spreadsheet program (e.g., Excel) to plot the electrostatic potential energy as a function of the distance.

Solution
The electrostatic potential energy between two charged particles is given by

$$V(r) = \frac{1}{4\pi\varepsilon_0}\frac{q_1 q_2}{r}$$

Solutions Manual for Principles of Physical Chemistry, Third Edition. Edited by Hans Kuhn, David H. Waldeck, and Horst-Dieter Försterling.

For an electron–proton pair, the magnitude of the charge is 1.602×10^{-19} C. This makes the energy at 100 pm equal to

$$V(r) = \frac{(-1.602 \times 10^{-19}\ \mathrm{C}) \cdot (+1.602 \times 10^{-19}\ \mathrm{C})}{4\pi(8.854 \times 10^{-12}\ \mathrm{C^2\ m^{-1}\ J^{-1}})\left(100. \times 10^{-12}\ \mathrm{m}\right)} = -2.307 \times 10^{-18}\ \mathrm{J}$$

This energy corresponds to

$$2.307 \times 10^{-18}\ \mathrm{J} \cdot \left(\frac{1\ \mathrm{eV}}{1.602 \times 10^{-19}\ \mathrm{J}}\right) = 14.40\ \mathrm{eV}$$

Similarly at 1.0 nm, $V = -2.307 \times 10^{-19}$ J and -1.440 eV, and at 10.0 nm, $V = -2.307 \times 10^{-20}$ J and -0.1440 eV.

The distance corresponding to an energy of $-13.6\ \mathrm{eV} = -21.79 \times 10^{-19}$ J is

$$r = \frac{q_1 q_2}{4\pi\varepsilon_0 V} = \frac{(-1.602 \times 10^{-19}\ \mathrm{C}) \cdot (+1.602 \times 10^{-19}\ \mathrm{C})}{4\pi\,(8.854 \times 10^{-12}\ \mathrm{C^2\ m^{-1}\ J^{-1}})\left(-21.79 \times 10^{-19}\ \mathrm{J}\right)} = 67.6 \times 10^{-12}\ \mathrm{m} = 67.6\ \mathrm{pm}$$

This distance is comparable to the 90 pm found for the average distance of the electron from the proton in the box model approximation for the hydrogen atom.

E2.3 Use the fact that the size of a He atom is about 220 pm in diameter to estimate the kinetic energy of the electrons in this atom. In particular, use the one-dimensional box model to estimate the electron's de Broglie wavelength and then its average kinetic energy. Report your answer in J and in eV. Also estimate the average speed of the electron. At what fraction of the speed of light is the electron traveling?

Solution

If a He atom in its electronic ground state is about 220 pm in diameter the wavelength describing the electronic state is about twice this size, or about 440 pm. Given this wavelength, the de Broglie relationship gives the momentum p as

$$p = mv = \frac{h}{\Lambda} = \frac{6.626 \times 10^{-34}\ \mathrm{J \cdot s}}{440 \times 10^{-12}\ \mathrm{m}} = 1.506 \times 10^{-24}\ \mathrm{kg\ m\ s^{-1}}$$

This momentum corresponds to an average kinetic energy $\overline{T}$ that is given by

$$\overline{T} = \frac{1}{2}mv^2 = \frac{1}{2}m\frac{p^2}{m^2} = \frac{1}{2}\frac{p^2}{m} = \frac{\left(1.506 \times 10^{-24}\ \mathrm{kg\ m\ s^{-1}}\right)^2}{2 \cdot 9.11 \times 10^{-31}\ \mathrm{kg}} = 1.245 \times 10^{-18}\ \mathrm{J} = 7.715\ \mathrm{eV}$$

The average speed can be obtained in a number of ways. One is from the momentum, i.e.,

$$v = \frac{p}{m} = \frac{1.506 \times 10^{-24}\ \mathrm{kg\ m\ s^{-1}}}{9.11 \times 10^{-31}\ \mathrm{kg}} = 1.653 \times 10^6\ \mathrm{m\ s^{-1}} = 0.0055\ c_0$$

where c_0 is the speed of light.

E2.4 Approximate an H_2 molecule as a one-dimensional box of length 216 pm (found from the bond length plus twice the atomic radius) which confines its valence electrons. Determine the kinetic energy of an electron that has a wavelength of twice the box length. Use this kinetic energy with the known ionization potential (or electron binding energy) of H_2, which is 15.43 eV, to estimate the potential energy of the electron.

Solution

Modeling the H_2 molecule as a one-dimensional box of length 216 pm gives the kinetic energy in the $n = 1$ level to be

$$E_{\mathrm{kin}} = \frac{h^2}{8mL^2}n^2 = \frac{h^2}{8mL^2}$$

Because the H_2 molecule has two electrons, both in the $n = 1$ level, the energy is

$$E_{\mathrm{kin}} = 2 \cdot \frac{h^2}{8mL^2} = \frac{2 \cdot \left(6.626 \times 10^{-34}\ \mathrm{J\ s}\right)^2}{8 \cdot 9.11 \times 10^{-31}\ \mathrm{kg} \cdot \left(216 \times 10^{-12}\ \mathrm{m}\right)^2} = 25.82 \times 10^{-19}\ \mathrm{J} = 16.12\ \mathrm{eV}$$

This kinetic energy, when combined with the known ionization energy E_{ion} (or electron binding energy), gives an estimate of the average potential energy $\overline{V}$ of the energy, namely

$$E_{\text{total}} = E_{\text{kin}} + \overline{V} = -E_{\text{ion}}$$

Solving for $\overline{V}$ we find

$$\overline{V} = -E_{\text{ion}} - E_{\text{kin}} = -15.43\text{ eV} - 16.12\text{ eV} = -31.55\text{ eV}$$

Note: This value is near to what one expects from the Virial Theorem, which states that $\overline{T} = -\frac{1}{2}\overline{V}$.

E2.5 Consider the three one-dimensional wavefunctions sketched in Figure E2.5. Each wavefunction (A, B, and C) is plotted versus the x coordinate whose scale is set by tic marks at zero and one. Order these wavefunctions by their kinetic energy.

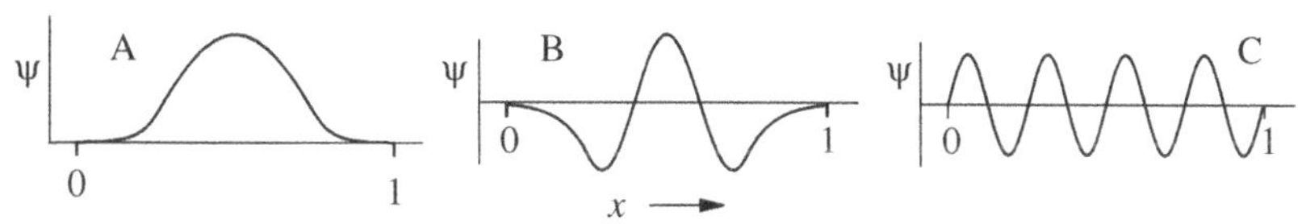

Figure E2.5 Each panel shows a sketch of a plausible wavefunction ψ versus a normalized coordinate x.

Solution
The energy of a wavefunction for a constant length box increases as the wavelength decreases, or as the number of nodes in the wavefunction increases. Thus the energy of wavefunction A is less that than for wavefunction B, which is less than that for wavefunction C.

E2.6 In Fig. 2.6 the wavefunction for an electron trapped in a one-dimensional box is plotted for various values of the total energy, which happens to be the kinetic energy. What observations can you make about the shape of the electron wave as the kinetic energy increases? Realizing that the "node" of a mathematical function occurs at a point where the function changes sign (from plus to minus, or minus to plus) make a quantitative hypothesis about the number of nodes and the kinetic energy of an electron wave.

Solution
The wavefunction has more nodes as the energy increases; hence, it has a smaller de Broglie wavelength. The decrease in de Broglie wavelength corresponds to a higher momentum and kinetic energy. For the simple particle-in-a-box wavefunction, the number of nodes k_{nodes} in the wavefunction is equal to $n - 1$ where n is the quantum number; hence, the kinetic energy is proportional to $(k_{\text{nodes}} + 1)^2$, it is strongly increasing with the number of nodes.

E2.7 Equation (2.13) gives a standing wave for a particle in a one-dimensional box with n as an integer. Plot this function for a box of length 100 pm and n values of 1, 2, 5, and 10. Also plot the function for five non-integer n values between 1 and 2. Describe the differences you observe for the functions. Describe the necessary characteristics for a standing wave of the box.

Solutions
The wavefunctions for integer values are plotted in the three panels of Fig. E2.7a. In each panel the $n = 1$ function is shown as a solid black curve. The grey curves show the wavefunction plots for each of the n values: $n = 2$ in the left panel, $n = 5$ in the center panel, and $n = 10$ in the right panel. For each of the wavefunctions we see that it has a zero at the edges of the box. We also see that the number of nodes for the wavefunctions is given by $n - 1$. In these plots the wavefunctions have been normalized by the factor $\sqrt{2/L}$.

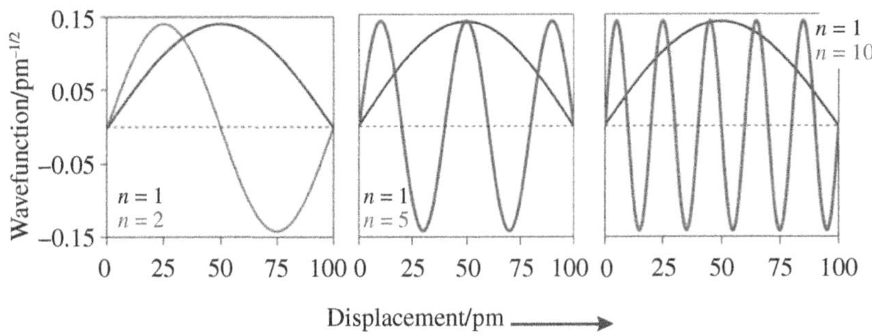

Figure E2.7a Wavefunctions for a particle in a one-dimensional box for $n = 2$, $n = 5$, and $n = 10$ and compared to that for $n = 1$, in each case.

Figure E2.7b shows the wavefunction plotted for five non-integer values of n. The black curve in each panel shows the case of $n = 1$, as a reference. In the left-most panel, the solid grey line is the case of $n = 1.1$ and the dashed grey line is the case of $n = 1.3$. In the center panel the solid grey line is the case of $n = 1.5$ and the dashed grey curve is the case of $n = 1.7$. The rightmost panel shows the case of $n = 1.9$ for the solid grey line. Note that in the cases of noninteger values of n the wavefunctions are not zero at the right border of the box, in contrast to the fact that the solutions of the Schrödinger equation must vanish at this point.

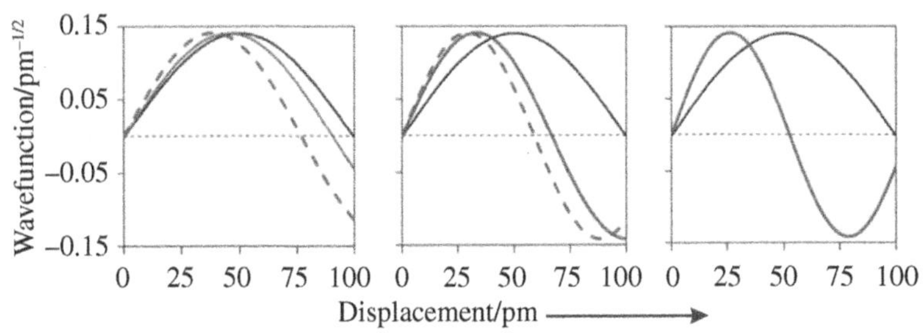

Figure E2.7b Wavefunctions for a particle in a one-dimensional box for $n = 1$, $n = 1.1$ and 1.3 (left), $n = 1$, $n = 1.5$, and $n = 1.7$ (center) and $n = 1$, $n = 1.9$ (right).

E2.8 Use the resonance condition of Equation (2.12) to derive the kinetic energy expression of Equation (2.14).

Solution

We begin by writing the expression for the kinetic energy classically, and then substitute for the speed using the de Broglie relationship

$$T_n = \frac{1}{2}m_e v^2 = \frac{1}{2}m_e\left(\frac{h}{m_e\Lambda}\right)^2$$

which gives on substitution of the resonance condition ($\Lambda = 2L/n$)

$$T_n = \frac{1}{2}m_e\left(\frac{h}{m_e(2L/n)}\right)^2 = \frac{1}{2}m_e\left(\frac{n^2h^2}{m_e^2 4L^2}\right)$$

or

$$T_n = \frac{1}{2}m_e\left(\frac{n^2h^2}{m_e^2 4L^2}\right) = \frac{h^2}{8m_eL^2}n^2$$

E2.9 In our box model of the hydrogen atom, we found that the ground-state electron wave corresponded to a box length of 376 pm. However, the proton is also contained in this volume of space. Are the wave characteristics of the proton important to consider in this case? Explain.

Solutions

The wavelength of the proton will be 1837 times smaller than that of the electron or (376 pm)/1837 = 0.205 pm because the mass of the proton is 1837 times larger. This value is small enough (i.e., much smaller than the "box size" of 376 pm) that the wave nature of the proton does not need to be considered.

E2.10 Reproduce Fig. 2.10 using a spreadsheet program, such as Excel, and find the value of L that minimizes the energy. Also determine the minimum value of L analytically, i.e., set the first derivative of the energy with respect to L equal to zero and solve for L. Compare your numerical and analytical results.

Solution

The energy is given by Equation (2.27) of the text and it is plotted in Fig. E2.10. The energy minimum is found to fall between 374 pm and 375 pm. See the Excel sheet "ch2_e10.xls."

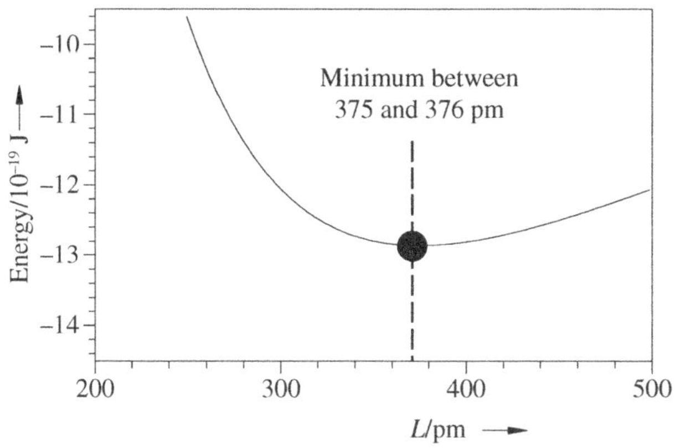

Figure E2.10 A plot of the variational energy for the H-atom, in which the electron wavefunction is approximated by a one-dimensional particle in box wavefunction, as a function of the box length L.

The analytical derivation of the box side length L that minimizes the energy is provided in Box 2.1 and is given here also. We begin with Equation (2.27), namely

$$\varepsilon = \frac{3h^2}{8m_eL^2} - \frac{1}{4\pi\varepsilon_0}\frac{4.18 \cdot e^2}{L}$$

To find the value of the parameter L that lies at the minimum of the energy function we take the derivative and set it equal to zero, and find

$$\left(\frac{d\varepsilon}{dL}\right)_{L=L_{\text{min}}} = -\frac{6h^2}{8m_eL_{\text{min}}^3} + \frac{1}{4\pi\varepsilon_0}\frac{4.18 \cdot e^2}{L_{\text{min}}^2} = 0$$

If we solve this equation for L_{min}, we find

$$L_{\text{min}} = 4\pi\varepsilon_0\frac{h^2}{m_ee^2}\frac{3}{4 \times 4.18} = 375 \text{ pm}$$

Formally we would need to take a second derivative and analyze it also, in order to prove that this value of L is a minimum. From the plot of the function above it is clear that the extremum must be a minimum.

E2.11 Use Equation (2.20) to determine the average potential energy of an electron in the H atom, $\overline{V}$. Use this average potential energy to determine the average value of $\overline{1/r}$. Compare your value for $\overline{1/r}$ to the value of $1/a_0$. Comment on your result.

Solutions

Equation (2.20) writes the energy of the ground state of the hydrogen atom as

$$E = -\frac{1}{4\pi\varepsilon_0}\frac{e^2}{2a_0} = -21.8 \times 10^{-19} \text{ J}$$

and using the Virial theorem we can write that

$$\overline{V} = 2E = -\frac{e^2}{4\pi\varepsilon_0}\frac{1}{a_0} = -43.6 \times 10^{-19} \text{ J}$$

Using Coulomb's law, the potential energy is given by

$$V = -\frac{e^2}{4\pi\varepsilon_0}\frac{1}{r}$$

so that

$$\overline{V} = -\frac{e^2}{4\pi\varepsilon_0} \cdot \overline{1/r} = -\frac{e^2}{4\pi\varepsilon_0}\frac{1}{a_0}$$

and we see that

$$\overline{1/r} = \frac{1}{a_0}$$

E2.12 To explore the properties of the particle in the box wavefunctions further, perform the MathCAD exercise, titled PinBwaves.mcd, and answer the questions:

a) For each of the above wavefunctions ($n = 1$ and $n = 2$), find the number of nodes.
b) Plot out the eigenfunctions corresponding to quantum numbers 2 and 3.
c) From the graph, read off the position at which the nodes can be found for the cases of $n = 1, 2, 3,$ and 4.
d) Plot the probability density, or the square of the wavefunctions for $n = 3$ and $n = 4$.

Solution

The solutions to the questions and activities requested in the MathCAD exercise are given in Figs. E2.12a and E2.12b. It is also available as PinBwaves_solutions.mcd.

a) The $n = 1$ wavefunction has no nodes and the $n = 2$ wavefunction has one node.
b) The MathCAD plots for $n = 2$ and $n = 3$ are

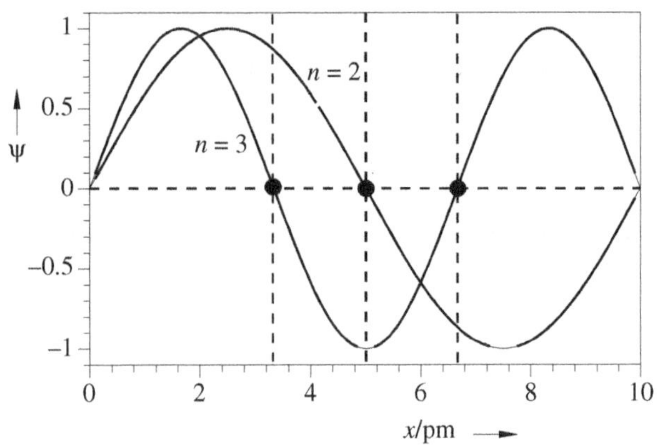

Figure E2.12a Plots of the wavefunction of a particle in a one-dimensional box for $n = 2$ and $n = 3$.

c) For $n = 1$ we have no node, for $n = 2$ a node occurs at $x = 5$ pm; for $n = 3$ nodes occur at $x = 3.333$ pm and $x = 6.666$ pm; and for $n = 4$ nodes occur at $x = 2.5$ pm, $x = 5.0$ pm, and $x = 7.5$ pm.
d) The probability density plots for $n = 3$ and $n = 4$ are

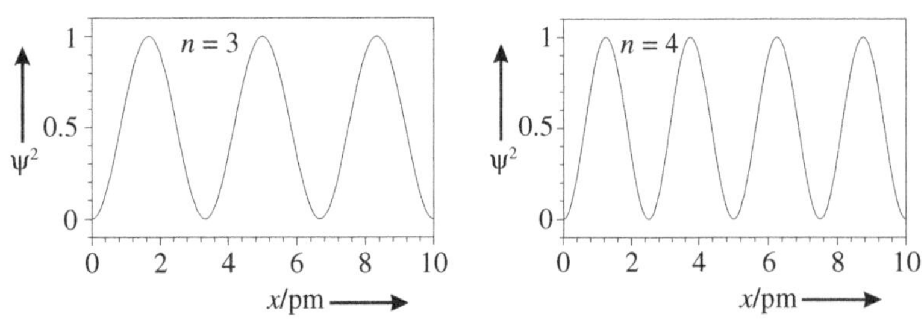

Figure E2.12b Plots of the probability density of a particle in a one-dimensional box for $n = 3$ and $n = 4$.

E2.13 Using Equation (2.1) find the energy that corresponds to the 589 nm Na emission line (see Fig. 2.1). Compare your result to the energy spacing shown in the electron energy loss spectrum of Fig. 2.2.

Solution

The energy of a 589 nm photon is

$$E = \frac{hc_0}{\lambda} = \frac{(6.626 \times 10^{-34}\ \mathrm{J \cdot s})\,(2.998 \times 10^{8}\ \mathrm{m \cdot s^{-1}})}{589 \times 10^{-9}\ \mathrm{m}} = 3.373 \times 10^{-19}\ \mathrm{J}\left(\frac{1\ \mathrm{eV}}{1.602 \times 10^{-19}\ \mathrm{J}}\right) = 2.105\ \mathrm{eV}$$

This value is exactly the spacing between lines in the energy loss spectrum, suggesting that similar energy levels are involved.

E2.14 When an electron is bound to a proton it makes a hydrogen atom, for which the accepted atomic diameter is 128 pm.

a) Estimate the largest wavelength that an electron could have if it were trapped inside of a box of this size; assume that the wave has to go to zero at the boundaries of the box. From this wavelength calculate the average kinetic energy for the electron.

b) Perform an estimation similar to that of part a) for an electron that is confined in i) a benzene ring, ii) a C_{60} molecule, or iii) a 10 nm diameter Au colloid. Estimate the diameters of benzene and C_{60} by using their density and the molecular weights.

c) Based on your calculations, discuss how you expect the kinetic energy of an electron to change as you change its ability to delocalize throughout a molecule or a nanometer-sized particle.

Solution

a) The appropriate box size would be approximately the diameter d of the atom or 128 pm; this distance corresponds to one-half of the de Broglie wavelength, which is then 256 pm. The average kinetic energy is given by

$$E = \frac{p^2}{2m} = \frac{(h/\Lambda)^2}{2m} = \frac{h^2}{8mL^2}$$

For $\Lambda = 2L = 256$ pm we find $L = 128$ pm; thus, the kinetic energy is 3.68×10^{-18} J.

b) The volume of one molecule of the substance is found from the density ρ and the molar mass $\mathbf{M}$ of the substance as

$$V_{\text{molecule}} = \frac{m}{\rho} = \frac{\mathbf{M}/\mathbf{N}_A}{\rho} = \frac{\mathbf{M}}{\rho \mathbf{N}_A}$$

The volume of the molecule is related to its diameter d (assuming the molecule is spherical) by

$$V_{\text{molecule}} = \frac{4}{3}\pi\left(\frac{d}{2}\right)^3$$

For the three species of interest here at 293 K

Species	$M/(\mathrm{g\ mol^{-1}})$	$\rho/(\mathrm{g\ mL^{-1}})$	$V_{\text{molecule}}/(\mathrm{nm^3})$	$d/(\mathrm{pm})$	$\Lambda/(\mathrm{nm})$	$E/(\mathrm{J})$
Benzene	78.11	0.8765	0.1480	656.2	1.312	3.500×10^{-20}
C60	720.6	1.197	0.9997	999.9	1.999	1.508×10^{-20}
10 nm Au	6.050×10^{6}	19.32	520	10,000	20.0	1.5×10^{-22}

c) As the size of the particle increases, the electron delocalizes more and its kinetic energy decreases. The lowest kinetic energy of the electron will correspond to a de Broglie wavelength of $2d$; $\Lambda = 2d$.

2.2 Problems

P2.1 Figure P2.1a shows an atomic emission spectrum from excited electronic states of the hydrogen atom to its lower energy electronic state with principal quantum number $n = 2$, the Balmer series. The first few lines correspond to: $H_\alpha = 656.28$ nm, $H_\beta = 486.13$ nm, $H_\gamma = 434.05$ nm, $H_\delta = 410.17$ nm. Use these wavelengths to determine the energy level spacing between the ground state and each of these excited states. Make an energy level diagram which shows the ground electronic state and these first few excited states. In your diagram, set the zero of energy to correspond to the separated electron and proton and use the fact that the ground electronic state lies 13.6 eV below this value. What do you think the H_∞ label indicates?

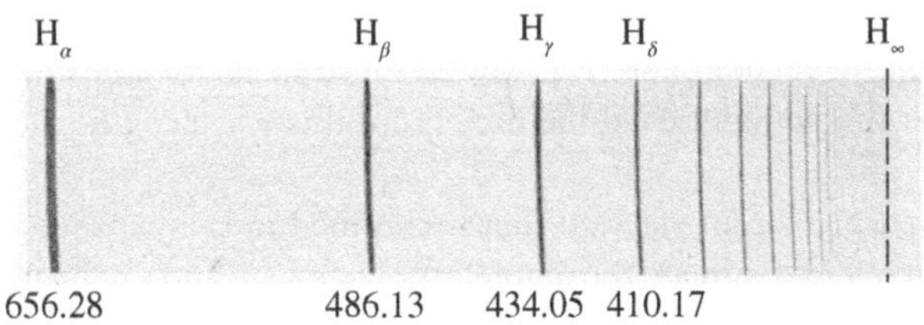

Figure P2.1a Emission lines of the Balmer series. The image is taken from G. Herzberg, *Atomic Spectra and Atomic Structure* (Prentice Hall, New York 1937). The numbers denote the wavelengths measured in units of nm.

Compare and contrast the energy level structure you found here for the H-atom to that calculated for an electron confined in a one-dimensional energy well whose potential energy does not change with displacement, see Fig. 2.6. Do you think that we will need to develop a more detailed model for the structure of the hydrogen atom, than the box potential? Please explain your answer in detail.

Solution

Begin by recognizing the quantum numbers and calculating the energies associated with each transition. For example, from the 656.28 nm wavelength, we find that

$$\Delta E = \frac{h\,c_0}{\lambda} = \frac{(6.62608 \times 10^{-34}\ \mathrm{J \cdot s})\,(2.9979 \times 10^{8}\ \mathrm{m \cdot s^{-1}})}{656.28 \times 10^{-9}\ \mathrm{m}}$$
$$= 3.0268 \times 10^{-19}\ \mathrm{J}$$

Proceeding in a like manner for the other wavelengths we find the values in the table below. The H_∞ label in the figure indicates a transition from $n_{\mathrm{upper}} = \infty$ to $n_{\mathrm{lower}} = 2$.

	n_{lower}	n_{upper}	λ/nm	$\Delta E/10^{-19}$ J
H_α	2	3	656.28	3.0268
H_β	2	4	486.13	4.0862
H_γ	2	5	434.05	4.5765
H_δ	2	6	410.17	4.8429

The energy level spacings, which can be deduced from these calculations, are shown in Fig. P2.1b; and they disagree strongly with those represented by Fig. 2.6 in the text. Namely, the energy levels get closer together as n increases for an H-atom, whereas a one-dimensional box model results in energy levels that get farther apart as the quantum number increases—a clear disagreement. Equation (2.14) and Fig. 2.6 describe the kinetic energy for a particle in a one-dimensional box and is not a model for the H-atom. A model for the H-atom must be three-dimensional and has to include a Coulomb potential energy contribution, as described in Section 2.42.

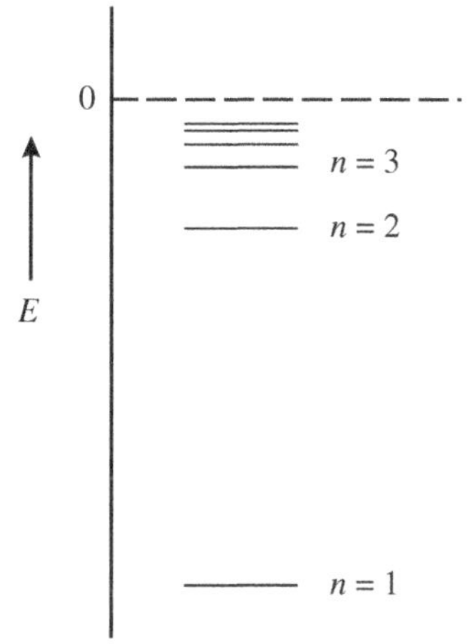

Figure P2.1b Plot of the H-atom energy levels. Note that the levels converge on the $E = 0$ limit as the principal quantum number becomes large.

P2.2 Figure P2.2 shows data that were measured using a Franck–Hertz apparatus with Ne gas rather than Na gas. It is reproduced from M.M. Kash and G.C. Shields, *J. Chem. Ed.* 1994, **71**, 466. Use these data to determine the resonance energy for excitation of Ne atoms. The data are collected for two values of the opposing voltage. You may use either data set.

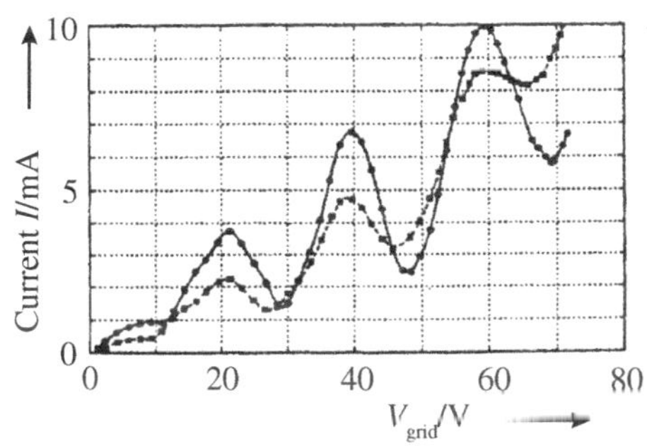

Figure P2.2 Plots of the current–voltage profiles observed in a Franck–Hertz experiment for Ne gas.

The authors describe how the gas emits a red glow when the atoms get excited. Using the energy difference that you found from the current–voltage curve, calculate the wavelength of a photon that an excited Ne atom can emit. Could such photons explain the red glow?

Solution

The three apparent maxima in the largest amplitude curve (shown in smoothed form) occur at 59.28, 39.39, and 20.93 V. The average voltage difference between successive peaks is thus 19.18 V. This value of the electric potential corresponds to an energy difference of 3.072×10^{-18} J. This energy difference then corresponds to photons of wavelength 64.6 nm, which would be in the far UV. The observed red glow in a tube of excited neon probably arises from transitions involving two excited states.

P2.3 Show that a standing wave with nodes at the positions $x = 0$ and $x = L$ is formed from the overlap of two traveling waves of the same wavelength moving in opposite directions. Note that a wave traveling from left to right with speed υ can be described by

$$\varphi_+ = A \sin\left(\frac{2\pi(\upsilon t - x)}{\Lambda}\right)$$

and when this wave is reflected at the right boundary the back traveling wave is described by

$$\varphi_- = -A \sin\left(\frac{2\pi(\upsilon t + x)}{\Lambda}\right)$$

Solution

The waves interfere. Using superposition, the displacement φ at a point x is given by

$$\varphi = \varphi_+ + \varphi_- = A\left[\sin\left(\frac{2\pi(vt-x)}{\Lambda}\right) - \sin\left(\frac{2\pi(vt+x)}{\Lambda}\right)\right]$$

Using the trigonometric relation, $\sin\alpha - \sin\beta = 2\cos\left(\frac{\alpha+\beta}{2}\right)\sin\left(\frac{\alpha-\beta}{2}\right)$, we find that

$$\varphi = 2A\cos\left(\frac{2\pi vt}{\Lambda}\right)\cdot\sin\left(\frac{2\pi x}{\Lambda}\right) = \cos\left(\frac{2\pi vt}{\Lambda}\right)\cdot\psi$$

with $\psi = 2A \cdot \sin(2\pi x/\Lambda)$ where ψ is the equation for a standing wave. With $\Lambda = 2L/n$ we obtain

$$\psi = 2A\cdot\sin\left(\frac{\pi x n}{L}\right), \quad n = 1,\ 2,\ \ldots$$

Note that φ changes in time; however, it is always zero at $x = 0$ and $x = L$.

P2.4 Since ψ^2 is a probability function, i.e., the probability of finding an electron in a region of space, we can use it to compute the statistical outcome of measurements. Explore this application, for a simple situation in which you know the answer, by computing the average value of an electron's position in the ground electronic state of a box. Use a spreadsheet program, to numerically determine the average position of an electron for the ground-state wavefunction of the particle in a box of length 100 pm. Take the wavefunction to be

$$\psi(x) = \sqrt{\frac{2}{L}}\sin\left(\frac{\pi x}{L}\right)$$

Remember that the average position $\bar{x}$ for a discrete set of values is given by

$$\bar{x} = \sum_{i=1}^{N} x_i\ \mathcal{P}_i,\ \mathcal{P}_i = \rho\cdot\Delta x_i$$

where x_i is the value of the displacement in the ith element and $\mathcal{P}_i$ is the probability in the ith element. In our case $\mathcal{P}_i$ is the electron probability density ρ times Δx_i.

Perform the same calculation analytically, for which the average value is given by

$$\bar{x} = \int_0^L x\,\psi^2\,\mathrm{d}x$$

and compare your findings in the two cases.

Solution

The Excel sheet titled "ch2_p5.xls" performs the numerical sum and finds an average value of 50 pm. This sheet is provided on Wiley's website. Hence the average displacement of the particle is at the center of the box, $L/2$.

To perform the result analytically, we first verify that $\psi(x)$ is normalized; hence, we must evaluate

$$\int_0^L \psi^2\,\mathrm{d}x = \frac{2}{L}\int_0^L \sin^2\left(\frac{\pi x}{L}\right)\mathrm{d}x$$

From Appendix B.1 we know that

$$\int \sin^2(z) = \frac{z}{2} - \frac{1}{4}\sin 2z$$

Making the substitution that $z = \pi x/L$, we find

$$\int_0^L \psi^2\,\mathrm{d}x = \frac{2}{\pi}\int_0^\pi \sin^2(z)\,\mathrm{d}z = \frac{2}{\pi}\left(\frac{z}{2} - \frac{1}{4}\sin 2z\right)_0^\pi = \frac{2}{\pi}\left(\frac{\pi}{2} - \frac{1}{4}\sin 2\pi\right) - \frac{2}{\pi}\left(\frac{0}{2} - \frac{1}{4}0\right) = 1$$

Having verified that the wavefunctions are normalized, we can evaluate $\bar{x}$ by

$$\bar{x} = \int_0^L x\,\psi^2\,\mathrm{d}x = \frac{2}{L}\int_0^L x\sin^2\left(\frac{\pi x}{L}\right)\mathrm{d}x$$

From Appendix B.1 we know that

$$\int z\sin^2(z) = \frac{z^2}{4} - \frac{z}{4}\sin 2z - \frac{\cos 2z}{8}$$

Making the substitution that $z = \pi x/L$, we find

$$\overline{x} = \int_0^L x\psi^2\,dx = \frac{2L}{\pi^2}\int_0^\pi \sin^2(z)\,dz = \frac{2L}{\pi^2}\left(\frac{z^2}{4} - \frac{z}{4}\sin 2z - \frac{\cos 2z}{8}\right)_0^\pi$$
$$= \frac{2L}{\pi^2}\left(\frac{\pi^2}{4} - 0 - \frac{1}{8}\right) - \frac{2L}{\pi^2}\left(0 - 0 - \frac{1}{8}\right) = \frac{L}{2}$$

The analytical and numerical results are in excellent agreement.

P2.5 Use Equation (2.25) in Section 2.4.1 to show that the mean potential energy $\overline{V}$ for an electron in the ground state of the H-atom (ρ is given by Equation (2.22)) is $\overline{V} = -e^2/(4\pi\varepsilon_0 a_0)$. Hint: Proceed by converting the integral over all space into spherical coordinates: i.e.,

$$\overline{V} = -\frac{e^2}{4\pi\varepsilon_0}\int_{\text{all space}} \frac{1}{r}\rho\,d\tau = -\frac{e^2}{4\pi\varepsilon_0}\int_0^{2\pi} d\phi \int_0^\pi \sin\theta\,d\theta \int_0^\infty \frac{1}{r}\rho\,r^2 dr$$

Solution

Because the Coulomb potential of the proton is isotropic in space, we proceed by using spherical coordinates to perform the integral. Thus we can write

$$\overline{V} = -\frac{e^2}{4\pi\varepsilon_0}\int_{\text{all space}} \frac{1}{r}\rho\,d\tau = -\frac{e^2}{4\pi\varepsilon_0}\int_0^{2\pi} d\phi \int_0^\pi \sin\theta\,d\theta \int_0^\infty \frac{1}{r}\rho\,r^2 dr$$

Because the probability density ρ depends only on r, the integration of the angles θ and ϕ is straightforward and gives the solid angle 4π. Using Equation (2.22) we can write

$$\overline{V} = -\frac{e^2}{4\pi\varepsilon_0}\int_0^\infty \frac{1}{r}\frac{1}{\pi a_0^3}\exp(-2r/a_0)\,4\pi r^2 dr = -\frac{e^2}{4\pi\varepsilon_0}\frac{4\pi}{\pi a_0^3}\int_0^\infty r\exp(-2r/a_0)\,dr$$

With the substitutions $x = 2r/a_0$ and $dx = 2dr/a_0$, this expression can be converted into

$$\overline{V} = -\frac{e^2}{4\pi\varepsilon_0}\frac{4\pi}{\pi a_0^3}\frac{a_0}{2}\frac{a_0}{2}\int_0^\infty x\cdot\exp(-x)\cdot dx$$

According to Appendix B the integral is

$$\int_0^\infty x\cdot\exp(-x)\cdot dx = 1! = 1$$

and for the mean potential energy we obtain

$$\overline{V} = -\frac{e^2}{4\pi\varepsilon_0}\frac{4\pi}{\pi a_0^3}\frac{a_0}{2}\frac{a_0 2}{\cdot}1 = -\frac{e^2}{4\pi\varepsilon_0 a_0}$$

P2.6 As we have discussed, the ionization energy of a hydrogen atom is the energy required to remove its electron. The table below provides the measured ionization energies for the hydrogen atom and six atomic ions, each of which contains only one electron. From Coulomb's law, we expect the potential energy of the electron to depend on the nuclear charge, or atomic number. Analyze the data and empirically determine this dependence. Given these ionization energies, estimate the kinetic energy of the electron and the electron's average speed. Do we need to account for relativistic effects?

Atomic Number (Z)	Atom/Ion	Ionization Energy (IE)
1	H	13.60 eV
2	He^+	54.40 eV
3	Li^{2+}	122.43 eV
4	Be^{3+}	217.67 eV
5	B^{4+}	340.11 eV
6	C^{5+}	489.81 eV

Solution

We begin by constructing a table which converts the ionization energy from eV into Joule, determines the electron speed, and compares that speed to the speed of light (c_0)

Z	Species	$IE/(10^{-18}$J)	$\overline{T}$ (10^{-18} J)	v/(m s^{-1})	v/c_0
1	H	2.179	2.179	2.187×10^6	0.00732
2	He^+	8.715	8.715	4.374×10^6	0.01463
3	Li^{2+}	19.61	19.61	6.562×10^6	0.02195
4	Be^{3+}	34.87	34.87	8.750×10^6	0.02927
5	B^{4+}	54.49	54.49	1.094×10^7	0.03658
6	C^{5+}	78.47	78.47	1.313×10^7	0.04390

By way of example, consider the calculations for He^+. In this case the energy conversion proceeds by

$$IE = 54.4 \text{ eV} \cdot \frac{1.602177 \times 10^{-19} \text{ J}}{\text{eV}} = 8.715 \times 10^{-18} \text{ J}$$

From Equation (2.30) of the text, we see that the average kinetic energy $\overline{T}$ is the same as this ionization energy. To find the average speed we use the basic formula $T = mv^2/2$ so that for He^+

$$v = \sqrt{\frac{2 \cdot 8.715 \times 10^{-18}}{9.1094 \times 10^{-31}}} \text{ m/s} = 4.374 \times 10^6 \text{ m/s}$$

Because each of the atomic ions is a one-electron system, the origin of the ionization energy difference must come from the nuclear charge differences; hence, we assume that

$$\text{Ionization Energy} = \text{constant} \cdot Z^m$$

To determine m, we plot $\ln(IE/10^{-18} \text{ J})$ versus $\ln(Z)$ and find that $m = 2.00$ provides the best fit to the data. The data for the ionization energy versus the atomic number are plotted in Figure P2.6 and clearly show that the ionization energy increases as the square of the atomic number.

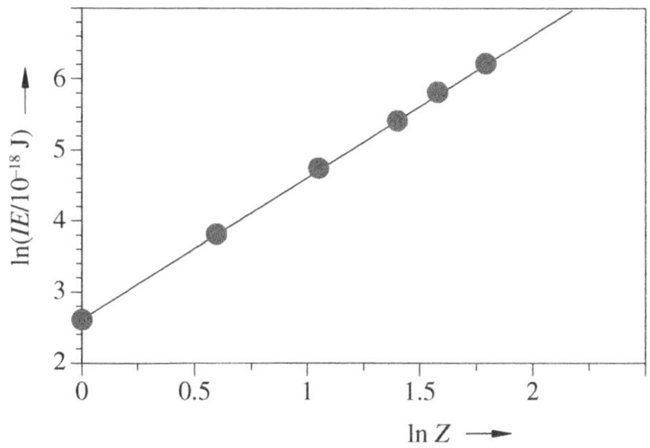

Figure P2.6 The natural logarithm of the ionization energy (IE) is plotted versus the natural logarithm of the atomic number Z.

The second question asks if relativity needs to be considered. From the data we see that the maximum speed for these ions is less than 5% of the speed of light; thus, relativistic effects do not need to be considered.

P2.7 Given the expression for the energy of an electron trapped in a one-dimensional box, $E_{kin} = [h^2/8m_0L^2)] \cdot n^2$, see Equation (2.14), describe how the energy changes with the size of the box. Make a plot/graph which shows this dependence for $n = 1$, $n = 2$, and $n = 3$. Based on your findings describe how you would expect the electronic energy levels of a chain of hydrogen atoms to change as the number of atoms in the chain is increased. Remember that as you increase the number of atoms in the chain, you are increasing the box length and the number of electrons.

Solution

The kinetic energy E_{kin} of the electron decreases as the box length becomes larger because it can delocalize more. In each case the dependence is a decrease that falls off as $1/L^2$. Figure P2.7 shows plots of the kinetic energy versus the length L for each of the cases: $n = 1$, $n = 2$, and $n = 3$.

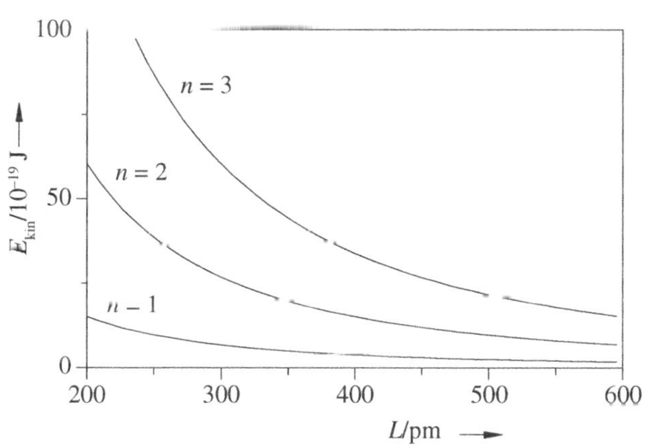

Figure P2.7 Kinetic energy versus the box length L for each of the cases: $n = 1$, $n = 2$, and $n = 3$.

While a chain of real H-atoms would spontaneously segregate into H_2 diatoms, we consider here the artificial condition in which the distance between the H-atoms in the chain is restricted to be fixed at the value $2a_0$. For a chain of N hydrogen atoms we can write the box length as $L = N \cdot 2a_0$. If we model the H-atom chain as a box, the electrons delocalize and the kinetic energy of each energy level n decreases as $1/L^2$; i.e., $1/N^2$. The number of electrons grows as N and the value of the quantum number for the most weakly bound electron is $n = N/2$ for N even and $n = (N + 1)/2$ for N odd. Consequently we expect that the kinetic energy of the most weakly bond electron will decrease as $1/N$.

P2.8 Using the probability density given by Equation (2.22), compute the average distance of the electron from the proton in a hydrogen atom.

Solutions

The average distance is given by

$$\bar{r} = \int r \cdot \rho(r) \cdot d\tau \text{ where the integral is taken over all space}$$

From Equation (2.22) we see that the probability density depends only on the distance r, not the angles so that

$$\bar{r} = 4\pi \int_0^{\infty} r \cdot \rho(r) \cdot r^2 dr = \frac{4}{a_0^3} \int_0^{\infty} r^3 \cdot \exp(-2r/a_0) \cdot dr$$

where we have substituted the expression for $\rho(r)$. If we substitute $x = 2r/a_0$, we find

$$\bar{r} = \frac{1}{a_0} \int_0^{\infty} x^3 \cdot \exp(-x) \cdot dx$$

Using the integral from Appendix B.1, namely $\int_0^{\infty} x^n \cdot \exp(-x) \cdot dx = n$, we find that

$$\bar{r} = \frac{a_0}{4} \int_0^{\infty} x^3 \cdot \exp(-x) \cdot dx = \frac{a_0}{4} \cdot 6 = \frac{3}{2} a_0$$

P2.9 The table shows the variational energy $\varepsilon/(10^{-19}\text{ J})$ as a function of the parameters L and b for the box model wavefunctions of H_2^+. The internuclear distance is set at the experimental value of 106 pm. Plot these data and show that they generate a well-defined minimum. Determine the optimized values for L and b. Comment on the results.

L/pm	$b = 1.20$	$b = 1.25$	$b = 1.30$	$b = 1.35$	$b = 1.40$	$b = 1.45$	$b = 1.50$
200	−11.197	−11.779	−12.223	−12.544	−12.753	−12.885	−12.937
220	−13.910	−14.236	−14.451	−14.5637	−14.601	−14.5742	−14.475
240	−15.218	−15.358	−15.405	−15.387	−15.297	−15.157	−14.973
260	−15.644	−15.656	−15.593	−15.466	−15.300	−15.079	−14.839
280	−15.530	−15.441	−15.300	−15.103	−14.878	−14.606	−14.323
300	−15.073	−14.914	−14.715	−14.468	−14.200	−13.898	−13.581
320	−14.403	−14.196	−13.950	−13.674	−13.372	−13.056	−12.714
340	−13.608	−13.368	−13.087	−12.793	−12.468	−12.136	−11.792

Solutions

Figure P2.9a shows plots of ε versus L for four different values of b.

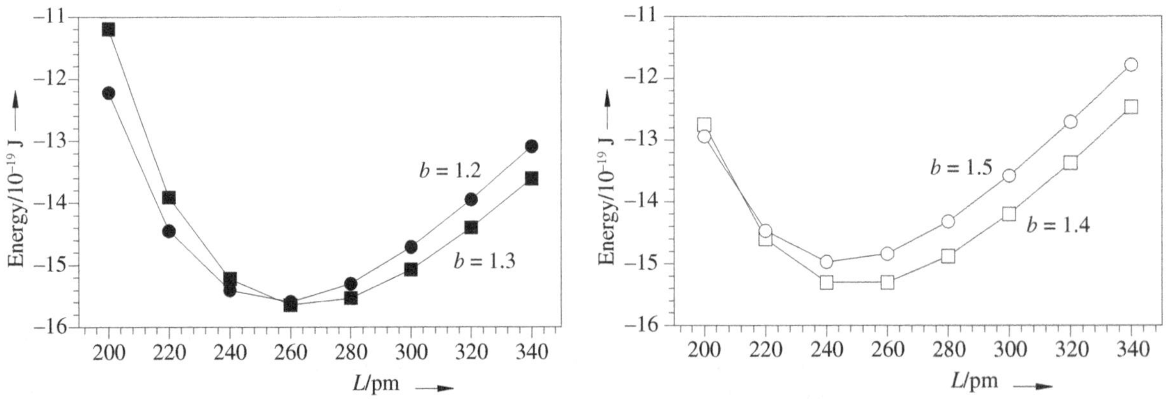

Figure P2.9a Variational energy ε of a particle in a three-dimensional box versus the box length L for four different values of the parameter b.

Using plots of this sort we find the value of L that minimizes the energy for each of the values of b; L_{min}. The values reported in the table below were obtained by fitting the curve to a cubic equation and then finding the minimum energy from the fit parameters (Fig. P2.9b).

b	1.20	1.25	1.30	1.35	1.40	1.45	1.50
L_{min}/pm	264	260	257	254	252	249	247
$\varepsilon/10^{-19}$ J	−15.241	−15.656	−15.611	−15.442	−15.299	−15.122	−14.926

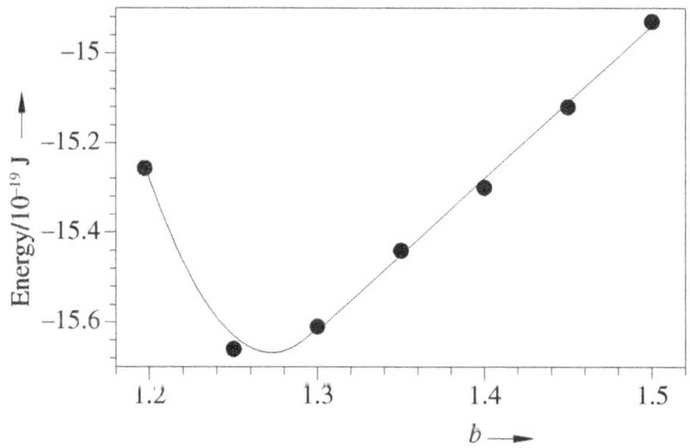

Figure P2.9b Variational energy of an electron in the field of two protons (H_2^+ model), with a three-dimensional box trial function, is plotted versus the box parameter b.

Figure P2.9b plots the minimum energy versus b and indicates that the optimized value of b lies between 1.2 and 1.3. The fact that b is greater than one indicates that the wavefunctions are more delocalized along the x-direction than along y or z.

3

Schrödinger Equation

3.1 Exercises

E3.1 Verify that Equation (3.38) is a solution of the Schrödinger Equation (3.31) for a particle in a three-dimensional box.

Solution

To verify this result we need to find the second partial derivatives of the wavefunction with respect to the variables x, y, and z, and then substitute into the Schrödinger Equation (3.31). To find the second partial derivative with respect to y, we first take the partial derivative with respect to y

$$\frac{\partial\psi}{\partial y} = A\left(\frac{n_y\pi}{L_y}\right)\sin\left(\frac{n_x\pi x}{L_x}\right)\cos\left(\frac{n_y\pi y}{L_y}\right)\sin\left(\frac{n_z\pi z}{L_z}\right)$$

and then take the partial derivative of this result, to find the second partial derivative with respect to y, namely

$$\frac{\partial^2\psi}{\partial y^2} = -A\left(\frac{n_y\pi}{L_y}\right)^2\sin\left(\frac{n_x\pi x}{L_x}\right)\sin\left(\frac{n_y\pi y}{L_y}\right)\sin\left(\frac{n_z\pi z}{L_z}\right) = -\left(\frac{n_y\pi}{L_y}\right)^2\psi.$$

We proceed in like fashion for the second partial derivative with respect to z. First we find the partial derivative with respect to z

$$\frac{\partial\psi}{\partial z} = A\left(\frac{n_z\pi}{L_z}\right)\sin\left(\frac{n_x\pi x}{L_x}\right)\sin\left(\frac{n_y\pi y}{L_y}\right)\cos\left(\frac{n_z\pi z}{L_z}\right)$$

and then the second partial derivative with respect to z

$$\frac{\partial^2\psi}{\partial z^2} = -A\left(\frac{n_z\pi}{L_z}\right)^2\sin\left(\frac{n_x\pi x}{L_x}\right)\sin\left(\frac{n_y\pi y}{L_y}\right)\sin\left(\frac{n_z\pi z}{L_z}\right) = -\left(\frac{n_z\pi}{L_z}\right)^2\psi.$$

Proceeding in a like fashion for the x variable, we find that

$$\frac{\partial^2\psi}{\partial x^2} = -A\left(\frac{n_x\pi}{L_x}\right)^2\sin\left(\frac{n_x\pi x}{L_x}\right)\sin\left(\frac{n_y\pi y}{L_y}\right)\sin\left(\frac{n_z\pi z}{L_z}\right) = -\left(\frac{n_x\pi}{L_x}\right)^2\psi.$$

Substituting these expressions for the second partial derivative into Equation (3.38) gives

$$-\frac{\hbar^2}{2m_e}\left[-\left(\frac{n_x\pi}{L_x}\right)^2\psi - \left(\frac{n_y\pi}{L_y}\right)^2\psi - \left(\frac{n_z\pi}{L_z}\right)^2\psi\right] = E\,\psi$$

or

$$E = \frac{\hbar^2}{2m_e}\left[\left(\frac{n_x\pi}{L_x}\right)^2 + \left(\frac{n_y\pi}{L_y}\right)^2 + \left(\frac{n_z\pi}{L_z}\right)^2\right] = \frac{h^2}{2m_e}\left[\left(\frac{n_x}{L_x}\right)^2 + \left(\frac{n_y}{L_y}\right)^2 + \left(\frac{n_z}{L_z}\right)^2\right]$$

which is Equation (3.39) of the text.

E3.2 In the spirit of the discussion in Section 3.2.3 consider an electron that is confined to a potential energy well with $V = 0$ in two dimensions and infinitely high walls $V = \infty$ outside of the box. Write the Schrödinger equation for

Solutions Manual for Principles of Physical Chemistry, Third Edition. Edited by Hans Kuhn, David H. Waldeck, and Horst-Dieter Försterling.

this case; show that the Schrödinger equation is separable; write an expression for the total wavefunction; and normalize the wavefunction.

Solution

Here we consider an electron confined to a well in the two dimensions: x and y. In this case, the potential energy depends only on x and y; and the particle has two kinetic energy terms so that the Laplacian is

$$\nabla^2 = \frac{\partial^2}{\partial x^2} + \frac{\partial^2}{\partial y^2}$$

As with the case considered in the text, the potential energy is zero inside of the box and is infinite outside of it. Thus the wavefunction is zero everywhere outside the box, and inside the box it is given by the Schrödinger equation,

$$\left(-\frac{\hbar^2}{2m_e}\nabla^2 + V\right)\psi = E\psi$$

to be

$$\left[-\frac{\hbar^2}{2m_e}\left(\frac{\partial^2}{\partial x^2} + \frac{\partial^2}{\partial y^2}\right)\right]\cdot\psi = E\psi$$

To demonstrate that this equation is separable, assume that $\psi(x,y) = \psi_x(x)\cdot\psi_y(y)$ and confirm that the terms are separable. By substituting $\psi(x,y)$ into the equation, we find

$$\begin{aligned} E\cdot\psi_x(x)\cdot\psi_y(y) &= \left[-\frac{\hbar^2}{2m_e}\left(\frac{\partial^2}{\partial x^2} + \frac{\partial^2}{\partial y^2}\right)\right]\cdot\psi_x(x)\cdot\psi_y(y) \\ &= \psi_y(y)\left[-\frac{\hbar^2}{2m_e}\frac{\partial^2}{\partial x^2}\psi_x(x)\right] + \psi_x(x)\left[-\frac{\hbar^2}{2m_e}\frac{\partial^2}{\partial y^2}\psi_y(y)\right] \end{aligned}$$

This can be rearranged to give

$$E = \frac{\left[-\frac{\hbar^2}{2m_e}\frac{\partial^2}{\partial x^2}\psi_x(x)\right]}{\psi_x(x)} + \frac{\left[-\frac{\hbar^2}{2m_e}\frac{\partial^2}{\partial y^2}\psi_y(y)\right]}{\psi_y(y)}$$

The two terms on the right are independent of each other, but each equal to a constant so that $E = E_x + E_y$ where

$$-\frac{\hbar^2}{2m_e}\frac{\partial^2}{\partial x^2}\psi_x(x) = E_x\cdot\psi_x(x) \qquad \text{and} \qquad -\frac{\hbar^2}{2m_e}\frac{\partial^2}{\partial y^2}\psi_y(y) = E_y\cdot\psi_y(y)$$

The solutions for the one-dimensional equations will be sine waves and are given by

$$\psi_x(x)\cdot\psi_y(y) = A\sin\left(\frac{n_x\pi x}{L_x}\right)\cdot\sin\left(\frac{n_y\pi y}{L_y}\right)$$

as found for the analogous cases in the text. That this wavefunction solves the Schrödinger equation can be confirmed by substituting it in and demonstrating the equality. We find

$$\begin{aligned} E\cdot\psi_x(x)\cdot\psi_y(y) &= \left[-\frac{\hbar^2}{2m_e}\left(\frac{\partial^2}{\partial x^2} + \frac{\partial^2}{\partial y^2}\right)\right]\cdot A\sin\left(\frac{n_x\pi x}{L_x}\right)\cdot\sin\left(\frac{n_y\pi y}{L_y}\right) \\ &= -\frac{\hbar^2}{2m_e}\left[-A\left(\frac{n_x\pi}{L_x}\right)^2\sin\left(\frac{n_x\pi x}{L_x}\right)\cdot\sin\left(\frac{n_y\pi y}{L_y}\right) - A\left(\frac{n_y\pi}{L_y}\right)^2\sin\left(\frac{n_y\pi y}{L_y}\right)\cdot\sin\left(\frac{n_x\pi x}{L_x}\right)\right] \\ &= \left[\frac{\hbar^2}{2m_e}\left(\frac{n_x\pi}{L_x}\right)^2 + \frac{\hbar^2}{2m_e}\left(\frac{n_y\pi}{L_y}\right)^2\right]A\sin\left(\frac{n_y\pi y}{L_y}\right)\cdot\sin\left(\frac{n_x\pi x}{L_x}\right) \\ &= \frac{h^2\pi^2}{(2\pi)^2 2m_e}\left(\frac{n_x^2}{L_x^2} + \frac{n_y^2}{L_y^2}\right)\cdot\psi_x(x)\cdot\psi_y(y) = \frac{h^2}{8m_e}\left(\frac{n_x^2}{L_x^2} + \frac{n_y^2}{L_y^2}\right)\cdot\psi_x(x)\cdot\psi_y(y) \end{aligned}$$

The probability of finding the electron in the area $dx\,dy$ is given by $dP = \psi^2(x,y)\,dx\,dy$. By normalizing the function we ensure that the probability of finding the particle somewhere in space is 1. Mathematically, this is performed by looking everywhere (integrating the probability over all space) or

$$\mathcal{P}_{total} = 1 = \int_{\text{all space}} dP = \int_{x=-\infty}^{x=\infty}\int_{y=-\infty}^{y=\infty}\psi^2(x,y)\cdot dxdy = \int_{x=-\infty}^{x=\infty}\left[\int_{y=-\infty}^{y=\infty}\psi^2(x,y)\cdot dy\right]dx$$

Substituting in the wavefunction, we get

$$1 = \int_{-\infty}^{\infty}\int_{-\infty}^{\infty} A^2 \sin^2\left(\frac{n_x\pi x}{L_x}\right)\sin^2\left(\frac{n_y\pi y}{L_y}\right) \mathrm{d}x\,\mathrm{d}y = A^2\int_{x=-\infty}^{x=\infty}\left[\int_{y=-\infty}^{y=\infty} \sin^2\left(\frac{n_x\pi x}{L_x}\right)\sin^2\left(\frac{n_y\pi y}{L_y}\right)\cdot \mathrm{d}y\right]\mathrm{d}x$$
$$= A^2\left[\int_{-\infty}^{\infty} \sin^2\left(\frac{n_x\pi x}{L_x}\right)\mathrm{d}x\right]\left[\int_{-\infty}^{\infty} \sin^2\left(\frac{n_y\pi y}{L_y}\right)\mathrm{d}y\right]$$

We recognize that each integral is of the same form,

$$\int \sin^2(bx)\,\mathrm{d}x = \frac{x}{2} - \frac{1}{4b}\sin(2bx)$$

which can be found in Appendix B. Upon substitution we find that

$$1 = A^2\left[\frac{x}{2} - \frac{L_x}{4n_x\pi}\sin\left(\frac{2n_x\pi x}{L_x}\right)\right]_0^{L_x}\left[\frac{y}{2} - \frac{L_y}{4n_y\pi}\sin\left(\frac{2n_y\pi y}{L_y}\right)\right]_0^{L_y}$$
$$= A^2\left(\left(\frac{L_x}{2} - 0\right) - (0-0)\right)\left(\left(\frac{L_y}{2} - 0\right) - (0-0)\right) = A^2\frac{L_x}{2}\frac{L_y}{2}$$

Upon rearrangement we find that

$$A = \sqrt{\frac{4}{L_xL_y}}$$

E3.3 Perform the integration in Equation (3.53) analytically and show that the integral is equal to zero.

Solution

The orthogonality condition is $\int_0^L \psi_1 \cdot \psi_2\,\mathrm{d}x = 0$ for the one-dimensional particle in a box. Upon substitution of the wavefunctions, we find that

$$\int_0^L \psi_1\cdot\psi_2\,\mathrm{d}x = \int_0^L \sqrt{\frac{2}{L}}\sin\left(\frac{n_1\pi x}{L}\right)\cdot\sqrt{\frac{2}{L}}\sin\left(\frac{n_2\pi x}{L}\right)\mathrm{d}x = \frac{2}{L}\int_0^L \sin\left(\frac{n_1\pi x}{L}\right)\sin\left(\frac{n_2\pi x}{L}\right)\mathrm{d}x$$

Using the integral table from Appendix B, this integral is

$$\int \sin ax\,\sin bx\,\mathrm{d}x = \frac{\sin((a-b)x)}{2(a-b)} + \frac{\sin((a+b)x)}{2(a+b)} \quad \text{for } a \neq b$$

Using this result for the integral gives

$$\int_0^L \psi_1\cdot\psi_2\,\mathrm{d}x = \left(\frac{2}{L}\right)\left[\frac{\sin((n_1-n_2)\pi x/L)}{2((n_1-n_2)\pi/L)} + \frac{\sin((n_1+n_2)\pi x/L)}{2((n_1+n_2)\pi/L)}\right]_0^L = \left[\frac{\sin((n_1-n_2)\pi)}{((n_1-n_2)\pi)} + \frac{\sin((n_1+n_2)\pi)}{((n_1+n_2)\pi)}\right]$$

Because n_1 and n_2 are integers, the sine terms are zero because the $\sin(n\pi) = 0$ for n an integer; hence, we find that

$$\int_0^L \psi_1\cdot\psi_2\,\mathrm{d}x = 0 \text{ for } n_1 \neq n_2$$

E3.4 By inspection, energy order the wavefunctions in Fig. 3.9. By inspection energy order the wavefunctions in Fig. 3.10.

Solution

The simplest way to energy order the wavefunctions is to count the number of nodes in the wavefunction, with a lower number of nodes being lower energy. Thus in Fig. 3.9 the energy increases from left to right because the leftmost figure has no nodal planes, the middle figure has 1 nodal plane, and the rightmost figure has two nodal planes. In Fig. 3.10, there are three wavefunctions, ψ_1, ψ_2, and ψ_3. Applying the same node counting principle, ψ_1 corresponds to the lowest energy state, ψ_2 the middle energy state, and ψ_3 the highest energy state.

Note that the total energy is a sum of the kinetic energy and potential energy; however, in these particular cases the potential energy is uniform and only the kinetic energy differs.

E3.5 Write an expression for the Schrödinger equation which needs to be solved for the electron in a box problem (i.e., consider the box to be one-dimensional and have infinitely high potential walls at x values of 0 and L). Be sure to write out the explicit form of the Hamiltonian operator $\mathcal{H}$. Show that $\sqrt{2/L}\sin(n\pi x/L)$ is a solution of this equation, where n is the quantum number.

Solution

The Schrödinger equation for the one-dimensional particle in a box is $\mathcal{H}\psi = E\psi$ where

$$\mathcal{H} = -\frac{\hbar^2}{2m}\frac{\partial^2}{\partial x^2} + V$$

so that

$$\left(-\frac{\hbar^2}{2m}\frac{d^2}{dx^2} + V\right)\psi = E\psi$$

In our case the potential energy is assumed to be $V = 0$ and the equation reduces to

$$-\frac{\hbar^2}{2m}\frac{d^2}{dx^2}\psi = E\psi$$

We are asked to show that $\psi(x) = \sqrt{2/L}\sin(n\pi x/L)$ is a solution of this equation over the range $0 \leq x \leq L$. The first derivative of ψ with respect to x is

$$\frac{d\psi(x)}{dx} = \sqrt{\frac{2}{L}}\left(\frac{n\pi}{L}\right)\cos\left(\frac{n\pi x}{L}\right)$$

and the second derivative is

$$\frac{d^2\psi(x)}{dx^2} = -\sqrt{\frac{2}{L}}\left(\frac{n\pi}{L}\right)^2\sin\left(\frac{n\pi x}{L}\right)$$

Upon substitution into the Schrödinger equation, we find that

$$-\frac{\hbar^2}{2m}\frac{d^2\psi(x)}{dx^2} = \frac{\hbar^2}{2m}\sqrt{\frac{2}{L}}\left(\frac{n\pi}{L}\right)^2\sin\left(\frac{n\pi x}{L}\right) = \frac{\hbar^2}{2m}\left(\frac{n\pi}{L}\right)^2\psi(x) = E\psi$$

Hence the equality holds for

$$E = \frac{\hbar^2}{2m}\left(\frac{n\pi}{L}\right)^2$$

E3.6 We can define the uncertainty in a particle's position as $\Delta x = \sigma_x = \sqrt{\left(\overline{x^2} - \overline{x}^2\right)}$. Find the uncertainty in the position of an electron localized in a one-dimensional box of length L and occupying the ground state of the box (i.e., $n = 1$). Evaluate $\overline{x}$ and $\overline{x^2}$ and then calculate Δx.

Solution

We are asked to evaluate the average value of x for the ground state of the particle in a one-dimensional box. We begin by writing the definition

$$\overline{x} = \frac{2}{L}\int_0^L \sin^2\left(\frac{\pi x}{L}\right)\cdot x \cdot dx$$

First we make use of the relation

$$\sin^2(ax) = \frac{1}{2}(1 - \cos 2ax)$$

Thus for the integral we obtain

$$\int_0^L \sin^2\left(\frac{\pi x}{L}\right)\cdot x\cdot dx = \int_0^L \frac{1}{2}\left[(1 - \cos\left(\frac{2\pi x}{L}\right)\cdot x\cdot dx\right] = \frac{1}{2}\int_0^L \left[x\cdot dx - \int_0^L \cos\left(\frac{2\pi x}{L}\right)\cdot x\cdot dx\right]$$

We calculate

$$\int_0^L x\cdot dx = \left|\frac{1}{2}x^2\right|_0^L = \frac{1}{2}L^2$$

Because of, see Appendix B.2,

$$\int x\cos ax = \frac{\cos ax}{a^2} + \frac{x\sin ax}{a}$$

we calculate the remaining integral as

$$\int_0^L \cos\left(\frac{2\pi x}{L}\right)\cdot x\cdot \mathrm{d}x = \left|\frac{\cos\frac{2\pi}{L}x}{\left(\frac{2\pi}{L}\right)^2} + \frac{x\sin\frac{2\pi}{L}x}{\frac{2\pi}{L}}\right|_0^L = \frac{\cos 2\pi}{\left(\frac{2\pi}{L}\right)^2} + \frac{L\sin 2\pi}{\frac{2\pi}{L}} - \frac{\cos 0}{\left(\frac{2\pi}{L}\right)^2} - \frac{0\cdot\sin 0}{\frac{2\pi}{L}}$$

$$= \frac{1}{\left(\frac{2\pi}{L}\right)^2} + \frac{L\cdot 0}{\frac{2\pi}{L}} - \frac{1}{\left(\frac{2\pi}{L}\right)^2} - 0 = 0$$

Thus we obtain

$$\int_0^L \sin^2\left(\frac{\pi x}{L}\right)\cdot x\cdot \mathrm{d}x = \frac{1}{2}\int_0^L \left[x\cdot\mathrm{d}x - \int_0^L \cos\left(\frac{2\pi x}{L}\right)\cdot x\cdot\mathrm{d}x\right] = \frac{1}{2}\cdot\frac{1}{2}L^2 = \frac{1}{4}L^2$$

and

$$\overline{x} = \frac{2}{L}\int_0^L \sin^2\left(\frac{\pi x}{L}\right)\cdot x\cdot\mathrm{d}x = \frac{2}{L}\cdot\frac{1}{4}L^2 = \frac{L}{2}$$

Accordingly we calculate

$$\overline{x^2} = \frac{2}{L}\int_0^L \sin^2\left(\frac{\pi x}{L}\right)\cdot x^2\cdot\mathrm{d}x = \frac{2}{L}\int_0^L \frac{1}{2}\left(1-\cos\frac{2\pi x}{L}\right)\cdot x^2\cdot\mathrm{d}x$$

$$= \frac{2}{L}\int_0^L \frac{1}{2}x^2\cdot\mathrm{d}x - \frac{2}{L}\int_0^L \frac{1}{2}\cos\left(\frac{2\pi x}{L}\right)\cdot x^2\cdot\mathrm{d}x$$

We calculate

$$\int_0^L \frac{1}{2}x^2\cdot\mathrm{d}x = \left|\frac{1}{6}x^3\right|_0^L = \frac{1}{6}L^3$$

From the integral table in Appendix B.2, we can write that

$$\int x^2\cos ax\,\mathrm{d}x = \frac{2x\cos ax}{a^2} + \left(\frac{x^2}{a} - \frac{2}{a^3}\right)\sin ax$$

we evaluate the remaining integral as

$$\int_0^L \cos\left(\frac{2\pi x}{L}\right)\cdot x^2\cdot\mathrm{d}x = \left|\frac{2x\cos\frac{2\pi}{L}x}{\left(\frac{2\pi}{L}\right)^2} + \left(\frac{x^2}{\frac{2\pi}{L}} - \frac{2}{\left(\frac{2\pi}{L}\right)^3}\right)\sin\frac{2\pi}{L}x\right|_0^L$$

$$= \frac{2L\cos 2\pi}{\left(\frac{2\pi}{L}\right)^2} + \left(\frac{L^2}{\frac{2\pi}{L}} - \frac{2}{\left(\frac{2\pi}{L}\right)^3}\right)\sin 2\pi - \frac{0\cdot\cos 0}{\left(\frac{2\pi}{L}\right)^2} - \left(\frac{0}{\frac{2\pi}{L}} - \frac{2}{\left(\frac{2\pi}{L}\right)^3}\right)\sin 0$$

$$= \frac{2L}{\left(\frac{2\pi}{L}\right)^2} = \frac{2L^3}{4\pi^2} = \frac{L^3}{2\pi^2}$$

Combining these results, we obtain

$$\overline{x^2} = \frac{2}{L}\int_0^L \sin^2\left(\frac{\pi x}{L}\right)\cdot x^2\cdot\mathrm{d}x = \frac{2}{L}\int_0^L \frac{1}{2}x^2\cdot\mathrm{d}x - \frac{2}{L}\int_0^L \frac{1}{2}\cos\left(\frac{2\pi x}{L}\right)\cdot x^2\cdot\mathrm{d}x$$

$$= \frac{2}{L}\left(\frac{1}{6}L^3 - \frac{L^3}{4\pi^2}\right) = \frac{1}{3}L^2 - \frac{L^2}{2\pi^2} = \frac{1}{3}L^2\left(1 - \frac{3}{2\pi^2}\right) = L^2\cdot 0.28267$$

Finally for the uncertainty we find

$$\overline{x^2} - \overline{x}^2 = L^2\cdot 0.283 - \frac{L^2}{4} = L^2\,(0.283 - 0.250) = 0.033\cdot L^2$$

This makes the uncertainty in the particle's position

$$\Delta x = \sigma_x = \sqrt{\left(\overline{x^2} - \overline{x}^2\right)} = \sqrt{\left(\frac{L^2}{3}\left(1 - \frac{3}{2\pi^2}\right) - \left(\frac{L}{2}\right)^2\right)}$$

$$= L\sqrt{\left(\frac{1}{12} - \frac{1}{2\pi^2}\right)} = 0.181 \cdot L$$

This is 18% of the box length. Also see Foundation 3.3.

E3.7 Consider a particle confined to a box whose walls are infinitely high and located at $x = 0$ and $x = L = 10.0$ nm. Find the probability that the particle has a spatial position between 3.00 and 7.00 nm for each of the energy levels $n = 1, n = 3$, and $n = 10$. In other words, explain how you expect the probability to behave as n becomes very large.

Solution

The probability that the particle is located between $x = 3.00$ nm and $x = 7.00$ nm is found as

$$\mathcal{P}_{3\text{ to }7\text{ nm}} = \frac{2}{10\text{ nm}} \int_{x=3\text{ nm}}^{x=7\text{ nm}} \sin^2\left(\frac{n\pi x}{L}\right) \mathrm{d}x$$

Using Appendix B, we find that $\int \sin^2 x\, \mathrm{d}x = \frac{1}{2}\left(x - \frac{1}{2}\sin(2x)\right)$, so that

$$\mathcal{P}_{3\text{ to }7\text{ nm}} = \frac{1}{5\text{ nm}}\frac{1}{2}\left[x|_{3\text{ nm}}^{7\text{ nm}} - \frac{\sin(2n\pi x/L)}{2n\pi/L}\bigg|_{3\text{ nm}}^{7\text{ nm}}\right] = \frac{1}{10\text{ nm}}\left(4\text{ nm} - \frac{10\text{ nm}}{2n\pi}\left(\sin\left(\frac{14n\pi}{10}\right)\right) - \sin\left(\frac{6n\pi}{10}\right)\right)$$

$$= \frac{2}{5} - \frac{1}{2n\pi}\left(\sin\left(\frac{7n\pi}{5}\right) - \sin\left(\frac{3n\pi}{5}\right)\right)$$

Thus for the different energy levels, we find that

n	1	3	10
$\mathcal{P}_{3\text{ to }7\text{ nm}}$	0.70	0.34	0.40

As n becomes larger, the particle is distributed more and more uniformly throughout the box, as expected from the ratio of 4 nm to the total space of 10 nm:

$$\mathcal{P}_{3\text{ to }7\text{ nm}} = \frac{7-3}{10} = 0.40$$

E3.8 Consider a particle confined to a box whose walls are infinitely high and located at $x = 0$ and $x = L = 10.0$ nm. Explicitly show how you can use the particle in a box wavefunction and the Schrödinger equation to find a general expression for the particle's energy and evaluate it for the first five energy levels.

Solution

The Schrödinger equation for the one-dimensional box is

$$-\frac{\hbar^2}{2m}\frac{\mathrm{d}^2\psi(x)}{\mathrm{d}x^2} = E\,\psi(x)$$

Upon substitution of the wavefunction $\psi(x) = \sqrt{\frac{2}{L}}\sin\left(\frac{n\pi x}{L}\right)$, we find that

$$-\frac{\hbar^2}{2m}\sqrt{\frac{2}{L}}\frac{\mathrm{d}^2(\sin(n\pi x/L))}{\mathrm{d}x^2} = E \cdot \sqrt{\frac{2}{L}}\sin\left(\frac{n\pi x}{L}\right)$$

so that

$$\frac{\hbar^2}{2m}\left(\frac{n\pi}{L}\right)^2\left(\sqrt{\frac{2}{L}}\sin\left(\frac{n\pi x}{L}\right)\right) = E \cdot \sqrt{\frac{2}{L}}\sin\left(\frac{n\pi x}{L}\right)$$

Solving for the energy, we find that

$$E = \frac{\hbar^2}{2m}\left(\frac{n\pi}{L}\right)^2 = \frac{n^2h^2}{8mL^2}$$

For the first five energy levels, the energies are (in units of $h^2/8mL^2$) 1, 4, 9, 16, and 25.

E3.9 Perform the integration for $\int_0^{2\pi} d\varphi \int_0^{\pi} \sin(\vartheta)\, d\vartheta$ and show that it has the value 4π.

Solution

The problem asks for the value of $\int_0^{2\pi} d\varphi \int_0^{\pi} \sin(\vartheta)\, d\vartheta$

$$\int_0^{2\pi} d\varphi \int_0^{\pi} \sin(\vartheta)\, d\vartheta = (2\pi) \cdot |-\cos\vartheta|_0^{\pi}$$
$$= (2\pi) \cdot (-(-1) - (-1)) = (2\pi) \cdot 2 = 4\pi$$

E3.10 Consider an electron that is confined to a potential energy well (or box) of length L which is enclosed on each side by a potential energy barrier of height V_0 (see Fig. 3.2). a) Sketch a wavefunction which could be a possible solution for this potential energy profile when $E < V_0$. (Note: Just focus on the qualitative behavior, rather than worrying about quantitative accuracy.) How does the wavefunction change as you decrease the mass of the particle? How does it change as you increase the mass of the particle? b) Sketch a wavefunction which could be a possible solution for this potential energy profile when $E > V_0$, but not much greater than V_0. How does the waveform change with the mass of the particle. Hint: The particle will not be confined to the well anymore, but its relative amounts of kinetic and potential energy will change in different regions of x.

Solution

a) Figure E3.10a shows a sketch of a wavefunction for a particle that is confined in a well (over the range $0 \leq x \leq L$) and is analogous to the $n = 3$ wavefunction in Fig. 3.2. Note that within the confines of the well, where $E > V$ the wavefunction oscillates, but outside of the well, where $E < V$, the wavefunction is damped and decays monotonically to zero.

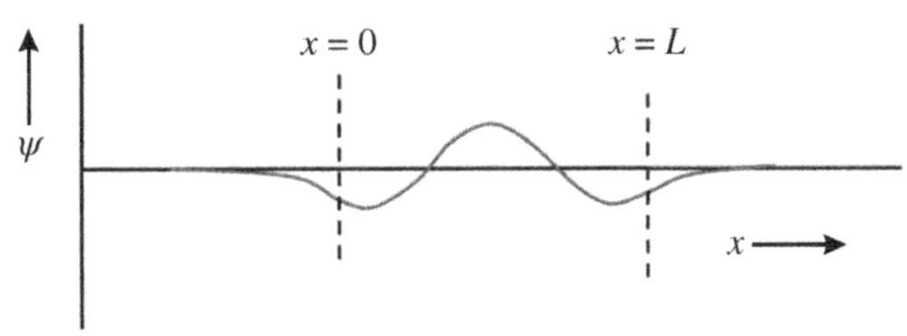

Figure E3.10a Sketch of a wavefunction for a particle confined in a well (kinetic energy smaller than the height of the walls).

We address the change in the wavefunction for the case of a fixed kinetic energy E and approximate the kinetic energy by

$$E = \frac{1}{2}mv^2 = \frac{p^2}{2m} = \frac{h^2}{2m\Lambda^2}$$

where the symbols have their usual meanings (m is the mass, h is Planck's constant, v is the speed, p is the momentum, and Λ is the de Broglie wavelength). If E is fixed, then as the mass m increases, the de Broglie wavelength Λ decreases in order to keep E constant. Hence as m increases, the wavefunction oscillates more and we would expect to see more nodes and a sharper curvature. Correspondingly, as the mass decreases, the wavelength increases; this means that the wavefunction has fewer nodes and spreads out more broadly in space.

b) Figure E3.10b shows a sketch of a wavefunction for a particle that is not confined to the well (over the range $0 \leq x \leq L$), but is affected by the well's presence. The wavefunction is sketched so that the wavelength is longer outside the well region and shorter inside the well region. This change is meant to illustrate the change in the particle's kinetic energy in the different regions; i.e., the kinetic energy is higher in the well region.

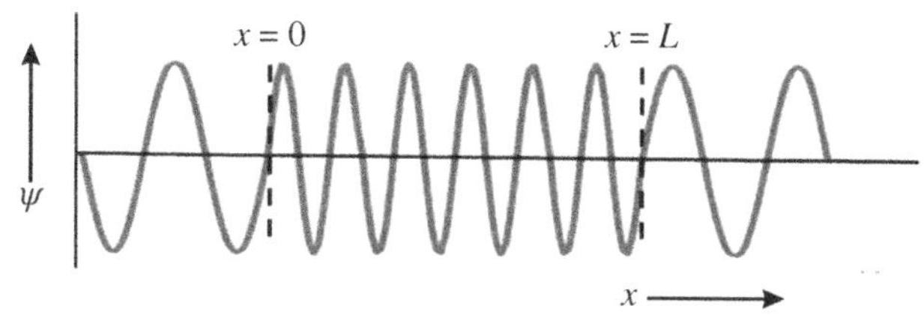

Figure E3.10b Sketch of a wavefunction for a particle that is not confined to the well (over the range $0 \le x \le L$), but is affected by the well's presence (kinetic energy greater than the height of the walls).

The change in the particle's de Broglie wavelength, for a fixed kinetic energy, will be the same as that discussed in part a). Higher mass leads to shorter de Broglie wavelength, and lower mass leads to longer de Broglie wavelength.

E3.11 Consider the situation where a particle of energy E and mass m is incident on a potential barrier of height V_0. Describe how the tunneling probability of a particle changes as you change a) the mass of the particle, b) the height of the barrier, as compared to the particle's energy (E/V_0), and c) the length of the barrier, L.

Solution

From Section 3.2.2.4 we know that the tunneling probability is given by

$$\frac{d\mathcal{P}(x = L)}{d\mathcal{P}(x = 0)} = \exp(-2bL)$$

where x is the displacement and L is the barrier width. The parameter b is given by

$$b = \frac{2\pi}{h}\sqrt{2m\left(V_0 - E\right)}$$

As the parameter b increases, the tunneling probability decreases exponentially. a) As the mass of the particle increases, the parameter b increases so that the tunneling probability decreases. b) As the energy E decreases with respect to V_0, then b increases and the tunneling probability decreases. c) As the width L of the barrier increases the tunneling probability decreases exponentially.

E3.12 Using the fact that the displacement along the circumference of a circle is $d\xi = r\,d\varphi$, transform the definition of the momentum operator to polar coordinates and show that the wavefunction $\psi = \frac{1}{\sqrt{2\pi}}\exp(+i\varphi)$ is one of its eigenfunctions with a momentum of h/Λ.

Solution

The displacement along the circumference of a circle is $d\xi = r\,d\varphi$; hence, the eigenvalue equation for p_{op} can be written as

$$p_{op}\psi = -i\hbar\frac{d\psi}{d\xi} = -\frac{i\hbar}{r}\frac{d\psi}{d\varphi} = p\cdot\psi$$

Substituting in the wavefunction $\psi = \frac{1}{\sqrt{2\pi}}\exp(+i\varphi)$ and using the circumference $L_c = 2\pi r$, we find

$$p_{op}\psi = -\frac{ih}{L_c}\frac{d\psi}{d\varphi} = -\frac{ih}{L_c}\frac{d\left(\frac{1}{\sqrt{2\pi}}\exp(+i\varphi)\right)}{d\varphi} = -\frac{ih}{L_c}\frac{1}{\sqrt{2\pi}}(i)\exp(i\varphi) = -\frac{ih}{L_c}(i)\psi = \frac{h}{L_c}\psi$$

Hence we see that the momentum eigenvalue is h/L_c. Making the correspondence with the de Broglie relation we find that the de Broglie wavelength Λ is L_c, as expected for this wavefunction. This wavefunction also corresponds to the doubly degenerate first excited state for which $\Lambda = L_c$ (see Section 3.5.2).

E3.13 Show that the wavefunction $\psi(\varphi) = Ae^{ik\varphi}$ is a solution of the Schrödinger Equation for an electron on a ring, namely

$$\mathcal{H}\psi = -\frac{\hbar^2}{2m_e r^2}\frac{d^2}{d\varphi^2}\psi = E\psi$$

Apply the boundary condition and show that the energy is quantized.

Solution
To show that a particular wavefunction is a solution to the Schrödinger equation, we show that the equality holds. Substituting $\psi(\varphi) = A\exp(ik\varphi)$ into the Schrödinger equation gives

$$\mathcal{H}\psi(\varphi) = -\frac{\hbar^2}{2m_e r^2}\frac{\partial^2(A\exp(ik\varphi))}{\partial\varphi^2} = E\cdot A\exp(ik\varphi)$$

or

$$E\cdot A\exp(ik\varphi) = -\frac{\hbar^2 ik}{2m_e r^2}\frac{\partial(A\exp(ik\varphi))}{\partial\varphi} = -\frac{\hbar^2 i^2 k^2}{2m_e r^2}A\exp(ik\varphi)$$

Solving for E, we find that

$$E = \frac{\hbar^2 k^2}{2m_e r^2}$$

Thus $\psi(\varphi) = A\exp(ik\varphi)$ is a solution to the differential equation. The boundary condition imposed for the particle on a ring is $\psi(\varphi) = \psi(\varphi + 2\pi)$ which on substitution is

$$\psi(\varphi) = A\exp(ik\varphi) = \psi(\varphi + 2\pi) = A\exp(ik(\varphi + 2\pi))$$

or $\exp(ik\varphi) = \exp(ik(\varphi + 2\pi))$ which can only be true if $k = 0, \pm 1, \pm 2, \ldots$
Upon substitution in the energy expression, we find that the energy

$$E = \frac{\hbar^2 k^2}{2m_e r^2}$$

is quantized since k is quantized.

E3.14 Normalize the wavefunction $\psi(\varphi) = A\sin(k\varphi)$, which is an energy eigenfunction of the Schrödinger Equation.

Solution
To normalize the wavefunction, we need to determine the value of the constant A in the wavefunction, $\psi(\varphi) = A\sin(k\varphi)$ by performing the integral $\int_{\text{all space}} \psi\cdot\psi\, d\tau$ and setting its value equal to unity. Thus,

$$1 = \int_0^{2\pi} A\sin(k\varphi)A\sin(k\varphi)\ d\varphi = A^2\int_0^{2\pi}\sin^2(k\varphi)\,d\varphi = A^2\left|\frac{1}{2}\varphi - \frac{1}{4k}\sin(2k\varphi)\right|_0^{2\pi} = \pi A^2$$

or $A = 1/\sqrt{\pi}$.

E3.15 Classically, the angular momentum $\vec{L}$ of a mass m moving on a circle of radius r with speed v has a magnitude of $\left|\vec{L}\right| = m\cdot v\cdot r$. Consider the case where the mass is the mass of an electron m_e and the speed is given by the de Broglie relation $v = h/(m_e\Lambda)$. By using the boundary condition that $\psi(\varphi + 2\pi) = \psi(\varphi)$, show that the angular momentum is quantized.

Solution
The boundary condition provides a constraint that only allows certain de Broglie wavelengths to form standing waves on the ring. As discussed in the text, one trivial example is Λ equal to a constant. In the other cases, the circumference $2\pi r$ must be a multiple of the wavelength Λ.

$$n\Lambda = 2\pi r \quad \text{where} \quad n = 1, 2, 3\ldots$$

Quantization of Λ implies quantization of $\left|\vec{L}\right|$. First, we show that this result for Λ implies that the speed is quantized.

$$v = \frac{h}{m_e\Lambda} = \frac{h}{m_e}\frac{n}{2\pi r}$$

Then we show that it leads to quantization of $\left|\vec{L}\right|$,

$$\left|\vec{L}\right| = m_e\cdot\left(\frac{h}{m_e}\frac{n}{2\pi r}\right)\cdot r = \frac{h}{2\pi}\cdot n = n\cdot\hbar \quad n = 1, 2, 3, \ldots$$

What about the case where Λ is equal to a constant? This corresponds to a very long wavelength, so that the velocity approaches zero. In this limit $\left|\vec{L}\right|$ approaches zero. We can incorporate this into our result above by letting $n = 0$ correspond to this special case, and we find that

$$\left|\vec{L}\right| = n \cdot \hbar \quad n = 0, 1, 2, 3, \ldots$$

E3.16 Use the orthogonality condition to claim that the integral for a product of cosines is zero when the odd integers n and j are not equal to each other; that is

$$\int_{-L/2}^{L/2} \cos\left(\frac{\pi}{L}jx\right) \cdot \cos\left(\frac{\pi}{L}nx\right) \, dx = 0$$

for $n \neq j$. Show that this integral is zero by explicitly evaluating it.

Solution

To begin we make the substitution that $u = \pi x/L$, which lets us write the integral as

$$\int_{-L/2}^{L/2} \cos\left(\frac{\pi}{L}jx\right) \cdot \cos\left(\frac{\pi}{L}nx\right) \, dx = \frac{L}{\pi}\int_{-\pi/2}^{\pi/2} \cos(ju) \cdot \cos(nu) \, du$$

From Appendix B, we use the integral

$$\int \cos nx \cos kx \cdot dx = \frac{1}{2} \cdot \left[\frac{\sin(n+k)x}{n+k} + \frac{\sin(n-k)x}{n-k}\right] \quad \text{for} \quad k^2 \neq n^2$$

to recast our integral as

$$\begin{aligned}\frac{L}{\pi}\int_{-\pi/2}^{\pi/2} \cos(ju) \cdot \cos(nu) \, du &= \frac{1}{2} \cdot \left[\frac{\sin(n+j)u}{n+j} + \frac{\sin(n-j)u}{n-j}\right]_{-\pi/2}^{\pi/2} \\ &= \frac{1}{2}\left(\frac{\sin\left(\frac{(n+j)\pi}{2}\right)}{n+j} + \frac{\sin\left(\frac{(n-j)\pi}{2}\right)}{n-j} - \left[-\frac{\sin\left(\frac{(n+j)\pi}{2}\right)}{n+j} - \frac{\sin\left(\frac{(n-j)\pi}{2}\right)}{n-j}\right]\right) \\ &\quad - \frac{\sin\left(\frac{(n+j)\pi}{2}\right)}{n+j} + \frac{\sin\left(\frac{(n-j)\pi}{2}\right)}{n-j}\end{aligned}$$

From the problem statement, we know that n and j are odd, which implies that $n+j$ is even and $n-j$ is even. If $n+j$ and $n-j$ are even then the arguments of the sine functions are a multiple of π which makes the sine functions equal to zero. Thus, we have that

$$\frac{L}{\pi}\int_{-\pi/2}^{\pi/2} \cos(ju) \cdot \cos(nu) \, du = 0$$

for j and n odd.

E3.17 For the potential well problem in Foundation 3.1 derive the result for the left hand side of the well.

Solution

In this case, we are considering the symmetric solutions of the Schrödinger equation. Recall that in the well we had

$$\psi_{\text{well}} = A \cdot \cos(ax) \text{ and } E = \frac{h^2}{8\pi^2 m_e}a^2$$

and on the right side of the well ($x > L/2$) we had

$$\psi_{\text{right}} = B e^{-bx} \text{ with } E = -\frac{h^2}{8\pi^2 m_e}b^2 + V_0 \text{ or } a^2 = \frac{8\pi^2 m_e}{h^2}V_0 - b^2$$

On grounds of symmetry, we use

$$\psi_{\text{left}} = B \cdot e^{bx}$$

for the wavefunction in the region $x < -L/2$ (left region). Our matching condition for the wavefunction at $x < -L/2$ (left region) is

$$\psi_{\text{well}}|_{x=-L/2} = \psi_{\text{leftt}}|_{x=-L/2}$$

and this gives the result that

$$A \cos\left(\frac{-aL}{2}\right) = B\mathrm{e}^{-bL/2}$$

Because the cosine is an even function, we see that our condition becomes

$$A \cos\left(\frac{aL}{2}\right) = B\mathrm{e}^{-bL/2}$$

which is identical to the corresponding equation in the Foundation. The matching condition for the wavefunction's slope is

$$\left(\frac{\mathrm{d}\psi_{\text{well}}}{\mathrm{d}x}\right)_{x=-L/2} = \left(\frac{\mathrm{d}\psi_{\text{left}}}{\mathrm{d}x}\right)_{x=-L/2}$$

and this gives the result that

$$-A\, a \sin\left(\frac{-aL}{2}\right) = B\, b\, \mathrm{e}^{-bL/2}$$

which rearranges to give Equation (3.105)

$$B = A\,\frac{a}{b} \sin\left(\frac{aL}{2}\right) \mathrm{e}^{bL/2}$$

Combining these two results we see that they give the same relationship between b and a, as we found for the matching on the right side of the well; namely

$$b = a \cdot \tan\frac{aL}{2}$$

E3.18 Verify that the wavefunction $\psi = \sqrt{\frac{2}{(n_{\max+1})}}\sqrt{\frac{2}{L}}\sum_{n=1}^{n_{\max}} \cos\left(\frac{\pi n x}{L}\right)$ is normalized for the case that $n_{\max} = 2$.

Solution

Equation (3.137) in Foundation 3.5 is

$$\psi = \sqrt{\frac{2}{(n_{\max}+1)}}\sqrt{\frac{2}{L}}\sum_{n=1}^{n_{\max}} \cos\left(\frac{\pi}{L}nx\right)$$

where $n = 1, 3, 5, \ldots, n_{\max}$. In the case that $n_{\max} = 2$ we write ψ as

$$\psi = \sqrt{\frac{2}{3}}\sqrt{\frac{2}{L}}\left[\cos\left(\frac{\pi}{L}x\right) + \cos\left(\frac{\pi}{L}2x\right)\right]$$

Thus

$$\int_{x=0}^{x=L} \psi^2 \cdot \mathrm{d}x = \frac{2}{3}\frac{2}{L}\int_{x=0}^{x=L}\left[\cos\left(\frac{\pi}{L}x\right) + \cos\left(\frac{\pi}{L}2x\right)\right]^2 \cdot \mathrm{d}x$$

$$= \frac{4}{3L}\left[\int_{x=0}^{x=L} \cos^2\left(\frac{\pi}{L}x\right) \cdot \mathrm{d}x + 2\int_{x=0}^{x=L} \cos\left(\frac{\pi}{L}x\right) \cdot \cos\left(\frac{\pi}{L}2x\right) \cdot \mathrm{d}x + \int_{x=0}^{x=L} \cos^2\left(\frac{\pi}{L}2x\right) \cdot \mathrm{d}x\right]$$

For the three integrals we obtain

$$\int_{x=0}^{x=L} \cos^2\left(\frac{\pi}{L}x\right) \cdot \mathrm{d}x = \frac{L}{\pi}\frac{\pi}{2} = \frac{L}{2}$$

$$\int_{x=0}^{x=L} \cos\left(\frac{\pi}{L}x\right) \cdot \cos\left(\frac{\pi}{L}2x\right) \cdot \mathrm{d}x = 0$$

$$\int_{x=0}^{x=L} \cos^2\left(\frac{\pi}{L}2x\right) \cdot \mathrm{d}x = \frac{L}{2\pi}\frac{\pi}{2} = \frac{L}{4}$$

and it follows that

$$\int_{x=0}^{x=L} \psi^2 \cdot dx = \frac{4}{3L} \cdot \left[\frac{L}{2} + \frac{L}{4}\right] = 1$$

E3.19 In this exercise, you are asked to combine basic wave–particle duality ideas from Chapters 1 and 2 with our results for the electron-in-a-box model to estimate the size of a sodium atom. While you are expected to use qualitative reasoning, you also need to perform mathematical calculations and compare values with each other. 1) The excited Na atom has a strong yellow emission wavelength of 589 nm. Assuming that this emission corresponds to relaxation of an electron from the lowest excited state to the ground state, estimate the energy gap between the lowest excited state and the ground state. 2) Using this energy gap and the one-dimensional (1D) electron-in-a-box model, estimate the size of a Na atom. Be sure to account for the fact that Na has 11 electrons and use the Aufbau principle to fill the box model's electronic energy levels. 3) Perform the estimates/calculations in part 2) for a three-dimensional (3D) electron-in-a-box model, also. Compare your results with each other and with literature estimates of the Na atom's size.

Solution

1) For the energy gap calculation we use the formula $\Delta E = h\nu = (hc_0)/\lambda$; and we find that

$$\Delta E = \frac{6.626 \times 10^{-34}\ \mathrm{J\ s} \cdot 2.998 \times 10^{8}\ \mathrm{m\ s^{-1}}}{589 \times 10^{-9}\ \mathrm{m}} = 3.37 \times 10^{-19}\ \mathrm{J} = 2.11\ \mathrm{eV}$$

2) For the one-dimensional electron-in-a-box model, the energy formula is $E = (n^2h^2)/(8m_eL^2)$. If we follow the Aufbau ("building-up") principle and place two electrons (one spin up and one spin down) in each energy level, then the first 5 energy levels are filled and there is one valence electron in the $n = 6$ energy level. Thus the transition giving rise to the emission would be from $n = 7$ to $n = 6$; and this energy will be given by

$$\Delta E = \frac{h^2}{8m_eL^2}(7^2 - 6^2) = \frac{13h^2}{8m_eL^2}$$

Assuming that this energy gap corresponds to the $\Delta E = 3.37 \times 10^{-19}$ J = 2.11 eV from above, we can solve this equation for L and find

$$L = \sqrt{\frac{13h^2}{8m_e\Delta E}} = \sqrt{\frac{13(6.626 \times 10^{-34}\ \mathrm{J\ s})^2}{8(9.109 \times 10^{-31}\ \mathrm{kg}) \cdot 3.37 \times 10^{-19}\ \mathrm{J}}} = 1.5 \times 10^{-9}\ \mathrm{m} = 1.5\ \mathrm{nm}$$

A length of 1.5 nm would correspond to a "radius" of 0.7 nm or 7 Angstrom. This value is of the same order of magnitude (albeit, higher by three times), as the value of 2.3 Angstroms reported as the van der Waals radius of Na.

3) For a three-dimensional cubic box, the energy level formula is given by

$$E = \frac{h^2}{8m_eL^2}(n_x^2 + n_y^2 + n_z^2) \text{ where } n_x, n_y, n_z = 1, 2, 3, 4, 5, \ldots.$$

Assuming that the electron transition occurs between the first set of unoccupied energy levels and the set of singly occupied energy levels; e.g., from $(n_x = 1, n_y = 1, n_z = 3)$ to $(n_x = 1, n_y = 2, n_z = 2)$. See Fig. 3.7. This transition has an energy gap of

$$\Delta E = \frac{h^2}{8m_eL^2}(11 - 9) = \frac{2h^2}{8m_eL^2}$$

so that

$$L = \sqrt{\frac{2h^2}{8m_e\Delta E}} = \sqrt{\frac{2(6.626 \times 10^{-34}\ \mathrm{J\ s})^2}{8 \cdot 9.109 \times 10^{-31}\ \mathrm{kg} \cdot 3.37 \times 10^{-19}\ \mathrm{J}}} = 0.58 \times 10^{-9}\ \mathrm{m} = 0.58\ \mathrm{nm}$$

which gives a "radius" of 0.29 nm or 2.9 angstroms; much closer to the van der Waals radius than the value obtained by the one-dimensional box model.

3.2 Problems

P3.1 Show that the wavefunction $\psi(x) = A\sin(kx)$ satisfies the Schrödinger equation for an electron in a one-dimensional box. Consider A and k to be constants that must be determined. Determine the constant k by requiring the wavefunction to have the value zero at $\psi(L) = 0$, and verify that the wavefunction has the value zero at $x = 0$. Determine the constant A by applying the normalization condition, Equation (3.44).

Solution

The Schrödinger equation for the one-dimensional particle in a box is

$$\left(-\frac{\hbar^2}{2m_e}\frac{d^2}{dx^2} + 0\right)\psi = E\psi$$

We are asked to show that $\psi(x) = A\sin(kx)$ is a solution. The first and second derivatives of ψ with respect to x are

$$\frac{d\psi(x)}{dx} = Ak\cos(kx) \qquad \text{and} \qquad \frac{d^2\psi(x)}{dx^2} = -Ak^2\sin(kx)$$

Upon substitution into the Schrödinger equation we find that

$$-\frac{\hbar^2}{2m_e}\frac{d^2\psi(x)}{dx^2} = -\frac{\hbar^2}{2m_e}\left[-Ak^2\sin(kx)\right] = \frac{\hbar^2}{2m_e}k^2 \cdot \psi(x)$$

with the energy identified by

$$E = \frac{\hbar^2}{2m_e}k^2$$

Thus the Schrödinger equation is satisfied by this wavefunction. Now we apply the boundary condition that $\psi(x)|_{x=L} = 0$, thus

$$\psi(L) = A\sin(kL) = 0$$

Because $A = 0$ is not an interesting solution; i.e., for $A = 0$ the wavefunction is zero everywhere and no particle is present, $A \neq 0$ and the boundary condition requires that $kL = n\pi$ or $k = n\pi/L$. Then the wavefunction and the energy become

$$\psi(x) = A\sin(kx) = A\sin\frac{n\pi}{L}kx \qquad \text{and} \qquad E = \frac{\hbar^2}{2m_e}k^2 = \frac{\hbar^2}{2m_e}\left(\frac{n\pi}{L}\right)^2 = \frac{h^2}{8mL^2}n^2$$

The value of A may be determined by applying the normalization condition. Doing so we find that

$$1 = \int_0^L A^2\sin^2\left(\frac{n\pi x}{L}\right)\,dx = A^2\int_0^L \sin^2\left(\frac{n\pi x}{L}\right)\,dx$$

Using the integral $\int \sin^2 x\,dx = \frac{1}{2}\left(x - \frac{1}{2}\sin(2x)\right)$ from Appendix B and making the substitution of $z = n\pi x/L$, we find that

$$1 = A^2\frac{L}{n\pi}\int_0^{n\pi}\sin^2(z)\,dz = A^2\frac{L}{n\pi}\left(\frac{z}{2} - \frac{1}{4}\cos(2z)\right)_0^{n\pi}$$
$$= A^2\frac{L}{n\pi}\left(\frac{n\pi}{2} - \frac{1}{4} - 0 + \frac{1}{4}\right) = A^2\frac{L}{2} \quad \text{so that } A = \sqrt{2/L}$$

Hence the normalized wavefunction may be written as

$$\psi(x) = \sqrt{\frac{2}{L}}\sin\left(\frac{n\pi x}{L}\right)$$

P3.2 Consider the electron-in-a-box system like that shown in Fig. 3.2. Consider the length of the box to be 250 pm and the potential energy V_0 to be 14 eV = 2.2×10^{-18} J. These numbers are chosen to mimic the case of a hydrogen atom. Consider the electron in this system to be in the ground-state energy level and imagine performing a photoemission experiment in which you measure the kinetic energy of the electron as a function of the excitation light source's frequency. a) Make a sketch that illustrates your observations; i.e., measured data. b) Draw a sketch of an experimental apparatus that shows the essential components needed in making such a measurement. c) Explain in words how this experiment can demonstrate the particle nature of light. d) Describe how your results would change if the electron was located in the $n = 2$ energy level, rather than the $n = 1$ energy level.

Solution

a) First we calculate/determine the energy of the bound electron, within the model defined in the problem. As shown in Fig. 3.2, we expect this energy to be shifted somewhat from that calculated for the infinitely high walls. Following the discussion in Box 3.1, we find this value by considering the function

$$y = w^2 + w^2 \cdot \tan^2\frac{w}{2} - \frac{8m_e\pi^2L^2}{h^2}V_0$$

and solve the transcendental equation

$$w_1^2 + w_1^2 \cdot \tan^2\frac{w_1}{2} - \frac{8m_e\pi^2L^2}{h^2}V_0 = 0$$

for w_1, see Fig. P3.2a.

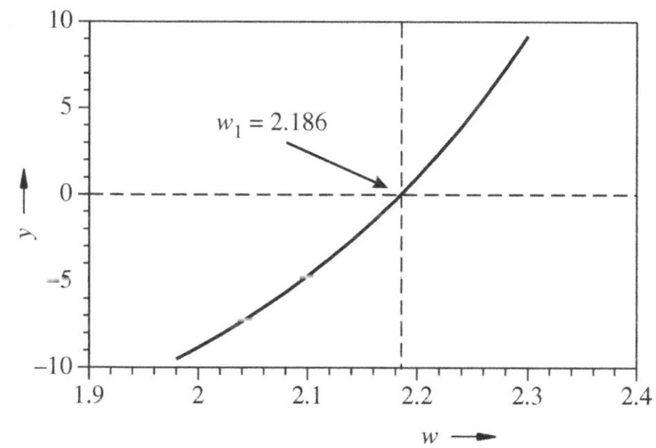

Figure P3.2a y as a function of w for $L = 250$ pm and $V_0 = 2.2 \times 10^{-18}$ J.

We use this value $w = 2.186$ in the energy expression

$$E = \frac{h^2}{8m_eL^2}\left(\frac{w_1}{\pi}\right)^2$$

to find that

$$E = \frac{(6.626 \times 10^{-34}\,\text{J-s})^2}{8\,(9.1094 \times 10^{-31}\,\text{kg})\,(250 \times 10^{-12}\,\text{m})^2}\left(\frac{2.186}{3.14159}\right)^2 = 4.67 \times 10^{-19}\,\text{J} = 2.92\,\text{eV}$$

Figure P3.2b shows two plots. The first panel gives a plot of the expected electron current as a function of electron kinetic energy if a photon energy of 21.2 eV is used. The second panel shows a plot of the observed electron kinetic energy versus the excitation photon wavelength that would be expected.

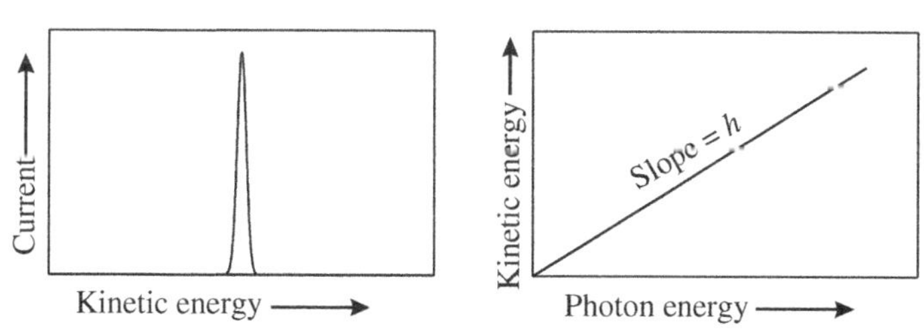

Figure P3.2b Electric current versus the kinetic energy of the photon (left) and kinetic energy versus the photon energy (right).

b) Figure P3.2c shows some of the essential elements of an apparatus. This sketch is similar to Fig. 11.39 of the text; however, the light source is tunable.

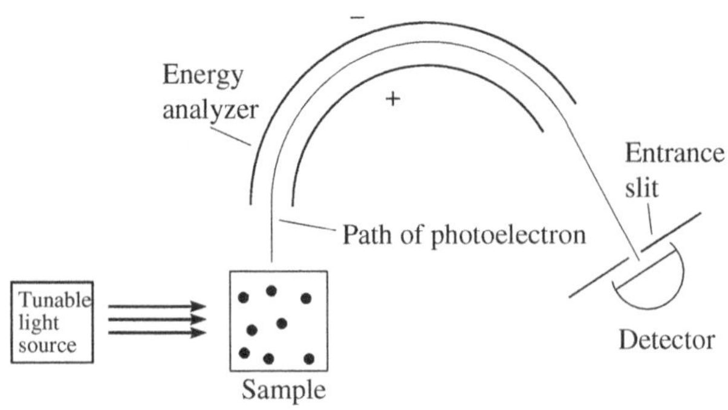

Figure P3.2c Apparatus for measuring the kinetic energy of an electron emitted by a collision with a photon.

c) The plot of measured electron kinetic energy versus the photon frequency shows that there is a direct proportionality between these two quantities, and the proportionality constant (slope of the plot) should be h. One could also change the intensity of the light source to illustrate that the electron kinetic energy does not change in this case. These findings are consistent with the hypothesis that light energy is absorbed in packets of $h\nu$.

d) If the electron were located in the $n = 2$ level, rather than the $n = 1$ level, we would see a linear shift of the kinetic energy; it would shift higher because the binding energy would be lower. The slope of the graph would still be h, however.
NOTE: To better mimic an H-atom, we should use 2×78 pm = 156 pm for the atomic diameter and choose a V_0 value of 21 eV so that the electron's binding energy is about 13.5 eV.

P3.3 Use a data sheet program (e.g., Excel) for this problem. a) Plot the ground electronic-state wavefunction for an electron confined to a box of length 100 pm. Plot the square of this wavefunction. b) Since the square of the wavefunction in a region of space represents the probability of detecting an electron in that region of space and our wavefunction is confined to the box, the probability of finding the electron somewhere in the box should be unity, or 100%. Determine what numerical factor you need to multiply your wavefunction by so that the total probability is unity. This numerical factor is called the normalization factor. c) In your data sheet program, change the box length and explore whether the normalization factor changes as you change the size of the box. d) You can also perform this procedure analytically by taking the integral of the square of the electron wavefunction and requiring it to be equal to one; namely

$$\int_0^L \psi^2(x)\,dx = 1$$

Perform this procedure for the wavefunction (the Mathematical Appendix may be useful) and determine the normalization factor. e) Show that your analytical result agrees with your calculated result.

Solution

a) Figure P3.3a plots the wavefunction (dashed line) and the square of the wavefunction (solid line) for an unnormalized ground-state wavefunction $f(x)$ representing an electron in a 100 pm box; namely

$$f(x) = \sin(\pi x/L) \quad \text{where} \quad L = 100\,\text{pm}$$

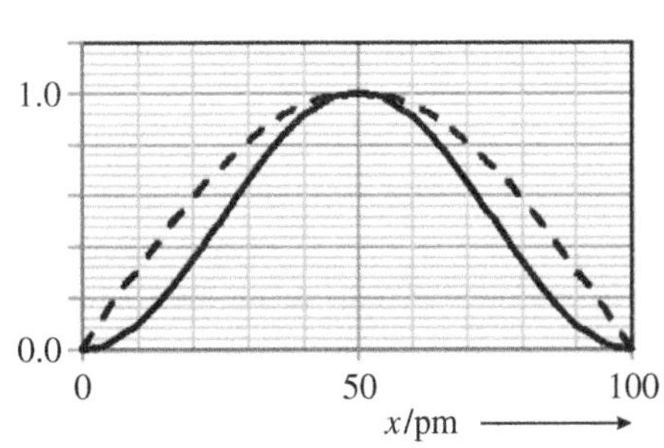

Figure P3.3a Plot of ground-state wavefunction (dashed curve) and its square (solid curve).

b) To determine the numerical factor for normalization, we first calculate the area under the curve for the square of the wavefunction (the probability density $f^2(x)$) and obtain a value of 50 pm. To normalize the probability to 1, we divide $f^2(x)$ by 50 pm. Hence our normalized wavefunction $\psi(x)$ will be

$$\psi(x) = \frac{f(x)}{\sqrt{50\,\text{pm}}} = 0.1414\,\text{pm}^{-1/2} \cdot \sin\frac{\pi x}{L}$$

This agrees with the expression

$$\psi(x) = \sqrt{\frac{2}{L}} \cdot \sin\frac{\pi x}{L} = \sqrt{\frac{2}{100\,\text{pm}}} \cdot \sin\frac{\pi x}{L} = 0.1414 \cdot \text{pm}^{-1/2} \cdot \sin\frac{\pi x}{L}$$

c) Using an Excel sheet we evaluate the area and determine the normalization for seven length values in the range of 100 to 1000 pm. The values are reported in the table. Figure P3.3b shows a plot of these factors and a best fit to a square root dependence (as predicted by the analytical result, see part d).

L /pm	100	200	300	400	600	800	1000
Normalization	0.1414	0.1000	0.0816	0.0707	0.0577	0.0500	0.0447

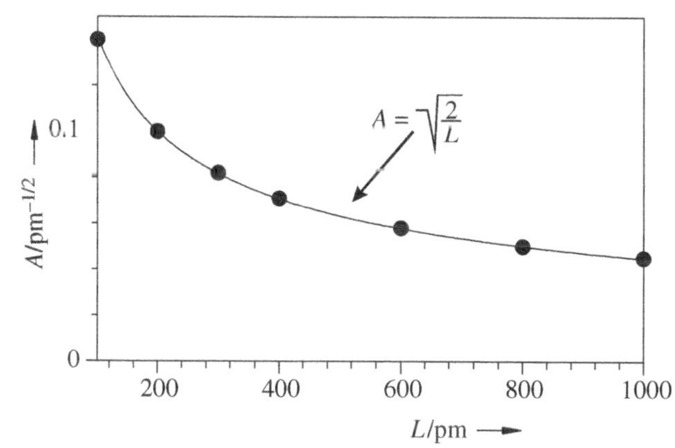

Figure P3.3b Normalization factor for one-dimensional box function. Full circles: Excel sheet evaluation, solid line: analytical calculation.

d) To normalize the wavefunction we require that

$$1 = \int_0^L A^2 \sin^2(\pi x/L)\,dx = A^2\frac{L}{\pi}\int_0^L \sin^2(z)\,dz$$

From Appendix B we know that $\int \sin^2 x\,dx = \frac{1}{2}\left(x - \frac{1}{2}\sin(2x)\right)$, so that

$$1 = A^2\frac{L}{\pi}\int_0^\pi \sin^2(z)\,dz = A^2\frac{L}{\pi}\left[\frac{1}{2}\left(z - \frac{1}{2}\sin(2z)\right)\right]_0^\pi$$
$$= A^2\frac{L}{\pi}\left[\frac{1}{2}\left(\pi - \frac{1}{2}\sin(2\pi)\right) - \frac{1}{2}\left(0 - \frac{1}{2}\sin(0)\right)\right] = A^2\frac{L}{2} \text{ or } A = \sqrt{2/L}$$

e) The graph in Fig. P3.3b shows that the square root dependence, predicted by the analytical result, matches the computational result.

P3.4 A scanning tunneling microscope (STM) uses the quantum-mechanical tunneling of electrons between a sample and a metallic tip to image the surface of a sample. With this device it is possible to achieve a spatial resolution which is of atomic dimension; *ca.* 100 pm.

a) If 100 pm is the uncertainty in an electron's position, what is the minimum uncertainty in its momentum?

b) Suppose you wish to energy resolve the electrons using your STM, in order to measure a spectrum and a spatial image simultaneously. How well can you determine the kinetic energy of 1 eV electrons before the spatial resolution is degraded to 500 pm?

To simplify, perform your estimates assuming a one-dimensional model.

Solution

a) The uncertainty principle states that $\Delta x\ \Delta p \geq \hbar/2$. Substituting for the uncertainty in the position as given in the problem (100 pm) gives

$$\Delta p = \frac{\hbar/2}{\Delta x} = \frac{\hbar}{2 \cdot \Delta x} = \frac{1.054 \times 10^{-34}\ \text{J s}}{2 \cdot 100 \times 10^{-12}\ \text{m}} = 5.27 \times 10^{-25}\ \text{kg m s}^{-1}$$

b) Similarly, Δp for a 500 pm resolution is 1.06×10^{-25} kg m s^{-1}. If we write the kinetic energy as $E = p^2/2m_e$, then

$$\sigma_E^2 = \left(\frac{p}{m_e}\right)^2 \sigma_p^2 = \frac{2E}{m_e} \cdot \sigma_p^2$$

so that

$$\Delta E \approx \sigma_E = \sqrt{\frac{2E}{m_e}} \cdot \sigma_p$$

and

$$\Delta E = \sqrt{\frac{2\left(1.602 \times 10^{-19}\ \text{J}\right)}{\left(9.11 \times 10^{-31}\ \text{kg}\right)}} \left(1.06 \times 10^{-25}\ \text{kg m s}^{-1}\right) = 0.65 \times 10^{-19}\ \text{J} = 0.39\ \text{eV}$$

P3.5 In the 1980s chemical physicists performed experiments in which Na atoms are trapped in a laser field. In this experiment, the Na atoms are "stopped" with photons so that their kinetic energy is very small, $\overline{E} \sim 0$. A typical value for the residual uncertainty in the kinetic energy is $\Delta E = 2.0 \times 10^{-28}$ J; i.e., $\overline{E} = 0 \pm 2.0 \times 10^{-28}$ J. Use this uncertainty in the kinetic energy to estimate the uncertainty in an atom's momentum Δp. Subsequently use this Δp to estimate the uncertainty in the atom's position. Express your answer in Bohr radii.
Given that the density of solid Na is 0.97 g/cm^3, use this number to estimate the size of a Na atom. Express your answer in Bohr radii.
Compare the uncertainty in the Na atom's position in the laser trapping experiment to its characteristic size in the solid.

Solution

The first part of the problem uses the uncertainty relation to determine an approximate size for the Na atom. Using the fact that $E = p^2/2m$, we can write the energy uncertainty as

$$\Delta E = \frac{\Delta\left(p^2\right)}{2m} \quad \text{or} \quad \sqrt{\Delta\left(p^2\right)} = \sqrt{2m \cdot \Delta E}$$

If we make the approximation that $\Delta p \simeq \sqrt{\Delta\left(p^2\right)}$, we can write that

$$\Delta p = \sqrt{2m \cdot \Delta E} = \sqrt{2\ \left(3.818 \times 10^{-26}\ \text{kg}\right)\ \left(2.0 \times 10^{-28}\ \text{J}\right)} = 3.9 \times 10^{-27}\ \text{kg m s}^{-1}$$

The uncertainty principle tells us that $\Delta p \cdot \Delta r = \hbar/2 = 5.27 \times 10^{-35}$ J s, which leads to

$$\Delta r \geq \frac{5.27 \times 10^{-35}\ \text{J s}}{3.9 \times 10^{-27}\ \text{kg m s}^{-1}} = 1.4 \times 10^{-8}\ \text{m} = 13.5\ \text{nm} = 255\ a_0$$

The second part of the problem uses the density of solid Na to determine the size of the Na atom. The volume associated with a single Na atom can be determined from the density as

$$V = \frac{m}{\rho} = \frac{\mathbf{M}}{\mathbf{N}_A \rho} = \frac{22.99\ \text{g mol}^{-1}}{6.022 \times 10^{23}\ \text{mol}^{-1} \cdot 0.97\ \text{g cm}^{-3}} = 3.9 \times 10^{-23}\ \text{cm}^3$$

we can approximate the atom as spherical to give a radius of

$$r = \sqrt[3]{\frac{3V}{4\pi}} = 0.21\text{ nm}$$

The uncertainty in the atom's position in the laser stopping experiment is almost 70 times larger than the physical size of the atom that is obtained from the solid's density.

P3.6 This problem requires a spreadsheet program. Use particle in the box wavefunctions with $L = 500$ pm. First, create plots of each wavefunction for the five lowest energy levels over the range of 0 to L. Second, create a superposition (i.e., sum them) of the two lowest wavefunctions and plot it. Third, create a superposition of the three lowest wavefunctions and plot it. Proceed in this manner until you have a superposition state that includes the first five wavefunctions. In addition to these wavefunction plots, plot the probability densities in each case. Describe how the probability density changes with the number of stationary state wavefunctions that you include in the superposition.

Solution

Figure P3.6a gives the normalized particle-in-a-box wavefunctions. In each panel the gray curve is for the $n = 1$ ground state. In the left panel the solid black curve is the $n = 2$ state and the dashed black curve is the $n = 3$ state. In the right panel the solid black curve is the $n = 4$ state and the dashed black curve is the $n = 5$ state.

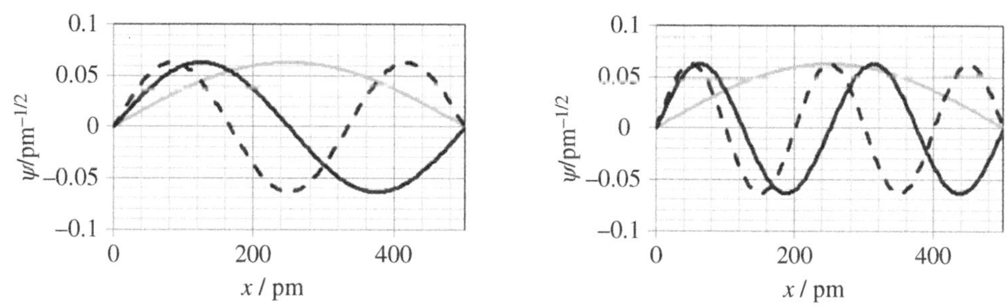

Figure P3.6a Normalized particle-in-a-box wavefunctions for $n = 1$, $n = 2$, and $n = 3$ (left) and $n = 1$, $n = 3$ and $n = 5$ (right).

Next we create a superposition state by summing the two wavefunctions and normalizing the sum (i.e., we divide the sum by the $\sqrt{2}$). Figure P3.6b shows a plot of the superposition function (curve 3) of the functions with $n = 1$ (curve 1) and $n = 2$ (curve 2). It is evident that the superposition state has enhanced amplitude on the left side of the box and decreased amplitude on the right side of the box; it is asymmetric.

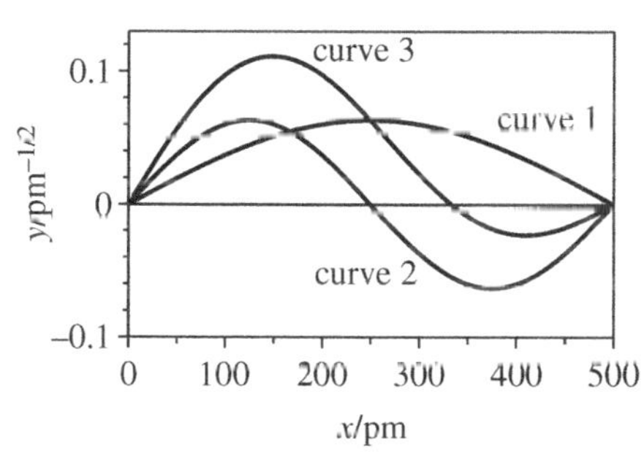

Figure P3.6b Plot of the superposition function (curve 3) of the functions with $n = 1$ (curve 1) and $n = 2$ (curve 2).

Figure P3.6c shows superposition plots that were created for the first five quantum states (functions with $n = 1$ to $n = 5$). The left panel shows a plot of the wavefunction amplitude for equally weighted superpositions of the first five quantum states. Note that the wavefunction is normalized. The right panel shows plots for the probability density of the superposition state. Notice that the probability density is localized mostly on the left side of the box.

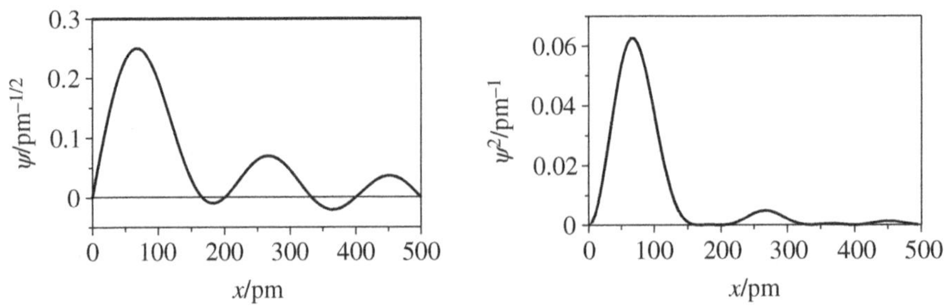

Figure P3.6c The left panel shows an amplitude plot for equally weighted superpositions of the first 5 quantum states (functions with $n = 1$ to $n = 5$). The right panel shows a plot for the corresponding probability density.

P3.7 Problem 2.3 shows that the superposition of two waves can give rise to a standing wave. It is common to describe a complicated electronic state as a superposition of the system's standing wave states. Use the MathCAD exercise, titled PinBsuperposition.mcdx, to explore how the superposition of a particle in the box standing waves can give rise to a diverse array of waveforms. Answer the following questions posed in the worksheet

a) How does the spread of the wavefunction in space relate to the number of contributing eigenstates in the four superposition states you are asked to investigate? How is the spread of the momentum influenced by the number of states?
b) Comment on how the spread of the wavefunction in position and momentum space changes for the superposition states. Is the Heisenberg Uncertainty principle ever violated? What is the uncertainty product for an eigenstate? You can check this by creating eigenstates rather than superposition states. Does it depend on the eigenstate's quantum number?
c) Comment on the appearance of the superposition states. Are the probability densities symmetric ? How does the symmetry of the superposition state depend on the coefficients?
d) Create a superposition state from only odd quantum numbers. Create a superposition state from only even quantum numbers. Plot the probability densities of each state. Plot the wavefunctions of each state. Verbally describe and explain the differences between these states.

Solution

The solution to this worksheet is given in the worksheet "PinBsuperposition_solution.mcd." Here we answer the questions asked in the worksheet for the default setting of the box length at $L = 1.0$ nm.

a) We see from the table that as the number of eigenstates contributing to the superposition state increases that the uncertainty in the position decreases; i.e., this means that the wavepacket becomes narrower. In contrast, the uncertainty in the momentum increases; i.e., our superposition state has more de Broglie wavelengths (more momenta) contributing as we increase the number of eigenstates in the superposition.
Now examine the results: for each of the above examples, we will look at $\Delta x \cdot \Delta p$. Notice how the product changes for our four example superposition states. The superposition states are defined by

$$\Psi_{\text{superposition}} = \sum_n c_n \cdot \sqrt{\frac{2}{L}} \sin\left(\frac{n\pi x}{L}\right)$$

State	c_1	c_2	c_3	c_4	c_5	c_6	c_7	c_8	c_9	c_{10}
1	$\sqrt{0.8}$	$\sqrt{0.1}$	$\sqrt{0.1}$							
2	$\sqrt{0.5}$	$\sqrt{0.2}$	$\sqrt{0.2}$	$\sqrt{0.1}$						
3	$\sqrt{0.182}$	$\sqrt{0.182}$	$\sqrt{0.136}$	$\sqrt{0.136}$	$\sqrt{0.091}$	$\sqrt{0.091}$	$\sqrt{0.091}$	$\sqrt{0.091}$		
4	$\sqrt{0.1}$	$\sqrt{0.1}$	$\sqrt{0.1}$	$\sqrt{0.1}$	$\sqrt{0.1}$	$\sqrt{0.1}$	$\sqrt{0.1}$	$\sqrt{0.1}$	$\sqrt{0.1}$	$\sqrt{0.1}$

The uncertainties and uncertainty products for these states are:

state1	$\Delta x_1 = 2.065 \cdot L$	$\Delta p_1 = 0.455 \cdot \hbar/L$	$\Delta x_1 \cdot \Delta p_1 = 0.94 \cdot \hbar$
state2	$\Delta x_2 = 1.616 \cdot L$	$\Delta p_2 = 0.681 \cdot \hbar/L$	$\Delta x_2 \cdot \Delta p_2 = 1.10 \cdot \hbar$
state3	$\Delta x_3 = 1.241 \cdot L$	$\Delta p_3 = 1.41 \cdot \hbar/L$	$\Delta x_3 \cdot \Delta p_3 = 1.75 \cdot \hbar$
state4	$\Delta x_4 = 1.170 \cdot L$	$\Delta p_4 = 1.95 \cdot \hbar/L$	$\Delta x_4 \cdot \Delta p_4 = 2.28 \cdot \hbar$

b) As stated in part (a) the position becomes better defined and the momentum less well defined for the superposition states used. For the $n = 1$ eigenstate the value of the uncertainty product is $0.568\ \hbar$, very near the theoretical limit of $\hbar/2$. The uncertainty relation is not violated.
The value of the uncertainty product increases as the quantum number of the eigenstate increases. In Chapter 3 of the text an analytical formula for its dependence on n is derived.

c) The superposition states, as given, do not appear to be symmetric. Rather they are localized on one side of the box. If only even quantum numbers are used or if only odd quantum numbers are used, then the superposition state has lobes that are either antisymmetric or symmetric with respect to the center of the box (see below).

d) For the case where the quantum number is odd, the wavefunction is even. This symmetric character is preserved in the superposition state if only odd quantum number states are used. Correspondingly when the quantum number is even the wavefunction is odd, and the superposition state wavefunction retains this property if coefficients of only one type are used.

A plot for an odd quantum number combination (note that the quantum number is 1 plus the index) is shown in Fig. P3.7a. This superposition state has a coefficient of 0.894 for the $n = 1$ eigenstate and a coefficient of 0.447 for the $n = 3$ eigenstate; i.e.,

$$\psi = 0.894\sqrt{\frac{2}{L}} \cdot \sin\left(\frac{\pi x}{L}\right) + 0.447\sqrt{\frac{2}{L}} \cdot \sin\left(\frac{3\pi x}{L}\right)$$

Both the wavefunction and the probability density are symmetric with respect to the center of the box, $x = 5$.

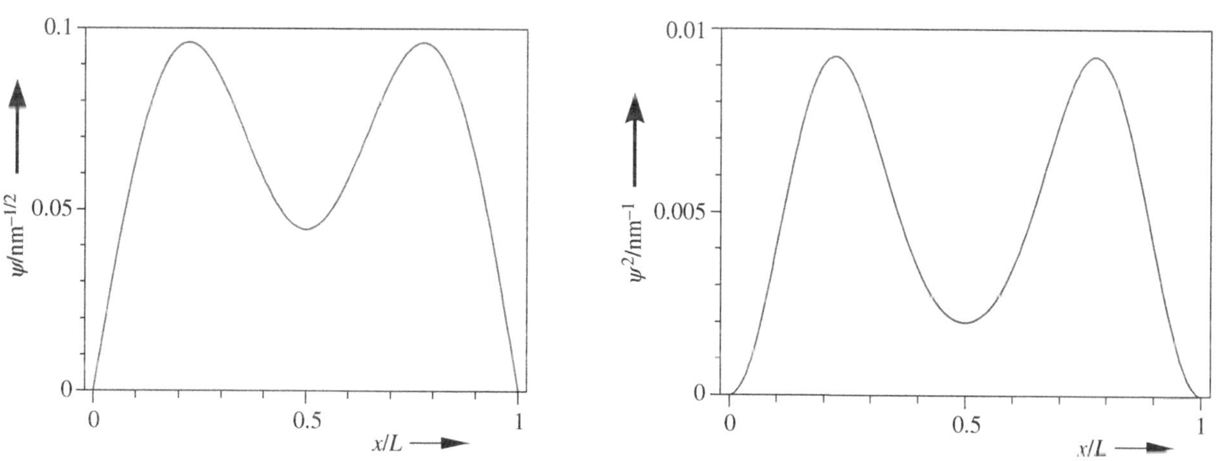

Figure P3.7a Plot of the wavefunction ψ (left) and the probability density ψ^2 (right) for a superposition of the $n = 1$ and $n = 3$ particle-in-a-box eigenstates with $L = 10$ nm.

A plot for an even quantum number combination

$$\psi = 0.447\sqrt{\frac{2}{L}} \cdot \sin\left(\frac{2\pi x}{L}\right) + 0.894\sqrt{\frac{2}{L}} \cdot \sin\left(\frac{4\pi x}{L}\right)$$

is shown in Fig P3.7b. This superposition state has a coefficient of 0.447 for the $n = 2$ eigenstate and a coefficient of 0.894 for the $n = 4$ eigenstate. In this case the wavefunction is antisymmetric (changes sign) with respect to the center of the box, $x = 0.5$; however, the probability density is symmetric because it is the square of the wavefunction.

These symmetry conditions are preserved for any of the other possible superpositions you might choose to create with the MathCAD worksheet.

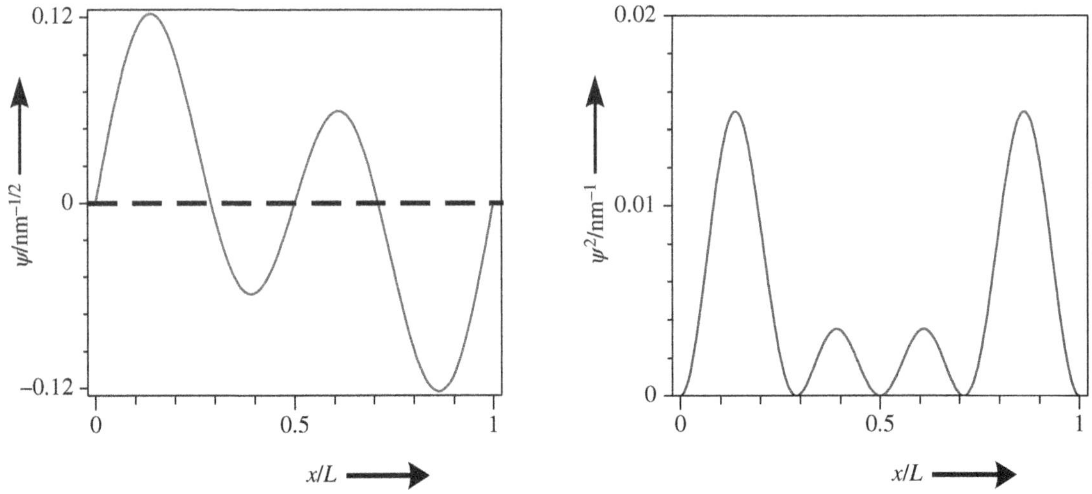

Figure P3.7b Plot of the wavefunction ψ (left) and the probability density ψ^2 (right) for a superposition of the $n = 2$ and $n = 4$ particle-in-a-box eigenstates with $L = 10$ nm.

P3.8 The MathCAD exercise tunneling.mcdx investigates features of quantum mechanical tunneling through an energy barrier. This exercise describes the tunneling electron as originating from a superposition state. Answer the questions posed in the worksheet; namely

a) Change the mass of the incoming particle. Take for example the mass of the reference particle to be 5 and compare it to the case where the mass is 50. Keep the incident energy E of the particles the same. Use at least four different masses and describe how the particles behavior changes with its mass.
b) Perform a similar series of studies, but keep the mass fixed and change the length (width) of the barrier. How does the barrier width change the tunneling probability ?
c) Perform a similar series of studies, but change the ratio of the incoming particle energy to the barrier height, E/V_0, and see how the wavefunctions change. Make sure that your incident energy E is still less than V_0 though, since we are concentrating on tunneling through a barrier.

Solution

Figure P3.8a shows a wave traveling from left to right (wave A). The wave is partly reflected from a barrier of height V_0 and width L and travels backward (wave B). Another part of the wave passes the barrier (wave C). Our task is to calculate the contributions of the reflected and the transmitted waves. To find the different contributions, we first apply the Schrödinger equation separately for each of the regions A, B, and C; and then we use wavefunction matching at the boundaries to find the reflection and transmission amplitudes.

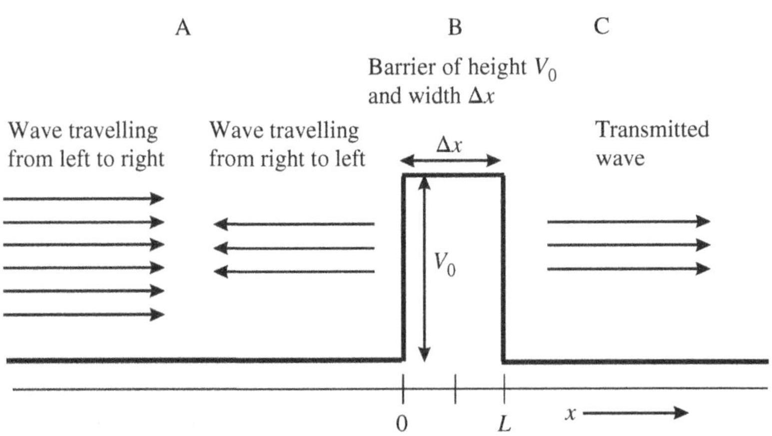

Figure P3.8a A matter wave traveling from left to right is partly reflected from a barrier of height V_0 and width L; part of the wave passes the barrier.

Description of Wavefunctions

In regions A and C the potential energy is zero, and in region B the potential energy equals V_0. Thus, the Schrödinger equation can be written as

$$\begin{array}{lll} \text{Region A:} & -\frac{\hbar^2}{2m}\frac{d^2}{dx^2}\psi_A = \quad E\cdot\psi_A & x \le 0 \\ \text{Region B:} & -\frac{\hbar^2}{2m}\frac{d^2}{dx^2}\psi_B + V_0 = E\cdot\psi_B & 0 \le x \le L \\ \text{Region C:} & -\frac{\hbar^2}{2m}\frac{d^2}{dx^2}\psi_C = \quad E\cdot\psi_C & x \ge L \end{array}$$

We describe the waves by the functions:

$$\text{A:}\ \psi_A = e^{ikx} + R\cdot e^{-ikx}$$

$$\text{B:}\ \psi_B = \alpha e^{iwx} + \beta\cdot e^{-iwx}$$

$$\text{C:}\ \psi_C = T\cdot e^{ikx}$$

where R is the reflection coefficient and T is the transmission coefficient. In region A we have one wave traveling from left to right and a reflected one traveling from right to left (ψ_A). In region B we have the wave ψ_B, and in region C we have the transmitted wave traveling from left to right (ψ_C). The coefficients R, T, α, and β are derived in Foundation 3.6 together with the expressions for the wavefunctions. The pictures displayed below show the real parts Re(ψ) of the complex wavefunctions. See Foundation 3.6.

a) The panel of four tunneling calculations in Fig. P3.8b shows the effect of the mass of the particle on the tunneling probability. In each graph the potential barrier is sketched in light gray and the wavefunction of the tunneling particle is plotted in gray or black. The mass of one of the particles is kept fixed at the value of 5 and is used as a reference for the other particle whose mass is varied in the different calculations. The reference wavefunction ($m_{ref} = 5$) is in black and the varied wavefunction is in gray. It is evident that as the mass of the particle increases, its wavefunction oscillates with a shorter de Broglie wavelength and it does not penetrate the barrier as well.

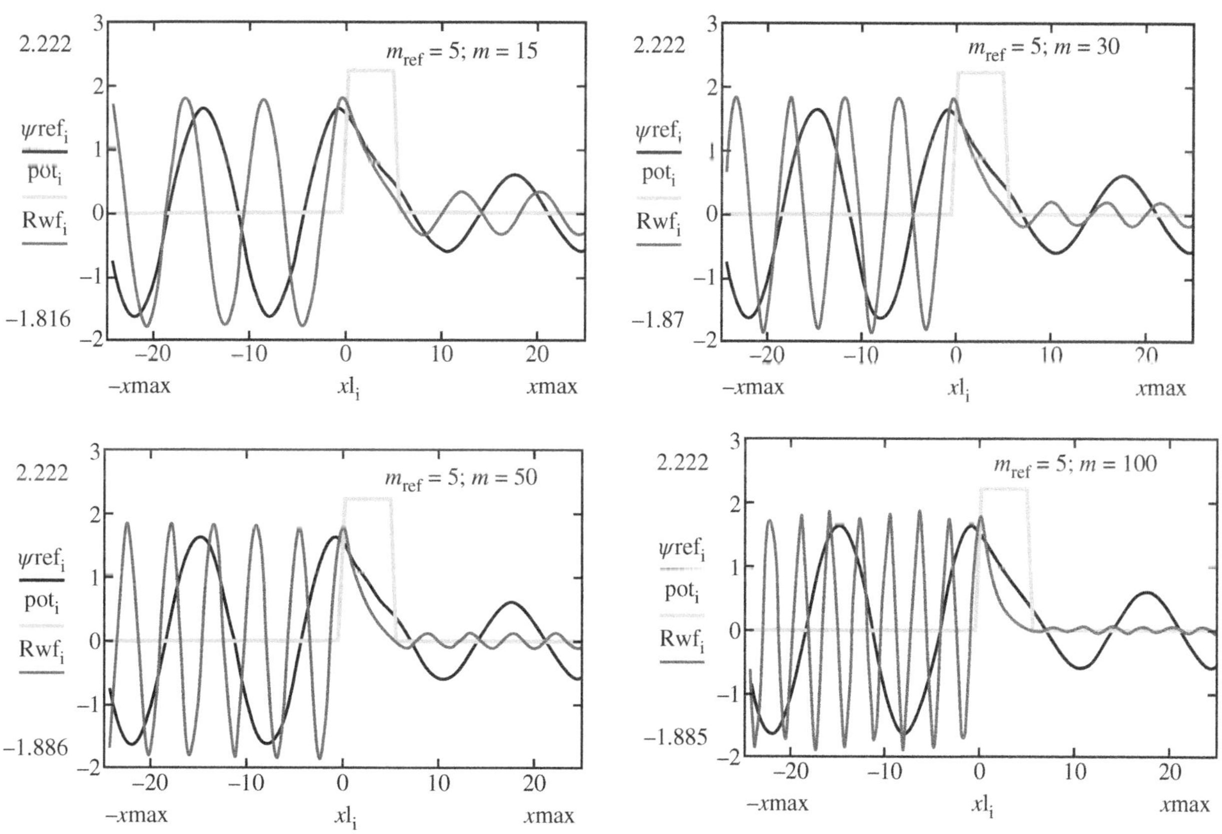

Figure P3.8b Outcomes of the MathCAD worksheet for tunneling through a potential barrier for different particle masses, but the same potential barrier properties and energy of the particle. The panels show plots of the wavefunction amplitudes versus the displacement coordinate. The barrier is represented by the light gray curve.

b) Figure P3.8c shows the effect of changing the width of the tunneling barrier on the transmission of the particle. In the top panel two calculations, in which the tunneling barrier's widths differ by a factor of two, are compared. It is evident that the thicker barrier attenuates the wavefunction more; note that the wavefunction has a lower oscillation amplitude to the right of the barrier in the case where it is thicker. The bottom two curves show a similar calculation for the case where the barrier is four times thicker.

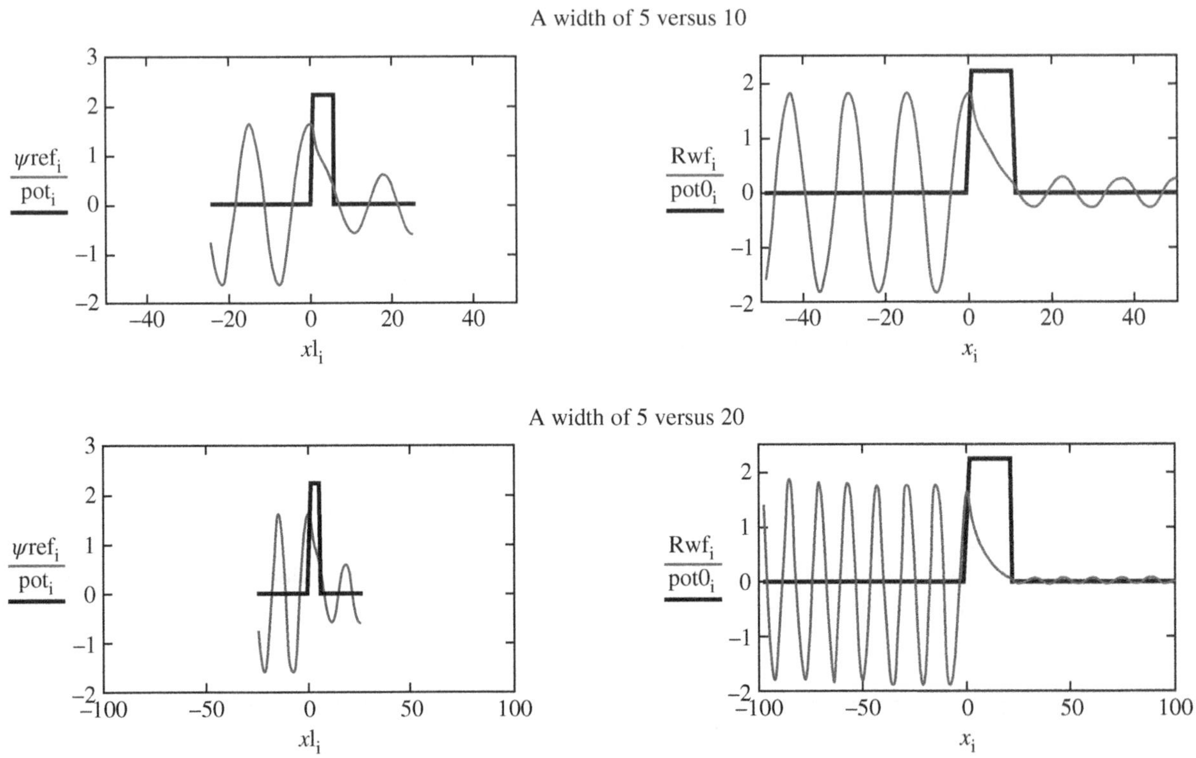

Figure P3.8c Outcomes of the MathCAD worksheet for tunneling through a potential barrier for different barrier widths, but the same mass and energy of the particle. The panels show plots of the wavefunction amplitudes versus the displacement coordinate. The barrier is represented by the black curve.

c) The reference value for E/V is 0.9. Figure P3.8d shows two calculations: one which compares an incident energy of $E/V = 0.5$ to that of 0.9 and another which compares an incident energy of $E/V = 0.25$ to that of 0.9. It is evident that as the incident energy of the tunneling particle drops, with respect to the height of the tunneling barrier, its transmission yield decreases.

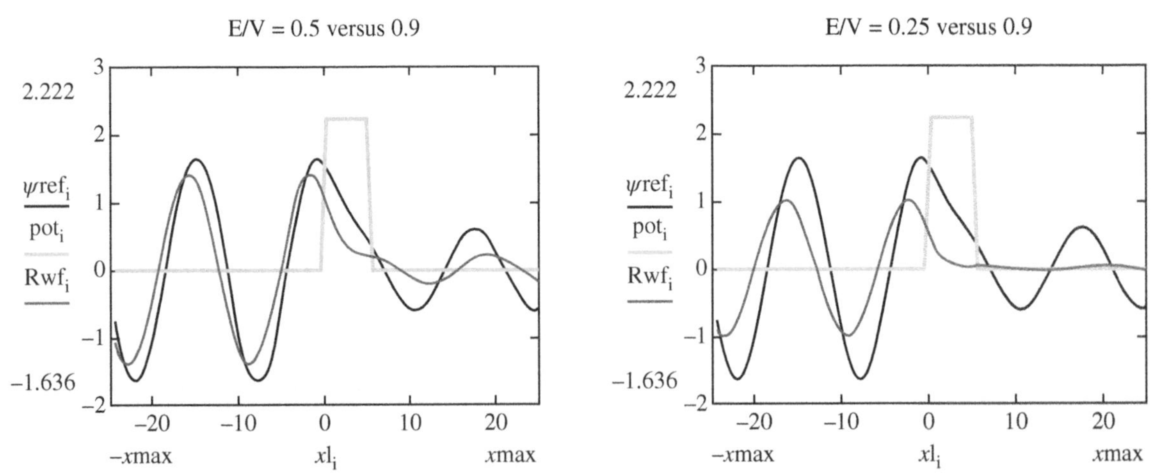

Figure P3.8d Outcomes of the MathCAD worksheet for tunneling through a potential barrier for different particle energies (relative to the barrier height), but the potential barrier width and the mass of the particle are fixed. The panels show plots of the wavefunction amplitudes versus the displacement coordinate. The barrier is represented by the light gray curve.

P3.9 a) Show that the wavefunction $\psi = \sin(kx)$ solves the Schrödinger Equation (3.16) with $V(x) = 0$, but it does not fulfill the condition that $p_{op}\psi = p\psi$.

b) Show that the wavefunction $\psi = Ce^{-ikx}$ solves the Schrödinger Equation (3.16) with $V(x) = 0$, and it fulfills the condition that $p_{op}\psi = p\psi$.

c) Compare and contrast the wavefunctions in parts a) and b). Describe how they describe different aspects of a physical system. Under what conditions might you choose one over the other?

Solution

a) To show that the wavefunction $\psi = \sin(kx)$ solves Equation (3.16) with $V(x) = 0$, it must be substituted into the equation

$$-\frac{\hbar^2}{2m_e}\frac{d^2\psi}{dx^2} = E\psi$$

Performing this operation gives

$$\frac{d\left[A\sin(kx)\right]}{dx} = k\left[A\cos(kx)\right] \qquad \frac{d^2\left[A\sin(kx)\right]}{dx^2} = -k^2\left[A\sin(kx)\right]$$

and

$$-\frac{\hbar^2}{2m_e}\frac{d^2\left[A\sin(kx)\right]}{dx^2} = -\frac{\hbar^2}{2m_e}\left[-k^2A\sin(kx)\right] = E\cdot A\sin(kx)$$

leading to

$$E = \frac{\hbar^2}{2m_e}k^2$$

Thus we see that the operation on the wavefunction generates a constant, the energy E in this case, times the function again. Thus it is an eigenfunction of the energy operator.
We analyze the case for the momentum operator in a similar manner, and we find

$$p_{op}\psi = -i\hbar\frac{d\psi}{dx} = -i\hbar\cdot Ak\cos(kx) \neq pA\sin(kx)$$

In this case, operation on the wavefunction (with the p_{op} operator) generates a new function; i.e., the result is not proportional to the original function. Thus it is not an eigenfunction of the momentum operator.

b) To show that the wavefunction $\psi = Ce^{-ikx}$ is an eigenfunction of the Equation (3.16) with $V(x) = 0$, in Section 3.5.2 we proceed similarly.

$$-\frac{\hbar^2}{2m_e}\frac{d^2\left(Ce^{-ikx}\right)}{dx^2} = E\cdot Ce^{-ikx}$$

In this case, the derivatives are

$$\frac{d\psi}{dx} = -C\cdot ike^{-ikx}, \quad \frac{d^2\psi}{dx^2} = -C\cdot k^2e^{-ikx}$$

in which C and k are constants and i is $\sqrt{-1}$. Thus we obtain

$$-\frac{\hbar^2}{2m_e}\left(-C\cdot k^2e^{-ikx}\right) = EC\cdot e^{-ikx}$$

and the energy E is

$$E = \frac{\hbar^2}{2m_e}\cdot k^2$$

as in the first case. Now consider the translational momentum of the electron. If we operate on the free electron's wavefunction with the operator p_{op} we find that

$$p_{op}\psi = -i\hbar\frac{d\psi}{dx} = -i\hbar\left(-C\cdot ike^{-ikx}\right) = -\hbar k\psi = p\psi$$

We see from this last equality that $p = -\hbar a$. Thus this wavefunction is an eigenfunction of both the energy and the linear momentum.

c) The sine function, sin(kx), describes standing waves whereas the exponential function, exp(ikx), describes traveling waves. The sine function can be represented as a combination of two traveling waves, one moving in the $+x$ direction and one moving in the $-x$ direction, with the same value of a. That is we can write

$$\sin(kx) = Const \cdot \left(e^{+ikx} - e^{-ikx}\right)$$

which shows that the standing wave is generated from counter propagating traveling waves.

The standing wave description is most useful for describing electrons in atoms and molecules, for which they are trapped in the potential well of the nuclei, and this type of electron wave is the one most commonly used to describe chemical bonding. The traveling wave description is most useful when describing electrons that have energies larger than any electron wells ("free electrons") or for describing the movement of electrons in metals and semiconductors (in which they are not strongly confined). An important exception is the case of traveling waves for describing the orbits of electrons in atoms and molecules, such as the electron on a ring model described in the chapter. In this latter case, we will find that the traveling wave description is useful for understanding the angular momenta and the energy state ordering in multielectron systems.

P3.10 Create a superposition of cosine waves and calculate the uncertainty product of a particle's momentum and position, $\Delta p \cdot \Delta x$, for a Gaussian distribution of momenta. Use wavefunctions

$$\cos(ax) \text{ where } a = \pm\frac{2\pi}{\Lambda} \text{ and } p = \pm\frac{h}{\Lambda} = \frac{h}{2\pi}a$$

in which the distribution of coefficients a is given by a Gaussian distribution; namely

$$\frac{dP}{da} = \frac{b}{\sqrt{\pi}} \cdot e^{-b^2 \cdot a^2}$$

The coefficients a are proportional to the momenta as shown above, and b is a parameter determining the full-width-at-half-maximum $a_{1/2}$ of the Gaussian distribution. The standard deviation σ_a for the spread in a values is given by $a_{1/2} = 2\sqrt{2 \cdot \ln 2} \cdot \sigma_a$ (see Appendix C).

Part 1: Use the expression for the Gaussian distribution of a to determine the distribution of a in the probability density. Then find the standard deviation of a in terms of the full-width-at-half-maximum of the distribution $a_{1/2}$ and use it to show that the uncertainty in the momenta is given by $\Delta p = \frac{h}{2\pi} \cdot \frac{1}{2b}$.

Part 2: Find the uncertainty in x, construct a superposition wavefunction of the form

$$\psi(x) = C \cdot \int_{-\infty}^{\infty} \frac{dP}{da} \cdot \cos(ax) \cdot da$$

where C is a normalization constant. Then use the probability density $\rho = \psi(x)^2$ to show that $\Delta x = b$.

Part 3: Combine your results to find the uncertainty product. In each case, use the standard deviation for the observable (i.e., x or p) to be its uncertainty.

Solution

Part 1: The distribution of momenta $p = \frac{h}{2\pi}a$ in the wavefunction is given by

$$\frac{dP}{da} = \frac{b}{\sqrt{\pi}} \cdot e^{-b^2 \cdot a^2}$$

so that the distribution of momenta in the probability density will be determined by

$$\left(\frac{dP}{da}\right)^2 = \frac{b^2}{\pi} \cdot e^{-2b^2 \cdot a^2}$$

which is also a Gaussian distribution. The value of a at half-height of the distribution (that is, where the distribution reaches one half of its maximum value) is denoted as a', and it can be calculated by the condition

$$\ln\left(\frac{1}{2}\right) = -2b^2 \cdot a'^2 \text{, thus } a' = \pm\frac{\sqrt{\ln 2}}{\sqrt{2}b}$$

and full-width-at-half-maximum $a_{1/2}$ is given by

$$a_{1/2} = \frac{2\sqrt{\ln 2}}{\sqrt{2}b} = \frac{\sqrt{2\ln 2}}{b}$$

Using $a_{1/2} = 2\sqrt{2 \cdot \ln 2} \cdot \sigma_a$, we find that

$$a_{1/2} = \frac{\sqrt{2 \cdot \ln 2}}{b} = 2\sqrt{2 \cdot \ln 2} \cdot \sigma_a$$

so that $\sigma_a = 1/(2b)$. Using σ_a for the uncertainty in the a values, we can write the uncertainty in the momentum as

$$\Delta p = \frac{h}{2\pi} \cdot \sigma_a = \frac{h}{2\pi} \cdot \frac{1}{2b}$$

Part 2: We consider the superposition

$$\psi(x) = C \cdot \int_{-\infty}^{\infty} \frac{d\mathcal{P}}{da} \cdot \cos(ax) \cdot da = \\ = \frac{C \cdot b}{\sqrt{\pi}} \cdot \int_{-\infty}^{\infty} e^{-b^2a^2} \cdot \cos(ax) \cdot da = 2 \cdot \frac{C \cdot b}{\sqrt{\pi}} \cdot \int_{0}^{\infty} e^{-b^2a^2} \cdot \cos(ax) \cdot da$$

where C is a constant. This expression for $\psi(x)$ has the form of the definite integral

$$\int_0^{\infty} \cos(r \cdot y) \cdot e^{-(p \cdot y)^2} \cdot dy = \frac{\sqrt{\pi}}{2p} e^{-r^2/(4p^2)}$$

which is given in Appendix B.2. We see that a corresponds to y, x corresponds to r, and b corresponds to p. Thus we can write that

$$\psi(x) = 2 \cdot \frac{C \cdot b}{\sqrt{\pi}} \cdot \left[\frac{\sqrt{\pi}}{2b} e^{-x^2/(4b^2)}\right] = C \cdot e^{-x^2/(4b^2)}$$

The probability distribution for the particle is proportional to the square of the wavefunction

$$\psi(x)^2 = \rho = C^2 \cdot e^{-x^2/(2b^2)}$$

This equation for ρ is a Gaussian, and we can find its standard deviation σ_x from its full-width-at-half-maximum $x_{1/2}$. That is $x_{1/2} = 2\sqrt{2 \cdot \ln 2} \cdot \sigma_x$, see Appendix C. The function ρ is at its half-height when

$$\ln\frac{1}{2} = -\frac{x'^2}{2b^2}, \text{ thus } x' = \pm\sqrt{2 \cdot \ln 2} \cdot b$$

and for the full-width-at-half-maximum $x_{1/2}$ we have

$$x_{1/2} = 2 \cdot \sqrt{2 \cdot \ln 2} \cdot b$$

Using this expression for $x_{1/2}$, we find

$$x_{1/2} = 2\sqrt{2 \cdot \ln 2} \cdot \sigma_x = 2 \cdot \sqrt{2 \cdot \ln 2} \cdot b$$

and hence that

$$\sigma_x = b = \Delta x$$

Part 3: Thus the uncertainty product $\Delta p \cdot \Delta x$ is

$$\Delta p \cdot \Delta x = \left(\frac{h}{2\pi} \cdot \frac{1}{2b}\right)(b) = \frac{h}{2\pi}\frac{1}{2} = \frac{\hbar}{2}$$

satisfying the Heisenberg uncertainty principle, which requires that $\Delta p \cdot \Delta x \geq \hbar/2$. Note that in this case, the uncertainty product matches the minimum uncertainty possible. Figure P3.10 plots the distribution of a values and the probability density of the $\psi(x)^2$ for two different values of b: $b = 3$ nm and $b = 1/3$ nm.

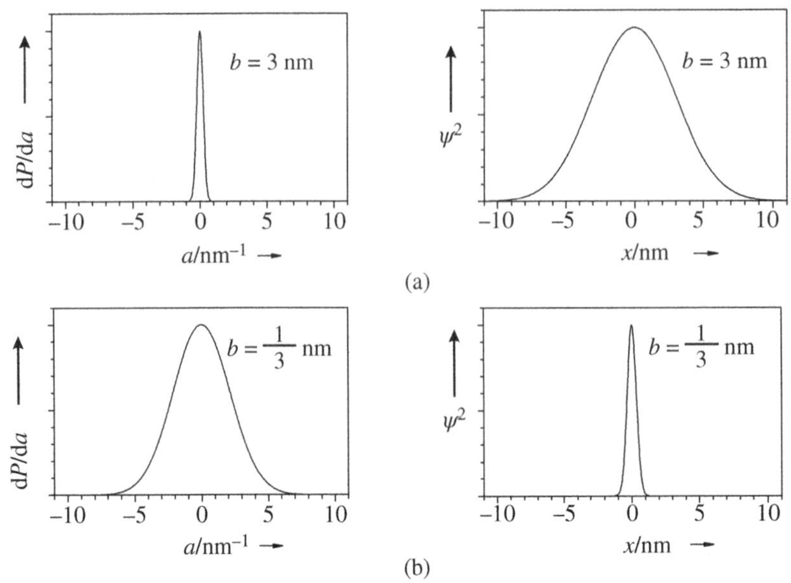

Figure P3.10 Gaussian distribution function dP/da for the momentum $p = \frac{h}{2\pi}a$ (left) and probability density ψ^2 for the location x (right) of a particle. (a) parameter $b = 3$ nm, (b) parameter $b = 1/3$ nm.

P3.11 Show that the angular momentum operator can be transformed from cartesian coordinates (Equation (3.69)) to spherical coordinates (Equation (3.70)). The angular momentum operator about the z-axis in cartesian coordinates x and y is written as

$$\mathcal{L}_{\text{op},z} = -\frac{ih}{2\pi}\left(x\frac{\partial}{\partial y} - y\frac{\partial}{\partial x}\right)$$

Hint: start with an equation where this operator acts on an operator $F(\varphi)$ depending on the spherical coordinate φ.

Solution

The spherical coordinates in two dimensions are r and φ:

$$r = \sqrt{x^2 + y^2}\, x = r\cos\varphi,\ y = \sin\varphi,\ v = \frac{x}{r},\ w = \frac{y}{r},\ \text{ and } \varphi = \arccos(v) = \arcsin(w)$$

First we calculate $\left(\frac{\partial\varphi}{\partial x}\right)_y$ and $\left(\frac{\partial\varphi}{\partial y}\right)_x$.

$$\left(\frac{\partial\varphi}{\partial x}\right)_y = \frac{\partial\varphi}{\partial v}\frac{\partial v}{\partial x} = \frac{-1}{\sqrt{1-v^2}}\cdot\frac{\partial}{\partial x}\frac{x}{\sqrt{x^2+y^2}} = -\frac{r}{y}\cdot\left(\frac{1}{r} - \frac{x^2}{r^3}\right) = -\frac{r}{y}\cdot\frac{r^2-x^2}{r^3} = -\frac{r}{y}\cdot\frac{y^2}{r^3} = -\frac{y}{r^2} = -\frac{\sin\varphi}{r}$$

$$\left(\frac{\partial\varphi}{\partial y}\right)_x = \frac{\partial\varphi}{\partial w}\frac{\partial w}{\partial x} = \frac{1}{\sqrt{1-w^2}}\cdot\frac{\partial}{\partial x}\frac{y}{\sqrt{x^2+y^2}} = \frac{\cos\varphi}{r}$$

We consider the operator $\left(x\frac{\partial}{\partial y} - y\frac{\partial}{\partial x}\right)$ depending on x and y acting on a function $F(\varphi)$ depending on the spherical coordinate φ.

$$\left(x\frac{\partial}{\partial y} - y\frac{\partial}{\partial x}\right)F(\varphi) = r\cos\varphi\cdot\left[\left(\frac{\partial F}{\partial\varphi}\right)\left(\frac{\partial\varphi}{\partial y}\right)_x\right] - r\sin\varphi\left[\left(\frac{\partial F}{\partial\varphi}\right)\left(\frac{\partial\varphi}{\partial x}\right)_y\right]$$

$$= r\cos\varphi\cdot\left[\frac{\partial F}{\partial\varphi}\frac{\cos\varphi}{r}\right] - r\sin\varphi\left[-\frac{\partial F}{\partial\varphi}\frac{\sin\varphi}{r}\right] = \cos^2\varphi\cdot\frac{\partial F}{\partial\varphi} + \sin^2\varphi\frac{\partial F}{\partial\varphi} = \frac{\partial}{\partial\varphi}F$$

Thus we can write the angular momentum as

$$\mathcal{L}_{\text{op},z} = -\frac{ih}{2\pi}\left(x\frac{\partial}{\partial y} - y\frac{\partial}{\partial x}\right) = -\frac{ih}{2\pi}\frac{\partial}{\partial\varphi}$$

We can also go the opposite way: to transform the angular momentum operator from polar coordinates to cartesian coordinates, we proceed by operating on the function $F(x(\varphi), y(\varphi))$

$$\frac{\partial}{\partial\varphi}F(x,y) = \left(\frac{\partial F}{\partial x}\right)_y\left(\frac{\partial x}{\partial\varphi}\right)_r + \left(\frac{\partial F}{\partial y}\right)_x\left(\frac{\partial y}{\partial\varphi}\right)_r$$

$$= -\left(\frac{\partial F}{\partial x}\right)_y r\sin\varphi + \left(\frac{\partial F}{\partial y}\right)_x r\cos\varphi = -y\left(\frac{\partial F}{\partial x}\right)_y + x\left(\frac{\partial F}{\partial y}\right)_x = \left(x\frac{\partial}{\partial y} - y\frac{\partial}{\partial x}\right)F$$

Thus we obtain

$$\mathcal{L}_{\text{op},z} = -\frac{ih}{2\pi}\frac{\partial}{\partial\varphi} = -\frac{ih}{2\pi}\left(x\frac{\partial}{\partial y} - y\frac{\partial}{\partial x}\right)$$

P3.12 Derive the equations for the antisymmetric solution of the potential energy well problem in Foundation 3.1.

Solution

The procedure for finding the antisymmetric solution is analogous to that we used to find the symmetric solution. Inserting the sine solution ($\psi_{\text{well}} = A'\sin ax$) into the Schrödinger equation

$$-\frac{h^2}{8\pi^2 m_e}\frac{d^2\psi}{dx^2} = E\psi$$

we find

$$-\frac{h^2}{8\pi^2 m_e}\left(-A'\,a^2\sin ax\right) = E\cdot\left(A'\sin ax\right)$$

The equality holds if the energy is given by

$$E = \frac{h^2}{8\pi^2 m_e}a^2 \qquad (3P12.1)$$

Now consider the case where the electron is in the region outside the energy well, and the solution to the equation is a damped function. We proceed by considering the right region, $x > L/2$, and the left region, $x < -L/2$, separately. For $x > L/2$, the wavefunction

$$\psi_{\text{right}} = B\,e^{-bx}$$

converges to zero for $x \to +\infty$. If we insert this wavefunction into the Schrödinger equation

$$-\frac{h^2}{8\pi^2 m_e}\frac{d^2\psi}{dx^2} + V_0\psi = E\psi$$

we find

$$E = -\frac{h^2}{8\pi^2 m_e}b^2 + V_0 \qquad (3P12.2)$$

The two expressions (3P12.1) and (3P12.2) for the electron's energy must be equivalent

$$E = \frac{h^2}{8\pi^2 m_e}a^2 = -\frac{h^2}{8\pi^2 m_e}\frac{d^2\psi}{dx^2} + V_0\psi$$

so

$$a^2 = \frac{8\pi^2 m_e}{h^2}V_0 - b^2 \quad \text{or} \quad a = \pm\sqrt{\frac{8\pi^2 m_e}{h^2}V_0 - b^2} \qquad (3P12.3)$$

which determines one of the conditions. For the calculation of the remaining constants b, A', and B we apply boundary conditions on our solution. The boundary conditions (matching conditions) on the wavefunction and its slope are

$$\psi_{\text{well}}|_{x=L/2} = \psi_{\text{right}}|_{x=L/2} \quad \text{and} \quad \left(\frac{d\psi_{\text{well}}}{dx}\right)_{x=L/2} = \left(\frac{d\psi_{\text{right}}}{dx}\right)_{x=L/2}$$

Matching the wavefunctions at the boundary results in the condition

$$A'\sin\left(\frac{aL}{2}\right) = B\,e^{-bL/2} \quad \text{or} \quad B = A'\sin\left(\frac{aL}{2}\right)e^{bL/2}$$

Matching the condition for the slopes at the boundary provides an additional relationship. In particular we find

$$A'\,a\,\cos\left(\frac{aL}{2}\right) = -B\,b\,e^{-bL/2} \text{ or } B = -A'\,\frac{a}{b}\,\cos\left(\frac{aL}{2}\right)\,e^{bL/2} \tag{3P12.4}$$

If we divide these two results for B, we find

$$1 = -\frac{b}{a}\,\tan\left(\frac{aL}{2}\right)$$

or

$$b = -a\,\cot\left(\frac{aL}{2}\right) \tag{3P12.5}$$

which provides a second relationship between a and b. Combining this result with the earlier result in Equation (3P12.3), we obtain

$$a^2 + a^2\,\cot^2\left(\frac{aL}{2}\right) - V_0\frac{8\pi^2 m_e}{h^2} = 0 \tag{3P12.6}$$

This equation determines a. If we let $w = aL$ and define a constant k through the relation

$$k = \frac{V_0}{h^2}8m_eL^2$$

we find that (3P12.6) becomes

$$w^2 + w^2 \cdot \cot^2\left(\frac{w}{2}\right) - \pi^2 k = 0$$

or using the relationship between the cotangent and the tangent we see that

$$w^2 + \frac{w^2}{\tan^2\left(\frac{w}{2}\right)} - \pi^2 k = 0$$

as written in the Foundation. For the energy Equation (3P12.1) we can write

$$E = \frac{h^2}{8\pi^2 m_e}\left(\frac{w}{L}\right)^2 = \frac{h^2}{8m_eL^2}\left(\frac{w}{\pi}\right)^2$$

The wavefunctions are given by

$$\psi(x) = \begin{cases} -B\cdot e^{bx} & (\text{for}\, x < -L/2) \\ A\cdot \sin(wx/L) & (\text{for} - L/2 < x < L/2) \\ B\cdot e^{-bx} & \text{for}\,(x > L/2) \end{cases}$$

where the parameter b (using 3.5) gives

$$b = -\frac{w}{L\cdot\tan\left(\frac{w}{2}\right)} = -\frac{w}{L}\cdot\cot\left(\frac{w}{2}\right)$$

and B (using 3.4) is given by

$$B = -A'\,\left(\frac{w}{bL}\right)\cos\left(\frac{w}{2}\right)\cdot\,e^{bL/2}$$

P3.13 Following the equation for $\overline{(\Delta A)^2}\cdot\overline{(\Delta B)^2} \leq$ in Foundation 3.4 prove the validity of the uncertainty relations

$$\Delta y\cdot\Delta p_y \geq \hbar/2 \qquad \Delta\varphi\cdot\Delta\mathcal{L}_z \geq \hbar/2$$

Solution

We start with the general relation

$$\overline{(\Delta A)^2}\cdot\overline{(\Delta B)^2} \geq \left[\frac{1}{2i}\left|\overline{[A_{op}, B_{op}]}\right|\right]^2$$
$$= \left[\frac{1}{2i}\int\psi^*(A_{op}B_{op} - B_{op}A_{op})\psi\cdot d\tau\right]^2$$

and set

$$y_{\text{op}} = y \qquad p_{y,\text{op}} = -\text{i}\hbar \frac{\text{d}}{\text{d}y}$$

Thus we obtain in analogy to $\overline{(\Delta p_x)^2}\ \overline{(\Delta x)^2}$

$$\overline{(\Delta p_y)^2}\ \overline{(\Delta y)^2} \geq \left[\frac{1}{2\text{i}}\left|\overline{[p_{y,\text{op}}, y_{\text{op}}]}\right|\right]^2 = \left[\frac{1}{2\text{i}}\int \psi^*(p_{y,\text{op}}y_{\text{op}} - y_{\text{op}}p_{y,\text{op}})\psi \cdot \text{d}y\right]^2$$
$$= \left[\frac{1}{2\text{i}}\int \psi^*(-\text{i}\hbar)\psi \cdot \text{d}y\right]^2 = \left[\frac{-\hbar}{2}\int \psi^*\psi \cdot \text{d}y\right]^2 = \left[\frac{-\hbar}{2}\right]^2 = \frac{\hbar^2}{4}$$

Next we set

$$\mathcal{L}_{z,\text{op}} = -\frac{\text{i}h}{2\pi}\frac{\text{d}}{\text{d}\varphi} \qquad \varphi_{\text{op}} = \varphi$$

and obtain

$$(\mathcal{L}_{z,\text{op}}\varphi_{\text{op}} - \varphi_{\text{op}}\mathcal{L}_{z,\text{op}})\psi = -\frac{\text{i}h}{2\pi}\frac{\text{d}}{\text{d}\varphi}(\varphi\psi) + \varphi\frac{\text{i}h}{2\pi}\frac{\text{d}}{\text{d}\varphi}\psi = -\frac{\text{i}h}{2\pi}\left[\psi + \varphi\frac{\text{d}}{\text{d}\varphi}\psi - \varphi\frac{\text{d}}{\text{d}\varphi}\psi\right]$$
$$= -\frac{\text{i}h}{2\pi}\left[\psi + \varphi\frac{\text{d}}{\text{d}\varphi}\psi - \varphi\frac{\text{d}}{\text{d}\varphi}\psi\right] = -\frac{\text{i}h}{2\pi}\psi$$

Thus we obtain in analogy to $\overline{(\Delta p_x)^2}\ \overline{(\Delta x)^2}$

$$\overline{(\Delta L_{z,\text{op}})^2}\ \overline{(\Delta\varphi)^2} \geq \left[\frac{1}{2\text{i}}\int_0^{2\pi} \psi^*\left(-\frac{\text{i}h}{2\pi}\right)\psi \cdot \text{d}\varphi\right]^2 = \left[\frac{1}{2\text{i}}\left(-\frac{\text{i}h}{2\pi}\right)\int_0^{2\pi} \psi^*\psi \cdot \text{d}\varphi\right]^2$$
$$= \left[\frac{1}{2}\left(-\frac{h}{2\pi}\right)\right]^2 = \frac{\hbar^2}{4}$$

P3.14 Consider an electron confined between two walls at a distance L (electron in a box). Imagine that a barrier of height V and width Δx is placed at the center of the box. Discuss qualitatively what will happen with the energies of the electron as the height of the barrier is increased from zero to infinity.

Solution

The energies E_n of the electron in the absence of the barrier are given by

$$E_n = \frac{h^2}{8mL^2}n^2,\ n = 1, 2, 3, \ldots$$

First we see that the wavefunctions with even n ($n = 2, 4, 6, \ldots$) have a node at the center of the box. Thus their energies will remain unchanged in the presence of the barrier. The wavefunctions with odd n ($n = 1, 3, 5, \ldots$) have antinodes at the center of the box, and they will be strongly affected by the barrier. With increasing height V their energies will increase. Finally, for infinite height of the barrier, the electron is confined within two boxes with box length $L/2$ and its energy will be described by

$$E_n = \frac{h^2}{8m(L/2)^2}n^2 = 4\frac{h^2}{8mL^2}n^2,\ n = 1, 3, 5, \ldots$$

4

Hydrogen Atom

4.1 Exercises

E4.1 Sketch the radial and angular parts of a 3*s* orbital's wavefunction. Identify the number of radial and angular nodes.

Solution

The unnormalized 3*s* radial part of the wavefunction is

$$R(r) = \left(27 - 18\frac{r}{a_0} + 2\frac{r^2}{a_0^2}\right) e^{-\frac{r}{3a_0}}$$

The radial part is displayed in Fig. E4.1a. The angular part is a sphere (Fig. E4.1a,b), and projected in two dimensions is a circle (Fig. E4.1c).

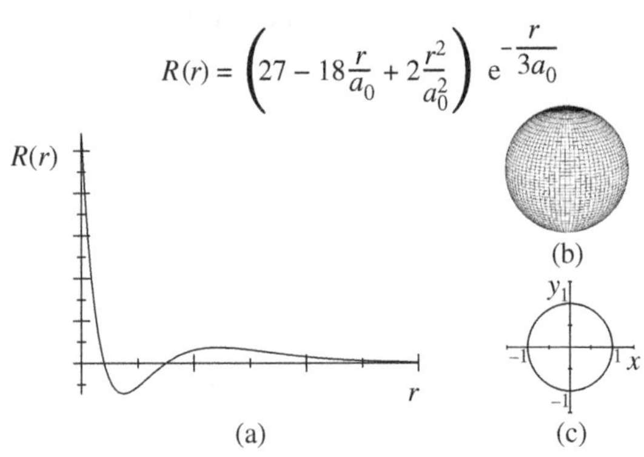

Figure E4.1 Radial (a) and angular (b and c) part of the 3*s*-wavefunction.

The radial part has two nodes and the angular part has zero nodes.

E4.2 Sketch the radial and angular parts of a d_{xy} orbital's wavefunction. Identify the number of radial and angular nodes.

Solution

a) The d_{xy} orbital is given by

$$\psi_{3d_{xy}} = \frac{\sqrt{2}}{81\sqrt{\pi a_0^3}} \frac{x}{a_0} \frac{y}{a_0} e^{-\frac{r}{3a_0}} = \frac{\sqrt{2}}{81\sqrt{\pi a_0^3}} \sin^2\theta \cdot \sin\varphi \cdot \cdot \cos\varphi \cdot \left(\frac{r}{a_0}\right)^2 e^{-\frac{r}{3a_0}}$$

where we have used

$$x \cdot y = (r \cdot \sin\theta \cdot \sin\varphi) \cdot (r \cdot \sin\theta \cdot \cos\varphi) = r^2 \sin^2\theta \cdot \sin\varphi \cdot \cos\varphi$$

Solutions Manual for Principles of Physical Chemistry, Third Edition. Edited by Hans Kuhn, David H. Waldeck, and Horst-Dieter Försterling.

The radial dependence is displayed in Fig. E4.2a. The angular portion of this orbital is

$$\frac{x}{a_0} \times \frac{y}{a_0}$$

(see Fig. E4.2b), and the angular part in two dimensions is sketched in Fig. E4.2c.

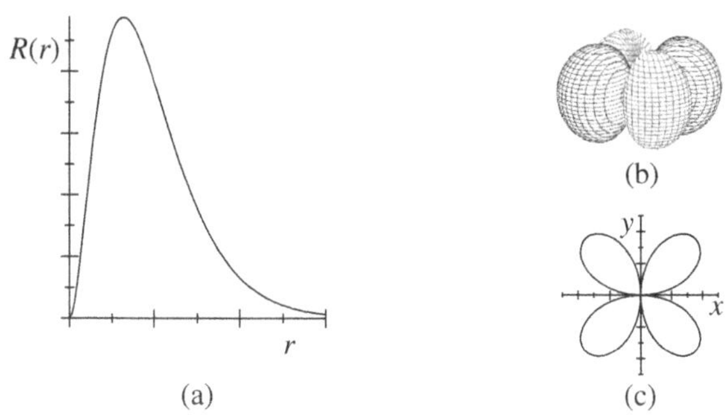

Figure E4.2 Radial (a) and angular (b and c) part of the $3d_{xy}$-wavefunction.

The radial part of the wavefunction does not have any nodes, but the angular part of the wavefunction has two nodal planes.

E4.3 Sketch the radial and angular parts of a $3d_{z^2}$ orbital's wavefunction. Identify the number of radial and angular nodes.

Solution

The radial part of the $3d_{z^2}$ orbital

$$R(r) = \left(\frac{r}{a_0}\right)^2 e^{-\frac{r}{3a_0}}$$

is displayed in Fig. E4.3a. The angular part of the $3d_{z^2}$ orbital is

$$3\cos(\theta)^2 - 1$$

(Fig. E4.3b) and the angular part is sketched in two-dimensions in Fig. E4.3c

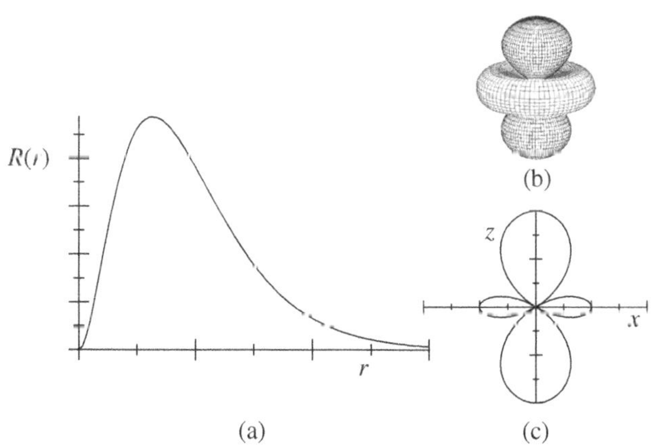

Figure E4.3 Radial (a) and angular (b and c) part of the $3d_{z^2}$-wavefunction.

The radial part of the wavefunction does not have any nodes. The angular part of the wavefunction has two nodal surfaces that are shaped like the surface of a cone with its apex at the origin and aligned along z and $-z$, respectively. In the two-dimensional sketch these surfaces appear as lines in each quadrant.

E4.4 Sketch the radial and angular parts of a $4p_z$ orbital's wavefunction. Identify the number of radial and angular nodes.

Solution

In general the radial part of the orbital is given as

$$R_{nl}(r) = \left\{ \frac{(n-l-1)!}{2n[(n+1)!]^3} \right\}^{1/2} \left(\frac{2}{na_0} \right)^{l+3/2} r^l e^{-r/na_0} L_{n+l}^{2l+1} \left(\frac{2r}{na_0} \right)$$

where $L_{n+1}^{2l+1}\left(\frac{2r}{na_0}\right)$ is the associated Laguerre polynomial. For the $4p_z$ radial part this becomes

$$R_{41}(r) = \frac{1}{32\sqrt{15}} \left(\frac{1}{a_0} \right)^{3/2} e^{-r/(4a_0)} \left[\left(20 - 10 \left(\frac{2r}{4a_0} \right) + \left(\frac{2r}{4a_0} \right)^2 \right) \right] \cdot \frac{2r}{4a_0}$$

(Fig. E4.4a).

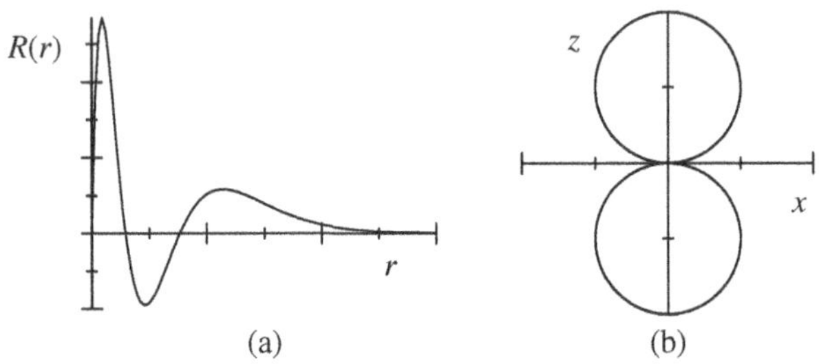

Figure E4.4 E4.4 Radial (a) and angular (b) part of the $4p_z$-wavefunction.

The unnormalized angular part is

$$Y(\theta.\phi) = \cos\theta$$

The radial part of the wavefunction has two nodes, and the angular part of the wavefunction has one nodal plane. Hence the total number of nodes in the wavefunction is three.

E4.5 Give the number of nodes for the radial and angular parts of the orbitals in E4.1–E4.4. Explain the trend observed.

Solution

The number of nodes for each of the orbitals is summarized in the table.

		Radial	Angular	Total
a	d_{xy}	0	2	2
b	$3s$	2	0	2
c	$3d_{z^2}$	0	2	2
d	$4p_z$	2	1	3

The number of radial nodes should be $n - l - 1$, and the number of angular nodes should be l. Hence the total number of nodes, which is their sum, is $n - 1$. The table satisfies these relations.

E4.6 Sketch the radial distribution function for a $3s$ orbital.

Solution

The radial distribution function for the $3s$ orbital is

$$P(r) = 4\pi r^2 |R_{30}(r)|^2 = 4\pi r^2 \left[\frac{1}{81\sqrt{3\pi \cdot a_0^3}} \left(27 - 18 \cdot \frac{r}{a_0} + 2 \cdot \frac{r^2}{a_0^2} \right) e^{-\frac{r}{3a_0}} \right]^2$$

and a sketch of this function is given in Figure E4.6.

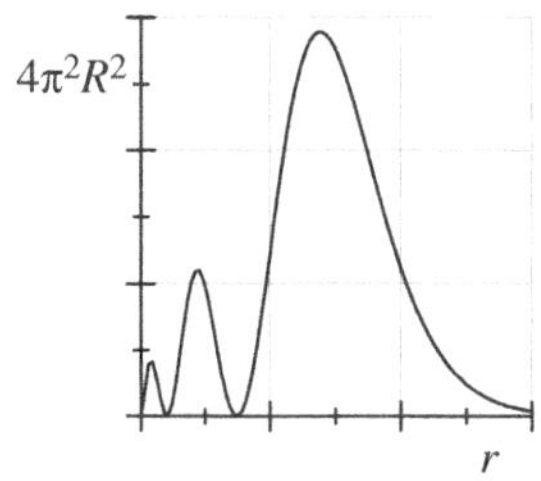

Figure E4.6 Probability function $\mathcal{P} = 4\pi^2R^2$ for the radial part of the 3s-wavefunction.

E4.7 Consider the 2s orbital of the hydrogen atom. Find the value(s) of r at which the electron probability density, ρ, is zero. Find the value(s) of r at which the electron probability density, ρ, has a maximum.

Solution

In this case we first plot the radial distribution function

$$\rho = |\psi_{2s}(r)|^2 = \left[\frac{1}{\sqrt{32\pi a_0^3}}\left(2-\frac{r}{a_0}\right)e^{-r/(2a_0)}\right]^2 = \frac{1}{32\pi a_0^3}\left(2-\frac{r}{a_0}\right)^2 e^{-r/a_0}$$

In Figure E4.7 we plot the probability density. However, to observe the zeroes and second lobe more clearly we expand the ordinate (Fig. E4.7b). Please note that ρ is given in units of pm^{-3}. From these plots we observe that the electron probability density has two zeroes, of which one occurs as $r \to \infty$. From inspection of the graph and the formula for the radial wavefunction, it is evident that a second zero appears at the value of $r = 2a_0$.

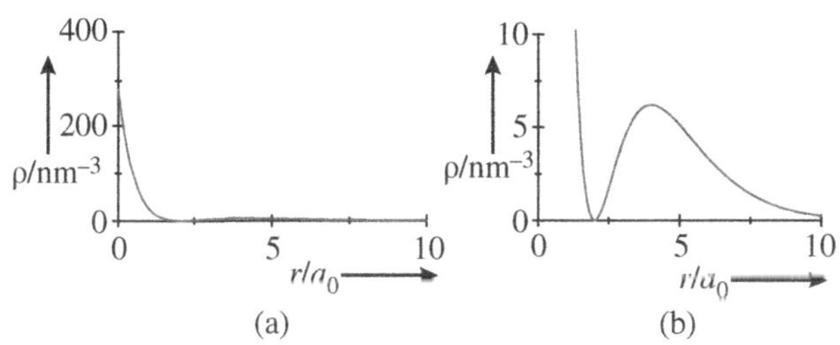

Figure E4.7 Density $\rho = \psi^2$ for the radial part of the 2s-wavefunction. (a) Total range of the wavefunction, (b) expansion of the ordinate by a factor of 40.

From inspection of the graph it is evident that a maximum occurs at $r = 0$ and at a value of r near $4a_0$. To find the value of r at which the maximum occurs, we take the derivative of the probability density and set it equal to zero

$$\frac{d}{dr}\left[\left(2-\frac{r}{a_0}\right)^2 e^{-r/a_0}\right] = 0$$

The solutions to this equation will determine the extrema of the function. Evaluating the derivative we have

$$0 - \frac{-1}{a_0}e^{-r/a_0}\left(2-\frac{r}{a_0}\right)^2 + 2\left(2-\frac{r}{a_0}\right)\left(\frac{-1}{a_0}\right)e^{-r/a_0} = \frac{-1}{a_0}e^{-r/a_0}\left(2-\frac{r}{a_0}\right)\left(4-\frac{r}{a_0}\right)$$

By inspection we see that $r = 2a_0$ and $r \to \infty$ both satisfy this equation, and they correspond to the zeroes of the wavefunction. The third solution is $r = 4a_0$, and it corresponds to the maximum of the wavefunction's outer lobe. The fact that this is a maximum could also be verified by taking the second derivative and analyzing its sign.

E4.8 Write the energy level formula for the hydrogen atom's electronic energy and identify the different terms in the equation.

Solution

The energy level formula is given by equation

$$E_n = -\frac{1}{4\pi\varepsilon_0}\frac{e^2}{2a_0}\cdot\frac{1}{n^2}$$

where ε_0 is the vacuum permittivity, e is the elementary charge, a_0 is the Bohr radius, and n is the principal quantum number.

E4.9 a) Consider the case where the electron can be described by the $2p_z$ wavefunction (or orbital). Sketch both the radial and angular parts of the wavefunction, and state the number of nodes in each part. b) Determine the distance from the nucleus at which it is most probable to find an electron; i.e., the peak of the radial distribution function.

Solution

a) The $2p_z$ orbital is

$$\psi_{2p_z} = \frac{1}{4\sqrt{2\pi a_0^3}}\frac{z}{a_0}\,e^{-r/(2a_0)} = \frac{\cos\theta}{4\sqrt{2\pi a_0^3}}\frac{r}{a_0}\,e^{-r/(2a_0)}$$

The unnormalized radial part, which has no nodes, is sketched in Fig. E4.9a.

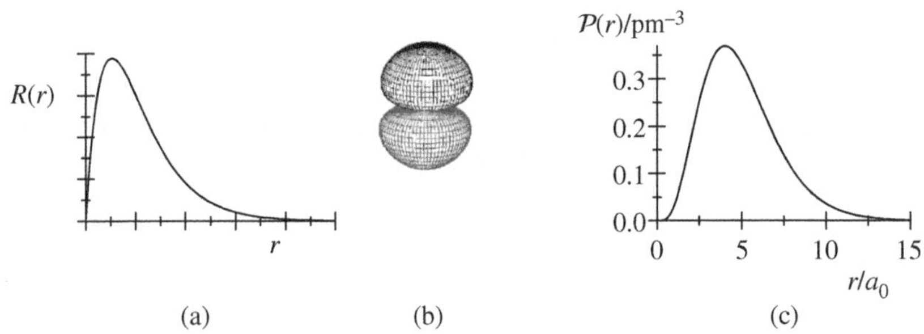

Figure E4.9 Radial and angular parts of the $2p_z$-wavefunction. (a) Radial part, (b) angular part of the wavefunction, (c) probability function $\mathcal{P}$ of the radial part.

The angular part is shown in Fig. E4.9b, and the two-dimensional angular part is sketched in Fig. E4.9c. It has one nodal plane (the *xy* plane).

b) The most probable distance of the electron from the nucleus is determined by taking the derivative of the radial probability distribution function, setting it equal to zero and finding the value of r that satisfies the equation.

$$\frac{d}{dr}\left(4\pi r^2\left(\frac{1}{4\sqrt{2\pi}}\frac{r}{a_0}e^{-\frac{r}{2a_0}}\right)^2\right) = 0 \text{ or } \frac{d}{dr}\left(r^4 e^{-\frac{r}{a_0}}\right) = 0$$

Taking the derivative, we find

$$4r^3\exp\left(-\frac{r}{a_0}\right) - \frac{r^4}{a_0}\exp\left(-\frac{r}{a_0}\right) = 0$$

or

$$r^3\exp\left(-\frac{r}{a_0}\right)\cdot(4 - r/a_0) = 0$$

Solving for r by inspection we have

$$r = 4a_0$$

We can check this result by plotting the radial probability distribution function

$$\frac{4\pi r^2}{32\pi a_0^3}\left(\frac{r}{a_0}e^{-\frac{r}{2a_0}}\right)^2$$

and locating the maximum (Fig. E4.9).

E4.10 Calculate the distance r at which the probability density ρ for the electron in the ground-state hydrogen atom has decreased to $1/1000$ of its maximum value.

Solution

The probability density for an electron in the ground-state hydrogen atom is given by

$$\rho = \frac{1}{\pi a_0^3}\, e^{-2r/a_0}$$

At $r = 0$, this function has the value $1/\pi a_0^3$, and we wish to find the value of r at which its value is equal to $1/1000$ of the value at zero; namely

$$\frac{1}{\pi a_0^3}\, e^{-2r/a_0} = \frac{1}{1000}\frac{1}{\pi a_0^3}$$

Solving for r, we find that

$$r = -\ln\left(\frac{1}{1000}\right)\left[\frac{a_0}{2}\right]$$

or

$$r = 3.454 \cdot a_0$$

E4.11 Show that the ψ_{1s} and the ψ_{2p_z} hydrogen atom wavefunctions are orthogonal to each other.

Solution

To demonstrate orthogonality we show that the integral over the product of the two wavefunctions is zero

$$\int_{-\infty}^{\infty} \psi_{1s}\ \psi_{2s}\ d\tau = 0$$

The hydrogen atom wavefunctions are $\psi_{1s} = c_{1s} \cdot e^{-r/a_0}$ and $\psi_{2\,p_z} = c_{2p} \cdot r \cdot \cos\theta \cdot e^{-r/2a_0}$, where we have used $z = r \cdot \cos\theta$ and do not explicitly write the normalization constants. The differential volume element $d\tau$ is given by $r^2 \sin\theta\ dr\ d\theta\ d\varphi$. Making the substitutions, the orthogonality expression becomes

$$\int_0^{\infty}\int_0^{\pi}\int_0^{2\pi} e^{-r/a_0} \cdot r\cos\theta \cdot e^{-r/2a_0}\ r^2 \sin(\theta)\ dr\ d\theta\ d\varphi = 0$$

or

$$\int_0^{\infty} e^{-r/a_0} \cdot e^{-r/2a_0} r^3 dr \cdot \int_0^{\pi} \sin(\theta)\cos(\theta)\ d\theta \cdot \int_0^{2\pi} d\varphi = 0$$

Note that we have neglected the overall normalization constants because we only wish to show that the integrals are zero. If any of these three integrals has the value zero, then the product will be zero and the orthogonality condition met. Performing the θ integral gives

$$\int_0^{\pi} \sin(\theta)\cos(\theta)\ d\theta = \frac{1}{2}\sin^2(\theta)\Big|_0^{\pi} = \frac{1}{2}\,(0-0) = 0$$

and we see that the orthogonality condition is satisfied.

Note that this result can also be obtained from symmetry considerations. The 1s-function has no nodes and is spherically symmetric, i.e., its amplitude is positive and depends only on the distance r, not the angles. In contrast, the 2p_z-function (see Fig. E4.9) has a node in the xy -plane, with its amplitude being of equal magnitude and opposite sign for the points (x, y, z) and $(x, y, -z)$. Thus the product of both functions has a node in the xy-plane, for which the positive and negative values of the product function cancel each other and the sum of all product values becomes zero.

E4.12 Plot the radial distribution functions for the first three s-orbitals of the hydrogen atom, and compare the most probable radius for the electron in each case. Comment on the trend you observe.

Solution

The three s-orbital radial distribution functions are

$$4\pi r^2 \rho_{1s} = 4\pi r^2 \frac{1}{\pi a_0^3} e^{-2r/a_0}$$

$$4\pi r^2 \rho_{2s} = 4\pi r^2 \frac{1}{32\pi a_0^3}\left(2 - \frac{r}{a_0}\right)^2 e^{-r/a_0}$$

and

$$4\pi r^2 \rho_{3s} = 4\pi r^2 \frac{1}{81^2 3\pi a_0^3}\left(27 - 18\frac{r}{a_0} + 2\frac{r^2}{a_0^2}\right)^2 e^{-2r/(3a_0)}$$

In Figure E4.12 we plot the functions ($\mathcal{P}(r)$ in units of pm^{-1}) versus the distance r/a_0

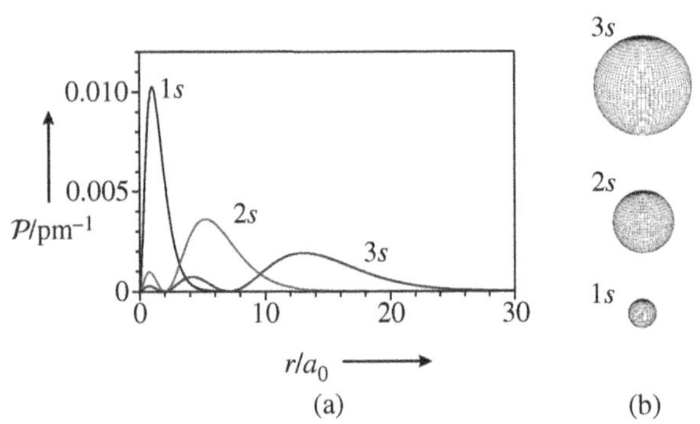

Figure E4.12 (a) Radial part of the 1s, 2s, and 3s-wavefunctions, the probability function $\mathcal{P}$ is shown. (b) Angular parts of the wavefunctions.

The most probable radius for these functions is the value of r at which the probability distribution has its maximum. By inspection of the graph we see that the most probable radius increases with the increase in the principal quantum number n; however, the functions show that the s-orbitals of higher quantum number have some electron density close to the nucleus as well. Considering the higher energy of the electron in the higher principal quantum number states, we expect a more diffuse electron cloud when the electron is excited.

E4.13 For $n = 3$, show that the size of the atom (as measured by the radial distribution function) decreases as the l quantum number increases. Provide a physical explanation for this trend.

Solution

One way to examine this trend is to plot the radial distribution functions (RDF) for $n = 3$; $l = 0$; $l = 1$; and $l = 2$; see Figure E4.13. For the 3s orbital, with $\psi(3s) = 1/(81\sqrt{3\pi a_0^3}) \cdot (27 - 18r/a_0 + 2r^2/a_0^2) \cdot \exp(-r/3a_0)$, we find that

$$\text{RDF}(3s) = \frac{1}{81^2 3\pi a_0}\left(\frac{r}{a_0}\right)^2\left(27 - 18\frac{r}{a_0} + 2\frac{r^2}{a_0^2}\right)^2 e^{-2r/(3a_0)} \int_0^{2\pi} d\phi \int_0^{\pi} \sin\theta \, d\theta$$

$$= 4\pi \frac{1}{81^2 3\pi a_0}\left(\frac{r}{a_0}\right)^2\left(27 - 18\frac{r}{a_0} + 2\frac{r^2}{a_0^2}\right)^2 e^{-2r/(3a_0)} = 4\pi r^2 \rho_{3s}$$

For the 3$p_z$orbital, with $\psi(3p_z) = \sqrt{2}/(81\sqrt{\pi a_0^3}) \cdot (r/a_0) \cdot (6 - r/a_0) \cdot \cos\theta \cdot \exp(-r/3a_0)$, we find that

$$\text{RDF}(3p_z) = \frac{2}{81^2\,\pi a_0}\left(\frac{r}{a_0}\right)^4\left(6 - \frac{r}{a_0}\right)^2 e^{-2r/(3a_0)} \int_0^{2\pi} d\phi \int_0^{\pi} \cos^2\theta \sin\theta \, d\theta$$

$$= \frac{4\pi}{3}\frac{2}{81^2\,\pi a_0}\left(\frac{r}{a_0}\right)^4\left(6 - \frac{r}{a_0}\right)^2 e^{-2r/(3a_0)}$$

For the d_{z^2} orbital, with $\psi(3d_{z^2}) = 1/(81\sqrt{6\pi a_0^3}) \cdot (r^2/a_0^2) \cdot [3\cos^2\theta - 1]\exp(-r/3a_0)$, we find that

$$\mathrm{RDF}(3d_{z^2}) = \frac{1}{81^2\, 6\pi a_0}\left(\frac{r}{a_0}\right)^6 e^{-2r/(3a_0)} \int_0^{2\pi} d\phi \int_0^{\pi} (3\cos^2\theta - 1)^2 \sin\theta\, d\theta$$

$$= \frac{16\pi}{5}\frac{1}{81^2\, 6\pi a_0}\left(\frac{r}{a_0}\right)^6 e^{-2r/(3a_0)}$$

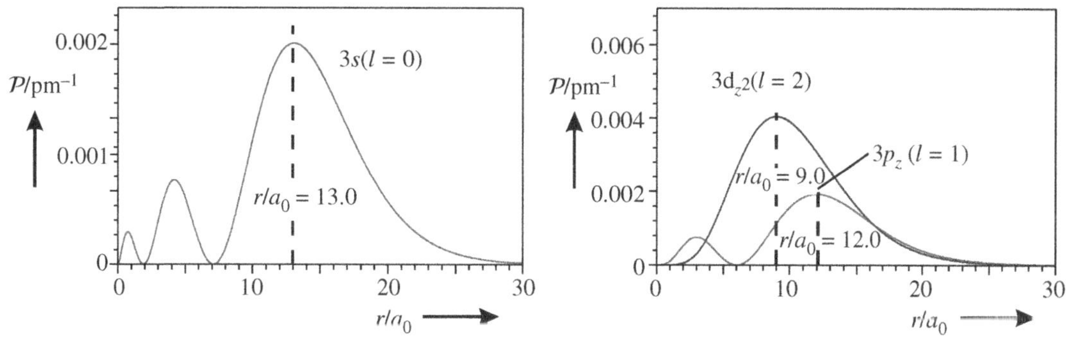

Figure E4.13 Probability function $\mathcal{P}$ (radial distribution function) of the wavefunctions $3s(l = 0)$ (left), $3p_x(l = 1)$, and $3d_{xy}(l = 2)$ (right).

From these plots, it is clear that as the quantum number l increases the maximum in the radial distribution function moves closer to the nucleus: from $r/a_0 = 13.0$ for $l = 0$ to $r/a_0 = 9.0$ for $l = 2$. This shift reflects the difference in the radial and angular contributions to the kinetic energy. The average total kinetic energy is the same for electrons with the same principal quantum number (see text and Virial Theorem). Hence, as the average contribution of the angular kinetic energy increases with increasing l, the average contribution of the radial kinetic energy decreases, and it is reflected in the smaller number of radial nodes and a contraction of the outer lobe of the wavefunction with the change in l.

E4.14 Calculate the normalization constant for the wavefunction of the hydrogen atom in its ground state, $\psi = A\, e^{-r/a_0}$. Remember that you must perform a three-dimensional integral. It is best to perform this integral in spherical coordinates (Fig. E4.14).

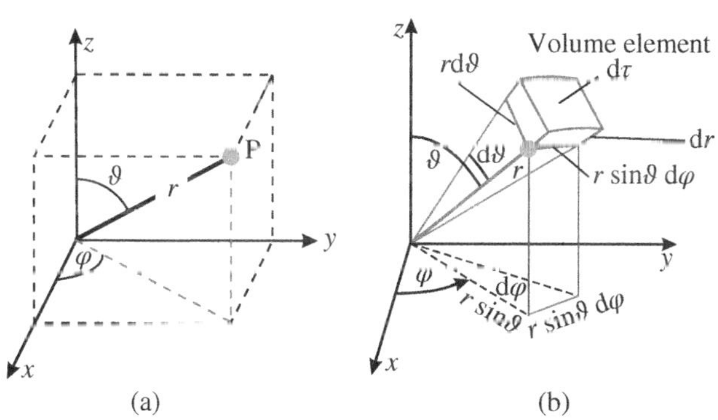

Figure E4.14 (a) Spherical coordinates r, ϑ, and φ of point P. From geometry we see that $z = r\cos\vartheta$; $x = r\sin\vartheta\cos\varphi$; $y = r\sin\vartheta\sin\varphi$. (b) Volume element $d\tau = (r\sin\vartheta\, d\varphi)(rd\vartheta)(dr) = r^2\sin\vartheta\, dr\, d\varphi\, d\vartheta$. Note that this relation is valid only for sufficiently small values of dr, $d\varphi$, and $d\vartheta$.

From Fig. E4.14 we find that the volume element is

$$d\tau = d\varphi \cdot \sin\vartheta \cdot d\vartheta \cdot r^2\, dr$$

Because ψ is a function of r, but has no explicit dependence on ϑ or φ, we can integrate over the angles to obtain

$$dV = \int_0^{2\pi} d\varphi \int_0^{\pi} \sin\vartheta \cdot d\vartheta \cdot r^2\ dr = |\varphi|_0^{2\pi} \cdot |-\cos\vartheta|_0^{\pi} \cdot r^2\ dr = 2\pi \cdot 2 \cdot r^2\ dr = 4\pi r^2\ dr$$

Solution

We start with

$$\int_0^{\infty} \left(Ae^{-r/a_0}\right)^2 4\pi r^2\ dr = 1$$

This leads to

$$4\pi A^2 \int_0^{\infty} e^{-2r/a_0} r^2\ dr = 1$$

evaluating the integral using $\int_0^{\infty} e^{-ax} x^n\ dx = \frac{n!}{a^{n+1}}$, $(a > 0,\ n$ positive integer) we have

$$\int_0^{\infty} e^{-2r/a_0} r^2\ dr = \frac{2}{(2/a_0)^3}$$

Using this value for the integral, we find that

$$4\pi A^2 \left(\frac{2a_0^3}{8}\right) = 1$$

Solving for A we obtain

$$A = \sqrt{\frac{1}{\pi a_0^3}}$$

Hence we see that

$$\psi = Ae^{-r/a_0} = \psi = \sqrt{\frac{1}{\pi a_0^3}} e^{-r/a_0}$$

which is Equation (4.3)

E4.15 Write the Schrödinger equation for He^+. The Schrödinger equation here is like that in Equation (4.2), but the nuclear charge is $2e$ rather than e. Given that the ground-state wavefunction of He^+ is $\psi = A\ e^{-2r/a_0}$ show that

$$E = -\frac{1}{4\pi\varepsilon_0}\frac{2e^2}{a_0}$$

Solution

The Hamiltonian operator for the He^+ ion is

$$\mathcal{H} = -\frac{\hbar^2}{2m_e}\left[\frac{1}{r^2}\frac{\partial}{\partial r}\left(r^2\frac{\partial}{\partial r}\right) + \frac{1}{r^2\sin\theta}\frac{\partial}{\partial\theta}\left(\sin\theta\frac{\partial}{\partial\theta}\right) + \frac{1}{r^2\sin^2\theta}\frac{\partial^2}{\partial\varphi^2}\right] - \frac{2e^2}{4\pi\varepsilon_0}\frac{1}{r}$$

Substituting the given wavefunction into the Schrödinger equation gives

$$\mathcal{H}\psi(r) = -\frac{\hbar^2}{2m_e}\left[\frac{1}{r^2}\frac{\partial}{\partial r}\left(r^2\frac{\partial}{\partial r}\right)\frac{1}{\sqrt{\pi}}\left(\frac{2}{a_0}\right)^{3/2} e^{-2r/a_0}\right] - \frac{2e^2}{4\pi\varepsilon_0}\frac{1}{r}\cdot\psi(r) = E\cdot\psi(r)$$

Note that the angular derivatives are equal to zero because the wavefunction only depends on r. Taking the first r derivative and factoring out the constants gives

$$\mathcal{H}\psi(r) = -\frac{\hbar^2}{2m_e}\frac{1}{\sqrt{\pi}}\left(\frac{2}{a_0}\right)^{3/2}\left[\frac{1}{r^2}\frac{\partial}{\partial r}\left(r^2 e^{-2r/a_0}\left(\frac{-2}{a_0}\right)\right)\right] - \frac{2e^2}{4\pi\varepsilon_0}\frac{1}{r}\cdot\psi(r) = E\cdot\psi(r)$$

Taking the second r derivative yields

$$\mathcal{H}\psi(r) = -\frac{\hbar^2}{m_e}\sqrt{\frac{2}{\pi a_0^3}}\left[\frac{-2}{a_0 r^2}\left(r^2 e^{-2r/a_0}\left(\frac{-2}{a_0}\right) + e^{-2r/a_0}(2r)\right)\right] - \frac{2e^2}{4\pi\varepsilon_0}\frac{1}{r}\psi(r) = E\cdot\psi(r)$$

which on factoring out the exponential is

$$\mathcal{H}\psi(r) = \frac{1}{\sqrt{\pi}}\left(\frac{2}{a_0}\right)^{3/2} e^{-2r/a_0}\left[\frac{\hbar^2}{2m_e}\frac{2}{a_0 r^2}\left(r^2\left(\frac{-2}{a_0}\right)+(2r)\right)\right] - \frac{2e^2}{4\pi\varepsilon_0}\frac{1}{r}\psi(r)$$

$$= \left[\frac{\hbar^2}{2m_e}\frac{2}{a_0 r^2}\left(r^2\left(\frac{-2}{a_0}\right)+(2r)\right)\right]\psi(r) - \frac{2e^2}{4\pi\varepsilon_0}\frac{1}{r}\psi(r) = E\cdot\psi(r)$$

From inspection we can write an expression for the energy as

$$E = \frac{\hbar^2}{2m_e}\left[\frac{2}{a_0 r^2}\left(r^2\left(\frac{-2}{a_0}\right)+(2r)\right)\right] - \frac{2e^2}{4\pi\varepsilon_0}\frac{1}{r} = -\frac{\hbar^2}{2m_e}\cdot\frac{4}{a_0^2} + \frac{\hbar^2}{2m_e}\cdot\frac{4}{a_0 r} - \frac{e^2}{2\pi\varepsilon_0 r}$$

In order to simplify the result, we substitute in the value of $a_0 = 4\pi\varepsilon_0\hbar^2/(m_e e^2)$ and write

$$E = -\frac{\hbar^2}{2m_e}\cdot\frac{4}{a_0\cdot\left(\frac{4\pi\varepsilon_0\hbar^2}{m_e e^2}\right)} + \frac{\hbar^2}{2m_e}\cdot\frac{4}{\left(\frac{4\pi\varepsilon_0\hbar^2}{m_e e^2}\right)r} - \frac{e^2}{2\pi\varepsilon_0 r}$$

$$= -\frac{1}{2}\cdot\frac{e^2}{a_0\cdot\pi\varepsilon_0} + \frac{e^2}{2\pi\varepsilon_0 r} - \frac{e^2}{2\pi\varepsilon_0 r} = -\frac{1}{2}\cdot\frac{e^2}{a_0\cdot\pi\varepsilon_0} = -\frac{1}{4\pi\varepsilon_0}\frac{2e^2}{a_0}$$

E4.16 In Example 4.1 in the text, we find the derivatives of ψ for the ground state of the H atom with respect to x. In an analogous way find the derivatives with respect to the variables y and z; and then show that

$$\nabla^2\psi = \frac{\partial^2\psi}{\partial x^2} + \frac{\partial^2\psi}{\partial y^2} + \frac{\partial^2\psi}{\partial z^2} = -\frac{1}{\sqrt{\pi a_0^3}}\cdot\frac{1}{a_0}\cdot\left(\frac{2}{r} - \frac{1}{a_0}\right)\cdot e^{-r/a_0}$$

Solution

We will start by writing $\psi(x, y, z)$ and then taking the derivatives.

$$\psi(x,y,z) = \frac{1}{\sqrt{\pi a_0^3}} e^{-\sqrt{(x^2+y^2+z^2)}/a_0}$$

Taking the first derivatives we find that

$$\frac{\partial\,\psi(x,y,z)}{\partial y} = \frac{1}{\sqrt{\pi a_0^3}}\frac{1}{a_0}\frac{(-y)}{\sqrt{(x^2+y^2+z^2)}} e^{-\sqrt{(x^2+y^2+z^2)}/a_0}$$

and

$$\frac{\partial\,\psi(x,y,z)}{\partial z} = \frac{1}{\sqrt{\pi a_0^3}}\frac{1}{a_0}\frac{(-z)}{\sqrt{(x^2+y^2+z^2)}} e^{-\sqrt{(x^2+y^2+z^2)}/a_0}.$$

The expressions for the second derivatives are thus

$$\frac{\partial^2\,\psi(x,y,z)}{\partial y^2} = -\frac{1}{\sqrt{\pi a_0^3}}\frac{1}{a_0}\cdot e^{-\sqrt{(x^2+y^2+z^2)}/a_0}\cdot\left[\frac{1}{\sqrt{(x^2+y^2+z^2)}} + \frac{-y^2}{(x^2+y^2+z^2)^{3/2}} + \frac{-y^2}{a_0\,(x^2+y^2+z^2)}\right]$$

$$= -\frac{1}{\sqrt{\pi a_0^3}}\frac{1}{a_0}\cdot e^{-\sqrt{(x^2+y^2+z^2)}/a_0}\cdot\left[\frac{r - y^2/r - y^2/a_0}{(x^2+y^2+z^2)}\right]$$

and

$$\frac{\partial^2\,\psi(x,y,z)}{\partial z^2} = -\frac{1}{\sqrt{\pi a_0^3}}\frac{1}{a_0}\cdot e^{-\sqrt{(x^2+y^2+z^2)}/a_0}\cdot\left[\frac{1}{\sqrt{(x^2+y^2+z^2)}} + \frac{-z^2}{(x^2+y^2+z^2)^{3/2}} + \frac{-z^2}{a_0\,(x^2+y^2+z^2)}\right]$$

$$= -\frac{1}{\sqrt{\pi a_0^3}}\frac{1}{a_0}\cdot e^{-\sqrt{(x^2+y^2+z^2)}/a_0}\cdot\left[\frac{r - z^2/r - z^2/a_0}{(x^2+y^2+z^2)}\right]$$

Combining the results together for x, y, and z, we find that

$$\nabla^2\psi = \frac{\partial^2\psi}{\partial x^2} + \frac{\partial^2\psi}{\partial y^2} + \frac{\partial^2\psi}{\partial z^2} = -\frac{1}{\sqrt{\pi a_0^3}} \cdot \frac{1}{a_0} \cdot \frac{3r - \frac{x^2}{r} - \frac{y^2}{r} - \frac{z^2}{r} - \frac{x^2}{a_0} - \frac{y^2}{a_0} - \frac{z^2}{a_0}}{x^2 + y^2 + z^2} \cdot e^{-r/a_0}$$

$$= -\frac{1}{\sqrt{\pi a_0^3}} \cdot \frac{1}{a_0} \cdot \frac{3r - \frac{r^2}{r} - \frac{r^2}{a_0}}{r^2} \cdot e^{-r/a_0} = -\frac{1}{\sqrt{\pi a_0^3}} \cdot \frac{1}{a_0} \cdot \frac{2r - \frac{r^2}{a_0}}{r^2} \cdot e^{-r/a_0}$$

$$= -\frac{1}{\sqrt{\pi a_0^3}} \cdot \frac{1}{a_0} \cdot \left(\frac{2}{r} - \frac{1}{a_0}\right) \cdot e^{-r/a_0}$$

E4.17 Show that the equality in Equation (4.12) holds. In particular work out the details for how the three-dimensional integral reduces to that used in Example 4.3 in the text.

Solution

The expression for the average potential energy is $\overline{V} = \int_{all\ space} V\ \rho\ d\tau$. For the ground state of the hydrogen atom $V = \frac{-e^2}{4\pi\varepsilon_0 r}$ and $\rho = \psi^2 = \frac{1}{\pi a_0^3} e^{-2r/a_0}$. Substitution of both these expressions into that for the average potential energy, and writing $d\tau$ out explicitly gives

$$\overline{V} = \int_{all\ space} V\ \rho\ d\tau = \int_0^{\infty}\int_0^{\pi}\int_0^{2\pi} \frac{-e^2}{4\pi\varepsilon_0 r} \cdot \frac{1}{\pi a_0^3} e^{-2r/a_0}\ r^2 \sin(\theta)\ dr\ d\theta\ d\varphi$$

Performing the θ and φ integrals and pulling the constants out gives

$$\overline{V} = \frac{-e^2}{4\pi\varepsilon_0} \cdot \frac{4\pi}{\pi a_0^3} \int_0^{\infty} e^{-2r/a_0}\ r\ dr$$

From the integral table $\int_0^{\infty} e^{-ax} x^n\ dx = \frac{n!}{a^{n+1}}$, ($a > 0$, $n$positive integer), so that the average potential energy is

$$\overline{V} = \frac{-e^2}{4\pi\varepsilon_0} \cdot \frac{4\pi}{\pi a_0^3} \cdot \frac{1!}{(2/a_0)^2} = \frac{-e^2}{4\pi\varepsilon_0} \cdot \frac{4\pi}{\pi a_0^3} \cdot \frac{a_0^2}{4} = \frac{-e^2}{4\pi\varepsilon_0} \cdot \frac{1}{a_0}$$

as given in the example.

E4.18 Calculate the average distance $\overline{r}$ for an electron in the $2p_z$ state of the hydrogen atom.

Solution

This calculation was performed in Section 4.3.3 of the text. For ψ_{2pz} we find

$$\psi_{2pz} = \frac{1}{4\sqrt{2\pi a_0^3}} \frac{z}{a_0} e^{-\frac{r}{2a_0}} = \frac{1}{4\sqrt{2\pi a_0^3}} \frac{r}{a_0} \cos\vartheta\ e^{-\frac{r}{2a_0}}$$

where we have replaced z by $r\cos\vartheta$, according to Fig. E4.14. In spherical coordinates

$$\overline{r} = \iiint \rho \cdot r\ d\tau = \frac{1}{32\pi a_0^3} \int_{r=0}^{\infty}\int_{\vartheta=0}^{\pi}\int_{\varphi=0}^{2\pi} \frac{r^2}{a_0^2} \cos^2\vartheta\ e^{-\frac{r}{a_0}} \cdot r \cdot r^2\ d\vartheta \cdot \sin\vartheta \cdot d\varphi \cdot dr$$

$$= \frac{1}{32\pi a_0^3} \int_{r=0}^{\infty} \frac{r^5}{a_0^2} e^{-\frac{r}{a_0}} \cdot dr \cdot \int_{\vartheta=0}^{\pi} \cos^2\vartheta \sin\vartheta\ d\vartheta \cdot \int_{\varphi=0}^{2\pi} d\varphi = \frac{1}{32\pi a_0^3}\left(a_0^4 \cdot 5! \cdot \frac{2}{3} \cdot 2\pi\right) = 5a_0 = 264\ \text{pm}$$

where we used the solutions of the integrals given in Appendix B.

E4.19 It is common for quantum chemists to work in a set of units called 'atomic units', which can be obtained by using the Bohr radius a_0 as the unit of distance. By transforming the coordinates (x, y, z) to $(a_0 \cdot x', a_0 \cdot y', a_0 \cdot z')$, show that the Schrödinger equation for the H-atom (Equation (4.2)) can be written as

$$-\frac{1}{2}\nabla^2\psi - \frac{1}{r'}\psi = E_{a.u} \cdot \psi \text{ where } E_{a.u} = \frac{1}{2} \cdot \frac{1}{n^2}$$

where $E_{a.u.}$ is the energy in atomic units.

Solution

First we make the substitution for the variables (x, y, z) and write

$$-\frac{\hbar^2}{2m_e}\frac{1}{a_0^2}\left(\frac{\partial^2}{\partial x'^2}+\frac{\partial^2}{\partial y'^2}+\frac{\partial^2}{\partial z'^2}\right)\psi-\frac{1}{a_0}\frac{e^2}{4\pi\varepsilon_0 r'}\psi=E\cdot\psi$$

Next we use the fact that $a_0 = 4\pi\varepsilon_0\frac{\hbar^2}{m_e e^2}$, to write

$$-\frac{\hbar^2}{2m_e}\frac{m_e^2e^4}{(4\pi\varepsilon_0)^2\hbar^4}\left(\frac{\partial^2}{\partial x'^2}+\frac{\partial^2}{\partial y'^2}+\frac{\partial^2}{\partial z'^2}\right)\psi-\frac{m_e e^2}{4\pi\varepsilon_0\hbar^2}\frac{e^2}{4\pi\varepsilon_0 r'}\psi=E\cdot\psi$$

which simplifies to

$$-\frac{1}{2}\frac{m_e e^4}{(4\pi\varepsilon_0)^2\hbar^2}\left(\frac{\partial^2}{\partial x'^2}+\frac{\partial^2}{\partial y'^2}+\frac{\partial^2}{\partial z'^2}\right)\psi-\frac{m_e e^4}{(4\pi\varepsilon_0)^2\hbar^2}\frac{1}{r'}\psi=E\cdot\psi$$

Factoring out the constants and rearranging, we find that

$$-\frac{1}{2}\left(\frac{\partial^2}{\partial x'^2}+\frac{\partial^2}{\partial y'^2}+\frac{\partial^2}{\partial z'^2}\right)\psi-\frac{1}{r'}\psi=\frac{(4\pi\varepsilon_0)^2\hbar^2}{m_e e^4}E\cdot\psi$$

or more compactly

$$-\frac{1}{2}\nabla^2\psi-\frac{1}{r'}\psi=E_{a.u}\cdot\psi \quad \text{where } E_{a.u}=\frac{(4\pi\varepsilon_0)^2\hbar^2}{m_e e^4}\cdot E$$

Using our solution, for the H-atom energy levels,

$$E=-\frac{m_e\ e^4}{2(4\pi\varepsilon_0)^2\ \hbar^2}\cdot\frac{1}{n^2}$$

we can substitute and show that

$$E_{a.u}=-\frac{(4\pi\varepsilon_0)^2\hbar^2}{m_e e^4}\cdot\frac{m_e\ e^4}{2(4\pi\varepsilon_0)^2\ \hbar^2}\cdot\frac{1}{n^2}=\frac{1}{2}\cdot\frac{1}{n^2}$$

E4.20 If a source of 58.5 nm wavelength photons (typical of that used in photoelectron spectroscopy) irradiates a sample of neutral hydrogen atoms, it is possible to eject electrons from the atoms and generate protons. The most stable hydrogen atoms (i.e., ground-state hydrogen atoms) bind the electron with about 13.6 eV of energy. Convert the 13.6 eV binding energy into Joules. Determine the energy of the 58.5 nm photon. Use energy conservation to deduce the kinetic energy of the photo-ejected electron. If you were trying to measure the electron's kinetic energy in an apparatus analogous to that illustrated in Fig. 1.2, what stopping voltage would you need to apply? Comment on why electron Volts (eV) might be a convenient energy unit for scientists doing photoelectron spectroscopy.

Solution

First we calculate the energy of a 58.5 nm photon and compare it to the binding energy of an electron in the H-atom. Using $E = h\nu$ we find that

$$E=h\nu=\frac{hc}{\lambda}=\frac{6.626\times10^{-34}\ \text{J s}\cdot2.998\times10^{8}\ \text{m s}^{-1}}{58.5\times10^{-9}\ \text{m}}=3.40\times10^{-18}\ \text{J for the photon energy.}$$

The binding energy of the electron to the proton in the H-atom is

$$13.6\ \text{eV}\cdot1.60\times10^{-19}\ \text{J/eV}=2.18\times10^{-18}\ \text{J.}$$

When the H-atom absorbs the light, the photon energy is converted into electronic energy, placing the electron above the ionization limit and it can escape from the atom. The kinetic energy of the photoejected electron is given by the difference between this absorbed energy and its initial state binding energy, hence the difference between these two energies, or

$$(3.40-2.18)\times10^{-18}\ \text{J}=1.22\times10^{-18}\ \text{J.}$$

To stop the electrons, the applied potential must be made at least as large as the electron kinetic energy. The kinetic energy computed above corresponds to an energy of $1.22\times10^{-18}\text{J}/(1.60\times10^{-19}\text{J /eV}) = 7.62$ eV. Hence the stopping potential would be 7.62 V.

The kinetic energies of the ejected electrons correspond to energies of 1 – 10 eV (a conveniently sized quantity because of its direct relation to the voltage difference used in detecting the photoelectrons).

E4.21 Consider the kinetic energy calculations in E1.3. Perform a similar calculation but use a photon wavelength of 23 nm. Perform a similar calculation but use a photon wavelength of 100 nm. Compare your three values of the kinetic energy and comment on them.
Assuming that the electron is bound by 13.6 eV, determine the largest wavelength that a photon may have if it is to eject an electron from a hydrogen atom.

Solution
For the 23 nm photon, we calculate

$$E = \frac{hc}{\lambda} = \frac{6.626 \times 10^{-34}\ \text{J s} \cdot 2.998 \times 10^{8}\ \text{m s}^{-1}}{23 \times 10^{-9}\ \text{m}} = 8.6 \times 10^{-18}\ \text{J per photon}$$

and for the 100 nm photon we find

$$E = \frac{hc}{\lambda} = \frac{6.626 \times 10^{-34}\ \text{J s} \cdot 2.998 \times 10^{8}\ \text{m s}^{-1}}{100 \times 10^{-9}\ \text{m}} = 1.99 \times 10^{-18}\ \text{J}$$

The kinetic energies of the photoejected electrons resulting from these wavelengths are -1.9×10^{-19} J (100 nm) and 6.4×10^{-18} J (23 nm). The negative value indicates that an electron cannot be ejected from an H atom by 100 nm light. The increase in kinetic energy with shorter wavelengths results because the shorter wavelength photons have a higher energy. The table shows how the photon energy and the kinetic energy of the electron increase with decreasing wavelength.

λ/nm	100	58.5	23
$E/10^{-18}$ J	1.99	3.40	8.6
kinetic energy/10^{-18} J	—	1.22	6.4

The longest wavelength photon that can photoionize an H-atom is that corresponding to a light energy of 13.6 eV, or

$$\lambda = \frac{hc}{E} = \frac{6.626 \times 10^{-34}\ \text{J s} \cdot 2.998 \times 10^{8}\ \text{m s}^{-1})}{13.6\ \text{eV} \cdot 1.602 \times 10^{-19}\ \text{J /eV}} = 91.2\ \text{nm}$$

4.2 Problems

P4.1 An early model of the atom, called the Bohr atom model, is of historical significance in the development of quantum mechanics. Before the introduction of wave mechanics, Bohr created an atomic model (Fig. P4.1).

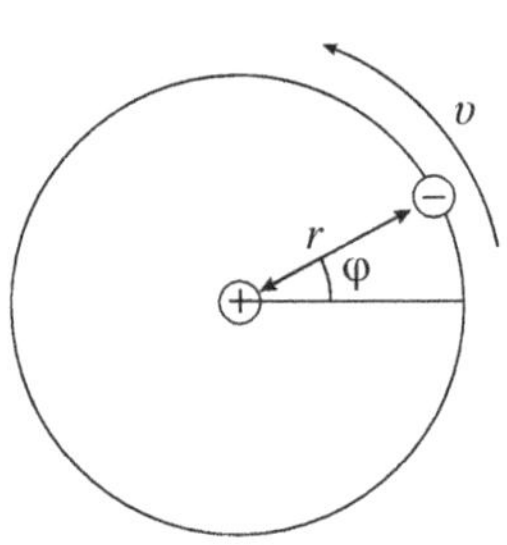

Figure P4.1 Bohr atom model; the electron circles at a distance r with the speed v around the nucleus.

This model views the electron as a particle that orbits about the proton at a distance r and has a velocity v. In this model, Bohr considered that the electrostatic attraction supplies the centripetal force ($m_e v^2/r$) which holds the electron in a circle. In addition, he postulated that the orbital angular momentum was quantized and given by the formula $m_e vr = h/(2\pi)$ in the ground electronic state. By using the balance of forces and the postulate for the angular momentum, show that the electron's distance from the nucleus is given by

$$r = 4\pi\varepsilon_0 \frac{h^2}{4\pi^2\, m_e e^2}$$

and the electron's speed is given by

$$v = \frac{4\pi^2}{4\pi\varepsilon_0}\frac{e^2}{h}$$

Note that the orbit's radius is the same as that we found for the most probable distance a_0; a_0 is called the *Bohr radius*. Proceed to calculate the kinetic and potential energy from v and r.

Solution

From the balance between the centripetal force, $m_e v^2/r$, and the Coulomb force, we can write

$$\frac{m_e v^2}{r} = \frac{1}{4\pi\varepsilon_0}\cdot\frac{e^2}{r^2}$$

which can be rearranged to write

$$m_e v^2 r = \frac{e^2}{4\pi\varepsilon_0}$$

Using Bohr's postulate that the angular momentum of the electron equals $\hbar = h/(2\pi)$ (i.e., $m_e v r = \hbar$), we can write that

$$\hbar v = \frac{e^2}{4\pi\varepsilon_0} \quad \text{or} \quad v = \frac{e^2}{4\pi\varepsilon_0\hbar}$$

Combining this result for the speed v and with the quantization postulate ($m_e v r = \hbar$), we can find r; namely

$$r = \frac{\hbar}{m_e v} = \frac{\hbar}{m_e}\frac{4\pi\varepsilon_0\hbar}{e^2} = 4\pi\varepsilon_0\frac{\hbar^2}{m_e e^2}$$

For the radius we obtain the same expression as for the most probable distance a_0; so a_0 is called the Bohr radius. We can calculate the kinetic and potential energy from v and r. For the kinetic energy, we find

$$T = \frac{1}{2}m_e v^2 = \frac{1}{2}m_e\left(\frac{1}{4\pi\varepsilon_0}\cdot\frac{e^2}{\hbar}\right)^2 = \frac{1}{4\pi\varepsilon_0}\cdot\frac{e^2}{2a_0}$$

and for the potential energy we find

$$V = -\frac{1}{4\pi\varepsilon_0}\cdot\frac{e^2}{a_0}$$

where we have used the fact that $r = a_0$. These results are the same as that found from solving the Schrödinger equation.

In spite of this agreement the Bohr theory leads to a false prediction; a nucleus and circling electron together make up a rotating dipole, and according to the theory of electromagnetism this dipole would have to emit electromagnetic radiation of frequency

$$\nu = \frac{v}{2\pi r} = \frac{1}{4\pi\varepsilon_0}\cdot\frac{2\pi e^2}{2\pi h a_0} = \frac{1}{4\pi\varepsilon_0}\cdot\frac{e^2}{h a_0} = 6\times10^{15}\ \text{s}^{-1}$$

and this does not occur. Moreover, the emission of radiation would lead to a decrease in the kinetic energy of the electron and finally to a crash of the electron into the nucleus. Nevertheless the agreement of the energies and of the Bohr radius with the most probable distance shows the close relation between classical and quantum mechanical considerations. For further information see A. Pais, Niels Bohr's *Times: In Physics, Philosophy and Polity* (Clarendon Press, Oxford, 1991).

P4.2 The energy of the H-atom was calculated with the assumption that the nucleus is at rest; actually, it is the center of mass which is stationary. Account for the movement of the nucleus (see Fig. P4.2) and calculate the energy. To calculate the kinetic and potential energy, you need to replace m_e by the reduced mass

$$m_{\text{red}} = \frac{m_e m_p}{m_e + m_p}$$

Determine the new energy expression by using the Bohr model of the atom. How much does the energy shift from our more approximate solution? Is this shift significant?

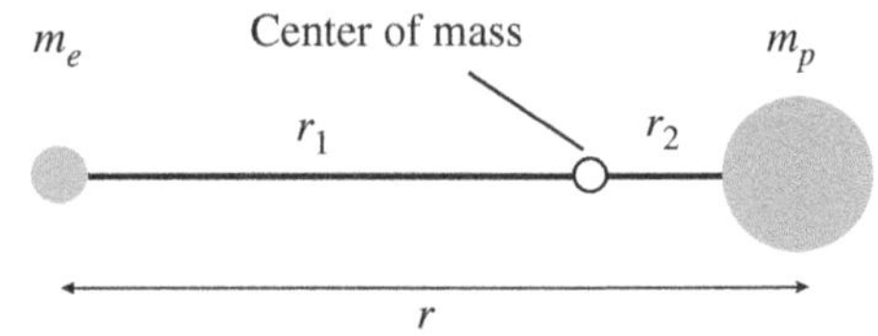

Figure P4.2 This diagram indicates the geometry for the center of mass.

Solution

The electron (mass m_e) and the proton (mass m_p) are rotating around the center of mass (distance r_1 for the electron, distance r_2 for the nucleus). Thus, the balance of the centripetal force and the Coulomb force leads to

$$\frac{m_e v^2}{r_1} = m_e r_1 \left(\frac{d\varphi}{dt}\right)^2 = \frac{1}{4\pi\varepsilon_0} \cdot \frac{e^2}{r^2}$$

With $r_1 + r_2 = r$ and $m_e r_1 = m_p r_2$ we find

$$m_e r_1 = \frac{m_e m_p}{m_e + m_p} r = m_{\text{red}} \cdot r$$

so that

$$m_{\text{red}} \cdot r \left(\frac{d\varphi}{dt}\right)^2 = \frac{1}{4\pi\varepsilon_0} \cdot \frac{e^2}{r^2}$$

The condition for the quantization of the angular momentum may be written as

$$m_e r_1^2 \frac{d\varphi}{dt} + m_p r_2^2 \frac{d\varphi}{dt} = \hbar \quad \text{or} \quad m_{\text{red}} \cdot r^2 \frac{d\varphi}{dt} = \hbar$$

To calculate the kinetic and potential energy we have to replace m_e by the reduced mass

$$m_{\text{red}} = \frac{m_e m_p}{m_e + m_p}$$

so that

$$T = \frac{1}{4\pi\varepsilon_0} \cdot \frac{e^2}{2a_0} \cdot \frac{m_p}{m_e + m_p}, \qquad V = -\frac{1}{4\pi\varepsilon_0} \cdot \frac{e^2}{a_0} \cdot \frac{m_p}{m_e + m_p}$$

The total energy is given by

$$E = T + V = \frac{1}{4\pi\varepsilon_0} \cdot \frac{e^2}{2a_0} \cdot \frac{m_p}{m_e + m_p} - \frac{1}{4\pi\varepsilon_0} \cdot \frac{e^2}{a_0} \cdot \frac{m_p}{m_e + m_p} = -\frac{1}{4\pi\varepsilon_0} \cdot \frac{e^2}{a_0} \cdot \frac{m_p}{m_e + m_p}$$

Using $a_0 = 4\pi\varepsilon_0 \frac{\hbar^2}{m_e e^2}$, we find

$$E = -\left(\frac{1}{4\pi\varepsilon_0}\right)^2 \cdot \frac{e^4}{\hbar^2} \cdot \frac{m_e m_p}{m_e + m_p} = -\left(\frac{e^2}{4\pi\varepsilon_0 \hbar}\right)^2 \cdot m_{\text{red}}$$

For the H atom, we obtain $m_{\text{red}} = m_e \times 0.99946$; the corresponding expression for the He atom is $m_{\text{red}} = m_e \times 0.99987$.

This small change in the energies can be detected by shifts in the frequency of the atomic emission lines (correction in Table 4.2).

P4.3 The MathCAD exercises included in the module "Hatom.mcd" asks you to plot and analyze features of the hydrogen atom orbitals.

Solution

The full solutions can be found in the file "Hatom_solutions.mcdx," and it consists of 6 parts, or Problems. Here we give only part of the solutions.

Problem 1 has three components. a) Figure P4.3a shows the radial wavefunction in the 1s, 2s, and 3s-state (left) and the squares (right). Note how the probability density for the electron near the nucleus decreases systematically as the principal quantum number increases.

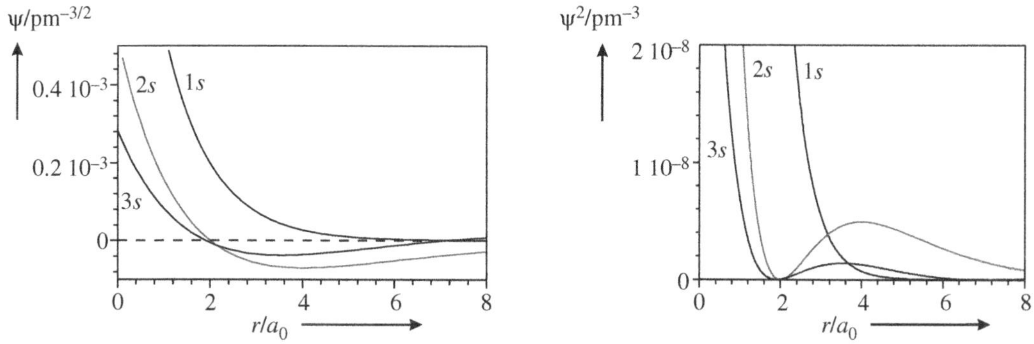

Figure P4.3a H-atom in the 1s, 2s, and 3s-state. Left: wavefunction ψ, right: square ψ^2.

In Fig. P4.3b we show the cases for the H-atom and the Li^{2+} ion with nuclear charge $Ze = 3e$:

$$\psi_{1s} = \frac{1}{\sqrt{\pi}}\left(\frac{Z}{a_0}\right)^{3/2} e^{-Zr/a_0}$$

$$\psi_{2s} = \frac{1}{\sqrt{32\pi}}\left(\frac{Z}{a_0}\right)^{3/2}\left(2 - \frac{Zr}{a_0}\right) e^{-Zr/(2a_0)}$$

$$\psi_{3s} = \frac{1}{81\sqrt{3\pi}}\left(\frac{Z}{a_0}\right)^{3/2}\left(27 - 18\frac{Zr}{a_0} + 2\left(\frac{Zr}{a_0}\right)^2\right) e^{-Zr/(3a_0)}$$

Note the contraction in the position of the s-orbitals outer lobes with the increase in the nuclear charge.

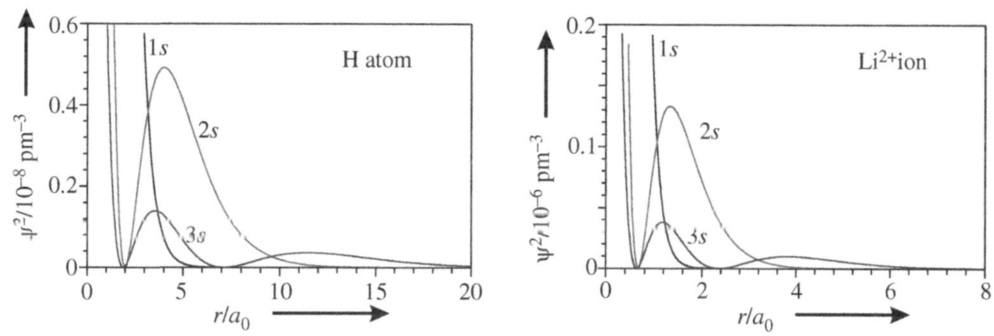

Figure P4.3b The figure shows a plot of the squares of the first three s-orbital wavefunctions for the H-atom on the left and the Li^{2+} atomic ion on the right.

In part b) we are asked to plot the radial distribution functions (1s, 2s, 3s). We show here the plots for the H-atom and the K^{+18} ion with $Z = 19$, see Fig. P4.3c. Note the change in scale and the strong contraction of the electron cloud for the K^{18+}.

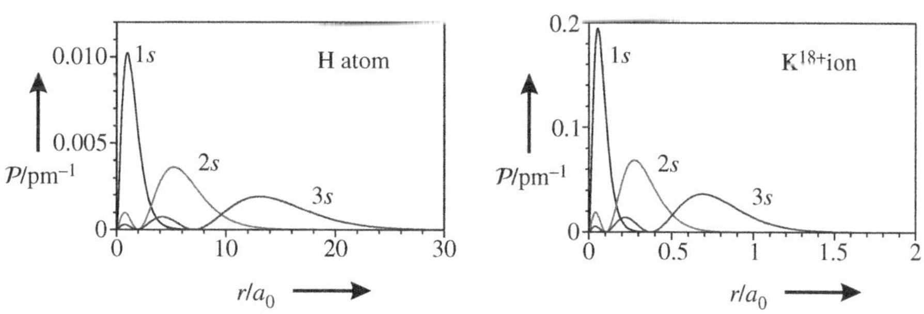

Figure P4.3c The radial distribution functions $\mathcal{P}$ are plotted for the first three s-orbitals of the H-atom (left) and the K^{18+}-atomic ion (right). Note the difference in the scaling of the r axes.

As the nuclear charge increases the wavefunction, and its square (labeled correspondingly as ($\rho1sq(r)$, $\rho2sq(r)$, and $\rho3s(r)$)), contract dramatically. Note the change in scale on the abscissa (r is given in units of Bohr radii). Correspondingly the amplitude of the wavefunctions increases, so that the area under the probability density is constant. In part c) we are asked to make contour plots of the electron density and radial distribution function for 1*s*, 2*s*, and 3*s* orbitals of an H-atom. In Figure P4.3d,e we show the cases for the 1*s* and 2*s* orbitals. The full color plots for all three orbitals can be found in the MathCAD solution file. The color code for the values is dark grey with the highest value and proceeding through the lighter shades of gray.

Problem 2 has three components. a) Here we plot the 3p (dark gray) and 3d (black) radial distribution functions for the H-atom (with $Z = 1$) and the Li^{2+} ion (with $Z = 19$).

$$\mathrm{RDF}(3p_z) = \frac{4\pi}{3}\frac{2Z}{81^2\pi a_0}\left(\frac{Zr}{a_0}\right)^4\left(6-\frac{Zr}{a_0}\right)^2 e^{-2Zr/(3a_0)}$$

$$\mathrm{RDF}(3d_{z^2}) = \frac{16\pi}{5}\frac{2Z}{81^2\,6\pi a_0}\left(\frac{Zr}{a_0}\right)^6 e^{-2Zr/(3a_0)}$$

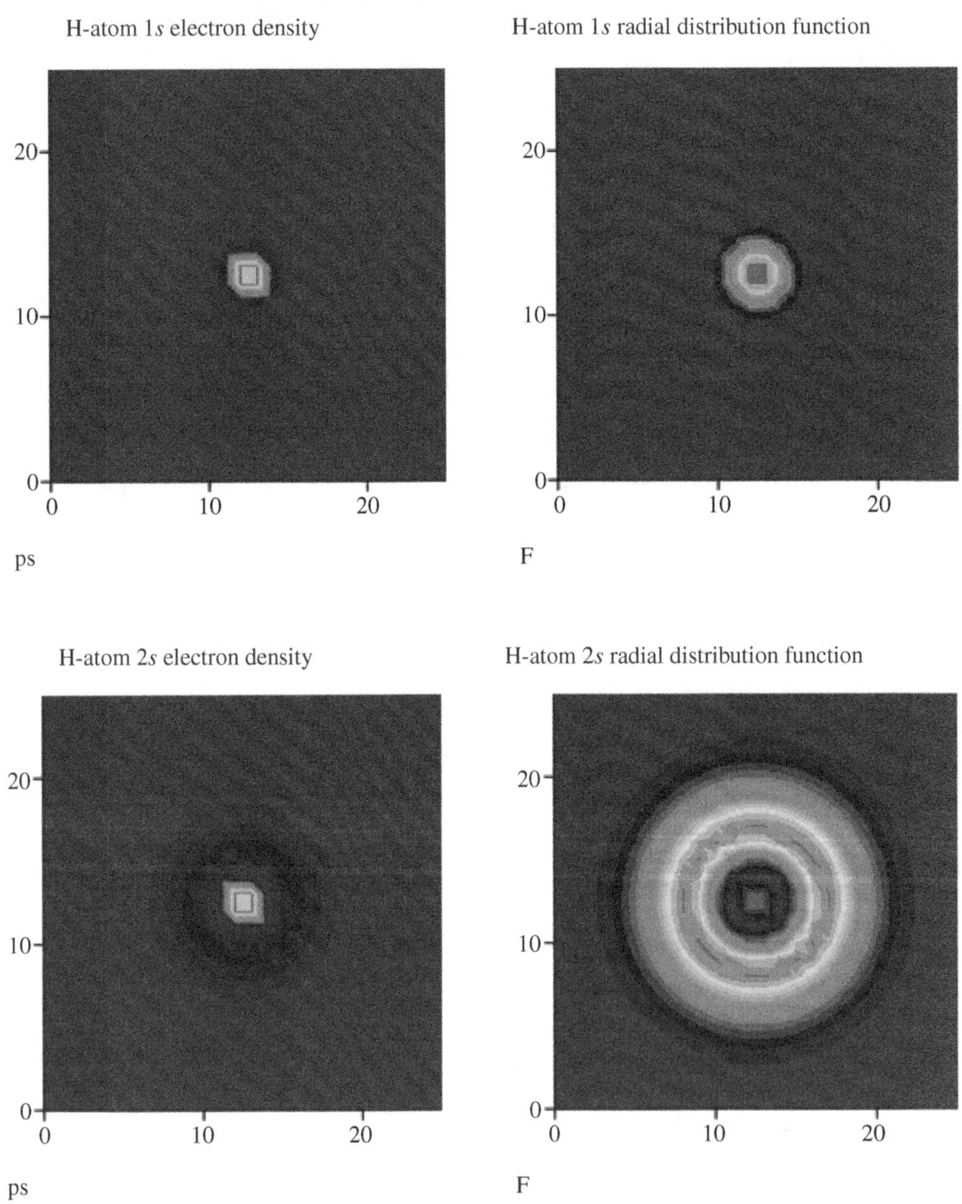

Figure P4.3d,e Contour plots are shown for the H-atom 1*s* orbital (top) and 2*s* orbital (bottom). The leftmost image in each case shows the probability density and the rightmost image shows the radial distributions function. The contours are distinguished by the darkness of the gray shading. The dark gray has the highest numerical value and proceeds systematicall to a lighter gray as it approaches the lowest value.

b) We find the maximum of the radial distribution function on the active MathCAD plots. The values of r where the maxima occur are reported in the table and displayed in Figure P4.3f.

	1s	**2s**	**2p**	**3s**	**3p**	**3d**
H	a_0	$5.24a_0$	$4a_0$	$13.1a_0$	$12.1a_0$	$9a_0$
Li^{+2}	$0.33a_0$	$1.75a_0$	$1.33a_0$	$4.36a_0$	$4a_0$	$3a_0$
K^{+18}	$0.053a_0$	$0.28a_0$	$0.21a_0$	$0.69a_0$	$0.63a_0$	$0.47a_0$

c) For the different wavefunctions we find the radial node distribution given in the table.

Radial Nodes	**1s**	**2s**	**2p**	**3s**	**3p**	**3d**
H	0	1	0	2	1	0
Li^{+2}	0	1	0	2	1	0
K^{+18}	0	1	0	2	1	0

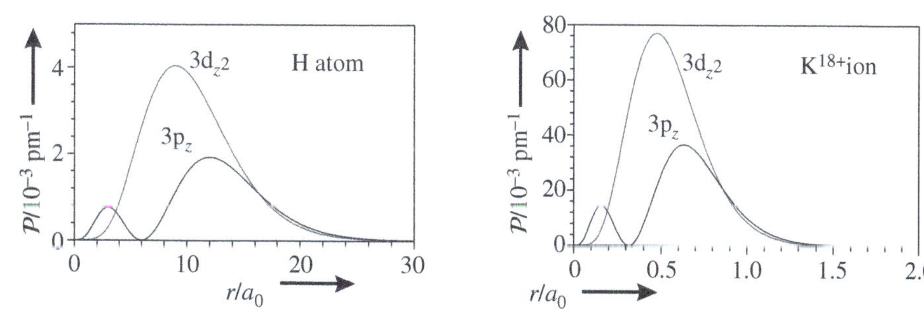

Figure P4.3f The figure shows plots of the radial distribution function$\mathcal{P}$ for the H-atom and the K^{+18} atomic ion. The dark gray curve shows the $3d_{z^2}$ radial distribution function and the light gray curve shows the $3p_z$ radial distribution function.

The number of nodes does not change with the nuclear charge but does change with the value of the principal quantum n and the angular momentum quantum number l. Moving across a row of the table we see that for a given value of l the number of radial nodes increases linearly with n. We also see that for a given n value, the number of radial nodes decreases linearly with l. Hence we can postulate that the number of radial nodes is equal to $n - l - 1$, and this is the formula borne out by more comprehensive comparisons and analysis.

Problem 3 has five components: a) Figure P4.3g we give contour plots of the $2p_x$ and $2p_y$ orbitals

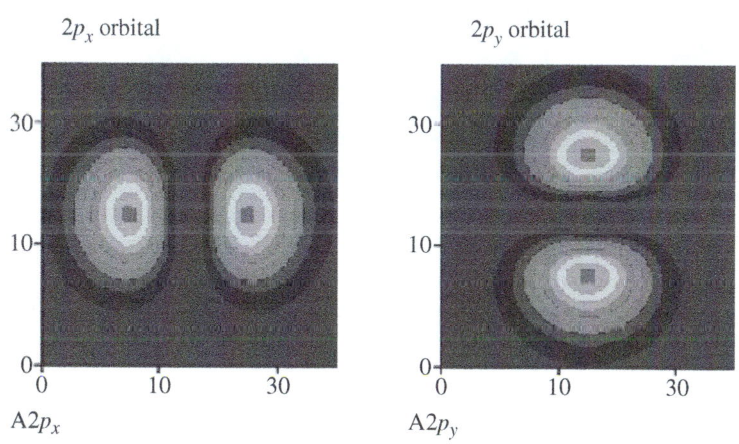

Figure P4.3g This image shows contour plots for the probability density of two of the three $2p$ orbitals. The contours are distinguished by the darkness of the gray shading. The dark gray has the highest numerical value and proceeds systematically to a lighter gray as it approaches the lowest value.

In Figure P4.3h we show angular plots for some of the 3d orbitals in Figure P4.3g.

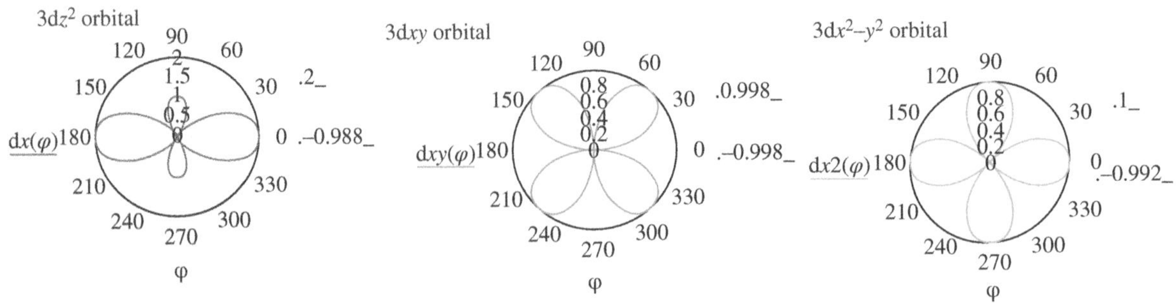

Figure P4.3h The image shows angular plots for three of the five d orbitals.

b) The $2p$ and $3p$ orbitals have one angular node. The $3d$ orbitals each have two angular nodes. c) and d) The total number of nodes for an orbital is the sum of its angular and radial pieces. These values are reported in the table. We see that the total number of nodes depends only on the principal quantum number and is equal to $n - 1$.

Total Nodes	1s	2s	2p	3s	3p	3d
H	0	1	1	2	2	2
Li^{+2}	0	1	1	2	2	2
K^{+18}	0	1	1	2	2	2

e) Note that the radial distribution function for the $3p$ orbital does have a zero originating from its zero, but it is difficult to see on this scale. Plotted for the x orbitals in Figure P4.3i we can see that the $3p_x$ is much more extended than the $2p_x$.

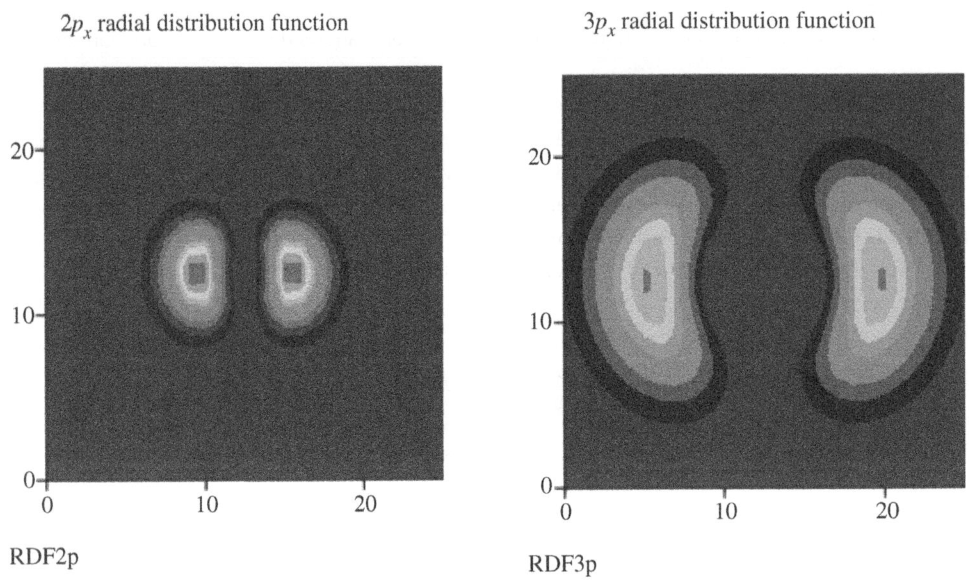

Figure P4.3i The image on the left shows a contour plot for the $2p_x$ orbital's probability density in the x, y plane, and it shows the probability density for the $3p_x$ probability density on the right. The contours are distinguished by the darkness of the gray shading. The dark gray has the highest numerical value and proceeds systematically to a lighter gray as it approaches the lowest value.

Problem 4 asks us to compute average values and uncertainty products. a) For some H-atom orbitals we calculate that

	2s	2p	3s	3p
$\overline{r}$	$6a_0$	$5a_0$	$13.5a_0$	$12.5a_0$
$\overline{r^2}$	$42a_0^2$	$30a_0^2$	$207a_0^2$	$180a_0^2$
$\Delta r = \sqrt{\overline{r^2} - \overline{r}^2}$	$2.45a_0$	$2.24a_0$	$4.98a_0$	$4.87a_0$

b) By way of comparison the formula $\overline{r} = \left[3n^2 - l(l+1)\right] a_0/2$ gives values of $6a_0$ for the 2*s* orbital, $5a_0$ for the 2*p* orbital, $13.5a_0$ for the 3*s* orbital, and $12.5a_0$ for the 3*p* orbital; in exact agreement with the integral evaluations. Note that this formula, which is provided in the MathCAD worksheet, is taken from Kjell Bockasten, Mean values of powers of the radius for hydrogenic electron orbits, *Phys Rev. A* 1974, **9**, 1087–1089.
c) Using the uncertainty relation $\Delta p = h/(4\pi\,\Delta r)$, we find that

$$\Delta v = \frac{h}{4\pi m_e\ \Delta r} = \frac{6.626\times10^{-34}\ \text{J-s}}{4\pi\left(9.109\times10^{-31}\ \text{kg}\right)}\frac{1}{\Delta r} = \frac{5.789\times10^{-5}}{\Delta r}\frac{\text{m}^2}{\text{s}}$$

Using the values of Δr in the table we find $\Delta v = 4.47\times10^5\ \text{m s}^{-1}$ for the 2*s* electron, $\Delta v = 4.89\times10^5\ \text{m s}^{-1}$ for the 2*p* electron, $\Delta v = 2.20\times10^5\ \text{m s}^{-1}$ for the 3*s* electron, and $\Delta v = 2.25\times10^5\ \text{m s}^{-1}$ for the 2p electron.
Problem 5 concerns the size of the H-atom and one electron ions. a) Using the code in the worksheet we find that 90% of the probability is contained within a radius of $2.67a_0$, that 95% of the probability is contained within a radius of $3.15a_0$, and that 99% is contained within a radius of $4.22a_0$ for the H-atom. b) For the H-atom and the three ions we find the values in the table. These values reveal the contraction of the electron cloud by the increased nuclear charge.

	90%	95%	99%
H	$2.67a_0$	$3.15a_0$	$4.22a_0$
He^+	$1.33a_0$	$1.58a_0$	$2.10a_0$
Li^{2+}	$0.89a_0$	$1.05a_0$	$1.40a_0$

Problem 6 asks us to find the most probable distance of an electron away from the nucleus for the H-atom orbitals and compare it to the prediction of the Bohr model. In the Bohr model of the atom, the electron remains in orbit because of a balance between the Coulomb force of attraction and the centripetal force from its motion about the nucleus. Hence,

$$\frac{e^2}{4\pi\varepsilon_0 r^2} = \frac{m_e v^2}{r}$$

Bohr assumed that the angular momentum, *l*, was quantized such that,

$$l = m_e v r = \frac{nh}{2\pi}$$

where *n* is the principal quantum number. We can solve these two equations to find the radius of the electron's orbit, namely

$$\frac{e^2}{4\pi\varepsilon_0} = \frac{m_e v^2 r^2}{r} = \frac{n^2h^2}{4\pi^2 m_e}\frac{1}{r}$$

so that

$$r = \frac{n^2h^2}{16\pi^2 m_e}\frac{4\pi\varepsilon_0}{e^2} = \frac{n^2h^2}{\pi m_e}\frac{\varepsilon_0}{e^2} = n^2 a_0$$

where we have used the definition of the Bohr radius a_0. For $n = 1, 2, 3\ldots$ we find exact agreement (see table).

	1s	**2p**	**3d**
Exact	a_0	$4a_0$	$9a_0$
Bohr Model	a_0	$4a_0$	$9a_0$

P4.4 In Section 4.3.3 we calculated the probability $\mathcal{P}(r < R)$ of finding a 2*s* electron inside a sphere with radius *R*. Show that the integration in Equation (4.18) leads to the result in Equation (4.19).

Solution

Equation (4.18) is written

$$\mathcal{P}(r < R) = \frac{4\pi}{32\pi a_0^3}\int_0^R \left(2 - \frac{r}{a_0}\right)^2 e^{-\frac{r}{a_0}}\ r^2 dr$$

To begin, we expand the polynomial in the integrand

$$\mathcal{P}(r < R) = \frac{1}{8a_0^3}\int_0^R \left(\frac{1}{a_0^2}r^4 - \frac{4}{a_0}r^3 + 4r^2\right) e^{-\frac{r}{a_0}}\ dr$$

Using the integral $\int x^n \cdot e^{-x} \cdot dx = -x^n \cdot e^{-x} + n \int x^{n-1} \cdot e^{-x} \cdot dx$ from Appendix B, we can integrate the separate terms. If we define $x = r/a_0$, the integral becomes

$$\mathcal{P}(r < R) = \frac{1}{8} \int_0^{R/a_0} (x^4 - 4x^3 + 4x^2)\, e^{-x}\, dx$$

For each of the different terms we find

$$\int x^2 \cdot e^{-x} \cdot dx = -x^2 \cdot e^{-x} - 2x \cdot e^{-x} - 2e^{-x}$$

$$\int x^3 \cdot e^{-x} \cdot dx = -x^3 \cdot e^{-x} - 3x^2 \cdot e^{-x} - 6x \cdot e^{-x} - 6e^{-x}$$

$$\int x^4 e^{-x}\, dx = -x^4 \cdot e^{-x} - 4x^3 \cdot e^{-x} - 12x^2 \cdot e^{-x} - 24x \cdot e^{-x} - 24e^{-x}$$

Hence we can write that

$$\int (x^4 - 4x^3 + 4x^2)\, e^{-x}\, dx = e^{-x} \begin{bmatrix} (-x^4 - 4x^3 - 12x^2 - 24x - 24) \\ + (4x^3 + 12x^2 + 24x + 24) + (-4x^2 - 8x - 8) \end{bmatrix} = e^{-x}(-x^4 - 4x^2 - 8x - 8)$$

Using this result for the indefinite form of the integral we find that

$$\mathcal{P}(r < R) = \frac{1}{8} \int_0^{R/a_0} (x^4 - 4x^3 + 4x^2)\, e^{-x}\, dx = \frac{1}{8} \left[e^{-x}(-x^4 - 4x^2 - 8x - 8) \right]_0^{R/a_0}$$

$$= \frac{1}{8} \left[e^{-R/a_0} \left(-\left(\frac{R}{a_0}\right)^4 - 4\left(\frac{R}{a_0}\right)^2 - 8\left(\frac{R}{a_0}\right) - 8 \right) + 8 \right]$$

This expression rearranges to

$$\mathcal{P}(r < R) = \left[1 - e^{-R/a_0} \left(\frac{1}{8}\left(\frac{R}{a_0}\right)^4 + \frac{1}{2}\left(\frac{R}{a_0}\right)^2 + \left(\frac{R}{a_0}\right) + 1 \right) \right]$$

which is the result we were asked to show.

P4.5 In Section 4.3.3 we calculated the probability $\mathcal{P}(r < R)$ of finding a $2s$ electron inside a sphere with radius R. Perform a corresponding calculation and obtain an expression for $\mathcal{P}(r < R)$ for a $2p$ electron. Evaluate these expressions for R values of $2a_0$, $4a_0$, $6a_0$, $8a_0$, and $10a_0$.

Solution

We start with

$$\mathcal{P}(r < R) = \frac{1}{32\pi a_0^3} \left[\int_0^R \left(\frac{r}{a_0}\right)^2 e^{-\frac{r}{a_0}} r^2\, dr \right] \left[\int_0^\pi (\cos\theta)^2 \sin\theta\, d\theta \right] \left[\int_0^{2\pi} d\varphi \right]$$

$$= \frac{1}{32\pi} \int_0^{R/a_0} x^4\, e^{-x}\, dx \left[-\frac{(\cos\theta)^3}{3} \right]_0^\pi 2\pi$$

where we have made the substitution for $x = r/a_0$ and performed the integrals over the angles. Simplifying the above result somewhat, we obtain

$$\mathcal{P}(r < R) = \frac{1}{24} \int_0^{R/a_0} x^4\, e^{-x}\, dx$$

Using the integral $\int x^n \cdot e^{-x} \cdot dx = -x^n \cdot e^{-x} + n \int x^{n-1} \cdot e^{-x} \cdot dx$ from Appendix B, we find

$$\int x^2 \cdot e^{-x} \cdot dx = -x^2 \cdot e^{-x} - 2x \cdot e^{-x} - 2e^{-x}$$

$$\int x^3 \cdot e^{-x} \cdot dx = -x^3 \cdot e^{-x} - 3x^2 \cdot e^{-x} - 6x \cdot e^{-x} - 6e^{-x}$$

$$\int x^4 e^{-x}\, dx = -x^4 \cdot e^{-x} - 4x^3 \cdot e^{-x} - 12x^2 \cdot e^{-x} - 24x \cdot e^{-x} - 24e^{-x}$$

Using this integral we can write

$$\mathcal{P}(r < R) = \frac{1}{24}\left[-x^4 \cdot e^{-x} - 4x^3 \cdot e^{-x} - 12x^2 \cdot e^{-x} - 24x \cdot e^{-x} - 24e^{-x}\right]_0^{R/a_0}$$

$$= 1 - e^{-R/a_0}\left[\frac{R^4}{24a_0^4} + \frac{R^3}{6a_0^3} + \frac{R^2}{2a_0^2} + \frac{R}{a_0} + 1\right]$$

Hence we have that

$$\mathcal{P}_{2s}(r < R) = 1 - e^{-R/a_0}\left(\frac{1}{8}\left(\frac{R}{a_0}\right)^4 + \frac{1}{2}\left(\frac{R}{a_0}\right)^2 + \left(\frac{R}{a_0}\right) + 1\right)$$

and

$$\mathcal{P}_{2p}(r < R) = 1 - e^{-R/a_0}\left[\frac{R^4}{24a_0^4} + \frac{R^3}{6a_0^3} + \frac{R^2}{2a_0^2} + \frac{R}{a_0} + 1\right]$$

Let us evaluate these probabilities for the different R/a_0 values. For $R/a_0 = 2$, we find

$$\mathcal{P}_{2s}(r < R) = 1 - e^{-2}\left(\frac{1}{8}(2)^4 + \frac{1}{2}(2)^2 + (2) + 1\right) = 1 - 7e^{-2} = 0.0526$$

and

$$\mathcal{P}_{2p}(r < R) = 1 - e^{-2}\left[\frac{2^4}{24} + \frac{2^3}{6} + \frac{2^2}{2} + 2 + 1\right] = 1 - 7e^{-2} = 0.0526$$

The numerical calculation is similar for the other R values, and we find

R/a_0	$\mathcal{P}_{2s}(r < R)$	$\mathcal{P}_{2p}(r < R)$
2	0.0526	0.0526
4	0.1758	0.4078
6	0.5365	0.7149
8	0.8145	0.9004
10	0.9405	0.9707

These results show that the $2p$ electron cloud is slightly more compact than that of the $2s$ electron.

P4.6 In Section 4.3.3 we calculated the average distance $\bar{r}$ of a $2p_z$ electron from the nucleus. Perform a corresponding calculation for a $2p_x$ and $2p_y$ electron.

Solution

The average distance $\bar{r}$ for a $2p_z$ electron is given by

$$\bar{r} = \int_0^{\pi} \sin\vartheta \, d\vartheta \int_0^{2\pi} d\varphi \int_0^{\infty} r\rho_{2p_z} r^2 dr$$

where

$$\rho_{2p_z} = \frac{1}{32\pi a_0^3}\frac{r^2}{a_0^2}e^{-r/a_0}(\cos\vartheta)^2$$

Using the expression for ρ_{2p_z} we have

$$\bar{r} = \frac{1}{32\pi a_0^3}\left[\int_0^{\pi}(\cos\vartheta)^2 \sin\vartheta \, d\vartheta\right]\left[\int_0^{2\pi} d\varphi\right]\left[\int_0^{\infty}\frac{r^5}{a_0^2}e^{-r/a_0}dr\right]$$

From the Table of Integrals in Appendix B we have

$$\int (\cos\vartheta)^2 \sin(\vartheta) \, d\vartheta = -\frac{(\cos\vartheta)^3}{3}$$

so that

$$\int_0^{\pi}(\cos\vartheta)^2 \sin\vartheta \, d\vartheta = -\frac{(\cos\vartheta)^3}{3}\bigg|_0^{\pi} = \frac{1}{3} + \frac{1}{3} = \frac{2}{3}$$

Given that

$$\int_0^{2\pi} d\varphi = 2\pi$$

we can perform the angular integrals to obtain

$$\bar{r} = \frac{1}{32\pi a_0^3}\left(\frac{2}{3}\right)(2\pi)\int_0^\infty \frac{r^5}{a_0^2} e^{-r/a_0} dr$$

To evaluate the radial part, we make the substitution $x = r/a_0$ and write

$$\bar{r} = \frac{a_0}{24}\int_0^\infty x^5 e^{-x} dx$$

Using the integral $\int_0^\infty e^{-x}\, x^n\, dx = n!$, ($n$ positive integer) from Appendix B, we find that

$$\bar{r} = \frac{a_0}{24}\int_0^\infty x^5 e^{-x}\, dx = \frac{5!}{24} a_0 = 5\, a_0$$

From symmetry, we expect the $2p_x$ and $2p_y$ orbitals to have the same average distance; however, let us show it. For the $2p_x$ wavefunction we start with

$$\rho = \frac{1}{64\pi a_0^3}\left(\frac{r}{a_0}\right)^2 (\sin\vartheta)^2\, e^{-r/a_0}$$

The $e^{i\varphi}$ does not appear because in writing ρ we multiply the complex conjugates of the wavefunction and $e^{i\varphi}e^{-i\varphi} = 1$. Substituting this expression into the $\bar{r}$ equation, we find that

$$\bar{r} = \frac{1}{64\pi a_0^5}\int_0^\pi (\sin\vartheta)^2 \sin\vartheta\, d\vartheta \int_0^{2\pi} d\varphi \int_0^\infty r^5 e^{-r/a_0} dr$$

Let us integrate the angular term over ϑ first. From the Tables of Integrals in Appendix B, we have

$$\int (\sin ax)^3\, dx = -\frac{1}{3a}\cos(ax)\left((\sin(ax))^2 + 2\right)$$

For this case we have

$$\int_0^\pi (\sin\vartheta)^3\, d\vartheta = -\frac{1}{3}(\cos\vartheta)\left[(\sin\vartheta)^2 + 2\right]\Big|_0^\pi = \frac{2}{3} + \frac{2}{3} = \frac{4}{3}$$

The second angular term is

$$\int_0^{2\pi} d\varphi = 2\pi$$

Hence we find that

$$\bar{r} = \frac{1}{24 a_0^5}\int_0^\infty r^5 e^{-r/a_0}\, dr$$

To evaluate the radial part, we make the substitution $x = r/a_0$ and write

$$\bar{r} = \frac{a_0}{24}\int_0^\infty x^5 e^{-x} dx = \frac{5!}{24} a_0 = 5\, a_0$$

which is the same as our intermediate result above; hence, the average electron distance is the same in the two orbitals.

Finally for the $2p_y$ we have the same ρ as we did for the $2p_x$ so the $\bar{r}$ will be the same.

Hence an electron in any one of the $2p$ states of a hydrogen atom has the same average distance from the proton.

P4.7 Consider the $2s$ orbital of the hydrogen atom. Find the value(s) of r at which the electron probability distribution $\mathcal{P} = 4\pi r^2 R^2$ is a maximum.

Solution

In this case we first plot the radial distribution function

$$\mathcal{P} = \frac{4\pi r^2}{8a_0^3}\left(2 - \frac{r}{a_0}\right)^2 e^{-r/(a_0)}$$

(see Fig. P4.7), and we observe that the electron probability density has three zeroes, of which two are at the extrema of $r = 0$ and $r \to \infty$ and the other occurs at $r = 2a_0$.

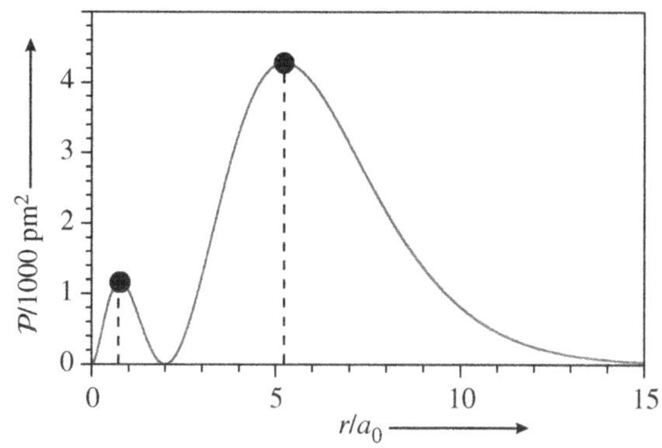

Figure P4.7 Probability $\mathcal{P}$ of the radial part of the 2s orbital of the H atom. The full black circles mark the maxima.

First we write the probability function in terms of $x = r/a_0$.

$$\mathcal{P} = \frac{4\pi r^2}{8a_0^3}\left(2 - \frac{r}{a_0}\right)^2 e^{-r/(a_0)} = \frac{4\pi}{8a_0}\frac{r^2}{a_0^2}\left(2 - \frac{r}{a_0}\right)^2 e^{-r/(a_0)} = \frac{4\pi}{8a_0}x^2(2 - x)^2 \cdot e^{-x}$$

From this equation we see that P becomes zero for $x = 2$. This means that the function has a minimum at this point. Further evaluation leads to

$$\mathcal{P} = \frac{4\pi}{8a_0}\left[x^2\left(4 - 4x + x^2\right)\right] \cdot e^{-x} = \frac{4\pi}{8a_0}\left[\left(4x^2 - 4x^3 + x^4\right)\right] \cdot e^{-x}$$

To find the value of x at which the maximum occurs, we take the derivative of the electron probability density and set it equal to zero to determine the extrema of the function.

$$\frac{d}{dx}\left[\left(4x^2 - 4x^3 + x^4\right) \cdot e^{-x}\right] = \left[8x - 12x^2 + 4x^3\right] \cdot e^{-x} - \left(4x^2 - 4x^3 + x^4\right) \cdot e^{-x}$$
$$= \left(-x^4 + 8x^3 - 16x^2 + 8x\right) \cdot e^{-x} = x \cdot \left(-x^3 + 8x^2 - 16x + 8\right) \cdot e^{-x}$$

By inspection we see that the derivative becomes zero at $x_1 = 0$ and $x_2 \to \infty$. Because the exponential is positive, it is sufficient to find the roots of the expression in parentheses to find the remaining extrema.

$$-x^3 + 8x^2 - 16x + 8 = 0$$

The cubic polynomial in the parentheses gives us three solutions for x. We concluded already that for $x = 2$ we have a minimum of P. This means that $x = 2$ is one solution of our cubic equation. If we divide the above expression by $(x - 2)$, we obtain:

$$-x^3 + 8x^2 - 16x + 8 = -\left(x - 2)(x^2 - 6x + 4\right)$$

and find a quadratic equation for x:

$$x^2 - 6x + 4 = 0$$

We use this quadratic formula to find the other two roots and obtain

$$x_\pm = 3 \pm \sqrt{-4 + 9} = 3 \pm \sqrt{5}$$

or

$$x_3 = x_+ = 3 + \sqrt{5} = 5.24 \qquad x_4 = x_- = 3 - \sqrt{5} = 0.76$$

We see that $x_3 = 5.24$ and $x_4 = 0.76$ are maxima, in accordance with the plot.
Note that one could also take the second derivative and determine its sign in order to distinguish the maxima from the minima.

P4.8 Plot the radial distribution functions for the first three *s*-orbitals of the hydrogen atom, and compare the most probable radius for the electron in each case. Analytically determine the value of the most probable radius for the 1*s* orbital and the 2*s* orbital.

Solution

The first part of this problem is identical to that of E4.13 and we refer you to that solution.

The most probable radius for these functions is the value of *r* at which the probability distribution has its maximum. By inspection of the graph we see that the most probable radius increases with the increase in the principal quantum number; however, the functions show that the *s*-orbitals of higher quantum number have electron density close to the nucleus as well.

For the 1*s* and 2*s* orbitals, we determine the functions extrema and compare with the plot to identify the maxima. We proceed by taking the derivative, setting it equal to zero, and solving for *r* in each case. For the 1*s* orbital we have

$$\frac{d}{dr}\left[4\pi r^2 \frac{1}{\pi a_0^3} e^{-2r/a_0}\right]$$

Taking the derivative we find that

$$\frac{4}{a_0^3}\cdot 2r\cdot e^{-2r/a_0} - 4r^2\frac{1}{a_0^3}\frac{2}{a_0}e^{-2r/a_0} = e^{-2r/a_0}\left(\frac{8}{a_0^3}r - r^2\frac{8}{a_0^4}\right) = e^{-2r/a_0}\frac{8}{a_0^4}\left(a_0 r - r^2\right) = \frac{8}{a_0^4} r\, e^{-\frac{2r}{a_0}}\left(a_0 - r\right)$$

By inspection, we find that the maximum occurs for $a_0 - r_{max} = 0$.

$$r_{max} = a_0 = 0.529$$

i.e., the Bohr radius is the most probable distance for an electron in a 1*s* orbital.

In Problem P4.7, we found the maxima for the 2*s* orbital to occur at $r = \left(3-\sqrt{5}\right)a_0$ and $r = \left(3+\sqrt{5}\right)a_0$. By comparison with the plot, we identify $r = \left(3+\sqrt{5}\right)a_0$ as the most probable radius, *r* value with the highest maximum in radial distribution function. This occurs at more than five times the value of the Bohr radius

$$r = \left(3+\sqrt{5}\right)a_0 = 5.236\cdot a_0 = 277.0\text{ pm}$$

Extension:

The case of the 3*s* wavefunction is more complicated. In this case we analyze the following derivative

$$\frac{d}{dr}\left[4\pi r^2\frac{1}{81^2 3\pi a_0^3}\left(27 - 18\frac{r}{a_0} + 2\frac{r^2}{a_0^2}\right)^2 e^{-2r/(3a_0)}\right]$$

which rearranges to (omitting the constant factors)

$$\frac{d}{dr}\left[r^2\left(27 - 18\frac{r}{a_0} + 2\frac{r^2}{a_0^2}\right)^2 e^{-2r/(3a_0)}\right]$$

With the abbreviation $x = r/a_0$ we obtain (again omitting the constant factors)

$$\frac{d}{dx}\left[x^2\left(27 - 18x + 2x^2\right)^2 e^{-2x/3}\right]$$

First we expand the polynomial

$$x^2\left(27 - 18x + 2x^2\right)^2 = 729x^2 - 972x^3 + 432x^4 - 72x^5 + 4x^6$$

Second we calculate the first derivative

$$\frac{d}{dx}\left[\left(729x^2 - 972x^3 + 432x^4 - 72x^5 + 4x^6\right)e^{-2x/3}\right] = \left(1458x - 2916x^2 + 1728x^3 - 360x^4 + 24x^5\right)e^{-2x/3}$$
$$+\left(729x^2 - 972x^3 + 432x^4 - 72x^5 + 4x^6\right)(-2/3)e^{-2x/3} = e^{-2x/3}\,.$$
$$(1458x - 2916x^2 + 1728x^3 - 360x^4 + 24x^5 - 486x^2 + 648x^3 - 288x^4 + 48x^5 - \frac{8}{3}x^6)$$

Next we order the polynomial according to ascending powers.

$$(1458x - 2916x^2 + 1728x^3 - 360x^4 + 24x^5 - 486x^2 + 648x^3 - 288x^4 + 48x^5 - \frac{8}{3}x^6)$$
$$= x(1458) + x^2(-2916 - 486) + x^3(1728 + 648) + x^4(-360 - 288) + x^5(24 + 48) + x^6(-\frac{8}{3})$$
$$= 1458x - 3402x^2 + 2376x^3 - 648x^4 + 72x^5 - \frac{8}{3}x^6$$

In order to find the extrema we consider the equation

$$y = (1458x - 3402x^2 + 2376x^3 - 648x^4 + 72x^5 - \frac{8}{3}x^6)e^{-2x/3}$$

and search for the values of x where $y = 0$. We compute values for this function. In a first run (low resolution of the data) we find the zeros for y at the x-values in Table P4.8 (second row):

Table P4.8 Calculating the Zeros of y.

i	x_i(First Run)	x_i(Final Run)	r_i/pm (Final Run)	Extremum
1	0.75	0.750	39.7	max
2	1.95	1.895	100.3	min
3	4.15	4.180	221.2	max
4	7.05	7.093	375.4	min
5	13.05	13.075	691.9	max

From Fig. E4.13 we know that the solutions with $i = 2$ and $i = 4$ are minima and that the solutions with $i = 1$, $i = 3$, and $i = 5$ are maxima. Next we perform a final calculation at a higher resolution of the computed data, see Table P4.8 (third row). From these values we obtain the corresponding r values (row 4). The radius for the outmost maximum is at $r_5 = 691.9$ pm. This value is about two and a half times larger than that for the 2s orbital.
From the different values for the most probable r and the plot, it is clear that the most probable distance of the electron from the nucleus increases with the increase in the principal quantum number.

P4.9 For a 3s orbital, evaluate the average value of the electron's distance from the proton and compare it to its most probable value of $r = 692$ pm (see solution to P4.8). Which value is larger and why?

Solution

To calculate the average value we must evaluate

$$\bar{r} = \int_{\text{all space}}^{2} \rho_{3s}^2\, r\, d\tau = \int_{\text{all space}} \psi_{3s}^2\, r\, d\tau$$

where

$$\rho_{3s} = \frac{1}{81^2 3\pi a_0^3}\left(27 - 18\frac{r}{a_0} + 2\frac{r^2}{a_0^2}\right)^2 e^{-2r/(3a_0)}$$

and $d\tau = r^2 \sin\theta\, dr\, d\theta\, d\varphi$. Making the substitutions, we can write

$$\bar{r} = \int_0^\infty \int_0^\pi \int_0^{2\pi} \frac{1}{81^2 3\pi a_0^3}\left(27 - 18\frac{r}{a_0} + 2\frac{r^2}{a_0^2}\right) e^{-2r/(3a_0)}\, r\, r^2 \sin\theta\, d\,r\, d\theta\, d\varphi$$
$$= \frac{1}{81^2 3\pi a_0^3}\left[\int_0^\infty \left(27 - 18\frac{r}{a_0} + 2\frac{r^2}{a_0^2}\right) e^{-2r/(3a_0)}\, r\, r^2\, dr\right]\left[\int_0^\pi \sin\theta\ d\theta\right]\left[\int_0^{2\pi} d\varphi\right]$$

By performing the angular integrals we can write

$$\bar{r} = \frac{1}{81^2 3\pi a_0^3}\left[\int_0^\infty \left(27 - 18\frac{r}{a_0} + 2\frac{r^2}{a_0^2}\right)^2 e^{-2r/(3a_0)} r^3 dr\right][-\cos\theta]_0^\pi\ [\varphi]_0^{2\pi}$$

$$= \frac{4\pi}{81^2 3\pi a_0^3}\left[\int_0^\infty \left(27 - 18\frac{r}{a_0} + 2\frac{r^2}{a_0^2}\right)^2 e^{-2r/(3a_0)} r^3 dr\right]$$

We proceed by expanding the radial integral so that

$$\bar{r} = \frac{4\pi}{81^2 3\pi a_0^3}\int_0^\infty \left(\frac{4}{a_0^4}r^7 - \frac{72}{a_0^3}r^6 + \frac{432}{a_0^2}r^5 - \frac{972}{a_0}r^4 + 729r^3\right) e^{-\frac{2r}{3a_0}}\ dr$$

From the integral table in the appendix we know that $\int_0^\infty e^{-ax} x^n\, dx = \frac{n!}{a^{n+1}}$, ($a > 0$, n positive integer). Thus we can perform the integral term by term; e.g.,

$$\int_0^\infty \frac{4}{a_0^4} r^7 e^{-\frac{2r}{3a_0}}\ dr = \frac{4}{a_0^4}\frac{7!}{(2/(3a_0))^8}$$

Performing this calculation for each term, we find that

$$\bar{r} = \frac{4\pi}{81^2 3\pi a_0^3}\cdot\left(\begin{matrix}\dfrac{4\cdot 7!\cdot(3a_0)^8}{2^8 a_0^4} - \dfrac{72\cdot 6!\cdot(3a_0)^7}{2^7 a_0^3} + \dfrac{432\cdot 5!\cdot(3a_0)^6}{2^6 a_0^2}\\ -\dfrac{972\cdot 4!\cdot(3a_0)^5}{2^5 a_0} + \dfrac{729\cdot 3!\cdot(3a_0)^4}{2^4}\end{matrix}\right)$$

or

$$\bar{r} = \frac{4}{(81^2)(3)}\left(\begin{matrix}\dfrac{7!\,(3^8)}{2^6} - \dfrac{(72)(6!)(3^7)}{2^7} + \dfrac{(432)(5!)(3^6)}{2^6}\\ -\dfrac{(972)(4!)(3^5)}{2^5} + \dfrac{(729)(3!)(3^4)}{2^4}\end{matrix}\right) a_0 = 13.5a_0$$

Using $a_0 = 52.9$ pm, we find that the average radius for the 3*s* orbital is $\bar{r} = 714$ pm.
Thus the average value is larger than the most probable value by 3%. This results from the exponential tail of the wavefunction giving a probability density that is asymmetric and extends out to large distances. We note however that the difference between these two measures of the electron cloud is much closer to each other for the 3*s* orbital than for the 1*s* orbital, for which they differ by 50% (see text).

P4.10 Begin with the general expression for the radial probability distribution function $\mathcal{P}(r < R) = \int_0^R \rho\, d\tau$ and derive a formula for $\mathcal{P}_{2s}$; namely, show that

$$\mathcal{P}(r < R) = 1 - e^{-R/a_0}\cdot\left[\frac{R^4}{8a_0^4} + \frac{R^2}{2a_0^2} + \frac{R}{a_0} + 1\right]$$

Solution

According to Section 4.3.3, Equation (4.18), the probability of finding the electron in the H atom in the 2*s* state inside a sphere with radius *R* is

$$\mathcal{P}(r < R) = \int_0^R \rho\, d\tau = \int_0^R \rho 4\pi r^2\ dr = \frac{1}{8a_0^3}\int_0^R \left(2 - \frac{r}{a_0}\right)^2 e^{-\frac{r}{a_0}}\cdot r^2\ dr$$

with the volume element $d\tau = 4\pi\, r^2 dr$. For the integral we find

$$I = \int_0^R \left(2 - \frac{r}{a_0}\right)^2 e^{-\frac{r}{a_0}}\cdot r^2\ dr = \int_0^R 4\, e^{-\frac{r}{a_0}}\cdot r^2\ dr - \int_0^R \frac{4r}{a_0} e^{-\frac{r}{a_0}}\cdot r^2\ dr + \int_0^R \left(\frac{r}{a_0}\right)^2 e^{-\frac{r}{a_0}} r^2\ dr$$

$$= 4\int_0^R e^{-\frac{r}{a_0}}\cdot r^2\ dr - \frac{4}{a_0}\int_0^R e^{-\frac{r}{a_0}}\cdot r^3\ dr + \frac{1}{a_0^2}\int_0^R e^{-\frac{r}{a_0}}\cdot r^4\ dr$$

We make the substitution $r/a_0 = x$, $dr/a_0 = dx$.

$$I = 4a_0^3 \int_0^{R/a_0} e^{-x}x^2 \, dx - 4a_0^3 \int_0^{R/a_0} e^{-x}x^3 \, dx + a_0^3 \int_0^{R/a_0} e^{-x}x^4 \, dx$$

The indefinite integrals are, according to Appendix B,

$$\int e^{-x}x^2 \, dx = e^{-x} \cdot \left(-x^2 - 2x - 2\right) + const$$

$$\int e^{-x}x^3 \, dx = e^{-x} \left(-x^3 - 3x^2 - 6x - 6\right) + const$$

$$\int e^{-x}x^4 \, dx = e^{-x} \cdot \left(-x^4 - 4x^3 - 12x^2 - 24x - 24\right) + const$$

Thus

$$\int_0^{R/a_0} e^{-x}x^2 \, dx = e^{-R/a_0} \cdot \left(-\frac{R^2}{a_0^2} - 2\frac{R}{a_0} - 2\right) + 2$$

$$\int_0^{R/a_0} e^{-x}x^3 \, dx = e^{-R/a_0} \cdot \left(-\frac{R^3}{a_0^3} - 3\frac{R^2}{a_0^2} - 6\frac{R}{a_0} - 6\right) + 6$$

$$\int_0^{R/a_0} e^{-x}x^4 \, dx = e^{-R/a_0} \cdot \left(-\frac{R^4}{a_0^4} - 4\frac{R^3}{a_0^3} - 12\frac{R^2}{a_0^2} - 24\frac{R}{a_0} - 24\right) + 24$$

and

$$I = 4a_0^3 \int_0^{R/a_0} e^{-x}x^2 \, dx - 4a_0^3 \int_0^{R/a_0} e^{-x}x^3 \, dx + a_0^3 \int_0^{R/a_0} e^{-x}x^4 \, dx$$

$$= e^{-R/a_0} \cdot a_0^3 \left[4\left(-\frac{R^2}{a_0^2} - 2\frac{R}{a_0} - 2\right) - 4\left(-\frac{R^3}{a_0^3} - 3\frac{R^2}{a_0^2} - 6\frac{R}{a_0} - 6\right)\right]$$

$$+ e^{-R/a_0} \cdot a_0^3 \left[+\left(-\frac{R^4}{a_0^4} - 4\frac{R^3}{a_0^3} - 12\frac{R^2}{a_0^2} - 24\frac{R}{a_0} - 24\right)\right] + 8a_0^3$$

$$= e^{-R/a_0} \cdot a_0^3 \cdot \left[-\frac{R^4}{a_0^4} - 4\frac{R^2}{a_0^2} - 8\frac{R}{a_0} - 8\right] + 8a_0^3$$

In this way for the probability we find

$$\mathcal{P}(r < R) = \frac{1}{8a_0^3} \int_0^R \left(2 - \frac{r}{a_0}\right)^2 e^{-\left(\frac{r}{a_0}\right)} r^2 \, dr = e^{-R/a_0} \cdot \frac{1}{8} \left[-\frac{R^4}{a_0^4} - 4\frac{R^2}{a_0^2} - 8\frac{R}{a_0} - 8\right] + 1$$

$$= +1 - e^{-R/a_0} \cdot \left[\frac{R^4}{8a_0^4} + \frac{R^2}{2a_0^2} + \frac{R}{a_0} + 1\right]$$

For example, for $R = a_0$, $R = 5a_0$, and $R = 10a_0$ we find

$$\mathcal{P}(r < a_0) = 1 - e^{-1} \cdot \left[\frac{1}{8} + \frac{1}{2} + 1 + 1\right] = 0.034$$

$$\mathcal{P}(r < 5a_0) = 1 - e^{-5} \cdot \left[\frac{625}{8} + \frac{25}{2} + 5 + 1\right] = 0.35$$

$$\mathcal{P}(r < 10a_0) = 1 - e^{-10} \cdot \left[\frac{10000}{8} + \frac{100}{2} + 10 + 1\right] = 0.94$$

These numbers are in accordance with our expectation from Fig. E4.12 in Exercise E4.12.

P4.11 Prove the identity

$$x\left(\frac{d^3\psi}{dx^3}\psi - \frac{d^2\psi}{dx^2} \cdot \frac{d\psi}{dx}\psi\right) = -2\psi\frac{d^2\psi}{dx^2} + \frac{d}{dx}\left[\psi^2 \frac{d}{dx}\left(\frac{x \cdot d\psi/dx}{\psi}\right)\right]$$

by performing the differentiation on the right-hand side. This is one step in deriving the Virial Theorem in Foundation 4.3.

Solution

We have to prove that

$$x\left(\frac{d^3\psi}{dx^3}\psi - \frac{d^2\psi}{dx^2}\cdot\frac{d\psi}{dx}\psi\right) = -2\psi\frac{d^2\psi}{dx^2} + \frac{d}{dx}\left[\psi^2\frac{d}{dx}\left(\frac{x\cdot d\psi/dx}{\psi}\right)\right]$$

We consider the expression on the right-hand side

$$y_1 = -2\psi^2\frac{d^2\psi}{dx^2} + \frac{d}{dx}\left[\psi^2\frac{d}{dx}\left(\frac{x\cdot d\psi/dx}{\psi}\right)\right]$$

and use the abbreviations

$$y_2 = \frac{d}{dx}\left(\frac{x\cdot d\psi/dx}{\psi}\right) \qquad y_3 = \frac{d}{dx}\left[\psi^2\cdot y_2\right]$$

First we evaluate y_2

$$y_2 = \frac{d}{dx}\left(\frac{x\cdot d\psi/dx}{\psi}\right) = \frac{\left(\frac{d\psi}{dx}+x\frac{d^2\psi}{dx^2}\right)\psi - \left(x\frac{d\psi}{dx}\cdot\frac{d\psi}{dx}\right)}{\psi^2}$$

and continue with y_3

$$y_3 = \frac{d}{dx}\left[\psi^2\cdot y_2\right] = \frac{d}{dx}\left[\left(\frac{d\psi}{dx}+x\frac{d^2\psi}{dx^2}\right)\psi - \left(x\frac{d\psi}{dx}\cdot\frac{d\psi}{dx}\right)\right] = \frac{d}{dx}\left(\frac{d\psi}{dx}\cdot\psi\right) + \frac{d}{dx}\left(x\frac{d^2\psi}{dx^2}\psi\right) - \frac{d}{dx}\left(x\frac{d\psi}{dx}\cdot\frac{d\psi}{dx}\right)$$
$$= \left(\frac{d^2\psi}{dx^2}\cdot\psi + \frac{d\psi}{dx}\cdot\frac{d\psi}{dx}\right) + \left[\left(\frac{d^2\psi}{dx^2}+x\frac{d^3\psi}{dx^3}\right)\psi + \left(x\frac{d^2\psi}{dx^2}\cdot\frac{d\psi}{dx}\right)\right] - \left[\left(\frac{d\psi}{dx}+x\frac{d^2\psi}{dx^2}\right)\cdot\frac{d\psi}{dx} + x\frac{d\psi}{dx}\cdot\frac{d^2\psi}{dx^2}\right]$$

leading to

$$y_3 = \frac{d^2\psi}{dx^2}\cdot\psi + \frac{d\psi}{dx}\cdot\frac{d\psi}{dx} + \frac{d^2\psi}{dx^2}\psi + x\frac{d^3\psi}{d\,x^3}\psi + x\frac{d^2\psi}{dx^2}\cdot\frac{d\psi}{dx} - \frac{d\psi}{dx}\cdot\frac{d\psi}{dx} - x\frac{d^2\psi}{dx^2}\cdot\frac{d\psi}{dx} - x\frac{d\psi}{dx}\cdot\frac{d^2\psi}{dx^2}$$
$$= 2\frac{d^2\psi}{dx^2}\cdot\psi + x\frac{d^3\psi}{dx^3}\psi - x\frac{d\psi}{dx}\cdot\frac{d^2\psi}{dx^2}$$

Finally we find for y_1

$$y_1 = -2\psi\frac{d^2\psi}{dx^2} + y_3 = x\frac{d^3\psi}{dx^3}\psi - x\frac{d^2\psi}{dx^2}\cdot\frac{d\psi}{dx}$$

Thus we have proved Equation (4.59) in Foundation 4F.3.

P4.12 In Foundation 4.2 we derived the Schrödinger Equation for the radial part R of the wavefunction for the H atom as

$$\frac{\hbar^2}{2m_e r^2}\cdot\frac{d}{dr}\left(r^2\frac{d}{dr}\right)R(r) + \left[E + \frac{e^2}{4\pi\varepsilon_0}\frac{1}{r} - \frac{\hbar^2}{2m_e r^2}l(l+1)\right]R(r) = 0$$

Here we apply this equation to the problem of standing waves of an electron in a spherical box with diameter D. In this case the potential energy is zero inside the box and the total energy equals the kinetic energy E_{kin}. Furthermore, we restrict our consideration to the case of spherical solutions ($l = 0$) and obtain:

$$\frac{\hbar^2}{2m_e r^2}\cdot\frac{d}{dr}\left(r^2\frac{d}{dr}\right)R(r) = -E_{kin}R(r)$$

Solve this equation using the separation ansatz

$$R(r) = \frac{1}{r}U(r), \qquad \frac{dR(r)}{dr} = \frac{1}{r}\frac{dU(r)}{dr} - \frac{1}{r^2}U(r)$$

Find the conditions for standing waves, determine the wavefunction $R(r)$, and find the energy E_{kin} for the lowest quantum state.

Solution

We start with

$$\frac{d}{dr}\left[r^2\frac{dR(r)}{dr}\right] = \frac{d}{dr}\left[r\frac{dU(r)}{dr} - U(r)\right] = \left[r\frac{d^2U(r)}{d^2r} + \frac{dU(r)}{dr} - \frac{dU(r)}{dr}\right] = r\frac{d^2U(r)}{d^2r}$$

and insert this result into the differential equation

$$\frac{\hbar^2}{2m_e} \cdot \frac{1}{r}\frac{d^2U(r)}{d^2r} = -E_{\text{kin}}R(r) = -E_{\text{kin}}\frac{1}{r}U(r)$$
$$\frac{\hbar^2}{2m_e} \cdot \frac{d^2U(r)}{d^2r} = -E_{\text{kin}}U(r)$$

We solve this equation by

$$U(r) = a \cdot \sin(cr) + b \cdot \cos(cr)$$

as we can show by insertion into the differential equation

$$\frac{\hbar^2}{2m_e} \cdot \left[-ac^2 \cdot \sin(cr) - bc^2 \cdot \cos(cr)\right] = -E_{\text{kin}}\left[a \cdot \sin(cr) + b \cdot \cos(cr)\right]$$
$$-\frac{\hbar^2}{2m_e} \cdot c^2\left[a \cdot \sin(cr) + b \cdot \cos(cr)\right] = -E_{\text{kin}}\left[a \cdot \sin(cr) + b \cdot \cos(cr)\right]$$
$$E_{\text{kin}} = \frac{\hbar^2}{2m_e} \cdot c^2$$

Then for the wavefunction $R(r)$ we obtain

$$R(r) = \frac{1}{r}U(r) = \frac{1}{r}\left[a \cdot \sin(cr) + b \cdot \cos(cr)\right]$$

To obtain the parameter c we use the fact that $R(r = 0)$ has to be finite. This is possible only for $b = 0$. Thus our solution reduces to

$$R(r) = a\frac{\sin(cr)}{r}$$

For $r = D/2$ (this is at the surface of the sphere) the wavefunction must be zero, so that

$$a\frac{\sin(cD/2)}{D/2} = 0$$

which is true if

$$\frac{cD}{2} = n\pi, \text{ with } n = 1, 2, 3...$$

Thus for the standing waves we obtain

$$R = a\frac{1}{r}\sin\left(\frac{2\pi n}{D}r\right), \text{with } n - 1, 2, 3...$$

and the allowed energies are

$$E_{\text{kin}} = \frac{\hbar^2}{2m_e} \cdot c^2 = \frac{\hbar^2}{2m_e} \cdot \frac{4n^2\pi^2}{D^2} = \frac{h^2}{2m_eD^2}n^2\ ,\ n = 1, 2, 3...$$

Note that D is the diameter of the sphere, comparable to the box length L in the case of an electron in a three-dimensional cubic box. For the cubic box with box length L the kinetic energy is

$$E_{\text{kin}}(\text{cubic box}) = \frac{3h^2}{8m_eL^2}n^2$$

Both quantities differ only slightly in the numerical factor $1/2 = 0.5$ for the sphere and $3/8 = 0.38$ for the cubic box. The remaining constant a in the equation for $R(r)$ can be fixed by using the normalization condition to obtain $a = \sqrt{1/(\pi D)}$. Figure P4.12a shows the square of the wavefunction in the lowest quantum state versus the coordinate r.

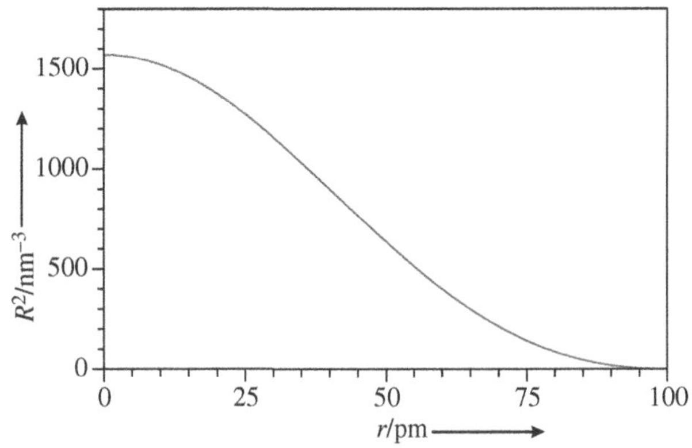

Figure P4.12a Square of wavefunction $R(r)$ versus the coordinate r inside a sphere of diameter 200 pm (lowest quantum state).

Extension: With the wavefunction $R(r)$ and the kinetic energy E_{kin} in hand, we can extend the consideration on a spherical box model of the H atom, as described in Section 2.4.2 for the cubic box model. We imagine a charge $+e$ at the center of the spherical box and calculate the average potential energy $\overline{V}$ as

$$\overline{V} = -\frac{e^2}{4\pi\varepsilon_0}\int_{r=0}^{r=D/2}\frac{1}{r}R^2 4\pi r^2 \mathrm{d}r = -\frac{e^2}{4\pi\varepsilon_0}\int_{r=0}^{r=D/2}\frac{1}{r}\frac{1}{\pi D}\frac{1}{r^2}\sin^2\left(\frac{2\pi}{D}r\right)4\pi r^2 \mathrm{d}r = -\frac{e^2}{4\pi\varepsilon_0}\frac{4}{D}\int_{r=0}^{r=D/2}\frac{1}{r}\sin^2\left(\frac{2\pi}{D}r\right)\mathrm{d}r$$

To solve the remaining integral we use the substitution

$$u = \frac{2\pi}{D}r \qquad \mathrm{d}u = \frac{2\pi}{D}\mathrm{d}r$$

and find

$$\int_{r=0}^{r=D/2}\frac{1}{r}\sin^2\left(\frac{2\pi}{D}r\right)\mathrm{d}r = \frac{2\pi}{D}\int_{u=0}^{u=\pi}\frac{1}{u}\sin^2 u\,\mathrm{d}u = \frac{2\pi}{D}\cdot I$$

Hence

$$\overline{V} = -\frac{e^2}{4\pi\varepsilon_0}\frac{4}{D}\cdot I$$

and for the energy E we obtain

$$E = E_{\mathrm{kin}} + \overline{V} = \frac{h^2}{2m_e}\frac{1}{D^2} - \frac{e^2}{4\pi\varepsilon_0}\frac{4}{D}\cdot I$$

The integral I is evaluated by numerical integration as $I = 1.214$. In Fig. P4.12b E is plotted as a function of D. We observe a minimum of $E = -13.0 \times 10^{-19}$ J at $D = 430$ pm. This is comparable with the result for the cubic box ($E = -12.9 \times 10^{-19}$ J at $L = 376$ pm).

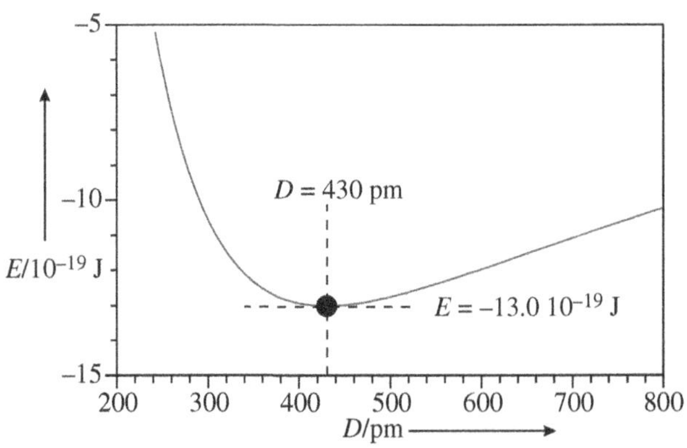

Figure P4.12b Energy E in the spherical box model for the H atom as a function of the diameter D of the sphere.

P4.13 In this problem you should use your knowledge of differential operators to show that in Foundation 4.3 (rigorous derivation of the Virial theorem) we obtain

$$\mathcal{T}(\mathcal{A}\psi) - \mathcal{A}(\mathcal{T}\psi) = (-i\hbar)\cdot 2\cdot \mathcal{T}\psi$$

where $\mathcal{T} = -\hbar^2/2m$ is the operator of the kinetic energy and the operator $\mathcal{A}$ is $\mathcal{A} = -i\hbar\left(\frac{\partial^2}{\partial x^2} + \frac{\partial^2}{\partial y^2} + \frac{\partial^2}{\partial z^2}\right)$ This is one step in deriving the Virial Theorem in Foundation 4.3.

Solution

We start with

$$\mathcal{T}(\mathcal{A}\psi) = (-\frac{\hbar^2}{2m})(-i\hbar)\left(\frac{\partial^2}{\partial x^2} + \frac{\partial^2}{\partial y^2} + \frac{\partial^2}{\partial z^2}\right)\left[\left(x\frac{\partial}{\partial x} + y\frac{\partial}{\partial y} + z\frac{\partial}{\partial z}\right)\psi\right]$$

First we consider the differentiation with respect to x.

$$\frac{\partial^2}{\partial x^2}\left[\left(x\frac{\partial\psi}{\partial x} + y\frac{\partial\psi}{\partial y} + z\frac{\partial\psi}{\partial z}\right)\right] = \frac{\partial}{\partial x}\left[\frac{\partial}{\partial x}\left(x\frac{\partial\psi}{\partial x} + y\frac{\partial\psi}{\partial y} + z\frac{\partial\psi}{\partial z}\right)\right]$$

$$= \frac{\partial}{\partial x}\left[\frac{\partial\psi}{\partial x} + x\frac{\partial^2\psi}{\partial x^2} + y\frac{\partial^2\psi}{\partial y\partial x} + z\frac{\partial^2\psi}{\partial z\partial x}\right] = \frac{\partial^2\psi}{\partial x^2} + \frac{\partial^2\psi}{\partial x^2} + x\frac{\partial^3\psi}{\partial x^3} + y\frac{\partial^3\psi}{\partial y\partial x^2} + z\frac{\partial^3\psi}{\partial z\partial x^2}$$

$$= 2\frac{\partial^2\psi}{\partial x^2} + x\frac{\partial^3\psi}{\partial x^3} + y\frac{\partial^3\psi}{\partial y\partial x^2} + z\frac{\partial^3\psi}{\partial z\partial x^2}$$

Correspondingly, we obtain for the differentiation with respect to y and z:

$$\frac{\partial^2}{\partial y^2}\left[\left(x\frac{\partial\psi}{\partial x} + y\frac{\partial\psi}{\partial y} + z\frac{\partial\psi}{\partial z}\right)\right] = 2\frac{\partial^2\psi}{\partial y^2} + y\frac{\partial^3\psi}{\partial y^3} + x\frac{\partial^3\psi}{\partial x\partial y^2} + z\frac{\partial^3\psi}{\partial z\partial y^2}$$

and

$$\frac{\partial^2}{\partial z^2}\left[\left(x\frac{\partial\psi}{\partial x} + y\frac{\partial\psi}{\partial y} + z\frac{\partial\psi}{\partial z}\right)\right] = 2\frac{\partial^2\psi}{\partial z^2} + z\frac{\partial^3\psi}{\partial z^3} + x\frac{\partial^3\psi}{\partial x\partial z^2} + y\frac{\partial^3\psi}{\partial y\partial z^2}$$

We continue with

$$\mathcal{A}(\mathcal{T})\psi = (-i\hbar)\cdot(-\frac{\hbar^2}{2m})\cdot\left(x\frac{\partial}{\partial x} + y\frac{\partial}{\partial y} + z\frac{\partial}{\partial z}\right)\left[\frac{\partial^2\psi}{\partial x^2} + \frac{\partial^2\psi}{\partial y^2} + \frac{\partial^2\psi}{\partial z^2}\right]$$

and consider again the differentiations with respect to x, y, and z.

$$x\frac{\partial}{\partial x}\left[\frac{\partial^2\psi}{\partial x^2} + \frac{\partial^2\psi}{\partial y^2} + \frac{\partial^2\psi}{\partial z^2}\right] = x\frac{\partial^3\psi}{\partial x^3} + x\frac{\partial^3\psi}{\partial y^2\partial x} + x\frac{\partial^3\psi}{\partial z^2\partial x}$$

$$y\frac{\partial}{\partial y}\left[\frac{\partial^2\psi}{\partial x^2} + \frac{\partial^2\psi}{\partial y^2} + \frac{\partial^2\psi}{\partial z^2}\right] = y\frac{\partial^3\psi}{\partial x^2\partial y} + y\frac{\partial^3\psi}{\partial y^3} + y\frac{\partial^3\psi}{\partial z^2\partial y}$$

$$z\frac{\partial}{\partial z}\left[\frac{\partial^2\psi}{\partial x^2} + \frac{\partial^2\psi}{\partial y^2} + \frac{\partial^2\psi}{\partial z^2}\right] = z\frac{\partial^3\psi}{\partial x^2\partial z} + z\frac{\partial^3\psi}{\partial y^2\partial z} + z\frac{\partial^3\psi}{\partial z^3}$$

An inspection of the expressions for $\mathcal{T}(\mathcal{A}\psi)$ and $\mathcal{A}(\mathcal{T})\psi$ reveals that most terms are identical and disappear if we calculate the difference $(\mathcal{T}\mathcal{A})\psi - (\mathcal{A}\mathcal{T})\psi$. In this way we obtain

$$(\mathcal{T}\mathcal{A})\psi - (\mathcal{A}\mathcal{T})\psi = (-\frac{\hbar^2}{2m})(-i\hbar)\left[2\frac{\partial^2\psi}{\partial x^2} + 2\frac{\partial^2\psi}{\partial y^2} + 2\frac{\partial^2\psi}{\partial z^2}\right]$$

$$= (-i\hbar)\cdot 2\cdot(-\frac{\hbar^2}{2m})\left[\frac{\partial^2\psi}{\partial x^2} + \frac{\partial^2\psi}{\partial y^2} + \frac{\partial^2\psi}{\partial z^2}\right] = (-i\hbar)\cdot 2\cdot\mathcal{T}\psi$$

Thus we have proved Equation (4.72).

5

Atoms and Variational Principle

5.1 Exercises

E5.1 Consider the wavefunction ψ for the He-atom that is a product of two 1s electron orbitals, $\psi = \phi_{1s}(x_1, y_1, z_1) \cdot \phi_{1s}(x_2, y_2, z_2)$. Show that this wavefunction solves Equation (5.1) if you ignore the electron–electron repulsion term, $e^2/(4\pi\varepsilon_0 r_{12})$.

Solution

Consider the wavefunction $\psi = \phi_{1s}(x_1, y_1, z_1) \cdot \phi_{1s}(x_2, y_2, z_2)$ for substitution into the Schrödinger equation

$$\left[-\frac{\hbar^2}{2m_e}\nabla_1^2 - \frac{\hbar^2}{2m_e}\nabla_2^2 - \frac{2e^2}{4\pi\varepsilon_0 r_1} - \frac{2e^2}{4\pi\varepsilon_0 r_2} + \frac{e^2}{4\pi\varepsilon_0 r_{12}}\right]\psi = E\psi$$

If we neglect the last term, this equation becomes

$$\left[-\frac{\hbar^2}{2m_e}\nabla_1^2 - \frac{\hbar^2}{2m_e}\nabla_2^2 - \frac{2e^2}{4\pi\varepsilon_0 r_1} - \frac{2e^2}{4\pi\varepsilon_0 r_2}\right]\phi_{1s}(1)\cdot\phi_{1s}(2) = E\cdot\left[\phi_{1s}(1)\cdot\phi_{1s}(2)\right]$$

or

$$E\cdot\left[\phi_{1s}(1)\cdot\phi_{1s}(2)\right] = \left[-\frac{\hbar^2}{2m_e}\nabla_1^2 - \frac{2e^2}{4\pi\varepsilon_0 r_1}\right]\phi_{1s}(1)\phi_{1s}(2) + \left[-\frac{\hbar^2}{2m_e}\nabla_2^2 - \frac{2e^2}{4\pi\varepsilon_0 r_2}\right]\phi_{1s}(1)\phi_{1s}(2)$$

for which each term in brackets only operates on one of the electrons. Hence the other electron's orbital factors through the operator, and we can write

$$E\cdot\left[\phi_{1s}(1)\cdot\phi_{1s}(2)\right] = \phi_{1s}(2)\left[-\frac{\hbar^2}{2m_e}\nabla_1^2 - \frac{2e^2}{4\pi\varepsilon_0 r_1}\right]\phi_{1s}(1) + \phi_{1s}(1)\left[-\frac{\hbar^2}{2m_e}\nabla_2^2 - \frac{2e^2}{4\pi\varepsilon_0 r_2}\right]\phi_{1s}(2)$$

Inspection of this equation shows that it is separable. This can be verified by dividing through by the wavefunction product and writing $E = E_1 + E_2$. This procedure yields two one-electron Schrödinger equations of the H-atom type.

Here we proceed explicitly. Each of the terms in brackets has the form of an H-atom Hamiltonian with a nuclear charge of 2e, and it operates on a 1s H-atom orbital. Hence the energy is the same as that for a He^+ ion, so that

$$\left[-\frac{\hbar^2}{2m_e}\nabla_1^2 - \frac{2e^2}{4\pi\varepsilon_0 r_1}\right]\phi_{1s}(1) = E_1\phi_{1s}(1) = -\frac{1}{4\pi\varepsilon_0}\frac{2e^2}{a_0}\phi_{1s}(1)$$

and

$$\left[-\frac{\hbar^2}{2m_e}\nabla_2^2 - \frac{2e^2}{4\pi\varepsilon_0 r_2}\right]\phi_{1s}(2) = E_2\phi_{1s}(2) = -\frac{1}{4\pi\varepsilon_0}\frac{2e^2}{a_0}\phi_{1s}(2)$$

see Equation (5.15). Hence, we find that

$$E_1 = -\frac{1}{4\pi\varepsilon_0}\frac{2e^2}{a_0} \quad \text{and} \quad E_2 = -\frac{1}{4\pi\varepsilon_0}\frac{2e^2}{a_0}$$

Solutions Manual for Principles of Physical Chemistry, Third Edition. Edited by Hans Kuhn, David H. Waldeck, and Horst-Dieter Försterling.

and

$$E = E_1 + E_2 = -\frac{1}{4\pi\varepsilon_0}\frac{4e^2}{a_0}$$

E5.2 Write the Hamiltonian for the lithium ion, Li^+, and for the lithium atom, Li. Identify the different terms in the Hamiltonian and discuss how the two Hamiltonians differ.

Solution

The Hamiltonian for the lithium ion, Li^+, is

$$\mathcal{H}_{Li^+} = -\frac{\hbar^2}{2m_e}\left[\nabla_1^2 + \nabla_2^2\right] - \frac{3e^2}{4\pi\varepsilon_0}\left(\frac{1}{r_1} + \frac{1}{r_2}\right) + \frac{e^2}{4\pi\varepsilon_0}\left(\frac{1}{r_{12}}\right)$$

The first term contains the kinetic energy operators for each of the electrons in Li^+. The negative $3e^2$ term is the potential energy for the Coulomb attraction between the two electrons and the nucleus, and the positive e^2 term, which is last, is the potential energy for the Coulomb repulsion between the two electrons.

For the lithium atom, Li, the Hamiltonian is

$$\mathcal{H}_{Li} = -\frac{\hbar^2}{2m_e}\left[\nabla_1^2 + \nabla_2^2 + \nabla_3^2\right] - \frac{3e^2}{4\pi\varepsilon_0}\left(\frac{1}{r_1} + \frac{1}{r_2} + \frac{1}{r_3}\right)$$
$$+\frac{e^2}{4\pi\varepsilon_0}\left(\frac{1}{r_{12}} + \frac{1}{r_{13}} + \frac{1}{r_{23}}\right)$$

The first term contains the kinetic energy operators for each of the three electrons. The negative $3e^2$ term is the potential energy for the Coulomb attraction between the three electrons and the nucleus. The last term, the positive e^2 term, is the potential energy for the electron–electron Coulomb repulsion between each pair of electrons.

Because the Li atom has one more electron than the Li^+ ion, we see that it has an extra kinetic energy operator (three versus two) and an extra Coulomb attraction contribution. In addition, the Li atom has three electron–electron repulsion contributions as compared to only one for the Li^+ ion.

In the general case of an n-electron atom or ion, the Hamiltonian will have n kinetic energy and Coulomb attraction contributions, one for each electron, and it will have $n\,(n-1)/2$ electron–electron repulsion contributions.

E5.3 The Schrödinger equation for He, given by Equation (5.1), does not include a kinetic energy operator for the He nucleus. Why is this reasonable? How can you account for the motion of the nucleus and the electrons?

Solution

The kinetic energy term for the helium nucleus can be neglected because the overall translational motion of the atom in space has a small effect on the electronic energy. To account for this effect properly, we would include the kinetic energy operator for the nucleus, and we would transform the Schrödinger equation to center-of-mass coordinates. Because the mass of the He nucleus is so much larger than that of the electrons, the center of mass would lie very close to the nucleus (in comparison to the size of the electron cloud). This mathematical procedure would generate a separable Schrödinger equation: one part which describes the atom's translation in space and a second part which describes the relative motion of the electron with respect to the nucleus. The electronic Schrödinger equation would be similar in structure to that written in Equation (5.1) except that the electron mass m_e would be replaced by the reduced mass μ (see Problem 4.2); where

$$\mu = \frac{m_e m_{He}}{m_e + m_{He}}$$

Since m_{He} is about 7400 times larger than m_e, it is reasonable to make the approximation that

$$\mu \simeq m_e$$

E5.4 Apply the normalization condition to the trial wavefunctions in Equations (5.24) and (5.25) to find the constant A. You may use the fact that the one-electron wavefunctions are normalized already.

Solution

The normalization condition requires that the probability of finding the particle somewhere in space be unity. For the purposes of this problem we will combine Equations (5.24) and (5.25) into

$$\phi_2 = A \cdot \left[\varphi_{1s}(1) \cdot \varphi_{2s}(2) \pm \varphi_{1s}(2) \cdot \varphi_{2s}(1)\right]$$

where the plus sign is for Equation (5.24) and the minus sign is for Equation (5.25).

The normalization condition for ϕ_2 takes the mathematical form

$$\iint \phi_2^2 \cdot \mathrm{d}\tau_1 \mathrm{d}\tau_2 = 1$$

or

$$\iint \left[\varphi_{1s}(1) \cdot \varphi_{2s}(2) \pm \varphi_{1s}(2) \cdot \varphi_{2s}(1)\right]^2 \cdot \mathrm{d}\tau_1 \mathrm{d}\tau_2 = 1/A^2$$

Multiplying out the terms in the integrand gives

$$\begin{aligned}&\left[\varphi_{1s}(1) \cdot \varphi_{2s}(2) \pm \varphi_{1s}(2) \cdot \varphi_{2s}(1)\right]^2 \\ &\quad = \varphi_{1s}^2(1) \cdot \varphi_{2s}^2(2) \pm 2\varphi_{1s}(1) \cdot \varphi_{2s}(2) \cdot \varphi_{1s}(2) \cdot \varphi_{2s}(1) + \varphi_{1s}^2(2) \cdot \varphi_{2s}^2(1)\end{aligned}$$

and with the abbreviations

$$\begin{aligned}I_1 &= \iint \varphi_{1s}^2(1) \cdot \varphi_{2s}^2(2) \cdot \mathrm{d}\,\tau_1 \mathrm{d}\tau_2 \\ I_2 &= \iint \varphi_{1s}(1) \cdot \varphi_{2s}(2) \cdot \varphi_{1s}(2) \cdot \varphi_{2s}(1) \cdot \mathrm{d}\tau_1 \mathrm{d}\tau_2 \\ I_3 &= \iint \varphi_{1s}^2(2) \cdot \varphi_{2s}^2(1) \cdot \mathrm{d}\,\tau_1 \mathrm{d}\tau_2\end{aligned}$$

we obtain

$$I_1 \pm 2I_2 + I_3 = 1/A^2$$

Now we evaluate the integrals I_1, I_2, and I_3. For the first integral we find

$$\begin{aligned}I_1 &= \iint \left[\varphi_{1s}^2(1) \cdot \varphi_{2s}^2(2)\right] \mathrm{d}\tau_1 \mathrm{d}\tau_2 \\ &= \left[\int \varphi_{2s}^2(2)\,\mathrm{d}\tau_2\right]\left[\int \varphi_{1s}^2(1)\,\mathrm{d}\tau_1\right] = 1\end{aligned}$$

where we have used the fact that the orbitals are each normalized. For the second integral we find

$$\begin{aligned}I_2 &= \iint \left[\varphi_{1s}(1) \cdot \varphi_{2s}(2) \cdot \varphi_{1s}(2) \cdot \varphi_{2s}(1)\right] \mathrm{d}\tau_1 \mathrm{d}\tau_2 \\ &= \left[\int \varphi_{1s}(1)\varphi_{2s}(1)\,\mathrm{d}\tau_1\right] \cdot \left[\int \varphi_{2s}(2)\varphi_{1s}(2)\,\mathrm{d}\tau_2\right] = 0\end{aligned}$$

where we have exploited the orthogonality of the individual one-electron wavefunctions. For the third integral we see that

$$\begin{aligned}I_3 &= \iint \varphi_{1s}^2(2) \cdot \varphi_{2s}^2(1)\,\mathrm{d}\tau_1 \mathrm{d}\tau_2 \\ &= \left[\int \varphi_{1s}^2(2)\,\mathrm{d}\tau_2\right]\left[\int \varphi_{2s}^2(1)\,\mathrm{d}\tau_1\right] = 1\end{aligned}$$

Hence we find that

$$\frac{1}{A^2} = I_1 \pm 2I_2 + I_3 = 2$$

or

$$A = \pm\frac{1}{\sqrt{2}}$$

E5.5 Verify that the wavefunctions in Equations (5.31) and (5.32) satisfy the condition of indistinguishability.

Solution

A wavefunction satisfies the requirement that the electrons be indistinguishable if the swapping of any two electron coordinates causes no change in the function, other than a change of sign. First we consider the wavefunction of Equation (5.31), which is

$$\psi(1,2) = \varphi_{1s}(1)\varphi_{1s}(2) \cdot \frac{1}{\sqrt{2}}\{\alpha(1)\beta(2) + \alpha(2)\beta(1)\}$$

If the wavefunction satisfies the indistinguishability requirement, then

$$\psi(1,2) = \pm\psi(2,1) \text{ so that } |\psi(1,2)|^2 = |\psi(2,1)|^2$$

Swapping the coordinates (or labels) for the electrons in our wavefunction expression, we write

$$\begin{aligned}\psi(2,1) &= \varphi_{1s}(2)\varphi_{1s}(1) \cdot \frac{1}{\sqrt{2}}\{\alpha(2)\beta(1) + \alpha(1)\beta(2)\}\\ &= \varphi_{1s}(1)\varphi_{1s}(2) \cdot \frac{1}{\sqrt{2}}\{\alpha(1)\beta(2) + \alpha(2)\beta(1)\}\\ &= \psi(1,2)\end{aligned}$$

Hence the two sides are equal; thus, it is symmetric and indistinguishability is satisfied.

Next we consider the wavefunction of Equation (5.32), which is

$$\psi(1,2) = \varphi_{1s}(1)\varphi_{1s}(2) \cdot \frac{1}{\sqrt{2}}\{\alpha(1)\beta(2) - \alpha(2)\beta(1)\}$$

Swapping the coordinates (or labels) for the electrons in our wavefunction expression, we write

$$\begin{aligned}\psi(2,1) &= \varphi_{1s}(2)\varphi_{1s}(1) \cdot \frac{1}{\sqrt{2}}\{\alpha(2)\beta(1) - \alpha(1)\beta(2)\}\\ &= -\varphi_{1s}(1)\varphi_{1s}(2) \cdot \frac{1}{\sqrt{2}}\{\alpha(1)\beta(2 - \alpha(2)\beta(1)\}\\ &= -\psi(1,2)\end{aligned}$$

Hence, the wavefunctions differ by a minus sign; thus, it is antisymmetric and indistinguishability is satisfied.

E5.6 Show that the wavefunctions in Equations (5.29) and (5.30) do not satisfy the Pauli Exclusion principle.

Solution

The Pauli Exclusion principle states that the sign of the total electronic wavefunction must change sign on interchange of a pair of electron coordinates. For the wavefunction of Equation (5.29)

$$\psi(1,2) = \varphi_{1s}(1)\varphi_{1s}(2) \cdot \alpha(1)\alpha(2)$$

we find that

$$\begin{aligned}\psi(2,1) &= \varphi_{1s}(2)\varphi_{1s}(1) \cdot \alpha(2)\alpha(1)\\ &= \varphi_{1s}(1)\varphi_{1s}(2) \cdot \alpha(1)\alpha(2) = \psi(1,2)\end{aligned}$$

Hence we see that the wavefunction is symmetric and thus does not obey the Pauli principle. Hence it is not an acceptable electronic wavefunction.

For the wavefunction of Equation (5.30)

$$\psi(1,2) = \varphi_{1s}(1)\varphi_{1s}(2) \cdot \beta(1)\beta(2)$$

we find that

$$\begin{aligned}\psi(2,1) &= \varphi_{1s}(2)\varphi_{1s}(1) \cdot \beta(2)\beta(1)\\ &= \varphi_{1s}(1)\varphi_{1s}(2) \cdot \beta(1)\beta(2) = \psi(1,2)\end{aligned}$$

This wavefunction is also symmetric and does not obey the Pauli principle.

An alternative, more formal approach: A process for analyzing this property of a wavefunction may be formalized by defining an operator $\mathcal{A}$ that interchanges two electrons' coordinates such that

$$\mathcal{A}\,\psi(1,2) = a\psi(2,1)$$

where a is an eigenvalue that is +1 for a symmetric wavefunction and −1 for an antisymmetric wavefunction.
For the wavefunction of Equation (5.29) we find that

$$\begin{aligned}\mathcal{A}\,\psi(1,2) &= \mathcal{A}\,\left[\varphi_{1s}(1)\varphi_{1s}(2)\cdot\alpha(1)\alpha(2)\right]\\ &= \left[\varphi_{1s}(2)\varphi_{1s}(1)\cdot\alpha(2)\alpha(1)\right]\\ &= \psi(1,2)\end{aligned}$$

Hence we see that the eigenvalue of the antisymmetrizer operator $\mathcal{A}$ is 1, which indicates that the wavefunction is symmetric and thus does not obey the Pauli principle. Hence it is not an acceptable electronic wavefunction.
For the wavefunction of Equation (5.30) the proof is similar

$$\begin{aligned}\mathcal{A}\,\psi(1,2) &= \mathcal{A}\,\left[\varphi_{1s}(1)\varphi_{1s}(2)\cdot\beta(1)\beta(2)\right]\\ &= \left[\varphi_{1s}(2)\varphi_{1s}(1)\cdot\beta(2)\beta(1)\right]\\ &= \psi(1,2)\end{aligned}$$

This wavefunction is also symmetric and does not obey the Pauli principle.

E5.7 Calculate the ionization energy for Li^{2+} and compare it with the experimental value.

Solution
Li^{2+} is a hydrogen-like ion. The ionization energies of the one-electron hydrogen-like systems are given by

$$E_{\text{ion}} = \frac{1}{4\pi\varepsilon_0}\cdot\frac{e^2}{2a_0}Z^2$$

For Li^{2+} this expression becomes

$$\begin{aligned}E_{\text{ion}} &= \frac{1}{4\pi\varepsilon_0}\cdot\frac{e^2}{2a_0}Z^2\\ &= E_{\text{ion},1}(H)\times 3^2\\ &= 13.6\ \text{eV}\times 9 = 122.4\ \text{eV}\end{aligned}$$

Table 5.1 gives the experimental ionization energy of Li^{2+} to be 122.45 eV, in excellent agreement with the calculated value.

E5.8 Consider the $2s$ orbital of the atoms in the second row of the periodic table. How does the binding energy of the electron in this orbital change as one moves from Li to Ne? Contrast this behavior with that of the first ionization energies of the atoms.

Solution
The binding energy for the $2s$ electron in the second row atomic elements shows a systematic increase as one proceeds from left to right across the row. This trend reflects the increasing nuclear charge (Z) with atomic number. This behavior differs from the first ionization energies (binding energy for the most weakly bound electron in each case) for the atoms which shows some structure (occasionally decreasing) depending on the electron configuration in the atom (see Exercise E5.9 for a detailed discussion of this latter trend).

E5.9 Consider Figure 5.12 and Table 5.1 for the first ionization potentials of the first 10 elements. Explain the trend in ionization potentials in the first column.

Solution
The increase in ionization potential from H to He reflects an increase in binding energy because of the nuclear charge acting on electrons in the $1s$ orbital. The drop in ionization potential from He to Li occurs because of the increase in principal quantum number for the outermost electron from $n = 1$ to $n = 2$. Electrons in the $n = 2$ shell are further from the nucleus (on average) and shielded by the $n = 1$ electrons.

The general trend of increasing first ionization potential from Li to Ne reflects the increasing nuclear charge and fixed principal quantum number, as one proceeds from left to right across the second row of the periodic table. The drop in ionization potential from Be to B occurs because of the change of electron occupation from 2s to $2p$, and the resulting increase in orbital energy. Although the $2s$ and $2p$ orbitals have the same energy for an H-atom, for multi-electron atoms the $2p$ orbital electrons are less stable. This decrease in stability can be explained by the difference in the radial part of the orbital wavefunctions; the s-wavefunctions penetrate more effectively to the nucleus and are not as well shielded by the $1s$ electrons; hence, the $2s$ electrons are somewhat more stable than the $2p$ electrons. The drop in ionization potential from N to O occurs because the electrons in the outermost orbital must now be paired, which decreases their average distance from one another, hence increasing their electron–electron repulsion energy.

E5.10 Consider Table 5.1 of ionization potentials for the first 10 elements. Explain the trend in ionization potential for the Be row.

Solution

The table reports the four ionization potentials for Be. The ionization energy increases monotonically as the electrons are removed. The second ionization energy, i.e., for Be^+, is larger than that for Be because only one $2s$ valence electron is in the $2s$ orbital of Be^+; in contrast to the electron–electron repulsion between the pair of $2s$ electrons in Be. The third ionization energy is dramatically larger than that of the first and second ionization potentials. This large increase occurs because the remaining two electrons (in Be^{2+}) are in the $1s$ orbital, hence closer to the nucleus and more highly stabilized. The subsequent increase in the ionization energy for Be^{3+} results because the remaining electron does not experience the destabilizing effect of electron–electron repulsion.

Note that the difference between the third and fourth ionization energies is much larger than the difference between the first and second ionization energies. This difference (of differences) exists because the electrons in the $1s$ shell are held more compactly (smaller average distance causes higher electron–electron repulsion) than are those in the $2s$ shell.

E5.11 Compare the trend in atomic radii with the trend in ionization energies for the first ten elements. Discuss why these trends might be correlated with each other.

Solution

As the ionization energy increases, the binding energy of the outermost electron has increased. The stronger binding energy results in a more compact electron cloud, creating a smaller atom.

E5.12 In Box 5.4 we computed an effective atomic number for Ca of 2.85. Use this effective atomic number to compute the ionization energy of a valence electron in the Ca atom. Compare your value with the literature value. Comment on your results.

Solution

The procedure in Box 5.4 gives $Z_{\text{eff,Ca}} = 2.85$. For hydrogen-like atoms we have

$$E = \frac{Z^2}{n^2} \cdot E_{\text{H-atom}}$$

where

$$E_{\text{H-atom}} = -13.599 \text{ eV} = -21.787 \times 10^{-19} \text{ J}$$

is the energy of an H-atom in the ground state. We estimate the ionization energy IE using this effective charge and the one-electron (hydrogen-like) approximation. We find that

$$IE = -E_{4s}(\text{Ca atom}) = -\frac{Z_{\text{eff}}^2}{n^2} \cdot E_{\text{H-atom}} = \frac{2.85^2}{4^2} \cdot 13.599 \text{ eV} = 6.90 \text{ eV}$$

The experimental value of 6.1132 eV differs significantly from this value. The small difference likely results from the energy associated with the pairing of the electron spins in the outermost orbital (remember calcium has a $4s^2$ valence electron occupation).

E5.13 Consider the He^+ atomic ion. Show that the average distance of the electron from the nucleus is half the value found for the H atom; i.e., $\overline{r} = \frac{3}{4}a_0$ rather than $\overline{r} = \frac{3}{2}a_0$. Provide a physical explanation for this finding.

Solution

For the He^+ ion with

$$\psi = A \cdot \exp\left(-2r/a_0\right) \quad \text{and} \quad A = \sqrt{8/\left(\pi a_0^3\right)}$$

we calculate the average distance of the electron from the nucleus by

$$\begin{aligned}\overline{r} &= \frac{8}{\pi a_0^3} \iiint \exp\left(-4r/a_0\right) \cdot r \cdot d\tau \\ &= \frac{8}{\pi a_0^3} \int_0^\infty \exp\left(-4r/a_0\right) \cdot r \cdot 4\pi r^2 \cdot dr \\ &= \frac{32}{a_0^3} \int_0^\infty \exp\left(-4r/a_0\right) \cdot r^3 \cdot dr\end{aligned}$$

where we have used spherical coordinates (see Fig. 4.6). The integration over the angles is trivial in this case and yields the value 4π. In Appendix B, we find the integral $\int_0^\infty x^n e^{-x}\,dx = n!$ and use it for our evaluation here. First we define $x = 4r/a_0$, $dx = 4r/a_0 \cdot dr$, so that

$$\begin{aligned}\overline{r} &= \int_0^\infty \exp\left(-4r/a_0\right) \cdot r^3 \cdot dr = \frac{32}{a_0^3} \int_0^\infty \exp(-x) \cdot \left(x\frac{a_0}{4}\right)^3 \cdot \frac{a_0}{4} dx \\ &= \frac{32}{a_0^3}\left(\frac{a_0}{4}\right)^4 \int_0^\infty \exp(-x) \cdot x^3 \; dx = \frac{1}{8}a_0\,[3!] = \frac{3}{4}a_0\end{aligned}$$

This value is half of the value $\overline{r} = \frac{3}{2}a_0$ found for the H-atom. This difference occurs because the nuclear charge in He^+ is twice the value in the H-atom.

E5.14 In Table 5.1 two ionization energies are reported for He. Assume that the difference in ionization energies is caused solely by electron–electron repulsion and estimate the magnitude of the electron–electron repulsion energy. Use this repulsion energy to estimate the average distance between the two electrons. You need only perform simple calculations here and make rough estimates.

Solution

The first and second ionization energies of He are reported to be 24.59 and 54.42 eV. The difference in their values is

$$(54.42 - 24.59)\;\;\text{eV} = 29.83\;\text{eV} = 4.78 \times 10^{-18}\;\;\text{J}$$

If we assign the difference in these two values to the electron–electron repulsion, we find that

$$\frac{e^2}{4\pi\varepsilon_0} \cdot \frac{1}{\overline{\overline{r_{12}}}} = 4.78 \times 10^{-18}\;\;\text{J}$$

and

$$\begin{aligned}\overline{\overline{r_{12}}} &= \frac{e^2}{4\pi\varepsilon_0} \cdot \frac{1}{4.78 \times 10^{-18}\;\;\text{J}} = \frac{(1.6021 \times 10^{-19})^2\;\text{C}^2}{4\pi \cdot 8.854 \times 10^{-12}\;\text{C}^2\;\text{J}^{-1}\;\text{m}^{-1}} \cdot \frac{1}{4.78 \times 10^{-18}\;\;\text{J}} \\ &= 48.3 \times 10^{-12}\;\;\text{m} = 48.3\;\text{pm}\end{aligned}$$

In Section 5.3.1 the mutual repulsion energy of the two electrons in He was roughly calculated as

$$\varepsilon_{\text{repulsion}} = \frac{1}{4\pi\varepsilon_0} \cdot \frac{5}{4} \cdot \frac{e^2}{a_0} = \frac{e^2}{4\pi\varepsilon_0} \frac{1}{\overline{\overline{r_{12}}}}$$

leading to

$$\overline{\overline{r_{12}}} = \frac{e^2}{4\pi\varepsilon_0} \cdot \frac{4\pi\varepsilon_0}{e^2} \cdot \frac{4}{5}a_0 = \frac{4}{5} \cdot 52.9\;\text{pm} = 42.3\;\text{pm}$$

which is not too different from the value estimated from the ionization energy. Hence we conclude that the average distance between the two 1s electrons in a helium atom is between 40 and 50 pm.

E5.15 Using a periodic table, identify the electron configurations of all the main group elements. Compare your identification to that you expect from analyzing Fig. 5.11.

Solution

The electron configurations for the main group elements are

Table E5.15 Periodic Table. Electron Configuration of the Highest Occupied Orbitals of the Main Group Elements.

H $1s^1$							He $1s^2$
Li $1s^2 2s^1$	Be $1s^2 2s^2$	B $1s^2 2s^2 2p^1$	C $1s^2 2s^2 2p^2$	N $1s^2 2s^2 2p^3$	O $1s^2 2s^2 2p^4$	F $1s^2 2s^2 2p^5$	Ne $1s^2 2s^2 2p^6$
Na $[Ne]3s^1$	Mg $[Ne]3s^2$	Al $[Ne]3s^2 3p^1$	Si $[Ne]3s^2 3p^2$	P $[Ne]3s^2 3p^3$	S $[Ne]3s^2 3p^4$	Cl $[Ne]3s^2 3p^5$	Ar $[Ne]3s^2 3p^6$
K $[Ar]4s^1$	Ca $[Ar]4s^2$	Ga $[Ar]4s^2 4p^1$	Ge $[Ar]4s^2 4p^2$	As $[Ar]4s^2 4p^3$	Se $[Ar]4s^2 4p^4$	Br $[Ar]4s^2 4p^5$	Kr $[Ar]4s^2 4p^6$
Rb $[Kr]5s^1$	Sr $[Kr]5s^2$	In $[Kr]5s^2 5p^1$	Sn $[Kr]5s^2 5p^2$	Sb $[Kr]5s^2 5p^3$	Te $[Kr]5s^2 5p^4$	I $[Kr]5s^2 5p^5$	Xe $[Kr]5s^2 5p^6$
Cs $[Xe]6s^1$	Ba $[Xe]6s^2$	Tl $[Xe]6s^2 6p^1$	Pb $[Xe]6s^2 6p^2$	Bi $[Xe]6s^2 6p^3$	Po $[Xe]6s^2 6p^4$	At $[Xe]6s^2 6p^5$	Rn $[Xe]6s^2 6p^6$

These configurations are exactly those predicted by the figure.

E5.16 Excited sodium atoms give rise to an intense yellow emission, called the sodium D-line, which results from an electronic transition from the $3p$ excited state configuration to the $3s$ ground-state configuration. What are the atomic term symbols for these configurations? The emission consists of two distinct lines at 588.995 and 589.592 nm. Use these emission lines to estimate the energy splitting that arises from spin–orbit interactions.

Solution

The filled inner shell portion of the electron configurations does not matter to the term symbol. Thus we need only consider the valence electrons, $3s$ for the ground state and $3p$ for the excited state. For both of these single-electron configurations, $S = s_1 = \frac{1}{2}$ and the multiplicity is then 2. The value of L for the ground state is $L = l_1 = 0$ and that for the excited state is $L = l_1 = 1$. Thus the term symbol for the ground state is $^2S_{1/2}$, and for the two different excited states $^2P_{1/2}$ and $^2P_{3/2}$.

One of the sodium D-lines is for the transition $^2P_{1/2} \rightarrow {}^2S_{1/2}$ and the other is for $^2P_{3/2} \rightarrow {}^2S_{1/2}$. Hence the difference in energy between the two sodium D-line transitions is equal to the difference between the $^2P_{3/2}$ and $^2P_{1/2}$. The two sodium atom lines occur at $\lambda_1 = 588.995$ nm and $\lambda_2 = 589.592$ nm, corresponding to the excitation energies $\Delta E_1 = hc_0/\lambda_1$ and $\Delta E_2 = hc_0/\lambda_1$. The energy difference $\Delta\Delta E$ between these two excitations is then

$$\Delta\Delta E = \Delta E_1 - \Delta E_2 = hc_0\left(\frac{1}{\lambda_1} - \frac{1}{\lambda_2}\right)$$

$$= 6.6260 \times 10^{-34}\ \text{J} \cdot \text{s} \cdot 2.997 \times 10^{8}\ \text{m s}^{-1}\left(\frac{1}{\lambda_1} - \frac{1}{\lambda_2}\right)$$

$$= 1.986 \times 10^{-25}\ \text{J} \cdot\ \text{m} \cdot \left(\frac{1}{588.995 \times 10^{-9}\ \text{m}} - \frac{1}{589.592\ \times 10^{-9}\ \text{m}}\right)$$

$$= 1.986 \times 10^{-25}\ \text{J} \cdot\ \text{m} \cdot 1.719 \times 10^{3}\ \text{m}^{-1} = 3.3726 \times 10^{-19}\ \text{J} - 3.3692 \times 10^{-19}\ \text{J}$$

$$= 3.41 \times 10^{-22}\ \text{J}$$

Hence the spin orbit energy splitting is 3.41×10^{-22} J.

E5.17 In order to excite an electron in the 1s orbital of a hydrogen atom to the 3p orbital of the hydrogen atom, what wavelength of light is needed? What are the spectroscopic term symbols for the electron in each of these states?

Solution

The wavelength of light for the hydrogen atom transitions is independent of the value of the l quantum number with the energies of the states given by Equation (4.15) to be

$$E_n = -\frac{1}{4\pi\varepsilon_0}\frac{e^2}{2a_0}\frac{1}{n^2}$$

thus the transition energies are given by

$$\Delta E = -\frac{1}{4\pi\varepsilon_0}\frac{e^2}{2a_0}\left(\frac{1}{1^2} - \frac{1}{3^2}\right)$$
$$= (21.80 \times 10^{-19}\ \text{J})\frac{8}{9} = 19.38 \times 10^{-19}\ \text{J}$$

and the corresponding wavelength is

$$\lambda = \frac{hc_0}{\Delta E} = \frac{(6.626 \times 10^{-34}\ \text{J s})\,(2.998 \times 10^{8}\ \text{m s}^{-1})}{(19.38 \times 10^{-19}\ \text{J})}$$
$$= 1.025 \times 10^{-7}\ \text{m} = 102.5\ \text{nm}$$

The term symbols are $1s$: $^2S_{1/2}$ for the ground state with $S = \frac{1}{2}$ (a single unpaired electron), $L = 0$ (an s electron with $l = 0$) and

$$J = |L + S|, \ldots, |L - S| = \frac{1}{2}$$

The excited state term symbol has $S = \frac{1}{2}$ (a single unpaired electron), $L = 1$ (a p electron with $l = 1$) and

$$J = |L + S|, \ldots, |L - S| = 3/2, 1/2$$

Thus a 3p electron configuration has the two energy levels $^2P_{3/2}$ and $^2P_{1/2}$.

E5.18 What is the ground electronic state configuration for the oxygen atom? What are the spectroscopic term symbols associated with this electronic configuration? Energy order these states using Hund's rules and state the rationale used to justify them.

Solution

The ground-state electron configuration for the oxygen, O, atom is $1s^2 2s^2 2p^4$. Using Table 5.3, we see that the spectroscopic term symbols associated with this configuration are $^3P_2$, $^3P_1$, $^3P_0$, $^1D_2$, and $^1S_0$.

When energy ordered using Hund's rules, the energies are $E_{^3P_2} < E_{^3P_1} < E_{^3P_0} < E_{^1D_2} < E_{^1S_0}$.

E5.19 Write the ground electronic state's configuration for fluorine, F. Write the excited electronic state configuration which results from promoting the most weakly bound electron in the ground state to the next higher energy orbital in the atom. Determine the spectroscopic term symbols for your ground-state electron configuration and for your excited state electron configuration. If it is required, energy order the states that occur in each of these configurations.

Solution

The ground-state electron configuration for F is $1s^2 2s^2 2p^5$. The lowest energy excited state will result by promoting one of the 2p electrons to the 3s orbital, resulting in an electron configuration of $1s^2 2s^2 2p^4 3s^1$. Using Table 5.3 the term symbol for the ground state is that of a single electron missing in the 2p orbital or $^2P_{3/2}$. Note that a slightly higher energy state with the term symbol $^2P_{1/2}$ also exists for this electron configuration.

The excited state term symbol has multiple possibilities because it arises from that of two electrons missing in the 2p orbital in addition to a 3s electron. Let us derive these using the method of microstates that is discussed in Foundation 5.5. The table below shows all of the electron arrangements (orbital occupancies and spin) for the $2p^4 3s$-subshells of fluorine. The first four columns of each row show the different electron arrangements that are possible. The fifth column of each row has the m_s values summed to give M_s. Note that ↑ is $m_s = 1/2$ and ↓ is $m_s = -1/2$. The sixth column reports M_L, which is obtained by summing the k values for the occupied orbitals of the row.

3*s* $k = 0$	2*p* $k = 1$	2*p* $k = 0$	2*p* $k = -1$	M_S	M_L
↑	↑↓	↑↓		$\frac{1}{2}$	2
↑		↑↓	↑↓	$\frac{1}{2}$	−2
↑	↑↓		↑↓	$\frac{1}{2}$	0
↑	↑	↑	↑↓	$\frac{3}{2}$	−1
↑	↑↓	↑	↑	$\frac{3}{2}$	1
↑	↑	↑↓	↑	$\frac{3}{2}$	0
↑	↓	↓	↑↓	$-\frac{1}{2}$	−1
↑	↑↓	↓	↓	$-\frac{1}{2}$	1
↑	↓	↑↓	↓	$-\frac{1}{2}$	0
↑	↑	↓	↑↓	$\frac{1}{2}$	−1
↑	↑	↑↓	↓	$\frac{1}{2}$	0
↑	↓	↑	↑↓	$\frac{1}{2}$	−1
↑	↓	↑↓	↑	$\frac{1}{2}$	0
↑	↑↓	↑	↓	$\frac{1}{2}$	1
↑	↑↓	↓	↑	$\frac{1}{2}$	1
↓	↑↓	↑↓		$-\frac{1}{2}$	2
↓		↑↓	↑↓	$-\frac{1}{2}$	−2
↓	↑↓		↑↓	$\frac{1}{2}$	0
↓	↑	↑	↑↓	$\frac{1}{2}$	−1
↓	↑↓	↑	↑	$\frac{1}{2}$	1
↓	↑	↑↓	↑	$\frac{1}{2}$	0
↓	↓	↓	↑↓	$-\frac{3}{2}$	−1
↓	↑↓	↓	↓	$-\frac{3}{2}$	1
↓	↓	↑↓	↓	$-\frac{3}{2}$	0
↓	↑	↓	↑↓	$-\frac{1}{2}$	−1
↓	↑	↑↓	↓	$-\frac{1}{2}$	0
↓	↓	↑	↑↓	$-\frac{1}{2}$	−1
↓	↓	↑↓	↑	$-\frac{1}{2}$	0
↓	↑↓	↑	↓	$-\frac{1}{2}$	1
↓	↑↓	↓	↑	$-\frac{1}{2}$	1

First we find the largest M_L, which is equal to 2. This M_L is associated with the $M_S = 1/2$ and $-1/2$ microstates. Clearly this value of M_L must be associated with an $L = 2$ orbital angular momentum, which has five-fold degeneracy; i.e., the $L = 2$ state will have values of $M_L = 0, \pm1, \pm2$. The values of $M_S = 1/2$ and $-1/2$ imply that $S = 1/2$, i.e., a doublet spin multiplicity. The term symbol for this value of L and S is ^2D. Now we find ten distinct electron arrangements, which are associated with this combination of M_L and M_S values. These electron arrangements are

3s $k=0$	2p $k=1$	2p $k=0$	2p $k=-1$	M_S	M_L
↑	↑↓	↑↓		$\frac{1}{2}$	2
↑		↑↓	↑↓	$\frac{1}{2}$	−2
↑	↑↓		↑↓	$\frac{1}{2}$	0
↑	↑	↓	↑↓	$\frac{1}{2}$	−1
↑	↑↓	↓	↑	$\frac{1}{2}$	1
↓	↑↓	↑↓		$-\frac{1}{2}$	2
↓		↑↓	↑↓	$-\frac{1}{2}$	−2
↓	↑↓		↑↓	$-\frac{1}{2}$	0
↓	↑	↓	↑↓	$-\frac{1}{2}$	−1
↓	↑↓	↓	↑	$-\frac{1}{2}$	1

Assigning these five electron arrangements to the ^2D term leaves the following electron arrangements to be accounted for.

3s $k=0$	2p $k=1$	2p $k=0$	2p $k=-1$	M_S	M_L
↑	↑	↑	↑↓	$\frac{3}{2}$	−1
↑	↑↓	↑	↑	$\frac{3}{2}$	1
↑	↑	↑↓	↑	$\frac{3}{2}$	0
↑	↓	↓	↑↓	$-\frac{1}{2}$	−1
↑	↑↓	↓	↓	$-\frac{1}{2}$	1
↑	↓	↑↓	↓	$-\frac{1}{2}$	0
↑	↓	↑	↑↓	$\frac{1}{2}$	−1
↑	↑	↑↓	↓	$\frac{1}{2}$	0
↑	↓	↑↓	↑	$\frac{1}{2}$	0
↑	↑↓	↑	↓	$\frac{1}{2}$	1
↓	↑	↑	↑↓	$\frac{1}{2}$	−1
↓	↑↓	↑	↑	$\frac{1}{2}$	1
↓	↑	↑↓	↑	$\frac{1}{2}$	0
↓	↓	↓	↑↓	$-\frac{3}{2}$	−1
↓	↑↓	↓	↓	$-\frac{3}{2}$	1

↓	↓	↑↓	↓	$-\frac{3}{2}$	0
↓	↑	↑↓	↓	$-\frac{1}{2}$	0
↓	↓	↑	↑↓	$-\frac{1}{2}$	−1
↓	↓	↑↓	↑	$-\frac{1}{2}$	0
↓	↑↓	↑	↓	$-\frac{1}{2}$	1

Once again we find the largest value of M_L, which is 1, because it corresponds to a high value of L. This M_L value is associated with $M_S = \pm\frac{3}{2}, \pm\frac{1}{2}$. This value of M_L implies an $L = 1$ state for which we have $M_L = 0, \pm 1$, and the possible M_S values indicate an $S = \frac{3}{2}$ state for which $M_S = \pm\frac{3}{2}, \pm\frac{1}{2}$. These values of L and S correspond to the spectroscopic term symbol ^4P.

The 12 microstates for the ^4P are listed below.

3*s* $k = 0$	2*p* $k = 1$	2*p* $k = 0$	2*p* $k = -1$	M_S	M_L
↑	↑	↑	↑↓	$\frac{3}{2}$	−1
↑	↑↓	↑	↑	$\frac{3}{2}$	1
↑	↑	↑↓	↑	$\frac{3}{2}$	0
↑	↓	↑	↑↓	$\frac{1}{2}$	−1
↑	↓	↑↓	↑	$\frac{1}{2}$	0
↑	↑↓	↑	↓	$\frac{1}{2}$	1
↓	↓	↓	↑↓	$-\frac{3}{2}$	−1
↓	↑↓	↓	↓	$-\frac{3}{2}$	1
↓	↓	↑↓	↓	$-\frac{3}{2}$	0
↓	↓	↑	↑↓	$-\frac{1}{2}$	−1
↓	↓	↑↓	↑	$-\frac{1}{2}$	0
↓	↑↓	↑	↓	$-\frac{1}{2}$	1

Now only eight microstates remain.

3*s* $k - 0$	2*p* $k - 1$	2*p* $k - 0$	2*p* $k - -1$	M_S	M_L
↑	↓	↓	↑↓	$-\frac{1}{2}$	−1
↑	↑↓	↓	↓	$-\frac{1}{2}$	1
↑	↓	↑↓	↓	$-\frac{1}{2}$	0
↑	↑	↑↓	↓	$\frac{1}{2}$	0
↓	↑	↑	↑↓	$\frac{1}{2}$	−1
↓	↑↓	↑	↑	$\frac{1}{2}$	1
↓	↑	↑↓	↑	$\frac{1}{2}$	0
↓	↑	↑↓	↓	$-\frac{1}{2}$	0

Once again we find the largest value of M_L, which is 1, because it corresponds to a high value of L. This M_L value is associated with $M_S = \pm\frac{1}{2}$. This value of M_L implies an $L = 1$ state for which we have $M_L = 0, \pm 1$, and the possible M_S values indicate an $S = \frac{1}{2}$ state for which $M_S = \pm\frac{1}{2}$. These values of L and S correspond to the spectroscopic term symbol ^2P.

The six microstates for the ^2P are listed below.

3s $k = 0$	2p $k = 1$	2p $k = 0$	2p $k = -1$	M_S	M_L
↑	↓	↓	↑↓	$-\frac{1}{2}$	−1
↑	↑↓	↓	↓	$-\frac{1}{2}$	1
↑	↓	↑↓	↓	$-\frac{1}{2}$	0
↓	↑	↑	↑↓	$\frac{1}{2}$	−1
↓	↑↓	↑	↑	$\frac{1}{2}$	1
↓	↑	↑↓	↑	$\frac{1}{2}$	0

This leaves only two microstates, for which $M_L = 0$ and $M_S = \pm\frac{1}{2}$ so that the term symbol is ^2S.

3s $k = 0$	2p $k = 1$	2p $k = 0$	2p $k = -1$	M_S	M_L
↑	↑	↑↓	↓	$\frac{1}{2}$	0
↓	↑	↑↓	↓	$-\frac{1}{2}$	0

Following Hund's rules we propose that the energy ordering for these terms is $^4P < {}^2D < {}^2P < {}^2S$.

Because both L and S are non-zero in a number of these cases, we must consider their vector sum to form J values. Since $\vec{J} = \vec{\mathcal{L}} + \vec{S}$, the magnitude $|J|$ ranges from $|\mathcal{L} - S|$ to $|\mathcal{L} + S|$. Hence, for ^4P we obtain three distinct term symbols, namely $^4P_{5/2}$, $^4P_{3/2}$, and $^4P_{1/2}$; for ^2D we obtain two distinct term symbols, namely $^2D_{5/2}$ and $^2D_{3/2}$; and for ^2P we obtain two distinct term symbols, namely $^2P_{3/2}$ and $^2P_{1/2}$—in which the subscript identifies the value of $|J|$.

E5.20 Write the ground electronic state's configuration for Mg. Write the excited electronic state configuration which results from promoting the most weakly bound electron in the ground state to the next higher energy orbital in the atom. Determine the spectroscopic term symbols for your ground-state electron configuration and for your excited state electron configuration. If it is required, energy order the states that occur in each of these configurations.

Solution

The ground-state electron configuration for Mg is $1s^2 2s^2 2p^6 3s^2$. The appropriate excited state configuration is $1s^2 2s^2 2p^6 3s^1 3p^1$. The ground-state configuration has all spins paired in filled orbitals; thus, the spectroscopic term symbol is $^1S_0$. The excited state has both $3s^1$ and $3p^1$ open subshells, for which the total spin can be 0 or 1 and the total orbital angular momentum is 1. Using Table 5.3, the spectroscopic term symbols are $^1P_1$ and $^3P_{2,1,0}$. The energy ordering of these states is $E_{^3P_0} < E_{^3P_1} < E_{^3P_2} < E_{^1P_1}$.

E5.21 In Section 5.6.2 the atomic term symbols for the excited states of He are discussed. Draw a vector model (analogous to Fig. 4.11) which shows the different orbital and spin angular momenta that are possible for the $1s2s$ configuration and for the $1s2p$ configuration.

Solution

From the discussion in the text, the $1s2s$ configuration has two different energy levels. For the ^1S level the orbital angular momentum is zero and the spin angular momentum is zero, so the vector model is the trivial result of zero length for both the spin and orbital parts. For the ^3S, $L = 0$; however, $S = 1$ and the total vector will have a length $\sqrt{2}\hbar$. It can be oriented up with a component of its angular momentum along the z-direction being $\hbar$, it can be oriented perpendicular to z (in the $x-y$ plane), or it can be pointed down with a component of its angular momentum along the z-direction being $-\hbar$, see Fig. E5.21a.

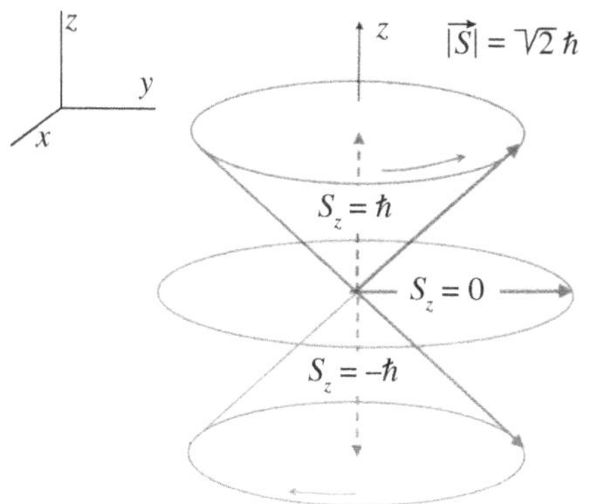

Figure E5.21a The spin angular momentum vector is drawn with its three possible z-components; i.e., oriented up, oriented perpendicular, and oriented down. The angular momentum vector has a total length of $\sqrt{2}\hbar$ and can have any orientation in the x, y-plane.

For the $1s2p$ configuration we found ^1P and ^3P levels. For ^1P, the spin angular momentum is zero and the vector representation has only three directions. In fact, the vector diagram for this state is like that for a 2p state, shown in Fig. 4.11. For the ^3P case, things are a bit more complicated. Because the spin and orbital angular momentum are each nonzero, we find that we get three J values $J = 2, 1, 0$ for the triplet state and obtain the terms symbols $^3P_2$, $^3P_1$, and $^3P_0$. The nine possible directions for its spin angular momentum are shown in shown in Fig. E5.21b.

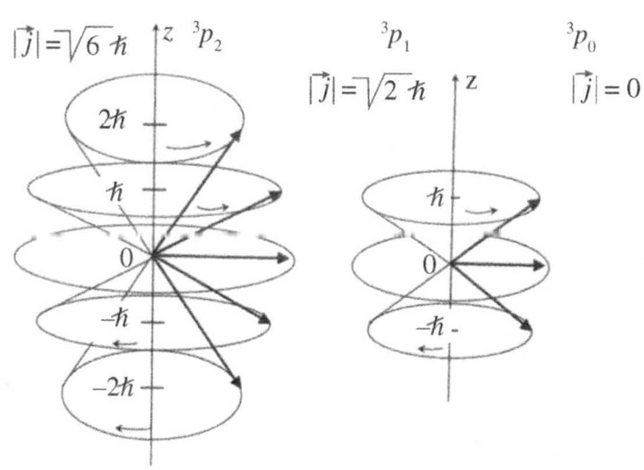

Figure 5.21b Nine possible combinations of the total angular momentum vector, $\vec{J} = \vec{L} + \vec{S}$. The left panel shows the case where the length of the total angular momentum vector is $\sqrt{6}\hbar$, which is possible for five relative orientations of the $\vec{L}$ and $\vec{S}$. vectors. The middle panel shows the case where the total length of the angular momentum vector is $\sqrt{2}\hbar$, which is possible for three relative orientations of the $\vec{L}$ and $\vec{S}$. vectors. The right panel shows the case where the total length of the angular momentum vector is zero; that is the $\vec{L}$ and $\vec{S}$. vectors are antiparallel and give no net $\vec{J}$.

5.2 Problems

P5.1 Given that the wavefunction of the lowest quantum state of the hydrogen atom is

$$\psi_1 = \frac{1}{\sqrt{\pi a_0^3}} e^{-\frac{r}{a_0}}$$

use Equation (5.4)

$$E = \int \psi \mathcal{H} \psi \cdot d\tau$$

to calculate the hydrogen atom's ground-state energy. To evaluate the average potential energy $\overline{V_1}$ follow the example in Section 4.2.5. In order to find the average kinetic energy, you should use the fact that

$$\left(\frac{\partial^2}{\partial x^2}+\frac{\partial^2}{\partial y^2}+\frac{\partial^2}{\partial z^2}\right)e^{-r/a_0}=-\frac{1}{a_0}\left(\frac{2}{r}-\frac{1}{a_0}\right)\cdot e^{-r/a_0}$$

which is discussed in Section 4.2.2 and you show in E4.16.

Solution

The Hamiltonian for the hydrogen atom is

$$\mathcal{H}=-\frac{\hbar^2}{2m_e}\left(\frac{\partial^2}{\partial x^2}+\frac{\partial^2}{\partial y^2}+\frac{\partial^2}{\partial z^2}\right)-\frac{1}{4\pi\varepsilon_0}\frac{e^2}{r}$$

so that

$$\begin{aligned}\mathcal{H}\psi_1&=-\frac{\hbar^2}{2m_e}\left(\frac{\partial^2}{\partial x^2}+\frac{\partial^2}{\partial y^2}+\frac{\partial^2}{\partial z^2}\right)\psi_1-\frac{1}{4\pi\varepsilon_0}\frac{e^2}{r}\psi_1\\&=\frac{\hbar^2}{2m_e}\frac{1}{a_0}\left(\frac{2}{r}-\frac{1}{a_0}\right)\psi_1-\frac{1}{4\pi\varepsilon_0}\frac{e^2}{r}\psi_1\end{aligned}$$

where we have used the relation given for the kinetic energy operator.

Substituting for the Hamiltonian and the wavefunction in Equation (5.4) gives

$$\begin{aligned}E&=\int\psi\mathcal{H}\psi\cdot d\tau\\&=\frac{1}{\pi a_0^3}\int e^{-\frac{r}{a_0}}\left[\frac{\hbar^2}{2m_e}\frac{1}{a_0}\left(\frac{2}{r}-\frac{1}{a_0}\right)-\frac{1}{4\pi\varepsilon_0}\frac{e^2}{r}\right]e^{-\frac{r}{a_0}}\cdot d\tau\end{aligned}$$

Using the facts that $d\tau = r^2\,dr\sin\theta\,d\theta\,d\varphi$ and

$$a_0=4\pi\varepsilon_0\hbar^2/(m_e e^2)$$

we can write

$$\begin{aligned}E&=\frac{1}{\pi a_0^3}\int_0^\infty e^{-\frac{2r}{a_0}}r^2\left[\frac{1}{2}\frac{e^2}{4\pi\varepsilon_0}\left(\frac{2}{r}-\frac{1}{a_0}\right)-\frac{1}{4\pi\varepsilon_0}\frac{e^2}{r}\right]dr\cdot\int_0^\pi\sin\theta\,d\theta\int_0^{2\pi}d\varphi\\&=\frac{e^2}{4\pi\varepsilon_0}\frac{1}{\pi a_0^3}\int_0^\infty e^{-\frac{2r}{a_0}}r^2\left[-\frac{1}{2a_0}\right]dr\cdot[-\cos\theta]_0^\pi\cdot[\varphi]_0^{2\pi}\\&=\frac{e^2}{\varepsilon_0}\frac{1}{2\pi a_0^4}\int_0^\infty e^{-\frac{2r}{a_0}}r^2\,dr\end{aligned}$$

If we make the substitution $x = 2r/a_0$, we obtain

$$E=-\frac{e^2}{2\pi\varepsilon_0}\frac{1}{a_0^4}\frac{a_0^3}{8}\int_0^\infty e^{-x}x^2dx$$

From Appendix B, we know that $\int_0^\infty e^{-x}x^n\,dx = n!$, so that

$$E=-\frac{e^2}{4\pi\varepsilon_0}\frac{1}{a_0^4}\frac{a_0^3}{4}(2!)=-\frac{e^2}{4\pi\varepsilon_0}\frac{1}{2a_0}$$

P5.2 How does the electronic energy of a one-electron atom depend on the atomic number? Perform the calculation in Problem P5.1 for the case where the nucleus has a charge of Ze, so that

$$V(r)=\frac{-Ze^2}{4\pi\varepsilon_0 r}$$

and the ground-state wavefunction is given by

$$\psi(r)=\sqrt{\frac{Z^3}{\pi a_0^3}}\,e^{-\frac{Zr}{a_0}}$$

Evaluate the energy for the case of $Z = 2$, He^+.

Solution

We consider an electron attracted by the nuclear charge Ze for which the Schrödinger equation becomes

$$-\frac{h^2}{8\pi^2 m_e}\nabla^2\psi - \frac{1}{4\pi\varepsilon_0}\cdot\frac{Ze\cdot e}{r}\psi = E\psi$$

With

$$\psi = A\cdot \mathrm{e}^{-Zr/a_0}$$

we find by analogy with the previous problem (P5.1) that

$$\mathcal{H}\psi = \frac{\hbar^2}{2m_e}\frac{Z}{a_0}\left(\frac{2}{r}-\frac{Z}{a_0}\right)\psi - \frac{1}{4\pi\varepsilon_0}\frac{Ze^2}{r}\psi$$

Using $a_0 = 4\pi\varepsilon_0\hbar^2/(m_e e^2)$ we can write

$$E = \frac{Z^3}{\pi a_0^3}\int_0^\infty \mathrm{e}^{-\frac{2Zr}{a_0}} r^2\left[\frac{1}{2}\frac{Ze^2}{4\pi\varepsilon_0}\left(\frac{2}{r}-\frac{Z}{a_0}\right) - \frac{1}{4\pi\varepsilon_0}\frac{Ze^2}{r}\right]\mathrm{d}r\cdot\int_0^\pi \sin\theta\,\mathrm{d}\theta\int_0^{2\pi}\mathrm{d}\varphi$$

$$= \frac{Z^5e^2}{4\pi\varepsilon_0}\frac{4}{a_0^3}\int_0^\infty \mathrm{e}^{-\frac{2Zr}{a_0}} r^2\left[\frac{1}{2a_0}\right]\mathrm{d}r = \frac{Z^5e^2}{4\pi\varepsilon_0}\frac{2}{a_0^4}\int_0^\infty \mathrm{e}^{-\frac{2Zr}{a_0}} r^2\,\mathrm{d}r$$

If we make the substitution $x = 2Zr/a_0$, we obtain

$$E = -\frac{Z^5e^2}{4\pi\varepsilon_0}\frac{2}{a_0^4}\frac{a_0^3}{8Z^3}\int_0^\infty \mathrm{e}^{-x}x^2\mathrm{d}x$$

From Appendix B, we know that $\int_0^\infty \mathrm{e}^{-x}x^n\,\mathrm{d}x = n!$, so that

$$E = -\frac{Z^2e^2}{4\pi\varepsilon_0}\frac{1}{a_o}\frac{1}{4}(2!) = -\frac{Z^2e^2}{4\pi\varepsilon_0}\frac{1}{2a_o}$$

In the case of He^+ we have $Z = 2$ and

$$\psi = A\cdot\mathrm{e}^{-2r/a_0},\ E = \frac{1}{4\pi\varepsilon_0}\cdot\frac{2e^2}{a_0}$$

The experimentally obtained ionization energy (87.184×10^{-19} J) is indeed four times as great for He^+ as for the H-atom. Actually 87.184×10^{-19} J$= 4.0016\times21.787\times10^{-19}$ J; the ionization energy of He^+ is thus somewhat greater than four times the ionization energy of the H-atom. The difference lies in the fact that the He nucleus is heavier, so the comovement of the He nucleus is less important.

P5.3 Here you will use the variational method to calculate the ground-state electronic energy for the H-atom with two different trial functions:

Part 1). Use a normalized Gaussian trial function

$$\phi_{1G} = \left(\frac{2C_{11}}{\pi}\right)^{3/4}\exp(-C_{11}r^2)\ \text{with}\ C_{11} > 0$$

where C_{11} is an adjustable constant and

Given that the variational energy for the single Gaussian case is

$$E_{1G} = \int \phi_{1G}\mathcal{H}\phi_{1G}\,\mathrm{d}\tau = \frac{3h^2}{8\pi^2 m_e}C_{11} - \frac{e^2}{4\pi\varepsilon_0}\sqrt{\frac{8C_{11}}{\pi}}$$

find the value of C_{11} that minimizes the variational energy. Calculate the percent error in this energy, as compared to that for the exact solution. Make a plot of the optimized Gaussian wavefunction and the exact wavefunction on the same graph. Comment on your findings.

Part 2). Use a sum of two Gaussians

$$\phi_{2G} = \frac{\left(\frac{2C_{21}}{\pi}\right)^{3/4}\exp(-C_{21}r^2) + w\left(\frac{2C_{22}}{\pi}\right)^{3/4}\exp(-C_{22}r^2)}{\sqrt{1+w^2+2\cdot w\cdot S}}\quad \text{with}\ C_{21},\quad C_{22} > 0$$

where C_{21} and C_{22} are adjustable constants, the parameter w adjusts the relative statistical weight of the Gaussian functions, and the quantity S accounts for the overlap of the two Gaussians

$$S = \sqrt{8}\cdot\frac{(C_{21}C_{22})^{3/4}}{(C_{21}+C_{22})^{3/2}}$$

Use the MathCAD program "Problem5_3" to calculate the variational energy for different values of C_{21} and C_{22}, with $w = 1$ and $w = 0.2$, and report the results for three cases in which the variational energy is equal to or lower than that you found in part 1) for a single Gaussian. As in part 1) calculate the percent error in these energies as compared to that for the exact solution. Make plots of the two Gaussian wavefunctions and compare them to the exact solution. Comment on your findings and discuss how the wavefunction properties might affect the energy values you obtained.

Solution

Part 1. Given the energy expression we can write that

$$E_{1G} = A_1 C_{11} - A_2 \sqrt{C_{11}}$$

with

$$A_1 = \frac{3h^2}{8\pi^2 m_e} \text{ and } A_2 = \frac{e^2}{4\pi\varepsilon_0}\sqrt{\frac{8}{\pi}}$$

Now we search for the minimum of the function $E = \overline{T} + \overline{V}$ analytically, by calculating its derivative with respect to C_{11}

$$\frac{\mathrm{d}E}{\mathrm{d}C_{11}} = A_1 - A_2 \frac{1}{2}\frac{1}{\sqrt{C_{11}}}$$

and solving this expression for $\mathrm{d}E/\mathrm{d}C_{11} = 0$. Hence we find that

$$\sqrt{C_{11,\min}} = \frac{A_2}{2 \cdot A_1} = \frac{e^2}{4\pi\varepsilon_0}\sqrt{\frac{8}{\pi}}\frac{1}{2}\frac{8\pi^2 m_e}{3h^2} = \frac{\sqrt{8\pi}}{3}\left(\frac{e^2 m_e}{\varepsilon_0 h^2}\right)$$

or

$$C_{11,\min} = \frac{8\pi}{9}\left(\frac{e^2 m_e}{\varepsilon_0 h^2}\right)^2 = \frac{8}{9\pi}\frac{1}{a_0^2}$$

where we used the fact that $a_0 = \varepsilon_0 h^2/(\pi m_e e^2)$. Hence we see that

$$E_{1G,\min} = \frac{3h^2}{8\pi^2 m_e}\frac{8}{9\pi}\frac{1}{a_0^2} - \frac{e^2}{4\pi\varepsilon_0}\sqrt{\frac{8}{\pi}} \cdot \sqrt{\frac{8}{9\pi}\frac{1}{a_0}}$$

$$= \frac{h^2}{3\pi^3 m_e}\frac{1}{a_0^2} - \frac{8}{3\pi}\frac{e^2}{4\pi\varepsilon_0}\frac{1}{a_0} = \frac{h^2}{3\pi^3 m_e a_0}\frac{1}{a_0} - \frac{2e^2}{3\pi^2\varepsilon_0}\frac{1}{a_0}$$

$$= \frac{h^2 \cdot \pi m_e e^2}{3\pi^3 m_e \varepsilon_0 h^2}\frac{1}{a_0} - \frac{2e^2}{3\pi^2\varepsilon_0}\frac{1}{a_0} = \frac{e^2}{3\pi^2\varepsilon_0}\frac{1}{a_0} - \frac{2e^2}{3\pi^2\varepsilon_0}\frac{1}{a_0}$$

$$= -\frac{1}{3\pi^2\varepsilon_0}\frac{e^2}{a_0}$$

compared to the exact energy

$$E_H = -\frac{1}{4\pi\varepsilon_0} \cdot \frac{e^2}{2a_0} = -\frac{1}{8\pi\varepsilon_0} \cdot \frac{e^2}{a_0}$$

Thus we have

$$\frac{E_{1G,\min}}{E_H} = \frac{8}{3\pi} = 0.849$$

As expected, the stabilization energy found with the trial function is less than the exact value. In this case it differs by 15%.

Now we plot the two wavefunctions. The optimized trial wavefunction is

$$\phi_{\min}(r) = \left(\frac{2C_{\min}}{\pi}\right)^{3/4} \exp\left(-C_{\min} r^2\right) = \left(\frac{16}{9\pi}\frac{1}{a_0^2}\right)^{3/4} \exp\left(-\frac{8r^2}{9\pi a_0^2}\right)$$

$$= 7.18 \times 10^{-4}\ \mathrm{pm}^{-3/2} \cdot \exp\left[-\frac{8}{9\pi} \cdot \left(\frac{r}{a_0}\right)^2\right]$$

The exact wavefunction is

$$\psi(r) = \frac{1}{\sqrt{\pi a_0^3}} \exp(-\frac{r}{a_0})$$

The plot of these functions is shown in Fig. P5.3a.

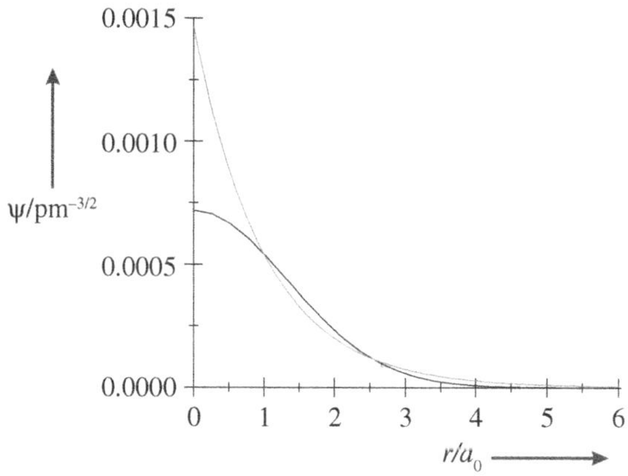

Figure P5.3a Plots of the exact wavefunction and the trial function ϕ_{min}.

The exact wavefunction has more amplitude closer to the origin than does the trial wavefunction. This features means that the average electron charge density is closer to the nucleus, leading to a greater electrostatic stabilization and a more stable atom.

Part 2. From the MathCAD Problem 5_3 worksheet, we obtain the results displayed in Table P5.3.

Table P5.3 Parameters w, C_{21}, and C_{22}, and Calculated Energies E_{2G}.

#	w	C_{21}	C_{22}	E_{2G}/eV	% error
1	1	$C_{11,min}$	$C_{11,min}$	−11.55	15
2	1	$C_{11,min}$	$2.4 \cdot C_{11,min}$	−11.67	14
3	1	$0.4 \cdot C_{11,min}$	$2.2 \cdot C_{11,min}$	−12.88	5.3
4	0.2	$C_{11,min}$	$7.5 \cdot C_{11,min}$	−13.03	4

where

$$C_{11,min} = \frac{8}{9\pi}\frac{1}{a_0^2}$$

as calculated in Part 1. The corresponding wavefunctions are

$$\phi_{2G} = \frac{\left(\frac{2C_{21}}{\pi}\right)^{3/4} \exp(-C_{21}r^2) + w \cdot \left(\frac{2C_{22}}{\pi}\right)^{3/4} \exp(-C_{22}r^2)}{\sqrt{1 + w^2 + 2w \cdot S}}$$

$$= \frac{\left(\frac{2C_{21}}{\pi}\right)^{3/4} \exp\left[-C_{21}a_0^2 \cdot (\frac{r}{a_0})^2\right] + w \cdot \left(\frac{2C_{22}}{\pi}\right)^{3/4} \exp\left[-C_{22}a_0^2 \cdot (\frac{r}{a_0})^2\right]}{\sqrt{1 + w^2 + 2w \cdot S}}$$

Using the values in Table P5.3, we obtain the plots in Fig. P5.3b.

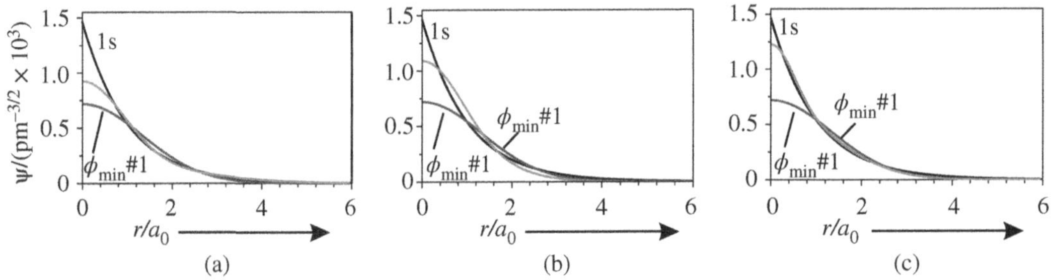

Figure P5.3b Radial dependence of the wavefunctions for the H-atom. Upper curves: 1s function, lower curves: trial function ϕ_{min}#1, Intermediate curves. (a) ϕ_{2G}#2, (b) ϕ_{2G}#3, (c) ϕ_{2G}#3.

P5.4 a) Insert the wavefunction given in Equation (5.23)

$$\phi_2(2,1) = \varphi_{1s}(2) \cdot \varphi_{2s}(1)$$

into the Schrödinger equation for the helium atom and show that the energy of the excited state is given by

$$E = E_{1s} + E_{2s} + \frac{e^2}{4\pi\varepsilon_0} \iint \frac{\varphi_{1s}^2 \varphi_{2s}^2}{r_{12}} \mathrm{d}\tau_1 \mathrm{d}\tau_2$$

Provide a physical interpretation for each of the terms in the energy equation. State what is wrong with this choice of a wavefunction.

b) Perform the same analysis for the wavefunction in Equation (5.24)

$$\phi_{2,\mathrm{sym}} = A\left[\varphi_{1s}(1)\varphi_{2s}(2) + \varphi_{1s}(2)\varphi_{2s}(1)\right]$$

with $A = 1/\sqrt{2}$. You should find that

$$E = E_{1s} + E_{2s} + \frac{e^2}{4\pi\varepsilon_0} \iint \frac{\varphi_{1s}^2 \varphi_{2s}^2}{r_{12}} \mathrm{d}\tau_1 \mathrm{d}\tau_2 + \frac{e^2}{4\pi\varepsilon_0} \iint \frac{\varphi_{1s}(1)\varphi_{2s}(2)\varphi_{1s}(2)\varphi_{2s}(1)}{r_{12}} \mathrm{d}\tau_1 \ \mathrm{d}\tau_2$$

Contrast your results in parts a) and b).

Solution

a) The wavefunction given by Equation (5.23) is

$$\phi_2(2,1) = \varphi_{1s}(2) \cdot \varphi_{2s}(1)$$

and the Schrödinger equation for the helium atom is given by

$$\mathcal{H}\phi_2(2,1) = E\phi_2(2,1)$$

so that

$$E = \int \phi_2(2,1)\mathcal{H}\phi_2(2,1)\,\mathrm{d}\tau$$

with

$$\mathcal{H} = -\frac{\hbar^2}{2m_e}\nabla_1^2 - \frac{\hbar^2}{2m_e}\nabla_2^2 - \frac{2e^2}{4\pi\varepsilon_0}\frac{1}{r_1} - \frac{2e^2}{4\pi\varepsilon_0}\frac{1}{r_2} + \frac{e^2}{4\pi\varepsilon_0}\frac{1}{r_{12}}$$

Substitution of the wavefunction and Hamiltonian into the energy expression gives

$$E = \int \varphi_{1s}(2)\cdot\varphi_{2s}(1)\begin{pmatrix} -\frac{\hbar^2}{2m_e}\nabla_1^2 - \frac{\hbar^2}{2m_e}\nabla_2^2 - \frac{2e^2}{4\pi\varepsilon_0}\frac{1}{r_1} \\ -\frac{2e^2}{4\pi\varepsilon_0}\frac{1}{r_2} + \frac{e^2}{4\pi\varepsilon_0}\frac{1}{r_{12}} \end{pmatrix}\varphi_{1s}(2)\cdot\varphi_{2s}(1)\,\mathrm{d}\tau_1 \ \mathrm{d}\tau_2$$

To simplify we separate the terms to obtain

$$
\begin{aligned}
E = &\int \varphi_{1s}(2)\cdot\varphi_{1s}(2)\,\mathrm{d}\tau_2 \int \varphi_{2s}(1)\left(-\frac{\hbar^2}{2m_e}\nabla_1^2\right)\varphi_{2s}(1)\,\mathrm{d}\tau_1 \\
&+\int \varphi_{2s}(1)\varphi_{2s}(1)\,\mathrm{d}\tau_1 \int \varphi_{1s}(2)\left(-\frac{\hbar^2}{2m_e}\nabla_2^2\right)\varphi_{1s}(2)\,\mathrm{d}\tau_2 \\
&+\int \varphi_{1s}(2)\varphi_{1s}(2)\,\mathrm{d}\tau_2\cdot\int \varphi_{2s}(1)\left(-\frac{2e^2}{4\pi\varepsilon_0}\frac{1}{r_1}\right)\varphi_{2s}(1)\,\mathrm{d}\tau_1 \\
&+\int \varphi_{2s}(1)\,\varphi_{2s}(1)\,\mathrm{d}\tau_1 \int \varphi_{1s}(2)\left(-\frac{2e^2}{4\pi\varepsilon_0}\frac{1}{r_2}\right)\varphi_{1s}(2)\,\mathrm{d}\tau_2 \\
&+\int \varphi_{1s}(2)\cdot\varphi_{2s}(1)\left(+\frac{e^2}{4\pi\varepsilon_0}\frac{1}{r_{12}}\right)\varphi_{1s}(2)\cdot\varphi_{2s}(1)\,\mathrm{d}\tau_1\mathrm{d}\tau_2
\end{aligned}
$$

The first integral in each term, except the last, is just the normalization integral; so the energy expression can be simplified to

$$
\begin{aligned}
E = &\int \varphi_{2s}(1)\left(-\frac{\hbar^2}{2m_e}\nabla_1^2\right)\varphi_{2s}(1)\,\mathrm{d}\tau_1 + \int \varphi_{1s}(2)\left(-\frac{\hbar^2}{2m_e}\nabla_2^2\right)\varphi_{1s}(2)\,\mathrm{d}\tau_2 \\
&+\int \varphi_{2s}(1)\left(-\frac{2e^2}{4\pi\varepsilon_0}\frac{1}{r_1}\right)\varphi_{2s}(1)\,\mathrm{d}\tau_1 + \int \varphi_{1s}(2)\left(-\frac{2e^2}{4\pi\varepsilon_0}\frac{1}{r_2}\right)\varphi_{1s}(2)\,\mathrm{d}\tau_2 \\
&+\int \varphi_{1s}(2)\cdot\varphi_{2s}(1)\left(+\frac{e^2}{4\pi\varepsilon_0}\frac{1}{r_{12}}\right)\varphi_{1s}(2)\cdot\varphi_{2s}(1)\,\mathrm{d}\tau_1\mathrm{d}\tau_2
\end{aligned}
$$

The first term in this expression is the kinetic energy of an electron in the $2s$ orbital; the second term is the kinetic energy of an electron in the $1s$ orbital; the third term is the nuclear-electron potential energy for a $2s$ electron; the fourth term is the nuclear-electron potential energy for a $1s$ electron; and the last term is the electron–electron repulsion energy. Combining the first and third terms and the second and fourth term gives

$$
E = E_{1s} + E_{2s} + \frac{e^2}{4\pi\varepsilon_0}\int \varphi_{1s}^2(2)\cdot\varphi_{2s}^2(1)\frac{1}{r_{12}}\mathrm{d}\tau_1\ \mathrm{d}\tau_2
$$

This wavefunction is not acceptable because the electrons are written as distinguishable particles.

b) In this part the wavefunction is given by Equation (5.24) to be

$$
\phi_{2,\mathrm{sym}} = \frac{1}{\sqrt{2}}\left[\varphi_{1s}(1)\varphi_{2s}(2) + \varphi_{1s}(2)\varphi_{2s}(1)\right]
$$

Substitution of the wavefunction and Hamiltonian into the energy expression given in the previous part of the problem gives

$$
E = \int \phi_{2,\mathrm{sym}}(2,1)\mathcal{H}\phi_{2,\mathrm{sym}}(2,1)\,\mathrm{d}\tau
$$

$$
= \int \left(\begin{array}{c}
\frac{1}{\sqrt{2}}\left[\varphi_{1s}(1)\varphi_{2s}(2) + \varphi_{1s}(2)\varphi_{2s}(1)\right] \\
\cdot\left(-\frac{\hbar^2}{2m_e}\nabla_1^2 - \frac{\hbar^2}{2m_e}\nabla_2^2 - \frac{2e^2}{4\pi\varepsilon_0}\frac{1}{r_1} - \frac{2e^2}{4\pi\varepsilon_0}\frac{1}{r_2} + \frac{e^2}{4\pi\varepsilon_0}\frac{1}{r_{12}}\right) \\
\cdot\frac{1}{\sqrt{2}}\left[\varphi_{1s}(1)\varphi_{2s}(2) + \varphi_{1s}(2)\varphi_{2s}(1)\right]
\end{array}\right)\mathrm{d}\tau_1\mathrm{d}\tau_2
$$

which on separation of the various terms gives

$$
\begin{aligned}
E = &\frac{1}{2}\int\left(\begin{array}{c}\left(\varphi_{1s}(1)\varphi_{2s}(2) + \varphi_{1s}(2)\varphi_{2s}(1)\right)\left(-\frac{\hbar^2}{2m_e}\nabla_1^2\right) \\ \cdot\left(\varphi_{1s}(1)\varphi_{2s}(2) + \varphi_{1s}(2)\varphi_{2s}(1)\right)\end{array}\right)\mathrm{d}\tau_1\mathrm{d}\tau_2 \\
&+\frac{1}{2}\int\left(\begin{array}{c}\left(\varphi_{1s}(1)\varphi_{2s}(2) + \varphi_{1s}(2)\varphi_{2s}(1)\right) \\ \cdot\left(-\frac{\hbar^2}{2m_e}\nabla_2^2\right)\left(\varphi_{1s}(1)\varphi_{2s}(2) + \varphi_{1s}(2)\varphi_{2s}(1)\right)\end{array}\right)\mathrm{d}\tau_1\mathrm{d}\tau_2 \\
&+\frac{1}{2}\int\left(\begin{array}{c}\left(\varphi_{1s}(1)\varphi_{2s}(2) + \varphi_{1s}(2)\varphi_{2s}(1)\right) \\ \cdot\left(-\frac{2e^2}{4\pi\varepsilon_0}\frac{1}{r_1}\right)\left(\varphi_{1s}(1)\varphi_{2s}(2) + \varphi_{1s}(2)\varphi_{2s}(1)\right)\end{array}\right)\mathrm{d}\tau_1\mathrm{d}\tau_2 \\
&+\frac{1}{2}\int\left(\begin{array}{c}\left(\varphi_{1s}(1)\varphi_{2s}(2) + \varphi_{1s}(2)\varphi_{2s}(1)\right) \\ \cdot\left(-\frac{2e^2}{4\pi\varepsilon_0}\frac{1}{r_2}\right)\left(\varphi_{1s}(1)\varphi_{2s}(2) + \varphi_{1s}(2)\varphi_{2s}(1)\right)\end{array}\right)\mathrm{d}\tau_1\mathrm{d}\tau_2 \\
&+\frac{1}{2}\int\left(\begin{array}{c}\left(\varphi_{1s}(1)\varphi_{2s}(2) + \varphi_{1s}(2)\varphi_{2s}(1)\right) \\ \cdot\left(\frac{e^2}{4\pi\varepsilon_0}\frac{1}{r_{12}}\right)\left(\varphi_{1s}(1)\varphi_{2s}(2) + \varphi_{1s}(2)\varphi_{2s}(1)\right)\end{array}\right)\mathrm{d}\tau_1\mathrm{d}\tau_2
\end{aligned}
$$

Combining the first and third terms (those involving only electron 1), and the second and fourth terms (those involving only electron 2) gives

$$E = \frac{1}{2}\int \begin{pmatrix} (\varphi_{1s}(1)\varphi_{2s}(2) + \varphi_{1s}(2)\varphi_{2s}(1)) \\ \cdot\left(-\frac{\hbar^2}{2m_e}\nabla_1^2 - \frac{2e^2}{4\pi\varepsilon_0}\frac{1}{r_1}\right)(\varphi_{1s}(1)\varphi_{2s}(2) + \varphi_{1s}(2)\varphi_{2s}(1)) \end{pmatrix} d\tau_1 d\tau_2$$
$$+\frac{1}{2}\int \begin{pmatrix} (\varphi_{1s}(1)\varphi_{2s}(2) + \varphi_{1s}(2)\varphi_{2s}(1)) \\ \cdot\left(-\frac{\hbar^2}{2m_e}\nabla_2^2 - \frac{2e^2}{4\pi\varepsilon_0}\frac{1}{r_2}\right)(\varphi_{1s}(1)\varphi_{2s}(2) + \varphi_{1s}(2)\varphi_{2s}(1)) \end{pmatrix} d\tau_1 d\tau_2$$
$$+\frac{1}{2}\frac{e^2}{4\pi\varepsilon_0}\int \begin{pmatrix} (\varphi_{1s}(1)\cdot\varphi_{2s}(2) + \varphi_{1s}(2)\cdot\varphi_{2s}(1)) \\ \cdot\left(\frac{1}{r_{12}}\right)(\varphi_{1s}(1)\cdot\varphi_{2s}(2) + \varphi_{1s}(2)\cdot\varphi_{2s}(1)) \end{pmatrix} d\tau_1 d\tau_2$$

Because the orbitals are eigenfunctions of the Hamiltonian, we can operate on them and extract the eigenvalues. One of the terms gives the energy of a 1*s* electron, and the other term gives the energy of a 2*s* electron. The third term involves the square of the wavefunction. Hence we can write the energy expression as

$$E = \frac{1}{2}\int \begin{pmatrix} (\varphi_{1s}(1)\varphi_{2s}(2) + \varphi_{1s}(2)\varphi_{2s}(1)) \\ \cdot(E_{1s}\varphi_{1s}(1)\varphi_{2s}(2) + E_{2s}\varphi_{1s}(2)\varphi_{2s}(1)) \end{pmatrix} d\tau_1 d\tau_2$$
$$+\frac{1}{2}\int \begin{pmatrix} (\varphi_{1s}(1)\varphi_{2s}(2) + \varphi_{1s}(2)\varphi_{2s}(1)) \\ \cdot(E_{2s}\varphi_{1s}(1)\varphi_{2s}(2) + E_{1s}\varphi_{1s}(2)\varphi_{2s}(1)) \end{pmatrix} d\tau_1 d\tau_2$$
$$+\frac{1}{2}\frac{e^2}{4\pi\varepsilon_0}\int\left(\frac{1}{r_{12}}\right)(\varphi_{1s}(1)\varphi_{2s}(2) + \varphi_{1s}(2)\varphi_{2s}(1))^2\, d\tau_1 d\tau_2$$

Expanding the first term gives four terms (two of which are zero by orthogonality); the other two are E_{1s} and E_{2s}, respectively. Similarly, the second term has two terms giving the same expression. The third term has two terms of the same value with the square of each wavefunction, and a third term containing all four single electron wavefunctions. Expanding the expression gives

$$E = \frac{1}{2}(E_{1s} + E_{2s}) + \frac{1}{2}(E_{1s} + E_{2s}) + \frac{e^2}{4\pi\varepsilon_0}\int\left(\frac{1}{r_{12}}\right)(\varphi_{1s}\cdot\varphi_{2s})^2\, d\tau_1 d\tau_2$$
$$+\frac{e^2}{4\pi\varepsilon_0}\int\left(\frac{1}{r_{12}}\right)(\varphi_{1s}(1)\cdot\varphi_{2s}(2)\cdot\varphi_{1s}(2)\cdot\varphi_{2s}(1))\, d\tau_1 d\tau_2$$

which simplifies to

$$E = E_{1s} + E_{2s} + \frac{e^2}{4\pi\varepsilon_0}\int\left(\frac{1}{r_{12}}\right)(\varphi_{1s}\cdot\varphi_{2s})^2\, d\tau_1 d\tau_2$$
$$+\frac{e^2}{4\pi\varepsilon_0}\int\left(\frac{1}{r_{12}}\right)(\varphi_{1s}(1)\cdot\varphi_{2s}(2)\cdot\varphi_{1s}(2)\cdot\varphi_{2s}(1))\, d\tau_1 d\tau_2$$

If we compare this energy expression with that found for the wavefunction of Equation (5.23), we see that it has the extra term

$$+\frac{e^2}{4\pi\varepsilon_0}\int\left(\frac{1}{r_{12}}\right)(\varphi_{1s}(1)\cdot\varphi_{2s}(2)\cdot\varphi_{1s}(2)\cdot\varphi_{2s}(1))\, d\tau_1 d\tau_2$$

Note that the orbitals in this expression are functions of different electron variables; i.e., one of the 2*s* orbitals depends on the coordinates of electron 1 and the other orbital depends on the coordinates of electron 2. This feature has its origin in the "electron indistinguishability" of the wavefunction of Equation (5.24). Note that this energy term is positive and destabilizes the He atom, compared to that found using Equation (5.23).

P5.5 Consider Equation (5.16) for the energy of a helium atom. Discuss how this energy expression would change if the nucleus had a charge of $Z = 1$, rather than $Z = 2$. Assuming that the electron–electron repulsion term is the same, would you expect the hydride ion H^- to be a stable ion? The hydride ion is known to have an ionization energy of 0.76 eV. Comment on this result, as compared to your rough estimations.

Solution

We follow the solution to Problem P5.2 for a one-electron atom with an arbitrary nuclear charge Ze. For the ground state of the atom we use a trial function ϕ that is a product of two 1s orbital wavefunctions φ_{1s}, namely

$$\phi(r_1, r_2) = \varphi_{1s}(r_1) \cdot \varphi_{1s}(r_2) = \sqrt{\frac{Z^3}{\pi a_0^3}} e^{-Zr_1/a_0} \cdot \sqrt{\frac{Z^3}{\pi a_0^3}} e^{-Zr_2/a_0}$$

where Z is the atomic number. By inserting this wavefunction into the approximate Schrödinger equation we obtain the energy E^0 as the sum of two atomic ground-state energies

$$E^0 = 2\left(-\frac{1}{4\pi\varepsilon_0}\frac{Z^2e^2}{2a_0}\right) = -\frac{1}{4\pi\varepsilon_0}\frac{Z^2e^2}{a_0}$$

As stated in the problem, we use the same expression for the electron–electron repulsion as given in Equation (5.16). By adding this correction term to the energy E^0, we find a corrected energy ε, namely

$$\varepsilon = E^0 + \overline{V}_e = \frac{1}{4\pi\varepsilon_0}\left(-\frac{Z^2e^2}{a_0} + \frac{5}{4}\frac{e^2}{a_0}\right)$$
$$= -\frac{1}{4\pi\varepsilon_0}\frac{11}{8}\frac{Ze^2}{a_0}$$

For the He atom ($Z = 2$) this provides exactly the same result as given in Equation (5.16), namely

$$\varepsilon_{He} = -\frac{1}{4\pi\varepsilon_0}\frac{11}{4}\frac{e^2}{a_0} = -119.9 \times 10^{-19}\ \text{J}$$

For the case of the hydride ion, $Z = 1$, and we find that

$$\varepsilon_{H^-} = \frac{1}{4\pi\varepsilon_0}\frac{1}{4}\frac{e^2}{a_0} = 10.9 \times 10^{-19}\ \text{J}$$

The energy found for the hydride ion is positive; hence, it is predicted to not be stable.

Two important caveats should be appreciated concerning this calculation. First the electron–electron repulsion term was taken to be the same as that for He. This assumption overestimates the electron–electron repulsion because it uses the more compact wavefunction of He ($Z = 2$) to describe the electron cloud, rather than the more diffuse wavefunction of H ($Z = 1$). Hence a more diffuse atomic orbital approximation would reduce the electron–electron repulsion and lower the energy. Second, this calculation neglects the evasion of the electrons (correlation effect). This effect can be taken into account by using two-electron trial functions which depend on the mutual distance r_{12} of the two electrons and would act to lower the energy.

P5.6 Consider the Na atom, which has a single valence electron.

a) Use the procedure in Box 5.4 to find the effective charge Z_{eff}.

b) Compare this charge to the effective charge needed for the Na atom to reproduce its measured ionization energy.

Atom	Ionization Energy
H	13.60 eV
Na	5.39 eV

c) Use your effective charges to determine the root mean square distance $\sqrt{\overline{r^2}}$ of the electron away from the nucleus. Comment on your results.

Solution

a) The procedure in Box 5.4 gives $\sigma = 8.80$, so that

$$Z_{eff,Na} = Z_{Na} - \sigma = 11 - 8.80 = 2.2''$$

b) For hydrogen-like atoms we have

$$E_n = \frac{Z^2}{n^2}E_{H\text{-atom}}$$

where $E_{H\text{-atom}}$ is the energy of the H-atom in its ground state.
For Hydrogen ($Z = 1$) in the 3s orbital, we have

$$E_{3s}(\text{H-atom}) = \frac{1}{9}E_{H\text{-atom}}$$

and for the H-atom in a 1s orbital we have

$$E_{1s}(\text{H-atom}) = E_{\text{H-atom}} = -13.599 \text{ eV}$$

therefore

$$E_{3s}(\text{H-atom}) = -\frac{13.599 \text{ eV}}{9} = -1.511 \text{ eV}$$

For the Na atom we expect the binding to be stronger because $Z_{\text{eff, Na}} > Z_H = 1$. In contrast to part a), we use the experimental ionization potential of Na to calculate the effective nuclear charge acting on its 3s electron. In particular, we use the ratio between the ionization energy for an electron in the 3s orbital of an H-atom to that in the 3s orbital of a Na; i.e.,

$$\frac{E_{3s}(\text{H-atom})}{E_{3s}(\text{Na-atom})} = \frac{Z^2(\text{H-atom})}{Z^2_{\text{eff}}(\text{Na-atom})}$$

Therefore,

$$Z^2_{\text{eff}}(\text{Na-atom}) = \frac{-5.139 \text{ eV}}{-1.511 \text{ eV}}(1)^2$$

and

$$Z_{\text{eff}}(\text{Na}) = (3.4)^{1/2} = 1.844$$

The effective charges calculated by the two methods are similar (they differ by 15–20%) and show that the shielding of the valence electron in Na, while quite effective, does not reduce the effective charge all the way to $+1e$.

c) The root mean square distance $\overline{(r^2)}$ of the electron away from the nucleus is given by the square root of the mean square distance calculated as

$$\overline{(r^2)} = \int_0^{2\pi} d\varphi \int_0^{\pi} \sin\theta \, d\theta \int_0^{\infty} \varphi^2_{3s} r^2 r^2 dr$$

with

$$\varphi_{3s} = \frac{1}{81\sqrt{3\pi}}\left(\frac{Z_{\text{eff}}}{a_0}\right)^{3/2}\left(27 - \frac{18Z_{\text{eff}}\, r}{a_0} + \frac{2Z^2_{\text{eff}}\, r^2}{a_0^2}\right) e^{-Z_{\text{eff}} r/3a_0}$$

Substituting the wavefunction into the integral and performing the integration over the angles gives

$$\overline{(r^2)} = 4\pi \int_0^{\infty} \frac{1}{81^2 3\pi}\left(\frac{Z_{\text{eff}}}{a_0}\right)^3 \left(27 - \frac{18Z_{\text{eff}} r}{a_0} + \frac{2Z^2_{\text{eff}} r^2}{a_0^2}\right)^2 e^{-2Z_{\text{eff}} r/3a_0} r^4 dr$$

By expanding the integrand we can write

$$\overline{(r^2)} = \frac{4}{81^2 3}\left(\frac{Z_{\text{eff}}}{a_0}\right)^3 \int_0^{\infty} \left(\begin{array}{c} 27^2 r^4 - 2(27)\frac{18Z_{\text{eff}}}{a_0} r^5 \\ +\left(\frac{18Z_{\text{eff}}}{a_0}\right)^2 r^6 + 2\cdot 27 \frac{2Z^2_{\text{eff}}}{a_0^2} r^6 \\ -2\frac{18Z_{\text{eff}}}{a_0}\frac{2Z^2_{\text{eff}}}{a_0^2} r^7 + \left(\frac{2Z^2_{\text{eff}}}{a_0^2}\right)^2 r^8 \end{array}\right) e^{-2Z_{\text{eff}} r/3a_0} dr$$

From Appendix B, we use the integral

$$\int_0^{\infty} x^n \, e^{-ax} dx = n!/a^{n+1}$$

and find

$$\overline{(r^2)} = \frac{2^2}{3^9}\left(\frac{Z_{\text{eff}}}{a_0}\right)^3 \left(\begin{array}{c} \frac{3^6\cdot 4!}{\left(\frac{2Z_{\text{eff}}}{3a_0}\right)^5} - 54\frac{18Z_{\text{eff}}}{a_0}\frac{5!}{\left(\frac{2Z_{\text{eff}}}{3a_0}\right)^6} + \frac{18^2 Z^2_{\text{eff}}}{a_0^2}\frac{6!}{\left(\frac{2Z_{\text{eff}}}{3a_0}\right)^7} \\ +108\frac{Z^2_{\text{eff}}}{a_0^2}\frac{6!}{\left(\frac{2Z_{\text{eff}}}{3a_0}\right)^7} - 72\frac{Z^3_{\text{eff}}}{a_0^3}\frac{7!}{\left(\frac{2Z_{\text{eff}}}{3a_0}\right)^8} + \frac{4Z^4_{\text{eff}}}{a_0^4}\frac{8!}{\left(\frac{2Z_{\text{eff}}}{3a_0}\right)^9} \end{array}\right)$$

or

$$\overline{(r^2)} = \frac{2^2}{3^9}\left(\frac{a_0}{Z_{\text{eff}}}\right)^2 \left(\begin{array}{c} \frac{3^5\cdot 3^6\cdot 4!}{2^5} - 54\cdot 18\cdot \frac{3^6\cdot 5!}{2^6} + 18^2\frac{3^7\cdot 6!}{2^7} + 108\frac{3^7\cdot 6!}{2^7} \\ -72\frac{3^8\cdot 7!}{2^8} + 4\frac{3^9\cdot 8!}{2^9} \end{array}\right)$$

This expression simplifies as

$$\overline{\langle r^2\rangle} = \left(\begin{array}{c}\frac{3^2\cdot 4!}{2^3} - \frac{2\cdot 27\cdot 2\cdot 5!}{3\cdot 2^4} + \frac{3^2\cdot 2^2\cdot 6!}{2^5} + \frac{2\cdot 6\cdot 6!}{2^5} \\ -\frac{2\cdot 36\cdot 7!}{2^6\cdot 3} + \frac{4\cdot 8!}{2^7}\end{array}\right)\left(\frac{a_0}{Z_{\text{eff}}}\right)^2$$

$$= \left(27 - \frac{27\cdot 5!}{12} + \frac{9\cdot 6!}{8} + \frac{6\cdot 6!}{16} - \frac{12\cdot 7!}{32} + \frac{8!}{32}\right)\left(\frac{a_0}{Z_{\text{eff}}}\right)^2$$

$$= (27 - 270 + 810 + 270 - 1890 + 1260)\left(\frac{a_0}{Z_{\text{eff}}}\right)^2$$

$$= 207\cdot\left(\frac{a_0}{Z_{\text{eff}}}\right)^2$$

Note that the expression has units of length squared, which is appropriate.

Now we evaluate the root mean square distance for each of the different effective charges. Using the above expression we see that

$$\sqrt{\langle r^2\rangle} = 14.3875\frac{a_0}{Z_{\text{eff}}} = \frac{761.35}{Z_{\text{eff}}}\ \text{pm}$$

For the effective charge of $Z_{\text{eff}} = 1.844$, we find that

$$\sqrt{\langle r^2\rangle} = \frac{761.35}{1.844}\ \text{pm} = 412.9\ \text{pm}$$

and for $Z_{\text{eff}} = 2.2$, we find

$$\sqrt{\langle r^2\rangle} = \frac{761.35}{2.2}\ \text{pm} = 350\ \text{pm}$$

These values appear reasonable and show that the root mean square distance decreases for the higher effective nuclear charge. However, the values are somewhat larger than the van der Waals radius reported for the atom, which is 230 pm.

P5.7 In this problem you will compare the Slater orbitals to the exact hydrogen atom orbitals. a) Plot the radial dependence of the Slater orbitals and hydrogen atom orbitals for the 2*s* orbitals on the same graph. Comment on how they compare. b) Show that the 1*s* Slater orbital is the same as the hydrogen atom 1*s* orbital. c) Show that the exact wavefunctions for the 1*s* and 2*s* orbitals are orthogonal to each other, but that the Slater orbitals are not orthogonal to each other.

Solution

a) The Slater 2*s* orbital is

$$\psi_{2s}^{\text{Slater}}(r) = A\times r\times \exp\left(-\frac{Z_{\text{eff}}r}{2a_0}\right)$$

First, we normalize the Slater 2*s* orbital

$$1 = 4\pi A^2\int_0^\infty r^4 e^{-Z_{\text{eff}}r/a_0}\,dr = 4\pi A^2\left(\frac{24a_0^5}{Z_{\text{eff}}^5}\right)$$

where we have used the integral $\int_0^\infty x^n e^{-ax}\,dx = n!/a^{n+1}$ from Appendix B. Rearranging, we find the normalization constant A is

$$A = \sqrt{\frac{Z_{\text{eff}}^5}{96\pi a_0^5}}$$

Now we set $Z_{\text{eff}} = 1$, so that the normalized Slater 2*s* orbital is

$$\psi_{2s}^{\text{Slater}}(r) = \sqrt{\frac{1}{96\pi a_0^3}}\times\frac{r}{a_0}\times\exp\left(-\frac{r}{2a_0}\right)$$

The hydrogen atom 2*s* orbital is

$$\psi_{2s}^{H}(r) = \frac{1}{4\sqrt{2\pi a_0^3}}\left(2-\frac{r}{a_0}\right)\times\exp\left(-\frac{r}{2a_0}\right)$$

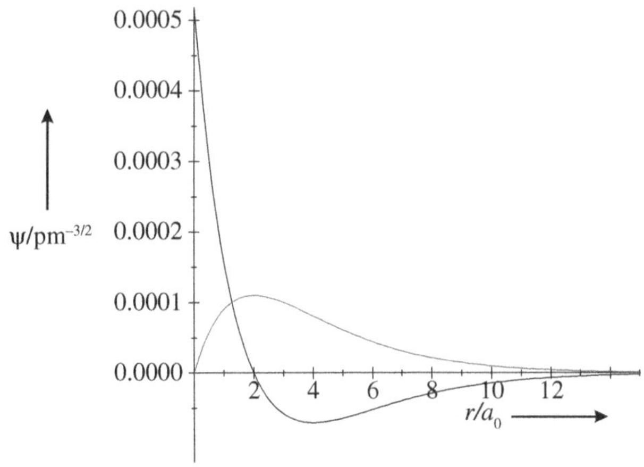

Figure P5.7 The curve starting at $\psi = 0.005\ \text{pm}^{-3/2}$ shows a plot of the H-atom 2*s* wavefunction and the curve starting at $\psi = 0.000\ \text{pm}^{-3/2}$ is a plot of the Slater 2*s* orbital.

Figure P5.7 compares both wavefunctions. Clearly these two representations of the 2*s* orbital's radial dependence are very different. Most notably the hydrogenic wavefunction has amplitude at the nucleus and a radial node. The wavefunctions are similar in that they both have amplitude at intermediate distances, albeit of opposite sign.

b) The Slater 1*s* orbital for the hydrogen is

$$\psi_{1s}^{\text{Slater}}(r) = A \times \exp\left(-\frac{Z_{\text{eff}} r}{a_0}\right)$$

and the hydrogen atom 1*s* orbital is

$$\psi_{1s}^{H}(r) = \frac{1}{\sqrt{\pi a_0^3}} \exp\left(-\frac{r}{a_0}\right)$$

For $\psi_{1s}^{\text{Slater}}(r) = \psi_{1s}^{H}(r)$, we find that

$$\frac{1}{\sqrt{\pi a_0^3}} \exp\left(-\frac{r}{a_0}\right) = A \times \exp\left(-\frac{Z_{\text{eff}} r}{a_0}\right)$$

which holds true as long as $Z_{\text{eff}} = 1$ and $A = 1/\sqrt{\pi a_0^3}$.

c) First we show that the $\psi_{1s}^{H}(r)$ and $\psi_{2s}^{H}(r)$ are orthogonal

$$\begin{aligned}
4\pi \int_0^\infty \psi_{1s}^{H} \psi_{2s}^{H} r^2 \, dr &= 4\pi \int_0^\infty \frac{1}{\sqrt{\pi a_0^3}} e^{-r/a_0} \frac{1}{4\sqrt{2\pi a_0^3}} \left(2 - \frac{r}{a_0}\right) e^{-r/(2a_0)} r^2 \, dr \\
&= 4\pi \frac{1}{\sqrt{\pi a_0^3}} \frac{1}{4\sqrt{2\pi a_0^3}} \int_0^\infty e^{-r/a_0} \left(2 - \frac{r}{a_0}\right) e^{-r/(2a_0)} r^2 \, dr \\
&= 4\pi \frac{1}{4\sqrt{2}\pi a_0^3} \left[2 \int_0^\infty r^2 \, e^{-3r/(2a_0)} \, dr - \frac{1}{a_0} \int_0^\infty r^3 e^{-3r/(2a_0)} \, dr\right] \\
&= \frac{1}{\sqrt{2} a_0^3} \left[2\left(\frac{16}{27} a_0^3\right) - \frac{1}{a_0}\left(\frac{32}{27} a_0^4\right)\right] \\
&= 0
\end{aligned}$$

Now we show that $\psi_{1s}^{\text{Slater}}(r)$ and $\psi_{2s}^{\text{Slater}}(r)$ are not orthogonal

$$\begin{aligned}
4\pi \int_0^\infty \psi_{1s}^{\text{Slater}} \psi_{2s}^{\text{Slater}} r^2 \, dr &= \frac{4\pi}{\sqrt{\pi a_0^3}} \int_0^\infty e^{-r/a_0} \frac{1}{\sqrt{96\pi a_0^5}} r e^{-r/(2a_0)} r^2 \, dr \\
&= 4\pi \frac{1}{\sqrt{\pi a_0^3}} \frac{1}{\sqrt{96\pi a_0^5}} \int_0^\infty r^3 e^{-r/a_0} e^{-r/(2a_0)} \, dr
\end{aligned}$$

$$= \frac{1}{6}\frac{\sqrt{6}}{a_0^4}\int_0^\infty r^3 e^{-3r/(2a_0)}\ \mathrm{d}r$$

$$= \frac{1}{6}\frac{\sqrt{6}}{a_0^4}\left(\frac{32}{27}a_0^4\right)$$

$$= \frac{1}{6}\sqrt{6}\left(\frac{32}{27}\right) = 0.483$$

Because it is not zero, there is net overlap and the orbitals are not orthogonal.

P5.8 Consider the Li atom, which has a single valence electron. In the text we approximated this valence electron as being hydrogenic and found $Z_{\text{eff}} = 1.26$. Using this value for Z_{eff}, find the values of r at which the radial distribution function of a 2*s* orbital has its maxima.

Solution

In this case we first plot the function to see where the radial distribution function has its maxima and its minima.

$$\mathcal{P} = 4\pi r^2 |R(r)|^2 = 4\pi r^2 \left|\sqrt{\frac{Z^3}{8a_0^3}}\left(2 - \frac{Zr}{a_0}\right)\exp\left(-\frac{Zr}{2a_0}\right)\right|^2$$

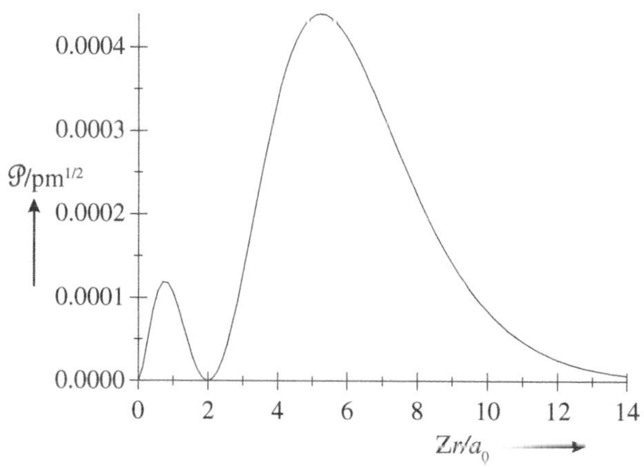

Figure P5.8 The 2s radial distribution function.

To find the maximum of the function

$$y = r^2\left(2 - \frac{Zr}{a_0}\right)\exp\left(-\frac{Zr}{2a_0}\right)^2$$

we take its derivative, set it equal to zero, and solve for r.

$$\frac{\mathrm{d}y}{\mathrm{d}r} = \frac{\mathrm{d}}{\mathrm{d}r}\left[r^2\left(2 - \frac{Zr}{a_0}\right)e^{-Zr/(2a_0)}\right]^2$$

or

$$\frac{\mathrm{d}y}{\mathrm{d}r} = \frac{\mathrm{d}}{\mathrm{d}r}\left[\left(4r^2 - 4\frac{Zr^3}{a_0} + \frac{Z^2r^4}{a_0^2}\right)e^{-Zr/a_0}\right]$$

Taking the derivative we find that

$$\frac{\mathrm{d}y}{\mathrm{d}r} = -\frac{Z}{a_0}\left(4r^2 - 4\frac{Zr^3}{a_0} + \frac{Z^2r^4}{a_0^2}\right)e^{-\frac{Zr}{a_0}} + \left(8r - \frac{12Zr^2}{a_0} + \frac{4Z^2r^3}{a_0^2}\right)e^{-\frac{Zr}{a_0}}$$

$$= r\left(8 - \frac{16Zr}{a_0} + \frac{8Z^2r^2}{a_0^2} - \frac{Z^3r^3}{a_0^3}\right)e^{-\frac{Zr}{a_0}}$$

$$= r\left(2 - \frac{Zr}{a_0}\right)\left(4 - \frac{6Zr}{a_0} + \frac{Z^2r^2}{a_0^2}\right)e^{-\frac{Zr}{a_0}}$$

By inspection of this expression and comparison to the graph, we see that $r_{min} = 0$ and $r_{min} = 2a_0/Z$ correspond to the minima (i.e., the zeros). We solve the quadratic form using the quadratic formula, so that

$$r_{max} = \frac{6Z/a_0 \pm \sqrt{36Z^2/a_0^2 - 16Z^2/a_0^2}}{2Z^2/a_0^2} = \frac{a_0}{Z}\left(3 \pm \sqrt{5}\right)$$

These latter values correspond to the maxima at $0.7639a_0/Z$ and $5.236a_0/Z$.

The maxima are at $r_{max} = 0.76a_0/Z$ and $5.24a_0/Z$, in accordance with Fig. P5.8. With the value of Z_{eff} given, the maxima in the radial distribution function occur at

$$r_{max} = 5.24a_0/1.26 = 5.24 \cdot 52.9 \text{ pm}/1.26$$
$$= 220 \text{ pm}$$

and

$$r_{max} = 31.9 \text{ pm}$$

P5.9 Consider the atomic and ionic radii given in the table. Use the electron shell structure of atoms to explain the trends in the radii. Look up the ionization energy for each of these species and plot it versus the radius. Identify any correlations between these data. Provide a physical explanation for the correlations you observe.

He, 128 pm	Li^+, 78 pm	Be^{2+}, 34 pm	B^{3+}, 23 pm
Ne, 160 pm	Na^+, 98 pm	Mg^{2+}, 78 pm	Al^{3+}, 57 pm
Ar, 174 pm	K^+, 133 pm	Ca^{2+}, 106 pm	Ga^{3+}, 62 pm
Kr, 198 pm	Rb^+, 149 pm	Sr^{2+}, 127 pm	In^{3+}, 92 pm
Xe, 218 pm	Cs^+, 165 pm	Ba^{2+}, 143 pm	Th^{3+}, 101 pm

Solution

Within the groups of data given in the problem, the first row has the electron configuration $1s^2$, the second row has the configuration $1s^2 2s^2 2p^6$, etc., and each species is closed shell. Going across the row, we expect that the increased nuclear charge will reduce the radius for constant n, and this is observed. Each column provides data for different species of similar charge but with one value of n larger. In this case, we expect that the increased n would increase the radius at constant Z_{eff}, and this trend is observed.

Figure P5.9 reveals that the measured ionization potentials IP of the atoms and ions are highly correlated with the corresponding atomic and ionic radii r.

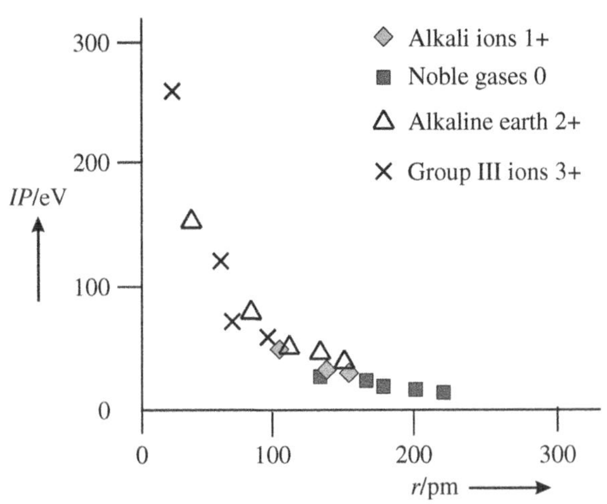

Figure P5.9 The ionization energies for the atoms and ions reported in the table are plotted versus the values reported in the table for their radii.

The ionization energies are taken from Herzberg's *Atomic Spectra and Atomic Structure* and from Emsley's *The Elements*.

He, 24.46 eV	Li^+, 75.26 eV	Be^{2+}, 153.1 eV	B^{3+}, 258.1 eV
Ne, 21.45 eV	Na^+, 47.06 eV	Mg^{2+}, 79.72 eV	Al^{3+}, 119.37 eV
Ar, 15.68 eV	K^+, 31.66 eV	Ca^{2+}, 50.96 eV	Ga^{3+}, 64 eV
Kr, 13.93 eV	Rb^+, 27.36 eV	Sr^{2+}, 43.4 eV	In^{3+}, 57.8 eV
Xe, 12.08 eV	Cs^+, 23.4 eV	Ba^{2+}, 37 eV	Th^{3+}, 51 eV

P5.10 In Foundation 5.2 (Fig. 5F.1) we consider an electron in a potential $V = 0$ (electron in a box with box length L) and estimated its change in energy for a change in the potential well of V_1 over the distance Δx. Use the same method to calculate the change in the electron's energy for the following cases: a) V increases by V_1 over the range Δx in the box center, and b) V changes as $V_0 \sin(3\pi x/L)$.

Solution

a) This derivation follows exactly that in Foundation 5.2, except V_1 is a small barrier, rather than a well. Hence we approximate the energy as

$$\varepsilon = E_0 + \Delta E \quad \text{with } \Delta E = \int \psi_0 V \psi_0 \cdot \mathrm{d}x = \int V \psi_0^2 \cdot \mathrm{d}x$$

where

$$E_0 = \frac{h^2}{8mL^2} n^2 \qquad \psi_0 = \sqrt{\frac{2}{L}} \sin\left(\frac{n\pi x}{L}\right) \tag{5P10.1}$$

for a particle confined to a box with infinitely high walls (Fig. P5.10a).

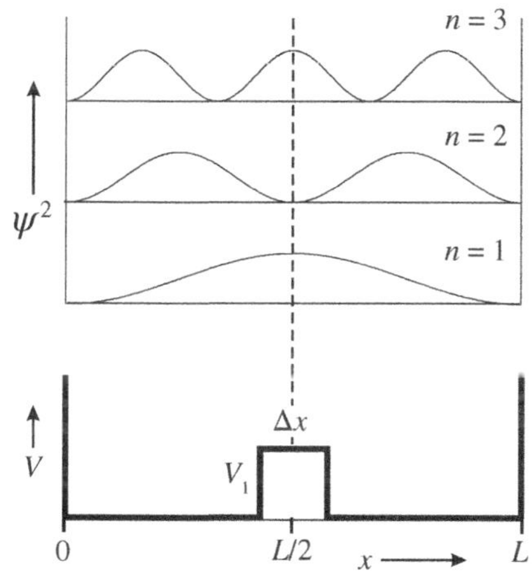

Figure P5.10a Bottom: Box of length L with infinite high walls; a potential energy barrier of height V_1 and width Δx is placed at the center of the box. Top: the square of the wavefunctions in Equation (5P10.1) for $n = 1$, $n = 2$, and $n = 3$.

The potential energy V equals V_1 between $x = x_1$ and $x = x_2$; otherwise, it is zero, hence we can write

$$\Delta E = \int V \psi_0^2 \cdot \mathrm{d}x = \int_{x_1}^{x_2} V_1 \psi_0^2 \cdot \mathrm{d}x = \int_{x_1}^{x_2} V_1 \frac{2}{L} \sin^2\left(\frac{n\pi x}{L}\right) \cdot \mathrm{d}x$$
$$= V_1 \frac{2}{L} \int_{x_1}^{x_2} \sin^2\left(\frac{n\pi x}{L}\right) \cdot \mathrm{d}x \tag{5P10.2}$$

If we restrict our solution to a small region at the center of the box, we can replace the sine function by its value at the position $x = L/2$.

$$\sin^2\left(\frac{n\pi x}{L}\right) = 1 \quad \text{for} \quad n = 1, 3, 5, \ldots \quad \text{and} \quad \sin^2\left(\frac{n\pi x}{L}\right) = 0 \quad \text{for} \quad n = 2, 4, 6, \ldots$$

This means that the correction term ΔE is zero for even values of n. For odd values of n we obtain

$$\Delta E = \int V \psi_0^2 \cdot \mathrm{d}x = V_1 \frac{2}{L} \int_{x_1}^{x_2} \cdot \mathrm{d}x = V_1 \frac{2}{L}(x_2 - x_1) = V_1 \frac{2}{L} \Delta x \tag{5P10.3}$$

The result of the calculation for $L = 1000$ pm, $V_1 = 2 \times 10^{-19}$ J, and $\Delta x = L/10$ is shown in Table P5.10a. Additionally, ΔE is calculated by numerically integrating Equation (5P10.2). The result is mainly the same; however, there are small differences in detail: the value for $n = 3$ is slightly smaller than that for $n = 1$, because this wavefunction decreases faster than the $n = 1$ function; the value for $n = 2$ is slightly different from zero, because the wavefunction is not exactly zero in the neighborhood of the center of the box.

Table P5.10a Result of the Calculation of ΔE with the Parameters $L = 1000$ pm, $V_1 = 2 \times 10^{-19}$ J, and $\Delta x = L/10$.

n	ΔE (Equation (5P10.3)) 10^{-19} J	ΔE (Equation (5P10.2)) 10^{-19} J
1	0.400	0.397
2	0.000	0.013
3	0.400	0.372

Column 2: result for the approximation given by Equation (5P10.3). Column 3: result for Equation (5P10.2) (numerical integration).

Finally we can write the energy as

$$\varepsilon = \frac{h^2 n^2}{8mL^2} + V_1 \frac{2}{L} \Delta x \quad \text{for} \quad n = 1, 3, 5, \ldots \quad \text{and} \quad \varepsilon = \frac{h^2 n^2}{8mL^2} \quad \text{for} \quad n = 2, 4, 6, \ldots$$

This result is identical to that in Foundation 5.2, except that V_1 is now positive rather than negative.

b) As in part a), we write the energy as

$$\varepsilon = E_0 + \Delta E \quad \text{with}\, \Delta E = \int \psi_0 V \psi_0 \cdot \mathrm{d}x = E_0 + \int V \psi_0^2 \cdot \mathrm{d}x$$

where

$$E_0 = \frac{h^2}{8mL^2} \quad \text{and} \quad \psi_0 = \sqrt{2/L} \sin\left(\frac{n\pi x}{L}\right) \tag{5P10.4}$$

for a particle confined to a box with infinitely high walls. The potential energy V is chosen to be

$$V = V_0 \sin(\frac{3\pi x}{L})$$

This situation is displayed in Fig. P5.10b.

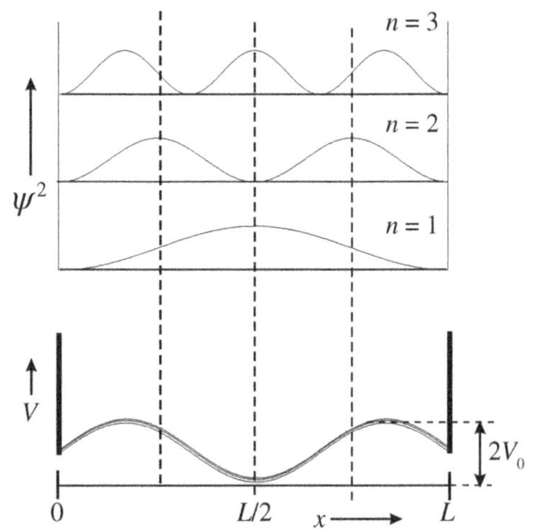

Figure P5.10b Bottom: Box of length L with infinite high walls; a sinus potential $V = V_0 \sin(3\pi x/L)$ is placed at the bottom of the box. Top: the square of the wavefunctions in Equation (5P10.4) for $n = 1$, $n = 2$, and $n = 3$ is displayed.

The first-order energy correction is

$$\Delta E = \int V\psi_0^2 \cdot \mathrm{d}x = \int_0^L V_0 \sin(3\pi x/L)\psi_0^2 \cdot \mathrm{d}x = V_0 \frac{2}{L}\int_0^L \sin(3\pi x/L)\sin^2\left(\frac{n\pi x}{L}\right) \cdot \mathrm{d}x \quad (5P10.5)$$

Let us first look qualitatively on the problem.

In Fig. P5.10b we recognize that the $n = 1$ probability density has a maximum at the center of the box, leading to a negative contribution to ΔE; on both sides there is a small overlap with the positive part of the potential energy curve, thus leading to a positive contribution to ΔE. Therefore for ΔE we expect a low negative value.

The $n = 2$ probability density is zero at the center of the box; thus, the corresponding contribution to ΔE will be zero; on both sides there is a big overlap with the positive part of the potential energy curve, leading to a high positive value of ΔE.

The $n = 3$ probability density has one maximum at the center of the box and two maxima on both sides. Therefore we expect a positive value for ΔE, but it should be smaller than that for the $n = 2$ case.

Finally we compare these qualitative results with a numeric solution of the integral in Equation (5P10.5). The result displayed in Table P5.10b shows that our qualitative prediction is in good agreement with the quantitative calculation.

Table P5.10b Result of the Calculation of ΔE According to Equation (5P10.5) with the Parameter $V_0 = 2 \times 10^{-19}$ J. The Integration Was Performed Numerically.

n	**ΔE (Equation (5P10.5))** 10^{-19} **J**
1	−0.339
2	0.979
3	0.566

As an example, we calculate the integral in Equation (5P10.5) for the case of $n = 1$ analytically. From Appendix B, we use the trigonometric relation

$$\sin(3x) = 3\sin x - 4\sin^3 x$$

to write

$$\int V\psi_0^2 \cdot \mathrm{d}x = V_0\frac{6}{L}\int_0^L \sin^3\left(\frac{\pi x}{L}\right) \cdot \mathrm{d}x - V_0\frac{8}{L}\int_0^L \sin^5\left(\frac{\pi x}{L}\right) \cdot \mathrm{d}x$$
$$= V_0\frac{6}{L}\int_0^L \sin\left(\frac{\pi x}{L}\right)\left(1-\cos^2\left(\frac{\pi x}{L}\right)\right) \cdot \mathrm{d}x$$
$$-V_0\frac{8}{L}\int_0^L \sin\left(\frac{\pi x}{L}\right)\left(1-\cos^2\left(\frac{\pi x}{L}\right)\right)^2 \cdot \mathrm{d}x$$

Expanding and collecting terms we can write

$$\Delta E = \int V\psi_0^2 \cdot \mathrm{d}x = -V_0\frac{2}{L}\int_0^L \sin\left(\frac{\pi x}{L}\right) \cdot \mathrm{d}x + 10\frac{V_0}{L}\int_0^L \sin\left(\frac{\pi x}{L}\right)\cos^2\left(\frac{\pi x}{L}\right) \cdot \mathrm{d}x$$
$$-V_0\frac{8}{L}\int_0^L \sin\left(\frac{\pi x}{L}\right)\cos^4\left(\frac{\pi x}{L}\right) \cdot \mathrm{d}x$$
$$= V_0\frac{2}{\pi}\left[\cos\left(\frac{\pi x}{L}\right)\right]_0^L - 10\frac{V_0}{\pi}\left[\frac{\cos^3\left(\frac{\pi x}{L}\right)}{3}\right]_0^L + 8\frac{V_0}{\pi}\left[\frac{\cos^5\left(\frac{\pi x}{L}\right)}{5}\right]_0^L$$
$$= V_0\frac{2}{\pi}\cdot(-2) - 10\frac{V_0}{\pi}\left(-\frac{2}{3}\right) + 8\frac{V_0}{\pi}\cdot\left(-\frac{2}{5}\right)$$
$$= \frac{V_0}{\pi}\left[-4 + \frac{20}{3} - \frac{16}{5}\right] = \frac{V_0}{\pi}(-0.533) = -0.170 \cdot V_0$$

and with $V_0 = 2 \times 10^{-19}$ J it follows that $\Delta E = -0.34 \times 10^{-19}$ J. This is identical with the numerical result in Table P5.10b.

P5.11 In this problem you use the perturbation theory method illustrated in Box 5.1 and Justification 5.1 to calculate the ground-state energy of the helium atom, using the trial wavefunction

$$\phi(r_1, r_2) = \psi_{\mathrm{He}^+}(r_1) \cdot \psi_{\mathrm{He}^+}(r_2) = \sqrt{\frac{C^3}{\pi a_0^3}} \mathrm{e}^{-Cr_1/a_0} \cdot \sqrt{\frac{C^3}{\pi a_0^3}} \mathrm{e}^{-Cr_2/a_0}$$

Proceed in three steps. In the first step, follow the procedure of Problem P5.1 for the case where the nucleus has a charge of $2e$ and find the energy for an electron in each of the orbitals ψ_{He^+} while neglecting the electron–electron repulsion; i.e., solve the Schrödinger equation

$$-\frac{h^2}{8\pi^2 m_e} \nabla^2 \psi - \frac{1}{4\pi\varepsilon_0} \cdot \frac{2e \cdot e}{r} \psi = E\psi$$

and show that

$$E(\mathrm{He}^+) = -\frac{e^2}{4\pi\varepsilon_0} \frac{[4C - C^2]}{2a_0}$$

In the second step, use the result from Justification 5.1 that

$$\overline{V}_{12} = \frac{e^2}{4\pi\varepsilon_0 a_0} \cdot \frac{5}{4} \frac{C}{2}$$

In part three, you should combine the two results from parts 1 and 2 and minimize the variational energy by minimizing the parameter C. You should find that

$$E = -\frac{e^2}{4\pi\varepsilon_0 a_0} \left[\frac{(27)^2}{(16)^2} \right] = -124.2 \times 10^{-19} \ \mathrm{J}$$

Compare this result to the calculation in the text for case of $C = 2$, and to the best Hartree Fock value of -124.8×10^{-19} J.

Solution

In the first part we are asked to solve the Schrödinger equation for He^+ using a $1s$ orbital form for the trial function but with an adjustable parameter C that determines the orbitals extent. This first step is very much like Problem P5.1. If we substitute the trial function $\psi = \sqrt{\frac{C^3}{\pi a_0^3}} \cdot \mathrm{e}^{-Cr/a_0}$ into the Schrödinger equation

$$\mathcal{H}\psi = -\frac{h^2}{8\pi^2 m_e} \nabla^2 \psi - \frac{1}{4\pi\varepsilon_0} \cdot \frac{2e \cdot e}{r} \psi = E\psi$$

we find

$$\mathcal{H}\psi = \frac{\hbar^2}{2m_e} \frac{C}{a_0} \left(\frac{2}{r} - \frac{C}{a_0} \right) \psi - \frac{1}{4\pi\varepsilon_0} \frac{2e^2}{r} \psi$$

(see Problem P5.1 for more details). Using $a_0 = 4\pi\varepsilon_0 \hbar^2/(m_e e^2)$ we can write

$$\begin{aligned} E(\mathrm{He}^+) &= \frac{C^3}{\pi a_0^3} \int_0^\infty \mathrm{e}^{-\frac{2Cr}{a_0}} r^2 \left[\frac{1}{2} \frac{e^2}{4\pi\varepsilon_0} \left(\frac{2C}{r} - \frac{C^2}{a_0} \right) - \frac{1}{4\pi\varepsilon_0} \frac{2e^2}{r} \right] \mathrm{d}r \cdot \int_0^\pi \sin\theta \, \mathrm{d}\theta \int_0^{2\pi} \mathrm{d}\varphi \\ &= \frac{C^3}{\pi a_0^3} \int_0^\infty \mathrm{e}^{-\frac{2Cr}{a_0}} r^2 \left[\frac{e^2}{4\pi\varepsilon_0} \left(\frac{(C-2)}{r} - \frac{C^2}{2a_0} \right) \right] \mathrm{d}r \cdot 4\pi \\ &= \frac{4C^3}{a_0^3} \frac{e^2}{4\pi\varepsilon_0} \left[(C-2) \int_0^\infty \mathrm{e}^{-\frac{2Cr}{a_0}} r \, \mathrm{d}r - \frac{C^2}{2a_0} \int_0^\infty \mathrm{e}^{-\frac{2Cr}{a_0}} r^2 \mathrm{d}r \right] \end{aligned}$$

If we make the substitution $x = 2Cr/a_0$, we obtain

$$E(\mathrm{He}^+) = \frac{4C^3}{a_0^3} \frac{e^2}{4\pi\varepsilon_0} \left[(C-2) \left(\frac{a_0}{2C} \right)^2 \int_0^\infty \mathrm{e}^{-x} x \, \mathrm{d}x - \frac{C^2}{2a_0} \left(\frac{a_0}{2C} \right)^3 \int_0^\infty \mathrm{e}^{-x} x^2 \mathrm{d}x \right]$$

From Appendix B, we know that $\int_0^\infty e^{-x}x^n\,dx = n!$, so that

$$E(\mathrm{He}^+) = \frac{4C^3}{a_0^3}\frac{e^2}{4\pi\varepsilon_0}\left[(C-2)\left(\frac{a_0}{2C}\right)^2(1) - \frac{C^2}{2a_0}\left(\frac{a_0}{2C}\right)^3(2)\right]$$
$$= \frac{C}{a_0}\frac{e^2}{4\pi\varepsilon_0}\left[(C-2) - C/2\right] = -\frac{e^2}{4\pi\varepsilon_0}\frac{\left[4C-C^2\right]}{2a_0}$$

In step 2, we use the trial function to calculate the first-order correction to the energy in perturbation theory. According to Justification 5.1 the perturbation energy is

$$\overline{V}_{12} = \frac{e^2}{4\pi\varepsilon_0 a_0}\cdot\frac{5}{4}\cdot\frac{C}{2}$$

In part 3 we are asked to combine these terms and find expression for the total energy. The energy is given by

$$E = 2\cdot E(\mathrm{He}^+) + \overline{V}_{12}$$
$$= 2\cdot\left(-\frac{e^2}{4\pi\varepsilon_0}\frac{\left[4C-C^2\right]}{2a_0}\right) + \frac{e^2}{4\pi\varepsilon_0 a_0}\cdot\frac{5}{4}\frac{C}{2}$$
$$= -\frac{e^2}{4\pi\varepsilon_0 a_0}\left[4C - C^2 - 5C/8\right] = -\frac{e^2}{4\pi\varepsilon_0 a_0}\frac{\left[27C-8C^2\right]}{8}$$

If we minimize the energy by taking the derivative with respect to C and setting it equal to zero, we find that

$$0 = -\frac{e^2}{4\pi\varepsilon_0 a_0}\frac{[27-16C]}{8} \quad \text{or } C = \frac{27}{16}$$

Thus the energy becomes

$$E = 2\cdot E(\mathrm{He}^+) + \overline{V}_{12} = -\frac{e^2}{4\pi\varepsilon_0 a_0}\left[\frac{(27)^2-(27)^2/2}{8\cdot 16}\right]$$
$$= -\frac{e^2}{4\pi\varepsilon_0 a_0}\left[\frac{(27)^2}{(16)^2}\right] = -124.2\times 10^{-19}\ \mathrm{J}$$

This energy is significantly lower than the value of -119.9×10^{-19} J obtained for the case of $C = 2$. The value of $C = 1.69$ implies a less compact electron cloud, which is consistent with the case of one electron partially screening the interaction of the other electron with the nucleus; however, the electrons are in the same spatial orbital. The increased spatial extent of the cloud also allows the electrons to be farther apart, on average, and this will reduce the electron–electron repulsion.

P5.12 Using the wavefunctions

$$\phi_2 = A\cdot\left[\varphi_{1s}(1)\cdot\varphi_{2s}(2) \pm \varphi_{1s}(2)\cdot\varphi_{2s}(1)\right]$$

calculate the energies for the symmetrical and antisymmetrical excited states of He.

Solution

Before we perform the calculation for the excited state, let us remember the calculation for the ground state of He. We write the ground-state wavefunction as

$$\phi_1(1,2) = \varphi_{1s}(1)\cdot\varphi_{1s}(2)$$

and for the repulsion energy we find

$$\overline{V}_{12,\mathrm{ground}} = \iint \phi_1^2(1,2)V_{12}\cdot d\tau_1 d\tau_2 = \frac{e^2}{4\pi\varepsilon_0}\iint \varphi_{1s}^2(1)V_{12}\varphi_{1s}^2(2)\cdot d\tau_1 d\tau_2$$

with

$$V_{12} = \frac{e^2}{4\pi\varepsilon_0}\frac{1}{r_{12}}$$

Now we consider the symmetrical excited state of He. In the text we calculated $A = \sqrt{2}/2$ for the normalization constant. Then in analogy to the ground state, we calculate the average repulsion energy for the symmetrical excited state as

$$\begin{aligned}
\overline{V}_{12,\text{excited,sym}} &= \iint \phi^2_{2,\text{sym}}(1,2)V_{12} \cdot \mathrm{d}\tau_1 \mathrm{d}\tau_2 \\
&= \frac{1}{2}\iint \left[\varphi_{1s}(1) \cdot \varphi_{2s}(2) + \varphi_{1s}(2) \cdot \varphi_{2s}(1)\right]^2 \cdot V_{12} \cdot \mathrm{d}\tau_1 \mathrm{d}\tau_2 \\
&= \frac{1}{2}\iint \left[\varphi_{1s}(1) \cdot \varphi_{2s}(2)\right]^2 \cdot V_{12} \cdot \mathrm{d}\tau_1 \mathrm{d}\tau_2 \\
&+ \frac{1}{2}\iint \left[\varphi_{1s}(2) \cdot \varphi_{2s}(1)\right]^2 \cdot V_{12} \cdot \mathrm{d}\tau_1 \mathrm{d}\tau_2 \\
&+ \frac{1}{2}\iint 2\left[\varphi_{1s}(1) \cdot \varphi_{2s}(2) \cdot \varphi_{1s}(2) \cdot \varphi_{2s}(1)\right] \cdot V_{12} \cdot \mathrm{d}\tau_1 \mathrm{d}\tau_2 \\
&= \frac{1}{2}\iint \varphi^2_{1s}(1)V_{12}\varphi^2_{2s}(2) \cdot \mathrm{d}\tau_1 \mathrm{d}\tau_2 \\
&+ \frac{1}{2}\iint \varphi^2_{1s}(2)V_{12}\varphi^2_{2s}(1) \cdot \mathrm{d}\tau_1 \mathrm{d}\tau_2 \\
&+ \iint \left[\varphi_{1s}(1) \cdot \varphi_{2s}(1) \cdot V_{12} \cdot \varphi_{2s}(2)\varphi_{1s}(2)\right] \cdot \mathrm{d}\tau_1 \mathrm{d}\tau_2
\end{aligned}$$

In the last expression the first two integrands are identical, because they differ only in the numbering of the electrons. Thus we obtain

$$\begin{aligned}
\overline{V}_{12,\text{excited,sym}} &= \iint \varphi^2_{1s}(1)V_{12}\varphi^2_{2s}(2) \cdot \mathrm{d}\tau_1 \mathrm{d}\tau_2 \\
&+ \iint \left[\varphi_{1s}(1) \cdot \varphi_{2s}(1) \cdot V_{12} \cdot \varphi_{2s}(2)\varphi_{1s}(2)\right] \cdot \mathrm{d}\tau_1 \mathrm{d}\tau_2 \\
&= J_{12} + K_{12}
\end{aligned}$$

The first integral is the Coulomb integral J_{12}, and the second integral is the exchange integral K_{12}.

Similarly, the result for the antisymmetrical excited state is

$$\begin{aligned}
\overline{V}_{12,\text{excited,anti}} &= \iint \varphi^2_{1s}(1)V_{12}\varphi^2_{2s}(2) \cdot \mathrm{d}\tau_1 \mathrm{d}\tau_2 \\
&- \iint \left[\varphi_{1s}(1) \cdot \varphi_{2s}(1) \cdot V_{12} \cdot \varphi_{2s}(2)\varphi_{1s}(2)\right] \cdot \mathrm{d}\tau_1 \mathrm{d}\tau_2 \\
&= J_{12} - K_{12}
\end{aligned}$$

In order to evaluate these integrals we write our trial functions as

$$\varphi_{1s}(1) = \sqrt{\frac{C^3}{\pi a_0^3}} \cdot \mathrm{e}^{-Cr/a_0} \quad \text{and } \varphi_{2s}(2) = \sqrt{\frac{C^3}{32\pi a_0^3}} \cdot \left(2 - \frac{Cr}{a_0}\right) \mathrm{e}^{-Cr/(2a_0)}$$

where C is the variational parameter, which was calculated in Problem P5.11 to be $C = 27/16$ in the case of the ground state of He, see Equation (5.18). The resulting energy E_{1s} is also calculated in Problem P5.11 as

$$E_{1s} = -\frac{1}{4\pi\varepsilon_0}\frac{e^2}{2a_0}\left(4C - C^2\right) = -\frac{1}{4\pi\varepsilon_0}\frac{e^2}{2a_0}3.902$$

Note that for $C = 2$ the last factor would be 4.0 instead of 3.902.

If we assume the same value of C for the excited states in He, we obtain

$$E_{2s} = -\frac{1}{4\pi\varepsilon_0}\frac{e^2}{2a_0}\left(4C - C^2\right)\frac{1}{4} = -\frac{1}{4\pi\varepsilon_0}\frac{e^2}{2a_0}3.902\frac{1}{4}$$

Thus it follows that

$$E_{1s} + E_{2s} = -\frac{1}{4\pi\varepsilon_0}\frac{e^2}{2a_0}3.902\left(1+\frac{1}{4}\right) = -\frac{1}{4\pi\varepsilon_0}\frac{e^2}{2a_0}4.878$$

Then for the symmetrical excited state of He we obtain

$$\begin{aligned}\varepsilon_{2,\mathrm{sym}} &= E_{1s} + E_{2s} + J_{12} + K_{12}\\ &= -\frac{1}{4\pi\varepsilon_0}\frac{e^2}{2a_0}4.878 + J_{12} + K_{12}\\ &= -106.2\times10^{-19}\ \mathrm{J} + J_{12} + K_{12}\end{aligned}$$

and

$$\begin{aligned}\varepsilon_{2,\mathrm{anti}} &= E_{1s} + E_{2s} + J_{12} - K_{12}\\ &= -106.2\times10^{-19}\ \mathrm{J} + J_{12} - K_{12}\end{aligned}$$

In Justification 5.2 we calculated J_{12} and K_{12} and found

$$J_{12} = 18.3\times10^{-19}\ \mathrm{J}\ \frac{C}{2} = 18.3\times10^{-19}\ \mathrm{J}\frac{27}{32} = 15.4\times10^{-19}\ \mathrm{J}$$

$$K_{12} = 1.91\times10^{-19}\ \mathrm{J}\ \frac{C}{2} = 1.91\times10^{-19}\ \mathrm{J}\frac{27}{32} = 1.61\times10^{-19}\ \mathrm{J}$$

Note that J_{12} for the excited state is only one-third of the value for the ground state (54.5×10^{-19} J). Thus the final result is

$$\begin{aligned}\varepsilon_{2,\mathrm{sym}} &= -106.2\times10^{-19}\ \mathrm{J} + 15.4\times10^{-19}\ \mathrm{J} + 1.61\times10^{-19}\ \mathrm{J} = -89.2\times10^{-19}\ \mathrm{J}\\ \varepsilon_{2,\mathrm{anti}} &= -106.2\times10^{-19}\ \mathrm{J} + 15.4\times10^{-19}\ \mathrm{J} - 1.61\times10^{-19}\ \mathrm{J} = -92.4\times10^{-19}\ \mathrm{J}\end{aligned}$$

6

A Quantitative View of Chemical Bonding

6.1 Exercises

E6.1 Calculate the bond energy for H_2 by using the bond energy of H_2^+, the H-atom ionization potential, and the experimental ionization potential of H_2 (15.426 eV). Justify your calculation by building an energy cycle to determine the electronic energy released in breaking the bond.

Solution

Using the cycle shown in Fig. E6.1 we can add up the energy changes in steps 1, 2, and 3 to obtain the net energy change between H_2 and the two separate H-atoms; namely

$$E_{\text{bond}} = \Delta E_1 + \Delta E_2 + \Delta E_3$$

Step 1 is the ionization energy of the H_2 molecule, 15.426 eV = 24.715×10^{-19} J. Step 2 is the bond energy of the H_2^+ molecule and is 4.48×10^{-19} J (see Table 6.1), and step 3 is the opposite of the ionization energy for an H-atom; namely

$$\Delta E_3 = -21.8 \times 10^{-19}\ \text{J}$$

Hence we find that the bond energy is

$$\begin{aligned} E_{\text{bond}} &= 24.715 \times 10^{-19}\ \text{J} + 4.48 \times 10^{-19}\text{J} - 21.8 \times 10^{-19}\ \text{J} \\ &= 7.40 \times 10^{-19}\ \text{J} \end{aligned}$$

It is also possible to calculate the total electronic energy of H_2 from the ionization potential of H_2 and the known electronic energy of H_2^+ (see Table 6.1); we find that

$$\begin{aligned} E_{\text{electronic}}(H_2) &= -24.715 \times 10^{-19}\ \text{J} - 26.28 \times 10^{-19}\ \text{J} \\ &= -51.00 \times 10^{-19}\ \text{J} \end{aligned}$$

These values are in reasonably good agreement with those in Table 6.1.

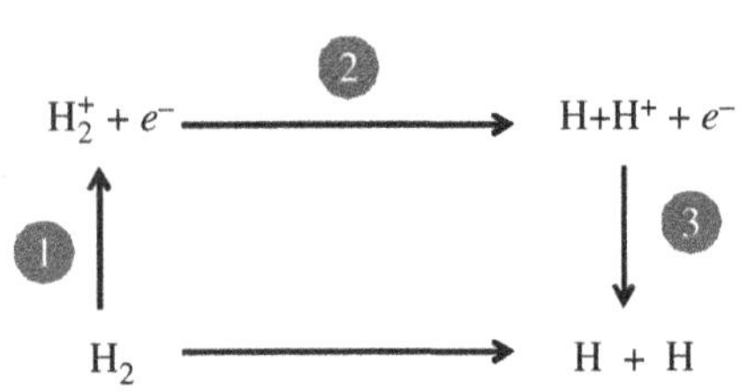

Figure E6.1 Energy cycle for calculating the bond energy of H_2.

Solutions Manual for Principles of Physical Chemistry, Third Edition. Edited by Hans Kuhn, David H. Waldeck, and Horst-Dieter Försterling.

E6.2 Use Fig. 6.1 and your knowledge of geometry to show that

$$r_a = \sqrt{(x + d/2)^2 + y^2 + z^2} \text{ and } r_b = \sqrt{(x - d/2)^2 + y^2 + z^2}$$

Solution

For Fig. 6.1 we choose the x-axis as the internuclear axis with negative x to the right. The positive y-axis is chosen to extend upward, and the positive z-axis is oriented out of the page toward the reader. Using this information, the position of the electron is specified by (x, y, z), the leftmost nucleus (a) is located at $(x - d/2, 0, 0)$, and the rightmost nucleus (b) is located at $(x + d/2, 0, 0)$. From this information and the three-dimensional version of the Pythagorean Theorem ($r^2 = x^2 + y^2 + z^2$), we readily obtain the desired formulae

$$r_a = \sqrt{(x + d/2)^2 + y^2 + z^2}$$

and

$$r_b = \sqrt{(x - d/2)^2 + y^2 + z^2}$$

E6.3 The particle-in-the-box wavefunctions are usually given as sine functions. In our analysis in Box 2.2, we used cosine functions for the box wavefunctions. Explain why this change is correct.

Solution

The cosine function is equivalent to a sine function that has a 90-degree phase shift; namely $\cos\theta = \sin(\theta + \pi/2)$. Hence the two functions are equivalent in shape possessing the same period, but displaced from each other. In Box 2.2, the edges of the box are offset from the positions of $x = 0$ and $x = L$ that were used to find the sine function solutions. Rather they occur at $x = -L/2$ and $x = +L/2$. For this reason the appropriate wavefunction is shifted by $L/2$ and should be written as

$$\sin\left[\frac{\pi}{L}\left(x + \frac{L}{2}\right)\right] = \sin\left(\frac{\pi x}{L} + \frac{\pi}{2}\right)$$
$$= \sin\left(\frac{\pi x}{L}\right)\cos\left(\frac{\pi}{2}\right) + \cos\left(\frac{\pi x}{L}\right)\sin\left(\frac{\pi}{2}\right)$$

Since the $\cos(\pi/2) = 0$ and $\sin(\pi/2) = 1$ we find that

$$\sin\left(\frac{\pi}{L}\left(x + \frac{L}{2}\right)\right) = \cos\left(\frac{\pi x}{L}\right)$$

The cosine function can then be high at the middle of this "offset" box in the ground state.

E6.4 Show in detail the mathematical steps needed to obtain Equation (6.11) from Equation (6.9).

Solution

Equation (6.9) is

$$\int \phi^2 d\tau = c^2 \cdot \left(\int \varphi_a^2 \, d\tau + 2\int \varphi_a\varphi_b \, d\tau + \int \varphi_b^2 \, d\tau\right) = 1$$

Using the fact that φ_a and φ_b are normalized gives

$$c^2 \cdot \left(1 + 2\int \phi_a\phi_b \, d\tau + 1\right) = 1$$

The definition of S_{ab} as $\int \varphi_a \, \varphi_b \, d\tau$ gives

$$2c^2(1 + S_{ab}) = 1$$

So that

$$c^2 = \frac{1}{2(1 + S_{ab})}$$

or

$$c = \sqrt{\frac{1}{2(1 + S_{ab})}}$$

which is Equation (6.11).

E6.5 Show in detail the mathematical steps needed to obtain Equation (6.12) from Equation (6.12).

Solution
Equation (6.12) is

$$\varepsilon = c^2\left(\int \varphi_a\,\mathcal{H}\,\varphi_a\,\mathrm{d}\tau + \int \varphi_a\,\mathcal{H}\,\varphi_b\,\mathrm{d}\tau + \int \varphi_b\,\mathcal{H}\,\varphi_a\,\mathrm{d}\tau + \int \varphi_b\,\mathcal{H}\,\varphi_b\,\mathrm{d}\tau\right)$$

Since φ_a and φ_b are identical functions that are centered on different nuclei we find that

$$\int \varphi_a\,\mathcal{H}\,\varphi_a\,\mathrm{d}\tau = \int \varphi_b\,\mathcal{H}\,\varphi_b\,\mathrm{d}\tau$$

and

$$\int \varphi_a\,\mathcal{H}\,\varphi_b\,\mathrm{d}\tau = \int \varphi_b\,\mathcal{H}\,\varphi_a\,\mathrm{d}\tau$$

So there are only two types of integrals that need to be examined. We write the integrals with orbitals on the same nuclear center as

$$H_{aa} = \int \varphi_a\,\mathcal{H}\,\varphi_a\,\mathrm{d}\tau = \int \varphi_b\,\mathcal{H}\,\varphi_b\,\mathrm{d}\tau$$

and those with orbitals on different centers as

$$H_{ab} = \int \varphi_a\,\mathcal{H}\,\varphi_b\,\mathrm{d}\tau = \int \varphi_b\,\mathcal{H}\,\varphi_a\,\mathrm{d}\tau$$

Substituting H_{aa} and H_{ab} for the integral expressions we find

$$\varepsilon = c^2(H_{aa} + H_{ab} + H_{ab} + H_{aa})$$

Using the result of E6.4, namely

$$c = \sqrt{\frac{1}{2(1+S_{ab})}}$$

we find that

$$\varepsilon = \frac{1}{2(1+S_{ab})}(2H_{aa} + 2H_{ab}) = \frac{H_{aa}+H_{ab}}{1+S_{ab}}$$

which is Equation (6.14).

E6.6 Evaluate the average kinetic energy of an electron in the ground state of a three-dimensional box with $b = 1.36$ and $L = 301$ pm; see Section 2.5. Compare this value to the optimized energy, given in the text, and show that the Virial Theorem is satisfied.

Solution
Equation (2.34) gives the average kinetic energy as

$$\overline{T} = \frac{h^2}{8m_eL^2}\left(\frac{1}{b^2}+2\right)$$

which on substituting the values from the problem statement gives

$$\overline{T} = \frac{(6.626\times10^{-34})^2\ \mathrm{J^2s^2}}{8(9.11\times10^{-31}\ \mathrm{kg})\,(301\times10^{-12}\ \mathrm{m})^2}\left(\frac{1}{(1.36)^2}+2\right)$$
$$= 1.689\times10^{-18}\,\mathrm{J}$$

The Virial Theorem states that $E = -\overline{T} = \frac{1}{2}\overline{V}$; comparison with the optimized energy $E = -1.689\times10^{-18}$J in Table 6.1 reveals that the Virial Theorem is followed.

E6.7 Write an explicit form for the electronic Schrödinger equation of HeH^{2+}. Discuss how it is different from that of H_2^+.

Solution

The molecular ion HeH^{2+} is a one-electron system with a proton and a He nucleus. Hence HeH^{2+} has a Hamiltonian operator of the form

$$\mathcal{H} = -\frac{\hbar^2}{2m_{\text{He}}}\nabla^2_{\text{He}} - \frac{\hbar^2}{2m_{\text{H}}}\nabla^2_{\text{H}} - \frac{\hbar^2}{2m_e}\nabla^2_e + \frac{2e^2}{4\pi\varepsilon_0}\frac{1}{R_{\text{HeH}}} - \frac{2e^2}{4\pi\varepsilon_0}\frac{1}{r_{\text{He}}} - \frac{e^2}{4\pi\varepsilon_0}\frac{1}{r_{\text{H}}}$$

which is very similar to that for H_2^+

$$\mathcal{H} = -\frac{\hbar^2}{2m_{\text{H}}}\nabla^2_{\text{H}(a)} - \frac{\hbar^2}{2m_{\text{H}}}\nabla^2_{\text{H}(b)} - \frac{\hbar^2}{2m_e}\nabla^2_e + \frac{e^2}{4\pi\varepsilon_0}\frac{1}{R_{\text{H-H}}} - \frac{e^2}{4\pi\varepsilon_0}\frac{1}{r_{\text{H}(a)}} - \frac{e^2}{4\pi\varepsilon_0}\frac{1}{r_{\text{H}(b)}}$$

The Hamiltonians have the same nuclear and electronic kinetic energy terms, as well as the potential energy terms from the nuclear–nuclear repulsion and the attraction of the electron to each nucleus. The differences are in the mass of one of the nuclei (4 for He instead of 1 for H) and in the terms involving the nuclear charge of He ($Z = 2$) rather than H ($Z = 1$).

E6.8 Draw a figure like that of Fig. 6.9 for the molecular ion HeH^+. Use this diagram to write the Hamiltonian for HeH^+. Be explicit when you write the Hamiltonian and provide a physical interpretation of the different terms. Compare this Hamiltonian to that given for H_2.

Solution

The only modification to Fig. 6.9 required in going from H_2 to HeH^+ is a change of one of the protons to a helium nucleus, and some renaming of the distances (see Fig. E6.8); namely

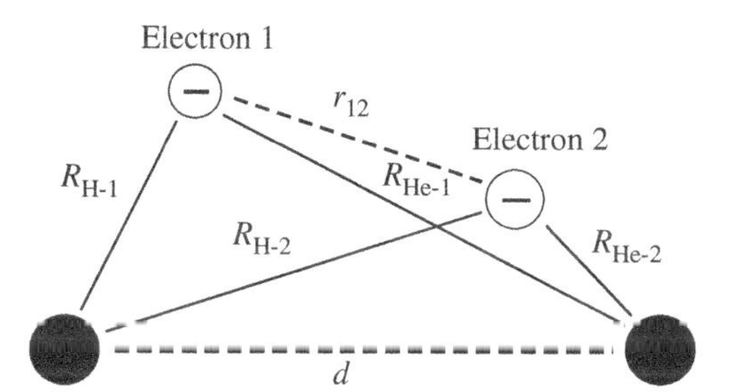

Figure E6.8 The HeH^+ molecule consists of two negatively charged electrons in the field of two positively charged nuclei, which are separated by a distance d. The solid lines represent attractive Coulomb interactions and the dashed lines represent repulsive Coulomb interactions.

The Hamiltonian is then

$$\mathcal{H} = -\frac{\hbar^2}{2m_{\text{He}}}\nabla^2_{\text{He}} - \frac{\hbar^2}{2m_{\text{H}}}\nabla^2_{\text{H}} - \frac{\hbar^2}{2m_{\text{e}}}\nabla^2_1 - \frac{\hbar^2}{2m_{\text{e}}}\nabla^2_2 + \frac{2e^2}{4\pi\varepsilon_0}\frac{1}{d}$$
$$- \frac{2e^2}{4\pi\varepsilon_0}\frac{1}{R_{\text{He-1}}} - \frac{2e^2}{4\pi\varepsilon_0}\frac{1}{R_{\text{He-2}}} - \frac{e^2}{4\pi\varepsilon_0}\frac{1}{R_{\text{H-1}}} - \frac{e^2}{4\pi\varepsilon_0}\frac{1}{R_{\text{H-2}}} + \frac{e^2}{4\pi\varepsilon_0}\frac{1}{r_{12}}$$

The first two terms correspond to the kinetic energy of the two nuclei, and the second two terms correspond to the kinetic energy of the two electrons. The fifth term corresponds to the Coulombic nuclear–nuclear repulsion. The next five times are Coulombic terms associated with the electrons; the sixth and sevenths terms are for the He-electron attraction, the eighth and ninth terms are for the H-electron attraction, and the last term is for the electron–electron repulsion.

The differences between this Hamiltonian and that for H_2 lie in the masses and nuclear charges only.

E6.9 Show that the Hamiltonian in Equation (6.21) can be written as the sum of two Hamiltonians for an H_2^+ molecular ion and an electron–electron repulsion term. Derive Equation (6.24).

Solution

The Hamiltonian given in Equation (6.21) is

$$\mathcal{H} = -\frac{\hbar^2}{2m_e}\nabla_1^2 - \frac{\hbar^2}{2m_e}\nabla_2^2 - \frac{e^2}{4\pi\varepsilon_0}\left(\frac{1}{r_{1a}} + \frac{1}{r_{1b}} + \frac{1}{r_{2a}} + \frac{1}{r_{2b}}\right) + \frac{e^2}{4\pi\varepsilon_0}\frac{1}{d} + \frac{e^2}{4\pi\varepsilon_0}\frac{1}{r_{12}}$$

Now we rewrite this expression with the terms for electron 1 combined, and those for electron 2 combined. In addition we associate a proton–proton repulsion term with each of these two pieces. Hence, we find

$$\mathcal{H} = \left[-\frac{\hbar^2}{2m_e}\nabla_1^2 + \frac{e^2}{4\pi\varepsilon_0}\left(-\frac{1}{r_{1a}} - \frac{1}{r_{1b}} + \frac{1}{d}\right)\right] + \left[-\frac{\hbar^2}{2m_e}\nabla_2^2 + \frac{e^2}{4\pi\varepsilon_0}\left(-\frac{1}{r_{2a}} - \frac{1}{r_{2b}} + \frac{1}{d}\right)\right] + \frac{e^2}{4\pi\varepsilon_0}\left(\frac{1}{r_{12}} - \frac{1}{d}\right)$$

By inspection we see that the term in brackets for each of the electrons has the same form as Equation (6.1) for the Hamiltonian of H_2^+, and we write

$$\mathcal{H} = \mathcal{H}_1(H_2^+) + \mathcal{H}_2(H_2^+) + \frac{e^2}{4\pi\varepsilon_0}\left(\frac{1}{r_{12}} - \frac{1}{d}\right)$$

where $\mathcal{H}_1$ is the H_2^+ Hamiltonian that we associate with electron 1 terms at internuclear distance d, and $\mathcal{H}_2$ is the H_2^+ Hamiltonian that we associate with the electron 2 terms at internuclear distance d. Lastly, we note that electrons are not distinguishable and this equation is

$$\mathcal{H} = 2\mathcal{H}(H_2^+) + \frac{e^2}{4\pi\varepsilon_0}\left(\frac{1}{r_{12}} - \frac{1}{d}\right)$$

which is the form of Equation (6.24).

E6.10 State the underlying physical premise of the Born–Oppenheimer approximation. Write the complete Hamiltonian of H_2 and identify its different terms. Contrast this Hamiltonian with the electronic Hamiltonian of H_2.

Solution

The underlying premise for the Born–Oppenheimer approximation is that the nuclei are much more massive than the electrons. Thus the electrons will rapidly adjust their charge distribution to accommodate the motion of the nuclei. For this reason the nuclear portion of the Hamiltonian does not need to explicitly involve the electron positions except as a parameterized averaged potential energy.

The complete Hamiltonian for H_2 is

$$\mathcal{H}_{\text{full}} = -\frac{\hbar^2}{2m_H}\nabla_{H_a}^2 - \frac{\hbar^2}{2m_H}\nabla_{H_b}^2 - \frac{\hbar^2}{2m_e}\nabla_1^2 - \frac{\hbar^2}{2m_e}\nabla_2^2 + \frac{e^2}{4\pi\varepsilon_0}\frac{1}{r_{H_a-H_b}} - \frac{e^2}{4\pi\varepsilon_0}\frac{1}{r_{H_{a1}}} - \frac{e^2}{4\pi\varepsilon_0}\frac{1}{r_{H_{a2}}} - \frac{e^2}{4\pi\varepsilon_0}\frac{1}{r_{H_{b1}}} - \frac{e^2}{4\pi\varepsilon_0}\frac{1}{r_{H_{b2}}} + \frac{e^2}{4\pi\varepsilon_0}\frac{1}{r_{12}}$$

The first two terms correspond to the kinetic energy of the two nuclei, the next two terms to the kinetic energy of the two electrons, and the fifth term to the nuclear–nuclear repulsion. In the second row, the next four terms (sixth through ninth) are the electrostatic attraction of the two nuclei for each of the electrons, and the last term is the electron–electron repulsion.

In writing the electronic Hamiltonian we neglect the first two terms for the kinetic energy of the nuclei and we keep the nuclear–nuclear distance set at a value $d_{H_a-H_b}$; it becomes a parameter used in the calculation rather than a variable. For the electronic Hamiltonian, we find

$$\mathcal{H}_{\text{electr}} = -\frac{\hbar^2}{2m_e}\nabla_1^2 - \frac{\hbar^2}{2m_e}\nabla_2^2 + \frac{e^2}{4\pi\varepsilon_0}\frac{1}{d_{H_a-H_b}} - \frac{e^2}{4\pi\varepsilon_0}\frac{1}{r_{H_{a1}}} - \frac{e^2}{4\pi\varepsilon_0}\frac{1}{r_{H_{a2}}} - \frac{e^2}{4\pi\varepsilon_0}\frac{1}{r_{H_{b1}}} - \frac{e^2}{4\pi\varepsilon_0}\frac{1}{r_{H_{b2}}} + \frac{e^2}{4\pi\varepsilon_0}\frac{1}{r_{12}}$$

where we have written the internuclear distance as $d_{H_a-H_b}$ rather than $r_{H_a-H_b}$ to emphasize that this distance is now treated as a parameter.

E6.11 Using the fact that the molecular orbitals φ and φ' are orthogonal to one another and each normalized, show that the two electron wavefunctions in Equations (6.28) and (6.29) are normalized.

Solution

To show that ϕ_{sym} is normalized, we look to see if

$$\iint \phi_{\text{sym}}^2 \, d\tau_1 d\tau_2 = 1$$

Substituting for the wavefunction from Equation (6.28) we write

$$\begin{aligned}
\iint \phi_{\text{sym}}^2 \, d\tau_1 d\tau_2 &= \iint \frac{1}{2}\left[\varphi(2)\varphi'(1) + \varphi(1)\varphi'(2)\right]^2 d\tau_1 d\tau_2 \\
&= \frac{1}{2}\left[\begin{array}{c}\iint \varphi^2(2)\varphi'^2(1)\cdot d\tau_1\, d\tau_2 + \iint \varphi^2(1)\varphi'^2(2)\cdot d\tau_1 d\tau_2 \\ +2\iint \varphi(2)\varphi'(1)\varphi(1)\varphi'(2)\cdot d\tau_1 d\tau_2\end{array}\right] \\
&= \frac{1}{2}\left[\begin{array}{c}\int \varphi^2(2)d\tau_2 \cdot \int \varphi'^2(1)d\tau_1 + \int \varphi^2(1)d\tau_1 \cdot \int \varphi'^2(2)d\tau_2 \\ +2\int \varphi'(1)\varphi(1)d\tau_1 \cdot \int \varphi(2)\varphi'(2)d\tau_2\end{array}\right] \\
&= \frac{1}{2}[1 + 1 + 2\cdot 0] = 1
\end{aligned}$$

To show that ϕ_{asym} is normalized, we look to see if

$$\iint \phi_{\text{asym}}^2 \, d\tau_1 d\tau_2 = 1$$

Substituting for the wavefunction from Equation (6.29) we write

$$\begin{aligned}
\iint \phi_{\text{antisym}}^2 \, d\tau_1 d\tau_2 &= \iint \frac{1}{2}\left[\varphi(2)\varphi'(1) - \varphi(1)\varphi'(2)\right]^2 d\tau_1 d\tau_2 \\
&= \frac{1}{2}\left[\begin{array}{c}\iint \varphi^2(2)\varphi'^2(1)\cdot d\tau_1\, d\tau_2 + \iint \varphi^2(1)\varphi'^2(2)\cdot d\tau_1 d\tau_2 \\ -2\iint \varphi(2)\varphi'(1)\varphi(1)\varphi'(2)\cdot d\tau_1 d\tau_2\end{array}\right] \\
&= \frac{1}{2}\left[\begin{array}{c}\int \varphi^2(2)d\tau_2 \cdot \int \varphi'^2(1)d\tau_1 + \int \varphi^2(1)d\tau_1 \cdot \int \varphi'^2(2)d\tau_2 \\ -2\int \varphi'(1)\varphi(1)d\tau_1 \cdot \int \varphi(2)\varphi'(2)d\tau_2\end{array}\right] \\
&= \frac{1}{2}[1 + 1 - 2\cdot 0] = 1
\end{aligned}$$

E6.12 Show that the spin wavefunctions in Equation (6.31) are normalized, by using the fact that

$$\int \alpha(\zeta)\alpha(\zeta)\, d\zeta = 1, \int \beta(\zeta)\beta(\zeta)\, d\zeta = 1, \int \beta(\zeta)\alpha(\zeta)\, d\zeta = 0$$

Solution

To show that the three components of Equation (6.31) are normalized we need to show that

$$1 = \iint \psi^2 \, d\zeta_1 d\zeta_2$$

For $\psi = \alpha(1)\alpha(2)$ we find

$$\iint \alpha^2(1)\alpha^2(2)\, d\zeta_1 d\zeta_2 = \left(\int \alpha^2(1)\, d\zeta_1\right)\left(\int \alpha^2(2)\, d\zeta_2\right) = 1\cdot 1 = 1$$

For $\psi = \beta(1)\,\beta\,(2)$ we find

$$\iint \beta^2(1)\beta^2(2)\, d\zeta_1 d\zeta_2 = \left(\int \beta^2(1) d\zeta_1\right)\left(\int \beta^2(2) d\zeta_2\right) = 1\cdot 1 = 1$$

For $\psi = [\alpha(1)\beta(2) + \alpha(2)\beta(1)]/\sqrt{2}$ the proof is similar but involves more algebra, namely

$$\begin{aligned}
\iint \psi^2 \, d\zeta_1 d\zeta_2 &= \frac{1}{2}\iint [\alpha(1)\beta(2) + \alpha(2)\beta(1)]^2 \, d\zeta_1 d\zeta_2 \\
&= \frac{1}{2}\left(\begin{array}{c}\iint \alpha^2(1)\beta^2(2)\, d\zeta_1 d\zeta_2 + \iint \alpha(1)\beta(2)\alpha(2)\beta(1)\, d\zeta_1 d\zeta_2 \\ + \iint \alpha^2(2)\beta^2(1)\, d\zeta_1 d\zeta_2\end{array}\right)
\end{aligned}$$

$$= \frac{1}{2}\left(\begin{array}{c} \int \alpha^2(1)\,d\zeta_1 \int \beta^2(2)d\zeta_2 + \int \alpha(1)\beta(1)d\zeta_1 \int \beta(2)\alpha(2)\,d\zeta_2 \\ + \int \beta^2(1)d\zeta_1 \int \alpha^2(2)d\zeta_2 \end{array} \right)$$

$$= \frac{1}{2}(1 \cdot 1 + 0 \cdot 0 + 1 \cdot 1) = 1$$

Thus all three wavefunctions are normalized.

E6.13 In Section 6.2.3 an LCAO wavefunction with an effective charge η was used to describe the H_2^+ molecule's ground-state energy. Provide a physically based discussion of why a value between $\eta = 1$ and $\eta = 2$ is reasonable.

Solution

The limit with $\eta = 1$ corresponds to the electron only interacting with one of the two protons, with the other very far away, or to well-separated nuclei (large d). The limit with $\eta = 2$ corresponds to the electron interacting with both nuclei together, much like with a helium nucleus, or to superimposed nuclei (small d). Thus it would be reasonable to expect that for intermediate internuclear distances, the value of η would be between these two values, or $1 \le \eta \le 2$.

E6.14 Provide a physically based explanation for why the total energy, obtained for H_2^+, with the box wavefunction is in worse agreement with the exact result than that obtained from the modified LCAO calculation.

Solution

The box trial wavefunction does not have peaks for the electron density at (or near) the nuclei, which we would expect from the Coulomb attraction force. In contrast, the modified LCAO method places more electron density near the nucleus, making it a more realistic description.

E6.15 Equation (2.35) gives an expression for the average potential energy of an electron. Show that it is equivalent to the expression

$$\overline{V} = \frac{-e^2}{4\pi\varepsilon_0} \int \left[\frac{1}{r_a} + \frac{1}{r_b} - \frac{1}{d}\right] \varphi^2 \, dx\, dy\, dz$$

Solution

Equation (2.35) gives the expression for the average potential energy of an electron as

$$\overline{V} = \frac{1}{4\pi\varepsilon_0}\frac{e^2}{d} - \frac{e^2}{4\pi\varepsilon_0}\int_{\text{all space}} \left(\frac{1}{r_a} + \frac{1}{r_b}\right) \rho \, d\tau$$

which is equivalent to

$$\overline{V} = -\frac{e^2}{4\pi\varepsilon_0}\left(-\frac{1}{d} + \int_{\text{all space}} \left(\frac{1}{r_a} + \frac{1}{r_b}\right) \rho \, d\tau\right)$$

Using the fact that $\rho = \varphi^2$ and $d\tau = dx\,dy\,dz$ in Cartesian coordinates, we can write that

$$\overline{V} = -\frac{e^2}{4\pi\varepsilon_0}\left(-\frac{1}{d} + \int_{\text{all space}} \left(\frac{1}{r_a} + \frac{1}{r_b}\right) \varphi^2 \, dx\, dy\, dz\right)$$

Using the fact that φ is normalized, we find

$$\overline{V} = -\frac{e^2}{4\pi\varepsilon_0}\left(-\frac{1}{d}\int_{\text{all space}} \varphi^2 \, dx\, dy\, dz + \int_{\text{all space}} \left(\frac{1}{r_a} + \frac{1}{r_b}\right) \varphi^2 \, dx\, dy\, dz\right)$$

$$\overline{V} = -\frac{e^2}{4\pi\varepsilon_0}\left(\int_{\text{all space}} -\frac{1}{d}\varphi^2 \, dx\, dy\, dz + \int_{\text{all space}} \left(\frac{1}{r_a} + \frac{1}{r_b}\right) \varphi^2 \, dx\, dy\, dz\right)$$

or

$$\overline{V} = -\frac{e^2}{4\pi\varepsilon_0}\left(\int_{\text{all space}} \left(\frac{1}{r_a} + \frac{1}{r_b} - \frac{1}{d}\right) \varphi^2 \, dx\, dy\, dz\right)$$

E6.16 Starting with Equation (6.18) for the anti-bonding orbital trial function of H_2^+ derive the expression for its energy that is given in Equation (6.20).

Solution

For describing the excited state we use the trial function

$$\phi = c \cdot (\varphi_a - \varphi_b)$$

where c is a normalization constant. According to the variational principle we use this trial wavefunction to evaluate the energy integral

$$\varepsilon = \int \phi \mathcal{H} \phi \cdot d\tau$$

to obtain

$$\varepsilon = \int \phi \mathcal{H} \phi \cdot d\tau = c^2 \int (\varphi_a - \varphi_b) \mathcal{H} (\varphi_a - \varphi_b) \cdot d\tau$$
$$= c^2 \left(\int \varphi_a \mathcal{H} \varphi_a \cdot d\tau - \int \varphi_a \mathcal{H} \varphi_b \cdot d\tau - \int \varphi_b \mathcal{H} \varphi_a \cdot d\tau + \int \varphi_b \mathcal{H} \varphi_b \cdot d\tau \right)$$

Again we use the shorthand notation,

$$\int \varphi_a \mathcal{H} \varphi_a \cdot d\tau = \int \varphi_b \mathcal{H} \varphi_b \cdot d\tau = H_{aa}$$
$$\int \varphi_a \mathcal{H} \varphi_b \cdot d\tau = \int \varphi_b \mathcal{H} \varphi_a \cdot d\tau = H_{ab}$$

to obtain

$$\varepsilon = \int \phi \mathcal{H} \phi \cdot d\tau = c^2 (H_{aa} - H_{ab} - H_{ab} + H_{aa})$$

and

$$\varepsilon = c^2 2 (H_{aa} - H_{ab}) \qquad (6E16.1)$$

Finally we normalize the wavefunction. In analogy to Equation (6.9) in Section 6.2.3 we start with

$$\int \phi^2 \cdot d\tau = c^2 \cdot \left(\int \varphi_a^2 \cdot d\tau - 2 \int \varphi_a \varphi_b \cdot d\tau + \int \varphi_b^2 \cdot d\tau \right) = 1 \qquad (6E16.2)$$

Using the normalization conditions of the atomic functions

$$\int \varphi_a^2 \cdot d\tau = \int \varphi_b^2 \cdot d\tau = 1$$

and the definition of the overlap integral

$$S_{ab} = \int \varphi_a \varphi_b \cdot d\tau$$

we can solve Equation (6E16.2) for c

$$\int \phi^2 \cdot d\tau = c^2 \cdot (1 - 2S_{ab} + 1) = 1$$

$$c^2 = \frac{1}{2(1 - S_{ab})} \qquad \text{and} \qquad c = \frac{1}{\sqrt{2(1 - S_{ab})}}$$

and with Equation (6E16.1) we find

$$\boxed{\varepsilon = \frac{H_{aa} - H_{ab}}{1 - S_{ab}}}$$

E6.17 In Problem P6.10 you are asked to calculate the energy ε_1 for the excited state of the H_2^+ molecular ion as a function of bond distance d. As part of this calculation you must evaluate the expression

$$F(d) = \frac{\eta}{a_0} + \frac{1}{d} \left[1 - \left(1 + \eta \frac{d}{a_0} \right) \cdot e^{-2\eta d / a_0} \right]$$

as the distance $d \rightarrow 0$. One must proceed carefully in the evaluation for small d in order to obtain a stable and accurate solution. a) Use L'Hopital's rule to show that $F(d) \rightarrow 2\eta/a_0$ as $d \rightarrow 0$. b) Expand the exponential to second order (using the Taylor's series method) and find a polynomial expression for $F(d)$ at small values of d.

Solution

a) If we write the expression for $F(d)$ as

$$F(d) = \frac{\eta \cdot d + \left[a_0 - \left(a_0 + \eta \cdot d\right) \cdot e^{-2\eta d/a_0}\right]}{a_0 \cdot d}$$

the need for using L'Hopital's rule becomes somewhat more self-evident; i.e., both the numerator and denominator approach zero as $d \rightarrow 0$. To proceed we take the derivative of the numerator and denominator with respect to d and evaluate the limit for that expression. We find

$$\lim_{d\to 0} F(d) = \lim_{d\to 0}\left(\frac{\eta + \left[-\eta \cdot e^{-2\eta d/a_0} + \frac{2\eta}{a_0}\left(a_0 + \eta \cdot d\right) \cdot e^{-2\eta d/a_0}\right]}{a_0}\right)$$

$$= \lim_{d\to 0}\left(\frac{\eta + \left[\eta \cdot e^{-2\eta d/a_0} + \frac{2\eta^2 d}{a_0} \cdot e^{-2\eta d/a_0}\right]}{a_0}\right)$$

$$= \lim_{d\to 0}\left(\frac{\eta}{a_0} + \left[\frac{\eta}{a_0} + \frac{2\eta^2 d}{a_0^2}\right] e^{-2\eta d/a_0}\right) = 2\frac{\eta}{a_0}$$

b) In performing the evaluation of the $F(d)$ expression for small values of d, numerical error (e.g., rounding error) can be minimized by simplifying the expression for $F(d)$ at small d values in order to eliminate having to evaluate the ratios of small numbers. To this end, we first expand the exponential term in a Taylor's series about the point $d = 0$. To second order, we find

$$e^{-2\eta d/a_0} = 1 - 2\eta d/a_0 + \frac{1}{2}(2\eta d/a_0)^2$$

Substitution of this expression into $F(d)$ gives

$$F(d)|_{\text{small } d} = \frac{\eta}{a_0} + \frac{1}{d}\left[1 - \left(1 + \eta\frac{d}{a_0}\right) \cdot e^{-2\eta d/a_0}\right]$$

$$= \frac{\eta}{a_0} + \left[\frac{1}{d} - \left(\frac{1}{d} + \eta\frac{1}{a_0}\right) \cdot \left(1 - \frac{2\eta d}{a_0} + \frac{1}{2}\left(\frac{2\eta d}{a_0}\right)^2\right)\right]$$

$$= \frac{\eta}{a_0} + \left[\frac{1}{d} - \frac{1}{d} + 2\frac{\eta}{a_0} - \frac{1}{2d}\left(\frac{2\eta d}{a_0}\right)^2 - \frac{\eta}{a_0} + \frac{2\eta^2 d}{a_0^2} - \frac{1}{2}\left(\frac{2\eta d}{a_0}\right)^2 \frac{\eta}{a_0}\right]$$

$$= \frac{\eta}{a_0} + \left[\frac{\eta}{a_0} - 2d\left(\frac{\eta}{a_0}\right)^2 + 2d\frac{\eta^2}{a_0^2} - 2d^2\left(\frac{\eta}{a_0}\right)^3\right] = 2\frac{\eta}{a_0} - 2d^2\left(\frac{\eta}{a_0}\right)^3$$

This result shows that $F(d)|_{\text{small } d}$ decreases with decreasing d parabolically, leading to $F(0) = 2\eta/a_0$, in agreement with our finding using L'Hopital's rule.

6.2 Problems

P6.1 In Equations (6.4) and (6.5) we calculated the average charge density of an electron at the protons of H_2^+ using the exact solution. Perform this calculation by using the optimized particle-in-a-box wavefunction ($d = 152$ pm, $b = 1.36$, and $L = 301$ pm; see Box 2.2). Also, perform this calculation using the optimized LCAO wavefunction $\phi(r) = \left[\varphi_a + \varphi_b\right] / \sqrt{2(1 + S_{ab})}$ with $\eta = 1.24$ and $d = 106$ pm; from Foundation 6.2 use the facts that

$$\varphi_a = \sqrt{\frac{\eta^3}{\pi a_0^3}} \cdot e^{-\eta r_a/a_0} \quad \text{and} \quad \varphi_b = \sqrt{\frac{\eta^3}{\pi a_0^3}} \cdot e^{-\eta r_b/a_0}$$

and

$$S_{ab} = \left(1 + \eta\frac{d}{a_0} + \frac{1}{3}\eta^2\frac{d^2}{a_0^2}\right) \cdot e^{-\eta d/a_0}$$

Solution

The optimized particle-in-a-box wavefunction (see Box 2.2) is

$$\varphi_a = \sqrt{\frac{8}{L_x L_y L_z}} \cos\left(\frac{\pi x_a}{L_x}\right) \cos\left(\frac{\pi y_a}{L_y}\right) \cos\left(\frac{\pi z_a}{L_z}\right)$$
$$= \sqrt{\frac{8}{bL^3}} \cos\left(\frac{\pi x_a}{bL}\right) \cos\left(\frac{\pi y_a}{L}\right) \cos\left(\frac{\pi z_a}{L}\right)$$

with $b = 1.36$, $L_x = bL = 409$ pm, and $L_y = L_z == L = 301$ pm. Because we are interested in the charge density ρ at the nucleus we can take $y = z = 0$, so that

$$\varphi_a|_{\substack{y=0\\z=0}} = \sqrt{\frac{8}{bL^3}} \cos\left(\frac{\pi x_a}{bL}\right)$$

and

$$\rho = \frac{8}{bL^3}\cos^2\left(\frac{\pi x_a}{bL}\right)$$

We have to calculate ρ at the positions of the nuclei, that is, $x_a = \pm d/2 = \pm 76$ pm. Then with $L = 301$ pm and $b = 1.36$ we obtain

$$\rho = \frac{8}{1.36 \cdot 301^3\ \text{pm}^3}\cos^2\left(\frac{\pi \cdot 76}{1.36 \cdot 301}\right)$$
$$= 2.16 \times 10^{-7}\ \text{pm}^{-3}\cos^2(0.583) = 2.16 \times 10^{-7}\ \text{pm}^{-3} \cdot 0.697 = 1.51 \times 10^{-7}\ \text{pm}^{-3}$$

To find the charge density, we multiply by the elementary charge $e = 1.602 \times 10^{-19}$ C, and find

$$\rho \cdot e = 1.51 \times 10^{-7}\ \text{pm}^{-3} \cdot 1.602 \times 10^{-19}\ \text{C} = 2.42 \times 10^{-26}\ \text{C pm}^{-3}$$

The optimized LCAO wavefunction is

$$\phi(r) = \frac{1}{\sqrt{2(1 + S_{ab})}}\sqrt{\frac{\eta^3}{\pi a_0^3}}\left[e^{-\eta r_a/a_0} + e^{-\eta r_b/a_0}\right]$$

and using our parameters above we find that $S_{ab} = 0.462$. Because we are interested in the average probability density at the nuclei we can write

$$\rho(r_a = 0) = \left[\frac{1}{2(1 + S_{ab})}\frac{\eta^3}{\pi a_0^3}\left[1 + e^{-\eta d/a_0}\right]^2\right]$$
$$= \left[\frac{1}{2(1 + 0.462)}\frac{(1.24)^3}{\pi a_0^3}[1 + 0.08335]^2\right]$$
$$= 0.244\, a_0^{-3} = 1.65 \times 10^{-6}\ \text{pm}^{-3}$$

where we have used the fact that nucleus a is at a distance d from nucleus b. To find the charge density, we multiply by the elementary charge $e = 1.602 \times 10^{-19}$ C, and find

$$\rho(r_a = 0) \cdot e = 1.65 \times 10^{-6}\ \text{pm}^{-3} \cdot 1.602 \times 10^{-19}\ \text{C} = 2.64 \times 10^{-25}\ \text{C pm}^{-3}$$

The two probability densities differ from each other significantly. The LCAO value is much closer to the exact value, whereas the particle-in-a-box wavefunction is much too low. See the discussion on this point in the text.

P6.2 In the first part of this problem, show that the Hamiltonian for the H_2 molecule's electronic Schrödinger equation can be written as a sum of two H_2^+ molecular ion Hamiltonians and an electron repulsion term. In the second part of this problem, use the wavefunction of Equation (6.23) in the electronic Schrödinger equation for H_2 and find the energy expression of Equation (6.25). This demonstration is formal, in the sense that you need not perform any derivatives explicitly. Rather substitute the expression for the wavefunction into the Schrödinger equation and show that the equation can now be written as a sum of two terms that correspond to the Schrö dinger equation for H_2^+ and a third term that is $-e^2/(4\pi\varepsilon_0 d)$.

Solution

In Equation (6.1) in Section 6.2.1 we considered the Hamiltonian of the H_2^+ molecular ion

$$\mathcal{H}(H_2^+) = -\frac{\hbar^2}{2m_e}\nabla^2 - \frac{e^2}{4\pi\varepsilon_0}\left(\frac{1}{r_a}+\frac{1}{r_b}\right) + \frac{e^2}{4\pi\varepsilon_0}\frac{1}{d}$$

where a and b refer to the two nuclei. According to Equation (6.22) in Section 6.3.1, the Hamiltonian for the H_2 molecule is

$$\mathcal{H}(H_2) = -\frac{\hbar^2}{2m_e}\nabla_1^2 - \frac{\hbar^2}{2m_e}\nabla_2^2 - \frac{e^2}{4\pi\varepsilon_0}\left(\frac{1}{r_{1a}}+\frac{1}{r_{1b}}+\frac{1}{r_{2a}}+\frac{1}{r_{2b}}\right) + \frac{e^2}{4\pi\varepsilon_0}\frac{1}{d} + \frac{e^2}{4\pi\varepsilon_0}\frac{1}{r_{12}}$$

where the indices 1 and 2 refer to the two electrons and the indices a and b refer to the two nuclei. We can recast this Hamiltonian as

$$\begin{aligned}\mathcal{H}(H_2) = &-\frac{\hbar^2}{2m_e}\nabla_1^2 - \frac{e^2}{4\pi\varepsilon_0}\left(\frac{1}{r_{1a}}+\frac{1}{r_{1b}}\right) + \frac{e^2}{4\pi\varepsilon_0}\frac{1}{d}\\ &-\frac{\hbar^2}{2m_e}\nabla_2^2 - \frac{e^2}{4\pi\varepsilon_0}\left(\frac{1}{r_{2a}}+\frac{1}{r_{2b}}\right) + \frac{e^2}{4\pi\varepsilon_0}\frac{1}{d}\\ &-\frac{e^2}{4\pi\varepsilon_0}\frac{1}{d} + \frac{e^2}{4\pi\varepsilon_0}\frac{1}{r_{12}}\end{aligned}$$

By inspection, we see that the terms for the kinetic energy and the attraction energy between electron 1 and the nuclei (first line) are equivalent to the Hamiltonian of an H_2^+ ion. Correspondingly we find the same result for the terms labeled as electron 2 (second line). Hence we can write that

$$\begin{aligned}\mathcal{H}(H_2) &= \mathcal{H}_1(H_2^+) + \mathcal{H}_2(H_2^+) - \frac{e^2}{4\pi\varepsilon_0}\frac{1}{d} + \frac{e^2}{4\pi\varepsilon_0}\frac{1}{r_{12}}\\ &= 2\cdot\mathcal{H}(H_2^+) - \frac{e^2}{4\pi\varepsilon_0}\frac{1}{d} + \frac{e^2}{4\pi\varepsilon_0}\frac{1}{r_{12}}\end{aligned}$$

where we have dropped the labels for electrons 1 and 2 on the Hamiltonian operator in the last line.

For the second part, we begin by writing the wavefunction as a product of two H_2^+ molecular orbitals φ

$$\phi_1 = \varphi(x_1, y_1, z_1)\cdot\varphi(x_2, y_2, z_2) = \varphi_1\varphi_2$$

Using the Hamiltonian

$$\mathcal{H}(H_2) = \mathcal{H}_1(H_2^+) + \mathcal{H}_2(H_2^+) - \frac{e^2}{4\pi\varepsilon_0}\frac{1}{d} + \frac{e^2}{4\pi\varepsilon_0}\frac{1}{r_{12}}$$

the Schrödinger equation becomes

$$\begin{aligned}\mathcal{H}\,\phi_1 &= \varphi_2\ \mathcal{H}_1(H_2^+)\,\varphi_1 + \varphi_1\ \mathcal{H}_2(H_2^+)\,\varphi_2 - \frac{e^2}{4\pi\varepsilon_0}\frac{1}{d}\varphi_1\varphi_2 + \frac{e^2}{4\pi\varepsilon_0}\frac{1}{r_{12}}\varphi_1\varphi_2\\ &= \left[2E(H_2^+) - \frac{e^2}{4\pi\varepsilon_0}\cdot\frac{1}{d} + \frac{e^2}{4\pi\varepsilon_0}\cdot\frac{1}{r_{12}}\right]\varphi_1\varphi_2 = E(H_2)\,\varphi_1\varphi_2\\ &= E(H_2)\phi_1\end{aligned}$$

Thus, the H_2 molecule energy $E(H_2)$ is given by

$$E(H_2) = 2E(H_2^+) - \frac{e^2}{4\pi\varepsilon_0}\cdot\frac{1}{d} + \frac{e^2}{4\pi\varepsilon_0}\cdot\frac{1}{r_{12}}$$

Identifying the first two terms with $E(H_2)_0$ and the last term as $E_{\text{repulsion}}$ gives Equation (6.25), and identifies

$$E(H_2)_0 = 2\cdot E(H_2^+) - \frac{e^2}{4\pi\varepsilon_0}\cdot\frac{1}{d}$$

P6.3 Figure 6.13 shows an electronic potential energy curve for the ground electronic state of H_2. The shape of this curve is characteristic of that for many diatomic molecules, and it is desirable to have an analytical expression that can reproduce such a shape. A widely used mathematical form for such electronic potential curves is the Morse potential $V(d - d_{eq})$ which is given by

$$V(d - d_{eq}) = \varepsilon_B[1 - \exp(-a(d - d_{eq}))]^2 \text{ , } a = \sqrt{\frac{k_f}{2\varepsilon_B}}$$

In this expression, ε_B is the well depth, d_{eq} is the equilibrium internuclear separation, and k_f is the force constant (see Chapter 7). Show that the Morse potential form is similar to that given in the figure by plotting it. For H_2 you should use $d_{eq} = 74.2$ pm, $\varepsilon_B = 7.60 \times 10^{-19}$ J, and $k_f = 575$ N/m.

Solution
The function to be plotted is

$$V\left(d - d_{eq}\right) = 7.60 \times 10^{-19}\ \text{J} \cdot \left[1 - \exp\left(-\sqrt{\frac{575}{2\left(7.60 \times 10^{-19}\right)}}(d - 74.2) \times 10^{-12}\right)\right]^2$$

It is displayed in Fig. P6.3 (solid line) together with the corresponding curve calculated by Kolos (full circles, see Fig. 6.13). The two curves are practically identical.

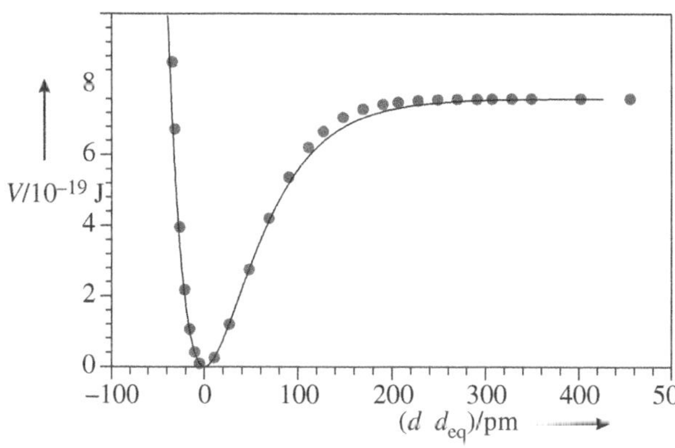

Figure P6.3 This plot compares the Morse function approximation to the numerical result of Kolos. Solid line: Morse function 6.3 with the parameters of the H_2 molecule. Full circles: electronic energy calculated by Kolos (see Fig. 6.13). $d - d_{eq}$ is the elongation from the equilibrium distance.

P6.4 In Chapter 2 the mean kinetic energy of the H_2^+ molecular ion for the box wavefunctions was calculated as $\overline{T} = 16.9 \times 10^{-19}$ J. For a hypothetical H_2^+ ion use the same box length L_x, but the different lengths $L_y = L_z = 376$ pm to calculate the average kinetic energy. These perpendicular lengths correspond to the box lengths for the H atom. Explain your result in comparison to the optimized value for H_2^+ and that found for the hydrogen atom in Section 2.3.

Solution
From Table 6.1 we obtain for the energy minimum $L_x = 1.36 \times 301$ pm $= 409$ pm. Using 409 pm for L_x and the given value for L_y and L_z in Equation (2.34), we obtain

$$\overline{T} = \frac{\hbar^2}{8m_e}\left(\frac{1}{L_x^2} + \frac{1}{L_y^2} + \frac{1}{L_z^2}\right) = 12.1 \times 10^{-19}\ \text{J}$$

This value is less than the mean kinetic energy of the electron in the H atom ($\overline{T} = 12.9 \times 10^{-19}$J) and significantly less than that given for the H_2^+ molecular ion. When forming the bond, the resulting box lengths $L_y = L_z = 301$ pm are smaller than in the H atom ($L_x = L_y = L_z = 376$ pm), and, as a consequence, the kinetic energy rises to $\overline{T} = 16.9 \times 10^{-19}$J, higher than the kinetic energy of the electron in the H atom. This contraction of the wavefunction reflects the increased Coulomb attraction of the electrons to the two protons, as compared to the single proton in the H-atom.

Note that in the simple LCAO model of H_2^+ the kinetic energy of the molecule at its equilibrium distance is smaller than that of the H atom (Fig. 6.5), because the extension of the electron cloud perpendicularly to the bond line remains unchanged. The variational LCAO method allows the wavefunction to contract however, and its kinetic energy is closer to the exact value (see Table 6.1).

P6.5 Consider the 1*s* orbital wavefunctions

$$\varphi_{1sa}(x, y, z) = \exp\left(-\eta\sqrt{(x - d/2)^2 + y^2 + z^2}\right)$$

and

$$\varphi_{1sb}(x, y, z) = \exp\left(-\eta\sqrt{(x + d/2)^2 + y^2 + z^2}\right)$$

Use a computer application to create a symmetric superposition of these two wavefunctions for $\eta = 1/a_0$ and $d = 74$ pm. Plot the superposition along the x axis ($y = 0, z = 0$). Also plot the value of the superposed wavefunction at $x = 0$ as a function of the internuclear distance d. Obtain an analytical expression that describes the d dependence you found.

Solution

The symmetric superposition is (unnormalized)

$$f(x, y, z; d) = \exp\left(-\frac{1}{a_0}\sqrt{(x - d/2)^2 + y^2 + z^2}\right) + \exp\left(-\frac{1}{a_0}\sqrt{(x + d/2)^2 + y^2 + z^2}\right)$$

A plot of this function along the x-axis (see Fig. P6.5a) is

$$f = \exp\left(-\frac{1}{52.9\text{ pm}}\sqrt{(x - 74/2)^2}\right) + \exp\left(-\frac{1}{52.9\text{ pm}}\sqrt{(x + 74/2)^2}\right) \quad (6P5.1)$$

Now we set $x = 0$ pm and plot f as a function of d (see Fig. P6.5b)

$$f = \exp\left(-\frac{1}{52.9\text{ pm}}\sqrt{(d/2)^2}\right) + \exp\left(-\frac{1}{52.9\text{ pm}}\sqrt{(d/2)^2}\right) \quad (6P5.2)$$

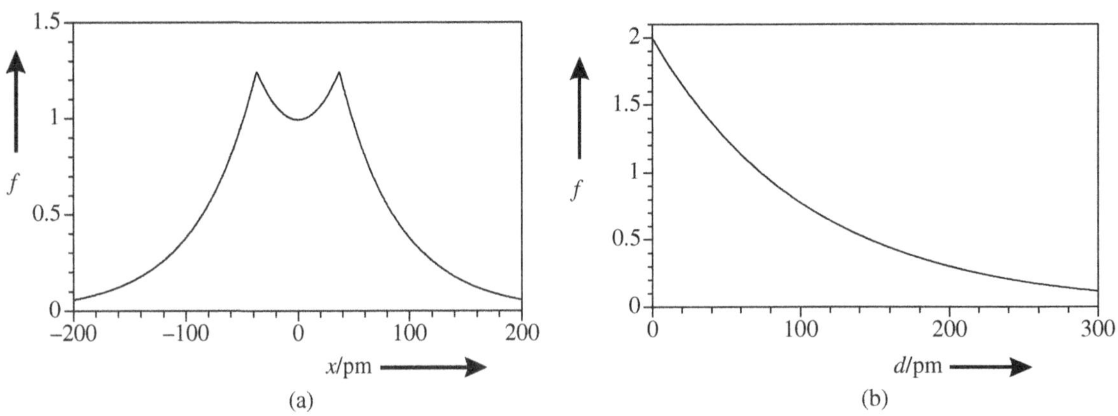

Figure P6.5 Plot of the function f. (a) f according to Equation (6P5.1) and (b) f according to Equation (6P5.2).

The analytical expression representing Fig. P6.5 is

$$f(d) = 2\exp\left(-\frac{1}{2 \times 52.9\text{ pm}}d\right)$$

P6.6 Perform Problem 6.5 for the antisymmetric combination of the 1*s* orbitals.

Solution

The antisymmetric superposition is (unnormalized)

$$f(x,y,z;d) = \exp\left(-\frac{1}{a_0}\sqrt{(x-d/2)^2+y^2+z^2}\right) - \exp\left(-\frac{1}{a_0}\sqrt{(x+d/2)^2+y^2+z^2}\right)$$

A plot of this function

$$f = \exp\left(-\frac{1}{52.9\ \text{pm}}\sqrt{(x-74/2)^2}\right) - \exp\left(-\frac{1}{52.9\ \text{pm}}\sqrt{(x+74/2)^2}\right) \quad (6P6.1)$$

is displayed in Fig. P6.6, in which we can see that the wavefunction changes sign at $x = 0$. Now we set $x = 0$ and plot as a function of d.

$$f = \exp\left(-\frac{1}{52.9\ \text{pm}}\sqrt{(-d/2)^2}\right) - \exp\left(-\frac{1}{52.9\ \text{pm}}\sqrt{(d/2)^2}\right) \quad (6P6.2)$$

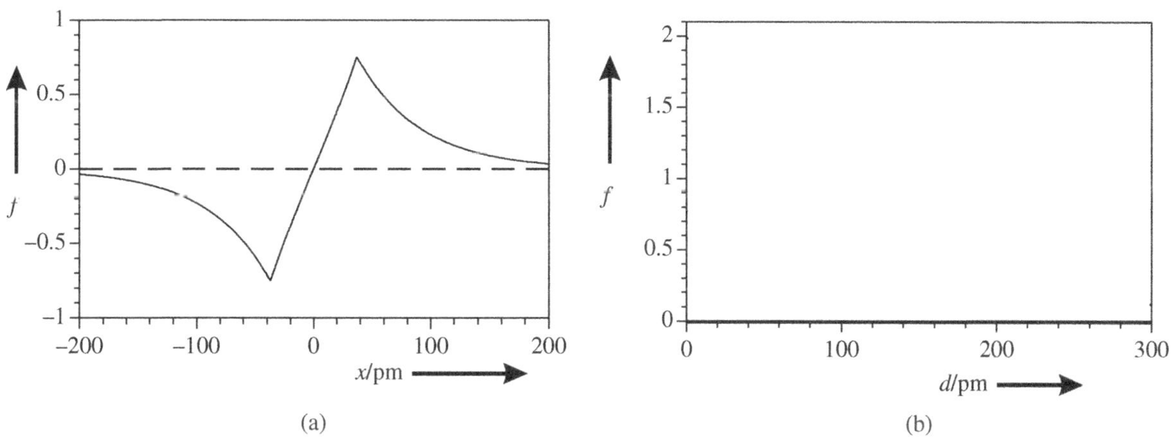

Figure P6.6 Plot of the function f. (a) f according to Equation (6P6.1) and (b) f according to Equation (6P6.2).

The analytical expression representing the above figure is

$$f(d) = 0$$

This reflects the fact that the nodal plane does not move; it remains exactly halfway between the two nuclei.

P6.7 For the hydrogen atom the energy of the electron is determined by the principal quantum number n. Hence the $n = 2$ energy level is four-fold degenerate because the 2*s* and the three 2*p* orbitals all have the same energy. Let us take a superposition of the 2*s* orbital and two of the 2*p* orbitals (the p_x and p_z). In particular we will restrict our discussion to the angular part of the wavefunctions and create three superposition wavefunctions that are each normalized and orthogonal to each other. These superposition states should have the angular shape of the sp^2 hybrid orbitals that you learned about in organic chemistry.

Given the sp^2 hybrid orbital

$$\psi_A = \sqrt{\frac{1}{3}}\psi_{2s} + \sqrt{\frac{2}{3}}\psi_{2p_z} = \sqrt{\frac{1}{3}}\psi_{2,0} + \sqrt{\frac{2}{3}}\psi_{2,1,0}$$

construct a second orbital of the form

$$\psi_B = \sqrt{\frac{1}{3}}\psi_{2s} + b_2\,\psi_{2p_z} + c_2\,\psi_{2px}$$

by requiring that ψ_A be orthogonal to ψ_B and that ψ_B be normalized. You should determine the numerical values for the b_2 and c_2 coefficients. Now derive the third hybrid orbital, namely

$$\psi_C = a_3\,\psi_{2s} + b_3\,\psi_{2p_z} + c_3\,\psi_{2px}$$

Hint: Each of these wavefunctions is orthogonal and normalized. If you use these properties, you will not need to perform an integration in solving this problem, rather you will be able to determine whether an integral is zero or unity by inspection.

Solution

In this problem we will designate ψ_{2s} as $2s$ and ψ_{2p} as $2p$ for simplicity. The two orbitals ψ_A and ψ_B must be orthogonal to each other, so

$$0 = \int \left(\frac{1}{\sqrt{3}}\, 2s + \sqrt{\frac{2}{3}}\, 2p_z\right)\left(\frac{1}{\sqrt{3}}\, 2s + b_2\, 2p_z + c_2\, 2p_x\right) \mathrm{d}\tau$$

Expanding the integrand gives

$$0 = \int \begin{pmatrix} \frac{1}{3}\,(2s)^2 + \frac{1}{\sqrt{3}}\, b_2\, 2s\, 2p_z + \frac{1}{\sqrt{3}}\, c_2\, 2s\, 2p_x \\ + \frac{\sqrt{2}}{3}\, 2s\, 2p_z + \sqrt{\frac{2}{3}}\, b_2\, (2p_z)^2 + c_2\sqrt{\frac{2}{3}}\, 2p_z\, 2p_x \end{pmatrix} \mathrm{d}\tau$$

The integrals are either 0 or 1 depending on the orbitals involved. Substituting in those values gives

$$0 = \frac{1}{3}\, 1 + \frac{1}{\sqrt{3}}\, b_2\, 0 + \frac{1}{\sqrt{3}}\, c_2\, 0 + \frac{\sqrt{2}}{3}\, 0 + \sqrt{\frac{2}{3}}\, b_2\, 1 + c_2\sqrt{\frac{2}{3}}\, 0 = \frac{1}{3} + \sqrt{\frac{2}{3}}\, b_2$$

so that $b_2 = -\sqrt{\frac{1}{6}}$.

In addition, ψ_B must be normalized, so that

$$1 = \int \left(\frac{1}{\sqrt{3}}\, 2s - \sqrt{\frac{1}{6}}\, 2p_z + c_2\, 2p_x\right)\left(\frac{1}{\sqrt{3}}\, 2s - \sqrt{\frac{1}{6}}\, 2p_z + c_2\, 2p_x\right) \mathrm{d}\tau$$

Multiplying out, and applying the orthogonality of the $2s$, $2p_z$, and $2p_x$ orbitals again gives

$$1 = \frac{1}{3} + \frac{1}{6} + c_2^2$$

so that $c_2 = \pm\sqrt{\frac{1}{2}}$. For convenience, we choose the negative root and write

$$\psi_B = \frac{1}{\sqrt{3}}\, 2s - \frac{1}{\sqrt{6}}\, 2p_z - \frac{1}{\sqrt{2}}\, 2p_x$$

While the choice of sign here is arbitrary, it will affect the coefficients that we find for ψ_C.

For the third orbital

$$\psi_C = a_3\, 2s + b_3\, 2p_z + c_3\, 2p_x$$

This orbital must be orthogonal to both ψ_A and ψ_B. The expression for orthogonality with ψ_A gives

$$0 = \int (a_3\, 2s + b_3\, 2p_z + c_3\, 2p_x)\left(\frac{1}{\sqrt{3}}\, 2s + \sqrt{\frac{2}{3}}\, 2p_z\right) \mathrm{d}\tau$$

or

$$0 = \sqrt{\frac{1}{3}} a_3 + \sqrt{\frac{2}{3}}\, b_3\,, \text{ so that } a_3 = -\sqrt{2} b_3$$

The expression for orthogonality with ψ_B gives

$$0 = \int (a_3\, 2s + b_3\, 2p_z + c_3\, 2p_x)\left(\frac{1}{\sqrt{3}}\, 2s - \frac{1}{\sqrt{6}}\, 2p_z - \frac{1}{\sqrt{2}}\, 2p_x\right) \mathrm{d}\tau$$

or

$$0 = \frac{1}{\sqrt{3}} a_3 - \frac{1}{\sqrt{6}}\, b_3 - \frac{1}{\sqrt{2}}\, c_3$$

By substituting our expression for a_3, we find that

$$0 = -\sqrt{\frac{3}{2}}\, b_3 - \frac{1}{\sqrt{2}}\, c_3 \text{ or } c_3 = -\sqrt{3}b_3$$

The final condition that we need to apply is that of normalization; i.e.,

$$\begin{aligned}1 &= \int \psi_C \cdot \psi_C \,\mathrm{d}\tau \\ &= \int \left(a_3\ 2s + b_3\ 2p_z + c_3\ 2p_x\right)\left(a_3\ 2s + b_3\ 2p_z + c_3\ 2p_x\right)\,\mathrm{d}\tau \\ &= \left(a_3^2 + 0 + 0 + 0 + b_3^2 + 0 + 0 + 0 + c_3^2\right)\end{aligned}$$

By substituting our expressions for c_3 and a_3 in terms of b_3, we find that

$$1 = 2b_3^2 + b_3^2 + 3b_3^2 \text{ or } b_3 = \frac{1}{\sqrt{6}}$$

and correspondingly we find that

$$c_3 = -\sqrt{3}b_3 = -\frac{1}{\sqrt{2}} \text{ and } a_3 = -\sqrt{2}b_3 = -\frac{1}{\sqrt{3}}$$

so that

$$\psi_C = -\frac{1}{\sqrt{3}}\,2s + \frac{1}{\sqrt{6}}\,2p_z - \frac{1}{\sqrt{2}}\,2p_x$$

Note that if we had chosen the negative root for our b_3 value that the overall sign of the wavefunction would change.

P6.8 Show that the trial wavefunctions for the H_2 molecule, ϕ_1 and ϕ_2 in Section 6.3.3, are orthogonal to each other.

Solution

The trial wavefunctions are·

$$\phi_1 = \frac{1}{\sqrt{2}} \cdot \left(\varphi(2)\varphi'(1) + \varphi(1)\varphi'(2)\right) \text{ and } \phi_2 = \frac{1}{\sqrt{2}} \cdot \left(\varphi(2)\varphi'(1) - \varphi(1)\varphi'(2)\right)$$

and the orbitals φ and φ' are orthogonal to each other and are each normalized. We need to show that

$$\iint \phi_1\phi_2 \,\mathrm{d}\tau_1\mathrm{d}\tau_2 = 0$$

We begin by substituting in the wavefunctions

$$\begin{aligned}\iint \phi_1\phi_2\,\mathrm{d}\tau_1\mathrm{d}\tau_2 &= \frac{1}{2}\iint \begin{bmatrix}\left(\varphi(2)\varphi'(1) + \varphi(1)\varphi'(2)\right) \\ \cdot\left(\varphi(2)\varphi'(1) - \varphi(1)\varphi'(2)\right)\end{bmatrix} \mathrm{d}\tau_1\mathrm{d}\tau_2 \\ &= \frac{1}{2}\iint \begin{bmatrix}\varphi^2(2)\varphi'^2(1) + \varphi(1)\varphi(2)\varphi'(2)\varphi'(1) \\ -\varphi(1)\varphi(2)\varphi'(1)\varphi'(2) - \varphi^2(1)\varphi'^2(2)\end{bmatrix} \mathrm{d}\tau_1\mathrm{d}\tau_2 \\ &= \frac{1}{2}\iint \left[\varphi^2(2)\varphi'^2(1) - \varphi^2(1)\varphi'^2(2)\right]\mathrm{d}\tau_1\mathrm{d}\tau_2 \\ &= \frac{1}{2}\int \varphi^2(2)\mathrm{d}\tau_2 \int \varphi'^2(1)\mathrm{d}\tau_1 - \frac{1}{2}\int \varphi^2(1)\mathrm{d}\tau_1 \int \varphi'^2(2)\mathrm{d}\tau_2 \\ &= \frac{1}{2}\cdot 1\cdot 1 - \frac{1}{2}\cdot 1\cdot 1 = 0\end{aligned}$$

P6.9 Using the trial function ϕ_2 for the excited state of the H_2^+ molecular ion in Section 6.2.5 derive Equations (6.19) and (6.20) for c_2 and for ε_2.

Solution

To determine c_2 we apply the normalization condition, which is

$$\int_{-\infty}^{\infty} \phi_2^*\phi_2 \,\mathrm{d}\tau = 1$$

so we have

$$1 = \int c_2 \cdot (\varphi_a - \varphi_b) \times c_2 \cdot (\varphi_a - \varphi_b)\ d\tau$$
$$= c_2^2 \int (\varphi_a - \varphi_b) \times (\varphi_a - \varphi_b)\ d\tau$$
$$= c_2^2 \int (\varphi_a^2 - 2\varphi_a\varphi_b + \varphi_b^2)\ d\tau$$

Since φ_a and φ_b are normalized, we find that

$$c_2^2\left(2 - 2\int \varphi_a\varphi_b\ d\tau\right) = 1$$

If we let

$$S_{ab} = \int \varphi_a\varphi_b\ d\tau$$

then our expression becomes

$$c_2^2\,(2 - 2S_{ab}) = 1$$

solving for c_2 we have

$$c_2 = \frac{1}{\sqrt{2 - 2S_{ab}}} = \frac{1}{\sqrt{2\,(1 - S_{ab})}}$$

To determine ε_2, we use

$$\varepsilon = \int \phi\mathcal{H}\phi \cdot d\tau$$

Upon substitution of ϕ we have

$$\varepsilon_2 = \frac{1}{2\,(1 - S_{ab})}\int (\varphi_a - \varphi_b)\,\mathcal{H}\,(\varphi_a - \varphi_b) \cdot d\tau$$
$$= \frac{\int \varphi_a\mathcal{H}\varphi_a \cdot d\tau - \int \varphi_a\mathcal{H}\varphi_b \cdot d\tau - \int \varphi_b\mathcal{H}\varphi_a \cdot d\tau + \int \varphi_b\mathcal{H}\varphi_b \cdot d\tau}{2\,(1 - S_{ab})}$$

Because the atomic functions φ_a and φ_b are identical (just centered on different nuclei), we have

$$\int \varphi_a\mathcal{H}\varphi_b \cdot d\tau = \int \varphi_b\mathcal{H}\varphi_a \cdot d\tau = H_{aa}$$

and

$$\int \varphi_a\mathcal{H}\varphi_a \cdot d\tau = \int \varphi_b\mathcal{H}\varphi_b \cdot d\tau = H_{ab}$$

which defines our shorthand notation for the matrix elements. Thus, we can write ε_2 as

$$\varepsilon_2 = \frac{H_{aa} - H_{ab} - H_{ab} + H_{aa}}{2\,(1 - S_{ab})}$$
$$= \frac{(2H_{aa} - 2H_{ab})}{2\,(1 - S_{ab})} = \frac{H_{aa} - H_{ab}}{1 - S_{ab}}$$

P6.10 In analogy to the procedure described in Foundation 6.2 for the calculation of the ground-state energy ε_0, calculate the energy ε_1 for the excited state of the H_2^+ molecular ion as a function of bond distance d.

Solution

According to Section 6.2.3 the excited state energy ε in the LCAO treatment of H_2^+ is

$$\varepsilon_1 = \overline{T_1} + \overline{V_1} = \frac{H_{aa} - H_{ab}}{1 - S_{ab}} = \frac{T_{aa} - T_{ab}}{1 - S_{ab}} + \frac{V_{aa} - V_{ab}}{1 - S_{ab}}$$

In Foundation 6.2 we find the following relations

$$T_{aa} = \frac{e^2}{4\pi\varepsilon_0} \cdot \frac{\eta^2}{2a_0}$$

and

$$T_{ab} = \frac{e^2}{4\pi\varepsilon_0} \cdot \frac{\eta^2}{2a_0} \cdot \left(2\sqrt{2}S'_{ab} - S_{ab}\right)$$

where the overlap integrals S'_{ab} and S_{ab} are

$$S'_{ab} = \frac{1}{\sqrt{2}}\left(1 + \eta\frac{d}{a_0}\right) \cdot \mathrm{e}^{-\eta d/a_0} \text{ and } S_{ab} = \left(1 + \eta\frac{d}{a_0} + \frac{1}{3}\eta^2\frac{d^2}{a_0^2}\right) \cdot \mathrm{e}^{-\eta d/a_0}$$

Hence we find that

$$\overline{T_1} = \frac{T_{aa} - T_{ab}}{1 - S_{ab}} = \frac{e^2}{4\pi\varepsilon_0}\frac{\eta^2}{2a_0}\frac{1 - 2\sqrt{2}S'_{ab} + S_{ab}}{1 - S_{ab}}$$

Correspondingly for V_{aa} and V_{ab} we find

$$V_{aa} = -\frac{e^2}{4\pi\varepsilon_0} \cdot \left\{+\frac{\eta}{a_0} + \frac{1}{d}\left[1 - \left(1 + \eta\frac{d}{a_0}\right) \cdot \mathrm{e}^{-2\eta d/a_0}\right]\right\} + \frac{e^2}{4\pi\varepsilon_0} \cdot \frac{1}{d}$$

$$V_{ab} = -\frac{e^2}{4\pi\varepsilon_0} \cdot \frac{2\sqrt{2} \cdot \eta \cdot S'_{ab}}{a_0} + \frac{e^2}{4\pi\varepsilon_0} \cdot \frac{1}{d} \cdot S_{ab}$$

$$\overline{V_1} = \frac{V_{aa} - V_{ab}}{1 - S_{ab}} = -\frac{e^2}{4\pi\varepsilon_0}\frac{\frac{\eta}{a_0} + \left(\frac{1}{d} - \frac{\eta}{a_0}\right) \cdot \mathrm{e}^{-2\eta d/a_0} - \left(2\sqrt{2} \cdot \eta \cdot S'_{ab}/a_0\right) + (S_{ab}/d)}{1 - S_{ab}}$$

with the same relations for S_{ab} and S'_{ab} as above. From these equations we calculate the energy ε_1 as

$$\varepsilon_1 = \overline{T_1} + \overline{V_1}$$

The calculation is performed in the program "excited." The result for the energy ε_1 is displayed in Fig. P6.10a. All three curves start at strongly positive values (due to the strong contribution of the repulsion energies of the nuclei) and decrease monotonically with increasing bond distance d. No energy minimum appears, indicating that the excited state is unbound.

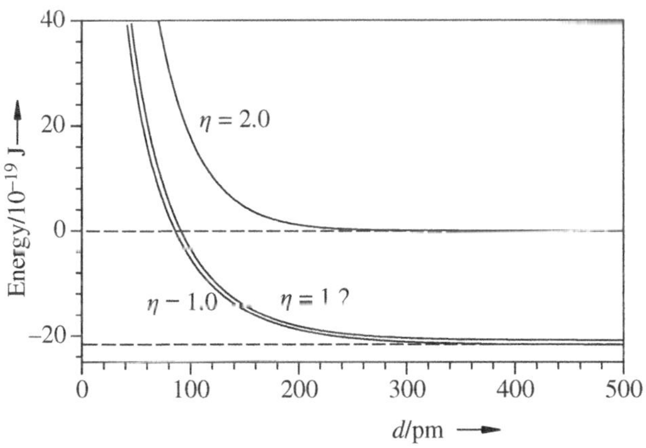

Figure P6.10a Energy ε_1 versus the bond distance d for the cases $\eta = 1.0$, $\eta = 1.2$, and $\eta = 2.0$.

For large distances, the energies are obtained by considering the limit as $d \to \infty$ in the above equations.

$$\lim_{d\to\infty} S_{ab} = 0 \quad \text{and} \quad \lim_{d\to\infty} S'_{ab} = 0$$

$$\lim_{d\to\infty} T_{aa} = \frac{e^2}{4\pi\varepsilon_0} \cdot \frac{\eta^2}{2a_0} \quad \text{and} \quad \lim_{d\to\infty} T_{ab} = 0$$

$$\lim_{d\to\infty} V_{aa} = -\frac{e^2}{4\pi\varepsilon_0} \cdot \frac{\eta}{a_0} \quad \text{and} \quad \lim_{d\to\infty} V_{ab} = 0$$

Thus we obtain

$$\lim_{d\to\infty} \overline{T}_1 = \lim_{d\to\infty} \frac{T_{aa} - T_{ab}}{1 - S_{ab}} = \frac{e^2}{4\pi\varepsilon_0} \cdot \frac{\eta^2}{2a_0}$$

$$\lim_{d\to\infty} \overline{V}_1 = \lim_{d\to\infty} \frac{V_{aa} - V_{ab}}{1 - S_{ab}} = -\frac{e^2}{4\pi\varepsilon_0} \cdot \frac{\eta}{a_0}$$

and

$$\lim_{d\to\infty} \varepsilon_1 = \lim_{d\to\infty} (\overline{T_1} + \overline{V_1}) = -\frac{e^2}{4\pi\varepsilon_0} \frac{1}{2a_0} = -21.8 \times 10^{-19} \text{ J for } \eta = 1.0$$

$$\lim_{d\to\infty} \varepsilon_1 = \lim_{d\to\infty} (\overline{T_1} + \overline{V_1}) = \frac{e^2}{4\pi\varepsilon_0} \frac{1.2^2 - 2 \cdot 1.2}{2a_0} = -0.96 \cdot 21.8 \times 10^{-19} \text{ J for } \eta = 1.2$$

$$\lim_{d\to\infty} \varepsilon_1 = \lim_{d\to\infty} (\overline{T_1} + \overline{V_1}) = -\frac{e^2}{4\pi\varepsilon_0} \frac{2^2 - 2 \cdot 2}{2a_0} = 0 \text{ for } \eta = 2.0$$

How can we understand the result that the limiting energy for $\eta = 2.0$ is zero? The limiting value of the energy for $d \to \infty$ is the energy of a hypothetical atom with a proton as a nucleus, but with a strongly compressed electron cloud. In the equations we see that the kinetic energy increases with the square of η, and the potential energy decreases linearly with η. Thus the kinetic energy of the electron increases more strongly than its potential energy decreases. Just for $\eta = 2.0$ both energies become identical, leading to a zero total energy.
For comparison, we calculate ε_0, the energy for the ground state. You will see immediately from the equations, that the result for $d \to \infty$ is the same as for the excited state energy. However, in the range in between, as shown in Fig. P6.10b, we observe an energy minimum, indicating a bonding state.

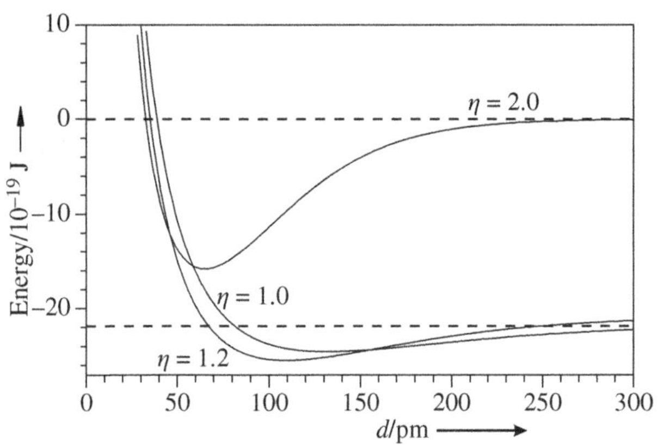

Figure P6.10b Energy ε_0 versus the bond distance d for the cases $\eta = 1.0$, $\eta = 1.2$, and $\eta = 2.0$.

Note that the limiting value for the $\eta = 1.2$ curve ($\varepsilon_0 = 0.96 \cdot \varepsilon_{\text{Hatom}}$) is a bit higher than that for the $\eta = 1.0$ curve ($\varepsilon_0 = 1.0 \cdot \varepsilon_{\text{Hatom}}$), whereas the limiting value for the corresponding curve ($\eta = 1.24$) in Fig. 6.8 equals $\varepsilon_0 = 1.0 \cdot \varepsilon_{\text{Hatom}}$. However, in this latter case, we used η-values as a function of the bond distance instead of a constant value for all bond distances.

P6.11 In Fig. 6.14 we consider the possibility that an electron in an H-atom can move to a proton over a distance d. Use the data given for the height and width of the barrier in Fig. 6.14 to calculate the probability $\mathcal{P}$ (see Eq. 3.22 in Chapter 3) for surmounting the barrier.

Solution
In Fig. 6.14 we find the data:

$$\text{a) } \Delta V = 17.2 \times 10^{-19} \text{ J}, \Delta x = 1.62 \times 10^{-9} \text{ m}, d = 2 \text{ nm}$$

$$\text{b) } \Delta V = 12.6 \times 10^{-19} \text{ J}, \Delta x = 0.63 \times 10^{-9} \text{ m}, d = 1 \text{ nm}$$

The probability $\mathcal{P}$ is

$$\mathcal{P} = \exp\left(-\frac{4\pi}{h}\sqrt{2m_e\left(V_0 - E\right)} \cdot \Delta x\right) = \exp\left(-\frac{4\pi}{h}\sqrt{2m_e\Delta V} \cdot \Delta x\right)$$
$$= \exp\left(-\frac{4\pi}{6.626 \times 10^{-34}\ \mathrm{J\,s}}\sqrt{2 \times 9.109 \times 10^{-31}\ \mathrm{kg}\,\Delta V} \cdot \Delta x\right)$$
$$= \exp\left(-1.89 \times 10^{34}\ \mathrm{J^{-1}\,s^{-1}}\sqrt{1.822 \times 10^{-30}\ \mathrm{J\,m^{-2}\,s^{2}}\ \Delta V} \cdot \Delta x\right)$$

Thus in case a) ($d = 2$ nm) we obtain

$$\mathcal{P} = \exp\left(-1.89 \times 10^{34}\ \mathrm{J^{-1}\,s^{-1}}\sqrt{1.822 \times 10^{-30}\ \mathrm{J\,m^{-2}\,s^{2}}\ 12.6 \times 10^{-19}\ \mathrm{J}} \cdot 0.63 \times 10^{-9}\ \mathrm{m}\right)$$
$$= \exp\left(-54.2\right) = 2.8 \times 10^{-24}$$

and the result for case b) ($d = 1$ nm) is

$$\mathcal{P} = \exp\left(-1.89 \times 10^{34}\ \mathrm{J^{-1}\,s^{-1}}\sqrt{1.822 \times 10^{-30}\ \mathrm{J\,m^{-2}\,s^{2}}\ 17.2 \times 10^{-19}\ \mathrm{J}} \cdot 1.62 \times 10^{-9}\ \mathrm{m}\right)$$
$$= \exp\left(-18.0\right) = 1.5 \times 10^{-8}$$

We see that the result is extremely sensitive to small changes of the distance d: a change of d from 2 to 1 nm results in a change of P by a factor of 10^{16}.

P6.12 In this problem you will explore the Equation

$$E_{\mathrm{kin,calc}} = -E - r \cdot \frac{\mathrm{d}E}{\mathrm{d}r} \tag{6P12.1}$$

You should run the program "h2plus_exact." This program calculates the exact wavefunctions and energies for the H_2^+ ion. As an output you will find the following files:

Energy.dat: this is the total energy $E/10^{-19}$ J
Vpot.dat: this is the potential energy $V_{\mathrm{pot}}/10^{-19}$ J
Ekin.dat: this is the kinetic energy $E_{\mathrm{kin}}/10^{-19}$ J
Virial_correction.dat: this is the correction term $r \cdot \mathrm{d}E/\mathrm{d}r$ considered in Equation (6P12.1)

These energies are given as a function of the bond distance r in the range from 20 to 450 pm.
Select a number of data points and check if the prediction of Equation (6P12.1) is correct.

Solution

We consider the H_2^+-ion. As an example, we choose values at bond distances r = 84.67, 105, 84, 149.93, and 199.32 pm and calculate $E_{\mathrm{kin,calc}}$ according to the equation:

$$E_{\mathrm{kin,calc}} = -E - r \cdot \frac{\mathrm{d}E}{\mathrm{d}r}$$

Exact Solution Calculated from Equation (6P12.1)

	Exact Solution			Calculated from Equation (6P12.1)		
r/pm	$E/10^{-19}$ J	$E_{\mathrm{kin}}/10^{-19}$ J	E/E_{kin}	$r \cdot \mathrm{d}E/\mathrm{d}r/10^{-19}$ J	$E_{\mathrm{kin,calc}}/10^{-19}$ J	$E_{\mathrm{kin}}/E_{\mathrm{kin,calc}}$
84.67	25.77	30.73	−0.84	−5.28	31.05	0.99
105.84	−26.28	26.27	−1.00	−0.13	26.41	1.00
149.93	−25.43	21.29	−1.19	4.13	21.30	1.00
199.32	−24.10	19.30	−1.25	4.83	19.27	1.00

Columns 2–3: result of the exact solution of the Schrödinger Equation for E and E_{kin}
Column 4: comparison of energy E and kinetic energy E_{kin}. Only in the case of r = 105.84 pm (this is the equilibrium distance) the ratio E/E_{kin} equals 1.00 as expected from the Virial Theorem. At smaller values of r this ratio is smaller than 1.00 and at greater values it is greater than 1.00.
Columns 5–7: Kinetic energy $E_{\mathrm{kin,calc}}$ calculated from the exact energy E using Equation (6P12.1). In all cases we find $E_{\mathrm{kin}}/E_{\mathrm{kin,calc}} = 1.0$, that is, the kinetic energy calculated from Equation (6P12.1) is identical with the value obtained directly from the Schrödinger Equation.

7

Bonding Described by Electron Pairs and Molecular Orbitals

7.1 Exercises

E7.1 Use the Lewis electron pair idea to rationalize the triple bond in N_2.

Solution

The N atom has five valence electrons, and by forming three electron pair bonds it can follow the "octet rule." In order for two N atoms to bind with each other and satisfy the "octet rule" they must form three-electron pair bonds, i.e. a triple bond, with each other.

E7.2 Use electron pair ideas to decide on the atom connectivity in the molecule NH_3. Explain why the structure you choose is expected to be the most stable, and comment on the geometry.

Solution

The diagrams below show three possible connectivity patterns for NH_3. Only the structure in which each H atom is bonded to the central N atom satisfies the octet rule for the nitrogen (see E7.1). Some other possible connectivities are shown in Figure E7.2 however, they are expected to be less stable. The decreased stability may be rationalized by the differences in the formal charges. For the case where each H atom is bonded to the N atom, each of the atoms in the molecule has a formal charge of zero. For the case where two H atoms are bonded to the N atom, we see that the N atom has a formal charge of +1e and one of the H-atoms has a formal charge of −1e. The structure with only one bond formed to the nitrogen is expected to be the least stable of the three because it has a formal charge of +2e on the nitrogen and a formal charge of −1e on two of the H-atoms. Since the arrangement with three bonds to the central N atom has the least amount of formal charge separation we expect it to be the most stable.

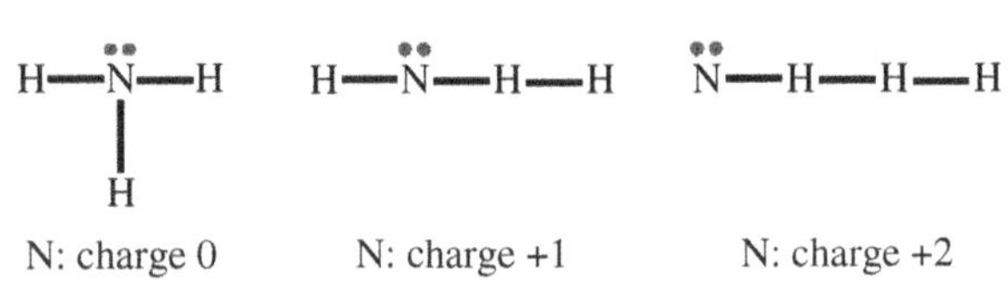

Figure E7.2 Three connectivities for the binding of three H atoms to a N atom. Formal charges 0, +1, and +2 on the N atom.

Here we apply the VSEPR rules to the most stable structure. In this structure, the N atom has four electron pairs in its valence shell (one lone pair and three electron pair bonds), and they are in a tetrahedral arrangement. The molecule's geometry is pyramidal.

E7.3 *Extension of* Example 7.2: Analyze the bonding requirements for an FCN ring compound and compare it to i).

Solution

Example 7.2 only considers linear connectivity patterns for FCN; however, it is also possible that a ring compound might be formed by the three atoms, as shown in Fig. E7.3. In this case, the electrons have been arranged in order to satisfy the octet rule on each of the atomic sites. Nevertheless this bonding pattern is less stable than the F-C≡N geometry (shown on the right in Fig. E7.3) deduced in Example 7.2.

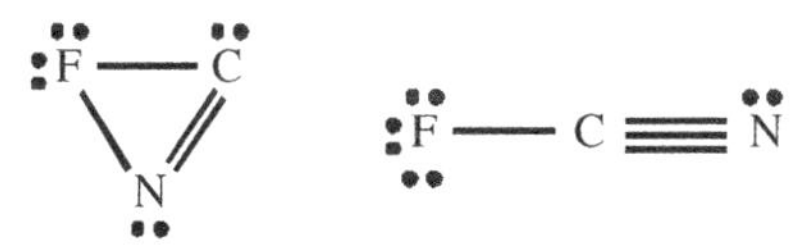

N: charge 0	N: charge 0
F: charge +1	F: charge 0
C: charge −1	C: charge 0

Figure E7.3 Connectivities for the binding in the FCN molecule.

The decreased stability can be rationalized by the consideration of formal charge. In the cyclic structure, the F-atom has a formal charge of +1e, the C-atom has a formal charge of −1e, and the N-atom has a zero formal charge. In the linear arrangement (shown on the right), each of the atoms has a formal charge of zero. We expect the linear geometry to be the most stable, because it has no separated formal charge.

E7.4 *Extension of* Example 7.3: A perfect tetrahedron has angles between its diagonals (bond axes for methane) of 109.5°. Assume that nonbonded electron pairs take up more space than bonded electron pairs and rationalize why the bond angle for water is 104.0° and the bond angle in ammonia is 106.0°.

Solution

If non-bonded electron pairs take up more space, they will force the bonded pairs closer together and decrease the angle between the bond axes. The electron pair structures of CH_4, NH_3, and H_2O all have four electron pairs around the central atom that are approximately in a tetrahedral arrangement. Methane has four equivalent bonds and we expect the bond angles to be 109.5°. Ammonia has three bonding electron pairs and one lone pair of electrons. Because the nonbonded electron pair is postulated to take up more space than the bonded electron pairs, then the H-N-H bond angle is somewhat less than 109.5°, in fact it is 106.0°. The water molecule has two bonding electron pairs and two non-bonded electron pairs; hence, we postulate that the H-O-H angle will be compressed even more than in ammonia; in fact it is 104.0°, significantly less than the H-N-H angle.

E7.5 Use the VSEPR method to predict the geometries of the molecules in Table 7.3.

Solution

The molecules BeH_2 and MgH_2 both have two bonding electron pairs about the central atom; hence, their geometry should be linear.

The molecules BH_2 and AlH_2 both have two bonding electron pairs and a single nonbonding electron around the central atom. If we had a full nonbonded electron pair on the central atom along with the two bonded pairs, we would predict a bent geometry with a bond angle somewhat less than 120°. Since the nonbonded electron is unpaired, we predict a bond angle near 120°. Note that the Al atom is somewhat larger than the B atom and we expect the AlH_2 bond angle to be smaller than that of NH_2, because the electrostatic repulsion between the bonding pairs is smaller for the larger atomic cores.

The molecules CH_2 and SiH_2 both have two bonding electron pairs and a nonbonded electron pair about the central atom. Hence we would predict a bent geometry that has a bond angle somewhat less than 120°. Note that the Si atom is somewhat larger than the C atom and we expect the SiH_2 bond angle to be smaller than that of H_2O because the electrostatic repulsion between the bonding pairs is smaller for the larger atomic cores.

The molecules NH_2 and PH_2 both have two bonding electron pairs, a nonbonding electron pair, and an unpaired electron in their valence shell. If we had two full nonbonded electron pairs on the central atom along with the two bonded pairs (e.g. water), we would predict a bent geometry with a bond angle that is somewhat less than 109.5°. Since we have one nonbonded pair and an unpaired electron we predict that the bond angle will not be quite as small and will lie closer to 109.5° than the corresponding case of water. Note that the P atom is somewhat larger than the N atom and we expect the H_2P bond angle to be larger than that of H_2N.

The molecules H_2O and H_2S each have two bonded electron pairs and two nonbonded electron pairs about the central atom; hence, we predict a bent geometry with a bond angle that is somewhat less than 109.5°. Note that the S atom is somewhat larger than the O atom and we expect the H_2S bond angle to be larger than that of H_2O.

Caveat: These arguments are given in the context of the VSEPR model, which is limited in a number of ways. See the text, especially with regard to Table 7.3 for an orbital model.

E7.6 Consider Equation (7.1), which is a linear combination of two wavefunctions that are each solutions to a Schrödinger equation. Show that if these two wavefunctions correspond to the same energy for that Schrödinger equation then any linear combination of them is a solution to that Schrödinger equation also. Show that this does not hold for a linear combination of wavefunctions belonging to different energies.

Solution

The Schrödinger equation is fulfilled if

$$\mathcal{H}(a\psi_{12} + b\psi_{21}) = E \cdot (a\psi_{12} + b\psi_{21})$$

We are told that ψ_{12} and ψ_{21} are each solutions to the Schrödinger equation, namely

$$\mathcal{H}\ \psi_{12} = E_{12}\ \psi_{12} \text{ and } \mathcal{H}\ \psi_{21} = E_{21}\ \psi_{21}$$

Hence, the superposition wavefunction gives

$$\mathcal{H}(a\psi_{12} + b\psi_{21}) = aE_{12}\psi_{12} + bE_{21}\psi_{21}$$

This means that in the general case ($E_{21} \neq E_{12}$) the trial function does not satisfy the Schrödinger equation, because

$$E \cdot (a\psi_{12} + b\psi_{21}) \neq aE_{12}\psi_{12} + bE_{21}\psi_{21}$$

Only in the special case that the energy levels are twofold degenerate ($E_{21} = E_{12} = E$) we find

$$\mathcal{H}(a\psi_{12} + b\psi_{21}) = E(a\psi_{12} + b\psi_{21})$$

in accordance with the requirements of the Schrödinger equation.

E7.7 Show that the wavefunctions in Equations (7.2) and (7.3) form an orthonormal set of hybrid wavefunctions.

Solution

The hybrid wavefunctions (7.2) and (7.3) are

$$\psi_{\text{I}} = a \cdot \psi_{p_x} + \sqrt{1 - a^2} \cdot \psi_{p_y}$$

and

$$\psi_{\text{II}} = \sqrt{1 - a^2} \cdot \psi_{p_x} - a \cdot \psi_{p_y}$$

To show orthonormality, we need to show that $\int \psi_i \psi_j \cdot \mathrm{d}\tau = \delta_{ij}$ where $\delta_{ij} = 1$ if $i = j$ and $\delta_{ij} = 0$ if $i \neq j$. To show that they are normalized we find

$$\begin{aligned}\int \psi_{\text{I}}^2 \mathrm{d}\tau &= a^2 \left(\int \psi_{p_x}^2 \mathrm{d}\tau\right) + 2a\sqrt{1-a^2}\left(\int \psi_{p_x}\psi_{p_y} \mathrm{d}\tau\right) + (1-a^2)\left(\int \psi_{p_y}^2 \mathrm{d}\tau\right)\\ &= a^2(1) + a\sqrt{1-a^2}(0) + (1-a^2)(1)\\ &= 1\end{aligned}$$

and

$$\begin{aligned}\int \psi_{\text{II}}^2 \mathrm{d}\tau &= (1-a^2)\left(\int \psi_{p_x}^2 \mathrm{d}\tau\right) - 2a\sqrt{1-a^2}\left(\int \psi_{p_x}\psi_{p_y} \mathrm{d}\tau\right) + a^2\left(\int \psi_{p_y}^2 \mathrm{d}\tau\right)\\ &= (1-a^2)(1) - a\sqrt{1-a^2}(0) + a^2(1)\\ &= 1\end{aligned}$$

To show that they are orthogonal, we find

$$\begin{aligned}\int \psi_{\text{I}}\psi_{\text{II}} \mathrm{d}\tau &= a\sqrt{1-a^2}\left(\int \psi_{p_x}^2 \mathrm{d}\tau\right) + (1-2a^2)\left(\int \psi_{p_x}\psi_{p_y} \mathrm{d}\tau\right)\\ &\quad - a\sqrt{1-a^2}\left(\int \psi_{p_y}^2 \mathrm{d}\tau\right)\\ &= a\sqrt{1-a^2}(1) + (1-2a^2)(0) - a\sqrt{1-a^2}(1)\\ &= 0\end{aligned}$$

Thus the wavefunctions (7.2) and (7.3) are orthogonal.

E7.8 In the Example 7.4 we considered the hybrid function

$$\psi = a\,\psi_{p_x} + b\,\psi_{p_y}$$

for which an infinite number of functions are possible, depending on the choice of a and b. By requiring that the two hybrid functions be orthogonal and normalized, show that we obtain

$$\psi_1 = a_1\,\psi_{p_x} + \sqrt{1-a_1^2}\,\psi_{p_y}, \quad \psi_2 = -\sqrt{1-a_1^2}\,\psi_{p_x} + a_1\,\psi_{p_y}$$

Solution

We require that any two functions in this set are normalized and orthogonal to each other. Let us write one possible combination as $a = a_1$, $b = b_1$ and another as $a = a_2$ and $b = b_2$, then we find

$$\psi_1 = a_1\psi_{p_x} + b_1\psi_{p_y}, \quad \psi_2 = a_2\psi_{p_x} + b_2\psi_{p_y}$$

First we normalize ψ_1 and find

$$\begin{aligned}\int \psi_1^2 \mathrm{d}\tau &= \int a_1^2\psi_{p_x}^2 \mathrm{d}\tau + 2\int a_1\psi_{p_x} b_1\psi_{p_y} \mathrm{d}\tau + \int b_1^2\psi_{p_y}^2 \mathrm{d}\tau \\ &= a_1^2 \int \psi_{p_x}^2 \mathrm{d}\tau + 2a_1b_1 \int \psi_{p_x}\psi_{p_y} \mathrm{d}\tau + b_1^2 \int \psi_{p_y}^2 \mathrm{d} \\ &= a_1^2 + 0 + b_1^2 = 1\end{aligned}$$

In the last step we used the normalization and the orthogonality of the functions ψ_{p_x} and ψ_{p_y}. Thus we obtain

$$a_1^2 + b_1^2 = 1$$

or

$$b_1 = \sqrt{1-a_1^2}$$

In the same manner, for b_2 we obtain

$$b_2 = \sqrt{1-a_2^2}$$

Inserting these results into our initial general form for the hybrid functions, we write

$$\psi_1 = a_1\psi_{p_x} + \sqrt{1-a_1^2}\psi_{p_y}, \quad \psi_2 = a_2\psi_{p_x} + \sqrt{1-a_2^2}\psi_{p_y}$$

Now we require the functions in the set to be orthogonal to each other, namely

$$\int \psi_1\psi_2 \cdot \mathrm{d}\tau = 0$$

Thus

$$\begin{aligned}&\int \left(a_1\psi_{p_x} + \sqrt{1-a_1^2}\psi_{p_y}\right)\left(a_2\psi_{p_x} + \sqrt{1-a_2^2}\psi_{p_y}\right) \cdot \mathrm{d}\tau \\ &= \int a_1\psi_{p_x} a_2\psi_{p_x} \cdot \mathrm{d}\tau + \int a_1\psi_{p_x} \sqrt{1-a_2^2}\psi_{p_y} \cdot \mathrm{d}\tau \\ &\quad + \int \sqrt{1-a_1^2}\psi_{p_y} a_2\psi_{p_x} \cdot \mathrm{d}\tau + \int \sqrt{1-a_1^2}\psi_{p_y} \sqrt{1-a_2^2}\psi_{p_y} \cdot \mathrm{d}\tau \\ &= a_1a_2 \int \psi_{p_x}^2 \cdot \mathrm{d}\tau + a_1\sqrt{1-a_2^2} \int \psi_{p_x}\psi_{p_y} \cdot \mathrm{d}\tau \\ &\quad + \sqrt{1-a_1^2}a_2 \int \psi_{p_y}\psi_{p_x} \cdot \mathrm{d}\tau + \sqrt{1-a_1^2}\sqrt{1-a_2^2} \int \psi_{p_y}\psi_{p_y} \cdot \mathrm{d}\tau \\ &= a_1a_2 + \sqrt{1-a_1^2}\sqrt{1-a_2^2} = 0\end{aligned}$$

leading to the relation

$$a_1a_2 = -\sqrt{\left(1-a_1^2\right)\left(1-a_2^2\right)}$$

We solve this equation for a_2.

$$a_1^2a_2^2 = \left(1-a_1^2\right)\left(1-a_2^2\right) = 1 - a_2^2 - a_1^2 + a_1^2a_2^2$$

leading to

$$1 - a_2^2 - a_1^2 = 0$$

or

$$a_2^2 = 1 - a_1^2, \quad a_2 = \pm\sqrt{1 - a_1^2}$$

Because the product of a_1 and a_2 is negative, they must be opposite in sign and we take the negative root

$$a_2 = -\sqrt{1 - a_1^2} \text{ for } a_1 > 0$$

Substituting into our expression for the hybrid functions we can eliminate a_2 and find that

$$\psi_1 = a_1\psi_{p_x} + \sqrt{1 - a_1^2}\psi_{p_y}, \quad \psi_2 = -\sqrt{1 - a_1^2}\psi_{p_x} + a_1\psi_{p_y}$$

E7.9 Use hybridization ideas to predict the structure of $B(CH_3)_3$ and Hg(phenyl)$_2$, and O(phenyl)$_2$. For the case of Hg, only consider the $n = 6$ valence electrons.

Solution

For the molecule $B(CH_3)_3$, one electron is promoted on the central B atom from the 2*s* orbital into the 2*p* orbital; formation of a trigonal planar hybrid orbital results. Then each orbital of B and each CH_3 provide one electron to form three-electron pair bonds. Tetrahedral hybridization would require more promotion energy to form three $s^{1/4}p^{3/4}$ hybrid orbitals.

In Hg, one electron is promoted from the 6*s* orbital into the 6*p* orbital; formation of a linear hybrid orbital results. Thus Hg(phenyl)$_2$ should be linear about the Hg atom. In the case of the O(phenyl)$_2$, the O atom has six valence electrons and no empty 2*p* orbitals. Hybridization of the O atom to place two sets of lone pair electrons in an sp^3 hybrid orbital will allow its other two sp^3 hybrids available to make a covalent bond with the phenyl groups and O(phenyl)$_2$ should have a bent structure about the O similar to that of H_2O.

E7.10 Qualitatively describe the nature of the bonding in amine oxide $(H_3)\overset{\oplus}{N} - \overset{\ominus}{\underline{\overline{O}}}|$.

Solution

The nonbonded electron pair in NH_3(donor) is donated to form a dative bond with an unoccupied 2*p* orbital of the O (acceptor). The N in NH_3 has a formal positive charge, the O has a formal negative charge. In this case the nitrogen atom donates both electrons to form the bond and lowers the overall energy of the system.

E7.11 Why are the energy levels of the linear combinations of the $2p_x$ orbitals of the two atoms ($\sigma(2p_x)$ and $\sigma^*(2p_x)$) in Fig. 7.11 more strongly separated than those of the linear combinations of the $2p_y$ and $2p_z$ orbitals (π and π^* orbitals)?

Solution

The numbering of the energy levels in Fig. 7.15 (same system as that of Fig. 7.11) is used for our description here. In σ orbital 3 there are antinodes of the wavefunction along the bond axis; in π orbitals 1 and 2, the antinodes are at some distance from the bond axis. Thus the attraction of the electrons in σ orbital 3 to the nuclei increases more strongly with increasing nuclear charge than does that for the π orbitals, leading to a stronger decrease in energy for σ orbital 3. Also the overlap of the electrons for σ orbital 3 is higher than that for the π orbitals 1 and 2.

E7.12 Make a molecular orbital energy diagram for the oxygen molecule in which the atomic orbitals used are considered to be the two equal *sp* hybrids, the p_y, and the p_z. Compare this energy diagram and its predictions to that considered in the text.

Solution

Figure E7.12 shows the energy level diagram for an oxygen molecule constructed from two *sp* hybrid orbitals and the p_y and p_z orbitals on each oxygen atom. In this scheme the *sp*-hybrid orbitals on one of the oxygen atoms that is oriented away from the other oxygen atom is considered to be non-bonding and is labeled σ_{nb}. Assuming that the bonding (σ_{sp}) and anti-bonding (σ_{sp}^*) interactions between the *sp*-hybrids are much larger than those between the p-orbitals on the O-atom sites gives the molecular energy diagram in Fig. E7.12. Thus, the molecule has six bonding electrons, two antibonding electrons, and four nonbonding electrons which gives it a bond order of two. Note that the degenerate HOMO orbitals each contain an unpaired electron, which agrees with the paramagnetism of the oxygen molecule.

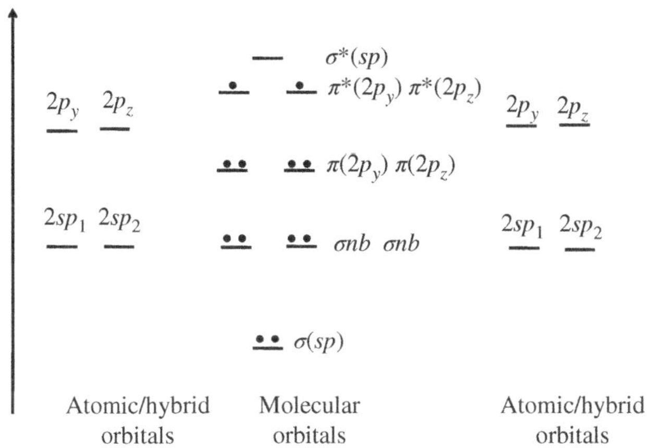

Figure E7.12 LCAO scheme of the O_2 molecule in which *sp* hybrids are used as basis orbitals on the O atoms. The subscript 'nb' is an abbreviation for non-bonding.

E7.13 Provide the molecular orbital energy diagrams for the three species C_2^+, NO, and Ne_2^+. Write out the electronic configurations for these diatomic molecules. Determine the bond order in each case.

Solution

The molecular orbital diagrams for C_2^+, NO, and Ne_2^+ are shown in Fig. E7.13. The electronic configuration for C_2^+ is $1\sigma^2 2\sigma^{*2} 1\pi^3$; that for NO is $1\sigma^2 2\sigma^{*2} 3\sigma^2 1\pi^4 2\pi^{*1}$; and that for Ne_2^+ is $1\sigma^2 2\sigma^{*2} 3\sigma^2 1\pi^4 2\pi^{*4} 4\sigma^{*1}$. The bond orders are given by

$$\frac{\text{number of bonding e}^- - \text{number of antibonding e}^-}{2}$$

In the case of C_2^+, we find

$$\frac{3-0}{2} = 1.5$$

Correspondingly, we find 2.5 for NO and 0.5 for Ne_2^+.

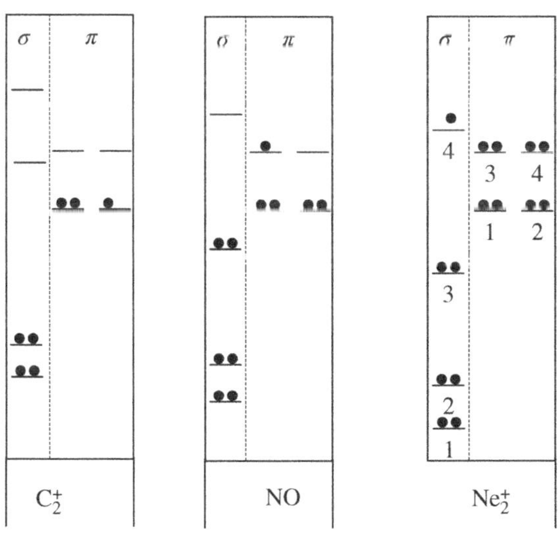

Figure E7.13 LCAO molecular orbital energy scheme for C_2^+, NO, and Ne_2^+.

E7.14 Draw a molecular orbital energy diagram for N_2^+. Use the valence shell electrons only (i.e. ignore the 1*s* electrons of each atom). Label the orbitals in your diagram. Using your diagram, determine the bond order of N_2^+. Write the electronic configuration of the ground state of N_2^+. Promote the most weakly bound electron to the next highest molecular orbital. What is the ion's bond order in this excited state configuration?

Solution

The molecular orbital energy diagram for the ground state of N_2^+ is shown in Fig. E7.14 and the orbitals are labeled by their designation. The 1σ and $2\sigma^*$ orbitals are the bonding and antibonding molecular orbitals formed by the 2*s* orbitals centered on each N atom, and the 3σ orbital is the bonding orbital formed by the $2p_x$ atomic orbitals centered on each N. The bond order for N_2^+ is thus

$$\frac{\text{number of bonding e}^- - \text{number of antibonding e}^-}{2} = \frac{5-0}{2} = 2.5$$

The complete electron configuration for the ground state is $1\sigma^2 2\sigma^{*2} 1\pi^4 3\sigma^1$. The excited state configuration is $1\sigma^2 2\sigma^{*2} 1\pi^4 2\pi^{*1}$, which has a bond order of

$$\frac{\text{number of bonding e}^- - \text{number of antibonding e}^-}{2} = \frac{4-1}{2} = 1.5$$

Note that we have neglected the 1*s* orbitals on each atom because their interaction is quite small.

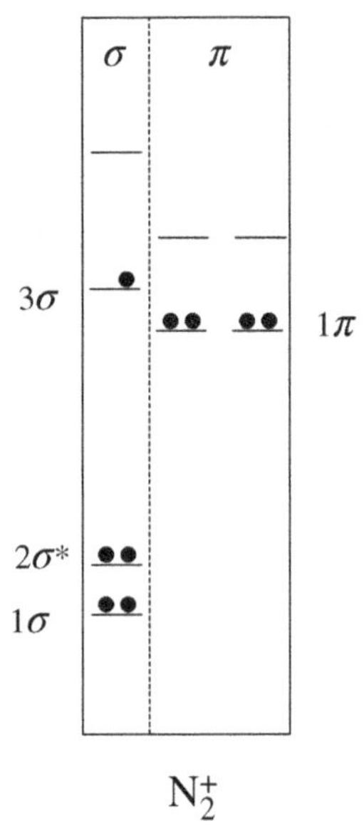

Figure E7.14 LCAO scheme for N_2^+..

E7.15 Draw the molecular orbital energy diagrams for the three molecules C_2, N_2, and O_2 and use them to explain the variation in their bond energy. Sketch the first σ^* antibonding orbital and one of the π bonding orbitals. The dissociation energies are 6.2 eV for C_2, 9.8 eV for N_2, and 5.1 eV for O_2.

Solution

The molecular orbital energy diagrams are shown in Fig. 7.15 of the text. Using this figure, the bond orders of C_2, N_2, and O_2 are found to be 2, 3, and 2, respectively. Thus the bond energy for N_2 should be greater than the other two molecules, which will have roughly equal bond energies. C_2 will have a greater bond energy than O_2 because it does not have the energy associated with the unpaired electrons. A sketch of the orbitals for the σ^* and π orbitals in any of these three molecules is approximated by the orbitals shown in Fig. E7.15.

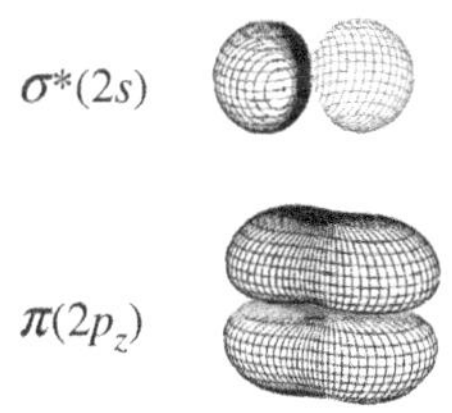

Figure E7.15 The renderings correspond to the $\pi(2p_z)$ orbital (lower image) and the $\sigma^*(2s)$ orbital upper image.

E7.16 The dipole moment of an HCl molecule is 3.4×10^{-30} C m = 1.03 Debye. Assume that the dipole moment can be represented by two partial charges, one on the H atom and one on the Cl atom, at a distance corresponding to the bond length 127 pm. Determine the magnitude of the partial charges.

Solution

We begin with the unnumbered first equation in Section 7.5.1, $\mu = Q \cdot a$, where Q is the partial charge and a is the distance between the charges. Rearrangement and substitution in the equation gives the partial charge as

$$Q = \frac{\mu}{d} = \frac{3.4 \times 10^{-30}\ \text{C m}}{127 \times 10^{-12}\ \text{m}} = 0.27 \times 10^{-19}\ \text{C} = 0.17 \cdot e$$

This is about 20% of the charge of an electron

E7.17 Explain the trend in the electronegativities of atoms in the periodic table (see Table 7.5) in terms of the trends in the atoms' electronic properties.

Solution

As first defined by Pauling, the electronegativity provides a measure of an atom's electron attracting power when it forms a bond. Here we use the Mulliken definition of the electronegativity as the average of an atom's ionization energy and its electron affinity. The ionization energy increases as you move to the right across a row of the periodic table because of the increased nuclear charge and the outermost electron having the same value of the principal quantum number, n. The electron affinity is a measure of the energy associated with an atom or ion gaining a single electron, and it also increases as one moves to the right across the periodic table (with the exception of the noble gases). For a review of electronegativity data see.[1] Thus the electronegativity increases as you move to the right across the periodic table in a given row. Because the ionization energies and the electron affinities both decrease as you move down the periodic table (because of the increasing principal quantum number, n, and the same effective nuclear charge), the electronegativity will also decrease as you move down a column in the periodic table.

E7.18 Consider the molecular orbital energy diagram for formaldehyde (Fig. 7.24). Draw this diagram in a way that shows the origin of the molecular orbitals from the two fragments C=O and the two H atoms, considered as a single unit.

Solution

The molecular orbital energy diagram drawn in Figure E7.18 considers the CO fragment to have the molecular orbital structure of a diatomic molecule (for homonuclear analogs see Figure 7.12) and to combine with symmetric and antisymmetric combination of the 1s orbitals on the two H atoms. The dashed lines are meant to indicate how the fragment orbitals correspond to the molecular orbitals. To simplify we only show the molecular orbitals that are filled.

Our convention is to choose the CO bond axis as the x-axis and the out-of-plane axis as the z-axis. In this case we see by symmetry that the π-orbitals on the CO fragment, originating from the p_z atomic orbitals, will be non-bonding and will correlate with the molecular orbital 5 of the formaldehyde (Fig. 7.12). In contrast, the π-orbitals that originate from the p_y atomic orbitals lie in the molecular plane and interact with the 1s orbitals of the H atoms. These orbitals form bonding (molecular orbital 3) and antibonding (molecular orbital 6) combinations with the antisymmetric superposition of the 1s orbitals of the H-atoms.

1 Hotop and Lineberger, *J. Phys. Chem. Ref. Data* 14, 731 (1985).

The other three molecular orbitals of the formaldehyde are formed from the interaction of the symmetric superposition of the H-atom 1*s* orbitals with the sigma orbitals of the CO fragment. The lowest energy 1σ orbital of the CO fragment forms a bonding interaction with the 1*s* orbitals of the H-atoms to generate molecular orbital 1. We find a somewhat weaker bonding interaction with the 2σ* fragment orbital and a somewhat antibonding interaction with the 3σ orbital fragment. The identification of each formaldehyde molecular orbital with a single CO fragment orbital is an oversimplification that is used here to show the methodology. In an actual calculation these latter three molecular orbitals of the CO are likely to have contributions from each of the different orbitals, rather than interact pairwise.

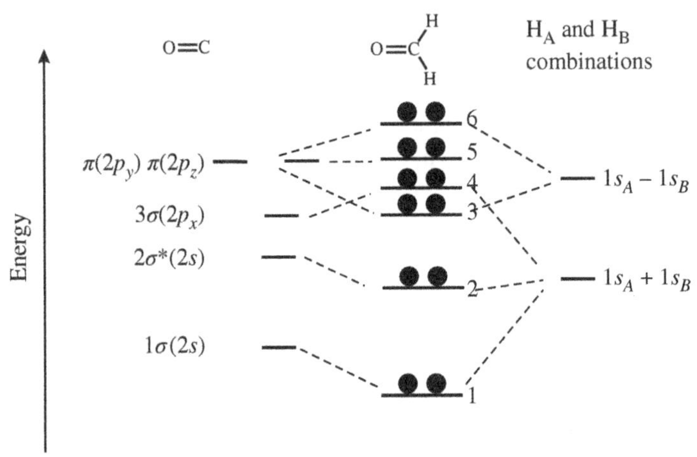

Figure E7.18 LCAO molecular orbital energy scheme for the formaldehyde molecule.

E7.19 Use the data in Table 7.6 to compute the force required to stretch the X–H bond in the homologous series of HF, HCl, and HI by a similar displacement *x*.

Solution

The equation to be used is

$$f = k_f \cdot x$$

where k_f is the force constant and x is the bond displacement as it is stretched. For HF we have $k_f = 966\ \text{N m}^{-1}$, and if we stretch the bond by 0.5 pm, the force will be

$$\begin{aligned} f &= 966\ \text{N m}^{-1} \times 0.5 \times 10^{-12}\ \text{m} \\ &= 4.83 \times 10^{-10} \frac{\text{m}}{\text{s}^2}\ \text{kg} \\ &= 4.83 \times 10^{-10}\ \text{N} = 0.483\ \text{nN} \end{aligned}$$

For HCl we find $f_{\text{HCl}} = 0.258$ nN and for HI we find $f_{\text{HI}} = 0.156$ nN.

E7.20 Given that the bending force constant for H_2O is 185 N m^{-1}, that for H_2S is 94 N m^{-1}, and that for H_2Se is 63 N m^{-1} compute the change in H atom displacement when a one nanoNewton force is applied.

Solution

The calculation in this exercise is virtually identical to that given under equation (7.15) namely

$$x = d_{\text{eq}} \cdot \Delta\alpha = \frac{f}{k_f'}.$$

The following table collects the data for the molecules in this exercise.

Molecule	$k_{f_r}/(\mathrm{N\,m^{-1}})$	x/pm
H_2O	185	5.41
H_2S	94	10.64
H_2Se	63	15.87

These displacements are 5% to 10% of the bond length in these molecules.

E7.21 Estimate the force needed to overcome the barrier to torsional motion of the methyl groups in ethane if the barrier height is 12 kJ mol^{-1}. Perform a similar calculation for the torsion of the OH group in methanol using a 4.5 kJ mol^{-1} barrier. In each case assume that the potential energy varies sinusoidally with torsion angle and is three-fold symmetric for the methyl rotation and two-fold symmetric for the OH rotation.

Solution

First we consider the methyl torsional potential, which we write as

$$V(\alpha) = \frac{1}{2}V_0 \cdot \sin(3\cdot\alpha - \pi/2) + \frac{1}{2}V_0$$
$$= \frac{1}{2}V_0 \cdot (1 - \cos(3\cdot\alpha))$$

The barrier height $V_0 = 1.99 \times 10^{-20}$ J is calculated from the 12 kJ/mol value reported in the problem statement. A plot of this function is shown in Fig. E7.21a.

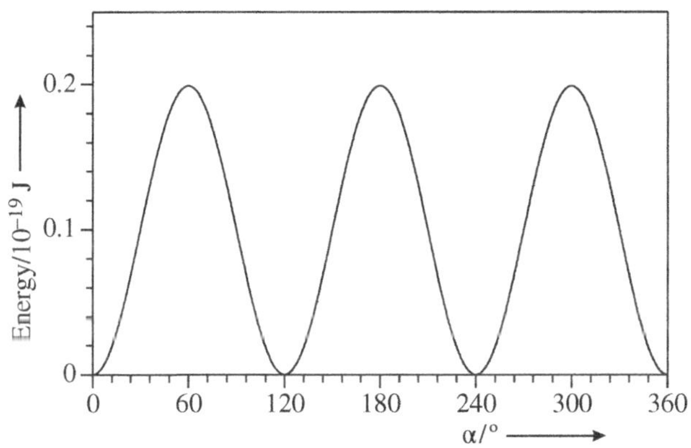

Figure E7.21a Torsion energy versus the torsion angle α for the methyl groups in ethane.

We consider the maximum force that must be applied to move over the barrier in the range of $0 \le \alpha \le 60°$. From the geometry illustrated by Fig. E7.21b we obtain $r_{eq} = d_{CH} \cdot \sin(70.5°) = 101.8$ pm. Hence the circumference of the circle is 639.7 pm, or 101.8 pm/radian.

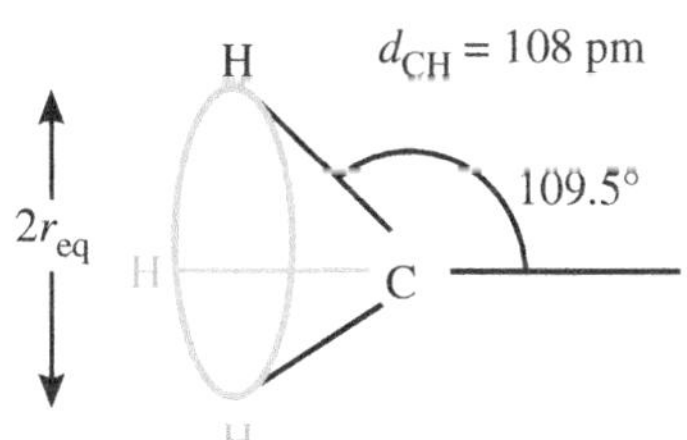

Figure E7.21b Geometry of the methyl group in ethane.

Now we use this geometry to write the potential energy in terms of the distance s that is traversed, namely

$$s = r_{eq} \cdot \alpha$$

This allows us to write the potential energy as

$$V(s) = \frac{1}{2}V_0 \cdot \left(1 - \cos\left(\frac{3 \cdot s}{r_{\mathrm{eq}}}\right)\right)$$

Thus, the restoring force f is given by

$$f = -\frac{\mathrm{d}V}{\mathrm{d}s} = -\frac{3}{2}\frac{V_0}{r_{\mathrm{eq}}} \cdot \sin\left(\frac{3 \cdot s}{r_{\mathrm{eq}}}\right)$$

By the symmetry of the sine function we can infer that the inflection point occurs at $\pi/6$, so that the maximum force is

$$f_{\max} = -\left.\frac{\mathrm{d}V}{\mathrm{d}\alpha}\right|_{\alpha=\pi/6} = -\frac{3}{2}\frac{V_0}{r_{\mathrm{eq}}}\frac{\mathrm{kJ/mol}}{\mathrm{pm}}$$

(NOTE: You can verify that the inflection occurs at $\pi/6$ by taking the derivative of the force, which is the second derivative of the potential energy, and setting it equal to zero). The value of $f_{\max}$ is

$$\begin{aligned} f_{\max} &= -\frac{3}{2} \cdot \frac{1.99 \times 10^{-20}\ \mathrm{J}}{101.8\ \mathrm{pm}} \\ &= -0.29 \times 10^{-9}\ \mathrm{N} = -0.29\ \mathrm{nN} \end{aligned}$$

We can proceed in a corresponding manner for the OH torsional potential energy, which we write as

$$V(\alpha) = \frac{V_0}{2} \cdot (1 - \cos(2 \cdot \alpha))$$

The barrier height $V_0 = 7.47 \times 10^{-21}$ J is calculated from the 4.5 kJ/mol value reported in the problem statement. Using methanol geometric parameters, we obtain $r_{eq} = d_{\mathrm{OH}} \cdot \sin(75^{\circ}) = 90.5$ pm (where $d_{\mathrm{OH}} = 93.7$ pm and a HOC bond angle of $105^{\circ}56'$ is used). Hence the circumference of the circle is 568.7 pm, or 90.5 pm/radian. The value of $f_{\max}$ becomes

$$\begin{aligned} f_{\max} &= \frac{V_0}{r_{\mathrm{eq}}} = \frac{7.47 \times 10^{-21}\ \mathrm{J}}{90.5\ \mathrm{pm}} \\ &= 0.083 \times 10^{-9}\ \mathrm{N} = 0.083\ \mathrm{nN} \end{aligned}$$

E7.22 For small displacements $x = d - d_{\mathrm{eq}}$, show that the a parameter in the Morse function

$$\varepsilon - \varepsilon_{\mathrm{eq}} = \varepsilon_{\mathrm{B}} \cdot \left(1 - \mathrm{e}^{-a \cdot (d - d_{\mathrm{eq}})}\right)^2$$

is given by $a = \sqrt{k_f/(2\varepsilon_{\mathrm{B}})}$, where k_f and ε_{B} are the parameters in the harmonic limit (see Section 7.5.3.2).

$$\Delta\varepsilon = \varepsilon - \varepsilon_{\mathrm{eq}} = \int_0^x f \cdot \mathrm{d}x = \int_0^x k_f x \cdot \mathrm{d}x = \frac{1}{2}k_f \cdot x^2$$

Solution

We start with the form of the Morse function

$$y = \varepsilon_{\mathrm{B}} \cdot \left(1 - \mathrm{e}^{-a \cdot x}\right)^2$$

with $x = d - d_{\mathrm{eq}}$ and $y = \varepsilon - \varepsilon_{\mathrm{eq}}$. For small enough x we can write the exponential function as

$$\mathrm{e}^{-a \cdot x} = 1 - a \cdot x$$

and for y we obtain

$$y = \varepsilon_{\mathrm{B}} \cdot (1 - 1 + a \cdot x)^2 = \varepsilon_{\mathrm{B}} \cdot a^2 \cdot x^2$$

Because

$$a = \sqrt{\frac{k_f}{2\varepsilon_{\mathrm{B}}}}$$

we find that

$$y = \varepsilon_B \cdot \frac{k_f}{2\varepsilon_B} \cdot x^2 = \frac{k_f}{2} \cdot x^2$$

E7.23 Using the Morse potential, derive the expression for f_{max} in Equation (7.13). Explain why the inflection point of the potential provides the maximum force.

Solution

Mathematically, the inflection point for the potential energy function corresponds to the point where the second derivative of the energy is zero. Because the first derivative of potential energy is the force, it follows that the inflection point corresponds to the point where the first derivative of the force is equal to zero. The points where the first derivative of the force is equal to zero are the extrema of the force function. Thus, the maximum in the force function occurs when the second derivative of the energy is zero. To find the expression for the maximum force f_{max}, we will first evaluate the inflection point for the Morse potential and then we will use it in the force function to find f_{max}.

The point of inflection occurs where the second derivative of a function changes sign (or alternatively where it equals zero). The Morse potential is given by

$$\varepsilon - \varepsilon_{eq} = \varepsilon_B \cdot \left(1 - e^{-\sqrt{k_f/(2\varepsilon_B)}\cdot(d-d_{eq})}\right)^2$$

To simplify the notation, we define $x = d - d_{eq}$ so that $dx = dd$; and we let $a = \sqrt{k_f/(2\varepsilon_B)}$. Thus we can write the Morse potential as

$$\varepsilon - \varepsilon_{eq} = \varepsilon_B \cdot \left(1 - e^{-ax}\right)^2$$

The first and second derivatives of ε with respect to x are then

$$\frac{d\varepsilon}{dx} = \varepsilon_B \cdot 2 \cdot \left(1 - e^{-ax}\right) \cdot \left(-e^{-ax}\right)(-a) = 2a \cdot \varepsilon_B \cdot \left(e^{-ax} - e^{-2ax}\right)$$

and

$$\frac{d^2\varepsilon}{dx^2} = 2a \cdot \varepsilon_B \cdot \left(e^{-ax}(-a) - e^{-2ax}(-2a)\right) = -2a^2 \cdot \varepsilon_B \cdot e^{-ax}\left(1 - 2e^{-ax}\right)$$

The points where the second derivative changes sign occur when

$$0 = \frac{d^2\varepsilon}{dx^2} = -2a^2\varepsilon_B \cdot e^{-ax} \cdot \left(1 - 2e^{-ax}\right) \quad \text{or} \quad e^{-ax}\left(1 - 2e^{-ax}\right) = 0$$

which are at $x = \infty$ and x such that

$$1 - 2e^{-ax}\big|_{x=x_{max}} = 0 \quad \text{or} \quad x_{max} = \frac{\ln 2}{a}$$

Upon substitution of the value of a from above we find that

$$x_{max} = \ln 2\sqrt{\frac{2\varepsilon_B}{k_f}}$$

The force f is defined by

$$f = -\left(\frac{d\varepsilon}{dx}\right)$$

which on substitution of the derivative found above is

$$f = -\frac{d\varepsilon}{dx} = -2a \cdot \varepsilon_B \cdot e^{-ax}\left(1 - e^{-ax}\right)$$

Using our expression for $x_{max} = \frac{\ln 2}{a}$, we find that

$$\begin{aligned} f_{max} &= f(x_{max}) = -2a \cdot \varepsilon_B \cdot e^{-a\cdot x_{max}}\left(1 - e^{-a\cdot x_{max}}\right) \\ &= -2a \cdot \varepsilon_B \cdot e^{-\ln 2}\left(1 - e^{-\ln 2}\right) \\ &= -2a \cdot \varepsilon_B \cdot \frac{1}{2}\left(1 - \frac{1}{2}\right) = -a \cdot \varepsilon_B \cdot \frac{1}{2} \end{aligned}$$

Upon substitution of the expression $a = \sqrt{k_f/(2\varepsilon_B)}$, we find that

$$f_{\max} = -\frac{1}{2} \cdot \varepsilon_B \cdot \sqrt{\frac{k_f}{2\varepsilon_B}} = -\frac{1}{2}\sqrt{\frac{k_f \varepsilon_B}{2}}$$

7.2 Problems

P7.1 Plot the radial wavefunctions for the 1*s*, 2*s*, and 2*p* orbitals on oxygen, using the Slater wavefunctions in Box 5.4. Consider two oxygen atoms at a distance *d* apart. What average distance would be needed to have an overlap integral that is $S_{ab} = 0.3$ for the 1*s* orbital (see Foundation 6.2)? Explain how these findings impact the role of the 1*s* electrons in bonding.

Solution

Both the 1*s* and 2*p* Slater wavefunctions have the functional form of their H-atom analogues, but with an effective nuclear charge; hence, we find that

$$\psi_{1s,\text{Slater}} = A\ \mathrm{e}^{-rZ_{\text{eff}}/a_0} \text{ and } \psi_{2p,\text{Slater}} = A \cdot \mathrm{e}^{-rZ_{\text{eff}}/(2a_0)}$$

The 2*s* radial wavefunction is not hydrogen-like, but is given by

$$\psi_{2s,\text{Slater}} = A \cdot r \cdot \mathrm{e}^{-rZ_{\text{eff}}/(2a_0)}$$

Now we follow the methodology specified in the Box 5.4 to determine the effective charges in each case. For the 1*s* wavefunction, we have $Z_{\text{eff}} = 8 - 0.30 = 7.70$ so that

$$\psi_{1s,\text{Slater}} = A\ \mathrm{e}^{-7.70\,r/a_0}$$

Applying the normalization condition

$$1 = \int \psi^2_{1s,\text{Slater}} \mathrm{d}\tau = A^2\ 4\pi \int_0^\infty \mathrm{e}^{-15.40\,r/a_0}\ r^2 \mathrm{d}r$$
$$= A^2\ 4\pi \left(\frac{a_0}{15.40}\right)^3 \left[\int_0^\infty \mathrm{e}^{-x}\ x^2 \mathrm{d}x\right] = A^2\ 4\pi \left(\frac{a_0}{15.40}\right)^3 [2!]$$

so that $A = 12.05 \cdot a_0^{-3/2}$. Hence

$$\psi_{1s,\text{Slater}} = \frac{12.05}{\sqrt{a_0^3}}\ \mathrm{e}^{-7.70\,r/a_0}$$

For the 2*p* wavefunction, we have $Z_{\text{eff}} = 8 - 5 \cdot 0.35 - 2 \cdot 0.85 = 4.55$ so that

$$\psi_{2p,\text{Slater}} = A \cdot \mathrm{e}^{-2.275r/a_0}$$

The normalization procedure gives

$$\psi_{2p,\text{Slater}} = \frac{1.936}{\sqrt{a_0^3}} \cdot \mathrm{e}^{-2.275r/a_0}$$

Similarly for the 2*s* wavefunction, we find

$$\psi_{2s,\text{Slater}} = \frac{2.543}{\sqrt{a_0^3}} \cdot \frac{r}{a_0} \cdot \mathrm{e}^{-2.275r/a_0}$$

Figure P7.1a plots these functions (black solid line is 1*s*, light grey solid line is 2*s*, and the dark grey line is 2*p*). The panel on the left shows the relative amplitude of the orbitals with distance over the entire ordinate range, and the right panel is expanded so that the $n = 2$ and $n = 1$ orbital functions can be better compared.

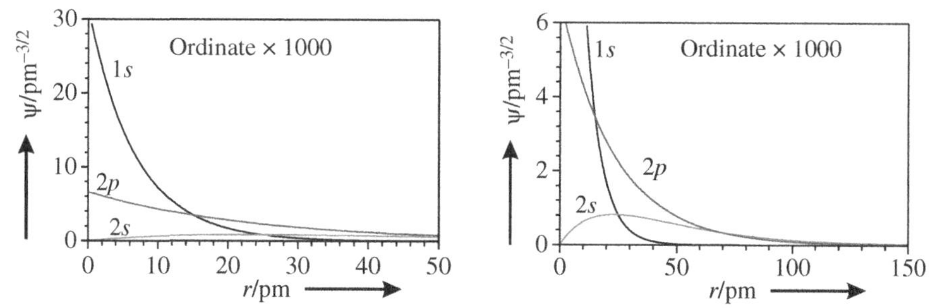

Figure P7.1a Slater functions1s, 2s, 2p. Note the different ordinate scaling and distance ranges in the two pictures.

From Foundation 6.2, we use the expression for the overlap integral S_{ab} with two 1s wavefunctions

$$S_{ab} = \left(1 + Z_{\text{eff}}\frac{d}{a_0} + \frac{1}{3}Z_{\text{eff}}^2\frac{d^2}{a_0^2}\right) \cdot \exp\left(-Z_{\text{eff}}d/a_0\right)$$

where d is the distance of the two nuclei. We solve this equation for d when $S_{ab} = 0.30$. To solve this transcendental equation we define the variable $y = Z_{\text{eff}}d/a_0$ to obtain

$$S_{ab} = \left(1 + y + \frac{1}{3}y^2\right) \cdot \exp(-y)$$

Figure P7.1b shows a plot of S_{ab} versus y.

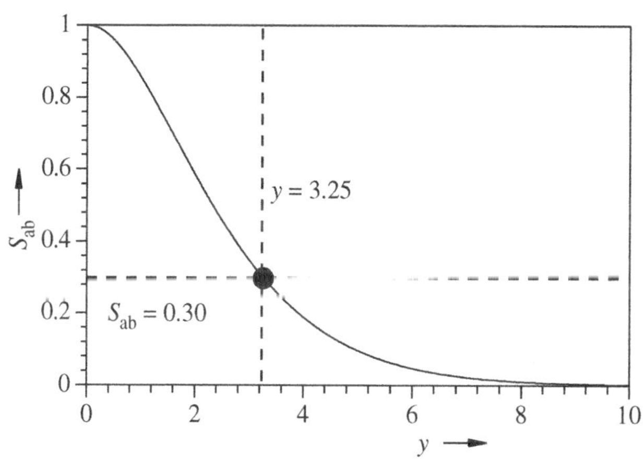

Figure P7.1b The overlap S_{ab} for the 1s function is plotted versus $y = Z_{\text{eff}}d/a_0$ with $Z_{\text{eff}} = 7.70$

From the plot we find that $y = 3.25$ is the solution for $S_{ab} = 0.30$. From this result we calculate d as

$$d = y\frac{a_0}{Z_{\text{eff}}} = 3.25\frac{52.9 \text{ pm}}{7.70} = 22.3 \text{ pm}$$

This value is only 40% of a Bohr radius and unrealistically small for an internuclear bond distance. For a more realistic bond distance of 121 pm (that for O_2), we find that the overlap integral has a value of 2.7×10^{-6}.

Thus, we conclude that the 1s orbital is so strongly contracted by the nuclear charge of the oxygen atoms it does not contribute significantly to the bonding, rather the bonding is dominated by the valence orbitals.

P7.2 Show that for a hybrid function $\phi = a\psi_{2s} + b\psi_{2p_x} + c\psi_{2p_y} + d\psi_{2p_z}$ the expression $\varepsilon_{\text{hybrid}} = E_{2s} + \left(b^2 + c^2 + d^2\right) \cdot \Delta E$ is obtained; calculate $\varepsilon_{\text{hybrid}}$ for the tetrahedral sp^3 hybrid state.

Solution

The energy of the hybrid state orbital is given by

$$\begin{aligned}
\varepsilon_{\text{hybrid}} &= \int \phi \mathcal{H} \phi \cdot d\tau \\
&= \int \begin{bmatrix} \left(a\psi_{2s} + b\psi_{2p_x} + c\psi_{2p_y} + d\psi_{2p_z}\right) \\ \cdot \mathcal{H}\left(a\psi_{2s} + b\psi_{2p_x} + c\psi_{2p_y} + d\psi_{2p_z}\right) \end{bmatrix} \cdot d\tau \\
&= \int \begin{bmatrix} \left(a\psi_{2s} + b\psi_{2p_x} + c\psi_{2p_y} + d\psi_{2p_z}\right) \\ \cdot \left(aE_{2s}\psi_{2s} + bE_{2p_x}\psi_{2p_x} + cE_{2p_y}\psi_{2p_y} + dE_{2p_z}\psi_{2p_z}\right) \end{bmatrix} \cdot d\tau \\
&= a^2 E_{2s} + b^2 E_{2p_x} + c^2 E_{2p_y} + d^2 E_{2p_z}
\end{aligned}$$

where we exploit the orthonormality of the H-atom orbitals in the last step. Since the *p*-orbitals have the same energy we can write that

$$E_{2p_x} = E_{2p_y} = E_{2p_z} = E_{2s} + \Delta E$$

which defines the quantity ΔE. Next we substitute into the expression for the hybrid orbitals energy and find

$$\varepsilon_{\text{hybrid}} = E_{2s} \cdot \left(a^2 + b^2 + c^2 + d^2\right) + \left(b^2 + c^2 + d^2\right) \cdot \Delta E$$

To evaluate the energy we must know the coefficients a, b, c, and d. By requiring that the hybrid orbitals ϕ each be normalized we obtain the condition $a^2 + b^2 + c^2 + d^2 = 1$. For the case of a tetrahedral sp^3 hybrid with equal contributions from each basis orbital, we have that $a^2 = b^2 = c^2 = d^2 = \frac{1}{4}$. In this limit, we obtain

$$\varepsilon_{\text{hybrid}} = E_{2s} + \frac{3}{4} \cdot \Delta E$$

P7.3 In strongly acidic solution the carbonium ion CH_5^+ and the carbenium ion CH_3^+ are formed according to

$$CH_4 + H^+ \rightarrow CH_5^+ \rightarrow CH_3^+ + H_2$$

Discuss the orbital bonding of these ions.

Solution

CH_3^+ can be formed from C^+ and three H atoms; in CH_3^+ a trigonal planar hybrid orbital offers the best bonding possibility, and thus CH_3^+ should be trigonal planar. CH_5^+ can be formed by addition of a proton to methane; one of the tetrahedral hybrid orbitals is shared by two protons (two-electron three-center bond similar to the case of polymer BeH_2).

P7.4 In Section 7.3.1.1 we discussed how symmetry considerations imply that the net overlap between σ and π orbitals is zero. Using $\psi_a = 2p_y$ and $\psi_b = 2s$ hydrogen atom wavefunctions, evaluate the overlap integral $S = \int \psi_a \psi_b \cdot d\tau$, and demonstrate that it is zero. Consider the bond axis to be the z-axis (see Fig. P7.4).

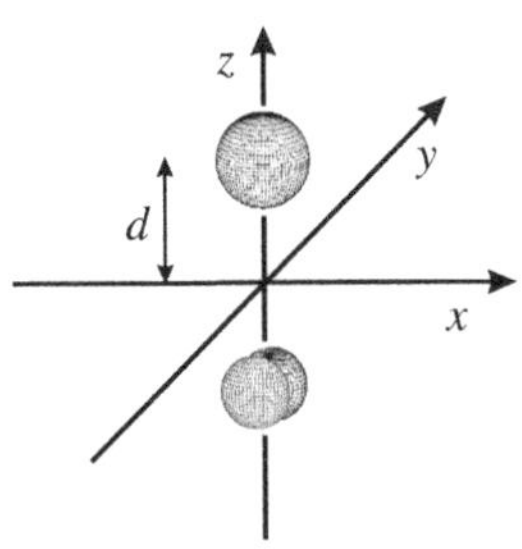

Figure P7.4 Orbital picture of a $2s$ (top) and a $2p_y$ (bottom) function of the H-atom.

Solution
The overlap integral for $\psi_a = 2p_y$ and $\psi_b = 2s$ is

$$S = \int_{-\infty}^{\infty}\int_{-\infty}^{\infty}\int_{-\infty}^{\infty} \psi_a(x, y, z)\,\psi_b(x, y, z + d)\,\mathrm{d}x\,\mathrm{d}y\,\mathrm{d}z$$

To evaluate the integrals we make the transformation to spherical coordinates, so that

$$\psi_a = A\frac{r}{a_0}\mathrm{e}^{-r/(2a_0)}\sin\vartheta\,\sin\varphi$$

and

$$\psi_b = A'\left(2 - \frac{r'}{a_0}\right)\mathrm{e}^{-r'/(2a_0)}$$

where

$$r' = \sqrt{x^2 + y^2 + (z + d)^2} = \sqrt{r^2\sin^2\vartheta + (r\cos\vartheta + d)^2}$$

Thus, we can write

$$S = \iiint A\frac{r}{a_0}\mathrm{e}^{-r/(2a_0)}\sin\vartheta\,\sin\phi\cdot A'\left(2 - \frac{r'}{a_0}\right)\mathrm{e}^{-r'/(2a_0)}r^2\sin\vartheta\,\mathrm{d}r\,\mathrm{d}\vartheta\,\mathrm{d}\varphi$$
$$= AA'\int_0^{\infty}\int_0^{\infty}\frac{r}{a_0}\mathrm{e}^{-r/(2a_0)}\sin\vartheta\cdot\left(2 - \frac{r'}{a_0}\right)\mathrm{e}^{-r'/(2a_0)}r^2\sin\vartheta\,\mathrm{d}r\,\mathrm{d}\vartheta\cdot\int_0^{2\pi}\sin\varphi\,\mathrm{d}\varphi$$

First, we perform the integral over φ and find

$$\int_0^{2\pi}\sin\varphi\cdot\mathrm{d}\varphi = -\cos\varphi|_0^{2\pi} = -1 + 1 = 0$$

Since this integral is multiplied by the others, we see that the net overlap integral is zero.

You arrive at the same result on grounds of a symmetry consideration. According to Fig. P7.4 the $2s$ function has an antinode in the xz -plane, whereas the $2p_y$ function has a node. Thus for the product of both functions the xz-plane is also a nodal plane and the sum over all points in space becomes zero.

P7.5 Our discussion of triatomic molecules considered hydride molecules, for which π-bonding is not important. Discuss how our bonding considerations must change to describe the bonding in CO_2. Sketch a molecular orbital energy diagram for CO_2.

Solution
Carbon dioxide is a linear triatomic molecule, and this information can be used to place important constraints on which orbitals can interact. In particular, we expect that the $2p_y$ orbitals on each atom center will only interact with each other, and the $2p_z$ orbitals on each center will interact with each other. Note that we take the bond axis to be along the x direction.

The three $2p_y$ orbitals (one on each atomic center) interact to form three molecular orbitals. These molecular combinations are shown in Fig. P7.5. Each of them is part of a doubly degenerate pair of molecular orbitals; and they correspond to the 1π bonding molecular orbital, the 2π nonbonding molecular orbital, and the 3π anti-bonding molecular orbital in the figure.

The analysis for the $2p_z$ orbital combinations is the same, and they correspond to the other orbitals in the doubly degenerate pairs.

The analysis for the σ framework is more complicated because the $2s$ orbitals and the $2p_x$ orbitals on the atoms can hybridize and/or interact; hence, we have six atomic orbitals combining to form six molecular orbitals (rather than two sets of three). The lowest two molecular orbitals comprise the symmetric and antisymmetric superpositions of the oxygen $2s$ orbitals, and we find that the symmetric superposition can make a bonding interaction with the carbon's $2s$ orbital and the antisymmetric superposition can make a bonding interaction with carbon's $2p_x$ orbital. These molecular orbitals are labeled 1σ and 2σ in the figure and are analogous to the lowest molecular orbitals discussed in relation to the hydrides (see Fig. 7.20 of text); however, the large energy shift between the $2s$ and $2p_x$ of the O and the $2s$ of the C leads to the amplitude of the C-atom's $2s$ orbital to be relatively small and the

O-atom $2s$ orbital combinations carry most of the electron probability density (hence, a dashed outline is placed on the C atom's $2s$ and $2p_x$ orbitals). In addition, we see that the symmetric and antisymmetric superpositions of the oxygen's $2p_x$ orbitals can form bonding interactions with the carbon $2p_x$ orbital; namely, the antisymmetric superposition forms a bonding interaction with the carbon $2s$ orbital (4σ) and the symmetric superposition forms a bonding interaction with the carbon $2p_x$ orbital (4σ). Two possible anti-bonding combinations are also shown. Note that the strict separation of interactions with the $2s$ and $2p_x$ orbitals is a simplification used here. In a quantitative treatment we would expect some mixing (hybridization) of the $2s$ and $2p_x$ orbitals.

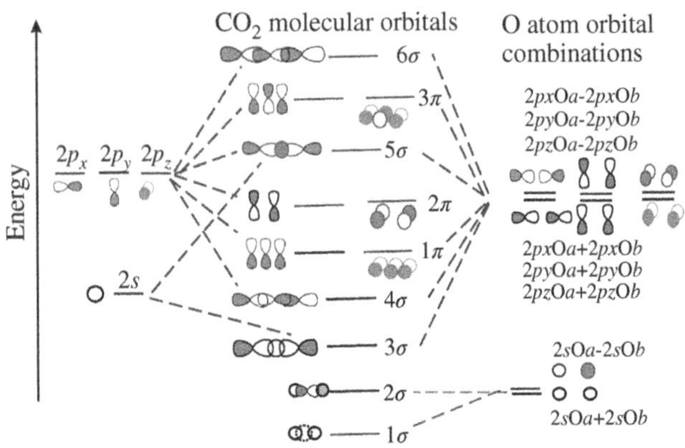

Figure P7.5 LCAO orbital scheme for the CO_2 molecule. A qualitative molecular orbital energy diagram is shown for CO_2, in which the C-atom orbitals are combined with symmetric and antisymmetric combinations of the O-atom orbitals.

P7.6 Make a qualitative molecular orbital energy diagram for methanol (CH_3OH). Sketch the molecular orbitals. Contrast this diagram with that of acetonitrile (Fig. 7.23).

Solution

A qualitative molecular orbital energy diagram for methanol is shown in Figure P7.6. This diagram is created by making an analogy with the acetonitrile case (Fig. 7.23). The two cases differ by the energetics of the fragment energy levels (OH versus CN) and by the fact that the OH system does not have the π orbitals that the CN system had. Rather it has p-orbitals localized on the O center; more like the HF case shown in Fig. 7.18. While some favorable interaction of the oxygen centered p-orbitals with the CH_3 fragment occurs, it is generally found to be weaker than a direct bonding interaction. The antibonding orbital combinations are not shown in this diagram; except for the second most stable orbital.

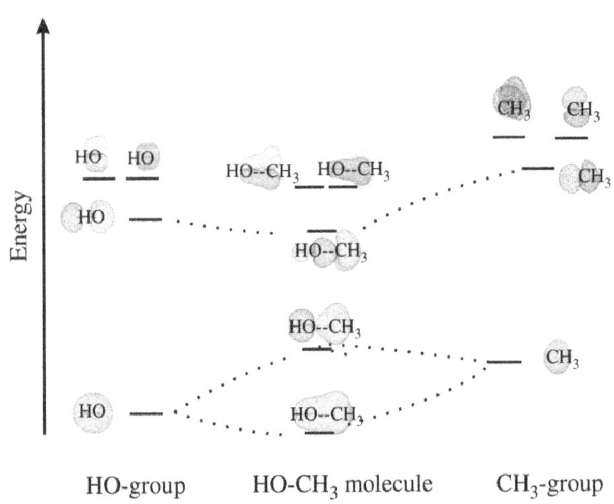

Figure P7.6 LCAO orbital scheme for the methanol molecule.

Extension: Use a Quantum Chemistry program (see P7.13) to examine this qualitative scheme and quantify it more.

P7.7 Discuss whether the electrostatic repulsion of the H atoms in the H_2O molecule is sufficient to cause a widening of the bond angle from 90^0 to 105^0. The dipole moment of H_2O is $\mu = 6.16 \times 10^{-30}$ C m = 1.85 Debye, the bending force constant k_f' is 68 N/m, and the bond length d is 96 pm.

Solution

Consider Fig. P7.7. The distance a between the O atom and the center of mass of the two H atoms is $a = d_1/\sqrt{2} = 67.9$pm. Exercise 7.16 provides an analogous problem for finding the partial charge Q; using that approach here we find $Q = \mu/a = 0.9 \times 10^{-19}$C. The force acting between the two hydrogen atoms is

$$f = \frac{1}{4\pi\varepsilon_0} \cdot \frac{Q^2}{(2a)^2} = 4 \times 10^{-9}\ \text{N}$$

We compare this force with the force required to increase the bond angle by $\Delta\alpha = 15°$ (for which $x = d \cdot \Delta\alpha =$ 25pm). In the Hooke's law limit we find

$$f = 68\ \text{N m}^{-1} \times 25 \times 10^{-12}\ \ \text{M} = 2 \times 10^{-9}\ \ \text{N}$$

which is of the same order of magnitude as the Coulomb repulsion force.

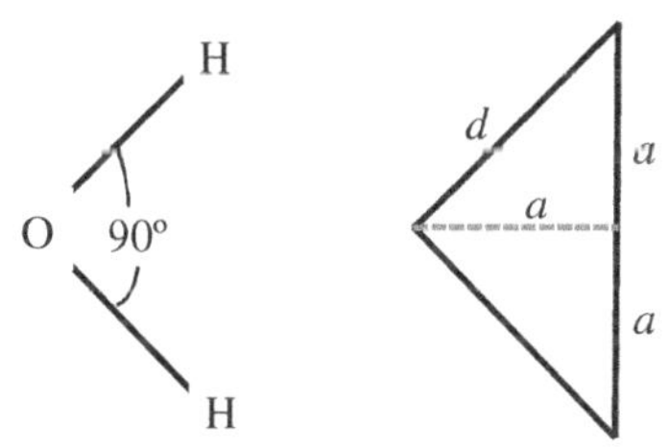

Figure P7.7 Geometry of the H_2O molecule.

P7.8 Use the Morse potential function in Equation (7.11) and determine how the force depends on distance. Plot the distance dependence of the force for the case of a hydrogen molecule, an oxygen molecule, and a nitrogen molecule. See Table 7.6 for values of the bond parameters.

Solution

The Morse function is

$$\varepsilon = \varepsilon_{\text{eq}} + \varepsilon_B \cdot \left(1 - e^{-a(d-d_{\text{eq}})}\right)^2$$

with $a = \sqrt{k_f/2\varepsilon_D}$. The force associated with a potential is the negative of the potential energy's first derivative. We found in Exercise 7.23 that the expression for the force from a Morse potential is

$$f = -\frac{d\varepsilon}{dx} = -2a\varepsilon_B \cdot \left(e^{-ax} - e^{-2ax}\right)$$

We plot this function in Fig. P7.8 (H_2 is the black solid line, O_2 is the light gray solid line, and N_2 is the dark gray line) by using the following data

Molecule	k_f, N/m	ε_B, kJ/mol	a, pm^{-1}	d_{eq}, pm
H_2	574	436	199	74
O_2	1177	498	267	121
N_2	2296	945	270	110

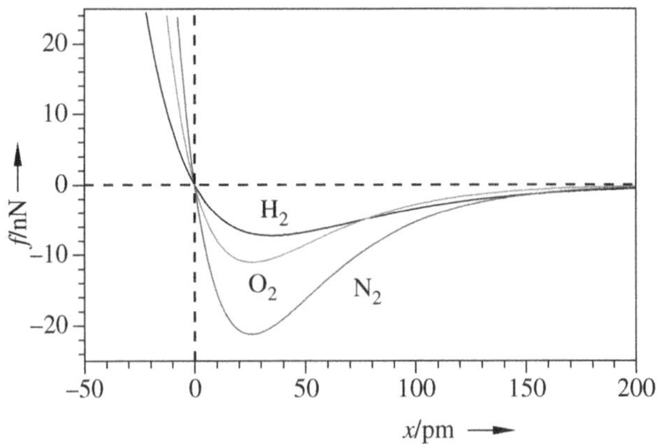

Figure P7.8 Restoring force f versus the elongation $x = d - d_{eq}$ of the bond for H_2, O_2, and N_2. At the equilibrium positions of the nuclei ($x = 0$) the restoring force is zero.

From the plot it is clear that when the bond distance is smaller than the equilibrium distance the force is positive and the atoms are pushed apart; whereas when the bond distance is greater than the equilibrium distance the force is negative and the atoms are pulled together. In addition, we note that the order of magnitude for the force is nN to tens of nN.

P7.9 Figure P7.9 shows a photoelectron spectrum of H_2S. Construct a qualitative molecular orbital energy diagram for H_2S and use it to explain the structure in the photoelectron spectrum.

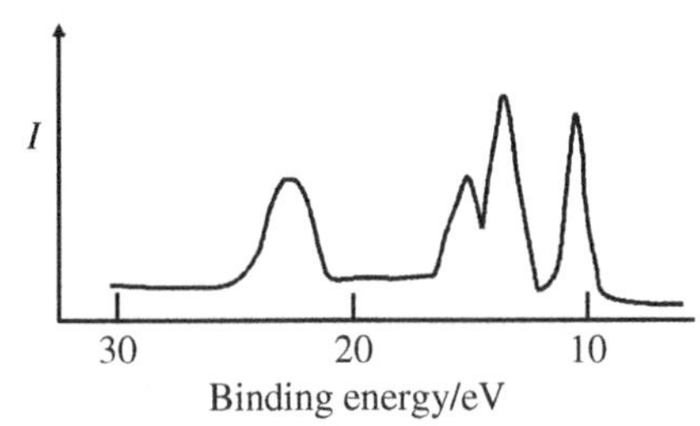

Figure P7.9 Photoelectron spectrum of H_2S.

Solution

The spectrum is drawn from that reported by Siegbahn (K. Siegbahn, *ESCA Applied to Free Molecules*, North Holland, 1969, Amsterdam), the valence shell binding energies occur at 22, 15.1, 13.2, and 10.3 eV. The binding energies for the sulfur's 2*p*-orbital electrons lie in the range of 170 to 172 eV, and those for the 2*s* -orbital electrons is at 234.5 eV.

The H_2S molecule has a geometry and structure similar to that of H_2O, and we can assign the energies by analogy to those of water (see Fig. 7.21). We expect that the most stable valence shell molecular orbital (22 eV) will be formed by a symmetric combination of the 3*s*-orbital and the symmetric superposition of the H-atom 1*s* orbitals. The second most stable valence orbitals (15.1 eV) are formed by the bonding combination of the $3p_y$ orbital of the S-atom, which lies in the molecular plane and are perpendicular to the molecule's symmetry axis, and the antisymmetric combination of the H-atom 1*s* orbitals. The third valence orbitals (13.2 eV) arise from a bonding combination of the symmetric superposition of the H-atom 1s orbitals and the $3p_z$ orbital on the S-atom, which is oriented along the molecule's symmetry axis and bisects the bond angle. The most weakly bound electrons (10.3 eV) correspond to the $3p_x$ orbital on S which is oriented perpendicular to the molecular plane and is nonbonding. Note that the 22 eV and 13.2 eV orbitals likely have a mixture of the 3*s* and $3p_z$ orbitals contributing to each orbital (i.e. it is hybridized).

P7.10 The table provides energy E versus internuclear distance d values that are calculated for HF at a sophisticated level of theory (see X. Li, J. Mol. Structure (Theochem) **2001**, 547, 69). Plot these data and then fit the Morse potential function to the data. Using your best fit parameters for the Morse potential calculate the force constant of the bond.

d/a_0	1.100	1.200	1.300	1.390	1.475	1.560	1.646	1.683
ε/eV	0.920	−0.727	−3.080	−4.426	−5.237	−5.714	−5.954	−6.003
d/a_0	1.693	1.703	1.713	1.723	1.733	1.743	1.753	1.763
ε/eV	−6.010	−6.016	−6.020	−6.022	−6.022	−6.020	−6.017	−6.013
d/a_0	1.773	1.790	1.870	1.950	2.030	2.110	2.166	2.310
ε/eV	−6.006	−5.995	−5.883	−5.705	−5.486	−5.238	−5.051	−4.546
d/a_0	2.460	2.600	2.800	3.000	3.300	3.600	4.000	4.500
ε/eV	−4.011	−3.524	−2.871	−2.287	−1.557	−1.004	−0.516	−0.199

Solution

Figure P7.10 compares the data (circles) and a best fit curve to the Morse potential energy expression

$$\varepsilon = \varepsilon_B \cdot \left(1 - e^{-a(d-d_{eq})}\right)^2$$

The best fit parameters are $\varepsilon_{eq} = -5.991$ eV, $d_{eq} = 1.750\ a_0$, $\varepsilon_B = 6.510$ eV, and $a = 1.113/a_0$.

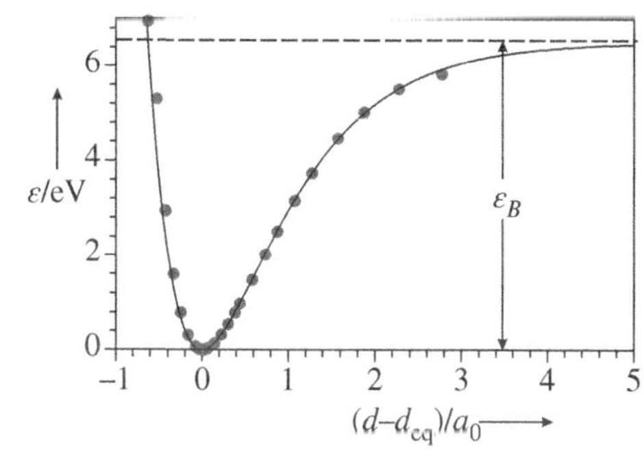

Figure P7.10 Energy ε of the HF molecule. Full circles: quantum chemical calculation. Solid line: Morse function with the parameters $\varepsilon_B = 6.510$ eV, $a = 1.113/a_0$, and $d_{eq} = 1.750\ a_0$.

The force constant for the bond is most easily found from the expression for

$$a = \sqrt{\frac{k_f}{2\varepsilon_B}} = \frac{1.113}{a_0} = 2.104 \times 10^{10}\ \text{m}^{-1}$$

so that

$$k_f = 2\varepsilon_B\left(\frac{1.113}{a_0}\right)^2$$

$$= 2 \cdot 6.5096\ \text{eV} \times \frac{1.602 \times 10^{-19}\text{J}}{\text{eV}}\left(\frac{1.113}{52.9177 \times 10^{-12}\ \text{m}}\right)^2$$

$$= 922.6\ \text{N m}^{-1}$$

P7.11 The table provides data for the energy ε of the H_2^+ ion as a function of the distance d between the nuclei (d_{eq} is the equilibrium distance, ε_{eq} is the energy at the equilibrium distance). Assume that the dependence of the energy on the displacement follows the cubic form

$$\varepsilon - \varepsilon_{eq} = \frac{1}{2}k_f \cdot x^2 + bx^3$$

where we have set $x = d - d_{eq}$. You should analyze the data for each of the three electronic structure models and extract a force constant, k_f, for the H_2^+ bond.

$\frac{d-d_{eq}}{pm}$	$\frac{\varepsilon-\varepsilon_{eq}}{10^{-19}\ J}$ Exact	Box model	LCAO model
−20	0.45	0.15	0.26
−15	0.24	0.08	0.14
−10	0.09	0.03	0.06
−5	0.02	0.00	0.01
0	0	0	0
5	0.02	0.01	0.01
10	0.07	0.03	0.04
15	0.15	0.07	0.09
20	0.24	0.12	0.15

a) Rearrange the equation to a linear form ($y = m \cdot x + b$), by showing that

$$y = \frac{\varepsilon - \varepsilon_{eq}}{x^2} = \frac{1}{2}k_f + bx$$

What is the slope and the intercept you expect?

b) Plot the data and extract a k_f value. Comment on the linearity of the data?
For each of the electronic energy curves, analyze the data for the case of $x \leq 0$ and for $x \geq 0$ and extract values for both b and k_f.
Compare and contrast the k_f values. Comment on the physical meaning of the parameter b.

Solution

a) If we divide both sides of the equation by x^2, we find that

$$y = \frac{\varepsilon - \varepsilon_{eq}}{x^2} = \frac{1}{2}k_f + bx$$

Thus a plot of $y = \left(\varepsilon - \varepsilon_{eq}\right)/x^2$ versus x should give a line with a slope of b and an intercept of $k_f/2$.

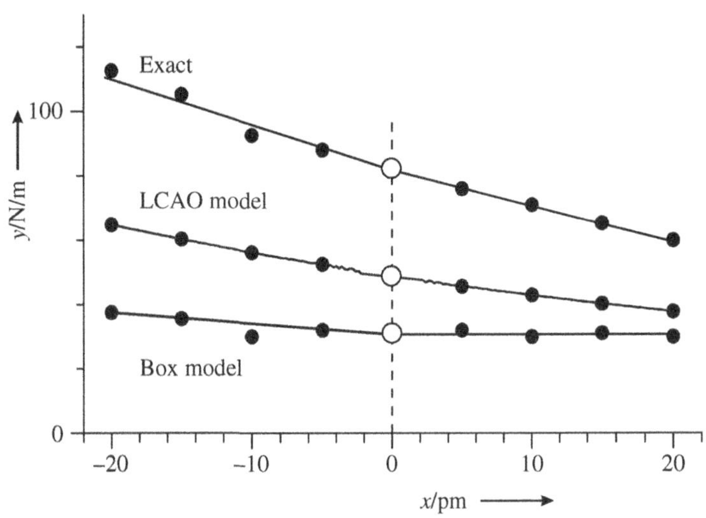

Figure P7.11 Evaluation of the force constant k_f of the H_2^+ ion (exact calculation), box model, and LCAO model. The quantity $y = (\varepsilon - \varepsilon_{eq})/x^2$ is plotted versus $x = (d - d_{eq})$. The intercept at point $x = 0$ equals one-half of the force constant. Full circles: data points, solid lines: interpolation of the data points to $x = 0$ (open circles).

The data for each of the three cases are shown in Figure P7.11. The data show a deviation from linearity when viewed over the whole range, but they appear to be quite linear on the two sides of $x = 0$. Figure 7.7 shows a plot of the data with linear fits on the two sides of $x = 0$, and the table reports the values for b and k_f.

When plotting y as a function of x, we expect a straight line with slope b and intersection $k_f/2$. For the three models we obtain the following results:

Model	$\frac{y(x=0)}{\text{N m}^{-1}}$ (N m^{-1})	k_f (N m^{-1})	b(left) (10^{10} N m^{-2})	b(right) (10^{10} N m^{-2})	b(mean value) (10^{10} N m^{-2})
Exact	82	164	−142	−112	−127
Box model	30	60	−80	−55	−68
LCAO model	49	98	−36	0	−18

b) The slopes of the solid lines in the figure are larger for negative elongations of the bond than those for positive ones; this indicates that the power series describing $(\varepsilon - \varepsilon_{eq})$ is not completely adequate. For this reason the force constant obtained for the exact solution is slightly larger than the value $k_f = 160$ N m^{-1} shown in Table 7.6. The b values are significantly larger on the left side ($x < 0$), which indicates that the electronic energy is changing more sharply when the bond is compressed as compared to when it is stretched than is accounted for by the cubic form of the equation.

While the force constant obtained from the Exact solution is in good agreement with the experiment, the force constants obtained from the LCAO model and the Box model are significantly smaller than what is found experimentally. This indicates that these models are underestimating how strongly the electronic energy changes with the distance.

P7.12 Consider the case of LiH which has an average bond distance of 159.5 pm. Given this constraint for the bond distance, evaluate the optimum geometry for the triatomic molecule Li_2H^+. More specifically, assume that the two LiH bond lengths are fixed and that the interactions between the ions $Li^+$and $H^-$are strictly electrostatic. Vary the angle of the $[\text{Li-H-Li}]^+$. Make a plot of the change in the energy as a function of the angle and find the most stable (lowest energy) geometry. Provide a physical explanation for the answer that you find.

Solutions

Figure P7.12 gives the arrangement of the three atoms. With the consideration that each Li atom has a single positive charge, and the hydrogen atom has a negative charge, the potential energy expression becomes

$$V = \frac{-e^2}{4\pi\varepsilon_0 r_{\text{Li-H}}} + \frac{-e^2}{4\pi\varepsilon_0 r_{\text{Li-H}}} + \frac{e^2}{4\pi\varepsilon_0 r_{\text{Li-Li}}}$$
$$= \frac{-e^2}{4\pi\varepsilon_0}\left(\frac{1}{r_{\text{Li-H}}} + \frac{1}{r_{\text{Li-H}}} - \frac{1}{2r_{\text{Li-H}}\sin(\theta/2)}\right)$$

where we have used the fact that $r_{\text{Li-Li}} = 2r_{\text{Li-H}}\sin(\theta/2)$; see the diagram in panel a). Simplifying, we can write that

$$V = \frac{e^2}{4\pi\varepsilon_0 r_{\text{Li-H}}}\left(\frac{1}{2\sin(\theta/2)} - 2\right)$$

A graph of this potential function is given as panel b) of Fig. P7.12. Because the energy is a minimum at an angle of 180°, we would predict that this molecular ion has a linear geometry. This geometry minimizes the electrostatic repulsion between the lithium ions while maintaining strong electrostatic attraction between each lithium ion and the hydride ion.

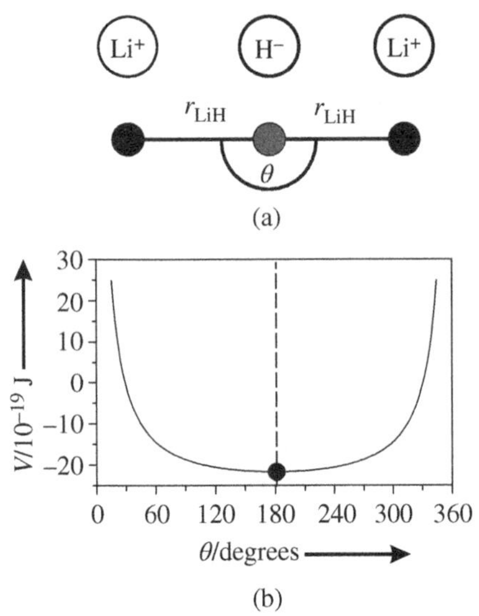

Figure P7.12 Panel (a) shows the geometry for the Li-H-Li ionic compound, and panel (b) shows a plot of how the electrostatic energy changes with the bond angle θ.

The minimum energy geometry at 180° minimized the electrostatic repulsion between the Li^+ ions.

P7.13 Perform electronic structure calculations for hydrogen fluoride, HF, with the three different basis sets STO-3G, 3-21G, and 6-31G (see Foundation 7.2 for a discussion of these basis sets); and compare the results to experimental data. Many universities provide access to software programs for such calculations, but "freeware" versions also exist. The solutions given for this problem in the solution manual use the WebMO program, which you can access at https://www.webmo.net/demoserver/cgi-bin/webmo/login.cgi. If you use this application, select the GAMESS software package, which is free to users. You can obtain experimental values for HF properties from the USA government website: http://www.nist.gov/data.

a) Determine the equilibrium internuclear bond distance for HF and compare it to the experimental value.
b) Determine the dipole moment of HF and compare it to the experimental value.
c) Determine the molecular orbital energies of HF for the HOMO-2, HOMO-1, HOMO, and LUMO orbitals. (You may find the energy conversion factor of 1 Hartree = 27.2114 eV to be useful).
d) Plot the molecular orbital energy diagram that includes the valence electrons; i.e. make a plot to scale that shows the molecular orbital energies and the corresponding atomic orbital energies for the 6-31G basis set, which should be the most accurate one. There is no need to include the 1s orbital of the F or its corresponding molecular orbital. Take the atomic orbital energies to be −13.6 eV for the H atom, −39 eV for the 2s atomic orbital of F[2], and −17.4 eV for the 2p orbitals of F.[3]

Finally, you should discuss and explain your results. Note that the calculations optimize the electronic structure for the total energy of the molecule, not the individual molecular orbitals, and you may see some variation in their energies.

Solutions

Parts a) and b): Values for the bond length and molecular dipole moment that are obtained from the HF-SCF quantum chemistry calculations with the STO-3G basis set, with the 3-21G basis set, and.with the 6-31G basis set are collated in the table, along with experimental values from the National Institute of Standards and Technology (NIST) website.

---

2 M. B. Trzhaskovskaya, V. I. Nefedov, and V. G. Yarzhemsky "Atomic Data and Nuclear Data Tables" 77, 97–159 (2001).
3 https://webbook.nist.gov/

	Bond distance /pm	Dipole moment /D
STO-3G	95.54	1.252
3-21G	93.74	2.174
6-31G	91.09	1.972
Experiment	91.68	1.820

(The experimental values taken from NIST: http://www.nist.gov/data, NIST CCCBDB).

The calculations show that the split valence basis sets (3-21G and 6-31G) show a better agreement of the dipole moment and bond distance with the experimental value. The best agreement between the experiment and calculation results for the 6-31G case. This calculation uses basis functions with a larger number of gaussian functions (6 for the core orbitals and 4 for the valence orbitals). See Foundation 7.2 for a discussion of these different basis set choices. Lastly, we note that more sophisticated approaches, such as those which account for correlation energy effects, provide even better agreement with experiment.[4] c) The molecular orbital energies that are found for HF by the different computational methods are collected in the table.[5]

	HOMO-2	HOMO-1	HOMO (degeneracy of 2)	LUMO
STO-3G	−39.86 eV	−15.76 eV	−12.61 eV	16.53 eV
3-21G	−42.09 eV	−19.01 eV	−16.26 eV	7.14 eV
6-31G	−43.07 eV	−20.3 eV	−17.1 eV	5.95 eV
Experiment	39.6	19.9	16.1 eV (ionization energy)	

Recall that Koopmans' theorem associates the orbital energies with the ionization energy, and this is an idealization which is known to lead to errors of the magnitude seen here; see https://en.wikipedia.org/wiki/Koopmans\LY1\textbackslash%27_theorem.

d) A plot of the orbital energy diagram in Fig. 7P.13 is the same as that shown in Fig. 7.18 of the text.

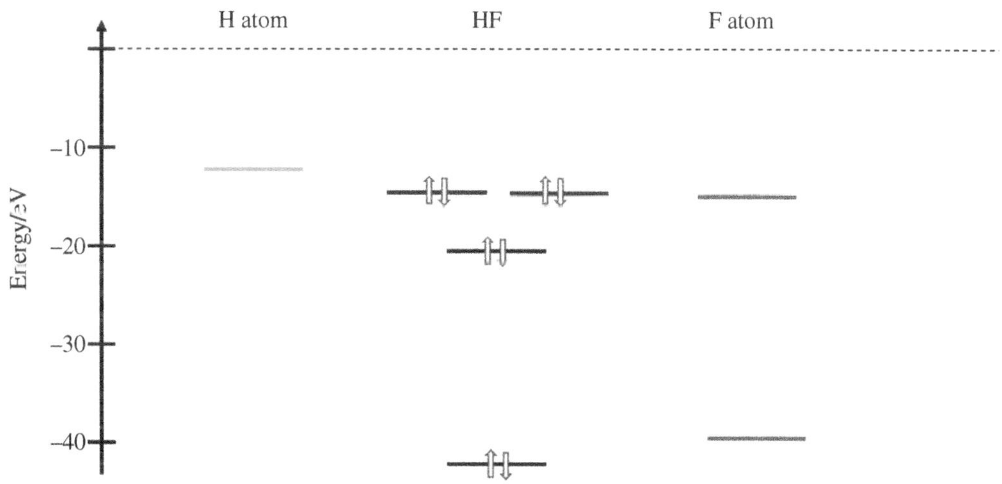

Figure 7P.13 Energy states of HF in comparison with those of the H atom and the F atom.

4 D. Feller and K. A. Peterson Hydrogen fluoride: a critical comparison of theoretical and experimental results. *J. Molec Structure (Theochem)* 1997, **400**, 69–92.

5 The experimental ionization energies are taken from M. S. Sanna, R. Malutzki and V. Schmidt, *J. Chem. Phys.* 1987, **87**, 1582.

8

Molecules with PI-Electron Systems

8.1 Exercises

E8.1 Explain the data in Table 8.1. In particular, discuss a) the correlation between bond energy and bond length and b) the correlation between bond energy and force constant.

Solution

a) As the bond length decreases one observes the bond strength increasing, as the C–C bond evolves from a single bond for ethane to a double bond for ethene and a triple bond for ethyne. We know from the Virial Theorem (Chapter 6) that the bond energy should correlate with the average potential energy. Hence the higher bond energies correspond to larger average potential energies which correspond to the electron cloud being held more tightly to the nuclear cores, hence a smaller bond length.

b) Here one observes that as the bond length decreases from ethane to ethyne the force constant increases. The force constant reflects the stiffness of the bond, which increases as we proceed from the C–C single bond of ethane to the C–C double bond of ethene and the C–C triple bond of ethyne. The force constant increases as the bond energy increases because the energy well becomes "tighter" (the well is deeper and narrower as one proceeds from the C–C single bond to a double bond and then a triple bond).

E8.2 Use Equation (8.3) to calculate the π orbital energy levels in the cases of ethene and butadiene.

Solution

UsingEquation (8.3) and evaluating the constants, we can write that

$$E_n = \frac{h^2}{8m_eL^2} \cdot n^2 = \frac{h^2}{8m_e} \cdot \frac{n^2}{L^2}$$
$$= \frac{(6.6260 \times 10^{-34}\ \text{J s})^2}{8 \times 9.109 \times 10^{-31}\ \text{kg}} \cdot \frac{n^2}{L^2} = 6.026 \times 10^{-38}\,\text{J m}^2 \cdot \frac{n^2}{L^2}$$

where L is the length of the box and n is the quantum number of the π orbitals. Ethene has two π electrons that occupy the $n = 1$ energy level. If we use the value $L = 3 \cdot 140$ pm $= 420$ pm, then for $n = 1$ we find that

$$E_1 = 6.026 \times 10^{-38}\,\text{J m}^2 \cdot \frac{1}{(420\,\text{pm})^2} = 3.42 \times 10^{-19}\ \text{J} = 2.13\,\text{eV}$$

Butadiene has four π electrons that occupy the $n = 1$ and $n = 2$ energy levels. Using a value for L of $5 \cdot (140\ \text{pm}) =$ 700pm, we find that the energy levels for butadiene are

$$E_1 = 6.026 \times 10^{-38}\,\text{J m}^2 \cdot \frac{1}{(700\,\text{pm})^2} = 1.23 \times 10^{-19}\ \text{J} = 0.78\,\text{eV}$$

and

$$E_2 = 6.026 \times 10^{-38}\,\text{J m}^2 \cdot \frac{4}{(700\,\text{pm})^2} = 4.92 \times 10^{-19}\ \text{J} = 3.07\,\text{eV}$$

The difference in the butadiene energy levels is 3.07 – 0.767 = 2.3 eV, which compares reasonably well with the difference in the experimental ionization energies, 2.5 eV.

Solutions Manual for Principles of Physical Chemistry, Third Edition. Edited by Hans Kuhn, David H. Waldeck, and Horst-Dieter Försterling.

E8.3 The experimental first ionization energy of ethene, 10.51 eV, is the energy required to remove a π-electron from the double bond, assuming Koopmans theorem applies:

$$IE = -E_{\text{orbital}}$$

see Section 5.5.4.1. Use this binding energy value and the energy found from the FEMO model to estimate the average potential energy for the electron.

Solution

The electron's binding energy is the sum of its average kinetic energy and potential energy $\overline{V}$. The energy from the FEMO model is the average kinetic energy for the π orbital in ethene. From Exercise 8.2 we take it to be 2.13 eV. So the average potential energy is

$$\overline{V} = -10.51 - 2.13\,\text{eV} = -12.64\,\text{eV}$$

E8.4 Use the FEMO model to calculate the resonance energy ε_R for butadiene. Given that the hydrogenation energy of ethene is -2.16×10^{-19} J and the hydrogenation energy of butadiene is -3.72×10^{-19} J, compute an experimental resonance energy for butadiene and compare it to your predicted value.[1]

Solution

Following the procedure outlined in Section 8.7, we use the experimental data to calculate a resonance energy for butadiene that is

$$\begin{aligned}\varepsilon_R &= \Delta E_{\text{butadiene}} - 2 \times \Delta E_{\text{ethene}} \\ &= -3.72 \times 10^{-19}\ \text{J} - 2 \times \left(-2.16 \times 10^{-19}\ \text{J}\right) \\ &= 0.60 \times 10^{-19}\ \text{J}\end{aligned}$$

In the FEMO model we calculated (see Exercise 8.2) ethene π-electron energies of 3.42×10^{-19} J and for butadiene we calculated π -electron energies of 1.23×10^{-19} J and 4.92×10^{-19} J. Using these results we calculate a resonance energy of

$$\begin{aligned}\varepsilon_{R,\text{FEMO}} &= 2 \times 2\left(3.42 \times 10^{-19}\ \text{J}\right) - 2\left(1.23 \times 10^{-19}\ \text{J}\right) - 2\left(4.92 \times 10^{-19}\ \text{J}\right) \\ &= 13.68 \times 10^{-19}\ \text{J} - 12.20 \times 10^{-19}\ \text{J} \\ &= 1.48 \times 10^{-19}\ \text{J}\end{aligned}$$

A result that is of the same order of magnitude as the experimental resonance energy, but about 2–3 times larger. See E8.19 for a comparison with the HMO model for the resonance energy.

E8.5 Figure 8.5a shows the π-orbital energy levels of benzene and their orbital shapes in the LCAO approximation. Compare the π -electron nodal patterns for the LCAO and FEMO wavefunctions of benzene.

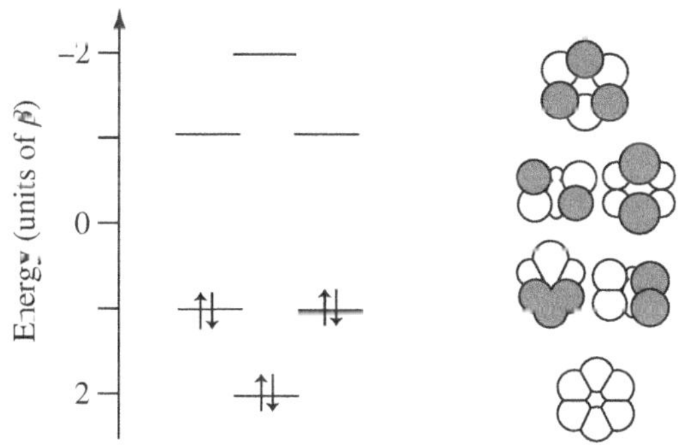

Figure E8.5a The LCAO energy levels and orbitals of benzene.

Solution

Plots of the first five FEMO wavefunctions for the first four ($n = 0, \pm1, \pm2,$ and ±3) energy levels of benzene are shown in Fig. E8.5b. The dashed lines define the zero for the shifted wavefunction plots.

1 Hydrogenation enthalpies are taken from M. Cowperthwaite, S.H. Bauer, *JCP* 1962, **36**, 1743.

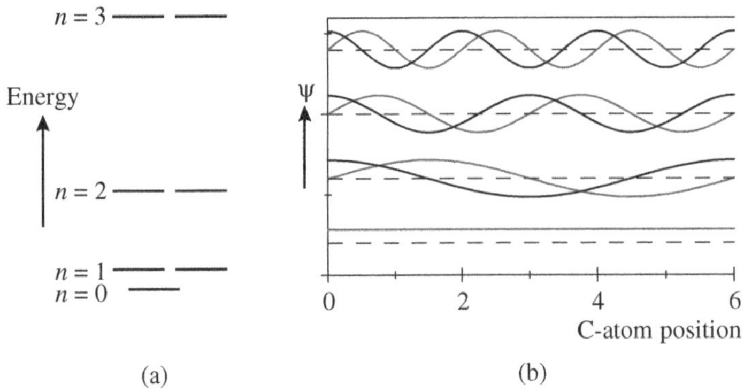

Figure E8.5b Energies (panel a) and wavefunctions (panel b) of benzene according to the FEMO model.

The FEMO wavefunction for the ground electronic level ($n = 0$) is nondegenerate, and it has equal amplitude on all of the C-atom sites. This wavefunction has a kinetic energy of zero along the s-direction and corresponds to the ground-state wavefunction with a π-bond energy of 2β in the HMO model.

The FEMO wavefunction for the next higher electronic state ($n = 1$) is doubly degenerate. One of the wavefunctions (shown in gray) has nodes at C-atom sites 3 and 6, and equal amplitude on the other four C-atom sites. The other wavefunction (shown in black) has equal amplitude on atom sites 1, 2, 4, and 5 and a somewhat higher amplitude on sites 3 and 6. The nodes for this wavefunction appear between sites 1 and 2, and between sites 4 and 5. These orbital patterns are the same as those found for the two degenerate orbitals in the HMO model with a π-bond energy of β.

The FEMO wavefunction for the next higher electronic state ($n = 2$) is doubly degenerate also. One of the wavefunctions (shown in gray) has nodes at C-atom site 3, site 6, between the sites 1 and 2, and between the sites 4 and 5. The other wavefunction (shown in black) has nodes between sites 1 and 6, between sites 2 and 3, between sites 3 and 4, and between sites 5 and 6; it also has a larger amplitude on sites 3 and 6 than it does on the other atom sites. These orbital patterns are the same as those found for the two degenerate orbitals in the HMO model with a π-bond energy of $-\beta$.

The FEMO wavefunction for the next higher electronic state ($n = 3$) is predicted to be doubly degenerate also. One of the wavefunctions (shown in red) has nodes on each C-atom site and the other wavefunction (shown in black) has nodes occurring between each C-atom site. In contrast, the HMO model with a π-bond energy of -2β predicts a nondegenerate energy level with nodes occurring between each C-atom site, which corresponds to one of the FEMO wavefunctions. The FEMO model fails in this situation; as discussed in the text it can be corrected by accounting for the stabilization of the π-cloud due to the proximity of the electron density to the C-atom cores.

We note that there is a direct correspondence between the nodal pattern and amplitudes of the $2p$ basis orbitals on each C-atom site in the HMO model and the nodal pattern and wavefunction amplitude in the free electron model.[2]

E8.6 In the FEMO model we confine the π-electrons for benzene to a circular box of circumference $6d_0$, where d_0 is the carbon–carbon bond length. Use a characteristic diameter for a carbon atom's $2p$ orbital of 400 pm and the experimental ionization energy of 9.40 eV = 1.506×10^{-18} J to estimate the extent to which the HOMO wavefunction tunnels outside of the box. You may wish to refer to Chapter 3.

Solution

The HOMO wavefunction is oscillatory along the ring and is given by

$$\psi_{1a} = \sqrt{\frac{2}{L_c}} \sin\left(\frac{2\pi s}{L_c}\right) \quad \text{and} \quad \psi_{1b} = \sqrt{\frac{2}{L_c}} \cos\left(\frac{2\pi s}{L_c}\right)$$

In addition the total wavefunction has a component F that decays exponentially with distance in directions perpendicular to the ring direction, see Section 8.2.2. If we take the C-atom's p-orbital to have a diameter of 400 pm (i.e., distance over which the electron's total energy is greater than its potential energy), then the wavefunction will

2 A.A. Frost, B. Musulin, A Mnemonic Device for Molecular Orbital Energetics, *J. Chem. Phys.* 1952, **20**, 1761.

decay like $F(x) = A\exp(-bx)$ for $x > 200$ pm and $b = 2\pi\sqrt{2m_e(V_0 - E)}/h$ (see Section 3.2.2). Since the ionization energy corresponds to the energy barrier for electron escape, we calculate b by

$$b = \frac{2\pi\sqrt{2 \cdot 9.109 \times 10^{-31} \times 1.506 \times 10^{-18}}}{6.626 \times 10^{-34}} \text{m}^{-1}$$
$$= 1.57 \times 10^{10}\ \text{m}^{-1}$$

The probability $\mathcal{P}$ of finding the electron outside of the box along the x-direction will be

$$\mathcal{P} = \frac{\int_{200\ \text{pm}}^{\infty} \psi^2(s) \cdot F^2(x)\,\text{d}x}{\int_0^{\infty} \psi^2(s) \cdot F^2(x)\,\text{d}x} = \frac{\psi^2(s) \cdot \int_{200\ \text{pm}}^{\infty} A^2 \exp(-2bx)\,\text{d}x}{\psi^2(s) \cdot \int_0^{\infty} A^2 \exp(-2bx)\,\text{d}x}$$
$$= \exp\left(-2 \cdot 1.57 \times 10^{10}\ \text{m}^{-1} \cdot 200\,\text{pm}\right) = 0.00187\,\text{or}\,0.19\%$$

As shown by the expression for the probability we might expect the actual amplitude of the wavefunction to vary along s; however, this is not the case for the HOMO orbital of benzene. Because the orbitals are filled and the charge density is constant, given by

$$\frac{\text{d}Q}{\text{d}s} = -2e \cdot \frac{1}{L_c} \cdot \left[2 \cdot \sin^2\frac{2\pi s}{L_c} + 2 \cdot \cos^2\frac{2\pi s}{L_c}\right] = -\frac{4e}{L_c}$$

Hence the charge density is uniform around the ring and the electron density that "leaks out" is uniform.

E8.7 If we replace one of the carbon atoms (and its associated H atom) in benzene by a nitrogen atom, we obtain the molecule pyridine which is also an aromatic six-membered ring. Use Fig. 8.7 to describe how you expect the energy levels of pyridine to differ from those of benzene.

Solution

Because of the larger effective charge of the nitrogen core we expect that charge density localized on the N-atom site to be somewhat higher than that on the C-atom sites. In general, this should cause the energy levels to shift downward, i.e., stabilize. In addition, we expect the amount of stabilization to depend on the local charge density near the N-atom site. This dependence will break the degeneracy of the orbitals for the FEMO $n - 1$ doubly degenerate state; i.e., the orbital with a node on the N-atom site will be higher in energy than the other π-orbital.

E8.8 Show that the two orthogonal wavefunctions ψ_B and ψ_D in Fig. 8.12 can be superposed to generate the wavefunctions ψ_A and ψ_C (Fig. E8.8).

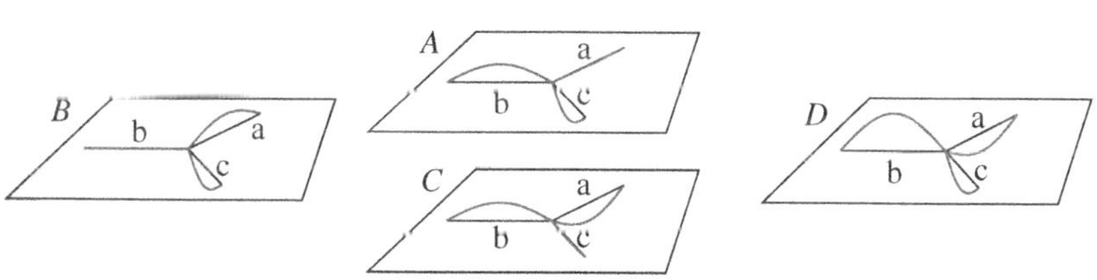

Figure E8.8 Degenerate wavefunctions for the guanidinium ion.

Solution

Let us take the intersection point of the branches to be the origin and the length of each branch to be $L/2$. We can write the wavefunction ψ_B (labeled B in the picture) as

$$\psi_B = \begin{cases} \sin(2\pi s_a/L) & 0 \le s_a \le L/2 \\ 0 & 0 \le s_b \le L/2 \\ -\sin(2\pi s_c/L) & 0 \le s_c \le L/2 \end{cases}$$

where s_a is the displacement along the a branch and s_c is the displacement along the c branch. This function provides a positive going lobe along the a branch and a negative lobe along the c branch. Similarly, for the wavefunction ψ_D (labeled D in the picture), we can write

$$\psi_D = \begin{cases} -\sin(2\pi s_a/L) & 0 \le s_a \le L/2 \\ 2\sin(2\pi s_b/L) & 0 \le s_b \le L/2 \\ -\sin(2\pi s_c/L) & 0 \le s_c \le L/2 \end{cases}$$

in which the amplitude along the b branch is twice the magnitude along that of a and c.

If we take the linear combination of ψ_B and ψ_D the wavefunction along branch b will have an up amplitude, branch a will have no amplitude, and branch c will have a down amplitude. This linear combination will produce wavefunction ψ_A. More specifically, we write

$$\psi_A = \psi_B + \psi_D$$

and sum the wavefunction components along each branch to find

$$\psi_A = \begin{cases} \sin(2\pi s_a/L) & 0 \le s_a \le L/2 \\ 0 & 0 \le s_b \le L/2 \\ -\sin(2\pi s_c/L) & 0 \le s_c \le L/2 \end{cases} + \begin{cases} -\sin(2\pi s_a/L) & 0 \le s_a \le L/2 \\ 2\sin(2\pi s_b/L) & 0 \le s_b \le L/2 \\ -\sin(2\pi s_c/L) & 0 \le s_c \le L/2 \end{cases}$$

$$= \begin{cases} 0 & 0 \le s_a \le L/2 \\ 2\sin(2\pi s_b/L) & 0 \le s_b \le L/2 \\ -2\sin(2\pi s_c/L) & 0 \le s_c \le L/2 \end{cases}$$

If we subtract ψ_B and ψ_D (i.e., the linear combination $\psi_B - \psi_D$), the wavefunction along branch b will have an up amplitude, along branch a the amplitude will be down and along branch c there will be zero amplitude. Thus, this linear combination will produce wavefunction ψ_C. More specifically, we write

$$\psi_C = \psi_D - \psi_B$$

and sum the wavefunction components along each branch to find

$$\psi_C = \begin{cases} -\sin(2\pi s_a/L) & 0 \le s_a \le L/2 \\ 2\sin(2\pi s_b/L) & 0 \le s_b \le L/2 \\ -\sin(2\pi s_c/L) & 0 \le s_c \le L/2 \end{cases} - \begin{cases} \sin(2\pi s_a/L) & 0 \le s_a \le L/2 \\ 0 & 0 \le s_b \le L/2 \\ -\sin(2\pi s_c/L) & 0 \le s_c \le L/2 \end{cases}$$

$$= \begin{cases} -2\sin(2\pi s_a/L) & 0 \le s_a \le L/2 \\ 2\sin(2\pi s_b/L) & 0 \le s_b \le L/2 \\ 0 & 0 \le s_c \le L/2 \end{cases}$$

Note that we have neglected to normalize the wavefunctions in this analysis.

E8.9 For the guanidinium ion, the HOMO orbital is doubly degenerate, and we chose a particular combination of orthogonal wavefunctions. Perform the same analysis, but choose a different orthogonal set of wavefunctions.

Solution

Rather than superpose ψ_A and ψ_C to obtain ψ_D, as discussed in the text, one could superpose ψ_A and ψ_B to obtain

$$\psi_E = c_{1A} \cdot \psi_A + c_{1B} \cdot \psi_B$$

where c_{1A} and c_{1B} are constants which have to be determined. Before proceeding formally, inspection of the wavefunction plots in Fig. E8.8 indicates that ψ_E has positive amplitude along branches a and b, and a negative amplitude along branch c for positive values of the constants c_{1A} and c_{1B}. To show that ψ_E is orthogonal to ψ_C one has to demonstrate that $\int \psi_C \psi_E \, ds = 0$. The product $\psi_C \psi_E$ has a positive amplitude in branch a, negative amplitude in branch b, and zero amplitude in branch c. Integrating this wavefunction leads to zero.

Now we give a formal proof. As discussed in the text, we take the length of each branch to be $L/2$. First we write ψ_A and ψ_B

$$\psi_A = \begin{cases} 0 & 0 \le s_a \le L/2 \\ C\sin\left(2\pi s_b/L\right) & 0 \le s_b \le L/2 \\ -C\sin\left(2\pi s_c/L\right) & 0 \le s_c \le L/2 \end{cases}$$

and

$$\psi_B = \begin{cases} C\sin\left(2\pi s_a/L\right) & 0 \le s_a \le L/2 \\ 0 & 0 \le s_b \le L/2 \\ -C\sin\left(2\pi s_c/L\right) & 0 \le s_c \le L/2 \end{cases}$$

where C is a normalization constant. For ψ_A, normalization requires that

$$1 = C^2\left(\int_0^{L/2} \sin^2\left(2\pi s_c/L\right)\ \mathrm{d}\ s_c + \int_0^{L/2} \sin^2\left(2\pi s_b/L\right)\ \mathrm{d}\ s_b\right)$$

We make the substitution that $x_c = 2\pi s_c/L$, so that

$$\begin{aligned} 1 &= C^2\left(\frac{L}{2\pi}\int_0^{\pi} \sin^2\left(x_c\right)\ \mathrm{d}\ x_c + \frac{L}{2\pi}\int_0^{\pi} \sin^2\left(x_b\right)\ \mathrm{d}\ x_b\right) \\ &= \frac{L}{2\pi}C^2\left(\frac{1}{2}\left(x_c - \frac{1}{2}\sin\left(2x_c\right)\right)_0^{\pi} + \frac{1}{2}\left(x_b - \frac{1}{2}\sin\left(2x_b\right)\right)_0^{\pi}\right) \\ &= \frac{L}{2\pi}C^2\left(\pi + \pi\right) = LC^2 \end{aligned}$$

and we find that $C = \sqrt{1/L}$ (see the integral table in Appendix B). A similar result holds for ψ_B, and the normalized wavefunctions are

$$\psi_A = \begin{cases} 0 & 0 \le s_a \le L/2 \\ \frac{1}{\sqrt{L}}\sin\left(2\pi s_b/L\right) & 0 \le s_b \le L/2 \\ \frac{-1}{\sqrt{L}}\sin\left(2\pi s_c/L\right) & 0 \le s_c \le L/2 \end{cases}$$

and

$$\psi_B = \begin{cases} \frac{1}{\sqrt{L}}\sin\left(2\pi s_a/L\right) & 0 \le s_a \le L/2 \\ 0 & 0 \le s_b \le L/2 \\ \frac{-1}{\sqrt{L}}\sin\left(2\pi s_c/L\right) & 0 \le s_c \le L/2 \end{cases}$$

Now we superpose the wavefunctions so that

$$\psi_E = c_{1A}\cdot\psi_A + c_{1B}\cdot\psi_B$$

and we require that

$$\int \psi_E^2\,\mathrm{d}\tau = 1 \text{ and } \int \psi_C\psi_E\ \mathrm{d}\tau = 0$$

where

$$\psi_C = \begin{cases} \frac{-1}{\sqrt{L}}\sin\left(2\pi s_a/L\right) & 0 \le s_a \le L/2 \\ \frac{1}{\sqrt{L}}\sin\left(2\pi s_b/L\right) & 0 \le s_b \le L/2 \\ 0 & 0 \le s_c \le L/2 \end{cases}$$

For the normalization we find that

$$\begin{aligned} 1 &= \int \psi_E^2\,\mathrm{d}\tau = c_{1A}^2\int \psi_A^2\mathrm{d}\tau + 2c_{1A}c_{1B}\int \psi_A\psi_B\mathrm{d}\tau + c_{1B}^2\int \psi_B^2\mathrm{d}\tau \\ &= c_{1A}^2\cdot(1) + c_{1B}^2\cdot(1) + 2c_{1A}c_{1B}\int \psi_A\psi_B\mathrm{d}\tau \end{aligned}$$

Using the normalized wavefunctions ψ_A and ψ_B, we find that

$$\int \psi_A\psi_B \mathrm{d}\tau = \frac{\int_0^{L/2} 0\cdot\sin\left(2\pi s_a/L\right)\mathrm{d}s_a}{\sqrt{L}} + \frac{\int_0^{L/2} 0\cdot\sin\left(2\pi s_b/L\right)\mathrm{d}s_b}{\sqrt{L}} + \frac{\int_0^{L/2}\sin^2\left(2\pi s_c/L\right)\mathrm{d}s_c}{L}$$

Making the substitution that $x_c = 2\pi s_c/L$, we find

$$\int \psi_A\psi_B \mathrm{d}\tau = \frac{1}{L}\frac{L}{2\pi}\int_0^{L/2}\sin^2\left(x_c\right)\mathrm{d}x_c = \frac{1}{2\pi}\frac{1}{2}\left(x_c - \frac{1}{2}\sin\left(2x_c\right)\right)_0^{\pi} = \frac{1}{4}$$

Inserting this value into the normalization condition gives

$$1 = c_{1A}^2 + c_{1B}^2 + c_{1A}c_{1B}/2$$

which is one condition on the constants c_{1A} and c_{1B}.

Now we apply the orthogonality condition to find the second algebraic equation for the coefficients

$$\begin{aligned}
0 &= \int \psi_C\psi_E\ \mathrm{d}\tau = c_{1A}\int \psi_C\psi_A\ \mathrm{d}\tau + c_{1B}\int \psi_C\psi_B\ \mathrm{d}\tau \\
&= c_{1A}\frac{1}{L}\int_0^{L/2}\sin^2\left(2\pi s_b/L\right)\ \mathrm{d}\ s_b - c_{1B}\frac{1}{L}\int_0^{L/2}\sin^2\left(2\pi s_a/L\right)\ \mathrm{d}\ s_a \\
&= c_{1A}\frac{1}{2\pi}\int_0^{\pi}\sin^2\left(x_b\right)\mathrm{d}x_b - c_{1B}\frac{1}{2\pi}\int_0^{\pi}\sin^2\left(x_a\right)\ \mathrm{d}\ x_a \\
&= c_{1A}\frac{1}{2\pi}\frac{1}{2}\left(x_b - \frac{1}{2}\sin\left(2x_b\right)\right)_0^{\pi} - c_{1B}\frac{1}{2\pi}\frac{1}{2}\left(x_a - \frac{1}{2}\sin\left(2x_a\right)\right)_0^{\pi} \\
&= c_{1A}\frac{1}{2\pi}\frac{\pi}{2} - c_{1B}\frac{1}{2\pi}\frac{\pi}{2} = \frac{1}{4}\left(c_{1A} - c_{1B}\right)
\end{aligned}$$

so that $c_{1A} = c_{1B}$. By substituting into the normalization condition we find that

$$1 = c_{1A}^2 + c_{1B}^2 + c_{1A}c_{1B}/2 = \frac{5}{2}c_{1A}^2$$

or

$$c_{1A} = c_{1B} = \sqrt{2/5}$$

Hence we see that

$$\psi_E = \sqrt{\frac{2}{5}}\left(\psi_A + \psi_B\right)$$

The product $\psi_C\psi_E$ has a positive amplitude in branch a, negative amplitude in branch b, and zero amplitude in branch c.

As was assumed in Exercise 8.8, we found that c_{1A} and c_{1B} must be equal and $\sqrt{\frac{2}{5}}$ is the normalization factor. This comparison shows that our initial choice of a basis orbital ψ along the branches is not unique; however, once it is chosen then the other degenerate wavefunction must be constructed through the requirements of orthogonality and normalization. This means that we can make the orbital choice in a way that is most convenient for the problem.

E8.10 Show that the energy expression in Equation (8.26)

$$\varepsilon_2 = 2c_2^2\cdot\left(H_{aa} - H_{ab}\right)$$

results from using the antisymmetric wavefunction

$$\phi_2 = c_2\cdot\left(\varphi_a - \varphi_b\right)$$

in the Schrödinger equation. Note: you may need to refer to Chapter 6.

Solution

We start with

$$\varepsilon_2 = \frac{\int \phi_2 \mathcal{H} \phi_2 \cdot d\tau}{\int \phi_2 \phi_2 \cdot d\tau}$$

Substituting $\phi_2 = c_2 \cdot (\varphi_a - \varphi_b)$ we find that

$$\varepsilon_2 = \frac{c_2^2 \int (\varphi_a - \varphi_b) \mathcal{H} (\varphi_a - \varphi_b) \cdot d\tau}{c_2^2 \int (\varphi_a - \varphi_b)(\varphi_a - \varphi_b) \cdot d\tau}$$

Expanding the integrands yields

$$\varepsilon_2 = \frac{\int (\varphi_a \mathcal{H} \varphi_a - \varphi_a \mathcal{H} \varphi_b - \varphi_b \mathcal{H} \varphi_a + \varphi_b \mathcal{H} \varphi_b) \cdot d\tau}{\int (\varphi_a \varphi_a - 2\varphi_a \varphi_b + \varphi_b \varphi_b) \cdot d\tau}$$

Let $H_{aa} = \int \varphi_a \mathcal{H} \varphi_a d\tau$, $H_{ab} = \int \varphi_a \mathcal{H} \varphi_b d\tau = \int \varphi_b \mathcal{H} \varphi_a d\tau$, and $H_{bb} = \int \varphi_b \mathcal{H} \varphi_b d\,\tau$, so that we find

$$\varepsilon_2 = \frac{(H_{aa} - 2H_{ab} + H_{bb})}{\int (\varphi_a \varphi_a - 2\varphi_a \varphi_b + \varphi_b \varphi_b) \cdot d\tau}$$

Now let $\int \varphi_a \varphi_a d\tau = \int \varphi_b \varphi_b \, d\tau = 1$ and $\int \varphi_a \varphi_b d\tau = S_{ab}$, so that we now have

$$\varepsilon_2 = \frac{(H_{aa} - 2H_{ab} + H_{bb})}{2(1 - S_{ab})}$$

Because $H_{aa} = H_{bb}$, we can write

$$\varepsilon_2 = \frac{2(H_{aa} - H_{ab})}{2(1 - S_{ab})} = \frac{1}{1 - S_{ab}} (H_{aa} - H_{ab})$$

E8.11 A HMO model treatment of naphthalene gives the π-electron energy level diagram shown in the table.

Table E8.11 List of Eigenvalues, $x_i = (\alpha - \varepsilon_i)/\beta$, which results from an HMO Calculation for Naphthalene.

Orbital i	1	2	3	4	5	6	7
x_i	−2.30	−1.62	−1.30	−1.00	−0.62	0.62	1.00

Use these x_i values to find the ε_i value for each molecular orbital i. Compare the energy gaps between the occupied HMO orbitals to those you obtain from the first four ionization energies: 8.15, 8.87, 10.08, and 10.83 eV of naphthalene and calculate a value for β.[3]

Solution

To compare with experiment we use the schematic energy diagram shown in Figure E8.11a. If we assume that the first four ionization energies correspond to the HMO molecular orbitals (Koopmans Theorem: $IE = -\varepsilon_{\text{orbital}}$, see Section 5.5.4.1), we can calculate the spacing between the levels. Figure E8.11b plots the energy of the ionization energy IP versus the HMO orbital energy $x = (\alpha - \varepsilon)/\beta$. According to Koopmans theorem we have

$$IE = -\varepsilon_{\text{orbital}} = -(\alpha - x\beta) = -\alpha + x\beta$$

In this equation β is the slope of the straight line in Fig. E8.11b. A linear fit to the data gives the slope as −2.07 eV; thus, $\beta = -2.07$ eV.

3 The ionization energies are taken from W. Schmidt, *JCP* 1977, **66**, 828.

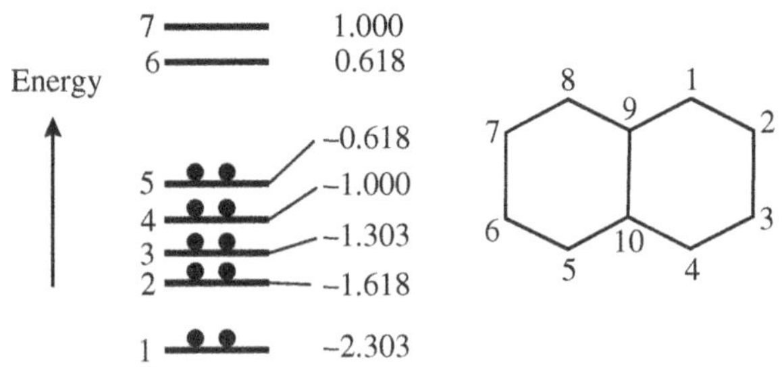

Figure E8.11a HMO eigenvalues $x_i = (\alpha - \varepsilon_i)/\beta$ for naphthalene.

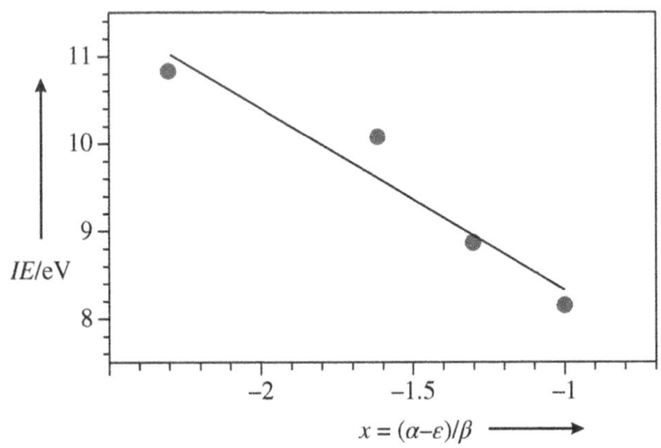

Figure E8.11b The first four ionization energies *IE* for naphthalene versus the HMO orbital eigenvalues $x = (\alpha - \varepsilon)/\beta$. Full circles: experimental data, solid line: regression curve.

E8.12 Use the FEMO model to determine the π-electron charge density of π electron rings with $(4n + 2)$ carbon atoms, $n = 1, 2, 3, \ldots$ (following the Hückel Rule). Compare to ring systems with $4n$ carbon atoms, $n = 1, 2, 3, \ldots$ Compare the degree of bond length alternation in these two cases.

Solution

According to Fig. 8.6 we find two-fold degenerate orbitals in cyclic π-electron systems for energy levels above $n = 0$. For $n = 0$, the energy level is non-degenerate and can be occupied by two electrons. In analogy to the benzene molecule the charge is equally distributed and the bond lengths are equal, if each of the degenerate orbitals is occupied by two electrons. This condition holds when the π -system has $4n + 2$ electrons ($n = 1$ for benzene and $n = 2$ for cyclodecapentaene).

For cyclic molecules with $4n$ electrons the charge cannot be equally distributed. For example, in cyclobutadiene two electrons are in the orbital with quantum number $n = 0$, and the two degenerate orbitals with $n = 1$ can be occupied with one electron each or the electrons could pair in one of the orbitals. As it turns out, the energy of the system is lower if two electrons occupy one of the degenerate orbitals and the second one remains empty. In this case the antinodes of the wavefunction are located between two C-atom sites, causing a contraction of the positively charged C-atom cores that stabilizes the system and leads to double bond formation. This effect leads to an alternation of the bonds as for the polyenes.

E8.13 Consider the cyclopropene radical C_3H_3. Use the HMO method to find the π-electron energy level diagram. Determine the net binding energy in terms of β, for the species $C_3H_3^+$, C_3H_3, or $C_3H_3^-$. Which of these species has the largest net π-bonding and which has the least amount of net π -bonding?

Solution

For cyclopropene the Hückel secular determinant is

$$\begin{vmatrix} (\alpha - \varepsilon) & \beta & \beta \\ \beta & (\alpha - \varepsilon) & \beta \\ \beta & \beta & (\alpha - \varepsilon) \end{vmatrix} = 0$$

By expanding the determinant one finds that

$$
\begin{aligned}
0 &= (\alpha-\varepsilon)\begin{vmatrix}(\alpha-\varepsilon) & \beta \\ \beta & (\alpha-\varepsilon)\end{vmatrix} - \beta\begin{vmatrix}\beta & \beta \\ \beta & (\alpha-\varepsilon)\end{vmatrix} + \beta\begin{vmatrix}\beta & (\alpha-\varepsilon) \\ \beta & \beta\end{vmatrix} \\
&= (\alpha-\varepsilon)\left[(\alpha-\varepsilon)(\alpha-\varepsilon)-\beta^2\right] - \beta\left[\beta(\alpha-\varepsilon)-\beta^2\right] + \beta\left[\beta^2-\beta(\alpha-\varepsilon)\right] \\
&= (\alpha-\varepsilon)(\alpha-\varepsilon)(\alpha-\varepsilon) - \beta^2(\alpha-\varepsilon) - \beta^2(\alpha-\varepsilon) + \beta^3 + \beta^3 - \beta^2(\alpha-\varepsilon) \\
&= (\alpha-\varepsilon)(\alpha-\varepsilon)(\alpha-\varepsilon) - 3\beta^2(\alpha-\varepsilon) + 2\beta^3 \\
&= \alpha^3 - 3\alpha^2\varepsilon - 3\alpha\beta^2 + 3\alpha\varepsilon^2 + 2\beta^3 + 3\beta^2\varepsilon - \varepsilon^3
\end{aligned}
$$

This polynomial can be factored to give

$$(\varepsilon - \alpha + \beta)^2 (\varepsilon - \alpha - 2\beta) = 0$$

By inspection, we see that it has the three roots $\alpha - \beta, \alpha - \beta$, and $\alpha + 2\beta$. Figure E8.13 shows the energy level structure for each of the species

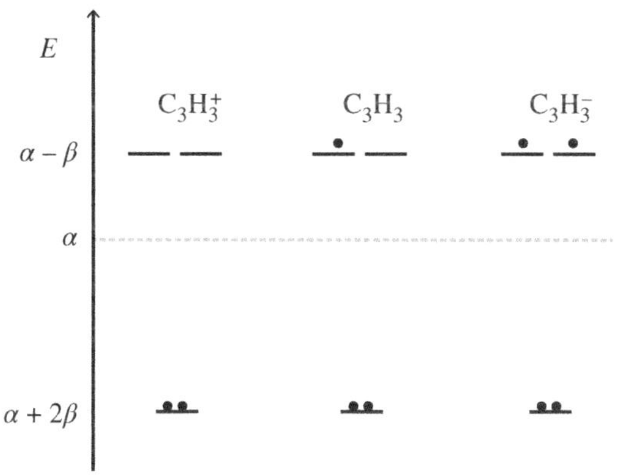

Figure E8.13 HMO orbital diagrams of $C_3H_3^+$, C_3H_3, and $C_3H_3^-$.

Because C_3H_3 has three π electrons, their binding energy is

$$2 \cdot (\alpha + 2\beta) + (\alpha - \beta) = 3\alpha + 3\beta,$$

or 3β below the self-energy of the three electrons. Because $C_3\ H_3^+$ has only two π electrons, the binding is

$$2 \times (\alpha + 2\beta) = 2\alpha + 4\beta$$

or 4β below the self-energy of the two electrons. Because $C_3H_3^-$ has four π electrons, the binding is

$$2 \times (\alpha + 2\beta) + 2 \times (\alpha - \beta) = 4\alpha + 2\beta$$

or 2β below the self-energy of the electrons. Hence we see that the cation has the largest π bond stabilization energy (4β), the neutral radical species has (3β) of π-bond stabilization energy, and the anion has the least amount of π-bond stabilization energy, with only 2β.

E8.14 Analyze the π-structure of cyclopropene radical using the FEMO model and compare your results to those found with the HMO model in E8.13.

Solution

The energy of the FEMO model is

$$E_n = \frac{h^2}{2m_e L_c^2} \cdot n^2$$

For cyclopropene we have $L_c = 3d_0 = 420$ pm where we have used $d_0 = 140$ pm The first two energy levels are given by the expression

$$E_n = \frac{\left(6.626 \times 10^{-34}\ \text{J s}\right)^2}{2 \times \left(9.109 \times 10^{-31}\ \text{kg}\right) \times \left(420 \times 10^{-12}\ \text{m}\right)^2} \cdot n^2$$

so that

$$E_0 = 0;\ \text{and}\ E_1 = 1.37 \times 10^{-18}\ \text{J}$$

This model gives the same orbital structure as that found in the HMO treatment, and we would expect a similar trend in energetics.

E8.15 Use the normalization condition to determine the normalization factor for the molecular wavefunctions ϕ_1 through ϕ_4 that we found for butadiene in Section 8.6.2 (Table 8.3).

Solution

The normalization condition for ϕ_1 is $\int \phi_1\phi_1 d\tau = 1$.

$$1 = N^2 \int \left(0.372\varphi_1 + 0.602\varphi_2 + 0.602\varphi_3 + 0.372\varphi_4\right)^2 d\tau$$
$$= N^2 \left[(0.372)^2 + (0.602)^2 + (0.602)^2 + (0.372)^2\right] = N^2\,[1.001\,6]$$

where we have used the fact that $\int \varphi_i\varphi_i d\tau = 1$ and $\int \varphi_i\varphi_j d\tau = 0$ in the HMO model. Solving for N we have

$$N = \frac{1}{\sqrt{1.0016}} = 0.9992$$

Note that this shows that ϕ_1 is normalized.

We can generalize the above process to show that the normalization constant is

$$N = \frac{1}{\sqrt{\sum_{i=1}^{4} c_i^2}}$$

In particular

$$1 = N^2 \int \left(c_1\varphi_1 + c_2\varphi_2 + c_3\varphi_3 + c_4\varphi_4\right)^2 d\tau$$
$$= N^2 \left[c_1^2 + c_2^2 + c_3^2 + c_4^2\right] = N^2 \sum_{i=1}^{4} c_i^2$$

which rearranges to the stated result.

Using this general expression, we find for ϕ_2 that

$$(0.602)^2 + (0.372)^2 + (-0.372)^2 + (-0.602)^2 = 1.001\,6 \approx 1$$

for ϕ_3 we have

$$(0.602)^2 + (-0.372)^2 + (-0.372)^2 + (0.602)^2 = 1.001\,6 \approx 1$$

and for ϕ_4 we have

$$(0.372)^2 + (-0.602)^2 + (0.602)^2 + (-0.372)^2 = 1.001\,6 \approx 1$$

Hence the wavefunctions given in Table 8.3 are normalized, with the small deviation given above resulting from rounding error in the numerical evaluations.

E8.16 Consider the orthogonality of the HMO wavefunctions found for butadiene. Note that the orthogonality for $\psi_1\psi_2$, $\psi_1\psi_4$, $\psi_2\psi_3$, and $\psi_3\psi_4$ follows directly from the symmetry of the wavefunctions, see Fig. 8.18. The orthogonality of the remaining two pairs of wavefunctions can be shown by evaluating the integral $\int \phi_i\phi_j d\tau$. Perform these calculations using the normalized molecular wavefunctions found for butadiene in Section 8.6.2 (Table 8.3).

Solution

To demonstrate orthogonality for the remaining cases $\phi_1\phi_3$ and $\phi_3\phi_4$ we require that

$$\int \phi_i \phi_j \mathrm{d}\tau = 0 \text{ for } i \neq j$$

For the case of ϕ_1 and ϕ_3 this requirement leads to

$$\int \phi_1 \phi_3 \mathrm{d}\tau = \int \left(c_{11}\varphi_1 + c_{12}\varphi_2 + c_{13}\varphi_3 + c_{14}\varphi_4\right) \cdot \left(c_{31}\varphi_1 + c_{32}\varphi_2 + c_{33}\varphi_3 + c_{34}\varphi_4\right) \mathrm{d}\tau$$

Here we are using the first subscript in the constant c for the molecular orbital and the second subscript for the atomic orbital. Next we expand the integrand.

$$\begin{aligned}
&\left(c_{11}\varphi_1 + c_{12}\varphi_2 + c_{13}\varphi_3 + c_{14}\varphi_4\right) \cdot \left(c_{31}\varphi_1 + c_{32}\varphi_2 + c_{33}\varphi_3 + c_{34}\varphi_4\right) \\
&\quad = \left(c_{11}c_{31}\varphi_1^2 + c_{11}c_{32}\varphi_1\varphi_2 + c_{11}c_{33}\varphi_1\varphi_3 + c_{11}c_{34}\varphi_1\varphi_4\right) \\
&\qquad + \left(c_{12}c_{31}\varphi_2\varphi_1 + c_{12}c_{32}\varphi_2^2 + c_{12}c_{33}\varphi_2\varphi_3 + c_{12}c_{34}\varphi_2\varphi_4\right) \\
&\qquad + \left(c_{13}c_{31}\varphi_3\varphi_1 + c_{13}c_{32}\varphi_3\varphi_2 + c_{13}c_{33}\varphi_3^2 + c_{13}c_{34}\varphi_3\varphi_4\right) \\
&\qquad + \left(c_{14}c_{31}\varphi_4\varphi_1 + c_{14}c_{32}\varphi_4\varphi_2 + c_{14}c_{33}\varphi_4\varphi_3 + c_{14}c_{34}\varphi_4^2\right)
\end{aligned}$$

In the integration we obtain terms of the form $\int c_{11}c_{3j}\varphi_1\varphi_j \mathrm{d}\tau$ with $j = 2,3,4$ and a term of the form $\int c_{11}c_{31}\varphi_1^2 \mathrm{d}\tau$ The terms of the form $\int c_{11}c_{31}\varphi_1\varphi_j \mathrm{d}\tau = c_{11}c_{31}\int \varphi_1\varphi_j \mathrm{d}\tau$ are zero because of the orthogonality of the atomic orbitals. The terms of the form $\int c_{11}c_{31}\varphi_1^2 \mathrm{d}\tau = c_{11}c_{31}\int \varphi_1^2 \mathrm{d}\tau$ are equal to $c_{11}c_{31}$ because the atomic orbitals are normalized. Inspection of the products involving c_{12}, c_{13},and c_{14} proceeds in an analogous fashion. Thus for the integral we obtain the following four contributions.

$$\int \phi_1 \phi_3 \mathrm{d}\tau = c_{11}c_{31} + c_{12}c_{32} + c_{13}c_{33} + c_{14}c_{34}$$

Lastly, we insert the constants c from Table 8.2.

$$\int \phi_1 \phi_3 \mathrm{d}\tau = 0.372 \cdot 0.602 - 0.602 \cdot 0.372 - 0.602 \cdot 0.372 + 0.372 \cdot 0.602 = 0$$

We repeat this procedure for the wavefunctions ϕ_2 and ϕ_4 to obtain

$$\int \phi_2 \phi_4 \mathrm{d}\tau = 0.602 \cdot 0.372 - 0.372 \cdot 0.602 - 0.372 \cdot 0.602 + 0.602 \cdot 0.372 = 0$$

Thus we have verified that all wavefunctions of butadiene are orthogonal to each other.

E8.17 Show that the energy ε_2 for the antibonding combination of the $2p_z$ orbitals in ethene (Equation (8.25)) can be written as

$$\varepsilon_2 = \int \phi_2 H \phi_2 \cdot \mathrm{d}\tau = 2c_2^2 \cdot \left(H_{aa} - H_{ab}\right)$$

Solution

The wavefunction ϕ_2 for the first excited state must be orthogonal to ϕ_1 and corresponds to the antisymmetric combination

$$\phi_2 = c_2 \cdot \left(\varphi_a - \varphi_b\right)$$

The proof here follows the same procedure that is used in Exercise 6.16 for the excited state of the H_2^+ ion. The normalization factor and the energy are,

$$c_2 = \frac{1}{\sqrt{2(1 - S_{ab})}}, \quad \varepsilon_2 = \frac{H_{aa} - H_{ab}}{1 - S_{ab}}$$

The only difference to the H_2^+ case is that the atomic orbitals are Slater-type p_z-orbitals rather than $1s$ orbitals.

E8.18 Using the FEMO model for the most weakly bound π-electron in butadiene, determine the uncertainty in the electron's momentum. Compare this uncertainty to that for a π electron in ethene. It may be helpful to refer to Section 3.5.5.

Solution

In Chapter 3 we define the uncertainty in the momentum Δp by

$$\Delta p = \sqrt{\overline{p^2} - \overline{p}^2}$$

where $\overline{p^2}$ is the average of the momentum squared and $\overline{p}^2$ is the square of the average momentum $\overline{p}$. In the FEMO model, the electron wavefunctions are particle-in-a-box wavefunctions, namely $\psi = \sqrt{2/L}\sin(\pi ns/L)$ and the energy eigenvalues correspond to the kinetic energy of the electron wave.

To find $\overline{p^2}$, we use the fact that the average kinetic energy $\overline{T}$ is

$$\overline{T} = \frac{1}{2m_e}\overline{p^2} \quad \text{and} \quad \overline{T} = \frac{h^2}{8m_e L^2}n^2$$

so that

$$\overline{p^2} = \frac{h^2}{4L^2}n^2$$

To find the average momentum $\overline{p}$ we use the fact that

$$\overline{p} = \int_0^L \psi_n \; p_{op} \; \psi_n \, \mathrm{d}s = -\left(i\frac{h}{2\pi}\right)\int_0^L \psi_n \; \frac{\mathrm{d}}{\mathrm{d}s} \; \psi_n \, \mathrm{d}s$$

so that

$$\begin{aligned}\overline{p} &= -\left(i\frac{h}{2\pi}\right)\frac{2}{L}\int_0^L \sin\left(\frac{\pi ns}{L}\right) \frac{\mathrm{d}}{\mathrm{d}s}\sin\left(\frac{\pi ns}{L}\right) \mathrm{d}s \\ &= -\left(i\frac{h}{2\pi}\right)\frac{2}{L}\left[\sin^2\left(\frac{\pi ns}{L}\right)\right]_0^L \\ &= -\left(i\frac{h}{\pi L}\right)\left[\sin^2(\pi n) - 0\right] = 0\end{aligned}$$

where the last equality holds because n is an integer. The average momentum of the electron is zero. This is what we expect for these linear ethene and butadiene systems where it is equally probable that the electron moves from left to right or from right to left.

Hence we find that

$$\begin{aligned}\Delta p &= \sqrt{\overline{p^2} - \overline{p}^2} = \sqrt{\frac{h^2}{4L^2}n^2 - 0} \\ &= \frac{h}{2L}n\end{aligned}$$

For ethene the most weakly bound electron has $n = 1$ and $L = 3d_0 = 420$ pm ($d_0 = 140$ pm), so that

$$\Delta p = \frac{6.626 \times 10^{-34}\ \text{J s}}{2 \cdot 420 \times 10^{-12}\ \text{m}} \cdot 1 = 7.89 \times 10^{-19}\ \text{kg m s}^{-1}$$

For butadiene the most weakly bound electron has $n = 2$ and $L = 5d_0 = 700$ pm ($d_0 = 140$ pm), so that

$$\Delta p = \frac{6.626 \times 10^{-34}\ \text{J s}}{2 \cdot 700 \times 10^{-12}\ \text{m}} \cdot 2 = 9.47 \times 10^{-19}\ \text{kg m s}^{-1}$$

Why is the uncertainty in the momentum higher for the π HOMO electrons of butadiene than those of ethene? The FEMO-derived energy of the π-electrons corresponds to kinetic energy, which implies that $\sqrt{\overline{p^2}} = p_{\text{rms}}$ increases with increasing orbital energy. Because the electron forms a standing wave along the C-skeleton (e.g., from left to right), it is equally likely that the electron is moving from left-to-right (p_{rms}) as it is to be moving from right-to-left ($-p_{\text{rms}}$). The difference in these values increases linearly with p_{rms}. Thus we can trace the increase of Δp for the HOMO of butadiene as compared to ethene to its HOMO electrons' increased kinetic energy.

E8.19 In Exercise 8.4 you use the FEMO model to find the resonance energy of butadiene. Perform the same calculation using the HMO model of butadiene. In order to determine a value for the β parameter, compare the π-electron energy levels to the experimental ionization energies of butadiene's π-electrons.

Solution

In Exercise 8.4 we found that

$$\varepsilon_R = \Delta E_{\text{butadiene}} - 2 \times \Delta E_{\text{ethene}} = 0.60 \times 10^{-19}\,\text{J}$$

from the experimental data, and that the FEMO model gives

$$\varepsilon_{R,\text{FEMO}} = 1.48 \times 10^{-19}\,\text{J}$$

This result is the same order of magnitude as the experimental resonance energy, but about 2 to 3 times larger.

Section 8.6 of the text uses the HMO model to calculate and compare the π-bond energetics of ethene and butadiene. For ethene we calculated a π -bond energy of 2 β and for butadiene we calculated π-bond energy levels of 1.618 β and 0.618 β. By convention β has a negative value. The energy difference of β between the butadiene levels should correspond roughly to the 2.5 eV difference observed in the photoelectron spectrum (see Section 8.3). Using the net π-bond energy for butadiene of 4.472 β gives a resonance energy of

$$\varepsilon_{R,\text{HMO}} = 2 \times 2\,\beta - 4.472\,\beta = -0.472\,\beta$$

Using $\beta = -2.5$ eV gives

$$\varepsilon_{R,\text{HMO}} = 0.472 \cdot 2.5\,\text{eV} \times \frac{1.602 \times 10^{-19}\,\text{J}}{\text{eV}} = 1.89 \times 10^{-19}\,\text{J}$$

Although of the same order as the experimental value, it is a factor of three larger.

E8.20 Using the bond orders in the HMO model and the bond lengths curve in Fig. 8.19, compute the bond lengths for anthracene (Figure E8.20a). Compare your result to the experimental data.[4]

Bond	Bond Order	Bond Length/pm (Experiment)
1–2	0.736	137
1–11	0.539	144
2–3	0.582	142
11–12	0.486	143
9–11	0.606	140

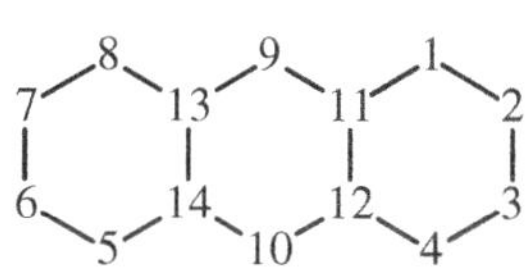

Figure E8.20a. The diagram shows the anthracene carbon framework and numbers the carbon atom sites.

Solution

The solution is obtained by using the data in Table 8.4 and reading the bond lengths from the curve. Figure E8.20b fits the data to a quadratic equation in which

$$d/\text{pm} = 30.525 \;\; (\text{bond order})^2 - 72.036 \cdot (\text{bond order}) + 174.31$$

and d is given in pm. Using this fit we find

Bond	Bond Order	*d*/pm (Experiment)	*d*/pm (HMO)
1–2	0.736	137	137.8
1–11	0.539	144	144.4
2–3	0.582	142	142.7
11–12	0.486	143	146.5
9–11	0.606	140	142.0

4 The bond lengths are taken from D.W.J. Cruickshank and R.A. Sparks, Experimental and theoretical determinations of bond lengths in naphthalene, anthracene and other hydrocarbons, *Proc. R. Soc. Lond. A Math. Phys. Sci.* 1960, **258**, 270–285.

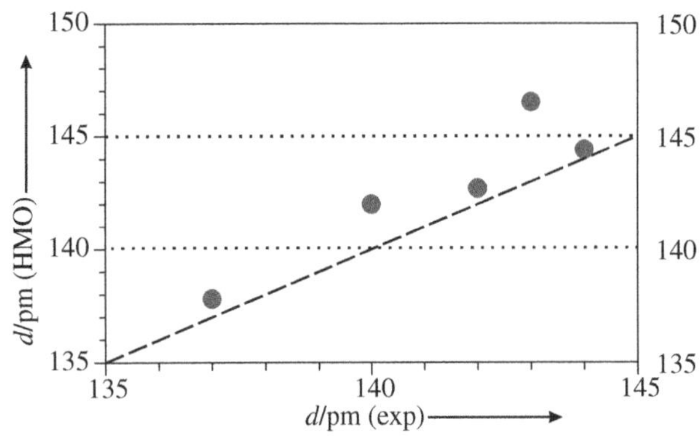

Figure E8.20b Bond length d (HMO) versus d (exp). The dashed line indicates a perfect agreement of both bond lengths.

While the predicted bond lengths are in a reasonable range, they are typically larger than the experimental values and the experimental bond lengths do not systematically follow the bond order trend.

E8.21 Example 8.1 provides the energy eigenvalues and the wavefunctions for the occupied π-electron levels of the amidinium ion. Draw the energy level diagram and plot the wavefunction amplitudes and the probability densities for each of the energy levels; i.e., make a figure like that of Fig. 8.5 but for the amidinium ion.

Solution

In Example 8.1 we showed that amidinium will have 4 π-electrons in the two lowest energy orbitals, at $E_1 = 1.92 \times 10^{-19}$ J and $E_2 = 7.68 \times 10^{-19}$ J, and we showed that the orbital wavefunctions are given by

$$\psi_1 = \sqrt{\frac{2}{L}} \sin\left(\frac{\pi s}{L}\right) \quad \text{and} \quad \psi_2 = \sqrt{\frac{2}{L}} \sin\left(\frac{2\pi s}{L}\right)$$

where L is 560 pm and s ranges from zero to 560 pm. Figure E8.21 shows the energy level plot, the wavefunctions, and the probability densities.

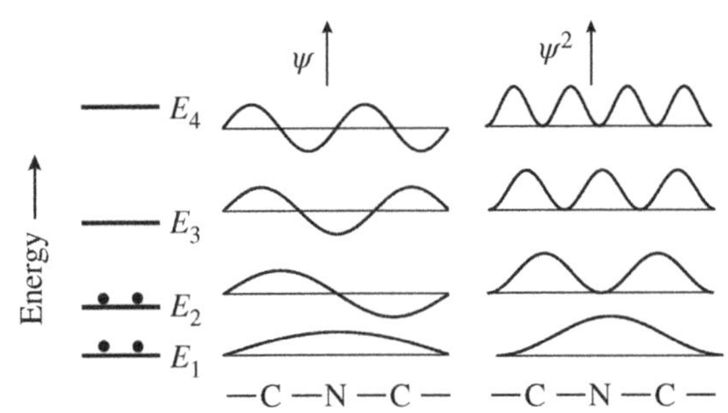

Figure E8.21 Energy level plot, the wavefunctions, and the probability densities for the amidinium ion.

E8.22 (Hückel rule.) Extend the argument in Section 8.3.3.1 to the case of equal bond lengths in rings with $(4n + 2)$ carbon atoms: e.g., benzene ($n = 1$), [10] annulene (decapentaene, $n = 2$), [14] annulene ($n = 3$).

Solution

According to Fig. 8.7 we find two-fold degenerate orbitals in cyclic π-electron systems. In analogy to the benzene molecule the charge is equally distributed and the bond lengths are equal, if each of the degenerate orbitals is occupied by two electrons. This is the case if there are $4n + 2$ electrons in the molecule ($n = 1$ for benzene, $n = 2$ for cyclodecapentaene).

For cyclic molecules with $4n$ electrons the charge cannot be equally distributed. For example, in cyclobutadiene two electrons are in the orbital with quantum number $n = 0$, and the two degenerate orbitals with $n = 1$ can be occupied with one electron each. However, the energy of the system is lower if two electrons occupy one of the degenerate orbitals and the second one remains empty. In this case, the antinodes of that orbital's wavefunction concentrate the electron density (a formal double bond) and attract the positively charged nuclei, shortening the bond distance in that region and stabilizing the system. This effect leads to an alternation of the bonds as for the polyenes.

8.2 Problems

P8.1 Use the HMO method to calculate the energy levels for fulvene's π electrons. Make an orbital energy diagram and show its filling with electrons. You can use fulvene.mcdx to facilitate this calculation.

Solution

This problem is solved by setting up the secular determinant for fulvene

$$\begin{vmatrix} (\alpha-\varepsilon) & \beta & 0 & 0 & 0 & 0 \\ \beta & (\alpha-\varepsilon) & \beta & 0 & 0 & \beta \\ 0 & \beta & (\alpha-\varepsilon) & \beta & 0 & 0 \\ 0 & 0 & \beta & (\alpha-\varepsilon) & \beta & 0 \\ 0 & 0 & 0 & \beta & (\alpha-\varepsilon) & \beta \\ 0 & \beta & 0 & 0 & \beta & (\alpha-\varepsilon) \end{vmatrix} = 0$$

If we define $x = (\alpha - \varepsilon)/\beta$, then we can write the determinant as

$$\begin{vmatrix} x & 1 & 0 & 0 & 0 & 0 \\ 1 & x & 1 & 0 & 0 & 1 \\ 0 & 1 & x & 1 & 0 & 0 \\ 0 & 0 & 1 & x & 1 & 0 \\ 0 & 0 & 0 & 1 & x & 1 \\ 0 & 1 & 0 & 0 & 1 & x \end{vmatrix} = 0$$

To find the resulting polynomial we expand the determinant

$$x\begin{vmatrix} x & 1 & 0 & 0 & 1 \\ 1 & x & 1 & 0 & 0 \\ 0 & 1 & x & 1 & 0 \\ 0 & 0 & 1 & x & 1 \\ 1 & 0 & 0 & 1 & x \end{vmatrix} - 1\begin{vmatrix} 1 & 1 & 0 & 0 & 1 \\ 0 & x & 1 & 0 & 0 \\ 0 & 1 & x & 1 & 0 \\ 0 & 0 & 1 & x & 1 \\ 0 & 0 & 0 & 1 & x \end{vmatrix} = 0$$

continuing

$$0 = x^2\begin{vmatrix} x & 1 & 0 & 0 \\ 1 & x & 1 & 0 \\ 0 & 1 & x & 1 \\ 0 & 0 & 1 & x \end{vmatrix} - x\begin{vmatrix} 1 & 1 & 0 & 0 \\ 0 & x & 1 & 0 \\ 0 & 1 & x & 1 \\ 1 & 0 & 1 & x \end{vmatrix} + x\begin{vmatrix} 1 & x & 1 & 0 \\ 0 & 1 & x & 1 \\ 0 & 0 & 1 & x \\ 1 & 0 & 0 & 1 \end{vmatrix}$$

$$-1\begin{vmatrix} x & 1 & 0 & 0 \\ 1 & x & 1 & 0 \\ 0 & 1 & x & 1 \\ 0 & 0 & 1 & x \end{vmatrix} + 1\begin{vmatrix} 0 & 1 & 0 & 0 \\ 0 & x & 1 & 0 \\ 0 & 1 & x & 1 \\ 0 & 0 & 1 & x \end{vmatrix} - 1\begin{vmatrix} 0 & x & 1 & 0 \\ 0 & 1 & x & 1 \\ 0 & 0 & 1 & x \\ 0 & 0 & 0 & 1 \end{vmatrix}$$

continuing

$$0 = x^3\begin{vmatrix} x & 1 & 0 \\ 1 & x & 1 \\ 0 & 1 & x \end{vmatrix} - x^2\begin{vmatrix} 1 & 1 & 0 \\ 0 & x & 1 \\ 0 & 1 & x \end{vmatrix} - x\begin{vmatrix} x & 1 & 0 \\ 1 & x & 1 \\ 0 & 1 & x \end{vmatrix} + x\begin{vmatrix} 0 & 1 & 0 \\ 0 & x & 1 \\ 1 & 1 & x \end{vmatrix}$$

$$+x\begin{vmatrix} 1 & x & 1 \\ 0 & 1 & x \\ 0 & 0 & 1 \end{vmatrix} - x^2\begin{vmatrix} 0 & x & 1 \\ 0 & 1 & x \\ 1 & 0 & 1 \end{vmatrix} + x\begin{vmatrix} 0 & 1 & 1 \\ 0 & 0 & x \\ 1 & 0 & 1 \end{vmatrix} - x\begin{vmatrix} x & 1 & 0 \\ 1 & x & 1 \\ 0 & 1 & x \end{vmatrix}$$

$$+\begin{vmatrix} 1 & 1 & 0 \\ 0 & x & 1 \\ 0 & 1 & x \end{vmatrix} - 1\begin{vmatrix} 0 & 1 & 0 \\ 0 & x & 1 \\ 0 & 1 & x \end{vmatrix} + x\begin{vmatrix} 0 & x & 1 \\ 0 & 1 & x \\ 0 & 0 & 1 \end{vmatrix} - 1\begin{vmatrix} 0 & 1 & 1 \\ 0 & 0 & x \\ 0 & 0 & 1 \end{vmatrix}$$

continuing

$$\begin{aligned}
0 = x^3 &\left(x \begin{vmatrix} x & 1 \\ 1 & x \end{vmatrix} - 1 \begin{vmatrix} 1 & 1 \\ 0 & x \end{vmatrix} \right) - x^2 \left(1 \begin{vmatrix} x & 1 \\ 1 & x \end{vmatrix} - 1 \begin{vmatrix} 0 & 1 \\ 0 & x \end{vmatrix} \right) \\
&-x \left(x \begin{vmatrix} x & 1 \\ 1 & x \end{vmatrix} - 1 \begin{vmatrix} 1 & 1 \\ 0 & x \end{vmatrix} \right) + x \left(-1 \begin{vmatrix} 0 & 1 \\ 1 & x \end{vmatrix} \right) \\
&+x \left(1 \begin{vmatrix} 1 & x \\ 0 & 1 \end{vmatrix} - x \begin{vmatrix} 0 & x \\ 0 & 1 \end{vmatrix} + 1 \begin{vmatrix} 0 & 1 \\ 0 & 0 \end{vmatrix} \right) - x^2 \left(-x \begin{vmatrix} 0 & x \\ 1 & 1 \end{vmatrix} + 1 \begin{vmatrix} 0 & 1 \\ 1 & 0 \end{vmatrix} \right) \\
&+x \left(-1 \begin{vmatrix} 0 & x \\ 1 & 1 \end{vmatrix} + 1 \begin{vmatrix} 0 & 0 \\ 1 & 0 \end{vmatrix} \right) - x \left(x \begin{vmatrix} x & 1 \\ 1 & x \end{vmatrix} - 1 \begin{vmatrix} 1 & 1 \\ 0 & x \end{vmatrix} \right) \\
&+ \left(1 \begin{vmatrix} x & 1 \\ 1 & x \end{vmatrix} - 1 \begin{vmatrix} 0 & 1 \\ 0 & x \end{vmatrix} \right) - 1 \left(-1 \begin{vmatrix} 0 & 1 \\ 0 & x \end{vmatrix} \right) \\
&+x \left(-x \begin{vmatrix} 0 & x \\ 0 & 1 \end{vmatrix} + 1 \begin{vmatrix} 0 & 1 \\ 0 & 0 \end{vmatrix} \right) - 1 \left(-1 \begin{vmatrix} 0 & x \\ 0 & 1 \end{vmatrix} + 1 \begin{vmatrix} 0 & 0 \\ 0 & 0 \end{vmatrix} \right)
\end{aligned}$$

and continuing

$$\begin{aligned}
0 = &\left(x^4 \left(x^2 - 1\right) - x^4\right) - x^2 \left(x^2 - 1\right) - 0 - x^2 \left(x^2 - 1\right) \\
&+x^2 + x + x - 0 + 0 - x^4 + x^2 + x^2 + 0 \\
&-x^2 \left(x^2 - 1\right) + x^2 + \left(x^2 - 1\right) - 0 - 0 - 0 + 0 - 0 + 0
\end{aligned}$$

Now we collect terms of like powers, and find

$$\begin{aligned}
0 &= x^6 - x^4 - x^4 - x^4 + x^2 - x^4 + x^2 + x^2 + x + x \\
&\quad -x^4 + x^2 + x^2 - x^4 + x^2 + x^2 + x^2 - 1 \\
&= x^6 - 6x^4 + 8x^2 + 2x - 1
\end{aligned}$$

The roots to this equation can be found by plotting the polynomial and finding the x values where it is zero; we find that the roots are

$$\begin{aligned}
&x = -1.0, \quad x = 0.2541, \quad x = -2.1149, \\
\text{and} \quad &x = -0.61803, \quad x = 1.618, \quad x = 1.608
\end{aligned}$$

To find the energies we use the roots and the definition $x = (\alpha - \varepsilon)/\beta$ to solve for ε. The energies are

$$-1.0 = \frac{(\alpha - \varepsilon)}{\beta}; \qquad \varepsilon = \alpha + \beta$$

$$0.2541 = \frac{(\alpha - \varepsilon)}{\beta}; \qquad \varepsilon = \alpha - 0.2541\beta$$

$$-2.1149 = \frac{(\alpha - \varepsilon)}{\beta}; \qquad \varepsilon = \alpha + 2.1149\beta$$

$$-0.61803 = \frac{(\alpha - \varepsilon)}{\beta}; \qquad \varepsilon = \alpha + 0.61803\beta$$

$$1.618 = \frac{(\alpha - \varepsilon)}{\beta}; \qquad \varepsilon = \alpha - 1.618\beta$$

$$1.8608 = \frac{(\alpha - \varepsilon)}{\beta}; \qquad \varepsilon = \alpha - 1.8608\beta$$

Figure P8.1 plots the four bound energy levels and fills the bottom three with fulvene's six π electrons.

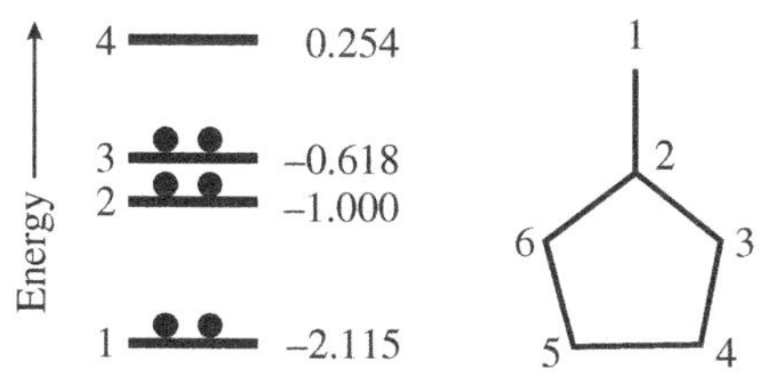

Figure P8.1 HMO energy scheme for fulvene. The numbers denote the eigenvalues $x = (\alpha - \varepsilon)/\beta$.

P8.2 Using the HMO coefficients listed below and the bond lengths curve in Fig. 8.19, calculate the bond lengths of fulvene from the HMO bond orders.

	c_1	c_2	c_3	c_4	c_5	c_6	Energy
ϕ_1	0.247	0.523	0.429	0.385	0.385	0.429	2.11β
ϕ_2	0.500	0.500	0.00	−0.500	−0.500	0.00	1.00β
ϕ_3	0.00	0.00	0.602	0.372	−0.372	−0.602	0.62β
ϕ_4	−0.749	0.190	0.351	−0.280	−0.280	0.351	-0.25β
ϕ_5	0.00	0.00	−0.372	0.602	−0.602	0.372	-1.62β
ϕ_6	0.357	−0.664	0.439	−0.153	−0.153	0.439	-1.86β

Solution

The HMO bond order for bonded atoms i and j is defined by Equation (8.30)

$$\text{HMO bond order} = \sum_k N_k c_{ik} c_{jk}$$

where c_{ik} and c_{jk} are the HMO coefficients (see Foundation 8.4) of the atomic orbitals on atoms i and j in molecular orbital k, and N_k is the number of π electrons in molecular orbital k.

Using Equation (8.30) for the bond between atoms $i = 1$ and $j = 2$, the HMO bond order is given by the sum over the first three molecular orbitals, namely

$$\text{HMO bond order} = 2c_{11}c_{12} + 2c_{12}c_{22} + 2c_{13}c_{23}$$
$$= 2 \cdot 0.247 \cdot 0.523 + 2 \cdot 0.500 \cdot 0.500) + 2 \cdot 0 \cdot 0 = 0.758$$

Note that molecular orbitals 4, 5, and 6 do not contribute because their electron populations are zero; i.e., $N_4 = 0$, $N_5 = 0$, and $N_6 = 0$. In a corresponding way we calculate the bond order for the 2–3 bond by

$$\text{HMO bond order} = 2 \cdot 0.523 \cdot 0.429 + 2 \cdot 0.500 \cdot 0 + 2 \cdot 0 \cdot 0.602 = 0.448$$

and the bond order for the 3–4 bond is

$$\text{HMO bond order} = 2 \cdot 0.429 \cdot 0.385 + 2 \cdot (-0.500) \cdot 0 + 2 \cdot \cdot 0.372 \cdot 0.602 = 0.778$$

From symmetry it is evident that the two sides of the molecule are equivalent so that the 3–4 and 5–6 bond orders are equal, as well as the 2–3 and 2–6 bond orders. Lastly, we calculate the 4–5 bond order in the same manner and obtain 0.520.

Using these values we obtain $d_{12} = 137$ pm, $d_{23} = 148$ pm, $d_{34} = 137$ pm and $d_{45} = 145$ pm from the curve constructed in Fig. 8.19.

P8.3 Use the HMO method to calculate the π electron energies (in terms of β) for cyclobutadiene in a square geometry and in a rectangular geometry of the carbon ring. To model the rectangular geometry, assume that the nearest neighbor coupling is β for the short bond length and it is β' for the long bond length. Comment on which structure you find to be more reasonable for cyclobutadiene.

Solution

The secular determinant for square cyclobutadiene is

$$\begin{vmatrix} (\alpha-\varepsilon) & \beta & 0 & \beta \\ \beta & (\alpha-\varepsilon) & \beta & 0 \\ 0 & \beta & (\alpha-\varepsilon) & \beta \\ \beta & 0 & \beta & (\alpha-\varepsilon) \end{vmatrix} = 0$$

The π-electron energies are found by finding the eigenvalues and eigenvectors to

$$\begin{vmatrix} x & 1 & 0 & 1 \\ 1 & x & 1 & 0 \\ 0 & 1 & x & 1 \\ 1 & 0 & 1 & x \end{vmatrix} = 0$$

so that

$$\begin{aligned} 0 &= x\begin{vmatrix} x & 1 & 0 \\ 1 & x & 1 \\ 0 & 1 & x \end{vmatrix} - \begin{vmatrix} 1 & 1 & 0 \\ 0 & x & 1 \\ 1 & 1 & x \end{vmatrix} - \begin{vmatrix} 1 & x & 1 \\ 0 & 1 & x \\ 1 & 0 & 1 \end{vmatrix} \\ &= x^2\begin{vmatrix} x & 1 \\ 1 & x \end{vmatrix} - x\begin{vmatrix} 1 & 1 \\ 0 & x \end{vmatrix} - \begin{vmatrix} x & 1 \\ 1 & x \end{vmatrix} + \begin{vmatrix} 0 & 1 \\ 1 & x \end{vmatrix} - \begin{vmatrix} 1 & x \\ 0 & 1 \end{vmatrix} + x\begin{vmatrix} 0 & x \\ 1 & 1 \end{vmatrix} - \begin{vmatrix} 0 & 1 \\ 1 & 0 \end{vmatrix} \\ &= x^4 - x^2 - x^2 - x^2 + 1 - 1 - 1 - x^2 + 1 \\ &= x^2\left(x^2 - 4\right) \end{aligned}$$

By inspection we see that the roots for this equation are 0,0, 2, and −2. Using the definition $x = (\alpha - \varepsilon)/\beta$, we find that the energies are

$$-2 = \frac{\alpha - \varepsilon_1}{\beta}; \qquad \varepsilon_1 = \alpha + 2\beta$$

$$0 = \frac{\alpha - \varepsilon_2}{\beta}; \qquad \varepsilon_2 = \alpha$$

$$0 = \frac{\alpha - \varepsilon_3}{\beta}; \qquad \varepsilon_3 = \alpha$$

$$2 = \frac{\alpha - \varepsilon_4}{\beta}; \qquad \varepsilon_4 = \alpha - 2\beta$$

Now we consider the rectangular form. In this case, we use a β_1 value to describe the interaction for the longer bonds and a β_2 value to describe the interaction for the shorter bonds, where $\beta_1 < \beta$ and $\beta_2 > \beta$. If we make the approximation that the increase in β_2 over β matches the magnitude of the decrease in β_1, we can write that $\beta_2 = \beta + \delta$ and $\beta_1 = \beta - \delta$. The secular determinant for rectangular cyclobutadiene is

$$\begin{vmatrix} (\alpha-\varepsilon) & \beta-\delta & 0 & \beta+\delta \\ \beta-\delta & (\alpha-\varepsilon) & \beta+\delta & 0 \\ 0 & \beta+\delta & (\alpha-\varepsilon) & \beta-\delta \\ \beta+\delta & 0 & \beta-\delta & (\alpha-\varepsilon) \end{vmatrix} = 0$$

Using our definition of $x = (\alpha - \varepsilon)/(\beta - \delta)$ and letting $r = (\beta + \delta)/(\beta - \delta)$ we can write the secular determinant as

$$\begin{vmatrix} x & 1 & 0 & r \\ 1 & x & r & 0 \\ 0 & r & x & 1 \\ r & 0 & 1 & x \end{vmatrix} = 0$$

so that

$$\begin{aligned} 0 &= x\begin{vmatrix} x & r & 0 \\ r & x & 1 \\ 0 & 1 & x \end{vmatrix} - \begin{vmatrix} 1 & r & 0 \\ 0 & x & 1 \\ r & 1 & x \end{vmatrix} - r\begin{vmatrix} 1 & x & r \\ 0 & r & x \\ r & 0 & 1 \end{vmatrix} \\ &= x^2\begin{vmatrix} x & 1 \\ 1 & x \end{vmatrix} - xr\begin{vmatrix} r & 1 \\ 0 & x \end{vmatrix} - \begin{vmatrix} x & 1 \\ 1 & x \end{vmatrix} + r\begin{vmatrix} 0 & 1 \\ r & x \end{vmatrix} - r\begin{vmatrix} r & x \\ 0 & 1 \end{vmatrix} + rx\begin{vmatrix} 0 & x \\ r & 1 \end{vmatrix} - r^2\begin{vmatrix} 0 & r \\ r & 0 \end{vmatrix} \end{aligned}$$

$$= x^4 - x^2 - x^2r^2 - x^2 + 1 - r^2 - r^2 - x^2r^2 + r^4$$
$$= x^4 - 2\left(1 + r^2\right)x^2 + \left(r^4 - 2r^2 + 1\right)$$

Note that $r \geq 1$ because $(\beta - \delta) \leq (\beta + \delta)$. If we define $y = x^2$, then we find

$$0 = y^2 - 2\left(1 + r^2\right)y + \left(r^2 - 1\right)^2$$

and we can use the quadratic equation to write that

$$y = \frac{2\left(1 + r^2\right) \pm \sqrt{4\left(1 + r^2\right)^2 - 4\left(r^2 - 1\right)^2}}{2}$$
$$= \left(1 + r^2\right) \pm \sqrt{\left(1 + r^2\right)^2 - \left(r^2 - 1\right)^2}$$
$$= \left(1 + r^2\right) \pm \sqrt{4r^2} = \left(1 + r^2\right) \pm 2r$$

Thus we find the two roots of $y = (r + 1)^2$ and $y = (r - 1)^2$. Using our definition of y, we find that

$$x = \pm\sqrt{y} = \pm(r + 1) \text{ and } x = \pm\sqrt{y} = \pm(r - 1)$$

For rectangular butadiene, $r > 1$ ($(\beta + \delta) > (\beta - \delta)$), and the four roots are

$$x = (r + 1), (r - 1), -(r - 1) \text{ and } -(r + 1)$$

Using the definitions of $r = (\beta + \delta)/(\beta - \delta)$ and $x = (\alpha - \varepsilon)/(\beta - \delta)$, we find that

$$(\alpha - \varepsilon) = 2\beta,\ 2\delta, -2\delta \text{ and } -2\beta$$

so that the energies are

$$\varepsilon = \alpha - 2\beta, \alpha - 2\delta,\ \alpha + 2\delta \text{ and } \alpha + 2\beta$$

We check our result in two ways. 1) For the limit $r = 1$ ($\delta = 0$), we find the four roots $x = 2, 0, 0$, and -2 which correspond to the square cyclobutadiene case, as expected. 2) Let us choose $\beta_1 = 0.8\beta$ for the longer bond and $\beta_2 = 1.2\beta$ for the shorter bond to obtain the secular determinant

$$\begin{vmatrix} x & 0.8 & 0 & 1.2 \\ 0.8 & x & 1.2 & 0 \\ 0 & 1.2 & x & 0.8 \\ 1.2 & 0 & 0.8 & x \end{vmatrix} = 0$$

This corresponds to the case of $\delta = 0.2\beta$, so that $\varepsilon = \alpha - 2\beta, \alpha - 0.4\beta,\ \alpha + 0.4\beta$, and $\alpha + 2\beta$. If we solve this determinant numerically (using the FEMO_HMO program), we find that

$$\begin{aligned} x_1 &= -2.0 \qquad & \varepsilon &= \alpha + 2\beta \\ x_2 &= -0.4 & \varepsilon &= \alpha + 0.4\beta \\ x_3 &= +0.4 & \varepsilon &= \alpha - 0.4\beta \\ x_4 &= 2.0 & \varepsilon &= \alpha - 2\beta \end{aligned}$$

and these values agree exactly with our analytical result.

Within the context of the HMO model, both structures are stable; however, the rectangular form is predicted to have a more stable π-electron energy than the square form. Figure P8.3 shows the HMO energy level diagram for the two calculations. Because of the alternating β-values, the rectangular form has a more stable HOMO orbital than does the square form. Note, however, that the sigma bonding energy also changes with bond length, and any firm conclusions about the most stable form of the molecule must consider those molecular orbital contributions as well. The rectangular form is in accordance with experiment.[5]

5 H. Irngartinger, N. Riegler, K.-D. Malsch, K.-A. Schneider, G. Maier, Structure of Tetra-tert-butylcyclobutadiene, *Angew. Chem. Int. Ed. Engl.* 1980, **19**, 211–212.

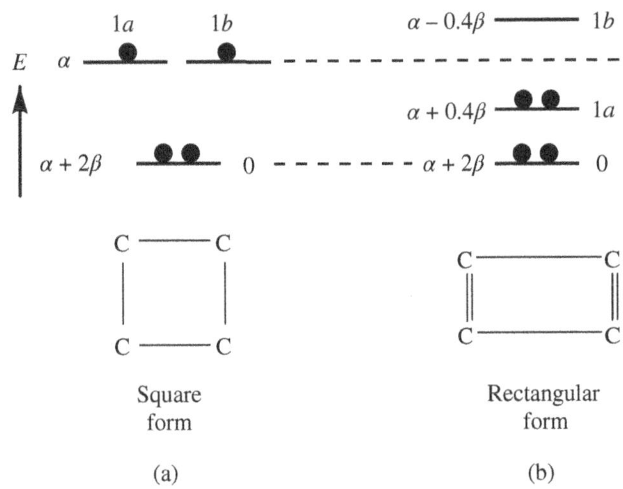

Figure P8.3 HMO energy scheme for the square and rectangular forms of cyclobutadiene (with 1.2β for the double bond and 0.8β for the single bond).

P8.4 Use the FEMO model to calculate the energies, wavefunctions, and charge densities in the centers of the bonds for cyclobutadiene. Perform your calculation for both a square and a rectangular shape of the molecule.

Solution

By analogy to benzene, the energies and the wavefunctions are

$$E = \frac{h^2}{2m_e L_C^2} \cdot n^2$$

and

$$\psi_0 = \frac{1}{\sqrt{L_C}}, \ \psi_{1a} = \frac{1}{\sqrt{L_C}} \cdot \sin\left(\frac{2\pi s}{L_C}\right), \quad \text{and} \quad \psi_{1b} = \frac{1}{\sqrt{L_C}} \cdot \cos\left(\frac{2\pi s}{L_C}\right)$$

for a square structure. Two electrons occupy the orbital with $n = 0$, and the remaining two electrons occupy the degenerate orbitals with $n = 1$. Thus cyclobutadiene should be a diradical, as shown in panel (a) of Fig. P8.4.

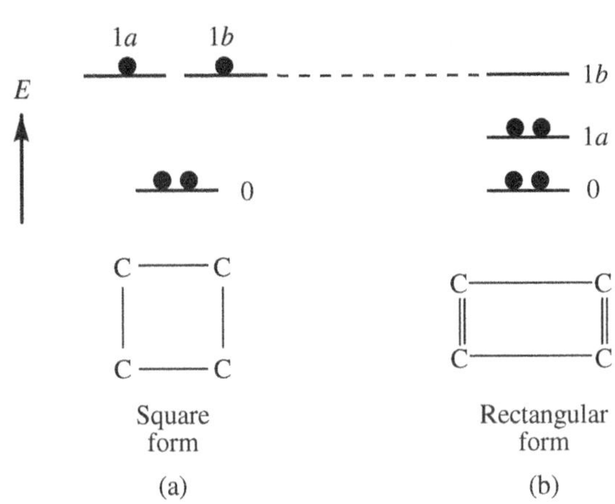

Figure P8.4 Energy scheme for cyclobutadiene using the FEMO model. Panel (a): square form and panel (b): rectangular form.

If we assume, however, that there are two single and two double bonds in cyclobutadiene (rectangular structure, panel (b) in Fig. P8.4), then the energies of the orbitals $1a$ and $1b$ are different. If the double bonds are at the place of the antinodes of the wavefunction $1a$, then $E_{1a} < E_{1b}$ and both electrons share the orbital $1a$.

We argue that the rectangular structure is more reasonable than the square structure. Let us start with a square-shaped cyclobutadiene; if, by chance, one of the bonds is somewhat decreased, then the electrons will be more strongly attracted by this bond. Consequently, more negative charge will accumulate in this bond, and the bond will be shortened further. This process continues until double and single bonds have been established. This can be demonstrated in a corresponding calculation as shown for dodecahexaene (see Box 8.1). In a more detailed consideration, the energy to change the σ bonds has to be taken into account, and the energy decrease of the π electrons is partially compensated by the energy increase of the σ electrons.[6]

Note that the implications of the FEMO model are similar to what is found in Problem 8.3 using the HMO model.

P8.5 Consider the molecular radical fluorocyclopropene (C_3H_2F) and describe the π-bonding in this system. Begin by assuming that each p_z-orbital has the same self-energy and calculate the energy level diagram and find the π molecular orbitals using Hückel theory; i.e., find the π-orbitals for cyclopropene. Next, you should describe (verbally) how you expect the fluorine to modify the molecular orbital energies.

Solution

The secular determinant for cyclopropene is

$$\begin{vmatrix} (\alpha-\varepsilon) & \beta & \beta \\ \beta & (\alpha-\varepsilon) & \beta \\ \beta & \beta & (\alpha-\varepsilon) \end{vmatrix} = 0$$

The π-electron energies are found by finding the eigenvalues and eigenvectors to

$$\begin{vmatrix} x & 1 & 1 \\ 1 & x & 1 \\ 1 & 1 & x \end{vmatrix} = 0$$

so that

$$\begin{aligned} 0 &= x\begin{vmatrix} x & 1 \\ 1 & x \end{vmatrix} - \begin{vmatrix} 1 & 1 \\ 1 & x \end{vmatrix} + \begin{vmatrix} 1 & x \\ 1 & 1 \end{vmatrix} \\ &= x^3 - x - x + 1 + 1 - x \\ &= x^3 - 3x + 2 \\ &= (x+2)(x^2 - 2x + 1) = (x+2)(x-1)^2 \end{aligned}$$

The roots for this equation are −2, 1, and 1. Using the definition $x = (\alpha - \varepsilon)/\beta$, we find that the energies are

$$-2 = \frac{\alpha - \varepsilon_1}{\beta}; \qquad \varepsilon_1 = \alpha + 2\beta$$

$$1 = \frac{\alpha - \varepsilon_2}{\beta}; \qquad \varepsilon_2 = \alpha - \beta$$

$$1 = \frac{\alpha - \varepsilon_3}{\beta}; \qquad \varepsilon_3 = \alpha - \beta$$

(see Fig. P8.5).

As described in Foundation 8.4, we can obtain the c_j coefficients and hence the molecular orbitals by inserting the solution for ε into the matrix equation

$$\begin{pmatrix} (\alpha-\varepsilon) & \beta & \beta \\ \beta & (\alpha-\varepsilon) & \beta \\ \beta & \beta & (\alpha-\varepsilon) \end{pmatrix} \begin{pmatrix} c_1 \\ c_2 \\ c_3 \end{pmatrix} = 0$$

and resolving it. For the eigenvalue $\varepsilon_1 = \alpha + 2\beta$, we find

$$\begin{pmatrix} -2\beta & \beta & \beta \\ \beta & -2\beta & \beta \\ \beta & \beta & -2\beta \end{pmatrix} \begin{pmatrix} c_1 \\ c_2 \\ c_3 \end{pmatrix} = 0$$

6 W.T. Borden, E.R. Davidson, P. Hart, The potential surfaces for the lowest singlet and triplet states of cyclobutadiene, *J. Am. Chem. Soc.* 1978, **100**, 388; *Angew. Chem.*, 1991, **103**, 1048; 1988, **100**, 317.

so that we have the three equations

$$
\begin{aligned}
-2c_1 + c_2 + c_3 &= 0 \\
c_1 - 2c_2 + c_3 &= 0 \\
c_1 + c_2 - 2c_3 &= 0
\end{aligned}
$$

whose solution gives $c_1 = c_2 = c_3$. Hence the ground-state molecular orbital ϕ_1 is

$$\phi_1 = c_1 (\varphi_1 + \varphi_2 + \varphi_3)$$

Now we normalize the orbital (require $\int \phi_1^2 \, d\tau = 1$) and find that

$$\phi_1 = \sqrt{\frac{1}{3}} (\varphi_1 + \varphi_2 + \varphi_3)$$

In the case of $\varepsilon_2 = \varepsilon_3 = \alpha - \beta$ we must solve the matrix equation

$$\begin{pmatrix} \beta & \beta & \beta \\ \beta & \beta & \beta \\ \beta & \beta & \beta \end{pmatrix} \begin{pmatrix} c_1 \\ c_2 \\ c_3 \end{pmatrix} = 0$$

which gives rise to three identical equations

$$
\begin{aligned}
c_1 + c_2 + c_3 &= 0 \\
c_1 + c_2 + c_3 &= 0 \\
c_1 + c_2 + c_3 &= 0
\end{aligned}
$$

from which we need to find two eigenvectors. In one case we choose the solution $c_1 = 0$ and $c_2 = -c_3$, and find

$$\phi_2 = \sqrt{\frac{1}{2}} (\varphi_2 - \varphi_3)$$

In the other case, we solve for c_1 to find

$$c_1 = -(c_2 + c_3)$$

so that

$$\phi_3 = (-(c_2 + c_3)\varphi_1 + c_2\varphi_2 + c_3\varphi_3)$$

We determine the two coefficients c_2 and c_3 by applying conditions of orthogonality and normalization. For orthogonality, we require that

$$
\begin{aligned}
0 = \int \phi_3\phi_2 d\tau &= \sqrt{\frac{1}{2}} \int (\varphi_2 - \varphi_3)(-(c_2 + c_3)\varphi_1 + c_2\varphi_2 + c_3\varphi_3) \, d\tau \\
&= \sqrt{\frac{1}{2}} \left[0 + c_2 \int \varphi_2^2 d\tau + 0 - 0 - 0 - c_3 \int \varphi_3^2 d\tau \right] = [c_2 - c_3] \quad \text{or } c_2 = c_3
\end{aligned}
$$

so that

$$\phi_3 = c_2 (-2\varphi_1 + \varphi_2 + \varphi_3)$$

Applying the normalization condition, we find that $c_2 = 1/\sqrt{6}$ and

$$\phi_3 = \frac{1}{\sqrt{6}} (-2\varphi_1 + \varphi_2 + \varphi_3)$$

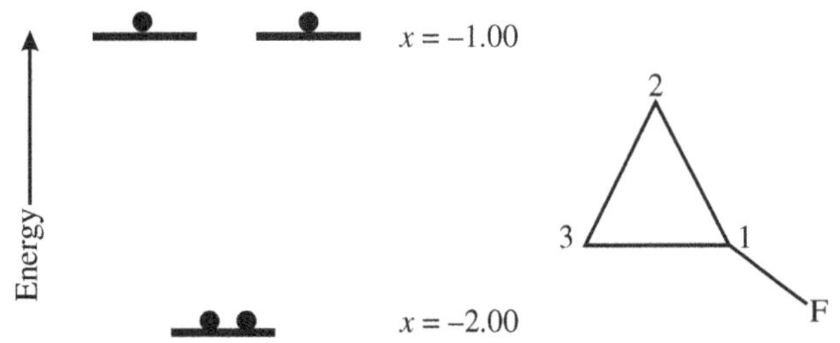

Figure P8.5 HMO energy scheme for cyclopropene with the eigenvalues $x = (\alpha - \varepsilon)/\beta$.

What happens as we shift the energy of one of the orbitals by attaching an F atom through a sigma bond to a carbon? This substitution would be expected to shift the self energy of the atomic orbitals on the F substituted carbon. Hence, the substitution should shift the energy of the lowest energy molecular orbital and break the degeneracy of the ϕ_3 and ϕ_2 molecular orbitals by shifting the ϕ_3 orbital lower and leaving the ϕ_2 orbital unchanged.

P8.6 The description of the π-orbital structure of benzene with the FEMO model gives rise to a degenerate HOMO orbital. Rather than the standing wave solutions used in the text, describe the HOMO orbital of benzene with traveling waves, rotating clockwise and counterclockwise. Refer to Section 3.5. Given that the angular momentum operator is $\hat{l}_z = -i\hbar\frac{\partial}{\partial\varphi}$ where φ is the angle about the ring, evaluate the angular momentum of an electron in the $k = -1$ orbital, where $|k| = n$. Given this value for the angular momentum determine the magnetic moment.

Solution

In benzene the standing wave HOMO molecular orbitals (Equations (8.6)) were found to be

$$\psi_{1a} = \sqrt{\frac{2}{L_c}} \cdot \sin\left(\frac{2\pi s}{L_c}\right) \quad \text{and} \quad \psi_{1b} = \sqrt{\frac{2}{L_c}} \cdot \cos\left(\frac{2\pi s}{L_c}\right) \quad \text{with } \Lambda_1 = L_c$$

These can be recast as traveling wave solutions. Because the wavefunctions are degenerate, we can create superpositions and find a new orthonormal pair of solutions. First, we consider a symmetric combination

$$\begin{aligned}
\psi_+ &= \frac{1}{\sqrt{2}}\left(\psi_{1b} + i\cdot\psi_{1a}\right) = \frac{1}{\sqrt{L_c}}\left(\cos\left(\frac{2\pi s}{L_c}\right) + i\cdot\sin\left(\frac{2\pi s}{L_c}\right)\right) \\
&= \frac{1}{\sqrt{L_c}}\left(\frac{\exp\left(i\frac{2\pi}{L_c}s\right) + \exp\left(-i\frac{2\pi}{L_c}s\right)}{2} + \frac{\exp\left(i\frac{2\pi}{L_c}s\right) - \exp\left(-i\frac{2\pi}{L_c}s\right)}{2}\right) \\
&= \frac{1}{\sqrt{L_c}}\exp\left(i\frac{2\pi}{L_c}s\right)
\end{aligned}$$

If we use the fact that $2\pi s = \varphi L_c$, we can write the solution as

$$\psi_+ = \frac{1}{\sqrt{2\pi}}\exp(i\varphi)$$

An analogous procedure shows that the antisymmetric solution is

$$\psi_- = \frac{1}{\sqrt{2}}\left(\psi_{1b} - i\cdot\psi_{1a}\right) = \frac{1}{\sqrt{L_c}}\exp\left(-i\frac{2\pi}{L_c}s\right)$$

in displacement or

$$\psi_- = \frac{1}{\sqrt{2\pi}}\exp(-i\varphi)$$

in angle variables. We can demonstrate that these wavefunctions are orthogonal to one another; namely that

$$\begin{aligned}
0 &= \int_0^{L_c} \psi_+^*\psi_- \mathrm{d}s = \frac{1}{L_c}\int_0^{L_\varphi} \exp\left(-i\frac{4\pi}{L_c}s\right)\mathrm{d}s \\
&= \frac{1}{L_c}\left(-i\frac{L_c}{4\pi}\right)\left[\exp\left(-i\frac{4\pi}{L_c}s\right)\right]_0^{L_c} = \frac{-i}{4\pi}[1-1] = 0
\end{aligned}$$

For the orbital ψ_-, we calculate the orbital angular momentum by

$$\begin{aligned}
l_- &= \int_0^{2\pi} \psi_-^*\hat{l}_z\psi_- \mathrm{d}\varphi = \frac{-i\hbar}{2\pi}\int_0^{2\pi} \exp(i\varphi)\frac{\partial}{\partial\varphi}\exp(-i\varphi)\,\mathrm{d}\varphi \\
&= \frac{-\hbar}{2\pi}\int_0^{2\pi} \mathrm{d}\varphi = -\hbar
\end{aligned}$$

To estimate the magnetic moment we perform considerations like those used in Box 5.2 for the H-atom's orbiting electron. We model the benzene ring as a current loop with an area A and calculate the magnetic moment μ by

$$\mu = IA$$

where I is the current. To estimate the area, we consider the circumference of the loop to be six times the bond length d_0, which allows us to calculate the radius r by $r = 6d_0/2\pi$ and the area A by

$$A = \pi r^2 = \frac{9d_0^2}{\pi}$$

We use the electron's orbital frequency ν to estimate the current, by way of the classical expression for the orbital angular momentum and the speed; namely

$$l_- = m_e v r = -\hbar$$

so that the speed is

$$v = \frac{-\hbar}{m_e r} = \frac{-\hbar\,\pi}{3d_0 m_e} = \frac{-h}{6d_0 m_e}$$

Hence the orbital frequency is

$$\nu = \frac{v}{6d_0} = \frac{-h}{36d_0^2 m_e}$$

Now we can calculate a current by

$$I = \nu e = \frac{-eh}{36d_0^2 m_e}$$

and a magnetic moment by

$$\mu = IA = \frac{-eh}{36d_0^2 m_e}\frac{9d_0^2}{\pi} = -\frac{eh}{4\pi m_e}$$

To get an intuition for the magnitude of the magnetic moment we compare it to the size of the Bohr magneton, which was found in Box 5.2 to be

$$\mu_{\text{Bohr}} = \frac{eh}{4\pi m_e}$$

and we see that they are of equal magnitude. Although the area of the loop in the benzene ring is significantly larger than the area of the orbit in the H-atom, the speed (hence current) of the electron motion is slower. These two effects cancel.

P8.7 In Example 8.4 we solved for the HMO energies of the amidinium ion but we assumed that $\alpha_C = \alpha_N$ and $\beta = \beta_{CN}$. Find expressions for the HMO energies without making these assumptions.

Solution

Here we follow the solution in Example 8.4, with the requested generalization. The structural formula of the amidinium ion is $\overset{1}{N}H_2 = \overset{2}{C}H\text{–}\overset{3}{N}\,H_2$, where the atoms that contribute p-orbitals are labeled 1, 2, and 3. By inspection we can write the secular determinant as

$$\begin{vmatrix} \alpha_N - \varepsilon & \beta_{CN} & 0 \\ \beta_{CN} & \alpha - \varepsilon & \beta_{CN} \\ 0 & \beta_{CN} & \alpha_N - \varepsilon \end{vmatrix} = 0$$

Expanding the determinant, we find

$$\begin{aligned} 0 &= (\alpha_N - \varepsilon)\begin{vmatrix} \alpha - \varepsilon & \beta_{CN} \\ \beta_{CN} & \alpha_N - \varepsilon \end{vmatrix} - \beta_{CN}\begin{vmatrix} \beta_{CN} & \beta_{CN} \\ 0 & \alpha_N - \varepsilon \end{vmatrix} \\ &= (\alpha_N - \varepsilon)\left[(\alpha_N - \varepsilon)(\alpha - \varepsilon) - \beta_{CN}^2\right] - \beta_{CN}^2(\alpha_N - \varepsilon) - 0 \\ &= (\alpha_N - \varepsilon)\left[\varepsilon^2 - (\alpha + \alpha_N)\varepsilon + (\alpha \cdot \alpha_N - 2\beta_{CN}^2)\right] \end{aligned}$$

By inspection we see that one of the roots for this equation is $\varepsilon = \alpha_N$, which corresponds to a nonbonding orbital on one of the N-atoms. The other two roots are obtained using the quadratic formula for the term in square brackets; namely

$$\varepsilon_\pm = \frac{(\alpha + \alpha_N) \pm \sqrt{(\alpha + \alpha_N)^2 - 4\left(\alpha \cdot \alpha_N - 2\beta_{CN}^2\right)}}{2}$$
$$= \frac{(\alpha + \alpha_N) \pm \sqrt{(\alpha - \alpha_N)^2 + 8\beta_{CN}^2}}{2}$$
$$= \frac{(\alpha + \alpha_N)}{2} \pm \sqrt{\frac{(\alpha - \alpha_N)^2}{4} + 2\beta_{CN}^2}$$

If we compare this result to that in Example 8.3 we see that it converges to $\varepsilon_\pm = \alpha \pm \sqrt{2}\beta$ as $\alpha \to \alpha_N$ and $\beta \to \beta_{CN}$. The leading term is the average self-energy of the electron on the C-site and the N-site, which will be α in the limit that $\alpha = \alpha_N$; the first term under the square root sign depends on the difference in the self-energies and goes to zero in the limit that $\alpha = \alpha_N$; and the last term β_{CN} is the energy from the overlap density.

P8.8 Use the MathCAD code "Problem8_8.mcdx" to calculate the HMO energies for naphthalene. Make an energy level diagram for naphthalene. Determine the total stabilization energy from the π-system, in terms of β. Compare your result to that found from the energy level diagram in E8.11.

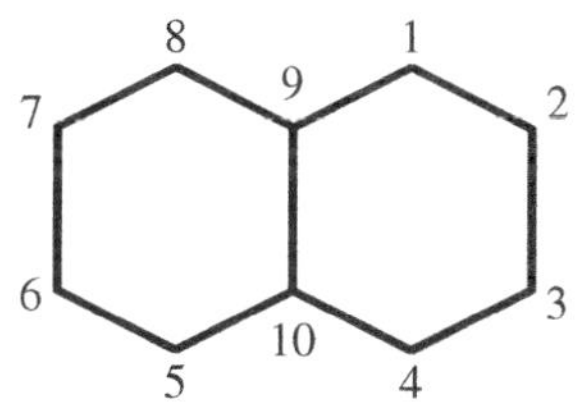

Figure P8.8a Molecular skeleton of naphthalene.

Solution

Using the accepted numbering scheme for naphthalene and the definition of $x = (\alpha - \varepsilon)/\beta$, the MathCAD sheet gives the characteristic polynomial

$$y = x^{10} - 11x^8 + 41x^6 - 65x^4 + 43x^2 - 9$$

and finds the roots $x = -2.303, -1.618, -1.303, -1.0, -0.618, 0.618, 1.0, 1.303, 1.618$, and 2.303. Figure P8.8b plots the polynomial and identifies the roots.

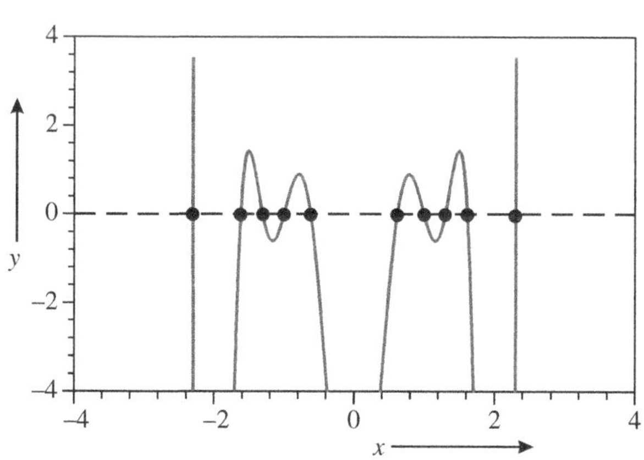

Figure P8.8b Determination of the roots of the characteristic polynomial for naphthalene.

Using the definition of x, we find the eigenvalues

$$\varepsilon_1 = \alpha + 2.303\beta;\ \varepsilon_2 = \alpha + 1.618\beta;\ \varepsilon_3 = \alpha + 1.303\beta;\ \varepsilon_4 = \alpha + 1.0\beta;$$
$$\varepsilon_5 = \alpha + 0.618\beta;\ \varepsilon_6 = \alpha - 0.618\beta;\ \varepsilon_7 = \alpha - 1.0\beta;\ \varepsilon_8 = \alpha - 1.303\beta;$$
$$\varepsilon_9 = \alpha - 1.618\beta;\ \text{and}\ \varepsilon_{10} = \alpha - 2.303\beta.$$

which is in excellent agreement with the energy level diagram in E8.11.
Please see the MathCAD worksheet "Problem8_8soln.mcdx" for more detail.

P8.9 Consider the FEMO model for benzene and compute the correction to the energies of the $n = 3$ states that arises from the carbon atom periodicity. Use a corrugated potential around the ring, where the attractive regions are centered on the atomic sites and their depths are $2b$.

Solution

According to Section 8.3.2, Equation (8.9) the FEMO energies of benzene are

$$E = \frac{h^2}{2m_e L_c^2} n^2\ ,\ n = 0,1,2,\dots.$$

where L_c is the circumference of the benzene ring. The energy levels with $n = 1,2,\ldots$are doubly degenerate. We focus on the case $n = 3$ where the degenerate wavefunctions are

$$\psi_{3a} = \sqrt{\frac{2}{L_c}} \cdot \sin\left(\frac{6\pi s}{L_c}\right) \quad \text{and} \quad \psi_{3b} = \sqrt{\frac{2}{L_c}} \cdot \cos\left(\frac{6\pi s}{L_c}\right)$$

These wavefunctions are shown in Fig. P8.9a as a function of the coordinate s along the ring system. In the derivation of the energy equation it was assumed that the potential energy of the π electrons is constant along the ring. Here we consider the fact that the electrons are more attracted to the positive charges at the carbon atoms than to the centers of the bonds. Figure P8.9a shows that wavefunction ψ_{3a} has nodes at the positions of the carbon atoms, whereas wavefunctions ψ_{3b} have antinodes. This means that the energy of state 3a is unchanged, but the energy of state $3b$ is lowered.

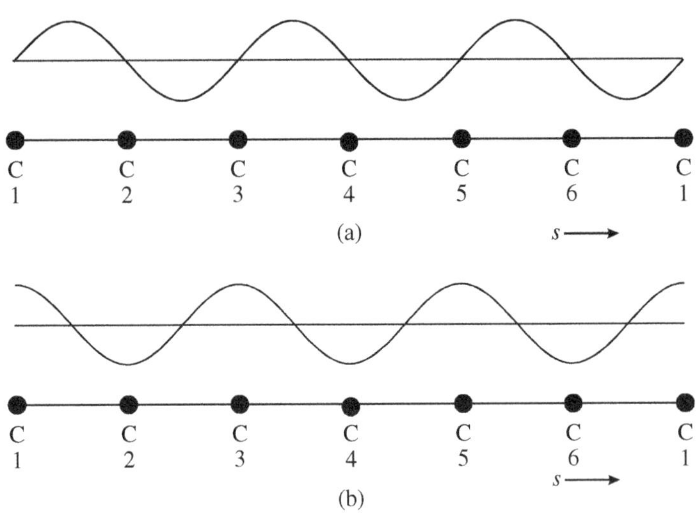

Figure P8.9a. FEMO wavefunctions of benzene for $n = 3$ as a function of the coordinate s along the ring system. The numbering corresponds to the six carbon atoms along the ring. (a) $\psi_{3a} = \sqrt{\frac{2}{L_c}} \cdot \sin\left(\frac{6\pi s}{L_c}\right)$ and (b) $\psi_{3b} = \sqrt{\frac{2}{L_c}} \cdot \cos\left(\frac{6\pi s}{L_c}\right)$.

We estimate this effect by using perturbation theory (see Foundation 5.2). Then the energy is

$$\varepsilon_3 = E_3 + \Delta E_3 = E_3 + \int \psi_3 V \psi_3 \cdot \mathrm{d}s$$

where E_3 is the energy; ψ_3 is the wavefunction of the original system; and V is the potential energy corresponding to the perturbation. In our case the integration is to be performed in the neighborhood of the carbon atoms. In this region wavefunction ψ_{3b} has antinodes exactly at the places of the carbons. Therefore, it is a good approximation of the integral if we replace the actual values of the wavefunction by its value at the maxima. Furthermore, we assume that V is constant around the location of the carbons (within a distance Δs) and zero otherwise. Thus we obtain

$$\begin{aligned}\Delta E_{3b} &= \int \psi_{3b} V \psi_{3b} \cdot \mathrm{d}s \\ &\approx \sum \psi_{3b}(s_{\text{carbon}}) V \psi_{3b\psi_{3b}} \psi_{3b}(s_{\text{carbon}}) \cdot \Delta s \\ &= 6 \cdot \psi_{3b}^2(s_{\text{carbon}}) \cdot V\Delta s\end{aligned}$$

s_{carbon} is the coordinate at a carbon atom and the factor of six takes account of the fact that the six carbon atoms contribute equally to the overall perturbation. Inserting the actual value of the wavefunction at its maximum ($\psi^2_{3b,\text{maximum}} = 2/L_c$) we obtain

$$\Delta E_{3b} = 6 \cdot \frac{2}{L_c} \cdot V\Delta s = \frac{12}{6d_0} \cdot V\Delta s = \frac{2}{d_0} \cdot V\Delta s$$

where $d_0 = 140$ pm is the bond length in benzene.

We use the potential trough shown in Section 8.2, Fig. 8.4 to calculate the quantity $V\Delta s$

$$V\Delta s = -28.8 \times 10^{-19}\ \text{J} \cdot 25\,\text{pm} = -7.2 \times 10^{-17}\ \text{J pm}$$

to estimate the perturbation energy as

$$\Delta E_{3b} = -\frac{2}{d_0} \cdot 7.2 \times 10^{-17}\ \text{J pm} = \ 10 \times 10^{-19}\ \text{J}$$

For comparison, we perform the same calculation for the quantum states with $n = 1$ and $n = 2$ (Figure P8.9b):

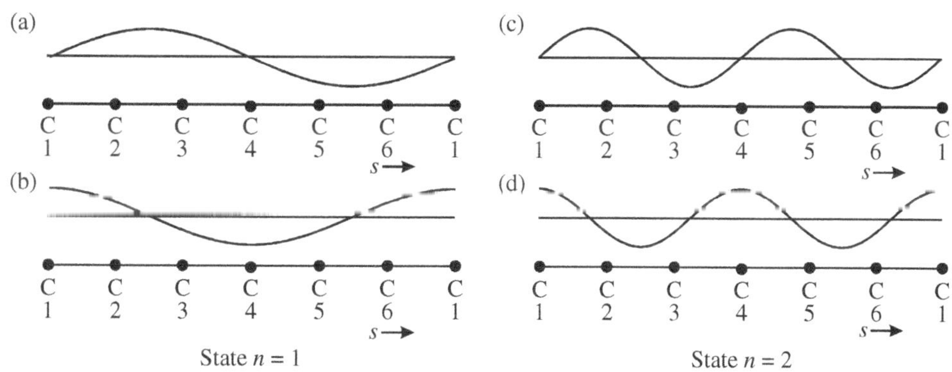

Figure P8.9b Wavefunctions for quantum states with (a, b) $n = 1$ and (c, d) $n = 2$.

In these cases the values of the wavefunction at the positions of the carbon atoms must be calculated explicitly.

Atom	$\sin^2 x$	$\cos^2 x$	$\sin^2 2x$	$\sin^2 2x$
1	0.00	1.00	0.00	1.00
2	0.75	0.25	0.75	0.25
3	0.75	0.25	0.75	0.25
4	0.00	1.00	0.00	1.00
5	0.75	0.25	0.75	0.25
6	0.75	0.25	0.75	0.25
$\sum_{\text{atoms}}$	3.00	3.00	3.00	3.00

In analogy we find

$$\Delta E_{1a} = \Delta E_{1b} = \Delta E_{2a} = \Delta E_{2b} = 3 \cdot \frac{2}{L_c} \cdot V\Delta s = 3 \cdot \frac{V\Delta s}{d_0} = -5 \times 10^{-19}\ \mathrm{J}$$

In summary, for the perturbation energy ΔE of all considered states we obtain

State	$\Delta E/10^{-19}$ J	State	$\Delta E/10^{-19}$ J
0	−5.0		
1*a*	−5.0	1*b*	−5.0
2*a*	−5.0	2*b*	−5.0
3*a*	0	3*b*	−10.0

This result shows that the degeneracy of the states with $n = 1$ and $n = 2$ is not affected by the perturbation; however, both energy levels decrease by the same amount, as does the level for $n = 0$. The resulting energy levels are shown schematically in Fig. P8.9c.

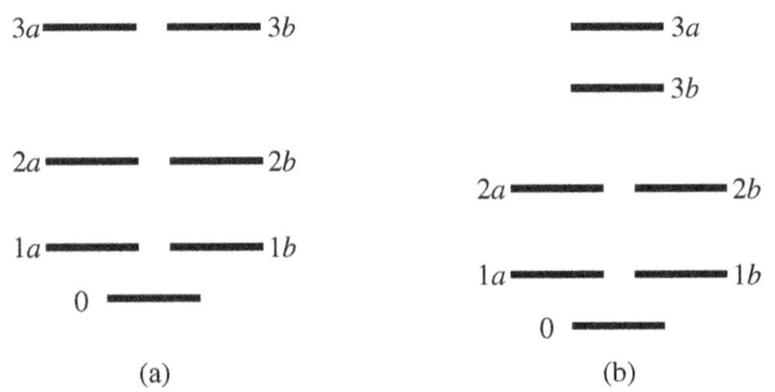

Figure P8.9c Energy levels of benzene according to the FEMO model (schematically). (a) Potential along the chain of carbon atoms assumed to be constant. (b) Attraction of the electron at the cores of the carbon atoms taken into account. In this case levels 3*a* and 3*b* are no more degenerate, whereas the degeneracies of levels 1*a*, 1*b* and 2*a*, 2*b* are not affected.

More Sophisticated Consideration

Now we repeat the consideration by calculating the perturbation terms more precisely.

We assume that the potential wells of depth $2b$ are located with their center on each carbon site and have a width of $d_0/2$ ($\pm d_0/4$ about the site). We use this potential with the FEMO wavefunctions and compute the energy shift (this corresponds to the first-order correction to energy in perturbation theory, see Foundation 5.2).

For the $n = \pm 3$ state, the wavefunction is

$$\psi_{3a} = \sqrt{\frac{2}{L_c}} \cdot \sin\left(\frac{6\pi s}{L_c}\right), \quad \psi_{3b} = \sqrt{\frac{2}{L_c}} \cdot \cos\left(\frac{6\pi s}{L_c}\right)$$

The corrected energy for the 3*a* state is

$$E_{3a} = \frac{9h^2}{2m_e L_c^2} - \frac{2}{L_c}\int_{-d_0/4}^{d_0/4} 2b\sin^2\left(\frac{6\pi s}{L_c}\right)\,ds - \frac{2}{L_c}\int_{d_0-d_0/4}^{d_0+d_0/4} 2b\sin^2\left(\frac{6\pi s}{L_c}\right)\,ds$$
$$-\frac{2}{L_c}\int_{2d_0-d_0/4}^{2d_0+d_0/4} 2b\sin^2\left(\frac{6\pi s}{L_c}\right)\,ds - \frac{2}{L_c}\int_{3d_0-d_0/4}^{3d_0+d_0/4} 2b\sin^2\left(\frac{6\pi s}{L_c}\right)\,ds$$
$$-\frac{2}{L_c}\int_{4d_0-d_0/4}^{4d_0+d_0/4} 2b\sin^2\left(\frac{6\pi s}{L_c}\right)\,ds - \frac{2}{L_c}\int_{5d_0-d_0/4}^{5d_0+d_0/4} 2b\sin^2\left(\frac{6\pi s}{L_c}\right)\,ds$$

Now we define $x = 6\pi s/L_c = \pi s/d_0$ and use

$$\int_g^f \sin^2(x)\,dx = \frac{1}{2}\left(x - \frac{1}{2}\sin 2x\right)\Big|_g^f = \frac{1}{2}\left(f - g - \frac{1}{2}\sin 2f + \frac{1}{2}\sin 2g\right)$$

to write

$$E_{3a} = \frac{9h^2}{2m_eL_c^2} - \frac{1}{3\pi}\int_{-\pi/4}^{\pi/4} 2b\sin^2(x)\ \mathrm{d}x - \frac{1}{3\pi}\int_{3\pi/4}^{5\pi/4} 2b\sin^2(x)\ \mathrm{d}x$$
$$-\frac{1}{3\pi}\int_{7\pi/4}^{9\pi/4} 2b\sin^2(x)\ \mathrm{d}x - \frac{1}{3\pi}\int_{11\pi/4}^{13\pi/4} 2b\sin^2(x)\ \mathrm{d}x$$
$$-\frac{1}{3\pi}\int_{15\pi/4}^{17\pi/4} 2b\sin^2(x)\ \mathrm{d}x - \frac{1}{3\pi}\int_{19\pi/4}^{21\pi/4} 2b\sin^2(x)\ \mathrm{d}x$$

Performing the integration we find that

$$E_{3a} = \frac{9h^2}{2m_eL_c^2} - \frac{2b}{3\pi}\left(x - \frac{1}{2}\sin 2x\right)_{-\pi/4}^{\pi/4} - \frac{2b}{3\pi}\left(x - \frac{1}{2}\sin 2x\right)_{3\pi/4}^{5\pi/4}$$
$$-\frac{2b}{3\pi}\left(x - \frac{1}{2}\sin 2x\right)_{7\pi/4}^{9\pi/4} - \frac{2b}{3\pi}\left(x - \frac{1}{2}\sin 2x\right)_{11\pi/4}^{13\pi/4}$$
$$-\frac{2b}{3\pi}\left(x - \frac{1}{2}\sin 2x\right)_{15\pi/4}^{17\pi/4} - \frac{2b}{3\pi}\left(x - \frac{1}{2}\sin 2x\right)_{7\pi/4}^{9\pi/4}$$

which simplifies to

$$E_{3a} = \frac{9h^2}{2m_eL_c^2} - \frac{2b}{3\pi}\left(\frac{\pi}{2} - 1\right) - \frac{2b}{3\pi}\left(\frac{\pi}{2} - 1\right) - \frac{2b}{3\pi}\left(\frac{\pi}{2} - 1\right)$$
$$-\frac{2b}{3\pi}\left(\frac{\pi}{2} - 1\right) - \frac{2b}{3\pi}\left(\frac{\pi}{2} - 1\right) - \frac{2b}{3\pi}\left(\frac{\pi}{2} - 1\right)$$
$$= \frac{9h^2}{2m_eL_c^2} - 2b + \frac{4b}{\pi}$$

In a corresponding way we find the corrected energy for the $3b$ state; namely

$$E_{3b} = \frac{9h^2}{2m_eL_c^2} - \frac{2}{L_c}\int_{-d_0/4}^{d_0/4} 2b\cos^2\left(\frac{6\pi s}{L_c}\right)\ \mathrm{d}s - \frac{2}{L_c}\int_{d_0-d_0/4}^{d_0+d_0/4} 2b\cos^2\left(\frac{6\pi s}{L_c}\right)\ \mathrm{d}s$$
$$-\frac{2}{L_c}\int_{2d_0-d_0/4}^{2d_0+d_0/4} 2b\cos^2\left(\frac{6\pi s}{L_c}\right)\ \mathrm{d}s - \frac{2}{L_c}\int_{3d_0-d_0/4}^{3d_0+d_0/4} 2b\cos^2\left(\frac{6\pi s}{L_c}\right)\ \mathrm{d}s$$
$$-\frac{2}{L_c}\int_{4d_0-d_0/4}^{4d_0+d_0/4} 2b\cos^2\left(\frac{6\pi s}{L_c}\right)\ \mathrm{d}s - \frac{2}{L_c}\int_{5d_0-d_0/4}^{5d_0+d_0/4} 2b\cos^2\left(\frac{6\pi s}{L_c}\right)\ \mathrm{d}s$$

Now we define $x = 6\pi s/L_c = \pi s/d_0$ and use

$$\int_g^f \cos^2(x)\,\mathrm{d}x = \frac{1}{2}\left(x + \frac{1}{2}\sin 2x\right)\Big|_g^f = \frac{1}{2}\left(f - g + \frac{1}{2}\sin 2f - \frac{1}{2}\sin 2g\right)$$

to write

$$E_{3b} = \frac{9h^2}{2m_eL_c^2} - \frac{1}{3\pi}\int_{-\pi/4}^{\pi/4} 2b\cos^2(x)\ \mathrm{d}x - \frac{1}{3\pi}\int_{3\pi/4}^{5\pi/4} 2b\cos^2(x)\ \mathrm{d}x$$
$$-\frac{1}{3\pi}\int_{7\pi/4}^{9\pi/4} 2b\cos^2(x)\ \mathrm{d}x - \frac{1}{3\pi}\int_{11\pi/4}^{13\pi/4} 2b\cos^2(x)\ \mathrm{d}x$$
$$-\frac{1}{3\pi}\int_{15\pi/4}^{17\pi/4} 2b\cos^2(x)\ \mathrm{d}x - \frac{1}{3\pi}\int_{19\pi/4}^{21\pi/4} 2b\cos^2(x)\ \mathrm{d}x$$

Performing the integration we find that

$$E_{3b} = \frac{9h^2}{2m_e L_c^2} - \frac{2b}{3\pi}\left(x + \frac{1}{2}\sin 2x\right)_{-\pi/4}^{\pi/4} - \frac{2b}{3\pi}\left(x + \frac{1}{2}\sin 2x\right)_{3\pi/4}^{5\pi/4}$$

$$-\frac{2b}{3\pi}\left(x + \frac{1}{2}\sin 2x\right)_{7\pi/4}^{9\pi/4} - \frac{2b}{3\pi}\left(x + \frac{1}{2}\sin 2x\right)_{11\pi/4}^{13\pi/4}$$

$$-\frac{2b}{3\pi}\left(x + \frac{1}{2}\sin 2x\right)_{15\pi/4}^{17\pi/4} - \frac{2b}{3\pi}\left(x + \frac{1}{2}\sin 2x\right)_{7\pi/4}^{9\pi/4}$$

which simplifies to

$$E_{3b} = \frac{9h^2}{2m_e L_c^2} - \frac{2b}{3\pi}\left(\frac{\pi}{2} + 1\right) - \frac{2b}{3\pi}\left(\frac{\pi}{2} + 1\right) - \frac{2b}{3\pi}\left(\frac{\pi}{2} + 1\right)$$

$$-\frac{2b}{3\pi}\left(\frac{\pi}{2} + 1\right) - \frac{2b}{3\pi}\left(\frac{\pi}{2} + 1\right) - \frac{2b}{3\pi}\left(\frac{\pi}{2} + 1\right)$$

$$= \frac{9h^2}{2m_e L_c^2} - 2b - \frac{4b}{\pi}$$

The energy difference between these two states will be

$$\Delta E = E_{3a} - E_{3b} = \frac{9h^2}{2m_e L_c^2} - 2b + \frac{4b}{\pi} - \left(\frac{9h^2}{2m_e L_c^2} - 2b - \frac{4b}{\pi}\right)$$

$$= 8b/\pi$$

This result differs from that obtained in the first part of this solution for the energy difference; namely

$$\Delta E = \frac{2}{d_0} V \Delta s = \frac{2}{d_0} 2b \frac{d_0}{2} = 2b$$

instead of $\Delta E = (8/\pi)b = 2.5b$. This result differs from the result in the first part (where we assume that the potential trough is infinitely small) by about 20%.

P8.10 Make a rough estimate of the ΔV value found for the calculation in Box 8.1, by comparing the electrostatic energy of an electron in a $2p_z$ Slater-type orbital in a C=C double bond to that in the case of a C–C single bond. You can assume that the electron is at a distance $\delta = 75$ pm above the molecular plane, which corresponds to the center of the p_z orbital's probability density, that $Z_{\text{eff}} = 3.25$, and that the bond lengths are $d_{\text{C=C}} = 133$ pm and $d_{\text{C-C}} = 148$ pm.

Solution

To estimate ΔV we calculate the potential energy of an electron in the middle of a bond in the electric field of the adjacent carbon atoms. The adjacent atoms are assumed to act as point charges $e \cdot (Z_{\text{eff}} - 1)$, since the Slater charge is approximately shielded by one electron. The remaining atoms are considered to be completely shielded. In the figure we draw a diagram, to find the distance of the electron from the two nuclei. The quantity ΔV should be the difference in the Coulomb attraction energy of the electron in the center of a double bond or in the center of a single bond, respectively:

$$\Delta V = \frac{2 \cdot (Z_{\text{eff}} - 1) \cdot e^2}{4\pi\varepsilon_0} \cdot \left(\frac{1}{r_{\text{C=C}}} - \frac{1}{r_{\text{C-C}}}\right)$$

In Fig. P8.10 the distances $r_{\text{C=C}}$ and $r_{\text{C-C}}$ are given by

$$r_{\text{C=C}} = \sqrt{\left(\frac{d_{\text{C=C}}}{2}\right)^2 + \delta^2} \qquad r_{\text{C-C}} = \sqrt{\left(\frac{d_{\text{C-C}}}{2}\right)^2 + \delta^2}$$

where $d_{\text{C=C}}$ and $d_{\text{C-C}}$ are the bond lengths of a double bond and a single bond. For the geometry shown and the bond parameters given in the problem statement, we find that

$$r_{\text{C=C}} = 100.4\,\text{pm}\; r_{\text{C-C}} = 105.3\,\text{pm}$$

Using these distance we find,

$$\Delta V = 4.8 \times 10^{-19}\ \text{J}$$

which is only slightly smaller than the value 6.59×10^{-19} J obtained in the step potential method, achieving self-consistency between bond lengths and π electron density.

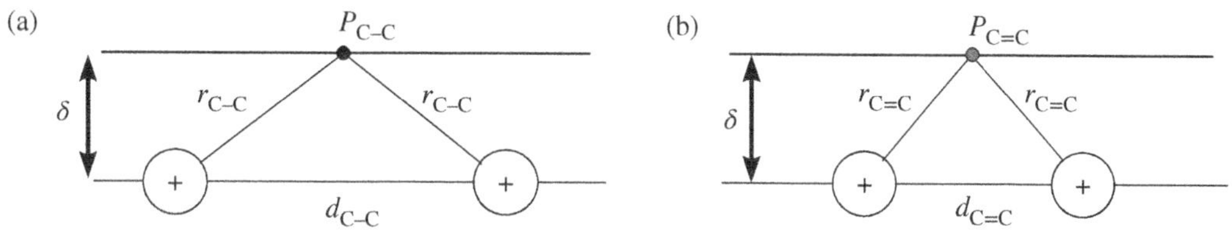

Figure P8.10. Estimating ΔV. The electron is considered to be in points P_{C-C} and $P_{C=C}$, respectively, exposed to the effective charges of the adjacent C atoms.

9

Absorption of Light

9.1 Exercises

E9.1 For the molecules shown in Figure E9.1 discuss whether the nitrogen atoms contribute to the number of π electrons in the conjugated system, and identify the number of π-electrons it contributes.

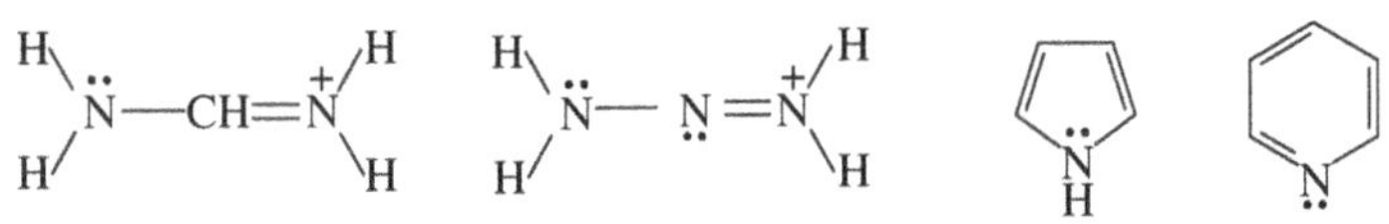

Figure E9.1 Contribution of nitrogen atoms to the π-electron system.

Solution

In contrast to a carbon atom (four electrons in the 2*s* and 2*p* orbitals), a nitrogen atom has five valence electrons (two of them in the 2*s* orbital and three in the 2*p* orbital). The N-atom hybridizes to distribute the valence electrons in two dominant characteristic ways to form a trigonal planar structure. In one case, three of the electrons are in the sp^2 hybrid orbitals that form three σ bonds and the remaining two electrons occupy a $2p_z$ orbital. In the second case, four of the electrons are in the sp^2 hybrid orbitals and they form two σ bonds and a lone electron pair (which extends in the plane of the σ electrons); the remaining electron occupies a $2p_z$ orbital. Using these facts, it follows that both of the atoms in the cyanine contribute two electrons to the π-system; note that the net positive charge implies that the molecule has only four π-electrons. The two end nitrogens on the azacyanine each contribute two electrons to the π-system and the center N-atom only contributes one electron. The lone pair is not part of the π-system. The nitrogen in the five-membered ring of pyrrole forms three σ bonds, and thus it contributes two electrons to the number of π electrons. The lone electron pair on the nitrogen of the six-membered ring (pyridine) does not participate in the π-system; in this case the N-atom contributes one π-electron.

E9.2 Discuss the color of the dyes in Table 9.1. Explain the shift in the absorbance maximum in terms of the FEMO model. Explain the changing color of the solution and how that relates to the molecule's absorption spectrum.

Table 9.1 Absorption of the Diethylthiacarbocyanine Dyes with Different Values of *j*.

j	λ_{max} /nm	Color of Absorbed Light	Color of Solution
0	420	Blue	Yellow
1	540	Green	Red
2	640	Orange	Blue
3	740	Red	Bluish green

Solutions Manual for Principles of Physical Chemistry, Third Edition. Edited by Hans Kuhn, David H. Waldeck, and Horst-Dieter Försterling.

Solution

The dyes differ in the number of CH=CH groups between the end structure. The entire length of the molecule is conjugated, with the π electrons moving along the length of the molecule. A π electron can then be modeled as a particle moving on a box of length corresponding to the length of the molecule. As the number of groups increases, the length of the box increases, and the number of π electrons increases. The increasing length of the box causes the energy level spacing to decrease. The decreasing energy level spacing causes the wavelength of the absorption bands to increase, correlating with the observed color changes for the compounds.

The color of the solution, as observed by your eye, is complementary to the color of the light absorbed by the dye molecules. For white light (a spectrum of colors from red to violet; red-orange-yellow-green-blue-indigo-violet, ROYGBIV)) incident on the solution, your eye (as the detector) only observes the photons that are not absorbed by the sample. Hence when the sample absorbs the bluish light (the case of $k = 0$), your eye will detect predominantly wavelengths in the red to green part of the spectrum; overall, it will be perceived as yellow. An analogous rationalization applies for the other solution colors.

E9.3 Explain the shift in absorbance maxima for the different sets of dye molecules shown in Figure E9.3.

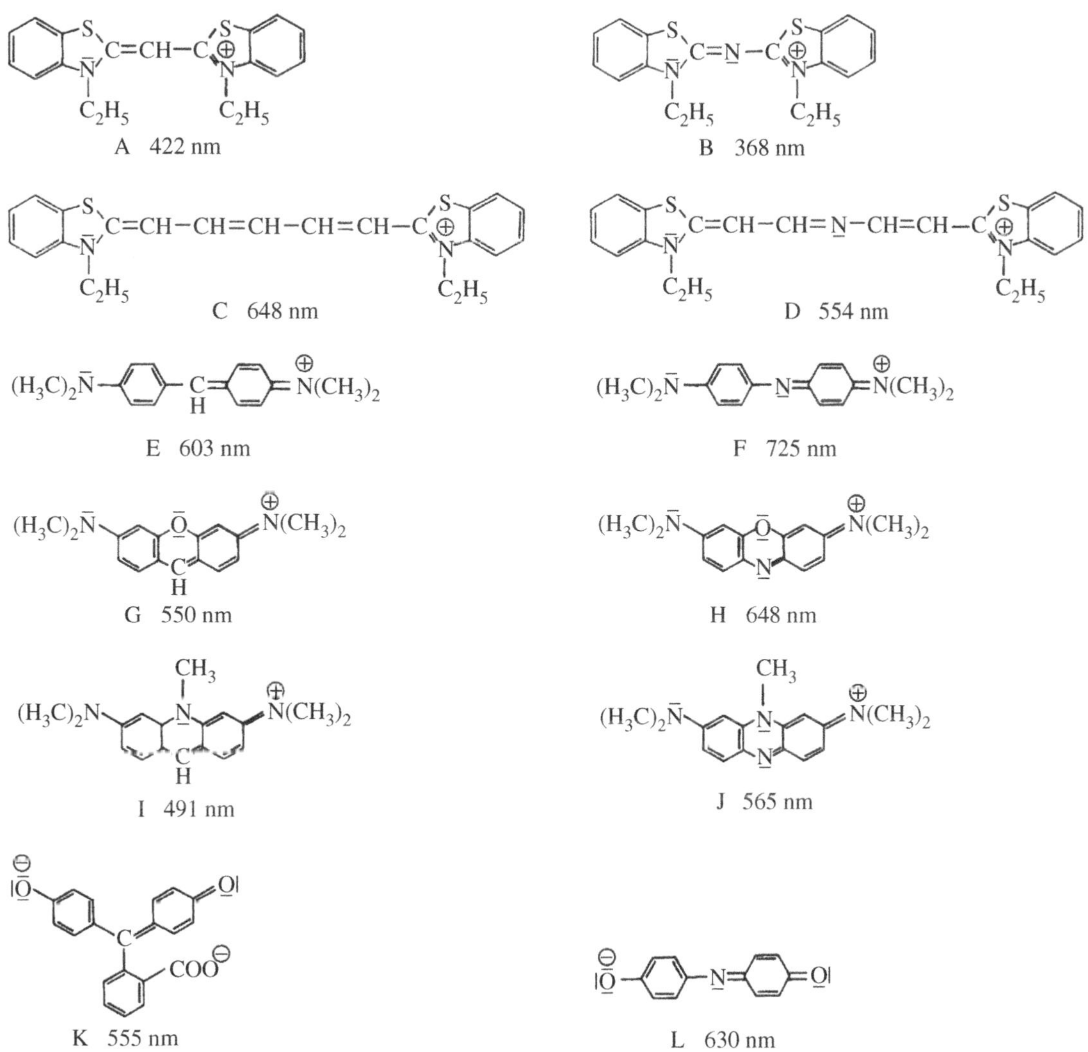

Figure E9.3 Absorption maxima of different sets of dyes.

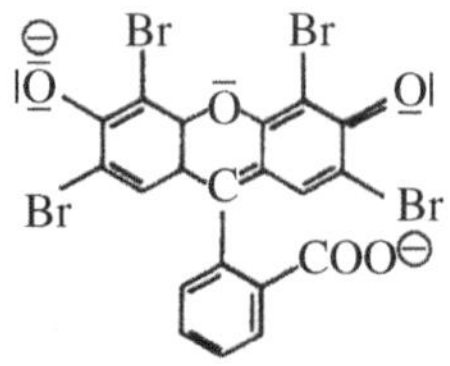

M 516 nm

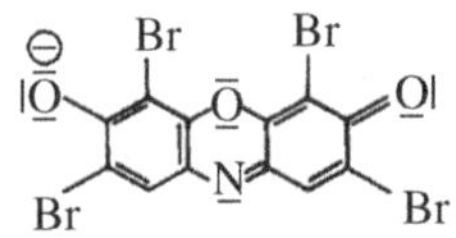

N 609 nm

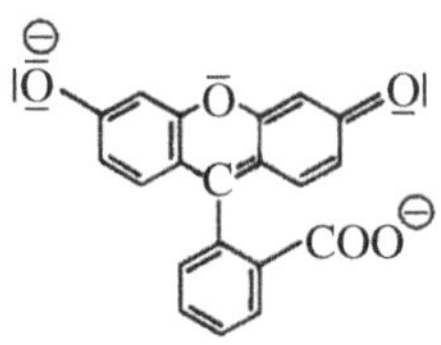

O 489 nm

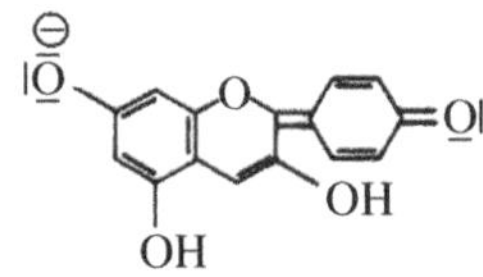

P 655 nm

Figure E9.3 *(Continued)*

Solution

The electronic structure of A corresponds to the case $N = 6$ in Table 9.2 (332 nm) and that of C corresponds to the case of $N = 10$ (587 nm); however, the polarizability of the benzene rings causes a shift to longer wavelengths in both cases. We expect that the azacyanine B will have an absorption maximum shifted to shorter wavelengths than the corresponding cyanine A, and that the azacyanine D will have an absorption maximum shifted to shorter wavelengths than the cyanine C. See the discussion in Section 9.2.4 for an explanation of how the aza substitution shifts the absorption wavelength. The λ_{max} of F, as compared to E; for J as compared to I; for L as compared to K; and for N as compared to M and O are shifted to longer wavelengths for a similar reason, see Section 9.2.4. In all these cases the number of π electrons along the chain is $N = 12$. P compares with the case $N = 14$ in Table 9.2 (735 nm).

E9.4 Use the FEMO model to calculate the λ_{max} for the two dye molecules A and B shown in Figure E9.4.

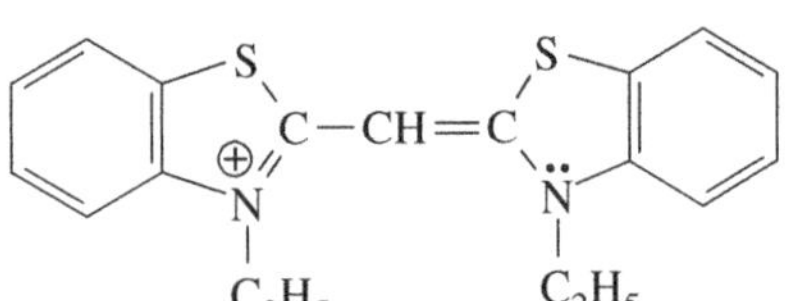

A

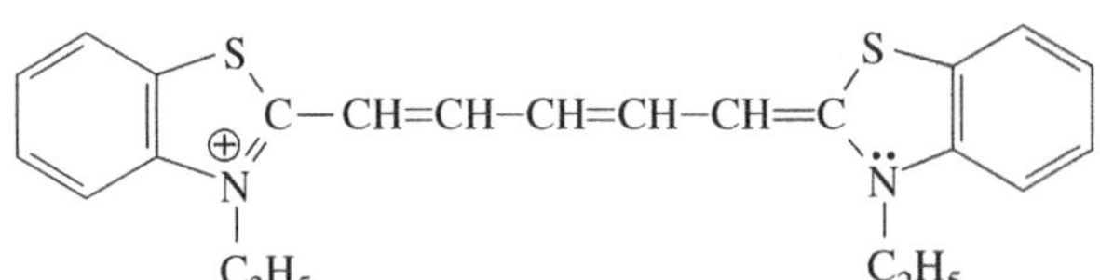

B

Figure E9.4 Two different dye molecules A and B.

Solution

Molecule A is a diethylthiacarbocyanine with $k = 1$, while molecule B corresponds to the case with $k = 3$. The expression for the transition energy found from Equation (9.17) is

$$\Delta E = \frac{h^2}{8m_e L^2}(N + 1)$$

where N is the number of π-electrons and

$$L = Nd_0$$

is the length of the π-electron system. Dye A has $N = 6$ π-electrons; thus,

$$L = 6d_0 \qquad \text{with } d_0 = 140 \text{ pm}$$

and

$$\Delta E = \frac{h^2}{8m_e(5d_0)^2} \cdot 7 = \frac{h^2}{8m_e {d_0}^2} \cdot \frac{7}{36} = 3.07 \times 10^{-18}\ \text{J} \cdot \frac{7}{36} = 5.97 \times 10^{-19}\ \text{J}$$

corresponding to a wavelength of

$$\lambda = \frac{hc_0}{\Delta E} = 332 \text{ nm}$$

Accordingly, for dye B with $N = 10$ we obtain

$$\Delta E = 3.07 \times 10^{-18}\ \text{J} \cdot \frac{11}{100} = 3.38 \times 10^{-19}\ \text{J}$$

and

$$\lambda = \frac{hc_0}{\Delta E} = 588\ \text{nm}$$

A comparison to Table 9.1 shows that the experimental values $\lambda = 420$ nm and $\lambda = 640$ are much larger than those calculated here. This discrepancy can be reduced by increasing the overall π-system length by $d_0/2$, which gives $L = 6.5d_0$ and $\lambda = 391$ nm for dye A and $L = 10.5d_0$ and $\lambda = 648$ nm for dye B. This is near to the experimental data.

How can one rationalize this extra distance increment when the last group on the nitrogen is an aromatic ring system versus a CH_3 unit? Because the π-system of the benzene can interact with the π-system in the dye, their partial overlap can extend the spatial extent of the dye's π-system (the length of the nitrogen π-system increases) which leads to an increase in the overall box length. An alternative, yet related explanation, is to consider the height of the potential barrier at the edge of the dye's π-system. The height of the potential wall for a benzene moiety (first ionization energy of 9.2 eV) is much lower than that for a methyl unit (first ionization energy of ethane is 11.5 eV); thus, we expect the dye's π-system to tunnel farther through the edge of the box in the benzene case than the methyl case. This lowering of the barrier causes shifts in the energy level spacings; for example see Section 3.2 (Fig. 3.2) of the text.

E9.5 Show that the energy gap calculated in Example 9.3 for Vitamin B12 corresponds to a λ_{max} of 580 nm.

Solution

The energy gap given in Example 9.3 is 3.35×10^{-19} J. The value of λ_{max} calculated using this energy gap is

$$\lambda_{max} = \frac{hc_0}{\Delta E} = \frac{(6.626 \times 10^{-34}\ \text{J} \cdot \text{s}^{-1})\,(2.998 \times 10^{8}\ \text{m})}{3.35 \times 10^{-19}\ \text{J}}$$
$$= 5.930 \times 10^{-7}\ \text{m} = 593.0\ \text{nm}$$

E9.6 In Example 9.5, we calculated energy gaps for transitions in phthalocyanine. Use these energy gaps to calculate the λ_{max} values.

Solution

In Example 9.5, the energy gaps for phthalocyanine were calculated to be $\Delta E_{4b\to5b} = \Delta E_{4b\to5a} = 2.57 \times 10^{-19}$J, and $\Delta E_{4a\to5a} = \Delta E_{4a\to5b} = 6.07 \times 10^{-19}$J. The wavelengths corresponding to these energies are

$$\lambda_{max} = \frac{hc_0}{\Delta E} = \frac{(6.626 \times 10^{-34}\ \text{J} \cdot \text{s})\,(2.998 \times 10^{8}\ \text{m} \cdot \text{s}^{-1})}{2.57 \times 10^{-19}\ \text{J}}$$
$$= 1.097 \times 10^{-25}\ \text{J m}\,\frac{1}{2.57 \times 10^{-19}\ \text{J}} = 7.729 \times 10^{-7}\ \text{m} = 772.9\ \text{nm}$$

and

$$\lambda_{max} = \frac{hc_0}{\Delta E} = 1.097 \times 10^{-25}\ \text{J m}\,\frac{1}{6.07 \times 10^{-19}\ \text{J}} = 3.273 \times 10^{-7}\ \text{m} = 327.3\ \text{nm}$$

E9.7 Analyze the data in Table 9.4. Plot the energy of the experimental absorption maximum (taken as ΔE(exp)) versus $1/(N+1)$ and fit the data. Is it linear? How does the slope compare with the prediction of Equation (9.43)?

Solution

The Figure E9.7 shows the desired graph of ΔE versus $(N+1)^{-1}$. The graph is linear and has an intercept of 3.6×10^{-19} J, indicating that a finite gap exists for an infinitely long polyene chain. The slope of the graph is 2.8×10^{-18}J. The predicted theoretical slope from Equation 9.43 is

$$\frac{h^2}{8m_e d_0^2} = \frac{(6.626 \times 10^{-34}\ \text{J} \cdot \text{s})^2}{8 \cdot 9.11 \times 10^{-31}\ \text{kg} \cdot (140 \times 10^{-12}\ \text{m})^2} = 3.074 \times 10^{-18}\ \text{J}$$

in reasonable agreement with the experimental slope.

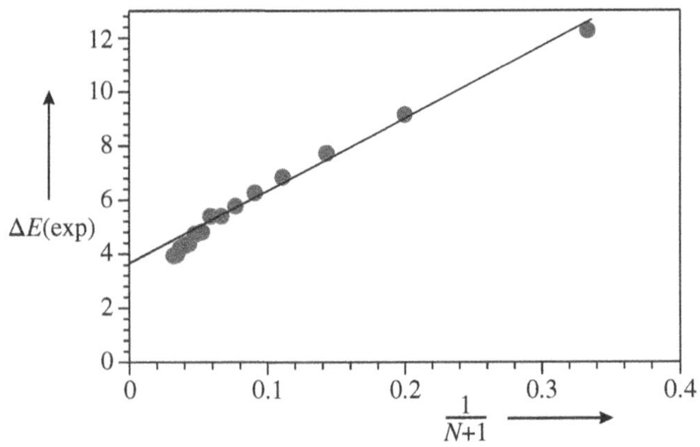

Figure E9.7 Polyenes. ΔE(exp) versus $1/(N + 1)$, where N is the number of π-electrons.

E9.8 In Example 9.5 we calculated the energies of the HOMO and LUMO electronic states in phthalocyanine using the FEMO model, and we corrected it for the self-energy of the N atom *p*-orbitals. Perform a similar calculation for porphyrin (see Figure E9.8).

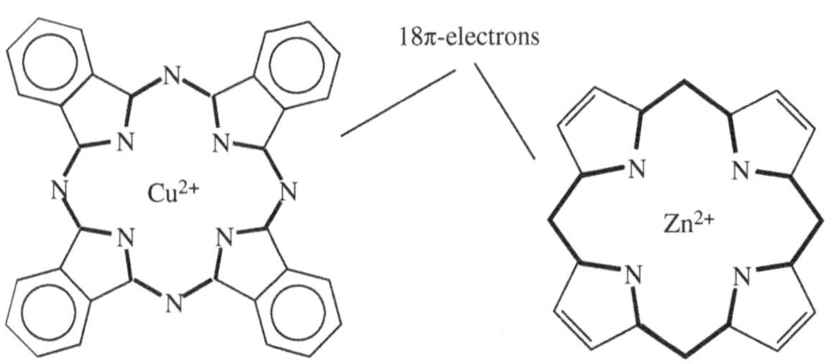

Figure E9.8 Molecular skeleton of phthalocyanine and porphyrin.

Solution

Porphyrin has an azacyanine ring system with 18 π-electrons. In principle, porphyrin should have the same π electron system as in phthalocyanine. However, the shift of the energy levels arising from the nitrogen atoms is smaller, because the four outermost nitrogen atoms are missing. Therefore the energy shift is just one-half of the shift in phthalocyanine.

The calculation follows that used for the phthalocyanine example very closely. From the electron count, the FEMO model gives $n = 4$ for the filled HOMO level and $n = 5$ for the empty LUMO level. As for phthalocyanine and the azacyanine dyes, we take the extra stabilization energy at a nitrogen atom to be given by $-a \cdot d_0 \cdot \rho$, where ρ is the probability density at the atomic center, $d_0 = 140\,\text{pm}$, and $a = 3.5 \times 10^{-19}$ J. In state 4*a* with the wavefunction

$$\psi_{4a} = \sqrt{\frac{2}{L_c}} \cdot \cos\left(4 \cdot \frac{2\pi s}{L_c}\right)$$

all four N atoms are placed at the antinodes of the wavefunction. The average potential energy change at a nitrogen atom site is given by

$$\overline{V} = -a \int_{s_1}^{s_2} \psi^2 \cdot ds = -a \cdot \left(\psi^2\right)_{\text{at center}} \cdot d_0$$

For state 4*a* the first antinode is at $s = 0$, thus $\left(\psi^2\right)_{\text{at center}} = 2/L_c$ and

$$\overline{V} = -a \cdot \frac{2}{L_c} \cdot d_0$$

In contrast to the case of phthalocyanine, only four such sites are present so that $\Delta E_{\text{nitrogen}} = 4\overline{V}$ and

$$E_{4a} = \frac{h^2}{2m_e L_c^2} \cdot 16 - \frac{8}{L_c} \cdot d_0 \cdot a$$

In state $4b$ all N atoms are at nodes of the wavefunction, so that

$$E_{4b} = \frac{h^2}{2m_e L_c^2} \cdot 16$$

In state $5a$ the wavefunction is

$$\psi_{5a} = \sqrt{\frac{2}{L_c}} \cdot \cos\left(5 \cdot \frac{2\pi s}{L_c}\right)$$

so that two N atoms are at an antinode (at $s = 0$ and $s = L_c/2$) and two are at a node (at $s = L_c/4$ and $s = 3L_c/4$); hence

$$E_{5a} = \frac{h^2}{2m_e L_c^2} \cdot 25 - 2a \cdot \frac{2}{L_c} \cdot d_0 = \frac{h^2}{2m_e L_c^2} \cdot 25 - \frac{4}{L_c} \cdot d_0 \cdot a$$

For state $5b$, the result is the same;

$$E_{5b} = \frac{h^2}{2m_e L_c^2} \cdot 25 - 2a \cdot \frac{2}{L_c} \cdot d_0 = \frac{h^2}{2m_e L_c^2} \cdot 25 - \frac{4}{L_c} \cdot d_0 \cdot a$$

however, the antinodes are located at $s = L_c/4$ and $s = 3L_c/4$ and the nodes are located at $s = 0$ and $s = L_c/2$. With $L_c = 16 \cdot d_0 = 2240$ pm ($d_0 = 140$ pm), we find that the transitions occurring from the $4b$ state occur at lower energy than those originating from the $4a$ state. The transition energies are

$$\begin{aligned}\Delta E_{4b\to5b} = \Delta E_{4b\to5a} &= \left(\frac{h^2}{2m_e L_c^2} \cdot 25 - \frac{4}{L_c} \cdot d_0 \cdot a\right) - \frac{h^2}{2m_e L_c^2} \cdot 16 \\ &= \frac{h^2}{512 m_e d_0^2} \cdot 9 - \frac{1}{4}a = (4.32 - 0.875) \times 10^{-19}\ \text{J} = 3.45 \times 10^{-19}\ \text{J}\end{aligned}$$

and

$$\begin{aligned}\Delta E_{4a\to5a} &= \Delta E_{4a\to5b} \\ &= \left(\frac{h^2}{2m_e L_c^2} \cdot 25 - \frac{4}{L_c} \cdot d_0 \cdot a\right) - \left(\frac{h^2}{2m_e L_c^2} \cdot 16 - \frac{8}{L_c} \cdot d_0 \cdot a\right) \\ &= \frac{h^2}{512 m_e d_0^2} \cdot 9 + \frac{1}{4}a = (4.32 + 0.875) \times 10^{-19}\ \text{J} = 5.20 \times 10^{-19}\ \text{J}\end{aligned}$$

These energies correspond to $\lambda_{max} = 576$ nm and $\lambda_{max} = 382$ nm and are in reasonable agreement with experiment.

E9.9 A new spectroscopic technique (Cavity Ring Down Polarimetry) can be used to measure the optical rotation resulting from a gas of propylene oxide. Use the optical rotation data in the table to compute the molar rotation for propylene oxide. Consider the propylene oxide to behave like an ideal gas, and take the path length of the optical cell to be 1.6 m.
Optical Rotation at $T = 323$ K and $\lambda = 355$ nm

P/bar	0.1	0.4	0.7	1.0
Rotation angle α	0.0006	0.0025	0.0043	0.0062

Solution

The data can be recalculated as the concentration c against rotation by using the ideal gas equation of state. When this is done the data becomes

P/bar	c/mol L^{-1}	Rotation/degrees
0.1000	0.0037	6.0000×10^{-4}
0.4000	0.0149	2.5000×10^{-3}
0.7000	0.0261	4.3000×10^{-3}
1.0000	0.0372	6.2000×10^{-3}

Figure E9.9 shows a plot of these data, and the slope of the graph is found to be $0.1665°/(\text{mol L}^{-1})$. The slope is associated with the value of the specific rotation times the path length; thus, the specific rotation of propylene oxide is

$$(0.1665°/\text{mol L}^{-1})/(1.6\text{ m}) = 0.1041°\text{ mol L}^{-1}\text{m}^{-1}$$

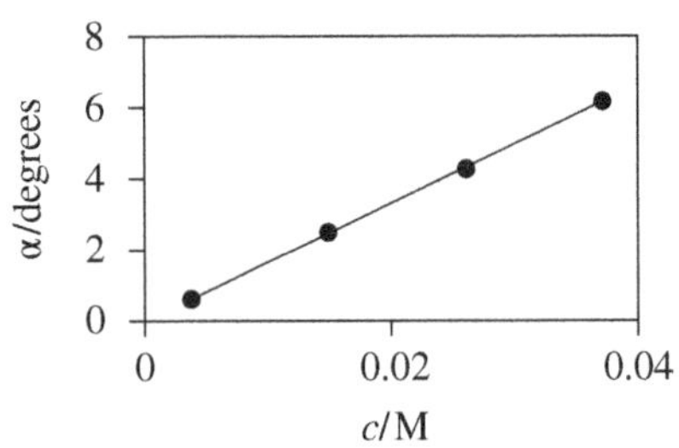

Figure E9.9 Optical rotation α versus the concentration c of propylene oxide.

E9.10 In the discussion of the azacyanines, we found the parameter $a = 3.5 \times 10^{-19}$ J to represent the difference in energy for a 2p-electron of a carbon atom versus a nitrogen atom. Compare this energy difference to the difference in first ionization potentials of carbon and nitrogen atoms (see Table 5.1).

Solution

From Table 5.1 we see that the first ionization energy of C is 11.26 eV and that for N is 14.53 eV. Taking the difference and converting to J we have

$$(14.53\text{ eV} - 11.26\text{ eV}) \times \frac{1.602 \times 10^{-19}\text{ J}}{\text{eV}} = 5.23 \times 10^{-19}\text{ J}$$

We compare this value to the value of a by computing the percent difference, and obtain

$$\frac{5.2 - 3.5}{5.2} \times 100 = 33\%$$

E9.11 Figure 9.19 diagrams the FEMO energy of Cu phthalocyanine, excluding the difference in self-energy of nitrogen and carbon atoms. Calculate the difference in energy between the $n = 4$ and $n = 5$ states using this model.

Solution

The energy level structure is that for a particle on a ring

$$E_n = \frac{h^2}{2m_e L_c^2} \cdot n^2$$

where $L_c = 16 \cdot d_0 = 2240$ pm ($d_0 = 140$ pm). So that the energy difference is

$$\begin{aligned}\Delta E = E_5 - E_4 &= \frac{h^2}{2m_e L_c^2} \cdot 25 - \frac{h^2}{2m_e L_c^2} \cdot 16 \\ &= \frac{\left(6.626 \times 10^{-34}\right)^2 \text{J}^2}{2 \times \left(9.11 \times 10^{-31}\text{ kg}\right)\left(2240 \times 10^{-12}\right)^2 \text{m}^2} \cdot 9 \\ &= 4.32 \times 10^{-19}\text{ J}\end{aligned}$$

This value is intermediate between the values found in Example 9.5, as one would expect.

E9.12 Imagine that you synthesize a free porphyrin which is supposed to have a molar extinction coefficient of $\varepsilon = 50,000\ \mathrm{M}^{-1}\ \mathrm{cm}^{-1}$ at $\lambda = 600$ nm. Describe how you can use electronic spectroscopy to check the purity of your sample.

Solution

The purity can be checked by measuring the absorbance of a sample of a known concentration. For example you may prepare a solution with a concentration of 0.400×10^{-6} M. The absorbance of such a solution in a $l = 1.00$ cm path length optical cell should be exactly 0.200.

$$A = \varepsilon \cdot c \cdot l$$
$$= 50,000\ \mathrm{M^{-1}cm^{-1}} \cdot 4.00 \times 10^{-6}\ \ \mathrm{M} \cdot 1.00\ \mathrm{cm} = 0.200$$

Any difference from this absorbance value can be attributed to impurities in the solution (which should cause the absorbance to be less than 0.200 unless their absorbance overlaps with that of the porphyrin).

Is this method very sensitive? To assess the sensitivity imagine that the sample is only 95% pure. Then the concentration would be 3.80×10^{-6} M and the absorbance would be 0.190. Such a difference is readily detected but requires a careful experiment. If the sample contained a 1% impurity however, the absorbance would be 0.198 and only differ by 0.002 which can be challenging to measure.

E9.13 Using a spreadsheet program plot the electric field found from Equation (9.52) with $a = 1$ and $b = 0.2$. Your plot should show that it traces out an ellipse.

Solution

The components along each of the axes are given by

$$F_z = a \cdot \cos \omega t \quad \text{and } F_y = b \cdot \sin \omega t$$

A plot of the field amplitudes versus each other is shown in Figure E9.13. For a given total electric field amplitude, each point on the ellipse (F_z, F_y) represents a particular value of ωt. As time proceeds ωt increases, however the electric field amplitude and direction retraces itself at multiples of 2π.

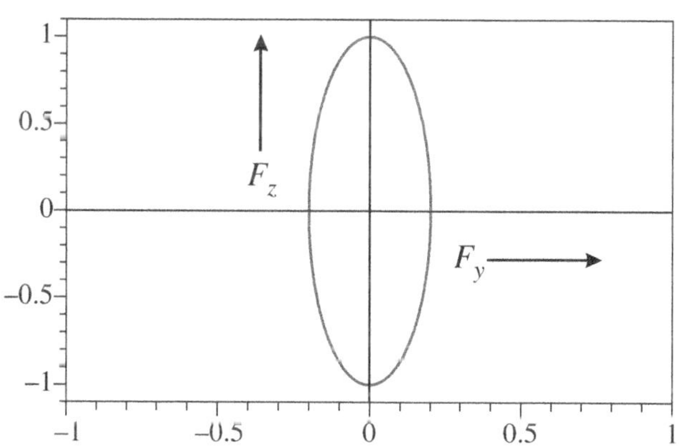

Figure E9.13 Vector sum F_z versus F_y of the two components of an elliptically polarized light wave.

E9.14 The equation for a right circularly polarized field amplitude is $\vec{F}_{\text{right}} = F_x \sin(\omega t) \cdot \hat{x} + F_y \cos(\omega t) \cdot \hat{y}$ where F_x and F_y are the components along the directions x and y, respectively, and are equal to each other in magnitude. Plot F_{right} using a spreadsheet program and show that it traces out a circle by rotating in a clockwise fashion. What is the formula for F_{left}? Show that the sum of F_{right} and F_{left} gives linearly polarized light.

Solution

From the formula for the vector $\vec{F}_{\text{right}}$ and the condition that $F_x = F_y = F$, we find that the magnitude of $\vec{F}_{\text{right}}$ is

$$\left|\vec{F}_{\text{right}}\right| = \sqrt{F^2 \sin^2(\omega t) + F^2 \cos^2(\omega t)} = F$$

that is, the total electric field amplitude is constant. This property is consistent with the claim that the electric field traces a circle, since a circle will be at a fixed distance from its origin at any point. To show that the circle is traced out in time by rotating in a clockwise direction, we plot the angle subtended between the electric field vector $\vec{F}_{\text{right}}$ and the y-axis versus the frequency time product, ωt; see Figure E9.14a. It is evident that the angle increases linearly in time until it reaches 180° and then it starts to decrease back to zero. The plot only goes from $\omega t = 0$ to $\omega t = 2\pi$; however, a continuation in time would lead to repetition of this cycle.

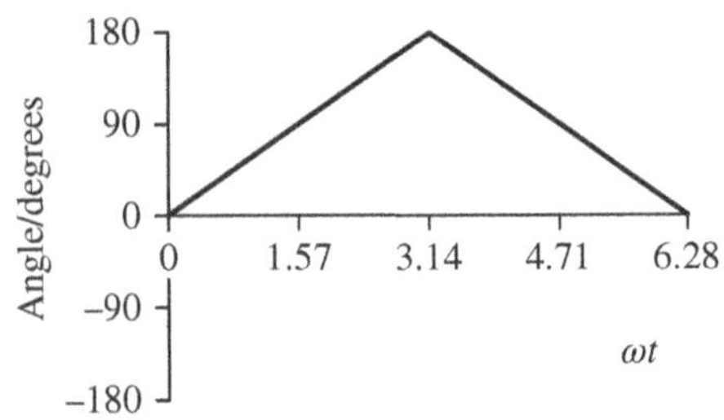

Figure E9.14a Right polarized light. Angle between the electric field vector $\vec{F}_{\text{right}}$ and the y-axis versus the time, ωt.

Left circularly polarized light has the same amplitude qualities; however, the angle subtended from the y-axis proceeds in the other direction. In this case we have

$$\vec{F}_{\text{left}} = F_x \sin \omega t\, \hat{x} - F_y \cos \omega t\, \hat{y}$$

and

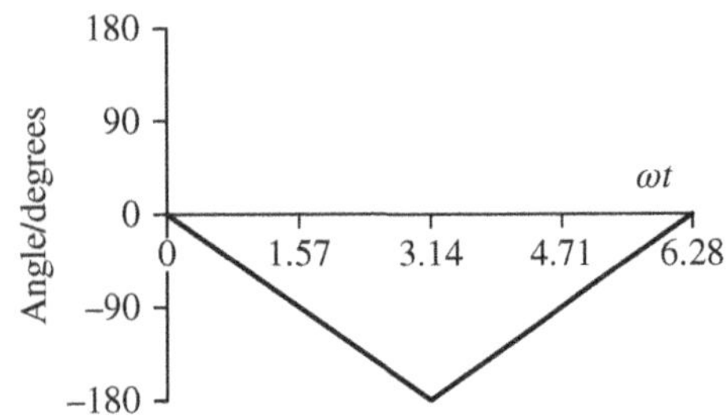

Figure E9.14b Left polarized light. Angle between the electric field vector $\vec{F}_{\text{right}}$ and the y-axis versus the time, ωt.

E9.15 Calculate the force on an electron in an electric field of $F = 10^6$ V/m (typical of an applied light field). Calculate the force on an electron in an applied magnetic field $B = F/c_0 = 3 \cdot 10^{-3}$ T, assuming that the force is given by $\mu B/a_0$ with a_0 the Bohr radius and μ the Bohr magneton. Compare the magnitudes of these two forces.

Solution

The force on an electron in an electric field of $F = 10^6$ V/m (typical of an applied light field) is

$$f = eF = \left(1.602 \times 10^{-19}\ \text{C}\right) \cdot \left(1 \times 10^{6}\ \text{V m}^{-1}\right) = 1.6 \times 10^{-13}\ \text{N}$$

The corresponding magnetic flux density of the light field is $B = F/c_0 = 3 \cdot 10^{-3}$ T. We assume that the force on an electron in this field is given by $\mu_B B/a_0$ with a_0 the Bohr radius and μ_B the Bohr magneton. Then the force acting on the electron is

$$f = \frac{\mu_B B}{a_0} = \frac{\left(9.27 \times 10^{-24}\ \text{J T}^{-1}\right) \cdot \left(3 \times 10^{-3}\ \text{T}\right)}{52.9\ \text{pm}} = 5.3 \times 10^{-16}\ \text{N}$$

This force is smaller than the force in the electric field by three orders of magnitude.

E9.16 An alternative description for the optical rotatory dispersion (ORD) signal is given by the difference of the index of refraction for left and right circularly polarized light, namely

$$\alpha = \frac{180 \cdot \left(n_l - n_r\right) \cdot l}{\lambda}$$

Using the difference in the rotation angle for saccharide, calculate the difference in refractive index.

Solution

In Section 9.3.1, the rotation angle of saccharide is 227° for a path length of 0.1 m and a wavelength of 589 nm. Substituting into the above equation, we find

$$
\begin{aligned}
(n_l - n_r) &= \frac{\alpha \cdot \lambda}{180 \cdot l} \\
&= \frac{227 \cdot 589 \times 10^{-9}\ \text{m}}{180 \cdot 0.1\ \text{m}} = 7.427 \times 10^{-6}
\end{aligned}
$$

The refractive index difference is 1 part in 10^5.

E9.17 In Section 9.2.6 we calculated the transition moment $M_x = 8.5\ \text{D} = 2.85 \times 10^{-29}$ C m for the transition between the 4a and 5a states of Cu phthalocyanine at $\lambda = 345$nm. Use this transition moment value, $\hat{n} = 1.6$, and Equation (9.31) to calculate the transition's oscillator strength.

Solution

Equation (9.31) is

$$
f = \frac{8\pi^2 m_e}{3h^2 e^2 \hat{n}} \cdot M^2 \cdot \Delta E
$$

We need the molecule-specific values of $\hat{n}$, M, and ΔE. $M_x = 2.85 \times 10^{-29}$ C m is given in the problem statement, as is $\hat{n} = 1.6$, and ΔE can be calculated from the wavelength at which the absorption spectrum has its maximum, namely

$$
\Delta E = \frac{hc}{\lambda} = \frac{(6.626 \times 10^{-34}\ \text{J s})\ (2.998 \times 10^{8}\ \text{m s}^{-1})}{345 \times 10^{-9}\ \text{m}} = 5.76 \times 10^{-19}\ \text{J}
$$

Given these parameters we can calculate f as

$$
\begin{aligned}
f &= \frac{8\pi^2\,(9.11 \times 10^{-31})\ \text{kg}}{3(6.626 \times 10^{-34}\ \text{J s})^2 (1.602 \times 10^{-19}\ \text{C})^2 \hat{n}} \cdot M^2 \cdot \Delta E \\
&= 2.13 \times 10^{75}\ \text{J}^{-1}\ \text{m}^{-2}\ \text{C}^{-2}\ \frac{1}{1.6} \cdot (2.85 \times 10^{-29}\ \text{C m})^2 \cdot 5.76 \times 10^{-19}\ \text{J} \\
&= 2.13 \times 10^{75}\ \text{J}^{-1}\ \text{m}^{-2}\ \text{C}^{-2}\ \frac{1}{1.6} \cdot 8.12 \times 10^{-50}\ \text{C m} \cdot 5.76 \times 10^{-19}\ \text{J} = 0.623
\end{aligned}
$$

Note that Equation (9.31) calculates the oscillatory strength for the excitation of one electron. In our case we have two electrons in the HOMO. Moreover, there are two different excitations possible, namely, in the x- and in the y-directions. Thus we obtain:

$$
f = 2\frac{8\pi^2 m_e}{3h^2 e^2 \hat{n}} \cdot (M_x^2 + M_y^2) \cdot \Delta E = 4 \cdot 0.623 = 2.5
$$

The value $f = 2.5$ is slightly smaller than the value $f = 2.8$ calculated in Table 9.5, because here we use the experimental value $\lambda = 345$ nm, whereas in the table we use the theoretical value $\lambda = 327$ nm. Both f-values are significantly higher than $f = 1.7$ found in the experiment.

E9.18 Consider the molecule anthanthrene dissolved in benzene. The first electronic transition in this molecule occurs near an energy of 23,090 cm^{-1}. A series of measurements of its absorbance A (at 23,090 cm^{-1}) as a function of concentration c gives

A	0.083	0.414	0.820	4.16
$c/10^{-6}$ mol L^{-1}	1.0	5.0	10.	50.

The path length of the optical cell was 1 cm. A) Determine the molar decadic absorption coefficient ε, in units of mol $\text{L}^{-1}\text{cm}^{-1}$, of the molecule at this wavelength. B) Determine the absorption cross-section σ of the molecule at this wavelength, where σ is given by

$$\sigma = \frac{\ln(10)}{\mathrm{N_A}} \cdot \varepsilon$$

C) Use bond increments and van der Waals radii to estimate the physical size of the molecule. Compare your absorption cross-section to the cross-sectional area of the molecule.

Solution

A) Figure E9.18 shows a plot of the absorbance versus the concentration and a fit of these data to a straight line. From the Lambert-Beer law the slope of the line is given by the product $\varepsilon \cdot b$. Hence we find that

$$\varepsilon = \frac{\text{slope}}{b} = \frac{0.0832 \cdot 10^6 \text{ mol L}^{-1}}{1 \text{ cm}} = 83,200 \text{ mol L}^{-1}\text{cm}^{-1}$$

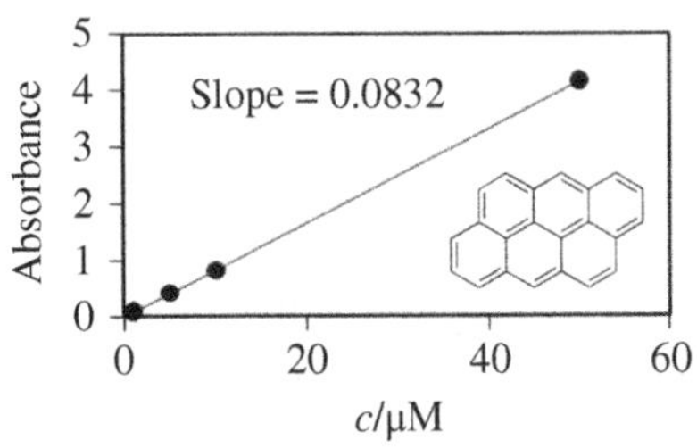

Figure E9.18 Absorbance of anthranthrene versus its concentration. The inset shows the C-skeleton of anthanthrene.

B) Using our value of ε and the formula given in the problem, we find that

$$\sigma = \frac{\ln(10)}{\mathrm{N_A}} \cdot \varepsilon = 3.82 \times 10^{-24} \text{ mol} \cdot 83,200 \text{ mol}^{-1} \text{ L cm}^{-1} =$$
$$= 3.18 \times 10^{-19} \text{ L cm}^{-1} = 3.18 \times 10^{-16} \text{ cm}^2 = 3.18 \times 10^{-20} \text{ m}^2$$
$$= 3.18 \times 10^4 \text{ pm}^2 = 3.18 \text{Å}^2$$

C) Here we estimate the molecule's cross-sectional area in its plane, hence providing an upper bound. Let us approximate the C–C bond lengths at 140 pm and the C–H bond lengths at 108 pm. Using these parameters, we dimensions of

$$2 \cdot (108 \text{ pm}) + (2 + 3 \cdot \cos(60))(140 \text{ pm}) = 706 \text{ pm}$$

by

$$2 \cdot \cos(30) \cdot (108 \text{ pm}) + (7 \cdot \cos(30))(140 \text{ pm}) = 1036 \text{ pm}$$

These values give the internuclear distances to the outermost H-atom nuclei. To account for the spatial extent of the H-atoms we add 120 pm on each end and find the molecule to have dimensions of 946 pm by 1276 pm; hence an area of

$$\text{area} = (946 \text{ pm}) \cdot (1276 \text{ pm}) = 1.21 \times 10^6 \text{ pm}^2 = 121 \text{ Å}^2$$

This calculation indicates that the molecule's absorption cross-section is about 2% of its physical planar cross-section.

E9.19 Consider the Beer's law derivation in Box 9.1 for the case of a solid material rather than a liquid. In this case, we can combine the product $\alpha \cdot c$ into a single parameter that is indicative of the solid, let us call it α'. In this case, show that

$$I = I_0 \exp(-\alpha' x)$$

where x is the distance into the solid sample. The table gives α' values for pure silicon at 300 K and four different photon energies; use these data to compute the distance $x_{1\%}$ at which 99% of the incident light would be absorbed.

α'/cm^{-1}	10^3	4×10^3	6×10^4	1×10^6
$h\nu/\text{eV}$	1.6	2.0	3.0	4.0

In your calculations, you can ignore the contribution of reflection at the surface on your results; however, you should comment on its effect.

Solution

Following the derivation in Box 9.1, we can write that the differential change in the light intensity $\mathrm{d}I$ between a position x' and $x' + \mathrm{d}x'$, is proportional to the incident light intensity I, and the solid material's absorption coefficient α', so that

$$\mathrm{d}I = -\alpha' \cdot I \cdot \mathrm{d}x'$$

To obtain the total light intensity at a depth x, we must integrate over the displacement from the surface of the solid at $x' = 0$ to the position $x' = x$

$$\int_{I_0}^{I(x)} \frac{\mathrm{d}I'}{I'} = -\alpha' \cdot \int_0^x \mathrm{d}x'$$

which gives

$$\ln\left(\frac{I(x)}{I_0}\right) = -\alpha' \cdot x \text{ or } I(x) = I_0 \cdot \mathrm{e}^{-\alpha' x}$$

The second part of the problem asks us to find the value of x, for which $I(x) = 0.01 \cdot I_0$; hence, we wish to find

$$\ln\left(\frac{0.01 \cdot I_0}{I_0}\right) = -\alpha' \cdot x_{1\%} \text{ or } x_{1\%} = \frac{4.605}{\alpha'}$$

Using the values given in the table, we find that

α'/cm^{-1}	10^3	4×10^3	6×10^4	1×10^6
$x_{1\%}/10^{-6}$ m	46	11.5	0.77	0.046
$h\nu/\text{eV}$	1.6	2.0	3.0	4.0

These values show that red light (near 1.6 eV) penetrates very deeply into silicon, whereas ultraviolet light (4 eV) is very strongly attenuated and only penetrates a few tens of nanometers.
Reflection from the front surface will reduce the light intensity that proceeds through the interface and will cause the penetration depth, as defined by $x_{1\%}$, to be reduced.

E9.20 Using the fundamental relation between the intensity I of a light beam and its electric field strength F, namely $I = c_0 \cdot \frac{1}{2}\varepsilon_0 F^2$, show that the force acting on an a valence electron in a carbon atom (use a screened charge of $q = 3.25e$ and a distance $d = 150$ pm for the electron from the carbon nucleus) is much larger than that from a light beam of intensity 100 W/cm ^{2}.

Solution

For $I = 100$ W/cm ^{2} we obtain

$$F = \sqrt{\frac{2 \cdot 100 \text{ W cm}^{-2}}{3 \times 10^8 \text{ m s}^{-1} \cdot 8.85 \times 10^{-12} \text{ C}^2 \text{ J}^{-1} \text{ m}^{-1}}}$$

$$= \sqrt{\frac{2 \cdot 100 \text{ J s}^{-1} \cdot 10^4 \text{ m}^{-2}}{3 \times 10^8 \text{ s}^{-1} \cdot 8.85 \times 10^{-12} \text{ J}^2 \text{ V}^{-2} \text{ J}^{-1}}}$$

where we used the relations

$$\mathrm{W} = \mathrm{J\,s^{-1}} \quad \text{and } \mathrm{J} = \mathrm{C\,V}$$

Thus

$$F = \sqrt{\frac{2\times 10^{6}}{3\times 10^{8}\ \cdot 8.85\times 10^{-12}}}\ \mathrm{V\,m^{-1}} = 2.7\times 10^{4}\ \mathrm{V\,m^{-1}}$$

The electric field experienced by a valence electron in a molecule can be estimated as

$$F_{\text{valence}} = -\frac{1}{4\pi\varepsilon_0}\frac{q}{d^2}$$

where d is the distance from the nucleus and q is the shielded nuclear charge, which the valence electron experiences. We use $q = 3.25e$ for a carbon atom and $d = 150$ pm for the distance to obtain

$$F_{\text{valence}} = -\frac{1}{4\pi\varepsilon_0}\frac{3.25e}{(150\ \mathrm{pm})^2} = -\frac{1}{4\pi\cdot 8.85\times 10^{-12}\ \mathrm{C^2\,J^{-1}\,m^{-1}}}\frac{3.25\cdot 1.602\times 10^{-19}\ \mathrm{C}}{2.25\times 10^{-20}\ \mathrm{m^2}}$$

$$= -\frac{1}{4\pi\cdot 8.85\times 10^{-12}}\frac{3.25\cdot 1.602\times 10^{-19}}{2.25\times 10^{-20}}\ \mathrm{V\,m^{-1}} = 2.1\times 10^{11}\ \mathrm{V\,m^{-1}}$$

Thus it is reasonable to assume that the light field is small compared to the molecular field.

E9.21 Evaluate the integral for the transition moment for phthalocyanine (Equation (9.45)) by approximating the phthalocyanine ring by a circle with a radius of r.

Solution

According to Equation (9.45) in Section 9.2.6 (cyclic systems) the transition moment for the $4a \rightarrow 5a$ transition in phthalocyanine is

$$M_x = e\frac{2}{L_c}\left(\frac{L_c}{2\pi}\right)^2 \cdot \int_0^{2\pi} \cos(4\varphi)\cos\varphi\cos(5\varphi)\,\mathrm{d}\varphi \tag{9E21.1}$$

where L_c is the circumference of the ring and e is the elementary charge.

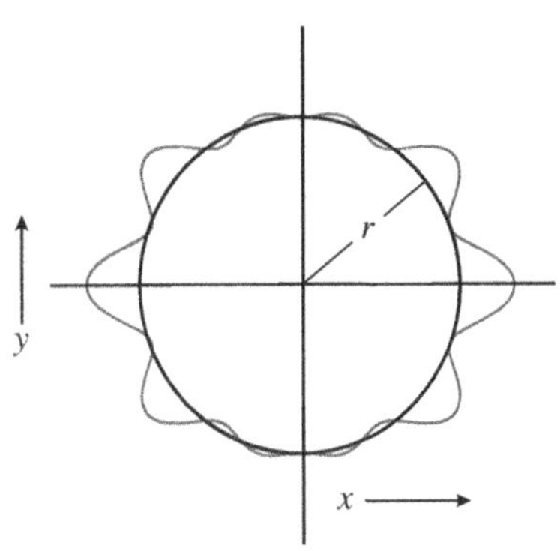

Figure E9.21 Integrand in Equation (9E21.1) along the circle approximating the ring in phthalocyanine.

The curve for the integrand is shown in Fig. E9.21 along a circle with radius r. From Figure E9.21 it is evident that the integrand is symmetric, with respect to the x-axis and to the y-axis.

The integral can be solved by numerical integration. The result is

$$\int_0^{2\pi} \cos(4\varphi)\cos\varphi\cos(5\varphi)\,\mathrm{d}\varphi = 1.5708$$

This result can also be obtained analytically. Using the trigonometric relations in Appendix B the integrand $I = \cos(4\varphi)\cos\varphi\cos(5\varphi)$ simplifies to

$$I = \cos\varphi \cdot \cos 4\varphi \cdot \cos 5\varphi = \cos\varphi \cdot \frac{1}{2}(\cos\varphi + \cos 9\varphi)$$
$$= \frac{1}{2}\cos^2\varphi + \frac{1}{2}\cos\varphi\cos 9\varphi = \frac{1}{2}\cos^2\varphi + \frac{1}{4}\cos 8\varphi + \frac{1}{4}\cos 10\varphi$$

Then for the integral we find

$$I = \int_0^{2\pi}\left[\frac{1}{2}\cos^2\varphi + \frac{1}{4}\cos 8\varphi + \frac{1}{4}\cos 10\varphi\right]\,d\varphi$$
$$= \left[\frac{1}{2}\left(\frac{\varphi}{2} - \frac{\sin 2\varphi}{4}\right)_0^{2\pi} + \frac{1}{4}\left(\frac{\sin 8\varphi}{8}\right)_0^{\pi} + \frac{1}{4}\left(\frac{\sin 10\varphi}{10}\right)_0^{\pi}\right]$$
$$= \left[\left(\frac{\pi}{2}\right) + \frac{1}{32}(0) - \frac{1}{40}(0)\right] = \frac{\pi}{2}$$

E9.22 Explain why the energy shift between the two low lying π-electron transitions in porphyrin is smaller than that in phthalocyanine. Using logic like that used in Section 9.2.6 for phthalocyanine estimate the absorption maxima for the porphyrin transitions.

Solution
Figure E9.22 compares the number of N-atom sites in porphyrin to that in phthalocyanine.

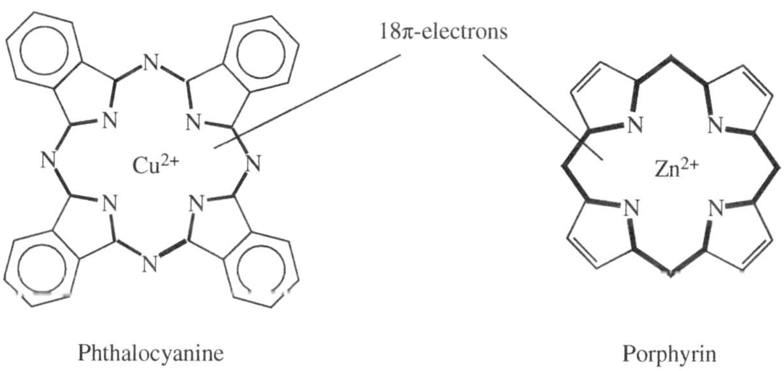

Figure E9.22 Comparison of phthalocyanine and porphyrin.

In principle, porphyrin should have the same π electron system as does phthalocyanine. However, the shift of the energy levels arising from the nitrogen atoms is smaller, because the four outermost nitrogen atoms are missing. Therefore the energy shift is just one-half of the shift in phthalocyanine. For the transitions we obtain

$$\Delta E_{4b\to5b} = \Delta E_{4b\to5a} = \frac{h^2}{512 m_e d_0^2} - \frac{1}{4}a = \left(4.32 - \frac{1.75}{2}\right) \times 10^{-19}\ \text{J} = 3.45 \times 10^{-19}\ \text{J}$$
$$\Delta E_{4a\to5a} = \Delta E_{4a\to5b} = \frac{h^2}{512 m_e d_0^2} + \frac{1}{4}a = \left(4.32 + \frac{1.75}{2}\right) \times 10^{-19}\ \text{J} = 5.20 \times 10^{-19}\ \text{J}$$

This corresponds to $\lambda_{max} = 576$ nm and $\lambda_{max} = 382$ nm.

9.2 Problems

P9.1 Show that the energy shift $\Delta E'$, when proceeding from a cyanine cation to the corresponding azacyanine cation, is given by $\Delta E' = \Delta E - a \cdot 2/N$ for $N/2 = 2, 4, 6, \ldots$ and $\Delta E' = \Delta E + a \cdot 2/N$ for $N/2 = 3, 5, 7, \ldots$. where a is a constant and N is the number of π-electrons. Use perturbation theory as discussed in Foundation 5.2 (on Wiley's website). Proceed by calculating the mean potential energy of a π electron in the electric field of the nitrogen atom's core using

$$\overline{V} = \int_0^L V_N \, \psi^2 \cdot \mathrm{d}s$$

V_N is the potential energy of a π electron in the field of the N atom (more precisely, the difference between the potential energies of N and C). As shown in Fig. 9.13 assume that $V_N = -a$ for $s_1 < s < s_2$ and $V_N = 0$ for s elsewhere.

Solution

We approximate V_N by assuming that the potential energy of an electron around an N atom is lower by a constant amount a compared to the constant potential energy of a chain of carbon atoms. According to Fig. 9.13 with $s_1 = L/2 - d_0/2$ and $s_2 = L/2 + d_0/2$ we have $V_N = -a$ for $s_1 < s < s_2$ and $V_N = 0$ for s elsewhere; so that

$$\overline{V} = \int_{s_1}^{s_2} -a\,\psi^2 \mathrm{d}s = -a \int_{s_1}^{s_2} \psi^2 \mathrm{d}s$$

For orbitals with a node at the center of the molecule, ψ^2 is practically zero for the whole region around the N atom, so that $\overline{V} \simeq 0$. For orbitals with an antinode at the center of the molecule, ψ^2 in the region of the N atom is only slightly different from its maximum value at the center, thus

$$\overline{V} \simeq -a \int_{s_1}^{s_2} \psi^2 \mathrm{d}s = -a(\psi^2)_{\text{at center}} \cdot d_0 = -a \cdot \frac{2}{L} \cdot d_0$$

This amount decreases with increasing chain length L. Because $L = N \cdot d_0$, we find that

$$\overline{V} = -a \cdot \frac{2}{L} \cdot d_0 = -a \cdot \frac{2}{N}$$

The energy shifts that are discussed with respect to Fig. 9.13, correspond to the HOMO to LUMO energy gap. For the case where $N/2$ is odd the HOMO orbital has an anti-node in the center and the LUMO has a node in the center; hence, the HOMO is stabilized by $-2a/N$ and the LUMO is not shifted. Consequently the energy gap increases from the cyanine value of ΔE to the corrected value

$$\Delta E' = \Delta E + a \cdot \frac{2}{N} \text{ for } \frac{N}{2} = 3,\ 5,\ 7,\ \ldots$$

For the case where $N/2$ is even the HOMO orbital has a node in the center and the LUMO has an anti-node in the center; hence, the HOMO is not shifted and the LUMO is stabilized by $-2a/N$. Consequently the energy gap decreases from the cyanine value of ΔE to the corrected value

$$\Delta E' = \Delta E - a \cdot \frac{2}{N} \text{ for } \frac{N}{2} = 2,\ 4,\ 6,\ \ldots$$

The constant a corresponds to the energy difference between an electron in a p-orbital of a carbon atom and that in a p-orbital of a nitrogen atom. From the ionization energies of C (18.0×10^{-19} J) and N (23.2×10^{-19} J) displayed in Table 5.1 it follows that $a = 5.2 \times 10^{-19}$ J (see E9.10), which is somewhat different from the value of $a = 3.5 \times 10^{-19}$ J that is used in the text.

P9.2 a) Calculate how the HOMO to LUMO gap energy shifts when proceeding from a hypothetical butadiene with equal bond lengths to a butadiene with alternating single and double bonds. In analogy to the considerations in Problem P9.1 the energy in orbital n should be

$$E' = E - 2 \cdot b \cdot \psi^2_{n,C=C} \cdot d_0 + b \cdot \psi^2_{n,C-C} \cdot d_0$$

where b is a constant, $\psi_{n,C=C}$ is the value of the wavefunction in orbital n at the place of the double bond, $\psi_{n,C-C}$ is the corresponding value at the place of the single bond, and d_0 is the bond length (see Fig. 9.16). You should find that $\Delta E' = \Delta E + \frac{6}{5}b$.

b) For longer polyenes explain why the difference between $\Delta E'$ and ΔE is approximately the same, independent of length.

Solution

a) By analogy with the consideration in Problem P9.1, the energy in orbital n should be

$$E' = E - 2 \cdot b \cdot \psi^2_{n,\text{C=C}} \cdot d_0 + b \cdot \psi^2_{n,\text{C-C}} \cdot d_0$$

where the factor of two arises because butadiene has two double bonds and one single bond; constant b is defined in Fig. 9.15. $\psi_{n,\text{C=C}}$ is the value of the wavefunction in orbital n at the place of the double bond, and $\psi_{n,\text{C-C}}$ is the corresponding value at the place of the single bond. From Fig. 9.16, we find that

$$\psi^2_{2,\text{C=C}} = \frac{2}{L} \quad \text{and} \quad \psi^2_{2,\text{C-C}} = 0$$

for the orbital with $n = 2$ and

$$\psi^2_{3,\text{C=C}} = 0 \quad \text{and} \quad \psi^2_{3,\text{C-C}} = \frac{2}{L}$$

for the orbital with $n = 3$. Using $L = 5d_0$ for butadiene, we obtain

$$\Delta E' = \Delta E + 2b\frac{2}{L}d_0 + b\frac{2}{L}d_0 = \Delta E + \frac{6}{5}b$$

b) For a polyene

$$\text{CH}_2(=\text{CH}-\text{CH})_k = \text{CH}_2, \quad k = 0, 1, 2, \ldots$$

the number of π-electrons is $2k + 2$ and $L = (2k + 3) \cdot d_0$. With bond alternation considered, we see that $k + 1$ double bond sites and k single bond sites are present. As the chain length of the polyene grows, in increments of two-carbon units and hence two π-electrons, the HOMO orbital retains a nodal structure in which each antinode lies on the center of a double bond and each node lies in the center of single bond. In a corresponding way, the LUMO orbital retains a pattern in which each antinode lies on the center of single bond and each node lies on the center of a double bond site. Generalizing our expression for the energy shift, we find

$$E'_n = E_n - (k+1) \cdot b \cdot \psi^2_{n,\text{C=C}} \cdot d_0 + k \cdot b \cdot \psi^2_{n,\text{C-C}} \cdot d_0$$

For the HOMO, $n = k + 1$ and

$$\psi^2_{\text{HOMO,C=C}} = \frac{2}{L}, \quad \psi^2_{\text{HOMO,C-C}} = 0; \text{ so that } E'_{k+1} = E_{k+1} - \frac{2\,(k+1) \cdot b}{L} \cdot d_0$$

For the LUMO, $n = k + 2$ and

$$\psi^2_{\text{LUMO,C=C}} = 0, \quad \psi^2_{\text{LUMO,C-C}} = \frac{2}{L}\text{ ; so that } E'_{k+2} = E_{k+2} + \frac{2k \cdot b}{L} \cdot d_0$$

Hence we find that

$$\Delta E' = E'_{k+2} - E'_{k+1} = \Delta E + \frac{(4k+2) \cdot b}{L} \cdot d_0$$
$$= \Delta E + \frac{(4k+2) \cdot b}{(2k+3)}$$

As the chain length becomes very long ($k \to \infty$), $\Delta E \to 0$ and $\Delta E' \to 2b$.

P9.3 a) For the porphyrin transitions in Exercise E9.8, calculate the transition dipole matrix elements $M_{x,\,4a\to5a}$ and $M_{x,\,4b\to5b}$ by using the free electron model and taking the porphyrin ring to be a circle with circumference $L_c = 16d_0$.

b) Use the facts $M_x = M_y$, $M_z = 0$, and your result in part a to evaluate the oscillator strength f where

$$f = \frac{8\pi^2 m_e}{3h^2 e^2 \hat{n}} \cdot M^2 \cdot \Delta E$$

and $\hat{n} = 1.5$.

Solution

We calculate the corresponding transition moments by approximating the 16-membered ring with circumference $L_C = 16d_0$ as a circle with radius

$$r = \frac{L_C}{2\pi} = \frac{16d_0}{2\pi}$$

Consider the FEMO model wavefunctions

$$\psi_{4a} = \frac{1}{\sqrt{\pi}} \cos(4\varphi) \quad \text{and} \quad \psi_{5a} = \frac{1}{\sqrt{\pi}} \cos(5\varphi)$$

where $x = r\cos\varphi$, so that

$$\begin{aligned} M_{x,4a\to 5a} &= e\int_0^{2\pi} \psi_{4a} \cdot x \cdot \psi_{5a} \cdot d\varphi \\ &= e\frac{16d_0}{2\pi^2}\int_0^{2\pi} \cos(4\varphi)\cdot\cos\varphi\cdot\cos(5\varphi)\cdot d\varphi \end{aligned}$$

Using the trigonometric relation $\cos\alpha\cdot\cos\beta = (\cos(\alpha+\beta)+\cos(\alpha-\beta))/2$, we find that

$$\begin{aligned} I &= \int_0^{2\pi} \cos(4\varphi)\cdot\cos\varphi\cdot\cos(5\varphi)\cdot d\varphi \\ &= \frac{1}{2}\int_0^{2\pi} (\cos(9\varphi)+\cos\varphi)\cdot\cos\varphi\cdot d\varphi \\ &= \frac{1}{2}\left[\int_0^{2\pi}\cos(9\varphi)\cos\varphi\cdot d\varphi + \int_0^{2\pi}\cos^2\varphi\cdot d\varphi\right] \\ &= \frac{1}{2}\left[\int_0^{2\pi}\frac{\cos(10\varphi)+\cos(8\varphi)}{2}\cdot d\varphi + \int_0^{2\pi}\cos^2\varphi\cdot d\varphi\right] \\ &= \frac{1}{4}\int_0^{2\pi}\cos(10\varphi)\cdot d\varphi + \frac{1}{4}\int_0^{2\pi}\cos(8\varphi)\cdot d\varphi + \frac{1}{2}\int_0^{2\pi}\cos^2\varphi\cdot d\varphi \\ &= \frac{1}{4}\frac{\sin(10\varphi)}{10}\bigg|_0^{2\pi} + \frac{1}{4}\frac{\sin(8\varphi)}{8}\bigg|_0^{2\pi} + \frac{1}{4}\left(x+\frac{1}{2}\sin 2x\right)\bigg|_0^{2\pi} \\ &= 0-0+0-0+\frac{1}{4}(2\pi+0-0-0) = \frac{\pi}{2} \end{aligned}$$

Hence we find that

$$\begin{aligned} M_{x,4a\to 5a} &= e\frac{16d_0}{2\pi^2}\int_0^{2\pi}\cos(4\varphi)\cdot\cos\varphi\cdot\cos(5\varphi)\cdot d\varphi \\ &= e\frac{16d_0}{2\pi^2}\cdot\frac{\pi}{2} = 2.85\times 10^{-29}\ \text{C m} \end{aligned}$$

The same value is obtained for $M_{y,4b\to 5b}$.

b) According to Equation (9.27) the oscillator strength for the transition of one electron is

$$f = \frac{8\pi^2 m_e}{3h^2e^2\hat{n}}\cdot M^2\cdot\Delta E$$

In the case of porphyrin, the transition energies are $\Delta E_{4b\to 5b} = 3.45\times 10^{-19}$J and $\Delta E_{4a\to 5a} = 5.20\times 10^{-19}$J. Using $\hat{n} = 1.5$ and the results for the transition dipoles we find

$$\begin{aligned} f_{4b\to 5b} &= \frac{8\pi^2(9.11\times 10^{-31}\ \text{kg})(2.85\times 10^{-29}\ \text{C m})^2}{3(6.626\times 10^{-34}\ \text{J s})^2(1.602\times 10^{-19}\ \text{C})^2(1.5)}(3.45\times 10^{-19}\text{J}) \\ &= 0.398 \end{aligned}$$

Through a similar calculation, we find that

$$f_{4a\to 5a} = 0.600$$

Note that these are the f values for the excitation of one electron. As outlined in Exercise E9.17, these values must be multiplied by a factor of four.

P9.4 Fit the data in Table 9.2 to a more general model of the energy shift, in which the box length at the ends of the π-system is an adjustable parameter γ and the bond length d_0 is an adjustable parameter. In particular analyze the data by

$$\Delta E = \frac{h^2}{8m_e d_0^2} \frac{2(k+2)+1}{4(k+1+\gamma)^2}$$

Comment on your findings.

Solution

We will start by writing the energy expression given in the problem statement.

$$\Delta E = \frac{h^2}{8m_e d_0^2} \cdot \frac{2(k+2)+1}{4(k+1+\gamma)^2}$$

Next we rearrange this expression to isolate the d_0 and γ terms. First we rearrange the expression to find

$$\frac{8m_e \Delta E}{h^2 \cdot (2(k+2)+1)} = \frac{1}{d_0^2} \cdot \frac{1}{4(k+1+\gamma)^2}$$

Then we invert this expression and take the square root of both sides to yield

$$\frac{1}{2}\sqrt{\frac{h^2 \cdot (2(k+2)+1)}{8m_e \Delta E}} = d_0(k+1+\gamma) = d_0(k+1) + d_0\gamma$$

Thus a plot of the left-hand side of this equation (which we label y in the graph) against $k+1$ should be linear with slope d_0 and intercept γd_0.

The following table presents the data for the plot, and the plot itself follows in Fig. P9.4.

k	$k+1$	λ/nm	$\Delta E/10^{-19}$J	$\frac{1}{2}\sqrt{\frac{h^2\cdot(2(k+2)+1)}{8m_e\Delta E}}/(10^{-10}\text{ m})$
0	1	224	8.87	2.91
1	2	313	6.35	4.08
2	3	416	4.78	5.33
3	4	519	3.83	6.58
4	5	625	3.18	7.85
5	6	735	2.70	9.14
6	7	848	2.34	10.5

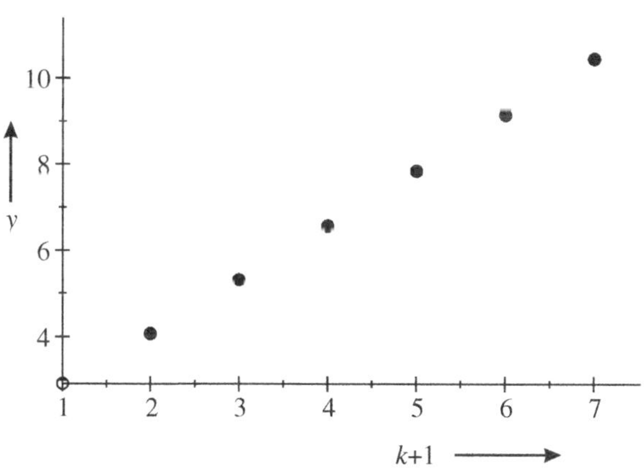

Figure P9.4 Evaluation of d_0 and γ for the cyanines in Table 9.2.

The slope is found to be 1.26×10^{-10} and the y-intercept (γd_0) is 1.58×10^{-10}, so that $d_0 = 1.26 \times 10^{-10}$ m and $\gamma = 1.25$. This value of d_0 is about 10% smaller than our commonly used value of 140 pm.

P9.5 Using the particle-in-a-box model, calculate the transition dipole moment for the excitation of an electron from the $n = 1$ level to the $n' = 2$ level. a) Determine the dependence of the transition dipole moment on the box length. b) Determine the transition dipole moment between an electron in the ground state $n = 1$ and an arbitrary excited sate n'. What general conclusion can you draw from this calculation; i.e., what is the selection rule?

Solution

a) We begin with our definition of the transition dipole moment (Equation (9.28))

$$M_{x,n\to n'} = e\int_0^L \psi_n \, x\psi_{n'} \cdot \mathrm{d}x$$

and use the fact that the particle-in-a-box wavefunctions are

$$\psi_n = \sqrt{\frac{2}{L}}\sin\left(\frac{n\pi x}{L}\right) \quad \text{and} \quad \psi_{n'} = \sqrt{\frac{2}{L}}\sin\left(\frac{n'\pi x}{L}\right)$$

Hence we can write the transition moment as

$$M_{x,n\to n'} = \frac{2e}{L}\int_0^L \sin\left(\frac{n\pi x}{L}\right)\, x \sin\left(\frac{n'\pi x}{L}\right)\cdot \mathrm{d}x$$

Now we use the trigonometric relation

$$2\cdot \sin a\cdot \sin b = \cos(a-b) - \cos(a+b)$$

(see Appendix B). Hence we can write that

$$\begin{aligned} M_{x,n\to n'} &= \frac{e}{L}\int_0^L \left[\cos\left(\frac{(n-n')\,\pi x}{L}\right) - \cos\left(\frac{(n+n')\,\pi x}{L}\right)\right] x\cdot \mathrm{d}x \\ &= \frac{e}{L}\left(\frac{L}{(n-n')\,\pi}\right)^2 \int_0^{(n-n')\pi} y\cos y\,\mathrm{d}y \\ &\quad -\frac{e}{L}\left(\frac{L}{(n+n')\,\pi}\right)^2 \int_0^{(n+n')\pi} y\cos y\cdot \mathrm{d}y \end{aligned}$$

Now we use the integral $\int x\cos x\,\mathrm{d}x = x\sin x + \cos x$ to write

$$\begin{aligned} M_{x,n\to n'} &= \frac{e}{L}\left(\frac{L}{(n-n')\,\pi}\right)^2 [y\sin y + \cos y]_0^{(n-n')\pi} \\ &\quad -\frac{e}{L}\left(\frac{L}{(n+n')\,\pi}\right)^2 [y\sin y + \cos y]_0^{(n+n')\pi} \\ &= \frac{eL}{\pi^2}\left(\frac{\left[\cos\left[(n-n')\,\pi\right]-1\right]}{(n-n')^2}\right) - \frac{eL}{\pi^2}\left(\frac{\left[\cos\left((n+n')\,\pi\right)-1\right]}{(n+n')^2}\right) \\ &= \frac{eL}{\pi^2}\left[\frac{\cos\left[(n-n')\,\pi\right]-1}{(n-n')^2} - \frac{\cos\left[(n+n')\,\pi\right]-1}{(n+n')^2}\right] \end{aligned}$$

This general result demonstrates that the transition moment grows linearly with the box length. For the particular case of $n = 1$ and $n' = 2$, we find that

$$\begin{aligned} M_{x,n\to n'} &= \frac{eL}{\pi^2}\left[\frac{\cos(-\pi)-1}{(-1)^2} - \frac{\cos(3\pi)-1}{(3)^2}\right] \\ &= \frac{eL}{\pi^2}\left[-2+\frac{2}{9}\right] = -\frac{16}{9}\frac{eL}{\pi^2} \end{aligned}$$

b) Now we consider the case $n = 1$ and n',

$$M_{x,n\to n'} = \frac{eL}{\pi^2}\left[\frac{\cos\left[(1-n')\,\pi\right]-1}{(1-n')^2} - \frac{\cos\left[(1+n')\,\pi\right]-1}{(1+n')^2}\right]$$

For n' odd, each of the arguments of the cosine function is a multiple of 2π and the transition moment integral is identically zero. For n' even, each of the arguments of the cosine function is an odd multiple of π and the transition moment integral is nonzero. Hence n' even is the selection rule for a transition from the $n = 1$ state.

P9.6 Explain why the terminal nitrogen atoms do not modulate the lowest energy electronic transitions of a cyanine dye, such as

$$(CH_3)_2\overset{-}{N} - CH = CH - CH = CH - CH = \overset{\oplus}{N}(CH_3)_2$$

as do nitrogen atoms in the center of the molecule, such as

$$(CH_3)_2\overset{-}{N} - CH = CH - \underset{-}{N} = CH - CH = \overset{\oplus}{N}(CH_3)_2$$

Solution

In Problem P9.1 we showed that the shift in energy of the orbital could be deduced using a perturbation theory treatment in which the mean potential energy $\overline{V}$ of a π electron in the electric field of the nitrogen atom's core was calculated.

$$\overline{V} = \int_0^L V_N \ \psi^2 \cdot ds$$

By making the approximation that $V_N = -a$ is the potential energy of a π electron in the field of the N atom, we were able to deduce how the energy of the HOMO to LUMO gap shifted between the azacyanine structure and the cyanine structure; the energy shift was given by $\Delta E' = \Delta E - a \cdot 2/N$ for $N/2 = 2, 4, 6, \ldots$ and by $\Delta E' = \Delta E + a \cdot 2/N$ for $N/2 = 3, 5, 7, \ldots.$ where N is the number of π-electrons.

What is the energy shift for the nitrogen atoms located at the end points of the chain? As discussed in the text, the HOMO and LUMO wavefunctions differ by whether they have a node or an antinode in the center of the molecule; however, the wavefunction near the end of the chain (remember that the chain length is d_0 farther than the final atom on each end) has about the same amplitude on the N atoms in both the HOMO and LUMO states. For a cyanine dye with N π-electrons in the chain the HOMO and LUMO wavefunctions are

$$\psi_{\text{HOMO}} = \sqrt{\frac{2}{L}} \sin\left(\frac{(N/2)\pi x}{L}\right) \quad \text{and} \quad \psi_{\text{LUMO}} = \sqrt{\frac{2}{L}} \sin\left(\frac{(N/2+1)\pi x}{L}\right)$$

and the chain length is $L = Nd_0$. We evaluate the probability density of the HOMO orbital on one of the N-atoms near the end of the chain and find that

$$\psi^2_{\text{HOMO}}(x = d_0) = \frac{2}{L}\sin^2\left(\frac{(N/2)\pi d_0}{Nd_0}\right) = \frac{2}{L}\sin^2\left(\frac{(N/2)\pi}{N}\right)$$
$$= \frac{2}{L}\sin^2\left(\frac{\pi}{2}\right) = \frac{2}{L}$$

Similarly for the LUMO orbital we find that

$$\psi^2_{\text{LUMO}}(x = d_0) = \frac{2}{L}\sin^2\left(\frac{(N/2+1)\pi d_0}{Nd_0}\right) = \frac{2}{L}\sin^2\left(\frac{(N/2+1)\pi}{N}\right)$$
$$= \frac{2}{L}\sin^2\left(\left[\frac{1}{2} + \frac{1}{N}\right]\pi\right) = \frac{2}{L}\sin^2\left(\left[\frac{\pi}{2} + \frac{1}{N}\pi\right]\right)$$

We see that the argument in the LUMO expression differs from that in the HOMO expression by the term π/N. If we use the trigonometric relation: $\sin(\alpha + \beta) = \sin\alpha \cdot \cos\beta + \cos\alpha \cdot \sin\beta$, we find that

$$\psi^2_{\text{LUMO}}(x = d_0) = \frac{2}{L}\left[\sin\left(\frac{\pi}{2}\right) \cdot \cos\left(\frac{\pi}{N}\right) + \cos\left(\frac{\pi}{2}\right) \cdot \sin\left(\frac{\pi}{N}\right)\right]^2$$
$$= \frac{2}{L}\cos^2\left(\frac{\pi}{N}\right)$$

Because $\cos^2(\pi/N) \leq 1$, the electron probability density on the nitrogen atoms at the end of the chain is somewhat higher for the HOMO orbital than for the LUMO orbital, and this difference decreases with increasing chain length. As N becomes large, $\cos^2(\pi/N) \to 1 - (\pi/N)^2$ so that the probability densities of the LUMO and HOMO wavefunctions on the terminal nitrogen atoms converge to the same value, as $1/N^2$. In contrast, the HOMO–LUMO gap decreases as $1/N$ as N becomes large (see Equation (9.19) of text), and this effect dominates for the decrease in HOMO–LUMO energy gap with increasing chain length.

For the cyanine dye considered in this problem we have $N = 8$ and $1/N = 0.125$. Thus

$$\psi^2_{\text{LUMO}}(x = d_0) = \frac{2}{L}\cos^2\left(\frac{1}{N}\pi\right) = \frac{2}{L}\cos^2(0.125\pi) = \frac{2}{L}0.85$$

and

$$\overline{V} = \int_{S_1}^{S_2} -a\psi^2 \mathrm{d}s = -a \int_{S_1}^{S_2} \psi^2 \mathrm{d}s$$

Hence the HOMO is more stabilized by 15% compared to the LUMO.

P9.7 (a) Show that the integrand in the transition moment integral, Equation (9.45), for the $4a \rightarrow 5a$ transition in phthalocyanine is symmetric with respect to the x and y directions (origin of the coordinate system at the center of the ring). Hint: plot the integrand as a function of the angle φ on the circumference of the ring. (b) Solve the transition moment integral, Equation (9.45).

Solution

According to Equation (9.45) in Section 9.2.6 the transition moment for the $4a \rightarrow 5a$ transition in phthalocyanine is

$$M_x = e\frac{2}{L_c}\left(\frac{L_c}{2\pi}\right)^2 \cdot \int_0^{2\pi} \cos(4\varphi)\cos\varphi\cos(5\varphi)\,\mathrm{d}\varphi$$

where L_c is the circumference of the ring and e is the elementary charge.

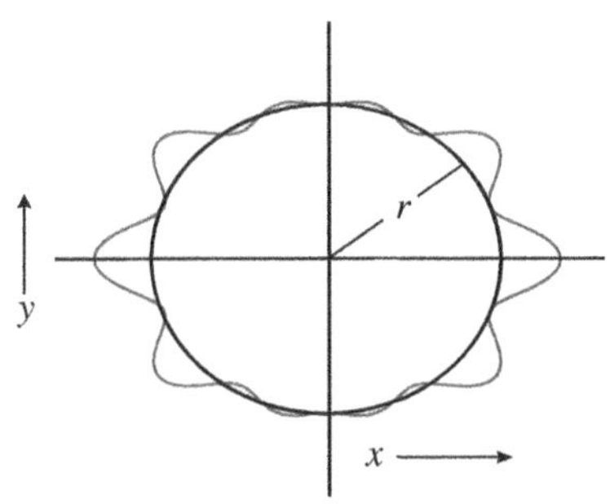

Figure P9.7 Integrand in Equation for M_x along the circle approximating the ring in phthalocyanine.

a) The curve for the integrand is shown in Fig. P9.7 along a circle with radius r. From the figure it is evident that the integrand is symmetric, with respect to the x-axis and to the y-axis. Furthermore, the positive values of the function are much larger than the negative ones. Thus the integral for M_x must be positive. The integral can be solved by numerical integration. The result is

$$\int_0^{2\pi} \cos(4\varphi)\cos\varphi\cos(5\varphi)\,\mathrm{d}\varphi = 1.5708 = \frac{\pi}{2}$$

This result can also be obtained analytically. Using the trigonometric relations in Appendix B the integrand $I = \cos(4\varphi)\cos\varphi\cos(5\varphi)$ simplifies to

$$I = \cos\varphi \cdot \cos 4\varphi \cdot \cos 5\varphi = \cos\varphi \cdot \frac{1}{2}(\cos\varphi + \cos 9\varphi)$$
$$= \frac{1}{2}\cos^2\varphi + \frac{1}{2}\cos\varphi\cos 9\varphi = \frac{1}{2}\cos^2\varphi + \frac{1}{4}\cos 8\varphi + \frac{1}{4}\cos 10\varphi$$

Then for the integral we find

$$I = \int_0^{2\pi} \left[\frac{1}{2}\cos^2\varphi + \frac{1}{4}\cos 8\varphi + \frac{1}{4}\cos 10\varphi\right] \mathrm{d}\varphi$$
$$= \left[\frac{1}{2}\left(\frac{\varphi}{2} - \frac{\sin 2\varphi}{4}\right)_0^{2\pi} + \frac{1}{4}\left(\frac{\sin 8\varphi}{8}\right)_0^{\pi} + \frac{1}{4}\left(\frac{\sin 10\varphi}{10}\right)_0^{\pi}\right]$$
$$= \left[\left(\frac{\pi}{2}\right) + \frac{1}{32}(0) - \frac{1}{40}(0)\right] = \frac{\pi}{2}$$

Thus for the transition moment we obtain

$$M_x = e\frac{2}{L_c}\left(\frac{L_c}{2\pi}\right)^2 \cdot \frac{\pi}{2} = e\frac{2L_c}{4\pi^2} \cdot \frac{\pi}{2} = e\frac{L_c}{4\pi}$$

P9.8 Show that the equation

$$\xi = \frac{Q}{m}F_0 \cdot \frac{1}{\sqrt{\left(\omega_0^2 - \omega^2\right)^2 + k_d^2\omega^2/m^2}} \cdot \cos(\omega t + \alpha)$$

in Foundation 9.1 is a solution of the differential equation

$$m \cdot \frac{d^2\xi}{dt^2} = -k_f \cdot \xi - k_d \cdot \frac{d\xi}{dt} + Q \cdot F_0 \cdot \cos(\omega t)$$

for a damped oscillator in an external electric field. Hint: In a first step solve the differential equation by using the general solution $\xi = A\cos(\omega t) + B\sin(\omega t)$ and obtain expressions for A and B. In a second step use the relation

$$\cos\alpha\cos\beta - \sin\alpha\sin\beta = \cos(\alpha + \beta)$$

with $\cos\alpha = A/x_0$, $\sin\alpha = -B/x_0$, and $\beta = \omega t$ and show that the general solution for ξ is equivalent to $\xi = x_0\cos(\omega t + \alpha)$. Finally, express x_0 and α in terms of the expressions for A and B obtained in the first step.

Solution

Step 1: We begin with the differential equation

$$m \cdot \frac{d^2\xi}{dt^2} = -k_f \cdot \xi - k_d \cdot \frac{d\xi}{dt} + Q \cdot F_0 \cdot \cos(\omega t)$$

and show that $\xi = A\cos(\omega t) + B\sin(\omega t)$ satisfies the equation. First we find the derivatives

$$\frac{d\xi}{dt} = -A\omega\sin(\omega t) + \omega B\cos(\omega t)$$

and

$$\frac{d^2\xi}{dt^2} = -A\omega^2\cos(\omega t) - \omega^2 B\sin(\omega t) = -\omega^2\xi$$

and then substitute the second derivative expression into the equation to find

$$0 = \left(m\omega^2 - k_f\right) \cdot \xi - k_d \cdot \frac{d\xi}{dt} + Q \cdot F_0 \cdot \cos(\omega t)$$

Substituting our expression for ξ and $d\xi/dt$, we find that

$$\begin{aligned} 0 &= \left(m\omega^2 - k_f\right) \cdot (A\cos(\omega t) + B\sin(\omega t)) \\ &\quad -k_d \cdot (-A\omega\sin(\omega t) + \omega B\cos(\omega t)\ ) + Q \cdot F_0 \cdot \cos(\omega t) \\ &= \left[A\left(m\omega^2 - k_f\right) - k_d\omega B + Q \cdot F_0\right]\cos(\omega t) \\ &\quad + \left[B\left(m\omega^2 - k_f\right) + k_d\omega A\right]\sin(\omega t) \end{aligned}$$

For this result to hold we must require that

$$\left[B\left(m\omega^2 - k_f\right) + k_d\omega A\right] = 0$$

and

$$\left[A\left(m\omega^2 - k_f\right) - k_d\omega B + Q \cdot F_0\right] = 0$$

The first of these equations gives

$$A = B\frac{\left(k_f - m\omega^2\right)}{k_d\omega}$$

which can be substituted into the second to find

$$\left[B\frac{\left(k_f - m\omega^2\right)}{k_d\omega}\left(m\omega^2 - k_f\right) - k_d\omega B + Q \cdot F_0\right] = 0$$

so that

$$\begin{aligned} B &= -Q \cdot F_0\frac{k_d\omega}{\left(k_f - m\omega^2\right)\left(m\omega^2 - k_f\right) - k_d^2\omega^2} \\ &= Q \cdot F_0\frac{k_d\omega}{\left(m\omega^2 - k_f\right)^2 + k_d^2\omega^2} \end{aligned}$$

Step 2: Given that $\cos\alpha\cos\beta - \sin\alpha\sin\beta = \cos(\alpha+\beta)$ we let $\cos\alpha = A/x_0$, $\sin\alpha = -B/x_0$, and $\beta = \omega t$ so that

$$\begin{aligned}\cos(\alpha+\beta) &= \cos\alpha\cos\beta - \sin\alpha\sin\beta \\ &= A/x_0\cos(\omega t) + B/x_0\sin(\omega t) \\ &= \cos(\alpha+\omega t) \text{ and } \alpha = \arccos\left(A/x_0\right) = \arcsin\left(-B/x_0\right)\end{aligned}$$

Hence we see that

$$A\cos(\omega t) + B\sin(\omega t) \;= \xi = x_0\cos(\omega t + \alpha)$$

Finally, we must express x_0 and α by the expressions for A and B obtained in the first step. Using our first relation between A and B, we find that

$$\tan\alpha = \frac{\sin\alpha}{\cos\alpha} = -\frac{B}{A} = -\frac{k_d\omega}{(k_f - m\omega^2)}$$

so that

$$\alpha = \arctan\left[-\frac{k_d\omega}{k_f - m\omega^2}\right]$$

Given the fact that $\cos^2\alpha + \sin^2\alpha = 1$ and our second condition, we find that

$$\begin{aligned}x_0^2 &= x_0^2\left(\sin^2\alpha + \cos^2\alpha\right) \\ &= x_0^2\left(\left(A/x_0\right)^2 + \left(B/x_0\right)^2\right) = A^2 + B^2 \\ &= B^2\left(\left[\frac{(k_f - m\omega^2)}{k_d\omega}\right]^2 + 1\right) \\ &= \left[Q\cdot F_0\frac{k_d\omega}{\left(m\omega^2 - k_f\right)^2 + k_d^2\omega^2}\right]^2\left(\frac{(k_f - m\omega^2)^2 + \left(k_d\omega\right)^2}{\left(k_d\omega\right)^2}\right) \\ &= \left(Q\cdot F_0\right)^2\left[\frac{1}{\left(m\omega^2 - k_f\right)^2 + k_d^2\omega^2}\right]\end{aligned}$$

If we define $\omega_0 = \sqrt{k_f/m}$, these equations become

$$\alpha = \arctan\left[-\frac{k_d\omega/m}{\omega_0^2 - \omega^2}\right]$$

and

$$x_0^2 = \left(\frac{Q\cdot F_0}{m}\right)^2\left[\frac{1}{\left(\omega^2 - \omega_0\right)^2 + k_d^2\omega^2/m^2}\right]$$

or

$$x_0 = \frac{Q\cdot F_0}{m}\frac{1}{\sqrt{\left(\omega^2 - \omega_0\right)^2 + k_d^2\omega^2/m^2}}$$

Putting the pieces together, we find the solution

$$\xi = \frac{Q}{m}F_0\cdot\frac{1}{\sqrt{\left(\omega_0^2 - \omega^2\right)^2 + k_d^2\omega^2/m^2}}\cdot\cos(\omega t + \alpha)$$

P9.9 According to Foundation 9.1, and derived in Problem P9.8, the phase shift α and the amplitude ξ of a damped driven oscillator are given by

$$\alpha = \arctan\left[-\frac{k_d\omega/m}{\omega_0^2 - \omega^2}\right]$$

and

$$\xi = \frac{Q}{m} F_0 \cdot \frac{1}{\sqrt{\left(\omega_0^2 - \omega^2\right)^2 + k_d^2 \omega^2 / m^2}} \cdot \cos(\omega t + \alpha)$$

where m is the mass of the charge, Q is the displaced charge, and ω_0 is the resonant frequency of the oscillator. The driving field has an amplitude F_0 and a frequency ω. Plot the amplitude ξ and the phase shift α of the damped oscillator versus the frequency ω of the external electric field, for $\omega_0 = 3.77 \times 10^{15}$ s^{-1}, $m = m_e = 9.109 \times 10^{-31}$kg, and damping constants $k_d = 2.5$, 5.0, and 20.0×10^{-16} N s m^{-1}.

Solution

The phase shift for the three different damping constants is plotted in Fig. P9.9a. Curve 1 is for $k_d = 2.5 \times 10^{-16}$ N s m^{-1}; curve 2 is for $k_d = 5.0 \times 10^{-16}$ N s m^{-1}; and curve 3 is for $k_d = 20 \times 10^{-16}$ N s m^{-1}. In each plot the zero crossing point of the phase occurs at the same value of $\omega = \omega_0 = 3.77 \times 10^{15}$ s^{-1}. The curves differ by their shape away from the resonance point. The plots with the higher damping coefficient have a "softer" decay of the phase angle with the deviation from the resonance point. Note that the oscillator is in phase with the external electrical field ($\alpha = 0$) at $\omega = 0$ (the oscillator follows a slowly changing field strength immediately). For $\omega = \omega_0$ (resonance) there is a phase shift of 90 °, and for $\omega \rightarrow \infty$ the oscillator and the field are out of phase ($\alpha = 180$ °).

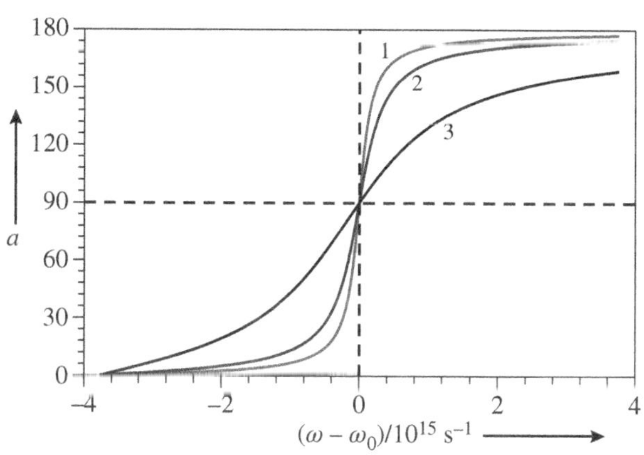

Figure P9.9a Phase shift α of a damped oscillator versus the frequency $\omega - \omega_0$. Damping constant $k_d = 2.5 \times 10^{-16}$ N s m^{-1} (curve 1), $k_d = 5.0 \times 10^{-16}$ N s m^{-1} (curve 2), and $k_d = 20 \times 10^{-16}$ N s m^{-1} (curve 3).

For the displacement, we plot the value scaled to the maximum displacement $\xi_{\max}$ which is given

$$\begin{aligned} \xi_{\max} &= \frac{Q}{m} F_0 \cdot \frac{1}{\sqrt{k_d^2 \omega^2 / m^2}} \cdot \cos(\omega t + \alpha) \\ &= \frac{Q}{m} F_0 \cdot \frac{1}{k_d \omega / m} \cdot \cos(\omega t + \alpha) \end{aligned}$$

Thus we plot the quantity

$$y = \frac{\xi}{\xi_{\max}} = \frac{k_d \omega / m}{\sqrt{\left(\omega_0^2 - \omega^2\right)^2 + k_d^2 \omega^2 / m^2}}$$

The function $y = \xi / \xi_{\max}$ is plotted in Fig. P9.9b. It is evident from this plot that all three curves display a resonance peak at ω_0. In addition, we note that the peaks have different widths; the case of higher damping constant corresponds to a broader resonance peak.

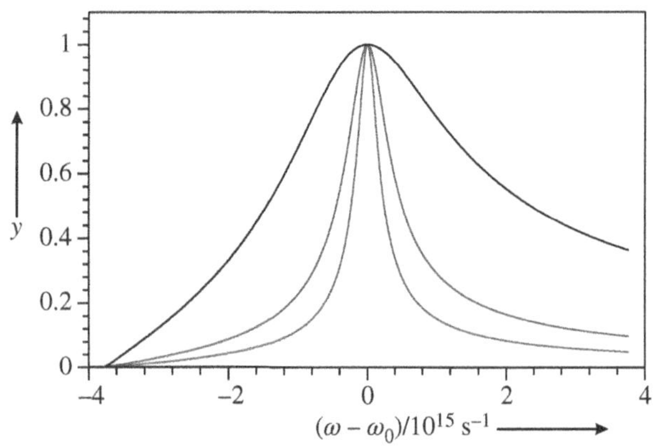

Figure P9.9b Displacement $y = \xi/\xi_{max}$ of the damped oscillator. Same parameters as in Fig. P9.9a.

P9.10 Here you are asked to estimate the mean potential energy $\overline{V}$ of a π electron in the electric field of the charge of the core of the aza nitrogen atom, by using the first-order perturbation theory result

$$\overline{V} = \int_0^L V_N\, \psi^2 \cdot \mathrm{ds}$$

where V_N is the potential energy of a π electron in the field of the N atom (more precisely: the difference between the potential energies of N and C). You can approximate V_N by assuming that the potential energy of an electron around a N atom is lower by a constant amount a compared to the constant potential energy of a chain of carbon atoms. You should show that

$$\overline{V} = -a \cdot \frac{2}{N}$$

Using this result explain the even–odd effect in the absorption spectra for the azacyanine dyes.

Solution

According to Fig. 9.12 with $s_1 = L/2 - d_0/2$ and $s_2 = L/2 + d_0/2$ we have

$$V_N = -a \quad \text{for} \quad s_1 < s < s_2$$

$$V_N = 0 \quad \text{for} \quad s \text{ elsewhere}$$

Then

$$\overline{V} = \int_{s_1}^{s_2} -a\psi^2 \cdot \mathrm{ds} = -a \int_{s_1}^{s_2} \psi^2 \cdot \mathrm{ds}$$

For orbitals with a node at the center of the molecule, ψ^2 is near zero for the whole region around the N atom, and $\overline{V} \approx 0$. For orbitals with an antinode at the center of the molecule, ψ^2 in the region of the N atom only deviates weakly from its maximum value at the center, thus

$$\overline{V} = -a \int_{x_1}^{x_2} \psi^2 \cdot \mathrm{dx} = -a \cdot \left(\psi^2\right)_{\text{at center}} \cdot d_0 = -a \cdot \frac{2}{L} \cdot d_0$$

This amount decreases with increasing chain length L. Because of $L = N \cdot d_0$, we find

$$\overline{V} = -a \cdot \frac{2}{N \cdot d_0} \cdot d_0 = -a \cdot \frac{2}{N}$$

Figure 9.13 shows the case that $N/2$ is odd: the HOMO level has an antinode at the location of the nitrogen, and the LUMO has a node; thus, the HOMO level is lowered by the amount $2a/N$ and consequently the excitation energy $\Delta E'$ becomes larger:

$$\Delta E' = \Delta E + a \cdot \frac{2}{N} \quad \text{for} \quad \frac{N}{2} = 3, 5, 7, \ldots$$

Figure 9.14 shows the case that $N/2$ is even: the HOMO level has a node at the location of the nitrogen, and the LUMO has an antinode; thus, the LUMO level is lowered by the amount $2a/N$ and consequently the excitation energy $\Delta E'$ becomes smaller:

$$\Delta E' = \Delta E - a \cdot \frac{2}{N} \quad \text{for} \ \frac{N}{2} = 2, 4, 6, \ldots$$

Extension: The constant a can be roughly estimated from the energy difference of an electron in a p orbital of a carbon atom and of a nitrogen atom. From the ionization energies displayed in Table 5.1 of C (11.26 eV= 18.04×10^{-19} J) and N (14.53 eV = 23.28×10^{-19} J), it follows that the energy difference is 5.3×10^{-19} J. A best fit to the experimental absorption data is $a = 3.5 \times 10^{-19}$ J. This is smaller than the difference of the ionization energies, but it is on the same order as estimated.

P9.11 In this problem you should calculate the HOMO to LUMO energy gap for polyenes, in which you include the effects of bond alternation. Use perturbation theory (in a manner analogous to that for Problem P9.10) to show that the HOMO-to-LUMO energy gap $\Delta E'$ in butadiene can be written as

$$\Delta E' = \Delta E + \frac{6}{5} b$$

where ΔE is the energy gap for the unperturbed FEMO model, $2b$ is the energy difference between the "double bond" and "single bond" sites (see Fig. 9.15), and d_0 is the bond length in the unperturbed model. Explain why you would expect this result to hold for longer polyenes.

Solution

In analogy to the considerations in Problem P9.10 the energy in orbital n should change by an amount

$$-2 \cdot b \cdot \psi^2_{n,C=C} \cdot d_0 + b \cdot \psi^2_{n,C-C} \cdot d_0$$

where the first term corresponds to the decrease of the potential energy at the location of the two double bonds, and the second term corresponds to the increase of the potential energy at the location of the single bond, and constant b is defined in Fig. 9.15. $\psi_{n,C=C}$ is the value of the wavefunction in orbital n at the place of the double bond, and $\psi_{n,C\text{-}C}$ is the corresponding value at the place of the single bond.

Using Fig. 9.16, to a good approximation we find that the orbital with $n = 2$ has

$$\psi^2_{2,C=C} = \frac{2}{L} \quad \text{and} \quad \psi^2_{2,C\text{-}C} = 0$$

and for the orbital with $n = 3$

$$\psi^2_{3,C=C} = 0 \quad \text{and} \quad \psi^2_{3,C\text{-}C} = \frac{2}{L}$$

Then the energies of levels $n = 2$ and $n = 3$ become

$$E'_2 = E_2 - 2 \cdot b \cdot \frac{2}{L} \cdot d_0 \quad \text{and} \quad E'_3 = E_3 + b \cdot \frac{2}{L} \cdot d_0$$

and for the excitation energy $\Delta E' = E'_3 - E'_2$ we obtain (using $\Delta E = E_3 - E_2$)

$$\Delta E' = E'_3 - E'_2 = \Delta E + b\frac{2}{L}d_0 - (-2b\frac{2}{L}d_0)$$

$$= \Delta E + b\frac{2}{L}d_0 + 2b\frac{2}{L}d_0$$

and with $L = 5d_0$ we find

$$\Delta E' = \Delta E + \frac{2}{5}b + \frac{4}{5}b = \Delta E + \frac{6}{5}b$$

Proceeding in a similar manner for longer polyenes we find that the difference between $\Delta E'$ and ΔE is approximately the same (on the one hand the number of correction terms increases, but on the other hand the length L increases to the same extent).

P9.12 Calculate the total probability density at the middle of the bonds and the transition energy for the amidinium cation $-\overline{\text{N}}\text{–CH=}\overset{+}{\text{N}}\text{–}$ for two cases: a) $d_0 = 140$ pm for the end length and b) $1.5d_0 = 210$ pm for the end length. Comment on the result.

Solution

According to Chapter 8 the total probability density is defined by

$$\rho_{\text{total}}(s) = \sum_k N_k \psi_k^2(s)$$

The amidinium cation has 4 π-electrons, thus two occupied orbitals, and $\rho_{\text{total}}(s)$ is

$$\rho_{\text{total}}(s) = 2\psi_1^2(s) + 2\psi_2^2(s) = 2\frac{2}{L}\left[\sin^2\left(\frac{\pi}{L}s\right) + \sin^2\left(\frac{2\pi}{L}s\right)\right]$$

a) Assuming an end length of $d_0 = 140$ pm, we have $L = 4d_0$ and the coordinates at the middle of the bonds are

$$s_{\text{bond1}} = \frac{3}{2}d_0 \quad \text{and} \quad s_{\text{bond2}} = \frac{5}{2}d_0$$

Thus

$$\begin{aligned}
\left[\rho_{\text{total}}(s)\right]_{\text{bond1}} &= \frac{4}{4d_0}\left[\sin^2\left(\frac{\pi}{4d_0}\frac{3}{2}d_0\right) + \sin^2\left(\frac{2\pi}{4d_0}\frac{3}{2}d_0\right)\right] \\
&= \frac{1}{d_0}\left[\sin^2\left(\frac{3\pi}{8}\right) + \sin^2\left(\frac{3\pi}{4}\right)\right] \\
&= 0.00714 \cdot [0.854 + 0.5]\ \text{pm}^{-1} = 9.67\ \text{nm}^{-1}
\end{aligned}$$

Accordingly, for $\left[\rho_{\text{total}}(s)\right]_{\text{bond2}}$ we obtain

$$\begin{aligned}
\left[\rho_{\text{total}}(s)\right]_{\text{bond2}} &= \frac{4}{4d_0}\left[\sin^2\left(\frac{\pi}{4d_0}\frac{5}{2}d_0\right) + \sin^2\left(\frac{2\pi}{4d_0}\frac{5}{2}d_0\right)\right] \\
&= \frac{1}{d_0}\left[\sin^2\left(\frac{5\pi}{8}\right) + \sin^2\left(\frac{5\pi}{4}\right)\right] \\
&= 0.00714 \cdot [0.854 + 0.5]\ \text{pm}^{-1} = 9.67\ \text{nm}^{-1}
\end{aligned}$$

This is the same value as for $\left[\rho_{\text{total}}(s)\right]_{\text{bond1}}$, as we expect according to the symmetry of the molecule. The transition energy for the amidinium cation is,

$$\Delta E = \frac{h^2}{8mL^2} \cdot \left[3^2 - 2^2\right] = \frac{h^2}{8m(4d_0)^2} \cdot 5 = 9.61 \times 10^{-19}\ \text{J}$$

corresponding to an absorbance maximum at

$$\lambda = \frac{hc_0}{\Delta E} = 207\ \text{nm}$$

b) Assuming an end length of $1.5d_0 = 210$ pm, we have $L = 5d_0$ and the coordinates at the middle of the bonds are

$$s_{\text{bond1}} = 2d_0 \quad \text{and} \quad s_{\text{bond2}} = 3d_0$$

Thus

$$\begin{aligned}
\left[\rho_{\text{total}}(s)\right]_{\text{bond1}} &= \frac{4}{5d_0}\left[\sin^2\left(\frac{\pi}{5d_0}2d_0\right) + \sin^2\left(\frac{2\pi}{5d_0}3d_0\right)\right] \\
&= \frac{4}{5d_0}\left[\sin^2\left(\frac{2\pi}{5}\right) + \sin^2\left(\frac{6\pi}{5}\right)\right] \\
&= 0.00571 \cdot [0.905 + 0.345]\ \text{pm}^{-1} = 7.14\ \text{nm}^{-1}
\end{aligned}$$

Because of symmetry, we obtain the same value for $\left[\rho_{\text{total}}(s)\right]_{\text{bond2}}$. The transition energy for the amidinium cation is,

$$\Delta E = \frac{h^2}{8mL^2} \cdot \left[3^2 - 2^2\right] = \frac{h^2}{8m(5d_0)^2} \cdot 5 = 6.15 \times 10^{-19}\ \text{J}$$

corresponding to an absorbance maximum at

$$\lambda = \frac{hc_0}{\Delta E} = 323\ \text{nm}$$

Comment:
In case a) we obtain $\left[\rho_{\text{total}}(s)\right]_{\text{bond1}} = 9.67\ \text{nm}^{-1}$. This is still more than the value 8.91 nm^{-1} for the double bond in butadiene, indicating a bond length alternation in the cyanine in contrast to our assumption that the bond lengths are as in benzene ($\rho_{\text{total}}(s) = 7.14\ \text{nm}^{-1}$). The calculated λ for the absorbance maximum is 207 nm, only slightly smaller than the experimental value $\lambda = 224$ nm.
In case b) we obtain $\left[\rho_{\text{total}}(s)\right]_{\text{bond1}} = 7.14\ \text{nm}^{-1}$, as we expect for benzene. The calculated λ for the absorbance maximum is 323 nm, considerably larger than the experimental value $\lambda = 224$ nm.
From these results we conclude that FEMO using an end length of 140 pm reproduces the absorbance maximum quite well, but the total probability density at the middle of the bonds is too large. On the other hand, FEMO with an end length of 210 pm predicts a much too high value for the absorbance maximum, but the total probability density at the middle of the bonds is in good accord with the FEMO assumptions for cyanine dyes.

P9.13 Calculate the total electron density and the transition energy for a cyanine cation with 8 π–electrons $-\overline{\text{N}}-\text{CH}=\text{CH}-\text{CH}=\text{CH}-\text{CH}=\overset{+}{\text{N}}-$ using a) $d_0 = 140$ pm for the end length, b) $1.5d_0 = 210$ pm for the end length. Compare with the amidinium cation. Comment the result.

Solution
The calculation is straightforward, in analogy to the calculation in Problem P9.12. The basic question is: is the conclusion we made in Problem P9.12 also valid for cyanines with a higher number of π–electrons than that in the case of the amidinum cation?
The total probability density for a cyanine with 8 π-electrons is

$$\begin{aligned}\rho_{total}(s) &= 2\psi_1^2(s) + 2\psi_2^2(s) + 2\psi_3^2(s) + 2\psi_4^2(s)\\ &= \frac{4}{L}\left[\sin^2\left(\frac{\pi}{L}s\right) + \sin^2\left(\frac{2\pi}{L}s\right) + \sin^2\left(\frac{3\pi}{L}s\right) + \sin^2\left(\frac{4\pi}{L}s\right)\right]\end{aligned}$$

a) Assuming the end length $d_0 = 140$ pm we have $L = 8d_0$ and the coordinates at the middle of the first 3 bonds (the densities at the last 3 bonds are given by symmetry)

$$s_{\text{bond1}} = \frac{3}{2}d_0, \quad s_{\text{bond2}} = \frac{5}{2}d_0, \quad \text{and} \quad s_{\text{bond3}} = \frac{7}{2}d_0$$

and

$$\begin{aligned}\left[\rho_{\text{total}}(s)\right]_{\text{bond1}} &= \frac{4}{8d_0}\left[\sin^2\left(\frac{\pi}{8}\frac{3}{2}\right) + \sin^2\left(\frac{2\pi}{8}\frac{3}{2}\right) + \sin^2\left(\frac{3\pi}{8}\frac{3}{2}\right) + \sin^2\left(\frac{4\pi}{8}\frac{3}{2}\right)\right]\\ &= 0.00357 \cdot 2.624\ \text{pm}^{-1} = 9.37\ \text{nm}^{-1}\end{aligned}$$

Accordingly, we obtain

$$\left[\rho_{\text{total}}(s)\right]_{\text{bond2}} = 7.44\ \text{nm}^{-1} \quad \text{and} \quad \left[\rho_{\text{total}}(s)\right]_{\text{bond3}} = 8.21\ \text{nm}^{-1}$$

The transition energy for the cyanine cation is,

$$\Delta E = \frac{h^2}{8mL^2} \cdot \left[5^2 - 4^2\right] = \frac{h^2}{8m(8d_0)^2} \cdot 9 = 4.323 \times 10^{-19}\ \text{J}$$

corresponding to an absorbance maximum at

$$\lambda = \frac{hc_0}{\Delta E} = 459\ \text{nm}$$

b) Assuming the end length $d_0 = 210$ pm we have $L = 9d_0$ and the coordinates at the middle of the first 3 bonds (the densities at the last 3 bonds are given by symmetry)

$$s_{\text{bond1}} = 2d_0, \quad s_{\text{bond2}} = 3d_0, \quad \text{and} \quad s_{\text{bond3}} = 4d_0$$

and

$$\begin{aligned}\left[\rho_{\text{total}}(s)\right]_{\text{bond1}} &= \frac{4}{9d_0}\left[\sin^2\left(\frac{\pi}{9}2\right) + \sin^2\left(\frac{2\pi}{9}2\right) + \sin^2\left(\frac{3\pi}{9}2\right) + \sin^2\left(\frac{4\pi}{9}2\right)\right]\\ &= 0.00318 \cdot 2.249\ \text{pm}^{-1} = 7.15\ \text{nm}^{-1}\end{aligned}$$

Accordingly, we obtain

$$[\rho_{\text{total}}(s)]_{\text{bond2}} = 7.17\ \text{nm}^{-1} \quad \text{and} \quad [\rho_{\text{total}}(s)]_{\text{bond3}} = 7.11\ \text{nm}^{-1}$$

The transition energy for the cyanine cation is,

$$\Delta E = \frac{h^2}{9mL^2} \cdot [5^2 - 4^2] = \frac{h^2}{8m(9d_0)^2} \cdot 9 = 3.42 \times 10^{-19}\ \text{J}$$

corresponding to an absorbance maximum at

$$\lambda = \frac{hc_0}{\Delta E} = 581\ \text{nm}$$

Comment:
In case a) we obtain $[\rho_{\text{total}}(s)]_{\text{bond1}} = 9.37\ \text{nm}^{-1}$, $[\rho_{\text{total}}(s)]_{\text{bond2}} = 7.44\ \text{nm}^{-1}$, and $[\rho_{\text{total}}(s)]_{\text{bond3}} = 8.21\ \text{nm}^{-1}$. This is still comparable with the bond length alternation in butadiene, and imply a bond length alternation in the cyanine in contrast to our assumption that the bond lengths are as in benzene ($\rho_{\text{total}}(s) = 7.14\ \text{nm}^{-1}$. The calculated λ for the absorbance maximum is 459 nm, only slightly larger than the experimental value $\lambda = 416$ nm.
In case b) we obtain $[\rho_{\text{total}}(s)]_{\text{bond1}} = 7.15\ \text{nm}^{-1}$, $[\rho_{\text{total}}(s)]_{\text{bond2}} = 7.17\ \text{nm}^{-1}$, and $[\rho_{\text{total}}(s)]_{\text{bond3}} = 7.11\ \text{nm}^{-1}$as we might expect for benzene. The calculated λ for the absorbance maximum is 583 nm, considerably larger than the experimental value $\lambda = 416$ nm.
From these results we conclude as in Problem P9.12 for the amidinium cation that FEMO using an end length of 140 pm reproduces the absorbance maximum quite well, but the total probability density at the middle of the bonds indicates bond alternation. On the other hand, FEMO with an end length of 210 pm predicts a much too high value for the absorbance maximum, however the total probability density at the middle of the bonds is in complete accordance with the FEMO assumptions for cyanine dyes.

P9.14 Using the numerical data for the experimental absorbance curves for the carbocyanines with $j = 0, j = 1, j = 2$, and $j = 3$ in Fig. 9.2 evaluate the oscillator strengths f. You can find the numerical data for these curves in Appendix A.6. The wavelengths are given in nanometers (in steps of $\Delta\lambda = 2$ nm) and the decadic absorbance coefficients ε in $10^5\ \text{L mol}^{-1}\ \text{cm}^{-1}$.

Solution
The oscillator strength f is defined as

$$f = \frac{4 \cdot \ln 10 \cdot c_0 \cdot \varepsilon_0 \cdot m_e}{\mathrm{N}_A \cdot e^2} \int_{\nu=-\infty}^{\nu=+\infty} \varepsilon(\nu) \cdot \mathrm{d}\nu$$

where $c_0 = 2.998 \times 10^8\ \text{m s}^{-1}$ is the speed of light, $\varepsilon_0 = 8.854 \times 10^{-12}\ \text{C}^2\ \text{J}^{-1}\ \text{m}^{-1}$ is the permittivity of vacuum. $m_e = 9.1094 \times 10^{-31}$ kg is the mass of the electron, $e = 1.6022 \times 10^{-19}$ C is the charge of the electron, and $N_A = 6.022 \times 10^{23}\ \text{mol}^{-1}$ is the Avogadro constant. The integral extends over the region of the absorbance band, and ν is the frequency of light. Because our experimental data describe ε as a function of wavelength λ, we have to rewrite the integral in terms of the wavelength λ. Note that

$$\nu\lambda = c_0 \quad \text{or} \quad \nu = \frac{c_0}{\lambda}$$

Thus for $\mathrm{d}\nu$ we find

$$\mathrm{d}\nu = -\frac{c_0}{\lambda^2} \cdot \mathrm{d}\lambda$$

and the integral becomes

$$\int_{\nu=-\infty}^{\nu=+\infty} \varepsilon(\nu) \cdot \mathrm{d}\nu = -\int_{\lambda=+\infty}^{\lambda=-\infty} \varepsilon(\lambda) \cdot \frac{c_0}{\lambda^2} \cdot \mathrm{d}\lambda = c_0 \int_{\lambda=-\infty}^{\lambda=+\infty} \varepsilon(\lambda) \cdot \frac{1}{\lambda^2} \cdot \mathrm{d}\lambda$$

Finally we replace the integration by a summation

$$\int_{\nu=-\infty}^{\nu=+\infty} \varepsilon(\nu) \cdot \mathrm{d}\nu = c_0 \sum_{\lambda=-\infty}^{\lambda=+\infty} \varepsilon(\lambda) \cdot \frac{1}{\lambda^2} \cdot \Delta\lambda = c_0 \cdot area$$

where *area* is the area between the absorbance curve and the abscissa.

Using the numerical data for the carbocyanine with $j = 0$ we obtain

$$area = 2.592 \times 10^{-4} \cdot 10^{5}\ \text{L mol}^{-1}\ \text{cm}^{-1}\ \text{nm}^{-1}$$

and the integral is

$$\int_{\nu=-\infty}^{\nu=+\infty} \varepsilon(\nu) \cdot d\nu = c_0 \cdot area = 2.998 \times 10^{8}\ \frac{\text{m}}{\text{s}} 2.592 \times 10^{-4} \cdot 10^{5}\ \text{L mol}^{-1}\ \text{cm}^{-1}\ \text{nm}^{-1}$$
$$= 7.77 \times 10^{18}\ \text{L mol}^{-1}\ \text{cm}^{-1}\ \text{s}^{-1}$$

Thus for the oscillator strength f we obtain

$$f_{\text{experimental}} = \frac{4 \cdot \ln 10 \cdot c_0 \cdot \varepsilon_0 \cdot m_e}{\mathbf{N}_A \cdot e^2} \cdot 7.77 \times 10^{18}\ \text{L mol}^{-1}\ \text{cm}^{-1}\ \text{s}^{-1}$$
$$= 1.441 \times 10^{-18}\ \text{s mol m}^{-2} \cdot 7.77 \times 10^{18}\ \text{L mol}^{-1}\ \text{cm}^{-1}\ \text{s}^{-1}$$
$$= 11.2\ \text{m}^{-2}\ \text{L cm}^{-1} = 1.12$$

Correspondingly, we obtain the f values for the carbocyanines with $j = 1$ to $j = 3$:

j	0	1	2	3
$f_{\text{experiment}}$	1.12	1.32	1.49	1.68
f_{FEMO}	1.0	1.3	1.5	1.8

We can compare these data to the FEMO f values calculated for the cyanines displayed in Table 9.2 (note that $j = 0$ corresponds to $N = 6$ in Table 9.2). However, note that the FEMO values are calculated for unbranched molecules, contrary to the carbocyanines considered here.

P9.15 Resonance frequency of two identical coupled oscillators. In Foundation 9.5 for the resonance frequency $\omega_{0,}$ of two coupled oscillators we found that

$$\omega_{0,\pm} = \sqrt{\frac{1}{2}\frac{k_{f,1} + k_{f,2}}{m} \pm \frac{\sqrt{k_{f,12}^2 + \frac{1}{4}\left(k_{f,1} - k_{f,2}\right)^2}}{m}}$$

where $k_{f,1}$ and $k_{f,2}$ are the force constants, m is the mass of the oscillators, and $k_{f,12}$ is the coupling constant. Consider the case that both oscillators are identical, and show that two new oscillation frequencies, at higher and lower frequencies, are generated

$$\omega_{0,1} = \sqrt{\frac{k_f}{m} + \frac{k_{f,12}}{m}} \quad \text{and} \quad \omega_{0,2} = \sqrt{\frac{k_f}{m} - \frac{k_{f,12}}{m}}$$

and use the results of Foundation 9.4 to show that the transition energies can be written as

$$\Delta E_{\text{coupled},1} = \sqrt{\Delta E^2 + 2\Delta E \cdot J_{12}} \qquad \Delta E_{\text{coupled},2} = \sqrt{\Delta E^2 - 2\Delta E \cdot J_{12}}$$

Solution

In this case the force constants are identical ($k_{f,1} = k_{f,2} = k_f$), and we have the solutions

$$\omega_{0,\pm} = \sqrt{\frac{1}{2}\frac{2k_f}{m} \pm \frac{\sqrt{k_{f,12}^2}}{m}} = \sqrt{\frac{1}{2}\frac{2k_f}{m} \pm \frac{k_{f,12}}{m}}$$

so that

$$\omega_{0,+} = \sqrt{\frac{k_f}{m} + \frac{k_{f,12}}{m}} \quad \text{and} \quad \omega_{0,-} = \sqrt{\frac{k_f}{m} - \frac{k_{f,12}}{m}}$$

That is, the coupled system can be in resonance at two equally higher or lower frequencies compared to the frequency

$$\omega = \sqrt{\frac{k_f}{m}}$$

of the uncoupled oscillators. In Foundation 9.4 we described coupled electronic transitions as coupled classical oscillators and we found the relations

$$\frac{k_f}{m} = \frac{4\pi^2}{h^2}\Delta E^2, \quad \frac{k_{f,12}}{m} = \frac{8\pi^2}{h^2}\Delta E \cdot J_{12}, \quad \text{and} \quad \omega_0 = \frac{2\pi}{h}\Delta E_{coupled}$$

between the classical force constant and coupling constant and the quantum mechanical transition energies ΔE and the quantum mechanical coupling integral J_{12}. Thus we can express our result for the classical coupled identical oscillators by the transition energies and the coupling integral of the corresponding electronic transitions

$$\frac{2\pi}{h}\Delta E_{\text{coupled},1} = \sqrt{\frac{4\pi^2}{h^2}\Delta E^2 + \frac{8\pi^2}{h^2}\Delta E \cdot J_{12}} \quad \text{and} \quad \frac{2\pi}{h}\Delta E_{\text{coupled},2} = \sqrt{\frac{4\pi^2}{h^2}\Delta E^2 - \frac{8\pi^2}{h^2}\Delta E \cdot J_{12}}$$

or

$$\Delta E_{\text{coupled},1} = \sqrt{\Delta E^2 + 2\Delta E \cdot J_{12}} \quad \text{and} \quad \Delta E_{\text{coupled},2} = \sqrt{\Delta E^2 - 2\Delta E \cdot J_{12}}$$

That is, instead of the transition energy ΔE of the two uncoupled electronic transitions we obtain transitions with two different transition energies $\Delta E_{\text{coupled},1}$ and $\Delta E_{\text{coupled},2}$.for the coupled system.

P9.16 Oscillator strength of two identical coupled oscillators. According to Foundation 9.3 the oscillator strength of a classical oscillator is

$$f = \frac{1}{3}\frac{m_e}{e^2 m} \cdot Q^2 \cdot \frac{1}{\hat{n}}$$

and according to Foundation 9.5 the oscillator strength of two coupled classical oscillators is

$$f = \frac{1}{3}\frac{m_e}{e^2 m} \cdot \frac{(Q_1\gamma_1 + Q_2\gamma_2)^2}{\gamma_1^2 + \gamma_2^2} \cdot \frac{1}{\hat{n}}$$

where Q_1 and Q_2 are the electrical charges of both oscillators and the amplitudes γ_1 and γ_2 of the oscillators are connected by

$$\frac{\gamma_2}{\gamma_1} = \frac{m\omega_0^2 - k_{f,1}}{k_{f,12}} \quad \text{and} \quad \frac{\gamma_2}{\gamma_1} = \frac{k_{f,12}}{m\omega_0^2 - k_{f,1}}$$

Consider the case that both oscillators are identical, and find expressions for the oscillator strength of the two different transitions. Discuss your result in terms of what you expect to observe in the experimental absorption spectrum.

Solution

With $k_{f,1} = k_{f,2} = k_f$ we find

$$\frac{\gamma_2}{\gamma_1} = \frac{m\omega_0^2 - k_f}{k_{f,12}} \quad \text{and} \quad \frac{\gamma_2}{\gamma_1} = \frac{k_{f,12}}{m\omega_0^2 - k_f}$$

or

$$\frac{\gamma_2}{\gamma_1} = \frac{\gamma_1}{\gamma_2}$$

leading to the condition

$$\gamma_2^2 = \gamma_1^2 \quad \text{and} \quad \gamma_2 = \pm\gamma_1$$

That is, we obtain two solutions for the amplitudes, namely,

$$\gamma_2 = +\gamma_1 \quad \text{and} \quad \gamma_2 = -\gamma_1$$

The first solution means that the two identical oscillators oscillate in phase, and the second solution means that the oscillators oscillate out of phase. Thus for the oscillator strength with $Q_1 = Q_2 = Q$ for the in phase oscillation we obtain

$$f_{\text{in-phase}} = \frac{1}{3}\frac{m_e}{e^2 m} \cdot \frac{(Q\gamma_1 + Q\gamma_1)^2}{\gamma_1^2 + \gamma_1^2} \cdot \frac{1}{\hat{n}} = \frac{1}{3}\frac{m_e}{e^2 m} \cdot \frac{4Q^2\gamma_1^2}{2\gamma_1^2} \cdot \frac{1}{\hat{n}}$$
$$= 2 \cdot \left(\frac{1}{3}\frac{m_e}{e^2 m} \cdot Q^2 \cdot \frac{1}{\hat{n}}\right) = 2f$$

This is just the double of the oscillatory strength of a single uncoupled oscillator. On the other hand, for the out-of-phase oscillation we obtain

$$f_{\text{out-of-phase}} = 0$$

This means that the two oscillators oscillating out of phase cannot absorb energy from the external field of the light wave.

P9.17 Coupling Effect in Cyanines. The HOMO energy level in a cyanine molecule is occupied by two electrons which can be excited to the LUMO energy level. Consider the transitions of both HOMO electrons to the LUMO as two coupled classical oscillators. Calculate the excitation energy considering the coupling effect.

Solution

From Problem P9.15 we know that the transition energy for the in-phase transition is

$$\Delta E_{\text{coupled},1} = \sqrt{\Delta E^2 + 2\Delta E \cdot J_{12}}$$

According to Table 9.2 for the cyanine cation with $N = 8\ \pi$ electrons we obtain for the HOMO→LUMO transition

$$\lambda_{\max} = 459\ \text{nm}, \Delta E = 4.33 \times 10^{-19}\ \text{J}$$

For the coupling integral we obtain $J_{12} = 0.5 \times 10^{-19}$ J (see Foundation 10.2), then

$$\Delta E_{\text{coupled,in-phase}} = \sqrt{4.33^2 + 2 \cdot 4.33 \times 0.5} \times 10^{-19}\ \text{J} = 4.80 \times 10^{-19}\ \text{J}$$

The corresponding absorption wavelength is $\lambda = 414$ nm, in excellent agreement with the experimental value 416 nm.

P9.18 Electron pairs in porphyrin. Calculate the shift of the transition energy for both separate transitions considering the coupling between the two electrons in each transition, in analogy to the cyanine considered in Problem P9.17.

Solution

For the transition energies of the electron pairs we obtain, as in the case of the cyanine,

$$\Delta E_{1,\text{pair}} = \sqrt{\Delta E^2 + 2 \cdot \Delta E \cdot J_{12}} = \sqrt{3.45^2 + 2 \cdot 3.45 \times 0.5} \times 10^{-19}\ \text{J}$$
$$= 3.92 \times 10^{-19}\ \text{J}$$

$$\Delta E_{2,\text{pair}} = \sqrt{\Delta E^2 + 2 \cdot \Delta E \cdot J_{12}} = \sqrt{5.20^2 + 5.20 \times 0.5}\ \text{J}$$
$$= 5.68 \times 10^{-19}\ \text{J}$$

P9.19 Coupling of the two electron pairs in porphyrin. Calculate the coupling between the two electron pairs (coupling integral $J_{12} = 0.6 \times 10^{-19}$ J), following the treatment in Foundation 9.4:

$$\frac{k_{f,\text{pair1}}}{m} = \frac{4\pi^2}{h^2}\Delta E^2_{\text{pair1}}, \quad \frac{k_{f,\text{pair2}}}{m} = \frac{4\pi^2}{h^2}\Delta E^2_{\text{pair2}}, \quad \text{and} \quad \frac{k_{f,12}}{m} = \frac{16\pi^2}{h^2}\sqrt{\Delta E_{\text{pair1}}\Delta E_{\text{pair2}}} \cdot J_{12}$$

Solution

For electron pairs 1 and 2 we find:

$$\frac{k_{f,\text{pair1}}}{m} = \frac{4\pi^2}{h^2}\Delta E^2_{\text{pair1}} = \frac{4\pi^2}{h^2}(3.92 \times 10^{-19}\ \text{J})^2 = 1.38 \times 10^{31}\ \text{s}^{-2}$$

$$\frac{k_{f,\text{pair2}}}{m} = \frac{4\pi^2}{h^2}\Delta E^2_{\text{pair2}} = \frac{4\pi^2}{h^2}(5.68 \times 10^{-19}\ \text{J})^2 = 2.90 \times 10^{31}\ \text{s}^{-2}$$

$$\frac{k_{f,12}}{m} = \frac{16\pi^2}{h^2}\sqrt{3.92 \times 5.68} \times 10^{-19}\ \text{J} \cdot 0.6 \times 10^{-19}\ \text{J} = 1.02 \times 10^{31}\ \text{s}^{-2}$$

Using these results we can find the resonance frequencies as (see Problem 9.15)

$$\omega_0^2 = \frac{1}{2}\left(\frac{k_{f,\text{pair1}}}{m} + \frac{k_{f,\text{ pair2}}}{m}\right) \pm \sqrt{\frac{k_{f,12}^2}{m^2} + \frac{1}{4}\left(\frac{k_{f,\text{pair1}}}{m^2} - \frac{k_{f,\text{pair2}}}{m^2}\right)^2}$$

$$= \frac{1}{2}(1.38 + 2.90) \times 10^{31}\ \text{s}^{-2} \pm \sqrt{(1.02 \times 10^{31}\ \text{s}^{-2})^2 + \frac{1}{4}(1.38 - 2.90)^2 \times 10^{62}\ \text{s}^{-4}}$$

$$= 2.14 \times 10^{31}\ \text{s}^{-2} \pm 1.27 \times 10^{31}\ \text{s}^{-2}$$

Thus we obtain

$$\omega_{0,\text{in-phase}}^2 = (2.14 + 1.27) \times 10^{31}\ \text{s}^{-2} = 3.41 \times 10^{31}\ \text{s}^{-2}$$

$$\omega_{0,\text{out-of-phase}}^2 = (2.14 - 1.27) \times 10^{31}\ \text{s}^{-2} = 0.87 \times 10^{31}\ \text{s}^{-2}$$

and

$$-\omega_{0,\text{in-phase}} = \sqrt{3.41 \times 10^{31}\ \text{s}^{-2}} = 5.84 \times 10^{15}\ \text{s}^{-1}$$

$$\omega_{0,\text{out-of-phase}} = \sqrt{0.87 \times 10^{31}\ \text{s}^{-2}} = 2.95 \times 10^{15}\ \text{s}^{-1}$$

The coupling increases the frequency of the high-frequency oscillator, and it decreases the frequency of the low-frequency oscillator. Using the relations

$$\omega_{0,\text{in-phase}}^2 = \frac{k_{f,\text{in-phase}}}{m} = \frac{4\pi^2}{h^2}\Delta E_{\text{in-phase}}^2$$

$$\omega_{0,\text{out-of-phase}}^2 = \frac{k_{f,\text{out-of-phase}}}{m} = \frac{4\pi^2}{h^2}\Delta E_{\text{out-of-phase}}^2$$

we calculate the excitation energies

$$\Delta E_{\text{in-phase}} = \frac{h}{2\pi}\omega_{0,\text{in-phase}} = \frac{6.626 \times 10^{-34}\ \text{J s}}{2\pi} 5.84 \times 10^{15}\ \text{s}^{-1} = 6.16 \times 10^{-19}\ \text{J}$$

$$\Delta E_{\text{out-of-phase}} = \frac{h}{2\pi}\omega_{0,\text{out-of-phase}} = \frac{6.626 \times 10^{-34}\ \text{J s}}{2\pi} 2.95 \times 10^{15}\ \text{s}^{-1} = 3.11 \times 10^{-19}\ \text{J}$$

and the wavelengths

$$\lambda_{\text{in-phase}} = \frac{hc_0}{\Delta E_{\text{in-phase}}} = 323\ \text{nm} \quad \text{and} \quad \lambda_{\text{out-of-phase}} = \frac{hc_0}{\Delta E_{\text{out-of-phase}}} = 639\ \text{nm}$$

Note that the originally calculated transition wavelengths (see data in Problem 9.18) are shifted by the coupling effect from 382 to 336 nm (shift by 46 nm to a shorter wavelength), and from 576 to 638 nm (shift by 62 nm to a longer wavelength).

P9.20 Oscillator strength f of porphyrin absorption bands. Calculate f for the coupled transitions using the transition moments $M_{x,1} = M_{x,2} = 2.85 \times 10^{-29}$ C m of transitions 1 and 2 and the excitation energies $\Delta E_1 = 3.92 \times 10^{-19}$ J, $\Delta E_2 = 5.68 \times 10^{-19}$ J.

Solution

The oscillator strength is defined as

$$f = \frac{1}{3}\frac{m_e}{e^2 m} \cdot Q^2 \cdot \frac{1}{\hbar}$$

According to Foundation 9.4 for the two uncoupled oscillators we have

$$\frac{Q_1^2}{m} = \frac{16\pi^2}{h^2}\Delta E_1 M_{x,1}^2 = \frac{16\pi^2}{h^2} 3.45 \times 10^{-19}\ \text{J} \cdot (2.85 \times 10^{-29}\ \text{C m})^2 = 10.08 \times 10^{-8}\ \text{C}^2\ \text{kg}^{-1}$$

$$\frac{Q_2^2}{m} = \frac{16\pi^2}{h^2}\Delta E_2 M_{x,2}^2 = \frac{16\pi^2}{h^2} 5.20 \times 10^{-19}\ \text{J} \cdot (2.85 \times 10^{-29}\ \text{C m})^2 = 15.19 \times 10^{-8}\ \text{C}^2\ \text{kg}^{-1}$$

Thus for the oscillatory strength of the uncoupled oscillators we obtain

$$f_1 = \frac{1}{3}\frac{m_e}{e^2}\cdot\frac{Q_1^2}{m}\cdot\frac{1}{\hat{n}} = 1.183\times10^{-8}\cdot 10.08\times10^{-8} = 1.19\cdot\frac{1}{\hat{n}}$$

$$f_2 = \frac{1}{3}\frac{m_e}{e^2}\cdot\frac{Q_2^2}{m}\cdot\frac{1}{\hat{n}} = 1.183\times10^{-8}\cdot 15.19\times10^{-8} = 1.79\cdot\frac{1}{\hat{n}}$$

Next we follow the treatment in Foundation 9.4 for doubly occupied orbitals and start as in Problem P9.16. The oscillator strength of two coupled oscillators is

$$f = \frac{1}{3}\frac{m_e}{e^2 m}\cdot\frac{(Q_1\gamma_1 + Q_2\gamma_2)^2}{\gamma_1^2+\gamma_2^2}\cdot\frac{1}{\hat{n}} = \frac{1}{3}\frac{m_e}{e^2}\frac{\left(Q_1/\sqrt{m} + Q_2/\sqrt{m}\cdot\frac{\gamma_2}{\gamma_1}\right)^2}{1+\left(\frac{\gamma_2}{\gamma_1}\right)^2}\cdot\frac{1}{\hat{n}}$$

From the expressions for Q_1^2/m and Q_2^2/m calculated above, we obtain

$$\frac{Q_1}{\sqrt{m}} = 3.175\times10^{-4}\ \mathrm{C\ kg^{-1/2}} \quad \text{and} \quad \frac{Q_2}{\sqrt{m}} = 3.898\times10^{-4}\ \mathrm{C\ kg^{-1/2}}$$

Furthermore, the ratio γ_2/γ_1 is

$$\frac{\gamma_2}{\gamma_1} = \frac{\omega_0^2 - k_{f,\mathrm{pair1}}/m}{k_{f,12}/m}$$

In Problem P9.19 we calculated

$$k_{f,\mathrm{pair1}}/m = 1.38\times10^{31}\ \mathrm{s^{-2}}\ ,\ k_{f,12}/m = 1.02\times10^{31}\ \mathrm{s^{-2}}$$
$$\omega_0^2 = (2.14\pm1.27)\times10^{31}\ \mathrm{s^{-2}}$$

For the in-phase oscillation we obtain

$$\omega^2_{0,\text{in-phase}} = (2.14+1.27)\times10^{31}\ \mathrm{s^{-2}} = 3.41\times10^{31}\ \mathrm{s^{-2}}$$

and

$$\left(\frac{\gamma_2}{\gamma_1}\right)_{\text{in-phase}} = \frac{3.41\times10^{31}\ \mathrm{s^{-2}} - 1.38\times10^{31}\ \mathrm{s^{-2}}}{1.02\times10^{31}\ \mathrm{s^{-2}}} = 1.99$$

Thus

$$f_{\text{in-phase}} = \frac{1}{3}\frac{m_e}{e^2}\frac{(3.178+3.898\cdot1.99)^2\times10^{-8}\ \mathrm{C^2kg^{-1}}}{4.96}\cdot\frac{1}{\hat{n}} = 2.85\cdot\frac{1}{\hat{n}}$$

Correspondingly, we obtain

$$\omega^2_{0,\text{out-of-phase}} = (2.14-1.27)\times10^{31}\ \mathrm{s^{-2}} = 0.87\times10^{31}\ \mathrm{s^{-2}}$$

$$\left(\frac{\gamma_2}{\gamma_1}\right)_{\text{out-of-phase}} = \frac{0.87\times10^{31}\ \mathrm{s^{-2}} - 1.38\times10^{31}\ \mathrm{s^{-2}}}{1.02\times10^{31}\ \mathrm{s^{-2}}} = -0.50$$

$$f_{\text{out-of-phase}} = \frac{1}{3}\frac{m_e}{e^2}\frac{(3.178-3.898\cdot0.50)^2\times10^{-8}\ \mathrm{C^2kg^{-1}}}{1.25}\cdot\frac{1}{\hat{n}} = 0.14\cdot\frac{1}{\hat{n}}$$

These are the oscillator strengths for the transitions with transition moment in the x-direction. The transitions in the y-direction contribute the same value; thus, the total oscillator strengths are $5.70/\hat{n} = 3.8$ and $0.28/\hat{n} = 0.19$, respectively, where we used the value $\hat{n} = 1.5$. Note that the sum of the oscillatory strengths ($f_1 + f_2 = 2.989/\hat{n}$) is not affected by the coupling.

The ratio of the oscillator strengths is

$$\frac{f_{\text{coupled},2}}{f_{\text{coupled},1}} = \frac{0.14}{2.85} = 0.049$$

The experimental data (see Table 9.5) are $\lambda_1 = 423$ nm, $\lambda_2 = 546$ nm, and $f_2/f_1 = 0.063$. The coupling effect on the intensities of the absorption bands is well represented by the model. However, the separation of the two bands ($\lambda_{\text{in-phase}} = 322$ nm, $\lambda_{\text{out-of-phase}} = 639$ nm) is overestimated; this is due to the neglect of branching as can be seen by refining the treatment.[1] Run the program "coupling.exe," which is available on Wiley's website, to see the result of the improved treatment.

1 H.D. Försterling, H.Kuhn, *Int. J. Quant. Chem.* 1968, **2**, 413.

P9.21 Coupling of transitions in phthalocyanine. Perform the same calculations as in Problems 9.18 to 9.20 for phthalocyanine. From the calculated absorption maxima $\lambda_1 = 772$ nm and $\lambda_2 = 327$ nm of phthalocyanine that are reported in Table 9.5 (uncoupled transitions), we obtain $\Delta E_1 = 2.57 \times 10^{-19}$ J, $\Delta E_2 = 6.07 \times 10^{-19}$ J. The coupling integrals for the electrons in the electron pairs ($J_{12} = 0.5 \times 10^{-19}$) J and between the two electron pairs ($J_{12} = 0.6 \times 10^{-19}$ J) and the transition moments ($M_{x,1} = M_{x,2} = 2.85 \times 10^{-29}$ C m) are the same as those for porphyrin. Thus for the transition energies of the electron pairs we obtain

$$\Delta E_1 = 3.03 \times 10^{-19}\ \text{J} \quad \text{and} \quad \Delta E_2 = 6.55 \times 10^{-19}\ \text{J}$$

Solution

We recognize that the data for phthalocyanine are the same as for porphyrin except those for the excitation energies. For electron pairs 1 and 2 we find:

$$\frac{k_{f,\text{pair1}}}{m} = \frac{4\pi^2}{h^2}\Delta E_{\text{pair1}}^2 = \frac{4\pi^2}{h^2}\left(3.03 \times 10^{-19}\ \text{J}\right)^2 = 0.83 \times 10^{31}\ \text{s}^{-2}$$

$$\frac{k_{f,\text{pair2}}}{m} = \frac{4\pi^2}{h^2}\Delta E_{\text{pair2}}^2 = \frac{4\pi^2}{h^2}\left(6.55 \times 10^{-19}\ \text{J}\right)^2 = 3.86 \times 10^{31}\ \text{s}^{-2}$$

$$\frac{k_{f,12}}{m} = \frac{16\pi^2}{h^2}\sqrt{3.03 \times 6.55} \times 10^{-19}\ \text{J} \cdot 0.6 \times 10^{-19}\ \text{J} = 0.96 \times 10^{31}\ \text{s}^{-2}$$

Using these results we can find the resonance frequencies as (see Problem P9.19)

$$\omega_0^2 = \frac{1}{2}\left(\frac{k_{f,\text{pair1}}}{m} + \frac{k_{f,\text{ pair2}}}{m}\right) \pm \sqrt{\frac{k_{f,12}^2}{m^2} + \frac{1}{4}\left(\frac{k_{f,\text{pair1}}}{m^2} - \frac{k_{f,\text{pair2}}}{m^2}\right)^2}$$

Thus we obtain

$$\omega_{0,\text{in-phase}} = \sqrt{3.41 \times 10^{31}\ \text{s}^{-2}} = 6.43 \times 10^{15}\ \text{s}^{-1}$$

$$\omega_{0,\text{out-of-phase}} = \sqrt{0.87 \times 10^{31}\ \text{s}^{-2}} = 2.33 \times 10^{15}\ \text{s}^{-1}$$

The coupling increases the frequency of the high-frequency oscillator, and it decreases the frequency of the low-frequency oscillator. Using the relation

$$\omega_{0,\text{in-phase}}^2 = \frac{k_{f,\text{in-phase}}}{m} = \frac{4\pi^2}{h^2}\Delta E_{\text{in-phase}}^2$$

$$\omega_{0,\text{out-of-phase}}^2 = \frac{k_{f,\text{out-of-phase}}}{m} = \frac{4\pi^2}{h^2}\Delta E_{\text{out-of-phase}}^2$$

we calculate the excitation energies

$$\Delta E_{\text{in-phase}} = 6.78 \times 10^{-19}\ \text{J} \quad \text{and} \quad \Delta E_{\text{out-of-phase}} = 2.46 \times 10^{-19}\ \text{J}$$

and the wavelengths

$$\lambda_{\text{in-phase}} = \frac{hc_0}{\Delta E_{\text{in-phase}}} = 293\ \text{nm} \quad \text{and} \quad \lambda_{\text{out-of-phase}} = \frac{hc_0}{\Delta E_{\text{out-of-phase}}} = 806\ \text{nm}$$

For the oscillator strengths we proceed as in Problem 9.20. For the two uncoupled oscillators we have

$$\frac{Q_1^2}{m} = \frac{16\pi^2}{h^2}\Delta E_1 M_{x,1}^2 = \frac{16\pi^2}{h^2} 2.57 \times 10^{-19}\ \text{J} \cdot \left(2.85 \times 10^{-29}\ \text{C m}\right)^2 = 7.51 \times 10^{-8}\ \text{C}^2\ \text{kg}^{-1}$$

$$\frac{Q_2^2}{m} = \frac{16\pi^2}{h^2}\Delta E_2 M_{x,2}^2 = \frac{16\pi^2}{h^2} 6.07 \times 10^{-19}\ \text{J} \cdot \left(2.85 \times 10^{-29}\ \text{C m}\right)^2 = 17.7 \times 10^{-8}\ \text{C}^2\ \text{kg}^{-1}$$

Thus for the oscillatory strengths of the uncoupled oscillators we obtain

$$f_1 = \frac{1}{3}\frac{m_e}{e^2} \cdot \frac{Q_1^2}{m} \cdot \frac{1}{\hbar} = 0.89 \cdot \frac{1}{\hbar}$$

$$f_2 = \frac{1}{3}\frac{m_e}{e^2} \cdot \frac{Q_2^2}{m} \cdot \frac{1}{\hbar} = 2.09 \cdot \frac{1}{\hbar}$$

The oscillator strength of two coupled oscillators is

$$f = \frac{1}{3}\frac{m_e}{e^2 m} \cdot \frac{(Q_1\gamma_1 + Q_2\gamma_2)^2}{\gamma_1^2 + \gamma_2^2} \cdot \frac{1}{\hat{n}} = \frac{1}{3}\frac{m_e}{e^2} \frac{\left(Q_1/\sqrt{m} + Q_2/\sqrt{m} \cdot \frac{\gamma_2}{\gamma_1}\right)^2}{1 + \left(\frac{\gamma_2}{\gamma_1}\right)^2} \cdot \frac{1}{\hat{n}}$$

From the expressions for Q_1^2/m and Q_2^2/m calculated above we obtain

$$\frac{Q_1}{\sqrt{m}} = 2.74 \times 10^{-4}\ \mathrm{C\ kg^{-1/2}} \quad \text{and} \quad \frac{Q_2}{\sqrt{m}} = 4.21 \times 10^{-4}\ \mathrm{C\ kg^{-1/2}}$$

Furthermore, the ratio γ_2/γ_1 is

$$\frac{\gamma_2}{\gamma_1} = \frac{\omega_0^2 - k_{f,\text{pair1}}/m}{k_{f,12}/m}$$

In analogy to Problem P9.20 we obtain

$$\omega^2_{0,\text{in-phase}} = 4.13 \times 10^{31}\ \mathrm{s^{-2}} \quad \text{and} \quad \omega^2_{0,\text{out-of-phase}} = 5.46 \times 10^{31}\ \mathrm{s^{-2}}$$

and

$$\left(\frac{\gamma_2}{\gamma_1}\right)_{\text{in-phase}} = 3.44 \quad \text{and} \quad \left(\frac{\gamma_2}{\gamma_1}\right)_{\text{out-of-phase}} = 0.29$$

Thus

$$f_{\text{in-phase}} = 2.73 \cdot \frac{1}{\hat{n}} \qquad f_{\text{out-of-phase}} = 0.25 \cdot \frac{1}{\hat{n}}$$

These are the oscillator strengths for the transitions with transition moment in the x-direction. The transitions in the y-direction contribute the same value, thus the total oscillator strengths are $5.46/\hat{n} = 4.0$ and $0.50/\hat{n} = 0.36$, respectively, with the value $\hat{n} = 1.36$ (value for the solvent used in the experiment)
The ratio of the oscillator strengths is

$$\frac{f_{\text{coupled},2}}{f_{\text{coupled},1}} = \frac{0.25}{2.73} = 0.092$$

The experimental data (see Table 9.5) are $\lambda_1 = 360$ nm, $\lambda_2 = 675$ nm, and $f_2/f_1 = 0.53$. The separation of the two bands $\lambda_{\text{in-phase}} = 292$ nm, $\lambda_{\text{out-of-phase}} = 806$ nm) is overestimated, as in the case of porphyrin. In contrast to porphyrin, the coupling effect on the intensities of the absorption bands is overestimated by a factor of five. A refined model including the branching of the electron gas (which results in smaller values for the coupling constants and for the higher energy transition moments)[2] gives a ratio $f_2/f_1 = 0.62$, in accordance with experiment. Run the program "coupling.exe," available on Wiley's website, to see the result of the improved treatment.

P9.22 Discuss the absorption spectrum of β-Carotene, distinguishing between an all-trans and a cis-configuration (Fig. P9.22a).

Solution
β-Carotene is a polyene with 22 π electrons (Fig. P9.22b).

---

2 H.D. Försterling, H.Kuhn, *Int. J. Quant. Chem.* 1968, **2**, 413.

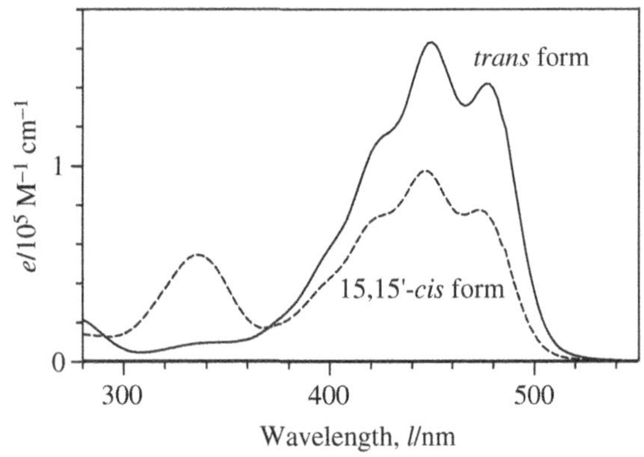

Figure P9.22a Absorption spectrum of β-Carotene. Solid line: trans-form, dashed line: cis-form.

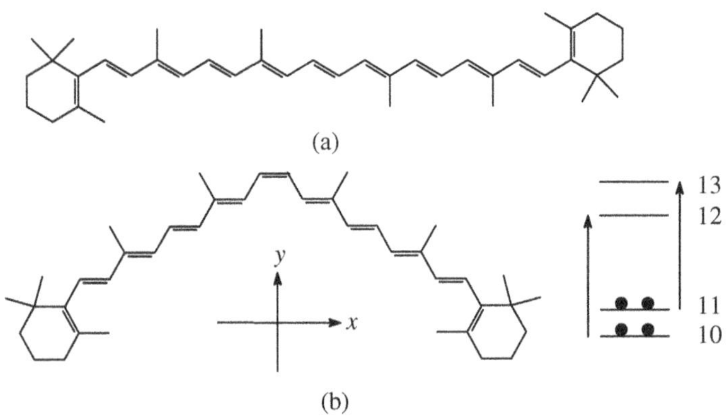

Figure P9.22b Structural formulae of β-carotene. (a) all-Trans-form, (b) $15,15'$-cis-form. The energy level diagram shows the $10 \rightarrow 12$ and $11 \rightarrow 13$ transitions.

For the HOMO–LUMO transition (11→12) we obtain $\lambda_{\text{max}} = 451$ nm in accordance with the experimental value $\lambda_{\text{max}} = 453$ nm (Fig. P9.22a). Thus we can understand that β-carotene has a yellow color. Without taking into account the bond-length alternation, an absorption maximum at $\lambda_{\text{max}} = 1400$ nm (this is in the infrared region) would be expected.

β-Carotene exists in different cis-trans isomers. In contrast to the all-trans β-carotene the 15,15 ′-cis β–carotene has, in addition to the absorption at 453 nm, another absorption band at 340 nm (Fig. P9.22a, dashed line: cis-peak) which is polarized in the y-direction, while the main absorption band (transition $11 \rightarrow 12$) is polarized in the x-direction. The cis peak can be considered as being due to a coupling of the transitions $11 \rightarrow 13$ ($\lambda_{max} = 386$ nm, $f = 0.44$) and $10 \rightarrow 12$ ($\lambda_{\text{max}} = 390$ nm, $f = 0.47$) leading to one absorption band shifted to higher frequency (cis peak, $\lambda_{\text{max}} = 320$ nm, $f = 0.91$) and a second band with small oscillator strength shifted to lower frequency ($\lambda_{\text{max}} = 425$ nm, $f = 0.001$).

Note that this bandis expected to be at 462 nm, which is a longer wavelength than the HOMO→LUMO peak. What is the reason for this unexpected behavior? The gap between the HOMO and the LUMO ($11 \rightarrow 12$) is much larger than the gap between the (HOMO−1) and the HOMO ($10 \rightarrow 11$) and the gap between the LUMO and the (LUMO+ 1), $12 \rightarrow 13$. Therefore, the transition energies (HOMO−1)→LUMO and HOMO→ (LUMO+1) are only slightly larger than the transition energy HOMO→LUMO; for this reason a small shift by coupling is sufficient to bring the out-of-phase coupled transition to longer wavelengths than the HOMO→LUMO transition. Experimentally, this effect can be observed in polyenes with four to six double bonds.

P9.23 Solve the coupling problem discussed for the cis-peak of β–carotene in Problem P9.22. Use the transition energies $\Delta E_{11\rightarrow13} = 5.15 \times 10^{-19}$ J and $\Delta E_{10\rightarrow12} = 5.09 \times 10^{-19}$ J, the coupling integral $J_{12} = 0.38 \times 10^{-19}$ J, and the transition moments $M_{y,1} = M_{y,2} = 2.29 \times 10^{-29}$ C m.

Solution

We proceed as in the case of porphyrin. From the transition energies $\Delta E_{11\to 13} = 5.15 \times 10^{-19}$ J and $\Delta E_{10\to 12} = 5.09 \times 10^{-19}$ J (corresponding to the wavelengths $\lambda_1 = 386$ nm and $\lambda_2 = 390$ nm) and the coupling integral $J_{12} = 0.38 \times 10^{-19}$ J we find

$$\Delta E_{\text{pair1}} = \sqrt{\Delta E^2_{11\to 13} + 2 \cdot J_{12} \cdot \Delta E^2_{11\to 13}} = 5.51 \times 10^{-19}\ \text{J}$$

$$\Delta E_{\text{pair2}} = \sqrt{\Delta E^2_{10\to 12} + 2 \cdot J_{12} \cdot \Delta E^2_{11\to 13}} = 5.46 \times 10^{-19}\ \text{J}$$

For electron pairs 1 and 2 we find:

$$\frac{k_{f,\text{pair1}}}{m} = 2.74 \times 10^{31}\ \text{s}^{-2} \quad \text{and} \quad \frac{k_{f,\text{pair2}}}{m} = 2.68 \times 10^{31}\ \text{s}^{-2}$$

Using these results we find the resonance frequencies (see Problem P9.15) to be

$$\omega_{0,\text{in-phase}} = 4.42 \times 10^{15}\ \text{s}^{-1} \quad \text{and} \quad \omega_{0,\text{out-of-phase}} = 2.95 \times 10^{15}\ \text{s}^{-1}$$

leading to the excitation energies

$$\Delta E_{\text{in-phase}} = 6.20 \times 10^{-19}\ \text{J} \quad \text{and} \quad \Delta E_{\text{out-of-phase}} = 4.67 \times 10^{-19}\ \text{J}$$

and to the wavelengths

$$\lambda_1 = 320\ \text{nm} \quad \text{and} \quad \lambda_2 = 426\ \text{nm}$$

From the transition moments $M_{y,1} = M_{y,2} = 2.29 \times 10^{-29}$ C m, we find

$$\frac{Q_1}{\sqrt{m}} = -3.11 \times 10^{-4}\ \text{C kg}^{-1/2} \quad \text{and} \quad \frac{Q_2}{\sqrt{m}} = -3.09 \times 10^{-4}\ \text{C kg}^{-1/2}$$

and

$$f_1 = \frac{1}{3}\frac{m_e}{e^2} \cdot \frac{Q_1^2}{m} \cdot \frac{1}{\hat{n}} = 1.15 \cdot \frac{1}{\hat{n}} \quad \text{and} \quad f_2 = \frac{1}{3}\frac{m_e}{e^2} \cdot \frac{Q_2^2}{m} \cdot \frac{1}{\hat{n}} = 1.13 \cdot \frac{1}{\hat{n}}$$

for the uncoupled oscillators.

The oscillator strengths f_{coupled} of the two coupled oscillators is

$$f_{\text{coupled}} = \frac{1}{3}\frac{m_e}{e^2 m} \cdot \frac{(Q_1\gamma_1 + Q_2\gamma_2)^2}{\gamma_1^2 + \gamma_2^2} \cdot \frac{1}{\hat{n}} = \frac{1}{3}\frac{m_e}{e^2} \frac{\left(Q_1/\sqrt{m} + Q_2/\sqrt{m} \cdot \frac{\gamma_2}{\gamma_1}\right)^2}{1 + \left(\frac{\gamma_2}{\gamma_1}\right)^2} \cdot \frac{1}{\hat{n}}$$

In order to simplify the problem we make the approximation that

$$\frac{Q_1}{\sqrt{m}} = \frac{Q_2}{\sqrt{m}} \quad \text{and} \quad \Delta E_{\text{pair1}} = \Delta E_{\text{pair2}}$$

In analogy to Problem 9.16, we obtain $\gamma_2/\gamma_1 = \pm 1$ for the ratio γ_2/γ_1. Thus the result for the oscillator strengths is

$$f_{\text{coupled}} = \frac{1}{3}\frac{m_e}{e^2} \frac{Q_1^2/m\left(1 + \frac{\gamma_2}{\gamma_1}\right)^2}{1 + \left(\frac{\gamma_2}{\gamma_1}\right)^2} \cdot \frac{1}{\hat{n}}$$

$$f_{1,\text{coupled}} = \frac{1}{3}\frac{m_e}{e^2}\frac{Q_1^2}{m} \cdot 2 \cdot \frac{1}{\hat{n}} \quad \text{for } \gamma_2/\gamma_1 = +1$$

and

$$f_{2,\text{coupled}} = 0 \text{ for } \gamma_2/\gamma_1 = -1$$

That is

$$f_{1,\text{coupled}} = 2 \cdot f_1 \qquad f_{2,\text{coupled}} = 0$$

The more rigorous treatment leads to

$$\frac{f_{1,\text{coupled}}}{f_1} = \frac{2.277}{1.15} = 1.98 \quad \text{and} \quad \frac{f_{2,\text{coupled}}}{f_2} = \frac{0.001}{1.13} = 0.0009$$

You can verify this result by running the program "coupling.exe."

P9.24 Using the step model and the Mathematica application "Cyanines.nb" show that in a long cyanine cation alternating bond lengths are obtained.

Solution

Run the application for a short cyanine and for a long cyanine. For example, for a cyanine with $k = 12$ (26 bonds) you should obtain a probability density like that displayed in Figure 9.6.

10

Emission of Light

10.1 Exercises

E10.1 The molecule 9,10-diphenylanthracene is a commonly used quantum yield standard in fluorescence spectroscopy. Its quantum yield is measured to be $\phi = 0.95$, and its fluorescence lifetime τ is 8.2 ns. Determine the radiative lifetime, the radiative rate constant, and the nonradiative rate constant for this molecule.

Solution

The radiative lifetime is given by

$$\tau_0 = \frac{\tau}{\phi} = \frac{8.2 \times 10^{-9}\ \text{s}}{0.95} = 8.6 \times 10^{-9}\ \text{s}$$

Hence the radiative rate constant k_{fl} is

$$k_{\text{fl}} = \frac{1}{\tau_0} = \frac{1}{8.6 \times 10^{-9}\ \text{s}} = 1.2 \times 10^{8}\ \text{s}^{-1}$$

and the nonradiative rate constant k' is

$$k' = \frac{1}{\tau} - k_{\text{fl}} = \frac{1}{8.2 \times 10^{-9}\ \text{s}} - 1.2 \times 10^{8}\ \text{s}^{-1} = 2.4 \times 10^{6}\ \text{s}^{-1}$$

E10.2 Single molecule spectroscopy relies on the ability to detect single photons. If a photomultiplier tube has a gain of 10^6 (i.e., generates a million electrons for each photon impinging on it), how many photons per second are needed to generate an average current of 10 pA from the photomultiplier tube. If a single photon strikes the photomultiplier tube it generates a short burst of electrons from the tube, estimate how short this burst of electrons is if it has an average current of 1 nA.

Solution

An ampere is the amount of charge that flows per second. For 10 pA that would mean we need 10×10^{-12} C per second. Each electron carries 1.602×10^{-19} C and so we would need

$$\frac{10 \times 10^{-12}\ \text{C}}{1.602 \times 10^{-19}\ \text{C}} = 6.2 \times 10^{7}$$

electrons. So if one photon can generate 10^6 electrons we would need

$$\frac{6.2 \times 10^{7}}{10^{6}} \approx 62$$

photons to provide 10 pA. To determine the time of the short burst we first note that

$$1\ \text{nA} = 1 \times 10^{-9}\ \text{C s}^{-1}$$

We divide that into the current generated from one photon to find the characteristic time

$$\frac{10^{6} \times 1.602 \times 10^{-19}\ \text{C}}{1 \times 10^{-9}\ \text{C s}^{-1}} \approx 2 \times 10^{-4}\ \text{s} = 0.2\ \text{ms}$$

Solutions Manual for Principles of Physical Chemistry, Third Edition. Edited by Hans Kuhn, David H. Waldeck, and Horst-Dieter Försterling.

E10.3 A He–Ne laser generates 1 mW of laser light at a wavelength of 632.8 nm. Calculate the number of photons per second that are being emitted by the laser.

Solution

We begin by recalling the definition of a milliwatt (mW) which is

$$Power = 1\ \text{mW} = 10^{-3}\ \text{J s}^{-1}$$

The energy of a 632.8 nm photon is

$$E = h\nu = \frac{hc}{\lambda}$$
$$= 6.626 \times 10^{-34}\ \text{J s} \times \frac{2.998 \times 10^{8}\ \text{m s}^{-1}}{632.8 \times 10^{-9}\ \text{m}} = 3.139 \times 10^{-19}\ \text{J}$$

Hence, the number of photons per second is

$$\frac{Power}{E} = \frac{10^{-3}\ \text{J s}^{-1}}{3.1391 \times 10^{-19}\ \text{J}} = 3.2 \times 10^{15}\ \text{photons per second}$$

E10.4 Using the results in Foundations 9.1 and 10.1 relate the radiative rate constant to the molecule's absorption spectrum; derive Equation 10.2, i.e.,

$$\frac{1}{\tau_0} = \ln 10 \cdot \frac{8\pi\nu_0^2}{3c_0^2\mathbf{N}_A} \cdot \int_0^\infty \varepsilon \cdot d\nu$$

which is Equation (10.2).

Solution

The primary result of Foundation 10.1 is the derivation of an expression for the radiative (or natural) lifetime τ_0

$$\tau_0 = \frac{3m\varepsilon_0 c_0^3}{2Q^2\pi\nu_0^2} \quad \text{or} \quad \tau_0 = \frac{3m_e\varepsilon_0 c_0^3}{2e^2\pi\nu_0^2} \cdot \frac{1}{f}$$

in terms of the oscillator strength, and in Foundation 9.1 (see Equation (9.91) for the special case of $\hat{n} = 1$) we showed that

$$\int_0^\infty \varepsilon \cdot d\nu = \frac{\mathbf{N}_A}{4 \cdot \ln 10 \cdot c_0\varepsilon_0} \frac{Q^2}{m}$$

If we rearrange the first of these relations for τ_0 and use the relation from Equation (9F2.6), $f = Q^2\ m_e/(m\ e^2)$, we can write that

$$\frac{1}{\tau_0} = \frac{2\pi\nu_0^2}{3\varepsilon_0 c_0^3} \frac{Q^2}{m}$$

Now we use our expression for the integrated absorption band to substitute for Q^2/m and find that

$$\frac{1}{\tau_0} = \frac{2\pi\nu_0^2}{3\varepsilon_0 c_0^3} \frac{4 \cdot \ln 10 \cdot c_0\varepsilon_0}{\mathbf{N}_A} \int_0^\infty \varepsilon \cdot d\nu$$
$$= \ln 10 \cdot \frac{8\pi\nu_0^2}{3c_0^2\mathbf{N}_A} \cdot \int_0^\infty \varepsilon \cdot d\nu$$

E10.5 The fluorescence excitation spectrum for terrylene (see Fig. 10.3) shows an intrinsic molecular linewidth of 0.1 GHz. Use this linewidth value to calculate the radiative lifetime of the molecule.

Solution

If we take the linewidth to result from the intrinsic radiative lifetime of the molecule, we can use Equation (10.3) to write that

$$\tau_0 = \frac{1}{2\pi\,\Delta\nu} = \frac{1}{2\pi\left(1 \times 10^{8}\ \text{s}^{-1}\right)} = 1.6 \times 10^{-9}\ \text{s} = 1.6\ \text{ns}$$

E10.6 Consider a 100 mW laser source that has a circular beam cross-section with a diameter d of 3 mm and a wavelength λ of 740 nm. Consider the light beam to consist of 10^8 light pulses per second and the light pulses to have a characteristic time width of 10^{-13} s (you may assume that the pulses have a square profile in time). a) Calculate the peak intensity of the light field and compare it to the corresponding instantaneous intensity you would find if the light beam were continuous wave (not pulsed). b) Justify why it is reasonable to assume that the peak field strength of the light pulse is small compared to a typical molecular field strength of 10^{11} V m^{-1}. (Hint: Use the fact that $I = c_0 \cdot \varepsilon_0 \cdot \hat{n} \cdot F_0^2$).

Solution

a) Given that the average power is 0.100 J s^{-1} and the beam has 100 million pulses per second, the energy per laser pulse (E_{pulse}) is

$$E_{\text{pulse}} = \frac{0.100\ \text{J s}^{-1}}{1 \times 10^8\ \text{s}^{-1}} = 1 \times 10^{-9}\ \text{J} = 1\ \text{nJ}$$

If the pulse persists for a time of 10^{-13} s and the energy is evenly distributed over this time (square shape), then the peak power of the pulse P_{pulse} is

$$P_{\text{pulse}} = \frac{E_{\text{pulse}}}{10^{-13}\ \text{s}} = \frac{1 \times 10^{-9}\ \text{J}}{10^{-13}\ \text{s}} = 1 \times 10^4\ \text{W}$$

To find the intensity of the pulse I_{pulse}, we must divide by its cross-sectional area A_{pulse} and we find that

$$I_{\text{pulse}} = \frac{P_{\text{pulse}}}{A_{\text{pulse}}} = \frac{P_{\text{pulse}}}{\pi(d/2)^2}$$

$$= \frac{1 \times 10^4\ \text{W}}{\pi\left(1.5 \times 10^{-3}\right)^2 \text{m}^2} = 1.4 \times 10^9\ \frac{\text{W}}{\text{m}^2}$$

By way of contrast, the same average laser power and beam cross-section for a continuous wave laser (not pulsed) has an intensity I_{cw} that is one-hundred thousand times lower

$$I_{\text{cw}} = \frac{P_{\text{avg}}}{A} = \frac{0.100\ \text{W}}{\pi\left(1.5 \times 10^{-3}\right)^2 \text{m}^2} = 1.4 \times 10^4\ \frac{\text{W}}{\text{m}^2}$$

b) We can write the field strength of the laser pulse F_{pulse} as

$$F_{\text{pulse}} = \sqrt{\frac{2 \cdot I_{\text{pulse}}}{c_0 \cdot \varepsilon_0}}$$

$$= \sqrt{\frac{2 \cdot 1.4 \times 10^9\ \text{W m}^{-2}}{3.0 \times 10^8\ \text{m s}^{-1} \cdot 8.85 \times 10^{-12}\ \text{C}^2\ \text{J}^{-1}\ \text{m}^{-1}}}$$

$$= 1.0 \times 10^6 \sqrt{\frac{\text{W m}^{-2}}{\text{m s}^{-1}\ \text{C}^2\ \text{J}^{-1}\ \text{m}^{-1}}} = 1.0 \times 10^6 \sqrt{\frac{\text{A V m}^{-2}}{\text{m s}^{-1}\ \text{A}^2\ \text{s}^2(\text{A V s})^{-1}\text{m}^{-1}}} = 1.0 \times 10^6\ \text{V m}^{-1}$$

This value is 10^5 times smaller than the molecular field strength of 10^{11} V m^{-1} that is quoted in the problem.

E10.7 In E10.6) you calculated the peak intensity of a 100 fs light pulse, assuming a diameter d of 3 mm which is typical of an unfocused laser source. a) Using the same parameters as in E10.6) calculate the intensity for a pulse that is focused to a 30 μm diameter spot. b) Calculate the intensity for the case where the beam is focused to near its diffraction limit, taken to be $\lambda/2$. b) Assuming that you can focus the beam to the diffraction limit and that you can change the laser power without changing its other parameters, calculate the power you would need for the peak field strength of the light pulse to be comparable to a typical molecular field strength of 10^{11} V m^{-1}. (Hint: Use the fact that $I = c_0 \cdot \varepsilon_0 \cdot \hat{n} \cdot F_0^2$).

Solution

a) In E10.6) we found that the power of the pulse was $P_{pulse} = 1 \times 10^4$ W and that the intensity of the pulse was $I_{pulse} = 1.4 \times 10^9$ W m^{-2} when its diameter was 3 mm. If we focus the laser beam to 30 μm, then the pulse intensity becomes

$$I_{pulse} = \frac{P_{pulse}}{\pi(d/2)^2} = \frac{1 \times 10^4 \text{ W}}{\pi\left(1.5 \times 10^{-5}\right)^2 \text{m}^2} = 1.4 \times 10^{13} \frac{\text{W}}{\text{m}^2}$$

That is it increases by 10,000 times because the area has decreased by 10,000 times.

b) If the laser beam is focused to a diameter of (740 nm/2 = 370 nm), then the intensity of the pulse becomes

$$I_{pulse} = \frac{1 \times 10^4 \text{ W}}{\pi\left(1.85 \times 10^{-7}\right)^2 \text{m}^2} = 9.3 \times 10^{16} \frac{\text{W}}{\text{m}^2}$$

c) From E10.6 part b, we can write that

$$F_{pulse} = \sqrt{\frac{2 \cdot I_{pulse}}{c_0 \cdot \varepsilon_0}} = \sqrt{\frac{2 \cdot P_{pulse}}{c_0 \cdot \varepsilon_0 \cdot A_{pulse}}}$$

Hence if we wish for $F_{pulse} = 10^{11}$ V m^{-1}, then we can solve the above equation for P_{pulse} and find

$$P_{pulse} = \frac{\left(10^{11} \text{ V m}^{-1}\right)^2 \cdot c_0 \cdot \varepsilon_0 \cdot A_{pulse}}{2}$$

$$= \frac{10^{22} \text{ V}^2\text{m}^{-2} \cdot 3.0 \times 10^8 \text{m s}^{-1} \cdot 8.85 \times 10^{-12} \text{ C}^2\text{J}^{-1} \text{ m}^{-1}\pi\left(1.85 \times 10^{-7}\right)^2 \text{m}^2}{2}$$

$$= 1.4 \times 10^6 \frac{\text{V}^2\text{m C}^2\text{m}^2}{\text{m}^2 \text{ s J m}} = 1.4 \times 10^6 \frac{\text{V}^2 \text{ (As)}^2}{\text{s (A V s)}} = 1.4 \times 10^6 \text{ A V} = 1.4 \times 10^6 \text{ W}$$

This value of the power is 140 times larger than the pulse power that was computed for the conditions in E10.6. By increasing the laser power from 100 mW to 14 W, this pulse power could be obtained. Alternatively, one could increase the laser power to 1.4 W and decrease the pulse width from 10^{-13} s to 10^{-14} s and obtain the same pulse power.

E10.8 Near U.V. chromophores are commonly found in detergents, because they make the clothing "brighter" and "whiter." Chemical derivatives of the stilbene chromophore (1,2-diphenylethylene) are one such fluorescent chromophore found in "whitening agents." The molecule *trans*-stilbene has a radiative rate constant $k_{rad} = 1/\tau = 3.6 \times 10^8$ s^{-1}. a) If radiative emission determines the lifetime of the excited state, what do you expect for the frequency width of the excited state (i.e., full width of the spectral line at half its maximum)? Assume you have a Lorentzian lineshape; namely

$$I(\omega - \omega_0) = \frac{I_0}{(\omega - \omega_0)^2 + (\gamma/2)^2}$$

and show that $\gamma = 1/\tau$. b) Take the transition energy of stilbene (i.e., from the ground vibrational, ground electronic state to the ground vibrational, excited electronic state) to be $32,234.74 \pm 0.02$ cm^{-1}, what precision must your spectrometer have in order to determine the bandwidth of the transition? Give your answer in terms of 1 part in 10^n, where you will determine n.

Solution

a) Using Equation (10.3), we find that

$$\Delta\nu = \frac{1}{2\pi}\frac{1}{\tau} = \frac{3.6 \times 10^8 \text{s}^{-1}}{2\pi} = 5.7 \times 10^7 \text{s}^{-1}$$

or

$$\Delta\omega = 2\pi \cdot \Delta\nu = \frac{1}{\tau} = 3.6 \times 10^8 \text{s}^{-1}$$

The Lorentzian function has its maximum I_{max} at $\omega = \omega_0$, so that

$$I_{max} = \frac{I_0}{(\gamma/2)^2}$$

From the definition of the half-width and assuming that the Lorentzian function applies, we can write that

$$\frac{I(\omega_{1/2}-\omega_0)}{I_{max}} = \frac{1}{2} = \frac{(\gamma/2)^2}{(\omega_{1/2}-\omega_0)^2+(\gamma/2)^2}$$

where $\omega_{1/2}$ is the value of the angular frequency when the intensity is at its half-height. Rearranging this expression we find that

$$(\omega_{1/2}-\omega_0)^2+(\gamma/2)^2 = 2\cdot(\gamma/2)^2$$

so that

$$(\omega_{1/2}-\omega_0) = (\gamma/2)$$

and

$$\Delta\omega = 2\cdot(\omega_{1/2}-\omega_0) = \gamma$$

Comparison with our result from Equation (10.3) shows that

$$\Delta\omega = \gamma = \frac{1}{\tau}$$

b) From part A, we know that $\Delta\nu = 5.7\times10^7\ s^{-1}$ and in units of wavenumbers it will be

$$\Delta\tilde{\nu} = \frac{\Delta\nu}{c_0} = \frac{(5.7\times10^7 s^{-1})}{(3.0\times10^8\ m\ s^{-1})} = 0.19\ m^{-1} = 0.0019\ cm^{-1}$$

To obtain 10 different intensity points along the line profile would require that we have a resolution of at least $0.00019\ cm^{-1}$. Hence for the transition, we require that

$$\frac{32,234.74\ cm^{-1}}{0.00019\ cm^{-1}} = 1.7\times10^8$$

that is, of 1 or 2 parts in 10^8, i.e., $n = 8$.

E10.9 Using a qualitative molecular orbital energy diagram, explain why you do not expect helium to form a stable diatomic molecule (He_2) in its electronic ground state, but that it could form stable bond in an excited electronic state.

Solution

In principle, it would have to be possible for two He nuclei and four electrons to form an He_2 molecule, analogous to H_2^+ and H_2. No such molecule has been experimentally observed (although two He atoms can weakly bind because of dispersion forces).

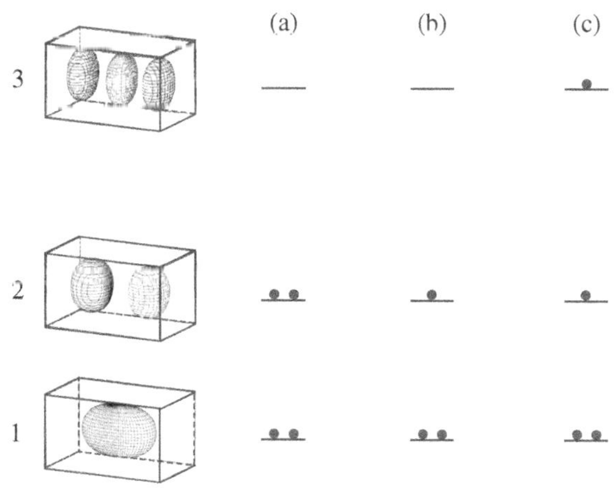

Figure E10.9 Energy scheme and box wavefunctions for systems consisting of two He atoms. (a) He_2 (two electrons in bonding, two electrons in antibonding state, unstable). (b) He_2^+ (two electrons in bonding, one electron in antibonding state, stable). (c) He_2^* (two electrons in bonding state 1, one electron in antibonding state 2, one electron in bonding state 3).

Let us assume that two He nuclei are fixed at a distance of, say, 100 pm and we add four electrons compensating for the nuclear charges. According to Fig. E10.9a, two electrons, like in H_2, occupy an electron cloud approximated by the lowest energy state which is a bonding orbital with an antinode between the He nuclei; however, the next two electrons, according to the Pauli exclusion principle, are excluded from the same orbital and they occupy the next higher energy orbital 2 (they occupy the antibonding orbital with a nodal plane between the nuclei). While the first two electrons in the bonding orbital stabilize the system, the other two electrons in the antibonding orbital destabilize it. The two effects cancel, so that no bond appears.

If there are two electrons in the bonding orbital 1 and only one in the antibonding orbital 2 (Fig. E10.9b), then the attraction dominates. In fact, He_2^+ is a known molecular ion. The binding energy is about the same as in H_2^+; this result could be estimated qualitatively because the binding effect of the first two electrons is about half compensated by the repulsion of the third.

A similar situation is realized if we consider He_2 in the excited state (Fig. E10.9c). Two electrons are in bonding state 1, one electron is in antibonding state 2, and one electron is in bonding state 3. Also in this case the attraction of the two nuclei dominates, and He_2^* exists as a bound excited state molecule.

E10.10 In Foundation 10.1, we write the equation of motion for a damped harmonic oscillator with a force constant k_f, a mass m, and a damping constant k_d as

$$\frac{d^2x}{dt^2} + \frac{k_d}{m}\frac{dx}{dt} + \frac{k_f}{m}x = 0$$

and claim that it has the solution

$$x(t) = Ce^{-(k_d/2m)t} \cdot \cos(\omega_0 t) \quad \text{and} \quad \omega_0 = \sqrt{\frac{k_f}{m}}$$

in the underdamped limit. More specifically, the underdamped limit corresponds to the case where the frictional decay rate k_d/m is small compared to the oscillation frequency ω_0. Show that $x(t)$ is a solution of the equation.

Solution

To show that the $x(t)$ is the solution, we insert it into the equation and solve. Taking the derivative of x with respect to t, we find

$$\begin{aligned}\frac{dx}{dt} &= C\left(-\frac{k_d}{2m}\right)e^{-(k_d/2m)t}\cos(\omega_0 t) - C\omega_0 e^{-(k_d/2m)t}\sin(\omega_0 t)\\ &= Ce^{-(k_d/2m)t}\left[-\frac{k_d}{2m}\cos(\omega_0 t) - \omega_0\sin(\omega_0 t)\right]\end{aligned}$$

In the underdamped limit ($k_d/(2m) \ll \omega_0$), we can write that

$$\frac{dx}{dt} = -Ce^{-(k_d/2m)t}\omega_0\sin(\omega_0 t)$$

For the second derivative we obtain

$$\begin{aligned}\frac{d^2x}{dt^2} &= -Ce^{-(k_d/2m)t}\frac{k_d}{2m}\left[-\frac{k_d}{2m}\cos(\omega_0 t) - \omega_0\sin(\omega_0 t)\right]\\ &\quad + Ce^{-(k_d/2m)t}\left[\frac{k_d}{2m}\omega_0\sin(\omega_0 t) - \omega_0^2\cos(\omega_0 t)\right]\\ &= Ce^{-(k_d/2m)t}\left[\left(\frac{k_d}{2m}\right)^2\cos(\omega_0 t) + \frac{k_d}{2m}\omega_0\sin(\omega_0 t) + \frac{k_d}{2m}\omega_0\sin(\omega_0 t) - \omega_0^2\cos(\omega_0 t)\right]\\ &= Ce^{-(k_d/2m)t}\left[\left(\frac{k_d}{2m}\right)^2\cos(\omega_0 t) + \frac{k_d}{m}\omega_0\sin(\omega_0 t) - \omega_0^2\cos(\omega_0 t)\right]\end{aligned}$$

In the underdamped limit ($k_d/(2m) \ll \omega_0$), we can write that

$$\frac{d^2x}{dt^2} = Ce^{-(k_d/2m)t}\left[\frac{k_d}{m}\omega_0\sin(\omega_0 t) - \omega_0^2\cos(\omega_0 t)\right]$$

Substituting these derivatives into the equation of motion, we find that

$$\begin{aligned}
&\frac{d^2x}{dt^2} + \frac{k_d}{m}\frac{dx}{dt} + \frac{k_f}{m}x \\
&= Ce^{-(k_d/2m)t}\left[\frac{k_d}{m}\omega_0 \sin(\omega_0 t) - \omega_0^2 \cos(\omega_0 t)\right] - \frac{k_d}{m}Ce^{-(k_d/2m)t}\omega_0 \sin(\omega_0 t) \\
&\quad + \frac{k_f}{m}Ce^{-(k_d/2m)t}\cos(\omega_0 t) \\
&= Ce^{-(k_d/2m)t}\left[\frac{k_d}{m}\omega_0 \sin(\omega_0 t) - \omega_0^2 \cos(\omega_0 t) - \frac{k_d}{m}\omega_0 \sin(\omega_0 t) + \frac{k_f}{m}\cos(\omega_0 t)\right] \\
&= Ce^{-(k_d/2m)t}\left[-\omega_0^2 \cos(\omega_0 t) + \frac{k_f}{m}\cos(\omega_0 t)\right] \\
&= Ce^{-(k_d/2m)t}\left[-\omega_0^2 \cos(\omega_0 t) + \omega_0^2 \cos(\omega_0 t)\right] = 0
\end{aligned}$$

E10.11 Compare Equation (10B1.2) in Box 10.1 to Equation (10.5) in the text and show that

$$B_{a\to b} = \frac{e^2}{12\varepsilon_0 m_e h\nu_{ab}} f$$

Solution

Equation (10.5) is

$$\left(\frac{dN}{dt}\right)_{a\to b} = B_{a\to b} \cdot \rho_\nu(\nu_{ab}) \cdot N_a$$

and we set it equal to Equation (10.21) in Box 10.1

$$\left(\frac{dN}{dt}\right)_{a\to b} = \left(\frac{dE}{dt}\right)_{a\to b}\frac{1}{\Delta E} = \frac{e^2}{12\varepsilon_0 m_e h\nu_{ab}} f \cdot \rho_\nu(\nu_{ab}) \cdot N_a$$

to find that

$$B_{a\to b} \cdot \rho_\nu(\nu_{ab}) \cdot N_a = \frac{e^2}{12\varepsilon_0 m_e h\nu_{ab}} f$$

10.2 Problems

P10.1 a) The intensity of a light beam I is the product of the light field's energy density and the velocity of light, namely

$$I = c_0 \cdot \rho_{\text{energy}} = c_0 \cdot \frac{1}{2}\varepsilon_0 F^2$$

where F is the light's electric field strength. Calculate the electric field strength for an intensity of 100 W cm^{-2}. b) The electric field experienced by a valence electron in a molecule can be estimated as

$$F_{\text{valence}} = -\frac{1}{4\pi\varepsilon_0}\frac{q}{d^2}$$

where d is the distance from the nucleus and q is the shielded nuclear charge, which the valence electron experiences. Use typical values for a π-electron on a carbon atom (let $d = 75$ pm and $q = -3.25\,e$; see Box 8.1, which are appropriate for the p_z-orbital on carbon) to calculate the field strength. c) Compare your two field strengths. Is it reasonable to assume that the light field is small compared to the molecular field?

Solution

a) We are given I the intensity of 100 W cm^{-2}. The electric field strength, F, is

$$F = \sqrt{\frac{2I}{c_0 \times \varepsilon_0}} = \sqrt{\frac{2 \times 100\ \text{W cm}^{-2} \times 10^4\ \text{cm}^2\ \text{m}^{-2}}{2.998 \times 10^8\ \text{m s}^{-1} \times 8.854 \times 10^{-12}\ \text{C}^2\text{J}^{-1}\text{m}^{-1}}}$$

$$= \sqrt{\frac{2}{2.998 \times 8.854}} \times 10^5 \sqrt{\frac{\text{A V s A V s}}{\text{m}^2\ (\text{A s})^2}} = 2.74 \times 10^4\ \frac{\text{V}}{\text{m}}$$

b) For a typical π-electron the distance from the nucleus, $d = 75$ pm, and a shielded nuclear charge of $q = 3.25(1.602 \times 10^{-19})$ C $= 5.21 \times 10^{-19}$ C, we find that

$$F_{\text{valence}} = \frac{1}{4\pi \times 8.854 \times 10^{-12}\ \text{C}^2\text{J}^{-1}\text{m}^{-1}} \frac{5.21 \times 10^{-19}\ \text{C}}{(75)^2 \times 10^{-24}\ \text{m}^2}$$
$$= \frac{5.21}{4\pi \times 8.854 \times (0.75)^2} \times 10^{13} \frac{\text{J}}{\text{C m}} = 8.3 \times 10^{11} \frac{\text{V}}{\text{m}}$$

c) The molecular field is 10 million times stronger than the optical field for these conditions. If one could increase the optical field to 1 GW cm^{-2}, then the two field strengths would be comparable.

P10.2 Calculate the electron–electron repulsion energy for S_1 and T_1, which are discussed in Section 10.2.4. You may use the MathCAD code which is provided, or you can write your own algorithm along the lines given in Foundation 10.2.

Solution

The electron–electron repulsion energies we calculate are 4.21×10^{-19} J for the singlet state (S_1) and 2.86×10^{-19} J for the triplet state. Details of the calculations can be found in the MathCAD worksheet "problem_10_2_solution." These values are in excellent agreement with the values reported in the text.

P10.3 Using the Einstein A coefficient (Equation (10B1.4)) and our expression for the oscillator strength (Equation (10B1.7)) calculate the lifetime for a $2p_z$ excited state of an H-atom, by assuming that it decays by the electronic transition for an electron in the $2p_z$ orbital of an H-atom to the $1s$ orbital of the H-atom.

Solution

Equation (10.23) is

$$A_{b\to a} = \frac{1}{\tau_0} = \frac{2\pi e^2 \nu_{ab}^2}{3m_e\varepsilon_0 c_0^3} f$$

and Equation (10.26) is

$$f = \frac{8\pi^2 m_e}{h^2 e^2} \cdot M^2 \cdot \Delta E$$

so that

$$\frac{1}{\tau_0} = \frac{2\pi e^2 \nu_{2p_z\to 1s}^2}{3m_e\varepsilon_0 c_0^3} \frac{8\pi^2 m_e}{h^2 e^2} \cdot M^2 \cdot \Delta E = \frac{16\pi^3 \nu_{2p_z\to 1s}^2}{3\varepsilon_0 c_0^3 h^2} \cdot M^2 \cdot \Delta E$$

Substituting $\Delta E = h\nu_{2p_z\to 1s}$, we find that

$$\frac{1}{\tau_0} = \frac{16\pi^3 \nu_{2p_z\to 1s}^3}{3\varepsilon_0 c_0^3 h} \cdot M^2$$

with

$$M = \int \psi_{1s} \cdot e \cdot z \cdot \psi_{2p_z} d\tau = e \cdot \int \psi_{1s} \cdot e \cdot z \cdot \psi_{2p_z} d\tau$$

Next, we substitute for the H-atom wavefunctions (see Chapter 4)

$$\psi_{1s} = \frac{1}{\sqrt{\pi a_0^3}} e^{-\frac{r}{a_0}} \qquad \psi_{2p_z} = \frac{1}{4\sqrt{2\pi a_0^3}} \frac{z}{a_0} e^{-\frac{r}{2a_0}}$$

and evaluate the integral

$$\int \psi_{1s} \cdot z \cdot \psi_{2p_z} d\tau = \frac{1}{4\pi a_0^4 \sqrt{2}} \int_0^\infty e^{-\frac{r}{a_0}} \cdot r\cos\theta \cdot r\cos\theta\, e^{-\frac{r}{2a_0}} \cdot r^2 \sin\theta dr d\theta d\varphi$$
$$= \frac{1}{4\pi a_0^4 \sqrt{2}} \int_0^\infty r^4 e^{-\frac{3r}{2a_0}} dr \cdot \int_0^\pi \cos^2\theta \cdot \sin\theta d\theta \cdot \int_0^{2\pi} d\varphi$$

Now we let $x = 3r/(2a_0)$, and find

$$\int \psi_{1s} \cdot z \cdot \psi_{2p_z} \mathrm{d}\tau = \frac{1}{4\pi a_0^4 \sqrt{2}} \left(\frac{2a_0}{3}\right)^5 \int_0^\infty x^4 \mathrm{e}^{-x} \mathrm{d}x \cdot \left| -\frac{\cos^3\theta}{3} \right|_0^\pi \cdot 2\pi$$
$$= \frac{1}{4\pi a_0^4 \sqrt{2}} \left(\frac{2a_0}{3}\right)^5 \int_0^\infty x^4 \mathrm{e}^{-x} \mathrm{d}x \cdot \frac{2}{3} \cdot 2\pi$$
$$= \frac{1}{4\pi a_0^4 \sqrt{2}} \left(\frac{2a_0}{3}\right)^5 \frac{4\pi}{3} \int_0^\infty x^4 \mathrm{e}^{-x} \mathrm{d}x$$
$$= \frac{2^5 a_0}{\sqrt{2} \cdot 3^6} \cdot 4! = \frac{2^8 a_0}{\sqrt{2} \cdot 3^5} = 0.745\, a_0$$

where we used the integral table in Appendix B. Now we substitute the definition of a_0, i.e., $a_0 = \varepsilon_0 h^2 / (\pi m_e e^2)$ and find

$$\frac{1}{\tau_0} = \frac{16\pi^3 e^2 \nu_{2p_z \to 1s}^3}{3\varepsilon_0 c_0^3 h} \cdot (0.745)^2 \frac{\varepsilon_0^2 h^4}{\pi^2 m_e^2 e^4}$$
$$= \frac{16\pi}{3} \cdot (0.745)^2 \cdot \frac{\varepsilon_0 \left(h\nu_{2p_z \to 1s}\right)^3}{m_e^2 e^2 c_0^3}$$

The product $h\nu_{2p_z \to 1s}$ is the energy for the transition and is given by

$$h\nu_{2p_z \to 1s} = E_{H-atom} \cdot \left(1 - \frac{1}{n^2}\right) = 21.80 \times 10^{-19}\ \mathrm{J} \cdot \left(1 - \frac{1}{4}\right) = 16.35 \times 10^{-19}\ \mathrm{J}$$

where we have used Equation (4.23) with $n = 2$. Hence we find the transition rate to be

$$\frac{1}{\tau_0} = \frac{16\pi}{3} \cdot (0.745)^2 \cdot \frac{\varepsilon_0}{m_e^2 e^2 c_0^3} \cdot 16.35 \times 10^{-19}\ \mathrm{J}$$
$$= 9.30 \cdot \frac{8.854 \times 10^{-12}\ \mathrm{C^2 J^{-1} m^{-1}} \cdot \left(16.35 \times 10^{-19}\ \mathrm{J}\right)^3}{\left(9.109 \times 10^{-31}\ \mathrm{kg}\right)^2 \cdot \left(1.602 \times 10^{-19}\ \mathrm{C}\right)^2 \cdot \left(2.998 \times 10^8\ \mathrm{m\ s^{-1}}\right)^3}$$
$$= 63.2 \times 10^7 \mathrm{s^{-1}} = 6.32 \times 10^8 \mathrm{s^{-1}}$$

so that the radiative lifetime is $\tau_0 = 1.58 \times 10^{-9}\ \mathrm{s} = 1.58$ ns. This value agrees with the experimental value.[1]

P10.4 Show that the transition moment integral between the singlet

$$\psi_{S_1} = \frac{1}{\sqrt{2}} \left[\phi_3(1) \cdot \phi_4(2) + \phi_3(2) \cdot \phi_4(1)\right]$$

and triplet states

$$\psi_{T_1} = \frac{1}{\sqrt{2}} \left[\phi_3(1) \cdot \phi_4(2) - \phi_3(2) \cdot \phi_4(1)\right]$$

given in Section 10.2 is zero; thus, the transition is "forbidden."

Solution

We calculate the transition moment in the x-direction

$$M_{x,S \to T} = e \iint \psi_{S_1}\ (x_1 + x_2)\ \psi_{T_1} \cdot \mathrm{d}\tau_1 \cdot \mathrm{d}\tau_2$$

where x_1 is the displacement of electron 1 and x_2 is the displacement of electron 2. The wavefunctions ψ_{S_1} and ψ_{T_1} considered in Section 10.2.4.1 are

$$\psi_{S_1} = \frac{1}{\sqrt{2}} \cdot \left[\phi_3(1) \cdot \phi_4(2) + \phi_3(2) \cdot \phi_4(1)\right]$$

1 W.S. Bickel, A.S. Goodman, *Phys. Rev.* 1966, **148**, 1.

and

$$\psi_{T_1} = \frac{1}{\sqrt{2}} \cdot [\phi_3(1) \cdot \phi_4(2) - \phi_3(2) \cdot \phi_4(1)]$$

Using these functions, the integrand I becomes

$$\begin{aligned} I = x_1 &\cdot [\phi_3(1) \cdot \phi_4(2) \cdot \phi_3(1) \cdot \phi_4(2) + \phi_3(2) \cdot \phi_4(1) \cdot \phi_3(1) \cdot \phi_4(2)] \\ -x_1 &\cdot [\phi_3(1) \cdot \phi_4(2) \cdot \phi_3(2) \cdot \phi_4(1) + \phi_3(2) \cdot \phi_4(1) \cdot \phi_3(2) \cdot \phi_4(1)] \\ +x_2 &\cdot [\phi_3(1) \cdot \phi_4(2) \cdot \phi_3(1) \cdot \phi_4(2) + \phi_3(2) \cdot \phi_4(1) \cdot \phi_3(1) \cdot \phi_4(2)] \\ -x_2 &\cdot [\phi_3(1) \cdot \phi_4(2) \cdot \phi_3(2) \cdot \phi_4(1) + \phi_3(2) \cdot \phi_4(1) \cdot \phi_3(2) \cdot \phi_4(1)] \end{aligned}$$

or

$$\begin{aligned} I = x_1 &\cdot [\phi_3^2(1) \cdot \phi_4^2(2) + \phi_3(1) \cdot \phi_4(1) \cdot \phi_3(2) \cdot \phi_4(2)] \\ -x_1 &\cdot [\phi_3(1) \cdot \phi_4(1) \cdot \phi_3(2) \cdot \phi_4(2) + \phi_3^2(2) \cdot \phi_4^2(1)] \\ +x_2 &\cdot [\phi_3^2(1) \cdot \phi_4^2(2) + \phi_3(1) \cdot \phi_4(1) \cdot \phi_3(2) \cdot \phi_4(2)] \\ -x_2 &\cdot [\phi_3(1) \cdot \phi_4(1) \cdot \phi_3(2) \cdot \phi_4(2) + \phi_3^2(2) \cdot \phi_4^2(1)] \end{aligned}$$

This expression simplifies to

$$\begin{aligned} I = x_1 &\cdot [\phi_3^2(1) \cdot \phi_4^2(2) - \phi_3^2(2) \cdot \phi_4^2(1)] \\ +x_2 &\cdot [\phi_3^2(1) \cdot \phi_4^2(2) - \phi_3^2(2) \cdot \phi_4^2(1)] \end{aligned}$$

Thus for the transition moment we obtain

$$\begin{aligned} M_{x,S\to T} = e &\iint x_1\phi_3^2(1) \cdot \phi_4^2(2) \cdot d\tau_1 \cdot d\tau_2 \\ -e &\iint x_1\phi_3^2(2) \cdot \phi_4^2(1) \cdot d\tau_1 \cdot d\tau_2 \\ +e &\iint x_2\phi_3^2(1) \cdot \phi_4^2(2) \cdot d\tau_1 \cdot d\tau_2 \\ -e &\iint x_2\phi_3^2(2) \cdot \phi_4^2(1) \cdot d\tau_1 \cdot d\tau_2 \end{aligned}$$

Consider the first term, which gives

$$\begin{aligned} &e \iint x_1\phi_3^2(1) \cdot \phi_4^2(2) \cdot d\tau_1 \cdot d\tau_2 \\ &= e \int \phi_4^2(2) \left[\int x_1\phi_3^2(1) \cdot d\tau_1\right] \cdot d\tau_2 \\ &= e \left[\int x_1\phi_3^2(1) \cdot d\tau_1\right] \left[\int \phi_4^2(2) \cdot d\tau_2\right] = 0 \end{aligned}$$

because the integrand in the left integral is antisymmetric with respect to x_1 and the right integral equals one. The same arguments hold for the remaining three terms; thus, the transition moment equals zero.

P10.5 Consider a two-state system (levels a and b) that is in thermal equilibrium with a radiation field having a spectral energy density of $\rho_\nu(\nu_{ab})$ and undergoes the three transitions absorption, stimulated emission, and spontaneous emission (see Fig. P10.5).

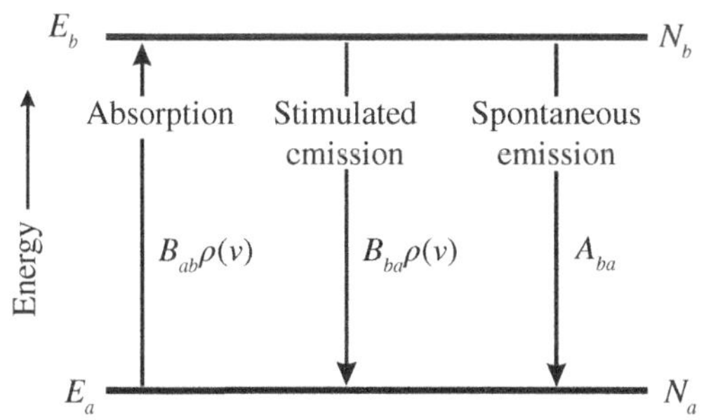

Figure P10.5 Absorption, spontaneous and stimulated emission in a two-level energy system with particle numbers N_a and $N_b = N_a \exp[-h\nu/(kT)]$.

First, show that the equilibrium condition implies that

$$\rho\left(\nu_{ab}\right) = \frac{A_{ba}}{\left(B_{ab} \cdot \exp\left(h\nu_{ab}/kT\right) - B_{ba}\right)}$$

Second, require that $\rho(\nu)$ be equal to the Planck distribution law $(\rho(\nu) = \left(8\pi\nu^2/c^3\right)(h\nu/(\exp(h\nu/kT) - 1)))$ and $B_{ab} = B_{ba}$, and then show that this leads to

$$A_{ba} = \frac{8\pi h\nu_{ab}^3}{c^3} B_{ba}$$

Solution

Part 1. At equilibrium the rates from state a to b and those from state b to a must be the same; thus, we can write that

$$\left[B_{ba}\ \rho(\nu) + A_{ba}\right] N_b = B_{ab}\ \rho(\nu) N_a$$

Also the thermal equilibrium implies that

$$\frac{N_b}{N_a} = \exp(-\Delta E/kT) = \exp\left(-h\nu_{ab}/kT\right)$$

Inserting these expressions for the populations into the equation for the rates, we find

$$\left[B_{ba}\ \rho(\nu) + A_{ba}\right] N_a \cdot \exp\left(-h\nu_{ab}/kT\right) = B_{ab}\ \rho(\nu) N_a$$

Next we collect the terms with $\rho(\nu)$ together, and write

$$B_{ab}\ \rho(\nu) - B_{ba}\ \rho(\nu) \exp\left(-h\nu_{ab}/kT\right) = A_{ba} \exp\left(-h\nu_{ab}/kT\right)$$

or

$$B_{ab}\ \rho(\nu) \exp\left(h\nu_{ab}/kT\right) - B_{ba}\ \rho(\nu) = A_{ba}$$

Isolating and solving for $\rho(\nu)$

$$\rho\left(\nu_{ab}\right) = \frac{A_{ba}}{\left(B_{ab} \cdot \exp\left(h\nu_{ab}/kT\right) - B_{ba}\right)}$$

Part 2. If we require that

$$\frac{A_{ba}}{\left(B_{ab} \cdot \exp\left(h\nu_{ab}/kT\right) - B_{ba}\right)} = \frac{8\pi h\nu^3}{c_0^3} \frac{1}{(\exp(h\nu/kT) - 1)}$$

and that $B_{ab} = B_{ba}$, we have that

$$\frac{A_{ba}}{B_{ba}\left(\exp\left(h\nu_{ab}/kT\right) - 1\right)} = \frac{8\pi h\nu^3}{c_0^3} \frac{1}{(\exp(h\nu/kT) - 1)},$$

which simplifies to

$$\frac{A_{ba}}{B_{ba}} = \left(\frac{8\pi h\nu^3}{c_0^3}\right) \text{ or } A_{ba} = \left(\frac{8\pi h\nu^3}{c_0^3}\right) B_{ba}$$

11

Nuclei: Particle and Wave Properties

11.1 Exercises

E11.1 Calculate the center of mass position in an HCl molecule, an HF molecule, and an H_2 molecule. Comment on the trend.

Solution

The center of mass is the balance point for the diatomic molecule, as illustrated in Fig. E11.1 and

$$r_1 = \frac{m_2 d}{m_1 + m_2}$$

where r_1 is the distance from the H-atom nucleus (proton). The center of mass for HCl is

$$r_1 = \frac{m_2 d}{m_1 + m_2} = \frac{35.45 \times 127 \text{ pm}}{1.008 + 35.4527} = 1.23 \times 10^{-10} \text{ m} = 123 \text{ pm}$$

the center of mass for HF is

$$r_1 = \frac{18.99 \times 92 \text{ pm}}{1.008 + 18.9984} = 8.7 \times 10^{-11} \text{ m} = 87 \text{ pm}$$

and the center of mass for H_2 is

$$r_1 = \frac{1.00794 \times 74 \text{ pm}}{1.00794 + 1.00794} = 3.7 \times 10^{-11} \text{ m} = 37 \text{ pm}$$

As the mass of the H-atom's bonding partner increases the center of mass shifts away from the H-atom nucleus; hence, the center of mass is only 4 pm from the Cl nucleus in HCl and 5 pm from the F nucleus in HF, whereas it is in the middle of the H_2 bond.

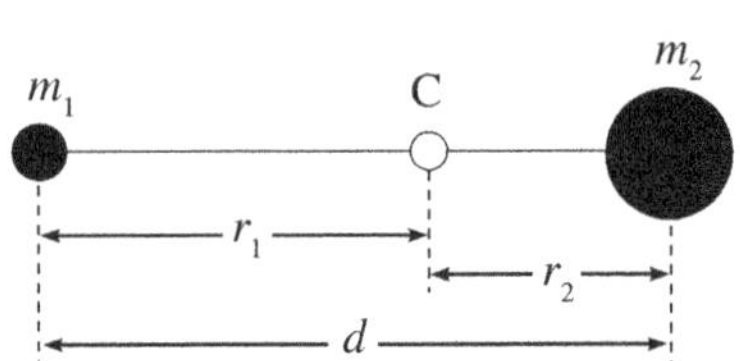

Figure E11.1 Calculation of center of mass.

Note: For the general case of a molecule with N nuclei, the center of mass can be found from the condition that

$$0 = \sum_{i=1}^{N} m_i \cdot \vec{r_i}$$

where m_i is the mass of nucleus i and $\vec{r_i}$ is the displacement of nucleus i from the center of mass.

Solutions Manual for Principles of Physical Chemistry, Third Edition. Edited by Hans Kuhn, David H. Waldeck, and Horst-Dieter Förstering.

E11.2 Given that the rotational constant of H_2 is $B = 1.209 \times 10^{-21}$ J, calculate the bond length of the molecule. If you use the mass of a hydrogen atom rather than the reduced mass, how large is the error in your calculation?

Solution

The rotational constant is

$$B = \frac{h^2}{8\pi^2 \mu_e d^2}$$

For the case of H_2, the reduced mass μ is given by

$$\mu = \frac{m_H}{2} = \frac{(1.008 \text{ g mol}^{-1})}{2 \times (6.022 \times 10^{23} \text{ mol}^{-1})} = 8.369 \times 10^{-25} \text{ g}$$

Using $\mu_e = \mu$ allows the calculation of d as

$$d = \sqrt{\frac{h^2}{8\pi^2 \; \mu \; B}}$$
$$= \sqrt{\frac{(6.626 \times 10^{-34} \text{ J·s})^2}{8\pi^2(8.369 \times 10^{-28} \text{ kg})(1.209 \times 10^{-21}\text{J})}}$$
$$= 7.441 \times 10^{-11} \; \mu_e = 74.4 \text{ pm}$$

If we use the mass of the H atom, m_H, rather than the reduced mass μ, then we find that

$$d = \sqrt{\frac{h^?}{8\pi^2 \; m_H \; B}}$$
$$= \sqrt{\frac{(6.626 \times 10^{-34} \text{ J·s})^2}{8\pi^2(1.6738 \times 10^{-27} \text{ kg})(1.209 \times 10^{-21}\text{J})}}$$
$$= 5.262 \times 10^{-11} \text{ m} = 52.62 \text{ pm}$$

A significant error is introduced by using the mass of a hydrogen atom, rather than the reduced mass.

E11.3 Given that the rotational constant of $^{12}C^{16}O$ is $B = 3.84 \times 10^{-23}$ J, calculate the bond length of the molecule. Compare your result to that found in Example 11.9.

Solution

The rotational constant is given by the expression

$$B = \frac{h^2}{8\pi^2 m \cdot d^2}$$

For CO the reduced mass is given by

$$m = \frac{m_C m_O}{m_C + m_O} = \frac{(12.000 \text{ g mol}^{-1})(15.995 \text{ g mol}^{-1})}{(12.000 \text{ g mol}^{-1} + 15.995 \text{ g mol}^{-1})(6.022 \times 10^{23} \text{ mol}^{-1})}$$
$$= 1.138 \times 10^{-26} \text{ kg}$$

This allows the calculation of d as

$$d = \sqrt{\frac{h^2}{8\pi^2 \; m \; B}}$$
$$= \sqrt{\frac{(6.626 \times 10^{-34} \text{ J·s})^2}{8\pi^2(1.138 \times 10^{-26} \text{ kg})(3.84 \times 10^{-23}\text{J})}}$$
$$= 1.13 \times 10^{-10} \text{ m} = 113 \text{ pm}$$

This result is identical to that in Example 11.8 within significant figures.

E11.4 Show that the wavefunction ψ_0 in Equation (11.26) satisfies the Schrödinger equation for the harmonic oscillator Equation (11.24).

Solution

We are asked to show that

$$\psi_0 = \sqrt[4]{\frac{\alpha}{\pi}} \cdot \exp\left(-\alpha x^2/2\right)$$

satisfies

$$-\frac{h^2}{8\pi^2\mu} \cdot \frac{d^2\psi}{dx^2} + \frac{1}{2}k_f\ x^2\ \psi = E\psi$$

We begin by writing the first and second derivatives of the wavefunction with respect to x, namely

$$\frac{d\psi_0}{dx} = \sqrt[4]{\frac{\alpha}{\pi}} \cdot (-\alpha x) \cdot \exp\left(-\alpha x^2/2\right)$$

and

$$\frac{d^2\psi_0}{dx^2} = \sqrt[4]{\frac{\alpha}{\pi}} \cdot (-\alpha) \cdot \exp\left(-\alpha x^2/2\right) \cdot [x(-\alpha x) + 1]$$
$$= \left[(\alpha^2 x^2) - \alpha\right]\ \psi_0$$

Upon substitution into the Schrödinger equation, we find that

$$-\frac{h^2}{8\pi^2\mu} \cdot \left[\alpha^2 x^2 - \alpha\right]\ \psi_0 + \frac{1}{2}k_f x^2\ \psi_0 = E\psi_0$$

By inspection, we obtain

$$E = -\frac{h^2}{8\pi^2\mu} \cdot \left[\alpha^2 x^2 - \alpha\right] + \frac{1}{2}k_f\ x^2 \tag{11E4.1}$$

First we rearrange Equation (11E4.1):

$$E = -\frac{h^2}{8\pi^2\mu} \cdot \alpha^2 x^2 + \frac{1}{2}k_f\ x^2 + \frac{h^2}{8\pi^2\mu}\alpha$$
$$= x^2\left[-\frac{h^2}{8\pi^2\mu} \cdot \alpha^2 + \frac{1}{2}k_f\right] + \frac{h^2}{8\pi^2\mu}\alpha \tag{11E4.2}$$

The expression on the right-hand side must be a constant because E is a constant.This is only possible if the expression in brackets becomes zero:

$$-\frac{h^2}{8\pi^2\mu} \cdot \alpha^2 + \frac{1}{2}k_f = 0 \tag{11E4.3}$$

Then the Equation (11E4.2) reduces to

$$E = \frac{h^2}{8\pi^2\mu}\alpha \tag{11E4.4}$$

From Equation (11E4.3) we calculate α

$$\alpha^2 = \frac{1}{2}k_f\frac{8\pi^2\mu}{h^2} \quad \text{or} \quad \alpha = \sqrt{\frac{1}{2}k_f\frac{8\pi^2\mu}{h^2}} = \frac{2\pi}{h}\sqrt{\mu k_f}$$

Using this value we obtain the energy E as

$$E = \frac{h^2}{8\pi^2\mu}\alpha = \frac{h^2}{8\pi^2\mu}\frac{2\pi}{h}\sqrt{\mu k_f}$$
$$= \frac{h}{4\pi}\sqrt{\frac{k_f}{\mu}} = \frac{1}{2}h\frac{1}{2\pi}\sqrt{\frac{k_f}{\mu}} = \frac{1}{2}h\nu_0$$

E11.5 Calculate the energy difference between the H^{35} Cl and the H^{37}Cl line in the high-resolution infrared spectrum of HCl (Fig. 11.16). Do you expect the energy spacing to change with the J quantum number?

Solution

The main influence of the isotopes on the transition is a change of the pure vibrational transition energy; namely $\Delta E^{35} = h\nu_0^{35}$ versus $\Delta E^{37} = h\nu_0^{37}$. The energy shift $\Delta\Delta E = (\Delta E^{35} - \Delta E^{37})$ is

$$\Delta\Delta E = h\nu_0^{35} - h\nu_0^{37} = \frac{h\sqrt{k_f}}{2\pi} \cdot \left(\frac{1}{\sqrt{\mu^{35}}} - \frac{1}{\sqrt{\mu^{37}}}\right)$$

With

$$\mu^{35} = \frac{m_H \cdot m_{^{35}Cl}}{m_H + m_{^{35}Cl}} = 1.627 \times 10^{-27} \text{ kg}$$

and

$$\mu^{37} = \frac{m_H \cdot m_{^{37}Cl}}{m_H + m_{^{37}Cl}} = 1.630 \times 10^{-27} \text{ kg}$$

we obtain $\Delta\Delta E = 5.3 \times 10^{-23}$ J, which corresponds to a shift of the center frequency by $\tilde{\nu} = 2.7$ cm^{-1} in accordance with the experiment (Fig. 11.16). Note that in this figure all ^{37}Cl lines are approximately equally shifted to smaller wavenumbers.

Correspondingly, the calculated shift associated with the difference in rotational energy is, according to Equation (11.13)

$$\begin{aligned}\Delta\Delta E_{rot} &= 2B^{35}(J+1) - 2B^{37}(J+1) = 2(J+1) \cdot (B^{35} - B^{37}) \\ &= 2(J+1) \cdot \frac{h^2}{8\pi^2 d^2}\left(\frac{1}{\mu^{35}} - \frac{1}{\mu^{37}}\right) \\ &= (6.3 \times 10^{-25}) \cdot 2(J+1)\end{aligned}$$

The shift does increase as J increases. For $J = 0$ the shift is only 1% of the shift calculated for the vibrational transition; even for $J = 10$ it is only 10%. Hence the most important contribution to the shift arises from the shift in the vibrational transition energy.

E11.6 Use the data given for O_2 and its ions to determine the force constants of their bonds, in the harmonic oscillator approximation. Explain the trend in the value of the force constant in terms of the bonding in the different species.

Molecular Species	$^{16}O_2$	$^{16}O_2^-$	$^{16}O_2^+$
Vibrational Frequency $\nu_0/10^{13}$ s^{-1}	4.74	3.27	5.71

Solution

For all these molecules, we need the reduced mass of the O-O pair. It is

$$\begin{aligned}\mu &= \frac{m_1 m_2}{m_1 + m_2} = \frac{m^2}{2m} = \frac{1}{2}m = \frac{1}{2}\frac{16 \times 10^{-3} \text{ kg mol}^{-1}}{6.022 \times 10^{23} \text{ mol}^{-1}} \\ &= 1.328 \times 10^{-26} \text{ kg}\end{aligned}$$

The frequency and force constant are related by

$$\nu_0 = \frac{1}{2\pi}\sqrt{\frac{k_f}{\mu}} \text{ or } k_f = (\nu_0 2\pi)^2 \mu$$

For O_2 we find that

$$k_f = [4.74 \times 10^{13}\text{s}^{-1} \cdot 2\pi]^2 \cdot 1.328 \times 10^{-26} \text{ kg} = 1178 \text{ N m}^{-1};$$

for O_2^-, we find that

$$k_f = 560.6 \text{ N m}^{-1};$$

and for O_2^+, we find that

$$k_f = 1709 \text{ N m}^{-1}$$

The trend in the force constant: O_2^+ the highest, O_2 intermediate, and O_2^- the smallest, follows the trend in bond order (BO): O_2^- (BO = 1.5) < O_2 (BO = 2) < O_2^+ (BO = 2.5), found from a molecular orbital energy diagram.

E11.7 The rotational period of a diatomic molecule is the time it takes for it to rotate through an angle of 2π; i.e., one full revolution. Consider the two diatomic molecules in the table below and the data associated with them. What is the rotational period of these two molecules, when they have an angular momentum of $\sqrt{2}\hbar$? What is the rotational period when they have an angular momentum of $\sqrt{110}\hbar$?

Molecule	$^{1}H_2$	$^{127}I_2$
I/kg m^2	4.59×10^{-48}	7.54×10^{-45}

Solution

The classical angular momentum L and moment of inertia I are related by $L = \omega I = 2\pi\nu I$, in which ν is the rotational frequency. Because the rotational period is $1/\nu$ we find that

$$\text{rotational period} = \frac{2\pi I}{L}$$

For the four cases considered here, the rotational period is

	$^{1}H_2$	$^{127}I_2$
$\sqrt{2}\hbar$	1.934×10^{-13} s	3.139×10^{-10} s
$\sqrt{110}\hbar$	2.608×10^{-14} s	4.283×10^{-11} s

Because the H_2 molecule is so light it has a very short rotational period (for a given angular momentum), and it ranges from 0.2 ps to 0.026 ps. In contrast, I_2 is quite massive and has a relatively long rotational period (for a given angular momentum) ranging from 314 ps to 43 ps.

E11.8 Consider the largest value of the rotational period found in E11.7. How many revolutions does this molecule make in one second? If we consider that human history has only been recorded for 5000 years, how many seconds in recorded history? Are there more seconds in recorded history, or are there more revolutions of a molecule in one second?

Solution

The largest value of the rotation period found in the previous problem is 3.139×10^{-10} seconds. The I_2 molecule would make $1/3.139 \times 10^{-10} = 3.185 \times 10^{9}$ rotations in one second. There are

$$5000\ \text{year} \cdot \frac{365.25\ \text{day}}{1\ \text{year}} \cdot \frac{24\ \text{hour}}{1\ \text{day}} \cdot \frac{3600\ \text{s}}{1\ \text{hr}} = 1.578 \times 10^{11}\ \text{s}$$

in recorded history. Thus there are about 50 times more seconds in recorded history.

Note: If you perform the same calculation for H_2, you find that there are more rotations of the molecule per second than there are seconds in recorded history.

E11.9 Consider the two potential energy curves (A and B) in Fig. E11.9 for a diatomic molecule.

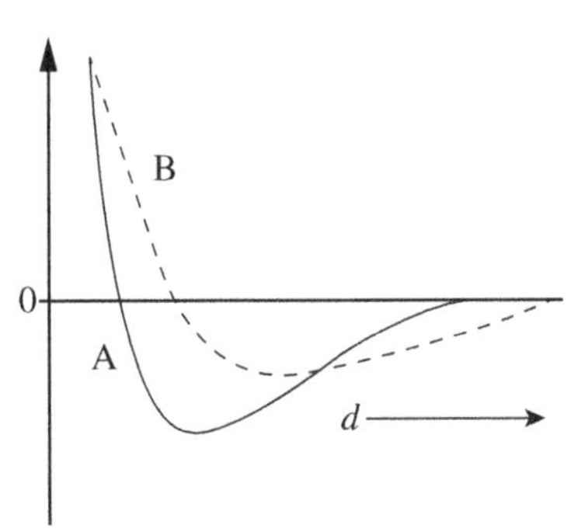

Figure E11.9 Potential energy curves. Energy V versus bond length d.

(Assume the same atomic nuclei for both curves). Which curve corresponds to the larger internuclear distance? Which curve corresponds to the higher moment of inertia? Which curve has a higher binding energy? Which curve has a higher vibrational energy level spacing? Which curve has a higher rotational energy level spacing?

Solution

We take the curves A and B to represent different electronic states of the same molecule, so that the reduced mass of the molecule does not change.

The equilibrium internuclear distance for curve B is larger than that for curve A. Taking the reduced mass to be the same would imply that the moment of inertia for curve B would be larger than for curve A.

Since the potential energy for large bond lengths is identical, the deeper potential well (curve A) has the greater binding energy.

The molecule with the deeper and narrower potential well (curve A) has the higher vibrational frequency and the larger vibrational energy spacing.

The molecule with the larger moment of inertia (curve B) will have the smaller rotational energy level spacing because of the inverse relationship between energy level spacing and moment of inertia for rotation.

E11.10 Given that the vibrational wavenumber of H_2 is $\tilde{\nu}_{0,H_2} = 4409\ \text{cm}^{-1}$, determine the vibrational frequencies of HD and D_2.

Solution

We can relate the vibrational wavenumber to the force constant and reduced mass by

$$\tilde{\nu}_{0,H_2} = \frac{1}{2\pi c_0}\sqrt{\frac{k_f}{\mu_{\text{H-H}}}}$$

Thus the ratio of the frequencies for H_2 and HD is

$$\frac{\tilde{\nu}_{0,\text{HD}}}{\tilde{\nu}_{0,H_2}} = \frac{\frac{1}{2\pi c_0}\sqrt{k_f/\mu_{\text{H-D}}}}{\frac{1}{2\pi c_0}\sqrt{k_f/\mu_{\text{H-H}}}} = \sqrt{\frac{\mu_{\text{H-H}}}{\mu_{\text{H-D}}}}$$

Similarly, we find that

$$\frac{\tilde{\nu}_{0,D_2}}{\tilde{\nu}_{0,H_2}} = \frac{\frac{1}{2\pi c_0}\sqrt{k_f/\mu_{\text{D-D}}}}{\frac{1}{2\pi c_0}\sqrt{k_f/\mu_{\text{H-H}}}} = \sqrt{\frac{\mu_{\text{H-H}}}{\mu_{\text{D-D}}}}$$

To perform calculations we need the reduced masses of H_2, HD, and D_2 which are

$$\mu_{\text{H-H}} = \frac{m_H m_H}{m_H + m_H} = \frac{(1.0078)\,(1.0078)}{1.0078 + 1.0078} \cdot \frac{10^{-3}\ \text{kg mol}^{-1}}{6.022\times\ 10^{23}\ \text{mol}^{-1}} = 8.368\times 10^{-28}\ \text{kg}$$

$$\mu_{\text{H-D}} = \frac{m_H m_D}{m_H + m_D} = \frac{(1.0078)\,(2.0141)}{1.0078 + 2.0141} \cdot \frac{10^{-3}\ \text{kg mol}^{-1}}{6.022\times\ 10^{23}\ \text{mol}^{-1}} = 1.115\times 10^{-27}\ \text{kg}$$

and

$$\mu_{\text{D-D}} = \frac{m_D m_D}{m_D + m_D} = \frac{(2.0141)\,(2.0141)}{2.0141 + 2.0141} \cdot \frac{10^{-3}\ \text{kg mol}^{-1}}{6.022\times\ 10^{23}\ \text{mol}^{-1}} = 1.672\times\ 10^{-27}\ \text{kg}$$

Now we can calculate the wavenumbers $\tilde{\nu}_0$ of HD and D_2, and find that

$$\tilde{\nu}_{0,\text{HD}} = \tilde{\nu}_{0,H_2}\sqrt{\frac{\mu_{\text{H-H}}}{\mu_{\text{H-D}}}} = 4409\ \text{cm}^{-1}\sqrt{\frac{8.368\times 10^{-28}}{1.115\times 10^{-27}}} = 3820\ \text{cm}^{-1}$$

and

$$\tilde{\nu}_{0,D_2} = \tilde{\nu}_{0,H_2}\sqrt{\frac{\mu_{\text{H-H}}}{\mu_{\text{D-D}}}} = 4409\ cm^{-1}\sqrt{\frac{8.368\times 10^{-28}}{1.672\times 10^{-27}}} = 3119\ \text{cm}^{-1}$$

E11.11 Plot the data in Table 11.1 and determine the best value of the parameter B using Equation (11.15). Use this B value to determine the HCl molecule's bond length. Compare your result with a value obtained from the literature.

Solution

We start with Equation (11.15)

$$E = B \cdot J(J+1) - \frac{2B^2}{k_f \cdot d_0^2} J^2(J+1)^2$$

Then we can calculate ΔE as

$$\Delta E = B[(J+1)(J+2) - J(J+1)] - \frac{2B^2}{k_f \cdot d_0^2}\left[(J+1)^2(J+2)^2 - J^2(J+1)^2\right]$$

which simplifies to

$$\Delta E = B(J+1)[(J+2) - J] - \frac{2B^2}{k_f \cdot d_0^2}(J+1)^2\left[(J+2)^2 - J^2\right]$$

or

$$\Delta E = 2B(J+1) - \frac{8B^2}{k_f \cdot d_0^2}(J+1)^3$$

If we rearrange this expression to write

$$\frac{\Delta E}{(J+1)} = 2B - \frac{8B^2}{k_f \cdot d_0^2}(J+1)^2$$

we see that a plot of $\Delta E/(J+1)$ versus $(J+1)^2$ should be a straight line and have an intercept of $2B$ and a slope of

$$-\frac{8B^2}{k_f \cdot d_0^2}$$

This plot is shown in Fig. E11.11; circles are the data and the line shows the best fit to a line that has an intercept of 41.46×10^{-23} J and a slope of -4.22×10^{-26} J. The intercept implies that $B = 20.73 \times 10^{-23}$ J or 10.43 cm^{-1}, which is in excellent agreement with the value found for $^1H^{35}Cl$ in the lowest vibrational state from the literature. Using this value of B we obtain a value of 128 pm for the bond length.

J	$1/\lambda$ (cm^{-1})	ΔE (10^{-23} J)	$\Delta(\Delta E)$ (10^{-23} J)
0	(20.93)	(41.58)	
1	41.78	82.99	41.4
2	62.60	124.3	41.3
3	83.35	165.6	41.3
4	104.02	206.6	41.0
5	124.74	247.8	41.2
6	145.36	288.7	40.9
7	165.83	329.4	40.7
8	186.25	370.0	40.6
9	206.54	410.3	40.3
10	226.71	450.3	40.0
11	246.82	490.3	40.0
12	266.70	529.8	39.5
13	286.50	569.1	39.3

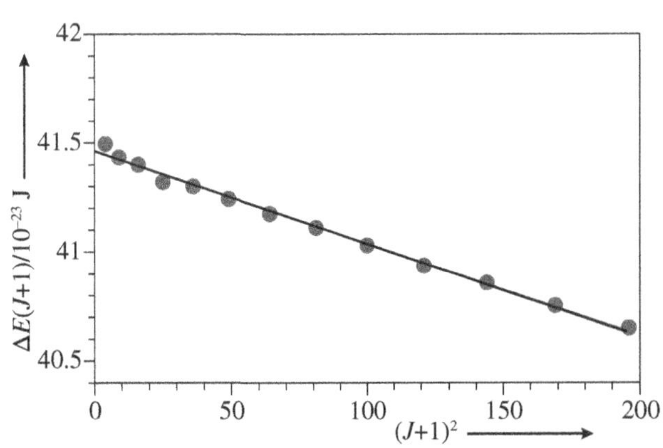

Figure E11.11 The plot shown here ($\Delta E/(J+1)$ versus $(J+1)^2$) allows the rotational constant B to be determined from the intercept.

E11.12 Using a force constant of 516 N m^{-1} calculate the vibrational frequencies for different isotopes of HCl, namely $D^{37}Cl$, $H^{37}Cl$, and $H^{35}Cl$. Comment on your findings.

Solution

In order to calculate the frequencies from the force constants, we need the values of the reduced mass for each of the isotopic variants. The reduced mass is defined as

$$\mu = \frac{m_1 m_2}{m_1 + m_2}$$

so that for $^1H^{35}Cl$ we find

$$\mu(^1H^{35}Cl) = \frac{1.0078 \times 34.965}{1.0078 + 34.965} \cdot \frac{10^{-3}\ \text{kg mol}^{-1}}{6.0225 \times 10^{23}\ \text{mol}^{-1}}$$
$$= 1.627 \times 10^{-27}\ \text{kg}$$

Similarly for the other isotopic variants, we find $\mu(^1H^{37}Cl) = 1.629 \times 10^{-27}$ kg; $\mu(^2H^{35}Cl) = 3.162 \times 10^{-27}$ kg; and $\mu(^2H^{37}Cl) = 3.172 \times 10^{-27}$ kg.

The relationship between force constant k_f, reduced mass μ, and frequency ν_0 is

$$\nu_0 = \frac{1}{2\pi}\sqrt{\frac{k_f}{\mu}}$$

Thus for $^1H^{35}Cl$, we find

$$\nu_0 = \frac{1}{2\pi}\sqrt{\frac{k_f}{\mu}} = \frac{1}{2\pi}\sqrt{\frac{516\ \text{N m}^{-1}}{1.627 \times 10^{-27}\ \text{kg}}} = 8.963 \times 10^{13}\ \text{s}^{-1}$$

and for the other isotopomers we find $\nu_0(^1H^{37}Cl) = 8.957 \times 10^{13}\text{s}^{-1}$; $\nu_0(^2H^{35}Cl) = 6.429 \times 10^{13}\text{s}^{-1}$; and $\nu_0(^2H^{37}Cl) = 6.419 \times 10^{13}\text{s}^{-1}$.

The frequencies only differ because of the reduced mass terms, with the ^{35}Cl and ^{37}Cl variants having only a small change in the reduced mass, and thus the frequency. The frequency for the ^2H variants is approximately 1.4 times smaller because of the larger change in the reduced mass.

E11.13 In Fig. 11.35 identify the P, Q, and R branches for the rotational fine structure in the infrared spectrum of methane.

Solution

The infrared spectrum for methane is shown in the top panel of Fig. 11.35. Peaks corresponding to the ν_3 mode are seen near 3000 cm^{-1}. The Q ($\Delta J = 0$) branch is seen as the strong single line at around 3000 cm^{-1}. The R ($\Delta J = 1$) branch extends to higher frequency from the narrow Q branch line, while the P ($\Delta J = -1$) branch extends to lower energies. A similar structure is seen, albeit less clearly for the ν_4 mode near 1300 cm^{-1}.

E11.14 Energy order the following spectroscopic transitions for the molecule CO: vibrational excitation of the CO bond, electronic excitation of CO (from HOMO to LUMO), rotational transition of CO, and photoionization of CO. Also, state the order of magnitude of the energy in each case.

Solution

The table lists the characteristic energies from lowest (at top) to highest (at bottom).

	Wavenumber	Energy
Rotational transition for CO	≈ 2 cm^{-1}	≈ 4×10^{-23} J
Vibrational transition for CO	≈ 2170 cm^{-1}	≈ 4.31×10^{-20} J
Electronic excitation of CO	≈ 49,000 cm^{-1}	≈ 9.7×10^{-19} J
Photoionization of CO	≈ 113,000 cm^{-1}	≈ 2.2×10^{-18} J

E11.15 Show that the kinetic energy of a classical oscillator is at a maximum when its potential energy is at a minimum and vice versa (see Box 11.1).

Solution

Box 11.1 gives the kinetic and potential energies of the classical oscillator to be

$$E_{kin} = \frac{A^2}{2} k_f \cos^2(2\pi \nu_0 t)$$

and

$$E_{pot} = \frac{A^2}{2} k_f \sin^2(2\pi \nu_0 t)$$

A plot of these two functions immediately shows that the kinetic energy and the potential energy are out of phase. Figure E11.15 shows the kinetic energy as the light gray curve and the potential energy as the dark gray curve. In the graph the energy has been rescaled so that it varies between 0 and 1 and the time is given in terms of the oscillation period $T = 1/\nu_0$.

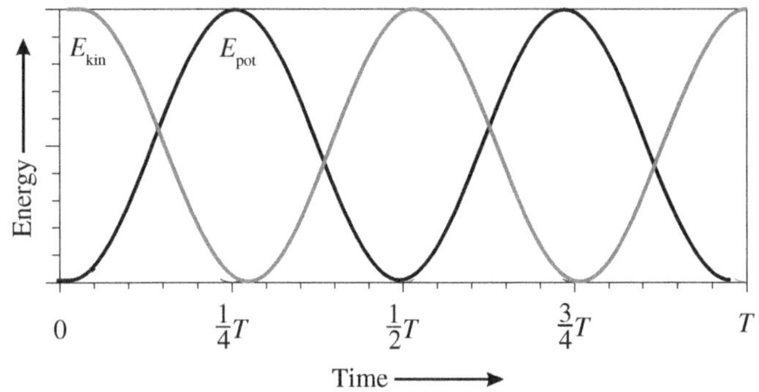

Figure E11.15 Potential (dark gray) and kinetic (light gray) energies of the harmonic oscillator.

For example, at $t = T/4$ the cosine function has a maximum, whereas the sine function is zero; vice versa, at $t = T/2$ the cosine function is zero, whereas the sine function has a maximum. Note that the sum of E_{pot} and E_{kin} is constant for all values of t, namely, $E_{pot} + E_{kin} = A^2\, k_f/2$.

E11.16 Use Fig. 11.14 to provide a physical explanation for how an infrared light wave excites the vibrational motion of a chemical bond.

Solution

As Fig. 11.14 indicates the applied electric field of the light couples with the changing dipole moment of the molecule for vibrational excitation; represented in the diagram by two differently signed charges on the ends of a spring. The applied electric field drives the separation and compression of the charges, depending upon its sign. Using the diagram, when the negative pole of the oscillating electric field is up and the positive pole is down, it stretches the spring because the positive charge is pulled upward and the negative charge is pulled downward. When the electric field is oriented in the other direction (positive pole up and negative pole down) it compresses the spring, pushing the positive charge down and the negative charge upward. When the oscillation period matches the natural period of motion for the spring, the system couples most strongly, i.e., the resonance condition. For fundamental vibrations, this condition is met by an infrared light wave.

E11.17 Explain why the bending motion and the asymmetric stretching motion of the CO_2 molecule generate a dipole moment but the symmetric stretching motion does not.

Solution

With its collinear geometry and equal C–O bond lengths, the carbon dioxide molecule in its ground state is symmetrical. Even though each of the C–O bonds has an asymmetric charge distribution (dipole moment), they are equal and oppositely oriented to give a net dipole moment of zero. Upon excitation of the bending mode, the collinear geometry of the molecule is broken as it vibrates, and the dipole moments no longer cancel each other. This gives rise to a time-dependent dipole moment that lies in the plane of the molecule and is oriented along the bisector of the O–C–O bond angle, which is also changing as the molecule vibrates. For the case of the asymmetric stretch motion, the dipole moment changes because unequal bond lengths occur as the molecule vibrates. These unequal bond lengths give rise to a time-dependent change in the effective dipole moments oriented along each C–O bond, giving rise to a net dipole moment that changes as the molecule vibrates. On the other hand, the symmetric stretching motion does not affect the molecular dipole moment. For this vibrational mode the C–O bond lengths are changing in phase with each other and by the same amounts so that the net dipole moment of the molecule is always zero.

One can also explain these facts from a symmetry argument. It is impossible for a molecule to have a dipole moment if it has a center of symmetry. The equilibrium geometry of the CO_2 molecule has a center of symmetry and its symmetrically stretched excited states maintain that center of symmetry. The bending and/or asymmetrically stretched molecule does not have a center of symmetry, so that it can have a dipole moment. Hence infrared light can excite the fundamental transitions from the ground state to either the bending or asymmetric stretch, but not the symmetric stretch.

E11.18 Compare the vibrational wavefunctions for a Morse potential to those for a harmonic oscillator potential by exploring the MathCAD exercise morse.mcdx.

Solution

Remember that the Morse potential is, according to Equation (11.40)

$$V = \varepsilon_B(1 - \exp(-ax))^2 \quad \text{with } a = \sqrt{\frac{k_f}{2\varepsilon_B}}$$

This curve is displayed in Fig. E11.18a for the parameters $k_f = 516$ N m^{-1}, $\varepsilon_B = 8.58 \times 10^{-19}$ J (these are the data for HCl), and $\varepsilon_B = 4.29 \times 10^{-19}$ J.

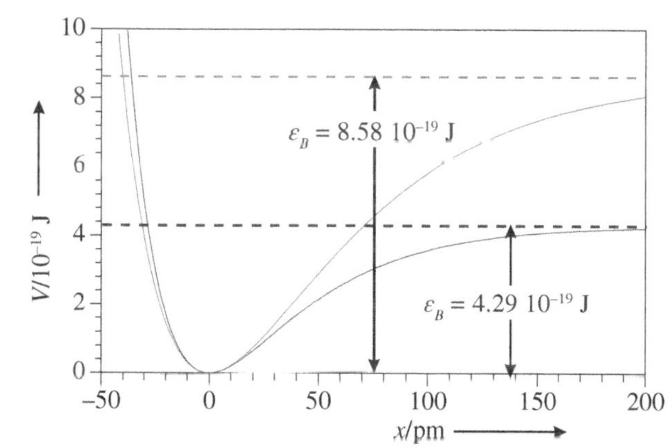

(a)

Figure E11.18a Morse functions with the parameters $k_f = 516$ N m^{-1}, $\varepsilon_B = 8.58 \times 10^{-19}$ J (dark gray solid line, these are the data for HCl), and $\varepsilon_B = 4.29 \times 10^{-19}$ J (light gray solid line).

The wavefunctions for the energy states look similar in terms of the number of nodes for each state. The largest differences are in the lack of symmetry about r_e for the Morse oscillator and the extent to which the wavefunction extends to large r. Both of these effects are more evident for higher quantum numbers.

Answers to questions posed in the MathCAD sheet are sketched here, but can be found in the file morse_solution.mcd.

Question 1: The a and ε_B parameters are correlated in the sense that they both affect the characteristic frequency of the vibration. The parameter ε_B controls the dissociation limit of the molecule, hence the bond energy.

A higher value of ε_B leads to a deeper well. In the case where all of the other parameters are the same, an increase in ε_B will cause an increase in the characteristic frequency of the vibration and an increase in the number of bound vibrational states. The parameter a changes the width of the well but does not change its depth. As a decreases the well becomes wider, and this could correspond to a smaller characteristic frequency. As a consequence the ladder of energy levels have a wider separation near the bottom of the well; however, they converge dramatically together as the vibrational energy approaches that of the dissociation limit.

Question 2:

a) Figures E11.18b–d shows plots of the squares of the Morse wavefunctions (light gray curves) and the harmonic oscillator wavefunctions (dark gray curves) for the cases of $n = 2,5$, and 8. It is evident that the wavefunction with the same quantum number has same number of nodes but that the Morse functions are asymmetric about the origin, having a lower amplitude for negative relative displacements and a higher amplitude for positive relative displacements than do the corresponding harmonic oscillator wavefunctions.

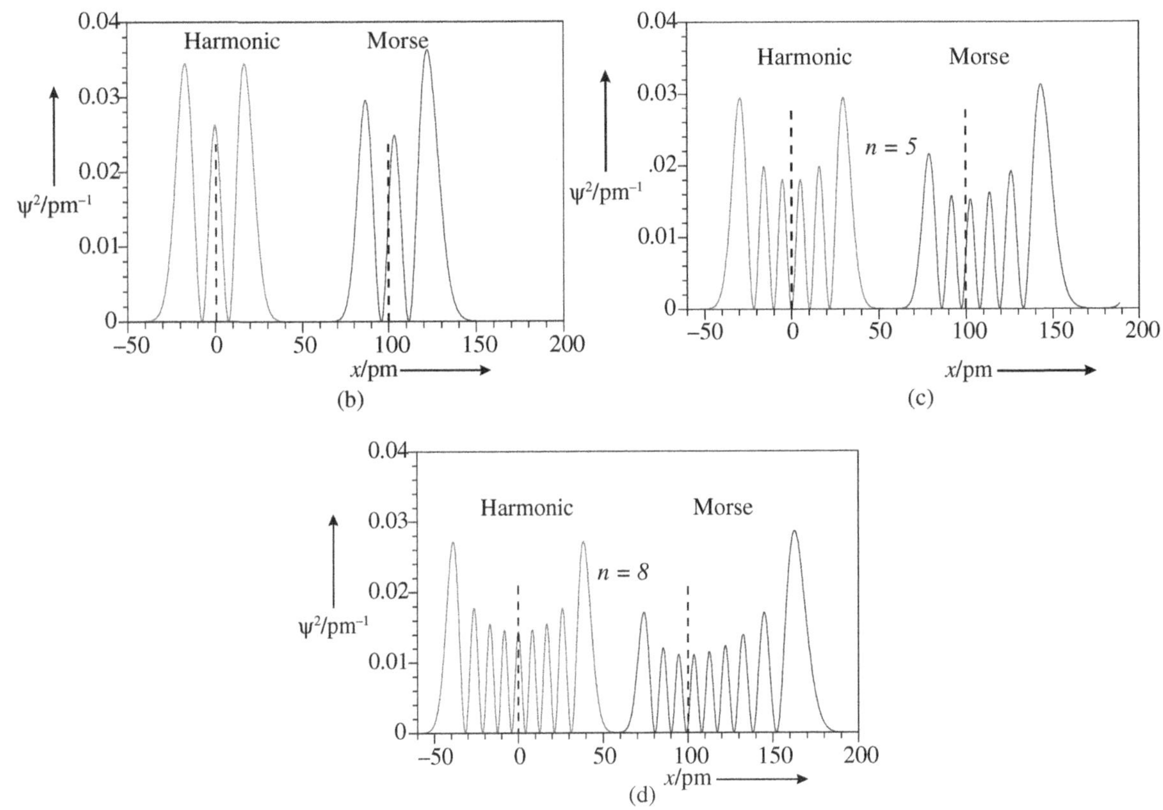

Figure E11.18b–d Squares of the wavefunctions of the harmonic (dark gray) and the anharmonic (Morse, light gray) oscillator for quantum numbers $n = 2,5$, and 8. The light gray curves are shifted in the x-direction by 100 pm to the right. The parameters are $k_f = 516\ \text{N m}^{-1}$ and $\varepsilon_B = 8.58 \times 10^{-19}$ J.

b) Figure E11.18e shows plots of the squares of the Morse wavefunctions for the case of $n = 5$ and the parameters $k_f = 516\ \text{N m}^{-1}$, $\varepsilon_B = 8.58 \times 10^{-19}$ J (dark gray), and $\varepsilon_B = 4.29 \times 10^{-19}$ J (light gray). We see that with decreasing ε_B the peaks become smaller and broader, due to the lower plateau of the Morse potential curve (see Fig. E11.18a). Also, the overall width increases slightly (from 105 pm for the left curve to 120 pm for the right curve).

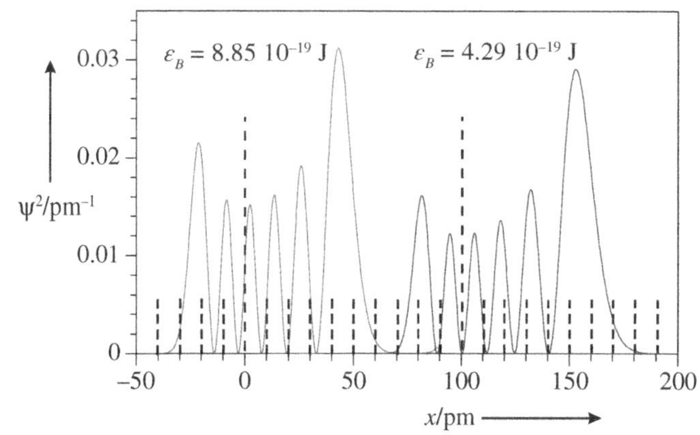

Figure E11.18e Squares of the wavefunctions of the anharmonic (Morse) oscillator for the quantum number $n = 5$ and for two different values of ε_B : $\varepsilon_B = 8.85 \times 10^{-19}$ J (dark gray) and $\varepsilon_B = 4.29 \times 10^{-19}$ J (light gray). The light gray curve is shifted in the x-direction by 100 pm to the right, for clarity.

E11.19 Use the MathCAD exercise FCfactors_displacement.xmcd to explore the vibronic structure of electronic transitions and its dependence on the displacement of the ground and excited electronic states. In this MathCAD excercise you assume that the ground-state electronic surface is parabolic with an energy

$$E_{gnd} = \frac{1}{2}\, k_f\, x^2$$

and the excited state surface is parabolic and has the same force constant as the ground state, namely

$$E_{exc} = \frac{1}{2}\, k_f\, (x - \delta)^2 + E_{00}$$

You are asked to evaluate the Franck–Condon factors (FCF) as a function of the displacement δ between the minima of the two parabolic curves. The absorption probability $\mathcal{P}$ in a spectroscopy experiment is proportional to

$$\mathcal{P} = \left| \int_{-\infty}^{\infty} \psi_g \psi_e \, d\tau \right|^2 = |M_{x,el}|^2 \cdot | \int \psi_{vib,g} \cdot \psi_{vib,e} \cdot dx|^2$$

where $|M_{x,el}|^2$ is the square of the electronic transition dipole moment integral and the vibrational overlap term is the FCF; namely

$$FCF = | \int \psi_{vib,g} \cdot \psi_{vib,e} \cdot dx|^2$$

You are asked to plot the FCF values as a function of the vibrational quantum number in each state and to rationalize the trends you observe.

Solution

The Franck–Condon factors depend on the differences between two potential energy curves. The two parameters which are important are the difference between the r_e values and the force constants k_f for the two electronic states. If the r_e's and k_f's are both similar, then the Franck–Condon factors with the most intense peaks will be those with the same quantum numbers in the upper and lower states, while if the values of r_e and k_f differ then the transition probability will be spread over multiple vibronic peaks, corresponding to different possible quantum number combinations (such as $0 \to 1, 0 \to 1, 0 \to 2, 1 \to 2, \ldots$).

In the MathCAD worksheet you can change the reduced mass of the oscillator, its force constant (by choosing different vibrational wavenumbers for the oscillators), and the displacement between the two minima (corresponding to a bond length change). Here we show some results for the case where the reduced mass is chosen to be 0.5 amu (1 amu= 1.6605×10^{-27} kg), the vibrational wavenumber is chosen to be 4000 cm^{-1}, and the E_{00} energy is taken to be 4.0 eV.

Note that this simplified model calculation assumes that all the molecules are in the ground vibrational state of the ground electronic state. The plots in Fig. E11.19 show the FCF factors (vibronic structure) for the electronic transition with two different displacements of the potential energy surfaces: 10 pm and 40 pm. For 10 pm most of the optical absorption will happen from the $n = 0$ vibrational state of the ground electronic state to the $n' = 0$, 1, and 2 vibrational states of the excited electronic state. Note that the maximum occurs for the transition from $n = 0$ to $n' = 1$. For the larger 100 pm displacement, it is evident that the absorption occurs to higher vibrational energy levels of the excited state pm. Also, we see that the probability of the transition to a particular n' state is significantly lower than the case of $\delta = 10$ pm, and that the absorption strength for the $n = 0$ to $n' = 0$ is very weak. As the displacement between the potential minima of the curves increases to large enough values the overlap of the vibrational wavefunctions will become exponentially small, however. These observations reveal how the vibronic structure of an electronic absorption spectrum is sensitive to the change in bond length between ground and excited states.

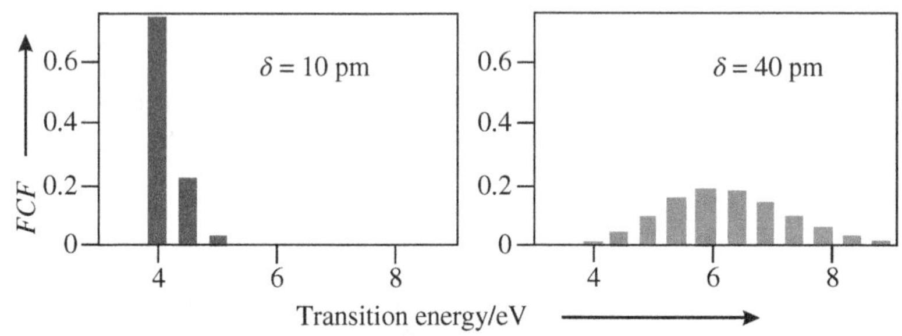

Figure E11.19a Plots of the Franck Condon (FCF) factors as a function of the transition energy (vibronic structure) are shown for harmonic oscillators displaced by $\delta = 10$ pm and $\delta = 40$ pm.

You can use the work sheet to explore other questions also. Such as, how does the spectrum change as $\delta \to 0$ pm? How does the vibronic structure change with the wavenumber? Is the overall shape correlated with the vibrational wavenumber chosen?

E11.20 In this exercise you extend the MathCAD-based calculations of Franck–Condon factors (see E11.19) to include changes in the force constants between the two harmonic oscillators, in addition to their displacement from each other. The calculations are a straightforward generalization of those used in E11.19, in which the ground-state vibrational energy spacing is given by

$$E_{\text{gnd},n} = \left(n + \frac{1}{2}\right)\, h\nu_g = \left(n + \frac{1}{2}\right)\, \frac{h}{2\pi}\sqrt{\frac{k_{fg}}{\mu}}$$

where n is the vibrational quantum number. The excited state surface, which is displaced by the electronic energy, E_{00}, has the vibrational spacing of

$$E_{\text{exc},n'} = \left(n' + \frac{1}{2}\right)\, h\nu_e + E_{00} = \left(n' + \frac{1}{2}\right)\, \frac{h}{2\pi}\sqrt{\frac{k_{fe}}{\mu}} + E_{00}$$

where n' is the vibrational quantum number in the excited state. In this case you should explore how the vibronic structure (FCFs) in the transition depends on the force constant differences. You are asked to plot the FCF values as a function of the vibrational quantum number in each state and to rationalize the trends you observe, for both the absorption spectrum and the fluorescence spectrum.

Solution

In the MathCAD worksheet you can change the reduced mass of the oscillator, its force constant (by choosing different vibrational wavenumbers), and the displacement between the minima. The plot in Fig. E11.20a shows a case in which we use two different force constants for the excited state, so that the excited state vibrational energy spacing corresponds to 1500 cm^{-1} (left panel) and 3000 cm^{-1} (right panel), and we keep the other parameters the same. Note that the ground-state vibrational wavenumber is 1500 cm^{-1} in each case. As the plots show, the vibronic transitions in the right panel are spaced apart more than those in the left panel, and the intensity distribution of the vibronic transitions is broader; that is, the peak intensity of the absorption spectrum shifts to a higher frequency, more vibronic transitions are evident, and the transitions are spaced farther apart. This comparison reveals that the vibronic structure in an electronic absorption spectrum can be used to probe the vibrational properties of the electronically excited state of a molecule.

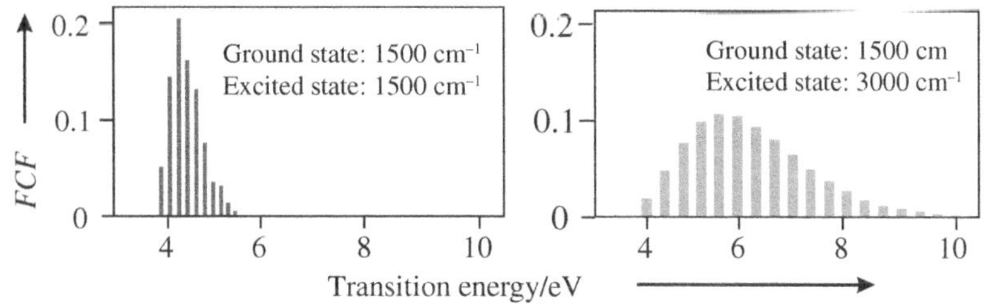

Figure E11.20a The plot shows the Franck–Condon factors (FCF) versus the transition energy (in eV) for two harmonic oscillator potentials that are displaced by 50 pm. The two panels show the absorption spectra for two different excited state vibrational wavenumbers; note that the ground-state oscillator has a fundamental vibrational wavenumber of 1500 cm^{-1}. Left panel: the excited state vibrational wavenumber is 1500 cm^{-1}; right panel: the excited state vibrational wavenumber is 3000 cm^{-1}.

Figure E11.20b shows the case for the fluorescence spectrum; i.e., the transition from the ground vibrational state of the excited electronic state to the different possible vibrational states in the ground electronic state. The same potential parameters ($\delta = 50$ pm, $\mu = 0.5$ amu, and $\tilde{\nu}_g = 1500$ cm^{-1}) as those used to generate Fig. E11.20a are shown here. The plots in Fig. E11.20b show two spectra; one is the case where the excited state vibrational wavenumber is 1500 cm^{-1} and the other has a vibrational wavenumber of 3000 cm^{-1}. Note that the spacing between the vibronic peaks is the same for these two spectra; this results because the ground electronic state's vibrational energy spacing determines the spectral line spacing. The intensities of the spectral lines do change somewhat between the two spectra, because of differences in the vibrational wavefunction overlap.

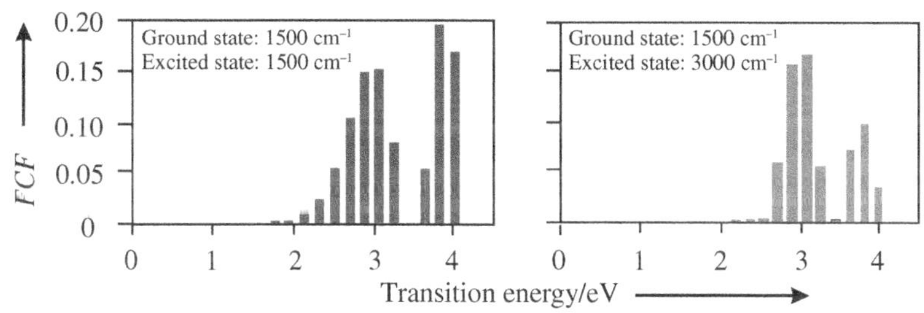

Figure E11.20b The plot shows the Franck–Condon factors (FCF) versus the transition energy (in eV) for two harmonic oscillator potentials that are displaced by 50 pm. The two panels show the fluorescence spectra for two different excited state vibrational wavenumbers. Left panel: the excited state vibrational wavenumber is 1500 cm^{-1}; right panel: the excited state vibrational wavenumber is 3000 cm^{-1}. Note that the ground-state oscillator has a fundamental vibrational wavenumber of 1500 cm^{-1}, and this accounts for the fact that the spectra are very similar – differing only in some detailed features.

E11.21 Using the diagram in Box 11.1 for a diatomic molecule, deduce the center of mass and show that the formula for the moment of inertia becomes $I = d^2 \cdot \mu$, where μ is the reduced mass $\left(\mu = m_A m_B / \left(m_A + m_B\right)\right)$. Use this more exact formula for I to determine the bond length of the $H^{35}Cl$ molecule, which has a rotational constant of $B = 10.439$ cm^{-1}

Solution

During the rotational motion the center of mass of the molecule stays at rest. Then according to Fig. B11.1 we have

$$am_A = bm_B \quad \text{and} \quad a + b = d$$

Thus

$$a = d \cdot \frac{m_B}{m_A + m_B} \quad \text{and} \quad b = d \cdot \frac{m_A}{m_A + m_B}$$

$I = md^2$ is the moment of inertia of a point mass rotating around a fixed point in distance d. This quantity must be replaced by the moment of inertia of a diatomic molecule:

$$I = m_A a^2 + m_B b^2 = m_A d^2 \cdot \frac{m_B^2}{(m_A + m_B)^2} + m_B d^2 \cdot \frac{m_A^2}{(m_A + m_B)^2}$$
$$= d^2 \cdot \frac{m_A m_B}{m_A + m_B} = d^2 \cdot \mu$$

with

$$\mu = \frac{m_A m_B}{m_A + m_B}$$

For the case of HCl, we can write the reduced mass as

$$\mu = \frac{m_{\mathrm{H}} \cdot m_{\mathrm{Cl}}}{m_{\mathrm{H}} + m_{\mathrm{Cl}}} = \frac{35}{36} \cdot m_{\mathrm{H}} = 0.972 \cdot m_{\mathrm{H}} = 1.627 \times 10^{-27}\ \mathrm{kg}$$

Equation (11.12) describes the rotational constant B of H_2 as

$$B = \frac{h^2}{8\pi^2 m_{\mathrm{H}} d^2}$$

Replacing m_{H} by μ and rearranging the Equation we find that

$$d = \sqrt{\frac{h^2}{8\pi^2 \mu} \cdot \frac{1}{B}} = 130\ \mathrm{pm}$$

which differs by less than 2% from that found in the text using m_{H}.

E11.22 Repeat the procedure in Equation (11.45), Table 11.4, and Fig. 11.25 using the calculated data in Table 11.3 for the anharmonic oscillator and compare the result with that obtained from the experimental data in Table 11.4.

Solution

The data in Table 11.4 were obtained from experimental values of the excitation energies. Here we use, instead, the data in Table 11.3 which are calculated using the anharmonic oscillator model. With these data, we construct the following table:

n	E_n 10^{-20} J	$\Delta E_{0\to n}$ 10^{-20} J	$n+1$	$\Delta E_{0\to n}/n$ 10^{-20} J
0	2.944			
1	8.678	5.734	2	5.734
2	14.206	11.262	3	5.631
3	19.528	16.584	4	5.528
4	24.644	21.700	5	5.425
5	29.556	26.612	6	5.322
6	34.261	31.317	7	5.220
7	38.763	35.819	8	5.117
8	43.065	40.121	9	5.015

According to Equation (11.45)

$$\frac{\Delta E_{0\to n}}{n} = h\nu_0 - C \cdot (n+1)$$

$\Delta E_{0\to n}/n$ is plotted versus $(n+1)$ in Fig. E11.22. From the intersection with the ordinate we obtain $h\nu_0 = 5.938 \times 10^{-20}$ J and from the slope $C = 0.1027 \times 10^{-20}$ J. These values are identical with the values obtained in the text. using experimental absorbance data. This means that the potential energy curve for the HCl molecule can be very well approximated by the Morse function.

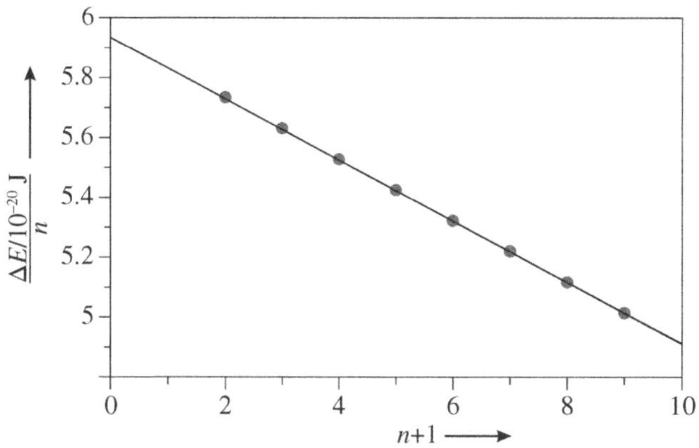

Figure E11.22 Evaluation of frequency ν_0 and the constant C. $\Delta E_{0\to n}/n$ is plotted versus $(n+1)$.

E11.23 Given that the probability $\mathrm{d}P_{\text{class}}$ of finding a classical oscillator between x and $x+\mathrm{d}x$ is inversely proportional to its classical speed v

$$\mathrm{d}P_{\text{class}} \sim \frac{1}{v}\cdot \mathrm{d}x$$

(i.e., the time it spends in the region of x to $x+\mathrm{d}x$), show that

$$\rho = \frac{\mathrm{d}P_{\text{class}}}{\mathrm{d}x} = \frac{1}{x_0\cdot\pi}\cdot\frac{1}{\sqrt{1-(x/x_0)^2}}$$

Proceed by using the fact that the time-dependent displacement $x(t)$ of a classical oscillator can be written as $x(t) = x_0 \cdot \sin(2\pi\nu_0 t)$, where ν_0 is the frequency of oscillation and x_0 is the amplitude of the oscillator.

Solution

First we find the speed by taking the derivative of the displacement with time.

$$\begin{aligned} v &= \frac{\mathrm{d}x}{\mathrm{d}t} = x_0 \cdot 2\pi\nu_0 \cdot \cos(2\pi\nu_0 t) \\ &= x_0 \cdot 2\pi\nu_0 \cdot \sqrt{1-\sin^2(2\pi\nu_0 t)} \\ &= x_0 \cdot 2\pi\nu_0 \cdot \sqrt{1-(x/x_0)^2} \end{aligned}$$

Thus, we can write the probability as

$$\mathrm{d}P_{\text{class}} = A\cdot\frac{1}{v}\cdot\mathrm{d}x = A\cdot\frac{1}{x_0\cdot 2\pi\nu_0\cdot\sqrt{1-(x/x_0)^2}}\cdot\mathrm{d}x$$

The constant A is determined by the condition that the probability of finding the oscillator between $-x_0$ and $+x_0$ equals 1; namely

$$\int_{-x_0}^{+x_0}\mathrm{d}P_{\text{class}}\cdot\mathrm{d}x = A\frac{1}{x_0\cdot 2\pi\nu_0}\int_{-x_0}^{+x_0}\frac{1}{\sqrt{1-(x/x_0)^2}}\cdot\mathrm{d}x = 1$$

We can solve the integral by using the substitution $u = x/x_0$ and $\mathrm{d}u = \mathrm{d}x/x_0$

$$\int\frac{1}{\sqrt{1-(x/x_0)^2}}\cdot\mathrm{d}x = x_0\int\frac{1}{\sqrt{1-u^2}}\cdot\mathrm{d}u$$

$$= x_0\cdot\arcsin u + const = x_0\cdot\arcsin\left(\frac{x}{x_0}\right) + const$$

where we used Appendix B. Inserting the limits we find

$$\int_{-x_0}^{+x_0} \frac{1}{\sqrt{1-(x/x_0)^2}} \cdot dx = x_0 \left| \arcsin \frac{x}{x_0} \right|_{-x_0}^{+x_0}$$

$$= x_0 (\arcsin 1 - \arcsin(-1))$$

$$= x_0 \left(\frac{\pi}{2} + \frac{\pi}{2} \right) = x_0 \pi$$

and for A we obtain

$$A = \frac{x_0 \cdot 2\pi v_0}{x_0 \pi} = 2v_0$$

Thus the result for the probability density is

$$\rho = \frac{dP_{class}}{dx} = \frac{1}{x_0 \cdot \pi} \cdot \frac{1}{\sqrt{1-(x/x_0)^2}}$$

E11.24 Consider the water molecule in its gaseous state. Given the experimental values for $I_B = 1.023 \times 10^{-47}$ kg m^2 and $I_C = 2.945 \times 10^{-47}$ kg m^2 [1] calculate the bond angle and the bond length, in analogy to the consideration in Foundation 11.1 for the NH_3 molecule.

Solution

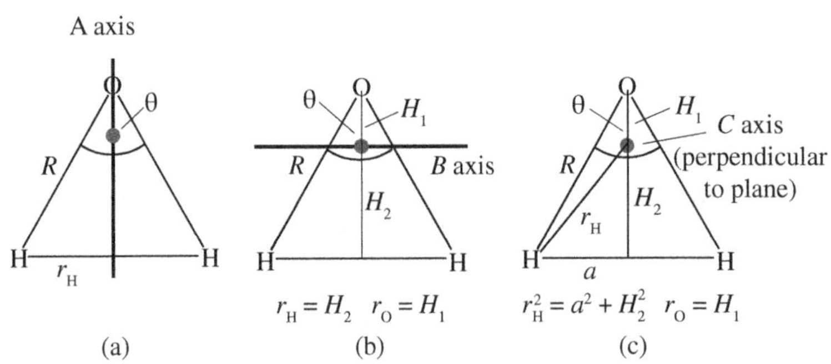

Figure E11.24 Rotation axes in the water molecule. (a) A-axis, (b) B-axis, (c) C-axis. The filled dark gray circle indicates the center of mass. r_O is the rotation radius of the oxygen atom, r_H is the rotation radius of the H atoms.

a) Calculation of I_A: We consider Fig. E11.24a. In this case the A-axis is the symmetry axis. The O atom is at rest and does not contribute to the inertial moment. The H atoms rotate at a distance r from the A axis. Thus

$$I_A = 2m_H \cdot r_H^2 = 2m_H \cdot R^2 \cdot \sin^2\frac{\theta}{2} = 2m_H \cdot R^2 \cdot \frac{1}{2}(1 - \cos\theta)$$

b) Calculation of I_B: We consider Fig. E11.24b. The O atom and the two H atoms rotate at distances $r_O = H_1$ and $r_H = H_2$ from the center of mass. Thus

$$I_B = m_O \cdot H_1^2 + 2m_H \cdot H_2^2$$

We calculate the distances H_1 and H_2 from the conditions

$$H_1 \cdot m_O = H_2 \cdot 2m_H \qquad H_1 + H_2 = H$$

where H is the distance between the O atom and the line connecting the two H atoms. Hence we have

$$H_1 \cdot m_O = (H - H_1) \cdot 2m_H = H \cdot 2m_H - H_1 \cdot 2m_H$$

so that

$$H_1 = H\frac{2m_H}{m_O + 2m_H} \quad \text{and} \quad H_2 = H\frac{m_O}{m_O + 2m_H}$$

1 Massachusetts Institute of Technology, Research Laboratory of Electronics, Technical Report No. 255, May 14, 1953, Microwave spectrum of the water molecule by Desmond Walter Posener.

and

$$I_B = m_O \cdot H^2 \left(\frac{2m_H}{m_O + 2m_H} \right)^2 + 2m_H \cdot H^2 \left(\frac{m_O}{m_O + 2m_H} \right)^2$$

$$= H^2 \left[m_O \cdot 2m_H \frac{2m_H}{(m_O + 2m_H)^2} + 2m_H \cdot m_O \frac{m_O}{(m_O + 2m_H)^2} \right]$$

$$= H^2 \left[m_O \cdot 2m_H \frac{2m_H + m_O}{(m_O + 2m_H)^2} \right] = H^2 \left[\frac{m_O \cdot 2m_H}{m_O + 2m_H} \right]$$

Finally we obtain H from the condition

$$H = R \cdot \cos \frac{\theta}{2}$$

where R is the O–H bond length. Thus

$$I_B = R^2 \cdot \cos^2 \frac{\theta}{2} \cdot \left[\frac{m_O \cdot 2m_H}{m_O + 2m_H} \right]$$

Using the relation (see Appendix C)

$$\cos^2 \frac{\theta}{2} = \frac{1}{2}(1 + \cos \theta)$$

we obtain

$$I_B = R^2 \cdot \frac{1}{2}(1 + \cos \theta) \left[\frac{m_O \cdot 2m_H}{m_O + 2m_H} \right]$$

$$= R^2 \cdot \frac{1}{2}(1 + \cos \theta) \cdot 2m_H \frac{16m_H}{16m_H + 2m_H}$$

$$= R^2 \cdot (1 + \cos \theta) \cdot m_H \frac{16}{18} = R^2 \cdot m_H \cdot \frac{8}{9} \cdot (1 + \cos \theta)$$

where we have made the approximation that $m_O = 16m_H$.

c) Calculation of I_C: To calculate I_C we consider Fig. E11.24c. Note that the C axis is perpendicular to the molecular plane. The O atom rotates at a distance H_1 from the center of mass. The H atoms rotate at a distance r_H from the center of mass, where r_H is given by the relation

$$r_H^2 = a^2 + H_2^2 - R^2 \sin^2 \frac{\theta}{2} + H_2^2$$

Thus

$$I_C = m_O \cdot H_1^2 + 2m_H \cdot \left(R^2 \sin^2 \frac{\theta}{2} + H_2^2 \right)$$

with

$$H_1 = H \frac{2m_H}{m_O + 2m_H} \qquad H_2 = H \frac{m_O}{m_O + 2m_H}$$

allows us to obtain

$$I_C = m_O \cdot H^2 \left[\frac{2m_H}{m_O + 2m_H} \right]^2 + 2m_H \cdot \left(R^2 \sin^2 \frac{\theta}{2} + H^2 \left[\frac{m_O}{m_O + 2m_H} \right]^2 \right)$$

$$= H^2 \left[\frac{m_O \cdot 4m_H^2}{(m_O + 2m_H)^2} + \frac{2m_H \cdot m_O^2}{(m_O + 2m_H)^2} \right] + 2m_H \cdot R^2 \sin^2 \frac{\theta}{2}$$

$$= H^2 \left[m_O \cdot 2m_H \frac{2m_H + m_O}{(m_O + 2m_H)^2} \right] + 2m_H \cdot R^2 \sin^2 \frac{\theta}{2}$$

$$= H^2 \frac{m_O \cdot 2m_H}{m_O + 2m_H} + 2m_H \cdot R^2 \sin^2 \frac{\theta}{2}$$

$$= R^2 \cdot \cos^2 \frac{\theta}{2} \cdot \frac{m_O \cdot 2m_H}{m_O + 2m_H} + 2m_H \cdot R^2 \sin^2 \frac{\theta}{2}$$

$$= R^2 \cdot \left(\cos^2 \frac{\theta}{2} \cdot \frac{m_O \cdot 2m_H}{m_O + 2m_H} + 2m_H \sin^2 \frac{\theta}{2} \right)$$

With (see Appendix C)

$$\cos^2\frac{\theta}{2} = (1+\cos\theta) \qquad \sin^2\frac{\theta}{2} = (1-\cos\theta)$$

we obtain

$$\begin{aligned}
I_C &= R^2\cdot\left[\frac{1}{2}(1+\cos\theta)\cdot\frac{m_O\cdot 2m_H}{m_O+2m_H} + 2m_H\frac{1}{2}(1-\cos\theta)\right]\\
&= R^2\cdot m_H\left[(1+\cos\theta)\cdot\frac{m_O}{m_O+2m_H} + (1-\cos\theta)\right]\\
&= R^2\cdot m_H\left[(1+\cos\theta)\cdot\frac{8}{9} + (1-\cos\theta)\right]\\
&= R^2\cdot m_H\left[\frac{8}{9} + \frac{8}{9}\cos\theta + 1 - \cos\theta\right]\\
&= R^2\cdot m_H\left[\frac{17}{9} - \frac{1}{9}\cos\theta\right]
\end{aligned}$$

d) Evaluation of bond length and bond angle. In order to eliminate the bond length we write that

$$\frac{I_C}{I_B} = \frac{R^2\cdot m_H\left[\frac{17}{9}-\frac{1}{9}\cos\theta\right]}{R^2\cdot m_H\cdot\frac{8}{9}\cdot(1+\cos\theta)} = \frac{\frac{17}{9}-\frac{1}{9}\cos\theta}{\frac{8}{9}\cdot(1+\cos\theta)}$$

so that

$$\frac{17}{9} - \frac{1}{9}\cos\theta = \frac{I_C}{I_B}\frac{8}{9}\cdot(1+\cos\theta) = \frac{I_C}{I_B}\frac{8}{9} + \frac{I_C}{I_B}\frac{8}{9}\cos\theta$$

To isolate the $\cos\theta$, we rearrange the above equation

$$\cos\theta\cdot\left[-\frac{1}{9} - \frac{I_C}{I_B}\frac{8}{9}\right] = \frac{I_C}{I_B}\frac{8}{9} - \frac{17}{9}$$

and obtain

$$\cos\theta = \left(\frac{I_C}{I_B}\frac{8}{9} - \frac{17}{9}\right) / \left(-\frac{1}{9} - \frac{I_C}{I_B}\frac{8}{9}\right)$$

With the experimental values for I_B and I_C we obtain

$$\frac{I_C}{I_B} = \frac{2.945\times10^{-47}\ \text{kg m}^2}{1.023\times10^{-47}\ \text{kg m}^2} = 2.879$$

and

$$\cos\theta = \frac{2.879\cdot\frac{8}{9} - \frac{17}{9}}{-\frac{1}{9} - 2.879\cdot\frac{8}{9}} = \frac{0.670}{-2.670} = -0.2510 \qquad \theta = 104.5°$$

Finally we obtain the bond length R:

$$I_B = 1.023\times10^{-47}\ \text{kg m}^2 = R^2\cdot m_H\cdot\frac{8}{9}\cdot(1+\cos\theta)$$

$$R^2 = \frac{1.023\times10^{-47}\ \text{kg m}^2}{m_H\cdot\frac{8}{9}\cdot(1+\cos\theta)} = \frac{1.023\times10^{-47}\ \text{kg m}^2}{1.01\cdot1.6605\times10^{-27}\ \text{kg}\cdot\frac{8}{9}\cdot(1-0.2510)}$$

$$= 9.174\times10^{-21}\ \text{m}^2 \quad \text{and} \quad R = 95.8\ \text{pm}$$

E11.25 Given that the rotational constant of CH_4 is $B = 5.336\times10^{-47}$ kg m^2 calculate the bond distance R of the molecule.

Solution

From the rotational constant we calculate the inertial moment I.

$$I = \frac{h^2}{8\pi^2 B} = \frac{(6{,}626\times10^{-34}\ \text{J s})^2}{8\pi^2 5.336\times10^{-47}\ \text{kg m}^2} = 5.336\times10^{-47}\ \text{kg m}^2$$

In order to express I by the bond length R we consider Fig. E11.25. During rotation the C atom and the upper H atom stay at rest. Thus for I we obtain

$$I = 3m_{\mathrm{H}} \cdot r^2$$

where r is the radius of rotation. From Fig. E11.25c we obtain

$$b^2 = 2r^2 - 2r^2 \cdot \cos\alpha = 2r^2\,(1 - \cos\alpha) = 2r^2\frac{3}{2} = 3r^2$$

with $\alpha = 120°$. From Fig. E11.25b we obtain

$$b^2 = 2R^2 - 2R^2 \cdot \cos\theta = 2R^2\,(1 - \cos\theta)$$

Combining both equations yields

$$r^2 = \frac{2}{3}R^2\,(1 - \cos\theta)$$

and

$$I = 3m_{\mathrm{H}} \cdot r^2 = 3m_{\mathrm{H}} \cdot \frac{2}{3}R^2\,(1 - \cos\theta) = 2m_{\mathrm{H}} \cdot R^2\,(1 - \cos\theta)$$

From the symmetry of the molecule we know that θ is the tetrahedral angle. In Appendix B4 we derive the relation

$$\cos\theta = -\frac{1}{3} \qquad \theta = 109.471°$$

Thus

$$I = 2m_{\mathrm{H}} \cdot R^2(1 + \frac{1}{3}) = \frac{8}{3}m_{\mathrm{H}} \cdot R^2$$

and the result for the bond length is

$$R = \sqrt{\frac{I}{\frac{8}{3}m_{\mathrm{H}}}} = \sqrt{\frac{5.336 \times 10^{-47}\ \mathrm{kg\ m^2}}{\frac{8}{3}1.01 \cdot 1.6605 \times 10^{-27}\ \mathrm{kg}}}$$
$$= \sqrt{1.193 \times 10^{-20}\ \mathrm{m^2}} = 1.092\ \mathrm{m} = 109.2\ \mathrm{pm}$$

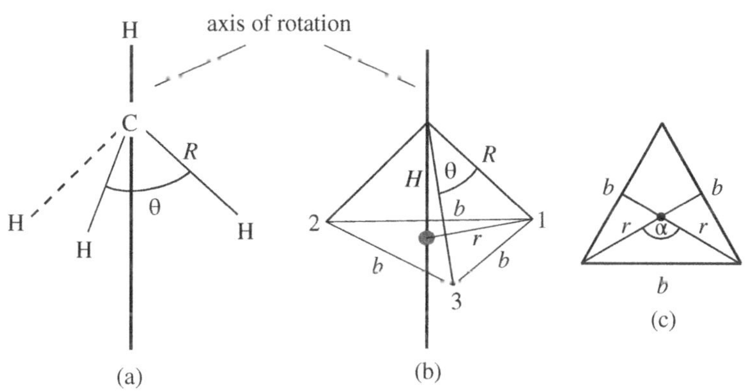

Figure E11.25 Rotation of the CH_4 molecule. (a) Molecular skeleton, (b) construction to calculate the rotational radius r of the three H atoms, (c) equal sided triangle 1-2-3 in (b) shown from top.

11.2 Problems

P11.1 Use Equation (11.19) to calculate the intensity pattern of the rotational absorption spectrum of HCl at 298 K. Compare your result with the experimental pattern shown in Fig. 11.7.

Solution

Figure 11.7 shows the experimental data for the intensity of the various J lines in a pure rotational spectrum, and Table 11.1 gives the transition energies and assignments to the lines; Table 11.1 is

J	$1/\lambda$ (cm^{-1})	ΔE (10^{-23} J)	$\Delta(\Delta E)$ (10^{-23} J)	J	$1/\lambda$ (cm^{-1})	ΔE (10^{-23} J)	$\Delta(\Delta E)$ (10^{-23} J)
0	(20.93)	(41.58)		7	165.83	329.4	40.7
1	41.78	82.99	41.4	8	186.25	370.0	40.6
2	62.60	124.3	41.3	9	206.54	410.3	40.3
3	83.35	165.6	41.3	10	226.71	450.3	40.0
4	104.02	206.6	41.0	11	246.82	490.3	40.0
5	124.74	247.8	41.2	12	266.70	529.8	39.5
6	145.36	288.7	40.9	13	286.50	569.1	39.3

Figure P11.14 gives a bar graph of the intensity predicted using Equation (11.19),

$$\text{absorbance} = const' \cdot \mu_{\text{perm}}^2 \cdot (J+1)^2 \cdot e^{-BJ(J+1)/(kT)}$$

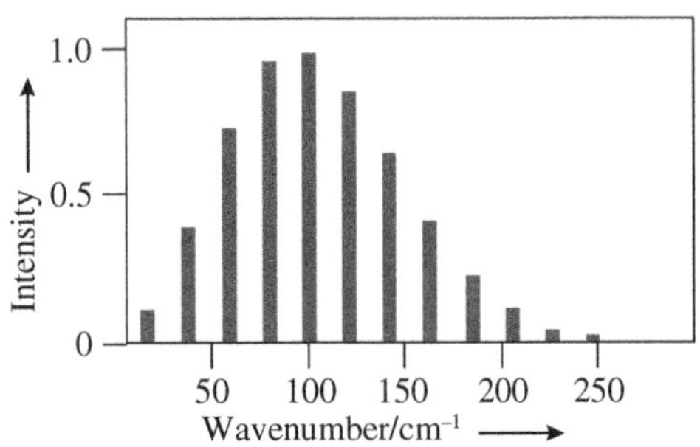

Figure P11.1 Intensity pattern of the absorption spectrum of HCl in the 10 to 250 cm^{-1} region.

Figure 11.7 shows the experimental data for the intensity of the various J lines in a pure rotational spectrum. Qualitative agreement is obtained, suggesting that the model is approximately correct in its prediction of the relative intensities.

P11.2 Show that the minimum of a potential energy curve is always harmonic. Consider an arbitrary potential energy function $V(x)$ and show that the leading term in its Taylor series expansion about the minimum is quadratic.

Solution

The Taylor series expansion for the function $V(x)$ about the point x_0 is

$$V(x) = V(x_0) + \frac{1}{1!}\left.\frac{dV(x)}{dx}\right|_{x=x_0}(x-x_0) + \frac{1}{2!}\left.\frac{d^2V(x)}{dx^2}\right|_{x=x_0}(x-x_0)^2 + \ldots$$

We limit our expansion to the second order term. If we take the point x_0 to be at the minimum of the function, then

$$\left.\frac{dV(x)}{dx}\right|_{x=x_0} = 0$$

so that

$$V(x) = V(x_0) + \frac{1}{2!}\left.\frac{d^2V(x)}{dx^2}\right|_{x=x_0}(x-x_0)^2$$

Next we choose our zero point of the energy scale to be $V(x_0) = 0$, and we find that

$$V(x) = \frac{1}{2!}\left.\frac{d^2V(x)}{dx^2}\right|_{x=x_0}(x-x_0)^2 = \frac{1}{2}k_f \cdot (x-x_0)^2$$

where k_f is the force constant. This expression has the form of the harmonic oscillator potential with

$$k_f = \frac{1}{2!}\frac{\mathrm{d}^2V(x)}{\mathrm{d}x^2}\bigg|_{x=x_0}$$

This last relation can also be understood from the relations

$$V = \frac{1}{2}k_f \cdot (x - x_0)^2$$
$$\frac{\mathrm{d}V}{\mathrm{d}x} = k_f \cdot (x - x_0)$$
$$\frac{\mathrm{d}^2V}{\mathrm{d}x^2} = k_f$$

P11.3 In Section 11.3.2 particle-in-a-box wavefunctions are used as trial functions for solving the Schrödinger equation of a harmonic oscillator. Use the variational principle with Equations (11.22) and (11.23) to calculate the energies ε_0 and ε_1.

Solution

For ε_0 we find that

$$\varepsilon_0 = \overline{T}_0 + \overline{V}_0 = \frac{h^2}{8\mu L_0^2} + \int_{-L_0/2}^{+L_0/2} \frac{1}{2}k_f x^2 \phi_0^2 \cdot \mathrm{d}x$$
$$= \frac{h^2}{8\mu L_0^2} + \frac{k_f}{L_0}\int_{-L_0/2}^{+L_0/2} x^2 \cos^2\left(\frac{\pi x}{L_0}\right) \cdot \mathrm{d}x$$

We solve the integral by the substitution

$$u = \frac{\pi}{L_0}x,\ \mathrm{d}u = \frac{\pi}{L_0}\mathrm{d}x$$

so that

$$\int x^2 \cdot \cos^2\left(\frac{\pi x}{L_0}\right) \cdot \mathrm{d}x = \left(\frac{L_0}{\pi}\right)^3 \cdot \int u^2 \cdot \cos^2 u \cdot \mathrm{d}u$$

Using Appendix B for the trigonometric functions we can write

$$\cos^2 u = \frac{1}{2} + \frac{1}{2}\cos 2u$$

and

$$\int u^2 \cdot \cos^2 u \cdot \mathrm{d}u = \frac{1}{2}\int u^2 \cdot \mathrm{d}u + \frac{1}{2}\int u^2 \cos(2u) \cdot \mathrm{d}u$$
$$= \frac{1}{2}\frac{1}{3}u^3 + \frac{1}{2}\frac{1}{8}\int \upsilon^2 \cos \upsilon \cdot \mathrm{d}\upsilon$$

where we used the substitution $\upsilon = 2u$. Appendix B gives

$$\int \upsilon^2 \cos \upsilon \cdot \mathrm{d}\upsilon = 2\upsilon \cdot \cos \upsilon + (\upsilon^2 - 2) \cdot \sin \upsilon + const$$
$$= 4u \cdot \cos(2u) + (4u^2 - 2) \cdot \sin(2u) + const$$

and we obtain with the limits $u = -\pi/2$ for $x = -L/2$ and $u = \pi/2$ for $x = L/2$

$$\int_{-\pi/2}^{\pi/2} u^2 \cos^2 u\ \mathrm{d}u = \left(\frac{u^3}{6} + \frac{u}{4}\cos(2u) + \left(\frac{u^2}{4} - \frac{1}{8}\right)\sin(2u)\right)\bigg|_{-\pi/2}^{\pi/2}$$
$$= \frac{1}{6}\left(\frac{\pi}{2}\right)^3 - \frac{1}{4}\frac{\pi}{2} + \frac{1}{6}\left(\frac{\pi}{2}\right)^3 - \frac{1}{4}\frac{\pi}{2} = \frac{1}{24}\pi^3 - \frac{1}{4}\pi$$

and

$$\frac{k_f}{L_0}\int_{-L_0/2}^{L_0/2} x^2 \cdot \cos^2\frac{\pi x}{L_0}\cdot dx = \left(\frac{L_0}{\pi}\right)^3 \frac{k_f}{L_0}\left(\frac{1}{24}\pi^3 - \frac{1}{4}\pi\right)$$
$$= \frac{L_0^2}{\pi^3}k_f\cdot\left(\frac{1}{24}\pi^3 - \frac{1}{4}\pi\right)$$
$$= L_0^2 k_f \cdot 0.0163$$

Thus the result for the energy $\varepsilon_{n=0}$ is

$$\varepsilon_{n=0} = \frac{h^2}{8\mu L_0^2} + k_f L_0^2\cdot a \quad \text{with}\, a = 0.0163$$

According to the variational principle, we optimize the energy by variation of the wavefunction (in this case by the parameter L_0); hence, we evaluate

$$\frac{d\varepsilon_0}{dL_0} = -\frac{h^2}{4\mu L_0^3} + 2a\cdot k_f\cdot L_0 = 0$$

so that

$$\frac{h^2}{4\mu L_0^3} = 2a\cdot k_f\cdot L_0 \text{ or } L_0^2 = \sqrt{\frac{h^2}{\mu k_f}}\sqrt{\frac{1}{8\,(0.0163)}} = 2.77\frac{h}{\sqrt{k_f\mu}}$$

and

$$\varepsilon_0 = \frac{h^2}{8\mu L_0^2} + k_f L_0^2\cdot a = \frac{h^2}{8\mu}\frac{\sqrt{k_f\mu}}{2.77\,h} + 2.77\frac{h}{\sqrt{k_f\mu}}k_f\cdot a$$

$$= \left(\frac{1}{4}\frac{\pi}{2.77} + 2.77\cdot 2\pi\cdot a\right)\frac{h}{2\pi}\sqrt{\frac{k_f}{\mu}} = 0.57\,\frac{h}{2\pi}\sqrt{\frac{k_f}{\mu}} = 0.57\cdot h\nu_0$$

which agrees with the result given in the text.

With the trial function φ_1 we obtain in a similar way (length $= L_1$)

$$\varepsilon_{n=1} = \frac{h^2}{2\mu L_1^2} + k_f L_1^2\cdot b \quad \text{with} \quad b = 0.0353$$

$$L_1^2 = \frac{h}{\sqrt{k_f\mu}}\cdot\beta \quad \text{with} \quad \beta = \frac{1}{\sqrt{2b}} = 3.76$$

and

$$\varepsilon_1 = 1.67\cdot\frac{h}{2\pi}\cdot\sqrt{\frac{k_f}{\mu}}$$

For ε_1 we find that

$$\varepsilon_1 = \overline{T}_1 + \overline{V}_1 = \frac{h^2}{8\mu L_1^2} + \int_{-L_1/2}^{+L_1/2}\frac{1}{2}k_f x^2\phi_1^2\cdot dx$$
$$= \frac{h^2}{2\mu L_1^2} + \frac{k_f}{L_1}\int_{-L_1/2}^{+L_1/2} x^2 \sin^2\left(\frac{2\pi x}{L_1}\right)\cdot dx$$

Using Appendix B for the trigonometric functions we can write

$$\sin^2 u = \frac{1}{2} - \frac{1}{2}\cos(2u)$$

and

$$\int u^2\cdot\sin^2 u\cdot du = \frac{1}{2}\int u^2\cdot du - \frac{1}{2}\int u^2\cos(2u)\cdot du$$
$$= \frac{1}{2}\frac{1}{3}u^3 - \frac{1}{2}\frac{1}{8}\int v^2\cos v\cdot dv$$

where we used the substitution $v = 2u$. Appendix B gives

$$\int v^2 \cos v \cdot dv = 2v \cos v + (v^2 - 2) \sin v + const$$
$$= 4u \cos (2u) + (4u^2 - 2) \sin (2u) + const$$

and we obtain with the limits $u = -\pi$ for $x = -L/2$ and $u = \pi$ for $x = L/2$

$$\int_{-\pi}^{\pi} u^2 \cdot \sin^2 u \cdot du = \left[\frac{u^3}{6} - \frac{u}{4}\cos (2u) - (\frac{u^2}{4} - \frac{1}{8}) \sin (2u)\right]_{-\pi}^{\pi}$$
$$= \frac{1}{6}(\pi)^3 - \frac{1}{4}\pi + \frac{1}{6}(\pi)^3 - \frac{1}{4}\pi = \frac{1}{3}\pi^3 - \frac{\pi}{2}$$

and

$$\frac{k_f}{L_1}\int_{-L_1/2}^{L_1/2} x^2 \cdot \sin^2\left(\frac{2\pi x}{L_1}\right) \cdot dx = \left(\frac{L_1}{2\pi}\right)^3 \frac{k_f}{L_1}\int_{-\pi}^{\pi} u^2 \cdot \sin^2 u \cdot du$$
$$= \left(\frac{L_1}{2\pi}\right)^3 \frac{k_f}{L_1}\left(\frac{\pi^3}{3} - \frac{\pi}{2}\right)$$
$$= L_1^2 k_f \cdot b \text{ where } b = 0.0353$$

Thus the result for the energy $\varepsilon_{n=1}$ is

$$\varepsilon_{n=1} = \frac{h^2}{2\mu L_1^2} + k_f L_1^2 \cdot b$$

According to the variation principle, we optimize the energy by variation of the wavefunction (in this case by the parameter L_1); hence, we evaluate

$$\frac{d\varepsilon_1}{dL_1} = -\frac{h^2}{\mu L_1^3} + 2b \cdot k_f L_1 = 0$$

so that

$$\frac{h^2}{\mu L_1^3} = 2b \cdot k_f L_1 \text{ or } L_1^2 = \sqrt{\frac{h^2}{\mu k_f}}\sqrt{\frac{1}{2 \cdot b}} = 3.76\frac{h}{\sqrt{k_f \mu}}$$

and

$$\varepsilon_1 = \frac{h^2}{2\mu L_1^2} + k_f L_1^2 \cdot b = \frac{h^2}{2\mu}\frac{\sqrt{k_f\mu}}{3.76\, h} + 3.76\frac{h}{\sqrt{k_f\mu}}k_f \cdot b$$
$$= \left(\frac{1}{2}\frac{2\pi}{3.76} + 3.76 \cdot 2\pi \cdot 0.0353\right)\frac{h}{2\pi}\sqrt{\frac{k_f}{\mu}} = 1.67\frac{h}{2\pi}\sqrt{\frac{k_f}{\mu}} = 1.67\, h\nu_0$$

P11.4 Given the bond length $d = 156 \times 10^{-12}$ m and force constant $k_f = 250\ \mathrm{N\ m^{-1}}$ for ^{6}LiF, use the rigid rotor/harmonic oscillator approximation to construct, to scale, an energy level diagram for the first four rotational levels in both the $n = 0$ and $n = 1$ vibrational states (i.e., a total of eight states). Indicate the allowed transitions in an infrared absorption experiment. Clearly label which transitions belong to the R branch and which belong to the P branch.

Solution

Using the given parameters and $\mu = \frac{6 \cdot 19}{25} 1.66 \times 10^{-27}\mathrm{kg} = 7.57 \times 10^{-27}\mathrm{kg}$, we calculate

$$\nu_0 = \frac{1}{2\pi}\sqrt{\frac{k_f}{\mu}} = \frac{1}{2\pi}\sqrt{\frac{259}{7.57 \times 10^{-27}}} = 2.94 \times 10^{13}\ \mathrm{s^{-1}}$$

and in wavenumbers we find

$$\tilde{\nu}_0 = c_0 \nu_0$$
$$= \left(2.998 \times 10^8 \frac{\mathrm{m}}{\mathrm{s}}\right)(2.94 \times 10^{13}\ \mathrm{s^{-1}})$$
$$= 98100\ \mathrm{m^{-1}} = 981\ \mathrm{cm^{-1}}$$

For the rotational constant, we find that

$$B = \frac{h^2}{8\pi^2 \mu d^2} = \frac{(6.626)^2 10^{-68}}{8\pi^2\left(7.57\times10^{-27}\right)(156)^2\times10^{-24}} \text{J}$$

$$= \frac{(6.626)^2}{8\pi^2(7.57)(156)^2} \times 10^{-17}\text{J} = 3.02\times10^{-23}\ \text{J}$$

Using the general energy level formula

$$E(n, J) = (n + 1/2)\cdot h\nu_0 + B\cdot J(J+1)$$

we find that

n, J	0,0	0,1	0,2	0,3	1,0	1,1	1,2	1,3
$J(J+1)$	0	2	6	12	0	2	6	12
$E(n,J)/10^{-21}$ J	9.744	9.803	9.924	10.11	29.22	29.29	29.41	29.59

The energy levels are plotted in Fig. P11.4. The dark gray arrows mark the R-branch transitions and the light gray arrows mark the P-branch transitions

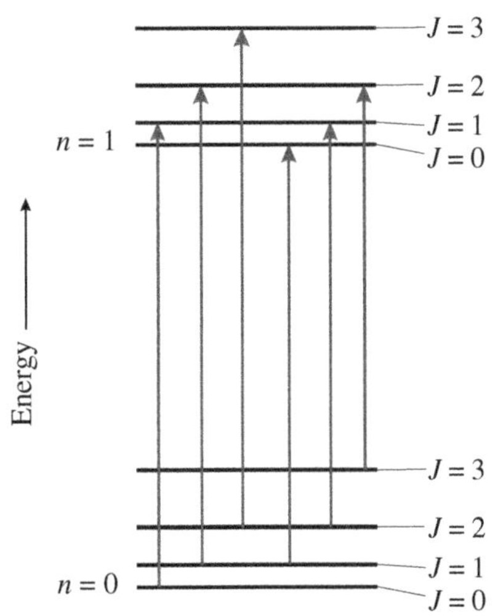

Figure P11.4 Energy scheme for the vibrational/rotational spectrum of LiF. Dark gray solid arrows: R-branch transitions, light gray solid arrows: P-branch transitions. Note that the energy difference between the levels $n = 0/J = 0$ and $n = 1/J = 0$ is greater by a factor of about 60 than the energy difference between the levels $n = 0/J = 0$ and $n = 0/J = 3$..

P11.5 From the different masses of ^{35}Cl and ^{37}Cl calculate the line splitting in the HCl spectrum shown in Fig. 11.16. Note that this problem is very similar to E11.5.

Solution

The different reduced mass for the two isotopomers affects both the vibrational and rotational energy levels; however, the main influence of the different isotopes is a change of the pure vibrational transition energy. For the vibrational transition, the energies are $\Delta E^{35} = h\nu_0^{35}$ and $\Delta E^{37} = h\nu_0^{37}$. The energy difference

$$\Delta E' = \Delta E^{35} - \Delta E^{37}$$

is

$$\Delta E' = h\nu_0^{35} - h\nu_0^{37} = \frac{h\sqrt{k_f}}{2\pi}\cdot\left(\frac{1}{\sqrt{\mu^{35}}} - \frac{1}{\sqrt{\mu^{37}}}\right)$$

Because of

$$\mu^{35} = \frac{m_{\mathrm{H}} \cdot m_{^{35}Cl}}{m_{\mathrm{H}} + m_{^{35}Cl}} = 1.627 \times 10^{-27}\ \mathrm{kg}, \mu^{37} = 1.630 \times 10^{-27}\ \mathrm{kg}$$

we obtain $\Delta E' = 5.3 \times 10^{-23}$ J corresponding to a shift of the lines by $\tilde{\nu} = 2.7\ \mathrm{cm}^{-1}$ in accordance with the experiment (Fig. 11.16). Note that in this figure all ^{37}Cl lines are mainly equally shifted to smaller wavenumbers.
As can be calculated correspondingly, the expected shift in the rotational energies is

$$\begin{aligned}\Delta E'_{rot} &= 2B^{35}(J+1) - 2B^{37}(J+1) = 2(J+1) \cdot \left(B^{35} - B^{37}\right) \\ &= 2(J+1) \cdot \frac{h^2}{8\pi^2 d^2}\left(\frac{1}{\mu^{35}} - \frac{1}{\mu^{37}}\right) = 1.26 \times 10^{-24}(J+1)\ \mathrm{J}\end{aligned}$$

For $J = 0$ this shift is only 1% of that found for vibration, and for $J = 10$ this shift is only 10% of the shift arising from the vibrational frequency.

P11.6 Consider a vibrational transition from quantum number $n = 0$ to $n = 3$. Using harmonic oscillator wavefunctions show that the corresponding transition moment M_x is zero. Use Equation 11.81; i.e.,

$$M_x = a \int_{-\infty}^{\infty} \psi_i x \psi_j \cdot \mathrm{d}x$$

from Foundation 11.6.

Solution
This problem asks you to find the value of the transition moment M_x for a harmonic oscillator transition from $n = 0$ to $n = 3$. There are several ways to consider for solving this problem. One is the brute force method of substituting the wavefunctions and doing all the calculus and algebra and calculating a value of zero. A second method sometimes applied to such problems is to use the symmetry properties of the wavefunction, and a third method is to use the mathematical properties of the wavefunction. We illustrate that method here.
The expression for the transition moment is

$$M_x = a \int_{-\infty}^{\infty} \psi_i\ x \psi_f \cdot \mathrm{d}x$$

or on substitution of the wavefunctions for this problem

$$\begin{aligned}M_x &= a \int_{-\infty}^{\infty} \psi_0\ x \psi_3 \cdot \mathrm{d}x \\ &= aA \int_{-\infty}^{\infty} H_0(x) x H_3(x) \exp\left(-\alpha x^2\right) \cdot \mathrm{d}x\end{aligned}$$

The recursion relation for the Hermite polynomials

$$xH_n(x) = \frac{1}{2}H_{n+1}(x) + nH_{n-1}(x)$$

allows us to write the integrand as

$$\begin{aligned}M_x &= aA \int_{-\infty}^{\infty} H_0(x) \cdot \left[\frac{1}{2}H_4(x) + 3H_2(x)\right] \exp\left(-\alpha x^2\right)\ \mathrm{d}x \\ &= \frac{aA}{2} \int_{-\infty}^{\infty} H_0(x) H_4(x) \exp\left(-\alpha x^2\right)\ \mathrm{d}x \\ &\quad + 3aA \int_{-\infty}^{\infty} H_0(x) H_2(x) \exp\left(-\alpha x^2\right)\ \mathrm{d}x \\ &= \frac{a}{2} \int_{-\infty}^{\infty} \psi_0 \psi_4 \cdot \mathrm{d}x + 3a \int_{-\infty}^{\infty} \psi_0 \psi_2 \cdot \mathrm{d}x\end{aligned}$$

Both of these integrals are equal to zero, because of the orthogonality of ψ_0 and ψ_4 or ψ_2, respectively; hence, their sum is zero. Thus $M_x = 0$ for the $n = 0$ to $n = 3$ transition.

P11.7 Consider the radical species OH and OD. Their vibrational energy level spectrum is given by Equation (11.42):

$$E_n = h\nu_0(n + \frac{1}{2}) - C(n + \frac{1}{2})^2 \quad \text{with} \quad C = \frac{(h\nu_0)^2}{4\varepsilon_B}$$

with the following parameters

Species	d/pm	ε_B/eV	$\frac{\nu_0}{c_0}$ (cm^{-1})	$\frac{\nu_0}{c_0}C$ (cm^{-1})
OH	97.06	4.35	3735.2	82.8
OD	96.99	4.39	2720.9	44.2

a) Graph the positions of the fundamental vibration and the first four overtones of OH; i.e., the infrared transitions. b) On the same graph plot the fundamental and the first four overtones of OD. c) Why is the anharmonic term ($\nu_0 C$) smaller for OD than OH? d) Sketch the vibrational spectrum you would expect to observe for OH.

Solution

a,b) First we calculate the fundamentals using the formula

$$\Delta E_{0\to 1} = E_1 - E_0 = h\nu_0 - 2C$$

Rather than convert the parameters from wavenumbers to energies, we perform the calculation in wavenumbers. Because $E = hc_0\tilde{\nu}$, the ordering of the energy level scheme will be the same in either set of units. Hence we find

$$\tilde{\nu}_{0\to 1}(\text{OH}) = 3735.2 - 2(82.8) = 3569.6\ \text{cm}^{-1}$$

$$\tilde{\nu}_{0\to 1}(\text{OD}) = 2720.9 - 2(44.2) = 2632.5\ \text{cm}^{-1}$$

To calculate the overtones, we use the formula

$$\tilde{\nu}_{0\to n} = \tilde{\nu}_n - \tilde{\nu}_0 = n\tilde{\nu}_0 - \left(n^2 + n\right)\tilde{\nu}_0 x$$

and find

	$\tilde{\nu}_{0\to n}$(OH)	$\tilde{\nu}_{0\to n}$ (OD)	
0 → 1	3569.6 cm^{-1}	2632.5 cm^{-1}	fundamental
0 → 2	6973.6 cm^{-1}	5176.6 cm^{-1}	
0 → 3	10212.0 cm^{-1}	7632.3 cm^{-1}	
0 → 4	13284.8 cm^{-1}	9999.6 cm^{-1}	
0 → 5	16192.0 cm^{-1}	12278.5 cm^{-1}	

The plot of the energy levels is given in Fig. P11.7a. We have chosen the zero of the scale to be the location of the electronic energy minimum so that the relative zero point vibrational energies would be apparent.

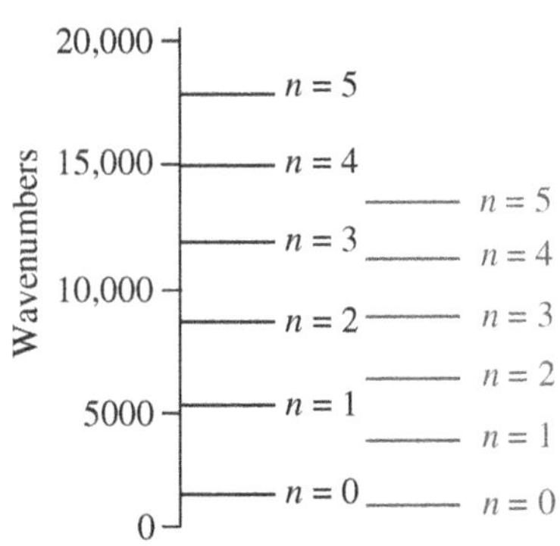

Figure P11.7a Fundamental vibration and the first four overtones of OH (black). Fundamental and the first four overtones of OD (dark gray).

c) The anharmonic term is higher for OH because its characteristic frequency is higher than that of OD (because of its reduced mass). For this reason OH samples higher energy regions of the electronic potential energy well than does OD, for a given value of n.

d) In Fig. 11.7b we plot the transition energies for OH (black lines) and OD (dark gray lines). The "stick spectrum" does not provide information on the relative intensities of the transitions. Only the position of the transition is significant in this plot, i.e., the heights are arbitrarily set to 1.

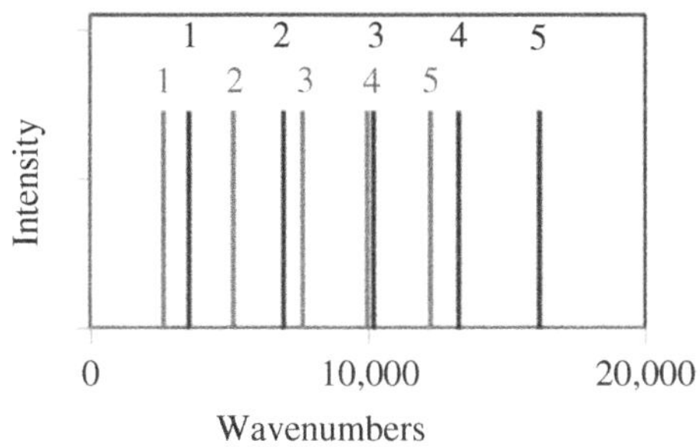

Figure P11.7b Transition energies for OH (black lines) and OD (dark gray lines). The numbers 1 to 5 denote transitions $0 \rightarrow 1, 0 \rightarrow 2, 0 \rightarrow 3, 0 \rightarrow 4$, and $0 \rightarrow 5$.

P11.8 The table provides transition wavelengths and absorption cross-sections $\sigma = 2.303\varepsilon/\mathbf{N}_A$, where ε is the molar decadic extinction coefficient, for vibronic transitions in the first allowed electronic transition of CO. Using these data make a qualitative sketch of the ground and excited state potential energy curves. Discuss how your potential energy curves can explain the relative transition intensities and energy spacings. Explain how you can determine the force constant for the excited state potential energy surface and do so for the lowest energy and highest energy spacings.

Ground State (X)	Excited State (A)	λ/nm	$\sigma/10^{-18}$ cm^2
$n = 0$	$n = 0$	154.45	350 ± 35
$n = 0$	$n = 1$	150.98	680 ± 68
$n = 0$	$n = 2$	147.77	820 ± 82
$n = 0$	$n = 3$	144.74	700 ± 70
$n = 0$	$n = 4$	141.91	460 ± 46
$n = 0$	$n = 5$	139.26	280 ± 28
$n = 0$	$n = 6$	136.77	172 ± 17
$n = 0$	$n = 7$	134.43	95 ± 10
$n = 0$	$n = 8$	132.22	45 ± 5
$n = 0$	$n = 9$	130.15	25.4 ± 3
$n = 0$	$n = 10$	128.19	9.5 ± 1.2
$n = 0$	$n = 11$	126.35	4.0 ± 0.4

Solution

Figure P11.8 sketches the ground and excited state potential energy curves. The horizontal lines (dashed in ground state and solid in excited state) demark the vibrational energy levels in the two electronic states. The minima of the two excited electronic state curves are displaced from one another slightly so that the maximum Franck–Condon overlap results for the transition from $n = 0$ in the ground electronic state to $n = 2$ in the excited electronic state.

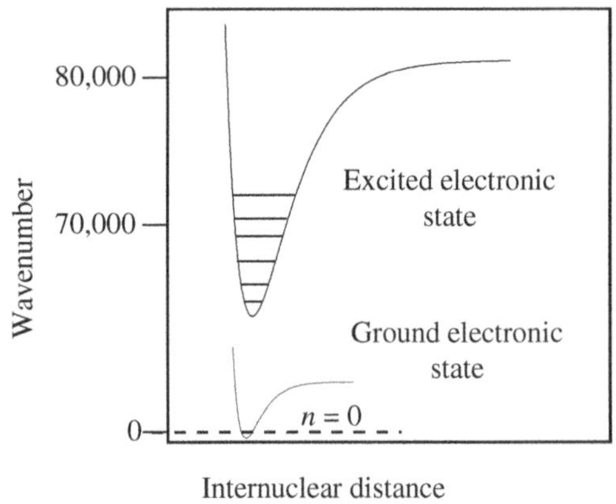

Figure P11.8 Potential energy curves for CO.

To calculate the force constant we find the difference in energy between two of the vibronic lines that have their initial state in the ground vibronic level. For example the difference in energy between the first two vibronic transitions corresponds to the difference in energy between $n = 0$ and $n = 1$ of the excited. Namely we find

Ground State (X)	Excited State (A)	λ (nm)	$\Delta E/10^{-18}$ J
$n = 0$	$n = 0$	154.45	1.2862
$n = 0$	$n = 1$	150.98	1.3157

so that ΔE(J) of $0 \rightarrow 1$ in the A-state is

$$\Delta E = 1.3157 \times 10^{-18}\ \text{J} - 1.2862 \times 10^{-18}\ \text{J} = 2.95 \times 10^{-20}\ \text{J}$$

and when converted to wavenumbers is 1500 cm^{-1}. In the harmonic oscillator approximation, this transition energy is given by

$$\Delta E = \frac{h}{2\pi}\sqrt{\frac{k_f}{\mu}}$$

Using $\mu = 1.138 \times 10^{-26}$ kg (see E11.3), we find that

$$\begin{aligned} k_f &= \mu\left(\frac{2\pi \cdot \Delta E}{h}\right)^2 \\ &= \left(1.138 \times 10^{-26}\ \text{kg}\right)\left(\frac{2\pi \cdot 2.95 \times 10^{-20}\text{J}}{6.626 \times 10^{-34}\ \text{J s}}\right)^2 \\ &= 891\ \text{kg s}^{-2} = 891\ \text{N m}^{-1} \end{aligned}$$

This value is significantly weaker than the 1903 N m^{-1} value found for the ground state of CO, and it is consistent with the promotion of a valence electron from a bonding orbital to an antibonding orbital thereby weakening the triple bond of CO.
Correspondingly for the highest energy transitions, we find ΔE of $10 \rightarrow 11$ in the A-state is

$$\Delta E = 1.5722 \times 10^{-18}\ \text{J} - 1.5496 \times 10^{-18}\ \text{J} = 2.26 \times 10^{-20}\ \text{J}$$

The change in the energy spacing indicates a significant anharmonicity in the potential. Although it is straightforward to use the harmonic oscillator approximation to calculate a force constant of 522 N m^{-1}, its relevance is not clear.

P11.9 a) Illustrate the Franck–Condon principle using the potential energy curves of ^{40}CaH that are given in Fig. P11.9a. Sketch a vibrationally resolved absorption spectrum associated with transitions from the ground rovibronic state of ^{40}CaH to the energy levels of the excited electronic state of ^{40}CaH, which is given in the diagram. Describe the important features of your sketch. b) Sketch the absorption spectrum you would expect to observe for transitions between the ground vibrational state of this molecule's ground electronic state to the first excited vibrational state of this molecule's ground electronic state. This transition is centered at 1299 cm^{-1}. Indicate clearly any features in your spectrum. c) State four molecular parameters that can be determined from the spectra in parts a and b. Describe in a detailed manner how you would go about determining them.

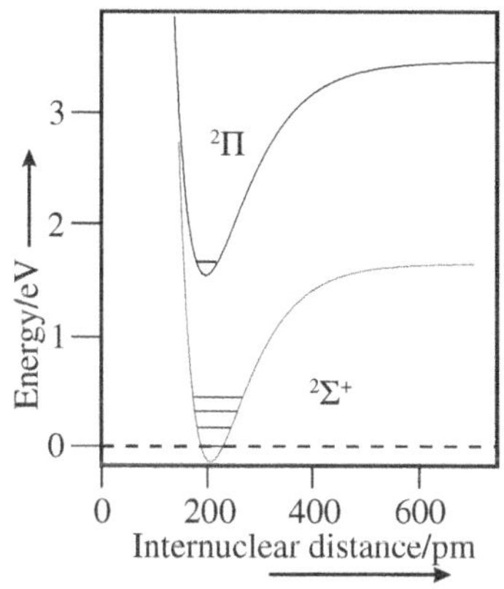

Figure P11.9a Potential energy curves of ^{40}CaH. This image is redrawn from the curve reported by G. Herzberg, *Molecular Spectra and Molecular Structure 1. Spectra of Diatomic Molecules*, 2nd edition (Van Nostrand, New York, 1950).

Solution

a) The minima of the potential energy curves in Fig. P11.9a are nearly coincident. Hence we expect that the highest intensity peak in the transition will be the 0–0 transition (if not then it is likely the 0–1 transition). The stick diagram illustrates this features (Fig. P11.9b). The spacing between the lines in this spectrum is determined by the energy spacing between the vibrational levels in the excited electronic state potential curve. In principle this spacing could be used to determine features of the excited state potential, such as its force constant. The intensity pattern is controlled by the overlap of the vibrational wavefunctions, and it could be used to determine the magnitude of the displacement between the minima of the two energy curves.

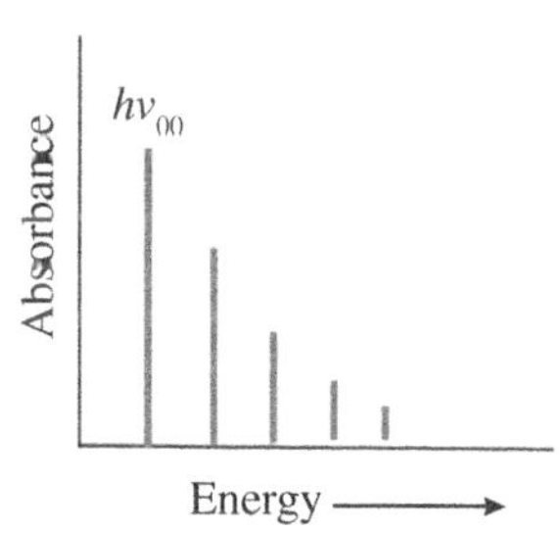

Figure P11.9b Transitions for the case that the minima of the potential energy curves in Fig. P11.10a are nearly coincident.

b) This transition occurs within the ground electronic state and shows rovibrational structure. The Q branch does not appear because the electronic state does not carry any angular momentum. The P and R branches will occur and the spacing of the lines can be used to determine features of the molecular geometry, such as the inertial moment (Fig. P11.9c).

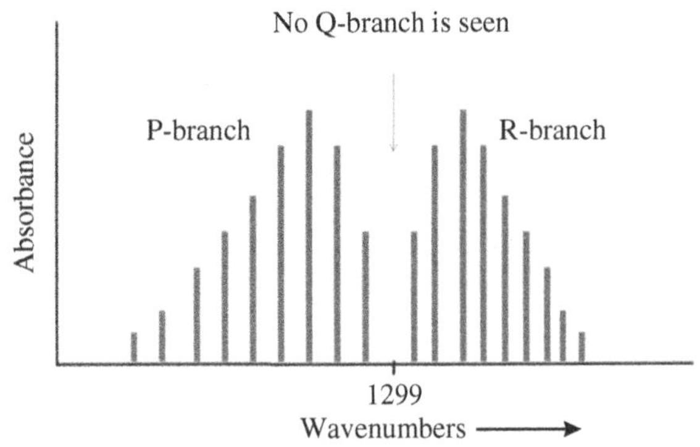

Figure P11.9c Rovibrational spectrum for ^{40}CaH near its vibrational fundamental

c) From the vibronic spectrum in part a), one can determine the force constant in the excited state from the vibronic energy level spacing. If the transitions are pronounced and intense enough it might be possible to determine the anharmonicity also. The difference in electronic energy between the two minima of the potential energy curves could also be determined since the force constant of the excited state can be found and the ground-state characteristic frequency is known. From part b), it should be possible to determine the inertial moment of the molecule and hence the bond length of the molecule.

P11.10 For the CO molecule the following wavenumbers for the fundamental and the overtone bands were measured: $0 \to 1 : 2144\ \text{cm}^{-1}, 0 \to 2 : 4266\ \text{cm}^{-1}, 0 \to 3 : 6357\ \text{cm}^{-1}$. Evaluate the force constant k_f and the bond energy ε_B, in analogy to the treatment of HCl in Section 11.3.4

Solution

As for HCl we use the approximate energy expression (Equation (11.42)),

$$E = h\nu_0\left(\frac{1}{2}+n\right) - C\cdot\left(\frac{1}{2}+n\right)^2 \text{ with } C = \frac{(h\nu_0)^2}{4\varepsilon_B}$$

Hence the excitation energies from quantum state $n' = 0$ to a state with quantum number n, is given by Equation (11.44),

$$\Delta E_{0\to n} = h\nu_0 \cdot n - C\cdot n(n+1) \text{ or } \frac{\Delta E_{0\to n}}{n} = h\nu_0 - C\cdot(n+1)$$

Using the data given we find for CO that

n	$\tilde{\nu}/\text{cm}^{-1}$	$\nu/10^{13}\text{s}^{-1}$	$\Delta E_{0\to n}/10^{-20}$ J	$(n+1)$	$\frac{\Delta E_{0\to n}/10^{-20}\text{ J}}{n}$
1	2144	6.432	4.262	2	4.262
2	4266	12.80	8.481	3	4.240
3	6357	19.07	12.636	4	4.212

In Fig. P11.10 we plot $\Delta E_{0\to n}/n$ versus $(n+1)$.

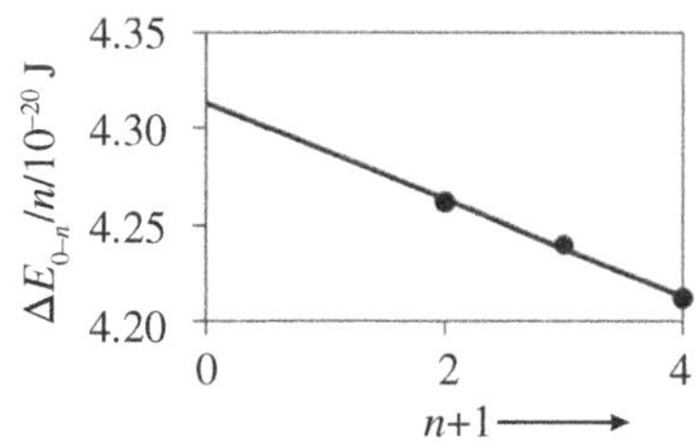

Figure P11.10a CO molecule. Excitation energy versus $n + 1$.

The intercept for this plot gives $h\nu_0 = 4.313 \cdot 10^{-20}$ J and the slope gives $C = 2.50 \cdot 10^{-22}$ J. Solving for the bond energy we find that

$$\varepsilon_B = \frac{(h\nu_0)^2}{4C} = \frac{(4.313 \cdot 10^{-20}\ \mathrm{J})^2}{4 \times 2.485 \cdot 10^{-22}\ \mathrm{J}} = 18.71 \cdot 10^{-19}\ \mathrm{J} = 11.68\ \mathrm{eV}$$

This value is much smaller than the value reported in Table 7.6 which is $\varepsilon_B = 1.788 \cdot 10^{-18}\mathrm{J} = 11.16$ eV. As stated in the text, this discrepancy results from the fact that the Morse function is only an approximation to the exact potential energy curve, and the small remaining deviations are sufficient to make an extrapolation to the bond energy a difficult problem.
To find the force constant we use the facts that

$$h\nu_0 = \frac{h}{2\pi}\sqrt{\frac{k_f}{\mu}} \text{ and } \mu = \frac{m_\mathrm{C} \cdot m_\mathrm{O}}{m_\mathrm{C} + m_\mathrm{O}} = \frac{(12)(16)}{(12+16)} \cdot 1.66 \cdot 10^{-27}\ \mathrm{kg}$$

so that

$$\begin{aligned} k_f &= \frac{(2\pi h\nu_0)^2}{h^2}\mu \\ &= \frac{4\pi^2(4.313 \cdot 10^{-20}\mathrm{J})^2}{(6.626 \cdot 10^{-34}\ \mathrm{J\ s})^2} \cdot 1.14 \cdot 10^{-26}\ \mathrm{kg} \\ &= 1905\ \mathrm{kg\ s^{-2}} = 1905\ \mathrm{N\ m^{-1}} \end{aligned}$$

which is in excellent agreement with the value 1903 N m^{-1} that is given in Table 7.6.

P11.11 Consider the ^{12}C^{14}N molecule. The potential energy curves of the first two electronic states are shown in Fig. P11.11a and the vibrational energy levels in each state are given by $E_n = h\nu_0(n + \frac{1}{2}) - C(n + \frac{1}{2})^2$. The molecular constants are given in the table. B is the rotational constant for the equilibrium bond length.

State	d/pm	$\left[\frac{B}{hc_0}\right]$/cm^{-1}	$\frac{\nu_0}{c_0}$ (cm^{-1})	$\frac{\nu_0}{c_0}C$ (cm^{-1})
X	117.18	1.8996	2068.71	13.14
A	123.27	1.7165	1814.43	12.83

a) Sketch the rovibrational spectrum of the molecule in the X state. Be sure to label the lines as quantitatively as possible. Assume the molecule is initially in $n = 0$. Perform a similar sketch for the A state. In what ways do these two sketches differ? b) Give an explicit expression for the Franck–Condon factor involved in the lowest energy allowed electronic transition between the X and A states. Assume harmonic oscillator wavefunctions are valid. Evaluate the integral.
(Note that the X state is $^2\Sigma^+$ and does not have electronic angular momentum, whereas the A state is $^2\Pi$ and does have angular momentum).

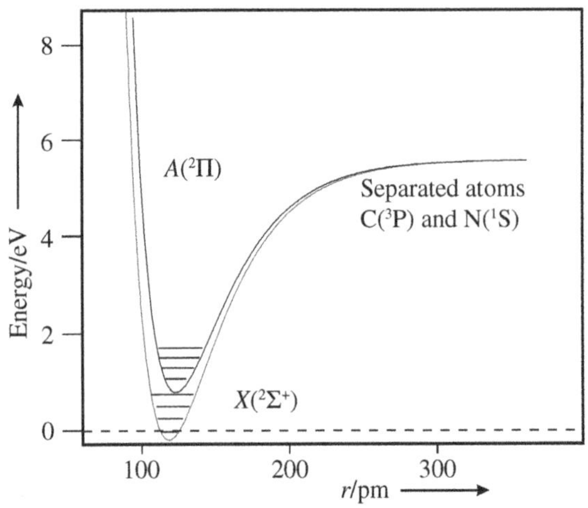

Figure P11.11a Potential energy curves of the first two electronic states of $^{12}C^{14}N$. This image is redrawn from the curve reported by G. Herzberg, *Molecular Spectra and Molecular Structure 1. Spectra of Diatomic Molecules*, 2nd edition (Van Nostrand, New York, 1950).

Solution

a) First we must determine the fundamental frequency that would be associated with the Q branch, and then we can determine the line positions for the P and R branches. The vibrational fundamental occurs at an energy of

$$\Delta E = E_1 - E_0 = h\nu_0 - 2h\nu_0 x$$

By converting to wavenumbers we find

$$\tilde{\nu} = \frac{\Delta E}{hc_0} = \frac{\nu_0}{c_0} - 2\frac{\nu_0}{c_0}x$$

For the X state we find that

$$\tilde{\nu} = (2068.71 - 2 \cdot 13.14)\ \text{cm}^{-1} = 2042.43\ \text{cm}^{-1}$$

is the position where the Q branch would occur (because the X-state has zero electronic angular momentum this transition does not occur); see Fig. P11.11b. A series of transitions will occur at higher wavenumber at $\tilde{\nu} + 2B/(hc_0)$, $\tilde{\nu} + 4B/(hc_0)$, $\tilde{\nu} + 6B/(hc_0)$, ... and so on corresponding to the R-branch. Also, a series of transitions will occur at lower wavenumber at $\tilde{\nu} - 2B/(hc_0)$, $\tilde{\nu} - 4B/(hc_0)$, $\tilde{\nu} - 6B/(hc_0)$, ... and so on corresponding to the P-branch.

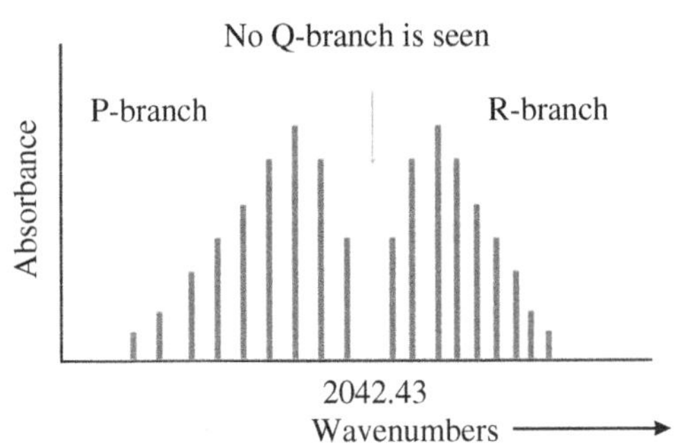

Figure P11.11b Rovibratonal spectrum for $^{12}C^{14}N$ (X-state) at its vibrational fundamental.

For the A state we find that

$$\tilde{\nu} = (1814.43 - 2 \cdot 12.83)\ \text{cm}^{-1} = 1788.77\ \text{cm}^{-1}$$

is the position where the Q branch occurs (because the A-state has electronic angular momentum this transition does occur); see Fig. P11.11b.

As with the X-state both an R- branch (transitions at $\tilde{\nu} + 2B/(hc_0)$, $\tilde{\nu} + 4B/(hc_0)$, $\tilde{\nu} + 6B/(hc_0)$, ...) and a P-branch (transitions at $\tilde{\nu} - 2B/(hc_0)$, $\tilde{\nu} - 4B/(hc_0)$, $\tilde{\nu} - 6B/(hc_0)$, ...) are seen.

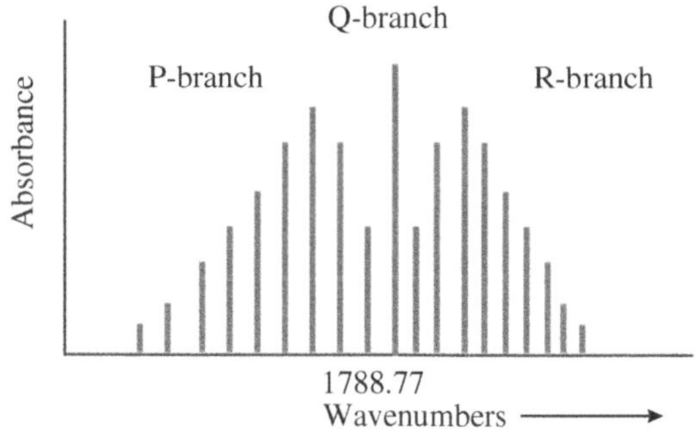

Figure P11.11c Rovibrational spectrum of $^{12}C^{14}N$ (A-state) near its vibrational fundamental. Note the presence of the Q branch.

The most dramatic difference between the two spectra is the appearance of the Q-branch for the A-state, which occurs because it has electronic angular momentum. In addition, the two spectra differ by the center frequency (it is lower for the A-state) and the spacing between the lines in the P and R branches (reflecting the change in bond length).

b) The Franck–Condon factor for the $0 - 0$ transition between the X and A state is

$$FCF = \left| \int_{-\infty}^{\infty} \psi_g \psi_e \, dx \right|^2$$

where

$$\int_{-\infty}^{\infty} \psi_g \psi_e \, dx = \sqrt[4]{\frac{a_g}{\pi}} \sqrt[4]{\frac{a_e}{\pi}} \int_{-\infty}^{\infty} \exp\left(-a_g x^2/2\right) \exp\left(-a_e (x-\delta)^2/2\right) \, dx$$

with

$$a_g = \frac{2\pi}{h}\sqrt{k_{f,g}\mu}; \; a_e = \frac{2\pi}{h}\sqrt{k_{f,e}\mu} \quad \text{and} \quad \delta = 123.27 - 117.18 = 6.09 \text{ pm}$$

We can simplify the integrand and write

$$\begin{aligned}
\int_{-\infty}^{\infty} \psi_g \psi_e \, dx &= \sqrt[4]{\frac{a_g}{\pi}} \sqrt[4]{\frac{a_e}{\pi}} \exp\left(-a_e \delta^2/2\right) \int_{-\infty}^{\infty} \exp\left(-\left(a_g + a_e\right) x^2/2 + \delta a_e x\right) dx \\
&= \sqrt[4]{\frac{a_g}{\pi}} \sqrt[4]{\frac{a_e}{\pi}} \exp\left(-a_e \delta^2/2\right) \int_{-\infty}^{\infty} \exp\left(-Ax^2 + Bx\right) dx
\end{aligned}$$

where $A = \left(a_g + a_e\right)/2$ and $B = \delta a_e$. Now we use the fact that

$$\int_{-\infty}^{\infty} \exp\left(-ax^2 + bx\right) dx = \sqrt{\frac{\pi}{a}} \exp\left(\frac{b^2}{4a}\right)$$

to write

$$\begin{aligned}
\int \psi_g \psi_e \, dx &= \sqrt{\frac{\sqrt{a_g a_e}}{\pi}} \exp\left(-a_e \delta^2/2\right) \sqrt{\frac{\pi}{A}} \exp\left(\frac{B^2}{4A}\right) \\
&= \sqrt{\frac{2\sqrt{a_g a_e}}{\left(a_g + a_e\right)}} \exp\left(-\frac{a_e \delta^2}{2} + \frac{\delta^2 a_e^2}{2\left(a_g + a_e\right)}\right) \\
&= \sqrt{\frac{2\sqrt{a_g a_e}}{\left(a_g + a_e\right)}} \exp\left(-\frac{\delta^2 a_e a_g}{2\left(a_g + a_e\right)}\right)
\end{aligned}$$

and we find that

$$FCF = \left|\int \psi_g \psi_e \, \mathrm{d}x\right|^2 = \frac{2\sqrt{a_g a_e}}{(a_g + a_e)} \exp\left(-\frac{\delta^2 a_e a_g}{(a_g + a_e)}\right)$$

Now we can evaluate this result for CN. Given that $a = \frac{2\pi}{h}\sqrt{k_f \mu}$ and $\nu_0 = \sqrt{k_f/\mu}/(2\pi)$, we can write that

$$a = \frac{2\pi}{h}\sqrt{k_f \mu} = \frac{(2\pi)^2}{h}\nu_0 \mu$$

so that

$$\frac{a_e a_g}{(a_g + a_e)} = \frac{(2\pi)^2}{h}\mu \frac{\nu_{0,e} \cdot \nu_{0,g}}{\nu_{0,e} + \nu_{0,g}} \text{ and } \frac{2\sqrt{a_g a_e}}{(a_g + a_e)} = 2\frac{\sqrt{\nu_{0,e} \cdot \nu_{0,g}}}{\nu_{0,e} + \nu_{0,g}}$$

Hence we can write that

$$FCF = \left|\int \psi_g \psi_e \, \mathrm{d}x\right|^2 = 2\frac{\sqrt{\nu_{0,e} \cdot \nu_{0,g}}}{\nu_{0,e} + \nu_{0,g}} \exp\left(-\delta^2 \frac{(2\pi)^2}{h}\mu \frac{\nu_{0,e} \cdot \nu_{0,g}}{\nu_{0,e} + \nu_{0,g}}\right)$$

Using $\delta = 6.09$ pm, $\nu_{0,g} = 6.20613 \times 10^{13}\ \mathrm{s}^{-1}$, $\nu_{0,e} = 5.44329 \times 10^{13}\ \mathrm{s}^{-1}$, $h = 6.635 \times 10^{-34}$ J s, and $\mu = 1.073 \times 10^{-26}$ kg, we find that

$$\begin{aligned} FCF &= 2\,(0.498927) \exp\left(-\frac{(6.09)^2 \times 10^{-24}(2\pi)^2}{6.635 \times 10^{-34}}\left(1.073 \times 10^{-26}\right)\left(2.89967 \times 10^{13}\right)\right) \\ &= 0.997854 \exp(-0.686) = 0.502 \end{aligned}$$

P11.12 Use Equation (11.15) to fit the experimental data in Table 11.1 for the infrared absorption lines in the rotational spectrum of HCl. Compare your result with the result from the rigid rotator calculation that is given in the text.

Solution

Equation (11.15) in Section 11.2.3 is

$$E_J = B \cdot J(J+1) - D \cdot J^2(J+1)^2 \quad \text{with} \quad D = \frac{2B^2}{k_f \cdot d_0^2}$$

Note that the term containing D is due to the centrifugal distortion. We calculate ΔE and $\Delta\Delta E$ and obtain

$$\Delta E = E_{J+1} - E_J = B \cdot 2\,(J+1) - 4D(J+1)^3$$

Subsequently, we find that

$$\begin{aligned} \Delta\Delta E &= \Delta E_{J+1 \to J+2} - \Delta E_{J \to J+1} \\ &= B \cdot 2\,(J+2) - 4D(J+2)^3 - B \cdot 2\,(J+1) + 4D(J+1)^3 \\ &= 2BJ + 4B - 4D\left[J^3 + 6J^2 + 12J + 8\right] \\ &\quad - 2BJ - 2B + 4D\left[J^3 + 3J^2 + 3J + 1\right] \\ &= 2B - 4D\left[3J^2 + 9J + 7\right] = 2B - 4D\,[3J\,(J+3) + 7] \\ &= 2B - 28D - 12D \cdot J\,(J+3) \end{aligned} \tag{11P12.1}$$

In Fig. P11.12 we plot $\Delta\Delta E$ versus $J(J+3)$, using the data in Table 11.1.

Table 11.1 Evaluation of the Absorption Lines in the Rotational Spectrum of HCl in Fig. 11.7.

J	$1/\lambda$	ΔE	$\Delta(\Delta E)$	J	$1/\lambda$	ΔE	$\Delta(\Delta E)$
	cm^{-1}	10^{-23} J			cm^{-1}	10^{-23} J	
0	(20.93)	(41.58)		7	165.83	329.4	40.7
1	41.78	82.99	41.4	8	186.25	370.0	40.6
2	62.60	124.3	41.3	9	206.54	410.3	40.3
3	83.35	165.6	41.3	10	226.71	450.3	40.0
4	104.02	206.6	41.0	11	246.82	490.3	40.0
5	124.74	247.8	41.2	12	266.70	529.8	39.5
6	145.36	288.7	40.9	13	286.50	569.1	39.3

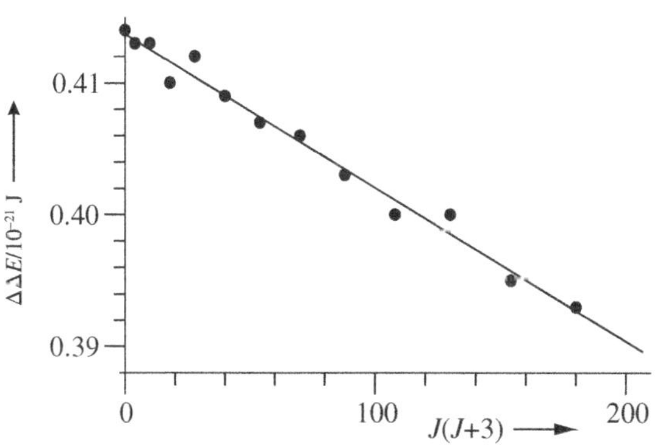

Figure P11.12 Plot of $\Delta\Delta E$ as a function of $J(J+3)$, according to Equation (11P12.1).

From the slope of the regression line, 1.194×10^{-25} J, we obtain

$$12D = 1.194 \times 10^{-25}\ \text{J} \quad \text{and} \quad D = 9.950 \times 10^{-27}\ \text{J}$$

From the intercept we obtain

$$2B - 28D = 0.414 \times 10^{-21}\ \text{J}$$

and

$$B = \frac{0.414 \times 10^{-21}\ \text{J} + 28D}{2}$$
$$= \frac{0.414 \times 10^{-21}\ \text{J} + 0.028 \times 10^{-23}\ \text{J}}{2} = 2.07 \times 10^{-22}\ \text{J}$$

This B value is slightly larger than the value $B = 2.029 \times 10^{-22}$ J calculated in Section 11.2.3 using an average for $\Delta\Delta E$.

For comparison, we calculate D from the Equation

$$D = \frac{2B^2}{k_f \cdot d_0^2}$$

with $B = 2.07 \times 10^{-22}$ J, $k_f = 516$ N m^{-1} (see Section 11.3.4), and $d_0 = 128$ pm:

$$D = \frac{2\left(2.07 \times 10^{-22}\ \text{J}\right)^2}{516\ \text{N m}^{-1} \cdot (128\ \text{pm})^2} = 1.01 \times 10^{-26}\ \text{J}$$

This value agrees excellently with the directly obtained value $D = 0.995 \times 10^{-26}$ J.

P11.13 Show that the anharmonicity of the potential energy curve causes the R-branch lines of HCl to come closer together with increasing J, whereas those for the P-branch move farther apart. Begin with the energy expressions for the two vibrational quantum states of

$$E_J(v=0) = B_0 J(J+1) + \frac{1}{2}h\nu_0 \ , \ J = 0,1,2,....$$
$$E_J(v=1) = B_1 J(J+1) + \frac{3}{2}h\nu_0 \ , \ J = 0,1,2,....$$

where $E_J(v = 0)$ is the energy of states originating from the lowest vibrational state and $E_J(v = 1)$ is the energy of states originating from the first excited vibrational state. Derive expressions for the difference in energy corresponding to the R-branch lines (a $\Delta E(J)$) and then an expression for the difference in energy of these ΔE (thus a $\Delta\Delta E$). Then perform a corresponding calculation for the P-branch lines and compare the formulas to each other.

Solution

The excitation energy of the allowed transitions (selection rule $\Delta J = \pm 1$) is

$$\text{(a)}\ \Delta J = +1(\text{R branch}), J = 0,1,2... :$$
$$\Delta E_{a,J} = E_{J+1}(v=1) - E_J(v=0)$$
$$= h\nu_0 + B_1(J+1)(J+2) - B_0 J(J+1)$$

$$\text{(b)}\ \Delta J = -1(\text{P branch}), J = 1,2,3...... :$$
$$\Delta E_{b,J} = E_{J-1}(v=1) - E_J(v=0)$$
$$= h\nu_0 + B_1(J-1)J - B_0 J(J+1)$$

The spacings between adjacent absorption lines are

$$\text{(a)}\ \Delta J = +1(\text{R branch}):$$
$$\Delta\Delta E_{a,J} = \Delta E_{a,J+1} - \Delta E_{a,J}$$
$$= B_1(J+2)(J+3) - B_0(J+1)(J+2) - B_1(J+1)(J+2) + B_0 J(J+1)$$
$$= 2B_1(J+2) - 2B_0(J+1) = 2(B_1 - B_0)J + (4B_1 - 2B_0)$$

$$\text{(b)}\ \Delta J = -1(\text{P branch}):$$
$$\Delta\Delta E_{b,J} = \Delta E_{b,J-1} - \Delta E_{b,J}$$
$$= B_1(J-2)(J-1) - B_0 J(J-1) - B_1 J(J-1) + B_0 J(J+1)$$
$$= B_1[(J-2) - J]\ (J-1) + 2B_0(J) = 2(B_0 - B_1)J + \ 2B_0$$

Because of the anharmonicity of the oscillator the equilibrium distance in the $v = 1$ state is larger than the equilibrium distance in the $n = 0$ state; thus, we obtain $B_1 < B_0$. Using the definition

$$B_0 = B_1 + \Delta B$$

we obtain

$$\text{(a)}\ \ \Delta J = +1(\text{R branch}): \Delta\Delta E_{a,J} = 2B_1 - 2\Delta B \cdot J - 2\Delta B = 2B_1 - 2\Delta B \cdot (J+1)$$

$$\text{(b)}\ \ \Delta J = -1(\text{P branch}): \Delta\Delta E_{b,J} = 2B_0 + 2\Delta B \cdot J' = 2B_1 + 2\Delta B \cdot (J+1)$$

In the special case $\Delta B = 0$, so that $B_1 = B_0$, these relations give

$$\Delta\Delta E_{a,J} = 2B_0 \ , \ \Delta\Delta E_{b,J} = 2B_0$$

as we derived for the harmonic oscillator. In the case of the anharmonic oscillator we see that the spacing of the lines in the R branch decreases with increasing quantum number J, and the spacing of the lines in the P branch increases with increasing quantum number J, in accordance with the absorption spectrum in Fig. 11.16.

In contrast, the effect of centrifugal distortion would lead to a decrease of the spacings with increasing J in both branches. Moreover, this effect is much smaller than the anharmonicity effect.

P11.14 Use the general form of the harmonic oscillator wavefunctions from Foundation 11.4 (on the book's website), namely

$$\psi_n(q) = N_n \cdot H_n(q) \cdot \exp(-q^2/2)$$

to derive the selection rule $\Delta n = \pm 1$ for a vibrational transition; i.e., show that

$$M_x = \int \psi_{n'}(q) \cdot q \cdot \psi_n(q) \, dq$$

is only nonzero for $n' = n \pm 1$.

Solution

We can write the transition moment integral as

$$M_x = A \int \psi_{n'}(q) \cdot q \cdot \psi_n(q) \, dq = A N_n N_{n'} \int H_{n'}(q) \cdot q \cdot H_n(q) \cdot \exp(-q^2/2) \, dq$$

where A is a constant (note that q is proportional to x). Using the recursion relation for the Hermite polynomials, $H_{n+1} = 2qH_n - 2nH_{n-1}$ (see Foundation 11.3), we can write

$$qH_n(q) = \frac{1}{2} H_{n+1}(q) + n \, H_{n-1}(q)$$

Now we substitute this expression into the integral for $qH_n(q)$ and it becomes

$$M_x = A N_n \; N_{n'} \int H_{n'}(q) \left[\frac{1}{2} H_{n+1}(q) + n \, H_{n-1}(q) \right] \exp(-q^2) dq$$

$$= A N_n \; N_{n'} \begin{bmatrix} \frac{1}{2} \int H_{n'}(q) H_{n+1}(q) \cdot \exp(-q^2) \, dq \\ + n \int H_{n'}(q) H_{n-1}(q) \cdot \exp(-q^2) \, dq \end{bmatrix}$$

This equation can be written in terms of the vibrational wavefunctions of the harmonic oscillator, namely

$$M_x = A N_n \left[\frac{1}{2N_{n+1}} \int \psi_{n'}(q) \; \psi_{n+1}(q) \, dq + \frac{n}{N_{n-1}} \int \psi_{n'}(q) \; \psi_{n-1}(q) \, dq \right]$$

From the orthonormality of the harmonic oscillator wavefunctions, we find that the first integral is zero unless $n' = n + 1$ and the second integral is zero unless $n' = n - 1$.
Thus, the selection rule for light absorption of a vibrating molecule is $\Delta n = \pm 1$ in the harmonic oscillator approximation.

P11.15 Use the infrared absorption spectrum of CO_2 in Fig. 11.19 to estimate the absorbance for the CO_2 bending mode in the atmosphere. The current pressure of CO_2 on sea level is about $P_{\text{atmosphere}} = 400$ ppm $\times 1$ bar $= 400 \times 10^{-6}$ bar. Assume that the atmosphere reaches up to 15km (troposphere). Compare your result to the CO_2 concentration of 280×10^{-6} bar in the pre-industrial period

Solution

From Fig. 11.19 we extract an absorbance of about $A_0 = 1.0$ for the absorption maximum of the bending mode at a path length of $d_0 = 10$ cm and a CO_2 pressure of $P_0 = 100$ mbar. Let us assume that the CO_2 pressure in the atmosphere decays linearly with height from 400×10^{-6} bar on sea level to about zero at a height of 15 km; that is, we assume an active path length of $d_{\text{atmosphere}} = 7.5$ km and an average CO_2 pressure of 200×10^{-6} bar. According to Box 9.1 the absorbance A is defined as

$$A = \varepsilon \, c \, d$$

where ε is the decadic absorption coefficient. Thus we obtain

$$A_{\text{atmosphere}} = A_0 \frac{\varepsilon \, c_{\text{atmosphere}} \; d_{\text{atmosphere}}}{\varepsilon \, c_0 \; d_0} = A_0 \frac{c_{\text{atmosphere}} \; d_{\text{atmosphere}}}{c_0 \; d_0}$$

with

$$\frac{c_{\text{atmosphere}} \; d_{\text{atmosphere}}}{c_0 \; d_0} = \frac{P_{\text{atmosphere}} \; d_{\text{atmosphere}}}{P_0 \; d_0}$$

Thus for the absorbance $A_{atmosphere}$ we obtain

$$A_{atmosphere} = 1.0\frac{200 \times 10^{-6}\ \text{bar}\, 7.5 \times 10^{3}\ \text{m}}{100 \times 10^{-3}\ \text{bar}\, 0.10\ \text{m}} = 150$$

Even if we assume that the mean absorbance within the absorption band of the bending mode is only 1/10 of the absorbance at the maximum we calculate a value of $A_{atmosphere} = 15$. Thus we conclude that all infrared radiation arriving within this absorption band is completely absorbed in the atmosphere. Using the preindustrial value of $P_{atmosphere} = 280 \times 10^{-6}$ bar we obtain $A_{atmosphere} = 10.5$. This means that even at this lower CO_2 pressure the infrared radiation is completely absorbed, and that there cannot be any additional absorbance within the bending mode band due to man-made CO_2 emissions.

Note that this does not mean that the global warming effect does not depend on man-made CO_2 emissions, because the overall warming depends not only on the infrared absorption of CO_2, but also on the infrared emission from the CO_2 molecules in the outer atmosphere into the interstellar space, see.[2]

2 https://skepticalscience.com/saturated-co2-effect.htm
J. Hansen, M. Sato, and R. Ruedy *J. Geophys. Res.* 1997, **102**, 6831–6864.
A. R. Ravishankara, Y. Rudich, and D. J. Wuebbles *Chem. Rev.* 2015, **115**, 3682–3703.
D. Archer, *Global Warming, Understanding the Forecast*, (Blackwell Publishing, 2007),
ISBN- 13: 978-1-4051-4039-9
ISBN-10: 1-4051-4039-9.
R. Wordsworth, J.T. Seeley, and K-P. Shine, Fermi Resonance and the Quantum Mechanical Basis of Global Warming, The Planetary Science Journal 5, 67 (2024).

12

Nuclear Spin

12.1 Exercises

E12.1 Show that the spacing of the lines in the Raman spectrum of O_2 (Fig. 12.4) is $8B$ instead of $2B$ as for heteronuclear diatomics. Use 120.752 pm for the bond length of O_2.

Solution

For heteronuclear diatomics $\Delta J = \pm 1$ while for homonuclear diatomics and CO_2 $\Delta J = \pm 2$. From the energy relationship $E = BJ(J+1)$, we can write

$$\Delta E_{J \to J+2} = B(J2)(J+3) - BJ(J+1)$$
$$= B(4J+6)$$

Hence the transition $J = 0 \to J = 2$ has a $\Delta E = 6B$, $J = 1 \to J = 3$ has a $\Delta E = 10B$, $J = 2 \to J = 4$ has a $\Delta E = 14B$, $J = 3 \to J = 5$ has a $\Delta E = 18B$, and so forth, giving a series of lines that are spaced by $4B$. A corresponding argument can be applied for the case of $\Delta E_{J \to J-2}$ transitions.

As discussed in the text, the nuclear spin quantum number of the O nucleus is zero; hence, it is always totally symmetric. Because the wavefunction for the molecule must be symmetric, only rotational states with symmetric wavefunctions are realized; these are the $J = 0, 2, 4, 6, \ldots$ energy levels. Combining this latter consideration and the selection rules for the transitions, we expect that transitions will occur at $J = 0 \to J = 2$ ($\Delta E = 6B$), $J = 2 \to J = 4$ ($\Delta E = 14B$), $J = 4 \to J = 6$ ($\Delta E = 22B$), and so forth. This means that the observed spectral transitions will be separated by $8B$.

In Fig. 12.4 we see that the spacing between adjacent lines is approximately 11–12 cm^{-1}. To show that the line spacing in the Raman spectrum of O_2 is $8B$ we begin with

$$B_{O_2} = \frac{h^2}{8\pi^2 \mu d^2}$$
$$= \frac{(6.626 \times 10^{-34}\ \text{J s})^2}{8 \times \pi^2 \times \frac{16 \times 16}{16+16} \times 1.672 \times 10^{-27}\ \text{kg} \times (120.752 \times 10^{-12})^2}$$
$$= 2.85 \times 10^{-23}\ \text{J}$$

then

$$8B_{O_2} = 8 \times 2.85 \times 10^{-23}\ \text{J} = 2.28 \times 10^{-22}\ \text{J}$$

This energy corresponds to a wavenumber of

$$\frac{1}{\lambda} = 2.28 \times 10^{-22}\ \text{J} \cdot \frac{1}{hc_0} = 11.48\ \text{cm}^{-1}$$

Also see Justification 12.2.

E12.2 Calculate the relative intensity of the Raman lines in the O_2 spectrum for the first eight transitions.

Solution

As discussed in the text, the nuclear spin quantum number of the O nucleus is zero; hence, it is always totally symmetric, and O is a boson. Because the wavefunction for the molecule must be symmetric, only rotational states with symmetric wavefunctions are realized; these are the $J = 0,2,4,6,\ldots$energy levels.

For oxygen we only need consider the one symmetric spin state. Hence the intensity of a transition is given directly by Equation (11.55)

$$I = const' \cdot (\Delta\alpha)^2 \cdot \frac{3\,(J+1)\,(J+2)}{2\,(2J+3)} \cdot e^{-BJ(J+1)/(kT)} \tag{12E2.1}$$

with (see Exercise 12.1)

$$B = 2.85 \times 10^{-23}\ \text{J}$$

For example, for $J = 0$ we obtain

$$I = const' \cdot (\Delta\alpha)^2 \cdot \frac{6}{6} \cdot e^{-2.85\times10^{-23}\cdot 0} = const' \cdot (\Delta\alpha)^2 \cdot 1 \cdot 1$$

Using this result we can construct Table E12.2.

Table E12.2 Evaluation of Equation (12E2.1).

J	$\frac{3\,(J+1)\,(J+2)}{2\,(2J+3)}$	$e^{-BJ(J+1)/(kT)}$	$\frac{I}{const' \cdot (\Delta\alpha)^2}$
0	1	1	1
2	2.57	0.96	2.47
4	4.09	0.87	3.56
6	5.60	0.75	4.19
8	7.11	0.61	4.31
10	8.61	0.47	4.02
12	10.1	0.34	3.43
14	11.6	0.23	2.71

The relative intensities computed in the table agree well with the observed spectrum; see Fig. 12.4.

E12.3 Following the procedure for D_2 explain the intensity distribution of the Raman lines of N_2 (Fig. 12.3).

Solution

^{14}N has a total spin of 1, so that it is a boson. Hence N_2 can be treated in the same way as D_2, and each nucleus can assume the spin values −1, 0, and +1. Accordingly, in this case we have three spin functions α, β, and γ which combine in the N_2 molecule in the way shown in Table 12.4.

Symmetric	Antisymmetric
$\alpha(1)\,\alpha(2)$	
$\beta(1)\,\beta(2)$	
$\gamma(1)\,\gamma(2)$	
$\alpha(1)\,\beta(2) + \alpha(2)\,\beta(1)$	$\alpha(1)\,\beta(2) - \alpha(2)\,\beta(1)$
$\alpha(1)\,\gamma(2) + \alpha(2)\,\gamma(1)$	$\alpha(1)\,\gamma(2) - \alpha(2)\,\gamma(1)$
$\beta(1)\,\gamma(2) + \beta(2)\,\gamma(1)$	$\beta(1)\,\gamma(2) - \beta(2)\,\gamma(1)$

Altogether, N_2 has six symmetric and three antisymmetric nuclear spin combinations. Since the total wavefunction must be symmetric, the symmetric spin functions combine with the symmetric rotational functions (J even) and the antisymmetric spin functions combine with the antisymmetric rotational functions (J odd). Therefore the states with $J = 0, 2, 4, \ldots$ will be sixfold degenerate, while the states with $J = 1, 3, 5, \ldots$ will be threefold degenerate. In the Raman spectrum we should expect to see alternating lines with differing heights with the stronger line twice the height of the line next to it.

See Justification 12.1 for a detailed calculation of the relative line intensities.

E12.4 Estimate the spacing of the spectral lines in the Raman spectrum of CO_2. Use this spacing to calculate the C–O bond length. Note that carbon dioxide will have a rotational spectrum like that found for oxygen.

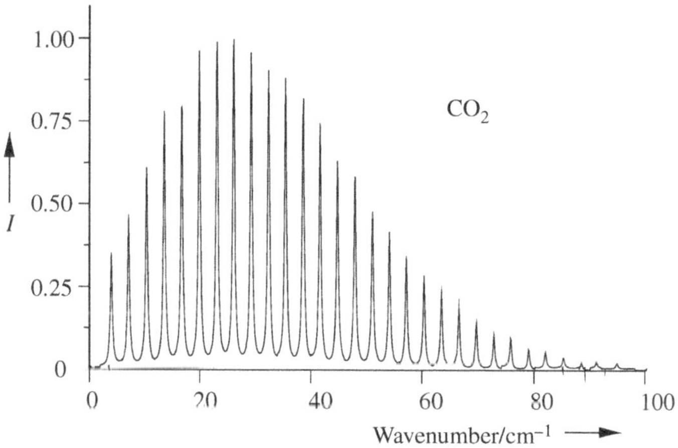

Figure E12.4 Rotational Raman spectrum of CO_2.

Solution

Using a ruler we estimate the mean distance between two Raman peaks in CO_2 to be 3.11 cm^{-1}, which corresponds to an energy of

$$\Delta\Delta E = hc_0\frac{1}{\lambda} = 6.626 \times 10^{-34}\ \text{J s} \cdot 2.998 \times 10^{8}\ \text{m s}^{-1} \cdot 3.13\ \text{cm}^{-1}$$
$$= 6.22 \times 10^{-23}\ \text{J}$$

The carbon atom lies at the center of mass and the two oxygen atoms rotate about this point. In analogy to the case of oxygen, the spectral lines are separated by $8B$; hence, we calculate

$$B = \frac{1}{8}\Delta\Delta E = \frac{1}{8} \cdot 6.22 \times 10^{-23}\ \text{J} = 7.78 \times 10^{-24}\ \text{J}$$

By combining this fact with the relation $B = h^2/(8\pi^2 I)$, we find

$$I = \frac{h^2}{8\pi^2 B} = \frac{(6.626 \times 10^{-34}\ \text{J s})^2}{8\pi^2 \cdot 7.78 \times 10^{-24}\ \text{J}} =$$
$$= 7.15 \times 10^{-46}\ \text{kg m}^2$$

The inertial moment of CO_2 is $I = 2m_O d^2$, where $m_O = 2.657 \times 10^{-26}$ kg; hence, we find that

$$d = \sqrt{\frac{I}{2m_O}} = \sqrt{\frac{7.15 \times 10^{-46}\ \text{kg-m}^2}{2\,(2.657 \times 10^{-26}\ \text{kg})}} = 116\ \text{pm}$$

E12.5 Calculate the magnetic field strength that satisfies the nuclear resonance condition for protons in a radiofrequency field of i) 60 MHz; ii) 300 MHz, and iii) 500 MHz.

Solution

We start with Equation (12.6)

$$\nu = \frac{g_N \cdot \mu_N \cdot B}{h}$$

Solving for B we have

$$B = \frac{h}{g_N \cdot \mu_N} \times \nu = \frac{6.626 \times 10^{-34}\ \text{J s}}{5.586 \times 5.051 \times 10^{-27}\ \text{J T}^{-1}} \times \nu$$
$$= 2.35 \times 10^{-8}\ \text{T s} \times \nu$$

where we have used $g_N = 5.586$ for the proton and $\mu_N = 5.051 \times 10^{-27}\ \text{J T}^{-1}$.
For $\nu = 60$ MHz we find

$$B = 2.35 \times 10^{-8}\ \text{T s} \times 60\ \text{MHz} = 1.41\ \text{T}$$

For 300 MHz we find

$$B = 2.35 \times 10^{-8}\ \text{T s} \times 300\ \text{MHz} = 7.05\ \text{T}$$

and for 500 MHz we find

$$B = 4.70 \times 10^{-8}\ \text{T s} \times 500\ \text{MHz} = 11.75\ \text{T}$$

E12.6 Calculate the nuclear resonance frequency for a ^{13}C nucleus in a 10 T magnetic field.

Solution
We begin in the same way as for E12.5; however, in this case we will use $g_N = 1.405$ for ^{13}C

$$\nu = \frac{g_N \cdot \mu_N \cdot B}{h} = \frac{1.405 \times 5.051 \times 10^{-27}\ \text{J T}^{-1}}{6.626 \cdot 10^{-34}\ \text{J s}} \cdot 10\ \text{T}$$
$$= 1.07 \times 10^8\ \text{T}^{-1}\ \text{s}^{-1} = 1.07 \times 10^8\ \text{s}^{-1} = 107\ \text{MHz}$$

E12.7 For the ethanol molecule draw the eight different spin combinations for the methyl protons (A) and draw a splitting pattern for the methylene protons (B) that arises from the methyl protons.

Solution
The eight different spin combinations for the methyl protons (A) are best seen using the following table of spins

Hydrogen 1	Hydrogen 2	Hydrogen 3
α	α	α
α	α	β
α	β	α
β	α	α
α	β	β
β	α	β
β	β	α
β	β	β

Hence we have one case where all the protons are α, three cases where two protons are α and one is β, three cases where one proton is α and two are β, and one case where all three of the protons are β. Consequently, the splitting pattern of protons B will be 1 : 3 : 3 : 1 which reflects the eight different spin combinations for the methyl protons.

E12.8 Some of the fine structure in the NMR spectrum of ethanol (Fig. 12.9) is lost because of rapid OH proton exchange. Draw the type of spectrum you would expect if the rapid OH proton exchange could be stopped.

Solution
The spectrum in Fig. 12.9 would change by the splitting patterns of the B and C protons. The splittings of the B protons would double, from a quartet to an octet, and the ethanolic proton would be split into a triplet by the methylene protons (Fig. E12.8).

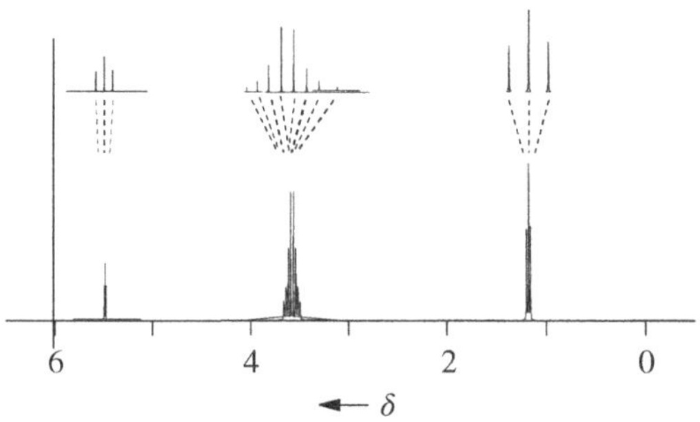

Figure E12.8 NMR spectrum of ethanol.

E12.9 Consider the cyanine cation

$$\overset{+}{-N} = \overset{1}{C}H-\overset{2}{C}H=\overset{3}{C}H-\overset{4}{C}\,H=\overset{5}{C}H-\overset{6}{C}H=\overset{7}{C}H-\overset{-}{N}-$$

The measured ^{13}C shielding parameters σ (see Equation (12.8)) at the C atoms 1–7 are[1]

C atom	C_1	C_2	C_3	C_4	C_5	C_6	C_7
σ/ppm	163.2	100.0	150.4	120.8	150.4	100.0	163.2

Use the FEMO model to compare these values to the total electron probability density at these atoms.

Solution

We calculate the total electron probability density ρ_{total} according to Section 8.3.3, Equation (8.10). You can use the program "E12.9.exe" to perform the calculation.

C atom	C_1	C_2	C_3	C_4	C_5	C_6	C_7
$\rho_{total}/\text{nm}^{-1}$	8.57	7.14	8.57	7.14	8.57	7.14	8.57

It is evident that the shielding parameter increases with ρ_{total}, that is, with increasing electron density.

12.2 Problems

P12.1 Energy order the following spectroscopic transitions for the molecule NO: vibrational excitation of the NO bond, nuclear spin "flip" of O nucleus, electronic excitation of NO (from HOMO to LUMO), electron spin "flip," rotational transition of NO, and photoionization of NO. Also, state the order of magnitude of the energy in each case.

Solution

Nuclear spin "flip" of O nucleus	$\approx 0.01\ \text{cm}^{-1}$	2.0×10^{-25} J
Electron spin "flip" in NO	$\approx 0.1\ \text{cm}^{-1}$	2.0×10^{-24} J
Rotational transition for NO	$\approx 2\ \text{cm}^{-1}$	4.0×10^{-23} J
Vibrational excitation of NO bond	$\approx 103\ \text{cm}^{-1}$	2.05×10^{-21} J
Electronic excitation of NO	$\approx 30{,}000\ \text{cm}^{-1}$	6.0×10^{-19} J
Photoionization of NO	$\approx 70{,}000\ \text{cm}^{-1}$	1.4×10^{-18} J

P12.2 In Section 12.3.3 we have considered the splitting of the NMR lines of two non-equivalent protons into two doublets. Perform a corresponding treatment for two equivalent protons (as the protons in dichloromethane).

1 H. Mustroph, *Phys. Sci. Rev.* 2020, **5**, 145.

Hint: equivalent protons cannot be distinguished; therefore, their spin functions must be either symmetric or antisymmetric. Transitions can only occur from a symmetric state to a symmetric state or from an antisymmetric state to an antisymmetric state.

Solution

Two nonequivalent protons can have four possible spin combinations ($\alpha\alpha$, $\alpha\beta$, $\beta\alpha$, and $\beta\beta$). For the case of two equivalent protons (the case of CH_2Cl_2), which are indistinguishable, the spin wavefunctions are three symmetric combinations

$$\alpha\alpha,\ \chi_+ = \frac{1}{\sqrt{2}}(\alpha\beta + \beta\alpha)\ ,\ \beta\beta$$

and one antisymmetric combination

$$\chi_- = \frac{1}{\sqrt{2}}(\alpha\beta - \beta\alpha)$$

The energies for these combinations are

Spin Function	Energies (Coupling Neglected)
$\alpha(1)\alpha(2)$	$E_\alpha = E_\alpha(1) + E_\alpha(2)$
χ_+	$E_+ = \left[E_\beta(1) + E_\alpha(2) + E_\alpha(1) + E_\beta(2)\right]/2$
$\beta(1)\beta(2)$	$E_\beta = E_\beta(1) + E_\beta(2)$
χ_-	$E_- = \left[E_\alpha(1) + E_\beta(2) - E_\beta(1) - E_\alpha(2)\right]/2$

Figure P12.2 plots these energy levels. A transition can only change the spin orientation of one of the protons. The two possible transitions are shown by the double arrow and have the same energy. Hence, only one spectral line is seen.

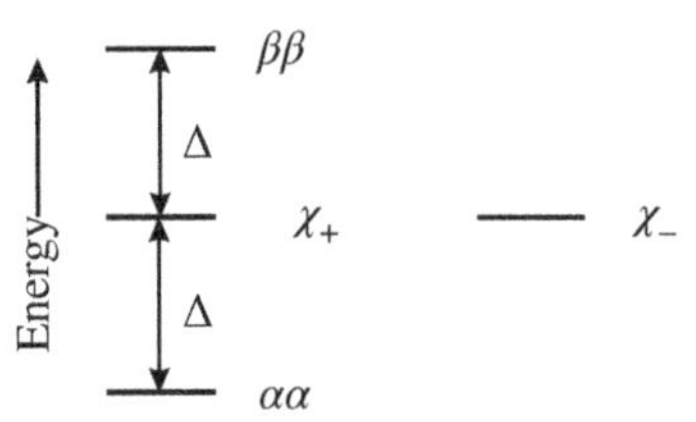

Figure P12.2 The sketch shows an energy level diagram for the case of two equivalent protons.

See Justification 12.3 for an alternative solution that also accounts for coupling between the protons.

13

Solids and Intermolecular Forces

13.1 Exercises

E13.1 Derive Equation (13.5) from Equation (13.4), and use this result to derive a formula for the energy of an ion pair in terms of the equilibrium bond distance.

Solution
Equation (13.4) is

$$E = -\frac{1}{4\pi\varepsilon_0} \cdot \frac{e^2}{d} + C \cdot \frac{1}{d^n}$$

The equilibrium bond distance corresponds to the distance at which the potential energy is at a minimum. Hence to find the equilibrium bond distance, d_{eq}, we take the derivative of the energy with respect to d

$$\frac{\mathrm{d}E}{\mathrm{d}d} = \frac{\mathrm{d}}{\mathrm{d}d}\left(-\frac{1}{4\pi\varepsilon_0} \cdot \frac{e^2}{d} + C \cdot \frac{1}{d^n}\right) = \frac{1}{4\pi\varepsilon_0} \cdot \frac{e^2}{d^2} - C \cdot n \cdot \frac{1}{d^{n+1}}$$

and set it equal to zero,

$$\left.\frac{\mathrm{d}E}{\mathrm{d}d}\right|_{d=d_{\text{eq}}} = 0 = \frac{1}{4\pi\varepsilon_0} \cdot \frac{e^2}{d_{\text{eq}}^2} - C \cdot n \cdot \frac{1}{d_{\text{eq}}^{n+1}}$$

and then we solve for d_{eq}. First we rearrange the equation as

$$\frac{1}{4\pi\varepsilon_0} \cdot \frac{e^2}{d_{\text{eq}}^2} = C \cdot n \cdot \frac{1}{d_{\text{eq}}^{n+1}}$$

and then we gather the d_{eq} terms on one side and the rest of the terms on the other, to find

$$\frac{d_{\text{eq}}^{n+1}}{d_{\text{eq}}^2} = d_{\text{eq}}^{n-1} = \frac{C \cdot n \cdot 4\pi\varepsilon_0}{e^2}$$

From this result, we see that the constant C can be expressed as

$$C = \frac{d_{\text{eq}}^{n-1} \cdot e^2}{n \cdot 4\pi\varepsilon_0}$$

Substituting this result back into Equation (13.4), we find

$$E = -\frac{1}{4\pi\varepsilon_0} \cdot \frac{e^2}{d} + \frac{d_{\text{eq}}^{n-1} \cdot e^2}{n \cdot 4\pi\varepsilon_0} \cdot \frac{1}{d^n}$$

At the equilibrium distance $E = E_{\text{eq}}$ and $d = d_{\text{eq}}$, so that

$$E_{\text{eq}} = -\frac{1}{4\pi\varepsilon_0} \cdot \frac{e^2}{d_{\text{eq}}} + \frac{d_{\text{eq}}^{n-1} \cdot e^2}{n \cdot 4\pi\varepsilon_0} \cdot \frac{1}{d_{\text{eq}}^n} = -\frac{1}{4\pi\varepsilon_0} \cdot \frac{e^2}{d_{\text{eq}}}\left[1 - \frac{1}{n}\right]$$

which is equivalent to Equation (13.8).

Solutions Manual for Principles of Physical Chemistry, Third Edition. Edited by Hans Kuhn, David H. Waldeck, and Horst-Dieter Försterling.

E13.2 Plot Equation (13.4) over the range of 190–1000 pm for the case where $C = e^2 d_{eq}^9 / (40\pi\varepsilon_0)$ and $d_{eq} = 250$ pm. Find the minimum and verify that it occurs at the distance 250 pm.

Solution

A plot of

$$E(d) = -\frac{1}{4\pi\varepsilon_0}\frac{e^2}{d} + \frac{e^2(250\ \text{pm})^9}{40\pi\varepsilon_0}\frac{1}{d^{10}}$$

is shown in Fig. E13.2. From the graph itself we can observe that the minimum occurs near 250 pm.

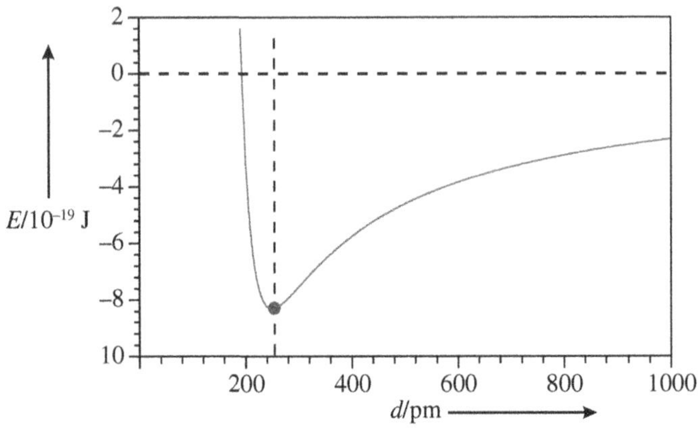

Figure E13.2 Energy E of an ion pair versus the distance d.

We can also locate and verify that the minimum occurs at 250 pm, analytically. We take the derivative of the energy expression with respect to d, set it equal to zero and solve for d.

$$0 = \left.\frac{dE(d)}{dd}\right|_{d=d_{min}} = \frac{d}{dd}\left(-\frac{1}{4\pi\varepsilon_0}\frac{e^2}{d} + \frac{e^2(250\ \text{pm})^9}{40\pi\varepsilon_0}\frac{1}{d^{10}}\right)\bigg|_{d=d_{min}}$$

$$= \frac{1}{4\pi\varepsilon_0}\frac{e^2}{d_{min}^2} - \frac{e^2(250\ \text{pm})^9}{40\pi\varepsilon_0}\frac{10}{d_{min}^{11}}$$

Solving for d_{min} we find that

$$\frac{1}{4\pi\varepsilon_0}\frac{e^2}{d_{min}^2} = \frac{e^2(250\ \text{pm})^9}{40\pi\varepsilon_0}\frac{10}{d_{min}^{11}}$$

which becomes

$$d_{min}^9 = \frac{10e^2(250\ \text{pm})^9}{40\pi\varepsilon_0}\frac{4\pi\varepsilon_0}{e^2} = (250\ \text{pm})^9$$

or

$$d_{min} = 250\ \text{pm}$$

E13.3 Use the experimental lattice energy of NaCl (equal to −769 kJ mol^{-1}) to estimate the equilibrium distance d_{eq}. Compare your result with that found from the ionic radii.

Solution

In this problem we use

$$E'_{eq} = -\frac{1}{4\pi\varepsilon_0} \times \frac{e^2}{d_{eq}} \times 1.748 \times \left[1 - \frac{1}{n}\right]$$

where $E'_{eq} = -769\ \text{kJ mol}^{-1}/\mathbf{N}_A = -12.8 \times 10^{-19}$ J is the energy per ion pair in the NaCl lattice. By rearranging this expression we write

$$d_{eq} = -\frac{e^2}{4\pi\varepsilon_0}\frac{1.748}{E'_{eq}}\left[1 - \frac{1}{n}\right]$$

Because n lies between 6 and 12 and $1/n$ is small compared to 1, in a first approximation we neglect this term and find

$$d_{eq} = \frac{(1.602 \times 10^{-19}\ \text{C})^2}{4\pi\,(8.854 \times 10^{-12}\ \text{C}^2\ \text{J}^{-1}\ \text{m}^{-1})} \cdot \frac{1.748}{12.8 \times 10^{-19}\ \text{J}}$$
$$= 3.16 \times 10^{-10}\ \text{m} = 316\ \text{pm}$$

The ionic radii are Na^+ (102 pm) and Cl^- (181 pm), which gives an equilibrium distance of 283 pm. This value differs by about 10% from 316 pm.
We use this result to estimate n.

$$\left[1 - \frac{1}{n}\right] = d_{eq}\frac{4\pi\varepsilon_0}{e^2}\frac{E'_{eq}}{1.748}$$

Then

$$\frac{1}{n} = 1 - d_{eq}\frac{4\pi\varepsilon_0}{e^2}\frac{E'_{eq}}{1.748} = 1 - 283\ \text{pm}\,\frac{4\pi \cdot 8.854 \times 10^{-12}\ \text{C}^2\ \text{J}^{-1}\ \text{m}^{-1}}{(1.602 \times 10^{-19}\ \text{C})^2}\frac{12.8 \times 10^{-19}\ \text{J}}{1.748}$$
$$= 1 - 283 \times 10^{-12}\ \text{m} \cdot 3.175 \times 10^{9}\ \text{m}^{-1} = 0.101$$

and

$$n = \frac{1}{0.101} = 10$$

E13.4 Use the experimental lattice energy of CsCl (equal to -657 kJ mol^{-1}) to estimate the equilibrium distance d_{eq}. Compare your result with that found from the ionic radii.

Solution
In this problem we use

$$E'_{eq} = -\frac{1}{4\pi\varepsilon_0} \times \frac{e^2}{d_{eq}} \times 1.763 \times \left[1 - \frac{1}{n}\right]$$

where $E'_{eq} = -657\ \text{kJ mol}^{-1}/\mathbf{N}_A = -10.9 \times 10^{-198}$ J is the energy per ion pair in the CsCl lattice. By rearranging this expression we write

$$d_{eq} = -\frac{e^2}{4\pi\varepsilon_0}\frac{1.763}{E'_{eq}}\left[1 - \frac{1}{n}\right]$$

Because n is between 6 and 12 and $1/n$ is small compared to 1, we neglect it in a first approximation and find

$$d_{eq} = \frac{(1.602 \times 10^{-19}\ \text{C})^2}{4\pi\,(8.854 \times 10^{-12}\ \text{C}^2\ \text{J}^{-1}\ \text{m}^{-1})} \cdot \frac{1.763}{1.09 \times 10^{-18}\ \text{J}}$$
$$= 3.73 \times 10^{-10}\ \text{m} = 373\ \text{pm}$$

The ionic radii are Cs^+ (167 pm) and Cl^- (181 pm), which gives an equilibrium distance of 348 pm. A result that is in reasonably good agreement with our estimate of the bond distance.
Note: A more precise calculation that includes the $1/n$ term with $n = 12$ gives a distance of 342 pm.

E13.5 Use the ionic radii of Na^+ and Br^- to estimate the equilibrium bond distance d_{eq} and calculate the lattice energy. Compare your result to the experimental value of -732 kJ mol^{-1}.

Solution
The ionic radii for Na^+(102 pm) and Br^-(196 pm) give a $d_{eq} = 298$ pm. If we insert this value into

$$E'_{eq} = -\frac{1}{4\pi\varepsilon_0} \times \frac{e^2}{d_{eq}} \times 1.748 \times \left[1 - \frac{1}{n}\right]$$

and neglect the term $1/n$, then we find

$$E'_{eq} = -\frac{(1.602 \times 10^{-19}\ \text{C})^2}{4\pi \cdot 8.854 \times 10^{-12}\ \text{C}^2\ \text{J}^{-1}\text{m}^{-1}} \cdot \frac{1.748}{298 \times 10^{-12}\ \text{m}}$$
$$= -13.5 \times 10^{-19}\ \text{J}$$

To compare with the experimental value we convert this energy per ion pair to a molar quantity; the lattice energy will be $\mathbf{N}_A \cdot E'_{\text{eq}} = -815$ kJ mol^{-1}. This value differs by 11% from the experimental value.
Typical values of n range from 6 to 12. For $n = 6$ the above formula gives -679 kJ mol^{-1}, and for $n = 12$ the above formula gives -747 kJ mol^{-1}. This range of values is in excellent agreement with the experimental value.

E13.6 Use the ionic radii of Li^+ and H^- to estimate the equilibrium bond distance d_{eq} and calculate the lattice energy. Compare your result to the experimental value of -858 kJ mol^{-1}.

Solution
The ionic radii for Li^+ (76 pm) and H^- (133 pm) give a $d_{\text{eq}} = 209$ pm. If we insert this value into

$$E'_{\text{eq}} = -\frac{1}{4\pi\varepsilon_0} \times \frac{e^2}{d_{\text{eq}}} \times 1.748 \times \left[1 - \frac{1}{n}\right]$$

and neglect the term $1/n$, then we find

$$E'_{\text{eq}} = -\frac{(1.6022 \times 10^{-19}\ \text{C})^2}{4\pi \cdot 8.8542 \times 10^{-12}\ \text{C}^2\ \text{J}^{-1}\ \text{m}^{-1}} \cdot \frac{1}{209 \times 10^{-12}\ \text{m}} = 19.3 \times 10^{-19}\ \text{J}$$

or, in molar terms,

$$E'_{\text{eq},m} = -19.3 \times 10^{-19}\ \text{J} \cdot \mathbf{N}_A = -1162\ \text{kJ mol}^{-1}$$

This value differs by 35% from the experimental value.
Typical values of n range from 6 to 12. For $n = 6$ the above formula gives -968 kJ mol^{-1}, and for $n = 12$ the above formula gives -1065 kJ mol^{-1}. This range of values is in better agreement with the experimental value but are consistently more negative than it.

E13.7 Calculate the electrostatic force that the Li^+ cation exerts on the H^- anion at their equilibrium bond distance, $d_{\text{eq}} = 209$ pm. Compare this force to the Coulomb force experienced by an electron that is one Bohr radius from a proton.

Solution
The Coulombic force f between two point charges at a distance d is given by

$$f = \frac{-1}{4\pi\varepsilon_0} \cdot \frac{e^2}{d^2}$$

Approximating Li^+ and H^- by point charges, we find that they exert a force of

$$f = \frac{-1}{4\pi\left(8.854 \times 10^{-12}\ \text{C}^2\ \text{J}^{-1}\text{m}^{-1}\right)} \cdot \frac{(1.602 \times 10^{-19}\ \text{C})^2}{(209 \times 10^{-12}\ \text{m})^2}$$
$$= -5.28 \times 10^{-9}\ \text{N}$$

The force between two point charges at the Bohr radius (a distance of 0.5292×10^{-10} m) is

$$f = \frac{-1}{4\pi \cdot 8.854 \times 10^{-12}\ \text{C}^2\ \text{J}^{-1}\ \text{m}^{-1}} \cdot \frac{(1.602 \times 10^{-19}\ \text{C})^2}{(0.5292 \times 10^{-10}\ \text{m})^2}$$
$$= -82.4 \times 10^{-9}\ \text{N}$$

The force between the atomic ions is smaller by a factor of 16 than that found for two point charges (electron and proton) at the Bohr radius.

E13.8 Calculate the binding energy of MgO for a Wurtzite structure and compare it to that for a cubic structure. In nature, MgO forms a cubic structure with an energy of -3795 kJ mol^{-1}. The ionic radius of O^{2-} is 140 pm and the ionic radius of Mg^{2+} is 72 pm.

Solution

The binding energy for MgO in a Wurtzite structure is

$$E'_{\text{eq,Wurtzite}} = -\frac{1}{4\pi\varepsilon_0}\frac{4e^2}{d_{\text{eq}}} \cdot 1.641 \cdot \left[1 - \frac{1}{n}\right]$$

Using an equilibrium ionic distance of $d_{\text{eq}} = 212$ pm and neglecting the $1/n$ term, we find that

$$\begin{aligned} E'_{\text{eq,Wurtzite}} &= -\frac{4 \cdot (1.602 \times 10^{-19}\ \text{C})^2 \cdot 1.641}{4\pi \cdot 8.854 \times 10^{-12}\ \text{C}^2\ \text{J}^{-1}\text{m}^{-1}} \cdot \frac{6.022 \times 10^{23}\ \text{mol}^{-1}}{212 \times 10^{-12}\ \text{m}} \\ &= -4300\ \text{kJ mol}^{-1} \end{aligned}$$

For a cubic lattice, we use

$$E'_{\text{eq,cubic}} = -\frac{1}{4\pi\varepsilon_0}\frac{4e^2}{d_{\text{eq}}} \cdot 1.748 \cdot \left[1 - \frac{1}{n}\right]$$

Neglecting the $1/n$ term and using $d_{\text{eq}} = \sqrt{2} \times 140$ pm (see Fig. 13.4), we find that

$$\begin{aligned} E'_{\text{eq,cubic}} &= -\frac{4 \cdot (1.602 \times 10^{-19}\ \text{C})^2 \cdot 1.748}{4\pi \cdot 8.854 \times 10^{-12}\ \text{C}^2\ \text{J}^{-1}\text{m}^{-1}} \cdot \frac{6.022 \times 10^{23}\ \text{mol}^{-1}}{198 \times 10^{-12}\ \text{m}} \\ &= -4905\ \text{kJ mol}^{-1} \end{aligned}$$

Comparing the two values we see that the cubic lattice is more stable by 605 kJ mol^{-1}. If we include the $1/n$ term in these calculations, they will not change our conclusion since both calculations will be affected in the same way. The calculated value for the cubic lattice differs from the experimental value by 29%. If we include the term $1/n$, we find that the calculated binding energy changes to −4088 kJ mol^{-1} for $n = 6$ and to −4496 kJ mol^{-1} for $n = 12$.

E13.9 Using the most common valency for the metals listed, calculate the Fermi energy for metallic Na and Al.

Solution

The Fermi energy for metals is given by Equation (13.18),

$$\begin{aligned} E_F &= \left(\frac{3}{\pi}\right)^{2/3} \cdot \frac{h^2}{8m_e} \cdot \left(\frac{N}{V}\right)^{2/3} \\ &= \left(\frac{3}{\pi}\right)^{2/3} \cdot \frac{(6.6260755 \times 10^{-34}\ \text{J s})^2}{8 \times (9.1093897 \times 10^{-31}\ \text{kg})} \cdot \left(\frac{N}{V}\right)^{2/3} \end{aligned}$$

For Na the valency is 1, and we have

$$\frac{N}{V} = \frac{\rho \mathbf{N}_A}{\mathbf{M}} = 1 \times \frac{971\ \text{kg m}^{-3} \times 6.022 \times 10^{23}\ \text{mol}^{-1}}{22.989\ \text{g mol}^{-1}} = 2.54 \times 10^{28}\ \text{m}^{-3}$$

which leads to

$$\begin{aligned} E_F &= \left(\frac{3}{\pi}\right)^{2/3} \cdot \frac{(6.626 \times 10^{-34}\ \text{J s})^2}{8 \cdot 9.109 \times 10^{-31}\ \text{kg}} \cdot (2.54 \times 10^{28}\ \text{m}^{-3})^{2/3} \\ &= 5.04 \times 10^{-19}\ \text{J} = 3.15\ \text{eV} \end{aligned}$$

For Al the valency is a filled *s*-subshell and a single *p* electron, so we have

$$\frac{N}{V} = \frac{\rho \mathbf{N}_A}{\mathbf{M}} = 1 \times \frac{2700\ \text{kg m}^{-3} \times 6.0221367 \times 10^{23}\ \text{mol}^{-1}}{26.98\ \text{g mol}^{-1}} = 6.027 \times 10^{28}\ \text{m}^{-3}$$

and

$$\begin{aligned} E_F &= \left(\frac{3}{\pi}\right)^{2/3} \times \frac{(6.6261 \times 10^{-34}\ \text{J s})^2}{8 \times (9.1094 \times 10^{-31}\ \text{kg})} \times (6.027 \times 10^{28}\ \text{m}^{-3})^{2/3} \\ &= 8.981 \times 10^{-19}\ \text{J} = 5.606\ \text{eV} \end{aligned}$$

E13.10 Given that the work function E_{work} of a metal (see Chapter 1) is the energy difference between the Fermi energy and the energy of a free electron (with zero kinetic energy), explain the trend in work function for the alkali metals (Li (2.9 eV), K (2.3 eV), and Cs (2.1 eV)) in terms of the metals' electronic structure.

Solution

From the problem description, we expect the trend in the work function to be mimicked by the Fermi energy. Since each of these alkali metals has a valency of one, their Fermi energies are given by Equation (13.18),

$$E_F = \left(\frac{3}{\pi}\right)^{2/3} \cdot \frac{h^2}{8m_e} \cdot \left(\frac{N}{V}\right)^{2/3}$$

$$= \left(\frac{3}{\pi}\right)^{2/3} \cdot \frac{(6.626 \times 10^{-34}\ \text{J s})^2}{8 \times (9.109 \times 10^{-31}\ \text{kg})} \cdot \left(\frac{N}{V}\right)^{2/3}$$

$$= 5.842 \times 10^{-38}\ \text{J m}^2 \cdot \left(\frac{N}{V}\right)^{2/3}$$

For Li we have

$$\frac{N}{V} = \frac{\rho \mathbf{N}_A}{\mathbf{M}} = 1 \times \frac{534\ \text{kg m}^{-3} \times 6.022 \times 10^{23}\ \text{mol}^{-1}}{6.941\ \text{g mol}^{-1}} = 4.63 \times 10^{28}\ \text{m}^{-3}$$

so that

$$\left(\frac{N}{V}\right)^{2/3} = 1.284 \times 10^{19}\ \text{m}^{-2}$$

which leads to

$$E_F = 5.842 \times 10^{-38}\ \text{J m}^2 \times 1.284 \times 10^{19}\ \text{m}^{-2} = 7.50 \times 10^{-19}\ \text{J} = 4.68\ \text{eV}$$

For K we have

$$\frac{N}{V} = \frac{\rho \mathbf{N}_A}{\mathbf{M}} = \frac{890\ \text{kg m}^{-3} \times 6.022 \times 10^{23}\ \text{mol}^{-1}}{39.09\ \text{g mol}^{-1}} = 1.37 \times 10^{28}\ \text{m}^{-3}$$

$$\left(\frac{N}{V}\right)^{2/3} = 0.572 \times 10^{19}\ \text{m}^{-2}$$

which leads to

$$E_F = 5.842 \times 10^{-38}\ \text{J m}^2 \times 0.572 \times 10^{19}\ \text{m}^{-2}$$
$$= 3.34 \times 10^{-19}\text{J} = 2.09\ \text{eV}$$

For Cs we have

$$\frac{N}{V} = \frac{\rho \mathbf{N}_A}{\mathbf{M}} = \frac{1930\ \text{kg m}^{-3} \times 6.0221367 \times 10^{23}\ \text{mol}^{-1}}{132.90545\ \text{g mol}^{-1}} = 0.8745 \times 10^{28}\ \text{m}^{-3}$$

$$\left(\frac{N}{V}\right)^{2/3} = 0.4226 \times 10^{19}\ \text{m}^{-2}$$

which leads to

$$E_F = 5.842 \times 10^{-38}\ \text{J m}^2 \times 0.4226 \times 10^{19}\ \text{m}^{-2}$$
$$= 2.47 \times 10^{-19}\ \text{J} = 1.54\ \text{eV}$$

Combining these results into a table gives

Element	E_{work}/eV	E_F/eV	r_+/pm	V_{eff}/eV
Li	2.9	4.68	76	−7.6
K	2.3	2.09	138	−4.4
Cs	2.1	1.54	167	−3.7

Given that E_F corresponds to the kinetic energy of the highest energy electron in the metal (see discussion around Equation (13.18)) and that the work function E_{work} is the difference between the highest energy electron and the vacuum level, we can calculate the effective potential V_{eff} binding the electron to the metal by

$$V_{eff} = -E_F - E_{work}$$

The binding energy of the electron to the metal ion cores changes systematically, being the strongest for Li^+ and the weakest for Cs^+. This trend is consistent with the size of the ionic core and the expected mean distance for the electron from the nuclear charge.

E13.11 For an energy difference ΔE between the two sides of a quantum wire and a change of g quantum states, show that $\Delta E = g\cdot(2n_z h^2)/(8m_e L^2)$.

Solution

Along the wire length (z-direction) the energy levels can be considered continuous, whereas they are quantized in the direction perpendicular to the wire's axis (x- and y-directions). We are only interested in the case where we have an energy difference along z that does not change the sub-band we are accessing (see discussion in Section 13.3.3). For a free electron moving along z the kinetic energy is

$$T = \frac{h^2}{8m_e L^2} n_z^2$$

Hence the energy difference for the electron on the two sides will be

$$\Delta E = E_{n_{z,j}} - E_{n_{z,i}}$$
$$= T_{n_{z,j}} - T_{n_{z,i}} = \frac{h^2}{8m_e L^2}\left(n_{z,j}^2 - n_{z,i}^2\right)$$

For this one-dimensional limit, when the quantum state changes by g on the two sides of the wire, the quantum number changes by g and we have that $n_{z,j} - n_{z,i} = g$. Hence we can write that

$$\Delta E = \frac{h^2}{8m_e L^2}\left(n_{z,j} - n_{z,i}\right)\left(n_{z,j} + n_{z,i}\right)$$
$$= g\frac{h^2}{8m_e L^2}\left(n_{z,j} + n_{z,i}\right)$$

Because we are in the quasi-continuous limit and ΔE is small enough, we can write that $\left(n_{z,j} + n_{z,i}\right) = 2n_z$ where n_z is the average quantum number for the electron moving across the wire. Hence we find that

$$\Delta E = g\frac{h^2}{8m_e L^2}\left(n_{z,j} + n_{z,i}\right) = g\frac{h^2}{8m_e L^2}2n_z$$

E13.12 The bulk solids, Si, Ge, and C (diamond), each have tetrahedral (sp^3) bonding. Why is diamond transparent, whereas Si and Ge have a silvery luster (much like a metal)?

Solution

The difference in the optical properties of these three solids arises from the difference in the energy (energy gap or bandgap) between their conduction band and their valence band, being largest for diamond (C), then Si, and finally Ge (see the table below). The magnitude of the bandgap for these elemental solids is determined by the strength of their bonding interaction. The strength of the bonding in the lattices of these three solids is shown by the bond energies reported in the table. If we use the sp^3 hybrid orbitals on each atomic site, then their interactions in these solids create a set of bands for their bonding and antibonding combinations. The bonding combinations generate a set of closely lying filled states, called the valence band, and the antibonding combinations generate a set of higher energy closely lying unfilled states, called the conduction band. The difference in energy between the highest filled state and the lowest unfilled state is called the band gap. Its magnitude increases as the interaction energy between overlapping hybrid orbitals increases. Hence, for these three elements it is largest for C and weakest for Ge.

Element	$\Delta H_{\text{atomization}}/\text{kJ mol}^{-1}$	Band Gap/eV	λ/nm
C (diamond)	717	5.47	227
Si	456	1.12	1108
Ge	377	0.66	1880

The band gaps for the three compounds and the lowest wavelength of a photon that can excite an electron from the valence band to the conduction band are shown in the table. For Ge and Si, the wavelength is longer than the wavelengths of photons in the visible region of the electromagnetic spectrum. Hence, when visible light is incident on the solid it can excite electrons. In contrast, the wavelength needed to excite an electron in diamond is 227 nm which is in the ultraviolet region of the spectrum and diamond appears to be transparent.

At room temperature, a small fraction of the electrons in Ge and Si are thermally excited from the valence band to the conduction band. These excited electrons are highly mobile/metal-like. Hence they give rise to some electrical conductivity and some optical properties (lustry/reflective) that are characteristic of metals.

E13.13 Using a semiconductor band diagram explain how impurity atoms in silicon (band gap of 1.12 eV) with an $E_{acceptor}$ energy 0.05 eV above the valence band edge can affect the conductivity. Use the fact that the probability to excite an electron from one electronic state to another (energy difference of ΔE) can be written as a probability $\sim \exp(-\Delta E/(kT))$.

Solution

First we consider the case of intrinsic silicon, i.e., no impurity/dopant ions. The density of silicon is 2330 kg m^{-3} so that its number density N/V is

$$\frac{N}{V} = \frac{2330\ \text{kg m}^{-3} \times 6.022 \times 10^{23}\ \text{mol}^{-1}}{0.0280855\ \text{kg mol}^{-1}} = 4.996 \times 10^{28}\ \text{m}^{-3}$$

Because each Si atom contributes four valence electrons to the bonding, we expect to have a density of electrons in the valence band of about $2 \times 10^{29}\,\text{m}^{-3}$ or $2 \times 10^{23}\,\text{cm}^{-3}$. Using the bandgap of silicon we estimate that only very few holes will be created by thermal excitation of electrons out of the valence band and into the conduction band; the hole density ρ_{hole} would be

$$\begin{aligned}\rho_{hole} &= \frac{N}{V} \times \exp\left(-\frac{\Delta E}{kT}\right)\\ &= \left(2 \times 10^{23}\ \text{cm}^{-3}\right) \times \exp\left(-\frac{1.12\ \text{eV} \times 1.602 \times 10^{-19}\ \text{J eV}^{-1}}{1.381 \times 10^{-23}\ \text{J K}^{-1}\ \ (298\ \text{K})}\right)\\ &= 5.0 \times 10^{4}\ \text{cm}^{-3}\end{aligned}$$

In contrast, consider having a dopant concentration of acceptors that is 0.001% of the Si concentration; namely $2 \times 10^{18}\,\text{cm}^{-3}$. If these acceptor states are 0.05 eV above the valence band, then on average one-seventh of them will be filled by electrons from the valence band, leaving that number of holes in the valence band; namely

$$\begin{aligned}\rho_{hole} &= \frac{N}{V} \times \exp\left(-\frac{\Delta E}{kT}\right)\\ &= \left(2 \times 10^{18}\ \text{cm}^{-3}\right) \times \exp\left(-\frac{0.05\ \text{eV} \times 1.602 \times 10^{-19}\ \text{J eV}^{-1}}{1.381 \times 10^{-23}\ \text{J K}^{-1}\ \ (298\ \text{K})}\right)\\ &= 2.9 \times 10^{17}\ \text{cm}^{-3}\end{aligned}$$

This hole density is 10 trillion times larger than that of the intrinsic material and allows for conduction through the valence band.

E13.14 Use Equation (13.35) to calculate the force between the two dipoles in Example 13.6.

Solution

The force between two dipoles is given by

$$f = \frac{dE}{dr} = \frac{1}{4\pi\varepsilon_0} \cdot \frac{6\mu_1\mu_2}{r^4}$$

the force between the two dipoles in Example 13.7 is

$$\begin{aligned}f &= \frac{1}{4\pi \cdot \left(8.854 \times 10^{-12}\ \text{C}^2\ \text{J}^{-1}\ \text{m}^{-1}\right)} \cdot \frac{6 \cdot \left(1.31 \times 10^{-29}\ \text{C m}\right)\left(1.31 \times 10^{-29}\ \text{C m}\right)}{(550\ \text{pm})^4}\\ &= 1.01 \times 10^{-10}\ \text{N} = 10.1\ \text{nN}\end{aligned}$$

E13.15 In Example 13.7 we considered two anti-parallel dipoles as a model for the acetonitrile dimer. Use the measured far-infrared bands at (80 cm^{-1} and 130 cm^{-1}) and calculate the force constant for the vibration in the harmonic oscillator approximation.

Solution

Here we use the harmonic oscillator result that

$$\nu = \frac{1}{2\pi}\sqrt{\frac{k_f}{\mu}} \text{ so that } k_f = 4\pi^2\mu\nu^2 = 4\pi^2\mu c_0^2\tilde{\nu}^2$$

and we estimate the reduced mass by

$$\mu = \frac{m_{CH_3CN}m_{CH_3CN}}{m_{CH_3CN} + m_{CH_3CN}} = m_{CH_3CN}/2$$
$$= \frac{0.041 \text{ kg}}{2 \cdot 6.022 \times 10^{23}} = 3.4 \times 10^{-26} \text{ kg}$$

Now we calculate the force constants. For $\tilde{\nu} = 80 \text{ cm}^{-1}$, we find

$$k_f = \frac{4\pi^2\mu}{c}\tilde{\nu} = 4\pi^2 \left(3.4 \times 10^{-26} \text{ kg}\right)\left(3.00 \times 10^8\right)^2 \text{ m}^2 \text{ s}^{-2}\left(8000 \text{ m}^{-1}\right)^2$$
$$= 7.7 \text{ kg s}^{-2} = 7.7 \text{ N m}^{-1}$$

For $\tilde{\nu} = 130 \text{ cm}^{-1}$, we find

$$k_f = \frac{4\pi^2\mu}{c}\tilde{\nu} = 4\pi^2 \left(3.4 \times 10^{-26} \text{ kg}\right)\left(3.00 \times 10^8\right)^2 \text{ m}^2 \text{ s}^{-2}\left(13{,}000 \text{ m}^{-1}\right)^2$$
$$= 20.4 \text{ kg s}^{-2} = 20.4 \text{ N m}^{-1}$$

These force constants are an order of magnitude smaller than those found for the stretching of covalent bonds, and somewhat smaller than those found for bending and torsional motions.

We note that for a 10 nN force f (calculated in E13.14) that the displacement would be

$$x = -f/k_f = \frac{10 \times 10^{-9} \text{ N}}{20.4 \text{ N m}^{-1}} = 490 \text{ pm}$$

E13.16 Calculate the attraction energy between two ICl dipoles arranged in the collinear geometry of Fig. 13.16a and the antiparallel geometry of Fig. 13.16b, with separation distances of 1.0 nm. The bond length of ICl is 232 pm and its dipole moment is 0.65 D.

Solution

For the collinear geometry we use Equation (13.34)

$$E = -\frac{1}{4\pi\varepsilon_0} \times \frac{2\mu^2}{r^3}$$

For ICl, we have that

$$\mu = 0.65 \text{ D} = 0.65 \times 3.3356 \times 10^{-30} \text{ C m}$$
$$= 2.17 \times 10^{-30} \text{ C m}$$

and the attraction energy is

$$E = -\frac{1}{4\pi\left(8.854 \times 10^{-12} \text{ C}^2 \text{ J}^{-1} \text{ m}^{-1}\right)}\frac{2\left(2.17 \times 10^{-30} \text{ C m}\right)^2}{(1 \text{ nm})^3}$$
$$= -8.46 \times 10^{-23} \text{ J}$$

For the antiparallel geometry we use the result that

$$E = -\frac{1}{4\pi\varepsilon_0}\frac{\mu^2}{r^3}$$

(see Box 13.4). That is, the energy is one-half of the first case, namely

$$E = -\frac{1}{2}8.46 \times 10^{-23} \text{ J} = -4.23 \times 10^{-23} \text{J}$$

We note that in both cases this energy is smaller than that associated with the product of Boltzmann's constant and the temperature, kT, at $T = 298$ K.

E13.17 Calculate the dispersion energy between two lithium atoms (take their polarizability to be 24.3×10^{-30} m^3 = $\alpha/(4\pi\varepsilon_0)$ and their ionization energy to be 5.39 eV) at distances of 350 pm (that in the solid) and 700 pm.

Solution

The dispersion energy between two atoms is given by Equation (13.50)

$$E_{\text{disp}} = -\frac{1}{(4\pi\varepsilon_0)^2} \cdot \frac{3}{2} \cdot \frac{\alpha_1 \cdot \alpha_2}{r^6} \frac{E_{\text{Ion},1} E_{\text{Ion},2}}{E_{\text{Ion},1} + E_{\text{Ion},2}}$$

Using the parameters given, we find

$$E_{\text{disp}} = -\frac{3}{2} \frac{\left(24.3 \times 10^{-30}\ \text{m}^3\right)^2}{(350\ \text{pm})^6} \frac{(5.39\ \text{eV})^2}{5.39\ \text{eV} + 5.39\ \text{eV}}$$
$$= -2.08 \times 10^{-19}\ \text{J}$$

at 350 pm, and we find

$$E_{\text{disp}} = -\frac{3}{2} \cdot \frac{24.3 \times 10^{-30}\ \text{m}^3 \times 24.3 \times 10^{-30}\ \text{m}^3}{(700\ \text{pm})^6} \frac{5.39\ \text{eV} \times 5.39\ \text{eV}}{5.39\ \text{eV} + 5.39\ \text{eV}}$$
$$= -3.25 \times 10^{-21}\ \text{J}$$

at 700 pm.

Compare these values to the characteristic thermal energy of $kT = 4.11 \times 10^{-21}$, at $T = 298$ K.

E13.18 Consider the dipole–dipole interaction between two ammonia molecules ($\mu = 1.5$ D) at 400 pm. In the point dipole approximation, what angle between the two dipoles has the most stable energy? How does this geometry compare to that you expect for hydrogen bond formation between the two ammonia molecules? (Refer to Box 13.4)

Solution

In Box 13.4 we show that the energy of interaction varies with angle as

$$\boxed{E = -\frac{1}{4\pi\varepsilon_0} \cdot \frac{\mu_1 \mu_2}{r^3} \cdot \left(3\cos^2\beta - 1\right)}$$

For a given μ and r, this function will have its most negative value (i.e., most stable geometry) when $\cos^2\beta = 1$, which occurs for $\beta = 0$ or π. Hence a collinear geometry is predicted to be the most stable.

This geometry is not what one would expect from a consideration of hydrogen bonding interactions, however. The typical geometry between two ammonia molecules is illustrated in Fig. E13.18

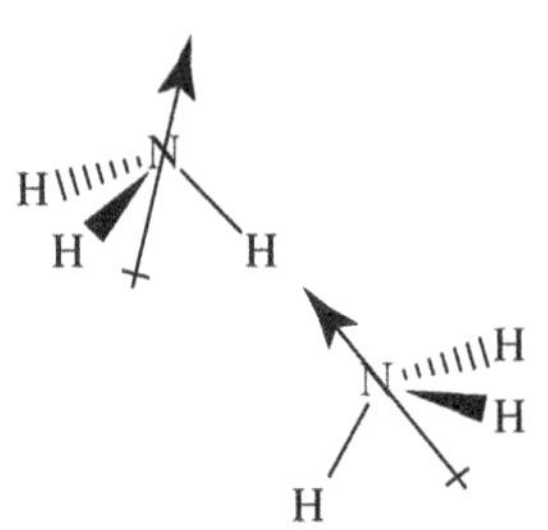

Figure E13.18 Dimer of two ammonia molecules.

E13.19 Consider an HCl molecule ($\mu = 1.18$ Debye) that is 400 pm from an Ar atom in a collinear geometry (Ar–HCl). Calculate the induced dipole moment in the Ar atom. Use the induced dipole to estimate the charge displacement (assume a unit charge) in units of the atom's radius.

Solution

The induced dipole moment is given by Equation (13.37), namely

$$\mu_{\text{ind}} = \alpha F$$

where F is the electric field strength from the dipole and α is the polarizability of the atom. The field strength that the dipole moment of HCl generates at a 400 pm distance is

$$F = \frac{1}{4\pi\varepsilon_0} \cdot \frac{2\mu}{r^3}$$
$$= \frac{1}{4\pi \cdot (8.854 \times 10^{-12}\ \mathrm{C^2\ J^{-1}\ m^{-1}})} \cdot \frac{2 \cdot (1.18 \times 3.33564 \times 10^{-30}\ \mathrm{C\ m})}{(400\ \mathrm{pm})^3}$$
$$= 1.11 \times 10^9 \frac{\mathrm{J}}{\mathrm{m\ C}} = 1.11 \times 10^9\ \frac{\mathrm{V}}{\mathrm{m}}$$

The polarizability of Ar can be approximated as

$$\alpha = 4\pi\varepsilon_0 \cdot r^3 = 4\pi \cdot (8.854 \times 10^{-12}\ \mathrm{C^2\ J^{-1}\ m^{-1}}) \cdot (182\ \mathrm{pm})^3$$
$$= 6.71 \times 10^{-40}\ \mathrm{C^2\ s^2\ kg^{-1}}$$

Combining these two results into Equation (13.37), we find that

$$\mu_{\text{ind}} = \alpha F = 6.71 \times 10^{-40}\ \mathrm{C^2}\frac{\mathrm{s^2}}{\mathrm{kg}} \cdot 1.11 \times 10^9 \frac{\mathrm{J}}{\mathrm{m\ C}} = 7.42 \times 10^{-31}\ \mathrm{C\ m}$$

The dipole moment is defined as

$$\mu = qd$$

Using our value of μ and the charge on an electron, we solve for d; namely

$$d = \frac{\mu}{q} = \frac{7.42 \times 10^{-31}\ \mathrm{C\ m}}{1.602 \times 10^{-19}\ \mathrm{C}} = 4.63 \times 10^{-12}\ \mathrm{m}$$

If we compare this result to the atomic radius of Ar

$$r_{\text{Ar}} = 182 \times 10^{-12}\ \mathrm{m}$$

we find that the fractional change is

$$\frac{4.63 \times 10^{-12}\ \mathrm{m}}{182 \times 10^{-12}\ \mathrm{m}} = 0.025$$

On a percentage basis it is small, roughly 2.5%.

E13.20 In Example 13.9 we calculated the bond energy of a proton with a water molecule. Perform a similar calculation for a sodium ion ($d_{\text{eq}} = 288$ pm).

Solution

Using the distances and partial charges, the electrostatic bond energy of the sodium ion to the water molecule is

$$E = \frac{e^2}{4\pi\epsilon_0}\left[\frac{-0.72}{288 \cdot 10^{-12}\ \mathrm{m}} + 2\frac{0.36}{507 \times 10^{-12}\ \mathrm{m}}\right]$$
$$= \frac{e^2}{4\pi\epsilon_0}\left[-1.09 \cdot 10^9\ \mathrm{m^{-1}}\right]$$

where the Na^+...H distance is

$$r = 2 \times 288\ \mathrm{pm} \times \sin\left(61.6 \times \frac{\pi}{180}\right) = 507 \times 10^{-12}\ \mathrm{m}$$

Putting everything together, we find that

$$E = \frac{(1.602 \times 10^{-19}\ \mathrm{C})^2}{4\pi\,(8.854 \times 10^{-12}\ \mathrm{C^2\ J^{-1} m^{-1}})}\left[-1.09 \cdot 10^9\ \mathrm{m^{-1}}\right]$$
$$= -2.49 \times 10^{-19}\ \mathrm{J} = -1.55\ \mathrm{eV}$$

E13.21 In the text we calculated the cohesion energy of Li metal by combining the experimental number density of Li atoms in the metal with the energy expression (Equations (13.24) and (13.25)) for the cohesive energy:

$$E = \overline{E}_{\text{pot}} + \overline{E}_{\text{kin}} = -A \cdot \left(\frac{N}{V}\right)^{1/3} + B \cdot \left(\frac{N}{V}\right)^{2/3}$$

where

$$A = \frac{e^2}{4\pi\varepsilon_0} \cdot \left(\frac{4\pi}{3}\right)^{1/3} = 3.27 \times 10^{-19}\ \text{J nm} \quad \text{and} \quad B = \frac{3}{5}\left(\frac{3}{\pi}\right)^{3/2} \frac{h^2}{8m_e} = 3.50 \times 10^{-20}\ \text{J nm}^2$$

In this problem, you are asked to estimate the optimum number density by minimizing E with respect to the number density ($x = N/V$), and then use this value to calculate the cohesion energy.

Solution

With the substitution $x = N/V$ we can write the cohesion energy as

$$E = \overline{E}_{\text{pot}} + \overline{E}_{\text{kin}} = -A \cdot x^{1/3} + B \cdot x^{2/3}$$

The minimum of E is reached if the first derivative equals zero.

$$\frac{\text{d}E}{\text{d}x} = -\frac{1}{3}A \cdot x^{-2/3} + \frac{2}{3}B \cdot x^{-1/3} = 0$$

Thus we obtain

$$\frac{1}{3}A \cdot x^{-2/3} = \frac{2}{3}B \cdot x^{-1/3}\text{, and } x^{1/3} = \frac{A}{2B}$$

and

$$\overline{E}_{\text{pot}} = -A \cdot x^{1/3} = -\frac{A^2}{2B}, \ \overline{E}_{\text{kin}} = B \cdot x^{2/3} = \frac{A^2}{4B}$$

Using the numerical data for A and B, we obtain

$$\left(\frac{N}{V}\right)_{\text{min}} = \left(\frac{A}{2B}\right)^3 = \left(\frac{3.27 \times 10^{-19}\ \text{J nm}}{2 \cdot 3.50 \times 10^{-20}\ \text{J nm}^2}\right)^3 = 101.9\ \text{nm}^{-3}$$

$$\overline{E}_{\text{pot}} = -\frac{A^2}{2B} = -\frac{(3.27 \times 10^{-19}\ \text{J nm})^2}{2 \cdot 3.50 \times 10^{-20}\ \text{J nm}^2} = -15.28 \times 10^{-19}\ \text{J}$$

$$\overline{E}_{\text{kin}} = -\frac{1}{2}\overline{E}_{\text{pot}} = 7.64 \times 10^{-19}\ \text{J}, E = \overline{E}_{\text{pot}} + \overline{E}_{\text{kin}} = -7.64 \times 10^{-19}\ \text{J}$$

E13.22 Consider the case of two electric dipoles arranged in a line (Fig. E13.22), for which the electrostatic energy is

$$E = \frac{1}{4\pi\varepsilon_0} \cdot \left[\frac{Q_1Q_2}{r} + \frac{Q_1Q_2}{r} - \frac{Q_1Q_2}{(r+a)} - \frac{Q_1Q_2}{(r-a)}\right]$$

Show that the energy expression simplifies to $2a^2/r^3$ in the limit where $a \ll r$.

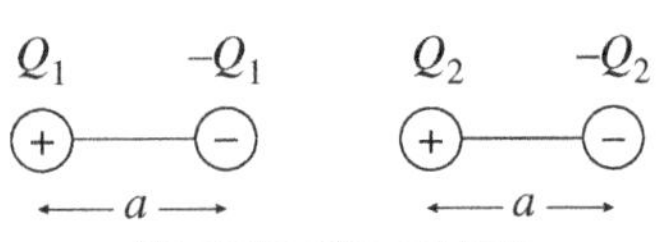

Figure E13.22 Two dipoles arranged in line.

Solution

We can rearrange the energy expression given and write

$$E = \frac{1}{4\pi\varepsilon_0} \cdot Q_1Q_2 \cdot \left[\frac{2}{r} - \frac{1}{(r+a)} - \frac{1}{(r-a)}\right]$$

If we analyze the term in square brackets, we find that

$$\frac{2}{r} - \frac{1}{r+a} - \frac{1}{r-a} = \frac{2r^2 - 2a^2 - r^2 + ra - r^2 - ra}{r(r+a)(r-a)} = -\frac{2a^2}{r^3} \cdot \frac{1}{1+(a/r)^2}$$

When $a \ll r$ the second term in the denominator can be neglected and we find that

$$\frac{2a^2}{r^3} \cdot \frac{1}{1+(a/r)^2} \approx -\frac{2a^2}{r^3}$$

Thus, in this limit we find that the energy E is

$$\boxed{E = \frac{1}{4\pi\varepsilon_0} \cdot \left(Q_1 \cdot a\right)\left(Q_2 \cdot a\right) \cdot \frac{2}{r^3} = -\frac{1}{4\pi\varepsilon_0} \cdot \frac{2\mu_1\mu_2}{r^3}}$$

with $\mu_1 = aQ_1$ and $\mu_2 = aQ_2$.

E13.23 Consider the hydrogen bonded aggregate

$(F \ldots H \ldots F)^-$

The H nucleus can be considered as a harmonic oscillator, which can be excited by infrared light. The infrared absorption bands for this aggregate occur at $1/\lambda = 1364\ \text{cm}^{-1}$ (excitation in the direction of the bond line: anti-symmetric vibration mode) and at $1217\ \text{cm}^{-1}$ (excitation perpendicular to the bond line: bending mode). Using a harmonic oscillator approximation for the vibrations, calculate the wavenumber shift in these transitions if the H atom is replaced by a deuterium atom. Compare your answer to the experimental value of 969 and $880\ \text{cm}^{-1}$.

Solution

In the harmonic oscillator approximation, the fundamental vibrational wavenumber $\overline{\nu}$ is given by

$$\overline{\nu} = \frac{1}{2\pi c}\sqrt{\frac{k}{\mu}}$$

where c is the speed of light, k is the force constant, and μ is the reduced mass for the vibrational mode. The use of deuterium changes the reduced mass for the vibrational mode, but it has a negligible effect on the force constant. To find an expression for the reduced mass of the asymmetric stretch and the bending mode, we can use the same procedure as that shown in Box 11.2 for the linear triatomic CO_2. For the anti-symmetric stretch mode of FDF, we find that

$$\mu_{\text{anti,FDF}} = \frac{m_D m_F}{m_D + 2m_F} = \frac{2 \cdot 19}{2+38}\ m_H = \frac{19}{20} u = 0.95\, m_H$$

whereas that for FHF $\mu_{\text{anti,FHF}} = 19/39\ m_H = 0.49\ m_H$. Using this change in the reduced mass and assuming that the force constant does not change, we find that

$$\frac{\overline{\nu}_{\text{anti,FDF}}}{\overline{\nu}_{\text{anti,FHF}}} = \sqrt{\frac{\mu_{\text{anti,FHF}}}{\mu_{\text{anti,FDF}}}} \text{ or } \overline{\nu}_{\text{anti,FDF}} = \sqrt{\frac{0.49}{0.95}} \cdot 1364\ \text{cm}^{-1} = 979\ \text{cm}^{-1}$$

For the bending mode, we find that

$$\mu_{\text{bend,FDF}} = \frac{m_D m_F}{2\left(m_D + 2m_F\right)} = \frac{2 \cdot 19}{2(40)}\ m_H = \frac{19}{40}\ m_H = 0.48\, m_H$$

and $\mu_{\text{bend,FHF}} = 19/78\ \text{u} = 0.24\ \text{u}$. In a manner similar to that used for the antisymmetric mode, we find that the wavenumber of the transition should shift to be $\overline{\nu}_{\text{bend,FDF}} = 860\ \text{cm}^{-1}$.

To compare with the experimental values, we calculate the percent error. For the antisymmetric stretch, we find that

$$\text{error} = \frac{979 - 969}{969} \cdot 100\% = 1.0\%$$

and that for the bending mode is

$$\text{error} = \frac{860 - 880}{880} \cdot 100\% = -2.3\%$$

The agreement is very good.

E13.24 Beginning with the energy expression for two dipoles at an angle β, given in Box 13.4, namely

$$E = \frac{Q_1 Q_2}{4\pi\varepsilon_0} \cdot \left[\frac{2}{r} - \frac{1}{r} \frac{1}{\sqrt{1 + \frac{a^2}{r^2} + 2\frac{a}{r}\cos\beta}} - \frac{1}{r} \frac{1}{\sqrt{1 + \frac{a^2}{r^2} - 2\frac{a}{r}\cos\beta}} \right]$$

expand each of the square root terms in brackets in a Taylor series and demonstrate that

$$E = \frac{1}{4\pi\varepsilon_0} \frac{Q_1 Q_2 a^2}{r^3} \cdot (1 - 3\cos^2\beta)$$

up to second order in a/r. Note that $\mu_1 = Q_1 a$ and $\mu_2 = Q_2 a$ are the dipole moments of the two dipoles.

Solution

The orientation of the dipoles is displayed in Fig. E13.24

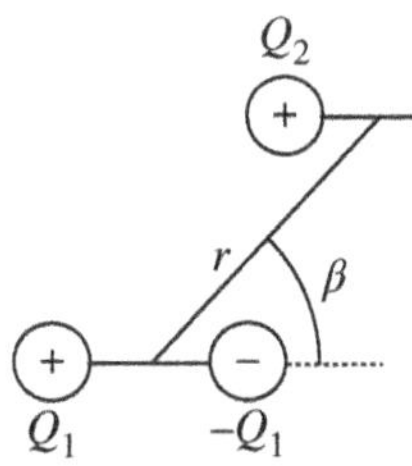

Figure E13.24 Orientation of the two dipoles at an angle β.

$$E = \frac{Q_1 Q_2}{4\pi\varepsilon_0} \cdot \left[\frac{2}{r} - \frac{1}{r} \frac{1}{\sqrt{1 + \frac{a^2}{r^2} + 2\frac{a}{r}\cos\beta}} - \frac{1}{r} \frac{1}{\sqrt{1 + \frac{a^2}{r^2} - 2\frac{a}{r}\cos\beta}} \right]$$

Expanding each of the square root terms in a Taylor series, see Appendix B

$$\frac{1}{\sqrt{1+x}} = 1 - \frac{1}{2}x + \frac{3}{8}x^2 - \cdots$$

gives

$$\begin{aligned}
&\frac{1}{\sqrt{1 + \frac{a^2}{r^2} + 2\frac{a}{r}\cos\beta}} \\
&= 1 - \frac{1}{2}\left(\frac{a^2}{r^2} + 2\frac{a}{r}\cos\beta\right) + \frac{3}{8}\left(\frac{a^2}{r^2} + 2\frac{a}{r}\cos\beta\right)^2 - \cdots \\
&= 1 - \frac{1}{2}\frac{a^2}{r^2} - \frac{a}{r}\cos\beta + \frac{3}{8}\frac{a^4}{r^4} + \frac{3}{2}\frac{a^3}{r^3}\cos\beta + \frac{3}{2}\frac{a^2}{r^2}\cos^2\beta - \cdots \\
&\approx 1 - \frac{1}{2}\frac{a^2}{r^2} - \frac{a}{r}\cos\beta + \frac{3}{2}\frac{a^2}{r^2}\cos^2\beta
\end{aligned}$$

where we neglected terms higher than second order in a/r. Correspondingly, we obtain

$$\frac{1}{\sqrt{1 + \frac{a^2}{r^2} - 2\frac{a}{r}\cos\beta}} \approx 1 - \frac{1}{2}\frac{a^2}{r^2} + \frac{a}{r}\cos\beta + \frac{3}{2}\frac{a^2}{r^2}\cos^2\beta$$

Combining these terms we find to second order in a/r that

$$\begin{aligned}
&\frac{2}{r} - \frac{1}{r_{14}} - \frac{1}{r_{23}} \\
&= \frac{2}{r} - \frac{1}{r}\left(1 - \frac{1}{2}\frac{a^2}{r^2} - \frac{a}{r}\cos\beta + \frac{3}{2}\frac{a^2}{r^2}\cos^2\beta\right) \\
&\quad - \frac{1}{r}\left(1 - \frac{1}{2}\frac{a^2}{r^2} + \frac{a}{r}\cos\beta + \frac{3}{2}\frac{a^2}{r^2}\cos^2\beta\right) \\
&= \frac{a^2}{r^3} - 3\frac{a^2}{r^3}\cos^2\beta = \frac{a^2}{r^3}(1 - 3\cos^2\beta)
\end{aligned}$$

and consider terms only up to second order in a/r, the energy E becomes

$$E = \frac{1}{4\pi\varepsilon_0} Q_1 Q_2 \cdot \left[\frac{2}{r} - \frac{1}{r_{14}} - \frac{1}{r_{23}}\right]$$
$$= \frac{1}{4\pi\varepsilon_0} Q_1 Q_2 \cdot \frac{a^2}{r^3}\left(1 - 3\cos^2\beta\right) = \frac{1}{4\pi\varepsilon_0}\frac{Q_1 Q_2 a^2}{r^3} \cdot \left(1 - 3\cos^2\beta\right)$$

E13.25 Beginning with the expressions for the distances r_{13} and r_{23} in the consideration of two dipoles at an angle β (see Box 13.4)

$$r_{23}^2 = r^2\left(1 - \frac{a}{r}\cos\beta + \frac{a^2}{4r^2}\right) \quad \text{and} \quad r_{13}^2 = r^2\left(1 + \frac{a}{r}\cos\beta + \frac{a^2}{4r^2}\right)$$

show that

$$\frac{1}{r_{13}} - \frac{1}{r_{23}} = -\frac{a}{r^2}\cos\beta$$

by expanding the term $1/r$ in a Taylor series and collecting terms up to second order in a/r.

Solution

Expanding the terms in a Taylor series and collect terms up to second order in a/r, gives

$$\frac{1}{r_{13}} = \frac{1}{r}\frac{1}{\sqrt{1 + \frac{a}{r}\cos\beta + \frac{a^2}{4r^2}}} \approx \frac{1}{r}\left(1 - \frac{a}{2r}\cos\beta \quad \frac{a^2}{8r^2}\right)$$
$$\frac{1}{r_{23}} = \frac{1}{r}\frac{1}{\sqrt{1 - \frac{a}{r}\cos\beta + \frac{a^2}{4r^2}}} \approx \frac{1}{r}\left(1 + \frac{a}{2r}\cos\beta - \frac{a^2}{8r^2}\right)$$

Thus, the difference is

$$\frac{1}{r_{13}} - \frac{1}{r_{23}} = -\frac{1}{r}\frac{a}{r}\cos\beta = -\frac{a}{r^2}\cos\beta$$

E13.26 Describe an experiment that demonstrates the wave nature of matter. Make a sketch that illustrates your observations; i.e., measured data. Draw a sketch of an experimental apparatus that shows the essential components needed in making such a measurement. Explain in words how this experiment demonstrates the wave nature of matter.

Solution

Low energy electron and/or neutron diffraction experiments both are examples of the wave nature of particles. In both cases a beam of the particles is focused onto a target, which diffracts the particles toward an angle scannable detector. The reflected particle intensity depends on angle because the wavelength of the particle is of the same magnitude as the spacing between atoms. The schematic diagram shows an electron beam impinging on a Ni crystal and diffracting (Fig. E13.26a). By scanning the detector over different angles the electron current (called the collector current below) can be measured.

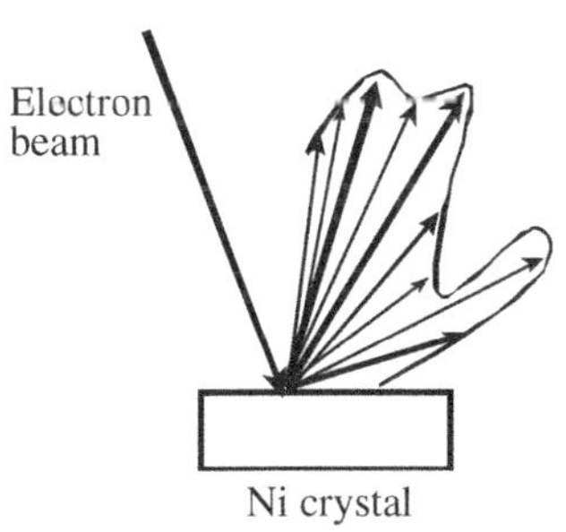

Figure E13.26a The diagram aims to illustrate that a beam of electrons impinges on a nickel crystal. Rather than simple specular reflection, the electron beam diffracts from the surface.

Davisson and Germer reflected a beam of electrons off of a crystalline Ni target and observed that the electrons diffracted; i.e., produced an intensity pattern that varied in space instead of falling off monotonically from the specular angle. Figure E13.26b shows data that they collected as a function of bias potential at different scattering angles. For the different detector angles, the peak intensity occurs at different incident electron wavelengths (hence the bombarding potential). Classical particles would not show such a dependence. Diffraction is a wave characteristic and this observation demonstrates that particles, namely electrons, can exhibit wave properties.

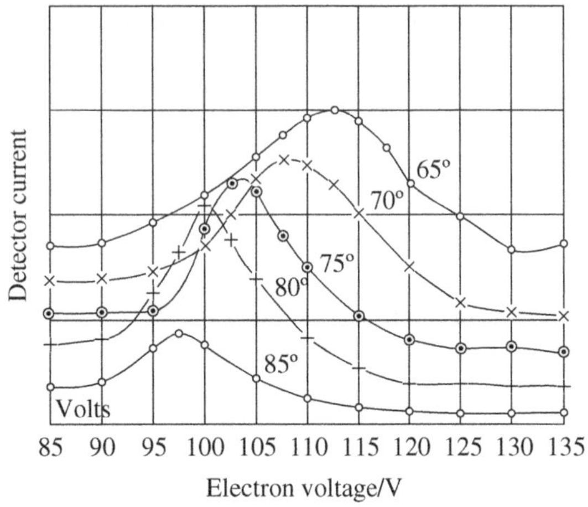

Figure E13.26b Collector current (detector current) versus bombarding potential (which determines the incident electron's wavelength) showing plane grating beams near grazing in {110}-azimuth. (Taken from C. Davisson, L.H. Germer, *Phys. Rev.* 1927, **30** 705.).

The experiment reported by C. Davisson and L.H. Germer, (*Phys. Rev.* 1927, **30**, 705) also provides an interesting example in serendipity. These initial investigations did not reveal much structure in the diffraction; however, they inadvertently heated the sample, which caused the Ni to recrystallize into larger crystallites upon cooling. They were observant enough to recognize what had occurred. Figure E13.26c shows the data that they obtained before and after the "accident"; i.e., before and after crystal growth has occurred.

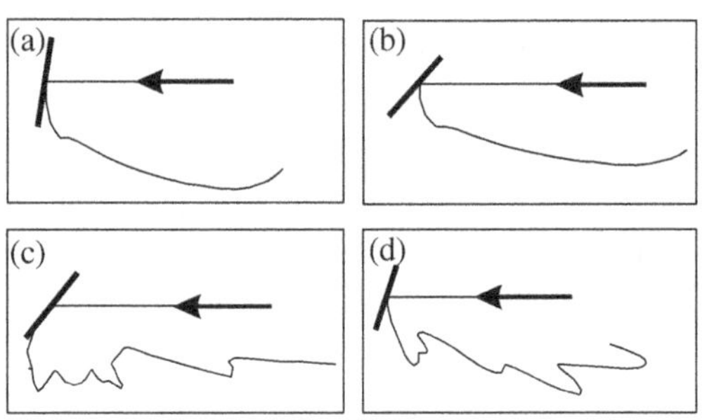

Figure E13.26c The scattered intensity pattern (75 eV electrons). (a,b): from a block of Ni (many small Ni crystals); (c,d) from several large Ni crystals.

13.2 Problems

P13.1 Estimate how much energy has to be supplied in order to increase or decrease the distance d of an ion pair by 1% and 10% ($d_{eq} = 100$ pm, $n = 10$).

Solution

We start with Equation 13.1 and use the relation

$$d = d_{eq}(1 + x)$$

Substituting our expression for d, we find

$$E = -\frac{1}{4\pi\epsilon_0}\frac{e^2}{d}\left[1-\frac{1}{n}\left(\frac{d_{eq}}{d}\right)^{n-1}\right] = -\frac{1}{4\pi\epsilon_0}\frac{e^2}{d_{eq}(1+x)}\left[1-\frac{1}{n}\left(\frac{d_{eq}}{d_{eq}(1+x)}\right)^{n-1}\right]$$

$$= -\frac{1}{4\pi\epsilon_0}\frac{e^2}{d_{eq}(1+x)}\left[1-\frac{1}{n}\left(\frac{1}{(1+x)}\right)^{n-1}\right] = -\frac{1}{4\pi\epsilon_0}\frac{e^2}{d_{eq}}\left[\frac{1}{(1+x)}-\frac{1}{n}\left(\frac{1}{(1+x)}\right)^{n}\right]$$

Using the table in Appendix B5, we find the expansions for the expressions $(1+x)^{-1}$ and $(1+x)^{-n}$,

$$\frac{1}{(1+x)} = (1+x)^{-1} = 1 - x + x^2$$

$$\left(\frac{1}{(1+x)}\right)^n = (1+x)^{-n} = 1 - nx + \frac{-n(-n-1)}{2}x^2 = 1 - nx + \frac{n^2+n}{2}x^2$$

Substituting these into our expression for E, we obtain

$$\left[\frac{1}{(1+x)}-\frac{1}{n}\left(\frac{1}{(1+x)}\right)^{n}\right] = -\left(1-x+x^2\right)-\left(\frac{1}{n}-x+\frac{n+1}{2}x^2\right)$$

$$= -1-\frac{1}{n}+\left(1-\frac{n+1}{2}\right)x^2 = 1-\frac{1}{n}+\left(\frac{1}{2}-\frac{n}{2}\right)x^2 = 1-\frac{1}{n}+\left(\frac{1-n}{2}\right)$$

and

$$E = -\frac{1}{4\pi\epsilon_0}\frac{e^2}{d_{eq}}\left[1-\frac{1}{n}+\left(\frac{1-n}{2}\right)x^2\right]$$

Thus, we see that E is increasing in proportion to nx^2. For $x = 1\%$ and $n = 10$ the change of the energy E is 0.05%; whereas for $x = 10\%$ it is 5%.

P13.2 The lattice energies and equilibrium ion distances for some simple halide salts (MCl) are provided in the table. Using these data (taken from CRC handbook), calculate the best value of n for Equation (13.8).

System	d/pm	U_m/kJ mol^{-1}	System	d/pm	U_m/kJ mol^{-1}
LiCl	257	−853	LiF	201	−1036
NaCl	282	−786	NaF	231	−923
KCl	314	−715	KF	266	−821
RbCl	329	−689	RbF	282	−785
CsCl	360	−659	CsF	300	−740

Solution

A number of different methods can be used to obtain n from Equation (13.8), which is

$$E'_{eq} = -\frac{1}{4\pi\epsilon_0}\times\frac{e^2}{d_{eq}}\times 1.748\times\left[1-\frac{1}{n}\right]$$

$$= \left(-\frac{1.748\ e^2}{4\pi\epsilon_0}\left[1-\frac{1}{n}\right]\right)\frac{1}{d_{eq}}$$

from which we write that

$$U_m = \mathbf{N}_A\left(-\frac{1.748\cdot e^2}{4\pi\epsilon_0}\left[1-\frac{1}{n}\right]\right)\frac{1}{d_{eq}}$$

In Fig. P13.2 we plot U_m versus $1/d_{eq}$ which should be a straight line and have a slope of

$$slope = -\mathbf{N}_A\frac{1.748\cdot e^2}{4\pi\epsilon_0}\left[1-\frac{1}{n}\right]$$

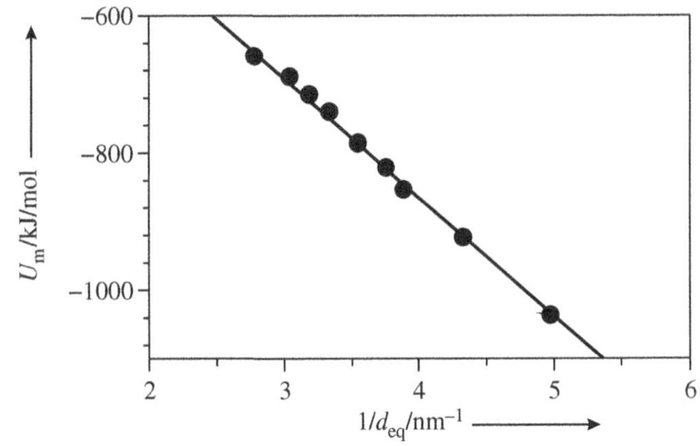

Figure P13.2 Energy U_m versus $1/d_{eq}$.

The slope is found to be −176.6 kJ nm mol^{-1} and if we solve the slope equation for $1/n$, we find that

$$\frac{1}{n} = 1 + \frac{4\pi\varepsilon_0}{N_A 1.748 \cdot e^2} slope$$
$$= 1 + \frac{4\pi\left(8.854 \times 10^{-12}\ C^2\ J^{-1}m^{-1}\right)}{1.748 \cdot \left(6.022 \times 10^{23}\ mol^{-1}\right)\left(1.602 \times 10^{-19}\ C\right)^2}\left(-1.766 \times 10^{-4}\ \frac{J\ m}{mol}\right)$$
$$= 0.2727$$

so that $n = 3.67$. Note that this value is somewhat smaller than the range of values ($6 \le n \le 12$) we typically find from other considerations.

P13.3 Using the crystal data for these different salts, which form cubic crystals with the sodium chloride lattice, and the fact that the radius of I^- is much greater than that of Li^+, determine "best values" for the ionic radii. Describe your reasoning.

Salt	d/pm	Salt	d/pm	Salt	d/pm	Salt	d/pm
LiF	201	LiCl	257	LiBr	275	LiI	300
NaF	231	NaCl	282	NaBr	297	NaI	323
KF	266	KCl	314	KBr	329	KI	353
RbF	282	RbCl	329	RbBr	343	RbI	363

Solution

One approach is to choose the system with the largest anion and the smallest cation, LiI. In this case we assume that the I^- anion is so much larger than the Li^+ that the iodide ions effectively "touch" each other. Hence we can use the geometry of the NaCl lattice to write that

$$r_{I^-} = \sqrt{(d/2)^2 + (d/2)^2} = 212\ \text{pm}$$

where $d = 300$ pm is the radius of the I^- anion. With this value in hand we can compute the cation radii from the other iodide salts, via $r_{M^+} = d - r_{I^-}$ and find

$$r_{Li^+} = 88\ \text{pm};\ r_{Na^+} = 111\ \text{pm};\ r_{K^+} = 141\ \text{pm};\ r_{Rb^+} = 151\ \text{pm}$$

By a similar procedure we can use these cation radii to find radii for the other anions; given here in pm.

Salt	r_{F^-}/pm	Salt	r_{Cl^-}/pm	Salt	r_{Br^-}/pm	Salt	r_{I^-}/pm
LiF	113	LiCl	169	LiBr	187	LiI	212
NaF	120	NaCl	171	NaBr	186	NaI	212
KF	125	KCl	173	KBr	188	KI	212
RbF	131	RbCl	178	RbBr	192	RbI	212

From the results it is evident that the anion radii change somewhat through the series. By way of comparison, the commonly accepted values of ionic radii for the metal cations and halide anions are

System	r_+/pm	System	r_-/pm
Li^+	76	F^-	133
Na^+	102	Cl^-	181
K^+	151	Br^-	196
Rb^+	161	I^-	220

Use of a more sophisticated method gives a more self-consistent set of radii; e.g., see T.C. Waddington, *Trans. Faraday Soc.* 1966, **62** 1482.

P13.4 The semiconductor Si has a bulk bandgap of 1.1 eV and a dielectric constant of 11.8. Graph the bandgap you expect for this material over a range of crystal sizes between 1 nm and 6 nm. At what size do you expect the material to have a bandgap of 2.5 eV (use $m_{e,\text{eff}} = 0.98m_e$ and $m_{h,\text{eff}} = 0.16m_e$ where $m_e = 9.109 \times 10^{-31}$ kg)?

Solution
Here we use Equation (13.32)

$$E_{\text{gap}} = E_{\text{bulk}} + \frac{3h^2}{8L^2}\left(\frac{1}{m_{e,\text{eff}}} + \frac{1}{m_{h,\text{eff}}}\right) - \frac{2e^2}{4\pi\varepsilon_0\varepsilon L}$$

and assume that $m_e = m_h = 9.109 \times 10^{-31}$ kg. Figure P13.4 plots this equation with the Si parameters given. The curve shows how the bandgap changes with the nanoparticle size L and the black filled circle shows the size (at $L = 2.33$ nm) where the bandgap is 2.5 eV.

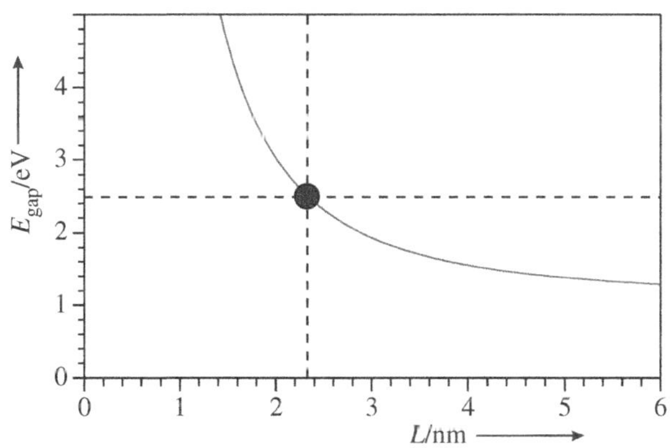

Figure P13.4 Energy gap versus L.

P13.5 A simple model for an impurity atom in a semiconductor is to approximate the impurity site as an effective hydrogen atom. Consider a phosphorous atom in silicon to use four of its valence electrons to participate in the sp^3 bonding of the silicon crystal and have its fifth valence electron to be loosely bound to the phosphorous cation. Approximate the interaction of this electron with the phosphorous cation as a hydrogen atom in a medium with the relative permittivity of silicon ($\varepsilon = 11.8$), and calculate 1) the energy of the ground electronic state and 2) the energy required to ionize this effective hydrogen atom. Also calculate and plot the ground-state wavefunction of the electron.

Solution
Here we refer to Chapter 4 and the electronic energy expression for the H-atom, namely

$$E_n = -\frac{m_e e^4}{8h^2\varepsilon_0^2}\frac{1}{n^2} \qquad n = 1, 2, 3, \ldots$$

For our model of the phosphorous impurity, we must include the dielectric constant $\varepsilon = 11.8$, rather than the dielectric constant of vacuum $\varepsilon = 1.0$. Hence the ground-state energy ($n = 1$) is

$$E_1 = -\frac{m_e e^4}{8h^2\varepsilon^2\varepsilon_0^2}$$

$$= -\frac{(9.109 \times 10^{-31}\ \text{kg}).(1.602 \times 10^{-19})^4\ \text{C}^4}{8 \cdot (6.626 \times 10^{-34})^2\ \text{J}^2\ \text{s}^2 \cdot (11.8)^2 \cdot (8.854 \times 10^{-12})^2\ \text{C}^4\ \text{J}^{-2}\ \text{m}^{-2}}$$

$$= -1.57 \times 10^{-20}\ \text{J} = -0.0975\ \text{eV}$$

Ionization occurs in the limit that $n \to \infty$, and corresponds to release of the electron into the conduction band of the silicon. As $n \to \infty$, $E_n \to 0$. Hence the ionization energy is $-E_1$ or 0.0975 eV.
For H, the ground-state wavefunction is

$$\frac{1}{\sqrt{\pi a_0^3}} \text{e}^{-\frac{r}{a_0}} \text{ with } a_0 = \frac{h^2\varepsilon_0}{\pi m_e e^2}$$

and in the dielectric solid it is

$$\frac{1}{\sqrt{\pi a_0^3}} \text{e}^{-\frac{r}{a_0}} \text{ with } a_0 = \frac{h^2\varepsilon\varepsilon_0}{\pi m_e e^2}$$

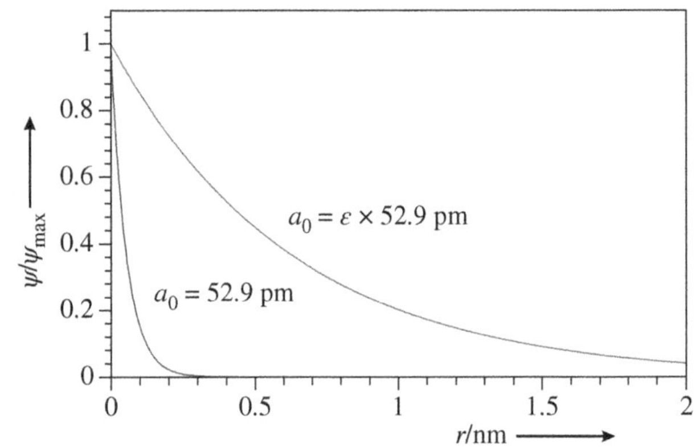

Figure P13.5 ψ/ψ_{max} versus distance r for 1s-wavefunction. $a_0 = 52.9$ pm (light gray) and $a_0 = \varepsilon \times 52.9$ pm (dark gray).

Note that the spatial extent of the electron wavefunction in Fig. P13.5 is significantly larger than that found for the H-atom in vacuum. See Fig. 4.1 of the text.

P13.6 The resistance of a semiconductor decreases with increasing temperature because it is possible to excite electrons through the bandgap; hence, the temperature dependence of the resistance R can be used to determine the bandgap of the material. In order to eliminate the effects of a solid's size one usually finds the resistivity $\rho = 1/\kappa$ reported, where $\kappa = l/(RA)$ is the conductivity. Thus $\rho = R \cdot A/l$ with A equal to the cross-sectional area of the sample and l equal to the sample length. Assume that the resistivity of germanium is given by

$$\rho = C \cdot \exp\left(E_{gap}/\left(2kT\right)\right) \qquad \text{(P13.6.1)}$$

where C is a constant and use the data in the table to determine its bandgap energy. Compare your result to the experimental value of 0.66 eV.

T/K	ρ/Ω cm	T/K	ρ/Ω cm
833	5.00×10^{-3}	357	4.55
700	1.24×10^{-2}	333	10.0
554	5.56×10^{-2}	312	25.0
500	1.43×10^{-1}	300	50.0
385	1.82	278	167

Solution

We convert Equation (P13.6.1) into

$$\ln\left(\frac{\rho}{\Omega\,\text{cm}}\right) = \ln C + \frac{E_{\text{gap}}}{2k}\frac{1}{T}$$

Figure P13.6 shows a plot of $\ln(\rho/(\Omega\text{ cm}))$ versus $1/T$ for the data given above. The data are fit to a line with a slope of 4348 K. Thus

$$\frac{E_{\text{gap}}}{2k} = 4348\text{ K}$$

Using the Boltzmann constant $k = 1.38 \times 10^{-23}\text{ J K}^{-1}$, we obtain

$$E_{\text{gap}} = 2k \cdot 4348\text{ K} = 1.20 \times 10^{-19}\text{ J} = 0.747\text{ eV}.$$

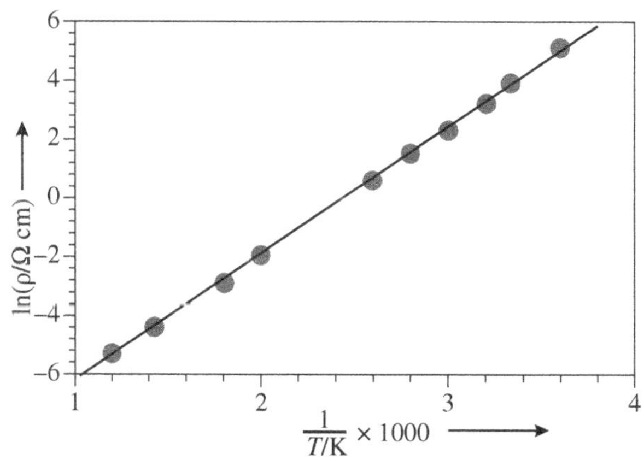

Figure P13.6 Evaluation of bandgap energy. $\ln(\rho/\Omega\text{ cm})$ versus $1/T$.

The bandgap obtained from this analysis is somewhat larger than the literature value reported for germanium.

P13.7 Consider two HCl molecules that are in a collinear geometry (Fig. 13.16a) and separated by a distance of $r = 600$ pm. Plot the dipole–dipole interaction energy as a function of the relative orientation of the dipoles; i.e., rotate one of the dipoles about its center of charge and compute the interaction energy for a series of angles. Note: Take the dipole moment for HCl to be 3.42×10^{-30} C m (See Table 13.4) and its bond length to be $a = 128$ pm.

Solution

The angle β is defined in Fig. P13.7a.

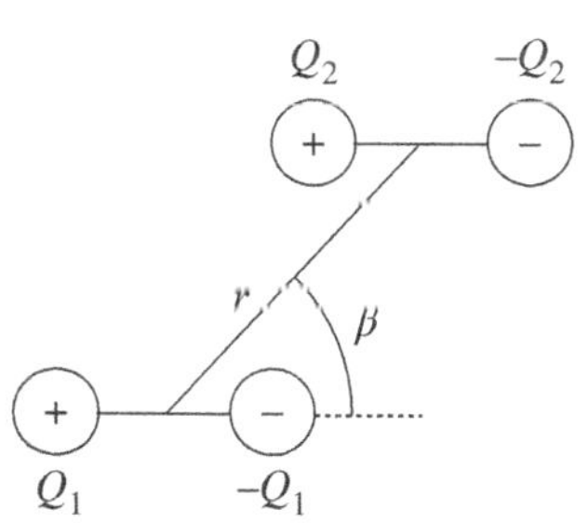

Figure P13.7a Orientation of two collinear dipoles by an angle β.

In Box 13.4 and in Exercise E13.24 we show that the energy of interaction varies with angle as

$$\boxed{E = -\frac{1}{4\pi\varepsilon_0} \cdot \frac{\mu_1\mu_2}{r^3} \cdot \left(3\cos^2\beta - 1\right)}$$

This function is plotted in Fig. P13.7b. Note that at $\beta = 0\,°$ and $\beta = 180\,°$ we have a maximum of attractive forces, and at $\beta = 90\,°$ we have a maximum of repulsive forces. With $\mu_1 = \mu_2 = 3.42 \times 10^{-30}$ C m, $r = 600$ pm, and $\beta = 0\,°$ for the dipole energy we obtain

$$E = -\frac{1}{4\pi \cdot 8.854 \times 10^{-12}\ \mathrm{C^2\ J^{-1}\ m^{-1}}} \cdot \frac{\left(3.42 \times 10^{-30}\ \mathrm{C\ m}\right)^2}{\left(600 \times 10^{-12}\ \mathrm{m}\right)^3} \cdot 2 = -9.26 \times 10^{-22}\ \mathrm{J}$$

This is about one-fourth the magnitude of the thermal energy kT ($E = 40 \times 10^{-22}$ J) at $T = 298$ K.

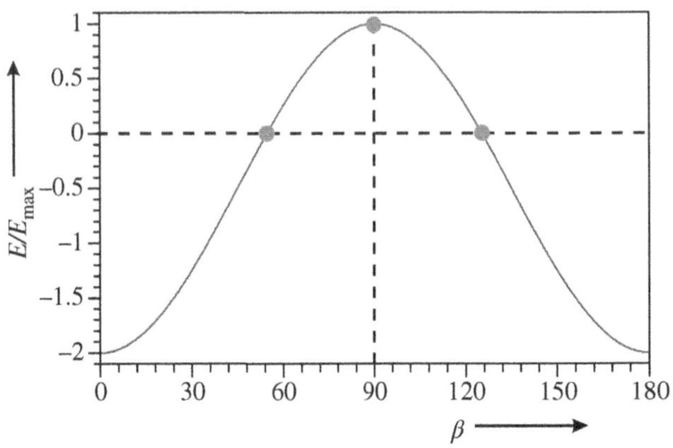

Figure P13.7b Dependence of energy E of two dipoles in collinear geometry on angle β. E_{max} is the energy at its maximum (at $\beta = 90\,°$).

Extension: Compare the point dipole approximation to the extended dipole approximation at 0°. First we use the dipole moment and bond length to calculate the charges on the atoms. Because we have identical molecules ($Q_1 = Q_2$), and we find

$$Q_1 = Q_2 = \frac{3.42 \times 10^{-30}\ \mathrm{C\ m}}{128 \times 10^{-12}\ \mathrm{m}} = 2.67 \times 10^{-20}\ \mathrm{C}$$

Beginning with the energy expression for two dipoles at an angle β (see Foundation 13.2)

$$E = \frac{Q_1 Q_2}{4\pi\varepsilon_0} \cdot \left[\frac{2}{r} - \frac{1}{r}\frac{1}{\sqrt{1 + \frac{a^2}{r^2} + 2\frac{a}{r}\cos\beta}} - \frac{1}{r}\frac{1}{\sqrt{1 + \frac{a^2}{r^2} - 2\frac{a}{r}\cos\beta}}\right]$$

$$= \frac{Q_1 Q_2}{4\pi\varepsilon_0} \cdot \left[\frac{2}{r} - \frac{1}{r}\left(\frac{1}{\sqrt{1 + \frac{a^2}{r^2} + 2\frac{a}{r}\cos\beta}} - \frac{1}{\sqrt{1 + \frac{a^2}{r^2} - 2\frac{a}{r}\cos\beta}}\right)\right]$$

For $\beta = 0°$, we find

$$E = -\frac{\left(2.67 \times 10^{-20}\right)^2\ \mathrm{C^2}}{4\pi \cdot 8.854 \times 10^{-12}\ \mathrm{C^2\ J^{-1} m^{-1}}} \cdot \left[\frac{2}{\left(600 \times 10^{-12}\ \mathrm{m}\right)} - \frac{1}{\left(600 \times 10^{-12}\ \mathrm{m}\right)}\left(\frac{1}{\sqrt{1 + \frac{128^2}{600^2} + 2\frac{128}{600}\cos(0)}} + \frac{1}{\sqrt{1 + \frac{128^2}{600^2} - 2\frac{128}{600}\cos 0}}\right)\right]$$

$$= 6.42 \times 10^{-30}\ \mathrm{J\ m} \cdot \frac{1}{600 \times 10^{-12}\ \mathrm{m}}\,[2 - (0.8242 + 1.2712)]$$

$$= -1.02 \times 10^{-21}\ \mathrm{J}$$

This value is only 10% larger than that found above, indicating that the point dipole approximation is reasonable under these conditions of $r \sim 4.7 \cdot a$.

P13.8 Consider two HCl molecules in the geometry of Fig. 13.16b. Plot the dipole–dipole energy, the induction energy, and the dispersion energy as a function of the distance between the HCl centers, ranging from van der Waals contact to 5 nm. (Take van der Waals contact to be 450 pm, the dipole moment to be 1.03 D = 3.44×10^{-40} C m, the polarizability α to be $4\pi\varepsilon_0 \left(2.63\,\text{Å}^3\right) = 2.93 \cdot 10^{-40} \text{C}^2\text{J}^{-1}\text{m}^{-1}$, and the ionization energy to be 21.5×10^{-19} J).

Solution

For the dipole interaction term we use

$$E_{\text{dipole}} = -\frac{1}{4\pi\varepsilon_0} \cdot \frac{\mu_1\mu_2}{r^3}$$

for the induction term we use

$$E_{\text{ind}} = -\frac{1}{(4\pi\varepsilon_0)^2} \cdot \frac{2\mu^2 \cdot \alpha}{r^6}$$

and for the dispersion term we use

$$E_{\text{disp}} = -\frac{1}{(4\pi\varepsilon_0)^2} \cdot \frac{3}{2} \cdot \frac{\alpha_1 \cdot \alpha_2}{r^6} \frac{E_{\text{Ion},1}E_{\text{Ion},2}}{E_{\text{Ion},1} + E_{\text{Ion},2}}$$

The plot is shown in Fig. P13.8. The dark gray curve is the dipole–dipole energy, the light gray curve is the induction energy, and the black curve is the dispersion energy.

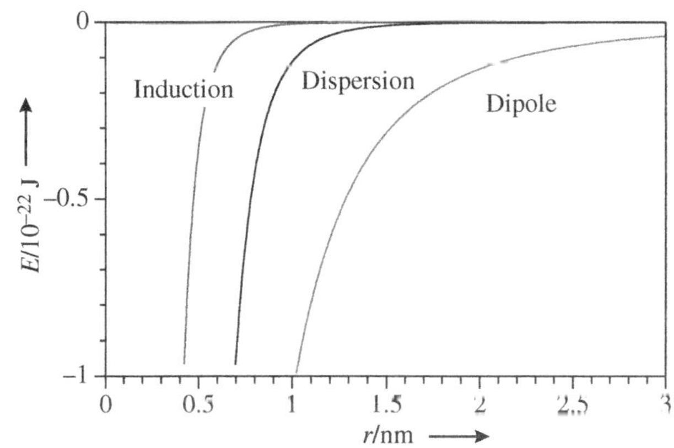

Figure P13.8 Energies of the dipole–dipole (dark gray), induction (light gray), and dispersion (black) interaction.

P13.9 Calculate the polarizability of an electrically conducting plate. Imagine a plate in a homogeneous electric field and calculate the induced dipole moment of the plate. Compare with the polarizability of a sphere of the same volume.

Solution

We consider a capacitor with area A and plate distance l (Fig. P13.9).

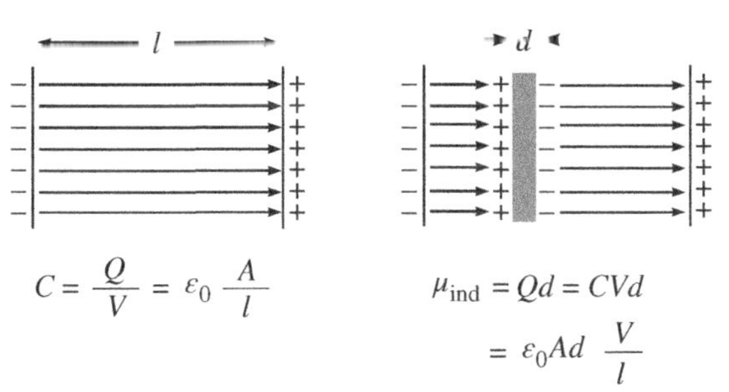

Figure P13.9 Determination of the polarizability of an electrically conducting plate.

Its capacitance is

$$C = \frac{Q}{V} = \varepsilon_0 \frac{A}{l}$$

where Q is the charge on the plates and V is the voltage. Now we place a metallic plate of the same area (thickness d) inside the capacitor. Then the charge Q is induced on the plate and the dipole moment of the plate is

$$\mu_{\text{ind}} = Q \cdot d = CV \cdot d = \varepsilon_0 \frac{A}{l} V \cdot d = \varepsilon_0 Ad \cdot F$$

where Ad is the volume of the plate, and $F = V/l$ is the electric field strength.
Then the polarizability is

$$\alpha = \frac{\mu_{\text{ind}}}{F} = \varepsilon_0 \cdot Ad = \varepsilon_0 \cdot \text{volume}$$

If we use this formula for a sphere, which has a volume of $(4\pi/3) \cdot r^3$, we find that the polarizability α would be

$$\alpha = \varepsilon_0 \cdot \frac{4\pi}{3} r^3$$

The actual α of a conducting sphere is larger by a factor of 3; namely

$$\alpha = \varepsilon_0 \cdot 4\pi r^3 = 3\varepsilon_0 \cdot \text{volume}$$

P13.10 Derive Equation (13.36), the attraction energy between a charge Q and a point dipole (dipole moment μ), for $\theta = 0°$.

Solution
The charge–dipole interaction between a point dipole μ and a charge Q is given by Equation (13.36), namely

$$E = \frac{1}{4\pi\varepsilon_0} \frac{Q\mu}{r^2} \cos\theta$$

where θ is the angle between the dipole and the line connecting it to the charge and r is their distance. For the case of $\theta = 0°$, it will be

$$E = \frac{1}{4\pi\varepsilon_0} \frac{Q\mu}{r^2}$$

The electrostatic field F created by a point dipole at a distance r is

$$F = \frac{1}{4\pi\varepsilon_0} \cdot \frac{2\mu}{r^3}$$

The electrostatic force between a point charge and this field is $Q \cdot F$; hence, we find that

$$\textit{force} = f = Q \cdot F = \frac{2}{4\pi\varepsilon_0} \frac{Q\mu}{r^3}$$

To find the energy we must integrate over space, namely

$$E = \int_{\infty}^{r} f \cdot \mathrm{d}r = \frac{2Q\mu}{4\pi\varepsilon_0} \int_{\infty}^{r} \frac{1}{r^3} \cdot \mathrm{d}r$$
$$= \frac{2Q\mu}{4\pi\varepsilon_0} \cdot \left(-\frac{1}{2} \frac{1}{r^2} \right|_{\infty}^{r} = -\frac{Q\mu}{4\pi\varepsilon_0} \frac{1}{r^2}$$

P13.11 We can use Equation (13.24) for the cohesion energy to find the number density for a metal. Calculate the minimum of the energy with respect to the number density by taking its derivative and setting it equal to zero.

Solution
Taking the derivative we find

$$\frac{\mathrm{d}E}{\mathrm{d}(N/V)} = -A\frac{1}{3}\left(\frac{N}{V}\right)_{\min}^{-2/3} + B\frac{2}{3}\left(\frac{N}{V}\right)_{\min}^{-1/3} = 0$$

so that

$$\left(\frac{N}{V}\right)_{\min}^{1/3} = \frac{A}{2B} = \frac{3.27 \times 10^{-19}\ \text{J nm}}{2 \cdot 3.50 \times 10^{-20}\ \text{J nm}^2} = 4.67\ \text{nm}^{-1}$$

or

$$\left(\frac{N}{V}\right)_{\min} = 101.9\ \text{nm}^{-3}$$

The typical radius about an atom in the metal is

$$l = \left(\frac{N}{V}\right)_{\min}^{-1/3} = (101.9\ \text{nm}^{-3})^{-1/3} = 0.214\ \text{nm} = 214\ \text{pm}$$

This result is on the order of the experimental value of 280 pm.
Then for the mean potential and kinetic energy we obtain

$$\bar{E}_{\text{pot}} = -A \cdot \left(\frac{N}{V}\right)^{1/3} = -A \cdot \frac{A}{2B} = -\frac{A^2}{2B}$$

and

$$\bar{E}_{\text{kin}} = B \cdot \left(\frac{N}{V}\right)^{2/3} = B \cdot \left(\frac{A}{2B}\right)^2 = \frac{A^2}{4B}$$

A result that is agreement with the Virial theorem. With the data given for A and B we find that

$$\overline{E}_{\text{pot}} = -\frac{A^2}{2B} = -\frac{\left(3.27 \times 10^{-19}\ \text{J nm}\right)^2}{2 \cdot 3.50 \times 10^{-20}\ \text{J nm}^2} = -15.28 \times 10^{-19}\ \text{J}$$

and

$$\overline{E}_{\text{kin}} = -\frac{1}{2}\overline{E}_{\text{pot}} = 7.64 \times 10^{-19}\ \text{J}, E = \overline{E}_{\text{pot}} + \overline{E}_{\text{kin}} = -7.64 \times 10^{-19}\ \text{J}$$

P13.12 Show that electron current traveling in the z-direction through a quantum wire in the $n_x = n_y = 1$ energy level is given by the formula

$$I = e\sum \frac{v_z}{L} = e\sum \frac{hn_z}{2m_eL^2}$$

Solution

Along the x- and y-directions the possible quantum states for the electron are constrained by the quantum numbers n_x and n_y, but in the z-direction a continuum of quantum states is available. The magnitude of the current I is given by the charge on the electron times the number of electrons moving in the positive z direction per unit time (the velocity v of the electrons moving to the right divided by the length of the quantum wire L). We can write this current as

$$I = e\sum \frac{v_z}{L}$$

where the sum is performed over the electronic states that lie between the Fermi levels on the left and right sides in Fig. 13.11 (since only these contribute to the net current). The velocity v_z is given by the de Broglie equation

$$v_z = \frac{h}{m_e\Lambda}$$

where

$$n_z \cdot \frac{\Lambda}{2} = L, \quad \Lambda = \frac{2L}{n_z}$$

Thus

$$v_z = \frac{hn_z}{2m_eL}$$

and the expression for the current becomes

$$I = e\sum \frac{v_z}{L} = e\sum \frac{hn_z}{2m_eL^2}$$

P13.13 Using Equation (13.7)

$$E = -\frac{1}{4\pi\varepsilon_0} \cdot \frac{e^2}{d} \cdot \left[1 - \frac{1}{n}\left(\frac{d_{eq}}{d}\right)^{n-1}\right]$$

calculate the energy E for an ion pair of NaCl for different values of the parameter n.

Solution

We use the same value for the equilibrium distance, $d_{eq} = 283$ pm, as we did in the text, but we perform the calculation for $n = 6$, $n = 8$, $n = 12$, and $n = 100$. The result for E is plotted in Fig. P13.13.

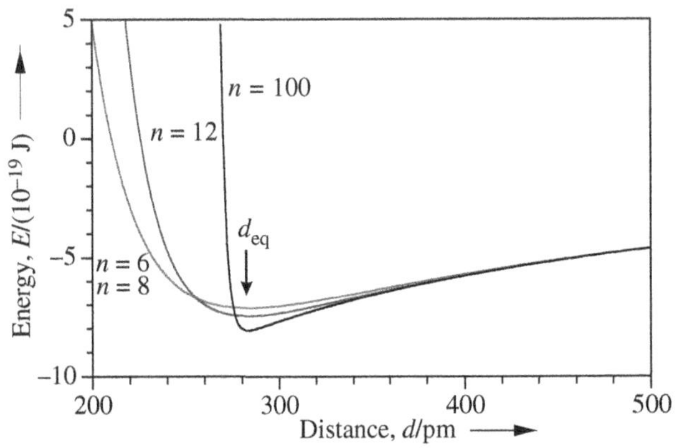

Figure P13.13 Energy of a NaCl ion pair calculated for $d_{eq} = 283$ pm and $n = 6$, $n = 8$ (dark gray solid line), $n = 12$ (light gray solid line), and $n = 100$ (black solid line).

We recognize that the curves for $n = 6$ and $n = 8$ coincide, the curve for $n = 12$ is slightly steeper, and the curve for $n = 100$ is nearly a vertical line in the range below the equilibrium distance. The equilibrium energy E_{eq} changes only slightly from -7.470×10^{-19} J at $n = 6$ and $n = 8$ to -7.471×10^{-19} J at $n = 12$ and -8.067×10^{-19} J at $n = 100$. According to Equation (13.6) the limiting value for $n \to \infty$ is

$$E_{eq} = -\frac{1}{4\pi\varepsilon_0} \cdot \frac{e^2}{d_{eq}} \cdot \left[1 - \frac{1}{n}\left(\frac{d_{eq}}{d_{eq}}\right)^{n-1}\right]$$

$$= -\frac{1}{4\pi\varepsilon_0} \cdot \frac{e^2}{d_{eq}} = -8.151 \times 10^{-19}\ \text{J}$$

P13.14 Often when constructing a theory, we begin with an idealization of the realistic system to see if the essential ideas apply. In this problem we consider the diffraction of X-ray light from a lattice of atoms. Let us view the crystalline solid as being a two-dimensional array of atoms as in Fig. P13.14a.

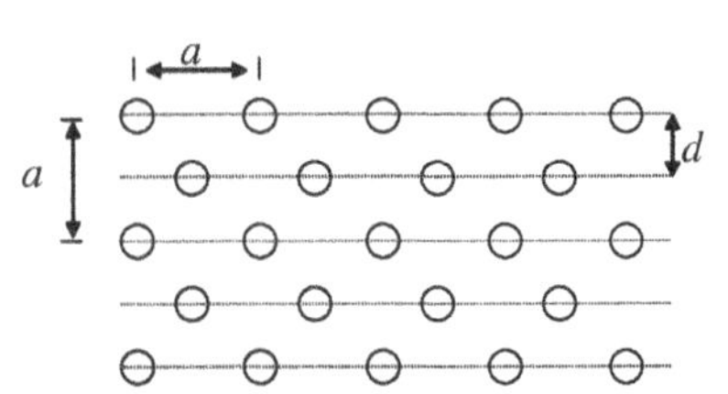

Figure P13.14a Two-dimensional lattice.

The distance a is called the lattice parameter, and let us take its value to be that of Au, $a = 408$ pm. If X-rays of wavelength λ are incident on this array of atoms from the top at an angle θ, it will be diffracted at some angle. You are asked to find this angle.

a) Consider the incident light field to be described by a sine wave and consider each layer of atoms (indicated by the dashed lines) to act as a partial mirror that reflects the X-rays. Using the superposition of light waves, show that the reflected waves add constructively when $n\lambda = 2d \cdot \sin(\theta)$ in which n is an integer and d is the distance between diffraction planes. This is called the Bragg condition.
b) One common source produces X-rays at a mean wavelength of 154 pm; using this value and the lattice constant given above determine the angles that satisfy the Bragg condition for $n = 1$ and $n = 2$.
c) The material C_{60} (fullerene) forms cubic crystals with a lattice constant of 1470 pm near room temperature. What value of the diffraction angles would you expect to find in this case?
d) Consider forming a solid lattice from micron-sized particles (sometimes called a colloidal crystal) that has a lattice parameter of 1.0 μm. What wavelength of light will your crystal diffract in first order ($n = 1$) at 15°? Diffraction peaks will appear at points where the wave amplitudes add in phase, and this occurs when the distances traveled by the different parts of the wave (e.g., that reflected from one layer of atoms versus that from the adjacent layer) differ by a multiple of the wavelength. Figure P13.14b shows that the difference in distance traveled by two light rays (dark gray and black) is $2l$, where $l = d \cdot \sin(\theta)$.

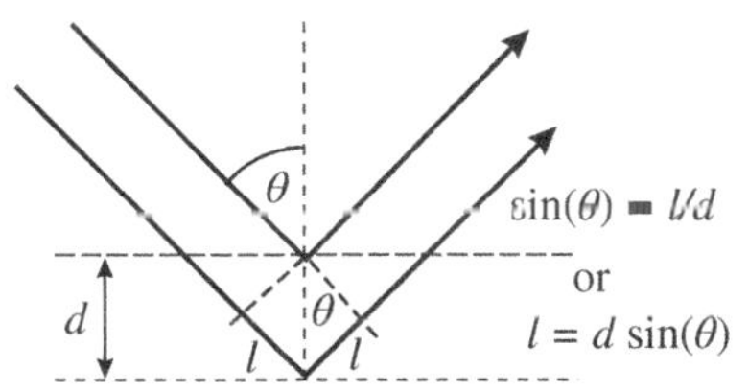

Figure P13.14b Diffraction: traveling distance.

Solution

a) In order for constructive interference to occur the extra distance that a wave travels must be a multiple of the wavelength, so that we find

$$n\lambda = 2l = 2d \cdot \sin(\theta)$$

which is Bragg's law. You can find a more detailed derivation in an introductory physics textbook.

b) Bragg's law can be rearranged as,

$$\sin(\theta) = \frac{n\lambda}{2d}$$

and substituting in values from the question we find

$$\sin(\theta) = \frac{n\ (154\ \text{pm})}{2\ (408\ \text{pm})} = n\ \cdot 0.189 \text{ where } n \text{ is an integer}$$

Hence we find that

n	1	2	3	4
$\sin(\theta)$	0.189	0.377	0.566	0.755
θ	10.8°	22.2°	34.5°	49.0°

c) Using the same wavelength of light with C60, the first two diffraction peaks would occur at

$$\sin(\theta) = \frac{n\lambda}{2d} = \frac{n\ (154\ \text{pm})}{2\ (1470\ \text{pm})} = n\ (0.0524)$$

so that for $n = 1$ we find $\sin(\theta) = 0.0524$ or $\theta = 3.0^\circ$ and for $n = 2$ we find $\sin(\theta) = 0.105$ or 6.0°.

d) Rearranging Bragg's law and substituting the values given, results in

$$\lambda = \frac{2d\ \sin(\theta)}{n} = \frac{2\left(1.0 \times 10^{-6}\ \text{m}\right)\ \sin(15^\circ)}{1} = 5.17 \times 10^{-7}\ \text{m} = 517\ \text{nm}$$

which is right in the middle of the visible spectrum.

P13.15 In this problem let us look at electron diffraction and consider the diffraction of electrons just from the top row of atoms in Fig. P13.14a with a spacing a.

a) Consider a situation in which electron waves of wavelength Λ are incident on the row of atoms at an angle normal to the surface and are scattered from the atoms at an angle θ. Show that the electron waves superpose constructively when $n\Lambda = a\ \sin(\theta)$, where n is an integer called the order of diffraction.

b) Consider the case when the row of atoms has a lattice parameter $a = 408$ pm (that of Au), what wavelength of electrons is needed for the first-order diffraction peak to appear at 15°? What must the kinetic energy of the electrons be for them to have the required wavelength?

c) A number of important experimental methods use electron diffraction to investigate the structural features of solid surfaces. In LEED (Low Energy Electron Diffraction) spectroscopy the energy of the electrons ranges from about 10–500 eV. What range of electron wavelengths are accessed in LEED spectroscopy and what characteristic structural features can you expect to study with LEED?

d) Another electron diffraction method, called RHEED (Reflection High-Energy Electron Diffraction), uses electron energies in the range of 5–50 keV. What range of electron wavelengths are accessed in this method?

Solutions

a) Diffraction peaks will appear at points where the electron wave amplitudes add in phase, and this occurs when the distances traveled by the different parts of the wave (e.g., that reflected from adjacent atoms) differ by a multiple of the electron wavelength Λ. Figure P13.15 shows that the difference in distance traveled by two rays (dark gray and black) is l, where $l = a \cdot \sin(\theta)$.

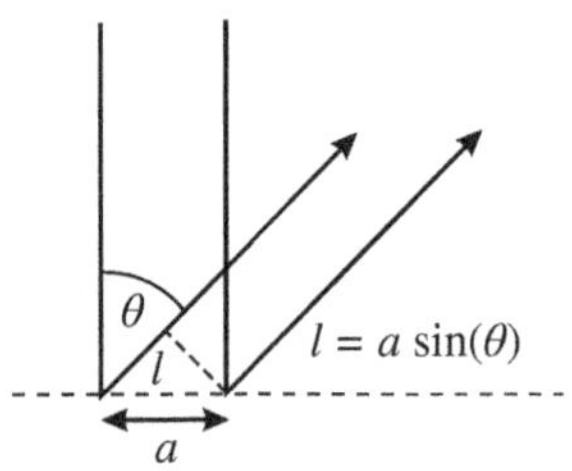

Figure P13.15 Diffraction: Traveling distance.

In order for constructive interference to occur $n\Lambda = l$, so that

$$n \cdot \Lambda = l = a \cdot \sin\theta$$

b) Rearranging this result and substituting the values given yields

$$\Lambda = \frac{a\ \sin(\theta)}{n} = \frac{(408\text{ pm})\ \sin(15°)}{1} = 106\text{ pm}$$

The kinetic energy can be found from the momentum, and thus the wavelength through the de Broglie relation as

$$\begin{aligned} E_{\text{kinetic}} &= \frac{p^2}{2\ m_e} = \frac{(h/\Lambda)^2}{2\ m_e} \\ &= \frac{\left(6.626 \times 10^{-34}\text{ J s}/106 \times 10^{-12}\text{ m}\right)^2}{2\ \left(9.11 \times 10^{-31}\text{ kg}\right)} \\ &= 21.4 \times 10^{-18}\text{ J} = 134\text{ eV} \end{aligned}$$

c) We can rearrange the previous expression for the kinetic energy in terms of the wavelength Λ and find

$$\begin{aligned} \Lambda &= \frac{h}{\sqrt{2m_e E_{\text{kinetic}}}} \\ &= \frac{6.626 \times 10^{-34}\text{ J s}}{\sqrt{2\left(9.11 \times 10^{-31}\text{ kg}\right)\left(10 \cdot 1.602 \times 10^{-19}\text{ J}\right)}} = 387.8\text{ pm} \end{aligned}$$

and

$$\Lambda = \frac{h}{\sqrt{2m_e E_{\text{kinetic}}}}$$
$$= \frac{6.626 \times 10^{-34}\ \text{J s}}{\sqrt{2\left(9.11 \times 10^{-31}\ \text{kg}\right)\left(500 \cdot 1.602 \times 10^{-19}\ \text{J}\right)}} = 54.84\ \text{pm}$$

This range is typical of interatomic spacings in solids, and LEED is commonly used to study the atomic arrangement of atoms on solid surfaces.

d) Performing the same type of calculation as in part c results in

$$\Lambda = \frac{h}{\sqrt{2m_e E_{\text{kinetic}}}}$$
$$= \frac{6.626 \times 10^{-34}\ \text{J s}}{\sqrt{2\left(9.11 \times 10^{-31}\ \text{kg}\right)\left(5 \times 10^{3} \cdot 1.602 \times 10^{-19}\ \text{J}\right)}} = 17.3\ \text{pm}$$

and

$$\Lambda = \frac{h}{\sqrt{2m_e E_{\text{kinetic}}}}$$
$$= \frac{6.626 \times 10^{-34}\ \text{J s}}{\sqrt{2\left(9.11 \times 10^{-31}\ \text{kg}\right)\left(50 \times 10^{3} \cdot 1.602 \times 10^{-19}\ \text{J}\right)}} = 5.48\ \text{pm}$$

14

Thermal Motion of Molecules

14.1 Exercises

E14.1 In Section 14.2 we calculated the de Broglie wavelength of H_2. Perform a similar calculation for N_2, CO_2, and He and compare to the H_2 calculation. As the molecular mass increases do you expect a molecule's translational motion to behave more classically or less classically? Of the five gases which would you expect to reveal quantum effects first?

Solution

The de Broglie wavelength is given by $\Lambda = h/(m\upsilon)$. In Section 14.3 we showed that the average speed of the molecules in a gas is given by $\overline{\upsilon} = \sqrt{8\boldsymbol{R}T/(\pi\boldsymbol{M})}$. Thus, for N_2 we find

$$\overline{\upsilon} = \sqrt{\frac{8\boldsymbol{R}T}{\pi\boldsymbol{M}}} = \sqrt{\frac{8\left(8.314\ \text{J mol}^{-1}\ \text{K}^{-1}\right) 300\ \text{K}}{\pi\left(0.02802\ \text{kg mol}^{-1}\right)}} = 476.3\ \text{m s}^{-1}$$

If we use this average speed in the calculation for the de Broglie wavelength we find that

$$\Lambda = \frac{h}{m\overline{\upsilon}} = \frac{6.626\times10^{-34}\ \text{J s}}{\left(0.02802\ \text{kg mol}^{-1}/6.022\times10^{23}\ \text{mol}^{-1}\right) 476.3\ \text{m s}^{-1}}$$
$$= 2.990\times10^{-11}\ \text{m} = 29.90\ \text{pm}$$

We proceed in a similar way for each of the gases and find the values given in the table.

	$\overline{\upsilon}$/m/s	Λ/pm
H_2	1782	112
He	1260	79.2
N_2	476.3	29.9
CO_2	380.0	23.9

It is evident that the de Broglie wavelength becomes progressively smaller as the mass increases. The wavelength is smaller than the size of any practical container, and hence we expect the gas to behave classically. Quantum effects are likely to become apparent for hydrogen or helium before they become apparent for other gases.

E14.2 Consider a "head-on" collision (one-dimensional) between a CO_2 molecule and an H_2 molecule that are each traveling at 500 m s^{-1} in opposite directions. Use the constraints that the linear momentum is conserved and the kinetic energy is conserved to determine the final speeds of the two molecules.

Solution

We begin by defining the initial conditions of the CO_2 moving to the right at 500 m s^{-1}, and the H_2 moving to the left at 500 m s^{-1}. We take the displacement along the collision direction to be x and consider CO_2 to be moving

Solutions Manual for Principles of Physical Chemistry, Third Edition. Edited by Hans Kuhn, David H. Waldeck, and Horst-Dieter Försterling.

in the positive x direction. The statement of conservation of linear momentum leads to the equation (the index i denotes "initial" and f denotes "final")

$$m_{CO_2}v_{CO_2,i} - m_{H_2}v_{H_2,i} = m_{CO_2}v_{CO_2,f} + m_{H_2}v_{H_2,f}$$

and the conservation of energy leads to

$$m_{CO_2}v^2_{CO_2,i} + m_{H_2}v^2_{H_2,i} = m_{CO_2}v^2_{CO_2,f} + m_{H_2}v^2_{H_2,f}$$

Writing $v_{CO_2,i} = v_{H_2,i} = 500\ \text{m s}^{-1}$, the conservation of momentum constraint gives

$$\left(m_{CO_2} - m_{H_2}\right)\ 500\ \text{m s}^{-1} = m_{CO_2}v_{CO_2,f} + m_{H_2}v_{H_2,f}$$

so that

$$\frac{\left(m_{CO_2} - m_{H_2}\right)\ 500\ \text{m s}^{-1} - m_{H_2}v_{H_2,f}}{m_{CO_2}} = v_{CO_2,f}$$

Now we can substitute this result into the equation that is obtained from the conservation of energy to give

$$\left(m_{CO_2} + m_{H_2}\right)\left(500\ \frac{\text{m}}{\text{s}}\right)^2 = \left[\begin{array}{c} m_{H_2}v^2_{H_2,f} \\ +\left[\left(m_{CO_2} - m_{H_2}\right)\ 500\ \frac{\text{m}}{\text{s}} - m_{H_2}v_{H_2,f}\right]^2 / m_{CO_2} \end{array}\right]$$

We want to solve this expression for the final speed of the H_2. To do so we expand the square and rearrange the equation into the form of a quadratic expression

$$\begin{aligned}\left(m_{CO_2} + m_{H_2}\right)\left(500\ \frac{\text{m}}{\text{s}}\right)^2 &= m_{H_2}v^2_{H_2,f} + \frac{\left(m_{CO_2} - m_{H_2}\right)^2}{m_{CO_2}}\left(500\ \frac{\text{m}}{\text{s}}\right)^2 + \frac{m^2_{H_2}}{m_{CO_2}}v^2_{H_2,f} \\ &\quad -2\left(\left(m_{CO_2} - m_{H_2}\right)\ (500\ \text{m s}^{-1})\right)\frac{m_{H_2}}{m_{CO_2}}v_{H_2,f}\end{aligned}$$

then we collect terms to find

$$\begin{aligned}0 = &\left[\frac{1}{m_{CO_2}}\left(m_{CO_2} - m_{H_2}\right)^2 - \left(m_{CO_2} + m_{H_2}\right)\right]\ (500\ \text{m s}^{-1})^2 \\ &-2\left[\left(m_{CO_2} - m_{H_2}\right)\ (500\ \text{m s}^{-1})\right]\frac{m_{H_2}}{m_{CO_2}}v_{H_2,f} + \left(\frac{m^2_{H_2}}{m_{CO_2}} + m_{H_2}\right)v^2_{H_2,f}\end{aligned}$$

Upon substitution of the masses of H_2 (0.3347×10^{-26} kg) and CO_2 (7.306×10^{-26} kg), we find

$$\begin{aligned}0 = &-0.9888 \times 10^{-26}\ \text{kg}\ (500\ \text{m s}^{-1})^2 - 0.6387 \times 10^{-26}\ \text{kg}\ (500\ \text{m s}^{-1})\ v_{H_2,f} \\ &+0.3500 \times 10^{-26}\ \text{kg}\ v^2_{H_2,f}\end{aligned}$$

which can be solved for $v_{H_2,f}$ by using the quadratic equation, namely

$$\begin{aligned}v_{H_2,f} &= \frac{319.4 \pm \sqrt{1.020 \times 10^5 + 3.461 \times 10^5}}{0.700}\frac{\text{m}}{\text{s}} = (456. \pm 956.)\frac{\text{m}}{\text{s}} \\ &= -500\frac{\text{m}}{\text{s}}\ \text{or}\ 1412\frac{\text{m}}{\text{s}}\end{aligned}$$

The value of $-500\ \text{m s}^{-1}$ is not physically possible because the speed v is defined to be a positive quantity. Hence we use the value $v_{H_2,f} = 1412\ \text{m s}^{-1}$ and substitute into the expression for the conservation of momentum to find

$$\begin{aligned}v_{CO_2,f} &= \frac{\left(m_{CO_2} - m_{H_2}\right)\ 500\ \text{m s}^{-1} - m_{H_2}v_{H_2,f}}{m_{CO_2}} \\ &= \left(1 - \frac{m_{H_2}}{m_{CO_2}}\right)\ 500\ \text{m s}^{-1} - \frac{m_{H_2}}{m_{CO_2}}v_{H_2} \\ &= \left(1 - \frac{3.354}{73.08}\right)\ 500\ \text{m s}^{-1} - \frac{3.354}{73.08}1412\ \text{m s}^{-1} = 412\ \text{m s}^{-1}\end{aligned}$$

You can check this result by running the program "collision."

E14.3 Two spheres (masses m_1 and m_2; initial speeds v_1 and v_2) collide head-on (central collision). Find the final speed for each of the particles if the initial conditions are $m_1 = m_2 = m$; $v_1 = 0$, $v_2 = v$.

Solution

We define the speeds after the collision as $v_{f,1}$ and $v_{f,2}$. According to the conservation of energy we can write

$$\frac{1}{2}mv_2^2 = \frac{1}{2}mv_{f,1}^2 + \frac{1}{2}mv_{f,2}^2$$

and from the conservation of momentum we can write

$$mv_2 = mv_{f,1} + mv_{f,2}$$

From the conservation of momentum constraint, we can find that $v_{f,2} = \left(v_2 - v_{f,1}\right)$, so that

$$v_{f,2}^2 = \left(v_2 - v_{f,1}\right)^2 = v_2^2 - 2v_{f,1}v_2 + v_{f,1}^2$$

By using the conservation of energy constraint to substitute for v_2^2, we find

$$v_{f,2}^2 = v_{f,1}^2 + v_{f,2}^2 - 2v_{f,1}v_2 + v_{f,1}^2$$

so that

$$0 = 2v_{f,1}^2 - 2v_{f,1}v_2 = 2v_{f,1}\left(v_{f,1} - v_2\right)$$

and

$$v_{f,1}(v_2 - v_{f,1}) = 0$$

This equation has the solution $v_{f,1} = v_2$ or $v_{f,1} = 0$. Only the former result is physically reasonable, since some momentum will be transferred to particle 1. Hence we choose the solution $v_{f,1} = v_2$ for which $v_{f,2} = 0$. This means that particle 2 transfers its kinetic energy completely to particle 1.

You can check this result by running the program "collision."

E14.4 Two spheres (masses m_1 and m_2; initial speeds v_1 and v_2) collide head-on (central collision). Take the displacement along the collision direction to be x and consider particle 1 to be moving in the positive x-direction. Find the final speed for each of the particles if the initial conditions are $m_1 \gg m_2$; $v_1 = v$, $v_2 = -v$.

Solution

Again we apply the laws of conservation of energy and momentum:

$$\frac{1}{2}m_1v^2 + \frac{1}{2}m_2v^2 = \frac{1}{2}m_1v_{f,1}^2 + \frac{1}{2}m_2v_{f,2}^2$$

and

$$m_1v - m_2v = m_1v_{f,1} + m_2v_{f,2}$$

From the energy conservation, we can write that

$$v^2 \cdot \left(m_1 + m_2\right) = m_1v_{f,1}^2 + m_2v_{f,2}^2$$

or

$$v_{f,1}^2 = v^2\left(1 + \frac{m_2}{m_1}\right) - v_{f,2}^2\frac{m_2}{m_1}$$

From the momentum conservation we can write that

$$v \cdot \left(m_1 - m_2\right) = m_1v_{f,1} + m_2v_{f,2}$$

If we solve it for $v_{f,1}$ and then square it, we find that

$$v_{f,1}^2 = v^2\left(1 - \frac{m_2}{m_1}\right)^2 - 2vv_{f,2}\left(1 - \frac{m_2}{m_1}\right)\frac{m_2}{m_1} + \left(\frac{m_2}{m_1}\right)^2 v_{f,2}^2$$

Now we can equate these equations to eliminate $v_{f,1}^2$ and find

$$v^2\left(1 + \frac{m_2}{m_1}\right) - v_{f,2}^2\frac{m_2}{m_1} = v^2\left(1 - \frac{m_2}{m_1}\right)^2 - 2vv_{f,2}\left(1 - \frac{m_2}{m_1}\right)\frac{m_2}{m_1} + \left(\frac{m_2}{m_1}\right)^2 v_{f,2}^2$$

$$v_{f,2}^2\left(1 + \frac{m_2}{m_1}\right) - 2vv_{f,2}\left(1 - \frac{m_2}{m_1}\right) = v^2\left(3 - \frac{m_2}{m_1}\right)$$

In the limit that $m_1 \gg m_2$, we obtain

$$v_{f,2}^2 - 2vv_{f,2} = 3v^2 \quad \text{or} \quad v_{f,2}^2 - 2vv_{f,2} - 3v^2 = 0$$

which has the solution

$$v_{f,2} = \frac{2v \pm \sqrt{4v^2 + 12v^2}}{2} = v \pm 2v \text{ so that } v_{f,2} = -v \text{ or } 3v$$

The $v_{f,2} = -v$ solution is not physically reasonable, because the speed v is defined as a positive quantity. Hence, we use the solution

$$v_{f,2} = 3v, \text{ and find that } v_{f,1} = v$$

That is, the heavy particle moves with the same velocity as before the collision, and the light particle moves with three times the velocity of the heavy particle.

The result can be easily understood. The light particle has a velocity of $-2v$ relative to the heavy particle; thus, it will be reflected with a velocity $+2v$ relative to the heavy particle. Since the heavy particle is moving at a velocity v, the velocity of the light particle becomes $+3v$.

You can check this result by running the program "collision."

E14.5 Calculate the mean free path for an H_2 molecule in a container of hydrogen gas at a pressure of 1 bar and a temperature of 300 K. Approximate the H_2 molecules as spheres of radius 138 pm.

Solution

With $(r_1 + r_2) = 2r_1 = 276$ pm, we obtain $\sigma = 2.39 \times 10^{-19}\,\text{m}^2$. Using this σ and $N/V = 2.41 \times 10^{25}\,\text{m}^{-3}$ in Equation (14.39), we obtain

$$\bar{\lambda}_{H_2} = \frac{1}{\sqrt{2} \cdot 2.39 \times 10^{-19}\ \text{m}^2 \cdot 2.41 \times 10^{25}\ \text{m}^{-3}} = 123\ \text{nm}$$

E14.6 Calculate the mean free path for a Br_2 molecule moving among a collection of N_2 molecules at a pressure of 1 bar and a temperature of 300 K. Approximate the molecules as spheres of radius 312 pm for Br_2 and 190 pm for N_2.

Solution

In this case $r_1 + r_2 = (312 + 190)$ pm $= 502$ pm, where the index 1 refers to the Br_2 molecule and the index 2 to the N_2 molecules so that $\sigma = \pi(r_1 + r_2)^2 = 7.92 \times 10^{-19}$ m^2. Using the ideal gas law to find N/V we obtain

$$\frac{N}{V} = \frac{P}{kT} = \frac{10^5\ \text{N m}^{-2}}{\left(1.38 \times 10^{-23}\ \text{J K}^{-1}\right) 300\ \text{K}} = 2.41 \times 10^{25}\ \text{m}^{-3}$$

With $N/V = 2.41 \times 10^{25}$ m^{-3} and $m_1/m_2 = 159.6/28.0 = 5.7$, we can use Equation (14.42) to obtain the mean free path of Br_2

$$\bar{\lambda}_{Br_2} = \frac{1}{7.92 \times 10^{-19}\ \text{m}^2 \cdot 2.41 \times 10^{25}\ \text{m}^{-3}} \cdot \frac{1}{\sqrt{1 + 5.7}} = 21\ \text{nm}$$

E14.7 Consider a container of I_2 gas molecules at a temperature of 298 K. What is the average translational energy of an I_2 molecule in the container? Consider two I_2 molecules, X and Y, to have the average energy. If molecule X collides with molecule Y and transfers all of its translational energy to molecule Y, is it possible for molecule Y to become vibrationally excited? If so, what is the maximum vibrational energy level to which Y could be excited? The I_2 molecule has a fundamental vibrational frequency ν_0 of 6.42×10^{12} s^{-1}.

Solution

The average translational energy of an I_2 molecule is

$$\begin{aligned}\bar{E}_{\text{kin}} &= \frac{3}{2}kT \\ &= \frac{3}{2}\left(1.3806 \times 10^{-23}\ \text{J K}^{-1}\right)(298\ \text{K}) = 6.17 \times 10^{-21}\ \text{J}\end{aligned}$$

The vibrational excitation energy is

$$\begin{aligned}\Delta E_{\text{vib}} &= h\nu_0 \\ &= \left(6.626 \times 10^{-34}\ \text{J s}\right)\ \left(6.42 \times 10^{12}\ \text{s}^{-1}\right) = 4.25 \times 10^{-21}\ \text{J}\end{aligned}$$

Thus sufficient translational energy is available to vibrationally excite I_2 in a single collision from the vibrational state with $n = 0$ to the vibrational state with $n = 1$. An excitation to the vibrational state with $n = 2$ would require an energy of $2h\nu_0 = 8.5 \times 10^{-21}$ J; this is more than the thermal energy.

E14.8 For practice you should compute the root mean square speeds of some gases; e.g., He, N_2, O_2, Ar, and Xe at $T = 298$ K and compare the results you find with the speed of sound in air, which is about 340 m s^{-1} at sea level.

Solution

The expression for the root mean square speed is

$$v_{\text{rms}} = \left(\frac{3kT}{m}\right)^{1/2} = \left(\frac{3RT}{\boldsymbol{M}}\right)^{1/2}$$

where m is the mass of one molecule and $\boldsymbol{M}$ is the molar mass in the appropriate units. For nitrogen we have that

$$v_{\text{rms}} = \sqrt{\frac{3RT}{\boldsymbol{M}}} = \sqrt{\frac{3\left(8.314\ \text{J mol}^{-1}\ \text{K}^{-1}\right)(298\ \text{K})}{0.02802\ \text{kg mol}^{-1}}} = 515\ \text{m s}^{-1}$$

The root-mean-square speed of He, N_2, O_2, Ar, and Xe at $T = 298$ K is reported in the table

Molecule	He	N_2	O_2	Ar	Xe
$v_{\text{rms}}/\left(\text{m s}^{-1}\right)$	1363	515	482	431	238

The values are roughly comparable to the speed of sound, 346 m s^{-1}; however, the speed of sound is somewhat less than the average speed found for the atmospheric gases.

E14.9 Calculate the collision frequency of a nitrogen molecule (the dominant component of air) at 1 bar pressure and 300 K. Compute the number of collisions that occur in one second. Compare this number to the number of seconds in recorded history, which you can estimate to be 5000 years.

Solution

The collision frequency for a single nitrogen molecule is given by Equations (14.45) and (14.43)

$$z_{\text{coll}} = \frac{\overline{v}}{\lambda} = \left(\sqrt{2}\sigma\ \frac{N}{V}\right)\ \overline{v}.$$

For N_2 we use the diameter $d = 375$ pm from Table 14.3 which gives $\sigma = 4.42 \times 10^{-19}$ m^2. At 300 K and 1 bar

$$\overline{v} = \sqrt{\frac{8RT}{\pi \boldsymbol{M}}} = 476\ \text{m s}^{-1}$$

with $\boldsymbol{M} = 0.02801$ kg mol^{-1}. Using the ideal gas law ($PV = NkT$) to find the number density we find that

$$\begin{aligned}\frac{N}{V} &= \frac{P}{kT} \\ &= \frac{10^5\ \text{N m}^{-2}}{\left(1.3806 \times 10^{-23}\ \text{J K}^{-1}\right)\ (300\ \text{K})} = 2.41 \times 10^{25}\ \text{m}^{-3}\end{aligned}$$

Combining these intermediate results in our expression for the collision frequency, we find that

$$\begin{aligned}z_{\text{coll}} &= \left(\sqrt{2}\sigma\ \frac{N}{V}\right)\ \overline{v} \\ &= \left(\sqrt{2}\left(4.42 \times 10^{-19}\text{m}^2\right)\ 2.41 \times 10^{25}\ \text{m}^{-3}\right)\ 476\ \text{m s}^{-1} \\ &= 7.17 \times 10^9\ \text{s}^{-1}\end{aligned}$$

So that in one second, 7.17×10^9 collisions occur.

By way of comparison, the number of seconds in recorded history

$$5000 \text{ year} \left(\frac{365.25 \text{ days}}{1 \text{ year}}\right)\left(\frac{24 \text{ hours}}{1 \text{ day}}\right)\left(\frac{3600 \text{ s}}{1 \text{ hour}}\right) = 1.578 \times 10^{11} \text{ s}$$

In order for the number of collisions to be equal to the number of seconds in recorded history, we would need to let the gas molecules collide for 20 to 25 seconds.

E14.10 Compute the mean free path and average speed of a He atom under the following conditions: a) 1 bar pressure and 300 K (typical of Earth surface), b) 0.01 bar pressure and 250 K (typical of Mars surface), c) 90 bar pressure and 700 K (typical of Venus surface), d) 1.4×10^{-5} bar pressure and 40 K (typical surface values for Triton, a moon of Neptune).

Solution

Helium has an atomic/molecular diameter of 218 pm from Table 14.3. This value gives a cross-section of $\sigma = \pi(218\text{pm})^2/4 = 3.73 \times 10^{-20}$ m^2. We can then do the calculations using the formulas in Equations (14.43) and (14.36) with $\boldsymbol{M} = 0.00400$ kg mol^{-1}, namely

$$\lambda = \frac{V}{\sqrt{2}\sigma N} = \frac{kT}{\sqrt{2}\sigma P} \quad \text{and} \quad \overline{v} = \sqrt{\frac{8\boldsymbol{R}T}{\pi \boldsymbol{M}}}$$

By way of example, we perform the calculation for Venus conditions and find

$$\lambda = \frac{\left(1.3806 \times 10^{-23} \text{ J K}^{-1}\right) 700 \text{ K}}{\sqrt{2}\left(3.73 \times 10^{-20} \text{m}^2\right)\left(90 \times 10^5 \text{ N m}^{-2}\right)} = 20.4 \text{ nm}$$

and

$$\overline{v} = \sqrt{\frac{8\boldsymbol{R}T}{\pi \boldsymbol{M}}} = \sqrt{\frac{8(8.314 \text{ J mol}^{-1} \text{ K}^{-1})700\text{K}}{\pi\left(0.00400 \text{ kg mol}^{-1}\right)}} = 1925 \text{ m s}^{-1}$$

Planet	$P/10^5$ Pa	T/K	$\overline{v}$/(m s^{-1})	λ
Earth	1.00	300	1260	788 nm
Mars	0.01	250	1150	65.6 μm
Venus	90	700	1925	20.4 nm
Triton	1.4×10^{-5}	40	460	7.48 mm

E14.11 Consider a gas of nitrogen at 273 K. Calculate the average time it takes for a nitrogen molecule to travel a distance of 5585 km (about the distance from London, UK, to New York, NY, USA). Compare this time to an average airline flight time of about 8 hours.

Solution

First we calculate the root mean square velocity of N_2 at 1 bar and 273 K and find

$$v_{\text{rms}} = \left(\frac{3\left(1.3806 \times 10^{-23} \text{ J K}^{-1}\right)(273 \text{ K})}{\left(4.65 \times 10^{-26} \text{ kg}\right)}\right)^{1/2} = 493 \text{ m s}^{-1}$$

Thus the time required to travel a distance of 5585 km would be

$$t = \frac{d}{v_{\text{rms}}} = \frac{5{,}585{,}000 \text{ m}}{493 \text{ m s}^{-1}} = 11326 \text{ s} = 3.15 \text{ hour}$$

This value is reasonably close to the time required for a supersonic jet to travel this distance. Of course, the nitrogen molecule undergoes many collisions and hence does not travel in a straight line. Rather the path of the molecule would be described by diffusion.

E14.12 Use the atmospheric distribution function in Box 14.5 to calculate the atmospheric pressure at the top of Mount Everest. Take the height of Mount Everest to be 8882 m. Use a temperature of $T = 273$ K as an average of the temperature at sea level $x = 0$ and at $x = 8882$ m.

Solution

The atmospheric distribution function is given as

$$P(h) = P(0)\ \exp(-mgh/(kT))$$

To find P on top of Mt. Everest, we use $P(0) = 1.01325$ bar, $m = 29$ g mol^{-1} or 4.82×10^{-26} kg, $g = 9.8$ m s^{-2}, $h = 8882$ m, $T = 273$ K, and $k = 1.3806 \times 10^{-23}$ J K^{-1} to yield

$$P = (1.01\text{ bar}) \cdot \exp\left(-\frac{(9.8\text{ m s}^{-2})(8882\text{ m})}{(1.3806 \times 10^{-23}\text{ J K}^{-1})(273\text{ K})}\right)$$
$$= (1.01325\text{ bar})\ (0.329) = 0.333\text{ bar}$$

E14.13 Analyze the following viscosity data as a function of temperature for Ar at 1 bar pressure ($\eta = 1.5951 \times 10^{-5}$ Pa s at 200 K, 2.2676×10^{-5} Pa s at 300 K, and 2.8627×10^{-5} Pa s at 400 K). Determine the effective diameter of an argon atom from these data. Comment on any temperature dependence you observe for the diameter that is obtained.

Solution

The viscosity is related to the mean speed and the mean free path through Equation (14.78)

$$\eta = \frac{5\pi}{32}\frac{N}{V}\, m \cdot \overline{v} \cdot \overline{\lambda} \quad \text{so that} \quad \frac{1}{\overline{\lambda}} = \frac{5\pi}{32}\frac{N}{V}\frac{m \cdot \overline{v}}{\eta}$$

Upon substitution of the relation for the mean free path $\overline{\lambda} = V/\left(N\sqrt{2}\sigma\right)$, we find

$$\sqrt{2}\sigma\frac{N}{V} = \frac{5\pi}{32}\frac{N}{V}\frac{m \cdot \overline{v}}{\eta} \quad \text{or} \quad \sigma = \frac{5\pi}{32\sqrt{2}}\frac{m \cdot \overline{v}}{\eta}$$

Next we substitute for $\overline{v}$, via $\sqrt{8kT/(\pi m)}$, so that

$$\sigma = \frac{5\pi}{32\sqrt{2}}\frac{m}{\eta}\sqrt{\frac{8kT}{\pi m}} = \frac{5\sqrt{\pi m k}}{16}\frac{\sqrt{T}}{\eta}$$
$$= \frac{5\sqrt{\pi\,(6.634 \times 10^{-26}\text{ kg})\,(1.3806 \times 10^{-23}\text{J K}^{-1})}}{16}\frac{\sqrt{T}}{\eta}$$
$$= 5.30 \times 10^{-25}\ \frac{\text{kg m}}{\text{s K}^{1/2}}\frac{\sqrt{T}}{\eta}$$

For the different temperatures and viscosities given above we find the values

T /K	η /Pa s	σ / m^2	d / pm
200	1.5951×10^{-5}	47.0×10^{-20}	387
300	2.2676×10^{-5}	40.5×10^{-20}	359
400	2.8627×10^{-5}	37.0×10^{-20}	343

where the diameter d is calculated from the relation $\sigma = \pi d^2$.

The diameter shows a systematic, but small (ca. 10%), decrease with increasing temperature. This temperature dependence may reflect nonidealities in the gas properties and/or the fact that higher temperature collisions will sample different regions of the interatomic potential, on average.

E14.14 In Section 14.3.3.5 we discussed the diffusion of Br_2 from a graduated cylinder and approximated the concentration gradient as linear. Why is this approximation reasonable?

Solution

The partial pressure of bromine near the liquid at the bottom of the cylinder is 100 mbar, and it decays to 0 mbar at the top of the cylinder (10 cm away). The total pressure in the cylinder is 1 bar (because of the air), and we can use Equation (14.42) to calculate the mean free path of a bromine molecule. In E14.6 we used Equation (14.42) to estimate the mean free path for a bromine molecule in nitrogen at 1 bar and room temperature, and we found that $\overline{\lambda}_{Br_2} = 21$ nm, which is much smaller than the length of the cylinder. Hence the bromine molecules travel a very short distance between collisions. This condition means that the bromine molecules will travel a very random path to the top of the cylinder, which means that they will mix with adjacent layers on each side causing a slowly changing concentration gradient. Hence the approximation of a linear pressure gradient for the partial pressure of bromine is appropriate.

E14.15 Here you consider the removal of excess thiosulfate from a photographic emulsion by diffusion. How long does the process take if the thickness of the emulsion is $d = 0.2$ mm? Consider the emulsion to be a gel with $\eta = 0.0089\ \mathrm{g\,s^{-1}cm^{-1}}$ and the thiosulfate to be a sphere of radius $r = 120$ pm.

Solution

We calculate the diffusion coefficient for the thiosulfate from the Stokes–Einstein equation (Equation (14.85)),

$$D = \frac{kT}{6\pi\eta r}$$

$$= \frac{\left(1.3806 \times 10^{-23}\ \mathrm{J\,K^{-1}}\right)(300\mathrm{K})}{6\pi\left(0.00089\ \mathrm{kg\,s^{-1}m^{-1}}\right)\left(120 \times 10^{-12}\ \mathrm{m}\right)} = 2.06 \times 10^{-9}\ \mathrm{m^2s^{-1}}$$

Then we can use the Einstein–Smoluchowski equation to find

$$\Delta t = \frac{\left(\Delta x^2\right)}{2D} = \frac{\left(2 \times 10^{-4}\right)^2\ \mathrm{m^2}}{2 \cdot 2.1 \times 10^{-9}\ \mathrm{m^2s^{-1}}} = 10\ \mathrm{s}$$

E14.16 Consider the formation of a bubble of methane gas (diameter 1 cm) at the bottom of a lake (depth 10 m, $P = 2$ bar and $T = 278$ K). What is the diameter of the bubble when it reaches the surface? Consider the lake to have a surface temperature of 298 K and a surface pressure of 1 bar.

Solution

Because the bubble is initially at a pressure of $P = 2$ bar at the bottom, its internal gas pressure will need to decrease to $P = 1$ bar and $T = 298$ K at the top, and we consider that this occurs by the expansion of the bubble as it rises. According to the ideal gas law the volume is proportional to T/P and the diameter is proportional to $V^{1/3}$; thus, for the diameter d_{surface} we obtain

$$d_{\mathrm{surface}} = d_{\mathrm{bottom}} \cdot \sqrt[3]{\frac{2\ \mathrm{bar}}{1\ \mathrm{bar}} \cdot \frac{298\ \mathrm{K}}{278\ \mathrm{K}}} = 1.3\ \mathrm{cm}$$

E14.17 The air inside a refrigerator (volume 100 L) is chilled from 20°C to 0°C after closing the door. What is the force needed to open the refrigerator door (area 1 m^2)? Suppose the refrigerator is totally airtight and neglect the friction between the door seal and the refrigerator.

Solution

P_1 and P_2 are the pressures at $T_1 = 293$ K and at $T_2 = 273$ K, respectively. Then

$$P_2 = P_1\frac{T_2}{T_1} = (1\ \mathrm{bar})\frac{273\ \mathrm{K}}{293\ \mathrm{K}} = 0.93\ \mathrm{bar}$$

The force acting on the door is $f = (P_1 - P_2) \cdot A = (1.00 - 0.93)\ \mathrm{bar} \cdot 1\ \mathrm{m^2} = 7000$ N.

E14.18 In a previous form of an automobile airbag (in use until 1995), NaN_3 and KNO_3 react to make metal oxides and nitrogen gas (1.6 mol for each mole of reactant). Consider the typical airbag to be 60 L in volume and fill to a pressure of 1.5 bar. How many grams of NaN_3 will be needed?

Solution

The chemical reactions involved are the decomposition of sodium azide to sodium metal and nitrogen gas, induced by an electrical ignition

$$2\ \mathrm{NaN_3}(s) \rightarrow 2\ \mathrm{Na}(s) + 3\ \mathrm{N_2}(g)$$

and the subsequent reaction of sodium metal with potassium nitrate to form sodium oxide and potassium oxide and additional nitrogen gas

$$10\ \mathrm{Na}(s) + 2\ \mathrm{KNO_3}(s) \rightarrow 5\ \mathrm{Na_2O}(s) + \mathrm{K_2O}(s) + \mathrm{N_2}(g)$$

Thus the overall reaction for production of gas is

$$10\ \mathrm{NaN_3}(s) + 2\ \mathrm{KNO_3}(s) \rightarrow 5\ \mathrm{Na_2O}(s) + \mathrm{K_2O}(s) + 16\ \mathrm{N_2}(g)$$

The amount of nitrogen gas required at 60 L and 1.5 bar and 298 K is just

$$n_{N_2} = \frac{PV}{RT} = \frac{(1.5 \text{ bar})\ (60 \text{ L})}{(0.08314 \text{ L bar mol}^{-1} \text{ K}^{-1})\ (298 \text{ K})} = 3.6 \text{ mol}$$

This amount of N_2 is produced from

$$3.6 \text{ mol } N_2 \times \left(\frac{10 \text{ mol NaN}_3}{16 \text{ mol N}_2}\right) \times \left(\frac{65.199 \text{ g NaN}_3}{1 \text{ mol NaN}_3}\right) = 148 \text{ g NaN}_3$$

Thus 148 g of NaN_3 are required in the air bag.

E14.19 In obtaining the rate of escape for effusion, we assume that the gas remains at equilibrium and the hole is smaller than the mean free path in the gas. Discuss how the flow of molecules from the container changes when the hole is larger than the mean free path.

Solution

If the hole is larger than the mean free path, the gas molecules collide with each other and exchange energy that leads to directed flow through the hole, which results in a non-equilibrium situation. The non-equilibrium situation results in a modified expression for the rate of molecules leaving the gas container (remember they were assumed to have a Boltzmann distribution of speeds among other assumptions), a non-equilibrated velocity distribution for the departed gas molecules (narrowed distribution at a peak velocity, which is often termed supersonic flow in extreme conditions) and a narrowed angular distribution of the departing gas molecules from the orifice.

E14.20 Consider a standard mixture of air (78.084% N_2, 20.9476% O_2, 0.934% Ar, 0.0314% CO_2, 0.001818% Ne, 0.000524% He, 0.000114% Kr, 0.0000087% Xe, 0.0002% CH_4, and 0.00005% H_2) at 1 bar pressure. Calculate the partial pressure of each of the atomic gases. What is the partial pressure arising from all of the atomic gases? What is the partial pressure arising from all of the molecular gases?

Solution

We proceed by constructing a table of the percentages and partial pressures. Note that in the ideal gas limit the pressure is directly proportional to the particle number. Hence for the atomic gases we find that

Atom	Ar	Ne	He	Kr	Xe
%	0.934	$1.818 \cdot 10^{-3}$	$5.24 \cdot 10^{-4}$	$1.14 \cdot 10^{-4}$	$8.7 \cdot 10^{-6}$
P_i bar	0.00934	$1.818 \cdot 10^{-5}$	$5.24 \cdot 10^{-6}$	$1.14 \cdot 10^{-6}$	$8.7 \cdot 10^{-8}$

and the total partial pressure of atomic gases is 0.0094 bar. For the molecular gases we find

Species	N_2	O_2	CO_2	CH_4	H_2
%	78.084	20.948	$3.14 \cdot 10^{-2}$	$2.0 \cdot 10^{-4}$	$5.0 \cdot 10^{-5}$
P_i /bar	0.78084	0.20948	$3.14 \cdot 10^{-4}$	$2.0 \cdot 10^{-6}$	$5.0 \cdot 10^{-7}$

and the total partial pressure of molecular gases is 0.9906 bar.

Thus the partial pressures of the monatomic gases total 0.0094 bar, while that of the molecular gases total 0.9906 bar.

E14.21 Use Equation (14.52) to derive Equation (14.53).

Solution

We begin with Equation (14.52)

$$P_j(x, y, z) \cdot \mathrm{d}x\mathrm{d}y\mathrm{d}z = const \cdot \exp\left[-\frac{3}{2j\lambda^2}r^2\right] \cdot \mathrm{d}x\mathrm{d}y\mathrm{d}z$$

and transform it into spherical coordinates so that

$$P_j(r, \theta, \phi) \cdot \sin(\theta)\, r^2\, \mathrm{d}r\, \mathrm{d}\theta\, \mathrm{d}\phi = const \cdot \exp\left[-\frac{3}{2j\lambda^2}r^2\right] \cdot \sin(\theta)\, r^2\, \mathrm{d}r\, \mathrm{d}\theta\, \mathrm{d}\phi$$

Equation (14.53) calculates the probability at a distance r, so we integrate over the angles θ and ϕ, and find that

$$\begin{aligned}
\mathcal{P}_j(r)\cdot r^2\mathrm{d}r &= \int_0^{\pi}\sin(\theta)\,\mathrm{d}\theta\int_0^{2\pi}\mathrm{d}\phi\mathcal{P}_j(r,\theta,\phi)\cdot r^2\mathrm{d}r \\
&= \int_0^{\pi}\sin(\theta)\,\mathrm{d}\theta\int_0^{2\pi}\mathrm{d}\phi\cdot\exp\left[-\frac{3}{2j\lambda^2}r^2\right]\cdot r^2\mathrm{d}r \\
&= const\cdot 2\pi\int_0^{\pi}\sin(\theta)\,\mathrm{d}\theta\cdot\exp\left[-\frac{3}{2j\lambda^2}r^2\right]\cdot r^2\mathrm{d}r \\
&= const\cdot 2\pi\,[-\cos(\theta)|_0^{\pi}\cdot\exp\left[-\frac{3}{2j\lambda^2}r^2\right]\cdot r^2\mathrm{d}r \\
&= const\cdot 4\pi\exp\left[-\frac{3}{2j\lambda^2}r^2\right]\cdot r^2\mathrm{d}r
\end{aligned}$$

which is Equation (14.53).

14.2 Problems

P14.1 A gas thermometer problem. The following molar volumes V_m are measured for SO_2 when the gas is examined at different pressures:

P/bar	0.253	0.506	0.660	1.013
$V_m/(\mathrm{L\ mol^{-1}})$	89.13	44.30	33.83	21.89

What is the temperature of the gas?

Solution

We calculate the product PV_m and plot PV_m as a function of P (see Fig. P14.1). From the figure we obtain

$$\lim_{P\to 0} PV_m = 22.67\ \mathrm{bar\ L\ mol^{-1}}$$

Thus the temperature is

$$T = \frac{\lim_{P\to 0} PV_m}{R} = \frac{22.67\ \mathrm{bar\ L\ mol^{-1}}}{0.08314\ \mathrm{bar\ L\ K^{-1}\ mol^{-1}}} = 272.7\ \mathrm{K}$$

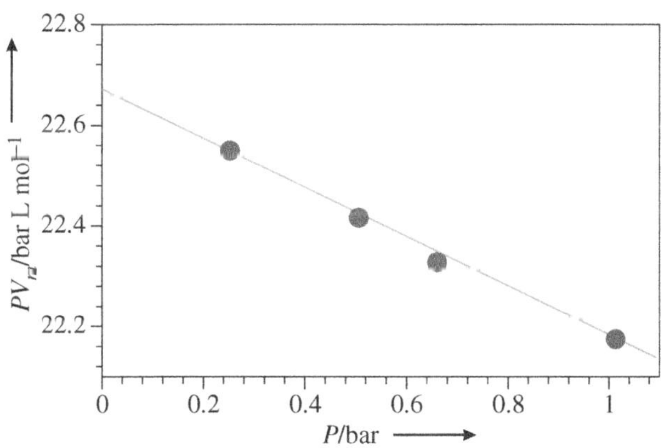

Figure P14.1 *PV of a SO_2 gas as a function of the pressure P.*

P14.2 In Box 14.5 we considered the pressure P of a gas in height h above the bottom of a container

$$P = P_0\cdot\exp\left[-mgh/(kT)\right]$$

where P_0 is the pressure at the bottom, m is the mass of a gas molecule, k is the Boltzmann constant, and T is the temperature. mgh is the potential energy of the molecule and kT is the thermal energy. At $T = 0$ all gas molecules are at the bottom, but with increasing temperature more and more molecules have the chance to take up energy from the temperature bath to reach a height h above the bottom. We also see that molecules with a lighter mass can reach greater heights than heavier molecules can. In principle we can use this effect to separate isotopes. However, the effect is too small in the gravity field, and an ultracentrifuge is used to apply a much greater force on the molecules.

Consider a gaseous mixture of $^{235}UF_6$ and $^{238}UF_6$ in an ultracentrifuge cylinder rotating with a speed of $v = \omega r = 350\,\mathrm{m\,s^{-1}}$ at a radius $r = 15$ cm. The force f acting on a particle of mass m in the centrifuge is given by $f = m\omega^2 r$. Consider that the partial pressure of $^{235}UF_6$ is 0.7% of the partial pressure of $^{238}UF_6$. What will be the enrichment of $^{235}UF_6$ in this process?

Solution

For the gas in a cylinder with radius r rotating at a frequency $\nu = \omega/(2\pi)$ around its vertical axis in an ultracentrifuge, the force acting on a molecule at distance r from the center is

$$force = m\omega^2 r$$

As a consequence the molecules are moving from the center of the cylinder ($r = 0$) to the outer wall (distance r from the center). Then in analogy to the case of gravity we obtain

$$P(r) = P(r = 0) \cdot \exp\left[m\omega^2 r \cdot r/(kT)\right] = P_0 \cdot \exp\left[m\omega^2 r^2/(kT)\right]$$

Note that in this case the gas pressure increases when we proceed from the center of the cylinder to its wall.

Now consider a mixture of two gases with masses m_1 and m_2 and partial pressures P_1 and P_2 at distance r and pressures $P_{0,1}$ and $P_{0,2}$ at the center:

$$P_1 = P_{0,1} \cdot \exp\left[m_1\omega^2 r^2/(kT)\right]$$
$$P_2 = P_{0,2} \cdot \exp\left[m_2\omega^2 r^2/(kT)\right]$$

Then the ratio of the partial pressures at the wall of the cylinder is

$$\frac{P_1}{P_2} = \frac{P_{0,1}}{P_{0,2}} \exp\left[(m_1 - m_2)\omega^2 r^2/(kT)\right]$$

Now let us calculate the effect for a natural mixture of $^{238}UF_6$ ($\boldsymbol{M}_1$=352 g mol^{-1}) and $^{235}UF_6$ ($\boldsymbol{M}_1$=349 g mol^{-1}). Then

$$\frac{m_1 - m_2}{kT} = \frac{\boldsymbol{M}_1 - \boldsymbol{M}_2}{\boldsymbol{R}T} = \frac{3\ \mathrm{g\,mol^{-1}}}{8.314\ \mathrm{J\,mol^{-1}\,K^{-1}} \cdot 300\ \mathrm{K}}$$
$$= 1.20 \times 10^{-6}\ \mathrm{kg\,N^{-1}\,m^{-1}} = 1.20 \times 10^{-6}\ \mathrm{s^2\,m^{-2}}$$

A typical value for the rotation speed of the cylinder is $\omega r = 350\ \mathrm{m\,s^{-1}}$ at a radius of 15 cm. Then we obtain

$$\frac{(m_1 - m_2)\omega^2 r^2}{kT} = 1.20 \times 10^{-6}\ \mathrm{s^2\,m^{-2}} \cdot (350\ \mathrm{m\,s^{-1}})^2$$
$$= 0.147$$

and

$$P_1 = P_2 \frac{P_{0,1}}{P_{0,2}} \exp(0.147) = \frac{P_{0,1}}{P_{0,2}} 1.16$$

for the partial pressure of the heavy molecules at the wall of the cylinder and

$$P_2 = P_1 \frac{P_{0,2}}{P_{0,1}} 0.86$$

for the partial pressure of the light molecules at the center of the cylinder. This means the enrichment in this first step equals 16%. In practice, the enriched gas is fed into a second ultracentrifuge, and the process is repeated. If we repeat 10 times the separation factor becomes

$$P_2 = P_1 \frac{P_{0,2}}{P_{0,1}} (1.16)^{10} = \frac{P_{0,2}}{P_{0,1}} 4.41$$

Given that $P_{0,2}/P_{0,1} = 0.7\%$ the lighter molecules are enriched to 3.1%, enough for the use in a nuclear plant.

P14.3 For a one-dimensional diffusion path (random walk with steps of $+\lambda$ and $-\lambda$) calculate the distribution function (i.e., probability versus position) after 10 diffusion steps; i.e., extend the results given in Section 14.3. Use your distribution function to calculate the average displacement of the particle and its mean square displacement.

Solution

The different stages of diffusion are shown in Table P14.3, which shows the distribution of the particle displacements for each step (the rows correspond to different steps and the columns correspond to the particle displacements in units of λ). The values in the body of the table give the number of trajectories that have an endpoint at that particular displacement for that number of steps.

Table P14.3 Displacement in a One-Dimensional Random Walk for 10 Steps.

Displacement in Units of λ	−9	−8	−7	−6	−5	−4	−3	−2	−1	0	1	2	3	4	5	6	7	8	9
1										1									
2									1		1								
3								1		2		1							
4							1		3		3		1						
5						1		4		6		4		1					
6					1		5		10		10		5		1				
7				1		6		15		20		15		6		1			
8			1		7		21		35		35		21		7		1		
9		1		8		28		56		70		56		28		8		1	
10	1		9		36		84		126		126		84		36		9		1

From the symmetry of the distribution it is apparent that the mean displacement is zero; i.e., it is just as likely to find the particle to the negative side or to the positive side of the zero position. Hence on average the particle is found at zero.

The mean square displacements after subsequent steps are collected in the second table. We start with $x = 0$. After one step the particle can be at $x = -\lambda$ or at $x = +\lambda$ with equal probability. Then the mean square of the displacement after this step is $\overline{x^2} = \frac{1}{2}(\lambda^2 + \lambda^2) = \lambda^2$. After two steps the particle can be found at $x = -2\lambda$, at $x = 0$ (two possibilities), or at $x = +2\lambda$, and we obtain $\overline{x^2} = \frac{1}{4}(4\lambda^2 + 0 + 4\lambda^2) = 2\lambda^2$. Proceeding in a like manner for the other steps we find the values shown in the table.

Step	1	2	3	4	5	6	7	8	9	10
Trajectories	2	3	8	16	32	64	128	256	512	1024
$\overline{x^2}$	λ^2	$2\lambda^2$	$3\lambda^2$	$4\lambda^2$	$5\lambda^2$	$6\lambda^2$	$7\lambda^2$	$8\lambda^2$	$9\lambda^2$	$10\lambda^2$

P14.4 Our development of the diffusion process is called the "random walk" model. In the limit that we allow the diffusion steps to be continuous, rather than discrete, we obtain a continuous distribution function which corresponds to a Gaussian distribution

$$\frac{dN}{dx} = a\,e^{-bx^2}$$

If you have N_0 total particles and a root mean square displacement of 4λ, determine the constants a and b in the distribution function.

Solution

We are asked to find the values of a and b for the Gaussian distribution

$$\frac{dN}{dx} = a\,e^{-bx^2}$$

for which N_0 is the total number of particles and the root mean square displacement is 4λ. The value of the constant a is most easily found by recognizing that

$$N_0 = \int_{-\infty}^{\infty} \frac{dN}{dx} dx = \int_{-\infty}^{\infty} a\, e^{-bx^2} dx = a \int_{-\infty}^{\infty} e^{-bx^2} dx = a\sqrt{\frac{\pi}{b}}$$

and the root mean square displacement is given by

$$4\lambda = \frac{1}{N_0} \int_{-\infty}^{\infty} \frac{dN}{dx} x^2 dx = \frac{1}{N_0} \int_{-\infty}^{\infty} ax^2 e^{-bx^2} dx$$

$$= \frac{a}{N_0} \int_{-\infty}^{\infty} x^2 e^{-bx^2} dx = \frac{2a}{N_0} \left(\frac{1}{2^2 b} \sqrt{\frac{\pi}{b}} \right) = \frac{1}{N_0} \left(\frac{a}{2b} \sqrt{\frac{\pi}{b}} \right)$$

Substitution of the value for N_0 from above gives

$$4\lambda = \frac{1}{a\sqrt{\frac{\pi}{b}}} \left(\frac{a}{2b} \sqrt{\frac{\pi}{b}} \right) = \frac{1}{2b}$$

so that $b = 1/2\lambda$. Hence we find that

$$a = \frac{N_0}{\sqrt{2\lambda\pi}}$$

and our distribution law becomes

$$\frac{dN}{dx} = \frac{N_0}{\sqrt{2\lambda\pi}} \exp\left(-x^2/(8\lambda)\right)$$

P14.5 Consider a measuring cylinder (inner diameter 3 cm, height 15 cm) with a layer (height 3 cm) of water on the bottom. Above the water surface gaseous water is in equilibrium with liquid water (vapor pressure of 30 mbar at room temperature). The water molecules escape from the cylinder by diffusion. How long does it take until 1 mL of liquid water has disappeared by diffusion? How long does it take if the cylinder height is 5 cm?

Solution

We apply Fick's law (Equations (14.68) and (14.69)) in the form

$$\frac{\Delta n}{\Delta t} = AD \cdot \frac{c_H}{H}$$

(see Fig. 14.20). The concentration c_H of the water vapor on the bottom of the cylinder is

$$c_H = \frac{n}{V} = \frac{P_{vap}}{RT} = \frac{30 \text{ mbar}}{8.314 \cdot 300 \text{ J mol}^{-1}} = 1.2 \times 10^{-3} \text{ mol L}^{-1}$$

We calculate the diffusion coefficient with

$$\sigma = \pi(r_{H_2O} + r_{N_2})^2 = \pi(231 + 190)^2 \times 10^{-24} \text{ m}^2 = 5.6 \times 10^{-19} \text{ m}^2$$

(we treat air as nitrogen gas) and the number density $N/V = 2.65 \times 10^{25}$ m^{-3} of the nitrogen molecules. The reduced mass is

$$\mu = \frac{18 \times 28}{18 + 28} \cdot \frac{1}{6.02 \times 10^{23}} \text{ g} = 1.8 \times 10^{-26} \text{ kg}$$

Then the diffusion coefficient is

$$D = \frac{1}{2} \cdot \frac{1}{\sigma \cdot N/V} \cdot \sqrt{\frac{k_B T}{\mu}} = 1.6 \times 10^{-5} \text{ m}^2 \text{ s}^{-1}$$

With $H = (15 - 3)$ cm = 12 cm and $A = \pi \cdot 1.52$ cm^2 = 7.07 cm^2, we calculate the time Δt needed for the evaporation of 1 mL of liquid water (this corresponds to an amount of substance of (1/18) mol = 0.056 mol) as

$$\Delta t = \Delta n \cdot \frac{H}{ADc_H} = 4.9 \times 10^5 \text{ s} = 5.7 \text{ days}$$

and it will take almost one week for 1 mL of liquid water to evaporate. If the cylinder height is 5 cm, it will take

$$\Delta t = \Delta n \cdot \frac{H}{A\,Dc_{\rm H}}$$
$$= (0.056\ \text{mol}) \cdot \frac{(5-3)\ \text{cm}}{(7.07\ \text{cm}^2)(1.6\times10^{-5}\ \text{m}^2\ \text{s}^{-1})(1.2\ \text{mol m}^{-3})}\frac{100\ \text{cm}}{\text{m}}$$
$$= 8.25\times10^{4}\ \text{s} = 0.95\ \text{days}$$

to evaporate, six times faster.

P14.6 It is possible to view small spherical particles undergoing Brownian motion on the surface of a liquid with a video camera and analyze the mean square displacement of the particles. The table contains data for such an experiment that was performed on 107 polystyrene spheres of 1.02 μm diameter. Analyze these data by plotting the root mean square $\Delta x_{\rm rms}$ displacement versus time t and also versus $\sqrt{t}$. After demonstrating that the motion is diffusive, determine the diffusion coefficient. From your diffusion constant and the size of the spherical particle determine the viscosity of the liquid in which the spheres are diffusing.

$\Delta R^2/\mu\text{m}^2$	t/s	$\Delta R^2/\mu\text{m}^2$	t/s	$\Delta R^2/\mu\text{m}^2$	t/s	$\Delta R^2/\mu\text{m}^2$	t/s
0	0	6.9	3.6	13.6	7.2	20.8	10.8
1.25	0.4	7.7	4	13.6	7.6	22.1	11.2
1.8	0.8	8.2	4.4	15	8	22.3	11.6
2.5	1.2	8.75	4.8	15.2	8.4	22.5	12
2.9	1.6	9.8	5.2	16.25	8.8		
4.2	2	10.8	5.6	17.9	9.2		
5	2.4	11.7	6	19.0	9.6		
6.0	2.8	11.9	6.4	19.0	10		
6.5	3.2	12.4	6.8	19.7	10.4		

Solution

We use Equation 14.6 for the square of the displacement, $\overline{x^2} = 2D\,\Delta t$ or $\sqrt{\overline{x^2}} = \sqrt{2D}\sqrt{\Delta t}$. Figure P14.6 shows a plot of the root mean square displacement $\Delta x_{\rm rms}$ versus the square root of time and its fit to a line, with a slope of 1.34 μm/s$^{1/2}$. Note that a plot of the root mean square displacement versus the time is curved.

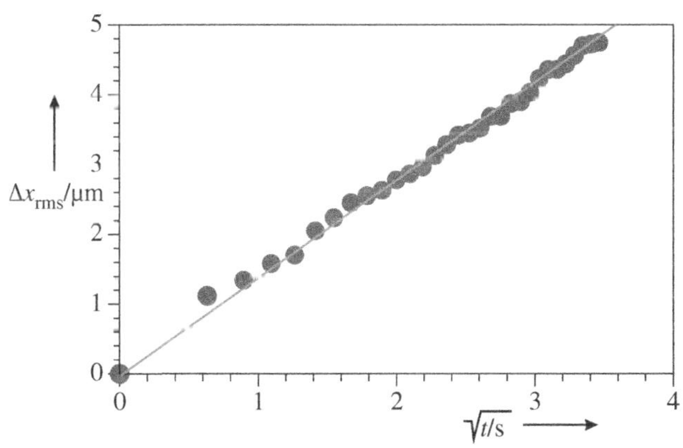

Figure P14.6 Root mean square displacement Δr versus the square root of time.

The slope of the plot is $\sqrt{2D}$. See Equation (14.60). Hence we find that

$$\sqrt{2D} = 1.34\,\mu\text{m/s}^{1/2} \quad \text{or} \quad D = 0.90\,\mu\text{m}^2\text{s}^{-1}$$

To find the effective viscosity, we use the Stokes–Einstein Equation (14.85), so that

$$\eta = \frac{kT}{6\pi Dr} = \frac{\left(1.3806 \times 10^{-23}\text{J K}^{-1}\right)(300\ \text{K})}{6\pi\left(0.90 \times 10^{-12}\ \text{m}^2\text{s}^{-1}\right)\left(0.51 \times 10^{-6}\text{m}\right)}$$
$$= 0.0029\ \text{kg m}^{-1}\ \text{s}^{-1}$$

P14.7 Use a statistical model to estimate the time it takes for an enzyme to assemble and interlock with a substrate of radius 1 nm that is present in a concentration of one molecule in a cubic micron of volume V. An enzyme **E** reacts with a substrate **S** if the point **s** on the substrate has approached point **e** on the enzyme within about $d = 100$ pm (see Figure P14.7). Then point **s′** must find point **e′**; this process, however, requires a short time compared to the first process. Analyze this phenomenon by considering the enzyme molecule to be immobilized and divide up the volume V into cells with an edge length of $d = 100$ pm. Consider point **s** of the substrate to be in one of these cells and diffuse from cell to cell until it reaches the cell with point **e**. Find the average time $\bar{t}$ required for this process. Why is the time you find so much longer than the time it takes the substrate to diffuse over a length of 1 μm?

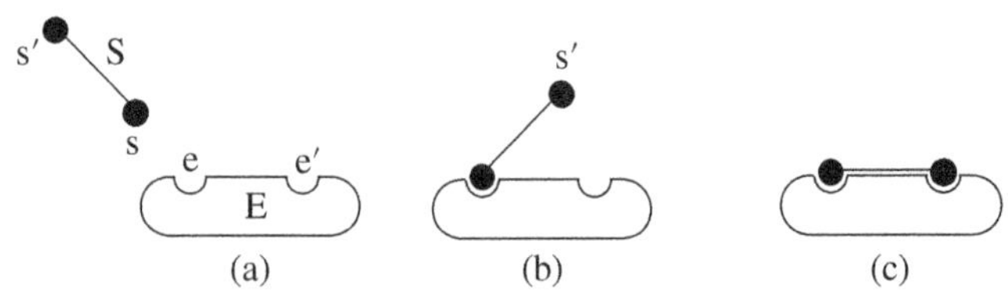

Figure P14.7 Enzyme interlocking with a substrate: (a) shows the substrate S approaching the enzyme E, (b) shows the substrate binding to a site on the enzyme, and (c) shows the substrate bound fully to the enzyme.

Solution
According to the equation of Einstein–Smoluchowski the average time τ needed by **S** to reach the next cell is

$$\tau = \frac{d^2}{2D}$$

(D = diffusion coefficient of **S**). Thus

$$\bar{t} = \frac{V}{d^3}\tau = \frac{V}{2dD}$$

If we assume that the substrate molecule has a mean radius of $R = 1$ nm, then

$$D = \frac{kT}{6\pi\eta R} = 2.5 \times 10^{-6}\ \text{cm}^2\ \text{s}^{-1}$$

for $\eta = 10^{-2}$ g cm^{-1}s^{-1} (water) and we find $\bar{t} = 20$ s. Note that this time is much larger than the time required for **S** to travel a distance of 1 μm by diffusion:

$$t' = \frac{(1\ \mu\text{m})^2}{2D} = 2\ \text{ms}$$

What is the reason for this difference? It is extremely improbable for the substrate molecule to meet the enzyme molecule on an arbitrary diffusion path of length 1 μm such that points **s** and **e** coincide. The substrate molecule must diffuse many times over a distance of 1 μm until it reaches the very particular location of the active site **e** on the enzyme.

P14.8 Consider the following data for water vapor as a function of temperature at 1 bar pressure. Compare the experimental values for the shear viscosity and the average kinetic energy with that you calculate using the kinetic theory of gases. Compare the molar volume with the value you calculate from the ideal gas law. Discuss any differences in terms of the molecular properties of water.

T/K	V_m/l/mol	E_{kin}/kJ/mol	η/μ Pa s
376.11	30.814	10.467	12.381
397.36	32.67	10.814	13.184
416.97	34.36	11.252	13.941
436.58	36.036	11.75	14.712
456.19	37.702	12.287	15.494
477.44	39.499	12.903	16.350
497.05	41.152	13.494	17.149
518.29	42.939	14.157	18.020
539.53	44.723	14.837	18.896
559.15	46.367	15.481	19.709
578.76	48.009	16.137	20.523
600.00	49.786	16.862	21.407

Solution

Construct the following table of the presented data, and that calculated using the following information

$$V_m = \boldsymbol{R}T/P$$

with $\boldsymbol{R} = 8.314 \times 10^{-2}$ L bar mol^{-1}K^{-1} and $P = 1.00$ bar

$$E_{\text{kin,calc}} = 3\boldsymbol{R}T$$

The viscosity

$$\eta = \frac{5}{64r^2}\sqrt{\frac{mkT}{\pi}}$$

with $m = 2.99 \times 10^{-26}$ kg, $k = 1.3806 \times 10^{-23}$ J molecule^{-1} K^{-1}, and $r = 0.209$ nm

T/K	V_m/M^{-1}	$V_{m,calc}$/M^{-1}	η/μPa s	η_{calc}/μPa s
376.11	30.814	31.270	12.381	12.56
397.36	32.67	33.037	13.184	12.91
416.97	34.36	34.667	13.941	13.22
436.58	36.036	36.297	14.712	13.53
456.19	37.702	37.928	15.494	13.83
477.44	39.499	39.694	16.350	14.15
497.05	41.152	41.325	17.149	14.44
518.29	42.939	43.091	18.020	14.74
539.53	44.723	44.857	18.896	15.04
559.15	46.367	46.488	19.709	15.31
578.76	48.009	48.118	20.523	15.58
600.00	49.786	49.884	21.407	15.86

Although some significant differences between the calculated and measured values exist, they are generally in reasonable agreement. The largest difference occurs for the calculated values of η, which agree well at low temperature but display a systematically weaker increase with temperature than do the data. A 10% to 15% error in the value of the collision diameter (value used here was taken from Table 14.4) can often account for the difference between the calculated and experimental values of η.

P14.9 A Fortran program is provided for performing computer simulations of diffusion of a particle. In the calculation the direction and the mean free path are chosen from a pseudorandom number. Use the program to run 200 simulations for the diffusion of a particle undergoing 200 collisions. Plot the endpoints of these trajectories and determine a) the average displacement, b) the mean square displacement, and c) the diffusion coefficient of the particle. (Note: To calculate the diffusion constant, you may assume that the collision frequency is 10^{13} collisions per second).

Solution

The end-point coordinates for twenty of the two hundred trajectories are provided in Table P14.9.

Table 14.9 Output file of (*x*, *y*, *z*) coordinates from simulation runs.

X-Direction/nm	*Y*-Direction/nm	*Z*-Direction/nm
−0.880866	−2.057393	0.377297
1.040614	1.122248	−2.703804
0.617357	−0.68581	−1.396362
−1.044772	−0.7706	−3.041617
0.059799	−1.22224	−0.749067
1.224998	−1.48534	0.158696
−0.147097	2.191663	−0.105567
−0.506266	1.987843	−0.852998
1.518813	0.692936	−0.4408
−0.48494	−0.634964	3.164001
0.527912	−0.319522	−2.025359
0.611253	−0.595726	2.725694
0.57117	−0.394574	−2.342295
1.425229	−0.532892	0.397156
−0.816497	−0.355017	0.169858
−0.579942	0.235192	−1.22824
−2.635055	1.955518	0.709946
0.104106	0.564878	−0.62538
0.113309	−0.144124	−0.68828
0.731946	1.380732	−0.116473

Figure P14.9 shows a plot of the *x* and *y* displacements of the particle after a trajectory with 200 steps. It is evident that the particle positions are centered around the origin. The dark gray circle that is plotted in the figure has a radius equal to the root-mean-square displacement of the particle.

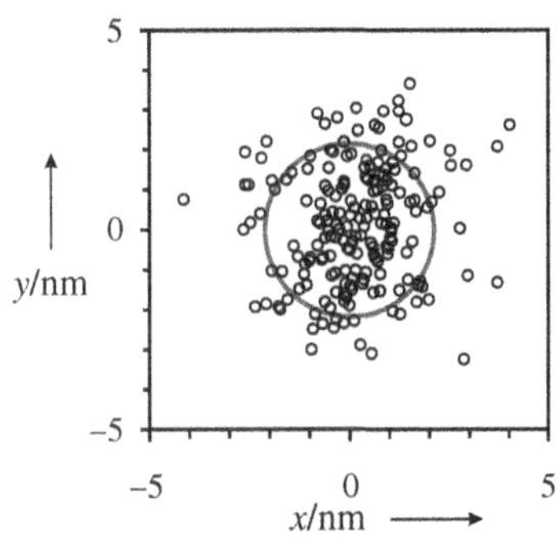

Figure P14.9 Diffusion path. *x*- and *y*-displacements of the particles after a trajectory with 200 steps. The dark gray circle indicates the mean displacement.

Because of the isotropy of space and the random nature of the diffusion, we expect the mean displacement to be zero. The mean displacement along the *x*−direction is computed by

$$\bar{x} = \frac{1}{200}\sum_{i=0}^{200} x_i = 0.179 \text{ nm}$$

In a similar manner we find that $\overline{y} = 0.153$ nm and $\overline{z} = -0.053$ nm. These values are much smaller than the typical displacements observed in the table; a result that reflects the cancelation of terms in the sum. As the number of trials (simulations) increases the mean displacement will approach its limiting value of zero.
The mean square displacement is computed as

$$\overline{r^2} = \frac{1}{200}\sum_{i=0}^{200}\left(x_i^2 + y_i^2 + z_i^2\right) = 5.35 \text{ nm}$$

and the root-mean-square displacement is

$$r_{\text{rms}} = \sqrt{\overline{r^2}} = 2.14 \text{ nm}$$

If we assume that we have 10^{-13} seconds per collision, then we can calculate the diffusion coefficient by way of Equation (14.60). Using 10^{-13} s per step, we find that 200 steps correspond to $\Delta t = 200 \cdot 10^{-13}\text{ s} = 2 \cdot 10^{-11}\text{s}$. Thus, we calculate a diffusion coefficient of

$$D = \frac{\overline{r^2}}{6\cdot} = \frac{(5.35)^2}{6 \cdot 2 \cdot 10^{-11}}\frac{\text{nm}}{\text{s}} = 2.39 \times 10^{11}\ \frac{\text{nm}^2}{\text{s}} = 2.39 \times 10^{-7}\ \frac{\text{m}^2}{\text{s}}$$

P14.10 Consider a plastic balloon that you fill to a pressure of 1.1 bar with 5 L of He gas at 300 K. Eventually the He escapes from the balloon through tiny "holes" between the polymer chains of the plastic. For a balloon that deflates to half its volume over a time of 24 hours, determine the effective area of the tiny "holes" (or microleaks) in the balloon. Model the escape using a model of effusion through the holes. You may neglect the presence of He outside the balloon and assume that the pressure in the balloon is 1.1 bar throughout its deflation. Perform your calculation by neglecting the rate of effusion of air into the balloon.

Solution
As outlined in Box 14.2 the number ΔN of He atoms that strike a unit area A of a wall in a time interval Δt can be written as

$$\Delta N = \frac{1}{4}\frac{N}{V}A\overline{v} \cdot \Delta t = \frac{1}{4}\frac{N}{V}A\sqrt{\frac{8k_BT}{\pi m}} \cdot \Delta t$$

If we use this equation for the conditions of the balloon and assume the ideal gas law ($PV = Nk_BT$), we find that

$$\Delta N = P \cdot A \cdot \sqrt{\frac{1}{2\pi m k_B T}} \cdot \Delta t$$

A balloon of 5 L and 1.1 bar at 300 K has

$$n = \frac{PV}{\boldsymbol{RT}} = \frac{1.1 \cdot 10^5 \text{ Pa}\left(5 \times 10^{-3}\text{m}^3\right)}{\left(8.3145 \text{ J mol}^{-1}\text{ K}^{-1}\right) 300 \text{ K}} = 0.220 \text{ mol}$$

or $N = n \cdot N_A = 1.32 \times 10^{23}$ atoms. Because it decreases by half in volume under the same pressure and temperature conditions, we find that $\Delta N = 0.660 \times 10^{23}$atoms in $\Delta t = 24$ hours $= 8.64 \times 10^4$ s. Using these values and the system parameters we can solve our equation for A, and find that

$$\begin{aligned} A &= \frac{\Delta N}{P \cdot \Delta t} \cdot \sqrt{2\pi m k_B T} \\ &= \frac{0.660 \times 10^{23}\sqrt{2\pi\left(6.642 \cdot 10^{-27}\text{ kg}\right)\left(1.3806 \times 10^{-23}\text{J K}^{-1}\right) 300 \text{ K}}}{\left(1.1 \cdot 10^5\text{N m}^{-2}\right) 8.64 \times 10^4\text{s}} \\ &= \frac{0.8677 \text{ kg m s}^{-1}}{9.545 \times 10^9 \text{ kg m}^{-1}\text{s}^{-1}} = 9.09 \times 10^{-11}\ \text{m}^2 = 90.9\ \mu\text{m}^2 \end{aligned}$$

This area represents the effective area for escape. For example, it could physically correspond to one million holes with an area of 90.9 nm^2 or some similar physical realization.

The effusion of nitrogen molecules into the balloon can also occur however that process will be slower by at least a factor of $\sqrt{28/4} = 2.6$ times. This process means that the deflated balloon was less than half full with He (contained some nitrogen) so that the effective area would need to be increased, correspondingly.

15

Energy Distribution in Molecular Assemblies

15.1 Exercises

E15.1 Calculate the characteristic vibrational temperature for HD ($\tilde{\nu} = 3813\,\text{cm}^{-1}$), HCl ($\tilde{\nu} = 2991\,\text{cm}^{-1}$), and CO ($\tilde{\nu} = 2169\,\text{cm}^{-1}$), where $\tilde{\nu} = 1/\lambda$ is the wavenumber. Over what temperatures are quantum effects important? At what temperatures is a classical description appropriate?

Solution

The characteristic temperature θ_{vib} is defined as $\theta_{\text{vib}} = \left(E_2 - E_1\right)/k$. For vibrational states, $E_2 - E_1 = h\nu = hc\tilde{\nu}$ and $\theta_{\text{vib}} = hc\tilde{\nu}/k$. Thus for HD we find

$$\theta_{\text{vib}} = \frac{hc\tilde{\nu}}{k} = \frac{6.626 \times 10^{-34}\ \text{J s} \cdot 2.998 \times 10^{10}\ \text{cm s}^{-1} \cdot 3813\ \text{cm}^{-1}}{1.3806 \times 10^{-23}\ \text{J K}^{-1}} = 5486\ \text{K}.$$

For HCl we find

$$\theta_{\text{vib}} = \frac{hc\tilde{\nu}}{k} = \frac{6.626 \times 10^{-34}\ \text{J s} \cdot 2.998 \times 10^{10}\ \text{cm s}^{-1} \cdot 2991\ \text{cm}^{-1}}{1.3806 \times 10^{-23}\ \text{J K}^{-1}} = 4302\ \text{K},$$

and for CO we find

$$\theta_{\text{vib}} = \frac{hc\tilde{\nu}}{k} = \frac{6.626 \times 10^{-34}\ \text{J s} \cdot 2.998 \times 10^{10}\ \text{cm s}^{-1} \cdot 2169\ \text{cm}^{-1}}{1.3806 \times 10^{-23}\ \text{J K}^{-1}} = 3120\ \text{K}.$$

From these characteristic temperatures, we can see that quantum mechanical effects will be important for describing the vibrational motion of these molecules up to temperatures above 1000 K (for example, see Fig. 15.11). Under the constraint that these molecules remain stable and do not react or decompose, a classical description would only be appropriate for stretching vibrations at very high temperatures, > 5000 K.

E15.2 Calculate the characteristic rotational temperature for HD ($d = 74$ pm), HCl ($d = 127$ pm), and CO ($d = 113$ pm). Over what temperatures are quantum effects important? At what temperatures is a classical description appropriate?

Solution

From the energy gap between the ground rotational energy level and the first excited rotational energy level

$$\Delta E_{0\to 1} = \frac{h^2}{4\pi^2 \mu d^2}$$

Solutions Manual for Principles of Physical Chemistry, Third Edition. Edited by Hans Kuhn, David H. Waldeck, and Horst-Dieter Försterling.

we calculate the corresponding characteristic rotational temperature

$$\theta_{\text{rot}} = \frac{\Delta E_{0\to1}}{2k}$$

By way of example, consider HD for which $\mu = 1.12 \times 10^{-27}$ kg and $d = 74$ pm. In this case, we find that

$$\Delta E_{0\to1} = \frac{\left(6.626 \times 10^{-34}\ \text{J s}\right)^2}{4\pi^2 \cdot 1.12 \times 10^{-27}\ \text{kg} \cdot \left(74 \times 10^{-12}\text{m}\right)^2} = 1.8 \times 10^{-21}\ \text{J}$$

and

$$\theta_{\text{rot}} = \frac{1.8 \times 10^{-21}\ \text{J}}{2 \cdot 1.3806 \times 10^{-23}\ \text{J K}^{-1}} = 66\ \text{K}$$

We proceed in a like manner for the other molecules and find the values given in the table.

	$\mu/10^{-27}$ kg	d/ pm	$\Delta E_{0\to1}/10^{-22}$ J	θ_{rot}/K
HD	1.12	74	18	66
HCl	1.63	127	4.23	15.5
CO	11.38	113	0.765	2.81

Thus quantum mechanical effects are important for temperatures less than those given in the table, and a classical description becomes appropriate for temperatures much greater than those in the table.

E15.3 Consider the I_2 molecule, with $\mu = 1.05 \times 10^{-25}$ kg and $d = 267$ pm, calculate the characteristic rotational temperature. Calculate the rotational energy level that has the highest population at $T = 300$ K.

Solution

For I_2, with $\mu = 1.05 \times 10^{-25}$ kg and $d = 267$ pm we obtain $B = 7.43 \times 10^{-25}$ J; thus, we calculate the characteristic temperature $\theta_{\text{rot}} = \Delta E/(2k) = B/k = 0.0538$ K. This characteristic temperature indicates that the equipartition principle is fulfilled at extremely low temperatures.

To find the rotational energy level with the highest population at $T = 300$ K we need to find the maximum of Equation (15.34), namely

$$N_J = \frac{N}{Z}(2J+1)\exp\left(-E_J/(kT)\right)$$

Because $\theta_{\text{rot}} << 300$ K, we use the high temperature limit to evaluate Z and find that

$$Z = \frac{kT}{B} = \frac{1.3806 \times 10^{-23}\text{J K}^{-1} \cdot 300\ \text{K}}{7.43 \times 10^{-25}\ \text{J}} = 5575$$

Using the fact that

$$\frac{B}{kT} = \frac{7.43 \times 10^{-25}\ \text{J}}{1.386 \times 10^{-23}\text{J K}^{-1} \cdot 300\ \text{K}} = 1.79 \times 10^{-4}$$

We calculate that the fractional population in level J will be

$$\frac{N_J}{N} = \frac{(2J+1)\exp\left(-BJ(J+1)/(kT)\right)}{Z}$$
$$= \frac{(2J+1)\exp\left[-J(J+1) \cdot 1.79 \times 10^{-4}\right]}{5575}$$

Figure E15.3 plots the function to see approximately where N_J/N is at a maximum.

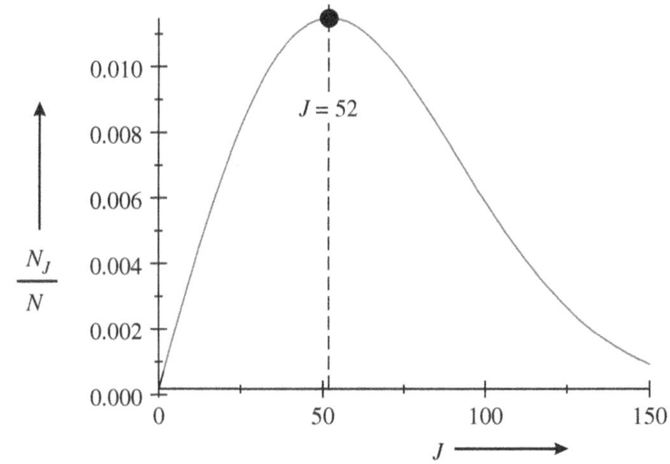

Figure E15.3 Fraction N_J/N as a function of the quantum number J.

To determine the maximum of the function analytically, we take its derivative and set it equal to zero; namely,

$$\frac{d}{dJ}\left(\frac{N_J}{N}\right)\bigg|_{J=J_{max}} = 0 = \frac{d}{dJ}\left(\frac{(2J+1)\exp\left(-BJ(J+1)/(kT)\right)}{Z}\right)\bigg|_{J=J_{max}}$$

$$= \left(\frac{2\cdot\exp\left(\frac{-BJ(J+1)}{kT}\right)}{Z} - \frac{(2J+1)^2 B\exp\left(\frac{-BJ(J+1)}{kT}\right)}{kT\cdot Z}\right)\Bigg|_{J=J_{max}}$$

$$= \frac{\exp\left(\frac{-BJ(J+1)}{kT}\right)}{Z}\cdot\left|2 - \frac{(2J+1)^2 B}{kT}\right|_{J=J_{max}}$$

By solving this expression for J_{max} we find

$$0 = 2 - \frac{\left(2J_{max}+1\right)^2 B}{kT}$$

so that

$$J_{max} = \sqrt{\frac{kT}{2B}} - \frac{1}{2} = \sqrt{\frac{1.38\times10^{-23}\,\text{J K}^{-1}\cdot 300\,\text{K}}{2\cdot 7.43\times10^{-25}\,\text{J}}} - \frac{1}{2} = 52.8 - 0.5 = 52.3$$

The rotational energy level with the highest population is $J = 52$.

E15.4 Obtain the fundamental vibrational frequencies of H_2, HI, and I_2; compute the energy difference between the ground vibrational state and the first excited vibrational state. Compute the ratio of this energy to $k_B T$ for temperatures of $T = 300$ K and $T = 1000$ K. Comment on the molecules abilities to store thermal energy in their vibrational degrees of freedom at these temperatures.

Solution

The fundamental vibrational wavenumbers for H_2, HI, and I_2 are 4401 cm^{-1}, 2230 cm^{-1}, and 213 cm^{-1} respectively. The energy difference between the ground vibrational state and first excited vibrational state is given by

$$\Delta E = h\nu = h\tilde{\nu}c_0$$

For H_2 we find that

$$\Delta E = \left(4401\text{cm}^{-1}\right)\left(2.998\times10^{10}\text{ cm/s}\right)\left(6.626\times10^{-34}\text{ J s}\right)$$
$$= 8.742\times10^{-20}\text{ J}$$

so that

$$\theta_{vib} = \frac{\Delta E}{k} = \frac{8.742\times10^{-20}\text{J}}{1.3806\times10^{-23}\text{J K}^{-1}} = 6332\text{ K}$$

The ratio of the vibrational excitation energy $\Delta E/(kT)$ is found as

$$\frac{\Delta E}{kT} = \frac{\theta_{\text{vib}}}{T} = \frac{6332\text{ K}}{300\text{ K}} = 21.1 \text{ for } 300\text{ K}$$

and

$$\frac{\Delta E}{kT} = \frac{\theta_{\text{vib}}}{T} = \frac{6332\text{ K}}{1000\text{ K}} = 6.332 \text{ for } 1000\text{ K}$$

Proceeding in a like manner for the other molecules we find the values reported in the table

	$\Delta E/10^{-20}$J	θ_{vib}/K	θ_{vib}/300 K	θ_{vib}/1000 K
H_2	8.742	6332	21.1	6.332
HI	4.430	3209	10.7	3.209
I_2	0.423	306	1.02	0.306

The characteristic vibrational temperature of H_2 is 21 times higher than that of I_2. The I_2 vibrational energy levels will be highly excited, even at moderate temperatures; whereas for H_2 the ground-state vibrational energy level will have the dominant population. This means that thermal energy can be stored in the vibrational mode of I_2 but not very readily in that of H_2 at moderate temperatures.

E15.5 Obtain the rotational constants for H_2, HI, and I_2 from the literature and compute the energy difference between the ground rotational state and the first excited rotational state. Compute the ratio of this energy to kT for temperatures of $T = 10$ K, $T = 100$ K, and $T = 300$ K. Comment on the molecules abilities to store thermal energy in their rotational degrees of freedom at these temperatures.

Solution

The rotational constants $\tilde{B}$ for H_2, HI, and I_2 are 60.85 cm^{-1}, 6.512 cm^{-1}, and 0.03737 cm^{-1}, respectively. The change in energy between the ground rotational state and first excited rotational state is given by

$$\Delta E = 2B = 2\tilde{B}hc_0$$

For H_2 we find that

$$\Delta E = 2\tilde{B}hc_0 = 2 \cdot 60.85\text{cm}^{-1} \cdot \frac{1.98645 \times 10^{-23}\text{J}}{\text{cm}^{-1}}$$
$$= 2.418 \times 10^{-21}\text{J}$$

The ratio of this energy to kT at 300 K for H_2 is

$$\frac{\Delta E}{kT} = \frac{2.418 \times 10^{-21}\text{J}}{1.3806 \times 10^{-23}\text{J K}^{-1} \times 300\text{ K}} = 0.5837$$

We proceed similarly and find for 100 K that

$$\frac{\Delta E}{kT} = \frac{2.418 \times 10^{-21}\text{J}}{1.3806 \times 10^{-23}\text{J K}^{-1} \times 100\text{ K}} = 1.751$$

and for 10 K that

$$\frac{\Delta E}{kT} = \frac{2.418 \times 10^{-21}\text{J}}{1.3806 \times 10^{-23}\text{J K}^{-1} \times 10\text{ K}} = 17.51$$

If we proceed in a likewise manner for the other molecules, we obtain the values in the table.

		$\Delta E/(kT)$		
	$\Delta E/10^{-22}$ J	$T = 300K$	$T = 100K$	$T = 10K$
H_2	24.176	0.58368	1.7511	17.511
HI	2.587	0.06151	0.1845	1.845
I_2	0.01485	0.0003584	0.001075	0.01075

We see that the thermal energy, kT, is already greater than the difference between the first two rotational states at readily accessible temperatures. Hence thermal energy can be stored in the rotational energy levels of the molecules. Only for the H_2 molecule does it appear that the energy level spacing will be high enough to inhibit thermal excitation, and even in this case we find that it happens at relatively low temperatures.

E15.6 Derive an expression for the internal energy U_{rot} of a molecule rotating in two dimensions, i.e., rotating on a circle (see the discussion for Fig. 11.2).

Solution

We will use the general Equation (15.41),

$$U_{\text{rot}} = NkT^2 \frac{\text{d} \ln Z_{\text{rot}}}{\text{d}T}$$

to find the internal energy from the rotational partition function. Given our analysis in E15.5, we proceed by assuming that the high temperature limit holds and find an expression for Z_{rot}. In analogy to the three-dimensional rotator (Section 15.5) we approximate the partition function by

$$Z_{\text{rot}} = 1 + 2 \cdot \sum_{n=1}^{\infty} \text{e}^{-Bn^2/kT} \simeq 1 + 2 \cdot \int_1^{\infty} \text{e}^{-Bn^2/kT} \cdot \text{d}n$$

(the factor of two occurs because of the twofold degeneracy of the energy levels for $n \geq 1$). Solving this expression, we find that

$$Z_{\text{rot}} = 1 + 2 \cdot \sqrt{\frac{kT}{B}} \cdot \int_0^{\infty} \text{e}^{-x^2} \cdot \text{d}x$$
$$= 1 + 2 \cdot \sqrt{\frac{kT}{B}} \cdot \frac{1}{2}\sqrt{\pi} = 1 + \sqrt{\frac{\pi kT}{B}}$$

In the case of the high temperature limit $kT \gg B$ this equation reduces to

$$Z_{\text{rot}} = \sqrt{\frac{\pi kT}{B}}$$

Using Equation (15.41), we can write that

$$U_{\text{rot}} = NkT^2 \frac{\text{d} \ln Z_{\text{rot}}}{\text{d}T} = NkT^2 \frac{1}{\sqrt{\pi kT/B}} \sqrt{\frac{\pi k}{B}} \frac{1}{2} T^{-1/2} = NkT^2 \frac{1}{2} \frac{1}{T} = \frac{NkT}{2}$$

This result coincides exactly with our considerations in Box 15.1 on the Equipartition Principle. A molecule confined to rotate in a plane has only one rotational degree of freedom (i.e., rotation angle) and in the classical (high temperature) limit it contributes $kT/2$ to the internal energy. Hence for N molecules we have an internal energy of $NkT/2$.

E15.7 Use the harmonic oscillator model to describe the vibrational energy levels of HCl (the fundamental vibration occurs at $\tilde{\nu} = 2991\ \text{cm}^{-1}$). What fraction of HCl molecules are in the 3rd excited state level at $T = 1000$ K? What fraction of HCl molecules are in the ground energy level at $T = 1000$ K?

Solution

The characteristic frequency θ_{vib} for HCl is

$$\theta_{\text{vib}} = \frac{\Delta E}{k} = \frac{h\nu_0}{k} = \frac{hc_0\tilde{\nu}}{k}$$
$$= \frac{(6.626 \times 10^{-34}\ \text{J s})\ (2.998 \times 10^{10}\ \text{cm s}^{-1})\ (2991\ \text{cm}^{-1})}{1.381 \times 10^{-23}\ \text{J K}^{-1}}$$
$$= 4302\ \text{K}$$

From Equation (15.23) for the vibrational partition function of a harmonic oscillator, we can write that

$$Z_{\text{vib}} = \frac{\text{e}^{-h\nu_0/(2kT)}}{1 - \text{e}^{-h\nu_0/(kT)}} = \frac{\text{e}^{-\theta_{\text{vib}}/(2T)}}{1 - \text{e}^{-\theta_{\text{vib}}/T}}$$
$$= \frac{\exp(-4302/2000)}{1 - \exp(-4302/1000)} = 0.1180$$

where we have performed the evaluation at $T = 1000$ K.

The fraction of the population in level n is given by

$$f_n = \frac{\frac{N}{Z_{vib}} \cdot \exp\left[-h\nu_0\left(\frac{1}{2}+n\right)/(kT)\right]}{N} = \frac{1}{Z_{vib}} \exp\left[-\theta\left(\frac{1}{2}+n\right)/T\right]$$

Hence the third excited state level ($n = 3$) contains

$$f_{n=3} = \frac{\exp\left(-7 \cdot \theta_{vib}/(2T)\right)}{Z_{vib}} = \frac{2.89 \times 10^{-7}}{0.118} = 2.449 \times 10^{-6}$$

and the fraction in the ground state is

$$f_{n=0} = \frac{\exp\left(-\theta_{vib}/(2T)\right)}{Z_{vib}} = 0.9862$$

Clearly, most of the population is in the ground vibrational energy level.

E15.8 Consider the diatomic molecules H_2, Cl_2, and I_2. At what temperature does the value of kT equal the energy level spacing between the lowest two energy levels for the vibrational levels, the rotational levels, and the translational levels in a cubic box of volume 1 mL? Compare trends between these three molecules and between the types of motions. Find literature values for the molecular parameters. (Note: Although these molecules are homonuclear you can neglect symmetry constraints arising from nuclear spin considerations for the purposes of this problem; i.e., treat the molecules as if they were heteronuclear diatomics.)

Solution

The energy relationships are

$$\Delta E_{vib} = h\nu = hc\tilde{\nu};\ \Delta E_{rot} = 2B = 2hc_0 \cdot \tilde{B};\ \text{and}\ \Delta E_{trans} = \frac{3h^2}{8mL^2}$$

The required data are given in the following table

	$\tilde{\nu}$/cm^{-1}	$\nu/10^{13}$s^{-1}	$\tilde{B}$/cm^{-1}	$B/10^{-23}$ J	M/g mol^{-1}	$m/10^{-26}$ kg
H_2	4401	13.19	60.85	120.9	2.0158	0.33518
Cl_2	554	1.66	0.2440	0.1817	70.904	11.790
I_2	213	0.629	0.03737	0.07423	253.808	42.2023

By way of example, we perform the calculations for H_2 and find that

$$\begin{aligned}\Delta E_{vib} &= h\nu \\ &= \left(6.626 \times 10^{-34}\ \text{J s}\right)\left(13.19 \times 10^{13}\text{s}^{-1}\right) \\ &= 8.740\ \times 10^{-20}\ \text{J}\end{aligned}$$

$$\begin{aligned}\Delta E_{rot} &= 2B \\ &= 2 \times \left(120.9 \times 10^{-23}\ \text{J}\right) = 2.418 \times 10^{-21}\ \text{J}\end{aligned}$$

and

$$\begin{aligned}\Delta E_{trans} &= \frac{3h^2}{8mL^2} \\ &= \frac{3\left(6.626 \times 10^{-34}\ \text{J s}\right)^2}{8\left(0.33518 \times 10^{-26}\ \text{kg}\right) \times \left(10^{-2}\ \text{m}\right)^2} = 4.883 \times 10^{-37}\ \text{J}\end{aligned}$$

The temperature at which kT will equal the energy spacing in each degree of freedom is

$$T = \frac{\Delta E}{k}$$

for vibrational

$$T = \frac{\Delta E}{k} = \frac{8.742\ \times 10^{-20}\ \text{J}}{1.3806 \times 10^{-23}\text{J K}^{-1}} = 6332\ \text{K}$$

for rotational

$$T = \frac{\Delta E}{k} = \frac{2.4176 \times 10^{-21}\ \text{J}}{1.3806 \times 10^{-23}\text{J K}^{-1}} = 175.1\ \text{K}$$

and for translational

$$T = \frac{\Delta E}{k} = \frac{4.883 \times 10^{-37}\ \text{J}}{1.3806 \times 10^{-23}\text{J K}^{-1}} = 3.537 \times 10^{-14}\ \text{K}$$

If we proceed in a similar manner for each of the three molecules we find the values reported in the table.

	$\Delta E_{vib}/10^{-20}$ J	θ_{vib}/K	$\Delta E_{rot}/10^{-23}$ J	θ_{rot}/K	$\Delta E_{trans}/10^{-37}$ J	$\theta_{trans}/10^{-14}$ K
H_2	87.42	6332	241.76	175.1	4.883	3.537
Cl_2	11.0	797	0.9694	0.7021	0.1388	0.1006
I_2	4.23	306	0.1485	0.1075	0.03878	0.02809

It is evident from this exercise that vibrational energy level spacings are the largest and translational energy level spacings are the smallest, with the rotational energy level spacings intermediate in magnitude. In addition, we see that H_2 which is the lightest molecule has the highest energy level spacings for a given degree of freedom; the next heavier molecule Cl_2 has smaller spacings; and I_2 has the smallest spacings.

E15.9 Using the harmonic oscillator approximation, compute the fraction of O_2 molecules ($\theta_{vib} = 2230$ K) in the first excited vibrational state at $T = 298$ K. Compute the fraction of O_2 molecules in all excited vibrational states at 298 K. Perform the same calculations for H_2 ($\theta_{vib} = 6332$ K). The bond strengths of these molecules are similar (in fact, the O_2 bond is a little stronger than H_2). Why is the H_2 population in the first excited level lower than that of O_2?

Solution

The fraction of molecules in a given vibrational state may be calculated by using Equation (15.24) (also see E15.7); namely

$$f_n = \frac{N_n}{N} = \frac{1}{Z_{vib}} \exp\left(-\frac{\theta_{vib}(n+1/2)}{T}\right)$$

where

$$Z_{vib} = \sum_{n=0}^{\infty} \exp\left(-\theta_{vib}(n+1/2)/T\right) = \frac{\exp\left(-\theta_{vib}/2T\right)}{1-\exp\left(-\theta_{vib}/T\right)}$$

The last equality holds in the case that the vibration can be described by a harmonic oscillator. For oxygen, we find that

$$Z_{vib} = \frac{\exp(-2230\ \text{K}/(2\cdot 298\ \text{K}))}{1-\exp(-2230\ \text{K}/298\ \text{K})} = 2.373 \times 10^{-2}$$

and

$$\frac{N_1}{N} = \frac{1}{Z_{vib}} \exp\left(-\frac{\theta_{vib}(1+1/2)}{T}\right)$$
$$= \frac{\exp(-2230\ \text{K}(1+1/2)/298\ \text{K})}{2.370 \times 10^{-2}} = 5.621 \times 10^{-4}$$

In a similar manner, we find for hydrogen that

$$Z_{vib} = 2.432 \times 10^{-5} \text{ and } \frac{N_1}{N} = 5.915 \times 10^{-10}$$

From Equation (15.24), we know that the probability $P_{n>0}$ of finding molecules in states $n > 0$ is given by

$$\mathcal{P}_{n>0} = \exp\left(-\frac{\theta_{vib}}{T}\right)$$

For oxygen we find that $\mathcal{P} = 5.624 \times 10^{-4}$ and for hydrogen we find that $\mathcal{P} = 5.915 \times 10^{-10}$.

The energy level spacing is not determined solely by the bond's force constant but also by the reduced mass of the nuclei involved in the vibrational motion. Because the hydrogen reduced mass is much lower than the oxygen reduced mass, the energy level spacing is larger and consequently it is more difficult to excite thermally than is that of oxygen.

E15.10 Determine the constants C and C' in Equations (15.67) and (15.69).

$$C = \frac{2N}{\sqrt{\pi}(kT)^{3/2}} \qquad C' = \frac{4N}{\sqrt{\pi}}\left(\frac{m}{2kT}\right)^{3/2}$$

which are the values given in the text.

Solution

Equation (15.67) is

$$dN(E, E + dE) = C \cdot \exp(-E/(kT)) \cdot \sqrt{E}\,dE$$

We use the normalization condition

$$\int_0^\infty dN(E, E + dE) = N$$

where N is the number of gas molecules in the container to determine the constants, so that the normalization condition becomes

$$N = C \cdot \int_0^\infty \exp(-E/(kT)) \cdot \sqrt{E}\,dE$$

If we make the substitution $x = E/(kT)$ so that $dx = dE/(kT)$, we find that

$$C = \frac{N}{(kT)^{3/2}\int_0^\infty \exp(-x) \cdot \sqrt{x}\,dx}$$

Using the Appendix B, we find $\int_0^\infty \exp(-x) \cdot \sqrt{x}\,dx = \sqrt{\pi}/2$, so that

$$C = \frac{2N}{\sqrt{\pi}(kT)^{3/2}}$$

which is the value given in the text.
Equation (15.69) is

$$dN(v, v + dv) = C' \cdot v^2 \cdot \exp\left(-mv^2/(2kT)\right) \cdot dv$$

We use the normalization condition

$$\int_0^\infty dN(v, v + dv) = N$$

so that

$$N = C' \int_0^\infty v^2 \exp\left(-mv^2/(2kT)\right) \cdot dv$$

Now we let $x^2 = mv^2/(2kT)$ so that $x\,dx = (m/(2kT))\,v\,dv$ and

$$N = C'\left(\frac{2kT}{m}\right)^{3/2} \int_0^\infty x^2 \exp\left(-x^2\right)\,dx$$

From the integral table in Appendix B, we find $\int_0^\infty x^2 \exp\left(-x^2\right) dx = \sqrt{\pi}/4$ and

$$N = C'\left(\frac{2kT}{m}\right)^{3/2} \frac{\sqrt{\pi}}{4}$$

so that

$$C' = \frac{4N}{\sqrt{\pi}}\left(\frac{m}{2kT}\right)^{3/2}$$

which is the same as the result given in the text.

E15.11 Calculate the root mean square speed v_{rms} and the average speed v for Ar gas at $T = 300$ K. Perform the same calculation for CH_4 gas at $T = 300$ K. Discuss your results.

Solution

The root mean square speed is given by

$$v_{\text{rms}} = \sqrt{\frac{3kT}{m}} = \sqrt{\frac{3RT}{\boldsymbol{M}}}$$

and the average speed is given by

$$\overline{v} = \sqrt{\frac{8kT}{\pi m}} = \sqrt{\frac{8RT}{\pi \boldsymbol{M}}}$$

where m is the mass of a molecule (atom) and $\boldsymbol{M}$ is the molar mass. For argon, we use $\boldsymbol{M}_{\text{Ar}} = 0.03995\text{kg mol}^{-1}$ and find that

$$v_{\text{rms,Ar}} = \sqrt{\frac{3RT}{\boldsymbol{M}_{\text{Ar}}}} = \sqrt{\frac{3 \times 8.314 \times 300}{0.03995}\frac{\text{m}}{\text{s}}} = 433\frac{\text{m}}{\text{s}}$$

and

$$\overline{v}_{\text{Ar}} = \sqrt{\frac{8RT}{\pi \boldsymbol{M}_{\text{Ar}}}} = \sqrt{\frac{8 \times 8.314 \times 300}{\pi \times 0.03995}\frac{\text{m}}{\text{s}}} = 398\frac{\text{m}}{\text{s}}$$

For methane we use $\boldsymbol{M}_{CH_4} = 0.01605\text{kg mol}^{-1}$ and find

$$v_{\text{rms},CH_4} = \sqrt{\frac{3RT}{\boldsymbol{M}_{CH_4}}} = \sqrt{\frac{3 \times 8.314 \times 300}{0.01605}\frac{\text{m}}{\text{s}}} = 684\frac{\text{m}}{\text{s}}$$

and

$$\overline{v}_{CH_4} = \sqrt{\frac{8RT}{\pi \boldsymbol{M}_{CH_4}}} = \sqrt{\frac{8 \times 8.314 \times 300}{\pi \times 0.01605}\frac{\text{m}}{\text{s}}} = 630\frac{\text{m}}{\text{s}}$$

The speeds for CH_4 are uniformly greater than those for Ar because CH_4 has a lower molar mass than Ar.

E15.12 Derive Equation (15.14) for the internal energy U using the approximations described in Section 15.2.3.

Solution

If a molecule's individual degrees of freedom (electronic, vibrational, rotational, and translational) are independent, then the energy $E_{k,l,m,n}$ of a particle which is in electronic quantum state k, vibrational quantum state l, rotational quantum state m, and translational quantum state n is

$$E_{k,l,m,n} = E_{\text{el},k} + E_{\text{vib,l}} + E_{\text{rot},m} + E_{\text{trans},n}$$

Accordingly, the molecular partition function Z can be written in terms of molecular partition functions for the electronic, vibrational, rotational, and translational contributions; i.e.,

$$\begin{aligned} Z &= \sum_{k,l,m,n=1}^{\infty} \text{e}^{-E_{k,l,m,n}/(kT)} \\ &= \sum_{k,l,m,n=1}^{\infty} \left(\text{e}^{-E_{\text{el},k}/(kT)} \cdot \text{e}^{-E_{\text{vib,l}}/(kT)} \cdot \text{e}^{-E_{\text{rot},m}/(kT)} \cdot \text{e}^{-E_{\text{trans},n}/(kT)}\right) \\ &= \sum_{k=1}^{\infty}\text{e}^{-E_{\text{el},k}/(kT)} \cdot \sum_{l=1}^{\infty}\text{e}^{-E_{\text{vib,l}}/(kT)} \cdot \sum_{m=1}^{\infty}\text{e}^{-E_{\text{rot},m}/(kT)} \cdot \sum_{n=1}^{\infty}\text{e}^{-E_{\text{trans},n}/(kT)} \end{aligned}$$

or

$$Z = Z_{\text{el}} \cdot Z_{\text{vib}} \cdot Z_{\text{rot}} \cdot Z_{\text{trans}}$$

Using Equation (15.13) for the internal energy from the text, we obtain

$$\begin{aligned}U &= NkT^2\frac{\mathrm{d}\ln Z}{\mathrm{d}T} = NkT^2\frac{\mathrm{d}\ln\left(Z_{\mathrm{el}}\cdot Z_{\mathrm{vib}}\cdot Z_{\mathrm{rot}}\cdot Z_{\mathrm{trans}}\right)}{\mathrm{d}T}\\ &= NkT^2\left[\frac{\mathrm{d}\ln\left(Z_{\mathrm{el}}\right)}{\mathrm{d}T}+\frac{\mathrm{d}\ln\left(Z_{\mathrm{vib}}\right)}{\mathrm{d}T}+\frac{\mathrm{d}\ln\left(Z_{\mathrm{rot}}\right)}{\mathrm{d}T}+\frac{\mathrm{d}\ln\left(Z_{\mathrm{trans}}\right)}{\mathrm{d}T}\right]\\ &= U_{\mathrm{el}}+U_{\mathrm{vib}}+U_{\mathrm{rot}}+U_{\mathrm{trans}}\end{aligned}$$

where U_{el}, U_{vib}, U_{rot}, and U_{trans} are the contributions of electronic, vibrational, rotational, and translational degrees of freedom to the internal energy of the gas, and are given by

$$U_{\mathrm{el}} = NkT^2\cdot\frac{\mathrm{d}\ln Z_{\mathrm{el}}}{\mathrm{d}T} = N\bar{E}_{\mathrm{el}}\,;\; U_{\mathrm{vib}} = NkT^2\cdot\frac{\mathrm{d}\ln Z_{\mathrm{vib}}}{\mathrm{d}T} = N\bar{E}_{\mathrm{vib}}$$

$$U_{\mathrm{rot}} = NkT^2\cdot\frac{\mathrm{d}\ln Z_{\mathrm{rot}}}{\mathrm{d}T} = N\bar{E}_{\mathrm{rot}}\,;\; U_{\mathrm{trans}} = NkT^2\cdot\frac{\mathrm{d}\ln Z_{\mathrm{trans}}}{\mathrm{d}T} = N\bar{E}_{\mathrm{trans}}$$

E15.13 The lowest electronically excited energy level of O_2 is $1.0\,\mathrm{eV} = 1.602\times10^{-19}$ J above the ground state. At what temperature $T_{1\%}$ would the population of excited molecules be 1.0% of that in the ground state? Take the degeneracy of the ground level to be $g_0 = 3$ and that of the excited levels to be $g_{i>0} = 1$. Assume that the energy of the ground state is $E_0 = 0$. Compute the electronic partition function for O_2 at the temperature $T_{1\%}$.

Solution

The fractional population in the first excited state, which we label 1, can be calculated by

$$\frac{N_1}{N} = \frac{1}{Z_{\mathrm{el}}}\exp\left[-E_1/(kT)\right]$$

where

$$\begin{aligned}Z_{\mathrm{el}} &= \sum_{i=0} g_i\exp\left[-E_i/(kT)\right]\\ &= 3\cdot\exp\left[-E_0/(kT)\right]+\exp\left[-E_1/(kT)\right]+\ldots\\ &= 3+\exp\left[-E_1/(kT)\right]+\ldots\end{aligned}$$

If we neglect the higher excited states (their energies are high enough that they contribute only weakly), then we find that

$$Z_{\mathrm{el}} = 3+\exp\left[-\frac{1.602\times10^{-19}\ \mathrm{J}}{\left(1.3806\times10^{-23}\ \mathrm{J/K}\right)\cdot T}\right] = 3+\exp\left[-\frac{11600\ \mathrm{K}}{T}\right]$$

The exponential function is $\ll 1$ for all available temperatures, thus

$$Z_{\mathrm{el}} \approx 3$$

For $N_1/N = 0.01$, we find

$$0.01 = \frac{\exp\left[-\frac{11600\ \mathrm{K}}{T}\right]}{3} \quad\text{or}\quad 0.03 = \exp\left[-\frac{11600\ \mathrm{K}}{T}\right]$$

Thus

$$\ln 0.03 = -\frac{11600\ \mathrm{K}}{T} \quad\text{and}\quad T = -\frac{11600\ \mathrm{K}}{\ln 0.03} = 3310\ \mathrm{K}$$

For comparison, we calculate the electronic partition function more precisely as

$$Z_{\mathrm{el}} = 3+\exp\left[-\frac{11600}{3310}\right] = 3.03$$

Thus, if we approximate the electronic partition function by $Z_{el} = 3$ for calculations with oxygen at temperatures below 3000, then the error is less than 1%.

E15.14 Figure 15.15 shows that the distribution of translational speeds broadens with temperature. Quantify how the width of the distribution changes with temperature by finding an expression for the standard deviation in the translational speeds; i.e., $\sigma_v = \sqrt{\overline{v^2} - \overline{v}^2}$.

Solution

The width of the distribution can be characterized by the standard deviation, σ, which can be calculated for the speed as

$$\sigma_v = \sqrt{\overline{v^2} - \overline{v}^2} = \sqrt{\frac{3RT}{M} - \left(\sqrt{\frac{8RT}{\pi M}}\right)^2} = \sqrt{\frac{3RT}{M} - \frac{8RT}{\pi M}}$$

This expression can be further simplified to

$$\sigma_v = \sqrt{\frac{RT}{M}}\sqrt{3 - \left(\frac{8}{\pi}\right)} = 0.673\sqrt{\frac{RT}{M}}$$

which increases as the square root of temperature.

E15.15 Analyze the expression for the most probable speed v_P of the Maxwell–Boltzmann distribution and determine how it changes with temperature. How does it change with molar mass?

Solution

The most probable speed (Equation (15.70)) is

$$v_P = \sqrt{\frac{2RT}{M}}$$

Figure E15.15 (left panel) shows a plot of the most probable speed v_p versus temperature T at a fixed mass (solid line). It is apparent that the most probable speed increases with temperature, as the square root.

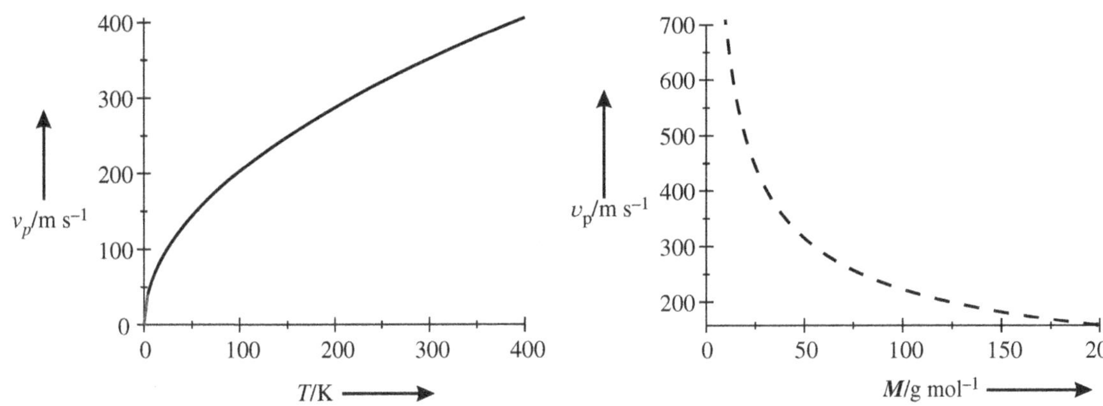

Figure E15.15 **Left panel:** Most probable speed v_p versus temperature T. **Right panel:** Most probable speed v_p versus molar mass $\boldsymbol{M}$.

Figure E15.15 (right panel) shows a plot of the most probable speed v_p versus molar mass $\boldsymbol{M}$ at a fixed temperature (dashed line); in this case it decreases as the mass increases with a square root dependence.

E15.16 Use the three-dimensional Maxwell–Boltzmann distribution of molecular speeds to find an expression for the average speed $\overline{v}$, Equation (15.71).

Solution

The average speed, or *mean speed*, is calculated by

$$\bar{v} = \frac{1}{N}\int_0^\infty v \cdot \mathrm{d}N$$
$$= 4\pi \cdot \left(\frac{m}{2\pi kT}\right)^{3/2} \int_0^\infty v \cdot v^2 \cdot \mathrm{e}^{-mv^2/2kT} \cdot \mathrm{d}v$$

If we substitute

$$x^2 = \frac{mv^2}{2kT}, \quad x\,\mathrm{d}x = \frac{mv}{2kT} \cdot \mathrm{d}v$$

then we obtain

$$\bar{v} = 4\pi \cdot \left(\frac{m}{2\pi kT}\right)^{3/2} \left(\frac{2kT}{m}\right)^2 \cdot \int_0^\infty x^3 \cdot \mathrm{e}^{-x^2} \cdot \mathrm{d}x$$
$$= 4\pi \cdot \left(\frac{m}{2\pi kT}\right)^{3/2} \cdot \left(\frac{2kT}{m}\right)^2 \cdot \frac{1}{2} = \sqrt{\frac{8kT}{\pi m}}$$

where we obtain the integral $\int_0^\infty x^3 \cdot \mathrm{e}^{-x^2} \cdot \mathrm{d}x = 1/2$ from Appendix B.

E15.17 Sketch the form of the three dimensional Maxwell–Boltzmann distribution law for molecular speeds. Perform this sketch for two different values of the temperature. On your graph mark the most probable speed, the average speed, and the root mean square speed of the gas particle.

Solution

Figure E15.17 shows a plot for He atoms at $T = 300$ K (light gray curve) and $T = 1200$ K (dark gray curve). The full circles mark the most probable speed v_p (1), the average speed $\bar{v}$ (2), and the root mean square speed v_{rms} (3).

T/K	v_p/m s^{-1}	$\bar{v}$/m s^{-1}	v_{rms}/m s^{-1}
300	1116	1259	1367
1200	2233	2519	2734

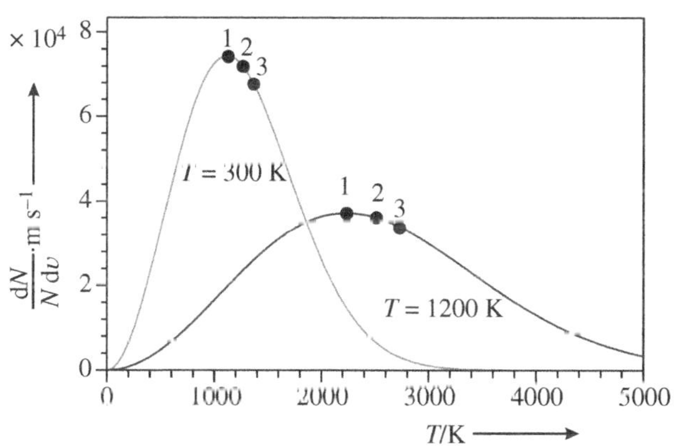

Figure E15.17 Distribution of speed s for the He atom at $T = 300$ K and $T = 1200$ K. The full circles mark the most probable speed v_p (1), the average speed $\bar{v}$ (2), and the root mean square speed v_{rms} (3).

E15.18 Sketch the Maxwell–Boltzmann distribution of molecular speeds for two monatomic gases of mass m_1 and m_2, at the same temperature. Assume that $m_1 > m_2$. Explain your sketch in words.

Solution

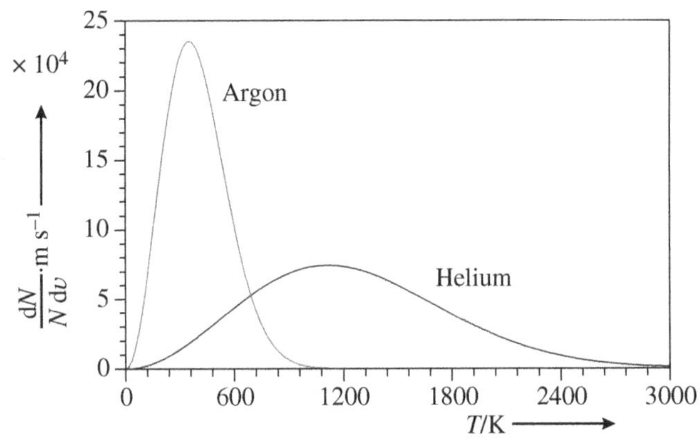

Figure E15.18 Distribution of speeds at $T = 300$ K for Ar atoms (light gray curve) and He atoms (dark gray curve).

The curves in Fig. E15.18 show that the distribution of molecular speeds is asymmetric with a tail toward higher speeds. The higher mass atoms are shifted to lower speed on average. This shift reflects the fact that the average kinetic energy of both gases is $3kT/2$. Hence the higher mass shows a correspondingly smaller average speed.

E15.19 An example of a two-state system is the case of a proton in an external magnetic field (as created in an NMR spectrometer). The energy splitting between the spin states of an H atom nucleus in a magnetic field is given by Equation (12.5)

$$\Delta\varepsilon = g_N \cdot \mu_N \cdot B$$

where B is the magnetic field strength. $g_N = 5.586$ is the nuclear g -factor for the proton and $\mu_N = 5.051 \times 10^{-27}$ J T^{-1} is the nuclear magneton. Ignore the translational degrees of freedom and compute the molecular partition function for the case of an H atom nucleus in a 12 Tesla magnetic field and $T = 77$ K. Also compute the relative populations of the two spin energy levels and their average energy in the magnetic field.

Solution

In a 12 T field (similar to that in a 600 MHz NMR), the energy splitting of the spin states is

$$\Delta\varepsilon = 5.586 \cdot 5.051 \times 10^{-27}\,\text{J T}^{-1} \cdot 12\,\text{T} = 3.38 \times 10^{-25}\,\text{J}$$

If we choose the zero point of energy to coincide with the ground spin energy level we find that the spin up level is

$$\varepsilon_\alpha = 0\,\text{J}$$

and the spin down state has an energy ε_β of

$$\varepsilon_\beta = 3.38 \times 10^{-25}\,\text{J}$$

The partition function Z is

$$\begin{aligned} Z &= \exp\left(-\frac{\varepsilon_\alpha}{kT}\right) + \exp\left(-\frac{\varepsilon_\beta}{kT}\right) \\ &= 1 + \exp\left(\frac{-3.38 \times 10^{-25}\,\text{J}}{1.3806 \times 10^{-23}\text{J K}^{-1} \cdot 77\,\text{K}}\right) \\ &= 1 + 0.99968 = 1.99968 \end{aligned}$$

Hence the fraction in the lower state f_α is

$$f_\alpha = \frac{\exp\left(-\varepsilon_\alpha/(kT)\right)}{Z} = \frac{1.0000}{1.99968} = 0.50008$$

and the fraction in the upper state f_β is

$$f_\beta = \frac{\exp\left(-\varepsilon_\beta/(kT)\right)}{Z} = \frac{0.99968}{1.99968} = 0.49992$$

The average energy may be written as

$$\begin{aligned}\overline{\varepsilon} &= f_\alpha \cdot \varepsilon_\alpha + f_\beta \cdot \varepsilon_\beta \\ &= (0.49992)\left(3.38 \times 10^{-25}\ \mathrm{J}\right) \\ &= 1.69 \times 10^{-25}\ \mathrm{J}\end{aligned}$$

E15.20 A molecular radical (e.g., methyl radical) has an unpaired electron. In an external magnetic field the ground electronic energy level splits into two different energy states associated with the two possible spin orientations. The energy splitting between the spin states of an unpaired electron in a magnetic field (without orbital angular momentum) is given by

$$\Delta\varepsilon = g \cdot \mu_{Bohr} \cdot B$$

where $g = 2.002$ is the g-factor of the electron and $\mu_{Bohr} = 9.274 \times 10^{-24}\ \mathrm{J\,T^{-1}}$ is the Bohr magneton. Compute the molecular partition function for the two different electron spin states in a 1.2 Tesla magnetic field at $T = 77$ K. Compute the relative populations of the two energy levels and the average energy of the electron in the magnetic field.

Solution

In a 1.2 Tesla field, the energy splitting of the unpaired electron states is

$$\Delta\varepsilon = 2.002 \cdot 9.274 \times 10^{-24}\mathrm{J\,T^{-1}} \cdot 1.2\ \mathrm{T} = 2.2 \times 10^{-23}\ \mathrm{J}$$

We choose the energy of the ground state (spin down state ε_β) to be zero

$$\varepsilon_\beta = 0$$

and the energy of the spin up state ε_α is given by

$$\varepsilon_\alpha = 2.2 \times 10^{-23}\ \mathrm{J}$$

In contrast to protons (see E15.19), the electron prefers to orient against the field direction. The partition function Z is

$$\begin{aligned}Z &= \exp\left(-\frac{\varepsilon_\beta}{kT}\right) + \exp\left(-\frac{\varepsilon_\alpha}{kT}\right) \\ &= 1 + \exp\left(\frac{-(2.2 \times 10^{-23}\ \mathrm{J})}{(1.3806 \times 10^{-23})(77)}\right) \\ &= 1 + 0.9795 = 1.9795\end{aligned}$$

Hence the fraction in the lower state f_β is

$$f_\beta = \frac{\exp\left(-\varepsilon_\beta/(kT)\right)}{Z} = \frac{1.0000}{1.9795} = 0.5052$$

and the fraction in the upper state f_β is

$$f_\beta = \frac{\exp\left(-\varepsilon_\beta/(kT)\right)}{Z} = \frac{0.9795}{1.9795} = 0.4948$$

Hence the populations differ by about 1%.

The average energy may be written as

$$\begin{aligned}\overline{\varepsilon} &= f_\beta \cdot \varepsilon_\beta + f_\alpha \cdot \varepsilon_\alpha \\ &= (0.4948)\left(2.2 \times 10^{-23}\ \mathrm{J}\right) \\ &= 1.09 \times 10^{-23}\ \mathrm{J}\end{aligned}$$

E15.21 Use the MathCAD exercise, "distribution.mcdx," to explore some fundamental properties of two different probability distributions: the Poisson distribution,

$$f(\lambda, k) = \exp(-\lambda) \cdot \frac{\lambda^k}{k!}$$

which often applies when the number of particles (or events) is small, and the Gaussian distribution,

$$f(x) = \frac{1}{\sigma\sqrt{2\pi}} \cdot \exp\left(-\frac{1}{2}\left(\frac{x-\mu}{\sigma}\right)^2\right)$$

which is very widely used and is appropriate when the particle (or event) number is large. In the first part of the exercise you are asked to plot the distributions with various parameter values and for various particle numbers N, to understand the meaning of the parameters for the distributions and to assess how the number of particles affects the shape of the distribution. In the second part of the exercise, you are asked to consider the probability distribution for the energy of a one-dimensional ideal gas, which is well described as a Gaussian distribution under common conditions. You should examine how the energy distribution for the gas changes with the number of particles N. In the last question for this part, you are given the relationship between the standard deviation σ of the distribution and the constant volume heat capacity C_V of the gas (i.e., $\sigma = \sqrt{kT^2C_V}$), and you are asked to comment on its implications for how the energy distribution must change near a phase transition.

Solution

Here we provide answers to the questions posed in the MathCAD sheet.

1. In the Poisson distribution, parameter b corresponds to the average value of the distribution. This can be verified by adjusting it in the MathCAD sheet and observing that the peak of the distribution shifts.

   In the Gaussian distribution, the parameter b corresponds to the average value and the parameter a corresponds to the variance. The meaning of b can be verified by changing its value and observing that the distribution shifts. The meaning of a can be verified by changing its value and observing that the distribution broadens.
2. The three plots in Fig. E15.21a show how the probability distribution changes with the number of particles; as N decreases the distribution broadens; the other parameters are held constant. The parameter a also changes the width of the distribution; however, the distribution broadens as a increases.

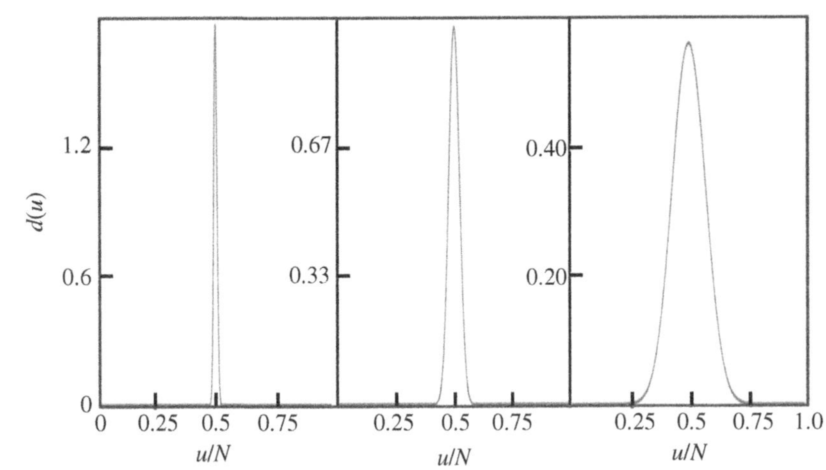

Figure E15.21a Change of the Gaussian distribution with the number of particles. From left to right, the three panels show the cases of 10,000 particles, 1000 particles, and 100 particles. In each case the distribution is centered at N/2.

3. The distribution shown in Fig. E15.21b narrows about the center point where $E = E_{av}$. As N becomes very large the distribution becomes so narrow that one can use the average value of the energy in place of the distribution. This approach is used in obtaining the thermodynamic limit.

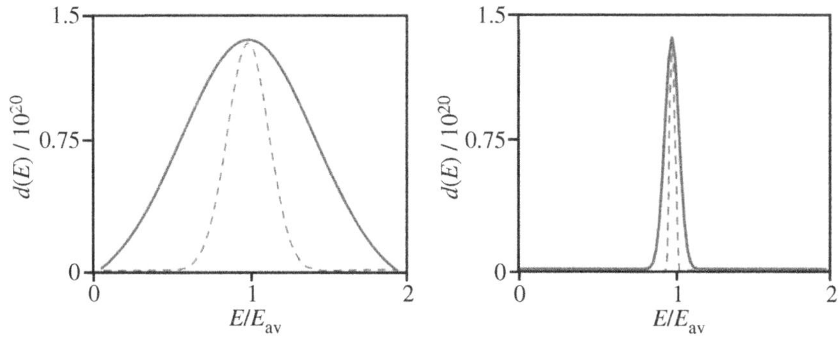

Figure E15.21b The Gaussian distributions shown here approach the mean value. The left panel shows the cases of $N = 10$ (solid curve) and $N = 100$ (dashed curve). The right panel shows the cases of $N = 1000$ (solid curve) and $N = 10{,}000$ (dashed curve).

4. As with the Gaussian distribution in part 1, σ is the standard deviation. As σ becomes small the distribution becomes very narrow. In the case of a phase transition where the heat capacity becomes extremely large, we would expect that the energy distribution of the particles would become very broad. Hence to properly describe a systems behavior we would need to use the entire distribution of energies rather than just the average value.

E15.22 Plot the following three functions on a single graph: $f_1(J) = 2J + 1$; $f_2(J) = \exp(-E_J/kT)$ with $E_J = 10^{-22}$ J; and their product $f_3(J) = f_1(J) \cdot f_2(J)$. Discuss your result in terms of the population distribution of rotational energy levels.

Solution

f_1 (J) is the solid line that increases with J. f_2 (J) is the dashed curve which has the shape of a Gaussian function. The product function f_3 (J) is plotted in Fig. E15.22 as the light gray curve.

Figure E15.22 Black solid line: $f_1(J) = 2J + 1$; Black dashed line: $f_2(J) = \exp(-E_J/kT)$ with $E_J = 10^{-22}$ J; Light gray line: their product $f_3(J) = f_1(J) \cdot f_2(J)$.

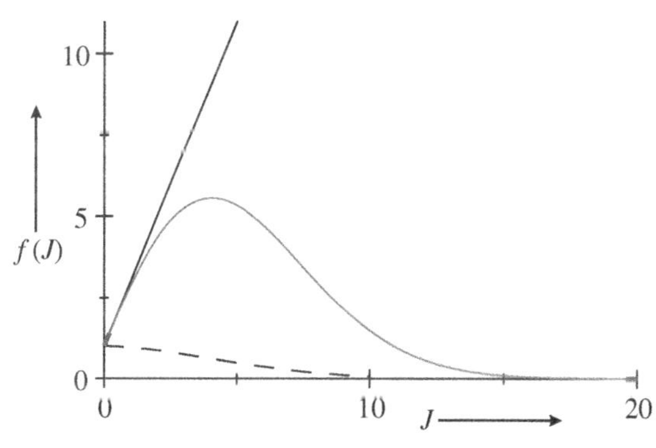

Initially f_3 (J) is dominated by the $2J + 1$ term and then after $J = 5$ the f_2 (J) term dominates. Hence at low J values we first see an increase in the population with the increasing energy level because the degeneracy is increasing; however, at larger J values the size of the gap becomes the dominant factor and the overall population of an energy level decreases with increasing J.

E15.23 Consider a two-state system with a ground-state energy of $E_1 = 0$ and an excited state energy of $E_2 = 1.6 \times 10^{-21}$ J. Write an expression for the molecular partition function and evaluate it at the temperatures 1 K, 10 K, 100 K, 1000 K, and 10,000 K. What value does the molecular partition function take as $T \to 0$? What value does the partition function take as $T \to \infty$? How does the partition function reflect the population of the states (refer to Fig. 15.3).

Solution

The partition function of a two-state system is given by

$$Z = \exp\left(-E_1/kT\right) + \exp\left(-E_2/kT\right)$$

Upon substitution of the data given we find that

$$Z = 1 + \exp\left(-\frac{1.6 \times 10^{-21}\ \text{J}}{(1.3806 \times 10^{-23}\ \text{J/K})\ T}\right) = 1 + \exp\left(\frac{-116\ \text{K}}{T}\right)$$

At 1, 10, 100, 1000, 10,000 K, the partition function is

T/K	1	10	100	1000	10,000
Z	1.00	1.00	1.28	1.88	1.99

As $T \to 0$, $Z = 1 \times 1 + 1 \times 0 = 1$; i.e., only the ground state is occupied at low enough temperature. As $T \to \infty$, $Z = 1 \times 1 + 1 \times 1 = 2$; i.e., at high temperatures where the thermal energy kT is greater than the energy level spacing it is equally likely to be in either state. The amount by which Z is greater than one reflects the relative importance (likelihood of being populated) of the excited state.

E15.24 A simple model for the lowest excited states of the Na atom is a ground energy level that is doubly degenerate ($E_1 = 0$, and $g_1 = 2$) and an excited energy level that is six-fold degenerate ($E_2 = 3.37 \times 10^{-19}$ J, $g_2 = 6$). (Note: the excited state configuration [Na]3p^1 is actually split into sublevels; however, they lie very close in energy as compared to the energy gap $E_2 - E_1$ so we ignore this difference.) Calculate the fraction of atoms in each of these levels at 1000 and 5000 K. For 0.1 mol of Na atoms, how many are in the excited level under these conditions?

Solution

The fraction of the population in each of the two states is given by

$$\frac{N_1}{N} = \frac{g_1 \exp(-E_1/kT)}{g_1 \exp(-E_1/kT) + g_2 \exp(-E_2/kT)}$$

and

$$\frac{N_2}{N} = \frac{g_2 \exp(-E_2/kT)}{g_1 \exp(-E_1/kT) + g_2 \exp(-E_2/kT)}$$

From the information in the problem statement ($g_1 = 2, E_1 = 0, g_2 = 6, E_2 = 3.37 \times 10^{-19}$ J), we find that

$$\frac{N_1}{N} = \frac{2}{2 + 6 \cdot \exp\left(-\frac{24410\ \text{K}}{T}\right)} \quad \text{and} \quad \frac{N_2}{N} = \frac{6 \cdot \exp\left(-\frac{24410\ \text{K}}{T}\right)}{2 + 6 \cdot \exp\left(-\frac{24410\ \text{K}}{T}\right)}$$

Thus the populations at the different temperatures are

T	N_1/N	N_2/N
1000	1.000	7.518×10^{-11}
5000	0.9978	0.0222

For 0.1 mol (6.022×10^{22} atoms) of Na we find that 4.528×10^{12} Na atoms are excited at 1000 K and 1.339×10^{21} atoms are excited at 5000 K.

E15.25 The molecular partition function Z_{trans} for a monatomic ideal gas is given by (see Equation (15.59))

$$Z_{\text{trans}} = V\left(\frac{2\pi mkT}{h^2}\right)^{3/2}$$

Evaluate Z_{trans} for a He atom at a temperature of 298 K and in a container of volume i) 1 cm^3; ii) 1 μm^3; iii) 10 nm^3.

Solution

We take the mass of a He atom to be 6.69×10^{-27} kg. The translational partition function for He at 298 K in a 1 cm^3 box is

$$Z_{trans} = V\left(\frac{2\pi mkT}{h^2}\right)^{3/2}$$
$$= 10^{-6}m^3\left(\frac{2\pi \cdot 6.69 \times 10^{-27}\ \text{kg} \cdot 1.3806 \times 10^{-23}\text{J K}^{-1} \cdot 298\ \text{K}}{(6.6260755 \times 10^{-34}\ \text{J s})^2}\right)^{3/2}$$
$$= 7.83 \times 10^{24}$$

We proceed in a like manner for the other volumes and find that

V	1 cm^3	1 μm^3	10 nm^3
Z_{trans}	7.83×10^{24}	7.83×10^{12}	7.83×10^{6}

E15.26 Calculate the molecular partition function for a mole of He atoms and for a mole of Ar atoms at 298 K and 1 bar. Assuming that Z_{trans} measures the number of available quantum states, calculate the average number of states available for each gas atom in the two cases. Compare your results with each other.

Solution

Equation (15.59) gives Z_{trans} as

$$Z_{trans} = V \cdot \left(\frac{2\pi mk}{h^2}\right)^{3/2} \cdot T^{3/2}$$

We can determine the volume using the ideal gas equation of state to be

$$V = \frac{nRT}{P}$$
$$= \frac{1.000\ \text{mol} \cdot 0.08314\ \text{bar L mol}^{-1}\text{K}^{-1} \cdot (298\ \text{K})}{1.000\ \text{bar}}$$
$$= 24.78\ \text{L} = 0.02478\ \text{m}^3$$

For He we take the mass to be 6.69×10^{-27} kg and the partition function becomes

$$Z_{trans}(\text{He}) = 0.02478\ \text{m}^3 \cdot \left(\frac{2\pi \cdot 6.69 \cdot 10^{-27}\ \text{kg} \cdot 1.381 \cdot 10^{-23}\ \text{J K}^{-1} \cdot 298\ \text{K}}{(6.626 \times 10^{-34}\ \text{J s})^2}\right)^{3/2}$$
$$= 1.92 \times 10^{29}$$

and for Ar we take the mass to be 6.64×10^{-26} kg and find the partition function

$$Z_{trans}(\text{Ar}) = 0.02478\ \text{m}^3 \cdot \left(\frac{2\pi \cdot 6.64 \cdot 10^{-26}\ \text{kg} \cdot 1.381 \cdot 10^{-23}\ \text{J K}^{-1} \cdot 298\ \text{K}}{(6.626 \times 10^{-34}\ \text{J s})^2}\right)^{3/2}$$
$$= 6.06 \times 10^{30}$$

Both values are large numbers indicating the classical nature of translational motion. The argon value is slightly larger because of its higher mass, which causes a somewhat smaller energy level spacing.

To find the average number of states available per gas atom in each case, we must divide by the number of moles. For He we find $\left(1.92 \times 10^{29}/\left(6.022 \times 10^{23}\right)\right) = 3.19 \times 10^5$ states per He atom, and for Ar we find $\left(6.06 \times 10^{30}/\left(6.022 \times 10^{23}\right)\right) = 1.01 \times 10^7$ states per Ar atom. Hence the gas is "dilute" in the sense that we are very unlikely to find two He atoms in the same quantum state.

E15.27 Show that for He in a volume of $V = 1000$ mL at $T = 300$ K and $P = 1$ bar the probability of finding an atom in the lowest translational quantum state is 1.3×10^{-28}.

Solution

According to the Boltzmann equation the number N_1 of particles in the lowest energy state, E_1, is

$$N_1 = \frac{N}{Z_{\text{trans}}} \exp\left(-E_1/kT\right), \text{ with } Z_{\text{trans}} = V\left(\frac{2\pi mk}{h^2}\right)^{3/2} \cdot T^{3/2}$$

Because the energy E_1 for a particle in a volume of macroscopic size is extremely small we can replace the exponential function by the value 1. Then

$$N_1 = \frac{N}{Z_{\text{trans}}}$$

With the data for He we obtain $Z_{\text{trans}} = 7.7 \times 10^{27}$ (see E15.26 for an example of how to perform this type of calculation) and the probability of finding an atom in the lowest translational quantum state is

$$\mathcal{P} = \frac{N_1}{N} = \frac{1}{Z_{\text{trans}}} = \frac{1}{7.7 \times 10^{27}} = 1.3 \times 10^{-28}$$

We calculate the total number of particles, N, from the ideal gas equation of state

$$N = \frac{PV}{kT} = 2.4 \times 10^{22}$$

to obtain the number of atoms in the lowest quantum state as

$$N_1 = 2.4 \times 10^{22}/7.7 \times 10^{27} = 3.1 \times 10^{-6} \approx 0.$$

E15.28 Consider a two-state system with an excitation energy of $\Delta E = 3.37 \times 10^{-19}$ J (e.g., the 3s to 3p electron promotion in Na is close to this value). Calculate the fraction of excited states at $T = 300$ K, 1000 K, and 2000 K.

Solution

We calculate the fraction at 300 K to be

$$\frac{N_2}{N_1} = \exp\left(-\frac{\Delta E}{kT}\right) = \exp\left(-\frac{3.37 \times 10^{-19}\ \text{J}}{1.38 \times 10^{-23} \text{J K}^{-1} \cdot 300\ \text{K}}\right) = 4 \times 10^{-36}$$

At 1000 K the fraction is 2×10^{-11}, and at 2000 K it is 5×10^{-6}. In Fig. 15.4, the population fractions N_1/N and N_2/N (see Equation (15.4)) in states 1 and 2 are plotted as a function of T. For this excitation energy, which is typical for electronic states of atoms and molecules, the temperature must be about 10,000 K to excite 10% of the particles to state 2.

E15.29 The fraction of molecules in a particular vibrational state depends on the energy spacing of the vibrational levels as well as the temperature. Use the data in Table E15.29 to plot the population probabilities $\mathcal{P}_n$ of molecules in vibrational level n for two different diatomic halogens at 300 K.

Table E15.29 Vibrational Parameters and Population Probabilities $\mathcal{P}_0$ and $\mathcal{P}_1$ for Some Interhalogens at 300 K.

System	k_f (N/m)	$\frac{1}{\lambda}$ (cm^{-1})	$\Delta E/10^{-21}$ J	$\mathcal{P}_0$	$\mathcal{P}_1$
$^{19}F^{35}Cl$	449	786	15.6	0.977	0.023
$^{19}F^{79}Br$	407	671	13.3	0.960	0.039
$^{127}I^{35}Cl$	239	384	7.63	0.842	0.133
$^{127}I^{79}Br$	208	269	5.34	0.725	0.200

Solution

According to Equation (15.24) the probability P_n of finding a molecule in quantum state n is

$$\begin{aligned}\mathcal{P}_n &= \frac{N_n}{N} = \left(1 - e^{-h\nu_0/(kT)}\right) \cdot e^{-nh\nu_0/(kT)} \\ &= \left(1 - e^{-\Delta E/(kT)}\right) \cdot e^{-n\Delta E/(kT)}\end{aligned}$$

Using the data for IBr and FBr the curves in Fig. E15.29 are obtained

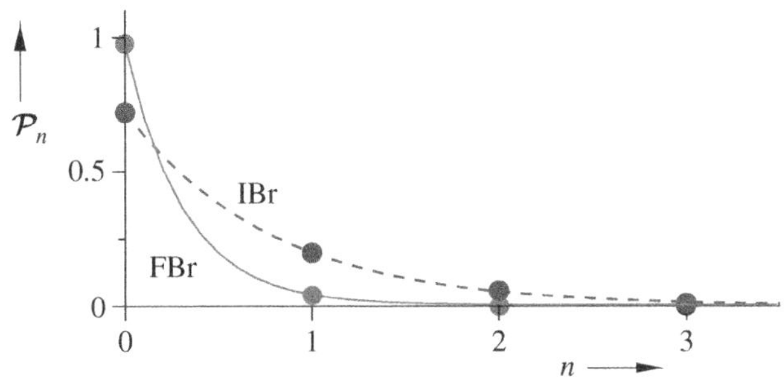

Figure E15.29 Population probability $P_n = N_n/N$ of molecules in a vibrational state n as a function of quantum number n for ${}^{19}F^{79}Br$ ($\Delta E = h\nu_0 = 15.6 \times 10^{-21}$ J, full light gray circles) and ${}^{127}I^{79}Br$ ($\Delta E = h\nu_0 = 5.34 \times 10^{-21}$ J, dark gray full circles) at a temperature of 300 K.

E15.30 Calculate the fraction of molecules that are in excited vibrational states at 300 K for HD, $h\nu_0 = 75 \times 10^{-21}$ J, and I_2, $h\nu_0 = 4.27 \times 10^{-21}$ J.

Solution

According to Equation (15.24) the probability P_n of finding a molecule in a vibrational state with quantum number n is

$$\mathcal{P}_n = \left(1 - e^{-h\nu_0/(kT)}\right) \cdot e^{-nh\nu_0/(kT)}$$

Then the probability of finding a molecule in a vibrational state with quantum number $n = 0$ is

$$\mathcal{P}_0 = 1 - e^{-h\nu_0/(kT)}$$

The probability of finding a molecule in states with any quantum number is one, thus

$$\mathcal{P}_{n>0} = 1 - \mathcal{P}_0 = e^{-h\nu_0/(kT)}$$

Evaluation of this expression for $T = 300$ K gives $P_{n>0} = 1.4 \times 10^{-8}$ for HD and $P_{n>0} = 0.36$ for I_2. At room temperature, 64% of all I_2 molecules are in the ground state, and 36% are distributed over vibrationally excited states, whereas only 14 in every billion HD molecules are in an excited state.

E15.31 Plot the fraction of molecules in each rotational energy level $\mathcal{P}_J$ as a function of the quantum number J for two interhalogen compounds ${}^{19}F^{35}Cl$ ($B = 1.02 \cdot 10^{-23}$ J; $r = 163$ pm) and ${}^{127}I^{79}Br$ ($B = 0.11 \cdot 10^{-23}$ J; $r = 247$ pm). Comment on your findings.

Solution

According to Equations (15.35) and (15.36)

$$\mathcal{P}_J = \frac{N_J}{N} = \frac{1}{Z_{\text{rot}}}(2J+1) \cdot \exp\left[-\frac{B \cdot J(J+1)}{kT}\right], \quad J = 0, 1, 2, 3, \ldots$$

and

$$Z_{\text{rot}} = \sum_{J=0}^{\infty}(2J+1) \cdot \exp\left[-\frac{B \cdot J(J+1)}{kT}\right]$$

we calculate $\mathcal{P}_J$ for different quantum numbers J. The data are plotted in Fig. E15.31. The rotational energy level spacing decreases from FCl to IBr – correlating with the increasing inertial moment for the molecules. The plot shows that higher J rotational levels are excited for the molecule with the smaller energy level spacing and that its distribution is broader.

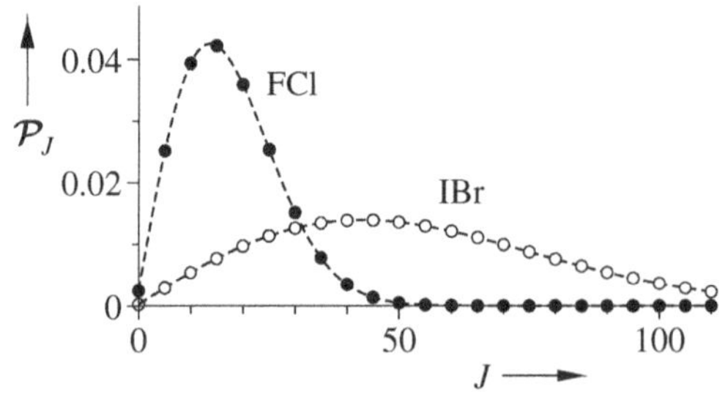

Figure E15.31 Population probability $\mathcal{P}_J = N_J/N$ of molecules in a rotational state J as a function of quantum number J for $^{19}F^{35}Cl$ ($B = 1.02 \times 10^{-23}$ J) and $^{127}I^{79}Br$ ($B = 0.11 \times 10^{-23}$ J) at a temperature of 300 K.

E15.32 Assuming that you can approximate the rotational partition functions for para and ortho hydrogen by

$$Z_{\text{rot,para}} = Z_{\text{rot,ortho}} = \frac{1}{2} Z_{\text{rot}} = \frac{1}{2} \sum_{J=0}^{\infty} g_J \cdot e^{-E_J/(kT)}, \quad J = 0, 1, 2, 3, 4, ..$$

at high temperature, show that $U_{\text{rot}} = NkT$.

Solution

According to Equation (15.49) we have

$$U_{\text{rot},m} = \frac{1}{4}\left[U_{\text{rot,para},m} + 3 \cdot U_{\text{rot,ortho},m}\right]$$

and

$$U_{\text{rot},m} = \frac{1}{4}\left[N_A kT^2 \frac{\mathrm{d}\ln Z_{\text{rot,para}}}{\mathrm{d}T} + 3 \cdot N_A kT^2 \frac{\mathrm{d}\ln Z_{\text{rot,ortho}}}{\mathrm{d}T}\right]$$

$$= \frac{1}{4}\left[N_A kT^2 \frac{\mathrm{d}\ln Z_{\text{rot}}/2}{\mathrm{d}T} + 3 \cdot N_A kT^2 \frac{\mathrm{d}\ln Z_{\text{rot}}/2}{\mathrm{d}T}\right]$$

$$= N_A kT^2 \frac{\mathrm{d}\ln Z_{\text{rot}}/2}{\mathrm{d}T} = N_A kT^2 \frac{\mathrm{d}\left(\ln Z_{\text{rot}} - \ln 2\right)}{\mathrm{d}T}$$

$$= N_A kT^2 \frac{\mathrm{d}\ln Z_{\text{rot}}}{\mathrm{d}T}$$

That is, for sufficiently high temperature the expression for the internal energy is the same as for heteronuclear diatomic molecules, namely

$$U_{\text{rot}} = NkT \quad \text{(high temperature limit)}$$

15.2 Problems

P15.1 Perform an analysis like that in Section 15.8 for a system with $N = 100$ and $U = 20a$. Calculate C and β in Equation (15.80).

Solution

For $N = 100$ and $U = 20a$ the following distributions on three energy levels with an energy spacing of a are possible:

	A	B	C	D	E	F	G	H	I	J	K
N_3	10	9	8	7	6	5	4	3	2	1	0
N_2	0	2	4	6	8	10	12	14	16	18	20
N_1	90	89	88	87	86	85	84	83	82	81	80
$\omega/10^{21}$	0	0	0	0	0.1	0.8	2.4	4.5	4.7	2.5	0.5

The maximum of ω is realized in the distribution I. Figure P15.1 shows a plot of $\ln N_i$ as a function of E_i for the values in column I. From the slope of the straight line we obtain $\beta = 1.86/a$, and from the ordinate intersection we obtain $\ln C = 4.48$ and $C = 88$.

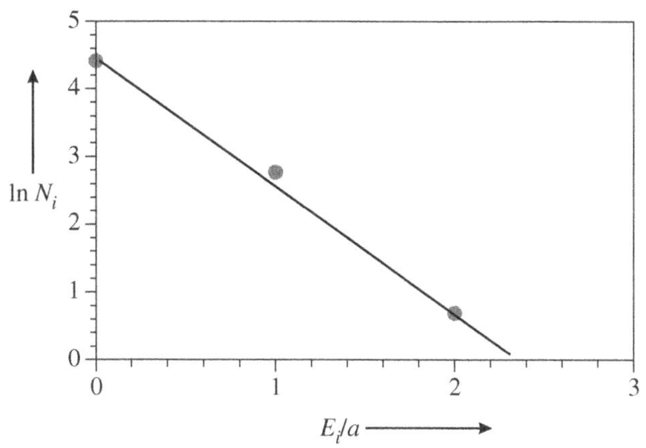

Figure P15.1 Plot of $\ln N_i$ as a function of E_i for the values in column I of the table.

P15.2 (adapted from Physical Chemistry, W. Moore) Consider the molecule N_2 ($\theta_{vib} = 3397$ K). A spectroscopic measurement of the population of molecules in excited vibrational states is

n	0	1	2	3
P_n	0.75	0.19	0.048	0.012

where $P_n = N_n/N$ is the probability of finding a molecule in energy level n. Describe how you would measure these populations spectroscopically; i.e., what spectroscopic method would you use and why (refer to Chapter 11)? Show that the gas is in thermodynamic equilibrium with respect to the distribution of vibrational energy. What is the temperature of the gas?

Solution

Raman spectroscopy can be used to probe the presence of excited state energy levels and their relative populations. Infrared spectroscopy is not appropriate because N_2 does not have a dipole moment and is thus infrared inactive. We assume that the energy levels of the N_2 atom are equidistant at an energy difference of $\Delta E = h\nu_0$. At equilibrium, the ratio of the fraction of molecules in state $n+1$ to that in state n is

$$\frac{P_{n+1}}{P_n} = \frac{N_{n+1}}{N_n} = \exp\left(-\frac{\Delta E}{kT}\right) = \exp\left(-\frac{h\nu_0}{kT}\right)$$

which is independent of n. Examining the ratios of the population numbers in the states given in the problem gives

N_1/N_0	0.25
N_2/N_1	0.25
N_3/N_2	0.25

The constant ratio suggests that the N_2 is at or near equilibrium with

$$0.25 = \exp\left(-\frac{h\nu_0}{kT}\right) = \exp\left(-\frac{\theta_{vib}}{T}\right)$$

so that

$$T = -\frac{3397\text{ K}}{\ln(0.25)} = 2540\text{ K}$$

P15.3 Use a computer to directly evaluate the rotational partition function Z_{rot} for HD and I_2 at 100 K, 300 K, and 1000 K. Compare your value to that obtained by considering the classical limit, for which $Z_{rot} = kT/B$.

Solution

The partition function to be evaluated is

$$Z_{rot} = \sum_{J=0}^{\infty} (2J+1)\, e^{-J(J+1)B/kT}$$

For HD the rotational constant B is

$$B = \frac{h^2}{8\pi^2 \mu d^2}$$

$$= \frac{\left(6.626 \times 10^{-34}\ \mathrm{J\ s}\right)^2}{8\pi^2 \left(1.12 \times 10^{-27}\ \mathrm{kg}\right)(74\ \mathrm{pm})^2} = 9.07 \times 10^{-22}\ \mathrm{J}$$

We find that $B/k = 65.7$ K and at 100 K we have $B/(kT) = 0.657$. Thus

$$Z_{rot}\,(100\ \mathrm{K}) = \sum_{J=0}^{\infty} (2J+1) \cdot e^{-J(J+1) \cdot B/(kT)} = \sum_{J=0}^{\infty} (2J+1) \cdot e^{-J(J+1)0.657}$$
$$= 1.00 + 0.806 + 0.0971 + 0.00264 + \ldots = 1.91$$

In a similar manner we find that $Z_{rot}\,(300\ \mathrm{K}) = 4.92$ and $Z_{rot}\,(1000\ \mathrm{K}) = 15.6$. For the classical limit (see Equation (15.38))

$$Z_{rot} = \frac{kT}{B}$$

we find

$$Z_{rot}(100\ \mathrm{K}) = 1.52;\ Z_{rot}(300\ \mathrm{K}) = 4.57;\ Z_{rot}(1000\ \mathrm{K}) = 15.2$$

It is evident that the partition function evaluations from the summation differ somewhat from that found by using the high temperature approximations. The values disagree the most for the low temperature of 100 K and converge toward better agreement as the temperature increases.
We proceed in a like manner for I_2. Using $B = 7.350 \times 10^{-25}$ J, we find the values in the table

	100 K	300 K	1000 K
Z from summation	1879	5635	18,784
$Z = kT/B$	1878	5635	18,784

In this case we find excellent agreement, within the rounding error, between the partition functions evaluated by the summation and those evaluated by the high temperature limiting expression. This good agreement results from the small energy spacing for I_2, which has a large moment of inertia, and should be contrasted to the case of HD, which has a large rotational energy spacing.

P15.4 Figure 11.16 shows an image of a high-resolution infrared spectrum for HCl. Calculate the populations of HCl molecules in the different rotational levels of the ground vibrational state and use them to explain the intensity profile in the spectrum.

Solution

The rotational populations in the ground vibrational state may be calculated by using the following expression

$$f_J = \frac{N_J}{N} = \frac{1}{Z}(2J+1)\, e^{-BJ(J+1)/(kT)}$$

where

$$Z = \sum_{J=0}^{\infty} (2J+1)\, e^{BJ(J+1)/(kT)}$$

For HCl we use $B = 2.104 \times 10^{-22}$ J and find that

$$Z = \sum_{J=0}^{\infty} (2J+1)\,\mathrm{e}^{-2.104\times10^{-22}\times J\times(J+1)/(1.3806\times10^{-23}\times298)} = 19.89$$

This leads to the following table of values for the fractional population $f_J = N_j/N$ in the first few energy levels

J	0	1	2	3	4	5	6	7	8
f_J	0.0503	0.136	0.185	0.191	0.163	0.119	0.0763	0.043	0.0215

We expect the absorbance of the different energy levels to reflect the populations of the molecules in those levels (from Beers law). Hence we would predict that the most strongly absorbing transitions would be those that originate in the $J = 2$ and $J = 3$ energy levels because they have the highest fractional population. To explain the intensity pattern of the HCl spectrum (Fig. 11.16), we simplify by considering only the R branch. The R branch shows a peak intensity for the third transition, which originates from the $J = 2$ state, and it decreases on either side, in agreement with the trend found for the populations. A similar comparison explains the trend of intensities for the P-branch transitions.
We note that this analysis does not account for the isotope splitting (see the text).

P15.5 Given that

$$U = NkT^2 \frac{\mathrm{d}\ln Z}{\mathrm{d}T}$$

Derive an expression for the internal energy U_{rot} of a nonlinear polyatomic molecule in the high temperature limit, using the rotational energy relation for a spherical top molecule

$$E_{\mathrm{rot}} = B \cdot J(J+1),\ J = 0,1,2,.. \quad \text{and} \quad g_J = (2J+1)^2$$

and for a symmetric top molecule:

$$E_{\mathrm{rot}} = B \cdot J(J+1) + (A-B)\cdot K^2 \quad \text{where} \quad J = 0,1,2,\ldots;\ K = 0, \pm1, \pm2, \ldots, \pm J$$

Compare your result with the equipartition principle. (Note: In your analysis, you can approximate the nuclei as having no spin. In the high temperature limit spin symmetry considerations will affect the partition function and the entropy, but not the energy. See Sections 15.5 and 17.8 of the text.)

Solution

Spherical Top: In this case we have

$$E_{\mathrm{rot}} = B \cdot J(J+1)$$

as in the case of a linear molecule, and we have

$$g_J = (2J+1)^2$$

in contrast to a linear molecule, where $g_J = 2J + 1$. Thus for the partition function we obtain

$$Z_{\mathrm{rot}} = \sum_{J=0}^{\infty} (2J+1)^2 \cdot \mathrm{e}^{-B\cdot J(J+1)/(kT)}$$

For sufficiently high temperature we can simplify $(2J+1)^2$ by $4J^2$ and $J(J+1)$ by J^2 and replace the summation by an integration

$$Z_{\mathrm{rot}} \approx 4 \int_{J=0}^{\infty} J^2 \cdot \mathrm{e}^{-B\cdot J^2/(kT)} \cdot \mathrm{d}J$$

With

$$x^2 = \frac{B}{kT} \cdot J^2 \qquad \mathrm{d}x = \sqrt{\frac{B}{kT}}\mathrm{d}J$$

we obtain

$$Z_{\text{rot}} = 4\left(\frac{kT}{B}\right)^{3/2} \int_{x=0}^{\infty} x^2 e^{-x^2} dx = 4\left(\frac{kT}{B}\right)^{3/2} \frac{\sqrt{\pi}}{4} = \sqrt{\pi}\left(\frac{kT}{B}\right)^{3/2}$$

and

$$U_{\text{rot}} = NkT^2 \frac{d \ln Z_{\text{rot}}}{dT} = NkT^2 \frac{3}{2}\frac{1}{T} = \frac{3}{2}NkT$$

as we expect from the equipartition principle.

Symmetric Top: In this case the partition function is

$$Z_{\text{rot}} = \sum_{J=0}^{\infty}\sum_{K=-J}^{J} (2J+1)\, e^{-E_{\text{rot}}/(kT)}$$

$$= \sum_{J=0}^{\infty} (2J+1)\, e^{-B \cdot J(J+1)/(kT)} \sum_{K=-J}^{J} e^{-(A-B)\cdot K^2/(kT)}$$

and we have the constraint that $A \neq B$; otherwise, it would be a spherical top. For the high temperature limit, we consider the energy level spacings to be very small relative to kT, and we approximate the summations by integrals; hence we write that

$$Z_{\text{rot}} = \int_0^{\infty} (2J+1)\, e^{-B\cdot J(J+1)/(kT)} dJ \int_{-J}^{J} e^{-(A-B)\cdot K^2/(kT)}\, dK$$

Because the integral of K is symmetric, we may write that

$$Z_{\text{rot}} = 2\int_0^{\infty} (2J+1)\, e^{-B\cdot J(J+1)/(kT)} \left[\int_0^{J} e^{-(A-B)\cdot K^2/(kT)}\, dK\right] dJ$$

Now we let $z^2 = (A-B)\cdot K^2/(kT)$ so that $z{\cdot}dz = (A-B)\cdot K/(kT)\, dK$ and the partition function becomes

$$Z_{\text{rot}} = 2\sqrt{\frac{kT}{(A-B)}} \int_0^{\infty} (2J+1)\, e^{-B\cdot J(J+1)/(kT)} dJ \int_0^{J\sqrt{(A-B)/(kT)}} e^{-z^2}\, dz$$

Next we use the definition of the error function, $\text{erf}(t) = \frac{2}{\sqrt{\pi}} \int_0^t e^{-z^2}\, dz$, to write that

$$Z_{\text{rot}} = \sqrt{\pi}\sqrt{\frac{(kT)}{(A-B)}} \int_0^{\infty} (2J+1)\, e^{-B\cdot J(J+1)/(kT)}\, \text{erf}\left(\sqrt{\frac{(A-B)}{kT}} J\right) dJ$$

(see Appendix C.6 for a discussion of the error function). If we let $y = J(J+1)$ and $dy = (2J+1)dJ$ then we can write that

$$Z_{\text{rot}} \simeq \sqrt{\pi}\sqrt{\frac{kT}{(A-B)}} \times \int_0^{\infty} dy e^{-By/(kT)} \cdot \text{erf}\left(\sqrt{\frac{(A-B)}{kT}}\sqrt{y}\right)$$

where we have used the approximation that $y \approx J^2$ in the expression for the error function. In this approximation, we can use the integral

$$\int_0^{\infty} e^{-at}\, \text{erf}\left(\sqrt{bt}\right) dt = \frac{1}{a}\sqrt{\frac{b}{a+b}}$$

(see Appendix B) to find that

$$Z_{\text{rot}} \simeq \sqrt{\pi}\sqrt{\frac{kT}{(A-B)}} \times \frac{kT}{B}\sqrt{\frac{\frac{(A-B)}{kT}}{\frac{(A-B)}{kT} + \frac{B}{kT}}} = \frac{\sqrt{\pi}}{\sigma}\sqrt{\frac{kT}{A}\frac{kT}{B}} = \frac{\sqrt{\pi}}{\sigma\sqrt{AB}} \cdot (kT)^{3/2}$$

We can compare this result to the more approximate result of Justification 15.1; namely

$$Z_{\text{rot}} = \sqrt{\frac{\pi kT}{A-B}} \cdot \frac{kT}{B} = \left[\sqrt{\frac{\pi}{A-B}} \cdot \frac{1}{B}\right] \cdot (kT)^{3/2} \text{ (approximate)}$$

These results differ only by a constant. Thus both methods lead to the same value of the internal energy U, namely,

$$U = NkT^2 \frac{d \ln Z}{dT} = NkT^2 \cdot \frac{3}{2}\frac{1}{T} = \frac{3}{2}NkT$$

where we used

$$Z_{\text{rot}} = aT^{3/2} \qquad \ln Z_{\text{rot}} = \ln\left(aT^{3/2}\right)$$

and

$$\frac{\mathrm{d}\ln Z_{\text{rot}}}{\mathrm{d}T} = \frac{1}{aT^{3/2}}\frac{3}{2}aT^{1/2} = \frac{3}{2}\frac{1}{T}$$

P15.6 Consider an atomic gas, which is constrained to move in two dimensions and deduce an expression for the normalized Maxwell–Boltzmann distribution of speeds. You should find that

$$\boxed{\mathrm{d}N = N\cdot\frac{m}{kT}\cdot\upsilon\cdot \mathrm{e}^{-m\upsilon^2/(2kT)}\cdot \mathrm{d}\upsilon}$$

Use the two-dimensional form of the Maxwell–Boltzmann distribution to show that the average speed of a gas atom is given by

$$\overline{|\upsilon|} = \sqrt{\frac{\pi kT}{2m}}$$

that the root mean square speed is given by

$$\sqrt{\overline{\upsilon^2}} = \sqrt{2kT/m}$$

and the most probable speed is given by

$$\upsilon_P = \sqrt{kT/m}$$

In addition, show that the average energy of a gas atom is given by $\overline{E} = kT$ and that the internal energy of a mole of that gas is given by $U_{\text{trans},m} = \boldsymbol{R}T$. (Hint: For the normalization use polar coordinates $\mathrm{d}\upsilon_x\,\mathrm{d}\upsilon_y = \upsilon\,\mathrm{d}\upsilon\,\mathrm{d}\varphi$).

Solution

We begin by recalling the expression for the one-dimensional velocity distribution (Equation (15.62))

$$\mathrm{d}N(\upsilon_x, \upsilon_x + \mathrm{d}\upsilon_x) \propto \mathrm{e}^{-m\upsilon_x^2/(2kT)}\cdot \mathrm{d}\upsilon_x$$

Because the translational motion of the gas particles is independent, we can write that

$$\mathrm{d}N(\upsilon_x, \upsilon_x + \mathrm{d}\upsilon_x; \upsilon_y, \upsilon_y + \mathrm{d}\upsilon_y) = const\cdot \mathrm{e}^{-m\upsilon_x^2/(2kT)}\mathrm{e}^{-m\upsilon_y^2/(2kT)}\cdot \mathrm{d}\upsilon_y\cdot \mathrm{d}\upsilon_x$$

Next we convert to polar coordinates, namely $\upsilon_x = \upsilon\cos\varphi$ and $\upsilon_y = \upsilon\sin\varphi$ where $\upsilon = \sqrt{\upsilon_x^2 + \upsilon_y^2}$ (see Appendix B), for which $\mathrm{d}\upsilon_y\cdot\mathrm{d}\upsilon_x = \upsilon\cdot\mathrm{d}\varphi\cdot\mathrm{d}\upsilon$. In polar coordinates, we find that

$$\begin{aligned}\mathrm{d}N(\upsilon, \upsilon + \mathrm{d}\upsilon; \varphi, \varphi + \mathrm{d}\varphi) &= \mathrm{d}N(\upsilon_x, \upsilon_x + \mathrm{d}\upsilon_x; \upsilon_y, \upsilon_y + \mathrm{d}\upsilon_y)\\ &= const\cdot \mathrm{e}^{-m\upsilon^2/(2kT)}\upsilon\cdot \mathrm{d}\upsilon\cdot \mathrm{d}\varphi\end{aligned}$$

Because we are only interested in the speeds and not the directions, we average of the angle variable φ and find that

$$\begin{aligned}\mathrm{d}N(\upsilon, \upsilon + \mathrm{d}\upsilon) &= \int_0^{2\pi} \mathrm{d}N(\upsilon, \upsilon + \mathrm{d}\upsilon; \varphi, \varphi + \mathrm{d}\varphi)\\ &= const\cdot \mathrm{e}^{-m\upsilon^2/(2kT)}\upsilon\cdot \mathrm{d}\upsilon\cdot \int_0^{2\pi} \mathrm{d}\varphi\\ &= const\cdot 2\pi\cdot \mathrm{e}^{-m\upsilon^2/(2kT)}\upsilon\cdot \mathrm{d}\upsilon\end{aligned}$$

To determine the normalization constant we require that each of the N particles has a speed, so that

$$\begin{aligned}N &= \int_0^{\infty} \mathrm{d}N(\upsilon, \upsilon + \mathrm{d}\upsilon)\\ &= const\cdot 2\pi\cdot \int_0^{\infty} \mathrm{e}^{-m\upsilon^2/(2kT)}\upsilon\cdot \mathrm{d}\upsilon\end{aligned}$$

To evaluate this integral, we make the substitution $z^2 = mv^2/(2kT)$ and $z\,dz = (m/(2kT))v\,dv$ so that

$$N = const \cdot 2\pi \frac{2kT}{m} \cdot \int_0^\infty e^{-z^2} z\,dz$$
$$= const \cdot 2\pi \frac{2kT}{m} \cdot \frac{1}{2}$$

where we have used the integral table in Appendix B. Solving for the constant we find that

$$const = N \frac{m}{2\pi kT}$$

and

$$dN(v, v + dv) = N \frac{m}{kT} \cdot e^{-mv^2/(2kT)} v \cdot dv$$

or

$$d\mathcal{P}(v, v + dv) = \frac{dN(v, v + dv)}{N} = \frac{m}{kT} \cdot e^{-mv^2/(2kT)} v \cdot dv$$

With the distribution of speeds in hand, it is straightforward to determine the averages that are requested. For the average speed we can write

$$\overline{v} = \int_0^\infty v\, d\mathcal{P}(v, v + dv)$$
$$= \left(\frac{m}{k_B T}\right) \int_0^\infty v^2\, e^{-mv^2/(2kT)}\, dv$$

If we make the substitution $z^2 = mv^2/(2kT)$ and $z\,dz = (m/(2kT))v\,dv$ we find that

$$\overline{v} = 2\sqrt{\frac{(2kT)}{m}} \int_0^\infty e^{-z^2} z^2\, dz$$
$$= 2\sqrt{\frac{(2kT)}{m}} \frac{\sqrt{\pi}}{4} = \sqrt{\frac{\pi kT}{2m}}$$

where we used the integral table in Appendix B to evaluate the integral. Correspondingly for the root-mean-square velocity we proceed by first finding the average speed squared:

$$\overline{v^2} = \int_0^\infty v^2\, d\mathcal{P}(v, v + dv)$$
$$= \left(\frac{m}{kT}\right) \int_0^\infty v^2\, e^{-mv^2/(2kT)}\, v\, dv$$

If we make the substitution $z^2 = mv^2/(2kT)$ and $z\,dz = (m/(2kT))v\,dv$ we find that

$$\overline{v^2} = \left(\frac{4kT}{m}\right) \int_0^\infty z^3\, e^{-z^2}\, dz = \left(\frac{4kT}{m}\right) \frac{1}{2} = \left(\frac{2kT}{m}\right)$$

where we have evaluated the integral by using the integral table in Appendix B. Hence we find that the root-mean-square speed v_{rms} is

$$v_{rms} = \sqrt{\overline{v^2}} = \sqrt{\frac{2kT}{m}}$$

To find the most probable speed v_p, we find the maximum of the distribution by setting its first derivative equal to zero; hence

$$0 = \left.\frac{d\left(\frac{m}{k_B T} \cdot e^{-mv^2/(2kT)} v\right)}{dv}\right|_{v=v_p}$$
$$= e^{-mv_p^2/(2kT)} - \frac{mv_p^2}{kT} e^{-mv_p^2/(2kT)} = 1 - \frac{mv_p^2}{kT}$$

so that $v_p = \sqrt{kT/m}$.

P15.7 Use a spreadsheet program (e.g., Excel) or MathCAD to plot the Maxwell–Boltzmann distribution for N_2 gas. Perform the plot for temperatures of 77 K, 300 K, and 1000 K. On the 300 K graph mark $\overline{v}$, v_{rms} and v_{max}.

Solution

The Maxwell–Boltzmann distribution for N_2 at 77 K is given as

$$\frac{\mathrm{d}N(v, v+\mathrm{d}v)}{N} = 4\pi\left(\frac{m}{2\pi kT}\right)^{3/2} \cdot v^2 \cdot \exp\left(-mv^2/(2kT)\right) \cdot \mathrm{d}v$$

If we use $m = 4.655 \times 10^{-26}$ kg, we find that

$$f(v) = \frac{\mathrm{d}N(v, v+\mathrm{d}v)}{N\,\mathrm{d}v}$$
$$= \frac{1.56 \times 10^{-4}\mathrm{K}^{3/2}}{T^{3/2}} \cdot v^2 \exp\left(-\frac{(1.696 \times 10^{-3})\,v^2}{T}\right)$$

This function is plotted in Fig. P15.7 for the temperatures 77 K (light gray solid curve), 300 K (dark gray solid curve), and 1000 K (black solid curve).

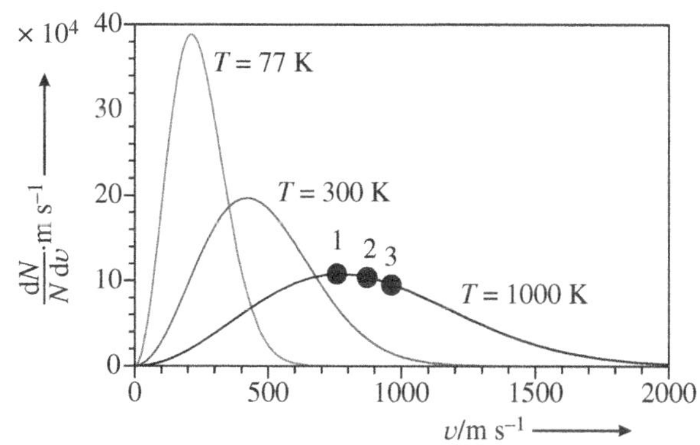

Figure P15.7 Distribution of speeds for the N_2 molecule at $T = 77$ K, $T = 300$ K, and $T = 1000$ K. The full circles mark the most probable speed v_p (1), the average speed $\overline{v}$ (2), and the root mean square speed v_{rms} (3) for $T = 1000$ K.

The full circles represent the average speed

$$\overline{v} = \sqrt{\frac{8 \times 1.3806 \times 10^{-23} \times 1000}{\pi \times 4.683 \times 10^{-26}}}\frac{\mathrm{m}}{\mathrm{s}} = 869\frac{\mathrm{m}}{\mathrm{s}}$$

the root-mean-square speed

$$v_{rms} = \sqrt{\frac{3 \times 1.3806 \times 10^{-23} \times 1000}{4.683 \times 10^{-26}}}\frac{\mathrm{m}}{\mathrm{s}} = 943\frac{\mathrm{m}}{\mathrm{s}}$$

and the speed v_p with the maximum probability

$$v_p = \sqrt{\frac{2 \times 1.3806 \times 10^{-23} \times 1000}{4.683 \times 10^{-26}}}\frac{\mathrm{m}}{\mathrm{s}} = 770\frac{\mathrm{m}}{\mathrm{s}}$$

P15.8 Plot the Maxwell–Boltzmann distribution of speeds for He, Ne, and Ar at a temperature of 300 K on the same graph. Perform a corresponding plot at a temperature of 1200 K. Discuss your findings.

Solution

As in P15.7, we plot

$$f(v) = 4\pi\left(\frac{m}{2\pi kT}\right)^{3/2} \cdot v^2 \cdot \exp\left(-mv^2/(2kT)\right)$$

Figure P15.8a shows the Ar distribution as a solid light gray curve, the Ne distribution as a solid dark gray curve, and the He distribution as a solid black curve. The temperature is 1200 K. It is evident that the He distribution is strongly shifted to higher speeds and is broader. These properties result from the smaller mass of the He atom. As expected from the form of the equation we see that the distributions show a systematic shift with the gas particle mass.

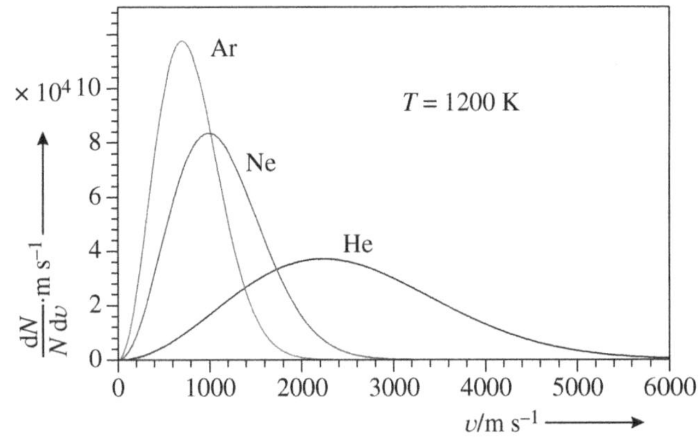

Figure P15.8a Distribution of speeds at $T = 1200$ K for Ar (light gray), Ne (dark gray), and He (black).

Figure P15.8b shows a similar plot for the three gases for the case where $T = 300$ K. The abscissa is chosen to have the same axis scaling as that for the higher temperature plot, in order to emphasize that distributions are shifted to smaller average velocities and are more narrow at lower temperatures. We also note that the relative ordering of the distributions follows the trend of their masses, as expected.

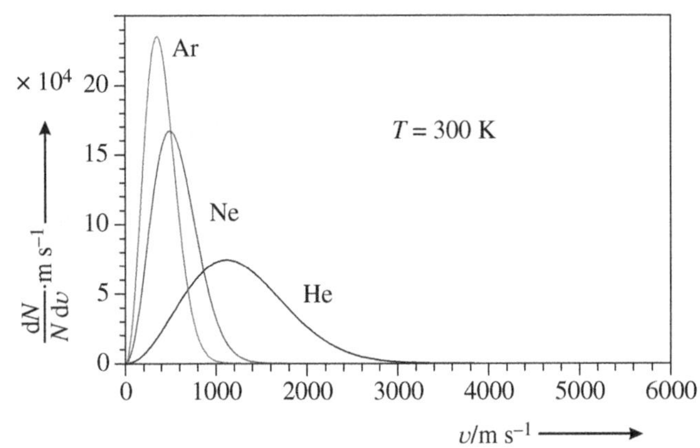

Figure P15.8b Distribution of speeds at $T = 300$ K for Ar (light gray), Ne (dark gray), and He (black).

P15.9 Find an expression for the number of available translational quantum states for a one-dimensional model gas. Compare your result with that we found for the corresponding three-dimensional gas (see Example 15.3).

Solution

The energy levels for a one-dimensional gas are

$$E_n = \frac{h^2}{8mL^2} \cdot n^2$$

In this case, no degeneracies exist and the number g of quantum states between $E = 0$ and $E = E_n$ equals the quantum number n, namely

$$g = n = \sqrt{\frac{8mL^2}{h^2}} \cdot \sqrt{E}$$

For a three-dimensional gas the number of quantum states is, according to (Equation (15.58)),

$$g_{xyz} = \frac{\pi}{6}\left(\frac{8mL^2}{h^2}\right)^{3/2} \cdot E^{3/2}$$

This means that $g_{xyz} \approx g^3$.

P15.10 Compare the vibrational energy levels for I_2 in the table[1] to those obtained from the harmonic oscillator approximation, with a characteristic frequency of $\nu_0 = 6.43 \times 10^{12} s^{-1}$, by plotting them on a graph. Calculate the vibrational partition function and the average vibrational energy of an iodine molecule at $T = 300$ K by using both the harmonic oscillator approximation and the values in the table. Comment on the differences between the two calculations.

n	$\tilde{\nu}$ (cm^{-1})	n	$\tilde{\nu}$ (cm^{-1})	n	$\tilde{\nu}$ (cm^{-1})	n	$\tilde{\nu}$ (cm^{-1})
0	107.120	6	1368.379	11	2384.577	16	3367.822
1	320.435	7	1574.194	12	2583.905	17	3560.397
2	532.515	8	1778.731	13	2781.904	18	3751.586
3	743.356	9	1981.980	14	2978.564	19	3941.377
4	952.952	10	2183.932	15	3173.874	20	4129.756
5	1161.296					21	4316.709

Solution

The frequency $\nu_0 = 6.43 \times 10^{12}$ s^{-1} corresponds to a wavenumber $\tilde{\nu} = 214$ cm^{-1}, so that the harmonic oscillator (HO) model will give an equally spaced ladder of states. The graph in Fig. P15.10 plots the predicted wavenumber from the HO model versus the experimental wavenumber; these are the circles. The solid line shows the idealized case in which the HO wavenumber and the experimental wavenumber would be equal; a line with unit slope. It is evident that the values deviate from the line in a manner that indicates that the experimental wavenumber is lower than the predicted value; hence, the potential must be anharmonic and the energy level spacing must decrease with increasing quantum number.

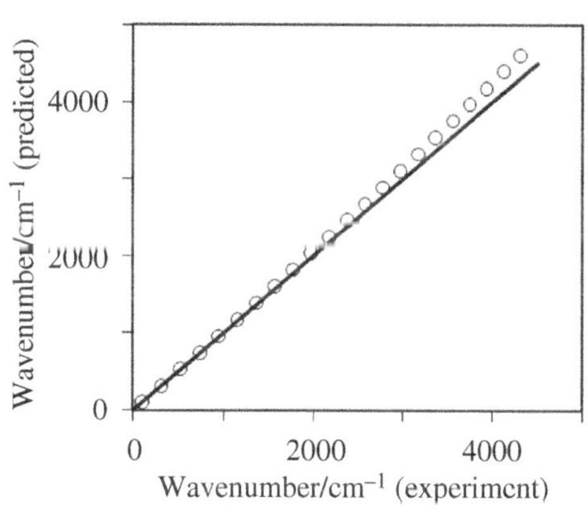

Figure P15.10 Iodine spectrum. Predicted wavenumbers versus experimental wavenumbers

To calculate the partition function we use

$$Z_{\text{vib}} = \exp\left(\frac{E_0}{kT}\right) \sum_{n=0}^{21} \exp\left(-E_n/(kT)\right)$$

For the experimental values we find $Z_{expt} = 1.5636$ and for the harmonic oscillator model we find $Z_{HO} = 1.558$. The average energies are calculated using the definition of the average energy as

$$\overline{E}_{\text{vib}} = \frac{\sum_{n=0}^{21} E_n \exp\left(-E_n/(kT)\right)}{\sum_{n=0}^{21} \exp\left(-E_n/(kT)\right)}$$

1 Values taken from R. J. LeRoy, *J. Chem. Phys.* 52 (1970) 2683.

For the experimental data we find that $\overline{E}_{\text{vib,expt}} = 4.521 \times 10^{-21}$ J and for the harmonic oscillator model we find $\overline{E}_{\text{vib,HO}} = 4.497 \times 10^{-21}$ J; values that are quite similar. The partition function and the average energies differ by less than 1%, indicating that the amount of anharmonicity is rather small. The Excel spreadsheet ch15_p10.xls provides details of the summations.

P15.11 Using a spreadsheet program (such as Excel) evaluate and plot the molar vibrational internal energy for CO_2 as a function of temperature from 10 K to 2000 K (see Equations (15.29) and (15.30) and Chapter 11). What is the molar vibrational internal energy at 300 K? at 2000 K? Compare these values to that expected in the high temperature, classical, limit.

Solution

From Box 11.2 we use the vibrational mode wavenumbers of $\tilde{\nu}_1 = 2349\ \text{cm}^{-1}$, $\tilde{\nu}_2 = 1388\ \text{cm}^{-1}$, $\tilde{\nu}_3 = 667\ \text{cm}^{-1}$, and $\tilde{\nu}_4 = 667\ \text{cm}^{-1}$ for the four modes of CO_2. Using the definition of the characteristic vibrational temperature $\theta_{\text{vib},i} = h\tilde{\nu}_i c/k_B$, we find $\theta_{\text{vib},1} = 3384$ K, $\theta_{\text{vib},2} = 1999$ K, $\theta_{\text{vib},3} = 961$ K, and $\theta_{\text{vib},4} = 961$ K. From Equation (15.30) we know that

$$U_{\text{vib},i} = Nh\nu_{0,i}\left(\frac{1}{2} + \frac{1}{e^{h\nu_{0,i}/(kT)} - 1}\right) = Nk\theta_{\text{vib},i}\left[\frac{1}{2} + \frac{1}{\exp\left(\theta_{\text{vib},i}/T\right) - 1}\right]$$

where we have used the relation $h\nu_{0,i}/k = \theta_{\text{vib},i}$. For the molar internal energy it follows that

$$U_{\text{vib},i,m} = \boldsymbol{N}_A k\theta_{\text{vib},i}\left(\frac{1}{2} + \frac{1}{e^{\theta_{\text{vib},i}/T} - 1}\right) = \boldsymbol{R}\theta_i\left(\frac{1}{2} + \frac{1}{e^{\theta_{\text{vib},i}/T} - 1}\right)$$

Here we focus on only the temperature-dependent part of the internal energy and ignore the zero point energy in each mode. Hence the total internal vibrational energy will be the sum of these three pieces, so that

$$U_{\text{vib},m} = \frac{\boldsymbol{R}\theta_{\text{vib},1}}{\exp\left(\theta_{\text{vib},1}/T\right) - 1} + \frac{\boldsymbol{R}\theta_{\text{vib},2}}{\exp\left(\theta_{\text{vib},2}/T\right) - 1} + \frac{\boldsymbol{R}\theta_{\text{vib},3}}{\exp\left(\theta_{\text{vib},3}/T\right) - 1} + \frac{\boldsymbol{R}\theta_{\text{vib},4}}{\exp\left(\theta_{\text{vib},4}/T\right) - 1}$$

This expression converges to the classical limit of

$$U_{\text{vib,m,classical}} = \boldsymbol{R}T + \boldsymbol{R}T + \boldsymbol{R}T + \boldsymbol{R}T = 4\boldsymbol{R}T$$

as T becomes very large. At 300 K, we find that $U_{\text{vib,m,1}} = 1692\ \boldsymbol{R}$ K, $U_{\text{vib,m,2}} = 1002\ \boldsymbol{R}$ K, $U_{\text{vib,m,3}} = 521\ \boldsymbol{R}$ K, and $U_{\text{vib,m,4}} = 521\ \boldsymbol{R}$ K, so that

$$\begin{aligned} U_{\text{vib,m}}(300\ \text{K}) &= 0.3551\ \text{J/mol} + 21.25\ \text{J/mol} + 338.3\ \text{J/mol} + 338.3\ \text{J/mol} \\ &= 698.2\ \text{J/mol} \end{aligned}$$

In a similar manner we find

$$\begin{aligned} U_{\text{vib,m}}(2000K) &= 6350\ \text{J/mol} + 9680\ \text{J/mol} + 12952\ \text{J/mol} + 12952\ \text{J/mol} \\ &= 41.93\ \text{kJ/mol} \end{aligned}$$

The classical limit of $4\boldsymbol{R}T$ is 9.977 kJ/mol at 300 K and is 66.51 kJ/mol at 2000 K. While CO_2 has not reached the classical limit at either of these temperatures, the low frequency bending mode has an internal energy of 12.95 kJ/mol which is 78% of the classical limit of $\boldsymbol{R}T = 16.62$ kJ/mol at 2000 K.

Figure P15.11 plots the molar vibrational internal energy (with the zero point energy removed) as a function of temperature.

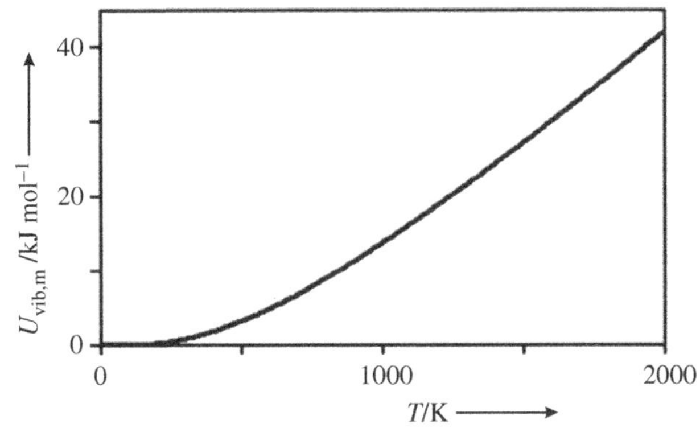

Figure P15.11 $U_{vib,m}$ of CO_2 versus the temperature T.

P15.12 In Chapter 14 (Box 14.2) we calculated the pressure of a gas and the number of collisions with a wall by including the different directions of motion of the gas molecules, but we assumed that the molecules are moving with equal speeds. Following the derivation in Box 14.2 and using the Maxwell–Boltzmann distribution (Equation (15.64)) show that in the expression for the pressure the square of the speed v^2 has to be replaced by its mean value $\overline{v^2}$ and in the expression for the number of collisions the speed v has to be replaced by its mean value $\overline{v}$.

Solution

First we consider the pressure calculation. In Box 14.2, we calculate the force $f(\vartheta)$ exerted on the wall from collisions with N gas particles moving at speed v and angle θ

$$f(\vartheta) = \left(\frac{N}{V}L^2 v\cos\vartheta\,\Delta t\right)\left(\frac{2v\cos\vartheta}{\Delta t}\right)$$

In our more sophisticated treatment, we replace N by $\mathrm{d}N(v, v+\mathrm{d}v)$ and find that

$$f(\vartheta, \mathrm{d}N) = \left(\frac{\mathrm{d}N(v, v+\mathrm{d}v)}{V}L^2 v\cos\vartheta\,\Delta t\right)\left(\frac{2v\cos\vartheta}{\Delta t}\right)$$

This force is that imparted to the wall by the subset of gas particles that are traveling with a speed between v and v+dv. From this expression we can determine the average force imparted by summing (integrating) over all the possible speeds; namely

$$\begin{aligned} f(\vartheta) &= \int_0^\infty f(\vartheta, \mathrm{d}N) \\ &= \frac{2\cos^2\vartheta}{V}L^2\int_0^\infty v^2\,\mathrm{d}N(v, v+\mathrm{d}v) \\ &= \frac{2\cos^2\vartheta}{V}L^2\,\overline{v^2} \end{aligned}$$

where we have used the definition of the mean square velocity. The rest of the derivation proceeds as in Box 14.2.

Next we consider the derivation for the collision frequency in Box 14.2. In this case the number of collisions with the wall in a time Δt and a particle density of N/V was written as

$$\Delta N = \int_0^{\pi/2}\frac{N}{V}L^2 v\cos\vartheta\,\Delta t\cdot\frac{1}{2}\sin\vartheta\cdot\mathrm{d}\vartheta = \frac{1}{4}\frac{N}{V}L^2 v\Delta t$$

Once again we can generalize this result to consider only the subpopulation of particles moving with a speed between v and v+dv and write

$$\begin{aligned} \Delta N(v, v+\mathrm{d}v) &= \int_0^{\pi/2}\frac{\mathrm{d}N(v, v+\mathrm{d}v)}{V}L^2 v\cos\vartheta\,\Delta t\cdot\frac{1}{2}\sin\vartheta\cdot\mathrm{d}\vartheta \\ &= \frac{1}{4}\frac{\mathrm{d}N(v, v+\mathrm{d}v)}{V}L^2 v\Delta t \end{aligned}$$

To account for all of the gas particles in the box, we must sum (integrate) over all possible speeds and find

$$\Delta N = \int_0^\infty \Delta N(v, v + dv)$$
$$= \frac{1}{4V} L^2 \Delta t \int_0^\infty v \, dN(v, v + dv) = \frac{1}{4V} L^2 \Delta t \, \overline{v}$$

where we have used the definition of the mean speed.

P15.13 Figure 15.4 plots the temperature dependence of the electronic internal energy of NO, using the two-state approximation. Actually the ground and first excited state levels are each doubly degenerate, perform the analysis for NO using this fact and show that the formulas for N_1, N_2, and U_{el} are the same as those given in the text. Calculate and plot the populations for the ground state N_1 and the excited state N_2 over the temperature range of 0 K to 1000 K for NO; i.e., a plot analogous to that shown for Na in Fig. 15.3. Perform a similar calculation for a system which is non-degenerate in the ground state, but three-fold degenerate in the excited state.

Solution

We begin with the general formula for the number of particles in a given energy level, namely

$$N_i = \frac{N}{Z} \cdot g_i \cdot \exp\left(-\frac{E_i}{kT}\right)$$

where

$$Z = \sum_i g_i \cdot \exp\left(-\frac{E_i}{kT}\right)$$

For the two-level system of NO we have that

$$N_1 = 2\frac{N}{Z} \cdot \exp\left(-\frac{E_1}{kT}\right) \quad \text{and} \quad N_2 = 2\frac{N}{Z} \cdot \exp\left(-\frac{E_2}{kT}\right)$$

where

$$Z = 2\left[\exp\left(-\frac{E_1}{kT}\right) + \exp\left(-\frac{E_2}{kT}\right)\right]$$

Hence we find that

$$N_1 = \frac{2N \cdot \exp\left(-\frac{E_1}{kT}\right)}{2\left[\exp\left(-\frac{E_1}{kT}\right) + \exp\left(-\frac{E_2}{kT}\right)\right]} = \frac{N}{\left[1 + \exp\left(-\frac{E_2 - E_1}{kT}\right)\right]}$$

and

$$N_2 = \frac{2N \cdot \exp\left(-\frac{E_2}{kT}\right)}{2\left[\exp\left(-\frac{E_1}{kT}\right) + \exp\left(-\frac{E_2}{kT}\right)\right]} = \frac{N}{\left[1 + \exp\left(\frac{E_2 - E_1}{kT}\right)\right]}$$

and

$$U_{\text{el}} = E_1 N_1 + E_2 N_2$$
$$= N\left[\frac{E_1}{\left[1 + \exp\left(-\frac{E_2 - E_1}{kT}\right)\right]} + \frac{E_2}{\left[1 + \exp\left(\frac{E_2 - E_1}{kT}\right)\right]}\right]$$

These equations are the same as those given in the text (e.g., see Equation (15.18)), with $\Delta E = E_2 - E_1$. For NO we find that $\Delta E = E_2 - E_1 = 2.40 \times 10^{-21}$ J. Figure P15.13 plots the internal energy versus temperature.

Figure P15.13 Internal energy U of a two-level system (NO molecules) versus temperature T. The dashed line marks the final value $U = 1.2 \times 10^{12}$ J for $T \to \infty$.

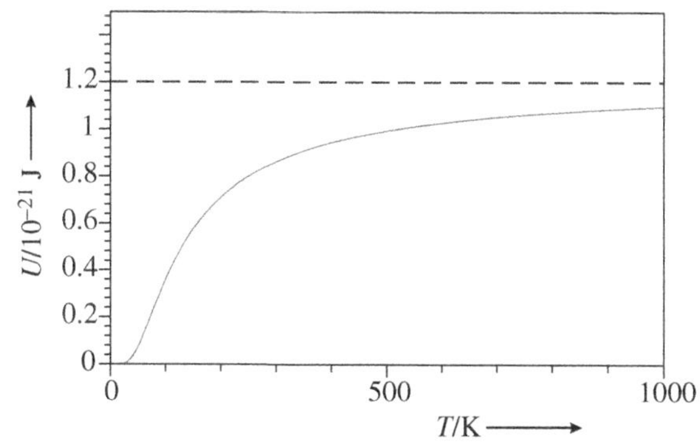

For a system that has a degeneracy of one in the ground state and three in the excited state, we proceed in a similar manner and find that

$$N_1 = \frac{N \cdot \exp\left(-\frac{E_1}{kT}\right)}{\left[\exp\left(-\frac{E_1}{kT}\right) + 3\exp\left(-\frac{E_2}{kT}\right)\right]} = \frac{N}{\left[1 + 3\exp\left(-\frac{E_2-E_1}{kT}\right)\right]}$$

and

$$N_2 = \frac{3N \cdot \exp\left(-\frac{E_2}{kT}\right)}{\left[\exp\left(-\frac{E_1}{kT}\right) + 3\exp\left(-\frac{E_2}{kT}\right)\right]} = \frac{3N}{\left[1 + 3\exp\left(\frac{E_2-E_1}{kT}\right)\right]}$$

and

$$U_{\text{el}} = E_1 N_1 + E_2 N_2$$

$$= N\left[\frac{E_1}{\left[1 + 3\exp\left(-\frac{E_2-E_1}{kT}\right)\right]} + \frac{3E_2}{\left[1 + 3\exp\left(\frac{E_2-E_1}{kT}\right)\right]}\right]$$

P15.14 Show that the expression for ω in Equation (15.79) (written in terms of the quantum states l) can be written for the rotational motion in terms of the quantum number J, namely

$$\omega = N! \cdot \frac{g_1^{n_1} \cdot g_2^{n_2} \cdot g_3^{n_3} \cdots}{N_1! \cdot N_2! \cdot N_3! \cdots}$$

where N is the total number of particles, g_J is the degeneracy of energy level with quantum number J, and N_J is the number of particles with quantum number J.

Solution

Let us distribute N particles on g energy states and let us start with $g = 2$. For $N = 1$ we obtain two representations ($\omega = 2$: particle 1 on left or on right energy state). For $N = 2$ we obtain $\omega = 4$, and for $N = 3$ we expect $\omega = 8$: (1|23), (2|13), (3|12),(12|3), (13|2), (23|1), (123|), (|123). Using the same logic we can find the number of representations for any value of N, and we deduce that the number of representations is

$$\omega = 2^N$$

Accordingly, if we have three degenerate energy states ($g = 3$), then we find $\omega = 3$ for $N = 1$, and $\omega = 9$ for $N = 2$, and we can describe the number of representations as

$$\omega = 3^N$$

for an arbitrary number of particles. By deduction, we see that for N particles occupying g degenerate states, we have

$$\omega = g^N$$

Finally, we generalize the consideration for the case that N_1 particles are in energy level 1 that has a degeneracy g_1, N_2 particles are in level 2 with degeneracy g_2, and so on. Then the total number of representations is described by

$$\begin{aligned}\omega &= \frac{N!}{N_1! \cdot N_2! \cdot N_3! \cdots} \cdot g_1^{N_1} \cdot g_2^{N_2} \cdot g_3^{N_3} \cdots \\ &= N! \cdot \frac{g_1^{N_1} \cdot g_2^{N_2} \cdot g_3^{N_3} \cdots}{N_1! \cdot N_2! \cdot N_3! \cdots}.\end{aligned}$$

This result is equivalent to Equation (15.79); whereas the numbering in Equation (15.79) is over the occupied quantum states, the numbering here is over the energy levels.

To demonstrate this equivalence we repeat the derivation of the Boltzmann law, starting with Equation (15.81) in Section 15.8.2.

$$\frac{\partial \ln \omega}{\partial N_i} - \alpha - \beta E_i = 0$$

Using our result for ω, we obtain

$$\begin{aligned}\ln \omega = &\ln N! + N_1 \ln g_1 + N_2 \ln g_2 + N_3 \ln g_3 + \cdots \\ &- \ln N_1! - \ln N_2! - \ln N_3! - \cdots\end{aligned}$$

Next we make use of Stirling's formula

$$\ln N! = N \cdot \ln N - N$$

and find that

$$\begin{aligned}\ln \omega = &(N \cdot \ln N - N) + N_1 \ln g_1 + N_2 \ln g_2 + N_3 \ln g_3 + \cdots \\ &- (N_1 \cdot \ln N_1 - N_1) - (N_2 \cdot \ln N_2 - N_2) - (N_3 \cdot \ln N_3 - N_3) - \cdots\end{aligned}$$

Evaluating the derivative with respect to N_1 we obtain

$$\begin{aligned}\frac{\partial \ln \omega}{\partial N_1} &= \ln g_1 - \ln N_1 - N_1 \cdot \frac{1}{N_1} + 1 \\ &= \ln g_1 - \ln N_1\end{aligned}$$

and with Equation (15.81) we can write

$$\ln g_1 - \ln N_1 - \alpha - \beta E_1 = 0$$

This result leads to

$$\ln \frac{N_1}{g_1} = \alpha + \beta E_1$$

and

$$\frac{N_1}{g_1} = e^{\alpha + \beta E_1}, \quad N_1 = g_1 \cdot e^{\alpha + \beta E_1}$$

As in Section 15.8.2 we generalize this equation for arbitrary N_i, g_i.

$$N_i = g_i \cdot e^{\alpha + \beta E_i}$$

Replacing β by $-1/(kT)$ we obtain

$$N_i = g_i \cdot e^{\alpha} e^{-E_i/(kT)}$$

This equation corresponds to Equation (15.34) in Section 15.5.1 when we applied the Boltzmann law to the degenerate energy levels of the rotational motion.

P15.15 Equation (15.112) for the number of photons emitted by a black body in the frequency range $\mathrm{d}\nu$ is written in terms of frequencies. Rewrite the equation in terms of wavelengths.

Solution

Equation (15.112) is

$$\mathrm{d}n(\nu) = \frac{8\pi}{c_0^3} V \frac{\nu^2}{\exp(h\nu/(kT) - 1} \mathrm{d}\nu$$

We use the relations

$$\nu = \frac{c_0}{\lambda} \qquad \frac{\mathrm{d}\nu}{\mathrm{d}\lambda} = -\frac{c_0}{\lambda^2}$$

to find that

$$\mathrm{d}n(\lambda) = \frac{8\pi}{c_0^3} V \frac{c_0^2/\lambda^2}{\exp(h\nu/(kT) - 1} \frac{c_0}{\lambda^2} \mathrm{d}\lambda$$
$$= 8\pi V \frac{1}{\exp(hc_0/(\lambda kT) - 1} \frac{1}{\lambda^4} \mathrm{d}\lambda$$

Accordingly, for the spectral energy density $\rho(\lambda)$ we obtain

$$\rho(\lambda) = \mathrm{d}n(\lambda)\, h\nu = \mathrm{d}n(\lambda)\, h\frac{c_0}{\lambda}$$
$$= 8\pi \frac{hc_0}{\exp(hc_0/(\lambda kT) - 1} \frac{1}{\lambda^5} \mathrm{d}\lambda$$

Figure P15.15 plots ρ as a function of the wavelength λ for different temperatures. Note that the maximum of the curves is shifted to higher frequencies with increasing temperature, in analogy to the curves for $\rho(\nu)$ in Fig. F15.26.

Note that the maximum for $T = 5800$ K is at $\lambda = 500$ nm. In the frequency representation of ρ in Fig. F15.26 the maximum for this temperature is at $\nu = 0.340 \times 10^{15}\ \mathrm{s}^{-1}$, that corresponds to a wavelength of

$$\lambda = \frac{c_0}{\nu} = \frac{2.998 \times 10^8 \mathrm{m\ s}^{-1}}{0.340 \times 10^{15} \mathrm{s}^{-1}} = 881 \mathrm{nm}$$

This difference is due to the fact that $\rho(\nu)$ measures the density in the range $\mathrm{d}\nu$ and $\rho(\lambda)$ measures in the range $\mathrm{d}\lambda$. Both ranges are different because of

$$\lambda = \frac{c_0}{\nu} \qquad \mathrm{d}\lambda = -\frac{c_0}{\upsilon^2}\, \mathrm{d}\nu$$

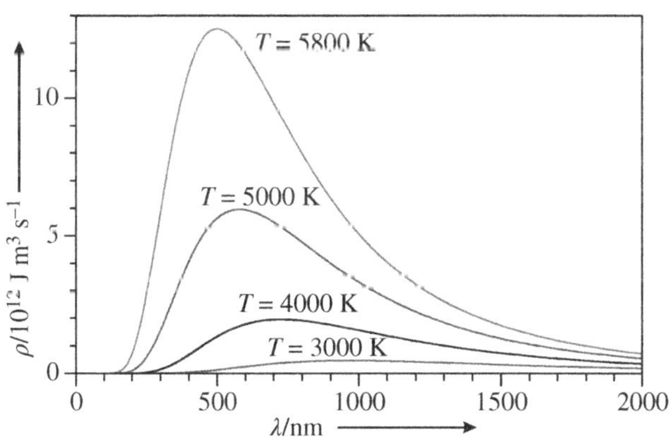

Figure P15.15 Planck's radiation law: spectral energy density $\rho(\lambda)$ versus the wavelength λ of the emitted radiation for different temperatures (5800 K is the temperature at the sun's surface).

P15.16 Generalize the one-dimensional wave equation (Equation 3.6) to three dimensions, in analogy to the Schrödinger Equation (3.25). Using this equation, derive the equation for the frequency ν of a three-dimensional standing wave in a cubic box with side length L,

$$\nu = \frac{\upsilon}{2L}\sqrt{n_x^2 + n_y^2 + n_z^2}$$

Solution

Equation (3.6) is

$$\frac{d^2\psi(x)}{dx^2} = -\frac{4\pi^2}{\Lambda^2}\psi(x)$$

Following the procedure in Chapter 3 for the three-dimensional Schrödinger equation we use the equation

$$\frac{\partial^2\psi(x,y,z)}{\partial x^2} + \frac{\partial^2\psi(x,y,z)}{\partial y^2} + \frac{\partial^2\psi(x,y,z)}{\partial z^2} = -\frac{4\pi^2}{\Lambda^2}\psi(x,y,z)$$

and solve it by the ansatz

$$\psi(x,y,z) = A\,\sin\left(\frac{n_x\pi}{L}x\right)\sin\left(\frac{n_y\pi}{L}y\right)\sin\left(\frac{n_z\pi}{L}z\right)$$

so that

$$\frac{\partial\psi(x,y,z)}{\partial x} = -A\,\cos\left(\frac{n_x\pi}{L}x\right)\left(\frac{n_x\pi}{L}\right)\sin\left(\frac{n_y\pi}{L}y\right)\sin\left(\frac{n_z\pi}{L}z\right)$$

$$\frac{\partial\psi(x,y,z)}{\partial y} = -A\,\sin\left(\frac{n_x\pi}{L}x\right)\cos\left(\frac{n_y\pi}{L}y\right)\left(\frac{n_y\pi}{L}\right)\sin\left(\frac{n_z\pi}{L}z\right)$$

$$\frac{\partial\psi(x,y,z)}{\partial z} = -A\,\sin\left(\frac{n_x\pi}{L}x\right)\sin\left(\frac{n_y\pi}{L}y\right)\cos\left(\frac{n_z\pi}{L}z\right)\left(\frac{n_z\pi}{L}\right)$$

Thus, the second derivatives become

$$\frac{\partial^2\psi(x,y,z)}{\partial x^2} = -A\,\sin\left(\frac{n_x\pi}{L}x\right)\left(\frac{n_x\pi}{L}\right)^2\sin\left(\frac{n_y\pi}{L}y\right)\sin\left(\frac{n_z\pi}{L}z\right) = -\left(\frac{n_x\pi}{L}\right)^2\psi(x,y,z)$$

and

$$\frac{\partial^2\psi(x,y,z)}{\partial y^2} = -\left(\frac{n_y\pi}{L}\right)^2\psi(x,y,z) \qquad \frac{\partial^2\psi(x,y,z)}{\partial z^2} = -\left(\frac{n_z\pi}{L}\right)^2\psi(x,y,z)$$

Thus we obtain

$$\frac{\partial^2\psi(x,y,z)}{\partial x^2} + \frac{\partial^2\psi(x,y,z)}{\partial y^2} + \frac{\partial^2\psi(x,y,z)}{\partial z^2} = \psi(x,y,z)\left[\left(\frac{n_x\pi}{L}\right)^2 + \left(\frac{n_y\pi}{L}\right)^2 + \left(\frac{n_z\pi}{L}\right)^2\right]$$

and

$$\left(\frac{n_x\pi}{L}\right)^2 + \left(\frac{n_y\pi}{L}\right)^2 + \left(\frac{n_z\pi}{L}\right)^2 = \frac{4\pi^2}{\Lambda^2}$$

$$\frac{\pi^2}{L^2}\left(n_x^2 + n_y^2 + n_z^2\right) = \frac{4\pi^2}{\Lambda^2}$$

$$\frac{1}{\Lambda^2} = \frac{1}{4L^2}\left(n_x^2 + n_y^2 + n_z^2\right)$$

We need an expression for the frequency $\nu = \upsilon/\Lambda$, where υ is the speed of the wave. Thus

$$\upsilon = \upsilon\frac{1}{\Lambda} = \frac{\upsilon}{2L}\sqrt{n_x^2 + n_y^2 + n_z^2}$$

P15.17 Use Equation (11.9) in Section 11.2.2 for the rotational energy levels of symmetric top polyatomic molecules.

$$E_{J,K} = B \cdot J(J+1) + (A-B)\cdot K^2$$

where $|K| \le J, K = 0, \pm1, \pm2, \dots \pm J$; and A and B are the rotational constants

$$A = \frac{h^2}{8\pi^2 I_A},\ B = \frac{h^2}{8\pi^2 I_B}$$

to derive a formula for the rotational partition function Z_{rot} in the high temperature limit and show that the internal energy $U_{rot} = N\frac{3}{2}kT$.

Solution

For the partition function we obtain

$$\begin{aligned} Z_{\text{rot}} &= \sum_{J,K} (2J+1) \cdot e^{-E_{J,K}/(kT)} \\ &= \sum_{J,K} (2J+1) \cdot \exp\left[-\frac{B \cdot J(J+1) + (A-B) \cdot K^2}{kT}\right] \\ &= \sum_{J,K} (2J+1) \cdot \exp\left[-\frac{B \cdot J(J+1)}{kT}\right] \cdot \exp\left[-\frac{(A-B) \cdot K^2}{kT}\right] \end{aligned}$$

For sufficiently high temperature T we can replace the summation by an integration

$$Z_{\text{rot}} = \int_{J=0}^{\infty} (2J+1) \cdot \exp\left[-\frac{B \cdot J(J+1)}{kT}\right] \cdot \mathrm{d}J \cdot \int_{K=-\infty}^{\infty} \exp\left[-\frac{(A-B) \cdot K^2}{kT}\right] \cdot \mathrm{d}K$$

Note the simplification that we integrate from $K = -\infty$ to $K = +\infty$ instead of caring for the condition that $|K| \leq J$. This means the integral over K is overestimated. The first integral is solved by the substitution

$$x = \frac{B}{kT} J(J+1),\ \mathrm{d}x = \frac{B}{kT}(2J+1) \cdot \mathrm{d}J$$

$$\begin{aligned} &\int_{J=0}^{\infty} (2J+1) \cdot \exp\left[-(B \cdot J(J+1))/(kT)\right] \cdot \mathrm{d}J \\ &= \int_{x=0}^{\infty} (2J+1) \cdot \exp[-x] \frac{kT}{B(2J+1)} \cdot \mathrm{d}x \\ &= \frac{kT}{B} \int_{x=0}^{\infty} \cdot \exp[-x] \cdot \mathrm{d}x = \frac{kT}{B} \end{aligned}$$

The second integral is solved by the substitution

$$x^2 = \frac{A-B}{kT} K^2,\ \mathrm{d}x = \sqrt{\frac{A-B}{kT}} \cdot \mathrm{d}K$$

$$\begin{aligned} &\int_{K=-\infty}^{\infty} \exp\left[-\left((A-B) \cdot K^2\right)/(kT)\right] \cdot \mathrm{d}K \\ &= \sqrt{\frac{kT}{A-B}} \int_{x=-\infty}^{\infty} \exp\left[-x^2\right] \cdot \mathrm{d}x = \sqrt{\frac{kT}{A-B}} \cdot \sqrt{\pi} \end{aligned}$$

where we used Appendix B. Thus for the partition function we obtain

$$Z_{\text{rot}} = \frac{kT}{B} \cdot \sqrt{\frac{kT\pi}{A-B}}$$

The internal energy U_{rot} can be determined by using Equation (15.13) in Section 15.2.3

$$\begin{aligned} U_{\text{rot}} &= NkT^2 \frac{\mathrm{d} \ln Z_{\text{rot}}}{\mathrm{d}T} = NkT^2 \frac{\mathrm{d} \ln \left(\frac{k}{B} \sqrt{\frac{k\pi}{A-B}} \cdot T^{3/2} \right)}{\mathrm{d}T} \\ &= NkT^2 \cdot \frac{3}{2}\frac{1}{T} = N\frac{3}{2}kT \end{aligned}$$

This is what we expect for a rotator with three degrees of freedom, namely, $3N\frac{kT}{2}$.

P15.18 Calculate the occupation probability for He gas in the lowest translational quantum state in a 1 L cubic box at $T = 300$ K at $P = 1$ bar

Solution

According to the Boltzmann distribution in Section 15.6.1 the occupation probability $\mathcal{P}_i$ of a gas molecule in quantum state i is

$$\mathcal{P}_i = \frac{1}{Z_{\text{trans}}} \cdot e^{-E_i/(kT)}$$

where Z is the partition function. For the lowest translational quantum state ($i = 1$) we can set $E_1 = 0$. Thus we obtain

$$\mathcal{P}_1 = \frac{1}{Z_{\text{trans}}} = \frac{1}{V}\left(\frac{h^2}{2\pi mkT}\right)^{3/2}$$

We consider He gas at $T = 300$ K at $P = 1$ bar in a volume $V = 1000$ mL. Then we find

$$\mathcal{P}_1 = \frac{1}{1000 \times 10^{-6}\,\text{m}^3} \cdot \left(\frac{(6.626 \times 10^{-34}\ \text{J s})^2}{2\pi \cdot 6.64 \times 10^{-27}\ \text{kg} \cdot 1.38 \times 10^{-23}\ \text{J} \cdot 300\ \text{K}}\right)^{3/2}$$
$$= 1.68 \times 10^{-26}$$

and the number of gas particles in the lowest quantum state is

$$N_1 = N \cdot \mathcal{P}_1$$

Because of

$$N = \frac{PV}{kT} = \frac{1\ \text{bar} \cdot 1\ \text{L}}{1.38 \times 10^{-23}\text{J K}^{-1} \cdot 300\ \text{K}}$$
$$= \frac{100\ \text{J}}{1.38 \times 10^{-23} \cdot 300\ \text{J}} = 2.4 \times 10^{22}$$

we find

$$N_1 = 2.4 \times 10^{22} \cdot 1.68 \times 10^{-26} = 4 \times 10^{-4}$$

This means that the lowest quantum state is practically unoccupied and most of the gas atoms occupy higher quantum states.

16

Work w, Heat q, and Internal Energy U

16.1 Exercises

E16.1 Calculate the relative populations of the ground electronic energy level and the first excited electronic energy level for He ($\Delta E = E_2 - E_1 = 31.7 \times 10^{-19}$ J) at the temperatures 300 and 1000 K. (Note: You may find it helpful to use the fact that $e^{-a} = 10^{-a/\ln 10}$, rather than your calculator directly.)

Solution

This exercise examines the Boltzmann distribution of the helium atom between two electronic states. For the population ratio (N_2/N_1) between the ground and excited energy levels, the Boltzmann distribution gives

$$\frac{N_2}{N_1} = \frac{g_2}{g_1} e^{-(E_2-E_1)/(kT)}$$

where g_1 and g_2 are the degeneracies of the ground and excited energy levels, respectively. For He the ground state is a singlet and the first excited state is a triplet so that $g_2/g_1 = 3$. At 300 K we find that

$$\frac{E_2 - E_1}{kT} = \frac{31.7 \times 10^{-19}\ \mathrm{J}}{1.3806 \times 10^{-23}\ \mathrm{J\ K^{-1}} \times 300\ \mathrm{K}} = 765.4$$

and

$$\frac{N_2}{N_1} = 3 \cdot e^{-765.4} = 3 \cdot 3.90 \times 10^{-333} = 1.17 \times 10^{-332}$$

where we have used the fact that

$$e^{-a} = 10^{-a/\ln 10}$$

and at 1000 K we have

$$\frac{N_2}{N_1} = 3 \cdot e^{-229.6} = 3 \cdot 1.91 \times 10^{-100} = 5.74 \times 10^{-100}$$

Hence it is extremely unlikely to find excited states of He atoms in an equilibrated He gas at these temperatures.

E16.2 Calculate the relative populations of the ground electronic state and the first excited singlet electronic state for H_2 ($\Delta E = 18.3 \times 10^{-19}$ J) at the temperatures 300 and 1000 K.

Solution

This exercise examines the Boltzmann distribution of the hydrogen molecule between two electronic states. Because both the ground state and the excited states are singlets, the Boltzmann distribution gives a population ratio of

$$\frac{N_2}{N_1} = e^{-(E_2-E_1)/(kT)}$$

Solutions Manual for Principles of Physical Chemistry, Third Edition. Edited by Hans Kuhn, David H. Waldeck, and Horst-Dieter Försterling.

At 300 K we find that

$$\frac{\Delta E}{kT} = \frac{18.3 \times 10^{-19}\ \mathrm{J}}{1.3806 \times 10^{-23}\ \mathrm{J\ K^{-1}} \times 300\ \mathrm{K}} = 442 \quad \text{and} \quad \frac{N_2}{N_1} = \mathrm{e}^{-442} = 1.10 \times 10^{-192}$$

and at 1000 K we find that

$$\frac{N_2}{N_1} = \mathrm{e}^{-18.3\times10^{-19}/(1.3806\times10^{-23}\times1000)} = 2.72 \times 10^{-58}$$

This ratio is much larger than that found for helium in Exercise 16.1. However, it is still extremely unlikely to find excited electronic states of H_2 in an equilibrated H_2 gas at these temperatures.

E16.3 Consider the NO radical whose ground electronic state and first excited electronic state are separated by $\Delta E = 2.4 \times 10^{-21}$ J. How much does the inclusion of electronic excitation modify the heat capacity of NO from that for a diatomic molecule that does not include electronic excitation, at 300 and 1000 K? Compute a percentage error.

Solution

This exercise examines the effect of electronic excitation on the heat capacity of NO. In the approximation that the thermal excitation of the vibrational and electronic excitation can be neglected, the heat capacity of NO would be

$$C_{V,m} = \frac{3}{2}\boldsymbol{R} + \boldsymbol{R} = \frac{5}{2}\boldsymbol{R}$$

where we find a contribution of $\frac{1}{2}\boldsymbol{R}$ for each of the three translational degrees of freedom and each of the two rotational degrees of freedom.

Now we consider the case where the electronic degrees of freedom are included. Using Equation (16.27),

$$C_{V,\mathrm{el},m} = \left(\frac{\mathrm{d}U_{\mathrm{el},m}}{\mathrm{d}T}\right)_V = \boldsymbol{N}_A \frac{(\Delta E)^2}{kT^2} \frac{\mathrm{e}^{-\Delta E/kT}}{\left(1 + \mathrm{e}^{-\Delta E/kT}\right)^2}$$

we can directly evaluate the electronic contribution to the molar heat capacity. Inserting the value for the NO molecule's energy gap, we find that $\Delta E/k = 174$ K, so that

$$C_{V,\mathrm{el},m} = \boldsymbol{R}\left(\frac{174\ \mathrm{K}}{T}\right)^2 \frac{\exp\left[-174\ \mathrm{K}/T\right]}{\left(1 + \exp\left[-174\ \mathrm{K}/T\right]\right)^2}$$

Thus for $T = 300$ K we find with $174\ \mathrm{K}/300\ \mathrm{K} = 0.58$

$$\begin{aligned} C_{V,\mathrm{el},m} &= \boldsymbol{R} \cdot (0.58)^2 \frac{\exp[-0.58]}{(1 + \exp[-0.58])^2} = \boldsymbol{R} \cdot (0.58)^2 \frac{\exp[-0.58]}{(1 + \exp[-0.58])^2} \\ &= \boldsymbol{R} \cdot 0.336 \frac{0.56}{(1.56)^2} = \boldsymbol{R} \cdot 0.077 \end{aligned}$$

Accordingly, for $T = 1000\ K$ we find $C_{V,\mathrm{el},m} = 0.0075\,\boldsymbol{R}$.

To find the error in neglecting the electronic contribution, we calculate

$$\frac{C_{V,\mathrm{el},m}}{\frac{5}{2}\boldsymbol{R} + C_{V,\mathrm{el},m}} = \frac{0.077}{2.5 + 0.077} = 0.03 = 3\% \text{ for } T = 300\ \mathrm{K}$$

and

$$\frac{C_{V,\mathrm{el},m}}{\frac{5}{2}\boldsymbol{R} + C_{V,\mathrm{el},m}} = \frac{0.0075}{2.5 + 0.0075} = 0.003 = 0.3\% \text{ for } T = 1000\ \mathrm{K}$$

From this calculation it appears that our approximate solution (neglect of the electronic excitation) is more accurate at 1000 K than at 300 K. This occurs because the energy levels are nearly equally populated at the higher temperatures.

E16.4 The molar constant volume heat capacities $C_{V,m}$ of Ne and Xe are the same at $T = 300$ K. What is the value of $C_{V,m}$? Why are their heat capacities the same?

Solution

The molar constant volume heat capacities $C_{V,m}$ of Ne and Xe are the same at $T = 300$ K because Ne and Xe are monatomic gases and electronic excitations are not accessible at typical temperatures. Hence, only the translational degrees of freedom contribute and their molar heat capacity is $C_{V,m} = \frac{3}{2}\boldsymbol{R}$.

E16.5 The molar heat capacity of a diatomic gas is always higher than that of a monatomic gas. Why is the molar heat capacity of a diatomic gas higher and what range of values may it take?

Solution

The molar heat capacity of a diatomic gas is higher because the diatomic molecules have rotational and vibrational degrees of freedom that can contribute to the heat capacity. If we neglect the contribution of electronic excitations for the monatomic and diatomic gases, then the molar heat capacity of a monatomic gas is $C_{V,m} = \frac{3}{2}\boldsymbol{R}$, whereas $C_{V,m}$ ranges from $\frac{5}{2}\boldsymbol{R}$ to $\frac{7}{2}\boldsymbol{R}$, depending on T and θ_{vib}.

E16.6 The heat capacity of molecules is temperature dependent, even for an ideal gas of molecules. Explain why this is so.

Solution

The temperature dependence of a molecule's heat capacity arises from the excitation of vibrational degrees of freedom and/or electronic degrees of freedom (note that in some cases at low temperatures the change in rotational energy level populations with temperature must be accounted for also). While it can sometimes be approximated as constant over a limited range of values, for wide temperature ranges its temperature dependence must be accounted for in accurate calculations.

E16.7 [Adapted from R.E. Dickerson, *Molecular Thermodynamics* (Benjamin, 1969)]. *The Setting*: You have two identical containers and they each contain 1 mol of a gas at the same pressure and temperature. In one of the containers you know that the gas is nitrogen, which behaves ideally. In addition, you know that the second container has a pure substance X which behaves as an ideal gas, but you need to determine its identity.
The Clues: a) At 100 K, when vibrational modes are not excited both nitrogen and the unknown gas have the same heat capacity. b) At very high temperatures (when vibrational modes are fully excited) it is found that the molar heat capacity of nitrogen is $C_{V,m} = 29.1\ \text{J K}^{-1}\ \text{mol}^{-1}$, whereas that of unknown gas is $C_{V,m} = 54.0\ \text{J K}^{-1}\ \text{mol}^{-1}$.
The Question: What can you say about the molecular structure and properties of unknown gas? Explain your reasoning.

Solution

We conclude that X must be a linear triatomic gas. Clue a) implies that gas X is a linear molecule like N_2 and at 100 K its molar heat capacity is $\frac{5}{2}\boldsymbol{R}$, of which $\frac{3}{2}\boldsymbol{R}$ is from translational degrees of freedom and $\boldsymbol{R}$ is from rotational degrees of freedom. The heat capacity is $20.78\ \text{J K}^{-1}\ \text{mol}^{-1}$ at this low temperature. Clue b) shows that N_2 has 1 fully excited vibrational mode, because $\left(29.1 - 20.78\ \text{J K}^{-1}\ \text{mol}^{-1}\right) = 8.3\ \text{J K}^{-1}\ \text{mol}^{-1} = \boldsymbol{R}$, which is the contribution to the heat capacity that we expect for a fully excited vibrational mode. Correspondingly, we conclude that the molecule X must have 4 modes, since

$$\frac{54.0 - 20.78\ \text{J K}^{-1}\ \text{mol}^{-1}}{\boldsymbol{R}} = 4.0$$

Therefore molecule X must be linear and have four vibrational modes; hence, it is a linear triatomic. For example, it might be CO_2 (symmetric and antisymmetric stretching modes, 2 bending modes).

E16.8 Consider the molecule HF which has a characteristic vibrational frequency of $1.24 \times 10^{14}\ \text{s}^{-1}$ and a rotational constant $B = 4.2 \times 10^{-22}$ J. Sketch the constant pressure molar heat capacity $C_{P,m}$ versus temperature of this gas from a temperature of 200 K to a temperature of 1000 K. On the same graph sketch the temperature dependence of the heat capacity of Ne. Explain your reasoning.

Solution

The characteristic temperature of rotation is given by

$$\theta_{\text{rot}} = \frac{B}{k} = \frac{4.2 \times 10^{-22}\ \text{J}}{1.3806 \times 10^{-23}\ \text{J K}^{-1}} = 30\ \text{K}$$

and the characteristic temperature of vibration is given by

$$\theta_{\text{vib}} = \frac{h\nu_0}{k} = \frac{\left(6.626 \times 10^{-34}\ \text{J s}\right)\left(1.24 \times 10^{14}\ \text{s}^{-1}\right)}{1.3806 \times 10^{-23}\ \text{J K}^{-1}} = 5951\ \text{K}$$

The characteristic temperature of rotation is small enough that we can consider the classical limit for rotational contributions to the heat capacity, and the constant volume molar heat capacity as a function of temperature is given by

$$C_{V,m}(T) = \frac{5}{2}\boldsymbol{R} + \left(\frac{\theta_{\text{vib}}}{T}\right)^2 \frac{\exp\left(-\theta_{\text{vib}}/T\right)}{\left(1 - \exp\left(-\theta_{\text{vib}}/T\right)\right)^2}\boldsymbol{R}$$

Refer to Equation (16.23) for the vibrational contribution. Assuming that the gas behaves ideally we can write the constant pressure molar heat capacity as

$$C_{P,m}(T) = \boldsymbol{R} + C_{V,m}(T) = \frac{7}{2}\boldsymbol{R} + \left(\frac{\theta_{\text{vib}}}{T}\right)^2 \frac{\exp\left(-\theta_{\text{vib}}/T\right)}{\left(1 - \exp\left(-\theta_{\text{vib}}/T\right)\right)^2}\boldsymbol{R}$$

A sketch of $C_{p,m}$ versus T for HF (solid line) and Ne (dashed line) is shown in Fig. E16.8. For Ne we assume that only translational modes are important, so that $C_{p,m}$ is 2.5$\boldsymbol{R}$. For HF we assume that the vibrations are only weakly excited at low temperature so that the $C_{p,m}$ is very near 3.5$\boldsymbol{R}$ (three translational modes and two rotational modes), but that the vibrational contribution starts to become important as the gas is warmed. Because the final temperature of 1000 K is still much lower than the characteristic temperature of vibrational we expect that the vibrational modes will still be only partly excited.

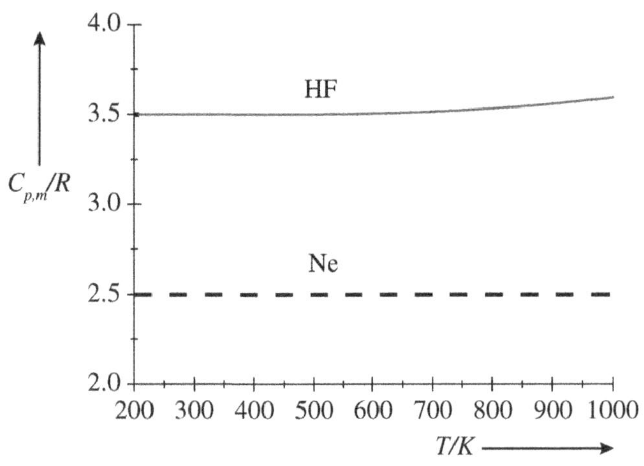

Figure E16.8 Temperature dependence of the molar heat capacity $C_{p,m}$ for HF and Ne.

E16.9 Why is the constant volume heat capacity always smaller than the constant pressure heat capacity?

Solution

When adding heat to a gas under constant pressure conditions, part of the energy must be used to perform the work of expanding the gas. For the constant volume case, no work is performed. For this reason (the additional energy required to expand the system) a gas at constant pressure has a higher heat capacity than the same gas held at constant volume.

E16.10 Consider the series of diatomic molecules HI, ICl, and I_2. Describe how the temperature-dependent constant volume heat capacities of these molecules differ under conditions where they behave as ideal gases. The vibrational frequencies are $\nu_0 = 6.92 \times 10^{13}\ s^{-1}$ for HI, $\nu_0 = 1.15 \times 10^{13}\ s^{-1}$ for ICl, and $\nu_0 = 0.64 \times 10^{13}\ s^{-1}$ for I_2; the rotational constants are $B = 128 \times 10^{-24}$ J for HI, $B = 2.26 \times 10^{-24}$ J for ICl, and $B = 0.74 \times 10^{-24}$ J for I_2. Use isotope 35 for Cl and 127 for I in your calculations.

Solution

At typical temperatures (300 K → 1000 K) the translational and rotational degrees of freedom will be fully excited; hence, they will each have a base value of $\frac{5}{2}\boldsymbol{R}$ for $C_{V,m}$. In addition, it is reasonable to neglect the electronic contribution. Given these considerations, the heat capacities of these gases are expected to differ primarily by differences in their vibrational contributions to $C_{V,m}$.

To assess the relative contributions of their vibrational mode to the molar heat capacity, we calculate their vibrational temperatures and compare them. For HI θ_{vib} is given by

$$\theta_{vib} = \frac{h\nu_0}{k} = \frac{6.626 \times 10^{-34}\ \text{J s} \cdot 6.92 \times 10^{13}\text{s}^{-1}}{1.3806 \times 10^{-23}\text{J K}^{-1}} = 3321\ \text{K}$$

In a corresponding way we find for ICI that θ_{vib} (ICl) = 552 K and for I_2 that $\theta_{vib}(I_2) = 307$ K. From this calculation we observe that

$$\theta_{vib}\,(\text{HI}) > \theta_{vib}\,(\text{ICl}) > \theta_{vib}\,(\text{I}_2)$$

Therefore we expect

$$C_{V,\text{vib},m}\,(\text{HI}) < C_{V,\text{vib},m}\,(\text{ICl}) < C_{V,\text{vib},m}\,(\text{I}_2)$$

at a given temperature, and hence

$$C_{V,m}\,(\text{HI}) < C_{V,m}\,(\text{ICl}) < C_{V,m}\,(\text{I}_2)$$

E16.11 Consider the molecules CH_4, CH_2Cl_2, and CCl_4. Describe how the temperature-dependent constant pressure heat capacities of these molecules differ under conditions where they behave as ideal gases.

Solution

These molecules are large enough that we expect the high temperature (classical) limit to apply for assessing the contributions of the translational and rotational degrees of freedom to the heat capacity. Assuming we can neglect the contributions from the electronic degrees of freedom, we can assess the differences in their heat capacities by the difference in their vibrational contributions to the heat capacity.

Because the molecules have similar structures and vibrational modes, we assess the difference in their vibrational modes by the differences in their bond strengths and their masses. For a given type of vibrational motion, we expect CCl_4 to have the lowest vibrational mode frequency, CH_2Cl_2 to be intermediate, and CH_4 to have the highest frequency. Using similar logic as in E16.10 we expect that

$$C_{P,\text{vib},m}\left(\text{CCl}_4\right) > C_{P,\text{vib},m}\left(\text{CH}_2\text{Cl}_2\right) > C_{P,\text{vib},m}\left(\text{CH}_4\right)$$

and

$$C_{P,m}\left(\text{CCl}_4\right) > C_{P,m}\left(\text{CH}_2\text{Cl}_2\right) > C_{P,m}\left(\text{CH}_4\right)$$

E16.12 Using $\Delta_f H^\ominus$ values at $T = 298$ K from Appendix A.1, calculate $\Delta_r H^\ominus$ for the combustion of diamond and of graphite. Use your result to determine the $\Delta_r H^\ominus$ for the formation of diamond from graphite. Perform the same calculation directly by using the heats of formation and compare your answers.

Solution

For the combustion of diamond ($C_{diamond} + O_2 \rightarrow CO_2$), we find that

$$\Delta_r H^\ominus(\text{C}_{\text{diamond}}) = \Delta_f H^\ominus(\text{CO}_2) - \Delta_f H^\ominus(\text{O}_2) - \Delta_f H^\ominus(\text{C}_{\text{diamond}}) = -393.5\ \text{kJ mol}^{-1} - 0 - 1.9\ \text{kJ mol}^{-1} = -395.4\ \text{kJ mol}^{-1}$$

For the combustion of graphite ($C_{graphite} + O_2 \rightarrow CO_2$), we find that

$$\begin{aligned}\Delta_r H^\ominus(C_{graphite}) &= \Delta_f H^\ominus(CO_2) - \Delta_f H^\ominus(O_2) - \Delta_f H^\ominus(C_{graphite}) \\ &= -393.5\ \text{kJ mol}^{-1} - 0 - 0 = -393.5\ \text{kJ mol}^{-1}\end{aligned}$$

Combining these two results we can calculate the heat of reaction for $C_{graphite} \rightarrow C_{diamond}$ as

$$\begin{aligned}\Delta_r H^\ominus &= \Delta_r H^\ominus(C_{graphite}) - \Delta_r H^\ominus(C_{diamond}) \\ &= -393.5\ \text{kJ mol}^{-1} + 395.4\ \text{kJ mol}^{-1} = 1.9\ \text{kJ mol}^{-1}\end{aligned}$$

If we use the heats of formation for $C_{graphite} \rightarrow C_{diamond}$ directly, we find that

$$\begin{aligned}\Delta_r H^\ominus &= \Delta_f H^\ominus(C_{diamond}) - \Delta_f H^\ominus(C_{graphite}) \\ &= 1.9\ \text{kJ mol}^{-1} - 0 = 1.9\ \text{kJ mol}^{-1}\end{aligned}$$

E16.13 Using $\Delta_f H^\ominus$ at $T = 298$ K calculate $\Delta_r H^\ominus$ for the hydrogenation of benzene and cyclohexene to cyclohexane. Use your result to determine the $\Delta_r H^\ominus$ for the hydrogenation of benzene to cyclohexene. Perform the same calculation directly by using the heats of formation and compare your answers.

Solution

We begin by determining $\Delta_r H^\ominus$ for the hydrogenation of benzene to cyclohexane using $\Delta_f H^\ominus$ at $T = 298$ K.

$$\begin{aligned}\Delta_r H^\ominus &= \Delta_f H^\ominus\,(\text{cyclohexane}) - \left[\Delta_f H^\ominus\,(\text{benzene}) + 3\Delta_f H^\ominus\,(H_2)\right] \\ &= -157.7\ \text{kJ mol}^{-1} - (48.95 + 3 \times 0)\ \text{kJ mol}^{-1} \\ &= -206.65\ \text{kJ mol}^{-1}\end{aligned}$$

Using $\Delta_f H^\ominus$ at $T = 298$ K we find for the hydrogenation of cyclohexene to cyclohexane that

$$\begin{aligned}\Delta_r H^\ominus &= \Delta_f H^\ominus\,(\text{cyclohexane}) - \left[\Delta_f H^\ominus\,(\text{cyclohexene}) + 3\Delta_f H^\ominus\,(H_2)\right] \\ &= -157.7\ \text{kJ mol}^{-1} - (-37.8 + 0)\ \text{kJ mol}^{-1} \\ &= -119.9\ \text{kJ mol}^{-1}\end{aligned}$$

The hydrogenation of benzene to cyclohexene is found by using Hess's Law

Reaction	$\frac{\Delta_r H^\ominus}{\text{kJ mol}^{-1}}$ at $T = 298$ K
Benzene + 3 H_2 $\rightarrow$ cyclohexane	−206.65
Cyclohexane $\rightarrow$ cyclohexene + H_2	119.9
Benzene + 2 H_2 $\rightarrow$ cyclohexene	−86.75

Using $\Delta_f H^\ominus$ at $T = 298$ K for cyclohexene and benzene, we find that

$$\begin{aligned}\Delta_r H^\ominus &= \Delta_f H^\ominus\,(\text{cyclohexene}) - \left[\Delta_f H^\ominus\,(\text{benzene}) + 2\Delta_f H^\ominus\,(H_2)\right] \\ &= -37.8\ \text{kJ mol}^{-1} - (48.95 + 2 \times 0)\ \text{kJ mol}^{-1} \\ &= -86.75\ \text{kJ mol}^{-1}\end{aligned}$$

As expected, the two different calculations agree exactly.

E16.14 Following an argument like that used for the internal energy, derive Equation (16.61) for the temperature dependence of the enthalpy.

Solution

We can use a cycle like that in Fig. 16.17 to quantify how the enthalpy of the reaction changes with temperature. Figure E16.14 shows this cycle.

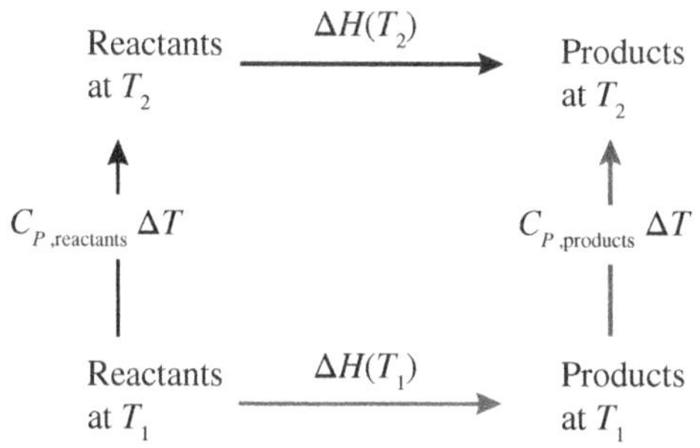

Figure E16.14 Cycle for determining reaction enthalpies.

To answer this question we imagine the reaction carried out at temperature T_1; after that the products are heated from T_1 to T_2. For this process the change of enthalpy is

$$\Delta H_{T_1} + \int_{T_1}^{T_2} C_{P,\text{products}} \cdot \mathrm{d}T$$

On the other hand, if the reactants are first heated from T_1 to T_2 and the reaction is carried out at temperature T_2, the change of enthalpy is

$$\Delta H_{T_2} + \int_{T_1}^{T_2} C_{P,\text{reactants}} \cdot \mathrm{d}T$$

Both these changes of enthalpy must be the same, so that

$$\Delta H_{T_1} + \int_{T_1}^{T_2} C_{P,\text{products}} \cdot \mathrm{d}T = \Delta H_{T_2} + \int_{T_1}^{T_2} C_{P,\text{reactants}} \cdot \mathrm{d}T$$

or

$$\boxed{\Delta H_{T_2} = \Delta H_{T_1} + \int_{T_1}^{T_2} \Delta C_P \cdot \mathrm{d}T}$$

where

$$\Delta C_P = C_{P,\text{products}} - C_{P,\text{reactants}}$$

E16.15 Use the bond enthalpies in Tables 16.5 and 16.6 to compute the heat of isomerization of cyclopropane to propylene. Comment on any difference between this value and the one found in E16.19.

Solution

Using bond enthalpies we have

Cyclopropane	Propylene
6 C–H	6 C–H
3 C–C	1 C–C
	1 C=C

hence we find that propylene has two fewer C–C bonds and one more C=C bond than does cyclopropane. So we have

$$\Delta_{\text{isom}}H = (614 - 2 \times 348)\ \text{kJ mol}^{-1} = -82\ \text{kJ mol}^{-1}$$

The method used in E16.19 is more accurate because it uses actual data for the specific molecule, whereas bond enthalpies are average values.

E16.16 Using $\Delta_f H^\ominus$ values at $T = 298$ K from Appendix A.2, calculate $\Delta_r H^\ominus$ for the combustion of methane and propane. Which substance generates the most heat per mol? Which generates the most heat per gram?

Solution

$\Delta_r H^\ominus$ for the combustion of methane is calculated as

$$\begin{aligned}\Delta_r H^\ominus &= \Delta_f H^\ominus (CO_2) + 2\Delta_f H^\ominus (H_2O) - [\Delta_f H^\ominus (CH_4) + 2\Delta_f H^\ominus (O_2)] \\ &= [-393.52 + 2(-241.83) - (-74.87 + 2 \times 0)] \text{ kJ mol}^{-1} \\ &= -802.31 \text{ kJ mol}^{-1}\end{aligned}$$

$\Delta_r H^\ominus$ for the combustion of propane is calculated as

$$\begin{aligned}\Delta_r H^\ominus &= 3\Delta_f H^\ominus (CO_2) + 4\Delta_f H^\ominus (H_2O) - [\Delta_f H^\ominus (C_3H_8) + 5\Delta_f H^\ominus (O_2)] \\ &= [3(-393.52) + 4(-241.83) - (-104.7 + 5 \times 0)] \text{ kJ mol}^{-1} \\ &= -2043.8 \text{ kJ mol}^{-1}\end{aligned}$$

we see that propane generates the most heat per mol.

To find the heat generated per gram we must use the molecular weights of the molecules. For methane we find that

$$\Delta H^\ominus = \frac{\Delta_r H^\ominus}{\boldsymbol{M}} = \frac{-802.31 \text{ kJ mol}^{-1}}{16 \text{ g mol}^{-1}} = -50.1 \text{ kJ g}^{-1}$$

and for propane

$$\Delta H^\ominus = \frac{\Delta_r H^\ominus}{\boldsymbol{M}} = -2043.8 \frac{\text{kJ}}{\text{mol}} \times \frac{\text{mol}}{44 \text{ g}} = -46.4 \text{ kJ g}^{-1}$$

We see that on a per gram basis methane generates more heat.

E16.17 Using $\Delta_f H^\ominus$ values at $T = 298$ K from Appendix A.2, calculate $\Delta_r H^\ominus$ for the combustion of n-butane and isobutane. Which substance generates the most heat per mol? Which generates the most heat per gram?

Solution

$\Delta_r H^\ominus$ for the combustion of n-butane, using $\Delta_f H^\ominus$ at $T = 298$ K, is found to be

$$\begin{aligned}\Delta_r H^\ominus &= 4\Delta_f H^\ominus (CO_2) + 5\Delta_f H^\ominus (H_2O) - \left[\Delta_f H^\ominus (n\text{-}C_4H_{10}) + \frac{26}{2}\Delta_f H^\ominus (O_2)\right] \\ &= \left[4(-393.52) + 5(-241.83) - \left(-125.6 + \frac{26}{2} \times 0\right)\right] \text{ kJ mol}^{-1} \\ &= -2657.6 \text{ kJ mol}^{-1}\end{aligned}$$

Correspondingly for isobutane we find that

$$\begin{aligned}\Delta_r H^\ominus &= 4\Delta_f H^\ominus (CO_2) + 5\Delta_f H^\ominus (H_2O) - \left[\Delta_f H^\ominus (i\text{-}C_4H_{10}) + \frac{26}{2}\Delta_f H^\ominus (O_2)\right] \\ &= \left[4(-393.52) + 5(-241.83) - \left(-134.2 + \frac{26}{2} \times 0\right)\right] \text{ kJ mol}^{-1} \\ &= -2649.0 \text{ kJ mol}^{-1}\end{aligned}$$

We see that n-butane will generate the most heat on a per mol basis. Because n-butane and *iso*-butane have the same molecular formula and thus the same molecular mass, n-butane will also generate the most heat on a per gram basis.

E16.18 Using $\Delta_f H^\ominus$ values at $T = 298$ K from Appendix A.1, calculate the heat of combustion for graphite. How many grams of graphite must be burned in order to warm the air in a 100 m^3 room from 0 to 25 m$^\circ$C? You may assume that the air has a heat capacity like that of pure nitrogen.

Solution

For the combustion of graphite ($C_{\text{graphite}} + O_2 \rightarrow CO_2$), we find that

$$\begin{aligned}\Delta_r H^\ominus(C_{\text{graphite}}) &= \Delta_f H^\ominus(CO_2) - \Delta_f H^\ominus(O_2) - \Delta_f H^\ominus(C_{\text{graphite}}) \\ &= -393.5 \text{ kJ mol}^{-1} - 0 - 0 = -393.5 \text{ kJ mol}^{-1}\end{aligned}$$

Because the constant pressure heat capacity of N_2 is

$$C_{P,m}\left(N_2\right) \approx C_{V,m}\left(N_2\right) + \boldsymbol{R} \approx \frac{7}{2}\boldsymbol{R}$$

we can calculate the required molar enthalpy for heating the room air as

$$\begin{aligned}\Delta H_{\text{required},m} &= C_{P,m}\left(N_2\right) \cdot \Delta T = \frac{7}{2}\boldsymbol{R} \cdot \Delta T \\ &= 3.5 \cdot \left(8.314\ \text{J mol}^{-1}\text{K}^{-1}\right)(25\ \text{K}) \ = 0.73\ \text{kJ mol}^{-1}\end{aligned}$$

To find the total enthalpy needed for the heating, we first find the amount of substance of gas molecules. In a $100\ \text{m}^3$ room and 1 bar $= 10^5\ \text{N m}^{-2}$ pressure at $T = 273$ K, we find that

$$\boldsymbol{n}_{N_2} = \frac{PV}{\boldsymbol{R}T} = \frac{10^5\ \text{N m}^{-2} \cdot 10^2\ \text{m}^3}{8.314\ \text{J K}^{-1}\ \text{mol}^{-1} \cdot 273\ \text{K}}\ \text{mol} = 4.4 \times 10^3\ \text{mol}$$

Hence the total enthalpy needed to warm the room is

$$\Delta H_{\text{required}} = 0.73\frac{\text{kJ}}{\text{mol}} \cdot 4.4 \times 10^3\ \text{mol} = 3216\ \text{kJ}$$

Now we determine the amount of graphite that can be combusted to yield this enthalpy, by

$$\boldsymbol{n}_{\text{graphite}} = \frac{3216\ \text{kJ}}{\left|\Delta_r H^{\ominus}(\text{C}_{\text{graphite}})\right|} = 8.2\ \text{mol}$$

and converting to weight we find that

$$m_{\text{graphite}} = 12\frac{\text{g}}{\text{mol}} \times 8.2\ \text{mol} = 98\ \text{g}$$

E16.19 The heat of combustion of cyclopropane (C_3H_6), graphite (C(s)), and H_2 is -2092 kJ mol^{-1}, -393.3 kJ mol^{-1}, and -285.83 kJ mol^{-1}, respectively. In addition, the heat of formation of propylene (CH_3-CH=CH_2) is $\Delta H_f^{\ominus} = 20.5$ kJ mol^{-1}. Calculate the heat of formation of cyclopropane. Calculate the heat of isomerization of cyclopropane to propylene. [(Adapted from R. E. Dickerson, Molecular Thermodynamics (Benjamin, 1969)].

Solution
Applying Hess's Law we have

Reaction	$\frac{\Delta_r H^{\ominus}}{\text{kJ mol}^{-1}}$ at $T = 298$ K
$3CO_2 + 3H_2O \rightarrow C_3H_6 + \frac{9}{2}O_2$	2092
$3H_2 + \frac{3}{2}O_2 \rightarrow 3H_2O$	−3 (285.83)
$3C + 3O_2 \rightarrow 3CO_2$	−3 (393.3)
$3H_2 + 3C \rightarrow C_3H_6$	54.61

This value represents $\Delta H_f^{\ominus}$ for cyclopropane.
To find the enthalpy of formation for the isomerization of cyclopropane to propylene is

$$\Delta_{\text{isom}}H = (20.5 - 54.6)\ \text{kJ mol}^{-1} = -34.1\ \text{kJ mol}^{-1}$$

E16.20 Compute the average translational kinetic energy of an argon atom at 300 K. If one converted the argon atom's kinetic energy to gravitational potential energy, how high could one raise the argon atom in the Earth's gravitational field (remember that $g = 9.81\ \text{m s}^{-2}$)? Perform the same calculation for a polymer molecule which has a molar mass of 100 kg mol^{-1}. Comment on your result.

Solution
The average translational kinetic energy of an argon atom at 300 K is

$$\begin{aligned}E_{\text{kin}} &= \frac{3}{2}kT \\ &= \frac{3}{2}\left(1.38 \times 10^{-23}\ \text{J K}^{-1}\right)(300\ \text{K})\ \ \text{J} = 6.21 \times 10^{-21}\ \text{J}\end{aligned}$$

Now we make use of energy balance

$$E_{\text{kin}} = \text{potential energy} = mgh$$

to find the height. Solving for h we have

$$h = \frac{\frac{3}{2}kT}{mg} = \frac{6.21 \times 10^{-21}\ \text{J}}{9.81\ \text{m s}^{-2} \cdot 0.039\ \text{kg mol}^{-1}\ /6.022 \times 10^{23}\ \text{mol}^{-1}}$$

$$= 9780 \frac{\text{N m}}{\text{m s}^{-2}\ \text{kg}} = 9780 \frac{\text{kg m s}^{-2}\ \text{m}}{\text{m s}^{-2}\ \text{kg}} = 9780\ \text{m or } 9.78\ \text{km}$$

For a polymer the kinetic energy would be the same, but the height would be only

$$h = \frac{6.21 \times 10^{-21}\ \text{J}}{9.81\ \text{m s}^{-2} \cdot 100\ \text{kg mol}^{-1}/6.022 \times 10^{23}\ \text{mol}^{-1}} = 3.82\ \text{m}$$

It is clear that the Ar gas atom can travel to very large heights in a gravitational field, but that the polymer molecule cannot.

E16.21 Make a pressure versus volume (PV) diagram that corresponds to the state change represented in Fig. 16.13. Provide a graphical interpretation of the work.

Solution

The process shown in Fig. 16.13 corresponds to a constant pressure expansion of a gas. The system evolves from the point (P, V_1) to the point (P, V_2) at constant P. From the definition of the work as

$$w = \int P \cdot \text{d}V$$

we see that the work is equivalent to the area under the line, as shown by the gray region in Fig. E16.21.

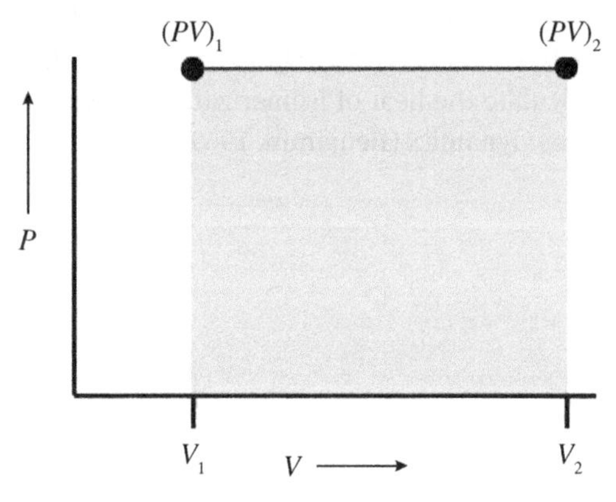

Figure E16.21 PV diagram.

E16.22 From the expression for the molar internal energy in Equation (16.26) calculate $C_V = \left(\text{d}U_{\text{el},m}/\text{d}T\right)_V = \text{d}U_{\text{el},m}/\text{d}T$ ($U_{\text{el},m}$ does not depend on V).

Solution

In Equation (16.26) in Section 16.3.6 we calculated the electronic contribution to the internal energy as

$$U_{\text{el},m} = \boldsymbol{N}_A \cdot \frac{E_1 + E_2\,\text{e}^{-\Delta E/kT}}{1 + \text{e}^{-\Delta E/kT}}$$

To obtain the heat capacity, we differentiate with respect to the temperature.

$$C_{V,\text{el},m} = \left(\frac{\partial U_{\text{el},m}}{\partial T}\right)_V$$

$$= \boldsymbol{N}_A \frac{\Delta E}{kT^2}\text{e}^{-\Delta E/kT}\left[\frac{E_2\left(1 + \text{e}^{-\Delta E/kT}\right)}{\left(1 + \text{e}^{-\Delta E/kT}\right)^2} - \frac{\left(E_1 + E_2\,\text{e}^{-\Delta E/kT}\right)}{\left(1 + \text{e}^{-\Delta E/kT}\right)^2}\right]$$

$$= \boldsymbol{N}_A \cdot \frac{\Delta E}{kT^2} \frac{E_2\, e^{-\Delta E/kT} - E_1\, e^{-\Delta E/kT}}{\left(1 + e^{-\Delta E/kT}\right)^2}$$

$$= \boldsymbol{N}_A \cdot \frac{\Delta E}{kT^2} \frac{\Delta E\, e^{-\Delta E/kT}}{\left(1 + e^{-\Delta E/kT}\right)^2}$$

E16.23 Using the equation of state for an ideal gas, show that the general expression for $C_P - C_V$, Equation (16.45), reduces to the simple form of Equation (16.44).

Solution
Starting with

$$C_P - C_V = TV\frac{\alpha_V^2}{\kappa_T}$$

and using

$$\kappa_T = \frac{1}{P} \text{ and } \alpha_V = \frac{1}{T}$$

we find that

$$C_P - C_V = TV\frac{P}{T^2} = \frac{PV}{T} = \boldsymbol{nR}$$

E16.24 Using Equation (16.45) calculate $C_{V,m}$ from $C_{P,m}$ data at 298 K for the data in the table; i.e., fill in the missing columns in the table. Comment on your results.

	α_V 10^{-6} K^{-1}	κ_T 10^{-6} bar^{-1}	V_m mL mol^{-1}	$C_{P,m}$ J K^{-1} mol^{-1}	$\frac{\alpha_V^2 V}{\kappa_T} T$ J K^{-1} mol^{-1}	$C_{V,m}$ J K^{-1} mol^{-1}
Benzene	1240	94	89.4	136.3		
Ethanol	1120	111	58.5	112.3		
Water	210	45.8	18	75.3		
Lead	86.7	2.42	17.9	26.4		
Aluminum	69	1.3	9.9	24.4		
Diamond	3.5	0.2	3.4	6.2		

Solution
Equation (16.45) is

$$C_P - C_V = T \cdot V\frac{\alpha_V^2}{\kappa_T}$$

As an example for how to proceed, consider the case of benzene. Rearranging Equation (16.45) gives

$$C_V = C_P - T \cdot V\frac{\alpha_V^2}{\kappa_T}$$

$$= 136.3\frac{\text{J}}{\text{K mol}} - 298\text{ K} \cdot \frac{\left(1240 \times 10^{-6}\text{ K}^{-1}\right)^2 \cdot 89.4 \times 10^{-6}\text{ m}^3\text{ mol}^{-1}}{94 \times 10^{-6}\text{ bar}^{-1}}$$

$$= 136.3\frac{\text{J}}{\text{K mol}} - 4.35 \times 10^{-4} \cdot 10^5\text{ K}^{-1}\text{ m}^3\text{ mol}^{-1}\text{ N m}^{-2}$$

$$= 136.3\frac{\text{J}}{\text{K mol}} - 43.5\text{ J K}^{-1}\text{ mol}^{-1} = 92.8\frac{\text{J}}{\text{K mol}}$$

In the table experimental data α_V, κ_T, V_m, and $C_{P,m}$ are given for some compounds at 298 K. $C_{V,m}$ is calculated according to Equation (16.45).

	α_V 10^{-6} K^{-1}	κ_T 10^{-6} bar^{-1}	V_m mL mol^{-1}	$C_{P,m}$ J K^{-1} mol^{-1}	$\frac{\alpha_V^2 V}{\kappa_T}T$ J K^{-1} mol^{-1}	$C_{V,m}$ J K^{-1} mol^{-1}
Benzene	1240	94	89.4	136.3	43.6	92.8
Ethanol	1120	111	58.5	112.3	23.3	89.0
Water	210	45.8	18	75.3	5.2	70.1
Lead	86.7	2.42	17.9	26.4	1.76	24.6
Aluminum	69	1.3	9.9	24.4	1.08	23.3
Diamond	3.5	0.2	3.4	6.2	0.005	6.2

The calculations show that the largest differences between C_P and C_V arise for the liquids: benzene, ethanol, and water. The liquids have much more significant changes in their volume with changes in the pressure and the temperature than do the solids. This sensitivity results in a work of expansion for the liquids, upon heating, that affects the measured heat capacity.

E16.25 What is the isobaric expansion coefficient of an ideal gas? Compare it to those reported for the substances in Table 16.3.

Solution

The isobaric expansion coefficient of an ideal gas is

$$\alpha_V = \frac{1}{V}\left(\frac{\partial V}{\partial T}\right)_P$$

For an ideal gas

$$V = \frac{nRT}{P}$$

so we find that

$$\alpha_V = \frac{1}{V}\left(\frac{nR}{P}\right) = \frac{1}{T}$$

If we evaluate this quantity at $T = 298$ K, we find that

$$\alpha_V = \frac{1}{298} = 3.35 \times 10^{-3} \text{K}^{-1}$$

Albeit somewhat larger, this value is similar to that of the volatile liquids reported in Table 16.3, but it is significantly larger than that of the solids reported in the table; as one might expect.

E16.26 What is the isothermal compressibility of an ideal gas? Compare it to those reported for the substances in Table 16.3.

Solution

The isothermal compressibility is given by

$$\kappa_T = -\frac{1}{V}\left(\frac{\partial V}{\partial P}\right)_T$$

For an ideal gas

$$V = \frac{nRT}{P}$$

so that the isothermal compressibility is

$$\kappa_T = -\frac{1}{V}\left(-\frac{nRT}{P^2}\right) = \frac{1}{P}$$

At 1 bar pressure κ would be 1 bar^{-1}. This value is much larger than the corresponding values in the table. Even the relatively volatile liquid ethanol is only 10^{-4} bar^{-1}. Given our picture that gas molecules are surrounded by mostly empty space, whereas molecules in a liquid are nearly always in van der Waals contact with another molecule, this difference should not be too surprising.

16.2 Problems

P16.1 The MathCAD exercise “workandheat.mcdx” asks you to explore our distinction of the heat and work from an atomistic perspective. You will use a one-dimensional particle in a box model for the energy levels of an ideal monatomic gas (pure translational energy), and you will examine how an adiabatic change in the length of the box affects the internal energy of the gas (this corresponds to “work”) and compare that to a change in the energy that can be realized at fixed length by changing the population of the energy levels (this corresponds to “heat”). For the “work” analysis, you are asked to show how the energy level structure changes for three different box lengths, to discuss the relationship between your calculated energy diagram and “macroscopic” energy for 10 gas atoms (assuming the populations of the energy levels is not changing), and to discuss how your model calculations are like performing an adiabatic compression (or expansion). For the “heat” analysis, you are asked to keep the box length fixed and to calculate the distribution of energy level populations; in particular, you compare three different temperatures (100, 400, and 1300 K) to the distribution at 300 K; and in the second part of this part you are asked to describe how the distributions change and what its implications might be. In the last part of the exercise, you are asked to calculate the total internal energy numerically, and to compare it with the value that you would find by integrating over the entire distribution.

Solution

Here we provide a sketch of the answers to the questions in the MathCAD worksheet. See the file “workandheat_solutions.mcdx” for solutions.

1. Figure P16.1a shows the energy

$$E = \frac{h^2}{8mL^2}n^2$$

of a particle in a one-dimensional box of length $L = 0.9$ nm, $L = 1.0$ nm and $L = 1.1$ nm. Only the first three energy levels are shown.

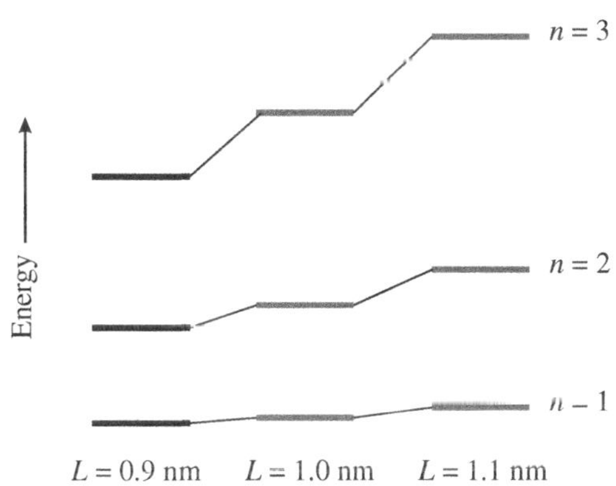

Figure P16.1a Energy for a particle in a one-dimensional box with length L.

2. The smaller box size corresponds to larger energy gaps between the energy levels. For a box with 10 atoms, we expect the smaller box to have a larger average energy; assuming that the population distribution of the energy levels is the same in each case. As the box becomes macroscopically sized however, this difference would become negligible. We note that this model does not have a temperature defined, so that the populations of the energy levels need to be specified in evaluating an average energy.
3. The change in the size of the container without changing the population distribution of the energy levels is analogous to the case of an adiabatic volume change. In such a case, work is performed, which can change the energy levels, but no heat is allowed to flow, which can change the population distribution of the energy levels (see P17.13).

4. The three plots are shown in Fig. P16.1b (the solid curve in each panel shows the 300 K population distribution, and the dashed curve shows the cases for the three other temperatures). For the decrease in temperature by 200 K we find that the population distribution is sharpened. For the increase in temperature we see that the population distribution of the energy levels becomes broader.

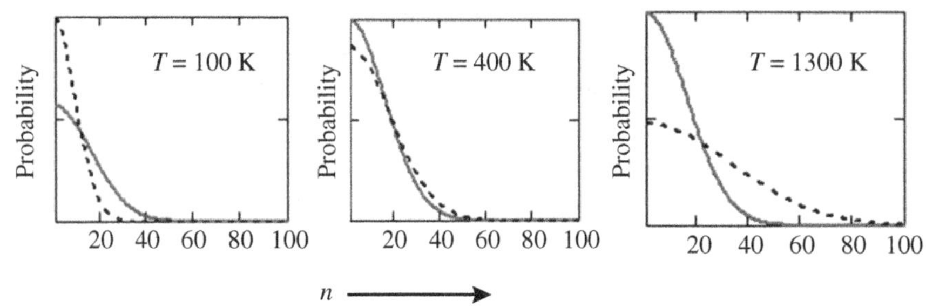

Figure P16.1b Population probability $N_n = (N/Z)\exp(-E_n/(kT))$ for particles in a one-dimensional box with length $L = 1.0$ nm for different temperatures. In each panel the solid curve shows the case for $T = 300$ K, and the dashed curves are for the temperatures shown on the graph. In each case the probability distribution is normalized to unity.

5. The calculations show that for higher temperatures it is more probable to have gas particles in higher energy levels; i.e., the tails of the distributions extend to higher n values. These more energetic gas particles should have larger collision energies and hence may have higher reaction probabilities.
6. By replacing the sum with an integral, we find that we must evaluate

$$\frac{\frac{h^2}{8mL^2}\int_0^\infty n^2 \exp\left(-\frac{h^2n^2}{8mL^2kT}\right)\,dn}{\int_0^\infty \exp\left(-\frac{h^2n^2}{8mL^2kT}\right)\,dn} = \frac{\sqrt{\frac{kT8mL^2}{h^2}}kT\int_0^\infty x^2\exp\left(-x^2\right)\,dx}{\sqrt{\frac{kT8mL^2}{h^2}}\int_0^\infty \exp\left(-x\right)\,dn}$$

$$= \frac{kT\left(\sqrt{\pi}/4\right)}{\sqrt{\pi}/2} = \frac{1}{2}kT$$

where we used $x^2 = h^2n^2/\left(8mL^2kT\right)$and $x\,dx = \left[h^2n/\left(8mL^2kT\right)\right]\,dn$. Figure P16.1c shows the numerically evaluated energy versus kT, and it is evident that this graph has a slope of 1/2.

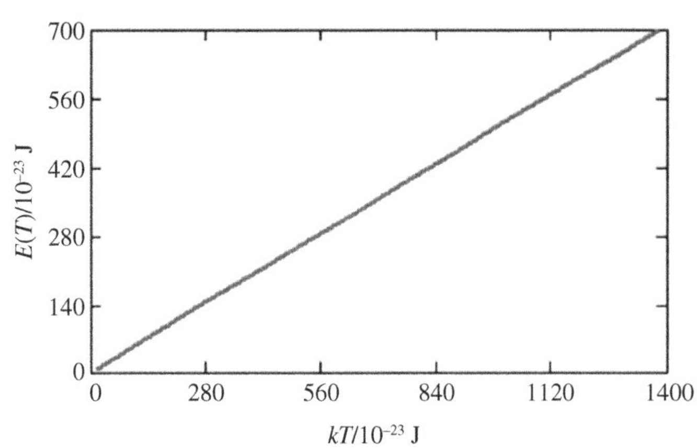

Figure P16.1c Numerically calculated energies for particles in a one-dimensional box with box length $L = 1.0$ nm versus the thermal energy kT.

P16.2 An ideal gas ($T = 300$ K, $P = 10$ bar, $V_1 = 1$ L) expands isothermally (see Fig. P16.2) by lifting a weight on a piston corresponding to the pressure $P_2 = 1$ bar. What is the volume V_2 of the gas after thermal equilibration, and what is the work w done on the gas and the heat q flowing in the gas during the process?

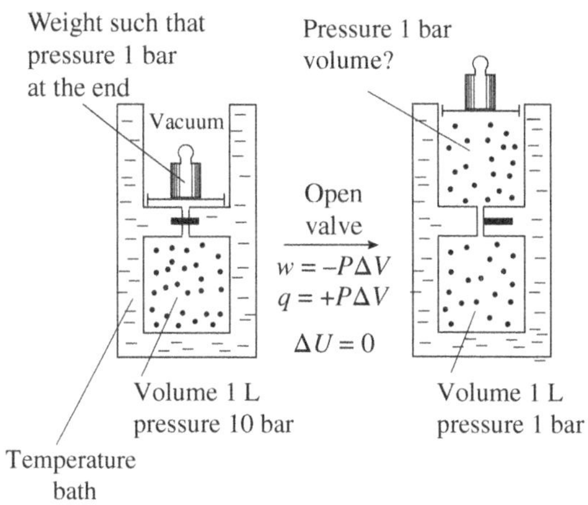

Figure P16.2 Isothermal expansion of an ideal gas.

Solution

Because of $P_1V_1 = P_2V_2$ we obtain

$$V_2 = V_1 \cdot \frac{P_1}{P_2} = 10 \text{ L}$$

The work w is

$$w = -P_2\Delta V = -P_2 \cdot (V_2 - V_1) = -1 \text{ bar} \cdot 9 \text{ L} = -900 \text{ J}$$

Because of $\Delta U = 0$ the heat is $q = -w = 900$ J.

P16.3 An ideal gas ($T_1 = 300$ K, $P_1 = 10$ bar, $V_1 = 1$ L) expands adiabatically to the pressure $P_2 = 1$ bar. Calculate T_2, V_2, and w.

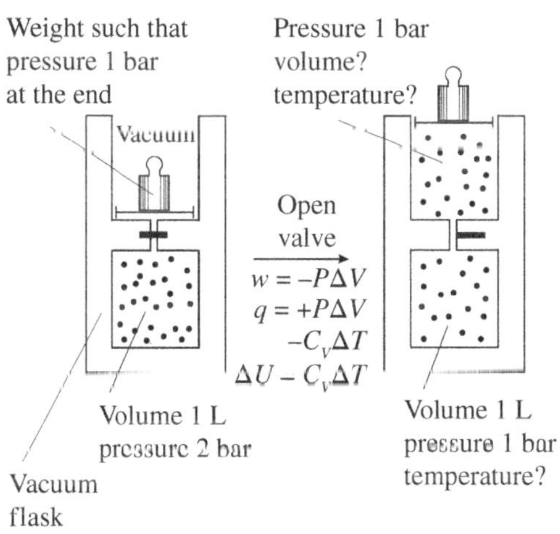

Figure P16.3 Adiabatic expansion of an ideal gas.

Solution

Figure P6.13 shows a diagram that illustrates the initial and final states for this process. In an adiabatic process we have $q = 0$; thus, we obtain

$$\Delta U = w$$

Because of

$$\Delta U = C_V \cdot (T_2 - T_1) \quad \text{and} \quad w = -P_2 \cdot (V_2 - V_1)$$

and

$$P_1V_1 = NkT_1, \quad P_2V_2 = NkT_2$$

we obtain

$$C_V \cdot (T_2 - T_1) = -P_2 \cdot \left(\frac{NkT_2}{P_2} - \frac{NkT_1}{P_1} \right) = Nk \cdot \left(-T_2 + \frac{P_2}{P_1} \cdot T_1 \right)$$

Then

$$T_2 = T_1 \cdot \frac{P_2/P_1 + C_V/Nk}{1 + C_V/Nk}$$

With the data given above and $C_V = \frac{3}{2}Nk$ for an atomic gas it follows

$$T_2 = 300\ \text{K} \cdot 0.64 = 192\ \text{K}$$
$$V_2 = V_1 \cdot \frac{P_1 T_2}{P_2 T_1} = 6.4\ \text{L}$$
$$w = -P_2 \Delta V = -1\ \text{bar} \cdot 5.4\ \text{L} = -540\ \text{J}$$

P16.4 What is the flame temperature of a Bunsen burner that operates with a methane–air mixture? Compare your result with that found for a hydrogen–air mixture in Box 16.2.

Solution

We consider the reaction

$$CH_4 + 2\,O_2 \rightarrow CO_2 + 2\,H_2O$$

As in Box 16.2 we start at $T_1 = 373$ K. Then we obtain

$$\Delta_r H^\ominus_{373} = \Delta_r H^\ominus_{298} + \int_{298}^{373} \Delta C^\ominus_{P,m} \cdot dT$$
$$= (-802 + 3.0)\ \text{kJ mol}^{-1} = -799\ \text{kJ mol}^{-1}$$

The heat capacity of the products is given by

$$C^\ominus_{P,m}(\text{products}) = C^\ominus_{P,m}(CO_2) + 2C^\ominus_{P,m}(H_2O) + 8C^\ominus_{P,m}(N_2)$$
$$= (40.3 + 2 \cdot 34.0 + 8 \times 29.2)\ \text{J K}^{-1}\ \text{mol}^{-1}$$
$$= 342\ \text{J K}^{-1}\ \text{mol}^{-1}$$

Then in analogy to the procedure in Box 16.2 we obtain

$$\Delta T = \frac{-\Delta_r H^\ominus_{373}}{C^\ominus_{P,m}(\text{products})} = \frac{799\ \text{kJ mol}^{-1}}{342\ \text{J K}^{-1}\ \text{mol}^{-1}} = 2340\ \text{K}$$

Remember that $\Delta_r H^\ominus$ and $C^\ominus_{P,m}$(products) refer to the reaction of one mol CH_4 and two mol O_2 to one mol CO_2 and two mol in the presence of eight mol N_2.
This value is 250 K cooler than that of the hydrogen flame.

P16.5 Consider the values for the experimental constant pressure heat capacity data of nitrogen as a function of temperature, given in the table. Use a spreadsheet program to plot these data. On the same graph plot the translational, the rotational, and the vibrational contribution to the heat capacity, and the sum of these three contributions (use the molecular parameters $k_f = 2296\ \text{N m}^{-1}$, $d_0 = 110$ pm).

T/K	300	400	500	600	700	800	900	1000	1100
$C_{P,m}/(\text{J K}^{-1}\ \text{mol}^{-1})$	28.88	29.35	29.91	30.47	31.02	31.55	32.06	32.54	32.99

Solution

Figure P16.5 shows the data plotted as filled circles. The dotted curve for the vibrational contribution is barely discernible from the axis. The dashed line is the rotational contribution; the solid black line is the translational contribution plus $\boldsymbol{R}$; and the thick solid line at the top is the total constant pressure heat capacity that is calculated; it is in excellent agreement with the data.

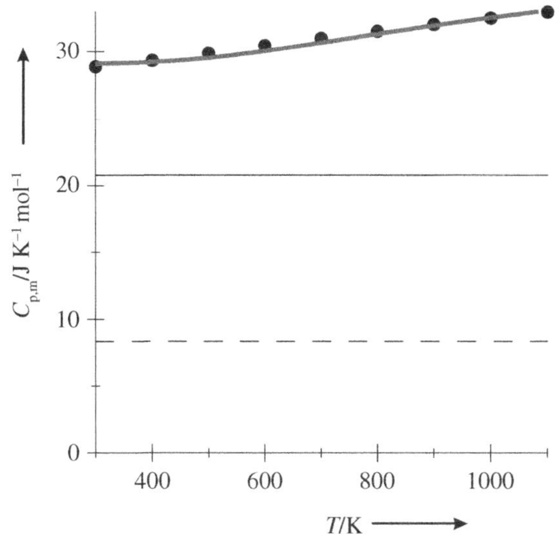

Figure P16.5 Molar heat capacity of N_2 versus temperature. Filled circles: experiment, thick solid line: calculation. Dashed line: calculated contribution of rotation, thin solid line: calculated contribution of translation.

In determining the vibrational contribution to the heat capacity we used the following quantities

$$\nu_0 = \frac{1}{2\pi}\sqrt{\frac{k}{\mu}}$$
$$= \frac{1}{2\pi}\sqrt{\frac{2296\ \text{N m}^{-1}}{1.1704\times10^{-26}\ \text{kg}}} = 7.0492\times10^{13}\ \text{s}^{-1}$$

where

$$\mu = \frac{14\times14}{14+14}\cdot1.672\times10^{-27}\ \text{kg} = 1.1704\times10^{-26}\ \text{kg}$$

The vibrational temperature is

$$\theta_{\text{vib}} = \frac{h\nu_0}{k} = \frac{6.6260755\times10^{-34}\ \text{J s}\cdot7.0492\times10^{13}\ \text{s}^{-1}}{1.3806568\times10^{-23}\ \text{J K}^{-1}} = 3383.1\ \text{K}$$

P16.6 Consider pure gases of each of the diatomic molecules H_2 ($k_f = 574\ \text{N m}^{-1}$), Cl_2 ($k_f = 328\ \text{N m}^{-1}$), and I_2 ($k_f = 173\ \text{N m}^{-1}$). Calculate the ratio of the number of molecules in the first excited vibrational level to the ground state vibrational energy level for each of these molecules at 298 K. Do you expect the vibrational contribution to the heat capacity of the gas to be large or small at this temperature?

Solution

From these force constants we can determine ν_0 from the formula

$$\nu_0 - \frac{1}{2\pi}\sqrt{\frac{k_f}{\mu}}$$

For H_2 with $\mu_{H_2} = 3.226\times10^{-27}$ kg we find that

$$\nu_0 = \frac{1}{2\pi}\sqrt{\frac{574\ \text{N m}^{-1}}{3.226\times10^{-27}\ \text{kg}}} = 6.71\times10^{13}\ \text{s}^{-1}$$

and the energy gap is

$$\Delta E = h\nu_0 = 6.626\times10^{-34}\ \text{J s}\cdot6.71\times10^{13}\ \text{s}^{-1}\ \text{J} = 4.453\times10^{-20}\ \text{J}$$

With this energy gap between ground state and first excited states the population ratio at $T = 298$ K for H_2 is

$$\frac{N_1}{N_0} = e^{-\Delta E/(kT)} = 4.5 \times 10^{-5}$$

We proceed in a similar manner for the other molecules and compile them in the table.

	$\mu/10^{-25}$ kg	$\nu_0/10^{13}$ s^{-1}	$\Delta E/10^{-21}$ J	n_1/n_0
H_2	0.03226	6.72	44.53	4.5×10^{-5}
Cl_2	1.136	8.56	5.672	0.276
I_2	4.67	3.06	2.028	0.624

From these population ratios it is evident that the vibrational contribution to the heat capacity for H_2 will be negligible; however, it will contribute significantly for Cl_2 and even more so for I_2.

P16.7 Calculate the vibrational contribution to the heat capacity of Cl_2 gas at 100, 298, 500, 1000, and 5000 K. Plot these values on a graph. What heat capacity would be expected if the vibrational mode was fully excited? [Adapted from R.E. Dickerson, *Molecular Thermodynamics* (Benjamin, 1969)].

Solution

The vibrational contribution to the heat capacity of Cl_2 gas is determined using

$$C_{V,\mathrm{vib},m} = \boldsymbol{R}\left(\frac{h\nu_0}{kT}\right)^2 \frac{\exp\left(h\nu_0/(kT)\right)}{\left(\exp\left(h\nu_0/(kT)\right) - 1\right)^2}$$

From P16.6 we have $\nu_0 = 8.56 \times 10^{13}$ s^{-1}, so that $h\nu_0 = 5.67 \times 10^{-21}$ J. At $T = 298$ K, we find

$$\frac{h\nu_0}{kT} = \frac{5.67 \times 10^{-21}\ \mathrm{J}}{1.38 \times 10^{-23}\ \mathrm{J\,K^{-1}} \cdot 298\ \mathrm{K}} = 1.38$$

$$C_{V,\mathrm{vib},m} = \boldsymbol{R} \cdot (1.38)^2 \frac{\exp(1.38)}{(\exp(1.38) - 1)^2} = 1.62 \times 10^{-3}\ \mathrm{J\,mol^{-1}K^{-1}}$$

Proceeding in the same manner for the other temperatures we find that

T/K	100	298	500	1000	5000
$C_{V,\mathrm{vib},m}$/J mol^{-1}K^{-1}	1.98×10^{-14}	1.62×10^{-3}	0.15	2.38	7.86

and plot them in Fig. P16.7.

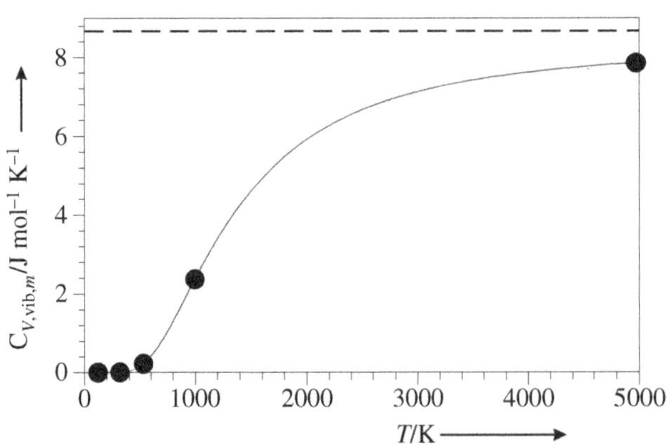

Figure P16.7 Temperature dependence of $C_{V,\mathrm{vib},m}$ for N_2 molecules (solid line) and maximum value $\boldsymbol{R}$ (dashed line). The full black circles mark the values displayed in the table.

P16.8 Consider the Debye function for the molar heat capacity $C_{V,m}$ of a solid, Equation (16.33). Find limiting expressions for this function as $T \to 0$, $T \to \infty$, and for small T.

Solution

Equation (16.33) in Section 16.3.8 gives the heat capacity of solids, using the Debye model (*Debye function of heat capacity*).

$$C_{V,m} = \frac{9h\boldsymbol{N}_A}{\nu_{\max}^3} \cdot \int_0^{\nu_{\max}} \frac{h\nu}{kT^2} \frac{\nu^3 e^{h\nu/kT}}{\left(e^{h\nu/kT} - 1\right)^2} \, d\nu$$

Limit $T \to 0$: In this case we can neglect the term -1 in the denominator of the integrand and we obtain

$$\begin{aligned} C_{V,m} &= \frac{9h\boldsymbol{N}_A}{\nu_{\max}^3} \cdot \int_0^{\nu_{\max}} \frac{h\nu}{kT^2} \frac{\nu^3 e^{h\nu/kT}}{\left(e^{h\nu/kT}\right)^2} \, d\nu \\ &= \frac{9h\boldsymbol{N}_A}{\nu_{\max}^3} \cdot \int_0^{\nu_{\max}} \frac{h\nu}{kT^2} \nu^3 e^{-h\nu/kT} \, d\nu \\ &= \frac{9h\boldsymbol{N}_A}{\nu_{\max}^3} \cdot \frac{h}{kT^2} \int_0^{\nu_{\max}} \nu^4 e^{-h\nu/kT} \, d\nu \end{aligned}$$

Using the substitution $h\nu/(kT) = x$ and $d\nu \cdot h/(kT) = dx$ we can write the integral as

$$\int_0^{\nu_{\max}} \nu^4 \, e^{-h\nu/kT} \, d\nu = \left(\frac{kT}{h}\right)^4 \frac{kT}{h} \int_0^{x_{\max}} x^4 \, e^{-x} \, dx$$

where $x_{\max} = h\nu_{\max} / (kT)$. In the limit $T \to 0$ we obtain $x_{\max} \to \infty$, thus

$$\int_0^{x_{\max}} x^4 \, e^{-x} \, dx \approx \int_0^{\infty} x^4 \, e^{-x} \, dx = 4! = 24$$

where we used Appendix B. Inserting this result into the expression for $C_{V,m}$, we find

$$\begin{aligned} \lim_{T\to 0} C_{V,m} &= \lim_{T\to 0} \frac{9h\boldsymbol{N}_A}{\nu_{\max}^3} \cdot \frac{h}{kT^2} \int_0^{\nu_{\max}} \nu^4 \, e^{-h\nu/kT} \, d\nu \\ &= \lim_{T\to 0} \frac{9h\boldsymbol{N}_A}{\nu_{\max}^3} \cdot \frac{h}{kT^2} \left(\frac{kT}{h}\right)^4 \frac{kT}{h} \int_0^{x_{\max}} x^4 \, e^{-x} \, dx \\ &= \lim_{T\to 0} \frac{9h\boldsymbol{N}_A}{\nu_{\max}^3} \cdot \frac{h}{kT^2} \left(\frac{kT}{h}\right)^4 \frac{kT}{h} \cdot 24 \\ &= \lim_{T\to 0} \frac{216 \cdot h\boldsymbol{N}_A}{\nu_{\max}^3} \cdot \frac{1}{T} \left(\frac{kT}{h}\right)^4 = 0 \end{aligned}$$

Limit $T \to \infty$: For large enough temperature T the denominator in the integrand of the Debye expression becomes

$$\left(e^{h\nu/kT} - 1\right)^2 = \left(1 + \frac{h\nu}{kT} - 1\right)^2 = \left(\frac{h\nu}{kT}\right)^2$$

and we obtain

$$\begin{aligned} C_{V,m} &= \frac{9h\boldsymbol{N}_A}{\nu_{\max}^3} \int_0^{\nu_{\max}} \frac{h\nu}{kT^2} \frac{\nu^3 \, e^{h\nu/kT}}{\left(\frac{h\nu}{kT}\right)^2} \, d\nu \\ &= \frac{9h\boldsymbol{N}_A}{\nu_{\max}^3} \cdot \frac{h}{kT^2} \left(\frac{kT}{h}\right)^2 \int_0^{\nu_{\max}} \nu^2 \, e^{h\nu/kT} \, d\nu \end{aligned}$$

In the limit $T \to \infty$ the exponential function approaches one, thus

$$\begin{aligned} \lim_{T\to\infty} C_{V,m} &= \lim_{T\to\infty} \frac{9h\boldsymbol{N}_A}{\nu_{\max}^3} \cdot \frac{h}{kT^2} \left(\frac{kT}{h}\right)^2 \int_0^{\nu_{\max}} \nu^2 \, d\nu \\ &= \lim_{T\to\infty} \frac{9h\boldsymbol{N}_A}{\nu_{\max}^3} \cdot \frac{h}{kT^2} \left(\frac{kT}{h}\right)^2 \cdot \frac{1}{3} \nu_{\max}^3 = 3\boldsymbol{N}_A \cdot k = 3\,\boldsymbol{R} \end{aligned}$$

C_V for Small Enough T: Using the above equation for the limit as T approaches zero, for sufficiently small temperature we obtain

$$C_{V,m} = \frac{216 \cdot h\mathbf{N}_A}{\nu_{max}^3} \cdot \frac{1}{T}\left(\frac{kT}{h}\right)^4$$
$$= \frac{216 \cdot h\mathbf{N}_A}{\nu_{max}^3} \cdot \frac{k^4}{h^4} \cdot T^3 = \frac{216 \cdot \mathbf{N}_A}{\nu_{max}^3} \cdot \frac{k^4}{h^3} \cdot T^3$$

P16.9 Consider the isomerization reaction of acetonitrile (H_3 C–C≡N) to isocyanomethane (H_3C–N≡C). The enthalpy of formation of acetonitrile is $\Delta_f H^{\ominus}_{298} = 31.4$ kJ mol^{-1} and the enthalpy of formation of isocyanomethane is $\Delta_f H^{\ominus}_{298} = 149.0$ kJ mol^{-1}. Compute the enthalpy change for the isomerization reaction at 298 K. If the heat capacity of acetonitrile is $C_{P,m} = 39.66$ J K^{-1} mol^{-1} + 0.242 J K^{-2} mol^{-1} $\cdot\, T$ and the heat capacity of isocyanomethane is $C_{P,m} = 51.05$ J K^{-1} mol^{-1} + 0.201 J K^{-2} mol^{-1} $\cdot\, T$, find the enthalpy change for the isomerization reaction at 750 K.

Solution

We first find

$$\Delta_r H_{298} = \Delta_f H^{\ominus}_{298}\,(\text{isocyanomethane}) - \Delta_f H^{\ominus}_{298}\,(\text{acetonitrile})$$
$$= (149.0 - 31.4)\text{ kJ mol}^{-1} = 117.6\text{ kJ mol}^{-1}$$

To calculate the enthalpy change at 750 K, we first calculate the enthalpies of the reactants and products at that temperature and subsequently compute their difference. Now we compute the change in enthalpy of acetonitrile that results from increasing its temperature to 750 K, namely

$$\Delta H_{CH_3\ CN} = \int_{298}^{750} C_{P,m}\, dT = \int_{298}^{750} \left(39.66\text{ J K}^{-1}\text{ mol}^{-1} + 0.242\text{ J K}^{-2}\text{ mol}^{-1} \cdot T\right)\, dT$$
$$= \left|39.66\text{ J K}^{-1}\text{ mol}^{-1} \cdot T\right|_{298}^{750} + \left|0.242\text{ J K}^{-2}\text{ mol}^{-1} \cdot \frac{1}{2}T^2\right|_{298}^{750}$$
$$= 39.66\text{ J K}^{-1}\text{ mol}^{-1}\,[750\text{ K} - 298\text{ K}] + 0.242\text{ J K}^{-2}\text{ mol}^{-1} \cdot \frac{1}{2}\left[750^2\text{ K}^2 - 298^2\text{ K}^2\right]$$
$$= 17.93\text{ kJ} + 57.32\text{ kJ} = 75.244\text{ kJ mol}^{-1}$$

so that

$$\Delta H_{CH_3\ CN}(750\text{ K}) = 31.4\text{ kJ mol}^{-1} + 75.24\text{ kJ mol}^{-1}$$
$$= 106.6\text{ kJ mol}^{-1}$$

Now we compute the change in enthalpy of isocyanomethane that results from increasing its temperature to 750 K, namely

$$\Delta H_{CH_3\ NC} = \int_{298}^{750} C_{P,m}\, dT = \int_{298}^{750} (51.05 + 0.201 \cdot T)\, dT$$
$$= 70{,}681\text{ J mol}^{-1} = 70.681\text{ kJ mol}^{-1}$$

so that

$$\Delta H_{CH_3\ NC}(750\text{ K}) = 149.0\text{ kJ mol}^{-1} + 70.681\text{ kJ mol}^{-1}$$
$$= 219.7\text{ kJ mol}^{-1}$$

Now we calculate the reaction enthalpy as

$$\Delta H_{750} = H_{CH_3\ NC}(750\text{ K}) - H_{CH_3\ CN}(750\text{ K})$$
$$= 219.7\text{ kJ mol}^{-1} - 106.6\text{ kJ mol}^{-1}$$
$$= 113.1\text{ kJ mol}^{-1}$$

This value only differs by a few percent from the value at 298 K.

P16.10 The heat of formation of gaseous isoprene, C_5H_8, is $\Delta_f H^{\ominus}_{298} = 75.5$ kJ mol^{-1}. Compute this heat of formation by using the bond enthalpies (the heat of atomization of graphite is 718 kJ mol^{-1}). Compare the heat of formation you calculate with the value given above. Use this energy difference to estimate the resonance energy of isoprene.

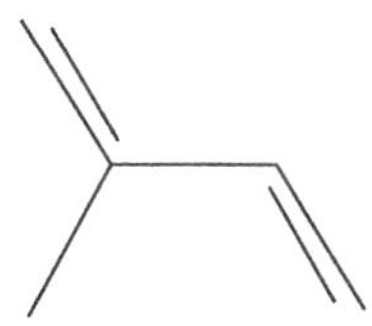

Solution

The enthalpy for the reaction

$$5C\,(s) + 4H_2\,(g) \rightarrow C_5H_8\,(g)$$

is the enthalpy of formation of isoprene. We construct this reaction through the following path. For the atomization of the graphite

$$5C\,(s) + 4H_2\,(g) \rightarrow 5C\,(g) + 4H_2\,(g)$$

we find an enthalpy of $\Delta H_1 = 5\,(718)$ kJ mol^{-1} = 3590.0 kJ mol^{-1}. The next step is to atomize $H_2(g)$

$$5C\,(g) + 4H_2\,(g) \rightarrow 5C\,(g) + 8H\,(g)$$

which has an enthalpy of $\Delta H_2 = 4\,(436)$ kJ mol^{-1} = 1744.0 kJ mol^{-1}. The final step is to form isoprene and calculate its enthalpy by the sum of bond enthalpies (2 C=C bonds, 2 C–C bonds, and 8 C–H bonds); for C_5H_8

$$5C\,(g) + 8H\,(g) \rightarrow C_5H_8\,(g)$$

The enthalpy for this process is $\Delta H_3 = -5228$ kJ. The overall enthalpy of formation is

$$\begin{aligned}\Delta_f H &= \Delta H_3 + \Delta H_1 + \Delta H_2 \\ &= (-5228 + 3590 + 1744)\ \text{kJ mol}^{-1} \\ &= 106\ \text{kJ mol}^{-1}\end{aligned}$$

The difference between $\Delta_f H$ and this value is

$$[106 - (75.5)]\ \text{kJ mol}^{-1} = 30.5\ \text{kJ mol}^{-1}$$

In part this difference may arise from error in the bond enthalpy approximation; however, part may also reflect the resonance stabilization for the isoprene that would not be manifest in the bond enthalpy approach but would contribute to the experimentally determined value.

P16.11 Consider the exchange reaction

$$Cl_2\,(g) + 2\,HBr\,(g) \rightleftarrows Br_2\,(g) + 2\,HCl(g)$$

A. Given that the heat of formation of HBr is $\Delta_f H^{\ominus}_{298} = -36.40$ kJ mol^{-1} and the heat of formation of HCl is $\Delta_f H^{\ominus}_{298} = -92.31$ kJ mol^{-1}, find the enthalpy change for this reaction at 298 K.
B. Find an expression for the enthalpy change in this exchange reaction in terms of its molecular constants. You may use the high temperature limit expressions for the rotational degrees of freedom and the translational degrees of freedom. For the vibrational partition functions of HBr and HCl you can assume that $\theta_{vib} \gg T$, but for Br_2 ($\theta_{vib} \sim 467$ K) and Cl_2 ($\theta_{vib} \sim 804$ K) you should include the vibrational partition function.
C. How would you expect the $\Delta_r H$ to change as the temperature increases from 298 to 500 K?

Solution

A. The enthalpy change for the above reaction at 298 K is

$$\Delta_r H = \Delta_f H\left(\mathrm{Br_2}\right) + 2\Delta_f H\,(\mathrm{HCl}) - \Delta_f H\left(\mathrm{Cl_2}\right) - 2\Delta_f H\,(\mathrm{HBr})$$
$$= [0 + 2\times(-92.31) - 0 - 2\times(-36.40)]\ \mathrm{kJ\ mol^{-1}}$$
$$= -111.82\ \mathrm{kJ\ mol^{-1}}$$

B. First find H of an ideal, diatomic gas

$$H = U + PV = U + NkT$$

Using

$$U = U_{\mathrm{trans}} + U_{\mathrm{rot}} + U_{\mathrm{vib}} + U_{\mathrm{elec}}$$

and $U_{\mathrm{trans}} + U_{\mathrm{rot}} = \frac{5}{2}NkT$ and $U_{\mathrm{elec}} = -\varepsilon_B$, where we only consider the electronic ground state. Therefore

$$H = \frac{7}{2}NkT - \varepsilon_B + U_{\mathrm{vib}}$$

Now we find

$$U_{\mathrm{vib}} = Nk\frac{\theta_{\mathrm{vib}}}{2} + Nk\frac{\theta_{\mathrm{vib}}\cdot\exp\left(-\theta_{\mathrm{vib}}/T\right)}{\left(1-\exp\left(-\theta_{\mathrm{vib}}/T\right)\right)}$$

Therefore

$$H_{\mathrm{diatomic}} = \frac{7}{2}NkT - \varepsilon_B + \frac{Nk\theta_{\mathrm{vib}}}{2} + \frac{Nk\theta_{\mathrm{vib}}\cdot\exp\left(-\theta_{\mathrm{vib}}/T\right)}{\left(1-\exp\left(-\theta_{\mathrm{vib}}/T\right)\right)}$$

If we assume that $\theta_{\mathrm{vib}} \gg T$ for the HBr and HCl cases, then we obtain

$$H_{\mathrm{diatomic}} = \frac{7}{2}NkT - \varepsilon_B + \frac{Nk\theta_{\mathrm{vib}}}{2}$$

If we let

$$-\varepsilon_{B,0} = -\varepsilon_B + \frac{Nk\theta_{\mathrm{vib}}}{2}$$

then

$$H_{\mathrm{diatomic}} = \frac{7}{2}NkT - \varepsilon_{B,0}$$

In this approximation the enthalpy of the reaction becomes

$$\Delta_r H = -\varepsilon_{B,0}\left(\mathrm{Br_2}\right) - 2\varepsilon_{B,0}\,(\mathrm{HCl}) + 2\varepsilon_{B,0}\,(\mathrm{HBr}) + \varepsilon_{B,0}\left(\mathrm{Cl_2}\right)$$
$$+\frac{Nk\theta_{\mathrm{vib,Br_2}}\times\exp\left(-\theta_{\mathrm{vib,Br_2}}/T\right)}{\left(1-\exp\left(-\theta_{\mathrm{vib,Br_2}}/T\right)\right)} - \frac{Nk\theta_{\mathrm{vib,Cl_2}}\times\exp\left(-\theta_{\mathrm{vib,Cl_2}}/T\right)}{\left(1-\exp\left(-\theta_{\mathrm{vib,Cl_2}}/T\right)\right)}$$

C. The $\varepsilon_{B,0}$ terms are temperature independent. The $\Delta_r H$ would show a temperature dependence that reflects the vibrational contributions of Br_2 and Cl_2. The Br_2 term will become important sooner, since $\theta_{\mathrm{vib,Br_2}} < \theta_{\mathrm{vib,Cl_2}}$. Hence the $\Delta_r H$ will become somewhat more positive as T increases from 298 K to 500 K.

P16.12 Consider the Otto Cycle (that used for the internal combustion engine) with the ideal gas as medium. The four steps in the Otto cycle are sketched in Figure P16.8 and may be stated as 1: adiabatic compression stage involving the compression and the expulsion of the gas, 2: isochoric heating at V_2 stage involving the ignition of the fuel, 3: adiabatic expansion stage involving the expansion of the piston, and 4: isochoric cooling at V_1 stage involving cooling at the end of the piston stroke.

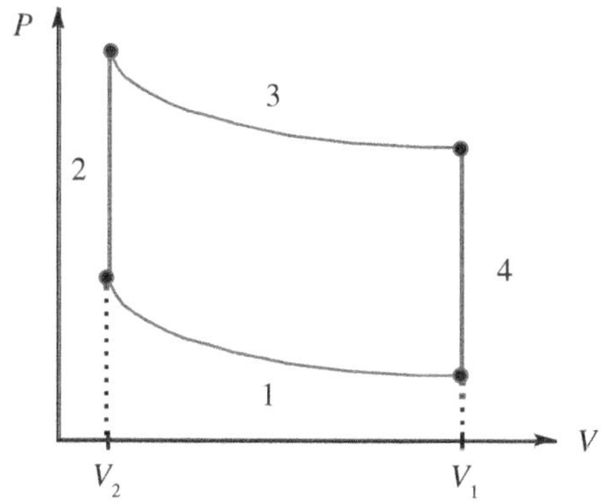

Figure P16.8 Otto cycle.

Assume that this cycle operates on a gas that follows the ideal gas law. Consider the gases formed in the combustion to have a heat capacity of $C_{V,m} = 2.5\boldsymbol{R}$. Take the gas which is combusted to be 0.001 moles of pure octane, such that

$$C_8H_{18} + 12.5\ O_2 \rightleftarrows 8\ CO_2 + 9\ H_2O$$

Assume your piston has a $V_2 = 50.0$ mL and a $V_1 = 500$ mL. Assume that your engine operates between 50°C at the low end, and the gas has a temperature of 1200°C at the high end. Assume that the gas has a temperature of 490°C after it is compressed but before it is ignited, and that the gas has a temperature of 310°C after it drives the expansion of the piston but before it is cooled. Find the work performed in each step of the cycle and the total work.

Solution
In step 1 we have

$$\begin{aligned} w_1 &= n \cdot C_V \cdot T_D \left[\left(\frac{V_1}{V_2} \right)^{R/C_V} - 1 \right] \\ &= (0.017)\,(2.5 \cdot 8.314)\,(323) \left(\left(\frac{500}{50} \right)^{1/2.5} - 1 \right) \text{ J} \\ &= 172.55 \text{ J} \end{aligned}$$

For step 2 $w_2 = 0$, because the volume is constant. For step 3 we have

$$\begin{aligned} w_3 &= n \cdot C_V \cdot T_B \left[\left(\frac{V_2}{V_1} \right)^{R/C_V} - 1 \right] \\ &= (0.017)\,(2.5 \times 8.314)\,(1473) \left(\left(\frac{50}{500} \right)^{1/2.5} - 1 \right) \text{ J} \\ &= -313.31 \text{ J} \end{aligned}$$

and for step 4 $w_4 = 0$. The net work is

$$\begin{aligned} w_{\text{net}} &= w_1 + w_2 + w_3 + w_4 \\ &= [172.55 + 0 + (-313.31) + 0] \text{ J} \\ &= -140.76 \text{ J} \end{aligned}$$

The negative sign tells us that the cycle performs work.

17

Reversible Work w_{rev}, Reversible Heat q_{rev}, and Entropy S

17.1 Exercises

E17.1 Consider the reversible isothermal expansion of 1 mol of an ideal gas at $T = 300$ K from a volume V_1 to a volume $V_2 = 2V_1$, and calculate the work performed. Using this value of the work, how many meters can this expansion lift a weight of 1 kg in earth's gravitational field 9.81 m s^{-2}?

Solution
Starting with the definition of PV work we find that

$$w = -\int_{V_1}^{V_2} P\,\mathrm{d}V = -\boldsymbol{n}RT\int_{V_1}^{V_2}\frac{\mathrm{d}V}{V} = -\boldsymbol{n}\boldsymbol{R}T\ln\left(\frac{V_2}{V_1}\right)$$
$$= -1\text{ mol}\cdot 8.314\text{ J K}^{-1}\text{ mol}^{-1}\cdot 300\text{ K}\cdot\ln(2) = -1729\text{ J}$$

Using this energy, we determine the height to be

$$h = \frac{-work}{m\cdot g} = \frac{1729\text{ J}}{1\text{ kg}\times 9.81\text{ m s}^{-2}} = 176\text{ J kg}^{-1}\text{ m}^{-1}\text{ s}^{2} = 176\,\frac{\text{kg m s}^{-2}\ \text{m}\cdot\ \text{s}^{2}}{\text{kg m}} = 176\text{ m}$$

E17.2 Calculate the work for an isothermal expansion of a gas from V_1 to $V_2 = 2V_1$ against a constant pressure P. Compare the work found in this case to that found for the isothermal expansion in E17.1.

Solution
Again starting with the definition of work

$$w = -\int_{V_1}^{V_2} P\,\mathrm{d}V = -P\int_{V_1}^{V_2}\mathrm{d}V = -P\left(V_2 - V_1\right)$$
$$= -\boldsymbol{n}RT\left(\frac{V_2}{V_1} - \frac{V_1}{V_1}\right) = -\boldsymbol{n}\boldsymbol{R}T\,(2-1) = -\boldsymbol{n}\boldsymbol{R}T$$

Hence we find that the work in a constant pressure expansion is smaller than that in an isothermal expansion by a factor of ln (2).

E17.3 In Section 17.5.2 we computed the entropy change for the mixing of two gases. Find this entropy by performing the reverse process. Compute the probability $\mathcal{P}$ that the mixture separates into gases A and B, and the change of the entropy.

Solution
Consider a container of volume $2V$ with a mixture of two gases A and B and calculate the probability that the gas A separates into a subvolume V and that the gas B separates into the other subvolume V, at pressure P, and temperature T (reverse of the process shown in Fig. 17.1b). We will calculate the entropy change in this irreversible separation process by counting the number of representations before and after the separation process. According

Solutions Manual for Principles of Physical Chemistry, Third Edition. Edited by Hans Kuhn, David H. Waldeck, and Horst-Dieter Försterling.

to the Sackur–Tetrode equation (17.49) the entropy of an atomic gas consisting of N atoms with mass m in a volume V is

$$S = Nk \cdot \ln\left(C \cdot \frac{m^{3/2}}{N} \cdot V\right) \quad \text{with} \quad C = e^{5/2}(2\pi)^{3/2}\frac{(k)^{3/2}}{h^3}T^{3/2}$$

We can write the initial entropy S_{initial} of the total system consisting of N_A atoms of mass m_A and N_B atoms with mass m_B as

$$\begin{aligned}
S_{\text{initial}} &= N_A k \cdot \ln\left(C\frac{m_A^{3/2}}{N_A} \cdot 2V\right) + N_B k \cdot \ln\left(C\frac{m_B^{3/2}}{N_B} \cdot 2V\right)\\
&= N_A k \cdot \left[\ln\left(C\frac{m_A^{3/2}}{N_A}V\right) + \ln 2\right] + N_B k \cdot \left[\ln\left(C\frac{m_B^{3/2}}{N_B}V\right) + \ln 2\right]\\
&= N_A k \cdot \ln\left(C\frac{m_A^{3/2}}{N_A}V\right) + N_A k \cdot \ln 2 + N_B k \cdot \ln\left(C\frac{m_B^{3/2}}{N_B}V\right) + N_B k \cdot \ln 2\\
&= N_A k \cdot \ln\left(C\frac{m_A^{3/2}}{N_A}V\right) + N_B k \cdot \ln\left(C\frac{m_B^{3/2}}{N_B}V\right) + [N_A k + N_B k] \cdot \ln 2
\end{aligned}$$

The final entropy S_{final} after they separate is the sum of the entropies of the two subvolumes, so that

$$S_{\text{final}} = N_A k \cdot \ln\left(C\frac{m_A^{3/2}}{N_A} \cdot V\right) + N_B k \cdot \ln\left(C\frac{m_B^{3/2}}{N_B} \cdot V\right)$$

For the entropy change we obtain

$$\Delta S = S_{\text{final}} - S_{\text{initial}} = -(N_A + N_B)\,k \cdot \ln 2$$

so that the entropy decreases upon separation.
The number of representations is given by $S = k \ln \Omega$, so that

$$\Omega_{\text{initial}} = \exp\left(\frac{S_{\text{initial}}}{k}\right) = \left(C\frac{m_A^{3/2}}{N_A} \cdot V\right)^{N_A}\left(C\frac{m_B^{3/2}}{N_B} \cdot V\right)^{N_B} \cdot 2^{(N_A+N_B)}$$

and

$$\Omega_{\text{final}} = \exp\left(\frac{S_{\text{final}}}{k}\right) = \left(C\frac{m_A^{3/2}}{N_A} \cdot V\right)^{N_A}\left(C\frac{m_B^{3/2}}{N_B} \cdot V\right)^{N_B}$$

Hence, we find that

$$\mathcal{P} = \frac{\Omega_{\text{final}}}{\Omega_{\text{initial}}} = \frac{1}{2^{(N_A+N_B)}} = \left(\frac{1}{2}\right)^{N_A+N_B}$$

Hence the final state is very improbable when the number of gas particles is large.
Aside: Note that this result is not valid for identical atoms A and B. In this case we have initially $2\,N_A$ atoms in volume $2\,V$

$$S_{\text{initial}} = 2N_A k \cdot \ln\left(C\frac{m^{3/2}}{2N_A}2V\right) = 2N_A k \cdot \ln\left(C\frac{m^{3/2}}{N_A}V\right)$$

and finally N_A atoms in each subvolume V

$$\begin{aligned}
S_{\text{final}} &= N_A k \cdot \ln\left(C\frac{m^{3/2}}{N_A}V\right) + N_A k \cdot \ln\left(C\frac{m^{3/2}}{N_A}V\right)\\
&= 2N_A k \cdot \ln\left(C\frac{m^{3/2}}{N_A}V\right)
\end{aligned}$$

Thus we obtain

$$\Delta S = S_{\text{initial}} - S_{\text{final}} = 0$$

E17.4 Perform an analysis for heat flow at constant volume for a system like that in Fig. 17.1a. Derive a mathematical expression for the entropy change in the upper container and for the final temperature in terms of the two bodies heat capacities, C_{upper} and C_{lower}, and initial temperatures, T_{upper} and T_{lower}. You may assume that the constant volume heat capacities are temperature independent. Evaluate the final temperature for the initial conditions of $T_{\text{lower}} = 273$ K, $T_{\text{upper}} = 373$ K, and 1 mol of a monatomic gas. Calculate how much the final temperature changes if the upper chamber has 2 mol of gas, rather than 1 mol.

Solution

When the systems reach thermal equilibrium they both have the final temperature T_f. For the case where the two gases are isolated from the external environment (but not from each other), we know that the net heat flow must be zero, so that

$$q_{\text{upper}} + q_{\text{lower}} = 0$$

Using this constraint, we can write

$$C_{\text{upper}}(T_f - T_{\text{upper}}) + C_{\text{lower}}(T_f - T_{\text{lower}}) = 0$$

Solving for T_f we find that

$$\frac{C_{\text{upper}} T_{\text{upper}} + C_{\text{lower}} T_{\text{lower}}}{C_{\text{upper}} + C_{\text{lower}}} = T_f$$

For 1 mol of a monatomic gas $C_{\text{upper}} = C_{\text{lower}} = \frac{3}{2}\boldsymbol{R} \cdot 1$ mol, so that

$$T_f = \frac{1 \text{ mol} \times \frac{3}{2}\boldsymbol{R} \times 373 \text{ K} + 1 \text{ mol} \times \frac{3}{2}\boldsymbol{R} \times 273 \text{ K}}{1 \text{ mol} \times 3\boldsymbol{R}} = 323 \text{ K}$$

whereas for 2 mols of gas in the upper chamber we find that

$$T_f = \frac{2 \text{ mol} \times \frac{3}{2}\boldsymbol{R} \times 373 \text{ K} + 1 \text{ mol} \times \frac{3}{2}\boldsymbol{R} \times 273 \text{ K}}{2 \text{ mol} \times \frac{3}{2}\boldsymbol{R} + 1 \times \frac{3}{2}\boldsymbol{R}} = 339 \text{ K}$$

For the change in entropy of the upper chamber, we find

$$\Delta S_{\text{upper}} = \int_{T_{\text{upper}}}^{T_f} \frac{C_{\text{upper}}}{T} dT = C_{\text{upper}} \ln\left(\frac{T_f}{T_{\text{upper}}}\right)$$

which is negative since $T_f < T_{\text{upper}}$ for the case described here. The decrease in entropy of the upper chamber is counteracted by the increase in entropy of the lower chamber, so that the total entropy of the system increases. You can show this by calculating the entropy in the lower chamber

$$\Delta S_{\text{lower}} = \int_{T_{\text{lower}}}^{T_f} \frac{C_{\text{lower}}}{T} dT = C_{\text{lower}} \ln\left(\frac{T_f}{T_{\text{lower}}}\right)$$

and adding the two entropy changes to find that $\Delta S_{\text{total}} = \Delta S_{\text{upper}} + \Delta S_{\text{lower}}$.

E17.5 Consider an isothermal and reversible expansion of an ideal gas to twice its initial volume. Determine whether the work w is positive, negative, or zero; the heat q is positive, negative, or zero; the internal energy change ΔU is positive, negative, or zero; and the entropy change ΔS is positive, negative, or zero.

Solution

The work for this process is negative; namely

$$w_{\text{rev}} = -\int_{V_1}^{V_2} P\, \text{d}V = -\boldsymbol{nRT}\int_{V_1}^{V_2} \frac{\text{d}V}{V} = -\boldsymbol{nR}T \ln\left(\frac{V_2}{V_1}\right)$$

Because $V_2 > V_1$ the logarithm term will be positive so that $w_{\text{rev}} < 0$.

Because this is an isothermal process, we have $\Delta U = 0$ for an ideal gas; i.e., the internal energy change is zero. From the first law (conservation of energy) we can write that

$$\Delta U = q_{\text{rev}} + w_{\text{rev}}$$

Because $\Delta U = 0$, we see that $q_{rev} = -w_{rev}$. Because $w_{rev} < 0$ we find that $q_{rev} > 0$.

The entropy change is given by

$$\Delta S = \int_{\text{initial}}^{\text{final}} \frac{dq_{rev}}{T}$$

and the entropy change in an isothermal process is given by

$$\Delta S = \frac{1}{T} \int_{\text{initial}}^{\text{final}} dq_{rev} = \frac{q_{rev}}{T}$$

Hence, we find that

$$\Delta S > 0$$

because $q_{rev} > 0$ and $T > 0$.

E17.6 In Equation (17.56) we calculated the entropy change for temperature equilibration as

$$\Delta S = \frac{3}{2} Nk \cdot \ln\left(\frac{T_{eq}^2}{T_1 T_2}\right) \quad \text{with } T_{eq} = \frac{T_1 + T_2}{2}$$

Show that ΔS is positive for any pair of temperatures T_1 and T_2.

Solution

First, we set

$$T_{eq} = T_1 + \Delta T = T_2 - \Delta T, \text{ thus } T_2 - T_1 = 2\Delta T$$

which allows us to write the temperature ratio in the argument of the logarithm as

$$\frac{T_{eq}^2}{T_1 T_2} = \frac{(T_1 + \Delta T)(T_2 - \Delta T)}{T_1 T_2} = \frac{T_1 T_2 + \Delta T (T_2 - T_1) - \Delta T^2}{T_1 T_2}$$
$$= \frac{T_1 T_2 + \Delta T (2\Delta T) - \Delta T^2}{T_1 T_2} = \frac{T_1 T_2 + \Delta T^2}{T_1 T_2} = 1 + \frac{\Delta T^2}{T_1 T_2}$$

This expression is always larger than 1, so that ΔS is always positive.

E17.7 Consider an ideal gas at 10 bar pressure and a temperature of 298 K in an isolated system. While remaining isolated, the gas is expanded into a vacuum from a volume $V_0 = 1$ L to a volume $V_f = 3$ L. Calculate ΔS for this process and the final temperature of the gas. Using your result, determine the heat q_{rev} that would be needed to perform this process reversibly.

Solution

We start with the Sackur–Tetrode equation

$$S = Nk \ln\left(C \frac{VT^{3/2}}{N}\right) \quad \text{where } C = e^{5/2} \frac{(2\pi mk)^{3/2}}{h^3}$$

Because the gas is ideal ($PV = NkT$) and the system is isolated $\Delta U = 0$, T is constant for the process, and we write the entropy change as

$$\Delta S = S_f - S_0 = Nk \ln\left(C \frac{V_f T^{3/2}}{N}\right) - Nk \ln\left(C \frac{V_0 T^{3/2}}{N}\right)$$
$$= Nk \ln\left(\frac{V_f}{V_0}\right) = Nk \cdot \ln(3)$$

If we performed this expansion reversibly and isothermally, the reversible heat would be

$$\Delta S = \int \frac{dq_{rev}}{T} = \frac{1}{T} \int dq_{rev} = \frac{q_{rev}}{T}$$

so that

$$q_{rev} = T\Delta S = NkT \cdot \ln(3)$$

E17.8 A piece of metal with heat capacity of 8 kJ K^{-1} and an initial temperature $T_1 = 273$ K is placed on top of another piece of metal with a heat capacity of 4 kJ K^{-1} and an initial temperature $T_2 = 373$ K. Assume that the two blocks are isolated from the environment and determine the final temperature and the total entropy change in the temperature equilibration between the two metal pieces.

Solution

We start with Fig. E17.8

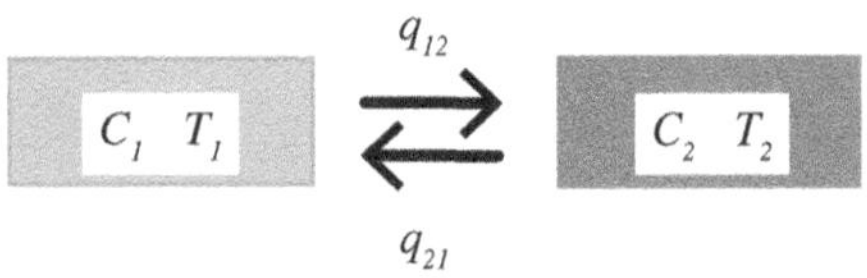

Figure E17.8 Two solids in temperature contact.

and noting that at equilibrium the change in heat going from 1 to 2 is q_{12} and is equal to $-q_{21}$; i.e., heat flows between the two metal blocks. Because the heat capacities are given as constants, we can write that $q_{12} = C_1 \Delta T_{1\to f}$ and $q_{21} = C_2 \Delta T_{2\to f}$. The final temperature is T_f can be found from the condition that

$$q_{12} = C_1 \Delta T_{1\to f} = -q_{21} = -C_2 \Delta T_{2\to f}$$

so that

$$C_1 (T_f - T_1) = -C_2 (T_f - T_2)$$

Solving for T_f we find that

$$T_f = \frac{C_1 T_1 + C_2 T_2}{C_1 + C_2} = \frac{2}{3} T_1 + \frac{1}{3} T_2 = 306.3 \text{ K}$$

With the final temperature determined, we can find the total entropy change. The total entropy change will be the sum of that for each block, so that

$$\begin{aligned} \Delta S_{\text{total}} &= \Delta S_1 + \Delta S_2 \\ &= \int_{T_1}^{T_f} \frac{q_{12}}{T} \mathrm{d}T + \int_{T_2}^{T_f} \frac{q_{21}}{T} \mathrm{d}T = \int_{T_1}^{T_f} C_1 \frac{\mathrm{d}T}{T} + \int_{T_2}^{T_f} C_2 \frac{\mathrm{d}T}{T} \end{aligned}$$

Upon integration (taking C_1 and C_2 as constant) we find that

$$\Delta S_{\text{total}} = C_1 \ln\left(\frac{T_f}{T_1}\right) + C_2 \ln\left(\frac{T_f}{T_2}\right)$$

and upon evaluation we obtain

$$\begin{aligned} \Delta S_{\text{total}} &= \left[8 \times \ln\left(\frac{306.3 \text{ K}}{273 \text{ K}}\right) + 4 \times \ln\left(\frac{306.3 \text{ K}}{373 \text{ K}}\right)\right] \text{ kJ K}^{-1} \\ &= (0.922 - 0.788) \text{ kJ K}^{-1} = 0.134 \text{ kJ K}^{-1} \end{aligned}$$

E17.9 Engineering tables list values of S for water. At 0°C, $S^{\ominus}_{273} = 3.9$ kJ K^{-1} for 1 kg of water. What is the number of representations Ω_{273} for water at 0 °C and atmospheric pressure? A kilogram of liquid water at 100°C and atmospheric pressure has the entropy $S^{\ominus}_{373} = S^{\ominus}_{273} + 1300$ J K^{-1}. Calculate the number of representations Ω_{373} for the water at 100 °C. Compare the two values.

Solution

At the temperature of 273 K we have $S^{\ominus}_{273} = k \cdot \ln \Omega$, so that

$$\Omega_{273} = \mathrm{e}^{S^{\ominus}_{273}/k} = \mathrm{e}^{2.8\times10^{26}}$$

At the temperature of 373 K, we find that

$$\Omega_{373} = \mathrm{e}^{S^{\ominus}_{373}/k} = \mathrm{e}^{3.7\times10^{26}} = \Omega_{273} \times 10^{9\times10^{25}}$$

Note that a small increase in the exponent (from 2.8×10^{26} to 3.7×10^{26}) corresponds to a very big increase of Ω.

E17.10 Consider the three-level system of energy $U = 40a$ in Table 15.3. What is the entropy of the system?

Solution

From Table 15.3 we know that

$$\Omega = 69.0 \times 10^{32}$$

so that the entropy is

$$\begin{aligned} S &= k \ln \Omega \\ &= \left(1.38 \times 10^{-23}\right) \ln \left(69.0 \times 10^{32}\right)\ \mathrm{J\,K^{-1}} = 1.25 \times 10^{-21}\ \mathrm{J\,K^{-1}} \end{aligned}$$

Why does this value seem so small? In part, it is small because this model system only has 100 particles.

E17.11 In Exercises E15.19 and E15.20 you computed the average energy and populations for spin sublevels in an applied magnetic field. Compute the entropy of each of those systems, for the cases of $N = 10$ and $N = 100$.

Solution

According to Equation (17.84)

$$\ln \Omega = N \ln Z + \frac{U}{kT} \quad \text{and} \quad S = Nk \ln Z + \frac{U}{T}$$

In E15.19 we considered the spins of a proton in a 12 Tesla magnetic field and found that $Z = 1.99968$ and $\bar{\varepsilon} = 1.69 \times 10^{-25}$ J at $T = 77$ K. Using $U = N\bar{\varepsilon}$, we find that

$$S = k \ln \left(Z^N\right) + \frac{N\bar{\varepsilon}}{T}$$

Substituting for $N = 100$, we find that $S = 9.56 \times 10^{-22}\ \mathrm{J\,K^{-1}}$ and for $N = 10$ we find $S = 9.56 \times 10^{-23}\ \mathrm{J\,K^{-1}}$.
In E15.20 we considered the spins of an electron in a 1.2 Tesla magnetic field and found that $Z = 1.9795$ and $\bar{\varepsilon} = 1.09 \times 10^{-23}$ J at $T = 77$ K. Using $U = N\bar{\varepsilon}$, we find that

$$S = k \ln \left(Z^N\right) + \frac{N\bar{\varepsilon}}{T}$$

Substituting for $N = 100$, we find that $S = 9.57 \times 10^{-22}\ \mathrm{J\,K^{-1}}$ and for $N = 10$ we find $S = 9.57 \times 10^{-23}\ \mathrm{J\,K^{-1}}$.

E17.12 Consider a gas that has expanded from the volume V_1 to the volume V_2 in Example 17.1. For the gas at V_2 calculate the probability of finding all of the gas molecules in the subvolume V_1. Calculate the entropy change for the gas.

Solution

Assuming ideal gas behavior, we can write that

$$\ln \Omega = const + N \ln V$$

where *const* is chosen to reproduce Equation (17.44) for an isothermal process. Thus we can write that

$$\ln \Omega_1 - \ln \Omega_2 - N \ln V_1 \quad N \ln V_2$$

This expression can be rearranged to give the probability of finding all particles in the subvolume V_1 when the gas is at the volume V_2; namely

$$\text{Probability} = \frac{\Omega_1}{\Omega_2} = \left(\frac{V_1}{V_2}\right)^N$$

and the entropy change can be written as

$$\Delta S = k \ln \Omega_2 - k \ln \Omega_1 = Nk \ln \left(\frac{V_2}{V_1}\right)$$

For the values given in the example, we have $N = (0.2\ \mathrm{mol})\left(6.022 \times 10^{23}\ \mathrm{mol^{-1}}\right) = 1.2 \times 10^{23}$, $V_1 = 7$ L, and $V_2 = 20$ L so that

$$\text{Probability} = \left(\frac{7}{20}\right)^{1.2\times 10^{23}} \approx \left(\frac{1}{3}\right)^{10^{23}}$$

which is miniscule. The entropy change is

$$\begin{aligned} \Delta S &= \left(1.2 \times 10^{23}\right)\left(1.38 \times 10^{-23}\ \mathrm{J\,K^{-1}}\right) \ln \left(\frac{20}{7}\right) \\ &= 1.7\ \mathrm{J\,K^{-1}} \end{aligned}$$

E17.13 Calculate the entropy for argon gas at $T = 300$ K and $P = 1$ bar in a container of volume V, where V is a) one milliliter, b) one microliter, and c) one nanoliter; assume that the gas behaves ideally.

Solution

We consider that Ar ($m = 6.63 \times 10^{-26}$ kg) behaves as an ideal monatomic gas and use the formula

$$S = Nk \ln\left[e^{5/2}(2\pi)^{3/2}\frac{(km)^{3/2}}{h^3}\frac{V}{N}T^{3/2}\right]$$

which can be written as

$$S = Nk \ln\left[e^{5/2}(2\pi)^{3/2}\frac{(kmT)^{3/2}}{h^3}\cdot\frac{V}{N}\right]$$
$$= Nk \ln\left(3.0 \times 10^{33}\ \text{m}^{-3}\cdot\frac{V}{N}\right)$$

Finally we calculate the particle number from the volume

$$PV = NkT \qquad N = \frac{PV}{kT}$$

so that

$$\frac{V}{N} = \frac{kT}{P} = \frac{1.38 \times 10^{-23}\ \text{J K}^{-1}\cdot 300\ \text{K}}{10^5\ \text{N m}^{-2}} = 4.14 \times 10^{-26}\ \text{m}^3$$

Thus for S we obtain

$$S = Nk \ln\left(3.0 \times 10^{33}\ \text{m}^{-3}\cdot\frac{V}{N}\right)$$
$$= Nk \ln\left(3.0 \times 10^{33}\ \text{m}^{-3}\cdot 4.14 \times 10^{-26}\ \text{m}^3\right)$$
$$= Nk \cdot \ln 1.24 \times 10^8 = Nk \cdot 18.6$$

and for the different volumes the result is

V	N	V/N	S/J K^{-1}
1 mL $= 10^{-6}$ m^3	2.42×10^{19}	4.14×10^{-26} m^3	5.8×10^{-3}
1 μL $= 10^{-9}$ m^3	2.42×10^{16}	4.14×10^{-26} m^3	5.8×10^{-6}
1 nL $= 10^{-12}$ m^3	2.42×10^{13}	4.14×10^{-26} m^3	5.8×10^{-9}

E17.14 Calculate the translational, rotational, and vibrational contributions to U, S, and C_V for 1 mol of ICl molecules at 300 and 1000 K. The fundamental vibrational frequency of ICl is $\nu_0 = 1.15 \times 10^{13}$ s^{-1} and the rotational constant of ICl is $B = 2.27 \times 10^{-24}$ J. You may consider the gas to behave ideally and have a pressure of 1 bar.

Solution

For the translational contributions we have an internal energy U_{trans} of

$$U_{\text{trans}} = \frac{3}{2}\boldsymbol{nRT}$$

which is

$$U_{\text{trans}} = \frac{3}{2} \times 1\ \text{mol} \times 8.314\ \text{J K}^{-1}\ \text{mol}^{-1} \times 300\ \text{K} = 3741.3\ \text{J}$$

at 300 K, and correspondingly 12, 471 J at 1000 K. For $C_{V,\text{trans}}$ we have that

$$C_{V,\text{trans}} = \frac{3\boldsymbol{n}}{2}\boldsymbol{R} = \frac{3}{2}\cdot 1\ \text{mol}\cdot 8.314\ \text{J K}^{-1} = 12.471\ \text{J K}^{-1}$$

and it is temperature independent, hence the same for 300 and 1000 K. To evaluate the entropy we need the mass ($m = 2.696 \times 10^{-25}$ kg) and volume in each case; for an ideal gas at 1 bar and 300 K we find

$$V = \frac{\boldsymbol{nRT}}{P} = \frac{1\ \text{mol} \times 8.314\ \text{J K}^{-1}\ \text{mol}^{-1} \times 300\ \text{K}}{10^5\ \text{N m}^{-2}} = 24.942\ \text{L}$$

and for 1 bar and 1000 K we find $V = 83.14$ L. For the entropy at 300 K we find

$$S_{\text{trans}} = \boldsymbol{nR} \ln \left[e^{5/2} (2\pi)^{3/2} \left(\frac{kTm}{h^2} \right)^{3/2} \frac{V}{N} \right]$$

$$= 1 \text{ mol} \cdot 8.314 \text{ J K}^{-1} \text{ mol}^{-1} \cdot$$

$$\ln \left[e^{5/2} \left[\frac{2\pi \cdot 1.38 \times 10^{-23} \text{ J K}^{-1} \cdot 300 \text{ K} \cdot 2.696 \cdot 10^{-25} \text{ kg}}{\left(6.626 \times 10^{-34} \text{ J s}\right)^2} \right]^{3/2} \frac{24.942 \times 10^{-3} \text{ m}^3}{6.022 \times 10^{23}} \right]$$

$$= 172.5 \text{ J K}^{-1}$$

at 1000 K we find $S_{\text{trans}} = 187.5 \text{ J K}^{-1}$.
For the rotational contribution, we have an internal energy contribution at $T = 300$ K of

$$U_{\text{rot}} = \boldsymbol{nRT} = 1 \text{ mol} \cdot 8.314 \text{ J K}^{-1} \text{ mol}^{-1} \cdot 300 \text{ K} = 2494.2 \text{ J}$$

and for $T = 1000$ K and $U_{\text{rot}} = 8314$ J. For C_V we find that

$$C_{V,\text{rot}} = \boldsymbol{nR} = 8.314 \text{ J K}^{-1}$$

For the entropy at $T = 300$ K we find that

$$S_{\text{rot}} = \boldsymbol{nR} \left[\ln \left(\frac{kT}{B} \right) + 1 \right]$$

$$= 8.314 \text{ J K}^{-1} \left[\ln \left(\frac{1.38 \times 10^{-23} \cdot 300}{2.27 \times 10^{-24}} \right) + 1 \right] = 70.74 \text{ J K}^{-1}$$

and for $T = 1000$ K we find that $S_{\text{rot}} = 80.75 \text{ J K}^{-1}$.
For the vibrational contribution, we will use the harmonic oscillator model which has a vibrational energy level spacing of

$$h\nu_0 = 6.626 \times 10^{-34} \text{ J s} \cdot 1.15 \times 10^{13} \text{ s}^{-1} = 7.62 \times 10^{-21} \text{ J}$$

Thus

$$\frac{h\nu_0}{kT} = \frac{7.62 \times 10^{-21} \text{ J}}{1.38 \times 10^{-23} \text{ J K}^{-1} \text{ 300 K}} = 1.84$$

For the internal energy contribution at 300 K we find

$$U_{\text{vib}} = N h\nu_0 \left(\frac{1}{2} + \frac{1}{\exp\left(h\nu_0/(kT)\right) - 1} \right)$$

$$= \left(6.022 \times 10^{23}\right) \left(7.62 \times 10^{-21} \text{ J}\right) \left(\frac{1}{2} + \frac{1}{\exp(1.84) - 1} \right)$$

$$= 4.59 \text{ kJ} \cdot 0.689 = 3.160 \text{ kJ}$$

and at $T = 1000$ K we find 8523 kJ. For C_V at $T = 300$ K we use

$$C_{V,\text{vib}} = \frac{N\left(h\nu_0\right)^2}{kT^2} \frac{\exp\left(h\nu_0/kT\right)}{\left(\exp\left(h\nu_0/kT\right) - 1\right)^2}$$

and find that $C_{V,\text{vib}} = 6.31 \text{ J K}^{-1}$. At $T = 1000$ K we find $C_{V,\text{vib}} = 8.10 \text{ J K}^{-1}$. For the entropy at $T = 300$ K we use

$$S_{\text{vib}} = \boldsymbol{nR} \left[\frac{h\nu_0/kT}{\exp\left(h\nu_0/kT\right) - 1} - \ln\left(1 - \exp\left(-h\nu_0/kT\right)\right) \right]$$

and find that $S_{\text{vib}} = 4.32 \text{ J K}^{-1}$. At $T = 1000$ K the entropy is $S_{\text{vib}} = 13.4 \text{ J K}^{-1}$.

E17.15 At 1 bar pressure and $T = 194.65$ K solid CO_2 sublimates to form CO_2 vapor. Compare the molar entropy of the solid (68.3 J $\text{mol}^{-1}\text{K}^{-1}$) and vapor (199.1 J $\text{mol}^{-1}\text{K}^{-1}$). Explain which motions contribute to each and how it explains the relative size of the entropies.

Solution

CO_2 vapor: Translational motion makes the largest contribution to the entropy. Rotational motion the second largest contribution to the entropy. The vibrational motion contributes weakly to the entropy and the electronic contribution can be neglected.

CO_2 solid: Rotation and translation are largely unavailable in the solid; however, the vibrational motion of CO_2 molecules about their positions in the solid will be important and make the largest contribution to the heat capacity; i.e., the solid has a large entropy contribution from intermolecular vibrational modes.

E17.16 Calculate the molar entropies of HF and HI from spectroscopic parameters (for HF $g_{el} = 1$, $\sigma = 1$, $\theta_{vib} = 5954$ K, $\theta_{rot} = 29.4$ K, and for HI $g_{el} = 1$, $\sigma = 1$, $\theta_{vib} = 3266$ K, $\theta_{rot} = 9.1$ K). Compare the entropy values and explain why they are different. You may assume that the gases behave ideally and are at standard conditions of 1 bar and 298.15 K.

Solution

The translational contribution to the molar entropy for HF ($m = 3.32 \times 10^{-26}$ kg) is

$$S_{\text{trans},m} = \boldsymbol{R} \ln \left[e^{5/2} \left(\frac{2\pi mkT}{h^2} \right)^{3/2} \frac{V}{N} \right]$$

With

$$\frac{V}{N} = \frac{kT}{P} = \frac{1.38 \times 10^{-23} \text{ J K}^{-1} \cdot 298.15 \text{ K}}{10^5 \text{ N m}^{-2}} = 4.11 \times 10^{-26} \text{ m}^3$$

we obtain

$$S_{\text{trans},m} = \boldsymbol{R} \ln \left[e^{5/2} \left[\frac{2\pi \left(3.32 \times 10^{-26} \text{ kg}\right) \left(1.38 \times 10^{-23} \text{ J K}^{-1}\right) (298.15 \text{ K})}{\left(6.626 \times 10^{-34} \text{ J s}\right)^2} \right]^{3/2} 4.11 \times 10^{-26} \text{ m}^3 \right]$$

$$= \boldsymbol{R} \ln \left[12.2 \cdot 8.64 \times 10^{31} \text{ m}^{-3} \cdot 4.11 \times 10^{-26} \text{ m}^3 \right] = 17.58 \cdot \boldsymbol{R} = 146.2 \text{ J K}^{-1} \text{ mol}^{-1}$$

and similarly for HI ($m = 2.12 \times 10^{-25}$ kg) we find that $S_{\text{trans},m} = 169.3 \text{ J K}^{-1}\text{mol}^{-1}$.

The rotational contribution to the molar entropy for HF is

$$S_{\text{rot},m} = \boldsymbol{R} \left[\ln \left(\frac{kT}{\sigma B} \right) + 1 \right] = \boldsymbol{R} \left[\ln \left(\frac{T}{\sigma \theta_{\text{rot}}} \right) + 1 \right]$$

$$= 8.314 \text{ J K}^{-1} \text{ mol}^{-1} \left[\ln \left(\frac{298.15 \text{ K}}{(1)(29.4 \text{ K})} \right) + 1 \right] \text{ J K}^{-1} \text{ mol}^{-1} = 27.57 \text{ J K}^{-1} \text{ mol}^{-1}$$

and similarly for HI we find that $S_{\text{rot},m} = 37.32 \text{ J K}^{-1}\text{mol}^{-1}$.

The vibrational contribution to the molar entropy for HF is

$$S_{\text{vib},m} = \boldsymbol{R} \left[\frac{\theta_{\text{vib}}/T}{\exp\left(\theta_{\text{vib}}/T\right) - 1} - \ln\left(1 - \exp\left(-\theta_{\text{vib}}/T\right)\right) \right]$$

$$= 8.314 \text{ J K}^{-1} \text{ mol}^{-1} \left[\left(\frac{5954/298.15}{\exp(5954/298.15) - 1} \right) - \ln\left(1 - \exp(-5954/298.15)\right) \right]$$

$$= 3.70 \times 10^{-7} \text{ J K}^{-1} \text{ mol}^{-1}$$

and similarly for HI we find that $S_{\text{vib},m} = 1.61 \times 10^{-3} \text{ J K}^{-1} \text{ mol}^{-1}$.

Because the degeneracies of the electronic ground states are 1 and the energy required to reach the first excited state is much greater than kT, the electronic contributions are taken to be zero.

The results are summarized in the table. Note that the total entropy is a sum of each contribution; i.e.,

$$S_{\text{total},m} = S_{\text{trans},m} + S_{\text{rot},m} + S_{\text{vib},m} + S_{\text{elec},m}$$

	$S_{\text{trans},m}$ J K^{-1} mol^{-1}	$S_{\text{rot},m}$ J K^{-1} mol^{-1}	$S_{\text{vib},m}$ J K^{-1} mol^{-1}	$S_{\text{elec},m}$ J K^{-1} mol^{-1}	$S_{\text{total},m}$ J K^{-1} mol^{-1}
HF	146.2	27.57	3.70×10^{-7}	0	173.8
HI	169.3	37.32	1.61×10^{-3}	0	206.6

To assess the origin of the difference in entropy values we compute

$$\Delta S_{\text{total},m} = S_{\text{total},m}(\text{HF}) - S_{\text{total},m}(\text{HI})$$
$$= \Delta S_{\text{trans},m} + \Delta S_{\text{rot},m} + \Delta S_{\text{vib},m}$$
$$= \frac{3}{2}\boldsymbol{R}\ln\left(\frac{m_{\text{HF}}}{m_{\text{HI}}}\right) + \boldsymbol{R}\ln\left(\frac{\theta_{\text{rot}}(\text{HI})}{\theta_{\text{rot}}(\text{HF})}\right) + \Delta S_{\text{vib},m}$$

Hence we see that the difference in entropy between these two gases results directly from the difference in their masses and in the value of their rotational energy level spacings. Because HI has the larger molecular mass and the larger inertial moment, it has a larger entropy. While it is also evident that the HI has a larger contribution from its vibrational mode (see table), it is evident that this effect is small at 298.15 K; i.e., the vibrations are largely frozen out at T = 298 K.

E17.17 Calculate the molar entropies of CO and N_2 from their spectroscopic parameters (for CO: $g_{\text{el}} = 1, \sigma = 1, \theta_{\text{vib}} = 3103$ K, $\theta_{\text{rot}} = 2.8$ K, and for N_2: $g_{\text{el}} = 1$, $\sigma = 2$, $\theta_{\text{vib}} = 3374$ K, $\theta_{\text{rot}} = 2.9$ K). Note that the difference in the symmetry factors σ is based on the fact that CO is a heteronuclear molecule and N_2 is homonuclear. Compare the entropy values and explain why they are different. You may assume that the gases behave ideally and are at standard conditions of 1 bar and 298.15 K.

Solution

The translational contribution to the molar entropy for CO ($m = 4.651 \times 10^{-26}$ kg) is

$$S_{\text{trans},m} = \boldsymbol{R}\ln\left[e^{5/2}\left(\frac{2\pi kTm}{h^2}\right)^{3/2}\frac{V}{N}\right]$$

With

$$\frac{V}{N} = \frac{kT}{P} = \frac{1.38 \times 10^{-23}\ \text{J K}^{-1} \cdot 298.15\ \text{K}}{10^5\ \text{N m}^{-2}} = 4.11 \times 10^{-26}\ \text{m}^3$$

we obtain

$$S_{\text{trans},m} = \boldsymbol{R}\ln\left[e^{5/2}\left(\frac{2\pi kTm}{h^3}\right)^{3/2}\frac{V}{N}\right]$$
$$= \boldsymbol{R}\ln\left[e^{5/2}\left[\frac{2\pi \cdot 1.38 \times 10^{-23}\ \text{J K}^{-1} \cdot 298.15\ \text{K} \cdot 4.65 \times 10^{-26}\ \text{kg}}{(6.626 \times 10^{-34}\ \text{J s})^2}\right]^{3/2} \cdot 4.11 \times 10^{-26}\ \text{m}^3\right]$$
$$= 18.09 \cdot \boldsymbol{R} = 150.4\ \text{J K}^{-1}\ \text{mol}^{-1}$$

and similarly for N_2 ($m = 4.653 \times 10^{-26}$ kg) we find that $S_{\text{trans},m} = 150.4\ \text{J K}^{-1}\text{mol}^{-1}$. The similarity of these two values results from the molecules very similar masses.

The rotational contribution to the molar entropy for CO is

$$S_{\text{rot},m} = \boldsymbol{R}\left[\ln\left(\frac{kT}{\sigma B}\right) + 1\right] = \boldsymbol{R}\left[\ln\left(\frac{T}{\sigma\theta_{\text{rot}}}\right) + 1\right]$$
$$= 8.314\left[\ln\left(\frac{298.15}{(1)(2.8)}\right) + 1\right]\ \text{J K}^{-1}\ \text{mol}^{-1} = 47.1\ \text{J K}^{-1}\ \text{mol}^{-1}$$

and similarly for N_2 (with $\sigma = 2$) we find that $S_{\text{rot},m} = 41.1\ \text{J K}^{-1}\text{mol}^{-1}$. Despite their similar rotational temperatures, nitrogen has a significantly smaller rotational contribution to its molar entropy because of its higher symmetry factor.

The vibrational contribution to the molar entropy for CO is

$$S_{\text{vib},m} = \boldsymbol{R}\left[\frac{\theta_{\text{vib}}/T}{\exp(\theta_{\text{vib}}/T) - 1} - \ln\left(1 - \exp\left(-\theta_{\text{vib}}/T\right)\right)\right]$$
$$= 8.314\left[\frac{3103/298.15}{\exp(3103/298.15) - 1} - \ln\left[1 - \exp\left(-\frac{3103}{298.15}\right)\right]\right]$$
$$= 2.64 \times 10^{-3}\ \text{J K}^{-1}\ \text{mol}^{-1}$$

and similarly for N_2 we find that $S_{\text{vib},m} = 1.16 \times 10^{-3}\ \text{J K}^{-1}\text{mol}^{-1}$.

Because the degeneracies of the electronic ground states are 1 and the energy required to reach the first excited state is much greater than kT, the electronic contributions are taken to be zero.
Note that the total entropy is a sum of each contribution; i.e.,

$$S_{\text{total},m} = S_{\text{trans},m} + S_{\text{rot},m} + S_{\text{vib},m} + S_{\text{elec},m}$$

The results are summarized in the table.

	$\frac{S_{\text{trans},m}}{\text{J K}^{-1}\text{ mol}^{-1}}$	$\frac{S_{\text{rot},m}}{\text{J K}^{-1}\text{ mol}^{-1}}$	$\frac{S_{\text{vib},m}}{\text{J K}^{-1}\text{ mol}^{-1}}$	$\frac{S_{\text{elec},m}}{\text{J K}^{-1}\text{ mol}^{-1}}$	$\frac{S_{\text{total},m}}{\text{J K}^{-1}\text{ mol}^{-1}}$
CO	150.4	47.1	2.64×10^{-3}	0	197.5
N_2	150.4	41.1	1.16×10^{-3}	0	191.5

To assess the origin of the difference in entropy values we compute

$$\begin{aligned}
\Delta S_{\text{total},m} &= S_{\text{total},m}(\text{CO}) - S_{\text{total},m}(\text{N}_2) \\
&= \Delta S_{\text{trans},m} + \Delta S_{\text{rot},m} + \Delta S_{\text{vib},m} \\
&= \frac{3}{2} \boldsymbol{R} \ln \frac{m_{\text{CO}}}{m_{\text{N}_2}} + \boldsymbol{R} \ln \frac{\sigma_{\text{N}_2} \theta_{\text{rot}}(\text{N}_2)}{\sigma_{\text{CO}} \theta_{\text{rot}}(\text{CO})} + \Delta S_{\text{vib},m} \\
&\approx \boldsymbol{R} \ln \frac{\sigma_{\text{N}_2}}{\sigma_{\text{CO}}} = \boldsymbol{R} \ln 2 = 5.8 \text{ J K}^{-1}\text{ mol}^{-1}
\end{aligned}$$

Hence we see that the difference in entropy between these two gases results almost solely from the difference in their symmetry factor.

E17.18 For a given substance, the entropy of the solid is less than that of the liquid which is less than that of the gas. Discuss how the different motions (translation, rotation, vibration, and electronic) contribute to each and explain this trend.

Solution
Solid: For a solid the translational and rotational contributions are not accessible (to a large extent). These degrees of freedom contribute significantly and are dominant for liquids and gases. Hence the entropy of a solid is least. Gas: For a gas the translational and rotational degrees of freedom are accessible and dominate the contributions to the entropy. In rare cases the rotations may not be fully accessible, but nevertheless the gas entropy is the largest. Liquid: The liquid is intermediate between the above two limits because the translational and rotational degrees of freedom are hindered but accessible.

E17.19 Consider the absolute entropies for the gases in Table 17.3. Explain the trend in the entropies for the monatomic series of gases. Explain the trend in the entropy for the diatomic gases. Explain the difference between the two sets of gases.

Solution
Monatomic Gases: The entropy of the monatomic gases is determined by the translational contribution to their entropy. Hence it correlates with the atomic mass, as the mass increases the entropy increases. Diatomic Gases: As with the monatomic gases the translational contribution to the entropy causes an increase with increasing mass. For the diatomic gases $F_2 \rightarrow I_2$, the inertial moment increases and the vibrational frequency decreases through the series. These two effects cause an increase in the rotational and vibrational contributions to their entropies through the series. Comparison: The diatomic entropies are much larger than the monatomic gases because of the contributions from their rotational and vibrational degrees of freedom.

E17.20 Consider the absolute entropies for the three gases Ne ($S^{\ominus}_{m,298} = 146.2\text{ J mol}^{-1}\text{K}^{-1}$), HF ($S^{\ominus}_{m,298} = 173.8\text{ J mol}^{-1}\text{K}^{-1}$) and D_2O ($S^{\ominus}_{m,298} = 198.3\text{ J mol}^{-1}\text{K}^{-1}$). These gases all have about the same total mass of 20 g mol^{-1}. Explain the difference in the entropies of the gases in terms of the gas molecule's molecular characteristics. Also consider the molecule ND_3, where do you expect its absolute entropy to lie with respect to Ne, HF, and D_2O?

Solution

Given that the total masses are the same we expect that the translational contributions to the entropy would be similar for the three gases. The increase of the molar entropies of HF and D_2O over that of Ne reflects the contributions from the rotational and vibrational degrees of freedom of these molecules. The increase in the molar entropy for D_2O, as compared to that of HF, reflects its additional rotational degree of freedom and the presence of lower frequency vibrational modes.

We would expect that the entropy of ND_3 would be larger than that of D_2O because of the presence of higher inertial moments and lower frequency vibrations.

E17.21 In Example 17.8 we compute the entropy change of an ideal gas in a two-step process: 1) heating at constant volume and 2) expansion at constant temperature. Find the entropy change between the same initial and final states by using a different two-step process: 1) heating at constant pressure and 2) expansion at constant temperature.

Solution

Step 1 (Isobaric heating): In this process the heat flow is given by

$$q_{\text{rev},1} = C_P \cdot \Delta T = \frac{5}{2}Nk \cdot \Delta T$$

To obtain the entropy change ΔS_1, we evaluate

$$\Delta S_1 = \int_{T_q}^{T_2} \frac{C_P}{T} \text{d}T = \frac{5}{2}Nk \ln\left(\frac{T_2}{T_1}\right)$$

The volume of the gas after the heating step is given by

$$V' = \frac{NkT_2}{P_1} = V_1\frac{T_2}{T_1}$$

Step 2 (Isothermal expansion): Because the process is isothermal and the gas is ideal we know that $\Delta U = 0$ and $q_{\text{rev}} = -w_{\text{rev}}$; hence, we can write that

$$\Delta S_2 = \frac{q_{\text{rev}}}{T_2} = \frac{-w_{\text{rev}}}{T_2}$$

Now we can evaluate the reversible work for the expansion from V' to V_2, and use it to find the entropy change; namely

$$\Delta S_2 = \frac{1}{T_2}\int_{V'}^{V_2} P\text{d}V = Nk \ln\frac{V_2}{V'} = Nk \ln\frac{V_2T_1}{V_1T_2}$$

Now we find the total entropy change as the sum of the two steps and obtain

$$\begin{aligned}\Delta S &= \Delta S_1 + \Delta S_2\\ &= \frac{5}{2}Nk \ln\frac{T_2}{T_1} + Nk \ln\frac{V_2T_1}{V_1T_2} = Nk \ln\left(\frac{V_2}{V_1}\frac{T_2^{3/2}}{T_1^{3/2}}\right)\end{aligned}$$

This result is identical to that found in Example 17.8.

E17.22 Consider two CO dipoles that are arranged side by side (take the CO dipole moment to be 0.1 D, its bond length to be 113 pm, and its diameter to be 350 pm). Using information from Chapter 13 calculate the difference in the dipole–dipole interaction energy for the antiparallel arrangement of the dipoles and for the parallel arrangement of the dipoles. At what temperature is this energy difference equal to kT?

Solution

For the first case in which the dipoles are parallel we have, in the point dipole approximation,

$$\begin{aligned}E_p &= \frac{1}{4\pi\varepsilon_0} \cdot \frac{\mu^2}{r^3}\\ &= \frac{1}{4\pi \cdot 8.854 \times 10^{-12}\ \text{C}^2\ \text{J}^{-1}\ \text{m}^{-1}} \cdot \frac{0.1 \cdot \left(3.3356 \times 10^{-30}\ \text{C m}\right)^2}{\left(350 \times 10^{-12}\ \text{m}\right)^3}\text{J}\\ &= 2.33 \times 10^{-23}\ \text{J}\end{aligned}$$

In the second case in which the dipoles are antiparallel we find that

$$E_a = -\frac{1}{4\pi\varepsilon_0} \cdot \frac{\mu^2}{r^3} = -2.33 \times 10^{-23} \text{ J}$$

The energy difference is

$$\Delta E = E_p - E_a$$
$$= \left(2.33 \times 10^{-23} + 2.33 \times 10^{-23}\right) \text{ J} = 4.66 \times 10^{-23} \text{ J}$$

To compare with kT, we determine the temperature for which

$$kT = 4.66 \times 10^{-23} \text{ J}$$

so that

$$T = \frac{4.66 \times 10^{-23} \text{ J}}{k} = \frac{4.66 \times 10^{-23} \text{ J}}{1.38 \times 10^{-23} \text{ J K}^{-1}} = 3.4 \text{ K}$$

Hence this interaction energy is relatively weak and not important for common temperatures.

E17.23 Two equal containers (each of volume V) contain two different ideal gases A and B (pressure P and temperature T). By connecting the containers the gases are mixed. What is the mixing entropy and what is the reversible work necessary to separate the gases?

Solution

From Section 17.5.2, we obtain for the entropy of a system of two different gases A and B

$$S_1 = N_A k \cdot \ln\left(C\frac{V}{N_A}\right) + N_B k \cdot \ln\left(C\frac{V}{N_B}\right)$$

with

$$C = e^{5/2}(2\pi)^{3/2}\frac{(km_A)^{3/2}}{h^3}T^{3/2}$$

After mixing, both gases occupy a volume $2V$, and thus the entropy is

$$S_2 = N_A k \cdot \ln\left(C\frac{2V}{N_A}\right) + N_B k \cdot \ln\left(C\frac{2V}{N_B}\right)$$
$$= N_A k \cdot \ln\left(C\frac{V}{N_A}\right) + N_B k \cdot \ln\left(C\frac{V}{N_B}\right) + N_A k \cdot \ln 2 + N_B k \cdot \ln 2$$
$$= N_A k \cdot \ln\left(C\frac{V}{N_A}\right) + N_B k \cdot \ln\left(C\frac{V}{N_B}\right) + (N_A k + N_B k) \cdot \ln 2$$

For the entropy change we obtain

$$\Delta S = S_2 - S_1 = (N_A + N_B)k \cdot \ln 2$$

This means that the entropy increases after mixing.

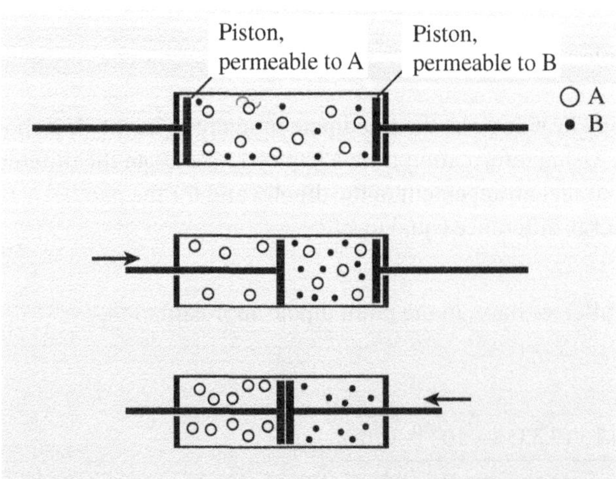

Figure E17.23 Separation of two gases using semipermeable pistons.

The reversible work w_{rev} and the reversible heat q_{rev} to separate the two gases are (see Fig. 17.2)

$$w_{rev} = -N_A kT \cdot \ln \frac{V}{2V} - N_B kT \cdot \ln \frac{V}{2V} = (N_A + N_B)kT \cdot \ln 2$$

and

$$q_{rev} = -w_{rev} = -(N_A + N_B)kT \cdot \ln 2$$

E17.24 Consider the evaporation of a liquid at constant pressure (as shown in Fig. 17.2). The entropy change for the process is

$$\Delta S = \frac{(q_{rev})_P}{T} = \frac{\Delta H}{T}$$

Given that the molar enthalpy of vaporization for water is 44.1 kJ mol^{-1} and its boiling point is 373 K at 1 bar pressure, calculate the molar entropy change for water when it vaporizes.

Solution

$$\Delta S_m = \frac{\Delta H_m}{T} = \frac{44.1 \text{ kJ mol}^{-1}}{373 \text{ K}} = 118 \text{ J K}^{-1} \text{ mol}^{-1}$$

E17.25 In our considerations of the entropy of subsystems, we wrote the number of representations of the total system as a product of that for the subsystems. Explain why this procedure is appropriate.

Solution

Because a rigid wall separates the subsystems their energy level spectrum and level occupations are independent of each other. Hence the possible configurations in one subsystem can be sampled for each and every configuration in the other subsystem. Consequently, the combined number of configurations is their product.

E17.26 Equation (17.12) relates the change in temperature for a reversible adiabatic expansion in terms of the ratio V_1/V_2. Rewrite the equation in terms of P_1/P_2. If you expand the gas from $P_1 = 10$ bar and $T_1 = 300$ K to $P_2 = 1$ bar, what is the final temperature?

Solution

Equation (17.12) is

$$T_2 = T_1 \cdot \left(\frac{V_1}{V_2}\right)^\gamma \quad \text{with } \gamma = \boldsymbol{nR}/C_{V,m}$$

Because of

$$P_1V_1 = \boldsymbol{nR}T_1 \qquad P_2V_2 = \boldsymbol{nR}T_2$$

we find

$$\frac{V_1}{V_2} = \frac{\boldsymbol{nR}T_1}{P_1}\frac{P_2}{\boldsymbol{nR}T_2} = \frac{P_2}{P_1}\frac{T_1}{T_2}$$

and

$$T_2 = T_1 \cdot \left(\frac{V_1}{V_2}\right)^\gamma = T_1 \cdot \left(\frac{P_2}{P_1}\right)^\gamma \left(\frac{T_1}{T_2}\right)^\gamma$$

or

$$T_2^{1+\gamma} = T_1^{1+\gamma} \cdot \left(\frac{P_2}{P_1}\right)^\gamma$$

and

$$T_2 = T_1 \cdot \left(\frac{P_2}{P_1}\right)^{\gamma/(1+\gamma)}$$

For example, for an ideal atomic gas we have

$$\gamma = \frac{\boldsymbol{nR}}{C_{V,m}} = \frac{\boldsymbol{nR}}{3\boldsymbol{nR}/2} = \frac{2}{3}$$

and

$$\frac{\gamma}{1+\gamma} = \frac{2}{3}\frac{1}{1+2/3} = \frac{2}{3}\frac{3}{5} = \frac{2}{5}$$

If we expand the gas from $P_1 = 10$ bar to $P_2 = 1$ bar from $T_1 = 300$ K we obtain

$$T_2 = 300\ \text{K} \cdot \left(\frac{1}{10}\right)^{2/5} = 300\ \text{K} \cdot 0.398 = 119\ \text{K}$$

17.2 Problems

P17.1 Figure 17.13 illustrates a process of temperature equilibration. Calculate the probability $\mathcal{P}$ that, by chance, the molecules are distributed as in State 1.

Solution

As illustrated in Fig. 17.13, only two energy levels are available to each molecule participating in the process. In the cold body, all four molecules occupy the lower energy level. In the hot body, two molecules occupy the lower energy level and two molecules occupy the upper energy level. Hence with $N = 4$ and $N_1 = 4$, we have

$$\Omega_{\text{cold}} = \frac{4!}{4!} = 1$$

Correspondingly, in the hot body we have $N = 4$, $N_1 = 2$, and $N_2 = 2$, so we obtain

$$\Omega_{\text{hot}} = \frac{4!}{2!2!} = 6$$

A first configuration of the hot body is: Particles 1 and 2 in the lower level, Particles 3 and 4 in the upper level (12|34). The five remaining configurations are (13|24), (14|23), (23|14), (24|13), and (34|12). The number Ω_1 of configurations in State 1 (cold/hot) is the product of Ω_{cold} and Ω_{hot}:

$$\Omega_1 = \Omega_{\text{cold}} \cdot \Omega_{\text{hot}} = 6$$

We bring the bodies in contact and wait for thermal equilibration (State 2). Because the internal energy U must remain constant, we find the following possibilities for State 2:

(a) left body: $N_1 = 4$, $N_2 = 0$; right body: $N_1 = 2$, $N_2 = 2$

(b) left body: $N_1 = 2$, $N_2 = 2$; right body: $N_1 = 4$, $N_2 = 0$

(c) left and right body: $N_1 = 3$, $N_2 = 1$

In all cases we have a total of six molecules in level 1 and two molecules in level 2. Then

$$\Omega_2 = \frac{8!}{6! \cdot 2!} = 28$$

So Ω_2 is greater than Ω_1.

In our numerical example, the probability $\mathcal{P}$ that after equilibration the system will be found accidentally in State 1 is

$$\mathcal{P} = \frac{\Omega_1}{\Omega_2} = \frac{6}{28} = 0.21$$

still relatively large.

If we increase the number of molecules, then $\mathcal{P}$ will decrease drastically. For example, for $N = 40$,

$$\Omega_1 = \frac{40!}{20! \cdot 20!} = 1.38 \times 10^{11}\ ,\ \Omega_2 = \frac{80!}{60! \cdot 20!} = 3.6 \times 10^{18}$$

so that we obtain

$$\mathcal{P} = \frac{1.38 \times 10^{11}}{3.6 \times 10^{18}} = 4 \times 10^{-8}$$

For arbitrary N we obtain

$$\Omega_1 = \frac{N!}{\left(\frac{N}{2}\right)! \cdot \left(\frac{N}{2}\right)!} \quad \text{and} \quad \Omega_2 = \frac{(2N)!}{\left(\frac{3N}{2}\right)! \cdot \left(\frac{N}{2}\right)!}$$

If we approximate $N!$ by Stirling's formula (see Appendix C.3)

$$N! \approx N^N \cdot \mathrm{e}^N$$

we find that

$$\mathcal{P} = \frac{\Omega_1}{\Omega_2} = \frac{2^N}{3.06^N} = (0.65)^N$$

If we set $N = 10^{23}$, then we can calculate

$$\mathcal{P} = (0.65)^{10^{23}} \approx 10^{-10^{22}}$$

P17.2 Consider an ideal atomic gas ($C_V = \frac{3}{2}Nk$) whose volume changes from V_1 to $V_2 = 2V_1$ in two different ways. In the first case the gas expansion proceeds isothermally and reversibly at the temperature $T_1 = 300$ K, and in the second case it proceeds adiabatically and reversibly. Find expressions for the work w, heat q, internal energy change ΔU, and entropy change ΔS for these two different processes and compare them.

Solution

Isothermal Process: Because the gas is ideal we can write that $\Delta U_T = 0$ and therefore $w_T = -q_T$, where we use the subscript T to emphasize that the process is isothermal. Hence we can write that

$$w_T = -q_T = -\int_{V_1}^{2V_1} P\mathrm{d}V = -\boldsymbol{nR}T_1 \ln\left(\frac{2V_1}{V_1}\right)$$
$$= -NkT_1 \ln(2)$$

The change in entropy of the gas is

$$\Delta S_T = \int \frac{\mathrm{d}q}{T} = \frac{q_T}{T_1} = Nk \ln(2)$$

Adiabatic Process: For an adiabatic process $q_a = 0$ so that the internal energy change and work are equal; namely

$$\Delta U_a = w_a = Nk\frac{3}{2} \cdot \Delta T$$

however we do not know the final temperature and hence ΔT. We can evaluate the work directly which we derived for an ideal gas (see Section 17.2.2.2), so that

$$w_a = -\int_{V_1}^{2V_1} P\mathrm{d}V = -P_1 \int_{V_1}^{2V_1} \left(\frac{V_1}{V}\right)^{5/3} \mathrm{d}V = -P_1 V_1^{5/3} \int_{V_1}^{2V_1} \frac{\mathrm{d}V}{V^{5/3}}$$
$$= -P_1 V_1^{5/3} \left(-\frac{3}{5}\right) \frac{1}{V^{2/3}}\bigg|_{V_1}^{2V_1} = \frac{3}{5} P_1 V_1 \left[\frac{V_1^{2/3}}{(2V_1)^{2/3}} - 1\right]$$
$$= \frac{3}{5} P_1 V_1 \left(\frac{1}{2^{2/3}} - 1\right) = -NkT_1 \frac{3}{5}\left(1 - \frac{1}{2^{2/3}}\right)$$

The change in entropy is

$$\Delta S = \int \frac{\mathrm{d}q}{T} = 0 \quad \text{since} \quad \mathrm{d}q = 0$$

To compare we first construct the table below.

	Work	Heat	ΔU	ΔS
Isothermal	$-\ln(2) \cdot NkT_1$	$NkT_1 \ln(2)$	0	$Nk \ln(2)$
Adiabatic	$-0.222 \cdot NkT_1$	0	$-0.222 \cdot NkT_1$	0

It is evident that the heat flow is higher and the worked performed is higher (because $\ln(2) > 0.222$) in the isothermal process. While the internal energy of the system decreases for the adiabatic process, the internal energy in the isothermal process does not change (because the heat flow into the system is used to perform work). The isothermal process increases the entropy of the system, whereas the adiabatic process does not.

P17.3 The molecular partition function Z for a monatomic ideal gas is given by

$$Z = V\left(\frac{2\pi mkT}{h^2}\right)^{3/2}$$

Use this expression for the molecular partition function to derive an expression for the internal energy U and for the entropy of an ideal monatomic gas.

Solution

From Chapter 15, Equation (15.13), we know that

$$U = NkT^2\frac{\mathrm{d}\ln Z}{\mathrm{d}T}$$

Substituting for Z, we find that

$$U = NkT^2\left[\frac{1}{Z_{\text{trans}}}\right]\cdot\frac{\mathrm{d}Z_{\text{trans}}}{\mathrm{d}T}$$

$$= NkT^2\left[\left(\frac{h^2}{2\pi mkT}\right)^{3/2}\frac{1}{V}\right]\left(\frac{2\pi mk}{h^2}\right)^{3/2}V\cdot\frac{3}{2}T^{1/2} = \frac{3}{2}NkT$$

To find the entropy, we use Equation (17.43) and the fundamental relation $S = k\ln\Omega$ to write that

$$S = Nk\ln Z + \frac{U}{T} + Nk - Nk\ln N$$

so that

$$S = Nk\ln\left[V\left(\frac{2\pi mkT}{h^2}\right)^{3/2}\right] + \frac{3}{2}Nk + Nk - Nk\ln N$$

$$= Nk\ln\left[\frac{V}{N}\left(\frac{2\pi mkT}{h^2}\right)^{3/2}\right] + \frac{5}{2}Nk = Nk\ln\left[\mathrm{e}^{5/2}\frac{V}{N}\left(\frac{2\pi mkT}{h^2}\right)^{3/2}\right]$$

which is the *Sackur–Tetrode equation.*

P17.4 Consider the following heat capacity data for the gases neon (Ne), water (H_2O) and hydrogen fluoride (HF). Plot these data with a spreadsheet program, such as Excel. Compare the data for the different substances and explain the different behavior from a molecular perspective.

T K	$C^{\ominus}_{P,m}$(Ne) J K^{-1}mol^{-1}	$C^{\ominus}_{P,m}$(HF) J K^{-1}mol^{-1}	$C^{\ominus}_{P,m}$(H_2O) J K^{-1}mol^{-1}
400	20.786	29.180	34.261
600	20.786	29.205	36.322
800	20.786	29.582	38.723
1000	20.786	30.142	41.627
1200	20.786	30.934	43.769
1400	20.786	31.832	46.055
1600	20.786	32.666	48.050
1800	20.786	33.396	49.747
2000	20.786	34.030	51.185

Use these data to find the enthalpy change at constant pressure for each gas upon heating it from $T = 400$ K to $T = 2000$ K. Use these data to find the entropy change at constant pressure for each gas upon heating it from 400 to 2000 K.

Solution

A plot of the data is shown in Fig. 17.3.

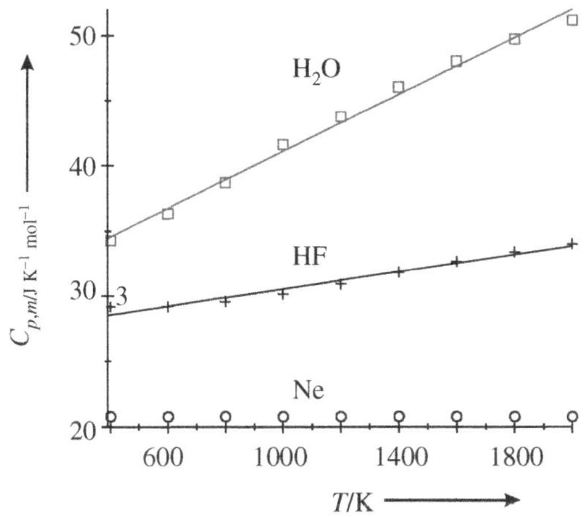

Figure P17.4 Heat capacity for Ne, HF, and H_2O versus temperature T.

The open circles show the data for Ne, which has a heat capacity that is not changing over this temperature range; i.e., $C_{P,m,\mathrm{Ne}}(T) = 20.786\ \mathrm{J\ mol^{-1}K^{-1}}$. The black crosses show the data for HF and these are fit to a line, such that $C_{P,m}(T) = \left[3.3193 \times 10^{-3}\ \mathrm{J\ mol^{-1}\ K^{-2}}T + 27.235\ \mathrm{J\ mol^{-1}\ K^{-1}}\right]$. The dark gray squares show the data for H_2O and these are fit by the equation of a line, such that $C_{P,m}(T) = \left[1.0921 \times 10^{-2}\ \mathrm{J\ mol^{-1}\ K^{-2}}T + 30.199\ \mathrm{J\ mol^{-1}\ K^{-1}}\right]$.

Because Ne is a monatomic gas its heat capacity is the smallest and independent of temperature over this range. The diatomic molecule HF has a larger heat capacity than Ne because of its two rotational degrees of freedom, and it shows a temperature dependence because of its vibrational degree of freedom. Water has the highest heat capacity because it has three translational degrees of freedom and three rotational degrees of freedom. The strong temperature dependence of water's heat capacity reflects the ability to access its vibrational degrees of freedom (especially the bending mode).

With this parameterization of the heat capacity data, it is straightforward to perform the integrals over temperature, namely

$$\Delta H_m = \int_{400}^{2000} C_{P,m} \mathrm{d}T$$

and find the enthalpy changes. For Ne we find that

$$\begin{aligned}\Delta H_m &= \int_{400}^{2000} C_{P,m} \mathrm{d}T \\ &= \left[20.786\ \mathrm{J\ mol^{-1}\ K^{-1}} \times (2000\ \mathrm{K} - 400\ \mathrm{K})\right] = 33.258\ \mathrm{kJ\ mol^{-1}}\end{aligned}$$

For HF we have

$$\begin{aligned}\Delta H_m &= \int_{400}^{2000} C_{P,m} \mathrm{d}T = \int_{400}^{2000} 3.3193 \times 10^{-3}\ \mathrm{J\ mol^{-1}\ K^{-2}} \cdot T + 27.325\ \mathrm{J\ mol^{-1}\ K^{-1}} \cdot \mathrm{d}T \\ &= \left(\frac{3.3193 \times 10^{-3}\ \mathrm{J\ mol^{-1}\ K^{-2}}}{2} T^2 + 27.235\ \mathrm{J\ mol^{-1}\ K^{-1}} \cdot T\right|_{400}^{2000} \\ &= 1.66 \times 10^{-3}\ \mathrm{J\ mol^{-1}\ K^{-2}} \cdot (1600\ \mathrm{K})^2 + 27.235\ \mathrm{J\ mol^{-1}\ K^{-1}} \cdot 1600\ \mathrm{K} \\ &= 4.25 \times 10^3\ \mathrm{J\ mol^{-1}} + 43.6 \times 10^3\ \mathrm{J\ mol^{-1}} \\ &= 47.8\ \mathrm{kJ\ mol^{-1}}\end{aligned}$$

and for H_2O we have

$$\Delta H_m = \int_{400}^{2000} C_{P,m} \mathrm{d}T$$
$$= \left(\frac{1.0921 \times 10^{-2}\ \mathrm{J\ mol^{-1}\ K^{-2}}}{2} T^2 + 30.199\ \mathrm{J\ mol^{-1}\ K^{-1}} \cdot T \right|_{400}^{2000} = 69.287\ \mathrm{kJ\ mol^{-1}}$$

The entropy changes may be found in an analogous manner; namely by

$$\Delta S_m = \int_{400}^{2000} \frac{C_{P,m}}{T} \mathrm{d}T$$

For Ne we find that

$$\Delta S_m = \int_{400}^{2000} \frac{C_{P,m,\mathrm{Ne}}}{T} \mathrm{d}T = C_{P,m,\mathrm{Ne}} \int_{400}^{2000} \frac{1}{T} \mathrm{d}T$$
$$= 20.786\ \mathrm{J\ K^{-1}\ mol^{-1}} \cdot \ln \frac{2000\ \mathrm{K}}{400\ \mathrm{K}} = 33.454\ \mathrm{J\ K^{-1}\ mol^{-1}}$$

For HF we find that

$$\Delta S_m = \int_{400}^{2000} \frac{C_{P,m}}{T} \mathrm{d}T$$
$$= \int_{400}^{2000} \frac{3.3193 \times 10^{-3}\ \mathrm{J\ K^{-2}\ mol^{-1}} T + 27.235\ \mathrm{J\ K^{-1}\ mol^{-1}}}{T} \mathrm{d}T$$
$$= 3.3193 \times 10^{-3}\ \mathrm{J\ K^{-2}\ mol^{-1}}\ (2000\ \mathrm{K} - 400\ \mathrm{K}) + 27.235\ \mathrm{J\ K^{-1}\ mol^{-1}} \ln \frac{2000\ \mathrm{K}}{400\ \mathrm{K}} = 49.144\ \mathrm{J\ K^{-1}\ mol^{-1}}$$

and for H_2O we find that

$$\Delta S_m = \int_{400}^{2000} \frac{C_{P,m}}{T} \mathrm{d}T$$
$$= \int_{400}^{2000} \frac{1.0921 \times 10^{-2}\ \mathrm{J\ K^{-2}\ mol^{-1}} T + 30.199\ \mathrm{J\ K^{-1}\ mol^{-1}}}{T} \mathrm{d}T$$
$$= 1.0921 \times 10^{-2}\ \mathrm{J\ K^{-2}\ mol^{-1}}\ (2000\ \mathrm{K} - 400\ \mathrm{K}) + 30.199\ \mathrm{J\ K^{-1}\ mol^{-1}} \ln \frac{2000\ \mathrm{K}}{400\ \mathrm{K}} = 66.077\ \mathrm{J\ K^{-1}\ mol^{-1}}$$

As shown by this calculation the differences in the entropy changes and the enthalpy changes can be traced to the differences in the molecules heat capacities, which were rationalized above.

P17.5 Consider a mole of CO_2 gas at 1 bar pressure and 194.67 K. The entropy of this gas was determined by measuring its heat capacity as a function of temperature and found to be 199.12 J mol^{-1} K^{-1} (see W.F. Giauque, C.J. Egan, *J. Chem. Phys.* 1937, **5**, 45). Compare this calorimetric value for the entropy to that you would calculate using the CO_2 molecule's spectroscopic parameters. Consider the molecule to be linear with $\theta_{rot} = 0.561$ K and $\sigma = 2$. Given that the molecule's normal modes have characteristic temperatures of $\theta_{vib} = 3360$ K, $\theta_{vib} = 1890$ K, $\theta_{vib} = 954$ K, $\theta_{vib} = 954$ K, the vibrational contribution to the entropy is small, 0.732 J mol^{-1} K^{-1}. Compute the percentage difference between the thermodynamic and spectroscopic values of the entropy.

Solution

We begin with the translational contribution to the molar entropy, From Equation (17.44) we can write

$$S_{\mathrm{trans},m} = \boldsymbol{R} \ln \left[\left(\frac{2\pi mkT}{h^2} \right)^{3/2} \frac{V}{N} \right] + \frac{5}{2} \boldsymbol{R}$$

If we use $V = NkT/P$, we can write this equation as

$$S_{\mathrm{trans},m} = \frac{5}{2} \boldsymbol{R} + \boldsymbol{R} \ln \left[\left(\frac{2\pi mkT}{h^2} \right)^{3/2} \frac{kT}{P} \right]$$
$$= 2.5\ \boldsymbol{R} + 15.20\ \boldsymbol{R} = 17.70\ \boldsymbol{R} = 147.17\ \mathrm{J\ K^{-1}\ mol^{-1}}$$

For the rotational contribution to the molar entropy, we find

$$S_{\text{rot},m} = \boldsymbol{R}\left[\ln\left(\frac{T}{\sigma\theta_{\text{rot}}}\right)+1\right]$$
$$= 8.314\ \text{J K}^{-1}\ \text{mol}^{-1}\left[\ln\left(\frac{194.67\ \text{K}}{2\times 0.561\ \text{K}}\right)+1\right]$$
$$= 51.18\ \text{J K}^{-1}\ \text{mol}^{-1}$$

The problem statement gives $S_{\text{vib},m} = 0.732$ J mol^{-1} K^{-1} and $S_{\text{elec},m} = 0$. The total molar entropy is

$$S_{\text{total},m} = S_{\text{trans},m} + S_{\text{rot},m} + S_{\text{vib},m} + S_{\text{elec},m}$$
$$= [147.17 + 51.18 + 0.732 + 0]\ \text{J K}^{-1}\ \text{mol}^{-1} = 199.08\ \text{J K}^{-1}\ \text{mol}^{-1}$$

Comparison of this value with the experimental value we have

$$\text{Percent error} = \frac{199.12 - 199.08}{199.12}\times 100\% = 0.02\%$$

P17.6 Consider a three-level system like that discussed in Section 15.8 with $N = 6$ particles, except it has a fixed temperature of $10a/k$ (a is the energy spacing and k is the Boltzmann constant), rather than a fixed energy. What is the average energy of the system? What is the entropy of the system?

Solution

In this problem we need to analyze the partition function

$$Z = \sum_{i=1}^{3}\exp\left(-\varepsilon_i/kT\right) = \exp(0) + \exp\left(-\frac{a}{kT}\right) + \exp\left(-\frac{2a}{kT}\right)$$
$$= 1 + \exp\left(-\frac{a}{kT}\right) + \exp\left(-\frac{2a}{kT}\right)$$

The internal energy U is

$$U = NkT^2\frac{\mathrm{d}\ln Z}{\mathrm{d}T}$$

Because of

$$\ln Z = \ln\left[1 + \exp\left(-\frac{a}{kT}\right) + \exp\left(-\frac{2a}{kT}\right)\right]$$

we find

$$\frac{\mathrm{d}\ln Z}{\mathrm{d}T} = \frac{\frac{a}{kT^2}\exp\left(-\frac{a}{kT}\right) + \frac{2a}{kT^2}\exp\left(-\frac{2a}{kT}\right)}{1 + \exp\left(-\frac{a}{kT}\right) + \exp\left(-\frac{2a}{kT}\right)}$$

For the case where $T = 10a/k$ and $N = 6$, we find

$$\frac{a}{kT} = 0.1$$

and

$$U = NkT^2\frac{\mathrm{d}\ln Z}{\mathrm{d}T} = NkT^2\cdot\frac{a}{kT^2}\frac{\exp\left(-\frac{a}{kT}\right) + 2\exp\left(-\frac{2a}{kT}\right)}{1 + \exp\left(-\frac{a}{kT}\right) + \exp\left(-\frac{2a}{kT}\right)}$$
$$= N\cdot a\cdot\frac{\exp(-0.1) + 2\exp(-0.2)}{1 + \exp(-0.1) + \exp(-0.2)}$$
$$= 6\cdot a\cdot\frac{2.54}{2.72} = 5.6\cdot a$$

Is this result reasonable? In our case ($kT = 10a$) the temperature is so high that a given particle is equally likely to be in any of the three states: 0, a, or 2a, as we remember from the simple model experiment displayed in Fig. 15.2. Then we find that the average energy per particle is

$$0\cdot\frac{1}{3} + a\cdot\frac{1}{3} + 2a\cdot\frac{1}{3} = a$$

and the average internal energy for 6 particles would be

$$U = 6a$$

This value is about 7% larger than the estimate made using the canonical partition function. This difference is reasonable because i) the second calculation overestimates the occupancy of the high energy states and ii) the particle number is small, so that the mean deviation in the energy calculation is, according to Box 15.2,

$$\overline{\Delta U} = \frac{5.6 \cdot a}{\sqrt{6}} = 2.3 \cdot a \tag{17.1}$$

Recall however that the canonical partition function in this form assumes that the number of states is sufficiently large and is clearly not right in this case.

For the entropy we have

$$S = Nk \ln Z + \frac{U}{T}$$

With our previous results

$$Z = 2.72 \qquad U = 5.6 \cdot a \qquad T = 10a/k \qquad N = 6$$

we obtain

$$S = 6 \cdot k \ln 2.72 + \frac{5.6 \cdot a}{10a/k} = k \cdot (6.00 + 0.56) = 6.56 \cdot k$$

We can check this result independently, if we calculate the number of configurations Ω starting with all configurations with $U = 6a$. The contribution ω of each configuration

$$\omega = \frac{N!}{N_1!N_2!N_3!}$$

is shown in the table:

	$U = 6a$			
N_3	0	1	2	3
N_2	6	4	2	0
N_1	0	1	2	3
ω	1	30	90	20

Then for Ω with $U = 6a$, we find

$$\Omega = 1 + 30 + 90 + 20 = 141$$

configurations, so that $S = k \cdot \ln(141) = 4.95 \cdot k$. This value is considerably smaller than the entropy calculated directly from the canonical partition function. As outlined in Box 15.2 we have to consider that the system is in connection with the temperature bath and fluctuations of the internal energy occur. This means that configurations with higher or lower energy than $U = 6a$ are possible. According to Equation (17.1) the average deviation in the internal energy is $\overline{\Delta U} = 2.3a$. Therefore let us also consider the configurations corresponding to $U = 7a$, $U = 5a$, $U = 6a$, $U = 4a$. We find

	$U = 7a$			$U = 5a$			$U = 8a$			$U = 4a$		
N_3	3	2	1	2	1	0	4	3	2	2	1	0
N_2	1	3	5	1	3	5	0	2	4	0	2	4
N_1	2	1	0	3	2	1	2	1	0	4	3	2
ω	60	60	6	60	60	6	15	10	15	15	10	15

These configurations give an additional

$$\Omega = 60 + 60 + 6 + 60 + 60 + 6 + 15 + 10 + 15 + 15 + 10 + 15 = 332$$

possible configurations. We continue with internal energies of $9a$, $3a$, $10a$, $2a$, $11a$, $1a$, $12a$, and $0a$ contributing additional $\Omega = 146$ configurations. Thus the total number of configurations is

$$\Omega_{\text{total}} = 141 + 332 + 146 = 619$$

and for the entropy we obtain

$$S = k \cdot \ln 619 = 6.19 \cdot k$$

comparable with the value calculated above.

Note that the number of configurations due to the energy fluctuations is essential in this case, because we have a small number of particles. Increasing the number of particles the distribution function of configurations becomes much smaller, and the fluctuations no longer play an important role. Compare Figs. 15.19 (six particles) and 15.20 (100 particles).

P17.7 Use the following calorimetric data for the molar constant pressure heat capacity of O_2 and oxygen phase transitions to determine the molar entropy of oxygen at 298 K and 1 bar pressure.

Solid I		Solid II		Solid III	
$\frac{T}{\text{K}}$	$\frac{C_{P,m}}{\text{J mol}^{-1}\text{K}^{-1}}$	$\frac{T}{\text{K}}$	$\frac{C_{P,m}}{\text{J mol}^{-1}\text{K}^{-1}}$	$\frac{T}{\text{K}}$	$\frac{C_{P,m}}{\text{J mol}^{-1}\text{K}^{-1}}$
12.97	4.60	25.02	22.68	45.90	46.13
14.14	6.36	26.75	24.07	47.76	46.34
15.12	7.07	28.0	25.32	48.11	46.08
15.57	7.49	29.88	27.67	48.97	46.00
16.66	9.75	30.63	29.05	50.55	46.08
16.94	9.42	33.05	31.48	51.68	46.17
18.13	11.18	34.41	33.82	52.12	46.29
18.45	11.68	35.77	35.54		
19.34	12.85	37.85	38.17		
20.26	14.65	38.47	41.02		
20.85	15.07	39.99	41.02		
21.84	17.58	40.67	41.52		
22.24	18.42	42.21	44.91		

Liquid		Gas	
$\frac{T}{\text{K}}$	$\frac{C_{P,m}}{\text{J mol}^{-1}\text{K}^{-1}}$	$\frac{T}{\text{K}}$	$\frac{C_{P,m}}{\text{J mol}^{-1}\text{K}^{-1}}$
56.95	53.41	91	30.75
57.95	53.24	100	29.93
60.97	53.20	140	29.50
65.57	53.20	180	29.30
69.12	53.37	220	29.25
74.95	53.79	260	29.30
81.13	53.91	298	29.44
84.79	54.12		
86.97	54.08		
90.33	54.37		

Phase transitions occur at a) $T = 23.66$ K with $\Delta H_m = 93.84$ J mol^{-1}, b) $T = 43.76$ K with $\Delta H_m = 743.4$ J mol^{-1}, c) $T = 54.39$ K with $\Delta H_m = 444.9$ J mol^{-1}, and d) $T = 90.13$ K with $\Delta H_m = 6817.6$ J mol^{-1}. The transitions at a) and b) are solid-to-solid phase transitions. The transition at c) is the solid-to-liquid phase transition and the transition at d) is the liquid-to-gas phase transition.

Solution

The plan is to first make a plot of the data, then perform fits of the data, followed by integration and finally sum all of the contributions.

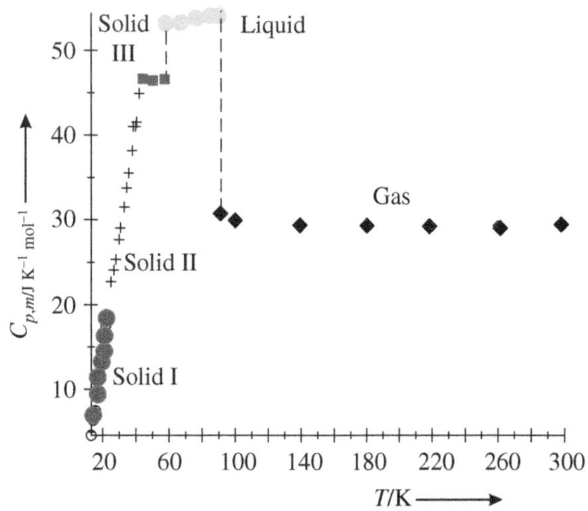

Figure P17.7 Heat capacity of O_2 versus temperature T. The dashed lines mark the phase transitions.

A regression fit of the heat capacity data for Solid I (dark gray circles) we find that

$$C_{P,m} = 1.4559 \text{ J K}^{-2} \text{ mol}^{-1} T - 14.800 \text{ J K}^{-1} \text{ mol}^{-1}$$

The regression fit for Solid II (crosses) gives

$$C_{P,m} = 1.3091 \text{ J K}^{-2} \text{ mol}^{-1} T - 11.021 \text{ J K}^{-1} \text{ mol}^{-1}$$

and the regression fit for Solid III (dark gray squares) is

$$C_{P,m} = \left[8.5898 \times 10^{-3} \text{ J K}^{-2} \text{ mol}^{-1} T + 45.732 \text{ J K}^{-1} \text{ mol}^{-1}\right]$$

For the heat capacity of the liquid phase (light gray circles) we find

$$C_{P,m} = \left[3.2995 \times 10^{-2} \text{ J K}^{-2} \text{ mol}^{-1} T + 51.265 \text{ J K}^{-1} \text{ mol}^{-1}\right]$$

and for the gas phase data (black diamonds) we find that

$$C_{P,m} = \left[30.531 \text{ J K}^{-1} \text{ mol}^{-1} - 4.8442 \times 10^{-3} \text{ J K}^{-2} \text{ mol}^{-1}\ T\right]$$

With this parameterization of the heat capacity data, we can compute the entropy change over these temperatures using

$$\Delta S_m = \int \frac{C_{P,m}}{T} dT$$

For Solid I we have

$$\Delta S_{\text{I},m} = \int_{12.97}^{23.66} \frac{1.4559 \text{ J K}^{-2} \text{ mol}^{-1} \cdot T - 14.800 \text{ J K}^{-1} \text{ mol}^{-1}}{T} dT$$

$$= 1.4559 \text{ J K}^{-2} \text{ mol}^{-1} (23.662 - 12.97) \text{ K} - 14.800 \text{ J mol}^{-1} \text{ K}^{-1} \cdot \ln \frac{23.66}{12.97} = 6.666 \text{ J mol}^{-1} \text{ K}^{-1}$$

for Solid II we find that

$$\Delta S_{\text{II},m} = \int_{23.66}^{43.76} \frac{1.3091 \text{ J K}^{-2} \text{ mol}^{-1} T - 11.021 \text{ J K}^{-1} \text{ mol}^{-1}}{T} dT$$

$$= 1.3091 \text{ J K}^{-2} \text{ mol}^{-1} (43.76 - 23.66) \text{ K} - 11.021 \text{ J K}^{-1} \text{ mol}^{-1} \ln \frac{43.76}{23.66} = 19.536 \text{ J mol}^{-1} \text{ K}^{-1}$$

and for Solid III we find that

$$\Delta S_{\text{III},m} = \int_{43.76}^{54.39} \frac{8.5898 \cdot 10^{-3}\ \text{J K}^{-2}\ \text{mol}^{-1} T + 45.732\ \text{J K}^{-1}\ \text{mol}^{-1}}{T} \text{d}T$$

$$= \left(8.5898 \cdot 10^{-3}\ \text{J K}^{-2}\ \text{mol}^{-1}\,(54.39 - 43.76)\ \ \text{K} + 45.732\ \text{J K}^{-1}\ \text{mol}^{-1} \ln \frac{54.39}{43.76}\right|_{43.76}^{54.39}$$

$$= 10.036\ \text{J mol}^{-1}\ \text{K}^{-1}$$

For the liquid phase we find

$$\Delta S_{\text{liq},m} = \int_{54.39}^{90.13} \frac{3.2995 \times 10^{-2}\ \text{J K}^{-2}\ \text{mol}^{-1} T + 51.265\text{J K}^{-1}\ \text{mol}^{-1}}{T} \text{d}T$$

$$= 3.2995 \cdot 10^{-2}\ \text{J K}^{-2}\ \text{mol}^{-1}\,(90.13 - 54.39)\ \ \text{K} + 51.265\ \text{J K}^{-1}\ \text{mol}^{-1} \ln \frac{90.13}{54.39}$$

$$= 27.072\ \text{J mol}^{-1}\ \text{K}^{-1}$$

and for the gas phase we find

$$\Delta S_{\text{gas},m} = \int_{90.13}^{298} \frac{30.531\ \text{J K}^{-1}\ \text{mol}^{-1} - 4.8442 \times 10^{-3}\ \text{J K}^{-2}\ \text{mol}^{-1}\ T}{T} \text{d}T$$

$$= 30.531\ \text{J K}^{-1}\ \text{mol}^{-1} \ln \frac{298}{90.13} - 4.8442 \times 10^{-3}\ \text{J K}^{-2}\ \text{mol}^{-1}\,(298 - 90.13)\ \ \text{K} = 35.503\ \text{J mol}^{-1}\ \text{K}^{-1}$$

Calculating the total entropy change requires that we sum each of the terms for the phases and add to them the entropy changes for each of the phase transitions; hence, we find that

$$\Delta S_{\text{total},m} = \Delta S_{\text{I},m} + \frac{\Delta H_{\text{I},m}}{T_{\text{I}}} + \Delta S_{\text{II},m} + \frac{\Delta H_{\text{II},m}}{T_{\text{II}}} + \Delta S_{\text{III},m} + \frac{\Delta H_{\text{fus},m}}{T_{\text{fus}}}$$

$$+ \Delta S_{\text{liq},m} + \frac{\Delta H_{\text{vap},m}}{T_{\text{bp}}} + \Delta S_{\text{gas},m}$$

$$= \left[\begin{matrix} 6.667 + \frac{93.84}{23.66} + 19.536 + \frac{743.4}{43.76} + 10.036 + \frac{444.9}{54.39} \\ +27.072 + \frac{6817.6}{90.13} + 35.503 \end{matrix}\right] \text{J mol}^{-1}\ \text{K}^{-1}$$

$$= 203.59\ \text{J K}^{-1}\ \text{mol}^{-1}$$

P17.8 In Problem P17.7 you should find that the calorimetrically determined entropy for diatomic oxygen is 203.6 J K^{-1}mol^{-1}. Calculate the entropy of oxygen from spectroscopic parameters and compare it to the calorimetric value. Note: In addition to the translational, rotational, and vibrational contributions to the entropy, you will also need to include the electronic contribution for oxygen. At room temperature the electronic ground state of diatomic oxygen may be considered to be a triplet, $g = 3$.

Solution

We start by noting that

$$S_{\text{total},m} = S_{\text{trans},m} + S_{\text{rot},m} + S_{\text{vib},m} + S_{\text{elec},m}$$

Now for the individual terms. For the translational contribution and using a mass of $m = 5.344 \times 10^{-26}$ kg, we find

$$S_{\text{trans},m} = \boldsymbol{R} \ln \left[\text{e}^{5/2} \left(\frac{2\pi mkT}{h^2} \right)^{3/2} \frac{V}{N} \right]$$

$$= \boldsymbol{R} \ln \left[\text{e}^{5/2} \left(\frac{2\pi\,(1.38 \times 10^{-23}\ \text{J K}^{-1})\,(5.344 \times 10^{-26}\ \ \text{kg})\ 298\ \text{K}}{(6.626 \times 10^{-34}\ \ \text{J s})^2} \right)^{3/2} 4.11 \times 10^{-26}\ \text{m}^3 \right]$$

$$= 152.13\ \text{J K}^{-1}\ \text{mol}^{-1}$$

where we used

$$\frac{V}{N} = \frac{kT}{P} = \frac{1.38 \times 10^{-23}\ \text{J K}^{-1} \cdot 298\ \text{K}}{1 \times 10^5\ \text{N m}^{-2}} = 4.11 \times 10^{-26}\ \text{m}^3$$

For the rotational contribution with $B = 2.88 \times 10^{-23}$ J and $\sigma = 2$, we find

$$S_{\mathrm{rot},m} = \boldsymbol{R}\left[\ln\left(\frac{kT}{\sigma B}\right)+1\right]$$
$$= 8.314\ \mathrm{J\,K^{-1}\,mol^{-1}} \times \left(\ln\left(\frac{(1.38\times10^{-23}\ \mathrm{J\,K^{-1}})\ 298\ \mathrm{K}}{2\left(2.88\times10^{-23}\right)\ \mathrm{J}}\right)+1\right)$$
$$= 43.80\ \mathrm{J\,K^{-1}\,mol^{-1}}$$

For the vibrational contribution with $h\nu_0 = 3.14 \times 10^{-20}$ J, we find

$$\frac{h\nu_0}{kT} = \frac{3.14\times10^{-20}\ \mathrm{J}}{1.38\times10^{-23}\ \mathrm{J\,K^{-1}}\ 298\ \mathrm{K}} = 7.64$$

$$S_{\mathrm{vib},m} = \boldsymbol{R}\left[\frac{h\nu_0/kT}{\exp\left(h\nu_0/kT\right)-1} - \ln\left(1-\exp\left(-h\nu_0/kT\right)\right)\right]$$
$$= \boldsymbol{R}\left[\frac{7.64}{\exp(7.64)-1} - \ln(1-\exp(-7.64))\right]$$
$$= \boldsymbol{R}\left[3.68\times10^{-3} + 4.81\times10^{-4}\right] = \boldsymbol{R}\cdot 4.16\times10^{-3}$$
$$= 3.47\times10^{-2}\ \mathrm{J\,K^{-1}\,mol^{-1}}$$

and finally for the electronic contribution we find

$$S_{\mathrm{elec}} = \boldsymbol{R}\ln(g) = \boldsymbol{R}\ln(3) = 9.134\ \mathrm{J\,K^{-1}\,mol^{-1}}$$

The total molar entropy change is found by summing these contributions, so that

$$S_{\mathrm{total},m} = S_{\mathrm{trans},m} + S_{\mathrm{rot},m} + S_{\mathrm{vib},m} + S_{\mathrm{elec},m}$$
$$= [152.13 + 43.80 + 0.0347 + 9.134]\ \mathrm{J\,K^{-1}\,mol^{-1}}$$
$$= 205.1\ \mathrm{J\,K^{-1}\,mol^{-1}}$$

This value is within 1% of the calorimetrically determined value in P17.7.

P17.9 We have calculated the rotational contribution S_{rot} for HD at 298 K with the assumption that the molecular partition function Z_{rot} can be evaluated by integration ($Z = kT/B$) and the internal energy by the high temperature value $U_{\mathrm{rot}} = NkT$. We learned in Chapter 15 that the internal energy U_{rot} for HD at 298 K is lower than the classical value (see Fig. 15.10). The numerical values at $T = 298$ K are $Z_{\mathrm{rot}} = 4.875$ and $U_{\mathrm{rot},m} = 2.295\ \mathrm{kJ\,mol^{-1}}$, whereas the integrated values used in Example 17.9 are $Z_{\mathrm{rot}} = 4.52$ and $U_{\mathrm{rot},m} = 2.48\ \mathrm{kJ\,mol^{-1}}$. Calculate S_{rot} for both data sets, using Equation (17.87).

Solution

From Equation (17.87) we obtain with the high temperature limit for U

$$S_{\mathrm{rot}} = Nk\ln\frac{Z_{\mathrm{rot}}}{\sigma} + \frac{U_{\mathrm{rot}}}{T} = Nk\ln\frac{Z_{\mathrm{rot}}}{\sigma} + \frac{NkT}{T} = Nk\left(\ln\frac{Z_{\mathrm{rot}}}{\sigma}+1\right)$$

and

$$S_{\mathrm{rot},m} = \boldsymbol{R}\left(\ln\frac{Z_{\mathrm{rot}}}{\sigma}+1\right)$$

a) Using the numerical value for Z_{rot} from Chapter 15 we find that

$$S_{\mathrm{rot},m} = \boldsymbol{R}\left(\ln\frac{Z_{\mathrm{rot}}}{\sigma}+1\right) = 8.314\ \mathrm{J\,K^{-1}\,mol^{-1}}\cdot\left(\ln\frac{4.875}{1}+1\right)$$
$$= 21.48\ \mathrm{J\,K^{-1}\,mol^{-1}}$$

b) Using the integrated (high temperature limit) value for Z_{rot} from Chapter 17 we find that

$$S_{\mathrm{rot},m} = \boldsymbol{R}\left[\ln\left(\frac{Z_{\mathrm{rot}}}{\sigma}\right)+1\right]$$
$$= 8.314\ \mathrm{J\,K^{-1}\,mol^{-1}}\cdot\left[\ln\left(\frac{4.52}{1}\right)+1\right] = 20.86\ \mathrm{J\,K^{-1}\,mol^{-1}}$$

In both cases we are using the high temperature limit of the internal energy.

c) For comparison, we calculate the entropy by

$$\begin{aligned}S_{\text{rot},m} &= \boldsymbol{R}\ln\frac{Z_{\text{rot}}}{\sigma} + \frac{U_{\text{rot},m}}{T}\\ &= \boldsymbol{R}\ln\frac{4.875}{1} + \frac{2.48\ \text{kJ mol}^{-1}}{298\ \text{K}}\\ &= (13.17 + 8.32)\ \text{J mol}^{-1}\ \text{K}^{-1} = 21.49\ \text{J mol}^{-1}\ \text{K}^{-1}\end{aligned}$$

This value is practically identical with that in case a). This means there is no net effect from the too low internal energy in the numerical calculation, because it is compensated for by the too high partition function.

P17.10 From Equation (17.77) show that the thermodynamic definition of temperature is equivalent to Equation (14.24), using the gas thermometer. Use the Carnot cycle with an ideal gas as a working medium.

Solution

Equation (17.77) in Section 17.7.1 relates the absolute temperature to the heat transferred in reversible processes:

$$\boxed{\frac{T_2}{T_1} = -\frac{q_{\text{rev},2}}{q_{\text{rev},1}}}$$

Here we show that this expression is equivalent to our former definition of the temperature, using the ideal gas law. If we use the Carnot cycle (Fig. 17.6) as a heat pump, then from Equation (17.24) in Section 17.2.2 we obtain

$$q_{\text{rev},2} = \boldsymbol{n}\boldsymbol{R}T_2 \cdot \ln\left(\frac{V_b}{V_a}\right)$$

and

$$q_{\text{rev},1} = -\boldsymbol{n}\boldsymbol{R}T_1 \cdot \ln\left(\frac{V_d}{V_c}\right)$$

Equation (17.21) in Section 17.2.2 states that

$$\frac{V_b}{V_a} = \frac{V_d}{V_c}$$

and we obtain

$$q_{\text{rev},2} = \boldsymbol{n}\boldsymbol{R}T_2 \cdot \ln\left(\frac{V_b}{V_a}\right), \quad q_{\text{rev},1} = -\boldsymbol{n}\boldsymbol{R}T_1 \cdot \ln\left(\frac{V_b}{V_a}\right)$$

to find

$$\frac{T_2}{T_1} = -\frac{q_{\text{rev},2}}{q_{\text{rev},1}} = \frac{\boldsymbol{n}\boldsymbol{R}T_2 \cdot \ln\left(\frac{V_b}{V_a}\right)}{\boldsymbol{n}\boldsymbol{R}T_1 \cdot \ln\left(\frac{V_b}{V_a}\right)} = \frac{\boldsymbol{n}\boldsymbol{R}T_2}{\boldsymbol{n}\boldsymbol{R}T_1} = \frac{(PV)_{T_2}}{(PV)_{T_1}}$$

P17.11 Reversible adiabatic compression. Consider a process with an ideal monatomic gas, which is thermally isolated from the surroundings. The gas (amount of substance $\boldsymbol{n} = 0.2$ mol) is reversibly compressed (*reversible adiabatic compression*) from volume $V_1 = 10$ L to volume $V_2 = 1$ L, while its temperature changes from $T_1 = 300$ K to T_2. Using the considerations on a reversible adiabatic expansion derive the final temperature T_2 and the reversible work w_{rev}.

Solution

We use Equation (17.13)

$$\Delta T = T_2 - T_1 = T_1\left[\left(\frac{V_1}{V_2}\right)^{\boldsymbol{n}\boldsymbol{R}/C_V} - 1\right]$$

For a monatomic gas we have

$$C_V = \frac{3}{2}\boldsymbol{n}\boldsymbol{R} \qquad \frac{\boldsymbol{n}\boldsymbol{R}}{C_V} = \frac{2}{3}$$

and

$$T_2 - T_1 = T_1 \left[\left(\frac{V_1}{V_2} \right)^{2/3} - 1 \right]$$
$$= 300\text{ K} \cdot \left[\left(\frac{10}{1} \right)^{2/3} - 1 \right] = 300\text{ K} \cdot 3.64 = 1092.5\text{ K}$$

Because of

$$\Delta U = q + w_{\text{rev}}$$

and $q = 0$, and the reversible work is

$$w_{\text{rev}} = \Delta U$$

Because the gas is monatomic and ideal, we know that $\Delta U = \boldsymbol{n}\frac{3}{2}\boldsymbol{R}\Delta T$, so that the work is

$$w_{\text{rev}} = \boldsymbol{n} \times \frac{3}{2}\boldsymbol{R}\Delta T$$
$$= 0.2\text{ mol} \cdot \frac{3}{2} \cdot 8.314\text{ J mol}^{-1}\text{ K}^{-1} \cdot (1092.5\text{ K}) = 2724.9\text{ J}$$

P17.12 In Section 17.8 we calculated the rotational contribution of the entropy S_{rot} of HD at 298 K and $P = 1$ bar from the molecular partition function Z_{rot}. Perform a similar calculation by describing S_{rot} as

$$S_{\text{rot}} = \int_0^{298} \frac{C_{P,\text{rot}}}{T} \text{d}T$$

Use the methods of Chapters 15 and 16 to calculate the values of $C_{P,\text{rot}}$ by direct summation. Compare the results of both treatments.

Solution

In Chapter 15 we calculated the rotational contribution to the molar internal energy of HD ($B = 0.91 \times 10^{-21}$ J or $\theta_{\text{rot}} = 65.6$) by

$$U_{\text{rot},m} = \frac{\boldsymbol{N_A}}{Z_{\text{rot}}} \sum_{J=1}^{\infty} g_J\, E_J \exp\left(-\frac{E_J}{kT}\right)$$

where $Z_{\text{rot}} = \sum_{J=1}^{\infty} g_J \exp\left(-E_J/(kT)\right)$. The constant pressure molar heat capacity will be

$$C_{P,\text{rot}} = \left(\frac{\partial \left(U_{\text{rot},m}\right)}{\partial T} \right)_P = \frac{\boldsymbol{N_A}}{Z_{\text{rot}}} \sum_{J=1}^{\infty} g_J \frac{E_J^2}{kT^2} \exp\left(-\frac{E_J}{kT}\right) - \frac{U_{\text{rot},m}}{Z_{\text{rot}}} \left(\frac{\partial Z_{\text{rot}}}{\partial T} \right)_P$$
$$= \frac{\boldsymbol{R}}{Z_{\text{rot}}} \sum_{J=1}^{\infty} g_J \left(\frac{E_J}{kT} \right)^2 \exp\left(-\frac{E_J}{kT}\right) - \frac{U_{\text{rot},m}}{Z_{\text{rot}}} \sum_{P\,J=1}^{\infty} g_J \frac{E_J}{kT^2} \exp\left(-\frac{E_J}{kT}\right)$$
$$= \frac{\boldsymbol{R}}{Z_{\text{rot}}} \sum_{J=1}^{\infty} g_J \left(\frac{E_J}{kT} \right)^2 \exp\left(-\frac{E_J}{kT}\right) - \frac{U_{\text{rot},m}^2}{\boldsymbol{R}T^2}$$

To calculate the molar entropy by integrating the heat capacity we will need to account for the phase transitions of HD at low temperature (we take the melting point to be 13.1 K and the boiling point to be 20.28 K). Because these transition temperatures are so low, we begin with the gas contribution to the molar entropy.

$$S_{\text{rot},m} = \int_{20.280}^{298} \left[\frac{\boldsymbol{R}}{TZ_{\text{rot}}} \sum_{J=1}^{\infty} g_J \left(\frac{E_J}{kT} \right)^2 \exp\left(-\frac{E_J}{kT}\right) - \frac{U_{\text{rot},m}^2}{\boldsymbol{R}T^3} \right] \text{d}T$$

A numerical evaluation (see the MathCAD sheet ch17_p12.mcdx for details) yields $S_{\text{rot},m} = 20.605\text{ J mol}^{-1}\text{K}^{-1}$. This value is somewhat less than the 20.9 J $\text{mol}^{-1}\text{K}^{-1}$ spectroscopic value computed in Example 17.9, as we might expect. Inclusion of the heat capacity changes for the solid and liquid at low temperatures would increase our calculated value.

P17.13 Consider a reversible adiabatic compression of an ideal gas from volume V_1 and temperature T_1 to volume V_2 and temperature T_2. Treating the gas as particles in a cubic box of side length L, show that the spacing of the energy levels increases, but the population probabilities remain the same.

Solution

According to Section 15.6 the energy levels of the gas particles are given by

$$E_{n_x,n_y,n_z} = \frac{h^2}{8mL^2}\left(n_x^2+n_y^2+n_z^2\right), \quad n_x,n_y,n_z = 1,2,3,\ldots$$

Because of

$$V = L^3$$

we can write for states 1 and 2 that

$$E(1)_{n_x,n_y,n_z} = \frac{\alpha}{V_1^{2/3}} \quad \text{and} \quad E(2)_{n_x,n_y,n_z} = \frac{\alpha}{V_2^{2/3}} \quad \text{with } \alpha = \frac{h^2}{8m}\left(n_x^2+n_y^2+n_z^2\right)$$

This result means that in the considered change of state the spacing of the energy levels increases with decreasing volume V.

The population probabilities in the two states are given by the Boltzmann law

$$\mathcal{P}(1)_{n_x,n_y,n_z} = \frac{1}{Z(1)_{\text{trans}}}\exp\left(\frac{-E(1)_{n_x,n_y,n_z}}{kT_1}\right)$$

and

$$\mathcal{P}(2)_{n_x,n_y,n_z} = \frac{1}{Z(2)_{\text{trans}}}\exp\left(\frac{-E(2)_{n_x,n_y,n_z}}{kT_2}\right)$$

with the molecular partition functions

$$Z(1)_{\text{trans}} = \left(\frac{2\pi kT_1 mL_1^2}{h^2}\right)^{3/2} = \left(\frac{2\pi km}{h^2}\right)^{3/2}\cdot T_1^{3/2}V_1$$

and

$$Z(2)_{\text{trans}} = \left(\frac{2\pi kT_2 mL_2^2}{h^2}\right)^{3/2} = \left(\frac{2\pi km}{h^2}\right)^{3/2}\cdot T_2^{3/2}V_2$$

In Section 17.2.2.2, for a reversible adiabatic compression, we derived the relation

$$T_2 = T_1\cdot\left(\frac{V_1}{V_2}\right)^{nR/C_V} = T_1\cdot\left(\frac{V_1}{V_2}\right)^{2/3}$$

(with $C_V = \frac{3}{2}\boldsymbol{nR}$ for an ideal atomic gas) which can be written as

$$T_2V_2^{2/3} = T_1V_1^{2/3} \text{ or } T_2^{3/2}V_2 = T_1^{3/2}V_1$$

Inserting these relations, we find that

$$Z(1)_{\text{trans}} = Z(2)_{\text{trans}}$$

Furthermore, we find that

$$\frac{-E(2)_{n_x,n_y,n_z}}{kT_2} = \frac{-\alpha/V_2^{2/3}}{kT_1\cdot\left(\frac{V_1}{V_2}\right)^{2/3}} = \frac{-\alpha/V_1^{2/3}}{kT_1} = \frac{-E(1)_{n_x,n_y,n_z}}{kT_1}$$

This result means that the population probabilities $\mathcal{P}(1)_{n_x,n_y,n_z}$ and $\mathcal{P}(2)_{n_x,n_y,n_z}$ remain unchanged in the considered change of state.

Note that this derivation assumes that the gas is an atomic gas. For a molecular gas the situation becomes more complicated because i) the heat capacity is different and ii) rotation and vibration contribute to the partition function.

P17.14 In this problem you find expressions for the internal energy and the entropy for a gas of N photons. Use the distribution function for photons in Equation (15.112) to show that the number N of photons is given by

$$N = \frac{8\pi}{c_0^3} V \left(\frac{kT}{h}\right)^3 \int_{x=0}^{\infty} \frac{x^2}{e^x - 1} dx = \frac{8\pi^4}{12.897 c_0^3} V \left(\frac{kT}{h}\right)^3$$

where $x = h\nu/kT$. The last equality results from evaluating the integral (see the Appendix B) which is

$$\int_{x=0}^{\infty} \frac{x^2}{e^x - 1} dx = 2.4041 = \frac{\pi^3}{12.897}$$

Next, you use the definition of the internal energy $U \equiv \int_{x=0}^{\infty} h\nu\, dN$, to show that

$$U = \frac{8\pi}{c_0^3} V \left(\frac{kT}{h}\right)^4 \int_{x=0}^{\infty} \frac{x^3}{e^x - 1} d\nu = \frac{8\pi}{c_0^3} V \left(\frac{kT}{h}\right)^4 \frac{\pi^4}{15}$$

In the last part of the problem, you will derive an expression for the entropy. Comparison of the distribution function for photons in the limit that $\exp(E_i/(kT)) \gg 1$, namely

$$N_i = \frac{g_i}{\exp(E_i/(kT)) - 1} \approx g_i e^{-E_i/(kT)}$$

with our Equation (15.9)[1]

$$N_i = \frac{N}{Z} g_i \, e^{-E_i/(kT)}$$

implies that

$$Z = N$$

Use this equality, and the general result that

$$S = Nk \ln Z + Nk(1 - \ln N) + \frac{U}{T}$$

to show that entropy S for a photon gas is

$$S = Nk + \frac{8\pi}{c_0^3} V \left(\frac{k}{h}\right)^4 \frac{\pi^4}{15} T^3$$

Using these formulae calculate the photon density (N/V), the energy density U/V, and the entropy S of the photon gas under standard conditions of 298 K and 1 atmosphere.

Solution

1) To find the number N of photons, we start with Equation (15.112) in the text

$$dN = \frac{8\pi}{c_0^3} V \frac{\nu^2}{\exp(h\nu/(kT)) - 1} d\nu$$

and calculate the number N of photons as

$$N = \int_0^{\infty} dN = \frac{8\pi}{c_0^3} V \int_{\nu=0}^{\infty} \frac{\nu^2}{\exp(h\nu/(kT)) - 1} d\nu$$

With the substitution

$$x = \frac{h}{kT}\nu \qquad dx = \frac{h}{kT} d\nu$$

1 Note that in Equation (15.9) the index i is related to the energy state i. In our procedure the index i is related to the quantum state i; therefore, we have to add the factor g_i.

we obtain

$$N = \frac{8\pi}{c_0^3} V \left(\frac{kT}{h}\right)^3 \int_{x=0}^{\infty} \frac{x^2}{e^x - 1} dx$$

which is the result that is requested. Using the integral evaluation in the Appendix B.3, namely

$$\int_{x=0}^{\infty} \frac{x^2}{e^x - 1} dx = 2.4041 = \frac{\pi^3}{12.897}$$

we can write that the number N of the photons is

$$N = \frac{8\pi}{c_0^3} V \left(\frac{kT}{h}\right)^3 \frac{\pi^3}{12.897} = \frac{8\pi^4}{12.897 c_0^3} V \left(\frac{kT}{h}\right)^3$$

Using this result, we can write the photon density N/V as

$$\frac{N}{V} = \frac{8\pi^4}{12.897 c_0^3} \left(\frac{kT}{h}\right)^3 \qquad \text{(equation for photon density)}$$

Note that $N/V = 0$ for $T = 0$ and that N/V increases proportional to T^3. For example, for $T = 298$ K we obtain

$$\frac{N}{V} = \frac{8\pi^4}{12.897(2.998 \times 10^8 \text{ m s}^{-1})^3} \left(\frac{1.38 \times 10^{-23} \text{ J K}^{-1} \; 298 \text{ K}}{6.626 \times 10^{-34} \text{ J s}}\right)^3$$
$$= 5.36 \times 10^{14} \text{ m}^{-3}$$

For comparison, the particle density of an ideal gas under standard conditions is

$$\frac{N}{V} = \frac{P}{kT} = \frac{1.013 \text{ bar}}{1.38 \times 10^{-23} \text{ J K}^{-1} \; 298 \text{ K}} \frac{10^5 \text{ Pa}}{1 \text{ bar}} = 2.46 \times 10^{25} \text{ m}^{-3}$$

2) Internal energy U

$$U = \int_{x=0}^{\infty} h\nu \, dn = \frac{8\pi}{c_0^3} V \int_{\nu=0}^{\infty} h\nu \frac{\nu^2}{\exp(h\nu/(kT)) - 1} d\nu$$
$$= \frac{8\pi h}{c_0^3} V \int_{\nu=0}^{\infty} \frac{\nu^3}{\exp(h\nu/(kT)) - 1} d\nu$$

If we make the substitution

$$x = \frac{h}{kT}\nu \qquad dx = \frac{h}{kT} d\nu$$

we obtain (see the Mathematical Appendix)

$$\int_{\nu=0}^{\infty} \frac{\nu^3}{\exp(h\nu/(kT)) - 1} d\nu = \left(\frac{kT}{h}\right)^4 \int_{x=0}^{\infty} \frac{x^3}{e^x - 1} d\nu = \left(\frac{kT}{h}\right)^4 \frac{\pi^4}{15}$$

Thus, we find that

$$U = \frac{8\pi h}{c_0^3} V \left(\frac{kT}{h}\right)^4 \frac{\pi^4}{15} = \frac{8\pi^5}{15} \frac{k^4 V}{c_0^3 h^3} T^4$$

This is the *Stefan–Boltzmann law*. By rearranging the equation for the photon density, which was found in part 1), we can write

$$(kT)^3 = \frac{12.897 c_0^3 h^3}{8\pi^4 V} N$$

and substituting it into our energy expression, we can write that

$$U = \frac{8\pi^5}{15}\frac{VkT}{c_0^3h^3}(kT)^3 = \frac{8\pi^5}{15}\frac{VkT}{c_0^3h^3}\frac{12.897c_0^3h^3}{8\pi^4 V}N$$
$$= \frac{12.897\pi}{15}NkT \simeq 2.70\,NkT$$

This value is almost a factor of two larger than the internal translational energy, $U = (3/2)NkT$, for an atomic gas. At 298 K, we find that

$$U/V = 2.70\frac{N}{V}kT$$
$$= 2.70\left(5.36 \times 10^{14}\ \text{m}^{-3}\right)\left(1.38 \times 10^{-23}\ \text{J K}^{-1}\right)(298\ \text{K})$$
$$= 5.95 \times 10^{-6}\ \text{J m}^{-3}$$

3) Entropy S

Using the result that $Z = N$ from the problem statement and Equation (17.83) for the translational entropy of an ideal gas

$$S = Nk\ \ln Z + k(N - N\ln N) + \frac{U}{T}$$
$$= Nk\ \ln Z + Nk(1 - \ln N) + \frac{U}{T}$$

we can write an expression for the entropy of the photon gas to be

$$S = Nk\ \ln N + Nk - Nk\ln N + \frac{U}{T} = Nk + \frac{U}{T}$$
$$= Nk + \frac{12.897\pi}{15}Nk \simeq Nk\,[1 + 2.70] = 3.70Nk$$

If we use our expression for the photon number, we can write that

$$S = 3.70\,Nk = 3.70\frac{8\pi^4}{12.897c_0^3}V\frac{k^4T^3}{h^3}$$
$$= 2.29\frac{\pi^4}{c_0^3}V\frac{k^4}{h^3}T^3$$

The entropy of the photon gas increases proportional to T^3, whereas the entropy of an atomic gas increases proportional to $\ln T$. The reason for this difference is that the number of particles is constant for an atomic gas, but for the photon gas the number of particles increases with temperature.

For example, for standard conditions ($V = 24.7$ L at $P = 1.013$ bar and $T = 298$ K) the entropy of the photon gas is

$$S^{\ominus}_{298} = 2.29\frac{\pi^4\ 24.7 \times 10^{-3}\ \text{m}^3(1.38 \times 10^{-23}\ \text{J K}^{-1})^4}{(2.998 \times 10^8\ \text{m s}^{-1})^3(6.626 \times 10^{-34}\ \text{J s})^3}298^3\ \text{K}^3 = 6.74 \times 10^{-10}\ \text{J K}^{-1}$$

$N = 5.36 \times 10^{14}$ photons are confined in a volume of 1 m^3 or $N = 1.32 \times 10^{13}$ in a volume of 24.7 L. Then the molar entropy is

$$S^{\ominus}_{m,298} = 6.74 \times 10^{-10}\ \frac{6.02 \times 10^{23}}{1.32 \times 10^{13}}\text{J K}^{-1}\ \text{mol}^{-1} = 30.7\ \text{J K}^{-1}\ \text{mol}^{-1}$$

compared to $S^{\ominus}_{m,298} = 154.1\ \text{J K}^{-1}\ \text{mol}^{-1}$ for an atomic gas consisting of argon atoms.

See also: H.S. Leff, Teaching the photon gas in introductory physics, *Am. J. Phys.* 2002, **70**, 792–797.

18

General Conditions for Spontaneity and its Application to Equilibria of Ideal Gases and Dilute Solutions

Many of the exercises and problems in this section require that you look up thermodynamic quantities of the substances. These facts may be found in Appendix A.

18.1 Exercises

E18.1 Calculate the Gibbs energy change and the equilibrium constant K for the reaction: n-butane → i-butane (n-$C_4H_{10} \rightarrow i$-C_4H_{10}) under standard conditions. Estimate the equilibrium constant at 1500 K. How does your estimated value compare to the more rigorous result of $K = 0.55$?

Solution
At 298 K the enthalpy of formation $\Delta_f H^{\ominus}_{gas}$ for n-butane is -125.6 kJ mol^{-1} and for i-butane it is -134.2 kJ mol^{-1}; so that $\Delta_r H = (-134.2 - (-125.6))$ kJ mol^{-1} = -8.6 kJ mol^{-1} at 298 K. The molar entropy, $S_{m,gas}$ for n-butane is 148.92 J K^{-1} mol^{-1} and for i-butane it is 145.98 kJ K^{-1}mol^{-1} at 298 K, so that $\Delta_r S = (145.98 - 148.92)$ J mol^{-1} = -2.94 J mol^{-1}. Assuming that these values do not change significantly with temperature we can use the relation

$$\Delta_r G = \Delta_r H - T\Delta_r S$$

to calculate $\Delta_r G$ at 1500 K; namely

$$\begin{aligned}\Delta_r G &= \left[(-8.6) - (1500) \cdot \left(-2.94 \times 10^{-3}\right)\right] \text{ kJ mol}^{-1} \\ &= -4.19 \text{ kJ mol}^{-1}\end{aligned}$$

To find the equilibrium constant we use the relation that $\Delta_r G = -\boldsymbol{R}T \ln K$, hence

$$\ln K = \frac{4.19 \text{ kJ mol}^{-1}}{8.314 \text{ J mol}^{-1} \text{ K}^{-1} \times 1500 \text{ K}} = 0.336$$

solving for K we have

$$K = \exp 0.336 = 1.40$$

The answer is off by a factor of nearly three times. A rigorous calculation that accounts for the temperature dependence of the enthalpy and entropy differences ($\Delta_r H$ and $\Delta_r S$) may be found in the work S. S. Chen, R. C. Wilhoit and B. J. Zwolinski; J. Chem. Phys. Ref. Data (1975) 4, 859–869.

E18.2 Consider the decomposition reaction for TNT (2,4,6-trinitrotoluene, $C_7H_5N_3O_6$).

$$2\text{ TNT} \rightarrow 3\text{ N}_2(\text{g}) + 7\text{ CO}(\text{g}) + 5\text{ H}_2\text{O}(\text{g}) + 7\text{ C}(\text{s})$$

Calculate the enthalpy change for this reaction at 298 K and 1 bar pressure. Compare your value to the enthalpy of combustion for TNT, which is -3410 kJ mol^{-1}. Use the value $\Delta_f H^{\ominus}_{298} = -63.2$ kJ mol^{-1} for TNT (these values are taken from the NIST Chemistry WebBook, 2008; http://webbook.nist.gov/).

Solutions Manual for Principles of Physical Chemistry, Third Edition. Edited by Hans Kuhn, David H. Waldeck, and Horst-Dieter Försterling.

Solution

The data for this problem are as follows:

Compound	TNT (s)	N_2(g)	CO (g)	H_2O (g)	C (s)
$\Delta_f H^\ominus_{298}/(\text{kJ mol}^{-1})$	−63.2	0.00	−110.53	−241.83	0

With these values we can calculate $\Delta_r H$ as

$$\begin{aligned}\Delta_r H &= \left[3\Delta_f H^\ominus_{298}(N_2) + 7\Delta_f H^\ominus_{298}(CO) + 5\Delta_f H^\ominus_{298}(H_2O) + 7\Delta_f H^\ominus_{298}(C) - 2\Delta_f H^\ominus_{298}(TNT)\right]\\ &= [3\cdot 0 + 7\cdot(-110.53) + 5\cdot(-241.83) + 7\cdot 0 - 2\cdot(-63.2)]\ \text{kJ mol}^{-1}\\ &= -1856.5\ \text{kJ mol}^{-1}\end{aligned}$$

This value is a factor of nearly two smaller than the enthalpy change for the combustion reaction (which includes the reactions of CO and C to CO_2).

E18.3 Calculate the Gibbs energy change and the equilibrium constant for the combustion of methane:

$$CH_4 + 2O_2 \rightarrow CO_2 + 2H_2O$$

Estimate the temperature at which this reaction is no longer spontaneous.

Solution

The data for this problem are

Compound	$\Delta_f H^\ominus_{298}$/kJ mol^{-1}	$S^\ominus_{m,298}$/J K^{-1} mol^{-1}
CH_4	−74.87	186.25
O_2	0.00	205.15
CO_2	−393.5	213.79
H_2O	−241.83	188.84

First we calculate $\Delta_r H^\ominus$ as

$$\begin{aligned}\Delta_r H^\ominus &= \Delta_f H^\ominus_{298}(CO_2) + 2\Delta_f H^\ominus_{298}(H_2O) - \Delta_f H^\ominus_{298}(CH_4) - 2\Delta_f H^\ominus_{298}(O_2)\\ &= (-393.5 + 2\times(-241.83) - (-74.87) - 2\times 0)\ \text{kJ mol}^{-1}\\ &= -802.3\ \text{kJ mol}^{-1}\end{aligned}$$

Next we calculate $\Delta_r S^\ominus$ as

$$\begin{aligned}\Delta_r S^\ominus &= S^\ominus_{m,298}(CO_2) + 2S^\ominus_{m,298}(H_2O) - S^\ominus_{m,298}(CH_4) - 2S^\ominus_{m,298}(O_2)\\ &= (213.79 + 2\times 188.84 - 186.25 - 2\times 205.15)\ \text{J K}^{-1}\ \text{mol}^{-1}\\ &= -5.08\ \text{J K}^{-1}\ \text{mol}^{-1}\end{aligned}$$

Lastly we calculate $\Delta_r G^\ominus$

$$\begin{aligned}\Delta_r G^\ominus &= \Delta_r H^\ominus - T\Delta_r S^\ominus\\ &= \left[-802.29 - 298\times\left(-5.08\times 10^{-3}\right)\right]\ \text{kJ mol}^{-1}\\ &= -800.78\ \text{kJ mol}^{-1}\end{aligned}$$

The temperature at which the reaction is no longer spontaneous corresponds to the temperature at which $\Delta_r G^\ominus = 0$. Assuming that $\Delta_r H^\ominus$ and $\Delta_r S^\ominus$ do not change with temperature allows us to calculate a value of T for which the reaction becomes nonspontaneous.

$$\Delta_r G^\ominus = \Delta_r H^\ominus - T\Delta_r S^\ominus = 0 \qquad \Delta_r H^\ominus = T\Delta_r S^\ominus$$

and

$$T = \frac{\Delta_r H^\ominus}{\Delta_r S^\ominus} = \frac{-802.29}{-5.08\times 10^{-3}}\ \text{K} = 1.58\times 10^5\ \text{K}$$

Hence we see that this reaction will remain spontaneous for all reasonable temperature values.

E18.4 Calculate the equilibrium constant for the reaction:

$$I_2(g) \rightarrow 2\,I(g)$$

at 1298 K. What is the fraction of iodine atoms at total pressures of 0.1, 1.0, and 10 bar?

Solution

From the data in Table A.1 we calculate

$$\Delta_r H^{\ominus}_{298} = [2 \cdot (106.6) - 62.2]\ \text{kJ mol}^{-1} = 151.0\ \text{kJ mol}^{-1}$$

and

$$\Delta_r S^{\ominus}_{m,298} = [2 \cdot (180.7) - 260.6]\ \text{J K}^{-1}\ \text{mol}^{-1} = 100.8\ \text{J K}^{-1}\ \text{mol}^{-1}$$

if we consider gaseous I_2. By using the data for the constant pressure heat capacity

$$C^{\ominus}_{P,m} = c_1 + c_2 T + c_3 T^2$$

	c_1 /J mol^{-1} K^{-1}	c_2 /J mol^{-1} K^{-2}	c_3 /J mol^{-1} K^{-3}
I	20.87	−0.00033	$0.025 \cdot 10^{-5}$
I_2	36.39	0.00233	$-0.077 \cdot 10^{-5}$

(c_1, c_2, and c_3 values from Foundation 16.2) we can calculate the temperature dependence of $\Delta_r H^{\ominus}$ and $\Delta_r S^{\ominus}$. Using $\Delta c_1 = 4.48$ J mol^{-1}K^{-1}, $\Delta c_2 = -0.00299$ J mol^{-1}K^{-2}, and $\Delta c_3 = 0.127 \cdot 10^{-5}$ J mol^{-1}K^{-3}, we find that

$$\begin{aligned}\Delta_r H^{\ominus}_{1298\ \text{K}} &= \Delta_r H^{\ominus}_{298\text{K}} + \int_{298}^{1298} \Delta C^{\ominus}_{P,m} \cdot dT \\ &= \left[\begin{array}{c}\Delta_r H^{\ominus}(_{298\ \text{K}}) + \Delta c_1\,(1298 - 298)\ \text{K} \\ + \frac{\Delta c_2}{2}(1298^2 - 298^2)\ \text{K}^2 + \frac{\Delta c_3}{3}(1298^3 - 298^3)\ \text{K}^3\end{array}\right] \\ &= (151.0 + 4.48 - 2.25 + 0.915)\ \text{kJ mol}^{-1} = 154.1\ \text{kJ mol}^{-1}\end{aligned}$$

and

$$\begin{aligned}\Delta_r S_{1298\ \text{K}} &= \Delta_r S^{\ominus}{}_{298\ \text{K}} + \int_{298}^{1298} \frac{\Delta C^{\ominus}_{P,m}}{T} \cdot dT \\ &= \left[\begin{array}{c}\Delta_r S^{\ominus}_{298\ \text{K}} + \Delta c_1 \cdot \ln\left(\frac{1298}{298}\right) + \Delta c_2 \cdot (1298 - 298)\ \text{K} \\ + \frac{\Delta c_3}{2} \cdot (1298^2 - 298^2)\ \text{K}^2\end{array}\right] \\ &= [100.8 + 6.59 - 2.99 + 1.01]\ \text{J K}^{-1}\text{mol}^{-1} = 105.4\ \text{J K}^{-1}\text{mol}^{-1}\end{aligned}$$

so that

$$\Delta_r G^{\ominus}_{1298} = \left(154.1 - \frac{1298 \cdot 105.4}{1000}\right)\ \text{kJ mol}^{-1} = 17.1\ \text{kJ mol}^{-1}$$

Hence the equilibrium constant is

$$\begin{aligned}K &= \exp\left(-\frac{\Delta_r G^{\ominus}}{RT}\right) \\ &= \exp\left(-\frac{17.1\ \text{kJ mol}^{-1}}{(8.314\ \text{J K}^{-1}\ \text{mol}^{-1})(1298\ \text{K})}\right) = 0.205\ \text{at 1298 K.}\end{aligned}$$

Given that

$$K = \frac{\widehat{P}_I^2}{\widehat{P}_{I_2}} \quad \text{and} \quad \widehat{P}_{\text{total}} = \widehat{P}_{I_2} + \widehat{P}_I$$

we can write that

$$\widehat{P}_{I_2} = \widehat{P}_{\text{total}} - \widehat{P}_I \quad \text{and} \quad K = \frac{\widehat{P}_I^2}{\widehat{P}_{\text{total}} - \widehat{P}_I}$$

and

$$\hat{P}_I^2 + \hat{P}_I K = K\hat{P}_{\text{total}}$$

Solving this equations for $\hat{P}_I$, we find that

$$\hat{P}_I = -\frac{1}{2}K + \sqrt{K\hat{P}_{\text{total}} + \frac{1}{4}K^2}$$

Substituting the values of K and $\hat{P}_{\text{total}}$ we can calculate $\hat{P}_I$. The fraction of I atoms is given by $\hat{P}_I/\hat{P}_{\text{total}}$. For the case of $\hat{P}_{\text{total}} = 0.1$, we find that

$$\hat{P}_I = -0.1025 + \sqrt{0.0205 + 0.0105} = 0.074$$

and

$$\hat{P}_I/\hat{P}_{\text{total}} = \frac{0.069}{0.1} = 0.74$$

$\hat{P}_{\text{total}}$	0.1	1.0	10
$\hat{P}_I$	0.074	0.362	1.33
$\hat{P}_I/\hat{P}_{\text{total}}$	0.74	0.362	0.133

With increasing pressure the reaction is shifted toward the reactant side; thus, the fraction of iodine atoms decreases in accordance with Le Chatelier's Principle.

E18.5 Calculate the equilibrium constant for the reaction:

$$H_2(g) + I_2(g) \rightarrow 2HI\,(g)$$

at 700 K. What is the fraction of HI at total pressures of 0.1, 1.0, and 10 bar?

Solution

From the data in Table A.1 we can calculate the reaction enthalpy and entropy as

$$\Delta_r H_{298}^{\ominus} = [2(25.9) - 0 - 62.2]\ \text{kJ mol}^{-1} = -10.4\ \text{kJ mol}^{-1}$$

and

$$\Delta_r S_{m,298}^{\ominus} = 2(206.3) - 260.6 - 130.6\ \text{J K}^{-1}\ \text{mol}^{-1} = 21.4\ \text{J K}^{-1}\ \text{mol}^{-1}$$

at 298 K for gaseous I_2. Then for $\Delta_r G^{\ominus}(298\ \text{K})$, we find that

$$\Delta_r G_{298}^{\ominus} = \Delta_r H^{\ominus} - T\cdot\Delta_r S^{\ominus}$$
$$= \left(-10.4 - \frac{298\times 21.4}{1000}\right)\ \text{kJ mol}^{-1} = -16.8\ \text{kJ mol}^{-1}$$

If we make the approximation that $\Delta_r H^{\ominus}$ and $\Delta_r S^{\ominus}$ do not change significantly with temperature, then we find that

$$\Delta_r G_{700}^{\ominus} = \left(-10.4\ \text{kJ mol}^{-1} - 700\ \text{K}\cdot 21.4\times 10^{-3}\ \text{kJ K}^{-1}\ \text{mol}^{-1}\right) = -25.4\ \text{kJ mol}^{-1}$$

Using Equation (18.53) we find that

$$K = \exp\left(-\frac{\Delta_r G^{\ominus}}{RT}\right) = \exp\frac{25.4\ \text{kJ mol}^{-1}}{8.314\ \text{J K}^{-1}\ \text{mol}^{-1}\cdot 700\ \text{K}} = e^{4.36} = 79$$

which should be compared to the directly measured value of $K = 55.2$. For a better approach to the experimental value we have to take account of the temperature dependence of $\Delta_r H^{\ominus}$ and $\Delta_r S^{\ominus}$.
At equilibrium, we can write that

$$K = \frac{\hat{P}_{HI}^2}{\hat{P}_{H_2}\hat{P}_{I_2}} \quad \text{and} \quad \hat{P}_{\text{total}} = \hat{P}_{H_2} + \hat{P}_{I_2} + \hat{P}_{HI}$$

If we restrict our discussion to a stoichiometric composition of the reactants:

$$\hat{P}_{H_2} = \hat{P}_{I_2}, \quad \hat{P}_{\text{total}} = 2\hat{P}_{H_2} + \hat{P}_{HI}$$

we can write that

$$K = \frac{\hat{P}_{\mathrm{HI}}^2}{\frac{1}{4}\left(\hat{P}_{\mathrm{total}} - \hat{P}_{\mathrm{HI}}\right)^2} \quad \text{and} \quad \frac{1}{2}\sqrt{K} = \frac{\hat{P}_{\mathrm{HI}}}{\left(\hat{P}_{\mathrm{total}} - \hat{P}_{\mathrm{HI}}\right)}$$

Solving for $\hat{P}_{\mathrm{HI}}$ we find that

$$\hat{P}_{\mathrm{xHI}} = \frac{\hat{P}_{\mathrm{total}}}{1 + 2/\sqrt{K}} \quad \text{and} \quad \frac{\hat{P}_{\mathrm{HI}}}{\hat{P}_{\mathrm{total}}} = \frac{1}{1 + 2/\sqrt{K}} = 0.82$$

This fraction does not depend on the total pressure, contrary to the situation in E18.4). This result is in accordance with Le Chatelier's Principle, because the number of product species is the same as the number of reactant species. *Extension.* Account for the temperature dependence of $\Delta_r H^\ominus$ and $\Delta_r S^\ominus$ and calculate K. By how much is the agreement with experiment improved?

E18.6 Use the standard enthalpy and entropies of formation for the reactants and products in the hydrogen and oxygen explosion to compute the $\Delta_r G$ for the reaction. Determine the temperature at which the reaction is no longer spontaneous.

Solution

The balanced chemical equation is

$$2H_2\,(g) + O_2\,(g) \rightarrow 2\,H_2O\,(g)$$

The thermodynamic data for the reaction species can be found in Table A.1 and are

	H_2	O_2	H_2O
$S_m^\ominus$ (J K^{-1} mol^{-1})	130.68	205.15	188.84
$\Delta_f H^\ominus$ (kJ mol^{-1})	0	0	−241.83

With these data we find that the reaction enthalpy is

$$\Delta_r H^\ominus = 2\Delta_f H^\ominus\left(H_2O\right) - 2\Delta_f H^\ominus\left(H_2\right) - \Delta_f H^\ominus\left(O_2\right)$$
$$= (2 \times (-241.83) - 2 \times 0 - 0)\ \mathrm{kJ\ mol^{-1}} = -483.66\ \mathrm{kJ\ mol^{-1}}$$

and the reaction entropy change is

$$\Delta_r S^\ominus = 2S^\ominus\left(H_2O\right) - 2S^\ominus\left(H_2\right) - S^\ominus\left(O_2\right)$$
$$= (2 \times 188.84 - 2 \times 130.68 - 205.15)\ \mathrm{J\ K^{-1}\ mol^{-1}}$$
$$= -88.83\ \mathrm{J\ K^{-1}\ mol^{-1}}$$

Hence the $\Delta_r G^\ominus$ for the reaction is

$$\Delta_r G^\ominus = \Delta_r H^\ominus - T\Delta_r S^\ominus$$
$$= \left[-483.66\ \mathrm{kJ\ mol^{-1}} - 298\ \mathrm{K} \cdot \left(-88.83 \times 10^{-3}\ \mathrm{kJ\ K^{-1}\ mol^{-1}}\right)\right]$$
$$= -457.19\ \mathrm{kJ\ mol^{-1}}$$

Assuming that the reaction enthalpy and entropy changes do not change too significantly with temperature, we find the temperature at which this reaction is no longer spontaneous to be

$$T = \frac{\Delta_r H^\ominus}{\Delta_r S^\ominus} = \frac{-483.66\ \mathrm{kJ\ mol^{-1}}}{-88.83 \times 10^{-3}\ \mathrm{kJ\ K^{-1}\ mol^{-1}}} = 5445\ \mathrm{K}$$

E18.7 Consider the molecular energy levels for the reactants and products in $2\,H_2 + O_2 \rightarrow 2\,H_2O$. Discuss how the reaction's enthalpy and entropy changes can be understood in terms of the molecular properties.

Solution

The enthalpy change is dominated by electronic and vibrational terms. The translations contribute a change of $\Delta\varepsilon_{\mathrm{trans}} = -3\boldsymbol{R}T/2$ and the rotations contribute a change of $\Delta\varepsilon_{\mathrm{rotn}} = 0$ for a net change of

$-3\boldsymbol{R}T/2 = -3.72$ kJ mol^{-1} at 298 K. From the following data on vibrations, it is evident that the vibrational modes are largely "frozen out" ($\theta_v \gg T$) at room temperature, so that only the zero point energy differences should be considered. Rather than consider the zero point energy difference explicitly, we combine it with the electronic contribution by way of the bond dissociation energies given in the table in a first step. In a second step we include the contributions of translation and rotation.

	H_2	O_2	H_2O
$h\nu_0/(10^{-20}\text{J})$	8.760	3.141	7.538, 7.258, 3.222
θ_v/K	6215	2256	5360, 5160, 2290
$\varepsilon_B/(10^{-19}\text{J})$	7.24	8.28	8.27 (H–OH) ; 7.1 OH

Hence we find that

$$\Delta\varepsilon_{\text{elec,vibration}} \sim 2\cdot\left[-\varepsilon_B\,(\text{H–OH}) - \varepsilon_B\,(\text{OH})\right] - 2\cdot\left[-\varepsilon_B\left(\text{H}_2\right)\right] + \varepsilon_B\left(\text{O}_2\right)$$
$$= (-2\cdot 15.37 + 2\cdot 7.24 + 8.28)\;\cdot 10^{-19}\text{ J} = -7.98\cdot 10^{-19}\text{ J}$$

which for the molar internal energy change for the formation of one mol of H_2O gives

$$\Delta U_{m,\text{elec,vibration}} = \frac{1}{2}\boldsymbol{N}_A\cdot\Delta\varepsilon_{\text{elec,vibration}} = \frac{1}{2}6.022\times 10^{23}\cdot(-7.98\cdot 10^{-19}\text{ J}) = -240.5\text{ kJ mol}^{-1}$$

Next we include the translational contribution and find that $\Delta U_m = -244.2$ kJ mol^{-1}. The difference between this estimate and the experimental value for the enthalpy of formation of gaseous water, given as $\Delta_f H = -241.3$ kJ mol^{-1} (see Appendix A), is quite small. If we convert $\Delta_f H$ to a ΔU by using the fact that $\Delta U = \Delta H - P\Delta V$ and $P\Delta V = P(V(H_2O) - V(H_2) - 1/2V(O_2)) = -1/2\boldsymbol{R}T = -1.24$ kJ/mol, we find $\Delta U_m = (-241.83 + 1.24)$ kJ/mol $= -240.59$ kJ/mol. Our calculation shows that the changes in bond energy dominate the energy changes for this reaction.

The entropy change for this reaction will have contributions from each of the electronic, vibrational, rotational, and translational degrees of freedom also. Because the electronic and vibrational energy levels of both the reactants and the products are widely spaced in energy (sparse), we expect that they will contribute only weakly to the entropy change. We note however that the oxygen molecule's ground electronic state is a triplet, and this change in degeneracy will make a contribution to the entropy change. We expect that the translational and rotational degrees of freedom will contribute most significantly to the entropy change (see Tables 17.2 and 17.3) because the number of gas particles is changing in the reaction; using the Sackur–Tetrode equation we can write that

$$\Delta S^{\ominus}_{\text{trans}} = Nk\left[2\cdot\ln\left(const\cdot m^{3/2}_{H_2O}\right) - 2\cdot\ln\left(const\cdot m^{3/2}_{H_2}\right) - \ln\left(const\cdot m^{3/2}_{O_2}\right)\right]$$
$$= Nk\left[\ln\left(const\cdot m^{3/2}_{H_2O}\right)^2 - \ln\left(const\cdot m^{3/2}_{H_2}\right)^2 - \ln\left(const\cdot m^{3/2}_{O_2}\right)\right]$$
$$= Nk\ln\frac{\left(const\cdot m^{3/2}_{H_2O}\right)^2}{\left(const\cdot m^{3/2}_{H_2}\right)^2\left(const\cdot m^{3/2}_{O_2}\right)}$$
$$= Nk\ln\left(\frac{const^2\cdot m^3_{H_2O}}{const^2\cdot m^3_{H_2}\,const\cdot m^{3/2}_{O_2}}\right) = Nk\ln\left(\frac{m^3_{H_2O}}{const\cdot m^3_{H_2}m^{3/2}_{O_2}}\right)$$

where

$$const = e^{5/2}(2\pi)^{3/2}\frac{(k)^{3/2}}{h^3}\cdot\frac{V}{N}T^{3/2}$$

for the translations. For $N = 6.022\cdot 10^{23}$, $V = 22.4\cdot 10^{-3}$ m^3, and $T = 298$ K, we find that

$$const = e^{5/2}(2\pi)^{3/2}\frac{\left(1.38\cdot 10^{-23}\text{ J/K}\right)^{3/2}}{\left(6.626\cdot 10^{-34}\text{ J-s}\right)^3}\cdot\frac{22.4\cdot 10^{-3}\text{ m}^3}{6.022\cdot 10^{23}}(298\text{ K})^{3/2}$$
$$= 6.47\cdot 10^{45}\frac{\left(\text{kg m}^2/\text{s}^2\right)^{3/2}}{\left(\text{kg m}^2/\text{s}\right)^3}\text{m}^3 = 6.47\cdot 10^{45}\;\frac{1}{(\text{kg})^{3/2}}$$

and

$$\Delta S^{\ominus}_{\text{trans}} = Nk \ln\left(\frac{m^3_{\mathrm{H_2O}}}{const \cdot m^3_{\mathrm{H_2}} m^{3/2}_{\mathrm{O_2}}}\right)$$

$$= Nk \ln\left(\frac{\left(18 \cdot 1.66 \cdot 10^{-27}\ \text{kg}\right)^3}{\left(6.47 \cdot 10^{45}\ \text{kg}^{-3/2}\right)\left(2 \cdot 1.66 \cdot 10^{-27}\ \text{kg}\right)^3\left(16 \cdot 1.66 \cdot 10^{-27}\ \text{kg}\right)^{3/2}}\right)$$

$$= Nk \ln\left(2.60 \cdot 10^{-5}\right) = -10.5 \cdot Nk$$

Thus we obtain

$$\Delta_r S^{\ominus}_{\text{trans}} = -10.5 \cdot \boldsymbol{N_A} k = -10.5 \cdot \boldsymbol{R} = -87.3\ \text{J K}^{-1}\ \text{mol}^{-1}$$

For comparison, we calculate the experimental value of the entropy change for this reaction, using the data in Appendix A:

$$\Delta_r S^{\ominus} = 2 \cdot S^{\ominus}_m(\mathrm{H_2O}) - 2 \cdot S^{\ominus}_m(\mathrm{H_2}) - S^{\ominus}_m(\mathrm{O_2})$$

$$= (2 \cdot 188.0 - 2 \cdot 130.6 - 205)\ \text{J K}^{-1}\ \text{mol}^{-1} = -90.2\ \text{J K}^{-1}\ \text{mol}^{-1}$$

This means that the entropy change is mainly due to the contribution of the translational degrees of freedom, and we can neglect the contribution of the rotational degrees of freedom. The entropy change is negative, because three diatomic gas molecules combine to form two gas molecules, with a subsequent loss of three translational degrees of freedom.

E18.8 Do the reactions written in Table E18.8 occur spontaneously under standard conditions (reactants and products in their standard states)?

Table E18.8 List of Chemical Reactions.

	Reaction
1	$Fe + S \rightarrow FeS$
2	$2\,AgI + Pb \rightarrow PbI_2 + 2\,Ag$
3	$2\,Al + Fe_2O_3 \rightarrow Al_2O_3 + 2\,Fe$
4	$C_2H_4 \rightarrow$ polyethylene
5	$CaCO_3 \rightarrow CaO + CO_2$
6	$H_2O\,(l) \rightarrow H_2O\,(g)$
7	$2Ag + PbI_2 \rightarrow 2AgI + Pb$

Solution

According to the data in Appendix A.1 the following values of $\Delta_r G^{\ominus}$ are obtained at 298 K:

Reaction	$\Delta_r G^{\ominus}$/kJ mol^{-1}
1	97.5
2	−40.5
3	−883.4
4	−54.0
5	129.7
6	8.7
7	40.8

Hence we see that reactions 1, 2, 3, and 4 are spontaneous but that reactions 5, 6, and 7 are not.

E18.9 What is the minimum temperature that is required for the decomposition of limestone on the top of Mount Everest (atmospheric pressure 0.34 bar) to be spontaneous?

Solution

The pressure of CO_2 must overcome the atmospheric pressure of 0.34 bar. In Fig. E18.9 ΔG is plotted versus temperature for $\hat{P}_{CO_2} = 0.34$:

$$\Delta_r G = \Delta_r G^{\ominus} + \boldsymbol{R}T \cdot \ln 0.34$$

From this plot we conclude that $\Delta_r G$ changes its sign at $T = 1079$ K. This is 66 K less than for atmospheric pressure at sea level.

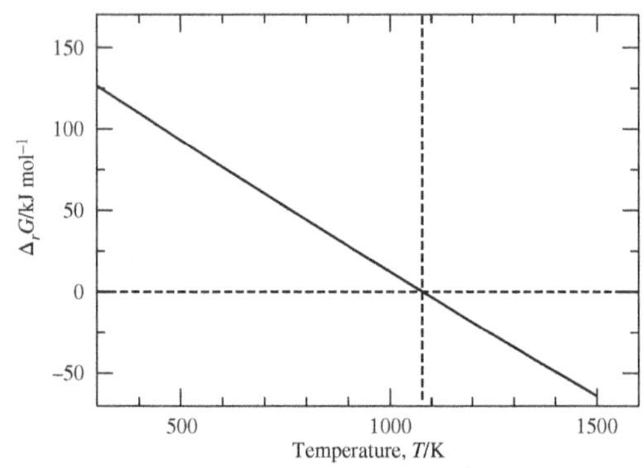

Figure E18.9 $\Delta_r G$ for the $CaCO_3 \rightarrow CaO + CO_2$ reaction versus temperature T.

E18.10 In the isotope exchange reaction in Example 18.11,

$$H_2 + D_2 \rightleftarrows 2HD$$

we considered the limit where the vibrational energy spacing is large compared to kT. Discuss how you expect the analysis to change as this condition is relaxed. In particular, comment on which motions contribute to the enthalpy and entropy changes in the reaction.

Solution

If we account for the thermal excitation of the vibrational modes of the molecules, then at high enough temperatures we expect the energy stored in vibrational modes to become similar. As a consequence the energy difference between the products and reactants (see Table 18.3) will become smaller, and we expect the equilibrium constant K to increase.

See E18.20 for a calculation at finite temperatures.

E18.11 Consider the gas reaction $H_2O \rightarrow H_2 + 1/2\ O_2$. Following Example 18.6 and Foundation 18.1 determine K at 1500 and 4000 K.

Solution

Using the fact that K for the reaction is

$$K = \frac{\hat{P}_{H_2} \cdot \hat{P}^{1/2}_{O_2,eq}}{\hat{P}_{H_2O,eq}},$$

Foundation 18.1 found the Gibbs energy for this reaction at 1500 K to be

$$\Delta_r G^{\ominus}_{1500} = +158 \text{ kJ mol}^{-1}$$

and

$$K_{1500} = \exp\left(-\frac{158 \text{ kJ mol}^{-1}}{8.3145 \text{ J K}^{-1} \text{ mol}^{-1} \cdot 1500 \text{ K}}\right) = \exp(-12.7) = 3.15 \times 10^{-6}$$

Foundation 18.1 showed that $\Delta_r H^{\ominus}(1500 \text{ K}) = 240.4 \text{ kJ mol}^{-1}$ and $\Delta_r S^{\ominus}(1500 \text{ K}) = 54.8 \text{ J K}^{-1} \text{ mol}^{-1}$.

At 4000 K we calculate $\Delta_r G^{\ominus}_{4000}$ by way of the fundamental relation $\Delta_r G^{\ominus} = \Delta_r H^{\ominus} - T\Delta_r S^{\ominus}$. First we find the enthalpy change at 4000 K, which may be written as

$$\Delta_r H^{\ominus}_{4000} = \Delta_r H^{\ominus}_{1500} + \int_{1500}^{4000} \Delta C^{\ominus}_p \cdot dT$$

Evaluating the integral (with the values $\Delta c_1 = 11.695$ J mol^{-1} K^{-1}, $\Delta c_2 = -0.0042$ J mol^{-1} K^{-2}, and $\Delta c_3 = -0.0925$ J mol^{-1} K^{-3} from the table in Foundation 16.2), we find that

$$\begin{aligned}\int_{1500}^{4000} \Delta C^{\ominus}_p \cdot dT &= \begin{bmatrix}\Delta c_1 \cdot (4000 - 1500)\ \mathrm{K} + \Delta c_2 \cdot \frac{1}{2}(4000^2 - 1500^2)\ \mathrm{K}^2 \\ + \Delta c_3 \cdot \frac{1}{3}(4000^3 - 1500^3)\ \mathrm{K}^3\end{bmatrix} \\ &= \begin{bmatrix}11.695 \cdot (2500) - \frac{1}{2} \cdot 0.0042 \cdot \left(1.375 \cdot 10^7\right) \\ - \frac{1}{3} \cdot 0.0925 \cdot 10^{-5} \cdot \left(6.0625 \cdot 10^{10}\right)\end{bmatrix} \frac{\mathrm{J}}{\mathrm{mol}} \\ &= -18.33\ \mathrm{kJ\ mol^{-1}}.\end{aligned}$$

so that

$$\Delta_r H^{\ominus}_{4000} = (240.4 - 18.33)\ \mathrm{kJ\ mol^{-1}} = 222.1\ \mathrm{kJ\ mol^{-1}}$$

For the entropy change at 4000 K, we can write that

$$\Delta_r S^{\ominus}_{4000} = \Delta_r S^{\ominus}_{1500} + \int_{1500}^{4000} \frac{\Delta C^{\ominus}_p\ (\mathrm{gas})}{T} \cdot dT$$

Evaluating the integral we find that

$$\begin{aligned}\int_{1500}^{4000} \frac{\Delta C^{\ominus}_p\ (\mathrm{gas})}{T} \cdot dT &= \begin{bmatrix}\Delta c_1 \cdot \ln\left(\frac{4000}{1500}\right) + \Delta c_2 \cdot (4000 - 1500)\ \mathrm{K} \\ + \frac{1}{2}\Delta c_3 \cdot (4000^2 - 1500^2)\ \mathrm{K}^2\end{bmatrix} \\ &= \begin{bmatrix}11.695\,(0.9808) - 0.0042 \cdot (2500) \\ - \frac{1}{2} \cdot 0.0925 \times 10^{-5} \cdot \left(1.375 \times 10^7\right)\end{bmatrix} \frac{\mathrm{J}}{\mathrm{K\ mol}} \\ &= -5.39\ \mathrm{J\ K^{-1}\ mol^{-1}}\end{aligned}$$

so that

$$\Delta_r S^{\ominus}_{4000} = (54.8 - 5.39)\ \mathrm{J\ K^{-1}\ mol^{-1}} = 49.4\ \mathrm{J\ K^{-1}\ mol^{-1}}$$

Combining these results we can calculate the Gibbs energy as

$$\begin{aligned}\Delta_r G^{\ominus}_{4000} &= \Delta_r H^{\ominus}_{4000} - T \cdot \Delta_r S^{\ominus}_{4000} \\ &= \left(222.1 - \frac{4000 \cdot (49.4)}{1000}\right)\ \mathrm{kJ\ mol^{-1}} = 24.5\ \mathrm{kJ\ mol^{-1}}\end{aligned}$$

The equilibrium constant at 4000 K is

$$K_{4000} = \exp\left(-\frac{24500\ \mathrm{J\ mol^{-1}}}{8.3145\ \mathrm{J\ K^{-1}\ mol^{-1}} \cdot 4000\ \mathrm{K}}\right) = \exp(-0.737) = 0.479$$

This result is in accordance with Le Chatelier's Principle, with increasing temperature the equilibrium of an endothermic reaction is pushed toward the products.

E18.12 Consider Example 18.9 for the ammonia synthesis. At each of the pressures in this example calculate the mole fraction of each component. Also calculate the entropy change that arises from the mixing of the different substances. Assume that the enthalpy change for mixing is zero, and use your entropy change to calculate the Gibbs energy change.

Solution

We calculate the the mole fractions from the partial pressures of the constituent gases

$$\begin{aligned}\widehat{P}_{\mathrm{total}} &= \widehat{P}_{N_2} + \widehat{P}_{H_2} + \widehat{P}_{NH_3} \\ &= \widehat{P}_{\mathrm{total}} \cdot x_{N_2} + \widehat{P}_{\mathrm{total}} \cdot x_{H_2} + \widehat{P}_{\mathrm{total}} \cdot x_{NH_3}\end{aligned}$$

or

$$\frac{\widehat{P}_{N_2}}{\widehat{P}_{\mathrm{total}}} = x_{N_2}, \quad \frac{\widehat{P}_{H_2}}{\widehat{P}_{\mathrm{total}}} = x_{H_2}, \quad \frac{\widehat{P}_{NH_3}}{\widehat{P}_{\mathrm{total}}} = x_{NH_3}$$

To find the entropy that arises from the mixing of gases, we use a thought experiment. Our experiment will have two steps. First we will imagine an isothermal expansion of each of the gases to the final volume. Second we will perform a reversible, adiabatic mixing of the gases. Step 1: For a reversible isothermal expansion of ideal gas i from V_i to V_{final}, $\Delta U = 0$. Because the process is reversible we find the entropy change arising from the volume expansion (see Equation (17.59)) of gas i to be

$$\Delta S_i = N_i k \ln\left(\frac{V_{\text{final}}}{V_i}\right)$$

We repeat this step for each of the gases i in the mixture, so that the entropy change in step 1 is

$$\Delta S_1 = \sum_i \Delta S_i = k \sum_i N_i \ln\left(\frac{V_{\text{final}}}{V_i}\right)$$

where i is an index for the gas type and the sum is performed over all gas types. Because step 2 is reversible and adiabatic $\Delta S_2 = 0$. The total entropy of mixing is

$$\Delta S = \Delta S_1 + 0 = k \sum_i N_i \ln\left(\frac{V_{\text{final}}}{V_i}\right)$$
$$= -k \sum_i N_i \ln\left(\frac{V_i}{V_{\text{final}}}\right)$$

It is convenient to express this equation in molar quantities; hence, we divide both sides by $\sum_i N_i$ and we multiply by $\boldsymbol{N}_A$ and find

$$\Delta_{\text{mix}} S = -\boldsymbol{R} \sum_i x_i \ln(x_i)$$

where x_i is the mole fraction of component i. For ideal gases, like the case considered here, the volume fraction of component i is the same as the mole fraction of component i when P and T are the same for the components.

For the entropy of mixing in the ammonia synthesis, we can write that

$$\Delta_{\text{mix}} S = -\boldsymbol{R}\left[x_{N_2} \ln x_{N_2} + x_{H_2} \ln x_{H_2} + x_{NH_3} \ln x_{NH_3}\right]$$

and because the enthalpy of mixing is taken to be zero ($\Delta_{\text{mix}} H$) we can write that

$$\Delta_{\text{mix}} G = -T \Delta_{\text{mix}} S$$

Using the values given in Table 18.2 for $\hat{P}_{\text{total}} = 3.95$ we find that

$$x_{N_2} = \frac{0.98}{3.95} = 0.248,\ x_{H_2} = \frac{2.93}{3.95} = 0.742,\ x_{NH_3} = \frac{0.05}{3.95} = 0.013$$

and

$$\Delta_{\text{mix}} S = -\boldsymbol{R}\,[(0.248)\ln(0.248) + (0.742)\ln(0.742) + (0.013)\ln(0.013)]$$
$$= 0.624 \cdot \boldsymbol{R} = 5.18 \text{ J mol}^{-1} \text{ K}^{-1}$$

so that

$$\Delta_{\text{mix}} G = -T \cdot \Delta_{\text{mix}} S$$
$$= -(700 \text{ K}) \cdot (5.18 \text{ J K}^{-1} \text{ mol}^{-1}) = -3.63 \text{ kJ mol}^{-1}$$

Proceeding in like fashion for the other entries we find the values in the table.

$\hat{P}_{N_2}$	$\hat{P}_{H_2}$	$\hat{P}_{NH_3}$	$\hat{P}_{\text{total}}$	x_{N_2}	x_{H_2}	x_{NH_3}	$\Delta_{\text{mix}} S$ J K^{-1} mol^{-1}	$\Delta_{\text{mix}} G$ kJ mol^{-1}
0.98	2.93	0.05	3.95	0.248	0.742	0.013	5.18	−3.63
20.1	60.2	19.9	100	0.201	0.602	0.199	7.89	−5.52
47	140	107	293	0.160	0.478	0.365	8.43	−5.90
70	210	260	540	0.130	0.389	0.481	8.18	−5.73

Comparison with Example 18.7 shows that the contribution to the Gibbs energy change from mixing is substantial.

E18.13 Calculate the partial pressure of ammonia for the ammonia synthesis reaction at $T = 298$ K and $P_{H_2} = P_{N_2} = 1$ bar. Perform the same calculation for $P_{H_2} = P_{N_2} = 100$ bar and compare your results.

Solution

Using the standard values for enthalpy and entropy we will determine the Gibbs energy and from that the equilibrium constant. The equilibrium constant will then be used to determine the partial pressures from the given data.

	H_2	N_2	NH_3
$\Delta_f H^\ominus$/ kJ mol^{-1}	0	0	−45.9
$S_m^\ominus$/J K^{-1} mol^{-1}	130.6	191.5	192.5

Using these data we have

$$\Delta_r H^\ominus = 2\Delta_f H^\ominus(\mathrm{NH_3}) - 3\Delta_f H^\ominus(\mathrm{H_2}) - \Delta_f H^\ominus(\mathrm{N_2})$$
$$= [2 \times (-45.9) - 0 - 0]\ \mathrm{kJ\ mol^{-1}} = -91.8\ \mathrm{kJ\ mol^{-1}}$$

and

$$\Delta_r S^\ominus = 2S^\ominus(\mathrm{NH_3}) - 3S^\ominus(\mathrm{H_2}) - S^\ominus(\mathrm{N_2})$$
$$= [2 \times 192.5 - 3 \times 130.6 - 191.5]\ \mathrm{J\ mol^{-1}\ K^{-1}} = -198.3\ \mathrm{J\ mol^{-1}\ K^{-1}}$$

so that

$$\Delta_r G^\ominus = [-91.8 \times 10^3 - 298 \times (-198.3)]\ \mathrm{J\ mol^{-1}}$$
$$= -32.7\ \mathrm{kJ\ mol^{-1}}$$

The equilibrium constant is

$$K = \exp\left(-\frac{-32.7\ \mathrm{kJ\ mol^{-1}}}{8.3145\ \mathrm{J\ K^{-1}\ mol^{-1}} \times 298\ \mathrm{K}}\right) = \exp(13.2) = 5.40 \times 10^5$$

The equilibrium constant expression for this reaction is

$$K = \frac{\hat{P}^2_{\mathrm{NH_3}}}{\hat{P}^3_{\mathrm{H_2}}\hat{P}_{\mathrm{N_2}}}$$

for the case $P_{H_2} = P_{N_2} = 1$ bar the pressure of ammonia is

$$\hat{P}_{\mathrm{NH_3}} = \sqrt{K \cdot \hat{P}^3_{\mathrm{H_2}}\hat{P}_{\mathrm{N_2}}}$$
$$= \sqrt{5.40 \times 10^5 \cdot 1^2 \cdot 1} = 734$$

That is, the contributions of the three molecules to the total pressure $\hat{P}_{\mathrm{total}} = (1 + 1 + 734) = 736$ are

$$\frac{\hat{P}_{\mathrm{NH_3}}}{\hat{P}_{\mathrm{total}}} = 0.997 \qquad \frac{\hat{P}_{\mathrm{N_2}}}{\hat{P}_{\mathrm{total}}} = \frac{\hat{P}_{\mathrm{H_2}}}{\hat{P}_{\mathrm{total}}} = 1.5 \times 10^{-3}$$

For the case $P_{H_2} = P_{N_2} = 100$ bar we obtain

$$\hat{P}_{\mathrm{NH_3}} = \sqrt{K \cdot \hat{P}^3_{\mathrm{H_2}}\hat{P}_{\mathrm{N_2}}}$$
$$= \sqrt{5.40 \times 10^5 \cdot 100^2 \cdot 100} = 734 \times 10^5$$

and

$$\frac{\hat{P}_{\mathrm{NH_3}}}{\hat{P}_{\mathrm{total}}} = 1.000 \qquad \frac{\hat{P}_{\mathrm{N_2}}}{\hat{P}_{\mathrm{total}}} = \frac{\hat{P}_{\mathrm{H_2}}}{\hat{P}_{\mathrm{total}}} = 1.4 \times 10^{-6}$$

This result shows that the equilibrium is strongly shifted toward the product, ammonia. Although the ammonia formation is strongly favored by the equilibrium, this reaction is kinetically hindered at room temperature and does not proceed significantly. The industrial Haber–Bosch process uses high temperatures and a catalyst to accelerate the reaction; and it accounted for 1 to 2% of the world's energy consumption in 2018.[1] Plants use enzymes to "fix" nitrogen at room temperature, however, and developing energy efficient methods to produce ammonia, which is essential as a fertilizer, is a current frontier in energy research.

E18.14 By using the heat capacity data given in Foundation 16.2 for the limestone decomposition reaction, calculate the Gibbs energy change for the decomposition of limestone $(CaCO_3 \rightarrow CaO + CO_2)$ at 1200 K.

Solution

The entropy change for the reaction at 1200 K may be found by using the result

$$\Delta_r S^{\ominus}(1200\ \text{K}) = \Delta_r S^{\ominus}(298\ \text{K}) + \int_{298}^{1200} \frac{\Delta C^{\ominus}_{P,m}}{T} \cdot dT$$

where

$$\Delta_r S^{\ominus}(298\ \text{K}) = 160.4\ \text{J K}^{-1}\ \text{mol}^{-1}$$

Using heat capacity data from Foundation 16.2, we can evaluate the integral as

$$\begin{aligned}&\int_{298}^{1200} \frac{\Delta C^{\ominus}_{P,m}}{T} \cdot dT\\ &= \left[2.67 \cdot \ln\left(\frac{1200}{298}\right) + (-0.027 \cdot 902) + \frac{1}{2} 0.74 \times 10^{-5} \cdot 1.35 \times 10^{6}\right]\ \text{J K}^{-1}\ \text{mol}^{-1}\\ &= -18.7\ \text{J K}^{-1}\ \text{mol}^{-1}\end{aligned}$$

Then

$$\Delta_r S^{\ominus}(1200\ \text{K}) = (160.4 - 18.7)\ \text{J K}^{-1}\ \text{mol}^{-1} = 144.7\ \text{J K}^{-1}\ \text{mol}^{-1}$$

In Foundation 16.2 we showed that $\Delta_r H^{\ominus}(1200\ \text{K}) = 165.7\ \text{kJ mol}^{-1}$, and we obtain

$$\Delta_r G^{\ominus}(1200\ \text{K}) = (165.7 \times 10^3 - 1200 \cdot 144.7)\ \text{J mol}^{-1} = -8.0\ \text{kJ mol}^{-1}$$

E18.15 Consider the reaction

$$CH_3CH_2OH\ (l) + CH_3\ COCl\ (l) \rightarrow CH_3CH_2O\text{-}C(O)CH_3\ (l) + HCl\ (g)$$

What is the equilibrium constant for this reaction at 298 K? Estimate the equilibrium constant at 370 K. How would the addition of a base, such as pyridine, affect this equilibrium?

	CH_3CH_2OH	CH_3COCl	$CH_3CH_2O(CO)CH_3$	HCl
$\Delta_f G^{\ominus}$ (kJ mol^{-1})	−174.8	−208.0	−337.2	−92.3
$\Delta_f H^{\ominus}$ (kJ mol^{-1})	−277.7	−273.8	−479.3	−95.3

Solution

The equilibrium constant at 298 K is

$$\begin{aligned}\Delta_r G^{\ominus} &= \begin{bmatrix} \Delta_f G^{\ominus}\ (CH_3CH_2OC(O)CH_3) + \Delta_f G^{\ominus}\ (HCl) \\ -\Delta_f G^{\ominus}\ (CH_3CH_2OH) - \Delta_f G^{\ominus}\ (CH_3COCl) \end{bmatrix}\\ &= (-337.2 - 92.3 + 174.8 + 208.0)\ \text{kJ mol}^{-1} = -46.7\ \text{kJ mol}^{-1}\end{aligned}$$

leading to

$$K = \exp\left(-\frac{-46.7\ \text{kJ mol}^{-1}}{8.3145\ \text{J K}^{-1}\ \text{mol}^{-1} \times 298\ \text{K}}\right) = \exp(18.8) = 1.5 \times 10^8$$

1 J. G. Chen et al., Beyond fossil fuel–driven nitrogen transformations, *Science* 2018, **360**, eaar6611. DOI: 10.1126/science.aar6611.

We can find the equilibrium constant at 370 K by way of the integrated form of the van't Hoff equation (see Equation (18.84)). First we find the reaction enthalpy, via

$$\Delta_r H^\ominus = \begin{bmatrix} \Delta_f H^\ominus \left(CH_3CH_2OC(O)CH_3\right) + \Delta_f H^\ominus \left(HCl\right) \\ -\Delta_f H^\ominus \left(CH_3CH_2OH\right) - \Delta_f H^\ominus \left(CH_3COCl\right) \end{bmatrix}$$

$$= (-479.3 - 95.3 + 277.7 + 273.8)\ \text{kJ mol}^{-1} = -23.1\ \text{kJ mol}^{-1}$$

Using Equation (18.84), we find that

$$\ln\left(\frac{K_2}{K_1}\right) = -\frac{\Delta_r H^\ominus}{\boldsymbol{R}}\left(\frac{1}{T_2} - \frac{1}{T_1}\right)$$

$$\ln\left(\frac{K_{370}}{1.5 \times 10^8}\right) = -\frac{-23.1\ \text{kJ mol}^{-1}}{8.3145\ \text{J K}^{-1}\ \text{mol}^{-1}}\left(\frac{1}{370\ \text{K}} - \frac{1}{298\ \text{K}}\right)$$

so that

$$\left(\frac{K_{370}}{1.5 \times 10^8}\right) = \exp(-1.81) = 0.163$$

Solving for K_{370}, we find that

$$K_{370} = 1.5 \times 10^8 \times 0.162\,97 = 2.44 \times 10^7$$

The addition of base to the reaction mixture would consume the acid and shift the equilibrium more toward the product form.

E18.16 Temperature dependence of ΔS. Use Equation (18.26) to derive Equation (18.27). Proceed in a manner similar to that we used in Section 16.5.3 for the temperature dependence of ΔU and ΔH.

Solution

Equation (18.26) is

$$\mathrm{d}S_i = \frac{C_{p,i} \cdot \mathrm{d}T}{T}$$

By integrating both sides

$$\int_{T_1}^{T_2} \mathrm{d}S_i = \int_{T_1}^{T_2} \frac{C_{p,i} \cdot \mathrm{d}T}{T}$$

we find that

$$S_{i,T_2} - S_{i,T_1} = \int_{T_1}^{T_2} \frac{C_{p,i} \cdot \mathrm{d}T}{T}$$

For a reaction we can write the reaction entropy change as

$$\Delta_r S = \sum_i^{\text{products}} S_i - \sum_i^{\text{reactants}} S_i$$

If we define the change in constant pressure heat capacity between the reactants and products as

$$\Delta_r C_p = \sum_i^{\text{products}} C_{p,i} \quad \sum_i^{\text{reactants}} C_{p,i}$$

then we can write that

$$\Delta_r S_{T_2} = \Delta_r S_{298} + \int_{298}^{T_2} \frac{\Delta_r C_p \cdot \mathrm{d}T}{T}$$

which is Equation (18.27).

E18.17 Reverse Osmosis. Consider sea water with a salt concentration of $c_1 = 1$ mol L^{-1} below the piston in Fig. 18.11. By exerting an external pressure on the piston, the sea water is concentrated, and pure water appears above the piston (reverse osmosis). What is the reversible work required to extract 0.5 L of pure water from 1 L of sea water at a temperature of 298 K?

Solution

To extract 0.5 L of pure water, the concentration in the remaining sea water must be increased by a factor of two, to $c_2 = 2$ mol L^{-1}. Thus with $n_{salt} = 1$ mol we obtain

$$w_{rev} = \boldsymbol{n}_{salt}\boldsymbol{R}T \cdot \frac{c_2}{c_1} = 1\ \text{mol} \cdot 8.314\ \text{J K}^{-1}\ \text{mol}^{-1} \cdot 298\ \text{K} \cdot \ln\frac{2}{1} = 1.7\ \text{kJ}$$

To purify the same amount of water by distillation, the heat of evaporation $\Delta H = \boldsymbol{n}_{water} \cdot 44$ kJ mol^{-1} = 1200 kJ is needed.

E18.18 Temperature Dependence of the Equilibrium Constant. Begin with Equation (18.53) for the equilibrium constant and derive the van't Hoff equation

$$\boxed{\frac{d\ln K}{dT} = \frac{\Delta_r H^\ominus}{\boldsymbol{R}T^2}}$$

Solution

To derive the temperature dependence of the equilibrium constant K we begin with Equation

$$\ln K = -\frac{\Delta_r G^\ominus}{\boldsymbol{R}T}$$

and take the derivative of both sides with respect to the temperature and find

$$\frac{d\ln K}{dT} = -\frac{1}{\boldsymbol{R}}\left(\frac{1}{T}\frac{d\Delta_r G^\ominus}{dT} - \Delta_r G^\ominus \frac{1}{T^2}\right)$$

From Box 18.1 we find for the temperature dependence of the Gibbs energy of reaction

$$\frac{d\Delta_r G^\ominus}{dT} = \left(\frac{\partial \Delta_r G}{\partial T}\right)_{P=P^\ominus} = -\Delta_r S^\ominus$$

Thus

$$\frac{d\ln K}{dT} = -\frac{1}{\boldsymbol{R}}\left(-\frac{1}{T}\Delta_r S^\ominus - \Delta_r G^\ominus \frac{1}{T^2}\right)$$

Using the fact that $\Delta_r G^\ominus = \Delta_r H^\ominus - T \cdot \Delta_r S^\ominus$ we obtain

$$\frac{d\ln K}{dT} = -\frac{1}{\boldsymbol{R}}\left(-\frac{1}{T}\Delta_r S^\ominus - (\Delta_r H^\ominus - T \cdot \Delta_r S^\ominus)\frac{1}{T^2}\right) = \frac{\Delta_r H^\ominus}{\boldsymbol{R}T^2}$$

(van't Hoff equation).

E18.19 Calculate the standard Gibbs energy change for the precipitation of calcium carbonate,

$$Ca^{2+}(aq) + CO_3^{2-}(aq) \rightarrow CaCO_3\ (s).$$

What is the concentration of the ions in a saturated solution?

Solution

We have the following data

	Ca^{2+}(aq)	CO_3^{2-}(aq)	$CaCO_3$(s)
$\Delta_f G^\ominus$/(kJ mol^{-1})	−553.0	−528.1	−1128.0

Using these data, the Gibbs energy of reaction is

$$\Delta_r G^\ominus = \Delta_f G^\ominus\,(CaCO_3) - \Delta_f G^\ominus\,(Ca^{2+}) - \Delta_f G^\ominus\,(CO_3^{2-})$$
$$= [-1128 - (-553.0) - (-528.1)]\ \text{kJ mol}^{-1} = -46.9\ \text{kJ mol}^{-1}$$

We can determine the concentration of ions by way of the equilibrium constant, which we calculate as

$$K = \exp\left(-\frac{\Delta_r G^\ominus}{\boldsymbol{R}T}\right)$$
$$= \exp\left(-\frac{-46.9\ \text{kJ mol}^{-1}}{8.3145\ \text{J K}^{-1}\ \text{mol}^{-1} \times 298\ \text{K}}\right) = \exp(18.93) = 1.66 \times 10^8$$

In this case

$$K = \frac{1}{\hat{c}_{Ca^{2+}} \cdot \hat{c}_{CO_3^{2-}}}$$

Because these are the only ions in solution we use the fact that $\hat{c}_{Ca^{2+}} = \hat{c}_{CO_3^{2-}} = x$, and calculate

$$x = \sqrt{\frac{1}{K}} = \sqrt{\frac{1}{1.66 \times 10^8}} = 7.76 \times 10^{-5}.$$

Hence we have

$$\hat{c}_{total} = \hat{c}_{Ca^{2+}} + \hat{c}_{CO_3^{2-}} = 1.55 \times 10^{-4}.$$

which indicates that calcium carbonate is only weakly soluble in water.

E18.20 From Equation (18.91) calculate the equilibrium constant K for the $H_2 + D_2 \rightleftharpoons 2$ HD equilibrium at $T = 83$ K, 195 K, and 383 K. Compare your result with the corresponding experimental values $K = 2.20$, 2.88, and 3.46.

Solution
Equation (18.91) is

$$K = \frac{2\left(m_H + m_D\right)}{\sqrt{m_H m_D}} \cdot \exp\left[-\frac{h\nu_{0,HD}}{kT}\left(1 - \frac{\sqrt{m_H} + \sqrt{m_D}}{\sqrt{2\left(m_H + m_D\right)}}\right)\right]$$

Plugging in the values for $T = 83$ K we find with

$$h\nu_{0,HD} = 6.626 \times 10^{-34}\ \text{J s} \cdot 1.14 \times 10^{14}\ \text{s}^{-1} = 7.55 \times 10^{-20}\ \text{J}$$

that

$$K = \frac{6}{\sqrt{2}} \cdot \exp\left[-\frac{7.55 \times 10^{-20}\ \text{J}}{1.380 \times 10^{-23}\ \text{J K}^{-1} \cdot 83\ \text{K}} \cdot \left(1 - \frac{1 + \sqrt{2}}{\sqrt{6}}\right)\right]$$
$$= 4.24 \cdot \exp\left[-65.9 \cdot 0.0144\right] = 4.24 \cdot 0.387 = 1.64$$

We proceed in a corresponding way for the other temperatures and find the values given in the table.

T/K	K_{calc}	K_{expt}
83	1.64	2.20
195	2.83	2.88
383	3.45	3.46

The calculated value deviates somewhat from the experimental value at low temperature but is in excellent agreement at higher temperatures. Some reasons for the discrepancy at 83 K might be our approximate treatment of the rotational degrees of freedom (we used the high T limit in obtaining Equation (18.91)) and/or nonidealities of the gas mixture at low temperature.

E18.21 Calculate the autoprotolysis constant K_w of water.

Solution
We consider the reaction

$$H_2O\ (\text{liq}) \rightarrow H^+\ (\text{aq}) + OH^-\ (\text{aq})$$

According to Table A.1 at 298 K we have

$$\Delta_f G^\ominus(H_2O) = -237.2\ \text{kJ mol}^{-1} \ \text{and} \quad \Delta_f G^\ominus_{aq}(OH^-) = -157.3\ \text{kJ mol}^{-1}$$

Then with $\Delta_f G^\ominus(H^+) = 0$ we obtain

$$\Delta_r G^\ominus = (-157.3 + 237.2)\ \text{kJ mol}^{-1} = 79.9\ \text{kJ mol}^{-1}$$

and

$$\ln K_w = -\frac{\Delta_r G^\ominus}{RT} = -32.2, \quad K_w = \hat{c}_{H^+,eq} \cdot \hat{c}_{OH^-,eq} = 1 \times 10^{-14}$$

In pure water at 298 K, H^+ and OH^- are present in concentrations of

$$c_{H^+,eq} = c_{OH^-,eq} = 10^{-7} \text{ mol L}^{-1}$$

E18.22 Zn^{2+} forms a complex with cyanide: $Zn^{2+} + 4\ CN^- \rightarrow \left[Zn(CN)_4\right]^{2-}$. For concentrations of Zn^{2+} and CN^- of 1×10^{-3}mol L^{-1}, calculate the equilibrium concentration of the complex.

Solution

With the data in Table A.1 we obtain

$$\Delta_r G^\ominus = [+446.9 - (-147.2 + 4 \times 1654.7)] \text{ kJ mol}^{-1}$$
$$= -68.7 \text{ kJ mol}^{-1}$$

so that

$$K = \frac{\hat{c}_{\text{complex,eq}}}{\hat{c}_{Zn^{2+},eq} \cdot \hat{c}^4_{CN^-,eq}} = 1.1 \times 10^{12}$$

In a solution with Zn^{2+} and CN^-, both in a concentration of 1×10^{-3} mol L^{-1}, the complex is present in a concentration of

$$c_{\text{complex,eq}} = 1.1 \times 10^{12} \cdot \left(10^{-3}\right)^5 \text{ mol L}^{-1}$$
$$= 1.1 \times 10^{-3} \text{ mol L}^{-1}$$

E18.23 Using the Clausius–Clapeyron equation in its integrated form, calculate the vapor pressure above liquid and solid water (a) in the range of $T = 350$ K to $T = 500$ K and (b) in the range of $T = 271$ K to $T = 275$ K. The molar enthalpies of evaporation and sublimation are $\Delta_{vap}H = 44.1$ kJ mol^{-1} and $\Delta_{sub}H = 50.1$ kJ mol^{-1}. The vapor pressure at $T_1 = 373$ K is $P_{vap,1} = 1.013$ bar, the vapor pressure at the triple point, $T_1 = 273.16$ K, is $P_{vap,1} = 6.11 \times 10^{-3}$ bar.

Solution

Inserting the given data in the integrated form of the Clausius–Clapeyron equation; namely

$$\ln \hat{P} = \ln \hat{P}_1 - \frac{\Delta_{vap} H^\ominus}{R} \left(\frac{1}{T} - \frac{1}{T_1}\right)$$

results in the plots that are shown in Fig. E18.23.

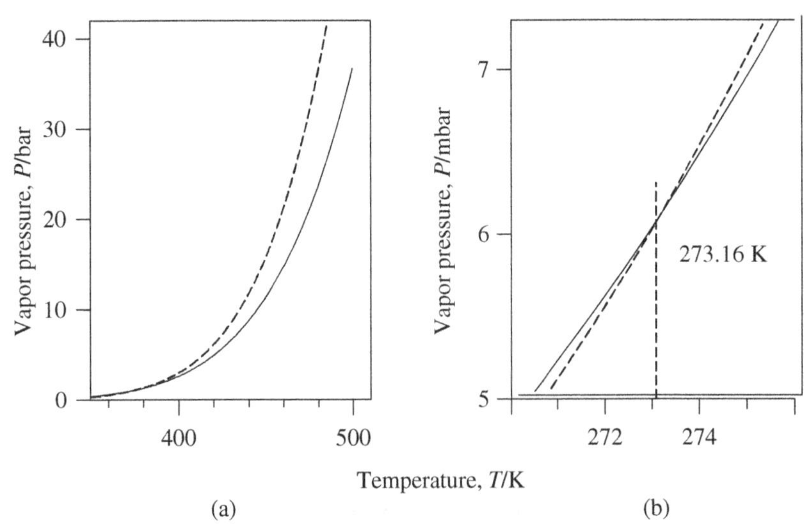

Figure E18.23 Vapor pressure P of water as a function of temperature calculated from Equation (18.41). Solid lines: vapor pressure above liquid water, dashed lines: vapor pressure above solid water. (a) Temperature range 350–500 K, (b) behavior at the triple point.

E18.24 The vapor pressure of water at 301 K is 0.0031 bar. Compute the vapor pressure at 301 K of a solution containing 68.0 g of sucrose, $C_{12}H_{22}O_{11}$ in 1000 grams of water. Assume the solution is ideal.

Solution

We expect that the sucrose will depress the vapor pressure of the water. To find out by how much, we use the relation $P_{H_2O} = x_{H_2O}P^{\ominus}$, in which $x_{H_2O} = 1 - x_{\text{sucrose}}$. To find the mole fraction of sucrose, we first calculate the molar mass $\boldsymbol{M}$ of sucrose as

$$\boldsymbol{M} = (12 \times 12.01 + 22 \times 1.01 + 11 \times 16.00)\ \text{g mol}^{-1} = 342.34\ \text{g mol}^{-1}$$

The mole fraction of sucrose in the solution is

$$x = \frac{68.0 \times \frac{1}{342.34}}{68.0 \times \frac{1}{342.34} + 55.5} = 3.57 \times 10^{-3}$$

so that the mole fraction of water is

$$x_{H_2O} = 1 - x = 0.996$$

Hence, we find that the vapor pressure is

$$\begin{aligned} P_{H_2O} &= xP^{\ominus} \\ &= 0.996 \cdot 0.0031\ \text{bar} = 0.00309\ \text{bar} \end{aligned}$$

E18.25 Given that the vapor pressure of pure water is $P_s = 0.0031$ bar at 298 K, and the vapor pressure of water above a sucrose solution is $P = 0.0030$ bar, determine the concentration of the sucrose solution.

Solution

This exercise is similar to E18.24, except we are asked to calculate the concentration of sucrose by way of the measured vapor pressure depression.

$$x_{H_2O} = \frac{P}{P^{\ominus}} = \frac{0.0030}{0.0031} = 0.97$$

If we recall the definition of the mole fraction

$$x_{H_2O} = \frac{\boldsymbol{n}_w}{\boldsymbol{n}_s + \boldsymbol{n}_w}$$

and take the concentration of H_2O in water to be 55.5 mol L^{-1}, we can write with $c_s = n_s/V$ that

$$x_{H_2O} = \frac{55.5\ \text{mol L}^{-1}}{c_s + 55.5\ \text{mol L}^{-1}} = 0.97$$

so that

$$c_s = \frac{\left(55.5\ \text{mol L}^{-1}\right) \cdot (1 - 0.97)}{0.97} = 1.72\ \text{mol L}^{-1}$$

E18.26 Given that its vapor pressure is 0.790 bar, determine the change of Gibbs energy for the evaporation of liquid Ar at 85.0 K.

Solution

Using Equation (18.35), for the change in Gibbs energy we find

$$\begin{aligned} \Delta_{\text{vap}}G^{\ominus} &= -\boldsymbol{R}T \cdot \ln \widehat{P} \\ &= \left[-\left(8.314\ \text{J K}^{-1}\ \text{mol}^{-1}\right) \cdot (85.0\ \text{K}) \cdot \ln 0.790\right]\ \text{J mol}^{-1} = 166.6\ \text{J mol}^{-1} \end{aligned}$$

E18.27 In Example 18.8 we found a positive entropy change for the reaction of one mole of H_2 with one mole of I_2 to form two moles of HI. Rationalize this entropy change from a molecular perspective.

Solution

We analyze the various contributions to

$$\Delta_r S^{\ominus} = 2S^{\ominus}_{HI} - S^{\ominus}_{H_2} - S^{\ominus}_{I_2}$$

Because each of the reactants and products is in singlet state and the electronic energy level spacings are large, we take the electronic contribution to the entropy change to be zero; i.e.,

$$\Delta S^{\ominus}_{\text{elec}} \approx 0$$

If we use the high temperature limit for the translational degrees of freedom, according to the Sackur–Tetrode equation we estimate an entropy change of

$$\Delta S^{\ominus}_{\text{trans}} = Nk\left[2 \cdot \ln\left(const \cdot m_{\text{HI}}^{3/2}\right) - \ln\left(const \cdot m_{\text{H}_2}^{3/2}\right) - \ln\left(const \cdot m_{\text{I}_2}^{3/2}\right)\right] = Nk \ln\left(\frac{m_{\text{HI}}^3}{m_{\text{H}_2}^{3/2} m_{\text{I}_2}^{3/2}}\right)$$

where

$$const = e^{5/2}(2\pi)^{3/2}\frac{(k)^{3/2}}{h^3} \cdot \frac{V}{N} T^{3/2}$$

Thus we obtain

$$\Delta S^{\ominus}_{\text{trans}} = Nk \ln\left(\frac{\left(const \cdot m_{\text{HI}}^{3/2}\right)^2}{const \cdot m_{\text{H}_2}^{3/2} const \cdot m_{\text{I}_2}^{3/2}}\right) = Nk \ln\left(\frac{m_{\text{HI}}^3}{m_{\text{H}_2}^{3/2} m_{\text{I}_2}^{3/2}}\right)$$

Inserting values for the molar masses gives

$$\Delta S^{\ominus}_{\text{trans}} = Nk \ln\left(\frac{\left(128 \text{ g mol}^{-1}\right)^3}{\left(2 \text{ g mol}^{-1}\right)^{3/2}\left(254 \text{ g mol}^{-1}\right)^{3/2}}\right) = Nk \cdot 5.21$$

which is greater than zero. For the rotational degrees of freedom we need the following data

	HI	H_2	I_2
I/kg pm^2	$4.27 \cdot 10^{-23}$	$4.55 \cdot 10^{-24}$	$7.52 \cdot 10^{-21}$
d/pm	161	74	267

and use the high temperature approximation Equation (17.87)

$$S_{\text{rot}} = Nk\left[\ln\left(\frac{8\pi^2 IkT}{\sigma h^2}\right) + 1\right] = Nk\left[\ln\left(const \cdot \frac{I}{\sigma}\right) + 1\right]$$

where

$$const = \frac{8\pi^2 kT}{h^2}$$

to write that

$$\begin{aligned}
\Delta S^{\ominus}_{\text{rot}} &= Nk\left[2 \cdot \left[\ln\left(\frac{const \cdot I_{\text{HI}}}{\sigma_{\text{HI}}}\right) + 1\right] - \left[\ln\left(\frac{const \cdot I_{\text{H}_2}}{\sigma_{\text{H}_2}}\right) + 1\right] - \left[\ln\left(\frac{const \cdot I_{\text{I}_2}}{\sigma_{\text{I}_2}}\right) + 1\right]\right] \\
&= Nk\left[\ln\left(\frac{const \cdot I_{\text{HI}}}{\sigma_{\text{HI}}}\right)^2 + 2 - \ln\left(\frac{const \cdot I_{\text{H}_2}}{\sigma_{\text{H}_2}}\right) - 1 - \ln\left(\frac{const \cdot I_{\text{I}_2}}{\sigma_{\text{I}_2}}\right) - 1\right] \\
&= Nk\left[\ln\left(\frac{I_{\text{HI}}^2 \cdot \sigma_{\text{H}_2} \cdot \sigma_{\text{I}_2}}{\sigma_{\text{HI}}^2 \cdot I_{\text{H}_2} \cdot I_{\text{I}_2}}\right)\right]
\end{aligned}$$

Inserting values for I and σ gives (note that H_2 and I_2 are homonuclear molecules, thus $\sigma = 2$)

$$\Delta S^{\ominus}_{\text{rot}} = Nk\left[\ln\left(\frac{\left(4.27 \cdot 10^{-23} \text{ kg pm}^2\right)^2 \cdot 2 \cdot 2}{1 \cdot 4.55 \cdot 10^{-24} \text{ kg pm}^2 \cdot 7.52 \cdot 10^{-21} \text{ kg pm}^2}\right)\right] = -Nk \cdot 1.55$$

which is greater than zero.

For the vibrational degrees of freedom we consider that the vibrational energy spacings (see the wavenumbers given in the table) of HI and H_2 are much larger than kT but that of I_2 is not.

	HI	H_2	I_2
$\tilde{\nu}/cm^{-1}$	2308.09	4400.39	214.57

$$\Delta S^{\ominus}_{vib} = 2 \cdot S^{\ominus}_{vib}(HI) - S^{\ominus}_{vib}(H_2) - S^{\ominus}_{vib}(I_2)$$
$$\approx 0 - 0 - S^{\ominus}_{vib}(I_2) \text{ which is less than zero}$$

Calculating $\Delta S^{\ominus}_{vib}$ explicitly, we find that

$$\Delta S^{\ominus}_{vib} = -Nk \cdot \left[\ln\left(\frac{kT}{h\nu_0}\right) + 1\right] = -Nk \cdot \left[\ln\left(\frac{4.14 \cdot 10^{-21}\ J}{4.27 \cdot 10^{-21}\ J}\right) + 1\right] = -Nk \cdot 0.969$$

Hence we see that the total entropy change is positive:

$$\Delta S^{\ominus} = Nk \cdot (5.21 - 1.55 - 0.969) = Nk \cdot 2.69$$

and

$$\Delta_r S^{\ominus} = \boldsymbol{R} \cdot 2.69 = 22.4\ J\ K^{-1}\ mol^{-1}$$

compared to the experimental value $\Delta_r S^{\ominus} = 21.4\ J\ K^{-1}\ mol^{-1}$ found in Exercise E18.5.

E18.28 Show that the Clausius–Clapeyron Equation (18.38) can be derived as a special case of the van't Hoff Equation (18.81).

Solution

The van't Hoff equation is

$$\ln\left(\frac{K_2}{K_1}\right) = -\frac{\Delta_r H^{\ominus}}{\boldsymbol{R}}\left[\frac{1}{T_2} - \frac{1}{T_1}\right]$$

If we write the following equilibrium

liquid $\rightleftarrows$ vapor

and recognize that

$$K = \hat{P}_{vapor}$$

then we can substitute into the van't Hoff equation and find that

$$\ln\left(\frac{P_2}{P_1}\right) = -\frac{\Delta_{vap} H^{\ominus}}{\boldsymbol{R}}\left[\frac{1}{T_2} - \frac{1}{T_1}\right]$$

18.2 Problems

P18.1 Find an expression for the equilibrium constant of the reaction: I_2 (g) → 2 I (g), in terms of the molecular and atomic parameters. Use the following data for I_2: $\theta_{vib} = 308$ K; $\theta_{rot} = 0.0537$ K; $\sigma = 2$; $\varepsilon_B = 2.51 \cdot 10^{-19}$ J and the ground electronic state of the I atom is $^2P_{3/2}$. Given your expression for the equilibrium constant, describe how this equilibrium will be affected by temperature. Evaluate the equilibrium constant at 1298 K and compare it to your result in E18.4.

Solution

From Equation (18.86), we can write

$$K = \frac{\left(\frac{Z_I^\ominus}{N_I}\right)^2}{\frac{Z_{I_2}^\ominus}{N_{I_2}}} = \frac{kT}{p^\ominus}\frac{\left(\frac{Z_I^\ominus}{V_I}\right)^2}{\frac{Z_{I_2}^\ominus}{V_{I_2}}}$$

For the I atom, we find, according to Equation (15.59), that

$$Z_I^\ominus = Z_{\text{trans}}^\ominus Z_{\text{elec}}^\ominus = V\left(\frac{2\pi m_I kT}{h^2}\right)^{3/2} \cdot 4 \cdot \exp\left(-\frac{E_I}{kT}\right)$$

where E_I is the energy of the I atom electronic state, and the factor of 4 arises from the degeneracy of the atom's electronic state, which can be calculated from $(2J + 1)$ and the $J = 3/2$ subscript of the atomic term symbol. Given that the zero of energy is arbitrary, we choose to place the zero of energy at the ground-state electronic energy of the I atom, thus $E_I = 0$. Hence we find that

$$\begin{aligned}\frac{Z_I^\ominus}{V} &= \left(\frac{2\pi \cdot 2.107 \cdot 10^{-25}\ \text{kg} \cdot 1.38 \cdot 10^{-23}\ \text{J K}^{-1} \cdot 1298\ \text{K}}{(6.626 \cdot 10^{-34}\ \text{J s})^2}\right)^{3/2} \cdot 4\ \text{m}^{-3} \\ &= 1.255 \cdot 10^{34} \cdot 4\ \text{m}^{-3} = 5.021 \cdot 10^{34}\ \text{m}^{-3}\end{aligned}$$

If we use the partition function for vibration, see Equation (15.23), and consider the high temperature limit for the rotational degrees of freedom for I_2, see Equation (15.38), we find

$$\begin{aligned}Z_{I_2}^\ominus &= Z_{\text{trans}}^\ominus Z_{\text{vib}}^\ominus \frac{Z_{\text{rot}}^\ominus}{\sigma} Z_{\text{elec}}^\ominus \\ &= V\left(\frac{2\pi m_{I_2} kT}{h^2}\right)^{3/2} \cdot \frac{\exp\left(-\frac{\theta_{\text{vib}}}{2T}\right)}{1-\exp\left(-\frac{\theta_{\text{vib}}}{T}\right)} \cdot \frac{T}{\sigma\theta_{\text{rot}}} \cdot g_{I_2} \cdot \exp\left(-\frac{E_{I_2}}{kT}\right) \\ &= V\left(\frac{2\pi m_{I_2} kT}{h^2}\right)^{3/2} \cdot \frac{\exp\left(-\frac{308\ \text{K}}{2 \cdot T}\right)}{1-\exp\left(-\frac{308\ \text{K}}{T}\right)} \cdot \frac{T}{2 \times 0.0537\ \text{K}} \cdot 1 \cdot \exp\left(-\frac{E_{I_2}}{kT}\right)\end{aligned}$$

where E_{I_2} is the electronic energy of the iodine molecule. Using the bond energy and the zero point vibrational energy of $h\nu/2 = k\theta_{\text{vib}}/2 = 2.126 \cdot 10^{-21}$ J, we can calculate that

$$\begin{aligned}E_{I_2} &= -\left(\varepsilon_B + \frac{h\nu_0}{2}\right) \\ &= -\left(2.51 \cdot 10^{-19} + 0.0213 \cdot 10^{-19}\right)\ \text{J} \\ &= -2.53 \cdot 10^{-19}\ \text{J}\end{aligned}$$

Hence we find that

$$Z_{\text{trans}} = V\left(\frac{2\pi \cdot 4.214 \times 10^{-25}\ \text{kg} \cdot 1.38 \times 10^{-23}\ \text{J K}^{-1} \cdot 1298\ \text{K}}{(6.626 \times 10^{-34}\ \text{J s})^2}\right)^{3/2} = V \cdot 3.55 \times 10^{34}\ \text{m}^{-3}$$

$$Z_{\text{rot}} = \frac{1298\ \text{K}}{2 \times 0.0537\ \text{K}} = 1.209 \times 10^4$$

$$Z_{\text{vib}} = \frac{\exp\left(-\frac{308\ \text{K}}{2 \cdot 1298\ \text{K}}\right)}{1-\exp\left(-\frac{308\ \text{K}}{1298\ \text{K}}\right)} = 4.204$$

$$Z_{\text{electronic}} = \exp\left(-\frac{E_{I_2}}{kT}\right) = \exp\left(-\frac{-2.53 \cdot 10^{-19}\ \text{J}}{1.38 \times 10^{-23}\ \text{J K}^{-1} \cdot 1298\ \text{K}}\right) = 1.36 \times 10^6$$

Thus we obtain

$$\frac{Z_{I_2}^\ominus}{V} = 3.55 \times 10^{34} \cdot 1.209 \times 10^4 \cdot 4.204 \cdot 1.36 \times 10^6 = 2.45 \times 10^{45}\ \text{m}^{-3}$$

Combining these results, we find that

$$K = \frac{kT}{P^{\ominus}} \frac{\left(Z_{\mathrm{I}}^{\ominus}/V\right)^2}{\left(Z_{\mathrm{I}_2}^{\ominus}/V\right)}$$

$$= \frac{(1.38 \cdot 10^{-23}\ \mathrm{J\,K^{-1}})(1298\ \mathrm{K})}{10^5\ \mathrm{N\,m^{-2}}} \frac{\left(5.021 \cdot 10^{34}\ \mathrm{m^{-3}}\right)^2}{2.45 \times 10^{45}\ \mathrm{m^{-3}}}$$

$$= \frac{(1.38 \cdot 10^{-23}\ )(1298\ )}{10^5} \frac{\left(5.021 \cdot 10^{34}\right)^2}{2.45 \times 10^{45}} \frac{\mathrm{N\,m}}{\mathrm{N\,m^{-2}}} \frac{\mathrm{m^{-6}}}{\mathrm{m^{-3}}}$$

$$= 0.184$$

This value is in reasonably good agreement with the value of 0.205 that was found in E18.4.

P18.2 In Example 18.7, we estimated the $\Delta_r G^{\ominus}$ and K for the reaction $3H_2 + N_2 \rightarrow 2\ NH_3$, at 700 K. Using the heat capacity data in the table, perform a rigorous calculation of the equilibrium constant's temperature dependence between 298 and 1000 K. The parameters in the table should be used in the formula

$$C^0_{P,m} = c_1 + c_2 \cdot T + c_3 \cdot T^2$$

Molecule	c_1 J /(K mol)	c_2 J /(K^2 mol)	c_3 J /(K^3 mol)
N_2	27.20	0.00535	−0.0121
H_2	29.04	−0.0008	0.197
NH_3	24.15	0.0392	−0.783

Solution

We will use the van't Hoff relation to find how the equilibrium shifts with temperature. We will evaluate the temperature dependence of the enthalpy by using

$$\Delta_r H\left(T_2\right) = \Delta_r H\left(T_1\right) + \int_{T_1}^{T_2} \Delta_r C_p\ \mathrm{d}T$$

and use the reaction enthalpy at 298 K as

$$\Delta_r H_{298} = 2\Delta_f H\left(\mathrm{NH_3}\right) - 3\Delta_f H\left(\mathrm{H_2}\right) - \Delta_f H\left(\mathrm{N_2}\right)$$
$$= [2 \cdot (-45.2) - 0 - 0]\ \mathrm{kJ\ mol^{-1}} = -90.4\ \mathrm{kJ\ mol^{-1}}$$

where $\Delta_r H_{298} = \Delta_r H\,(298)$. Now we calculate the change in the heat capacity between the two sides as

$$\Delta C_p = \Delta c_1 + \Delta c_2 \cdot T + \Delta c_3 \cdot T^2$$

From the data in the table we calculate

$$\Delta c_1 = [2 \times 24.15 - 3 \times 29.04 - 27.20]\ \mathrm{J\,K^{-1}\ mol^{-1}} = -66.02\ \mathrm{J\,K^{-1}\ mol^{-1}}$$

$$\Delta c_2 = [2 \times 0.0392 - 3 \times (-0.0008) - 0.00535]\ \mathrm{J\,K^{-2}\ mol^{-1}} = 0.0754\ \mathrm{J\,K^{-2}\ mol^{-1}}$$

and

$$\Delta c_3 = [2 \times (-0.783) - 3 \times 0.197 - (-0.012)] \cdot 10^{-5}\ \mathrm{J\,K^{-3}\ mol^{-1}}$$
$$= -2.144 \cdot 10^{-5}\ \mathrm{J\,K^{-3}\ mol^{-1}}$$

Using these results we can write that

$$\Delta_r H\,(T) = \Delta_r H_{298} + \int_{298}^{T} \Delta_r C_p\ \mathrm{d}T$$
$$= \Delta_r H_{298} + \int_{298}^{T} \left(\Delta c_1 + \Delta c_2 \cdot T + \Delta c_3 \cdot T^2\right) \cdot \mathrm{d}T$$

$$= \left[\begin{array}{c} \Delta_r H_{298} + \Delta c_1 \cdot (T - 298\ \text{K}) + \Delta c_2 \cdot \frac{1}{2}(T^2 - 298^2\ \text{K}^2) \\ + \Delta c_3 \cdot \frac{1}{3}(T^3 - 298^3\ \text{K}^3) \end{array}\right]$$

$$= Const + \Delta c_1 \cdot T + \Delta c_2 \cdot \frac{1}{2}T^2 + \Delta c_3 \cdot \frac{1}{3}T^3$$

where $Const = \Delta_r H_{298} - \Delta c_1 \cdot 298\ \text{K} - \Delta c_2 \cdot \frac{1}{2}298^2\ \text{K}^2 - \Delta c_3 \cdot \frac{1}{3}298^3\ \text{K}^3 = -73.9\ \text{kJ mol}^{-1}$.
From the van't Hoff equation (Equation (18.81)) we can write that

$$\ln\left(\frac{K(T)}{K(298)}\right) = \frac{1}{R}\int_{298}^{T} \frac{\Delta_r H(T)}{T^2}\, dT$$

By substituting into our expression for the heat capacity difference we find

$$\ln\left(\frac{K(T)}{K_{298}}\right) = \frac{1}{R}\int_{298}^{T} \frac{Const + \Delta c_1 \cdot T + \Delta c_2 \cdot \frac{1}{2}T^2 + \Delta c_3 \cdot \frac{1}{3}T^3}{T'^2}\, dT'$$

$$= \frac{1}{R}\left(\begin{array}{c} -Const \cdot \left(\frac{1}{T} - \frac{1}{298\ \text{K}}\right) + \Delta c_1 \cdot \ln\left(\frac{T}{298\ \text{K}}\right) \\ + \frac{\Delta c_2}{2}(T - 298\ \text{K}) + \frac{\Delta c_3}{9}(T^2 - 298^2\ \text{K}^2) \end{array}\right)$$

where $K_{298} = K(298)$.

Though somewhat cumbersome, this expression describes how the equilibrium constant changes with temperature.

If, we consider the range from 298 to 1000 K then we find that

$$\ln\left(\frac{K_{1000}}{K_{298}}\right) = \frac{1}{R}\left(\begin{array}{c} -Const \cdot \left(\frac{1}{1000\ \text{K}} - \frac{1}{298\ \text{K}}\right) + \Delta c_1 \cdot \ln\left(\frac{1000}{298}\right) \\ + \frac{\Delta c_2}{2}(1000 - 298)\ \text{K} + \frac{\Delta c_3}{9}(1000^2 - 298^2)\ \text{K}^2 \end{array}\right)$$

$$= \frac{(-174.09 - 79.93 + 26.47 - 2.17)}{8.314} = -27.63$$

and using $\ln K_{298} = 13.2$ (see Example 18.7), we find

$$\ln(K_{1000}) = \ln(K_{298}) - 27.63 = -14.43$$

$$K_{1000} = \exp(-14.43) = 5.41 \cdot 10^{-7}$$

P18.3 Temperature Regulation in the Human Body. A daily intake of 300 g glucose is sufficient to cover the human energy demand.

a) What is the heat produced by the oxidation of 300 g glucose to CO_2 and H_2O?
b) What would be the increase of temperature in the body in one day in the case of complete heat isolation? For simplicity, use the heat capacity of water.
c) The temperature of the body is kept constant by evaporation of water; what amount of water is necessary?
d) The temperature of the body ($T_2 = 100\ °\text{F} = 311\ \text{K}$) is kept constant by thermal conduction to the surroundings (temperature T_1) through clothing. Model the clothing by a bolster of air of thickness 1 cm (thermal conductivity $\kappa = 25 \times 10^{-5}\ \text{J K}^{-1}\text{cm}^{-1}\ \text{s}^{-1}$). Calculate T_1.
e) In analogy to part (d) calculate the demand of glucose for the case that $T_1 = -40\ °\text{F} = 233\ \text{K}$.
f) What power P, wattage, of a continuously burning electrical bulb could be operated by the energy equivalent of 300 g of glucose per day?

Solution

a) Because of the reaction

$$C_6H_{12}O_6\ (\text{glucose}) + 6\ O_2 \rightarrow 6\ H_2O + 6\ CO_2$$

we find

$$\Delta_r H^{\ominus} = -2802\ \text{kJ mol}^{-1} \text{ at } 298\ \text{K}$$

300 g glucose correspond to 1.67 mol; thus, the daily heat production is 4679 kJ.

b) The increase of temperature is $\Delta T = q/C_P$. We approximate C_P by the heat capacity of water (the body consists mainly of water). For a mass of 80 kg we obtain $\boldsymbol{n} = 4444$ mol of H_2O, thus

$$C_P = 75.3 \cdot 4444 \text{ J K}^{-1} = 334 \text{ kJ K}^{-1} \quad \text{and} \quad \Delta T = 14 \text{ K}$$

c) The heat of evaporation of water is $\Delta_{vap}H = 44 \text{ kJ mol}^{-1} \cdot \boldsymbol{n}_{H_2O}$. Then the amount of water needed for cooling the body is

$$\boldsymbol{n}_{H_2O} = \frac{4679 \text{ kJ}}{44 \text{ kJ mol}^{-1}} = 106 \text{ mol}$$

This corresponds to 2 L of water.

d) The surface of the human body is about $A = 1 \text{ m}^2 = 10^4 \text{ cm}^2$. The heat q passing the air bolster in one day is

$$\begin{aligned} q &= \kappa \cdot \frac{A}{\text{thickness}} \cdot (T_2 - T_1) \cdot t \\ &= 25 \times 10^{-5} \text{ J K}^{-1} \text{ cm}^{-1} \text{ s}^{-1} \cdot \frac{10^4 \text{ cm}^2}{1 \text{ cm}} \cdot (T_2 - T_1) \times 10^5 \text{ s} \\ &= 25 \times 10^4 \cdot (T_2 - T_1) \text{ J K}^{-1} \end{aligned}$$

This heat is compensated by 4679 kJ due to the combustion of 300 g glucose. Then

$$T_2 - T_1 = \frac{4679 \text{ kJ}}{25 \times 10^4 \text{ J}} \cdot \text{K} = 18 \text{ K}$$

and $T_1 = (311 - 18) \text{ K} = 293 \text{ K} = 68\,^\circ\text{F}$.

e) For $T_1 = 233$ K the energy demand is $(311 - 233) \cdot 25 \times 10^4 \text{ J} = 1.95 \times 10^4$ kJ. This corresponds to a daily intake of 1250 g glucose.

f) We assume that the electrical energy equals $\Delta G^\ominus$. Then, according to Appendix A.1 and Section 18.4.1,

$$P = \frac{\Delta G^\ominus}{24 \text{ h}} = \frac{1.67 \text{ mol} \cdot 2802 \text{ kJ mol}^{-1}}{86400 \text{ s}} = 55 \text{ W}$$

P18.4 Phosphorus pentachloride dissociates according to the reaction

$$PCl_5(g) \rightarrow PCl_3\,(g) + Cl_2\,(g)$$

Find an expression for the equilibrium constant in terms of the fraction ξ of PCl_5 that is dissociated at equilibrium and the total pressure P. Given that the equilibrium constant is 1.78 bar at 523 K, calculate ξ for total pressures of 0.01, 0.10, and 1.00 bar. Calculate and plot $\Delta_r G$ as a function of ξ at a pressure of 1 bar and a temperature of 523 K. (Adapted from Dickerson, Molecular Thermodynamics 1969).

Solution

If we begin with PCl_5 (amount of substance n), then at equilibrium we will have $n \cdot (1 - \zeta)$ of PCl_5, $\zeta \cdot n$ of PCl_3, and $\zeta \cdot n$ of Cl_2. From these considerations we can write expressions for the mole fractions at equilibrium and the partial pressures. At equilibrium the total amount of substance is $n \cdot (1 + \zeta)$ and we obtain the values in the table

	PCl_5	PCl_3	Cl_2
x	$\frac{(1-\xi)}{(1+\xi)}$	$\frac{\xi}{(1+\xi)}$	$\frac{\xi}{(1+\xi)}$
$\widehat{P}$	$\frac{(1-\xi)}{(1+\xi)}\widehat{P}_{total}$	$\frac{\xi}{(1+\xi)}\widehat{P}_{total}$	$\frac{\xi}{(1+\xi)}\widehat{P}_{total}$

Hence we can write the equilibrium constant as

$$K_P = \frac{\widehat{P}_{Cl_2} \cdot \widehat{P}_{PCl_3}}{\widehat{P}_{PCl_5}} = \frac{\frac{\xi^2}{(1+\xi)^2}\widehat{P}^2_{total}}{\frac{(1-\xi)}{(1+\xi)}\widehat{P}_{total}} = \frac{\xi^2}{(1-\xi^2)}\widehat{P}_{total}$$

If we solve this equation for ξ we find that

$$\xi = \sqrt{\frac{K_P}{\widehat{P}_{\text{total}} + K_P}}$$

At a total pressure of 0.01 bar and $K_P = 1.78$ we find that $\xi = 0.997$. Correspondingly we find 0.973 for a total pressure of 0.10 bar and 0.800 for a total pressure of 1 bar.

The reaction Gibbs energy is

$$\Delta_r G = \Delta_r G^{\ominus} + \boldsymbol{R}T \ln K_p$$
$$= \Delta_r G^{\ominus} + \boldsymbol{R}T \ln \left(\frac{\xi^2}{(1-\xi^2)} \widehat{P}_{\text{total}} \right)$$

For the situation where $\widehat{P}_{\text{total}} = 1$ and $T = 523$ K, we find that $\Delta_r G^{\ominus} = -2.51$ kJ mol^{-1} and

$$\Delta_r G = -2.51 \text{ kJ mol}^{-1} + 4.35 \cdot \ln \left(\frac{\xi^2}{1-\xi^2} \right) \text{ kJ mol}^{-1}$$

and it is plotted in Fig. P18.4.

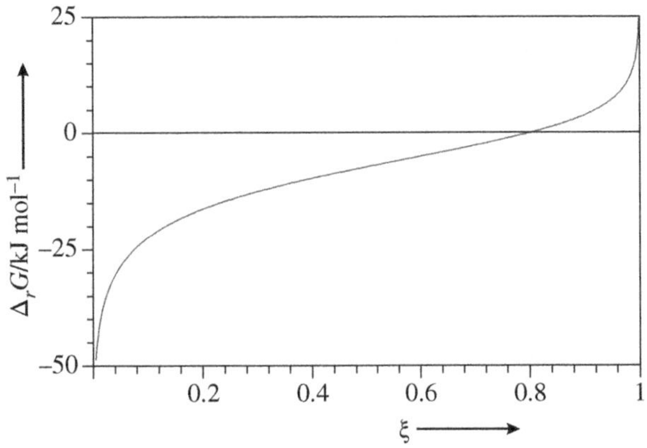

Figure P18.4 $\Delta_r G$ for the reaction PCl_5(g) → PCl_3 (g)+ Cl_2 (g) versus the fraction ξ of PCl_5 that is dissociated at equilibrium.

P18.5 Consider that 0.050 mol I_2 vapor is combined with 0.050 mol of Br_2(g) in a high temperature 10.0 L reaction vessel, and the equilibrium for the reaction is

$$I_2\text{ (g)} + Br_2\text{ (g)} \rightarrow 2\text{ IBr (g)}$$

The equilibrium is studied by extracting gas samples and using gas chromatography. At 1000 K the mole ratio of $\boldsymbol{n}_{IBr}/\boldsymbol{n}_{Br_2}$ is found to be 5.41, while at 1400 K it is 4.62. a) Determine the $\Delta_r G^{\ominus}$ for the reaction at these two temperatures, assuming that the gases behave ideally. b) Find $\Delta_r H^{\ominus}$ and $\Delta_r S^{\ominus}$ for the reaction, assuming that they are independent of temperature. Check the sign of your answer by using LeChatelier's principle. c) Provide a justification for the assumption that the enthalpy change is independent of temperature. Describe in detail what molecular properties will contribute to the temperature dependence of the enthalpy and which molecules will contribute most to the enthalpies temperature dependence. d) Combine the result of part B with the following data to estimate the bond energy of IBr. e) How would decreasing the volume of the reaction vessel have affected the results of the equilibrium measurements?

Substance	Br_2 (g)	I_2 (g)	Br (g)	I (g)
$\Delta_f H^{\ominus}$ (kJ mol^{-1})	30.7	62.2	111.8	106.6

Solution

a) From the reaction stoichiometry, we know that $n_{Br_2} = n_{I_2}$, and we can write the equilibrium constant as

$$K = \frac{c_{IBr}^2}{c_{I_2} \cdot c_{Br_2}} = \frac{c_{IBr}^2}{\left(c_{Br_2}\right)^2} = \frac{\left(\boldsymbol{n}_{IBr}/V\right)^2}{\left(\boldsymbol{n}_{Br_2}/V\right)^2} = \frac{\boldsymbol{n}_{IBr}^2}{\boldsymbol{n}_{Br_2}^2}$$

For $T = 1000$ K,

$$K_{1000} = \frac{n^2_{\text{IBr}}}{n^2_{\text{Br}_2}} = (5.41)^2 = 29.27$$

so at 1000 K we find that

$$\begin{aligned}\Delta_r G^{\ominus}_{1000} &= -\left(8.3145\ \text{J K}^{-1}\ \text{mol}^{-1}\right)(1000\ \text{K}) \cdot \ln(29.27)\ \text{J mol}^{-1} \\ &= -28.1\ \text{kJ mol}^{-1}\end{aligned}$$

and at 1400 K we find $K = (4.62)^2 = 19.91$ which implies that

$$\begin{aligned}\Delta_r G^{\ominus}_{1400} &= -\left(8.3145\ \text{J K}^{-1}\ \text{mol}^{-1}\right)(1400\ \text{K}) \ln(19.91)\ \text{J mol}^{-1} \\ &= -35.6\ \text{kJ mol}^{-1}\end{aligned}$$

b) If the reaction enthalpy and reaction entropy do not change with temperature, we can use the fundamental relation $\Delta_r G = \Delta_r H - T\Delta_r S$ to find their values. First we find the entropy change via

$$\Delta_r G(T_1) - \Delta_r G(T_2) = (T_2 - T_1)\,\Delta_r S$$

which upon rearrangement gives

$$\begin{aligned}\Delta_r S^{\ominus} &= \frac{\Delta_r G^{\ominus}(T_1) - \Delta_r G^{\ominus}(T_2)}{(T_2 - T_1)} \\ &= \left[\frac{-28.1 - (-35.6)}{1400 - 1000}\right]\ \text{kJ K}^{-1}\ \text{mol}^{-1} = 18.9\ \text{J K}^{-1}\ \text{mol}^{-1}\end{aligned}$$

To find the enthalpy change we use

$$\begin{aligned}\Delta_r H^{\ominus} &= \Delta_r G^{\ominus} + T\Delta_r S^{\ominus} \\ &= [-28100 + 1000 \times 18.9]\ \text{J mol}^{-1} = -9.2\ \text{kJ mol}^{-1}\end{aligned}$$

The negative enthalpy of reaction is consistent with LeChatelier's principle which states that for an exothermic reaction an increase in the temperature pushes the equilibrium toward the reactants.

c) For a diatomic molecule

$$H_{\text{diatomic}} = \frac{3}{2}RT + RT + H_{\text{vib}} + \varepsilon_e$$

For $\Delta_r H$ the translational and rotational enthalpies of the reactants and products will cancel. In addition the $\varepsilon_{B,m}$ term is well approximated as temperature independent. Only the vibrational part of H will have a strong temperature dependence and it will be a small correction to the $\varepsilon_{B.m}$ terms.

d) The reaction enthalpy can be calculated as

$$\begin{aligned}\Delta_r H^{\ominus} &= 2\Delta_f H^{\ominus}(\text{IBr}) - \Delta_f H^{\ominus}(\text{I}_2) - \Delta_f H^{\ominus}(\text{Br}_2) \\ &- 2\Delta_f H^{\ominus}(\text{IBr}) \quad 30.9\ \text{kJ mol}^{-1} - 62.2\ \text{kJ mol}^{-1} \\ &= -9.35\ \text{kJ mol}^{-1}\end{aligned}$$

Solving for $\Delta_f H$ (IBr) we find that

$$\Delta_f H^{\ominus}(\text{IBr}) = \frac{1}{2}\left(83.7\ \text{kJ mol}^{-1}\right) - 41.9\ \text{kJ mol}^{-1}$$

Next we find $\Delta_r H$ to dissociate Br_2 and I_2. For Br_2 we find that

$$\text{Br}_2(\text{g}) \rightarrow 2\text{Br}(\text{g}) \quad (2(111.8) - 30.7)\ \text{kJ mol}^{-1} = 192.9\ \text{kJ mol}^{-1}$$

and for I_2 we find that

$$\text{I}_2(\text{g}) \rightarrow 2\text{I}(\text{g}) \quad (2(106.6) - 62.2)\ \text{kJ mol}^{-1} = 151\ \text{kJ mol}^{-1}$$

With these values we can calculate a bond enthalpy by way of the reaction

$$2\text{I} + 2\text{Br} \rightarrow 2\text{IBr}$$

so that

$$2\Delta_r H_{\text{IBr}} = 2\Delta_f H_{\text{IBr}} - 2\,(151)\ \text{kJ mol}^{-1} - 2\,(192.9)\ \text{kJ mol}^{-1}$$

so that

$$\begin{aligned} 2\Delta_r H_{\text{IBr}} &= ((41.9) - (151) - (192.9))\ \text{kJ mol}^{-1} \\ &= -302\ \text{kJ mol}^{-1} \end{aligned}$$

or a bond enthalpy of 302 kJ ($\Delta_r H_{\text{IBr}} = -302$ kJ mol^{-1}).

e) There would have been no effect.

P18.6 Consider the isomerization reaction of methyl isocyanide to acetonitrile,

$$\text{CH}_3\text{NC} \rightleftarrows \text{CH}_3\text{CN}$$

a) Given the following thermochemical data, compute the $\Delta_r G^\ominus$ and the equilibrium constant K for this reaction at 298 K. Take both compounds to be in the gas phase.

Molecule	$\Delta_f H^\ominus$ (kJ mol^{-1})	$\Delta_f G^\ominus$ (kJ mol^{-1})	$S_m^\ominus$ (J mol^{-1} K^{-1})
CH_3NC	149.0	165.7	246.9
CH_3CN	64.3	81.7	245.1

b) Given the following spectroscopic data, compute the $\Delta_r G^\ominus$ and the equilibrium constant K for this reaction at 298 K.

Molecule	CH_3NC	CH_3CN
$\varepsilon_{B,m}$ (kJ mol^{-1})	2346	2476
$\theta_{\text{rot,A}}$ (K)	7.2	7.2
$\theta_{\text{rot,B}}$ (K)	0.48	0.45
$\theta_{\text{rot,C}}$ (K)	0.48	0.45

For this analysis, ignore the contribution associated with the thermal population of all vibrational modes with wavenumbers above 1000 cm^{-1}. The vibrational modes with wavenumbers less than 1000 cm^{-1} are 444 cm $^{-1}$ (2) for acetonitrile (corresponding to a characteristic vibrational temperature of $\theta_{\text{vib}} = 639$ K) and 343 cm^{-1} (2) for methylisocyanide (corresponding to a characteristic vibrational temperature of $\theta_{\text{vib}} = 343$ K). The number in parentheses after the vibrational wavenumber is the degeneracy of the normal mode.

c) Provide a justification for why it is reasonable to not consider the vibrational modes with wavenumbers of 1000 cm^{-1} and higher.

Solution

a)

$$\Delta_r G^\ominus = (81.7 - 165.7)\ \text{kJ mol}^{-1} = -84.0\ \text{kJ mol}^{-1}$$

and

$$\begin{aligned} K &= \exp\left(-\frac{\Delta_r G^\ominus}{RT}\right) \\ &= \exp\left(-\frac{-84.0\ \text{kJ mol}^{-1}}{8.3145\ \text{J K}^{-1}\ \text{mol}^{-1} \times 298\ \text{K}}\right) = \exp(33.902) = 5.28 \times 10^{14} \end{aligned}$$

b) First we calculate K from the molecular partition functions; namely

$$K = \frac{Z^{\ominus}_{CH_3CN}/V}{Z^{\ominus}_{CH_3NC}/V}$$
$$= \left[\frac{Z^{\ominus}_{trans}(CH_3CN)}{Z^{\ominus}_{trans}(CH_3NC)} \cdot \frac{Z^{\ominus}_{rot}(CH_3CN)}{Z^{\ominus}_{rot}(CH_3NC)} \cdot \frac{Z^{\ominus}_{vib}(CH_3CN)}{Z^{\ominus}_{vib}(CH_3NC)} \cdot \exp\left(\frac{\varepsilon_{B,m}(CH_3CN) - \varepsilon_{B,m}(CH_3NC)}{\boldsymbol{R}T}\right)\right]$$
$$= \left(\frac{m_{CH_3CN}}{m_{CH_3NC}}\right)^{3/2} \cdot \left(\frac{0.48}{0.45}\right) \cdot \left[\frac{1-\exp(-494/298)}{1-\exp(-639/298)}\right]^2 \cdot \exp(52.47)$$
$$= 1 \cdot 1.07 \cdot \left(\frac{0.809}{0.883}\right)^2 \cdot 6.13 \times 10^{22}$$
$$= 5.5 \times 10^{22}$$

Next we calculate $\Delta_r G$ as

$$\Delta_r G^{\ominus} = -\boldsymbol{R}T \ln K$$
$$= \left[-8.3145 \times 298 \times \ln\left(6.01 \times 10^{22}\right)\right] \text{ J mol}^{-1} = -130 \text{ kJ mol}^{-1}$$

This value differs by a considerable amount from that reported in part a). We note that a 2% error for the difference in the bonding energy of the atoms ε_B corresponds to about 50 kJ mol^{-1} and could account for this difference.

c) The vibrational degrees of freedom contribute in two ways. (1) The zero point energies modify the energy difference between reactants and products. This term is included in the difference of ε_B values. (2) The thermal excitations of the vibrational modes modify the partition functions. These contributions are small because θ_{vib}/T is big causing $\exp\left(-\theta_{vib}/T\right)$ to be small.

P18.7 Consider the decomposition of hydrogen peroxide

$$H_2O_2(l) \rightarrow H_2O(l) + \frac{1}{2}O_2(g)$$

where $\Delta_r G^{\ominus} = -123.1$ kJ mol^{-1}. Calculate $\Delta_r G$ under atmospheric conditions, $\widehat{P}_{O_2} = 0.209$, and decide whether the reaction is spontaneous. Calculate the equilibrium pressure of O_2 in a closed system held at 323 K and fixed volume, which initially contained only H_2O_2(l). Account for the temperature dependence of $\Delta_r H^{\ominus}$ by using $\Delta_r H^{\ominus} = -98.2$ kJ mol^{-1} at 298 K, and the heat capacities $C_{P,m}(H_2O_2(l)) = 89.1$ J K^{-1} mol^{-1}, $C_{P,m}(H_2O(l)) = 75.3$ J K^{-1} mol^{-1}, and $C_{P,m}(O_2(g)) = 29.4$ J K^{-1} mol^{-1}. Assume the heat capacities are independent of temperature over the temperature range of interest.

Solution

The standard state for the $\Delta_r G^{\ominus} = -123.1$ kJ mol^{-1} corresponds to $P_{O_2} = 1.0$ bar, so that for atmospheric conditions we find

$$\Delta_r G = \Delta_r G^{\ominus} + \boldsymbol{R}T \ln\left(\widehat{P}^{1/2}_{O_2}\right) = \Delta_r G^{\ominus} + \frac{1}{2}\boldsymbol{R}T \ln\left(\widehat{P}_{O_2}\right)$$
$$= \left[-123.1 \text{ kJ mol}^{-1} + \frac{1}{2} 8.314 \text{ J K}^{-1} \text{ mol}^{-1} \cdot 298 \text{ K} \cdot \ln(0.209)\right] \text{ J mol}^{-1} = -125.0 \text{ kJ mol}^{-1}$$

Hence the reaction is spontaneous.

For equilibrium in a closed container we can write that the total number of moles is given by $(1 + \alpha/2)\, n_0$, where α is the fraction of H_2O_2 that has dissociated and n_0 is the initial number of moles of hydrogen peroxide. Because the hydrogen peroxide and water are both liquids, and their molar volumes are not too different (0.0180 L mol^{-1} versus 0.0236 L mol^{-1}) the total volume of liquid in the container is approximated as constant. Thus we can write that

$$\widehat{c}_{H_2O_2} = \widehat{c}_{0,H_2O_2}(1-\alpha) \; ; \; \widehat{c}_{H_2O} = \widehat{c}_{0,H_2O_2}(\alpha) \text{ and } \widehat{c}_{O_2} = \widehat{c}_{0,H_2O_2}\left(\frac{\alpha}{2}\right)$$

so that the equilibrium constant is

$$K = \frac{\alpha \widehat{c}_{0,H_2O_2} \cdot \widehat{c}^{1/2}_{O_2}}{(1-\alpha)\widehat{c}_{0,H_2O_2}} = \frac{\alpha \cdot \widehat{c}^{1/2}_{O_2}}{(1-\alpha)}$$

Using $\alpha = (2\hat{c}_{O_2})/\hat{c}_{0,H_2O_2}$, we find that

$$K = \frac{2 \cdot \hat{c}_{O_2}^{3/2}}{\left(\hat{c}_{0,H_2O_2} - 2\hat{c}_{O_2}\right)}$$

and rearrangement gives

$$2 \cdot \hat{c}_{O_2}\left(\hat{c}_{O_2}^{1/2} + K\right) = \hat{c}_{0,H_2O_2}K$$

Hence if we know K, then we can find the concentration (hence the pressure) of oxygen as a function of the initial peroxide concentration.

To find K, we use the relationship

$$K = \exp\left(-\Delta_r G^\ominus / (\boldsymbol{R}T)\right)$$

and find $\Delta_r G^\ominus$ by

$$\Delta_r G^\ominus = \Delta_r H^\ominus - T\Delta_r S^\ominus$$

For the Gibbs energy change at 323 K, we can write that

$$\begin{aligned}\Delta_r G(323\text{ K}) &= \Delta_r G^\ominus(298\text{ K}) + \int_{298\text{K}}^{323\text{K}} \Delta C_{P,m}\mathrm{d}T - (323\text{ K})\int_{298\text{K}}^{323\text{K}} \frac{\Delta C_{P,m}}{T}\mathrm{d}T \\ &= \Delta_r G^\ominus(298\text{ K}) + \Delta C_{P,m}\cdot(323-298)\text{ K} - (323\text{ K})\,\Delta C_{P,m}\ln\left(\frac{323}{298}\right) \\ &= \left[-123,100 + (0.9)(25) - (323)(0.9)\ln\left(\frac{323}{298}\right)\right]\text{ J mol}^{-1} \\ &= -123.1\text{ kJ mol}^{-1}\end{aligned}$$

so that

$$K = \exp\left(-\frac{123100}{(8.314)(298)}\right) = 3.79\cdot 10^{18}$$

where we have used the fact that $\Delta C_{P,m} = 0.9\text{ J K}^{-1}\text{mol}^{-1}$. Hence we see that

$$2\cdot\hat{c}_{O_2}\left(\hat{c}_{O_2}^{1/2} + 3.79\cdot 10^{18}\right) = \hat{c}_{0,H_2O_2}3.79\cdot 10^{18}$$

Given that the molar density of hydrogen peroxide is $4.24\cdot 10^{-2}$ mol L^{-1} under standard conditions, we see that this equation has the approximate solution

$$\hat{c}_{O_2} \approx \frac{\hat{c}_{0,H_2O_2}}{2}$$

That is, nearly all of the peroxide decomposes to generate water and oxygen, at equilibrium. Using $\hat{c}_{0,H_2O_2} = 42.4$ we find $\hat{c}_{O_2} = 21.2$ so that its pressure would be

$$\begin{aligned}P_{O_2} &= \boldsymbol{R}T\frac{n_{O_2}}{V} = \boldsymbol{R}T\cdot\hat{c}_{O_2} \\ &= (0.08314)(298)(21.2)\text{ bar} = 525\text{ bar}\end{aligned}$$

This pressure is quite high and would require a reaction vessel that could contain such a pressure if it were to remain closed; i.e., not leak or explode.

P18.8 Consider the following reaction:

$$2\text{ XO (g)} \rightarrow 2\text{ X (g)} + O_2\text{ (g)}$$

where X is a halogen: Cl, Br, or I. Relevant information is given here for the different XO species. Find the other information you need in the Appendix

Table of data for the XO species.

Species XO	$\Delta_f H^\ominus$ kJ mol^{-1}	$S_m^\ominus$ J mol^{-1} K^{-1}	$\varepsilon_{B,m}$ kJ mol^{-1}	θ_{vib} K	θ_{rot} K
ClO	101.84	226.52	264.90 ± 0.12	1431	2.8
BrO	125.77	237.44	232.6 ± 2.4	1025	1.9
IO	175.06	245.39	176 ± 20	980	0.65

The ground electronic state for these XO species is a $^2\Pi_{3/2}$ (a degeneracy of 2) and the symmetry parameter σ is equal to 1 since they are heteronuclear diatomics. The ground electronic state of O_2 is a $^3\Sigma_g$ (a degeneracy of 3) and its symmetry parameter σ is equal to 2 since it is a homonuclear diatomic.

First, use thermochemical data to compute the $\Delta_r G^\ominus$ for these reactions at 298 K.

Second assume that the gases behave ideally and use the molecular parameters given above to find an expression for the $\Delta_r G^\ominus$ of these reactions and for their equilibrium constants. Evaluate your $\Delta_r G^\ominus$ expression at the temperature 298 K and compare your results with the thermochemical ones. Analyze your expression for $\Delta_r G^\ominus$ and explain (verbally) how you expect it to change with increasing temperature. Explain what terms contribute to the change in $\Delta_r G^\ominus$ through this series of three reactions. Which terms are most important and which terms are least important?

Solution

a) The Gibbs energy of reaction can be calculated from the fundamental relationship $\Delta_r G = \Delta_r H - T \cdot \Delta_r S$, For the reaction enthalpy we use

$$\Delta_r H^\ominus = \Delta_f H^\ominus\left(O_2\right) + 2\Delta_f H^\ominus(X) - 2\Delta_f H^\ominus(XO)$$
$$= 2\left[\Delta_f H^\ominus(X) - \Delta_f H^\ominus(XO)\right]$$

and for the reaction entropy we use

$$\Delta_r S^\ominus = S_m^\ominus\left(O_2\right) + 2S_m^\ominus(X) - 2S_m^\ominus(XO)$$

Using the data from the appendix

	O_2	Cl	Br	I
$\Delta_f H^\ominus$ (kJ mol^{-1})	0	121.1	96.4	106.6
$S_m^\ominus$ (J mol^{-1} K^{-1})	205.14	165.1	174.9	180.7

we find the following reaction enthalpies, entropies, and Gibbs energy changes.

Reaction	$\Delta_r H^\ominus$ (kJ mol^{-1})	$\Delta_r S^\ominus$ (J mol^{-1} K^{-1})	$\Delta_r G^\ominus$ (kJ mol^{-1})
$2ClO \rightarrow 2Cl + O_2$	38.5	82.3	14.0
$2BrO \rightarrow 2Br + O_2$	−58.8	80.1	−82.7
$2IO \rightarrow 2I + O_2$	−136.9	75.8	−159.5

b) The equilibrium constant expression is

$$K_p(T) = \frac{\hat{P}_{O_2}\hat{P}_X^2}{\hat{P}_{XO}^2} = \frac{kT}{P^\ominus}\frac{\left(Z_{O_2}^\ominus/V\right)\left(Z_X^\ominus/V\right)^2}{\left(Z_{XO}^\ominus/V\right)^2}$$
$$= \left(\frac{kT}{P^\ominus}\left[\frac{\left(Z_{\text{trans},O_2}^\ominus/V\right)\left(Z_{\text{trans},X}^\ominus/V\right)^2}{\left(Z_{\text{trans},XO}^\ominus/V\right)^2}\right]\left[\frac{Z_{\text{rot},O_2}^\ominus}{\left(Z_{\text{rot},XO}^\ominus\right)^2}\right]\left[\frac{Z_{\text{vib},O_2}^\ominus}{Z_{\text{vib},XO}^{\ominus 2}}\right]\cdot\left[\frac{Z_{\text{elec},X}^{\ominus 2} Z_{\text{elec},O_2}^\ominus}{Z_{\text{elec},XO}^{\ominus 2}}\right]\right)$$

The first term corresponds to the contribution from the translational degrees of freedom and simplifies to

$$\frac{kT}{p^{\ominus}}\left[\frac{\left(\frac{Z^{\ominus}_{\text{trans},O_2}}{V}\right)\left(\frac{Z^{\ominus}_{\text{trans,X}}}{V}\right)^2}{\left(\frac{Z^{\ominus}_{\text{trans,XO}}}{V}\right)^2}\right] = \frac{kT}{p^{\ominus}}\left[\left(\frac{2\pi m_{O_2}kT}{h^2}\right)^{3/2}\left(\frac{m_X^2}{m_{XO}^2}\right)^{3/2}\right]$$

$$= 4.11\times10^{-26}\left[\left(3.313\times10^{21}\right)^{3/2}\left(\frac{m_X}{m_O}\right)^3\right]$$

Upon substitution of the masses, we find the values given below in the table. The second term corresponds to the contributions from the rotational degrees of freedom and simplifies to

$$\left[\frac{Z^{\ominus}_{\text{rot},O_2}}{\left(Z^{\ominus}_{\text{rot,XO}}\right)^2}\right] = \frac{\theta^2_{\text{rot,XO}}}{2\theta_{\text{rot},O_2}T}$$

For $\theta_{\text{rot},O_2} = 2.07$ K and $T = 298$ K we find the values in the table. The last two terms correspond to the vibrational and electronic contributions and correspond to

$$\left[\frac{Z^{\ominus}_{\text{vib},O_2}}{Z^{\ominus 2}_{\text{vib,XO}}}\right]\left[\frac{Z^{\ominus 2}_{\text{elec,X}}Z^{\ominus}_{el,O_2}}{Z^{\ominus 2}_{el,XO}}\right] = \exp\left(\frac{\left(\varepsilon_B\left(O_2\right) - 2\varepsilon_B\left(XO\right)\right)}{kT}\right)\frac{\left(1-\exp\left(-\frac{\theta_{\text{vib,XO}}}{T}\right)\right)^2}{\left(1-\exp\left(-\frac{\theta_{\text{vib},O_2}}{T}\right)\right)}$$

We note that the separated atoms is chosen as the zero of energy here. The following table summarizes the results

X	$Z^{\ominus}_{\text{trans}}$ term	$Z^{\ominus}_{\text{rot}}$ term	$Z^{\ominus}_{\text{vib/elec}}$ term	K_p(298 K)
Cl	2.35×10^{6}	6.35×10^{-3}	4.40×10^{-7}	5.91×10^{-2}
Br	4.16×10^{6}	2.93×10^{-3}	8.84×10^{4}	9.72×10^{9}
I	5.04×10^{6}	3.42×10^{-4}	6.09×10^{24}	9.45×10^{28}

The last column in the table shows the equilibrium constant, which is calculated as the product. Note that the primary difference in the equilibrium constant arises from the vibrational/electronic terms.

With the equilibrium constants in hand, we can calculate $\Delta_r G^{\ominus}$ by way of

$$\Delta_r G^{\ominus} = -RT\ln\left(K_p\right)$$

The table summarizes the results

	$\Delta_r G^{\ominus}$ / kJ mol^{-1}
$2ClO \rightarrow 2Cl + O_2$	7.0
$2BrO \rightarrow 2Br + O_2$	−57.0
$2IO \rightarrow 2I + O_2$	−165

These values differ somewhat from the empirically determined values in part a). The most likely place of error is in the $\varepsilon_{B,m}$ values. Note however that the trend in the $\Delta_r G^{\ominus}$ is captured by the calculation.

The rotational and translational contributions combine to cause an increase in K_p with T; hence, $\Delta_r G^{\ominus}$ will become more negative, for all three systems. The electronic and vibrational pieces indicate that K_p will increase with T for the Cl case, and decrease with T for the Br and I cases.

P18.9 Consider the following data for the osmotic pressure of aqueous solutions of the protein α-chymotrypsin (298 K and pH = 8.25). Analyze these two datasets and determine the molecular weight of the protein. Data are taken from C. A. Haynes et al. J. Phys. Chem. 96 (1992) 905.

	0.1 M K_2SO_4		0.1 M Na_3PO_4
m/V	^{osm}P	m/V	^{osm}P
g/L	Pa	g/L	Pa
0.902	63.6	0.957	75.80
1.811	125.6	2.209	167.2
3.659	259.6	5.677	269.9
5.391	377.4	7.209	416.1
7.185	501.1	9.366	513.3
8.932	625.3		

Solution

The osmotic pressure is given by

$$^{osm}P = RT \cdot c \text{ where } c = \frac{n}{V}$$

Because the concentrations are given as m (g/L) rather than c (mols/L) we transform this equation as

$$^{osm}P = \left(\frac{RT}{M}\right) \cdot \frac{m}{V}$$

In Fig. 19.8 we show a plot of the osmotic pressure ^{osm}P versus m/V, so that

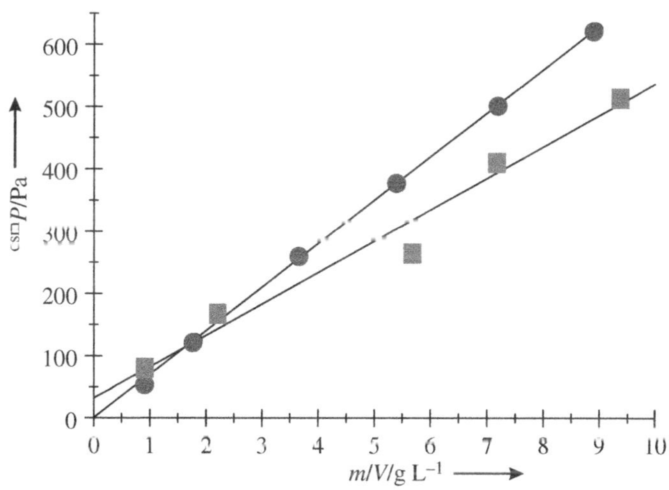

Figure P18.9 Osmotic pressure ^{osm}P of α-chymotrypsin versus m/V for Na_3PO_4 (filled squares) and K_2SO_4 (filled circles).

The slope for the K_2SO_4 solution is 69.9 Pa L g^{-1} = 69.9×10^{-3} N m^{-2} m^3 g^{-1} = 69.9×10^{-3} J g^{-1} and that for the Na_3PO_4 solution's data plot is 50.5 Pa L g^{-1} = 50.5×10^{-3} J g^{-1}. Using this slope we can calculate the molecular weight of the α-chymotrypsin in the K_2SO_4 solution and find

$$M = \frac{RT}{slope}$$

$$= \left[\frac{8.314 \text{ J K}^{-1} \text{ mol}^{-1} \times 298 \text{ K}}{69.9 \times 10^{-3} \text{ J g}^{-1}}\right] = 35.4 \times 10^3 \text{ g mol}^{-1} = 35.4 \text{ kg mol}^{-1}.$$

For the Na_3PO_4 solutions, we find a molecular weight of

$$M = \frac{RT}{slope} = \frac{8.314 \times 298}{50.5 \times 10^{-3}} \frac{\text{g}}{\text{mol}} = 49.1 \text{ kg mol}^{-1}$$

The actual molecular weight of α-chymotrypsin is 31.4 kg mol^{-1}.

P18.10 Calculate the enthalpy change, the entropy change, the Gibbs energy change, and the equilibrium constant for the autoionization of water, $H_2O \rightarrow H^+$ (aq) + OH^- (aq), at 298 K. Using a molecular analysis, estimate how the equilibrium constant changes upon replacing the protons by deuterons (i.e., the autoionization of D_2O). Compare your result to the experimental value of 1.0×10^{-15} M. Lastly, comment on the relative importance of enthalpic and entropic terms in the isotope effect.

Solution

The data to be used is

	H^+	OH^-	H_2O
$\Delta_f H^\ominus$ / kJ mol^{-1}	0	−229.95	−285.84
$S_m^\ominus$ / J K^{-1} mol^{-1}	0	−10.54	69.94
$\Delta_f G^\ominus$ / kJ mol^{-1}	0	−157.3	−237.2

For the water autoionization, the reaction enthalpy is

$$\Delta_r H^\ominus = (-229.95 + 285.84)\ \text{kJ mol}^{-1} = 55.89\ \text{kJ mol}^{-1}$$

and the reaction entropy is

$$\Delta_r S^\ominus = (-10.54 - 69.94)\ \text{J K}^{-1}\ \text{mol}^{-1} = -80.48\ \text{J K}^{-1}\ \text{mol}^{-1}$$

The Gibbs energy change

$$\Delta_r G^\ominus = (-157.3 + 237.2)\ \text{kJ mol}^{-1} = 79.9\ \text{kJ mol}^{-1}$$

and the equilibrium constant is

$$K = \exp\left(-\frac{\Delta_r G^\ominus}{RT}\right)$$
$$= \exp\left(-\frac{79.9 \times 10^3}{8.314 \times 298}\right) = \exp(-32.25) = 9.87 \times 10^{-15} \approx 10^{-14}$$

For the deuterium oxide autoionization, namely

$$D_2O \rightarrow D^+\ (\text{aq}) + OD^-\ (\text{aq})$$

we expect the major enthalpic difference to arise from the zero point energy of the bond that is broken O–D versus O–H; however, we note that the hydrogen bonds between the deuterium ion and the OD^- ion with the D_2O solvent molecules will be different than the corresponding ones of the proton and hydroxide ion with H_2O. The data needed for this comparison is given in the table, and we infer that the vibrational frequencies of the OH and OD can be used as reasonable estimates for a bond frequency.

	ν_1 / s^{-1}	ν_2 / s^{-1}	ν_3 / s^{-1}
H_2O	$11.47 \cdot 10^{13}$	$4.958 \cdot 10^{13}$	$11.80 \cdot 10^{13}$
D_2O	$8.268 \cdot 10^{13}$	$3.628 \cdot 10^{13}$	$8.650 \cdot 10^{13}$
OH	$11.20 \cdot 10^{13}$		
OD	$8.157 \cdot 10^{13}$		

Hence we estimate the H_2O autoionization to be more exothermic than the D_2O autoionization by about

$$\frac{1}{2}\left(h\nu_{\text{O-D}} - h\nu_{\text{O-H}}\right) \boldsymbol{N_A}$$
$$= \frac{1}{2} \cdot 6.626 \times 10^{-34} \cdot (8.157 - 11.20) \times 10^{13} \cdot 6.022 \times 10^{23}\ \text{J mol}^{-1}$$
$$= -6.07\ \text{kJ mol}^{-1}$$

So that the D_2O autoionization has a reaction enthalpy of about

$$\Delta_r H^{\ominus}(D_2O) = 55.89\ \text{kJ mol}^{-1} + 6.07\ \text{kJ mol}^{-1} = 61.96\ \text{kJ mol}^{-1}$$

From the reported autoionization constant for D_2O we calculate a Gibbs energy change of

$$\Delta_r G^{\ominus}(D_2O) = -\boldsymbol{R}T\ln K$$
$$= -\left[(8.314)(298)\ln\left(10^{-15}\right)\right]\ \text{J mol}^{-1} = 85.6\ \text{kJ mol}^{-1}$$

A value which is about 6 kJ mol^{-1} more endothermic than that of H_2O.
Because of the lack of experimental data for determining the entropy of OD^- (and D^+) in solution, we cannot directly calculate the entropy change in a meaningful way. However, if we combine our Gibbs energy of reaction with our estimate for the enthalpy of reaction we can obtain an estimate of the entropy of reaction for the autoionization of D_2O. We find

$$\Delta_r S^{\ominus}(D_2O) = -\frac{\Delta_r G^{\ominus}(D_2O) - \Delta_r H^{\ominus}(D_2O)}{T}$$
$$= -\frac{(85.6 - 61.96)\ \text{kJ mol}^{-1}}{298\ \text{K}} = -79.3\ \text{J K}^{-1}\ \text{mol}^{-1}$$

which is 1.14 J K^{-1}mol^{-1} less negative than that of water. At 298 K this contributes only about 0.3 kJ mol^{-1} to the Gibbs energy change for the reaction. Hence we expect that the enthalpy effect is the major reason for the shift in the autoionization constant.

P18.11 Depression of Melting point. Using Fig. 18.13 of the text, show that the depression of melting point as a function of mole fraction x of the solute is given by

$$\boxed{\Delta T_{\text{fus}} = -\frac{\boldsymbol{R}T^2}{\Delta_{\text{fus}}H}\cdot x}$$

Solution

From Fig. P18.11 (which is identical with Fig. 18.13 in the text) we obtain

$$\frac{\Delta P + \Delta P'}{\Delta T_{\text{fus}}} = -\left(\frac{dP}{dT}\right)_{\text{solid}} \quad\text{and}\quad \frac{\Delta P'}{\Delta T_{\text{fus}}} = -\left(\frac{dP}{dT}\right)_{\text{liquid}}$$

where $\Delta P = P_s - P$ and $\Delta T_{\text{fus}} = T_{\text{fus}} - T_{\text{fus},s}$ (where T_{fus} is the melting point of the solution and $T_{\text{fus},s}$ is the melting point of the pure solvent) and $\Delta P'$ is explained in the figure. Combining these results, we obtain

$$\frac{\Delta P}{\Delta T_{\text{fus}}} = -\left(\frac{dP}{dT}\right)_{\text{solid}} + \left(\frac{dP}{dT}\right)_{\text{liquid}}$$

$(dP/dT)_{\text{solid}}$ is the slope of the vapor pressure/temperature curve above the solid and $(dP/dT)_{\text{liquid}}$ is that above the liquid. According to the equation of Clausius and Clapeyron,

$$\left(\frac{dP}{dT}\right)_{\text{liquid}} = P\frac{\Delta_{\text{vap}}H}{\boldsymbol{R}T^2} \quad\text{and}\quad \left(\frac{dP}{dT}\right)_{\text{solid}} = P\frac{\Delta_{\text{sub}}H}{\boldsymbol{R}T^2}$$

where $\Delta_{\text{sub}}H$ is the enthalpy of sublimation, and we find

$$\frac{\Delta P}{\Delta T_{\text{fus}}} = -P\frac{\Delta_{\text{vap}}H}{\boldsymbol{R}T^2} + P\frac{\Delta_{\text{sub}}H}{\boldsymbol{R}T^2} = -\frac{P}{\boldsymbol{R}T^2}\left(\Delta_{\text{sub}}H - \Delta_{\text{vap}}H\right)$$

Using Raoult's law, Equation (18.102) in Section 18.6.2,

$$\Delta P = P_s \cdot x$$

we find

$$\frac{P_s\cdot x}{\Delta T_{\text{fus}}} = -\frac{P}{\boldsymbol{R}T^2}\left(\Delta_{\text{sub}}H - \Delta_{\text{vap}}H\right)$$

If ΔP is small enough, then $P_s/P \approx 1$, and we obtain

$$\Delta T_{\text{fus}} = \frac{-\boldsymbol{R}T^2}{\Delta_{\text{sub}}H - \Delta_{\text{vap}}H}\cdot x$$

and with $\Delta_{\text{sub}}H - \Delta_{\text{vap}}H = \Delta_{\text{fus}}H$ (the enthalpy of melting)

$$\boxed{\Delta T_{\text{fus}} = -\frac{RT^2}{\Delta_{\text{fus}}H} \cdot x}$$

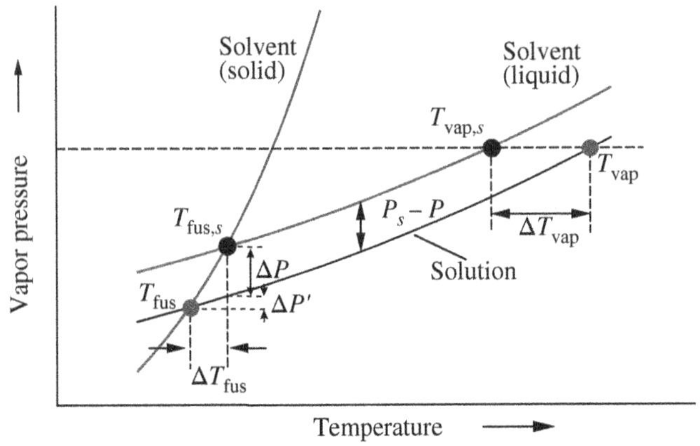

Figure P18.12 Depression ($P_s - P$) of vapor pressure, elevation ΔT_{vap} of boiling point, and depression ΔT_{fus} of melting point. Note that $\Delta T_{\text{vap}} = T_{\text{vap}} - T_{\text{vap},s}$ is positive, while $\Delta T_{\text{fus}} = T_{\text{fus}} - T_{\text{fus},s}$ is negative.

19

Formal Thermodynamics and Its Application to Phase Equilibria

19.1 Exercises

E19.1 Show that the relations

$$\Delta G = \int V\, \mathrm{d}P \text{ and } \Delta G = w_{\mathrm{rev}} = -\int P\, \mathrm{d}V$$

are equivalent for the isothermal expansion of an ideal gas.

Solution
By substituting for P with the ideal gas equation of state,

$$PV = \boldsymbol{nR}T$$

we find that

$$-\int P\, \mathrm{d}V = -\boldsymbol{nR}T \int_{V_1}^{V_2} \frac{1}{V} \mathrm{d}V = -\boldsymbol{nR}T \ln\left(\frac{V_2}{V_1}\right)$$

Correspondingly, we find

$$\int V\, \mathrm{d}P = \boldsymbol{nR}T \int_{P_1}^{P_2} \frac{1}{P} \mathrm{d}P = \boldsymbol{nR}T \ln\left(\frac{P_2}{P_1}\right) = -\boldsymbol{nR}T \ln\left(\frac{V_2}{V_1}\right)$$

Thus we verify that for an ideal gas

$$\int V\, \mathrm{d}P = -\int P\, \mathrm{d}V$$

E19.2 Use Equation (19.3) and the combined form of the first and second laws to find a general expression for the differential entropy in terms of directly measurable quantities.

Solution
Equation (19.3) is

$$\mathrm{d}U = \left(\frac{\partial U}{\partial T}\right)_V \mathrm{d}T + \left(\frac{\partial U}{\partial V}\right)_T \mathrm{d}V$$

and the combined form of the first and second laws is

$$\mathrm{d}U = T\mathrm{d}S - P\mathrm{d}V$$

From the combined form of the first and second laws, we can write that

$$\mathrm{d}S = \frac{1}{T} \cdot \mathrm{d}U + \frac{P}{T} \cdot \mathrm{d}V$$

Substituting for $\mathrm{d}U$, we find that

$$\mathrm{d}S = \frac{1}{T} \cdot C_V\, \mathrm{d}T + \left(\frac{\partial U}{\partial V}\right)_T \mathrm{d}V + \frac{P}{T} \cdot \mathrm{d}V$$

Solutions Manual for Principles of Physical Chemistry, Third Edition. Edited by Hans Kuhn, David H. Waldeck, and Horst-Dieter Försterling.

where we have used the definition of the constant volume heat capacity. Using the Thermodynamic Equation of State (Equation 19.16), we can write that

$$\begin{aligned} dS &= \frac{1}{T} \cdot C_V \, dT + \frac{1}{T}\left[T\left(\frac{\partial P}{\partial T}\right)_V - P\right] dV + \frac{P}{T} \cdot dV \\ &= \frac{1}{T} \cdot C_V \, dT + \left(\frac{\partial P}{\partial T}\right)_V dV \end{aligned}$$

This result expresses dS in terms of the temperature, the constant volume heat capacity, and the temperature dependence of the pressure, which are all measurable.

One way to assess our result is to compare it with our statistical mechanical evaluation of the entropy for an ideal monatomic gas. In this approximation, our general expression for dS becomes

$$\begin{aligned} dS &= \frac{1}{T} \cdot \left(\frac{3}{2} \, Nk\right) dT + \left(\frac{\partial}{\partial T}\left(\frac{NkT}{V}\right)\right)_V dV \\ &= \left(\frac{3Nk}{2T}\right) dT + \left(\frac{Nk}{V}\right) dV \end{aligned}$$

We can compare this result to the differential form of the Sackur–Tetrode equation for an ideal atomic gas (see Equation (17.49)). Taking the differential of Equation (17.49),

$$S = Nk \ln \left[e^{5/2}\left(\frac{2\pi km}{h^2}\right)^{3/2} \frac{V}{N} T^{3/2}\right]$$

we find

$$\begin{aligned} dS &= \left(\frac{\partial S}{\partial T}\right)_V dT + \left(\frac{\partial S}{\partial V}\right)_T dV \\ &= \frac{3}{2} Nk \cdot \frac{1}{T} dT + Nk \cdot \frac{1}{V} \, dV = \left(\frac{3Nk}{2T}\right) dT + \left(\frac{Nk}{V}\right) dV \end{aligned}$$

Hence they are equivalent.

E19.3 a) What is the criterion for a spontaneous process at constant T and P? b) What is the criterion for a spontaneous process at constant S and V?

Solution

The natural variables of G are T and P. Therefore

$$\Delta G < 0 \quad T, P \ \text{constant}$$

is the criterion for spontaneity. See the discussion in Section 19.3.2.

The natural variables of U are S and V. Therefore

$$\Delta U < 0 \quad S, V \ \text{constant}$$

is the criterion for spontaneity. See the discussion regarding Equation (19.34)

E19.4 Beginning with the Sackur–Tetrode Equation (17.49) and $U = C_V T$, show that the internal energy is

$$U(S, V) = c_1 \cdot V^{-R/C_V} \exp\left(S/C_V\right)$$

for an ideal gas. In the first part of your derivation, eliminate T from the Sackur–Tetrode equation by replacing it with U, and in the second part of the equation rearrange your results to write U as a function of S and V.

In the next part of this Exercise, use your knowledge of thermodynamics and this $U(S, V)$, to derive a) the ideal gas equation of state and b) show that the internal energy U is a function of T only.

Solution

The Sackur–Tetrode equation can be written as

$$\begin{aligned} S &= Nk \cdot \ln\left(T^{3/2}\left(\frac{V}{N}\right) \cdot c'\right) = \frac{3}{2} Nk \cdot \ln\left(T\left(\frac{V}{N}\right)^{2/3} \cdot c'^{2/3}\right) \\ &= C_V \cdot \ln\left(\frac{C_V \cdot T}{C_V}\left(\frac{V}{N}\right)^{2/3} \cdot c'^{2/3}\right) \end{aligned}$$

Writing this result in terms of molar quantities, we find that

$$\frac{S_m}{C_{V,m}} = \ln\left(\frac{U_m}{C_{V,m}}(V_m)^{2/3}\cdot c'^{2/3}\right) = \ln\left(U_m\cdot V_m^{\boldsymbol{R}/C_{V,m}}\right) + \ln\left(\frac{c'^{2/3}}{C_{V,m}}\right)$$

which rearranges to be

$$\ln\left(U_m\cdot V_m^{\boldsymbol{R}/C_{V,m}}\right) = \frac{S_m}{C_{V,m}} - \ln\left(\frac{c'^{2/3}}{C_{V,m}}\right)$$

or

$$U_m = V_m^{-\boldsymbol{R}/C_{V,m}}\cdot\exp\left(\frac{S_m}{C_{V,m}}\right)\cdot\left(\frac{c'^{2/3}}{C_{V,m}}\right) = V_m^{-\boldsymbol{R}/C_{V,m}}\cdot\exp\left(\frac{S_m}{C_{V,m}}\right)\cdot c_1$$

where $C_V = \frac{3}{2}Nk$ and *const.* is a constant.
For items a) and b), we start with the combined form of the first and second laws

$$dU = T\cdot dS - P\cdot dV$$

and compare it to the total differential of $U(S, V)$

$$dU = \left(\frac{\partial U}{\partial S}\right)_V\cdot dS + \left(\frac{\partial U}{\partial V}\right)_S\cdot dV$$

By inspection we can write that

$$\begin{aligned} T &= \left(\frac{\partial U}{\partial S}\right)_V \\ &= c_1\cdot V^{-\boldsymbol{R}/C_V}\frac{\partial}{\partial S}\exp\left(\frac{S}{C_V}\right) \\ &= c_1\cdot V^{-\boldsymbol{R}/C_V}\exp\left(\frac{S}{C_V}\right)\frac{1}{C_V} = \frac{U}{C_V} \end{aligned}$$

where we used the definition of $U(S, V)$ in the last equality. Rearrangement of this expression gives

$$U = C_V\cdot T$$

for the case that C_V is constant. This is the result desired for part b), and it was used in the derivation of the $U(S, V)$ expression from the Sackur–Tetrode equation.

Next we use the fact that

$$\begin{aligned} P &= -\left(\frac{\partial U}{\partial V}\right)_S \\ &= c_1\cdot\frac{\boldsymbol{R}}{C_V}V^{-\frac{R}{C_V}-1}\exp\left(\frac{S}{C_V}\right) = \frac{\boldsymbol{R}}{V}\cdot\frac{U}{C_V} \end{aligned}$$

where we used the definition of $U(S, V)$ in the last equality. If we substitute $U = C_V\cdot T$, as found above, we find

$$P = \frac{\boldsymbol{R}T}{V}$$

which is the ideal gas equation of state.

E19.5 In Section 19.2.6 we derived the Maxwell relation

$$\left(\frac{\partial T}{\partial P}\right)_S = \left(\frac{\partial V}{\partial S}\right)_P$$

Derive the other three Maxwell relations

$$\left(\frac{\partial V}{\partial T}\right)_P = -\left(\frac{\partial S}{\partial P}\right)_T \quad \left(\frac{\partial P}{\partial T}\right)_V = \left(\frac{\partial S}{\partial V}\right)_T \quad \left(\frac{\partial T}{\partial V}\right)_S = -\left(\frac{\partial P}{\partial S}\right)_V$$

from their corresponding energy functions: G, A, and U.

Solution

First, we consider the differential dH, Equation (19.20), and write

$$dH = TdS + VdP = \left(\frac{\partial H}{\partial S}\right)_P dS + \left(\frac{\partial H}{\partial P}\right)_S dP$$

Then by inspection we can write that

$$\left(\frac{\partial H}{\partial S}\right)_P = T \text{ and } \left(\frac{\partial H}{\partial P}\right)_S = V$$

and for the second cross derivatives it follows that

$$\left(\frac{\partial}{\partial P}\left(\frac{\partial H}{\partial S}\right)_P\right)_S = \left(\frac{\partial T}{\partial P}\right)_S \quad \text{and} \quad \left(\frac{\partial}{\partial S}\left(\frac{\partial H}{\partial P}\right)_S\right)_P = \left(\frac{\partial V}{\partial S}\right)_P$$

Because these cross derivatives must be equal, we find the Maxwell relation

$$\boxed{\left(\frac{\partial T}{\partial P}\right)_S = \left(\frac{\partial V}{\partial S}\right)_P}$$

Next we consider the change of the differential Helmholtz energy dA, Equation (19.38), and write

$$\begin{aligned} dA &= dU - TdS - SdT = TdS - PdV - TdS - SdT \\ &= -PdV - SdT = \left(\frac{\partial A}{\partial V}\right)_T dV + \left(\frac{\partial A}{\partial T}\right)_V dV \end{aligned}$$

Then

$$\left(\frac{\partial A}{\partial V}\right)_T = -P \quad \text{and} \quad \left(\frac{\partial A}{\partial T}\right)_V = -S$$

and for the second cross derivatives it follows that

$$\left(\frac{\partial}{\partial T}\left(\frac{\partial A}{\partial V}\right)_T\right)_V = -\left(\frac{\partial P}{\partial T}\right)_V \quad \text{and} \quad \left(\frac{\partial}{\partial V}\left(\frac{\partial A}{\partial T}\right)_V\right)_T = -\left(\frac{\partial S}{\partial V}\right)_T$$

Because these cross derivatives must be equal, we find the Maxwell relation

$$\boxed{\left(\frac{\partial P}{\partial T}\right)_V = \left(\frac{\partial S}{\partial V}\right)_T}$$

Finally we consider the change of the Gibbs energy dG and combine it with the equation for dH

$$dG = VdP - SdT = \left(\frac{\partial G}{\partial P}\right)_T dV + \left(\frac{\partial G}{\partial T}\right)_P dT$$

Then

$$\left(\frac{\partial G}{\partial P}\right)_T = V \quad \text{and} \quad \left(\frac{\partial G}{\partial T}\right)_P = -S$$

and for the second cross derivatives it follows that

$$\left(\frac{\partial}{\partial T}\left(\frac{\partial G}{\partial P}\right)_T\right)_P = \left(\frac{\partial V}{\partial T}\right)_P \quad \text{and} \quad \left(\frac{\partial}{\partial P}\left(\frac{\partial G}{\partial T}\right)_P\right)_T = -\left(\frac{\partial S}{\partial P}\right)_T$$

Because these cross derivatives must be equal, we find the Maxwell relation

$$\boxed{\left(\frac{\partial V}{\partial T}\right)_P = -\left(\frac{\partial S}{\partial P}\right)_T}$$

E19.6 Imagine a vertical tube filled with mercury (height h = 10 m). What is the freezing point of the mercury at the bottom when the freezing point at the top is 234.3 K?

Solution

The mercury exerts a pressure of

$$\begin{aligned} P = hg\rho &= (10 \text{ m})\left(9.81 \text{ m s}^{-2}\right)\left(13.6 \cdot 10^3 \text{ kg m}^{-3}\right) \\ &= 13.3 \cdot 10^5 \text{ Pa} = 13.3 \text{ bar} \end{aligned}$$

Next we use the Clapeyron equation and the facts $\Delta_{\text{fus}}H$ = 2.29 kJ mol $^{-1}$ and $\Delta_{\text{fus}}V = 0.517$ cm^3 mol^{-1} to estimate the change in freezing point

$$\begin{aligned} \delta T &= \frac{T \cdot \Delta_{\text{fus}}V}{\Delta_{\text{fus}}H} \cdot \delta P \\ &= \frac{(234.3)\left(0.517 \times 10^{-6}\right)}{2.29 \times 10^3}\left(13.3 \cdot 10^5\ \right) \text{K} = +0.07 \text{ K} \end{aligned}$$

The freezing point at the bottom is about 0.1 K higher than that at the top; this finding means that freezing of the mercury starts at the bottom of the tube.

E19.7 What is the boiling point of water on the top of Mount Everest (8882 m above sea level and a pressure of 0.34 bar)? You may take the enthalpy change for the vaporization of water $\Delta_{vap}H$ to be 32.83 kJ mol^{-1}.

Solution

Using the Clausius–Clapeyron equation

$$\ln\left(\frac{P_2}{P_1}\right) = -\frac{\Delta_{vap}H}{\boldsymbol{R}}\left(\frac{1}{T_2} - \frac{1}{T_1}\right)$$

we have

$$\ln\left(\frac{0.34}{1}\right) = -\frac{32,830}{8.314}\left(\frac{1}{T_2/\mathrm{K}} - \frac{1}{373}\right)$$

Solving for T_2, we find $T_2 = 338.5$ K

E19.8 Sketch and label a P versus T phase diagram for a pure substance which has a liquid phase at $P = 1$ bar. Indicate the normal melting point, boiling point, triple point, critical point, and label the phases. Using the triple point (T_3, P_3) as a reference, give equations describing each of the phase boundaries, assuming ideal behavior for the gas phase and neglecting the temperature dependence of the enthalpies.

Solution

Figure E19.8 shows a sketch of a phase diagram; the solid black curves are called "coexistence" curves and give the (P, T) points where two phases of a substance can coexist.

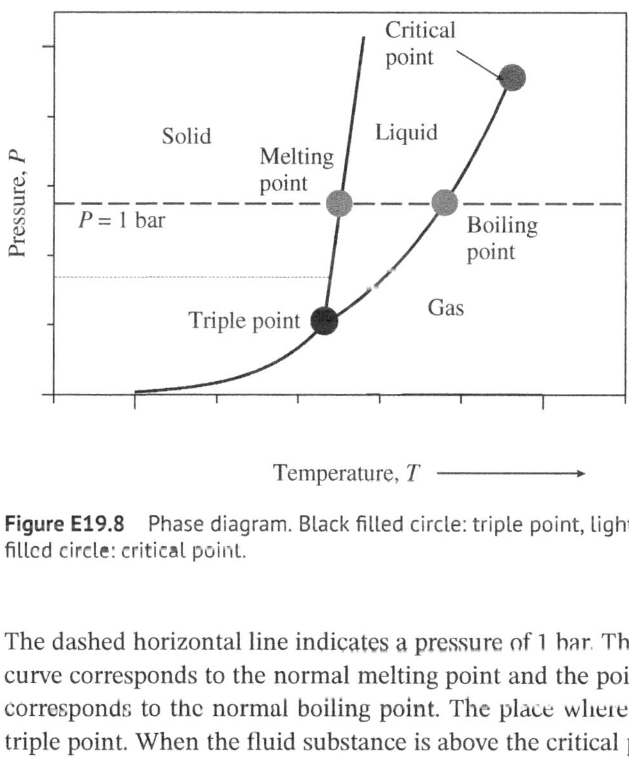

Figure E19.8 Phase diagram. Black filled circle: triple point, light gray filled circles: melting and boiling points, dark gray filled circle: critical point.

The dashed horizontal line indicates a pressure of 1 bar. The point where it crosses the solid liquid coexistence curve corresponds to the normal melting point and the point where it crosses the liquid–gas coexistence curve corresponds to the normal boiling point. The place where the three coexistence curves intersect is called the triple point. When the fluid substance is above the critical point, no distinction between the gas and liquid can be identified; thus, the critical point is the value of T and P where the liquid–gas coexistence curve terminates.

(b) The phase boundaries are defined by the relation

$$\mathrm{d}G_\alpha = \mathrm{d}G_\beta$$

where the subscripts α and β denote the phases. Using the fundamental relation that $\mathrm{d}G_\alpha = -S_\alpha \cdot \mathrm{d}T + V_\alpha \cdot \mathrm{d}P$ we find

$$\frac{\mathrm{d}P}{\mathrm{d}T} = \frac{\Delta S}{\Delta V} = \frac{\Delta H_{\alpha\to\beta}}{T \cdot \Delta V}$$

where $\Delta S = S_\beta - S_\alpha$ and $\Delta V = V_\beta - V_\alpha$. For the transition from the solid to the liquid we can write

$$\frac{\mathrm{d}P}{\mathrm{d}T} = \frac{\Delta H_{s\to l}}{T \cdot \Delta V}$$

If we assume that $\Delta H_{s\to l}$ and ΔV do not change too strongly along the coexistence curve, then we can readily integrate this equation from the triple point to another point on the solid/liquid curve; namely

$$\int_{P_3}^{P'} \mathrm{d}P = \frac{\Delta H_{s\to l}}{\Delta V} \int_{T_3}^{T'} \mathrm{d}T$$

and find

$$P' - P_3 = \frac{\Delta H_{s\to l}}{\Delta V} \ln\left(\frac{T'}{T_3}\right)$$

equation defines the solid–liquid coexistence curve.

For the solid-to-gas and the liquid-to-gas transition we also use the Clapeyron equation, but we include the volume change by way of the ideal gas law. For the solid-to-gas transition we write

$$\frac{\mathrm{d}P}{\mathrm{d}T} = \frac{\Delta H_{s\to g}}{T \cdot \Delta V} \approx \frac{\Delta H}{T \cdot V_{\text{gas}}} = \frac{\Delta H}{\boldsymbol{n}RT^2} P$$

which can be rearranged to

$$\frac{\mathrm{d}P}{P} = \frac{\Delta H}{\boldsymbol{n}\boldsymbol{R}} \frac{\mathrm{d}T}{T^2}$$

By assuming that the enthalpy of sublimation does not change very strongly with temperature, we can integrate this equation and find that

$$\int_{P_3}^{P'} \frac{\mathrm{d}P}{P} = \frac{\Delta H_{s\to g}}{\boldsymbol{n}\boldsymbol{R}} \int_{T_3}^{T'} \frac{\mathrm{d}T}{T^2}$$

which implies that

$$\ln\left(\frac{P'}{P_3}\right) = \frac{\Delta H_{s\to g}}{\boldsymbol{n}\boldsymbol{R}} \left(\frac{1}{T_3} - \frac{1}{T'}\right)$$

Lastly, we change the notation by replacing P' by P and T' by T, and find that

$$\left(\frac{P}{P_3}\right) = \frac{\Delta H_{s\to g}}{\boldsymbol{n}\boldsymbol{R}} \left(\frac{1}{T_3} - \frac{1}{T}\right)$$

This is the Clausius–Clapeyron equation.

We proceed similarly for the liquid-to-gas transition (which is the same as the steps used for the solid-to-gas transition except we use $\Delta H_{l\to g}$)

$$\ln\left(\frac{P}{P_3}\right) = \frac{\Delta H_{l\to g}}{\boldsymbol{n}\boldsymbol{R}} \left(\frac{1}{T_3} - \frac{1}{T}\right)$$

E19.9 Using the following experimental data, make a rough sketch of the phase diagram of acetic acid. Solid acetic acid may exist in two phases, α and β. The two solid phases α and β are in equilibrium at a temperature of 328 K and a pressure of 200 bar. The high pressure form β is more dense than α, and both solids are more dense than the liquid. The normal boiling point of the liquid is 391 K. The low pressure form α melts at 289 K under its own vapor pressure of 0.012 bar. The critical pressure and critical temperature are 57.9 bar and 593 K. (Adapted from: "Molecular Thermodynamics" by Dickerson).

Solution

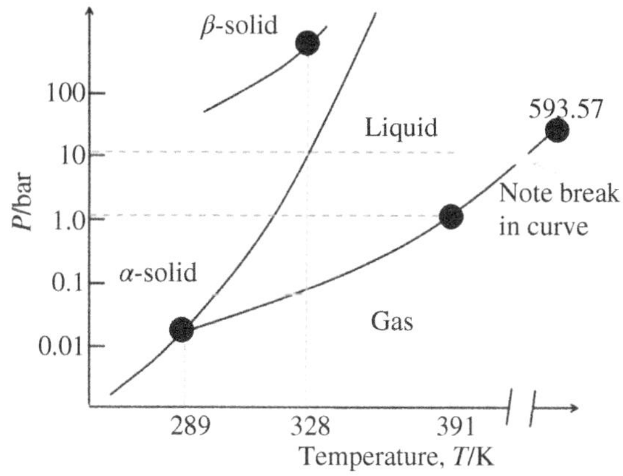

Figure E19.9 Phase diagram of acetic acid.

E19.10 Sketch the Gibbs energy for a gas and a liquid versus temperature to explain the phenomenon of boiling. What is the slope of the Gibbs energy curve with temperature of a substance in a given state? How does this slope change with temperature?

Solution

Figure E19.10 shows two sketches: one at the top which is G versus T for a gas and a liquid at a particular pressure P and the lower graph shows a plot of the pressure P versus T. The vertical line shows that the point on the P versus T coexistence curve corresponds to the temperature at which the Gibbs energy of the gas and the Gibbs energy of the liquid are equal. For different pressures, the crossing point of Gibbs energy curves of the two phases will shift; measurement of this shift can be used to find the coexistence curve.

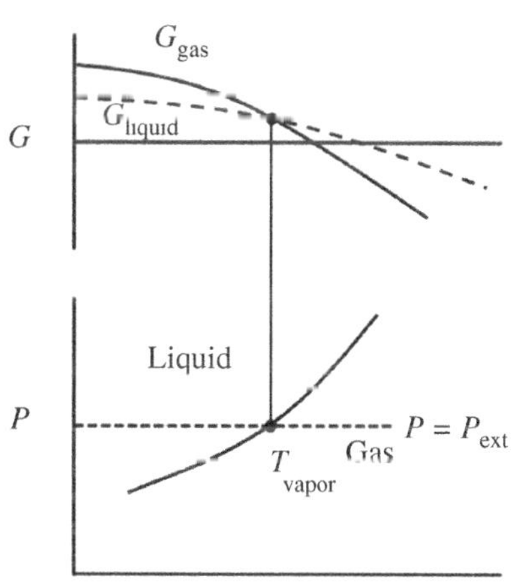

Figure E19.10 G versus T for a gas and a liquid at a particular pressure P (top) and the pressure P versus T (bottom).

The diagram at the top shows that the slope of the Gibbs energy for the gas has a higher negative value than that of the liquid. The slope of the G versus T curve is $-S$, i.e.,

$$\left(\frac{\partial G}{\partial T}\right)_P = -S$$

because $S_{gas} > S_{liquid}$ the gas has the steeper slope. Because of intermolecular attractive interactions ΔH_{liquid} dominates at low T, but as T increases the entropic contribution to $\Delta G = \Delta H - T\Delta S$ becomes more important and makes the gas phase more stable. As shown in the diagram $G_{liquid} < G_{gas}$ at low temperatures; however, at high temperatures $G_{gas} < G_{liquid}$.

E19.11 At 298 K and 1 bar, the molar entropies of liquid and gaseous water are 69.9 and 188.8 J $K^{-1}mol^{-1}$, respectively. The enthalpy of vaporization for water $\Delta_{vap}H^{\ominus}$ is 44.0 kJ mol^{-1} at 298 K. a) Will the phase change, $H_2O(l) \rightarrow H_2O(g, 1$ bar), occur spontaneously at 298 K? b) Will the phase change, $H_2O(l) \rightarrow H_2O(g, 0.01$ bar), occur spontaneously at 298 K? c) Estimate the normal boiling point of water from the data given above. State what assumptions or approximations you make. ((Adapted from R.E. Dickerson, *Molecular Thermodynamics* (Benjamin, 1969)).

Solution

a) We proceed by evaluating $\Delta_{vap}G^{\ominus} = \Delta_{vap}H^{\ominus} - T \cdot \Delta_{vap}S^{\ominus}$ and find

$$\Delta_{vap}G^{\ominus} = \left(44.0 \times 10^3 - (298)(188.7 - 69.9)\right) \text{ J mol}^{-1}$$
$$= 8.6 \text{ kJ mol}^{-1} > 0$$

Hence the vaporization of the liquid is not spontaneous because $\Delta_{vap}G^{\ominus} > 0$. Note that this analysis applies to a "closed" thermodynamic system.

b) For this case, we can use Equation (18.32) to write

$$\Delta_{vap}G_m = \Delta_{vap}G^{\ominus} + RT \ln\left(\frac{P}{P^{\ominus}}\right)$$

so that

$$\Delta_{vap}G_m\,(0.01 \text{ bar}) = 8.6 \text{ kJ mol}^{-1} + (8.3145)(298) \cdot \ln\left(\frac{0.01}{1}\right)$$
$$= -2.8 \text{ kJ mol}^{-1}$$

Because $\Delta_{vap}G_m\,(0.01 \text{ bar}) < 0$, we conclude that this process is spontaneous.

c) To estimate the normal boiling point of water, we use the relation

$$T_{BP} = \frac{\Delta_{vap}H^{\ominus}}{\Delta_{vap}S^{\ominus}} = \frac{44,000 \text{ J mol}^{-1}}{188.8 - 69.9 \text{ J K}^{-1} \text{ mol}^{-1}} = 370 \text{ K}$$

We have assumed that the ratio of $\Delta_{vap}H^{\ominus}$ to $\Delta_{vap}S^{\ominus}$ does not change much between 298 K and T_{BP}, even though the individual values may change.

Extension: Account for the temperature dependence of the entropy and enthalpy changes in the phase transition and determine the boiling point at 1 bar pressure. Comment on how much it improves the correspondence of the predicted boiling point temperature with the experimental value of 373.15 K.

E19.12 Take the molar volume of solid benzene to be 74.4 cm^3 mol^{-1} at 70.5 MPa and 298. K, its melting temperature. The molar volume of the liquid at this temperature and pressure is 83.4 cm^3mol^{-1}. If the melting temperature changes to 313 K at 128 MPa, calculate the molar enthalpy and entropy of fusion of the solid.

Solution

We can estimate the molar enthalpy and entropy of fusion by way of the Clapeyron equation

$$\frac{dP}{dT} = \frac{\Delta_{fus}S}{\Delta_{fus}V}$$

To make our estimation, we assume that the change of the pressure and temperature is small enough that we can write

$$\frac{dP}{dT} \sim \frac{\delta P}{\delta T} = \frac{(128 - 70.5) \text{ MPa}}{(313 - 298) \text{ K}} = 3.83 \text{ MPa K}^{-1}$$

From the Clapeyron equation, we can write that

$$\Delta_{fus}S \sim \frac{\delta P}{\delta T} \cdot \Delta_{fus}V$$
$$= (3.83) \text{ MPa K}^{-1} \cdot (83.4 - 74.4)\frac{\text{cm}^3}{\text{mol}} \times 10^{-6} \frac{\text{m}^3}{\text{cm}^3} = 34.5 \text{ J K}^{-1} \text{ mol}^{-1}$$

With $\Delta_{fus}S$ in hand and using T_{mp} = 298 K we can estimate $\Delta_{fus}H$ at 705 bar by the condition that $\Delta_{fus}G = 0$ to find

$$\Delta_{fus}H = T_{mp} \cdot \Delta_{fus}S$$
$$= (298)(34.5) \text{ J mol}^{-1} = 10.3 \text{ kJ mol}^{-1}$$

These values are in excellent agreement with the experimental values; see the NIST Chemistry WebBook.

E19.13 Using the Clausius–Clapeyron equation in its integrated form calculate and plot the vapor pressure above liquid and solid water (a) in the range of $T = 350$ K to $T = 500$ K and (b) in the range of $T = 271$ K to $T = 275$ K. The molar enthalpies of evaporation and sublimation are $\Delta_{vap}H = 44.1$ kJ mol^{-1}, and $\Delta_{sub}H = (6.0 + 44.1)$ kJ mol^{-1} $= 50.1$ kJ mol^{-1}. The vapor pressure at 373 K is 1.013 bar, and the vapor pressure at 273.16 K (triple point) is 6.11×10^{-3} bar.

Solution

The Clausius–Clapeyron equation is

$$\ln\left(\frac{P_{vap}}{P_{vap,1}}\right) = -\frac{\Delta H}{\boldsymbol{R}}\left(\frac{1}{T} - \frac{1}{T_1}\right)$$

a) Solving for the vapor pressure above the liquid P_{vap}, we have

$$P_{vap} = P_{vap,1} \cdot \exp\left[-\frac{\Delta_{vap}H}{\boldsymbol{R}}\left(\frac{1}{T} - \frac{1}{T_1}\right)\right]$$
$$= 1.013\ \text{bar} \cdot \exp\left[-\frac{44,100\ \text{J mol}^{-1}}{8.314\ \text{J mol}^{-1}\ \text{K}^{-1}}\left(\frac{1}{T} - \frac{1}{373\ \text{K}}\right)\right]$$

Figure E19.13a shows a plot of this equation.

b) Solving for the vapor pressure above the solid P_{vap}, we have

$$P_{vap} = P_{vap,1} \cdot \exp\left[-\frac{\Delta_{sub}H}{\boldsymbol{R}}\left(\frac{1}{T} - \frac{1}{T_1}\right)\right]$$
$$= 6.11 \times 10^{-3}\ \text{bar} \cdot \exp\left[-\frac{50,100\ \text{J mol}^{-1}}{8.314\ \text{J mol}^{-1}\ \text{K}^{-1}}\left(\frac{1}{T} - \frac{1}{273.16\ \text{K}}\right)\right]$$

Figure E19.13b shows a plot of this function.

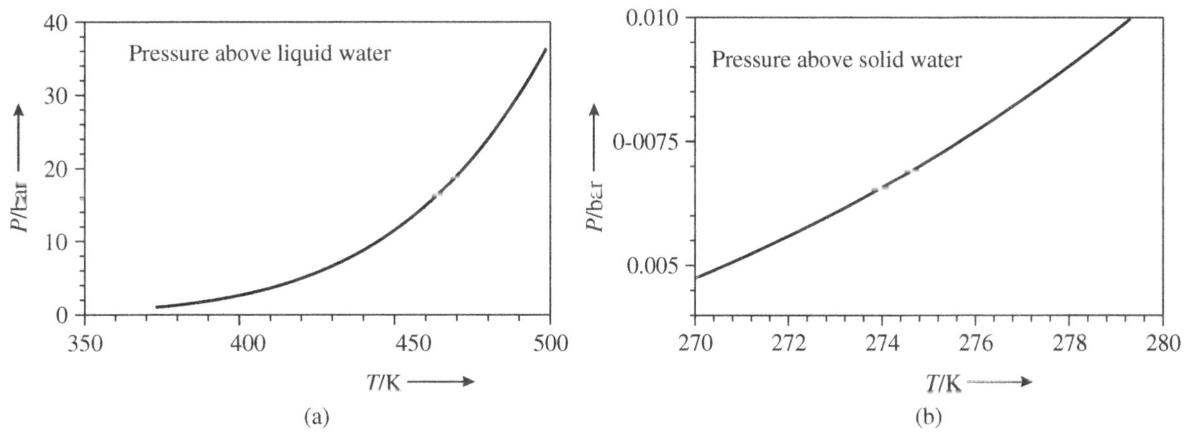

Figure E19.13 (a) Vapor pressure above liquid water versus temperature. (b) Vapor pressure above solid water versus temperature.

E19.14 Using equation

$$dS = \frac{1}{T}\left(\frac{\partial U}{\partial T}\right)_V \cdot dT + \frac{1}{T}\left[\left(\frac{\partial U}{\partial V}\right)_T + P\right] \cdot dV$$

and the Thermodynamic Equation of State, show that

$$dS = C_V\frac{dT}{T} + \left(\frac{\partial P}{\partial T}\right)_V dV$$

Solution

We start with

$$dS = \frac{1}{T}\left(\frac{\partial U}{\partial T}\right)_V \cdot dT + \frac{1}{T}\left[\left(\frac{\partial U}{\partial V}\right)_T + P\right] \cdot dV$$

and we can use the Thermodynamic equation of state to substitute for $(\partial U/\partial V)_T$ and obtain

$$\begin{aligned} dS &= \frac{1}{T}\left(\frac{\partial U}{\partial T}\right)_V \cdot dT + \frac{1}{T}\left[\left[T\left(\frac{\partial P}{\partial T}\right)_V - P\right] + P\right] \cdot dV \\ &= \frac{1}{T}\left(\frac{\partial U}{\partial T}\right)_V \cdot dT + \left(\frac{\partial P}{\partial T}\right)_V \cdot dV \end{aligned}$$

Using the definition of the constant volume heat capacity, we find

$$dS = C_V \frac{dT}{T} + \left(\frac{\partial P}{\partial T}\right)_V dV$$

E19.15 A butane cigarette lighter contains a biphasic (liquid–gas) mixture of butane. The normal boiling point of butane occurs at 272.6 K and 1 bar. The enthalpy of vaporization of butane is 23.8 kJ mol^{-1}K^{-1}. Use the Clausius–Clapeyron equation to estimate the vapor pressure of the gaseous butane in the cigarette lighter at 300 K.

Solution
By substituting into the Clausius–Clapeyron equation, we find

$$\begin{aligned} P_{\text{vap}} &= P_{\text{vap},1} \cdot \exp\left[-\frac{\Delta_{\text{vap}}H}{\boldsymbol{R}}\left(\frac{1}{T} - \frac{1}{T_1}\right)\right] \\ &= 1\ \text{bar} \cdot \exp\left[-\frac{23,800\ \text{J mol}^{-1}}{8.314\ \text{J mol}^{-1}\ \text{K}^{-1}}\left(\frac{1}{300\ \text{K}} - \frac{1}{272.6\ \text{K}}\right)\right] \\ &= 2.61\ \text{bar} \end{aligned}$$

E19.16 Use the equation $(dH = TdS + VdP)$ to derive the equation

$$dH = C_P \cdot dT + \left[V - T\left(\frac{\partial V}{\partial T}\right)_P\right] \cdot dP$$

by writing dS as a function of T and P and using a Maxwell relation.

Solution
We begin with $dH = TdS + VdP$ and substitute

$$dS = \left(\frac{\partial S}{\partial T}\right)_P dT + \left(\frac{\partial S}{\partial P}\right)_T dP$$

to find

$$dH = T\left(\frac{\partial S}{\partial T}\right)_P dT + T\left(\frac{\partial S}{\partial P}\right)_T dP + VdP$$

From the second law and the definition of C_P we know that $C_P = T\left(\frac{\partial S}{\partial T}\right)_P$, which gives

$$dH = C_P dT + T\left(\frac{\partial S}{\partial P}\right)_T dP + VdP$$

Lastly, we substitute the Maxwell relation $\left(\frac{\partial V}{\partial T}\right)_P = -\left(\frac{\partial S}{\partial P}\right)_T$ into this expression for dH, and we find

$$\begin{aligned} dH &= C_P \cdot dT - T\left(\frac{\partial V}{\partial T}\right)_P dP + VdP \\ &= C_P \cdot dT + \left[V - T\left(\frac{\partial V}{\partial T}\right)_P\right] dP \end{aligned}$$

E19.17 Use the fact that dV/V is an exact differential to relate the pressure dependence of the expansivity to the temperature dependence of the compressibility, namely prove that

$$\left(\frac{\partial \alpha_V}{\partial P}\right)_T = -\left(\frac{\partial \kappa_T}{\partial T}\right)_P$$

Solution

First, we write the differential volume as

$$\mathrm{d}V = \left(\frac{\partial V}{\partial P}\right)_T \mathrm{d}P + \left(\frac{\partial V}{\partial T}\right)_P \mathrm{d}T$$

Dividing by V and using our definitions of κ_T and α_V, we find

$$\frac{\mathrm{d}V}{V} = -\kappa_T\ \mathrm{d}P + \alpha_V\ \mathrm{d}T$$

Because $\mathrm{d}V/V$ is an exact differential the cross derivatives must be equal, hence

$$\left(\frac{\partial \alpha_V}{\partial P}\right)_T = -\left(\frac{\partial \kappa_T}{\partial T}\right)_P$$

E19.18 For a first-order phase transition, the molar entropy, the molar enthalpy, and the molar volume are discontinuous. What happens to the molar internal energy of the system at such a phase transition?

Solution

Using our definition of H_m we see that $U_m = H_m - PV_m$. For a phase transition at constant pressure from phase α to phase β, we find that

$$\begin{aligned}\Delta U_m &= U_{m,\alpha} - U_{m,\beta}\\ &= (H_{m,\alpha} - PV_{m,\alpha}) - (H_{m,\beta} - PV_{m,\beta}) = \Delta H_m - P\Delta V_m\end{aligned}$$

The internal energy changes discontinuously and the magnitude of its change is determined by the heat it absorbs and the work performed. Note that even though the molar energy, molar volume, molar enthalpy, and molar entropy changes between the phases are discontinuous, the total energy, volume, enthalpy, and entropy change smoothly in the biphasic region of the phase diagram.

E19.19 Prove that the volume dependence of the heat capacity at constant volume is equal to the second derivative of the pressure, namely

$$\left(\frac{\partial C_V}{\partial V}\right)_T = T\left(\frac{\partial^2 P}{\partial T^2}\right)_V$$

Use this result to show that the constant volume heat capacity of an ideal gas does not depend on the volume.

Solution

We begin with Equation (19.17),

$$\mathrm{d}U = C_V\ \mathrm{d}T + \left[T\left(\frac{\partial P}{\partial T}\right)_V - P\right]\ \mathrm{d}V$$

Because $\mathrm{d}U$ is an exact differential, the cross derivatives must be equal, namely

$$\begin{aligned}\left(\frac{\partial C_V}{\partial V}\right)_T &= \frac{\partial}{\partial T}\left[T\left(\frac{\partial P}{\partial T}\right)_V - P\right]_V\\ &= \left(\frac{\partial P}{\partial T}\right)_V + T\left(\frac{\partial^2 P}{\partial T^2}\right)_V - \left(\frac{\partial P}{\partial T}\right)_V = T\left(\frac{\partial^2 P}{\partial T^2}\right)_V\end{aligned}$$

For an ideal gas $P = \boldsymbol{nR}T/V$, and we see that

$$\left(\frac{\partial C_V}{\partial V}\right)_T = T\left(\frac{\partial^2 P}{\partial T^2}\right)_V = T\left(\frac{\partial}{\partial T}\left(\frac{\boldsymbol{nR}}{V}\right)\right)_V = 0$$

hence C_V does not depend on V for an ideal gas.

E19.20 The table contains vapor pressure data for water, taken from O.C. Bridgeman, E.W. Aldrich, *J. Heat Transf*. 1964, **86** 279. Plot these data. Compare the data to the result of the Clausius–Clapeyron equation and of Equation (19.62) using the values

$$\Delta_{\text{vap}}H^{\ominus} = 44.1\ \text{kJ mol}^{-1}\ ,\ \Delta C_P^{\ominus} = -41.7\ \text{J K}^{-1}\ \text{mol}^{-1}$$
$$T_1 = 373\ \text{K},\quad \text{and}\quad P_{\text{vap}}1 = 1\ \text{bar}$$

P (bar)	0.005948	0.022762	0.030844	0.071851	0.19406	0.46130	0.98678	1.9334	3.5189
T (K)	273.15	293.15	298.15	313.15	333.15	353.15	373.15	393.15	413.15
P (bar)	6.0189	9.7653	15.1451	22.592	32.6051	45.7124	62.5133	83.6708	109.9467
T (K)	433.15	453.15	473.15	493.15	513.15	533.15	553.15	573.15	593.15

Solution

Equation (19.62) is

$$\ln \hat{P}_{\text{vap}} = \ln \hat{P}_{\text{vap},1} - \frac{\Delta_{\text{vap}} H_1}{R}\left(\frac{1}{T} - \frac{1}{T_1}\right) + \frac{\Delta C_{P,m}}{R}\left[\ln \frac{T}{T_1} + T_1\left(\frac{1}{T} - \frac{1}{T_1}\right)\right]$$

with $\hat{P}_{\text{vap},1} = 1$ at $T_1 = 373$ K, and using the values of $\Delta_{\text{vap}} H^{\ominus} = (-241.8 + 285.9)$ kJ mol^{-1} = 44.1 kJ mol^{-1} and $\Delta C_P = (33.6 - 75.3)$ J K^{-1} mol^{-1} = -41.7 J K^{-1} mol^{-1} from Example 19.6. Figure E19.20 shows the experimental data points (black circles) together with calculated data: Clausius–Clapeyron equation (dark gray line), and Equation (19.62) (light line).

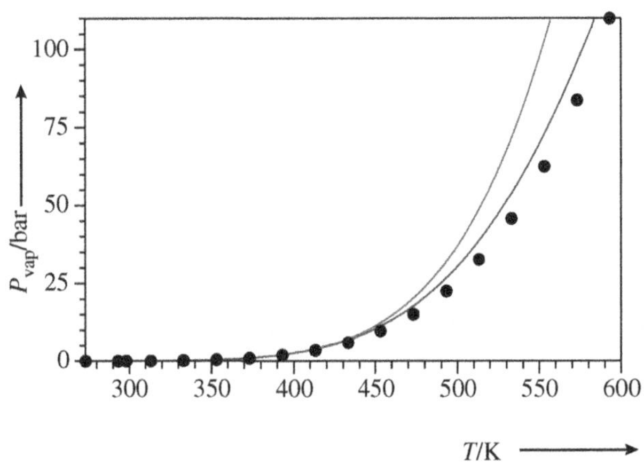

Figure E19.20 Vapor pressure of water versus temperature T. Experimental data points (black full circles), Clausius–Clapeyron equation (dark gray solid line) and Equation (19.62) (light gray solid line).

Up to a temperature of 450 K ($P_{\text{vap}} \approx 10$ bar) both calculations reproduce the experimental data well. At higher temperatures both calculations are significantly above the experimental data. As expected, Equation (19.62) gives values closer to the experiment than does the Clausius–Clapeyron equation. At 550 K we find 97 bar for Clausius–Clapeyron and 69 bar for Equation (19.62) compared to 62 bar in the experiment. These remaining deviations arise from the fact that the Clausius–Clapeyron equation, as well as Equation (19.62), assume that the vapor can be treated as an ideal gas.

E19.21 Using equation

$$dS = \frac{1}{T}\left(\frac{\partial U}{\partial T}\right)_V \cdot dT + \frac{1}{T}\left[\left(\frac{\partial U}{\partial V}\right)_T + P\right] \cdot dV$$

and the Thermodynamic Equation of State, show that

$$dS = C_V \frac{dT}{T} + \left(\frac{\partial P}{\partial T}\right)_V dV$$

Solution

We start with the equation

$$dS = \frac{1}{T}\left(\frac{\partial U}{\partial T}\right)_V \cdot dT + \frac{1}{T}\left[\left(\frac{\partial U}{\partial V}\right)_T + P\right] \cdot dV$$

and we can use the Thermodynamic equation of state to substitute for $(\partial U/\partial V)_T$ and obtain

$$dS = \frac{1}{T}\left(\frac{\partial U}{\partial T}\right)_V \cdot dT + \frac{1}{T}\left[\left[T\left(\frac{\partial P}{\partial T}\right)_V - P\right] + P\right] \cdot dV$$
$$= \frac{1}{T}\left(\frac{\partial U}{\partial T}\right)_V \cdot dT + \left(\frac{\partial P}{\partial T}\right)_V \cdot dV$$

Using the definition of the constant volume heat capacity, we find

$$dS = C_V \frac{dT}{T} + \left(\frac{\partial P}{\partial T}\right)_V dV$$

19.2 Problems

P19.1 The following data are found for Cl_2:

T/K	P_{vap}/bar	ρ_{liquid}/g cm^{-3}	ρ_{vapor} /g cm^{-3}
273	3.636	1.468	0.0128
283	4.993	1.438	0.0175
293	6.706	1.408	0.0226

Calculate the enthalpy of vaporization of Cl_2 at 283 K. If Cl_2 were assumed to behave as an ideal gas and the volume of the liquid is neglected, how would it change your results? (Adapted from R.E. Dickerson, *Molecular Thermodynamics* (Benjamin, 1969)).

Solution

We can analyze these data by way of the Clapeyron equation; namely

$$\frac{dP}{dT} = \frac{\Delta_{vap}S}{\Delta_{vap}V} = \frac{\Delta_{vap}H}{T \cdot \Delta_{vap}V}$$

First we plot the data and fit it to a polynomial. Using the data in the table

T/K	273	283	293
P_{vap}/bar	3.636	4.993	6.706

We find a good fit to a quadratic form with the formula

$$P_{vap}/\text{bar} = 0.001\,515(T/\text{K})^2 - 0.704\,(T/\text{K}) + 82.946;$$

see Fig. P19.1.

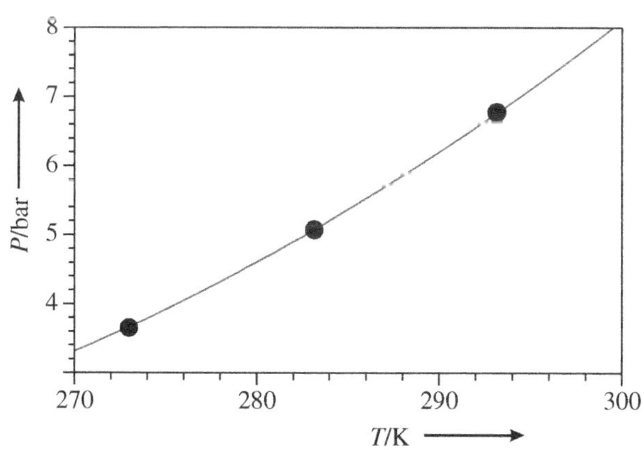

Figure P19.1 Vapor pressure of Cl_2 versus temperature T. Experiment (filled circles) and quadratic fit (black line).

The slope is given by

$$\frac{dP}{dT} = \frac{d}{dT}\left(0.001\,515(T/K)^2 - 0.704\ (T/K)\right) = 0.00303\ (T/K) - 0.704$$

and if we evaluate it at 283 K we find that

$$\left.\frac{dP}{dT}\right|_{283\ K} = \left[0.00303\ (T/K) - 0.704\right|_{T=283\ K} = 0.1535\ \text{bar K}^{-1}$$

Hence we can write that

$$0.1535\ \text{bar K}^{-1} = \frac{\Delta H}{T\Delta V}$$

From the data given we can calculate the specific volume of the vapor and the liquid as

$$\rho_{vap} = \frac{m}{V_{vap}} = \frac{\boldsymbol{M}}{V_{vap,m}}, \quad \rho_{liq} = \frac{m}{V_{liq}} = \frac{\boldsymbol{M}}{V_{liq,m}}$$

Thus we calculate the change in specific volume upon evaporation to be

$$\begin{aligned}\Delta_{vap}V = V_{vap,m} - V_{liq,m} &= \frac{\boldsymbol{M}}{\rho_{vap}} - \frac{\boldsymbol{M}}{\rho_{liq}} = \boldsymbol{M}\left(\frac{1}{\rho_{vap}} - \frac{1}{\rho_{liq}}\right)\\ &= 70.9\ \text{g mol}^{-1}\left(\frac{1}{0.0175\ \text{g cm}^{-3}} - \frac{1}{1.438\ \text{g cm}^{-3}}\right)\\ &= 70.9\ \text{g mol}^{-1}\cdot 56.45\ \text{g}^{-1}\ \text{cm}^3 = 4.00\ \text{L mol}^{-1}\end{aligned}$$

With these results we can evaluate the enthalpy of vaporization at 283 K and find that

$$\begin{aligned}\Delta_{vap}H &= (283\ \text{K})\left(0.1535\ \text{bar K}^{-1}\right)\left(4.00\ \text{L mol}^{-1}\right)\\ &= 174\ \text{bar L mol}^{-1} = 17.4\ \text{kJ mol}^{-1}\end{aligned}$$

If we assume that Cl_2 behaves as an ideal gas and $V_{gas} \gg V_{liq}$, we find that

$$\frac{dP}{dT} = \frac{P\cdot\Delta_{vap}H}{\boldsymbol{R}T^2} \text{ so that } \Delta_{vap}H = \frac{dP}{dT}\cdot\left(\frac{\boldsymbol{R}T^2}{P}\right)$$

If we evaluate this expression at 283 K, we find

$$\begin{aligned}\Delta_{vap}H &= \frac{0.1535\ \text{bar K}^{-1}\cdot 8.314\ \text{J K}^{-1}\ \text{mol}^{-1}\cdot(283\ \text{K})^2}{4.96\ \text{bar}}\\ &= 20.6\ \text{kJ mol}^{-1}\end{aligned}$$

Comparison with the previous value gives an error of about 18%.

P19.2 Figure P19.2a shows a sketch of the phase diagram of iodine. Some data on I_2 are a) the normal boiling point of iodine is 456 K; b) the vapor pressure of liquid I_2 at 390 K is 100 torr; c) the molar enthalpy of sublimation is 61.05 kJ mol^{-1}; d) the vapor pressure of solid I_2 is 1.00 torr at 312 K.

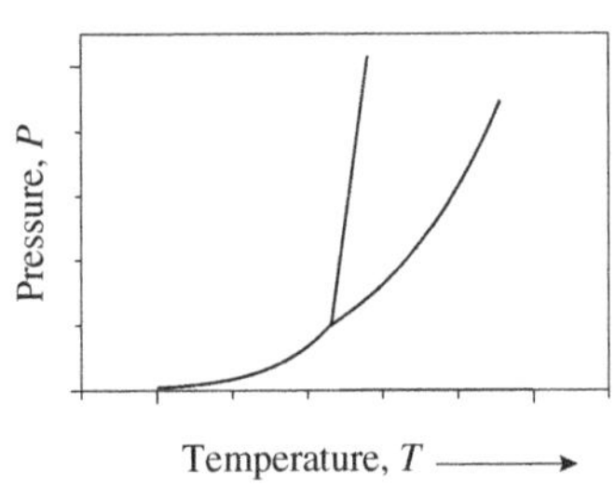

Figure P19.2a Sketch of phase diagram of I_2.

Identify the three phases on the diagram and label the points indicated with the appropriate values.

Using the data given above and your knowledge of phase transitions, calculate the molar enthalpy of vaporization of iodine.

Using the data given above and your knowledge of phase transitions, calculate the triple point temperature and pressure.

Sketch the Gibbs energy curves for gaseous, liquid, and solid I_2 as a function of temperature and at the triple point pressure. Label any of the temperatures you know.

On a different graph sketch the Gibbs energy curves for gaseous, liquid, and solid I_2 as a function of temperature and at a pressure of 1 bar. Label any temperatures that you know.

Solution

The relevant data are included in Fig. P19.2b.

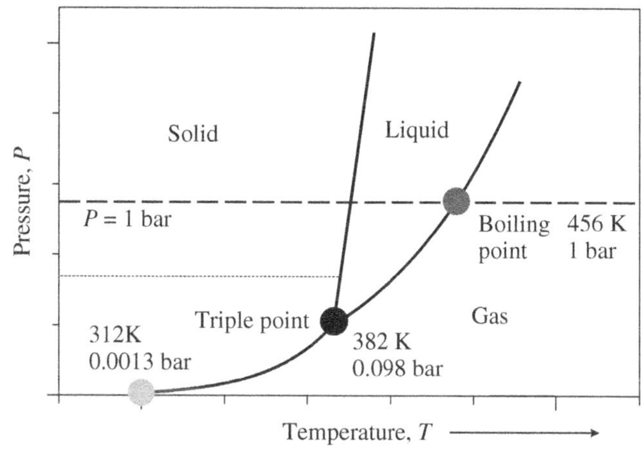

Figure P19.2b Phase diagram of I_2.

Molar Enthalpy: The molar enthalpy of vaporization is calculated using

$$\ln\left(\frac{P_2}{P_1}\right) = -\frac{\Delta_{vap}H}{\boldsymbol{R}}\left(\frac{1}{T_2} - \frac{1}{T_1}\right)$$

Solving for so $\Delta_{vap}H$ we find that

$$\Delta_{vap}H = -\boldsymbol{R}\ln\left(\frac{P_2}{P_1}\right)\left[\frac{1}{T_2} - \frac{1}{T_1}\right]^{-1}$$

$$= 8.314\ \text{J K}^{-1}\text{mol}^{-1} \cdot \ln\left(\frac{760}{100}\right)\left[\frac{1}{456\ \text{K}} - \frac{1}{390\ \text{K}}\right]^{-1}\ \text{J mol}^{-1} = 45.4\ \text{kJ mol}^{-1}$$

Triple point: The temperature and pressure can be determined by applying the Clausius–Clapeyron equation for both the sublimation equilibrium and the vaporization equilibrium; namely

$$\ln\left(\frac{P_{tp}}{1.0\ \text{torr}}\right) = -\frac{\Delta_{sub}H}{\boldsymbol{R}}\left(\frac{1}{T_{tp}} - \frac{1}{312\ \text{K}}\right)$$

and

$$\ln\left(\frac{P_{tp}}{100\ \text{torr}}\right) = -\frac{\Delta_{vap}H}{\boldsymbol{R}}\left(\frac{1}{T_{tp}} - \frac{1}{390\ \text{K}}\right)$$

If we subtract these two equations, we find that

$$\ln(100) = -\frac{\Delta_{sub}H}{\boldsymbol{R}}\left(\frac{1}{T_{tp}} - \frac{1}{312\ \text{K}}\right) + \frac{\Delta_{vap}H}{\boldsymbol{R}}\left(\frac{1}{T_{tp}} - \frac{1}{390\ \text{K}}\right)$$

which simplifies to

$$\ln(100) = \frac{\Delta_{sub}H}{\boldsymbol{R}\cdot 312\ \text{K}} - \frac{\Delta_{vap}H}{\boldsymbol{R}\cdot 390\ \text{K}} - \frac{1}{T_{tp}}\left(\frac{\Delta_{sub}H - \Delta_{vap}H}{\boldsymbol{R}}\right)$$

and can be solved to find

$$T_{tp} = 384\ \text{K}$$

Now that we know T_{tp}, we can evaluate the pressure by using

$$\ln\left(\frac{P_{tp}}{1.0\ \text{torr}}\right) = -\frac{\Delta_{sub}H}{\boldsymbol{R}}\left(\frac{1}{T_{tp}} - \frac{1}{312\ \text{K}}\right)$$

and $\Delta_{sub}H = 62.4$ kJ mol^{-1} (calculated from the data in Appendix A). so that

$$P_{tp} = 1.0 \text{ torr} \cdot \exp\left[-\frac{62.4 \text{ kJ mol}^{-1}}{8.314 \text{ J K}^{-1} \text{ mol}^{-1}}\left(\frac{1}{384 \text{ K}} - \frac{1}{312 \text{ K}}\right)\right]$$
$$= 1.0 \text{ torr} \cdot \exp\left[-7.51 \times 10^3 \cdot \left(-6.009 \times 10^{-4}\right)\right] = 1.0 \text{ torr} \cdot \exp[4.51] = 91.1 \text{ torr}$$

Figure P19.2c shows a plot of the Gibbs energy curves as a function of temperature at the triple point pressure of iodine. At this pressure the three Gibbs energy curves all cross at a single temperature value; at this temperature and pressure the system will be triphasic.

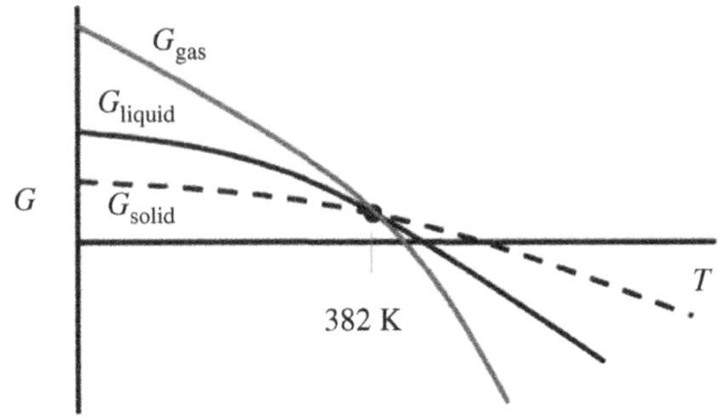

Figure P19.2c Triple point curves of I_2.

Figure P19.2d shows a plot of the Gibbs energy versus temperature for iodine at a pressure of 1 bar. In this case the liquid and gas curves cross at a temperature of 456 K, and the liquid and solid curves cross at a lower temperature. The crossing at 456 K corresponds to the normal boiling point of the liquid, and the lower crossing point corresponds to the normal melting point of the solid.

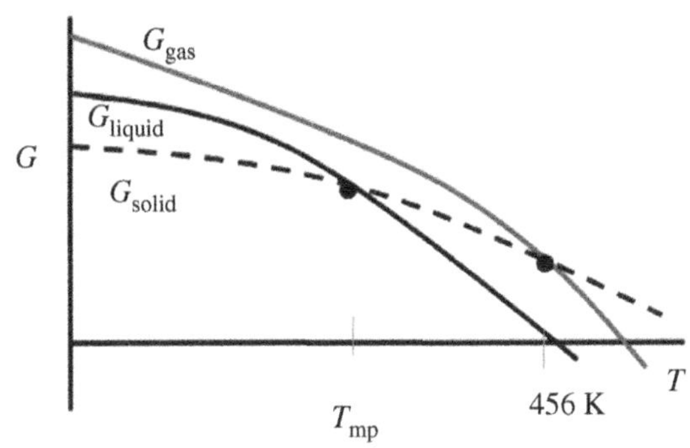

Figure P19.2d Gibbs energy G versus temperature for I_2.

P19.3 The table gives phase equilibrium data for CO_2.

Sublimation	Curve								
T (K)	130	140	155	170	185	194.7	205	216.658	
P (bar)	0.00032	0.00187	0.01674	0.0987	0.4402	1.013	2.271	5.185	
Vaporization	Curve								
T (K)	216.658	220	230	240	250	260	280	300	304.14
P (bar)	5.185	5.991	8.929	12.825	17.85	24.19	41.61	67.13	73.75

Liquid–Solid	Curve	
T (K)	216.55	236.45
P (bar)	5.185	1013

a) Use the data in this table to make a P versus T plot that shows the different phases of CO_2. The critical point occurs at $T = 304.14$ K and $P = 73.75$ bar, and the triple point occurs at $T = 216.658$ K and $P = 5.185$ bar. Mark the different phase regions on the graph and indicate the position of the triple point and critical point.
b) Use these data to determine $\Delta_{vap}H$ and $\Delta_{sub}H$ for CO_2 by making the assumption that $V_{gas} \gg V_{liq}$ and $V_{gas} \gg V_{solid}$. c) Use the density data given below for the liquid and gas[1] to determine $\Delta_{vap}H$ using the Clapeyron equation but accounting for $\Delta_{vap}V$ in a more realistic manner.

T (K)	233.15	243.15	253.15	263.15	273.15	283.15	293.15
ρ_{liquid} (kg m^{-3})	1115.9	1075.0	1031.0	980.97	925.07	859.99	772.02
ρ_{gas} (kg m^{-3})		36.78	52.03	72.31	97.46	132.	190.

Compare the enthalpies of vaporization with each other and with the literature value.

Solution

a) Figure P19.3a plots the data given in the form of a P versus T phase diagram.

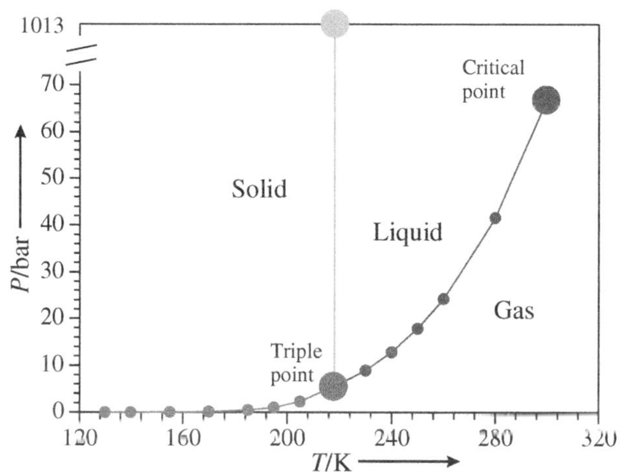

Figure P19.3a Phase diagram of CO_2. Vapor pressure P of CO_2 versus temperature T above the solid (dark gray circles) and above the liquid (light gray circles). Phase transition from solid to liquid (medium gray circle and medium gray line). Calculation using the Clausius–Clapeyron equation with $\Delta_{sub}H = 26.1$ kJ mol^{-1} and $\Delta_{vap}H = 16.5$ kJ mol^{-1} (dark gray and light gray solid lines).

b) From the Clausius–Clapeyron equation

$$\ln\left(\hat{P}\right) = -\frac{\Delta_{sub}H}{\boldsymbol{R}}\frac{1}{T} + const$$

we expect that a plot of the sublimation data in the form $\ln(\hat{P})$ versus $1/T$ should be linear. Such a plot of the data is shown in Fig. P19.3b and we fit the data by a line of the form

$$\ln(\hat{P}) = 16.175 - \frac{3144.6}{T/\text{K}}$$

A comparison with the Clausius–Clapeyron equation and its slope gives the sublimation enthalpy as

$$\Delta_{sub}H = (3144 \times 8.314)\ \text{J mol}^{-1} = 26.1\ \text{kJ mol}^{-1}$$

1 The density data are from the International Critical Tables, Volume 3 (McGraw Hill, 1928)

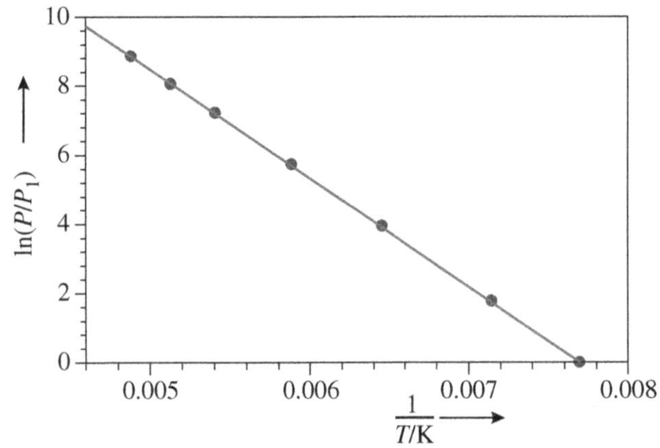

Figure P19.3b $\ln(P/P_1)$ (where P is the vapor pressure) versus $1/T$ for the sublimation data of CO_2. Experiment (light gray circles) and regression line (dark gray solid line).

For the vaporization data we proceed in a corresponding way. We write the Clausius–Clapeyron equation as

$$\ln\left(\hat{P}\right) = -\frac{\Delta_{\text{vap}}H}{\boldsymbol{R}}\frac{1}{T} + const$$

and plot the vaporization data in the form $\ln(\hat{P})$ versus $1/T$, which should be linear. Such a plot of the data is shown in Fig. P19.3c and we fit the data by a line of the form

$$\ln(\hat{P}) = 10.831 - \frac{1988.0}{T/\text{K}}$$

A comparison with the Clausius–Clapeyron equation and its slope gives the vaporization enthalpy as

$$\Delta_{\text{vap}}H = (1988 \times 8.314)\ \text{J mol}^{-1} = 16.5\ \text{kJ mol}^{-1}$$

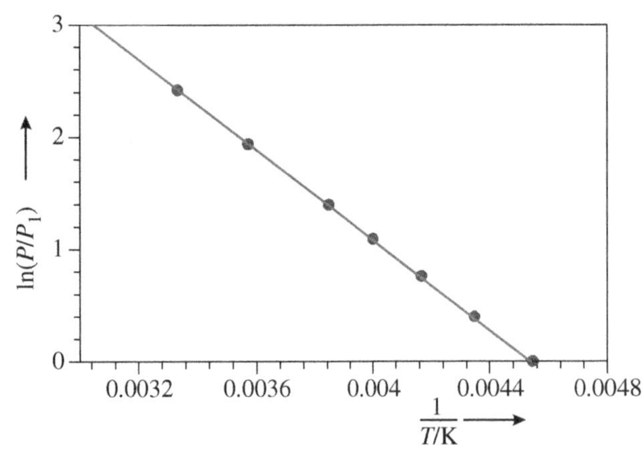

Figure P19.3c $\ln(P/P_1)$ (where P is the vapor pressure) versus $1/T$ for the vaporization data of CO_2. Experiment (light gray circles) and regression line (dark gray solid line).

c) Without the assumption that $V_{\text{gas}} \gg V_{\text{liq}}$, we use the data given to evaluate the volume change at each temperature. For $\boldsymbol{M}(CO_2) = 0.04401$ kg mol^{-1}, we can evaluate $\Delta_{\text{vap}}V$ at the temperatures specified by

$$\Delta_{\text{vap}}V = V_{m,\text{gas}} - V_{m,\text{liq}}$$
$$= \left(\frac{1}{\rho_{\text{gas}}} - \frac{1}{\rho_{\text{liquid}}}\right) \cdot \boldsymbol{M}$$

For example, at 243.15 K we find that

$$\Delta_{vap}V = V_{m,gas} - V_{m,liq}$$
$$= 0.04401\ \text{kg mol}^{-1} \cdot \left(\frac{1}{37.1} - \frac{1}{1066.9}\right)\frac{\text{m}^3}{\text{kg}} = 11.45 \cdot 10^{-4}\frac{\text{m}^3}{\text{mol}}$$

Proceeding in like manner for the other temperatures we obtain the values in the table.

T/K	243.15	253.15	263.15	273.15	283.15	293.15
$\Delta_{vap}V/10^{-4}\frac{\text{m}^3}{\text{mol}}$	11.45	8.156	5.828	4.103	2.797	1.760

If we fit these data to a quadratic form (Fig. P19.3d)

$$\Delta_{vap}V = 2.74518 \cdot 10^{-7}x^2 - 1.6615310^{-4}x + 2.53031 \cdot 10^{-2}$$

then we can interpolate and find $\Delta_{vap}V$ at the temperatures given for the phase transition.

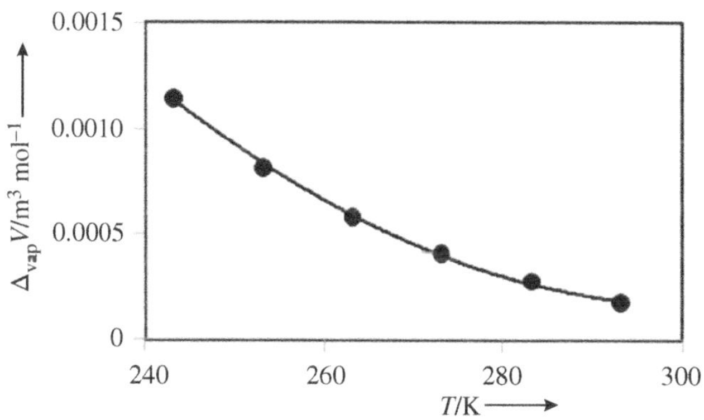

Figure P19.3d $\Delta_{vap}V$ versus temperature T for CO_2.

In a corresponding way we use our fit of the ln ($\hat{P}$) versus T data; to determine the value of the slope as

$$\frac{d\hat{P}}{dT} = \frac{1988.0}{(T/\text{K})^2}\hat{P}$$

In this way we can directly evaluate $\Delta_{vap}H$ from the data, via

$$\Delta_{vap}H = \left(T \cdot \Delta_{vap}V\right)\frac{dP}{dT}$$
$$= \left(\Delta_{vap}V/\text{m}^3\ \text{mol}^{-1}\right) \cdot \frac{1988.0}{(T/\text{K})} \cdot P$$

For 240 K, we find $\Delta_{vap}V = 1.239 \cdot 10^{-3}\text{m}^3\ \text{mol}^{-1}$ and $P = 12.825$ bar, thus

$$\Delta_{vap}H = \left|1.239 \cdot 10^{-3}\ \text{m}^3\ \text{mol}^{-1} \cdot \frac{1988.0}{240} \cdot 12.825\ \text{bar}\right|$$
$$= 0.1316\frac{\text{m}^3\ \text{bar}}{\text{mol}}10^5\frac{\text{Pa}}{\text{bar}} = 13.16\ \text{kJ mol}^{-1}$$

Proceeding in a like manner for the other temperatures, we find the values in the table.

Vaporization				
T (K)	240	250	260	280
P/bar	12.825	17.85	24.19	41.61
$\Delta_{vap}V/10^{-4}\frac{\text{m}^3}{\text{mol}}$	12.39	9.222	6.607	3.025
$\Delta_{vap}H$/kJ mol^{-1}	13.16	13.09	12.22	8.94

These values for the enthalpy of vaporization are in excellent agreement with the literature values of 10.32 kJ mol^{-1} at 273 K and 13.26 kJ mol^{-1} at 244 K.

P19.4 You observe that a certain liquid has a vapor pressure of 1 bar at 300 K; i.e., it boils. In addition the liquid has a molar volume of $V_{m,\text{liquid}} = 0.150\ \text{L mol}^{-1}$, the compression factor of the vapor above the liquid is $\Phi = PV/RT = 0.95$ and $dP/dT = 0.026\ \text{bar K}^{-1}$. Determine the enthalpy of vaporization by assuming that the volume of the liquid may be neglected. Determine the enthalpy of vaporization of the liquid without any assumptions.

Solution

We start with

$$\frac{dP}{dT} = \frac{\Delta_{\text{vap}}H}{T \cdot \Delta_{\text{vap}}V}$$

For $V_{\text{gas}} \gg V_{\text{liquid}}$ we have $\Delta_{\text{vap}}V \approx V_{\text{gas}} = \Phi RT/P$ which leads to

$$\frac{dP}{dT} = \frac{\Delta_{\text{vap}}H}{T \cdot \Phi \boldsymbol{R}T/P} = \frac{\Delta_{\text{vap}}H \cdot P}{T^2 \cdot \Phi \boldsymbol{R}}$$

Solving this equation for the enthalpy of vaporization gives

$$\begin{aligned}\Delta_{\text{vap}}H &= \frac{dP}{dT}\frac{\Phi \boldsymbol{R}T^2}{P} = 0.026\ \text{bar K}^{-1}\frac{0.95 \cdot 8.314\ \text{J K}^{-1}\ \text{mol}^{-1} \cdot 300^2\ \text{K}^2}{1.0\ \text{bar}} \\ &= 18.48\ \text{kJ mol}^{-1}\end{aligned}$$

For the second part we have

$$V_{m,\text{gas}} = \frac{\Phi \boldsymbol{R}T}{P} = \frac{0.95 \cdot 8.314 \times 10^{-2}\ \text{bar L} \cdot 300}{1.0\ \text{bar}} = 23.69\ \text{L mol}^{-1}$$

so that

$$\Delta_{\text{vap}}V = V_{m,\text{gas}} - V_{m,\text{liquid}} = (23.69 - 0.15)\ \text{L mol}^{-1} = 23.54\ \text{L mol}^{-1}$$

Hence we can directly calculate the enthalpy of vaporization and find

$$\begin{aligned}\Delta_{\text{vap}}H &= \frac{dP}{dT}T \cdot \Delta_{\text{vap}}V \\ &= 0.026\ \text{bar K}^{-1} \cdot 300\ \text{K} \cdot 23.54\ \text{L mol}^{-1} \cdot 10^2\ \text{J bar}^{-1}\ \text{L}^{-1} = 18.36\ \text{kJ mol}^{-1}\end{aligned}$$

P19.5 You plan to measure the spectrum of I_2 vapor in order to calibrate your spectrometer. You wish to have an absorbance of 1.0 or less for the I_2 vapor in your optical cell. If the maximum absorption cross-section[2] of I_2 is $\sigma = 4.84 \times 10^{-18}\text{cm}^2$ and your optical cell's path length is $d = 10$ cm, what number density of I_2 will you need to have? Assuming the gas is ideal, what pressure of I_2 will you need? Given that I_2 is a solid under standard conditions, what temperature is necessary to provide your desired vapor pressure ($\Delta_f G(I_2(\text{g})) = 19.3$ kJ mol^{-1} and $\Delta_{\text{sub}}H = 62.4$ kJ mol^{-1} at 298 K and 1 bar)? Assume that I_2 vapor behaves ideally and that $\Delta_f H$ is independent of T.

Solution

Recall that the absorbance A is given by

$$A = \sigma \cdot \rho \cdot d \text{ where } \rho = \frac{N}{V} \text{ is the number density}$$

Solving for the number density and using the parameters above, we find that

$$\begin{aligned}\rho &= \frac{A}{\sigma \cdot d} \\ &= \frac{1.0}{4.84 \times 10^{-18}\ \text{cm}^2 \times 10\ \text{cm}} = 2.07 \times 10^{16}\ \text{cm}^{-3} = 2.07 \times 10^{22}\ \text{m}^{-3}\end{aligned}$$

That is, a number density of $2.07 \times 10^{22}\ \text{m}^{-3}$ I_2 gas phase molecules should give an absorbance of 1.0 for a 10 cm cell.

2 This cross-section is determined for a wavelength of 533 nm (a Fourier spectrometer measurement at a resolution of 0.1 nm) and is taken from A. Saiz-Lopez, R.W. Saunders, D.M. Joseph, S.H. Ashworth, J.M.C. Plane *Atmos. Chem. Phys. Discuss.* 2004, **4** 2379.

From the ideal gas law, we can relate this density to a pressure as a function of the temperature T, i.e., we use $P = NkT/V = \rho kT$ to write

$$P = 2.07 \times 10^{22}\ \text{m}^{-3} \cdot 1.38 \times 10^{-23}\ \text{J K}^{-1} \cdot T = 0.286\ \text{J K}^{-1}\text{m}^{-3} \cdot T$$
$$= 0.286\ \text{N m}^{-2}\ \text{K}^{-1} \cdot T = 0.286 \times 10^{-5}\ \text{bar K}^{-1} \cdot T$$

which is the pressure at temperature T that will have the correct density of molecules. For example, at room temperature $T = 298$ K we obtain

$$P = 0.286 \times 10^{-5}\ \text{bar K}^{-1} \cdot 298\ \text{K} = 0.85\ \text{mbar}$$

The vapor pressure can be evaluated from the Gibbs energy data. At 298 K we find that

$$\hat{P}_{\text{vap}} = \exp\left(-\Delta_f G/\boldsymbol{R}T\right) = \exp\left(\frac{-19.3 \times 10^3}{8.3145 \times 298}\right) = 4.13 \times 10^{-4}$$

which corresponds to $P_{\text{vap}}(298\ \text{K}) = 4.13 \times 10^{-4}$ bar $= 0.413$ mbar.
If we use the Clausius–Clapeyron equation, we can find how the vapor pressure changes with temperature; namely

$$\ln\left(\frac{P_2}{P_1}\right) = -\frac{\Delta_{\text{sub}}H}{\boldsymbol{R}}\left(\frac{1}{T_2} - \frac{1}{T_1}\right)$$

we have

$$\ln\left(\frac{0.286 \times 10^{-5}\ \text{bar K}^{-1} \cdot T}{4.13 \times 10^{-4}\ \text{bar}}\right) = -\frac{62,400\ \text{J mol}^{-1}}{8.314\ \text{J mol}^{-1}\ \text{K}^{-1}}\left(\frac{1}{T} - \frac{1}{298\ \text{K}}\right)$$
$$\ln(0.00693 \cdot T/\text{K}) = -7505\ \text{K} \cdot \frac{1}{T} + 25.2$$

and

$$\ln(T/\text{K}) = -7505\ \text{K} \cdot \frac{1}{T} + 30.2$$

which reduces to the equation

$$y = \ln(T/\text{K}) + \frac{7505}{T/\text{K}} - 30.16 = 0$$

This equation can be solved using Excel or Mathcad. Another way is to solve it graphically by plotting y versus T (Fig. P19.5) and identifying the temperature at which the curve crosses the zero line.

Figure P19.5 $y = \ln(T/\text{K}) + \frac{7505}{T/\text{K}} - 30.16$ versus T.

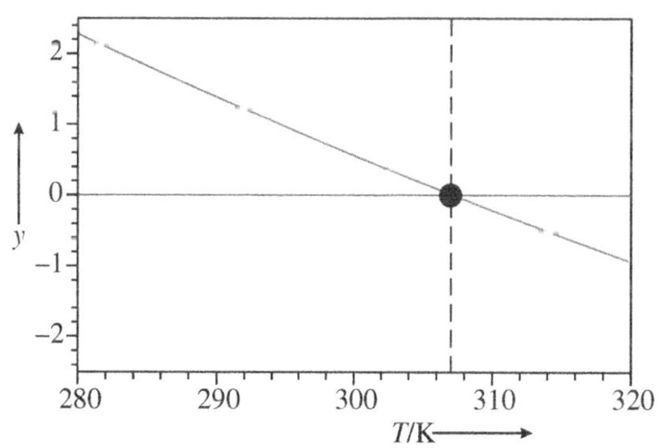

The two lines cross near 307 K = 34 °C.
At this temperature we obtain

$$P = 0.286 \times 10^{-5}\ \text{bar K}^{-1} \cdot 307\ \text{K} = 0.87\ \text{mbar}$$

P19.6 Use a Maxwell relation to show that a phase transition that has a discontinuity in the volume must also have a discontinuity in its entropy, and hence have a latent heat.

Solution

We use the Maxwell relation

$$\left(\frac{\partial S}{\partial V}\right)_T = \left(\frac{\partial P}{\partial T}\right)_V$$

which upon rearrangement gives

$$(\mathrm{d}S)_T = \left(\frac{\partial P}{\partial T}\right)_V (\mathrm{d}V)_T \text{ or } \Delta S_T = \left(\frac{\partial P}{\partial T}\right)_V \cdot \Delta V_T$$

The first term on the right-hand side of the equation is continuous; however, the second term corresponds to the volume change and is discontinuous. Hence, ΔS_T must be discontinuous.

Using the fact that

$$\mathrm{d}S = \frac{\mathrm{d}q}{T}$$

and T is fixed at the phase transition, we find that

$$(\Delta S)_T = \int \frac{\mathrm{d}q}{T} = \frac{1}{T}\int \mathrm{d}q = \frac{\Delta H}{T}$$

Hence the discontinuity in the entropy gives rise to a latent heat, enthalpy for the transition.

P19.7 Because the enthalpies $H_{m,c}$ and $H_{m,v}$ of two phases are state functions, their difference is also. Use this fact to show that the heat of vaporization of a substance may be written as

$$\frac{\mathrm{d}\left(\Delta_{\text{vap}}H\right)}{\mathrm{d}T} = \Delta C_{P,m} + \frac{\Delta_{\text{vap}}H}{T} - \frac{\Delta_{\text{vap}}H}{\Delta V_m}\left(\frac{\partial \Delta V_m}{\partial T}\right)_P$$

where $\Delta V_m = V_{v,m} - V_{c,m}$ and $\Delta C_{P,m} = C_{P,m,v} - C_{P,m,c}$

Solution

The differential $\mathrm{d}\left(\Delta_{\text{vap}}H\right)$ is

$$\begin{aligned}\mathrm{d}\left(\Delta_{\text{vap}}H\right) &= \left(\frac{\partial\left(H_{m,v} - H_{m,c}\right)}{\partial T}\right)_P \mathrm{d}T + \left(\frac{\partial\left(H_{m,v} - H_{m,c}\right)}{\partial P}\right)_T \mathrm{d}P \\ &= \left(C_{P,m,v} - C_{P,m,c}\right)\mathrm{d}T + \left[\left(\frac{\partial H_{m,v}}{\partial P}\right)_T - \left(\frac{\partial H_{m,c}}{\partial P}\right)_T\right]\mathrm{d}P\end{aligned}$$

Substituting for the pressure derivative with Equation (19.22), we find that

$$\mathrm{d}\left(\Delta_{\text{vap}}H\right) = \Delta C_{P,m} \cdot \mathrm{d}T + \left[\Delta V_m - T\left(\frac{\partial \Delta V_m}{\partial T}\right)_P\right]\mathrm{d}P$$

Now we use the Clapeyron equation, $\mathrm{d}P = \frac{\Delta_{\text{vap}}H}{T\cdot\Delta V_m}\mathrm{d}T$, and substitute for $\mathrm{d}P$ to find

$$\begin{aligned}\mathrm{d}\left(\Delta_{\text{vap}}H\right) &= \Delta C_{P,m} \cdot \mathrm{d}T + \left[\Delta V_m - T\left(\frac{\partial \Delta V_m}{\partial T}\right)_P\right]\frac{\Delta_{\text{vap}}H}{T\cdot\Delta V_m}\mathrm{d}T \\ &= \left[\Delta C_{P,m} + \frac{\Delta_{\text{vap}}H}{T} - \left(\frac{\partial \Delta V_m}{\partial T}\right)_P \frac{\Delta_{\text{vap}}H}{\Delta V_m}\right]\mathrm{d}T\end{aligned}$$

Simplification of this result gives

$$\frac{\mathrm{d}\left(\Delta_{\text{vap}}H_m\right)}{\mathrm{d}T} = \Delta C_{P,m} + \frac{\Delta_{\text{vap}}H}{T} - \frac{\Delta_{\text{vap}}H}{\Delta V_m}\left(\frac{\partial \Delta V_m}{\partial T}\right)_P$$

which is the requested result.

P19.8 Explain what happens to the constant pressure heat capacity of a system undergoing a first-order phase transition. Provide graphs of the Gibbs energy versus T (as in Fig. 19.2), H versus T, and C_P versus T, and a verbal explanation.

Solution

The constant pressure heat capacity is well-defined before the phase transition and after the phase transition, but is not well-defined at the phase transition. Because

$$C_P = \left(\frac{\partial H}{\partial T}\right)_P$$

and the fact that the enthalpy of the system changes at the phase transition while its temperature does not, we expect that the constant pressure heat capacity will become infinite at the phase transition This conclusion is evident from the diagrams shown in Fig. P19.8. The first diagram corresponds to Fig. 19.2 but for generic phases and we have drawn the physically realized Gibbs energy curve in dark gray. That is the Gibbs energy curve for the liquid is followed at low temperature, whereas the Gibbs energy curve for the gas is followed at high temperature. We draw the corresponding enthalpy as a function of temperature for the two phases in the middle diagram. At the phase transition temperature the enthalpy changes discontinuously; it jumps from one value to another and this event is marked by the two black points in the figure. The final diagram shows the heat capacity which can be found from the slope of the H versus T plot at each T. The slope is well-defined at each point other than the phase transition, which corresponds to the vertical arrow and would have an infinite slope.

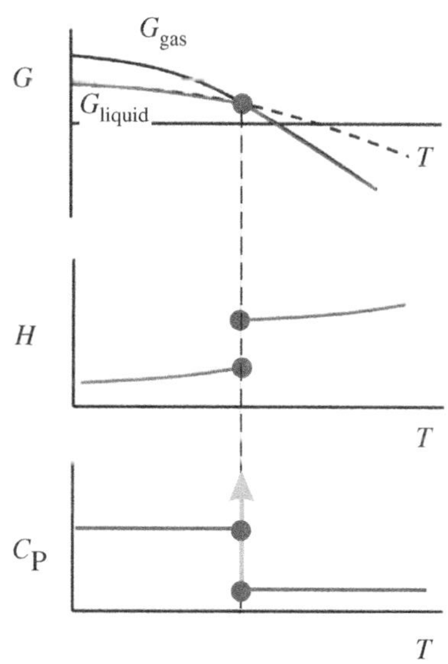

Figure P19.8 Top: Gibbs energy G for the liquid and the gas versus temperature T (black), experimentally realized energy curve (dark gray). Middle: Enthalpy H for the liquid and the gas versus temperature T (dark gray). Bottom: Heat capacity C_P of the liquid and the gas versus temperature T (dark gray). Note the jump of C_P to minus infinity at the phase transition (medium gray arrow).

P19.9 Use the fundamental expression for the Helmholtz free energy (Equation (19.34))

$$A = U - TS$$

and the results for U and S from Problem P17.14 to show that A for a photon gas is

$$A = -\frac{8\pi^1}{12.897}\frac{Vk^1}{c_0^3h^3}T^4$$

Then use this expression for A to find the radiation pressure P of the photon gas and deduce the equation of state for the photon gas. Lastly, calculate the radiation pressure created by an electric filament (incandescent light bulb filament) at $T = 1000$ K.

Solution

In Problem P17.14 it is shown that

$$S = Nk + \frac{8\pi}{c_0^3}V\left(\frac{k}{h}\right)^4\frac{\pi^4}{15}T^3 \quad \text{and} \quad U = \frac{8\pi}{c_0^3}V\left(\frac{kT}{h}\right)^4\frac{\pi^4}{15}$$

From Equation (19.34) we can write that

$$A = U - TS$$
$$= \frac{8\pi}{c_0^3} V \left(\frac{kT}{h}\right)^4 \frac{\pi^4}{15} - T\left[Nk + \frac{8\pi}{c_0^3} V \left(\frac{k}{h}\right)^4 \frac{\pi^4}{15} T^3\right] = -NkT$$

Also in Problem 17.14, we showed that the number N of photons is

$$N = \frac{8\pi}{c_0^3} V \left(\frac{kT}{h}\right)^3 \frac{\pi^3}{12.897} = \frac{8\pi^4}{12.897 c_0^3} V \left(\frac{kT}{h}\right)^3$$

Thus for the Helmholtz free energy of the photon gas we obtain

$$A = -NkT = -\frac{8\pi^4}{12.897 c_0^3} V \left(\frac{kT}{h}\right)^3 kT$$
$$= -\frac{8\pi^4}{12.897} \frac{Vk^4}{c_0^3 h^3} T^4$$

To find the radiation pressure P, we use the fact that, according to Equation (19.38),

$$\mathrm{d}A = \left(\frac{\partial A}{\partial T}\right)_V \mathrm{d}T + \left(\frac{\partial A}{\partial V}\right)_T \mathrm{d}V = -S\,\mathrm{d}T - P\mathrm{d}V$$

so that we can write

$$P = -\left(\frac{\partial A}{\partial V}\right)_T = \frac{\partial}{\partial V}\left[\frac{8\pi^4}{12.897} \frac{Vk^4}{c_0^3 h^3} T^4\right]$$
$$= \frac{8\pi^4}{12.897} \frac{k^4}{c_0^3 h^3} T^4 = \frac{8\pi^4}{12.897\, c_0^3} kT \left(\frac{kT}{h}\right)^3$$

If we use our result for the photon number N, namely

$$\frac{N}{V} = \frac{8\pi^4}{12.897\, c_0^3} \left(\frac{kT}{h}\right)^3 \qquad \left(\frac{kT}{h}\right)^3 = \frac{N}{V} \frac{12.897\, c_0^3}{8\pi^4}$$

then P becomes

$$P = \frac{8\pi^4}{12.897\, c_0^3} kT \cdot \frac{N}{V} \frac{12.897\, c_0^3}{8\pi^4} = \frac{NkT}{V}$$

This result is identical with the corresponding expression for an ideal atomic gas. While photons interact with matter, they do not interact directly with each other and the ideal gas equation of state should apply.

For an electric bulb ($T = 1000$ K) we calculate

$$P = \frac{8\pi^4}{12.897} \frac{\left(1.38 \times 10^{-23}\ \mathrm{J\ K^{-1}}\right)^4}{\left(2.998 \times 10^8\ \mathrm{m\ s^{-1}}\right)^3 \left(6.626 \times 10^{-34}\ \mathrm{J\ s}\right)^3} 298^4\ \mathrm{K}^4 = 2.20 \times 10^{-6}\ \mathrm{Pa}$$

This is an extremely low pressure.

For comparison, let us compare this value to the pressure exerted by an industrial CO_2 laser ($\lambda = 5\ \mu$m) with a mean power of 8 kW on a wall with area A. First we calculate the force f exerted by the photons on the wall

$$f = \Delta m \frac{\mathrm{d}v}{\mathrm{d}t} = \Delta N\, m_{\mathrm{photon}} \frac{\Delta v}{\Delta t} = \Delta N\, m_{\mathrm{photon}} \frac{2c_0}{\Delta t} = 2c_0 m_{\mathrm{photon}} \frac{\Delta N}{\Delta t}$$

where we have used the fact that $\Delta v = c_0 - (-c_0) = 2c_0$ because of the reflection of the photons on the wall. The pressure P is

$$P = \frac{f}{A} = \frac{2c_0 m_{\mathrm{photon}}}{A} \frac{\Delta N}{\Delta t}$$

where A is the area of the wall. According to Equation (1.9) the photon mass is

$$m_{\mathrm{photon}} = \frac{h}{c_0 \lambda} = \frac{h\nu}{c_0^2}$$

Thus

$$P = \frac{2c_0}{A}\frac{h\nu}{c_0^2}\frac{\Delta N}{\Delta t} = \frac{2}{A}\frac{h\nu}{c_0}\frac{\Delta N}{\Delta t}$$

Because the laser beam power is given by

$$power = \frac{\Delta N}{\Delta t}h\nu$$

we can write the pressure P as

$$P = \frac{2}{A}\frac{h\nu}{c_0}\frac{power}{h\nu} = \frac{2}{c_0}\frac{power}{A}$$

Using the power of 8 kW= 8×10^3 J s^{-1} and $A = 1$ mm^2 $= 1 \times 10^{-6}$ m^2, we find

$$P = \frac{2}{2.998 \times 10^8 \text{ m s}^{-1}}\frac{8 \times 10^3 \text{ J s}^{-1}}{1 \times 10^{-6} \text{ m}^2} = 53 \text{ N m}^{-2} = 53 \text{ Pa}$$

We see that even under these conditions the radiation pressure is rather low.

See also: H.S. Leff, Teaching the photon gas in introductory physics, *Am. J. Phys.* 2002, **70**, 792–797.

P19.10 Beginning with the definition of the enthalpy and the combined form of the first and second laws of thermodynamics, derive an expression for $(\partial H/\partial P)_T$.

Solution

Given Equation (19.20) in Section 19.2.4

$$\mathrm{d}H = T\mathrm{d}S + V\mathrm{d}P$$

we arrange it to solve for dS and find

$$\mathrm{d}S = \frac{1}{T}\mathrm{d}H - \frac{V}{T}\mathrm{d}P$$

Inserting the definition of the differential enthalpy, we find

$$\begin{aligned}\mathrm{d}S &= \frac{1}{T}\left(\frac{\partial H}{\partial T}\right)_P \mathrm{d}T + \frac{1}{T}\left(\frac{\partial H}{\partial P}\right)_T \mathrm{d}P - \frac{V}{T}\mathrm{d}P \\ &= \frac{1}{T}\left(\frac{\partial H}{\partial T}\right)_P \cdot \mathrm{d}T + \frac{1}{T}\left[\left(\frac{\partial H}{\partial P}\right)_T - \frac{V}{T}\right] \cdot \mathrm{d}P\end{aligned}$$

The first term in the expression for dS corresponds to $(\partial S/\partial T)_P$ and the second term corresponds to $(\partial S/\partial P)_T$; namely

$$\left(\frac{\partial S}{\partial T}\right)_P = \frac{1}{T}\left(\frac{\partial H}{\partial T}\right)_P \text{ and } \left(\frac{\partial S}{\partial P}\right)_T = \frac{1}{T}\left[\left(\frac{\partial H}{\partial P}\right)_T - V\right]$$

Now we evaluate the cross-derivatives to obtain

$$\frac{\partial^2 S}{\partial P \partial T} = \frac{1}{T}\left(\frac{\partial^2 H}{\partial P \partial T}\right)$$

$$\frac{\partial^2 S}{\partial T \partial P} = -\frac{1}{T^2}\left[\left(\frac{\partial H}{\partial P}\right)_T - V\right] + \frac{1}{T}\left[\frac{\partial^2 H}{\partial T \partial P} - \left(\frac{\partial V}{\partial T}\right)_P\right]$$

Since S is a state function we know that the cross-derivatives must be equal (dS is an exact differential), hence

$$\frac{\partial^2 S}{\partial P \partial T} - \frac{\partial^2 S}{\partial T \partial P} = \frac{1}{T}\frac{\partial^2 H}{\partial P \partial T} + \frac{1}{T^2}\left[\left(\frac{\partial H}{\partial P}\right)_T - V\right] - \frac{1}{T}\left[\frac{\partial^2 H}{\partial T \partial P} - \left(\frac{\partial V}{\partial T}\right)_P\right] = 0$$

H is a state function, thus

$$\frac{\partial^2 H}{\partial P \partial T} = \frac{\partial^2 H}{\partial T \partial P}$$

and it follows that

$$\frac{1}{T^2}\left[\left(\frac{\partial H}{\partial P}\right)_T - V\right] + \frac{1}{T}\left(\frac{\partial V}{\partial T}\right)_P = 0$$

or

$$\left(\frac{\partial H}{\partial P}\right)_T = -T\left(\frac{\partial V}{\partial T}\right)_P + V$$

P19.11 Calculate the vapor pressure of water in the range of $T = 373$ K to 500 K using the refined Equation (19.62) (data see Example 19.6 and 19.7).

Solution

Equation (19.62) is

$$\ln \hat{P}_{\text{vap}} = \ln \hat{P}_{\text{vap},1} - \frac{\Delta_{\text{vap}} H_1}{R}\left(\frac{1}{T} - \frac{1}{T_1}\right) + \frac{\Delta C_{P,m}}{R}\left[\ln \frac{T}{T_1} + T_1\left(\frac{1}{T} - \frac{1}{T_1}\right)\right]$$

with $\hat{P}_{\text{vap},1} = 1$ at $T_1 = 373$ K, and using the values of $\Delta_{\text{vap}}H^{\ominus} = (-241.8 + 285.9)$ kJ mol^{-1} = 44.1 kJ mol^{-1} and $\Delta C_P = (33.6 - 75.3)$ J K^{-1} mol^{-1} = -41.7 J K^{-1}mol^{-1} from Example 19.6 and 19.7. The data are plotted in Fig. P19.11. The calculated values reproduce the experimental data much better than in the simpler approximation (Equation (19.61)), even though the two high temperature values are significantly lower than calculated. This deviation is due to the fact that in Equation (19.62) the vapor is treated as an ideal gas.

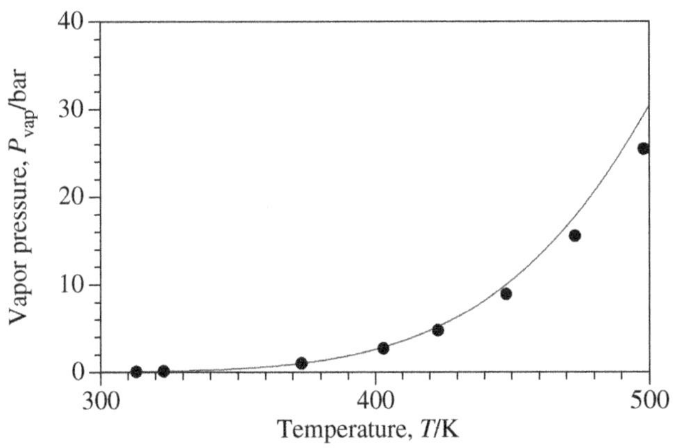

Figure P19.11 Vapor pressure of water in the range 300–500 K. Dark gray solid line: Calculated using Equation (19.62), full circles: experiment (data see Appendix A.5).

P19.12 Derive the equation for the vapor pressure curve from the Clausius–Clapeyron equation in Section 19.4.2 by accounting for the temperature dependence of $\Delta_{\text{vap}}H$. In particular, assume that the $\Delta_{\text{vap}}H$ at any temperature T is described by the linear equation

$$\Delta_{\text{vap}}H = \Delta_{\text{vap}}H_1 + \Delta C_P \cdot (T - T_1)$$

where $\Delta_{\text{vap}}H_1$ is the molar enthalpy of vaporization at the temperature T_1 and is the difference in the molar constant pressure heat capacities at T and T_1.

Solution

We begin with the Clausius–Clapeyron equation

$$\frac{\mathrm{d}\ln \hat{P}}{\mathrm{d}T} = \frac{\Delta_{\text{vap}}H}{RT^2}$$

and solve for d ln $\hat{P}$.

$$\mathrm{d}\ln \hat{P} = \frac{\Delta_{\text{vap}}H}{RT^2} \cdot \mathrm{d}T$$

Inserting the expression for $\Delta_{\text{vap}}H(T)$, we find.

$$\mathrm{d}\ln \hat{P} = \frac{\Delta_{\text{vap}}H_1 + \Delta C_P \cdot (T - T_1)}{RT^2} \cdot \mathrm{d}T$$
$$= \frac{\Delta_{\text{vap}}H_1}{RT^2} \cdot \mathrm{d}T + \frac{\Delta C_P \cdot T}{RT^2} \cdot \mathrm{d}T - \frac{\Delta C_P \cdot T_1}{RT^2} \cdot \mathrm{d}T$$

We integrate the three terms:

$$\int_{T_1}^{T} \frac{\Delta_{\text{vap}}H_1}{RT^2} \cdot dT = -\frac{\Delta_{\text{vap}}H_1}{RT} + \frac{\Delta_{\text{vap}}H_1}{RT_1} = -\frac{\Delta_{\text{vap}}H_1}{R}\left(\frac{1}{T} - \frac{1}{T_1}\right)$$

$$\int_{T_1}^{T} \frac{\Delta C_P \cdot T}{RT^2} \cdot dT = \frac{\Delta C_P}{R} \ln \frac{T}{T_1}$$

$$\int_{T_1}^{T} \frac{\Delta C_P \cdot T_1}{RT^2} \cdot dT = -\frac{\Delta C_P \cdot T_1}{RT} + \frac{\Delta C_P \cdot T_1}{RT_1} = \frac{\Delta C_P \cdot T_1}{R}\left(-\frac{1}{T} + \frac{1}{T_1}\right)$$

Thus we obtain

$$\ln \hat{P} = \ln \hat{P}_1 - \frac{\Delta_{\text{vap}}H_1}{R}\left(\frac{1}{T} - \frac{1}{T_1}\right) + \frac{\Delta C_P}{R} \ln \frac{T}{T_1} + \frac{\Delta C_P \cdot T_1}{R}\left(\frac{1}{T} - \frac{1}{T_1}\right)$$

and

$$\ln \hat{P} = \ln \hat{P}_1 - \frac{\Delta_{\text{vap}}H_1}{R}\left(\frac{1}{T} - \frac{1}{T_1}\right) + \frac{\Delta C_P}{R} \cdot \left[\ln \frac{T}{T_1} + T_1\left(\frac{1}{T} - \frac{1}{T_1}\right)\right]$$

P19.13 In Example 19.2 we calculate the change in enthalpy for the ammonia gas undergoing a transition between the equilibrium state of 1 bar and 400 K to 51 bar and 500 K. In that calculation we used the mean value of the heat capacity, molar volume, and temperature dependence of the molar volume. In this problem you will account for the full temperature and pressure dependence of those quantities to find the enthalpy change. Over this pressure range, the molar volume can be described by

$$\overline{V}_m = \left(\frac{41.573}{P/\text{bar}} - 0.0638\right) \text{ L mol}^{-1}$$

and its temperature dependence by

$$\left(\overline{dV_m/dT}\right)_P = \left(\frac{0.0831}{P/\text{bar}} + 5.83 \times 10^{-4}\right) \text{ L K}^{-1} \text{ mol}^{-1}$$

The temperature dependence of the constant pressure heat capacity can be described by a linear equation from 400 to 500 K.

$$C_{P,m} = (26.661 + 0.0307 \cdot T/\text{K}) \text{ J mol}^{-1} \text{ K}^{-1}$$

Compare your result to that reported in Example 19.2.

Solution

Similar to Example 19.2, we follow a two-step path. In the first step, we heat the gas at constant pressure from 400 to 500 K to find ΔH_1, and in the second step we perform an isothermal compression of the liquid from 1 bar to 51 bar at 500 K to find ΔH_2.

Step 1: Because the first step occurs at constant pressure $(dH)_P = C_P \cdot dT$, so that

$$\Delta H_{1,m} = \int_{400\text{ K}}^{500\text{ K}} C_{P,m} \cdot dT = \int_{400\text{ K}}^{500\text{ K}} (26.661 \text{ J mol}^{-1}\text{ K}^{-1} + 0.0307 \text{ J mol}^{-1}\text{ K}^{-2} \cdot T) \cdot dT$$

$$= \left[26.661 \cdot (100) + \frac{0.0307}{2}\left(500^2 - 400^2\right)\right] \text{ J mol}^{-1} = 4.048 \text{ kJ mol}^{-1}$$

This value is nearly identical to that found in Step 1 for Example 19.2.

Step 2: Because this step occurs at constant temperature, $(dH) = \left[V - T\left(\frac{\partial V}{\partial T}\right)_P\right] dP$, so that

$$\Delta H_{2,m} = \int_{1\text{ bar}}^{51\text{ bar}} V_m \, dP - T \int_{1\text{ bar}}^{51\text{ bar}} \left(\frac{\partial V_m}{\partial T}\right)_P dP$$

$$= \int_{1\text{ bar}}^{51\text{ bar}} \left(\frac{a_1}{P} - b_1\right) dP - T \int_{1\text{ bar}}^{51\text{ bar}} \left(\frac{a_2}{P} + b_2\right) dP$$

$$= |a_1 \ln P - b_1 P|_{1\text{bar}}^{51\text{bar}} - T\,|a_2 \ln P + b_2 P|_{1\text{bar}}^{51\text{bar}}$$

$$= \left[a_1 \ln \frac{51}{1} - b_1 (51 - 1) \text{ bar}\right] - T\left[a_2 \ln \frac{51}{1} + b_2 (51 - 1) \text{ bar K}^{-1}\right]$$

$$
\begin{aligned}
&= [41.573 \cdot 3.930 - 0.0638 \cdot 50] \text{ bar L mol}^{-1} \\
&\quad -T \left[0.0831 \cdot 3.930 + 5.83 \times 10^{-4} \cdot 50\right] \text{ bar L mol}^{-1} \\
&= [163.4 - 3.19] \text{ bar L mol}^{-1} - 500 \cdot [0.3266 + 0.0292] \text{ bar L mol}^{-1} \\
&= 160.2 \text{ bar L mol}^{-1} - 500 \cdot 0.3558 \text{ bar L mol}^{-1} \\
&= [160.2 - 177.9] \text{ bar L mol}^{-1} = -17.70 \text{ bar L mol}^{-1} = -1.77 \text{ kJ mol}^{-1}
\end{aligned}
$$

Combining these two contributions, we find that the total enthalpy change ΔH_{total} is

$$
\begin{aligned}
\Delta H_{\text{total},m} &= \Delta H_{1,m} + \Delta H_{2,m} \\
&= 4.048 \text{ kJ mol}^{-1} + 1.77 \text{ kJ mol}^{-1} = 2.28 \text{ kJ mol}^{-1}
\end{aligned}
$$

This value is in better agreement with the experimental result than is the value found by the more approximate method of Example 19.2. Note that the calculation of $\Delta H_{2,m}$ generates a small difference of two big numbers, so that a relatively small (<10%) error in one of the large numbers might result in a great relative error for the small difference.

20

Real Gases

20.1 Exercises

E20.1 Use the Thermodynamic Equation of State,

$$\boxed{\left(\frac{\partial U}{\partial V}\right)_T = T\left(\frac{\partial P}{\partial T}\right)_V - P,}$$

to show that the internal energy of a hard sphere gas depends only on T. Discuss your result and compare it to that for an ideal gas.

Solution

Generally, we can write the internal energy as a function of T and V, namely $U(T, V)$ such that

$$\mathrm{d}U = \left(\frac{\partial U}{\partial V}\right)_T \mathrm{d}V + \left(\frac{\partial U}{\partial T}\right)_V \mathrm{d}T = \left(\frac{\partial U}{\partial V}\right)_T \mathrm{d}V + C_V\,\mathrm{d}T$$

Hence if we can show that $\left(\frac{\partial U}{\partial V}\right)_T$ is zero, then will have shown that the internal energy is a function of T alone, $U(T)$. We begin with the hard sphere gas equation of state

$$P\left(V - V_1\right) = \boldsymbol{nR}T$$

for which we can evaluate the derivative

$$\left(\frac{\partial P}{\partial T}\right)_V = \frac{\boldsymbol{nR}}{\left(V - V_1\right)}$$

If we insert this derivative into the Thermodynamic Equation of State, we find that

$$\left(\frac{\partial U}{\partial V}\right)_T = T\frac{\boldsymbol{nR}}{\left(V - V_1\right)} - P = P - P = 0$$

This result shows that U is not a function of V and is only a function of T.

E20.2 Integrate the differential form of the entropy which you found in E19.2 and show that the entropy change of a monatomic hard sphere gas ($C_V = 1.5\,\boldsymbol{nR}$) is

$$\Delta S = \frac{3}{2}\boldsymbol{nR}\ln\left(\frac{T_2}{T_1}\right) + \boldsymbol{nR}\ln\left[\frac{\left(V_{m,2} - b\right)}{\left(V_{m,1} - b\right)}\right]$$

Solution

In Exercise E19.2, we showed that

$$\mathrm{d}S = \frac{C_V}{T}\mathrm{d}T + \left(\frac{\partial P}{\partial T}\right)_V \mathrm{d}V$$

If we use the hard sphere gas equation of state, namely

$$P = \frac{\boldsymbol{nR}T}{(V - \boldsymbol{n}b)}$$

Solutions Manual for Principles of Physical Chemistry, Third Edition. Edited by Hans Kuhn, David H. Waldeck, and Horst-Dieter Försterling.

we can find an explicit expression for the temperature derivative of the pressure

$$\left(\frac{\partial P}{\partial T}\right)_V = \frac{\boldsymbol{nR}}{(V - \boldsymbol{n}b)}$$

If we insert this form of the derivative into our equation for the differential entropy, we find that

$$dS = \frac{3}{2}\boldsymbol{nR}\frac{dT}{T} + \frac{\boldsymbol{nR}}{(V - \boldsymbol{n}b)}dV$$

In this form, we can integrate to solve this equation and find

$$\int_1^2 dS = \frac{3}{2}\boldsymbol{nR}\int_{T_1}^{T_2}\frac{dT}{T} + \boldsymbol{nR}\int_{V_1}^{V_2}\frac{dV}{(V - \boldsymbol{n}b)}$$

which becomes

$$\Delta S = \frac{3}{2}\boldsymbol{nR}\ln\left(\frac{T_2}{T_1}\right) + \boldsymbol{nR}\ln\left(\frac{(V_2 - \boldsymbol{n}b)}{(V_1 - \boldsymbol{n}b)}\right)$$

$$= \frac{3}{2}\boldsymbol{nR}\ln\left(\frac{T_2}{T_1}\right) + \boldsymbol{nR}\ln\left(\frac{(V_{2,m} - b)}{(V_{1,m} - b)}\right)$$

E20.3 Given that the molar Helmholtz energy A_m of a gas is

$$A_m = -\frac{a}{V_m} - \boldsymbol{R}T\ln\frac{(V_m - b)}{V_m^{\ominus}} + f(T)$$

where V_m is the molar volume of the gas, $V_m^{\ominus}$ is the molar volume of the gas under standard state conditions, $\boldsymbol{R}$ is the gas constant (8.314 J mol^{-1} K^{-1}), a and b are empirical constants, and $f(T)$ is an arbitrary function of T, find the equation of state for the gas. Interpret the parameters a and b. What features of the intermolecular interactions do they capture?

Solution

From Equation (19.38) we can write dA as

$$dA = -SdT - PdV$$

and use the definition of the differential to write

$$dA = \left(\frac{\partial A}{\partial T}\right)_V dT + \left(\frac{\partial A}{\partial V}\right)_T dV$$

By inspection of these two equations we can write that

$$-P = \left(\frac{\partial A}{\partial V}\right)_T$$

for which we can evaluate the derivative; namely

$$-P = \left(\frac{\partial A}{\partial V}\right)_T = \frac{a}{V_m^2} - \boldsymbol{R}T\frac{1}{V_m - b}$$

which leads to the van der Waals equation of state

$$\left(P + \frac{a}{V_m^2}\right) = \frac{\boldsymbol{R}T}{V_m - b}$$

The parameter b accounts for the physical size of the molecules. If $a = 0$ we obtain the hard sphere gas equation of state $P(V - \boldsymbol{n}b) = \boldsymbol{nR}T$.

The parameter a accounts for the attractive interactions between gas particles. One may show that

$$\left(\frac{\partial U}{\partial V}\right)_T = \frac{a}{V^2}$$

See Exercise E20.9 for this proof.

E20.4 Use the van der Waals equation of state $P = \frac{RT}{V_m - b} - \frac{a}{V_m^2}$ to find an expression for dU of a gas. Contrast your result with that for an ideal gas.

Solution

We begin with Equation (19.17) for the differential internal energy dU,

$$dU = C_V\, dT + \left[T\left(\frac{\partial P}{\partial T}\right)_V - P\right] dV$$

To find an explicit expression for dU, we evaluate the derivative $\left(\frac{\partial P}{\partial T}\right)_V$ and substitute it into this equation. Evaluating the derivative of the pressure with respect to the temperature, we find

$$\left(\frac{\partial P}{\partial T}\right)_V = \left(\frac{\partial}{\partial T}\left[\frac{RT}{V_m - b} - \frac{a}{V_m^2}\right]\right)_V = \frac{R}{V_m - b}$$

so that upon substitution we find

$$dU = C_V dT + \left[\frac{RT}{V_m - b} - P\right] dV = C_V\, dT + \frac{a}{V_m^2}\, dV$$

This result shows that the internal energy change for a van der Waals gas depends on both the volume and the temperature. This result differs from an ideal gas for which the internal energy depends only on the temperature T (see Chapter 14).

E20.5 Consider 1 mol of Ar gas in a volume $V = 1$ L. Compute the pressure P in the ideal gas approximation and the hard sphere gas approximation at 298 K. Perform the same calculation for methane. Comment on your findings.

Solution

If we use the ideal gas approximation, under these conditions we find a pressure of

$$\begin{aligned} P &= \frac{nRT}{V} \\ &= \frac{1\text{ mol} \cdot 8.314 \times 10^{-2}\text{ L bar mol}^{-1}\text{ K}^{-1} \cdot 298\text{ K}}{1\text{ L}} = 24.8\text{ bar} \end{aligned}$$

for both argon and methane. In contrast, if we use the hard sphere gas approximation for argon, (Ar: $b = 3.219 \times 10^{-2}$ L mol^{-1}, methane: $b = 4.278 \times 10^{-2}$ L mol^{-1}) we find that

$$\begin{aligned} P_{Ar} &= \frac{nRT}{V_m - b} \\ &= \frac{1\text{ mol} \cdot 8.314 \times 10^{-2}\text{ L bar mol}^{-1}\text{ K}^{-1} \cdot 298\text{ K}}{(1 - 3.219 \times 10^{-2})\text{ L mol}^{-1}}\text{ bar} = 25.3\text{ bar} \end{aligned}$$

and correspondingly we find for methane that

$$\begin{aligned} P_{CH_4} &= \frac{nRT}{V_m - b} \\ &= \frac{(1)\left(8.314 \times 10^{-2}\right)(298)}{(1 - 4.278 \times 10^{-2})\text{ L mol}^{-1}}\text{ bar} = 25.5\text{ bar} \end{aligned}$$

The hard sphere approximation leads to higher pressure with the larger pressure change being for the case of methane.

E20.6 Using the hard sphere gas equation of state, $P(V_m - b) = RT$, and assuming that $C_V = \frac{3}{2}nR$, find mathematical expressions for the internal energy, the entropy, and the enthalpy of the gas.

Solution

We begin with the total differential of the internal energy, namely

$$dU = \left(\frac{\partial U}{\partial T}\right)_V dT + \left(\frac{\partial U}{\partial V}\right)_T dV$$

In E20.1 we showed that

$$\left(\frac{\partial U}{\partial V}\right)_T = 0$$

and from the definition of the constant volume heat capacity, we can write that

$$\left(\frac{\partial U}{\partial T}\right)_V = C_V$$

Hence the differential internal energy may be written as

$$dU = C_V \, dT$$

Using the fact that $C_V = \frac{3}{2}\boldsymbol{nR}$, we can integrate this equation and find

$$\Delta U = \int dU = \int C_V \, dT = \frac{3}{2}\boldsymbol{nR} \int dT = \frac{3}{2}\boldsymbol{nR} \cdot \Delta T$$

For the enthalpy we can use Equation (20.34), namely

$$dH = C_P \, dT - \left[T\left(\frac{\partial V}{\partial T}\right)_P + V\right] dP$$

If we use the hard sphere gas equation of state to evaluate the second term, we find that

$$-\left[T\left(\frac{\partial V}{\partial T}\right)_P + V\right] = -T\left(\frac{\boldsymbol{nR}}{P}\right) + V = \boldsymbol{nb}$$

Thus we obtain

$$dH = C_P \cdot dT + \boldsymbol{nb} \cdot dP$$

Using Equation (19.24) for the constant pressure heat capacity, namely

$$C_P = C_V + T\left(\frac{\partial P}{\partial T}\right)_V\left(\frac{\partial V}{\partial T}\right)_P$$

we find that

$$C_P = C_V + T\left(\frac{\boldsymbol{nR}}{V - \boldsymbol{nb}}\right)_V\left(\frac{\boldsymbol{nR}}{P}\right)_P = C_V + \boldsymbol{nR}$$

so that

$$\Delta H = \int dH = (C_V + \boldsymbol{nR}) \cdot \Delta T + \boldsymbol{nb} \cdot \Delta P$$

For the entropy change see the derivation in E20.2, where we show that

$$\Delta S = \frac{3}{2}\boldsymbol{nR} \ln\left(\frac{T_2}{T_1}\right) + \boldsymbol{nR} \ln\left(\frac{(V_2 - \boldsymbol{nb})}{(V_1 - \boldsymbol{nb})}\right)$$

E20.7 Extend the calculations in E20.6 to determine the change in Gibbs energy and Helmholtz energy of the gas along an isotherm

Solution

For the Gibbs energy we start with

$$\Delta G = \Delta H - T\Delta S$$

Using the results from the previous exercise we can write that

$$(\Delta H)_T = \boldsymbol{nb} \cdot \Delta P$$

and

$$(\Delta S)_T = \boldsymbol{nR} \ln\left(\frac{(V_2 - \boldsymbol{nb})}{(V_1 - \boldsymbol{nb})}\right)$$

Inserting these results in our expression for ΔG, we find that

$$(\Delta G)_T = \boldsymbol{nb} \cdot \Delta P - \boldsymbol{nR}T \cdot \ln \frac{V_2 - \boldsymbol{nb}}{V_1 - \boldsymbol{nb}}$$

For the Helmholtz energy we start with

$$\Delta A = \Delta U - T\Delta S$$

and using the results for $(\Delta H)_T$ and $(\Delta S)_T$ we find that

$$(\Delta A)_T = -\boldsymbol{nR}T \cdot \ln \frac{V_2 - \boldsymbol{nb}}{V_1 - \boldsymbol{nb}}$$

E20.8 Consider an isothermal and reversible expansion of a hard sphere gas from an initial volume V_0 to a final volume V_f. Find expressions for the work and the heat in the process. Compare the reversible work to the change in Helmholtz energy between the initial and final states.

Solution

Using the definition of the work and the hard sphere gas equation of state, we can write that

$$w = -\int_{V_0}^{V_f} P\,\mathrm{d}V = -\int_{V_0}^{V_f} \frac{\boldsymbol{n}RT}{(V - \boldsymbol{n}b)}\,\mathrm{d}V$$

Performing the integration, we find that

$$w = -\boldsymbol{n}RT \cdot \ln \frac{V_f - \boldsymbol{n}b}{V_0 - \boldsymbol{n}b}$$

In E20.1 we proved that the internal energy of a hard sphere gas depends only on the temperature; hence, for an isothermal process we know that

$$(\Delta U)_T = 0$$

From the First Law of Thermodynamics we can write that

$$(\Delta U)_T = q + w$$

so that

$$q = -w = \boldsymbol{n}RT \cdot \ln \frac{V_f - \boldsymbol{n}b}{V_0 - \boldsymbol{n}b}$$

For a hard sphere gas, we showed in E20.7 that the change in the Helmholtz energy for an isothermal process is

$$(\Delta A)_T = \boldsymbol{n}RT \cdot \ln \frac{V_f - \boldsymbol{n}b}{V_0 - \boldsymbol{n}b}$$

By inspection we see that $(\Delta A)_T = -w$.

E20.9 Show that the change in internal energy with an isothermal volume change is $\boldsymbol{n}^2a/V^2$ for a van der Waals gas. Evaluate the reversible work to compress CO_2 isothermally from 1 bar to 66 bar.

Solution

We begin with the Thermodynamic Equation of State, namely

$$\left(\frac{\partial U}{\partial V}\right)_T = T\left(\frac{\partial P}{\partial T}\right)_V - P$$

Using the van der Waals equation of state in the form

$$P = \frac{\boldsymbol{n}RT}{(V - \boldsymbol{n}b)} - \frac{\boldsymbol{n}^2a}{V^2}$$

we can write the temperature derivative of the pressure as

$$\left(\frac{\partial P}{\partial T}\right)_V = \frac{\boldsymbol{n}R}{V - \boldsymbol{n}b}$$

Upon substitution, we find that

$$\left(\frac{\partial U}{\partial V}\right)_T = \frac{\boldsymbol{n}RT}{V - \boldsymbol{n}b} - P = \frac{\boldsymbol{n}^2a}{V^2}$$

The isothermal compression work of CO_2 (amount of substance $\boldsymbol{n}$) from P_0 to P_f is

$$w = -\int_{V_0}^{V_f} P\,\mathrm{d}V = -\int \frac{\boldsymbol{n}RT}{V - \boldsymbol{n}b}\,\mathrm{d}V + \int_{V_0}^{V_f} \frac{\boldsymbol{n}^2a}{V^2}\,\mathrm{d}V$$

$$= -\boldsymbol{n}RT \ln\left(\frac{V_f - \boldsymbol{n}b}{V_0 - \boldsymbol{n}b}\right) + \boldsymbol{n}^2a\left(\frac{1}{V_f} - \frac{1}{V_0}\right)$$

where V_0 and V_f depend on P_0 and P_f.

Next we need to find a relation between these quantities. To do this, we need to find V_0 and V_f by solving the van der Waals equation of state, which is cubic in the volume:

$$P = \frac{\boldsymbol{n}RT}{V - \boldsymbol{n}b} - \frac{\boldsymbol{n}^2 a}{V^2}$$

We do this numerically for the case that the gas is compressed from $P_0 = 1$ bar to $P_f = 66$ bar by plotting

$$y = \frac{\boldsymbol{n}RT}{V - \boldsymbol{n}b} - \frac{\boldsymbol{n}^2 a}{V^2} - P$$

as a function of V with $T = 313$ K, $\boldsymbol{n} = 1$ mol, $b = 4.267 \times 10^{-2}$ L mol^{-1}, and $a = 3.66$ bar L^2 mol^{-2} for the cases a) $P = P_0 = 1$ bar and b) $P = P_f = 66$ bar. The roots are the volumes at the location $y = 0$. Fig. E20.9a, panel a) shows the plot of

$$y = \frac{8.3145 \times 10^{-2}\ \text{L bar K}^{-1} \cdot 313\ \text{K}}{V - 4.267 \times 10^{-2}\ \text{L}} - \frac{3.66\ \text{bar L}^2}{V^2} - 1\ \text{bar}$$

for the $P = 1$ bar case, in which we find the volume at which $y = 0$ to be $V = 25.94$ L.

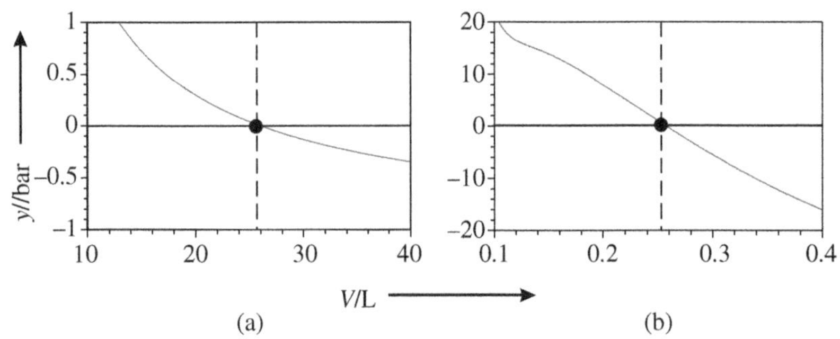

Figure E20.9 Determination of the roots of the van der Waals equation for CO_2 at T = 313 K. Plot of $y = \frac{nRT}{V-nb} - \frac{n^2a}{V^2} - P$ versus the volume V for $P = 1$ bar (panel a) and $P = 66$ bar (panel b).

Correspondingly for the $P = 66$ bar, case we plot (see Fig. E20.9, panel b)

$$y = \frac{8.3145 \times 10^{-2}\ \text{L bar K}^{-1} \cdot 313\ \text{K}}{V - 4.267 \times 10^{-2}\ \text{L}} - \frac{3.66\ \text{bar L}^2}{V^2} - 66\ \text{bar}$$

and find $V = 0.257$ L.

Using these values, the work is

$$w = -8.314\ \text{J K}^{-1}\ \text{mol}^{-1} \cdot 313\ \text{K} \cdot \ln\left(\frac{0.257 - 4.267 \cdot 10^{-2}}{25.94 - 4.267 \cdot 10^{-2}}\right) + 3.66\ \text{bar L}^2 \left(\frac{1}{0.257\ \text{L}} - \frac{1}{25.94\ \text{L}}\right)$$
$$= (124.8 + 14.1)\ \text{L bar} = 138.9\ \text{L bar} = 13.9\ \text{kJ}$$

E20.10 Show that the equation

$$G(T, P) = G^{\ominus}(T) + NkT \cdot \ln\left(\frac{P}{P^{\ominus}}\right)$$

can be used as an alternative definition of the ideal gas, by showing that it implies (a) $PV = NkT$ and (b) U is a function of only T.

Solution

We begin with the basic relation

$$dG = -SdT + VdP$$

from Section 19.3.2 from which we found that

$$\left(\frac{\partial G}{\partial T}\right)_P = -S \text{ and } \left(\frac{\partial G}{\partial P}\right)_T = V$$

Using the second relation, we can write that

$$V = \left(\frac{\partial G}{\partial P}\right)_T = \left(\frac{\partial}{\partial P}\left[G^{\ominus}(T) + NkT \cdot \ln P\right]\right)_T = 0 + NkT \cdot \frac{1}{P}$$

This result is the ideal gas equation of state

$$PV = NkT$$

To show that U is a function only of T, we use the fundamental relation

$$G = U - TS + PV \text{ or } U = G + TS - PV$$

so that

$$U = G^{\ominus}(T) + NkT \cdot \ln\left(\frac{P}{P^{\ominus}}\right) + TS - PV$$

Now we use the fact that

$$\left(\frac{\partial G}{\partial T}\right)_P = -S$$

and write

$$\begin{aligned} U &= G^{\ominus}(T) + NkT \cdot \ln\left(\frac{P}{P^{\ominus}}\right) + TS - PV \\ &= G^{\ominus}(T) + NkT \cdot \ln\left(\frac{P}{P^{\ominus}}\right) - T\left(\frac{\partial G}{\partial T}\right)_P - PV \\ &= G^{\ominus}(T) + NkT \cdot \ln\left(\frac{P}{P^{\ominus}}\right) - T\left(\frac{\partial G^{\ominus}(T)}{\partial T}\right)_P - NkT \cdot \ln\left(\frac{P}{P^{\ominus}}\right) - PV \\ &= G^{\ominus}(T) - T\left(\frac{\partial G^{\ominus}(T)}{\partial T}\right)_P - NkT \end{aligned}$$

Hence U is a function of T only; i.e., $U(T)$.

From this proof we see that the equation given for the Gibbs energy in the problem statement can be used as an alternative definition of the ideal gas.

E20.11 What is the Law of Corresponding States?

Solution

The Law of Corresponding States points out that all substances obey the same equation of state in terms of reduced variables. In the text, we gave the example where we rescale the intermolecular potential and plot the second Virial, see Fig. 20.9. Alternatively, we could rescale the pressure as $\pi = P/P_C$, the volume as $\phi = V/V_C$ and the temperature as $\tau = T/T_C$; where P_c, V_c, and T_c are the values of the pressure, volume, and temperature of the fluid at the critical point. These definitions allow us to write the van der Waals equation of state as

$$\left(\pi + \frac{3}{\phi^2}\right)(3\phi - 1) = 8\tau$$

which is independent of the parameters a and b. Hence if we use these reduced variables, all van der Waals gases may be plotted on the same graph – independent of their a and b values. The validity of this approach reflects the similarity in the intermolecular potentials for different molecules.

E20.12 Show that the Joule–Thomson coefficient for an ideal gas is zero.

Solution

We start with Equation (20.36) for the Joule–Thomson coefficient

$$\mu_{JT} = \frac{1}{C_P}\left(T\left(\frac{\partial V}{\partial T}\right)_P - V\right)$$

Using the ideal gas law, we find that

$$\left(\frac{\partial V}{\partial T}\right)_P = \frac{\boldsymbol{nR}}{P}$$

Inserting this result into the expression for μ_{JT} we find that

$$\mu_{JT} = \frac{1}{C_P}\left(T\left(\frac{nR}{P}\right) - V\right) = \frac{1}{C_P}(V - V) = 0$$

E20.13 Find an expression for the Joule–Thomson coefficient of a hard sphere gas. Assume the gas is monatomic and that $C_{P,m} = \frac{5}{2}\boldsymbol{R}$.

Solution

We start with Equation (20.36) for the Joule–Thomson coefficient

$$\mu_{JT} = \frac{1}{C_P}\left(T\left(\frac{\partial V}{\partial T}\right)_P - V\right)$$

Using the hard sphere gas equation of state,

$$P(V - \boldsymbol{n}b) = \boldsymbol{nR}T$$

we can evaluate the temperature derivative of the volume and find that

$$\left(\frac{\partial V}{\partial T}\right)_P = \frac{\boldsymbol{nR}}{P}$$

Inserting this result into the expression for μ_{JT} we find that

$$\mu_{JT} = \frac{1}{C_P}\left(\frac{\boldsymbol{nR}T}{P} - V\right) = \frac{1}{C_P}\left(\frac{\boldsymbol{nR}T}{\frac{\boldsymbol{nR}T}{(V-\boldsymbol{n}b)}} - V\right) = -\frac{\boldsymbol{n}b}{C_P} = -\frac{2}{5}\frac{b}{\boldsymbol{R}}$$

Hence this model predicts that the gas warms upon expansion at all temperatures, contrary to experiment.

E20.14 In Example 20.7 we have calculated the temperature jump when expanding a diatomic gas in the Joule–Thomson experiment from pressure P_1 to P_2, using Equation (20.38) for the Joule–Thomson coefficient. Repeat the calculation by using the more general Equation (20.37).

Solution

Equation (20.37) is

$$\mu_{JT} = \frac{1}{C_{P,m}}\left[\frac{2a}{\boldsymbol{R}T} - b - \frac{3ab}{\boldsymbol{R}^2T^2}\cdot P\right]$$

Using the van der Waals parameters for CO_2 $a = 3.66$ bar L^2 mol^{-2}, $b = 0.043$ L mol^{-1}, and $C_{P,m} = 7/2\,\boldsymbol{R}$ we calculate that

$$\mu_{JT} = \frac{2}{7\cdot \boldsymbol{R}}\cdot \left(\frac{2\cdot 3.66\text{ bar L}^2\text{ mol}^{-2}}{0.08314\text{ bar L K}^{-1}\text{ mol}^{-1}\cdot 300\text{ K}} - 0.043\text{ L mol}^{-1} - \frac{3\cdot 3.66\text{ bar L}^2\text{ mol}^{-2}\cdot 0.043\text{ L mol}^{-1}}{(0.08314\text{ bar L K}^{-1}\text{ mol}^{-1})^2\cdot(300\text{ K})^2}\cdot 11\text{ bar}\right)$$
$$= 3.44\text{ bar}^{-1}\text{ L}^{-1}\text{ K mol}\cdot\left(0.293\text{ mol}^{-1}\text{ L} - 0.043\text{ L mol}^{-1} - 0.0083\text{ L mol}^{-1}\right)$$
$$= 3.44\cdot\text{bar}^{-1}\text{ L}^{-1}\text{ K mol}\cdot 0.242\text{ L mol}^{-1} = 0.83\text{ K bar}^{-1}$$

for an initial pressure of 11 bar. Then we find that

$$\Delta T \approx \mu_{JT}\Delta P = 0.69\times(-10)\text{ K} = -8.3\text{ K}$$

A value similar to that found in Example 20.7 (−8.6 K).

E20.15 Use the fact that $\mathrm{d}U$ is an exact differential to show that the constant volume molar heat capacity of a van der Waals gas depends only on the temperature.

Solution

From the solution to E20.4 we know that the differential internal energy of a van der Waals gas can be written as

$$\mathrm{d}U_m = C_{V,m}\,\mathrm{d}T - \frac{a}{V_m^2}\,\mathrm{d}V_m$$

Because dU_m is exact, the cross-derivatives must be equal, hence we find that

$$\left(\frac{\partial C_{V,m}}{\partial V_m}\right)_T = \left(\frac{\partial\left(\frac{-a}{V_m^2}\right)}{\partial T}\right)_{V_m} = 0$$

which proves that the molar heat capacity is volume independent.

E20.16 Using the data provided in Table 20.1, determine the fugacity of CO_2 at 101 bar and 50.6 bar; i.e., perform the analysis of the text.

Solution

From Equation (20.26), we obtain

$$\int_{P_0}^{P} V_m \cdot dP = \boldsymbol{RT} \cdot \ln\left(\frac{P}{P_0}\right) + \boldsymbol{RT} \cdot \ln\phi$$

$$= \int_{P_0}^{P} V_{m,\text{ideal}} \cdot dP + \boldsymbol{RT} \cdot \ln\phi$$

where P_0 is a small enough pressure such that $\phi_0 = 1$ and V_m is the molar volume. Then

$$\ln\phi = \frac{1}{\boldsymbol{RT}} \int_{P_0}^{P} \left(V_m - V_{m,\text{ideal}}\right) \cdot dP$$

where

$$V_{m,\text{ideal}} = \frac{\boldsymbol{RT}}{P}$$

The molar volume V_m is obtained from the experimental PV data (Table 20.1, for $\boldsymbol{n}_{CO_2} = 1$ mol and $T = 313$ K). In Fig. 20.1, $V_m - V_{m,\text{ideal}}$ is plotted versus the pressure P. The shaded area in the figure corresponds to

$$\int_{0\text{ bar}}^{100\text{ bar}} \left(V_m - V_{m,\text{ideal}}\right) \cdot dP = -14 \text{ bar L}$$

Then

$$\ln\phi = -\frac{1}{\boldsymbol{RT}} \cdot 14 \text{ bar L} = -0.54, \quad \phi = 0.58$$

E20.17 Analyze the PV_m data[1] for hydrogen at 273 K to obtain values for the fugacity at 200 bar and 1000 bar.

P (bar)	1	50	100	200	300	400	600	800	1000
V_m (L mol^{-1})	22.712	0.46818	0.24146	0.12844	0.09096	0.072300	0.053670	0.044327	0.038670

Solution

We need to determine the quantity $\left(V_m - V_{m,\text{ideal}}\right)$ versus P. To do this we plot the data (see Fig. E20.17a) and numerically integrate it to obtain the area under the curve. For each pressure along the isotherm we calculate the quantity

$$V_{\text{diff}} = \left(V_m - V_{m,\text{ideal}}\right)$$

For example, at 1 bar pressure we find

$$V_{\text{diff}} = 22.712 \text{ L mol}^{-1} - \frac{\boldsymbol{RT}}{P} = 22.712 \text{ L mol}^{-1} - \frac{0.08314 \text{ bar L mol}^{-1} \cdot 273 \text{ K}}{1 \text{ bar}}$$

$$= 0.01478 \text{ L mol}^{-1}$$

1 E.W. Lemmon, M.O. McLinden, and D.G. Friend, "Thermophysical Properties of Fluid Systems" in NIST Chemistry WebBook, NIST Standard Reference Database Number 69, Eds. P.J. Linstrom and W.G. Mallard, National Institute of Standards and Technology, Gaithersburg MD, 20899, http://webbook.nist.gov.

Proceeding in a like manner we find the values in the table

P/bar	1	50	100	200	300	400	600	800	1000
V_{diff}/(L mol^{-1})	0.01478	0.014244	0.014488	0.014954	0.015309	0.015557	0.015841	0.015955	0.015973

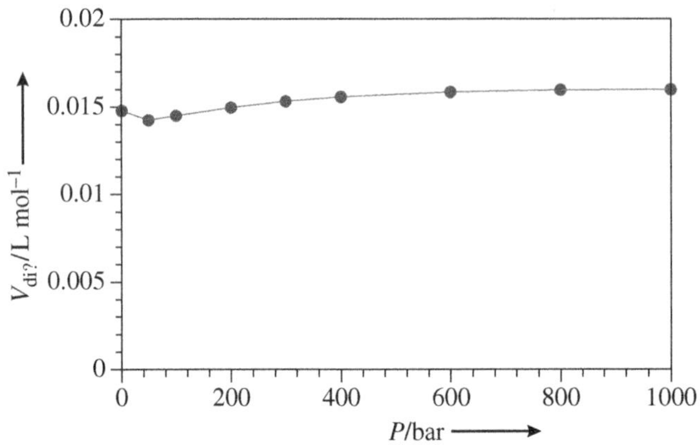

Figure E20.17a $V_{diff} = \left(V_m - V_{m,ideal}\right)$ versus pressure P $V_m = V_{m,ideal}$ for H_2 at 273 K.

To calculate the fugacity coefficient ϕ

$$\ln \phi = \frac{1}{RT} \int_{P_0}^{P} \left(V_m - V_{m,ideal}\right) \cdot dP$$

we need to determine the area under the curve in Fig. E20.17a from $P_0 = 0$ bar to P. However, the smallest experimental value is for $P_1 = 1$ bar. Therefore we start at $P_1 = 1$ bar. From the figure it can be seen that V_{diff} is an only slightly changing function. Therefore, in a first attempt, we replace the function by its mean value $V_{diff,mean}$.

$$\ln \phi = \frac{1}{RT} V_{diff,mean} \cdot (P - P_1)$$

In the range from $P = 1$ bar to $P = 200$ bar we find $V_{diff,mean} = 0.0145$ L mol^{-1} and it follows that

$$\ln \phi = \frac{1}{RT} \cdot 0.0145 \text{ L mol}^{-1} \cdot (200 - 1) \text{ bar} = \frac{1}{RT} \cdot 2.89 \text{ bar L mol}^{-1} = 0.127$$

and $\phi = 1.14$.

In the range from $P = P_0$ to $P = 1000$ bar we find $V_{diff,mean} = 0.0152$ L mol^{-1} and it follows that $\ln \phi = \frac{1}{RT} \cdot 0.0153$ L mol^{-1} $\cdot$ (1000 – 1) bar $= \frac{1}{RT} \cdot 15.19$ bar L mol^{-1} $= 0.669$ and $\phi = 1.95$. Thus for the fugacity $f = \phi \times P$ we obtain $f = 228$ bar for $P = 200$ bar and $f = 1950$ bar for $P = 1000$ bar

Now we refine our procedure by a step-by-step calculation of the area, see Fig. E20.17b, where we consider the first 4 data points in the table.

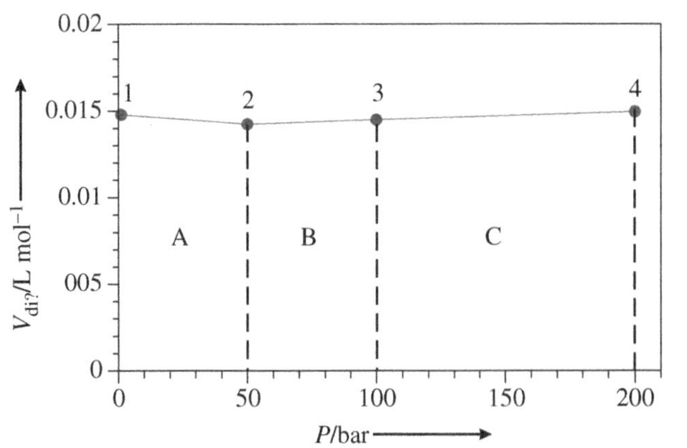

Figure E20.17b Step-by-step calculation of the area under a curve.

We calculate the total area by adding the areas A, B, and C.

$$area = A + B + C = \frac{(V_{\text{diff},1} + V_{\text{diff},2})}{2}\Delta P_A + \frac{(V_{\text{diff},2} + V_{\text{diff},3})}{2}\Delta P_B + \frac{(V_{\text{diff},3} + V_{\text{diff},4})}{2}\Delta P_C$$

with

$$\Delta P_A = P_2 - P_1 \quad \Delta P_B = P_3 - P_2 \quad \Delta P_C = P_4 - P_3$$

Inserting the data from the table results in

$$A = 0.7111 \text{ bar L mol}^{-1}; \quad B = 0.7183 \text{ bar L mol}^{-1}; \quad \text{and} \quad C = 1.4721 \text{ bar L mol}^{-1}$$

Thus we obtain

$$|area|_1^{200} = A + B + C = 2.901 \text{ bar L mol}^{-1}$$

This value differs from the estimated one by 0.3%. Correspondingly, for $P = 1000$ bar we obtain $|area|_1^{200} = 15.47$ bar L mol^{-1}, this differs from the estimated value by 2%.

In practice you will apply this method by using a computer application.

However, our result is not yet completely correct, because we calculated the area from $P_1 = 1$ bar to P instead of starting at $P_0 = 0$ bar. We can estimate the correction as

$$|area|_0^{1\text{ bar}} = \frac{1}{2}V_{\text{diff},1\text{ bar}} \cdot 1 \text{ bar} = \frac{1}{2}0.01478 \text{ L mol}^{-1} \cdot 1 \text{ bar} = 0.00739 \text{ bar L mol}^{-1}$$

The correction is 0.3% compared to $|area|_1^{200}$ and 0.04% compared to $|area|_1^{1000}$.

E20.18 Analyze the data given for nitrogen at 273.15 K[2] to obtain values for the fugacity at 202.65 bar and 1013.25 bar (Φ is the compression factor). To proceed, assume that the gas behaves ideally (i.e., the compression factor $\Phi = 1$, see Equation (20.8)) for pressures at 1.01325 bar and lower.

P (bar)	1.01325	50.6625	101.325	202.650	303.975	405.300	607.95	810.6	1013.25
Φ	1.0000	0.9839	0.9840	1.0330	1.1332	1.2540	1.5211	1.7985	2.0659

Use your result to determine the change in Gibbs energy of the gas between 202.65 bar and 1013.25 bar. Compare your result to that expected for an ideal gas.

Solution

To proceed, we determine the quantity V_m from Equation (20.8)

$$\Phi = \frac{PV_m}{RT} \quad V_m = \frac{RT}{P}\Phi$$

and calculate

$$V_{\text{diff}} = (V_m - V_{m,\text{ideal}}) = \left(\frac{RT}{P}\Phi - \frac{RT}{P}\right) = \frac{RT}{P}(\Phi - 1)$$

at the different pressures P and plot them (see Fig. E20.18). For example, at 50.6625 bar pressure we find

$$V_{\text{diff}} = (0.9839 - 1) \cdot \frac{0.083145 \cdot 273.15}{50.6625} \text{ L mol}^{-1} = -0.007217 \text{ L mol}^{-1}$$

Proceeding in a like manner we find the values in the table

P/bar	V_{diff}/(L mol^{-1})	P/bar	V_{diff}/(L mol^{-1})
1.01325	0.00	405.3	0.01423
50.6625	−50.6625	607.95	0.01947
101.325	−0.003586	810.6	0.02237
202.65	−0.003586	1013.25	0.02389
303.975	0.009951		

2 E. P. Bartlett, *J. Am. Chem. Soc* 1927, **49**, 687.

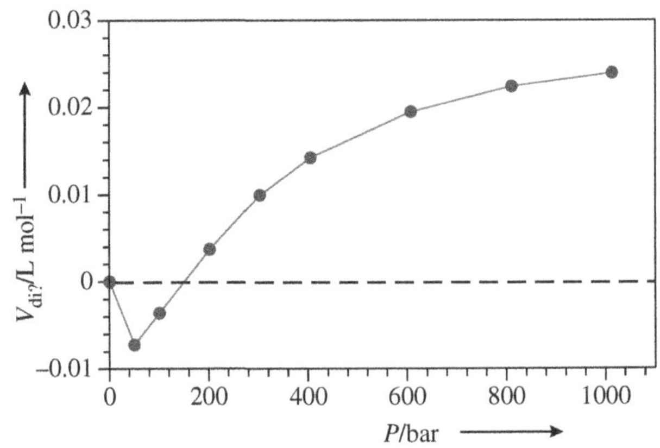

Figure E20.18 $V_{\text{diff}} = (V_m - V_{m,\text{ideal}})$ versus pressure P for N_2 at 273.15 K.

The area under the curve can be approximated by numerical integration using the method of trapezoids as outlined in Exercise E20.17. Thus we obtain $|area|_1^{202.65} = -0.447$ L bar mol^{-1} and $|area|_1^{1013.25} = 13.81$ L bar mol^{-1}.

The fugacity coefficient is

$$\ln \phi = \frac{1}{RT} \int_{P_0}^{P} \left(V_m - V_{m,\text{ideal}}\right) \cdot \mathrm{d}P = \frac{1}{RT} |area|_{P_0}^{P}$$

Thus we obtain for $P = 202.65$ bar

$$\ln \phi = \frac{-0.447}{0.08314 \times 273.15} = -0.0197 \Longrightarrow \phi = \exp(-0.0162) = 0.980$$

and the fugacity will be

$$f = \phi \times P = (0.980 \times 202.65) \text{ bar} = 198.7 \text{ bar}$$

In the 1013.25 bar case, the fugacity coefficient is

$$\ln \phi = \frac{13.81}{0.08314 \times 273.15} = 0.597 \Longrightarrow \phi = \exp(0.611) = 1.80$$

and the fugacity will be

$$f = \phi \times P = (1.80 \times 1013.25) \text{ bar} = 1823 \text{ bar}$$

The change in the Gibbs energy between 202.65 bar and 1013.25 bar is

$$\Delta G_m = RT \ln\left(\frac{f_2}{f_1}\right) = (8.314)(273.15) \ln\left(\frac{1823}{198.7}\right) = 5.03 \text{ kJ mol}^{-1}$$

and based upon the ideal gas approximation

$$\Delta G_m = RT \ln\left(\frac{P_2}{P_1}\right) = (8.314)(273.15) \ln\left(\frac{1013.25}{202.65}\right) = 3.66 \text{ kJ}$$

which is

$$\frac{5.03 - 3.66}{5.03} \times 100 = 27\%$$

in error.

E20.19 In Fig. 20.5 we compared the pressure dependence of the compression factor for CO_2 at 500 K to the prediction by the van der Waals equation of state model. Perform a similar comparison using the Beattie–Bridgman equation of state, which is

$$P = \frac{RT \cdot (1 - c/(V_m T^3))}{V_m^2}(V_m + B_0(1 - b/V_m)) - \frac{A_0 \cdot (1 - a/V_m)}{V_m^2}$$

where the parameters are $A_0 = 5.072836$ bar L^2/mol^2, $a = 0.07132$ L mol^{-1}, $B_0 = 0.10476$ L mol^{-1}, $b = 0.07235$ L mol^{-1}, and $c = 6.600 \cdot 10^5$ K^3 L mol^{-1}. The data for CO_2 are

P (bar)	1	50	100	150	200	250	300
V_m (L mol^{-1})	41.55	0.8091	0.389	0.253	0.187	0.149	0.125
Φ_{expt}	0.999	0.965	0.936	0.914	0.900	0.896	0.901

P (bar)	400	500	600	700	800	900	1000
V_m (L mol^{-1})	0.0972	0.0824	0.0733	0.0673	0.0630	0.0597	0.0571
Φ_{expt}	0.935	0.990	1.06	1.13	1.21	1.29	1.37

Solution

To compare the data with the Beattie–Bridgman equation of state we use the parameters given, and calculate the pressure and compression factor for a range of different molar volumes. For the pressure, we use

$$P_{\text{BB}} = \frac{RT \cdot (1 - c/(V_m T^3))}{V_m^2}(V_m + B_0(1 - b/V_m)) - \frac{A_0 \cdot (1 - a/V_m)}{V_m^2}$$

and for the compression factor, we use

$$\Phi_{\text{BB}} = \frac{P_{\text{BB}} V_m}{RT}$$

The resulting P/Φ_{BB} data are collected in the table:

P (bar)	1	50	100	150	200	250	300
V_m (L mol^{-1})	42.00	0.808	0.393	0.255	0.187	0.147	0.121
Φ_{BB}	0.999	0.973	0.947	0.922	0.902	0.886	0.875

P (bar)	400	500	600	700	800	900	1000
V_m (L mol^{-1})	0.0909	0.0746	0.0647	0.0582	0.0538	0.0502	0.0475
Φ_{BB}	0.874	0.896	0.933	0.983	1.034	1.091	1.151

The data are plotted in Fig. E20.19. It is evident that the equation of state does not quantitatively capture the behavior; however, it displays good agreement below 200 bar and has the correct qualitative behavior. Compared to the van der Waals equation the Beattie–Bridgman equation results in a better agreement at low pressure, but there are equally significant deviations at high pressure.

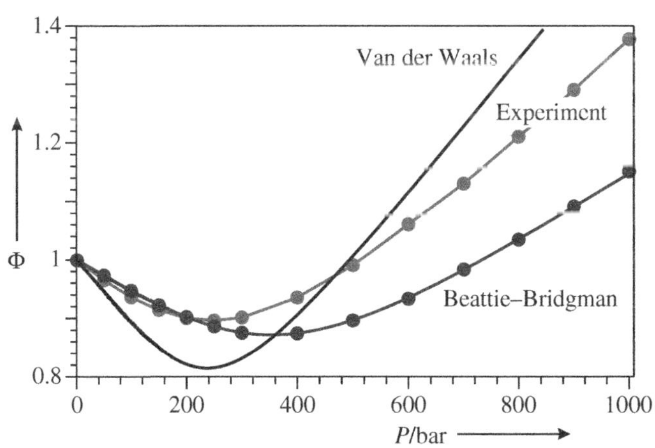

Figure E20.19 Compression factor Φ for CO_2 at $T = 500$ K. Solid lines with filled circles: Experiment and calculation according to the Beattie–Bridgman equation. In detail, in a first step P is calculated for different values of V_m; from these P/V_m data the product PV_m and the compression factor $\Phi = PV_m/(RT)$ are calculated. Filled circles: the data points corresponding to the experimental pressures are displayed. Solid line: result for the van der Waals equation, for comparison.

E20.20 The Dieterici equation of state is

$$P(V_m - b) \cdot \exp\left(\frac{a}{V_m \boldsymbol{R}T}\right) = \boldsymbol{R}T$$

where a and b are constants (parameters). a) Show that the Dieterici equation of state reduces to the ideal gas equation of state as the temperature T becomes large and the molar volume V_m becomes large. b) Use this equation of state and find an expression for the volume dependence of the gas's internal energy.

Solution

(a) As the temperature becomes large the exponential term in the Dieterici equation goes to 1, leaving

$$P(V_m - b) = \boldsymbol{R}T$$

and as the molar volume becomes large the $(V_m - b) \approx V_m$ leaving

$$PV_m = \boldsymbol{R}T$$

which is the ideal gas equation of state.

(b) We start with the thermodynamic equation of state, namely

$$\left(\frac{\partial U}{\partial V}\right)_T = T\left(\frac{\partial P}{\partial T}\right)_V - P$$

If we take the derivative of the pressure with respect to temperature, we find that

$$\begin{aligned}\left(\frac{\partial P}{\partial T}\right)_V &= \frac{\partial}{\partial T}\left(\frac{\boldsymbol{R}T}{(V_m - b)} \cdot \exp\left(-\frac{a}{V_m \boldsymbol{R}T}\right)\right)\\ &= \frac{\boldsymbol{R}}{(V_m - b)} \cdot \exp\left(-\frac{a}{V_m \boldsymbol{R}T}\right) + \frac{\boldsymbol{R}T}{(V_m - b)} \frac{a}{V_m \boldsymbol{R}T^2} \cdot \exp\left(-\frac{a}{V_m \boldsymbol{R}T}\right)\\ &= P\left(\frac{1}{T} + \frac{a}{V_m \boldsymbol{R}T^2}\right)\end{aligned}$$

By substituting this form for the derivative into the Thermodynamic Equation of State, we find that

$$\begin{aligned}\left(\frac{\partial U}{\partial V}\right)_T &= \left[P + \frac{a}{V_m \boldsymbol{R}T} \cdot P\right] - P\\ &= \frac{a}{V_m \boldsymbol{R}T} \cdot P = \frac{\boldsymbol{n}aP}{\boldsymbol{R}T}\frac{1}{V}\end{aligned}$$

Thus

$$(\mathrm{d}U)_T = \frac{\boldsymbol{n}aP}{\boldsymbol{R}T}\frac{1}{V} \cdot (\mathrm{d}V)_T$$

and substituting for P by

$$P = \frac{\boldsymbol{n}\boldsymbol{R}T}{(V - \boldsymbol{n}b)} \cdot \exp\left(-\frac{\boldsymbol{n}a}{V\boldsymbol{R}T}\right)$$

we find

$$(\mathrm{d}U)_T = \frac{\boldsymbol{n}}{(V - \boldsymbol{n}b)} \frac{\boldsymbol{n}a}{V} \cdot \exp\left(-\frac{\boldsymbol{n}a}{V\boldsymbol{R}T}\right) \cdot (\mathrm{d}V)_T$$

If we integrate along the isotherm from V_1 to V_2, we find that

$$\Delta U = \boldsymbol{n}^2 a \int_{V_1}^{V_2} \frac{\exp\left(-\frac{\boldsymbol{n}a}{V\boldsymbol{R}T}\right)}{V(V - \boldsymbol{n}b)} \cdot \mathrm{d}V$$

In the approximation that $V \gg \boldsymbol{n}b$, we can write this integral as

$$\Delta U \approx \boldsymbol{n}^2 a \int_{V_1}^{V_2} \frac{\exp\left(-\frac{\boldsymbol{n}a}{V\boldsymbol{R}T}\right)}{V^2} \cdot \mathrm{d}V$$

Making the substitution $x = \boldsymbol{n}a/(V\boldsymbol{R}T)$ and $dx = -\boldsymbol{n}a/(V^2\boldsymbol{R}T)\,dV$ we find that

$$\begin{aligned}\Delta U &= -\boldsymbol{nR}T\int_{x_1}^{x_2}\exp(-x)\ dx\\ &= \boldsymbol{nR}T\left[\exp(-x_2) - \exp(-x_1)\right]\\ &= \boldsymbol{nR}T\left[\exp\left(-\frac{\boldsymbol{n}a}{V_2\boldsymbol{R}T}\right) - \exp\left(-\frac{\boldsymbol{n}a}{V_1\boldsymbol{R}T}\right)\right]\end{aligned}$$

E20.21 A simple application of thermodynamic equilibria to chemically significant systems is the description of the phase transitions which pure substances undergo. In the MathCAD worksheet (sublimation.mcdx), you investigate the biphasic equilibrium between a solid and its monatomic gas, see Foundation 20.1. To construct the phase diagram you will calculate the Gibbs free energy for the solid ($G_{\text{solid}}(T)$), by using the Einstein model (i.e., one normal mode frequency) and for the gas ($G_{\text{gas}}(T)$), by using an ideal gas approximation. You will find the temperature and pressures at which the Gibbs energies of the two phases are equal and plot the coexistence curve.

Solution

See the file sublimation_solution.mcdx.

Question 1. Compare the temperature dependence of $G_{\text{gas}}(T,p)$ and $G_{\text{solid}}(T)$ as plotted above at 1 bar pressure. Which slope is larger and why?

The equations for $G_{\text{gas}}(T,p)$ and $G_{\text{solid}}(T)$ are those derived in Foundation 20.1:

$$G_{\text{solid}} = \frac{3}{2}Nk\theta_E + 3NkT\ln\left[1 - \exp\left(\frac{\theta_E}{T}\right)\right] + \frac{N\phi}{2}$$

and

$$G_{\text{gas}} = NkT\ln\left[\frac{p^{\ominus}}{a\cdot T^{5/2}}\right] + NkT\ln\hat{P}$$

with $\hat{P} = P/P^{\ominus}$, where $P^{\ominus}$ is the standard pressure, ϕ is the electronic interaction energy per pair of atoms, $\theta_E = h\nu_0/k$ is the characteristic temperature in the Einstein model, and the constant a is defined as

$$a = \frac{(2\pi k\cdot m)^{3/2}}{h^3}k = 6.70\times 10^5\,\text{N m}^{-2}$$

It is more convenient to replace G by the molar quantity G_m:

$$G_{m,\text{solid}} = \frac{3}{2}\boldsymbol{R}\theta_E + 3\boldsymbol{R}T\ln\left[1 - \exp\left(-\frac{\theta_E}{T}\right)\right] + \frac{\phi_m}{2}$$

$$G_{m,\text{gas}} = \boldsymbol{R}T\ln\left[\frac{p^{\ominus}}{a\cdot T^{5/2}}\right] + \boldsymbol{R}T\ln\hat{P}$$

with $\phi_m = \boldsymbol{N}_A\phi$. It is evident from Fig. E20.21a that the slope of the gas is more strongly negative than the slope of the solid. The slope of the Gibbs energy versus T plot is the negative of the entropy (i.e., $-S$); hence, the model gives an entropy for the gas that is higher than that of the solid – in accordance with our expectation.

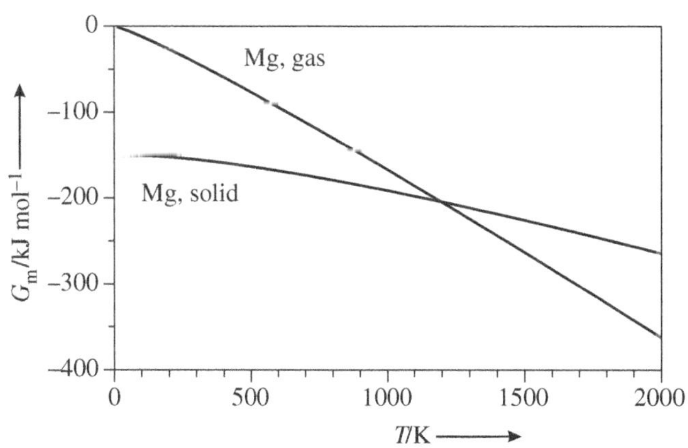

Figure E20.21a The molar Gibbs energy for Mg. $G_{m,\text{gas}}$ and $G_{m,\text{solid}}$ are plotted as a function of T (in K) at a pressure of $\hat{P} = 0.407$, with $\phi_m = -306$ kJ mol^{-1} and $\theta_E = 215$ K.

Question 2. Find the temperature of the crossing point for a number *f* different pressure. This provides the pressures and temperatures on the sublimation curve. Make a graph of the sublimation curve using this model; i.e., a *P* versus *T* diagram. [Note: Stay below the triple point of the substance.] Perform this procedure for a system which mimics the sublimation of Ar and for a system which mimics the sublimation of Mg.

We use the equation

$$\ln \hat{P} = \frac{3}{2}\frac{\theta_E}{T} + 3\ln\left[1 - \exp\left(-\frac{\theta_E}{T}\right)\right] + \frac{\phi}{2kT} - \ln\left[\frac{P^{\ominus}}{a \cdot T^{5/2}}\right] \tag{20E21.1}$$

derived in Foundation 20.1 with a, θ_E, and ϕ as defined above. For Ar let $\phi_m = -14.4$ kJ mol^{-1} or $\phi = \phi_m/N_A = -2.39 \times 10^{-20}$ J and $\theta_E = 30$ K. For argon the triple point is at 83.8 K and 0.680×10^5 Pa. Under these conditions we calculate the following data for the dependency of the pressure on the temperature.

T / K	56	59.5	67	71	78	81.5	83.4
$\hat{P}$	0.005	0.01	0.05	0.10	0.30	0.50	0.60

Figure E20.21b plots $\hat{P}$ in the temperature range from 50 to 90 K and compares the calculation to experiment.

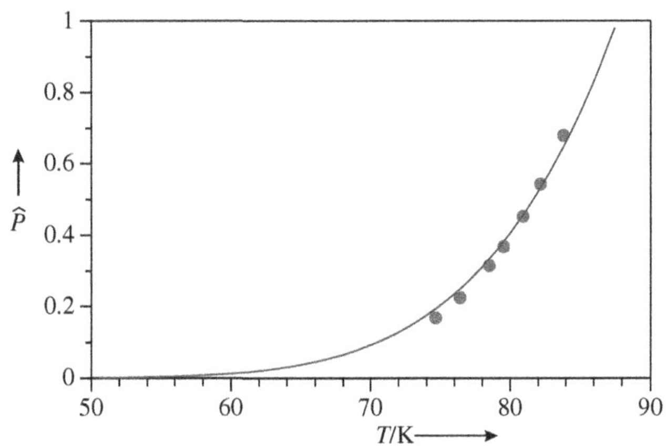

Figure E20.21b The pressure $\hat{P} = P/P^{\ominus}$ versus temperature T curve for Ar gas/solid coexistence is plotted. Dark gray filled circles: Experiment.[3] Light gray solid line: data calculated from Equation (20E21.1), using $\phi = -2.39 \times 10^{-20}$ J and $\theta_E = 30$ K.

For Mg we use $\phi_m = -306$ kJ mol^{-1} and $\theta_E = 215$ K For magnesium the triple point is at 922 K. From the MatchCAD simulation we find the following data.

T / K	500	550	600	650	700	750
$\hat{P}$	$3.45 \cdot 10^{-10}$	$8.50 \cdot 10^{-9}$	$1.29 \cdot 10^{-7}$	$1.29 \cdot 10^{-6}$	$8.95 \cdot 10^{-6}$	$5.63 \cdot 10^{-5}$

Figure E20.21c plots these simulation values along with the experimental data shown in the table below.

T / K	500	525	550	575	600	625	650
$\hat{P}$	$4.01 \cdot 10^{-10}$	$2.11 \cdot 10^{-9}$	$8.50 \cdot 10^{-9}$	$3.76 \cdot 10^{-8}$	$1.33 \cdot 10^{-7}$	$4.24 \cdot 10^{-7}$	$1.24 \cdot 10^{-6}$

3 H. H. Chen, C. C. Lim, and R. A. Aziz *J. Chem. Thermodynamics* 1978, **10**, 649–659.

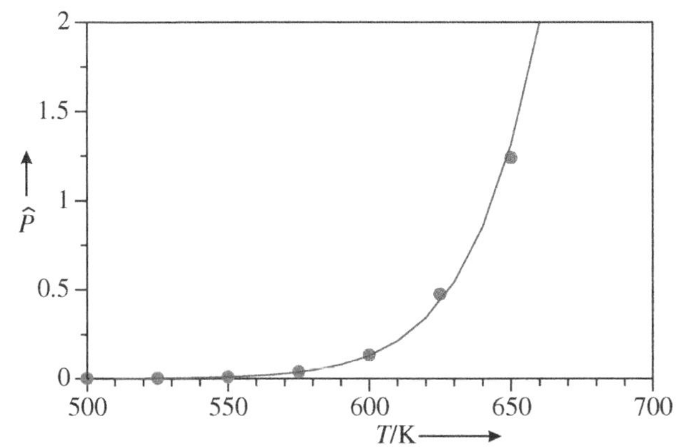

Figure E20.21c The pressure $\widehat{P} = P/P^{\ominus}$ versus temperature T curve for Mg gas/solid coexistence is plotted. Dark gray filled circles: Experiment.[4] Light gray solid line: data calculated from Equation (20E21.1), using $\phi = -5.08 \times 10^{-19}$ J and $\theta_E = 215$ K.

We see that the shape of the behavior is similar in the two cases, however, the temperature range at which these vapor pressures are achieved is quite different. Note also that we can plot vapor pressure curves over a wider temperature range than is available from experiment.The use of a microscopic model for the sublimation curve for this purpose makes such extrapolations more reliable than an empirical fit with no underlying model, such as a polynomial.

20.2 Problems

P20.1 Consider hydrogen to be a hard sphere gas $[P(V - \boldsymbol{n}b) = \boldsymbol{nR}T]$ with constant $b = 0.02661$ L mol^{-1}. a) Show that $\mathrm{d}H = C_P\,\mathrm{d}T + \left(V - T\left(\frac{\partial V}{\partial T}\right)_P\right)\mathrm{d}P$. b) Provide a molecular-based explanation as to why it is reasonable to estimate the molar heat capacity at constant pressure for hydrogen to be 29.100 J mol^{-1}K^{-1}. c) Use the information given in parts a) and b) to determine the enthalpy change for 10 moles of hydrogen that is compressed from 1 bar pressure at 298 K to 500 bar pressure at 298 K. d) Use the enthalpy expression in part a) to derive an expression for the internal energy of the hydrogen gas. Evaluate the internal energy change for compressing the gas from 1 bar to 500 bar at 298 K.

Solution

(a) We start with

$$\mathrm{d}H = \left(\frac{\partial H}{\partial T}\right)_P \mathrm{d}T + \left(\frac{\partial H}{\partial P}\right)_T \mathrm{d}P = C_P\,\mathrm{d}T + \left(\frac{\partial H}{\partial P}\right)_T \mathrm{d}P$$

and use Equation (19.22)

$$\left(\frac{\partial H}{\partial P}\right)_T = V - T\left(\frac{\partial V}{\partial T}\right)_P$$

to write that

$$\mathrm{d}H = C_P\,\mathrm{d}T + \left[V - T\left(\frac{\partial V}{\partial T}\right)_P\right]\mathrm{d}P$$

(b) H_2 is a weakly interacting gas. The three translational degrees of freedom contribute $\frac{3}{2}\boldsymbol{R}T$ to the energy and the two rotational degrees of freedom contribute $\boldsymbol{R}T$ to the energy, at high enough temperatures. The vibrational and electronic degrees of freedom have large energy spacing between their energy levels; hence, their temperature dependence may be neglected. Since $H = U + \boldsymbol{R}T$ for an ideal gas,

$$H = \frac{3}{2}\boldsymbol{R}T + \boldsymbol{R}T + \boldsymbol{R}T$$

4 W. P. Gilbreath "The Vapor Pressure of Magnesium between 223° and 385 °C" NASA Technical Note TN d-2723 (1965).

we find that

$$C_P = \left(\frac{\partial H}{\partial T}\right)_P = \frac{7}{2}\boldsymbol{R} = 29.1\ \mathrm{J\ K^{-1}\ mol^{-1}}$$

(c) Because the compression is performed isothermally, we can write that

$$(\Delta H)_T = \int (\mathrm{d}H)_P = \int_{P_0}^{P_f} \left(\frac{\partial H}{\partial P}\right)_T \mathrm{d}P$$

Using

$$\left(\frac{\partial H}{\partial P}\right)_T = \left[V - T\left(\frac{\partial V}{\partial T}\right)_P\right]$$

and the hard sphere gas equation of state $P(V - \boldsymbol{n}b) = \boldsymbol{nR}T$ we find that

$$\left(\frac{\partial V}{\partial T}\right)_P = \left(\frac{\partial}{\partial T}\left(\frac{\boldsymbol{nR}T}{P} + \boldsymbol{n}b\right)\right)_P = \frac{\boldsymbol{nR}}{P}$$

to get

$$\left(\frac{\partial H}{\partial P}\right)_T = \left[V - T\left(\frac{\boldsymbol{nR}}{P}\right)\right] = \boldsymbol{n}b$$

so that

$$\begin{aligned}(\Delta H)_T &= \int_{P_0}^{P_f} \boldsymbol{n}b\ \mathrm{d}P \\ &= (10\ \mathrm{mol})\left(0.02661\ \mathrm{L\ mol^{-1}}\right)(499\ \mathrm{bar}) \\ &= 132.8\ \mathrm{L\ bar} = 13.46\ \mathrm{kJ}\end{aligned}$$

(d) Starting with

$$H = U + PV = U + \boldsymbol{nR}T + \boldsymbol{n}bP$$

we can solve for the internal energy and find

$$U = H - \boldsymbol{nR}T - \boldsymbol{n}bP$$

In differential form this relation takes the form

$$\mathrm{d}U = \mathrm{d}H - \boldsymbol{nR}\ \mathrm{d}T - \boldsymbol{n}b\ \mathrm{d}P$$

Using the result in part (a) we can write that

$$\begin{aligned}\mathrm{d}U &= \boldsymbol{n}C_P\ \mathrm{d}T + \left[V - T\left(\frac{\partial V}{\partial T}\right)_P\right]\ \mathrm{d}P - \boldsymbol{nR}\ \mathrm{d}T - \boldsymbol{n}b\ \mathrm{d}P \\ &= \boldsymbol{n}C_P\ \mathrm{d}T + \boldsymbol{n}b\ \mathrm{d}P - \boldsymbol{nR}\ \mathrm{d}T - \boldsymbol{n}b\ \mathrm{d}P = \boldsymbol{n}C_P\ \mathrm{d}T - \boldsymbol{nR}\ \mathrm{d}T\end{aligned}$$

Because the process is isothermal ($\mathrm{d}T = 0$), we find that $\mathrm{d}U = 0$.

P20.2 The differential Gibbs energy change of a gas is

$$\mathrm{d}G = V\ \mathrm{d}P = \boldsymbol{nR}T\ \frac{\mathrm{d}f}{f}$$

under isothermal conditions. P is the pressure and f is the fugacity. a) Show that the fugacity is related to the compressibility factor Φ of the gas by

$$\ln\left(\frac{f}{P}\right) = \int_{P_0}^{P} \mathrm{d}P'\left(\frac{\Phi - 1}{P'}\right)$$

(b) Using the Berthelot equation of state for a gas, namely

$$PV_m = \boldsymbol{R}T + \frac{9\boldsymbol{R}T_c P}{128P_c}\left(1 - 6\frac{T_c^2}{T^2}\right)$$

find an expression for the fugacity of the gas. Note that T_c is the critical temperature and P_c is the critical pressure of the gas. c) Evaluate the fugacity of carbon dioxide at 298 K and 1 bar pressure, using your result from b) and the facts $T_c = 304$ K and $P_c = 74$ bar for CO_2.

Solution

(a) We begin with the relation given in the problem statement, and write

$$\mathrm{d}\ln f = \frac{V}{\boldsymbol{nRT}}\mathrm{d}P = \frac{\Phi}{P}\mathrm{d}P$$

Next we integrate this expression

$$\int_{P_0}^{f_1} \mathrm{d}\ln f = \ln f_1 - \ln P_0 = \int_{P_0}^{P_1} \frac{\Phi}{P}\mathrm{d}P$$

and find that

$$\begin{aligned}\ln f_1 - \ln P_0 &= \int_{P_0}^{P_1} \frac{\Phi}{P}\mathrm{d}P - \int_{P_0}^{P_1} \frac{\mathrm{d}P}{P} + \int_{P_0}^{P_1} \frac{\mathrm{d}P}{P}\\ &= \int_{P_0}^{P_1} \frac{\Phi - 1}{P}\mathrm{d}P + \int_{P_0}^{P_1} \frac{\mathrm{d}P}{P}\\ &= \int_{P_0}^{P_1} \frac{\Phi - 1}{P}\mathrm{d}P + \ln\left(P_1\right) - \ln P_0\end{aligned}$$

which simplifies to

$$\ln f_1 = \ln P_1 + \int_{P_0}^{P_1} \frac{\Phi - 1}{P'}\mathrm{d}P'$$

If we omit the index 1, we find that

$$\ln f = \ln P + \int_{P_0}^{P} \frac{\Phi - 1}{P'}\mathrm{d}P'$$

which is the requested result.

(b) Using the Berthelot equation of state, we can write

$$\Phi = \frac{PV_m}{\boldsymbol{RT}} = 1 + \frac{9}{128}\frac{T_c}{T}\frac{P}{P_c}\left(1 - 6\frac{T_c^2}{T^2}\right)$$

and hence

$$\ln f = \ln P + \int_{P_0}^{P} \frac{9}{128}\frac{T_c}{T}\frac{1}{P_c}\left(1 - 6\frac{T_c^2}{T^2}\right)\mathrm{d}P'$$

Integrating this expression, we find that

$$f = P \cdot \exp\left(\frac{9}{128}\frac{T_c}{T}\left(1 - 6\frac{T_c^2}{T^2}\right)\frac{P - P_0}{P_c}\right)$$

(c) In the limit that $P_0 \to 0$, we find

$$\begin{aligned}f &= 1\text{ bar} \cdot \exp\left[\frac{9}{128}\frac{304}{298}\left(1 - 6\frac{304^2}{298^2}\right)\left(\frac{1\text{ bar}}{74\text{ bar}}\right)\right]\\ &= 0.995\text{bar}\end{aligned}$$

which indicates that the gas is nearly ideal.

P20.3 Consider the thermodynamic properties of a "real" gas with a molar Gibbs energy of

$$G_m = G^{\ominus}(T) + \boldsymbol{R}T\ln\left(P/P^{\ominus}\right) + bP$$

where $G^{\ominus}(T)$ is the standard state value of the Gibbs energy, i.e., at 1 bar pressure. $\boldsymbol{R}$ is the gas constant, and b is an empirical constant. a) Find the equation of state for the gas. Interpret the parameter b. b) Assume that $G^{\ominus}(T) = c_1 T \cdot \ln\left(c_2 T\right)$, where c_1 and c_2 are constants, and find an expression for the entropy of the gas. c) Using your results from parts a) and b), write expressions for the internal energy of the gas, the enthalpy of the gas, and the Helmholtz energy of the gas. d) Find an expression for the maximum work that can be performed by isothermally expanding the gas from an initial volume $V_{m,1}$ to a final volume $V_{m,2}$.

Solution

(a) We begin with the relations

$$dG = VdP - SdT = \left(\frac{\partial G}{\partial P}\right)_T dP + \left(\frac{\partial G}{\partial T}\right)_P dT$$

By inspection, we can write that

$$V_m = \left(\frac{\partial G_m}{\partial P}\right)_T = \frac{\boldsymbol{R}T}{P} + b$$

which upon rearrangement gives

$$P\left(V_m - b\right) = \boldsymbol{R}T$$

This equation of state is that for the hard sphere gas. The parameter b is the excluded volume; i.e., volume of space excluded to gas particles by virtue of their finite size.

(b) To find the molar entropy, we take the derivative of the molar Gibbs energy with respect to the temperature and find that

$$\left(\frac{\partial G_m}{\partial T}\right)_P = -S_m = \boldsymbol{R}\ln P + c_1 \ln\left(c_2 T\right) + c_1$$

which can be rewritten as

$$S_m = -\boldsymbol{R}\ln P - c_1 \ln\left(c_2 T\right) - c_1$$

or

$$S_m = -\boldsymbol{R}\ln P - \frac{G^{\ominus}(T)}{T} - c_1$$

(c) With results of parts a) and b) in hand, the other energy functions are found by simple algebra. For the Helmholtz energy, we write that

$$\begin{aligned} A_m &= G_m - PV_m \\ &= \boldsymbol{R}T\ln\hat{P} + bP + G^{\ominus}(T) - \boldsymbol{R}T - bP \\ &= \boldsymbol{R}T\ln\hat{P} + G^{\ominus}(T) - \boldsymbol{R}T \end{aligned}$$

For the internal energy, we write that

$$\begin{aligned} U_m &= A_m + TS_m \\ &= \boldsymbol{R}T\ln\hat{P} + G^{\ominus}(T) - \boldsymbol{R}T - \boldsymbol{R}T\ln\hat{P} - G^{\ominus}(T) - c_1 T = -\boldsymbol{R}T - c_1 T \end{aligned}$$

For the enthalpy, we write that

$$\begin{aligned} H_m &= U_m + PV \\ &= -\boldsymbol{R}T - c_1 T + \boldsymbol{R}T + bP = bP - c_1 T \end{aligned}$$

(d) The maximum work in an isothermal expansion is given by the change in the Helmholtz energy; hence, we can write that

$$\begin{aligned} w_{\max,T} &= -(\Delta A)_T = -\boldsymbol{n}\left[\boldsymbol{R}T\ln\left(\hat{P}_2\right) - \boldsymbol{R}T\ln\left(\hat{P}_1\right)\right] \\ &= \boldsymbol{nR}T\ln\left(\frac{\hat{P}_1}{\hat{P}_2}\right) \\ &= \boldsymbol{nR}T\ln\left(\frac{\boldsymbol{R}T/\left(V_{m,1} - b\right)}{\boldsymbol{R}T/\left(V_{m,2} - b\right)}\right) \\ &= \boldsymbol{nR}T\ln\left(\frac{V_{m,2} - b}{V_{m,1} - b}\right) \end{aligned}$$

P20.4 In Problem P20.2, you showed that the fugacity f is related to the compression factor of the gas by

$$\ln\left(\frac{f}{P}\right) = \int_{P_0}^{P} dP' \left(\frac{\Phi - 1}{P'}\right)$$

Find the fugacity and the compression factor for a hard sphere gas whose equation of state is given by $P(V - \boldsymbol{n}b) = \boldsymbol{nR}T$. Sketch the dependence of Φ on P and the dependence of f on P at three different temperatures. Lastly, sketch the dependence of Φ on P for an ideal gas.

Solution

For the hard sphere gas we start with $P(V - \boldsymbol{n}b) = \boldsymbol{nR}T$, which may be written as

$$\frac{PV}{\boldsymbol{nR}T} - \frac{Pb}{\boldsymbol{R}T} = 1$$

or

$$\Phi = \frac{PV}{\boldsymbol{nR}T} = 1 + \frac{Pb}{\boldsymbol{R}T}$$

By substituting in our expression for the fugacity, we find that

$$\begin{aligned}\ln(f) &= \ln(P) + \int_{P_0}^{P} \frac{1}{P'}\frac{P'b}{\boldsymbol{R}T} dP' \\ &= \ln(P) + \int_{P_0}^{P} \frac{b}{\boldsymbol{R}T} dP' \\ &= \ln(P) + \frac{b}{\boldsymbol{R}T}(P - P_0)\end{aligned}$$

By exponentiating both sides, we find that

$$f = P\exp\left(\frac{b}{\boldsymbol{R}T}(P - P_0)\right)$$

(c) Using $b = 0.02661$ mol L^{-1} we plot Φ versus P at different temperatures in Fig. P20.4a. $T = 100$ K (black solid curve), 200 K (dark gray solid curve), 300 K (black dashed curve), and 400 K (dark gray dashed curve).

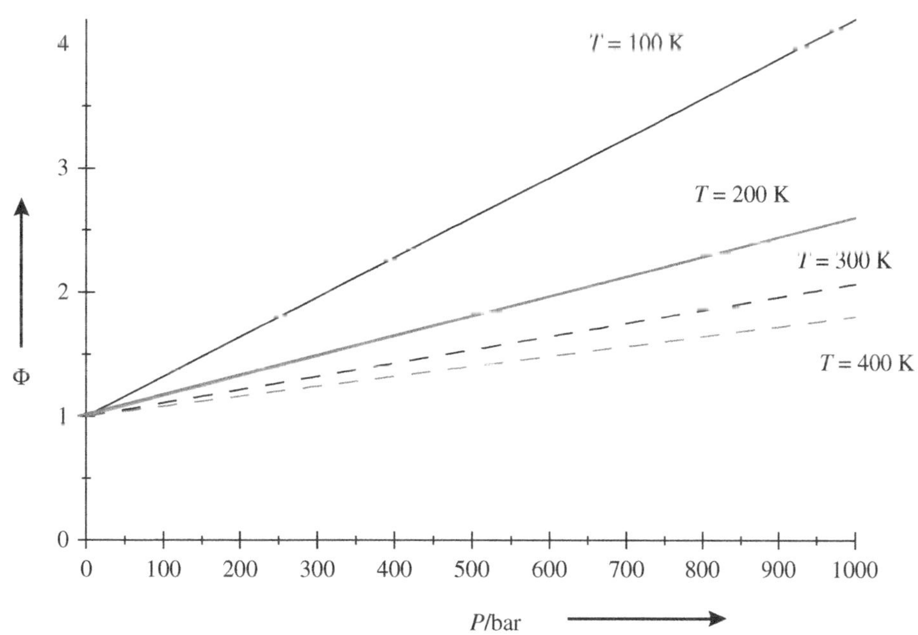

Figure P20.4a Compression factor Φ versus the pressure P for a hard sphere gas at different temperatures.

In Fig. P20.4b we plot f versus P at the temperatures $T = 100$ K (solid black curve), 200 K (solid dark gray curve), and 300 K (dashed black curve).

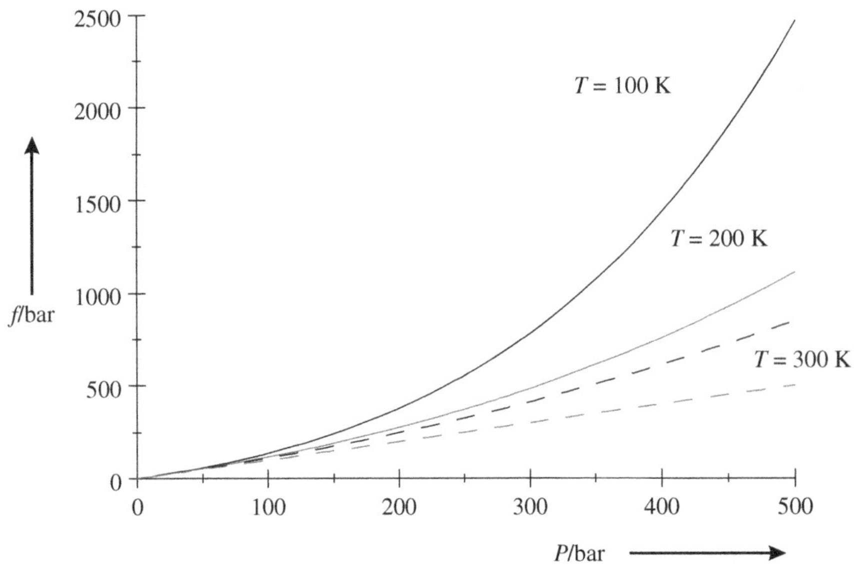

Figure P20.4b Fugacity f for a hard sphere gas versus the pressure P at different temperatures T. The dark gray dashed line represents ideal gas behavior, for which $f = P$ at all temperatures.

P20.5 Consider our definition of the isothermal compressibility κ_T. Derive a mathematical expression for this quantity in the case of an ideal gas equation of state. Derive a mathematical expression for the isothermal compressibility for a hard sphere gas equation of state $P(V - nb) = nRT$. Compare your results and find an expression for the deviation $\delta\kappa_T$ in the isothermal compressibility, defined as $\delta\kappa_T = \kappa_T(\text{ideal}) - \kappa_T(\text{hard sphere})$. Provide a physical interpretation for why the isothermal compressibility differs in the way it does.

Solution

For an ideal gas the isothermal compressibility is given by

$$\kappa_T = -\frac{1}{V}\left(\frac{\partial V}{\partial P}\right)_T = -\frac{1}{V}\left(\frac{\partial}{\partial P}\left[\frac{\boldsymbol{nRT}}{P}\right]\right)_T$$
$$= -\frac{1}{V}\left(\frac{-\boldsymbol{nRT}}{P^2}\right) = \frac{\boldsymbol{nRT}}{PV}\frac{1}{P} = \frac{1}{P}$$

For a hard sphere gas, it is given by

$$\kappa_T = -\frac{1}{V}\left(\frac{\partial V}{\partial P}\right)_T = -\frac{1}{V}\left(\frac{\partial}{\partial P}\left[\frac{\boldsymbol{nRT}}{P} + \boldsymbol{n}b\right]\right)_T$$
$$= -\frac{1}{V}\left(\frac{-\boldsymbol{nRT}}{P^2}\right) = \frac{1}{PV}\frac{\boldsymbol{nRT}}{P} = \frac{1}{PV}(V - \boldsymbol{n}b) = \frac{1}{P} - \frac{\boldsymbol{n}b}{PV}$$

(c) The difference between the two is

$$\delta\kappa_T = \frac{\boldsymbol{n}b}{PV}$$

which means that the compressibility of a hard sphere gas is smaller than that of an ideal gas. This difference arises from the finite size of the gas atoms, restricting the volume change.

P20.6 The molar Helmholtz free energy of a van der Waals gas may be written as

$$A_m = -\frac{a}{V_m} - \boldsymbol{R}T\ln\left[(V_m - b)/V_m^{\ominus}\right] - \boldsymbol{R}T - 3\boldsymbol{R}T\ln\left(C_1T^{1/2}\right)$$

where a, b, and C_1 are constants. a) Use this expression to derive an expression for the molar entropy of the van der Waals gas. b) Use the previous results to derive an expression for the molar internal energy of the van der Waals gas, U_m. c) Compare your expression for the molar internal energy of the van der Waals gas to that for one mole of a monatomic ideal gas. Determine how they differ mathematically and provide a physically based explanation for this difference. d) Find an expression for the molar constant volume heat capacity $C_{V,m}$ of the van der Waals gas. Compare your finding to that of a monatomic ideal gas and provide a physically based explanation for your findings.

Solution

(a) We begin with the differential form of A

$$\mathrm{d}A = -P\,\mathrm{d}V - S\,\mathrm{d}T = \left(\frac{\partial A}{\partial V}\right)_T \mathrm{d}V + \left(\frac{\partial A}{\partial T}\right)_V \mathrm{d}T$$

By inspection we can write that

$$\begin{aligned}\left(\frac{\partial A_m}{\partial T}\right)_{V_m} &= -S_m\\ &= 0 - \boldsymbol{R}T\ln\left(\frac{V_m - b}{V_m^{\ominus}}\right) - \boldsymbol{R} - 3\boldsymbol{R}\ln\left(C_1T^{1/2}\right) - \frac{3\boldsymbol{R}T}{2T}\\ &= -\boldsymbol{R}\ln\left(\frac{V_m - b}{V_m^{\ominus}}\right) - \frac{5}{2}\boldsymbol{R} - 3\boldsymbol{R}\ln\left(C_1T^{1/2}\right)\end{aligned}$$

or

$$S_m = \boldsymbol{R}\ln\left(\frac{V_m - b}{V_m^{\ominus}}\right) + \frac{5}{2}\boldsymbol{R} + 3\boldsymbol{R}\ln\left(C_1T^{1/2}\right)$$

(b) To find the molar internal energy we use the fundamental relation

$$U_m = A_m + TS_m$$

Substituting the results from part a) we find that

$$\begin{aligned}U_m &= \left[\begin{array}{c} -\frac{a}{V_m} - \boldsymbol{R}T\ln\left[(V_m - b)/V_m^{\ominus}\right] - \boldsymbol{R}T - 3\boldsymbol{R}T\ln\left(C_1T^{1/2}\right)\\ +T\left(\boldsymbol{R}\ln\left(\frac{V_m-b}{V_m^{\ominus}}\right) + \frac{5}{2}\boldsymbol{R} + 3\boldsymbol{R}\ln\left(C_1T^{1/2}\right)\right)\end{array}\right]\\ &= -\frac{a}{V_m} + \frac{3}{2}\boldsymbol{R}T\end{aligned}$$

(c) The difference between this form of the internal energy and that of an ideal gas is

$$\begin{aligned}\delta U_m &= U_{m,\text{ideal}} - U_{m,vdw}\\ &= \frac{3}{2}\boldsymbol{R}T - \left[\frac{3}{2}\boldsymbol{R}T - \frac{a}{V_m}\right] = \frac{a}{V_m}\end{aligned}$$

As discussed in the text, the factor $-a/V_m$ accounts for the attractive forces between the atoms.

(d) The molar heat capacity is

$$\left(\frac{\partial U_m}{\partial T}\right)_v = \left(\frac{\partial}{\partial T}\left[-\frac{a}{V_m} + \frac{3}{2}\boldsymbol{R}T\right]\right)_V = \frac{3}{2}\boldsymbol{R}$$

The difference between an ideal gas and van der Waals gas molar heat capacities is zero; namely

$$C_{V,m,\text{ideal}} - C_{V,m,\text{van der Waals}} = 0$$

The value of the heat capacity arises from the translational degrees of freedom of the atoms. As long as the forces between the atoms are weak enough, they do not influence the heat capacity.

P20.7 If the equation of state for a liquid is

$$V_m = V_{m,0}\left[1 + a\cdot T - b\cdot P\right]$$

where P is the pressure, $V_{m,0}$ is the molar volume of the liquid at $T = 298$ K and $P = 10^5$ Pa, and a and b are empirical constants, show that

$$\left(\frac{\partial S}{\partial P}\right)_T = -aV_{m,0}\,;\ \left(\frac{\partial S}{\partial V_m}\right)_T = -\frac{a}{b}\,;\text{ and }\left(\frac{\partial H}{\partial P}\right)_T = V_{m,0}\,(1 - bP)$$

Calculate ΔS and ΔH when 1 mol of the liquid is compressed at $T = 293$ K from $P = 10^5$ Pa to $P = 2.5 \times 10^6$ Pa. Assume that $V_{m,0} = 58.62$ mL mol^{-1}, $a = 1.432 \cdot 10^{-3}$ K^{-1}, and $b = 1.120 \times 10^{-9}$ Pa^{-1} typical of ethanol (adapted from R. E. Dickerson, Molecular Thermodynamics).

Solution

To show these results we will make use of the following Maxwell relations

$$\left(\frac{\partial S}{\partial P}\right)_T = -\left(\frac{\partial V_m}{\partial T}\right)_P \quad \text{and} \quad \left(\frac{\partial S}{\partial V_m}\right)_T = \left(\frac{\partial P}{\partial T}\right)_V$$

and for the last expression we use

$$\left(\frac{\partial H}{\partial P}\right)_T = -T\left(\frac{\partial V_m}{\partial T}\right) + V$$

For the first case, we find that

$$\begin{aligned}\left(\frac{\partial S}{\partial P}\right)_T &= -\left(\frac{\partial V_m}{\partial T}\right)_P \\ &= -\left(\frac{\partial}{\partial T}\left[V_{m,0}\left[1 + a \cdot T - b \cdot P\right]\right]\right)_P \\ &= -aV_{m,0}\end{aligned}$$

For the second one we rearrange the equation of state to the form

$$P = \frac{V_m}{V_{m,0}b} - \frac{1}{b} - \frac{a}{b}T$$

and then evaluate the derivative to find

$$\begin{aligned}\left(\frac{\partial S}{\partial V_m}\right)_T &= \left(\frac{\partial P}{\partial T}\right)_V \\ &= \left(\frac{\partial}{\partial T}\left[\frac{V_m}{V_{m,0}b} - \frac{1}{b} - \frac{a}{b}T\right]\right)_V = -\frac{a}{b}\end{aligned}$$

For the last expression we have

$$\begin{aligned}\left(\frac{\partial H}{\partial P}\right)_T &= -T\left(\frac{\partial V_m}{\partial T}\right) + V_m \\ &= -T\left(\frac{\partial}{\partial T}\left[V_{m,0}\left[1 + a \cdot T - b \cdot P\right]\right]\right) + V_m \\ &= -T\left(aV_{m,0}\right) + V_{m,0}\left[1 + a \cdot T - b \cdot P\right] = V_{m,0}(1 - bP)\end{aligned}$$

When compressing the liquid the change in entropy will be

$$\Delta S_m = -aV_{m,0}\int_{P_0}^{P_f} dP = -aV_{m,0}\left(P_f - P_0\right)$$

Inserting the values given in the problem statement ($V_{m,0} = 5.862 \times 10^{-5}$ m^3/mol, $a = 1.432 \cdot 10^{-3}$ K^{-1}, and $b = 1.120 \times 10^{-9}$ Pa^{-1}), we find that

$$\begin{aligned}\Delta S_m &= -aV_{m,0}\left(P_f - P_0\right) \\ &= -\left(1.432 \times 10^{-3}\ \text{K}^{-1}\right) \cdot V_{m,0} \cdot \left(25 \times 10^5\ \text{Pa} - 1 \times 10^5\ \text{Pa}\right) \\ &= -\left(5.862 \times 10^{-5}\text{m}^3/\text{mol}\right) \cdot \left(3.437 \cdot 10^3\ \text{Pa K}^{-1}\right) \\ &= -0.2015\ \text{J mol}^{-1}\ \text{K}^{-1}\end{aligned}$$

The molar enthalpy change is

$$\begin{aligned}\Delta H_m &= \int_{P_0}^{P_f} V_{m,0}(1 - bP)\, dP \\ &= V_{m,0} \cdot \left(P_f - P_0\right) - V_{m,0} \cdot b\int_{P_0}^{P_f} P\, dP \\ &= V_{m,0} \cdot \left(P_f - P_0\right) - \frac{V_{m,0} \cdot b}{2}\left(P_f^2 - P_0^2\right) \\ &= V_{m,0} \cdot \left(P_f - P_0\right)\left(1 - b(P_f + P_0)/2\right)\end{aligned}$$

Inserting the values given in the problem statement we find that

$$\Delta H_m = V_{m,0} \cdot \left(25 \times 10^5 \text{ Pa} - 1 \times 10^5 \text{ Pa}\right) \left(1 - \left(1.120 \times 10^{-9}\text{Pa}^{-1}\right) \cdot (25 \times 10^5 \text{ Pa} + 1 \times 10^5 \text{ Pa})/2\right)$$
$$= \left(5.862 \times 10^{-5} \text{ m}^3/\text{mol}\right) \cdot 2.393 \cdot 10^6 \text{ Pa} = 140.3 \text{ J mol}^{-1}$$

P20.8 In the MathCAD exercise, sublimation.mcdx, you study the phase equilibria of Ar between its solid and its vapor. For the molar Gibbs energy of Ar gas you use

$$G_{m,\text{gas}}(T,P) = \boldsymbol{R}T \ln(P/P^{\ominus}) - \boldsymbol{R}T \ln\left[\frac{kT}{P^{\ominus}}\left(\frac{2\pi mkT}{h^2}\right)^{3/2}\right]$$

and for the molar Gibbs energy of Ar solid you use

$$G_{m,\text{solid}}(T,P) = \frac{3}{2}\boldsymbol{R}\theta + \frac{\phi_m}{2} + 3\boldsymbol{R}T\ln\left(1 - e^{-\theta/T}\right)$$

where θ is 30 K and ϕ_m is -14.44 kJ mol^{-1}.[5] These equations are derived in Foundation 20.1. Use the Gibbs energy expression to determine the molar entropy of the gas, $S_{m,\text{gas}}$. Given that the entropy of the solid is

$$S_{m,\text{solid}} = -3\boldsymbol{R}\ln\left[1 - e^{-\theta/T}\right] + 3\boldsymbol{R}\frac{\theta/T}{(\exp(\theta/T) - 1}$$

find an expression for the entropy change in the phase transition. Use these entropy expressions and the Gibbs energy expressions to determine the enthalpy change in the phase transition. Simplify your answer and evaluate it for $T = 80$ K. Compare your result with that reported by H. H. Chen et al., *J. Chem. Thermodynamics* 1978, **10**, 649–659.

Solution

First we determine $S_{m,\text{gas}}$ as

$$-S_{m,\text{gas}} = \left(\frac{\partial G_m}{\partial T}\right)_P$$
$$= -\boldsymbol{R}\ln\left[\frac{kT}{P^{\ominus}}\left(\frac{2\pi mkT}{h^2}\right)^{3/2}\right] - \boldsymbol{R}T\left(\frac{5}{2T}\right) + \boldsymbol{R}\ln\left(\frac{P}{P^{\ominus}}\right)$$
$$= \boldsymbol{R}\ln\left[\frac{kT}{P^{\ominus}}\left(\frac{2\pi mkT}{h^2}\right)^{3/2}\right] + \frac{5}{2}\boldsymbol{R} - \boldsymbol{R}\ln\left(\frac{P}{P^{\ominus}}\right)$$

The entropy change for the phase transition is

$$\Delta S_m = S_{m,\text{gas}} - S_{m,\text{solid}}$$
$$= \left(\begin{array}{l} \boldsymbol{R}\ln\left[\frac{kT}{P^{\ominus}}\left(\frac{2\pi mkT}{h^2}\right)^{3/2}\right] + \frac{5}{2}\boldsymbol{R} - \boldsymbol{R}\ln\left(\frac{P}{P^{\ominus}}\right) \\ -\left[-3\boldsymbol{R}\ln\left[1 - e^{-\theta/T}\right] + 3\boldsymbol{R}\frac{\theta/T}{\exp(\theta/T)-1}\right] \end{array}\right)$$

Next we find the enthalpy change for the transition as

$$\Delta H = \Delta G + T\Delta S$$

For the molar Gibbs energy change we find

$$\Delta G_m = G_{m,\text{gas}} - G_{m,\text{solid}}$$
$$= \left(\begin{array}{l} \boldsymbol{R}T\ln(P/P^{\ominus}) - \boldsymbol{R}T\ln\left[\frac{kT}{P^{\ominus}}\left(\frac{2\pi mkT}{h^2}\right)^{3/2}\right] \\ -\left[\frac{3}{2}\boldsymbol{R}\theta + \frac{\phi}{2} + 3\boldsymbol{R}T\ln\left(1 - e^{-\theta/T}\right)\right] \end{array}\right)$$

5 a) H. H. Chen, R. Aziz and C. C. Lim *Canadian J. Phys.*, 1971, **49**, 1569.
b) H. H. Chen, C. C. Lim, and R. Aziz J. Chem. *Thermodynamics* 1978, **10**, 649–659.

so that the molar enthalpy change is

$$\Delta H_m = \left(\begin{array}{c} RT\ln\left(\frac{P}{P^\ominus}\right) - RT\ln\left[\frac{kT}{P^\ominus}\left(\frac{2\pi mkT}{h^2}\right)^{3/2}\right] \\ -\left[\frac{3}{2}R\theta + \frac{\phi}{2} + 3RT\ln\left(1-e^{-\theta/T}\right)\right] + RT\ln\left[\frac{kT}{P^\ominus}\left(\frac{2\pi mkT}{h^2}\right)^{3/2}\right] \\ +\frac{5}{2}RT - RT\ln\left(\frac{P}{P^\ominus}\right) - \left[-3RT\ln\left[1-e^{-\theta/T}\right] + 3RT\frac{\theta/T}{\exp(\theta/T)-1}\right] \end{array} \right)$$

By canceling the first two terms with the fourth and sixth terms, this expression can be written as

$$\Delta H_m = \left(\begin{array}{c} -\frac{3}{2}R\theta - \frac{\phi}{2} - 3RT\ln\left[1-e^{-\theta/T}\right] + \frac{5}{2}RT \\ +3RT\ln\left[1-e^{-\theta/T}\right] - 3RT\frac{\theta/T}{\exp(\theta/T)-1} \end{array} \right)$$

Inspection shows that the third and fifth terms also cancel, and we find that

$$\Delta H_m = -\frac{3}{2}R\theta - \frac{\phi}{2} + \frac{5}{2}RT - 3\frac{R\theta}{\exp(\theta/T)-1}$$

To evaluate the enthalpy we use the parameter values given in the problem statement and find that

$$\begin{aligned} \Delta H_m &= -\frac{3}{2}R\theta - \frac{\phi}{2} + \frac{5}{2}RT - 3\frac{R\theta}{\exp(\theta/T)-1} \\ &= -\frac{3}{2}\cdot 8.3145\ \text{J K}^{-1}\ \text{mol}^{-1}\cdot 30\ \text{K} - \frac{-14.44\times 10^3\ \text{J mol}^{-1}}{2} + \frac{5}{2}\cdot 8.3145\ \text{J K}^{-1}\ \text{mol}^{-1}\cdot 80\ \text{K} \\ &\quad -3\frac{30\cdot 8.3145\ \text{J K}^{-1}\ \text{mol}^{-1}}{(\exp(30/80)-1)} \\ &= \left[-374.0 + 7.22\times 10^3 + 1.663\times 10^3 - 1644\right]\ \text{J mol}^{-1} \\ &= 6.87\times 10^3\ \text{J mol}^{-1} = 6.87\ \text{kJ mol}^{-1} \end{aligned}$$

This value is within 12% of the experimental value 7.7894 kJ mol $^{-1}$ reported by Chen et al.[6]

P20.9 Given the data in the table and the fact that the constant pressure heat capacity at 5 bar follows the empirical equation $C_{P,m} = a\cdot T^2 - b\cdot T + c$ with $a = 1.680\cdot 10^{-7}$ J mol^{-1} K^{-3}, $b = 0.1043$ J mol^{-1} K^{-2}, and $c = 91.458$ J mol^{-1} K^{-1}, calculate the change in enthalpy of 1 mol of liquid water if it is compressed from 5 bar and 300 K to 100 bar and 400 K.

P/bar	ρ/kg m^{-3}	V_m/10^{-5} m^3 mol^{-1}	$\frac{\partial V_m}{\partial T}$/10^{-8} m^3 mol^{-1} K^{-1}
5	937.617	1.92139	1.140
7.5	937.745	1.92113	1.139
10	937.873	1.92086	1.138
20	938.385	1.91982	1.136
30	938.895	1.91877	1.134
40	939.404	1.91773	1.131
50	939.91	1.9167	1.129
60	940.415	1.91567	1.127
70	940.919	1.91465	1.124
80	941.42	1.91363	1.122
100	942.418	1.9116	1.118

6 H. H. Chen, C. C. Lim, and R. Aziz J. Chem. *Thermodynamics* 1978, **10**, 649–659.

Solution

To solve this problem we break it into two paths. The first is to calculate the enthalpy change going from a pressure of 5 bar and 300 K to 5 bar and 400 K. The second step is to determine the enthalpy change from 5 bar and 400 K to 100 bar and 400 K. The first step is

$$
\begin{aligned}
\Delta H_1 &= \int_{300}^{400} C_{P,m}(T)\ \mathrm{d}T \\
&= \int_{300}^{400} \left(a \cdot T^2 - b \cdot T + c\right)\ \mathrm{d}T = \left| a \cdot \frac{1}{3}T^3 - b \cdot \frac{1}{2}T^2 + c \cdot T \right|_{300}^{400} \\
&= a \cdot \frac{1}{3}\left(400^3 - 300^3\right)\,\mathrm{K}^3 - b \cdot \frac{1}{2}\left(400^2 - 300^2\right)\,\mathrm{K}^2 + c \cdot (400 - 300)\,\mathrm{K} \\
&= 1.680 \cdot 10^{-7}\ \mathrm{J\ mol^{-1}\ K^{-3}} \cdot \frac{1}{3}(3.70 \times 10^7)\ \mathrm{K}^3 - 0.1043\ \mathrm{J\ mol^{-1} K^{-2}} \cdot \frac{1}{2}(7.0 \times 10^4)\ \mathrm{K}^2 \\
&\quad + 91.458\ \mathrm{J\ mol^{-1}\ K^{-1}} \cdot 100.0)\mathrm{K} \\
&= \left(2.07 \times 10^3 - 3.65 \times 10^3 + 9.15 \times 10^3\right)\ \mathrm{J\ mol^{-1}} = 7.54 \times 10^3\ \mathrm{J\ mol^{-1}}
\end{aligned}
$$

For the second part we need to evaluate

$$
\left(\frac{\partial H}{\partial P}\right)_T = -T\left(\frac{\partial V_m}{\partial T}\right) + V_m
$$

which upon integration becomes

$$
\Delta H_2 = -T \int_5^{100} \left(\frac{\partial V_m}{\partial T}\right)\ \mathrm{d}P + \int_5^{100} V_m\ \mathrm{d}P
$$

We observe in the table that the derivative of the molar volume with respect to the temperature is a weak function of pressure. Inspection of the data in the table shows that V_m differs by only 0.5% in the full range; thus, we use the average value of $1.9165 \times 10^{-5}\ \mathrm{m^3\ mol^{-1}}$ without much loss of accuracy. Correspondingly $\frac{\partial V_m}{\partial T}$ differs by only 2% and we can take the average value to be $1.129 \times 10^{-8}\ \mathrm{m^3\ mol^{-1}\ K^{-1}}$. Using these values, we calculate

$$
\begin{aligned}
\Delta H_2 &= -400\ \mathrm{K} \cdot 1.129 \times 10^{-8}\ \mathrm{m^3\ mol^{-1}\ K^{-1}} \cdot 95\ \mathrm{bar} \\
&\quad + 1.9165 \times 10^{-5}\ \mathrm{m^3\ mol^{-1}} \cdot 95\ \mathrm{bar} \\
&= \left(-4.290 \times 10^{-4} + 1.821 \times 10^{-3}\right)\ \mathrm{m^3\ mol^{-1}\ bar} \\
&= 1.392 \times 10^{-3}\ \mathrm{m^3\ mol^{-1}\ bar} \\
&= 1.392\ \mathrm{L\ bar\ mol^{-1}} = 139.2\ \mathrm{J\ mol^{-1}}
\end{aligned}
$$

The overall enthalpy change is

$$
\begin{aligned}
\Delta H_1 + \Delta H_2 &= (7567.3 + 139.18)\ \mathrm{J\ mol^{-1}} \\
&= 7706.5\ \mathrm{J\ mol^{-1}}
\end{aligned}
$$

Note that a more rigorous evaluation of the integrals, realized by fitting the data to linear forms $[(\partial V_m/\partial T) = 1.1407 \times 10^{-8} - 2.3133 \times 10^{-12}\,P$ and $V_m = 1.9219 \times 10^{-5} - 10.322 \times 10^{-10}P]$ and then performing the integrals, gives the same result.

P20.10 In Section 20.2 we claim that an ideal gas can be defined as a substance whose Gibbs energy is given by

$$
G(T,P) = G^{\ominus}(T) + NkT \cdot \ln \hat{P}
$$

where $G^{\ominus}(T)$ is the Gibbs energy under standard conditions and $\hat{P} = P/P^{\ominus}$. Using your knowledge of thermodynamics, prove that this is true by deriving an expression for the ideal gas equation of state and showing that the internal energy depends only on T.

Solution

Equation of state: Consider the basic relation

$$
\mathrm{d}G = -S\mathrm{d}T + V\mathrm{d}P
$$

from Section 19.3.2 from which we found that

$$\left(\frac{\partial G}{\partial T}\right)_P = -S \text{ and } \left(\frac{\partial G}{\partial P}\right)_T = V$$

Using this second relation, we can write that

$$V = \left(\frac{\partial G}{\partial P}\right)_T = \left(\frac{\partial}{\partial P}\left[G^{\ominus}(T) + NkT \cdot \ln \hat{P}\right]\right)_T$$
$$= 0 + NkT \cdot \frac{1}{P}$$

This result is the ideal gas equation of state

$$PV = NkT$$

$U(T)$: From the general definitions of energy functions we can write that

$$G = H - TS = U + PV - TS$$

or

$$U = G - PV + TS = G - NkT + TS$$
$$= G^{\ominus}(T) + NkT \cdot \ln \hat{P} - NkT + TS$$

Next we find S, through the relation

$$S = -\left(\frac{\partial G}{\partial T}\right)_P = -\left(\frac{\partial}{\partial T}\left[G^{\ominus}(T) + NkT \cdot \ln \hat{P}\right]\right)_P$$
$$= -\left(\frac{\partial G^{\ominus}(T)}{\partial T}\right)_P - Nk \cdot \ln \hat{P}$$

and we substitute it back into the equation for U to find

$$U = G^{\ominus}(T) + NkT \cdot \ln \hat{P} - NkT - T\left[\left(\frac{\partial G^{\ominus}(T)}{\partial T}\right)_P + Nk \cdot \ln \hat{P}\right]$$
$$= G^{\ominus}(T) - NkT - T\left(\frac{\partial G^{\ominus}(T)}{\partial T}\right)_P$$

This result shows that U is a function of only T, not P; i.e., $U(T)$.

P20.11 Find mathematical expressions (Equation (20.18)) for the van der Waals parameters a and b in terms of the critical constants, T_c, P_c, and V_c, of a gas.

Solution

By differentiation of Equation (20.15)

$$P = \frac{RT}{V_m - b} - \frac{a}{V_m^2}$$

we obtain

$$\left(\frac{\partial P}{\partial V_m}\right)_T = -\frac{RT}{(V_m - b)^2} + \frac{2a}{V_m^3}$$
$$\left(\frac{\partial^2 P}{\partial V_m^2}\right)_T = \frac{2RT}{(V_m - b)^3} - \frac{6a}{V_m^4}$$

Because of the conditions in Equation (20.16)

$$\frac{\partial P}{\partial V} = 0 \quad \text{and} \quad \frac{\partial^2 P}{\partial V^2} = 0$$

we obtain

$$\frac{RT_c}{(V_{m,c} - b)^2} = \frac{2a}{V_{m,c}^3} \quad \text{and} \quad \frac{RT_c}{(V_{m,c} - b)^3} = \frac{3a}{V_{m,c}^4}$$

We divide the left equation by the right one and obtain

$$\left(V_{m,c} - b\right) = 2V_{m,c}/3$$

which rearranges to

$$V_{m,c} = 3b$$

Inserting this result into the left equation, we obtain

$$\frac{\boldsymbol{R}T_c}{(2b)^2} = \frac{2a}{(3b)^3} \quad \text{and} \quad T_c = \frac{8a}{27\boldsymbol{R}b}$$

Finally we insert $V_{m,c}$ and T_c in Equation (20.15)

$$P_c = \frac{\boldsymbol{R}T_c}{2b} - \frac{a}{(3b)^2} = \frac{\boldsymbol{R}\cdot 8a}{2b\cdot 27\boldsymbol{R}b} - \frac{a}{(3b)^2} = \frac{a}{27b^2}$$

P20.12 Derive an expression for the Joule–Thomson coefficient, and its inversion temperature, of a van der Waals gas. To simplify the analysis, you should write the van der Waals equation of state in the approximate form

$$V_m \simeq \frac{\boldsymbol{R}}{P}T + b - \frac{a}{\boldsymbol{R}T} + \frac{abP}{\boldsymbol{R}^2T^2}$$

Explain the reasoning behind this approximate form.

Solution

We start with the relation derived in Section 20.4.2 for the Joule–Thomson coefficient, μ_{JT},

$$\mu_{JT} = \left(\frac{\partial T}{\partial P}\right)_H = \frac{1}{C_{P,m}}\cdot\left[T\left(\frac{\partial V_m}{\partial T}\right)_P - V_m\right]$$

and the van der Waals equation derived in Section 20.3.2, Equation (20.12),

$$\left(P + \frac{a}{V_m^2}\right)\cdot\left(V_m - b\right) = \boldsymbol{R}T$$

By expanding the left side of this equation we find that

$$PV_m - Pb + \frac{a}{V_m} - \frac{ab}{V_m^2} = \boldsymbol{R}T$$

and we can rearrange it to write

$$V_m = \frac{\boldsymbol{R}}{P}T + b - \frac{a}{PV_m} + \frac{ab}{PV_m^2}$$

This form shows that the leading term for V_m is the ideal gas relation and that the last two terms on the right-hand side, which depend on V_m, are proportional to a and correspond to corrections to the ideal gas law. If we assume that these terms are significantly smaller than the leading term, then we can approximate V_m in the two correction terms by $V_m \approx \boldsymbol{R}T/P$. Making this substitution we find that

$$V_m \simeq \frac{\boldsymbol{R}}{P}T + b - \frac{a}{\boldsymbol{R}T} + \frac{abP}{\boldsymbol{R}^2T^2} \tag{20P12.1}$$

Taking the derivative with respect to T at constant P, we obtain

$$\left(\frac{\partial V_m}{\partial T}\right)_P = \frac{\boldsymbol{R}}{P} + \frac{a}{\boldsymbol{R}T^2} - \frac{2abP}{\boldsymbol{R}^2T^3} \tag{20P12.2}$$

In order to eliminate the term $\boldsymbol{R}/P$ in Equation (20P12.2) we solve Equation (20P12.1) for $\boldsymbol{R}/P$

$$\frac{\boldsymbol{R}}{P} = \frac{V_m}{T} - \frac{b}{T} + \frac{a}{\boldsymbol{R}T^2} - \frac{abP}{\boldsymbol{R}^2T^3} \tag{20P12.3}$$

By inserting Equation (20P12.3) in Equation (20P12.2) we obtain

$$\left(\frac{\partial V_m}{\partial T}\right)_P = \frac{a}{\boldsymbol{R}T^2} - \frac{2abP}{\boldsymbol{R}^2T^3} + \frac{V_m}{T} - \frac{b}{T} + \frac{a}{\boldsymbol{R}T^2} - \frac{abP}{\boldsymbol{R}^2T^3}$$

$$= \frac{2a}{\boldsymbol{R}T^2} - \frac{3abP}{\boldsymbol{R}^2T^3} + \frac{V_m}{T} - \frac{b}{T} \tag{20P12.4}$$

By substituting into our expression for the Joule–Thomson coefficient we obtain

$$\mu_{JT} = \left(\frac{\partial T}{\partial P}\right)_H = \frac{1}{C_{P,m}} \cdot \left[T\left(\frac{\partial V_m}{\partial T}\right)_P - V_m\right]$$
$$= \frac{1}{C_{P,m}} \cdot \left[\frac{2a}{\mathbf{R}T} - b - \frac{3ab}{\mathbf{R}^2T^2} \cdot P\right] \tag{20P12.5}$$

The condition for the inversion temperature T_i is

$$\mu_{JT} = 0$$

If we set μ_{JT} equal to zero, we find

$$\frac{2a}{\mathbf{R}T_i} - b - \frac{3ab}{\mathbf{R}^2T_i^2} \cdot P = 0$$

and can solve it for T_i to find

$$T_i = \frac{1}{b\mathbf{R}}\left[a \pm \sqrt{a^2 - 3ab^2 \cdot P}\right]$$

21

Real Solutions

21.1 Exercises

E21.1 Consider a 1.50 molal aqueous HCl solution and assume that the HCl is not volatile. When the external pressure is 355.1 Torr, the enthalpy of vaporization $\Delta_{vap}H$ for water is 40.7 kJ mol^{-1} and its entropy of vaporization is $\Delta_{vap}S$ is 115 J mol^{-1} K^{-1}. Find the boiling point of pure water and of the HCl solution under these conditions. Provide a physical explanation for the boiling point elevation, or vapor pressure lowering, of the water. In your explanation discuss the Gibbs energy and provide a sketch of its concentration dependence.

Solution

To find the boiling point of water at a 355.1 Torr pressure, we use the fact that $\Delta G = 0$ at the phase transition so that $\Delta_{vap}H = T_s \cdot \Delta_{vap}S$ and

$$T_s = \frac{\Delta_{vap}H}{\Delta_{vap}S}$$

Hence we find that

$$T_s = \frac{40{,}700 \text{ J mol}^{-1}}{115 \text{ J mol}^{-1}\text{K}^{-1}} = 354 \text{ K}$$

The boiling point elevation (ΔT) of the HCl solution is found by solving Equation (21.62), namely

$$\Delta T = -\frac{\boldsymbol{R}T_s^2}{\Delta_{vap}H} \cdot \ln\left(a_s\right)$$

where a_s is the activity of the solvent. If we make the approximation that $a_s \approx x_s$, where x_s is the mole fraction of the solvent we can proceed readily. At this molality the mole fraction of solute is

$$x_{\text{HCl}} = \frac{2 \times 1.50 \text{ molal}}{2 \times 1.50 \text{ molal} + 55.5 \text{ molal}} = 5.13 \times 10^{-2}$$

and the mole fraction of the solvent is

$$x_s = 1 - 0.051 = 0.949$$

Hence, the boiling point elevation is

$$\Delta T = \left[-\frac{8.314 \text{ J K}^{-1} \text{ mol}^{-1} \cdot (354 \text{ K})^2}{40{,}700 \text{ kJ mol}^{-1}} \cdot \ln(0.949)\right] \text{K} = 1.34 \text{ K}$$

and so the boiling point is

$$(354 + 1.34) \text{ K} = 355.3 \text{ K}$$

Because the dissolution of HCl in water is spontaneous, we know that $G_{mix} < G_{pure}$, and part of this arises from the fact that ΔS of mixing is positive. For HCl we also have a considerable enthalpy of mixing term from solvation. $HCl \rightarrow H^+_{(aq)} + Cl^-_{(aq)}$. At equilibrium, the molar Gibbs energy of the vapor (μ_{gas}) and that of the mixture ($\mu_{mixture}$) are equal, and we can write that

$$\mu_{gas} = \mu_{mixture} = \mu^{\ominus}_{H_2O} + \boldsymbol{R}T \ln\left(a_{H_2O}\right)$$

In the approximation of an ideal solution $a_{H_2O} = x_{H_2O} = 1 - x_{HCl}$, and we can write that

$$\mu_{gas} \approx \mu^{\ominus}_{H_2O} + RT \ln\left(1 - x_{HCl}\right)$$

This formula implies that μ_{gas} decreases as x_{HCl} increases. Because

$$\mu_{gas} = \mu^{\ominus}_{gas} + RT \ln \hat{P}$$

we find that $\hat{P}$ becomes lower as μ_{gas} becomes lower. A sketch of μ_{gas} versus concentration is given in Fig. E21.1.

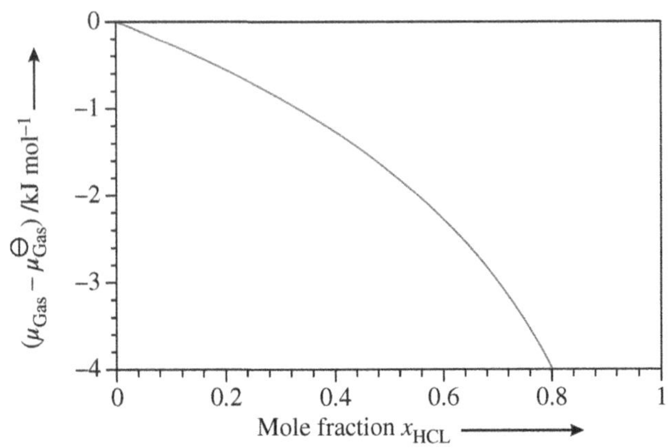

Figure E21.1 Chemical potential $\mu_{gas} - \mu^{\ominus}_{gas}$ in a solution of HCl in water versus the mole fraction x of the solute.

E21.2 The vapor pressure of pure water at 298 K is 23.76 Torr. Assume that the water vapor above an HCl solution acts ideally and compute the activity of the water in solution. If the vapor pressure of water above the solution at 298 K is 22.28 Torr, what is its activity?

Solution

Given that

$$P_{H_2O} = 23.76 \text{ Torr}$$

then

$$a_{H_2O} = \frac{P_{solution}}{P_{H_2O}} = \frac{22.28 \text{ Torr}}{23.76 \text{ Torr}} = 0.9377$$

E21.3 Charge Distribution $\rho(r)$: Prove that the total charge of the cloud of counter ions in the treatment of a spherical distribution of ions equals $-e$. Refer to Foundation 21.2.

Solution

From the result in Foundation 21.2, Equation (21F2.17), we write the spherical charge distribution of counterions around an ion as

$$\rho(r) = -2\frac{N}{V} \cdot \frac{e^2}{kT} \cdot \frac{e}{4\pi\varepsilon_0\varepsilon} \cdot \frac{1}{r} \cdot \cdot e^{-\kappa r}$$

where the charge density $\rho(r)$ is

$$\rho(r) = \left(\frac{N_+(r)}{V} - \frac{N_-(r)}{V}\right) \cdot e$$

N/V is the number density of the ions in the bulk and $N_+(r)/V$ and $N_-(r)/V$ are the number densities of the ions in the cloud of the counterions at the position r. κ is

$$\kappa = \sqrt{\frac{2e^2}{\varepsilon_0\varepsilon kT}\frac{N}{V}}$$

so that

$$\text{Total charge} = \int_0^\infty \rho(r) \cdot 4\pi r^2 \cdot \, dr$$
$$= -2\frac{N}{V} \cdot \frac{e^2}{kT} \cdot \frac{e}{\varepsilon_0 \varepsilon} \cdot \int_0^\infty r \cdot e^{-\kappa r} \cdot dr$$

We solve the integral by the substitution $x = \kappa r$ and $dx = \kappa \, dr$

$$\int_0^\infty r \cdot e^{-\kappa r} \cdot dr = \frac{1}{\kappa^2} \int_0^\infty x \cdot e^{-x} \cdot dx$$
$$= \frac{1}{\kappa^2} \cdot |-(x+1) \cdot e^{-x}|_0^\infty = \frac{1}{\kappa^2}$$

to obtain

$$\text{Total charge} = -2\frac{N}{V} \cdot \frac{e^2}{kT} \cdot \frac{e}{\varepsilon_0 \varepsilon} \cdot \frac{1}{\kappa^2}$$
$$= -2\frac{N}{V} \cdot \frac{e^2}{kT} \cdot \frac{e}{\varepsilon_0 \varepsilon} \cdot \frac{\varepsilon_0 \varepsilon kT}{2e^2} \frac{V}{N} = -e$$

E21.4 Consider an ideal mixture of two miscible liquids, and derive a formula for the composition that results in the largest entropy of mixing. Now consider a non-ideal mixture of two miscible liquids and derive a formula for the composition that results in the largest entropy of mixing. Assume that $a_i = \gamma_i x_i$ where γ_i is a constant.

Solution

We begin by considering the ideal mixture and write the entropy of mixing as

$$\Delta S = -\boldsymbol{R}\,[x \ln x + (1-x) \ln (1-x)]$$

For a derivation of this formula in the case of an ideal mixture, see the solution to Exercise E18.12.
To find the extremum (in this case a maximum) we take the derivative with respect to x

$$\frac{\partial (\Delta S)}{\partial x} = -\boldsymbol{R}\left[\ln x + x\frac{1}{x} - \ln(1-x) - (1-x)\frac{1}{1-x}\right]$$
$$= -\boldsymbol{R}\,[\ln x + 1 - \ln(1-x) - 1] = -\boldsymbol{R}\,[\ln x - \ln(1-x)]$$
$$= -\boldsymbol{R} \cdot \ln\left(\frac{x}{1-x}\right)$$

and set it equal to zero:

$$0 = \ln\left(\frac{x_1}{1-x_1}\right) \quad \text{or} \quad \frac{x_1}{1-x_1} = 1$$

Solving for x_1, we find that

$$x_1 = \frac{1}{2}$$

For the case of a nonideal solution, we start with

$$\Delta S = -\boldsymbol{R}\,(x \ln a_1 + x_2 \ln a_2)$$
$$-\boldsymbol{R}\,[x \ln(\gamma_1 x) + (1-x) \ln(\gamma_2 (1-x))]$$

Again we find the extremum by taking the derivative of ΔS with respect to x

$$\frac{\partial (\Delta S)}{\partial x} = -\boldsymbol{R}\left[\ln(\gamma_1 x) + x\gamma_1 \frac{1}{\gamma_1 x} - \ln(\gamma_2(1-x)) - (1-x)\frac{\gamma_2}{\gamma_2(1-x)}\right]$$
$$= -\boldsymbol{R}\left[\ln(\gamma_1 x) + 1 - \ln(\gamma_2(1-x)) - 1\right] = -\boldsymbol{R}\left[\ln\frac{\gamma_1 x}{\gamma_2(1-x)}\right]$$

and setting it equal to zero. In this case, we find that

$$0 = \ln\left(\frac{\gamma_1 x_1}{\gamma_2(1-x_1)}\right) \quad \text{or} \quad \frac{\gamma_1 x_1}{\gamma_2(1-x_1)} = 1$$

which we rearrange to

$$\gamma_2 \left(1 - x_1\right) = \gamma_1 x_1$$

and

$$x_1 = \frac{\gamma_2}{\gamma_2 + \gamma_1}$$

For example, if the activity coefficients of the two liquids are equal ($\gamma_2 = \gamma_1$) we obtain $x_1 = 1/2$ as for the ideal solution.

E21.5 Starting with Equation (21.60), namely

$$\frac{\Delta_{\text{vap}} H}{\boldsymbol{R}} \left(\frac{1}{T} - \frac{1}{T_s}\right) = \ln a_s$$

derive the approximate relation

$$\boxed{\Delta T = -\frac{\boldsymbol{R} T_s^2}{\Delta_{\text{vap}} H} \cdot \ln a_s \quad \text{(elevation of boiling point)}}$$

for ΔT.

Solution

For the temperature difference $\Delta T = T - T_s$ we can write

$$\frac{1}{T} - \frac{1}{T_s} = \frac{T_s - T}{T T_s} = -\frac{\Delta T}{\left(T_s + \Delta T\right) T_s}$$

For the case that $\Delta T \ll T_s$, we find that

$$\frac{1}{T} - \frac{1}{T_s} \approx -\frac{\Delta T}{T_s^2}$$

Substituting this expression into Equation (21.60), we obtain

$$-\frac{\Delta_{\text{vap}} H}{\boldsymbol{R}} \frac{\Delta T}{T_s^2} = \ln a_s$$

which rearranges to the desired result.

E21.6 An aqueous solution of KCl at 298 K, which is 1.2455 molar, has a freezing point depression of 4.048 K. Use these data to calculate the activity coefficient for the water.

Solution

We begin with the result found for the freezing point depression (Equation (21.64)) and write that

$$\ln\left(a_{H_2O}\right) = \frac{\Delta_{\text{fus}} H}{\boldsymbol{R}} \frac{\Delta T}{T_s^2}$$

$$= \frac{6000 \text{ J mol}^{-1}}{8.3145 \text{ J mol}^{-1} \text{ K}^{-1}} \frac{-4.048 \text{ K}}{(273 \text{ K})^2} = -0.039$$

where we have used 6.0 kJ mol^{-1} for the enthalpy of fusion. Hence the activity of the water solvent is

$$a_{H_2O} = \exp\left(-0.039\right) = 0.962$$

E21.7 The data in the table show the volume of a KCl solution as a function of its molality at 298 K and 1 bar pressure. Plot these data and use it to determine the partial molar volume of KCl at 0.5 mol of KCl. (Note that the molality is moles of solute per kilogram of solvent.)

m_{KCl}, molal	0.0	0.1656	0.3342	0.6689	1.6694	3.3481
V, mL	1000.	998.318	996.781	993.967	986.791	977.973

Solution

From the data given and the use of 1 kg of water solvent, we can write the data as

n_{KCl}, mole	0.0	0.1656	0.3342	0.6689	1.6694	3.3481
V, mL	1000.	998.318	996.781	993.967	986.791	977.973

A plot of the data is shown in Fig. E21.7 (dark gray full circles). The light gray solid curve shows a fit of the data to a quadratic equation:

$$V = \left[0.80598 \cdot (n/\text{mol})^2 - 9.2312 \cdot (n/\text{mol}) + 999.86\right] \text{ mL}$$

where $n = n_{KCl}$ are the amount of substance of KCl.

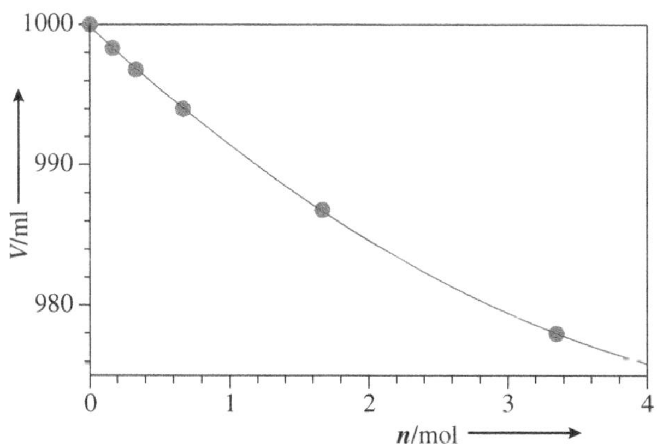

Figure E21.7 Volume V of a KCl solution as a function of the amount of substance n of KCl at 298 K and 1 bar.

The partial molar volume can be found as the derivative of this function, evaluated at $n = 0.5$ mol. The derivative is

$$\frac{dV}{dn} = \left[\frac{d}{dn}\left(0.80598 \cdot (n/\text{mol})^2 - 9.2312 \cdot (n/\text{mol}) + 999.86\right)\right] \text{mL mol}^{-1}$$
$$= (1.6120 \cdot (n/\text{mol}) - 9.2312) \text{ mL mol}^{-1}$$

Performing the evaluation at $n = 0.5$ mol, we find that

$$\left(\frac{\partial V}{\partial n}\right)_{n=0.5} = (1.6120 \cdot (0.5) - 9.2312) \text{ mL mol}^{-1}$$
$$= -8.4 \text{ mL mol}^{-1} = -8.4 \times 10^{-3} \text{ L mol}^{-1}$$

E21.8 The data in the table show the measured activity coefficient for HNO_3 as a function of molality. Plot these data versus the solution molality and fit it to the Debye–Hückel model. Comment on the quality of the fit and your best fit value for the radius R of the cloud of the counterions.

m, mol kg^{-1}	0.001	0.002	0.005	0.01	0.02
$\gamma_\pm$	0.965	0.951	0.927	0.902	0.871

m, mol kg^{-1}	0.05	0.1	0.2	0.5	1.0
$\gamma_\pm$	0.823	0.785	0.748	0.715	0.720

Solution

Figure E21.8 shows a plot of the data (dark gray filled circles) and a comparison to the Debye–Hückel model

$$\ln \gamma_\pm = -\frac{A \cdot z^2 \sqrt{I}}{1 + B \cdot R\sqrt{I}}$$

where $A = 1.175 \text{ L}^{1/2} \text{ mol}^{-1/2}$ and $B = 0.330 \times 10^{10} \text{ m}^{-1} \text{ L}^{1/2} \text{ mol}^{-1/2}$, and radius parameter R was optimized and found to be 660 pm. The quality of the fit (light gray solid line) is very good at low concentrations; however, at concentrations greater than 0.05 molal the fit is poor.

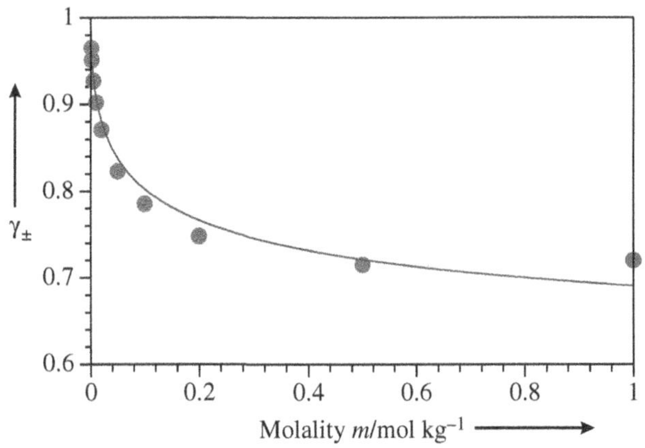

Figure E21.8 Activity coefficient γ versus molality m for an aqueous solution of HNO_3.

E21.9 The data in the table[1] show the measured activity coefficient for Na_2SO_4 as a function of concentration. Plot these data versus the solution molality and fit it to the Debye–Hückel model. Comment on the quality of the fit and your best fit value for R.

m, mol kg^{-1}	0.001	0.002	0.005	0.01	0.02
$\gamma_\pm$	0.8846	0.8505	0.7808	0.6889	0.6468

m, mol kg^{-1}	0.05	0.1	0.2	0.5	1.0
$\gamma_\pm$	0.5439	0.4342	0.3923	0.2996	0.2335

Solution

The sodium sulfate does not form a 1:1 electrolyte, but rather a 1:2 electrolyte. In this case we must calculate the ionic strength as

$$I = \frac{1}{2}\left[m_{Na^+}\cdot\left(z_{Na^+}\right)^2 + m_{SO_4^{2-}}\cdot\left(z_{H_2SO_4^{2-}}\right)^2\right]$$

with

$$m_{Na^+} = 2m,\ z_{Na^+} = 1 \quad m_{SO_4^{2-}} = m,\ z_{H_2SO_4^{2-}} = 2$$

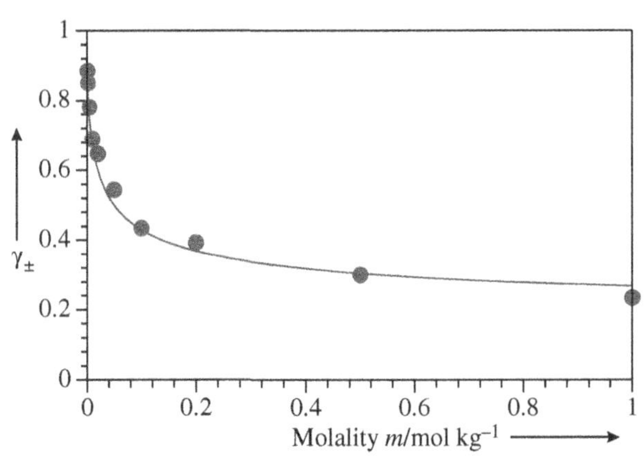

Figure E21.9 Activity coefficient $\gamma_\pm$ versus molality m for an aqueous solution of Na_2SO_4 at T = 298 K. Dark gray circles: experimental data, light gray solid line: Debye–Hückel model with R = 500 pm.

1 I. D. Zaytsev and G. G. Aseyev, eds *Properties of Aqueous Solutions of Electrolytes* (CRC, Boca Raton, 1992).

Figure E21.9 shows a plot of the data. A comparison to the Debye–Hü ckel model, with $\gamma_\pm = \sqrt{\gamma_{Na^+} \cdot \gamma_{H_2SO_4^{2-}}}$,

$$\ln \gamma_{Na^+} = -\frac{A \cdot (z_{Na^+})^2 \sqrt{I}}{1 + B \cdot R\sqrt{I}}, \quad \text{and} \quad \ln \gamma_{H_2SO_4^{2-}} = -\frac{A \cdot (z_{H_2SO_4^{2-}})^2 \sqrt{I}}{1 + B \cdot R\sqrt{I}}$$

where $A = 1.175$ L$^{1/2}$ mol$^{-1/2}$ and $B = 0.330 \times 10^{10}$ m^{-1} L$^{1/2}$ mol$^{-1/2}$ were used. The radius parameter R was varied to obtain a good fit and was found to be 500 pm.

E21.10 Beginning with Equation (21.83) prove that the van't Hoff equation,

$$\frac{\mathrm{d}(\ln K)}{\mathrm{d}T} = \frac{\Delta_r H^\ominus}{RT^2}$$

holds; i.e., it is a general result that does not depend on approximations about a system's ideality.

Solution

Equation (21.83) is

$$\Delta_r G^\ominus = -RT \ln K$$

First we rearrange the equation to isolate the $\ln K$ term, and then we take the derivative with respect to temperature and find

$$\frac{\mathrm{d}\ln K}{\mathrm{d}T} = \frac{\mathrm{d}}{\mathrm{d}T}\left(-\frac{\Delta_r G^\ominus}{RT}\right) = \frac{\mathrm{d}}{\mathrm{d}T}\left(-\frac{\Delta_r H^\ominus/T - \Delta_r S^\ominus}{R}\right) = \frac{\Delta_r H^\ominus}{RT^2}$$

The only approximation made here is that the change in enthalpy and change in entropy are independent of temperature, provided that we calculate the equilibrium constant in terms of activities.

E21.11 In Example 21.7 the K_{sp} of AgCl in water is given to be $K_{sp} = 1.77 \cdot 10^{-10}$. Using this knowledge and the Debye–Hückel model predict what the solubility of AgCl will be in a 0.100 molar solution of KCl.

Solution

The solution to this problem begins by determining the mean activity coefficients of the solution which is dominated by the KCl in this case. The Debye–Hückel model is

$$\ln \gamma_\pm = -\frac{A \cdot z^2 \sqrt{I}}{1 + B \cdot R\sqrt{I}}$$

First, we calculate the ionic strength by the formula

$$I = \frac{1}{2}\left(0.100\,(1)^2 + 0.100\,(-1)^2\right)\ \text{mol L}^{-1} = 0.1\ \text{mol L}^{-1}$$

Then we use the values $A = 1.175\ (\text{mol L}^{-1})^{-1/2}$, $B = 0.330 \times 10^{10}\ \text{m}^{-1}(\text{mol L}^{-1})^{-1/2}$, and $R_{Ag^+} = R_{Cl^-} = 500$ pm in the Debye–Hückel model to find

$$\ln \gamma_\pm = -\frac{1.175\,(\text{mol L}^{-1})^{-1/2} \cdot 1^2 \sqrt{0.1\ (\text{mol L}^{-1})}}{1 + \left(0.330 \times 10^{10}\ \text{m}^{-1}(\text{mol L}^{-1})^{-1/2}\right) \cdot (500 \times 10^{-12}\ \text{m}) \sqrt{0.1\ (\text{mol L}^{-1})}}$$
$$= -0.244$$

which gives a mean activity coefficient of

$$\gamma_\pm = 0.783$$

For the solubility product, we write that (using the fact that $c_{Cl^-} = 0.1$ mol L^{-1} and $\hat{c}_{Cl^-} = 0.1$)

$$K_{sp} = \gamma_\pm^2 \cdot \hat{c}_{Ag^+} \cdot \hat{c}_{Cl^-} = \gamma_\pm^2 \cdot \hat{c}_{Ag^+} \cdot 0.1$$

which can be expanded to give

$$\hat{c}_{Ag^+} = \frac{K_{sp}}{\gamma_\pm^2 \cdot 0.1} = \frac{1.77 \times 10^{-10}}{0.783^2 \cdot 0.1} = 2.89 \times 10^{-9}$$

E21.12 By far the major component of seawater is NaCl and it is about 0.55 molar. If sea water were an ideal solution with this concentration of NaCl, what would be the vapor pressure of the water in sea water at 298 K? Compare your value for the vapor pressure of seawater to that of pure water at 298 K. State whether it is higher or lower and explain why. If the actual vapor pressure of such a solution is 0.027 bar, what is the activity of the solution? Consider the vapor pressure of pure water to be 0.035 bar at 298 K, and take the density of 0.55 molar salt water to be $\rho = 0.98\ \text{g mL}^{-1}$.

Solution

The mole fraction of water, x_{H_2O} is

$$x_{H_2O} = \frac{54.4\ \text{moles H}_2\text{O}}{0.55\ \text{moles Na}^+ + 0.55\ \text{moles Cl}^- + 54.4\ \text{moles H}_2\text{O}} = 0.98$$

For an ideal solution we find a vapor pressure of

$$\begin{aligned} P_{\text{solution}} &= x_{H_2O} P^{\ominus}_{H_2O} \\ &= (0.98)\,(0.035\ \text{bar}) = 0.034\ \text{bar} \end{aligned}$$

The vapor pressure for sea water is lower than that for pure water. This decrease in the vapor pressure occurs because the Gibbs energy of the mixture is lower than that of the pure solvent. From the measured value of the vapor pressure, the activity of water in the solution is

$$a = \frac{0.034\ \text{bar}}{0.035\ \text{bar}} = 0.97$$

Provided that the vapor pressure of an unknown salt solution is 0.27 bar, we obtain an activity of

$$a = \frac{0.027\ \text{bar}}{0.035\ \text{bar}} = 0.77$$

E21.13 Use Equations (21.8) and (21.22) to write a general expression for the partial molar enthalpy of a component in a solution. Subsequently, use your enthalpy expression to show that

$$\left(\frac{\partial \ln \hat{a}_i}{\partial T}\right)_{P,\boldsymbol{n}_{i\neq i}} = \frac{h_i^{\ominus} - h_i}{\boldsymbol{R}T^2}$$

Solution

Using our expressions for μ_i and s_i along with the definition $G = H - TS$, we can write that

$$h_i = \mu_i + Ts_i = \mu_i - T\left(\frac{\partial \mu_i}{\partial T}\right)_{P,\boldsymbol{n}_i,\boldsymbol{n}_{j\neq i}}$$

Substituting $\mu_i = \mu_i^{\ominus} + \boldsymbol{R}T \cdot \ln \hat{a}_i$, we find that

$$\begin{aligned} h_i &= \mu_i^{\ominus} + \boldsymbol{R}T \cdot \ln \hat{a}_i - T\left[\frac{\partial \mu_i^{\ominus}}{\partial T} + \boldsymbol{R} \cdot \ln \hat{a}_i + \boldsymbol{R}T \cdot \frac{\partial \ln \hat{a}_i}{\partial T}\right] \\ &= \left[\mu_i^{\ominus} - T\frac{\partial \mu_i^{\ominus}}{\partial T}\right] - \boldsymbol{R}T^2 \cdot \frac{\partial \ln \hat{a}_i}{\partial T} \\ &= h_i^{\ominus} - \boldsymbol{R}T^2 \cdot \frac{\partial \ln \hat{a}_i}{\partial T} \end{aligned}$$

which defines $h_i^{\ominus} = \mu_i^{\ominus} - T\left(\frac{\partial \mu_i^{\ominus}}{\partial T}\right)$.

E21.14 Derive the expression for the depression of the freezing point, namely

$$\Delta T = \frac{\boldsymbol{R}T_s^2}{\Delta_{\text{vap}}H} \cdot \ln a_s \quad \text{(depression of freezing point)}$$

Solution

In analogy to the consideration of the elevation of the boiling point in Section 21.4.1 we consider the solid/liquid equilibrium, $\mu_{s,\text{solid}} = \mu_{s,\text{liq}}$ where $\mu_{s,\text{solid}}$ is the chemical potential of the solvent s in the solid state and $\mu_{s,\text{liq}}$ is the chemical potential of the solvent in the liquid state (i.e., in the solution). We may write that

$$\frac{\mu_{s,\text{solid}}}{T} = \frac{\mu_{s,\text{liq}}}{T}, \text{ so that } \mathrm{d}\left(\frac{\mu_{s,\text{solid}}}{T}\right) = \mathrm{d}\left(\frac{\mu_{s,\text{liq}}}{T}\right)$$

for a differential change of the chemical potential. For a solution we have

$$\mu_{s,\text{liq}} = \mu^{\ominus}_{s,\text{liq}} + \boldsymbol{R}T \cdot \ln a_s \quad \text{or} \quad \frac{\mu_{s,\text{liq}}}{T} = \frac{\mu^{\ominus}_{s,\text{liq}}}{T} + \boldsymbol{R} \cdot \ln a_s$$

where $a_s = x_s \cdot \gamma_s$ is the activity of the solvent in the solution (x_s = mole fraction of the solvent, γ_s is the activity coefficient of the solvent) and the standard state is defined for $a_s = 1$ (pure solvent). Thus we obtain

$$\mathrm{d}\left(\frac{\mu_{s,\text{solid}}}{T}\right) = \mathrm{d}\left(\frac{\mu^{\ominus}_{s,\text{liq}}}{T}\right) + \mathrm{d}\left(\boldsymbol{R} \ln a_s\right)$$

For a pure substance $\mu = G_m = H_m - TS_m$ and

$$\frac{\partial(\mu/T)}{\partial T} = \frac{\partial}{\partial T}\left(\frac{H_m}{T} - S_m\right) = \frac{\partial H_m}{\partial T}\frac{1}{T} - H_m\frac{1}{T^2} - \frac{\partial S_m}{\partial T}$$
$$= \frac{C_P}{T} - \frac{H_m}{T^2} - \frac{C_P}{T} = -\frac{H_m}{T^2}$$

Then

$$-\frac{H^{\ominus}_{m,s,\text{solid}}}{T^2}\mathrm{d}T = -\frac{H^{\ominus}_{m,s,\text{liq}}}{T^2}\mathrm{d}T + \mathrm{d}\left(\boldsymbol{R} \ln a_s\right)$$

where $H^{\ominus}_{m,s,\text{liq}}$ is the molar enthalpy of the pure liquid solvent in its standard state, and $H^{\ominus}_{m,s,\text{solid}}$ is the molar enthalpy of the pure solvent in the solid state. Then $\Delta_{\text{fus}}H = H^{\ominus}_{m,s,\text{liq}} - H^{\ominus}_{m,s,\text{solid}}$ is the molar enthalpy of melting of the pure solvent and we obtain

$$\boxed{\frac{\Delta_{\text{fus}}H}{\boldsymbol{R}T^2}\mathrm{d}T = \mathrm{d}\left(\ln a_s\right)}$$

This relation is exact.

For practical use we have to integrate this equation. Then

$$\int_{T=T_s}^{T} \frac{\Delta_{\text{fus}}H}{\boldsymbol{R}T'^2}\,\mathrm{d}T' = \int_{\ln a_s=0}^{\ln a_s} \mathrm{d}\left(\ln a_s'\right)$$

where T_s is the boiling temperature of the pure solvent and T is the boiling temperature of the solution with activity a_s. As an approximation we assume that i) $\Delta_{\text{fus}}H$ and ii) the activity coefficient γ_s do not depend on temperature. Often these assumptions are appropriate for the small range of temperature associated with the freezing point depression. In this approximation, we find that

$$-\left(\frac{\Delta_{\text{fus}}H}{\boldsymbol{R}T'}\right)\Bigg|_{T_s}^{T} - \left(\ln a_s'\right)\Big|_{0}^{\ln a_s}$$

or

$$-\frac{\Delta_{\text{vap}}H}{\boldsymbol{R}}\left(\frac{1}{T} - \frac{1}{T_s}\right) = \ln a_s$$

Because of

$$\frac{1}{T} - \frac{1}{T_s} = \frac{T_s - T}{TT_s} = -\frac{\Delta T}{TT_s} \approx -\frac{\Delta T}{T_s^2}$$

for sufficiently small temperature shifts we obtain

$$\frac{\Delta_{\text{fus}}H}{\boldsymbol{R}}\left(\frac{\Delta T}{T_s^2}\right) = \ln a_s$$

with $\Delta T = T - T_s$. Solving for ΔT we find

$$\boxed{\Delta T = \frac{\boldsymbol{R}T_s^2}{\Delta_{\text{fus}}H} \cdot \ln a_s \qquad \text{(depression of freezing point)}}$$

Because a_s is smaller than one, $\ln a_s$ is negative and ΔT is negative, the freezing point is lowered.

E21.15 Consider a solution of two volatile components, A and B, at fixed composition in equilibrium with its vapor at a total pressure P of 1 bar. By assuming that the vapor behaves like an ideal gas and that $\Delta_{\text{vap}}H$ is independent of temperature, derive the relationship

$$\ln\left(\frac{x_A}{\hat{a}_A}\right) = \frac{\Delta_{\text{vap}}H_A}{\boldsymbol{R}}\left(\frac{1}{T_{BP,A}} - \frac{1}{T}\right)$$

where $T_{BP,A}$ is the boiling temperature, x_A is the mole fraction of A in the vapor, and a_A is the activity of A. (Hint: Refer to the derivation in Section 21.4.)

Solution

We begin with the equilibrium condition $\mu_{A,\text{vap}} = \mu_{A,\text{liq}}$ from which we may write that

$$\mathrm{d}\left(\frac{\mu_{A,\text{vap}}}{T}\right) = \mathrm{d}\left(\frac{\mu_{A,\text{liq}}}{T}\right)$$

see Section 21.4. By substituting for the chemical potential, we find that

$$\mathrm{d}\left(\frac{\mu^{\ominus}_{A,\text{vap}}}{T} + \boldsymbol{R}\cdot\ln\hat{P}_A\right) = \mathrm{d}\left(\frac{\mu^{\ominus}_{A,\text{liq}}}{T} + \boldsymbol{R}\cdot\ln\hat{a}_A\right)$$

or

$$\mathrm{d}\left(\frac{\mu^{\ominus}_{A,\text{vap}}}{T} - \frac{\mu^{\ominus}_{A,\text{liq}}}{T}\right) = \mathrm{d}\left(\boldsymbol{R}\cdot\ln\hat{a}_A - \boldsymbol{R}\cdot\ln\hat{P}_A\right)$$

As proven in Section 21.4, the term on the left gives $-\frac{\Delta_{\text{vap}}H_A}{T^2}\mathrm{d}T$, so that

$$-\frac{\Delta_{\text{vap}}H_A}{\boldsymbol{R}T^2}\mathrm{d}T = \mathrm{d}\left(\ln\hat{a}_A - \ln\hat{P}_A\right)$$

We integrate from the boiling point temperature of pure A to a temperature T'

$$-\int_{T_{BP,A}}^{T'} \frac{\Delta_{\text{vap}}H_A}{\boldsymbol{R}T^2}\mathrm{d}T = \int_0^{\ln\hat{a}_A - \ln\hat{P}_A} \mathrm{d}y$$

and find that

$$\frac{\Delta_{\text{vap}}H_A}{\boldsymbol{R}}\left(\frac{1}{T'} - \frac{1}{T_{BP,A}}\right) = \ln\hat{a}_A - \ln\hat{P}_A$$

Because the vapor behaves ideally we can write that $P_A = x_A P$ so that $\hat{P}_A = x_A$ (Note that $P = 1$ bar) and the previous equation becomes

$$\ln\left(\frac{x_A}{\hat{a}_A}\right) = \frac{\Delta_{\text{vap}}H_A}{\boldsymbol{R}}\left(\frac{1}{T_{BP,A}} - \frac{1}{T}\right)$$

where we have changed the notation from T' to T.

E21.16 Derive the general relationship

$$\left(\frac{\partial\ln\hat{a}_i}{\partial P}\right)_{T,n_j} = \frac{v_i}{\boldsymbol{R}T}$$

which describes the pressure dependence of the activity for component i.

Solution

We begin with the definition of the chemical potential for a solution

$$\mu_i = \mu_i^{\ominus} + \boldsymbol{R}T\cdot\ln\hat{a}_A$$

If we take the partial derivative of each side with respect to the pressure P, we find that

$$\left(\frac{\partial\mu_i}{\partial P}\right)_{T,n_j} = 0 + \boldsymbol{R}T\cdot\left(\frac{\partial\ln\hat{a}_i}{\partial P}\right)_{T,n_j}$$

because $\mu_i^{\ominus}$ is a function of T only. Using the relation between the chemical potential and the partial molar volume, we can write

$$\left(\frac{\partial \mu_i}{\partial P}\right)_{T,\boldsymbol{n}_{j\neq i}} = v_i$$

so that

$$\left(\frac{\partial \ln \hat{a}_i}{\partial P}\right)_{T,\boldsymbol{n}_j} = \frac{v_i}{\boldsymbol{R}T}$$

E21.17 Show that

$$\left(\frac{\partial \ln \hat{a}_i}{\partial T}\right)_{p,\boldsymbol{n}_i,\boldsymbol{n}_{j\neq i}} = \frac{\Delta H_{m,i}}{\boldsymbol{R}T^2}$$

where $\Delta H_{m,i}$ is the difference in enthalpy between the solution and the standard state.

Solution

From the problem statement we can write that

$$\Delta H_{m,i} = H_{m,i} - H_{m,i}^{\ominus}$$

By using the fact that $H_{m,i} = G_{m,i} + TS_{m,i} = \mu_i + TS_{m,i}$, we can write that

$$H_{m,i} = \mu_i + TS_{m,i} = \mu_i - T\left(\frac{\partial \mu_i}{\partial T}\right)$$

so that

$$H_{m,i} - H_{m,i}^{\ominus} = \mu_i - \mu_i^{\ominus} - \left[T\left(\frac{\partial \mu_i}{\partial T}\right) - T\left(\frac{\partial \mu_i^{\ominus}}{\partial T}\right)\right]$$

or

$$\Delta H_{m,i} = \Delta\mu_i - T\left(\frac{\partial \Delta\mu_i}{\partial T}\right)$$

Substituting $\Delta\mu_i = \mu_i - \mu_i^{\ominus} = \boldsymbol{R}T \cdot \ln \hat{a}_i$, we find that

$$\begin{aligned}\Delta H_{m,i} &= \boldsymbol{R}T \cdot \ln \hat{a}_i - \boldsymbol{R}T \cdot \ln \hat{a}_i + \boldsymbol{R}T^2\left(\frac{\partial \ln \hat{a}_i}{\partial T}\right)\\ &= \boldsymbol{R}T^2\left(\frac{\partial \ln \hat{a}_i}{\partial T}\right)\end{aligned}$$

which rearranges to give

$$\left(\frac{\partial \ln \hat{a}_i}{\partial T}\right) = \frac{\Delta H_{m,i}}{\boldsymbol{R}T^2}$$

E21.18 The table shows data for the solubility x_{O_2} of oxygen in water at 1.013 bar pressure. Use these data to calculate the enthalpy of solution for O_2 in water.

x_{O_2}	$2.756 \cdot 10^{-5}$	$2.501 \cdot 10^{-5}$	$2.293 \cdot 10^{-5}$	$2.122 \cdot 10^{-5}$	$1.982 \cdot 10^{-5}$
T / K	288.15	293.15	298.15	303.15	308.15

Solution

If we make the approximation that $\hat{a}_i = x_i$, we can use the result of E21.17 to analyze these data; namely

$$\left(\frac{\partial \ln x_i}{\partial T}\right) = \frac{\Delta H_{m,i}}{\boldsymbol{R}T^2}$$

so that

$$\int \mathrm{d}\ln x_i = \int \frac{\Delta H_{m,i}}{\boldsymbol{R}T^2}\mathrm{d}T \simeq \frac{\Delta H_{m,i}}{\boldsymbol{R}}\int \frac{\mathrm{d}T}{T^2}$$

and

$$\ln x_i = -\frac{\Delta H_{m,i}}{R}\frac{1}{T} + const$$

Figure E21.18 shows a plot of the data given as a function of the inverse temperature (circles), and a best fit to a line with a slope of

$$-\frac{\Delta H_{m,i}}{R} = 1464\ \text{K}$$

which rearranges to give

$$\Delta H_{m,i} = -12.17\ \text{kJ mol}^{-1}$$

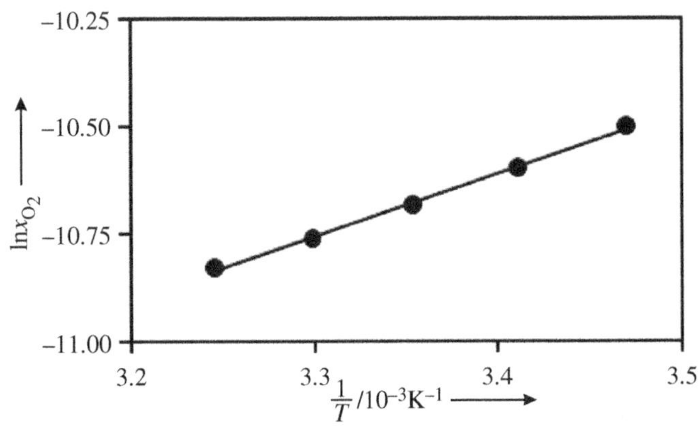

Figure E21.18 $\ln x_{O_2}$ versus $1/T$ for a solution of oxygen in water.

E21.19 The table shows data for the lowering of the vapor pressure of water as a function of NaCl concentration at a temperature of 373 K and pressure of 760 Torr. Using these data calculate how much NaCl must be added to 1 L of water in order to raise its boiling point by 5 °C.

$c_{\text{NaCl}}/\text{mol L}^{-1}$	0.50	1.00	2.00	3.00	4.00	5.00	6.00
$\Delta P/\text{Torr}$	12.3	25.2	52.1	80.0	111.0	143.0	176.5

Solution

Using Equation (21.62) with a ΔT of 5° degrees and $\Delta_{\text{vap}}H = 40.65\ \text{kJ mol}^{-1}$, we find

$$\ln \hat{a}_s = -\frac{\Delta_{\text{vap}}H}{RT_s^2}\Delta T$$

$$= -\frac{40{,}650\ \text{J mol}^{-1}}{8.314\ \text{J mol}^{-1}\ \text{K}^{-1} \cdot (373.15\ \text{K})^2} \cdot 5 = -0.1756$$

so that $\hat{a}_s = 0.8390$.

Here we calculate the activity from the vapor pressure lowering

$$\hat{a}_s = \frac{760 - \Delta P/\text{Torr}}{760}$$

$c_{\text{NaCl}}/\text{mol L}^{-1}$	0.50	1.00	2.00	3.00	4.00	5.00	6.00
$\Delta P/\text{torr}$	12.3	25.2	52.1	80.0	111.0	143.0	176.5
$\hat{a}_s$	0.984	0.967	0.929	0.895	0.854	0.812	0.768

Hence we see that the concentration of NaCl must lie between 4 and 5 molar; circa 4.36 M.

E21.20 Consider the Debye–Hückel equation for the mean activity coefficient of a 1:1 electrolyte (e.g., KCl) and show that it may be written in terms of the concentration c as

$$\ln \gamma_\pm = -\frac{A\sqrt{c}}{1 + B \cdot R\sqrt{c}}$$

For the salts KCl and NaCl estimate the value of R from the average of their ionic radii and calculate the concentration c at which the two terms in the denominator are equal. Determine the molarity at which the denominator is 1.01. Under what conditions would you expect to be able to use the more approximate equation

$$\ln \gamma_\pm = -A\sqrt{c}$$

Solution

The Debye–Hückel equation is written as

$$\ln \gamma_\pm = -\frac{A \cdot z^2\sqrt{I}}{1 + B \cdot R\sqrt{I}}$$

and for a 1:1 electrolyte (M^+X^-) with $z = 1$ the ionic strength is

$$I = \frac{1}{2}\left(c(1)^2 + c(-1)^2\right) = c$$

so that

$$\ln \gamma_\pm = -\frac{A \cdot \sqrt{c}}{1 + B \cdot R\sqrt{c}}$$

The ionic radii reported for the ions in the solid are Cl^- is 181 pm, K^+ is 133 pm, and Na^+ is 98 pm; however, we consider that the ions (especially the cations) will have a considerably larger radius in water because of their hydration shells. Hence we estimate that $R \approx 300$ pm. If we use the value $B = 0.330 \times 10^{10}\ \text{m}^{-1}\ \text{L}^{1/2}\ \text{mol}^{-1/2}$, then for the condition where the terms in the denominator are equal we find that

$$B \cdot R\sqrt{c} = 1$$

and

$$c = \left(\frac{1}{B \cdot R}\right)^2$$

$$= \left[\frac{1}{(0.330 \times 10^{10}\ \text{m}^{-1}(\text{mol L}^{-1})^{-1/2}) \cdot (3.00 \times 10^{-10}\ \text{m})}\right]^2 = 1.02\ \text{mol L}^{-1}$$

For the condition where the denominator is 1.01, we have a concentration such that

$$B \cdot R\sqrt{c} = 0.01$$

and

$$c = \left(\frac{0.01}{B \cdot R}\right)^2$$

$$= \left[\frac{0.01}{(0.330 \times 10^{10}\text{m}^{-1}(\text{ mol L}^{-1})^{-1/2}) \cdot (3.00 \times 10^{-10}\ \text{m})}\right]^2 \text{M}$$

$$= 1.02 \times 10^{-4}\ \text{mol L}^{-1}$$

The approximate expression will have an error of <1% for concentrations $<10^{-4}$ M.

E21.21 Starting with Equation (21.60), show that it reduces to Equation (21.61) for a small temperature shift.

Solution

Starting with Equation (21.60),

$$\frac{\Delta_{\text{vap}}H}{\boldsymbol{R}}\left(\frac{1}{T} - \frac{1}{T_s}\right) = \ln a_s$$

and the temperature shift $\Delta T = T - T_s$, we can write that

$$\ln a_s = \frac{\Delta_{\text{vap}}H}{R}\left(\frac{1}{\Delta T + T_s} - \frac{1}{T_s}\right)$$
$$= \frac{\Delta_{\text{vap}}H}{R}\left(\frac{T_s - (\Delta T + T_s)}{(\Delta T + T_s)\cdot T_s}\right) = -\frac{\Delta_{\text{vap}}H}{R}\left(\frac{\Delta T}{(\Delta T + T_s)\cdot T_s}\right)$$

For $\Delta T \ll T_s$. we can write that $(\Delta T + T_s)\cdot T_s \simeq T_s^2$, which gives

$$-\frac{\Delta_{\text{vap}}H}{R}\left(\frac{\Delta T}{T_s^2}\right) = \ln a_s$$

Solving for ΔT we find that

$$\boxed{\Delta T = -\frac{RT_s^2}{\Delta_{\text{vap}}H}\cdot \ln a_s}$$

(elevation of the boiling point).

21.2 Problems

P21.1 The data in the table report masses and volumes for preparing NaCl solutions with water. In each case 14.881 g NaCl with a volume of 6.902 mL was used.

m mol kg^{-1}	Mass$_{\text{solution}}$ g	Mass$_{H_2O}$ g	V_{solution} mL	V_{H_2O} mL
5.1420	58.836	43.955	49.504	44.084
2.5638	108.842	93.961	99.286	94.237
1.0273	257.213	242.332	247.779	243.045
0.5110	506.986	492.105	498.163	493.551
0.2543	1008.572	993.691	1001.094	996.610

Fit these data and write an empirical expression for the partial molar volume over this range of concentration. Evaluate your expression for the case of 2 molal NaCl solution.

Solution

Using the given volume of NaCl and the volume of water in each trial it is possible to calculate the total volume for ideal mixing of the two components. Scaling this value to 1 L for the ideal solution $V_{\text{ideal}}^{\text{scaled}}$ and assuming that the density of water is exactly 1 kg = 1 L, it is possible to scale the measured solution volume $V_{\text{solution}}^{\text{scaled}}$.

m molal	V_{ideal} L	V_{solution} L	$V_{\text{ideal}}^{\text{scaled}}$ L	$V_{\text{solution}}^{\text{scaled}}$ L
5.1420	0.050986	0.049504	1.000000	0.970933
2.5638	0.101139	0.099286	1.000000	0.981679
1.0273	0.249947	0.247779	1.000000	0.991326
0.5110	0.500453	0.498163	1.000000	0.995424
0.2543	1.003512	1.001094	1.000000	0.997590

A plot of these solution volumes versus the amount of substance of NaCl is shown in Fig. P21.1 and is fit to a quadratic equation,

$$V = \left[5.645 \cdot 10^{-4} \cdot (\boldsymbol{n}/\text{mol})^2 - 8.417 \cdot 10^{-3} \cdot (\boldsymbol{n}/\text{mol}) + 0.9996\right] \text{ L}$$

where **n** is the amount of NaCl (determined by $\boldsymbol{n} = (m/(\text{molal}) \cdot 1\text{ L})$.

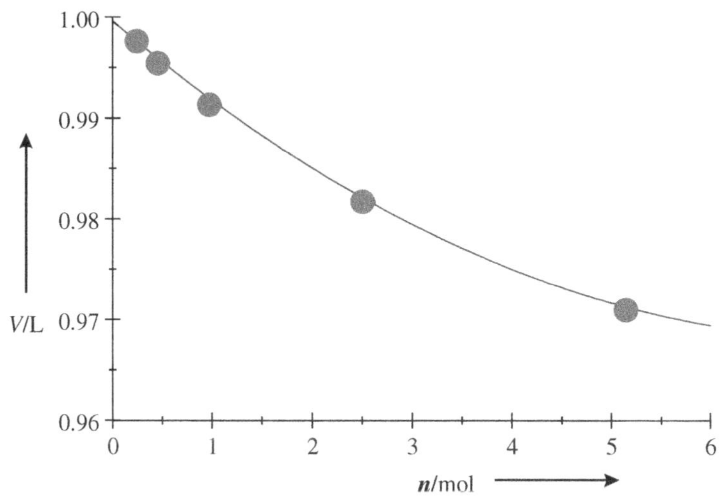

Figure P21.1 Volume V of a NaCl solution in water versus the amount of substance **n**.

To determine the partial molar volume, we take the derivative with respect to $\boldsymbol{n}$ and find

$$\left(\frac{\partial V}{\partial \boldsymbol{n}}\right) = \upsilon_{\text{NaCl}} = \left[1.1290 \cdot 10^{-3} \cdot (\boldsymbol{n}/\text{mol}) - 8.417 \cdot 10^{-3}\right] \text{ L mol}^{-1}$$

At $\boldsymbol{n} = 2$ mol, we find that $\upsilon_{\text{NaCl}} = -6.159 \text{ L mol}^{-1}$.

P21.2 Consider the following set of data for mixtures of acetone and water.

	$T = 298$ K		$T = 333$ K	
x_{acetone}	$\frac{P_{\text{water}}}{\text{Torr}}$	$\frac{P_{\text{acetone}}}{\text{Torr}}$	$\frac{P_{\text{water}}}{\text{Torr}}$	$\frac{P_{\text{acetone}}}{\text{Torr}}$
0.0	23	0	149	0
0.0333	27	28	149	190
0.0720	27	77	143	342
0.117	25	105	134	443
0.171	24	125	145	495
0.236	22	146	129	553
0.318	28	152	126	588
0.420	23	164	116	624
0.554	16	182	102	672
0.737	17	192	97	711
1.00	0	229	0	860

Plot the vapor pressures of each component and the total vapor pressure as a function of the mole fraction. Perform this plot for each temperature. Explain, in words, how the observed vapor pressure data differs from that which one would predict for an ideal solution.

Compute the activity of each component at each of the mole fractions and temperatures. Present a table of the activities and give a representative calculation in detail. Use the activities to determine the $\Delta_{\text{mix}}G$ at each of the intermediate mole fractions and for each temperature. Compute the $\Delta_{mix}G$ that you would expect if acetone and water formed an ideal solution. Present your values in tabular form and plot the data. Compare your two results and try to explain it in terms of the intermolecular interactions between acetone and water.

Solution

The plots for $T = 298$ K and $T = 333$ K are shown in Fig. P21.2a. The dark gray circles represent the acetone vapor pressures, the light gray circles represent the water vapor pressures, and the black circles represent the total vapor pressure.

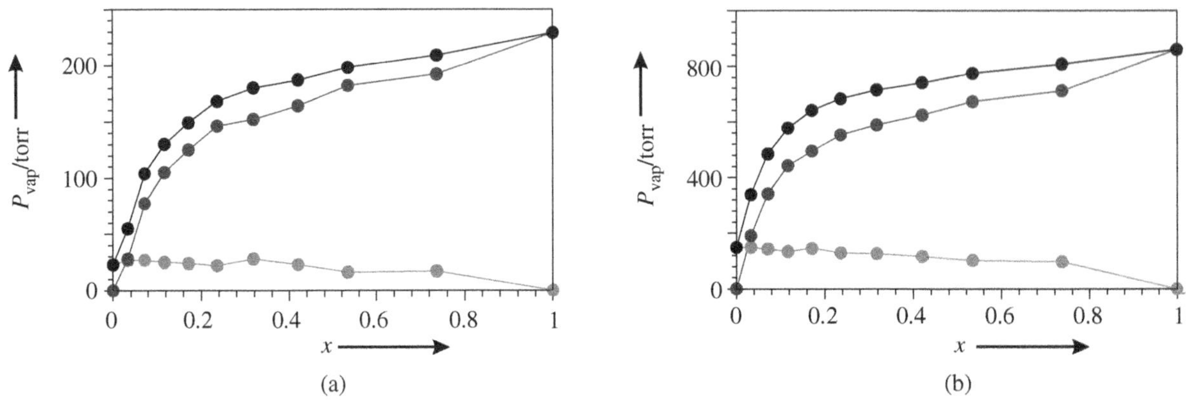

Figure P21.2a Vapor pressure P_{vap} versus the mole fraction x of acetone. Dark gray filled circles: acetone; light gray filled circles: water; black filled circles: total pressure. (a) Temperature 298 K, (b) temperature 333 K.

In an ideal solution one expects the vapor pressure to follow Raoult's Law, which says the vapor pressure is a linear function of the composition (for an example see Fig. 18.12). It is evident that the acetone/water mixture does not follow Raoult's law.

Here we show example calculations for the activity of each component, the Gibbs energy of mixing for an ideal solution, and the actual Gibbs energy of mixing. Given that the standard vapor pressure at 298 K for acetone is $P^{\ominus}_{H_2CO} = 229$ Torr and that of water is $P^{\ominus}_{H_2O} = 23$ Torr, we can calculate the activities. For an acetone mole fraction of 0.236 in the solution we find that

$$\hat{a}_{H_2CO} = \frac{P_{H_2CO}}{P^{\ominus}_{H_2CO}} = \frac{146 \text{ Torr}}{229 \text{ Torr}} = 0.638$$

and for water we find that

$$\hat{a}_{H_2O} = \frac{P_{H_2O}}{P^{\ominus}_{H_2O}} = \frac{22 \text{ Torr}}{23 \text{ Torr}} = 0.956$$

For an ideal solution the Gibbs energy of mixing is

$$\Delta_{mix}G^{ideal} = \boldsymbol{RT}\left(x_{H_2CO} \cdot \ln\left(x_{H_2CO}\right) + \left(1 - x_{H_2CO}\right)\ln\left(1 - x_{H_2CO}\right)\right)$$

For the case of $x_{H_2CO} = 0.236$ and $T = 298$ K, we find that

$$\begin{aligned}\Delta_{mix}G^{ideal} &= \boldsymbol{RT}\,(0.236 \cdot \ln(0.236) + 0.764 \cdot \ln(0.764)) \\ &= 8.314 \text{ J mol}^{-1}\text{K}^{-1} \cdot 298 \text{ K} \cdot (-0.564) = -1.35 \text{ kJ mol}^{-1}\end{aligned}$$

The values for each of the temperatures and mole fractions are reported in the table given below. For a real solution we use the activity to calculate the Gibbs energy of mixing, namely

$$\Delta_{mix}G = \boldsymbol{RT}\left(x_{H_2CO} \cdot \ln\left(a_{H_2CO}\right) + \left(1 - x_{H_2CO}\right)\ln\left(a_{H_2O}\right)\right)$$

For the case of $x_{H_2CO} = 0.236$ and $T = 298$ K, we have $\hat{a}_{H_2CO} = 0.64$ and $\hat{a}_{H_2O} = 0.96$, so that

$$\begin{aligned}\Delta_{mix}G &= \boldsymbol{RT}\,(0.236 \cdot \ln(0.64) + (0.764)\ln(0.96)) \\ &= 8.314 \text{ J mol}^{-1}\text{K}^{-1} \cdot 298 \text{ K} \cdot (-0.137) = -0.339 \text{ kJ mol}^{-1}\end{aligned}$$

The values for each of the temperatures and mole fractions are reported in the tables, one for 298 K and the other for 333K.

			T = 298 K	
$x_{acetone}$	$\hat{a}_{water}$	$\hat{a}_{acetone}$	$\Delta_{mix}G$ kJ mol^{-1}	$\Delta_{mix}G^{ideal}$ kJ mol^{-1}
0.0	1.00	0.00		
0.0333	1.17	0.12	0.201	−0.362
0.0720	1.17	0.34	0.169	−0.641
0.117	1.09	0.46	−0.0366	−0.894
0.171	1.04	0.55	−0.173	−1.133
0.236	0.96	0.64	−0.338	−1.354
0.318	1.22	0.66	0.00863	−1.549
0.420	1.00	0.72	−0.342	−1.686
0.554	0.70	0.79	−0.718	−1.703
0.737	0.74	0.84	−0.515	−1.428
1.00	0.00	1.00		

			T = 333 K	
$x_{acetone}$	a_{water}	$a_{acetone}$	ΔG_{mix} kJ mol^{-1}	ΔG_{mix}^{ideal} kJ mol^{-1}
0.0	1.00	0.00		
0.0333	1.00	0.22	−0.140	−0.404
0.0720	0.96	0.40	−0.288	−0.717
0.117	0.90	0.52	−0.469	−0.999
0.171	0.97	0.58	−0.328	−1.267
0.236	0.87	0.64	−0.586	−1.513
0.318	0.85	0.68	−0.646	−1.731
0.420	0.78	0.73	−0.765	−1.884
0.554	0.68	0.78	−0.857	−1.903
0.737	0.65	0.83	−0.694	−1.595
1.00	0.00	1.00		

A plot of the results at T = 298 K (A) and at T = 333 K (B) are shown in Fig. P21.2b. In each plot the circles represent the experimental $\Delta_{mix}G$ and the smooth curves represent $\Delta_{mix}G^{ideal}$).

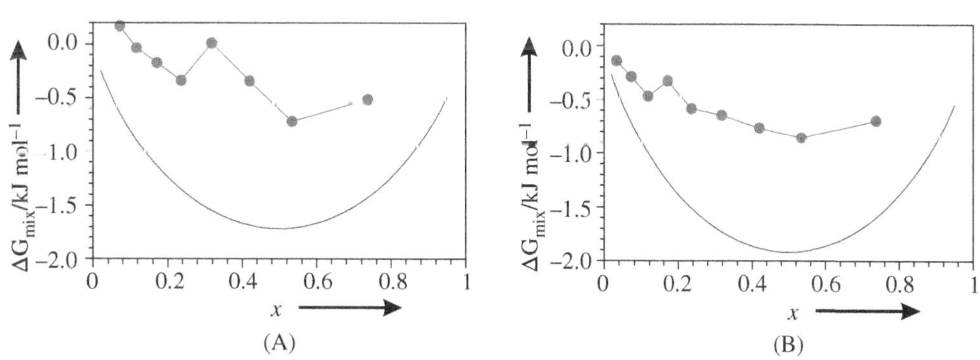

Figure P21.2b $\Delta_{mix}G$ for a mixture of water and acetone versus the mole fraction x of acetone. Dark gray: $\Delta_{mix}G$; light gray: $\Delta_{mix}G^{ideal}$. (a) Temperature 298 K, (b) Temperature 333 K.

It is evident that the model of an ideal solution captures the qualitative features of the behavior, but that the actual Gibbs energy of mixing is smaller than that predicted by the ideal model.

P21.3 Using Equation (21.38) and the data for the activity coefficients of the solvent γ_s in Table 21.1, calculate the corresponding activity coefficients γ of the solute.

Solution

In Section 21.3.1 we calculated the activity coefficients γ_s for solutions of sucrose in water, depending on the mole fraction x_s of the solvent (see data in Table 21.1, columns 2 and 6). According to Equation (21.38) in Section 21.3.1 we can calculate the activity coefficient γ of the solute

$$\ln \gamma_1 = -\int_{\ln \gamma_s = 0}^{\ln \gamma_{s,1}} \frac{x_s}{1 - x_s} \cdot \mathrm{d} \ln \gamma_s$$

where γ_1 is the activity coefficient of the solute with mole fraction $x_{s,1}$. Here we solve the integral by numerical integration and calculate the data given in column 7 of Table P21.3.

Table P21.3 Relative Change of Vapor Pressure $\Delta P/P_s$ for Solutions of Sucrose in Water at 0 °C. x is the Mole Fraction of Sucrose, x_s is the Mole Fraction of Water.

				Calculated		
	Calculated		Experiment	Equation (21.36)		Equation (21.38)
x	$x_s = 1 - x$	$\frac{\Delta P}{P_s}$	$\frac{\Delta P}{P_s}$	$\hat{a}_s$	γ_s	γ
0.000	1.000	0.000	0.000	1.000	1.000	1.00
0.0036	0.996	−0.0036	−0.00371	0.996	1.000	1.00
0.0089	0.991	−0.0089	−0.00939	0.991	1.000	1.00
0.0177	0.982	−0.0177	−0.0197	0.980	0.998	1.13
0.059	0.941	−0.059	−0.0839	0.916	0.973	2.18
0.075	0.925	−0.075	−0.1124	0.888	0.960	2.69
0.082	0.918	−0.082	−0.1278	0.872	0.949	2.93
0.098	0.902	−0.098	−0.1555	0.845	0.937	3.41

First we calculate the coefficients of a cubic regression curve, using the data in Table 21.1, column 6.

$$\gamma_s = b_0 + b_1 \cdot x_s + b_2 \cdot x_s^2 + b_3 \cdot x_s^3$$

with

$$b_0 = 0.99975,\ b_1 = 0.141589,\ b_2 = -13.3632,\ b_3 = 54.3498$$

This curve is shown in Fig. P21.3a together with the data points. Using the regression curve, we calculate the quantities

$$\ln \gamma_s \ \text{and} \ \frac{x_s}{1 - x_s}$$

and plot $x_s/(1 - x_s)$ as a function of $\ln \gamma_s$. We approximate the integral by a summation

$$\int_{\ln \gamma_s = 0}^{\ln \gamma_{s,1}} \frac{x_s}{1 - x_s} \cdot \mathrm{d} \ln \gamma_s = \sum_{\ln \gamma_s = 0}^{\ln \gamma_{s,1}} \frac{x_s}{1 - x_s} \cdot \Delta \ln \gamma_s$$

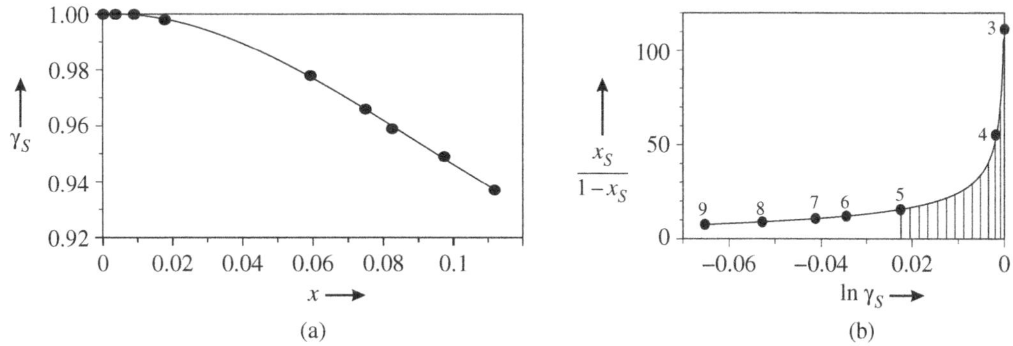

Figure P21.3 Panel (a) Activity coefficient γ_s for solutions of sucrose in water at 0 °C versus the mole fraction x of sucrose. Panel (b) The graph plots x_s $(1 - x_s)$ as a function of $\ln \gamma_s$. The numbers correspond to the data points in lines 3 to 8 in Table 21.1.

For example, for data point 5 the integral corresponds to the dashed area in Fig. P21.3b.

$$\sum_{\ln \gamma_s=0}^{\ln \gamma_{s,5}} \frac{x_s}{1-x_s} \cdot \Delta \ln \gamma_s = -0.7789$$

Thus we obtain

$$\ln \gamma_5 = -\int_{\ln \gamma_s=0}^{\ln \gamma_{s,1}} \frac{x_s}{1-x_s} \cdot \mathrm{d} \ln \gamma_s = 0.779 \quad \text{or} \quad \gamma_5 = \mathrm{e}^{0.779} - 2.18$$

Accordingly, the result for all data points in Table P21.3, column 7, is

Data Point	3	4	5	6	7	8
$\ln \gamma$	–	0.126	0.779	0.991	1.075	1.227
γ	1.00	1.13	2.18	2.69	2.93	3.41

P21.4 Consider a solution of glycol $(CH_2OH)_2$) in water (glycol is used in the water cooling system of cars). Calculate the mole fraction x of glycol that is necessary to decrease the freezing point by 3, 10, and 20 K. How much glycol must be added to 1 L of water?

Solution

The freezing point depression expression to be used is

$$\Delta T = \frac{RT_s^2}{\Delta_{\text{fus}} H} \cdot \ln \hat{a}_s$$

In the approximation that $\hat{a}_s = x_s = 1 - x_{\text{glycol}}$ we can write that

$$\ln\left(1 - x_{\text{glycol}}\right) = \frac{\Delta_{\text{fus}} H}{RT_s^2} \cdot \Delta T$$

so that

$$x_{\text{glycol}} = 1 - \exp\left(\frac{\Delta_{\text{fus}} H}{RT_s^2} \cdot \Delta T\right)$$

where $\Delta_{\text{fus}} H = 6020 \text{ J mol}^{-1}$ and $T_s = 273.15$ K. For the case of $\Delta T = -1$ K, we find that

$$x_{\text{glycol}} = 1 - \exp\left(-\frac{6020 \text{ J mol}^{-1}}{(8.314 \text{ J mol}^{-1}\text{K}^{-1})(273.15 \text{ K})^2} \cdot 3 \text{ K}\right)$$
$$= 2.87 \times 10^{-2} \simeq 3\%$$

The amount of glycol to be added to 1 L of water can be found from the relation

$$2.87 \times 10^{-2} = \frac{n_{\text{glycol}}/\text{mol}}{n_{\text{glycol}}/\text{mol} + 55.5 \text{ mol}}$$

Solving this relation for the moles of glycol, we find $\boldsymbol{n}_{\text{glycol}} = 1.64$ mol corresponding to

$$1.64 \text{ mol} \times 62.068 \text{ g mol}^{-1} = 101.8 \text{ g}$$

We proceed in a similar manner for the other cases and find that

$\Delta T/\text{K}$	−3	−10	−20
x_{glycol}	0.0287	0.0925	0.176
$\boldsymbol{n}_{\text{glycol}}/\text{mol}$	1.64	5.66	11.9

Note that we expect our approximation that $\hat{a}_s = x_s = 1 - x_{\text{glycol}}$, may not hold at the higher mole fractions of glycol. The depression by 3 degrees for $x_{\text{glycol}} = 2.87 \times 10^{-2}$ corresponds to a volume fraction between 9 and 10% glycol in water, which compares favorably with the experimental value of 3 degrees for a solution that is 10% glycol by volume.

P21.5 Lead iodide, PbI_2, has a solubility product of 9.49×10^{-9}. Calculate the solubility of Pb^{2+} in a 0.01 molar KNO_3 solution, in a 0.01 molar KI solution, and a 0.1 molar KI solution. You can assume that the KNO_3 and KI are strong electrolytes and infinitely soluble. Comment on your findings.

Solution

Given the solubility product of 9.49×10^{-9}, we would expect the solubility of PbI_2 in an ideal solution to be given by

$$K_{sp} = \left(\hat{c}_{\text{Pb}^{2+}}\right)\left(\hat{c}_{I^-}^2\right) = 4\hat{s}^3$$

where we have used the stoichiometry to write that $\hat{c}_{\text{Pb}^{2+}} = \hat{c}_{I^-}/2 \equiv \hat{s}$. Thus we find that $\hat{s} = 4.87 \times 10^{-5}$.

The Case of 0.01 M KNO$_3$: We begin by using the Debye–Hückel formula to calculate the activity coefficients for 0.01 M KNO_3. If we assume that $R = 500$ pm, we find that

$$\ln \gamma_\pm = -\frac{A \cdot z^2\sqrt{I}}{1 + B \cdot R\sqrt{I}}$$

$$= -\frac{\left(1.175 \text{ M}^{-1/2}\right) \cdot (1)^2\sqrt{0.01 \text{ M}}}{1 + \left(0.330 \times 10^{10} \text{ M}^{-1/2}\text{m}^{-1}\right) \cdot \left(500 \times 10^{-12}\text{m}\right)\sqrt{0.01 \text{ M}}} = -0.101$$

so that $\gamma_\pm = 0.904$. For the solubility product of the lead iodide, we write that

$$K_{sp} = \left(\gamma_{\text{Pb}^{2+}} \cdot \hat{c}_{\text{Pb}^{2+}}\right)\left(\gamma_{I^-}^2 \cdot \hat{c}_{I^-}^2\right)$$

Using the fact that $\hat{c}_{\text{Pb}^{2+}} = \hat{c}_{I^-}/2 \equiv \hat{s}$, we find

$$K_{sp} = \left(4\gamma_\pm^3 \cdot \hat{s}^3\right)$$

If we assume that the ion activity is controlled by the KNO_3, then we can make the approximation that $\gamma_{\text{Pb}^{2+}} = \gamma_\pm$ and $\gamma_{I^-} = \gamma_\pm$ and calculate

$$\hat{s} = \sqrt[3]{\frac{K_{sp}}{4\gamma_\pm^3}}$$

For the case of 0.01 M KNO_3, we find that

$$\hat{s} = \sqrt[3]{\frac{K_{sp}}{4\gamma_\pm^3}} = \sqrt{\frac{9.49 \cdot 10^{-9}}{4 \cdot (0.904)^3}} = 5.67 \times 10^{-5}$$

Note that the presence of the KNO_3 acts to enhance the solubility of PbI_2 from 4.87×10^{-5} to 5.67×10^{-5}. From a thermodynamic perspective, we see that this increase results from a change in the activity coefficient. From a more molecular perspective, we can rationalize this result as arising from a change in the solvation about the Pb^{2+}; i.e., the likelihood of nitrate ions in the solvation shell (ionic atmosphere) of the Pb^{2+} is large compared to that of iodide ions. Correspondingly, the ion atmosphere of iodide ions has a higher likelihood of containing potassium ions than lead ions. The phenomenon here is called the "ionic strength effect on solubility."

The Case of KI: In contrast to KNO_3, the iodide ions from KI can participate in the equilibrium of the Pb^{2+} ions. If we assume that KI is a strong electrolyte, then it contributes an initial amount of I^- to the solution, which will be either 0.01 M or 0.1 M; we call it $\hat{c}_{K+}$. We also have I^- that can arise from dissolution of PbI_2, via $\hat{c}_{Pb^{2+}} = \hat{c}_{I^-}/2 \equiv \hat{s}$; thus, the total concentration of the I^- is

$$\hat{c}_{I^-} = \hat{c}_{K^+} + 2 \cdot \hat{c}_{Pb^{2+}} = \hat{c}_{K^+} + 2 \cdot \hat{s}$$

Inserting this result into the K_{sp} expression gives

$$K_{sp} = \left(\gamma_{Pb^{2+}} \cdot \hat{s}\right) \left(\gamma_{I^-}^2 \cdot \left(\hat{c}_{K^+} + 2 \cdot \hat{s}\right)^2\right)$$

If we make the assumption that KI controls the ionic strength and replace the activity coefficients with the mean activity coefficient, we find that

$$K_{sp} = \gamma_\pm^3 \cdot \hat{s} \cdot \left(\hat{c}_{K^+} + 2 \cdot \hat{s}\right)^2$$

While this cubic equation can be solved exactly, we can find an accurate result by making the approximation that $2 \cdot \hat{s} \ll \hat{c}_{K^+}$, which is based on our finding in the first part that $\hat{s} < 10^{-4}$. In this approximation, we find that

$$\hat{s} \approx \frac{K_{sp}}{\gamma_\pm^3 \cdot \hat{c}_{K^+}^2}$$

Given this approximation, we first consider the case of 0.1 molar KI. At 0.1 M KI, we find a mean activity coefficient $\gamma_\pm$ of 0.783 and hence

$$\hat{s} = \frac{9.49 \cdot 10^{-9}}{(0.783)^3 \cdot (0.1)^2} = 1.98 \times 10^{-6}$$

Thus, we see that the iodide concentration pushes the PbI_2 equilibrium toward the precipitate and reduces the Pb^{2+} solubility. Note however that the activity coefficient is suppressed by the KI, and it still acts to increase the solubility. If we assumed that the activity coefficient were unity, then the solubility would be even lower; i.e., $9.49 \cdot 10^{-7}$. In contrast to the ionic strength effect (described in the first part), the phenomenon shown here is called the "common ion effect."

Using this same approximation for the case of 0.01 M KI, we find a lead solubility of

$$\hat{s} = \frac{9.49 \cdot 10^{-9}}{(0.904)^3 \cdot (0.01)^2} = 1.28 \times 10^{-4}$$

This value is significantly higher than that found for the 0.1 M KI case, and it is a factor of 2 to 3 more than that found for the KNO_3 electrolyte, which does not participate in the equilibrium. This latter result underscores the competition between the "ionic strength effect" and the "common ion effect."

P21.6 One important feature of a biological membrane is its ability to maintain concentration gradients of ions. Consider the idealized situation in which a membrane separates two solutions, one with a concentration of 0.15 molal KCl and the other with a concentration of 5×10^{-3} molal KCl, typical differences for the inside and outside of a cell. Assume that both solutions have a density of $\rho = 1\ \text{g mL}^{-1}$ and are at 298 K. a) Find the molar Gibbs energy difference between the two sides of the membrane, assuming that the solutions are ideal. b) Using the Debye–Hückel limiting law given by

$$\ln \gamma_\pm = -1.141 \cdot z_+ z_- \sqrt{I} \text{ with } I = \frac{1}{2}\sum_i \hat{c}_i z_i^2$$

find the activity for each of the two solutions. c) Using your activities from part b) find the molar Gibbs energy difference between the two sides. d) Compute the percentage error in the two molar Gibbs energy differences that you found in parts a) and c).

Solution

(a) The molar Gibbs energy difference between the two sides of the membrane is

$$\Delta G_m = -\boldsymbol{RT} \ln Q$$

where

$$Q = \frac{(\hat{a}_{\text{KCl}})_{\text{in}}}{(\hat{a}_{\text{KCl}})_{\text{out}}}$$

For ideal solutions $\gamma = 1$, we have

$$Q = \frac{(0.15)}{(0.005)} = 30$$

and therefore

$$\Delta G_{m,\text{ideal solution}} = \left[-\left(8.314 \text{ J mol}^{-1}\text{K}^{-1}\right)(298 \text{ K}) \ln(30)\right] \text{ J mol}^{-1} = -8.4 \text{ kJ mol}^{-1}$$

(b) For the case in which the solutions are not ideal we begin by evaluating the Debye–Hückel limiting law

$$\ln \gamma_{\pm} = -A \cdot |z_+| \cdot |z_-| \sqrt{I} \text{ with } A = 1.175 \text{ L}^{1/2}\text{mol}^{-1/2} \text{ at } 298 \text{ K}$$

The ionic strength of the inside is with $z_i^2 = 1$

$$I = \frac{1}{2}\sum_i c_i z_i^2$$
$$= \frac{1}{2}\left(0.15 \text{ mol L}^{-1} + 0.15 \text{ mol L}^{-1}\right) = 0.15 \text{ mol L}^{-1}$$

leading to (with $|z_+| = |z_-| = 1$)

$$\ln \gamma_{\pm} = -1.175 \cdot \sqrt{0.15} = -0.455 \text{ or } \gamma_{\pm} = \exp\left(-3.72 \times 10^{-3}\right) = 0.634$$

and for the outside of the cell the ionic strength is

$$\ln \gamma_{\pm} = -1.175 \cdot \sqrt{0.005} = -0.0831 \text{ or } \gamma_{\pm} = \exp\left(-3.72 \times 10^{-2}\right) = 0.920$$

The activities become, inside the cell

$$\hat{a}_{\text{inside}} = \gamma_{\pm}\hat{m} = 0.0951$$

and for outside the cell

$$\hat{a}_{\text{outside}} = \gamma_{\pm}\hat{m} = 0.0046$$

(c) With the activities known, we calculate the molar Gibbs energy difference as

$$\Delta G_{\text{real solution}} - 8.314 \text{ J mol}^{-1}\text{K}^{-1} \cdot 298 \text{ K} \cdot \ln\left(\frac{0.0951}{0.0046}\right) = -7.50 \text{ kJ mol}^{-1}$$

(d) The percentage difference between real and ideal solution is

$$\text{percent error} = \frac{\Delta G_{m,\text{ideal solution}} - \Delta G_{\text{real solution}}}{\Delta G_{m,\text{ideal solution}}} \frac{-8.4 - (-7.50)}{-7.50} \times 100 = 12\%$$

P21.7 Show that the expression for the chemical potential of an ideal solution in terms of the mole fraction (e.g., Equation (21.24) of the text) and the expression in terms of the concentration (e.g., Equation (21.16) of the text) differ from each other only in terms of the reference state.

Solution

First we show the connection for an ideal solution. Although Equation (21.16) is written for an ideally diluted solution, it has the same form for the more stringent requirement of an ideal solution that is discussed here. While Equation (21.16) is written for the solute, it can be generalized and written for a component i, taking the form

$$\mu_i = \mu_i^{\ominus} + \boldsymbol{RT} \cdot \ln \hat{c}_i \text{ with } \mu_i^{\ominus} = G_{m,i}^{\ominus}$$

Generalizing Equation (21.24) in an analogous manner, we can write that

$$\mu_i = \mu_i^{\ominus\prime} + \boldsymbol{RT} \cdot \ln x_i$$

where we indicate the change in the reference state by the use of the apostrophe symbol.

In order to create a consistent description, we require that the chemical potential of species i be the same for the two cases, hence we find that

$$\mu_i = \mu_i^{\ominus} + \boldsymbol{R}T \cdot \ln \hat{c}_i = \mu_i^{\ominus\prime} + \boldsymbol{R}T \cdot \ln x_i$$

so that

$$\begin{aligned}
\mu_i^{\ominus\prime} &= \mu_i^{\ominus} + \boldsymbol{R}T \cdot \ln \hat{c}_i - \boldsymbol{R}T \cdot \ln x_i \\
&= \mu_i^{\ominus} + \boldsymbol{R}T \cdot \ln \left(\frac{c_i}{c_i^{\ominus}} \frac{1}{x_i} \right) = \mu_i^{\ominus} + \boldsymbol{R}T \cdot \ln \left(\frac{n_i}{n_i^{\ominus}} \frac{n_{\text{total}}}{n_i} \right) \\
&= \mu_i^{\ominus} + \boldsymbol{R}T \cdot \ln \left(\frac{n_{\text{total}}}{n_i^{\ominus}} \right) = \mu_i^{\ominus} - \boldsymbol{R}T \cdot \ln \left(x_i^{\ominus} \right)
\end{aligned}$$

where $x_i^{\ominus}$ is the mole fraction of species i in a solution having the standard state concentration of $c_i^{\ominus}$.
If we use the convention that the standard state for the chemical potential of component i is the chemical potential of the pure component ($x_i^{\ominus} = 1$) whenever the mole fractions are used, then we see that

$$\mu_i^{\ominus\prime} = \mu_i^{\ominus}$$

P21.8 Show that

$$\phi(x) = \frac{1}{2} a_0 \left(e^{\kappa x} + e^{-\kappa x} \right) + \frac{1}{2} a_1 \frac{1}{\kappa} \left(e^{\kappa x} - e^{-\kappa x} \right) \tag{P21.8.1}$$

is a solution of the differential equation

$$\frac{d^2 \phi(x)}{dx^2} = 2n \frac{e^2}{\varepsilon_0 \varepsilon k T} \cdot \phi(x) \tag{P21.8.2}$$

in Foundation 21.2, and that for small values of x we have

$$\phi(x) = a_0 + a_1 \cdot x$$

Solution

We begin by finding the second derivative of $\phi(x)$. For the first derivative we find

$$\frac{d\phi(x)}{dx} = \frac{\kappa}{2} a_0 \left(e^{\kappa x} - e^{-\kappa x} \right) + \frac{1}{2} a_1 \left(e^{\kappa x} + e^{-\kappa x} \right)$$

and for the second derivative we find

$$\begin{aligned}
\frac{d^2 \phi(x)}{dx^2} &= \frac{\kappa^2}{2} a_0 \left(e^{\kappa x} + e^{-\kappa x} \right) + \frac{\kappa}{2} a_1 \left(e^{\kappa x} - e^{-\kappa x} \right) \\
&= \kappa^2 \cdot \left[\frac{1}{2} a_0 \left(e^{\kappa x} + e^{-\kappa x} \right) + \frac{1}{2} a_1 \frac{1}{\kappa} \left(e^{\kappa x} - e^{-\kappa x} \right) \right] \\
&= \kappa^2 \cdot \phi(x)
\end{aligned}$$

Inserting this result into the differential equation gives

$$\frac{d^2 \phi(x)}{dx^2} = 2n \frac{e^2}{\varepsilon_0 \varepsilon k T} \cdot \phi(x) = \kappa^2 \phi(x)$$

Thus we find that Equation (P21.8.1) is a solution of the differential Equation (P21.8.2), if κ is identified by

$$\kappa = \sqrt{\frac{2ne^2}{\varepsilon_0 \varepsilon k T}}$$

Next we consider the expression for $\phi(x)$ for small values of x. Making the approximation

$$e^{\kappa x} = 1 + \kappa x \qquad e^{-\kappa x} = 1 - \kappa x$$

we find that

$$\phi(x) = \frac{1}{2} a_0 \cdot 2 + \frac{1}{2} a_1 \frac{1}{\kappa} \cdot 2\kappa x = a_0 + a_1 x \qquad \text{for small values of} x.$$

22

Reaction Equilibria in Aqueous Solutions and Biosystems

22.1 Exercises

E22.1 The pK_a of benzoic acid in water at 298 K is 4.213. Calculate the pH of a 1:1 mixture of benzoic acid (C_6H_5COOH, HA) and benzoate ion ($C_6H_5COO^-$, A^-). Use the pK_a to calculate the Gibbs energy change for the acid ionization reaction.

Solution
The acid ionization reaction is

$$HA(aq) \rightleftarrows A^-(aq) + H^+(aq)$$

Assuming that the solution behaves ideally, we can write that, according to Equation (22.7),

$$pH = pK_a + \log\left(\frac{\hat{c}_{A^-}}{\hat{c}_{HA}}\right)$$

and taking $\hat{c}_{A^-} = \hat{c}_{HA}$, we find that $pH = pK_a = 4.213$. Note that this analysis ignores the autoionization of water.
Using the fact that,

$$\Delta_r G^\ominus = 2.303 \cdot \boldsymbol{R}T \cdot pK_a$$

and taking $T = 298$ K, we find that

$$\Delta_r G^\ominus = 2.303 \cdot 8.314\ \text{J mol}^{-1}\text{K}^{-1} \cdot 298\ \text{K} \cdot 4.213$$
$$= 24.04\ \text{kJ mol}^{-1}$$

E22.2 Determine the pK_a of lactic acid by using Gibbs energies of formation data and compare it to the experimental value of 3.86 at $T = 298$ K. Calculate the pH of a 1:1 mixture of lactic acid (CH_3 C(OH)HCOOH) and lactate ion ($CH_3C(OH)HCOO^-$).

Solution
The acid ionization reaction is

$$CH_3C(OH)HCOOH(aq) \rightleftarrows CH_3C(OH)HCOO^-(aq) + H^+(aq)$$

From Appendix A we use the facts that $\Delta_f G(CH_3C(OH)HCOOH) = -538.77\ \text{kJ mol}^{-1}$, $\Delta_f G(CH_3C(OH)HCOO^-) = -516.72\ \text{kJ mol}^{-1}$, and $\Delta_f G(H^+) = 0\ \text{kJ mol}^{-1}$, to calculate

$$\Delta_r G^\ominus = [0 - 516.72 + 538.77]\ \text{kJ mol}^{-1} = 22.05\ \text{kJ mol}^{-1}$$

Using this value and the relation $\Delta_r G^\ominus = 2.303 \cdot \boldsymbol{R}T \cdot pK_a$, we find that

$$pK_a = \frac{\Delta_r G^\ominus}{2.303 \cdot \boldsymbol{R}T}$$
$$= \frac{22{,}050\ \text{J mol}^{-1}}{2.303 \cdot 8.314\ \text{J mol}^{-1}\text{K}^{-1} \cdot 298\ \text{K}} = 3.864$$

Solutions Manual for Principles of Physical Chemistry, Third Edition. Edited by Hans Kuhn, David H. Waldeck, and Horst-Dieter Försterling.

which is in excellent agreement with the experimental value given. For the case of $\hat{c}_{CH_3C(OH)HCOOH} = \hat{c}_{CH_3C(OH)HCOO^-}$, we find that the pH = $pK_a = 3.864$.

E22.3 Calculate the pH of acetic acid CH_3COOH ($K_a = 2.5 \times 10^{-5}$), a weak acid, different total concentrations of $0.010\ \text{mol L}^{-1}$, $1.0 \times 10^{-3}\ \text{mol L}^{-1}$, and $1.0 \times 10^{-4}\ \text{mol L}^{-1}$?

Solution

We calculate $\hat{c}_{H^+}$ according to Equation (22.12),

$$\hat{c}_{H^+} = -\frac{1}{2}K_a + \sqrt{\hat{c}_{\text{total}} \cdot K_a + \frac{1}{4}K_a^2}$$

With this relation a difficulty arises for $\hat{c}_{\text{total}} = 0$, because it yields $\hat{c}_{H^+} = 0$, a clearly wrong result. The reason is that we neglected the dissociation of water in its derivation. This neglect affects $\hat{c}_{H^+}$ at very low acid concentration. We account for the dissociation of water by setting $\hat{c}_{H^+} = 1.0 \times 10^{-7}$ for $\hat{c}_{\text{total}} = 0$ and starting with the evaluation of Equation (22.12) at $c_{\text{total}} = 3 \times 10^{-6}$ (for this value we calculate $\hat{c}_{H^+} = 30 \times 10^{-7}$, and the dissociation of water can be neglected). The resulting pH = $-\log(\hat{c}_{H^+})$ is displayed in Fig. E22.3.

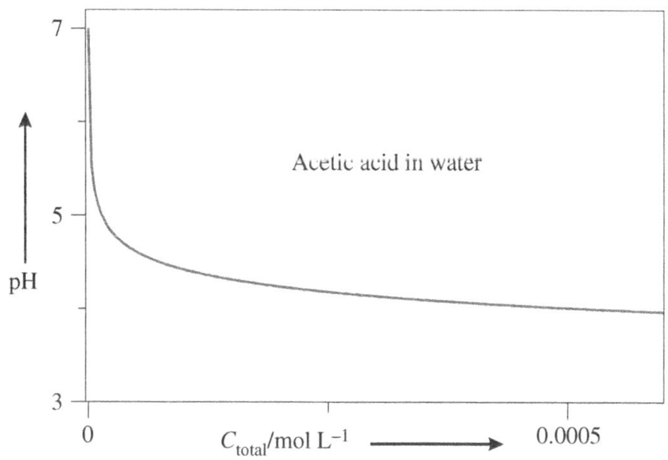

Figure E22.3 pH of acetic acid as a function of its concentration c_{total}.

E22.4 Test the approximation about complete dissociation of the acid HCl by calculating its acid dissociation constants (using Gibbs energies of formation) and comparing it to K_w. Contrast your result to that of acetic acid with its pK_a of 4.62.

Solution

The acid ionization reaction is

$$HCl(g) \rightleftarrows Cl^-(aq) + H^+(aq)$$

From Appendix A, we use the facts that $\Delta_f G(HCl) = -95.3\ \text{kJ mol}^{-1}$, $\Delta_f G(Cl^-) = -131.2\ \text{kJ mol}^{-1}$, and $\Delta_f G(H^+) = 0\ \text{kJ mol}^{-1}$ and calculate a reaction Gibbs energy of

$$\Delta_r G^\ominus = [0 - 131.2 + 95.3]\ \text{kJ mol}^{-1} = -35.9\ \text{kJ mol}^{-1}$$

With this value in hand we can calculate the pK_a as

$$pK_a = \frac{\Delta_r G^\ominus}{2.303 \cdot \boldsymbol{RT}} = \frac{-35.9\ \text{kJ mol}^{-1}}{2.303 \cdot 8.314\ \text{J mol}^{-1}\text{K}^{-1} \cdot 298\ \text{K}} = -6.292$$

or a $K_a = 1.96 \times 10^6$. This value is 10^{20} times larger than the value of K_w, and nearly 10^{10} to 10^{11} times larger than the $K_a = 2.30 \times 10^{-5}$ value of acetic acid.

E22.5 In Section 22.3.2 the pH of amino acid solutions is calculated. Equation (22.37) was simplified to Equation (22.38) by considering large enough c_{total}. Can the dissociation of water and K_1 be neglected?

Solution

Consider the equilibria for the amino acid to be

$$AH^+ \rightarrow BH + H^+ \rightarrow C^- + 2H^+$$

We will use our pH value from the approximate solution to estimate the magnitude of the AH^+ and C^- forms to assess our treatment.

The analysis in the text calculates a value of pH = 6.1 and reports a value of $K_1 = 10^{-2.4}$. If we use this value in Equation (22.36) and take $\hat{c}_{\text{total}} = 0.1$, we find that

$$\hat{c}_{AH^+} = \frac{\hat{c}_{H^+} \cdot \hat{c}_{\text{total}}}{K_1} = \frac{10^{-6.1} \cdot 10^{-1}}{10^{-2.4}} = 10^{-4.7}$$

This $\hat{c}_{AH^+}$ value is much smaller than that of $\hat{c}_{\text{total}}$, indicating that the equilibrium is strongly shifted away from the protonated form of the amino acid.

Accordingly, from Equation (22.31), we find

$$\hat{c}_{C^-} = K_2 \frac{\hat{c}_B}{\hat{c}_{H^+}} \approx K_2 \frac{\hat{c}_{\text{total}}}{\hat{c}_{H^+}} = 10^{-4.7}$$

Hence, the $\hat{c}_{OH^-} = 10^{-7.9}$ given by the dissociation equilibrium of water can be neglected. K_1 can be neglected in the denominator of Equation (22.37) until $\hat{c}_{\text{total}} > 0.01$.

E22.6 In Section 22.3.2 the pH of a glycine solution is calculated. First calculate the pH by assuming that the reaction from B to A^+ does not happen. Second, calculate the pH by assuming that the reaction from B to C^- does not take place.

Solution

We consider the equilibria of the amino acid glycine discussed in Section 22.3.2.

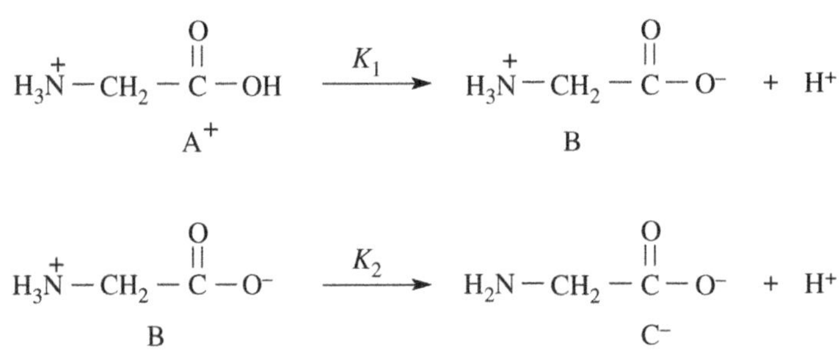

and calculate the pH for two limiting cases:

1) Equilibrium $A^+ \rightleftharpoons B$ is neglected.

   In this case we have to consider the dissociation reaction $B \rightarrow C^-$, and the pH is obtained in analogy to the procedure for acetic acid in Section 22.2.2, Equation (22.12)

$$\hat{c}_{H^+} = -\frac{1}{2}K_2 + \sqrt{\hat{c}_{\text{total}}\, K_2 + \frac{1}{4}K_2^2}$$

   With $K_2 = 10^{-9.8}$ and $\hat{c}_{\text{total}} = 0.1\ \text{mol L}^{-1}$ we obtain

$$\begin{aligned}\hat{c}_{H^+} &= -\frac{1}{2} \cdot 10^{-9.8} + \sqrt{0.1 \cdot 10^{-9.8} + \frac{1}{4}10^{-19.6}}\\ &= -7.93 \times 10^{-11} + \sqrt{1.58 \times 10^{-11} + 6.29 \times 10^{-21}}\\ &= -7.93 \times 10^{-11} + 3.97 \times 10^{-6} = 3.97 \times 10^{-6}\end{aligned}$$

   This concentration corresponds to a pH of

$$\text{pH} = -\log\left(3.97 \times 10^{-6}\right) = 5.40$$

2) Equilibrium $B \rightleftharpoons C^-$ is neglected
In this case B reacts with one H^+ to form one A^+. In this process the H^+ is taken from the water, and one OH^- is being formed. Thus

$$\hat{c}_{A^+} = \hat{c}_{OH^-}$$

Then we obtain

$$K_1 = \frac{\hat{c}_{H^+} \cdot \hat{c}_B}{\hat{c}_{A^+}} = \frac{\hat{c}_{H^+} \cdot \hat{c}_B}{\hat{c}_{OH^-}} = \frac{\hat{c}_{H^+} \cdot \hat{c}_{total}}{K_w / \hat{c}_{H^+}}$$

and

$$\hat{c}_{H^+} = \sqrt{\frac{K_1 K_w}{\hat{c}_{total}}}$$

With $K_1 = 10^{-2.4}$ and $\hat{c}_{total} = 0.1$ mol L^{-1} we obtain

$$\hat{c}_{H^+} = \sqrt{\frac{10^{-2.4} \cdot 1 \times 10^{-14}}{0.1}} = 1.99 \times 10^{-8}$$

This corresponds to a pH of

$$pH = -\log\left(1.99 \times 10^{-8}\right) = 7.70$$

E22.7 The pK_a values for the dissociation of H_2O and D_2O at two different temperatures are:

T(K)	pK_a (H_2O)	pK_a (D_2O)
283	14.528	15.527
313	13.542	14.462

Determine $\Delta_r H$ and $\Delta_r S$ for the autoionization of H_2O and D_2O. Provide a molecular level explanation for any differences you find between the enthalpies and entropies.

Solution
First we consider H_2O and use the fact that

$$\Delta_r G = -\boldsymbol{R}T \cdot \ln K_a = -\boldsymbol{R}T \cdot \log K_a = 2.303 \cdot \boldsymbol{R}T \cdot pK_a$$

and find

$$\begin{aligned}\Delta_r G(283\text{ K}) &= 2.303 \cdot 8.314\text{ J mol}^{-1}\text{ K}^{-1} \cdot 283\text{ K} \cdot 14.528\\ &= 78{,}722\text{ J mol}^{-1}\\ \Delta_r G(313\text{ K}) &= 2.303 \cdot 8.314\text{ J mol}^{-1}\text{ K}^{-1} \cdot 313\text{ K} \cdot 13.542\\ &= 81{,}158\text{ J mol}^{-1}\end{aligned}$$

Next, we combine these two values with the fundamental relationship

$$\Delta_r G = \Delta_r H - T\Delta_r S$$

to find the entropy change; namely

$$\begin{aligned}\Delta_r S^{\ominus} &= \frac{\Delta_r G(T_2) - \Delta_r G(T_1)}{T_1 - T_2}\\ &= \frac{(81{,}158 - 78{,}722)\text{ J mol}^{-1}}{(283 - 313)\text{ K}} = -81\text{ J mol}^{-1}\text{ K}^{-1}\end{aligned}$$

Note that this analysis neglects any temperature change in $\Delta_r S$ over this 30 degree temperature range. With $\Delta_r S$ in hand we can calculate $\Delta_r H$ via

$$\Delta_r H = \Delta_r G + T\Delta_r S$$

so that

$$\Delta_r H(283\text{ K}) = 55.80\text{ kJ mol}^{-1} \text{ and } \Delta_r H(313\text{ K}) = 55.81\text{ kJ mol}^{-1}$$

which indicates only a slight change over this temperature range.

We can proceed in a like manner for the case of D_2O, and we find that

$$\Delta_r S = \frac{\Delta_r G(T_2) - \Delta_r G(T_1)}{T_1 - T_2}$$
$$= \frac{(86.7 - 84.13)\text{ kJ mol}^{-1}}{(283 - 313)\text{ K}} = -86\text{ J mol}^{-1}\text{K}^{-1}$$

and a $\Delta_r H = 59.7\text{ kJ mol}^{-1}$.

The enthalpy change of the D_2O case is expected to be larger than that of H_2O because of the larger reduced mass of the OD bond (hence a lower zero point vibrational energy) and the larger energy barrier for the bonds dissociation.

E22.8 In analogy to K_a

$$\text{HA} \rightleftarrows \text{H}^+ + \text{A}^- \text{ for which } K_a = \frac{\hat{c}_{\text{A}^-}\,\hat{c}_{\text{H}^+}}{\hat{c}_{\text{HA}}},$$

we can define an equilibrium constant K_b for the reaction of an acid's conjugate base

$$\text{H}_2\text{O} + \text{A}^- \rightleftarrows \text{HA} + \text{OH}^- \text{ for which } K_b = \frac{\hat{c}_{\text{HA}}\,\hat{c}_{\text{OH}^-}}{\hat{c}_{\text{A}^-}}$$

Show that $K_a K_b = K_w$.

Solution

We proceed by substituting our expressions for K_a and K_b and then simplifying the result and comparing it to K_w. Substituting for K_a and K_b we find that

$$K_a K_b = \left(\frac{\hat{c}_{\text{H}^+}\,\hat{c}_{\text{A}^-}}{\hat{c}_{\text{HA}}}\right)\left(\frac{\hat{c}_{\text{HA}}\,\hat{c}_{\text{OH}^-}}{\hat{c}_{\text{A}^-}}\right)$$
$$= \hat{c}_{\text{H}^+}\hat{c}_{\text{OH}^-} = K_w$$

E22.9 Using the information in E22.1 and E22.2, write the equilibria for benzoate, the conjugate base of benzoic acid, and lactate, the conjugate base of lactic acid. Determine their pK_b values.

Solution

For benzoate we find that

$$\text{H}_2\text{O} + \text{C}_7\text{H}_5\text{O}_2^-(\text{aq}) \rightleftarrows \text{C}_7\text{H}_6\text{O}_2(\text{aq}) + \text{OH}^-\ (\text{aq})$$

for which

$$K_b = \frac{\hat{c}_{\text{C}_7\text{H}_6\text{O}_2}\,\hat{c}_{\text{OH}^-}}{\hat{c}_{\text{C}_7\text{H}_5\text{O}_2^-}}$$

For lactate we find that

$$\text{H}_2\text{O} + \text{CH}_3\text{C(OH)HCOO}^-(\text{aq}) \rightleftarrows \text{CH}_3\text{C(OH)HCOOH(aq)} + \text{OH}^-(\text{aq})$$

for which

$$K_b = \frac{\hat{c}_{\text{CH}_3\text{C(OH)HCOOH}}\,\hat{c}_{\text{OH}^-}}{\hat{c}_{\text{CH}_3\text{C(OH)HCOO}^-}}$$

Using the fact that $K_a K_b = K_w$(see E22.8) we find that

$$pK_b = pK_w - pK_a$$

Using 14.00 for pK_w and 4.213 for the pK_a of benzoic acid, we find that its pK_b is

$$pK_b = 14.00 - 4.213 = 9.79$$

Using 14.00 for pK_w and 3.864 for the pK_a of lactic acid, we find that its pK_b is

$$pK_b = 14.00 - 3.864 = 10.14$$

E22.10 Calculate the pH of an equimolar mixture of NH_4NO_3 and NH_3 which acts as a basic buffer. What relative concentrations are needed to generate a pH of 9.50? Use a value of 9.25 for the pKa of ammonia at 298 K.

Solution
The acid ionization equilibrium is

$$NH_4^+(aq) \rightleftarrows NH_3(aq) + H^+(aq)$$

with

$$pH = pK_a + \log\left(\frac{\hat{c}_{NH_3}}{\hat{c}_{NH_4^+}}\right)$$

For $\hat{c}_{NH_4^+} = \hat{c}_{NH_3}$ we find that $pH = pK_a = 9.25$. For the case of $pH = 9.5$, we find that

$$9.5 - 9.25 = \log\left(\frac{\hat{c}_{NH_3}}{\hat{c}_{NH_4^+}}\right)$$

so that

$$\frac{\hat{c}_{NH_3}}{\hat{c}_{NH_4^+}} = 10^{0.25} = 1.78$$

Nearly twice as much ammonia as ammonium ion is needed to shift the pH by a quarter or so of a pH unit.

E22.11 Human blood plasma is a complex mixture of many important biochemicals; however, it also contains simple buffering compounds (HCO_3^-/H_2CO_3 and $HPO_4^{2-}/H_2PO_4^-$) that maintain its pH at 7.4. Calculate the ratio of concentrations of HCO_3^-/H_2CO_3 that create a pH of 7.4. Also perform the calculation for $HPO_4^{2-}/H_2PO_4^-$. Use pK_a values of 6.35 and 10.33 for carbonic acid at 298 K, and use pK_a values of 2.16, 7.21, and 12.32 for phosphoric acid.

Solution
For carbonic acid we see that the pK_a of 6.35 for the first acid dissociation is significantly smaller than that of the bicarbonate ionization, hence we consider only this first equilibrium; namely,

$$H_2CO_3(aq) \rightleftarrows HCO_3^-(aq) + H^+(aq)$$

Using the relation

$$pH = pK_a + \log\left(\frac{\hat{c}_{HCO_3^-}}{\hat{c}_{H_2CO_3}}\right)$$

we can solve for the concentration ratio by

$$\log\left(\frac{\hat{c}_{HCO_3^-}}{\hat{c}_{H_2CO_3}}\right) = 7.4 - 6.35 = 1.05$$

or

$$\frac{\hat{c}_{HCO_3^-}}{\hat{c}_{H_2CO_3}} = 10^{1.05} = 11.2$$

Hence the bicarbonate concentration is 11 times that of the carbonic acid concentration at a pH of 7.4.

For phosphoric acid we see that the pK_a of 7.21 for the second acid dissociation is significantly different from that of the other two; hence, we only consider

$$H_2PO_4^-(aq) \rightleftarrows HPO_4^{2-}(aq) + H^+(aq)$$

Using the relation

$$\mathrm{pH} = \mathrm{p}K_a + \log\left(\frac{\hat{c}_{\mathrm{HPO_4^{2-}}}}{\hat{c}_{\mathrm{H_2PO_4^-}}}\right)$$

we can solve for the concentration ratio by

$$\log\left(\frac{\hat{c}_{\mathrm{HPO_4^{2-}}}}{\hat{c}_{\mathrm{H_2PO_4^-}}}\right) = 7.4 - 7.21 = 0.19$$

or

$$\frac{\hat{c}_{\mathrm{HPO_4^{2-}}}}{\hat{c}_{\mathrm{H_2PO_4^-}}} = 10^{0.19} = 1.5$$

Hence the ion concentrations are nearly equal.

Extension: Show that inclusion of the second equilibrium leads to a small correction.

E22.12 Perform the Gibbs energy calculation for the reaction of Equation (22.39) at a pH of 7 rather than the 1 molar solution.

Solution

Reaction 22.39 is

$$\frac{1}{2}\,\mathrm{Zn} + \mathrm{H^+} \rightleftarrows \frac{1}{2}\mathrm{Zn^{2+}} + \frac{1}{2}\mathrm{H_2}$$

and has a reaction Gibbs energy of $\Delta_r G^{\ominus} = -73.6\,\mathrm{kJ\,mol^{-1}}$ under standard conditions. To find how the Gibbs energy changes with pH we write

$$\Delta_r G = \Delta_r G^{\ominus} + \boldsymbol{RT}\ln\left(\frac{\hat{c}_{\mathrm{Zn^{2+}}}^{1/2}\cdot\hat{P}_{\mathrm{H_2}}^{1/2}}{\hat{c}_{\mathrm{H^+}}}\right)$$

Under the conditions of $\hat{c}_{\mathrm{Zn^{2+}}} = 1, \hat{P}_{\mathrm{H_2}} = 1$, and $\hat{c}_{\mathrm{H^+}} = 10^{-7}$, we find that

$$\begin{aligned}\Delta_r G &= \left[-73.6\,\mathrm{kJ\,mol^{-1}} + 8.314\,\mathrm{J\,mol^{-1}K^{-1}}\cdot 298\,\mathrm{K}\cdot\ln\left(\frac{1\cdot 1}{10^{-7}}\right)\right]\\ &= -33.7\,\mathrm{kJ\,mol^{-1}}\end{aligned}$$

Hence we expect the reaction to still be spontaneous.

E22.13 The amino acid alanine has a $\mathrm{p}K_{\mathbf{a}}$ of 2.3 for its carboxylic acid group. What is the pH of a solution that is 0.1 molar HCl and 0.1 molar alanine?

Solution

Because HCl is such a strong acid we consider it as a proton source, and here we consider the equilibrium

$$\mathrm{Ala^+} \rightleftarrows \mathrm{Ala} + \mathrm{H^+} \quad \text{with } \mathrm{p}K_a = 2.3$$

First we let us make the approximation that $\hat{c}_{\mathrm{H^+}}$ is given by the concentration of the added HCl, $\hat{c}_{\mathrm{H^+}} = 0.1$ (i.e., we neglect the contribution from $\mathrm{Ala^+}$ dissociation). Because of $\hat{c}_{\mathrm{Ala}} + \hat{c}_{\mathrm{Ala^+}} = 0.1$ we can write that $\hat{c}_{\mathrm{Ala^+}} = 0.1 - \hat{c}_{\mathrm{Ala}}$. Then, we can write

$$K_a = \frac{\hat{c}_{\mathrm{H^+}}\hat{c}_{\mathrm{Ala}}}{\hat{c}_{\mathrm{Ala^+}}} = \frac{\hat{c}_{\mathrm{H^+}}\cdot\hat{c}_{\mathrm{Ala}}}{0.1 - \hat{c}_{\mathrm{Ala}}}$$

By solving for $\hat{c}_{\mathrm{Ala}}$, we find with $K_a = 10^{-\mathrm{pK}} = 10^{-2.3} = 5.01\times 10^{-3}$

$$\hat{c}_{\mathrm{Ala}} = \frac{K_a\cdot 0.1}{K_a + \hat{c}_{\mathrm{H^+}}} = \frac{5.01\times 10^{-3}\cdot 0.1}{5.01\times 10^{-3} + 0.1} = 4.77\times 10^{-3}$$

and the pH is

$$\mathrm{pH} = -\log\hat{c}_{\mathrm{H^+}} = -\log 0.1 = 1.0$$

In a more rigorous calculation in which we account for the Ala^+ dissociation, we have to consider that $\hat{c}_{H^+} = 0.1 + \hat{c}_{Ala}$, because of the additional protons formed by the dissociation of the amino acid. Then we obtain

$$\hat{c}_{Ala} = \frac{K_a \cdot 0.1}{K_a + \hat{c}_{H^+}} = \frac{5.01 \times 10^{-3} \cdot 0.1}{5.01 \times 10^{-3} + 0.10477} = 4.56 \times 10^{-3}$$

This is 4% less than our first estimate. Then the final pH of the solution is

$$-\log \hat{c}_{H^+} = -\log \hat{c}_{H^+}(0.1 + \hat{c}_{Ala}) = -\log \hat{c}_{H^+} \cdot 0.10456) = 0.98$$

E22.14 The amino acid alanine (NH_2-$CH(CH_3)CO_2H$) has a p$K_\mathbf{a}$ of 9.9 for its amino group and a pK_a of 2.3 for its acid group. What is the pH of a solution that is 0.1 molar NaOH and 0.1 molar alanine?

Solution

In this case, the NaOH is a strong base and we expect it to be fully dissociated; see Figure E22.14. Given the low pK_a of the carboxylic group and the basic solution we consider that the hydroxide reacts with the zwitterion form of the amino acid (Ala) to generate its conjugate base (Ala^-)

Figure E22.14 The figure shows the species involved in the chemical equilibrium.

We consider the equilibrium

$$Ala^- \rightleftarrows Ala + OH^- \quad \text{with} \quad pK_b = 4.1$$

where pK_b = 14.0–p$K_\mathbf{a}$ = 14 – 9.9 = 4.1.

First we assume that the concentration of the hydroxyl ions is controlled by the concentration of the strong base, $\hat{c}_{OH^-} = 0.1$. Because of $\hat{c}_{Ala} + \hat{c}_{Ala^-} = 0.1$ we can write that $\hat{c}_{Ala^-} = 0.1 - \hat{c}_{Ala}$ so that

$$K_b = \frac{\hat{c}_{Ala} \cdot \hat{c}_{OH^-}}{\hat{c}_{Ala^-}} = \frac{\hat{c}_{Ala} \cdot \hat{c}_{OH^-}}{0.1 - \hat{c}_{Ala}}$$

and we can resolve this equation for $\hat{c}_{Ala}$ to obtain with $K_b = 10^{-pK_b} = 7.94 \times 10^{-5}$.

$$\hat{c}_{Ala} = \frac{K_b \cdot \hat{c}_{OH^-}}{K_b + 0.1} = \frac{7.94 \times 10^{-5} \cdot 0.1}{7.94 \times 10^{-5} + 0.1} = 7.93 \times 10^{-5}$$

Then the pH is

$$pH = 14 - pOH = 14 - 1 = 13$$

In a more rigorous calculation we have to consider that $\hat{c}_{OH^-} = 0.1 - \hat{c}_{Ala}$, because hydroxyl ions are consumed by the deprotonation of the amino acid. Then we obtain

$$\hat{c}_{Ala} = \frac{K_b \cdot 0.1}{K_b + \hat{c}_{OH^-}} = \frac{7.94 \times 10^{-5} \cdot 0.1}{7.94 \times 10^{-5} + 0.1 - 7.93 \times 10^{-5}} = 7.93 \times 10^{-5}$$

That is same result as before, because of the much smaller value of $K_b = 7.94 \times 10^{-5}$ compared to $K_a = 5.01 \times 10^{-3}$ in Exercise E22.13.

E22.15 In Foundation 22.2 we derived equations for the concentrations c_{H_2A}, c_{HA^-}, and $c_{A^{2-}}$ for the coupled equilibria in Equation (22.24). Show that these equations simplify to Equation (22.17) for $K_{a2} = 0$.

Solution

In Foundation 22.2 we find that

$$\hat{c}_{H_2A} = \hat{c}_{total} \frac{(\hat{c}_{H^+})^2}{(\hat{c}_{H^+})^2 + K_{a1}\hat{c}_{H^+} + K_{a1}K_{a2}}$$

that

$$\hat{c}_{\text{HA}^-} = \hat{c}_{\text{total}} \frac{K_{a1}\hat{c}_{\text{H}^+}}{\hat{c}^2_{\text{H}^+} + K_{a1}\hat{c}_{\text{H}^+} + K_{a1}K_{a2}}$$

and that

$$\hat{c}_{\text{A}^{2-}} = \hat{c}_{\text{total}} \frac{K_{a1}K_{a2}}{\hat{c}^2_{\text{H}^+} + K_{a1}\hat{c}_{\text{H}^+} + K_{a1}K_{a2}}$$

If we set $K_{a2} = 0$ (i.e., we 'turn off' the second equilibrium), we find by inspection that $\hat{c}_{\text{A}^{2-}} = 0$ because it is directly proportional to K_{a2}. This makes sense because our choice of K_{a2} eliminates the possibility of the second ionization event. For the other two expressions we substitute $K_{a2} = 0$, which eliminates one of the terms in the denominator and we find that

$$\hat{c}_{\text{H}_2A} = \hat{c}_{\text{total}} \frac{(\hat{c}_{\text{H}^+})^2}{(\hat{c}_{\text{H}^+})^2 + K_{a1}\hat{c}_{\text{H}^+} + 0} = \hat{c}_{\text{total}} \frac{\hat{c}_{\text{H}^+}}{\hat{c}_{\text{H}^+} + K_{a1}}$$

and that

$$\hat{c}_{\text{HA}^-} = \hat{c}_{\text{total}} \frac{K_{a1}\hat{c}_{\text{H}^+}}{\hat{c}^2_{\text{H}^+} + K_{a1}\hat{c}_{\text{H}^+} + 0} = \hat{c}_{\text{total}} \frac{K_{a1}}{\hat{c}_{\text{H}^+} + K_{a1}}$$

If we draw the correspondence of $\hat{c}_{\text{H}_2A}$ with the acid form in Equation (22.17) and $\hat{c}_{\text{HA}^-}$ with the conjugate base form, we see that these results agree exactly.

E22.16 Determine $\Delta_r G$ for the reaction

$$\text{NADH} \rightarrow \text{NAD}^+ + \text{H}^+ + 2\text{e}^-$$

under standard conditions at pH = 7, at pH = 5, and at pH = 9. Use $\Delta_r G^\ominus = 21.8\,\text{kJ mol}^{-1}$ for the standard conditions of $\hat{c}_{H^+} = 1$ (pH = 0).

Solution

From the equilibrium, we can write that

$$\begin{aligned} \Delta_r G &= \Delta_r G^\ominus + \boldsymbol{R}T \ln \frac{\hat{c}_{\text{NAD}^+} \cdot \hat{c}_{\text{H}^+}}{\hat{c}_{\text{NADH}}} \\ &= \Delta_r G^\ominus + \boldsymbol{R}T \ln \frac{\hat{c}_{\text{NAD}^+}}{\hat{c}_{\text{NADH}}} + \boldsymbol{R}T \ln \hat{c}_{\text{H}^+} \end{aligned}$$

For the standard state conditions $\hat{c}_{\text{NAD}^+} = 1, \hat{c}_{\text{NADH}} = 1$, we find that

$$\Delta_r G = \Delta_r G^\ominus + \boldsymbol{R}T \ln \hat{c}_{\text{H}^+}$$

and using a proton concentration of $\hat{c}_{\text{H}^+} = 10^{-7}$ (pH = 7), we find that

$$\begin{aligned} \Delta_r G &= \left[21.8\,\text{kJ mol}^{-1} + 8.314\,\text{J mol}^{-1}\text{K}^{-1} \cdot 298\,\text{K} \cdot \ln\left(10^{-7}\right)\right] \\ &= -18.1\,\text{kJ mol}^{-1} \end{aligned}$$

The calculations at other pH values proceed in a similar manner. For pH = 5 we find that $\Delta_r G = -6.7\,\text{kJ mol}^{-1}$, and for pH = 10 we find that $\Delta_r G = -35.2\,\text{kJ mol}^{-1}$.

E22.17 Use Equations (22.15) and (22.16) to derive an expression for $\hat{c}_{HA}$ in terms of $\hat{c}_{\text{total}}$, $\hat{c}_{H^+}$, and K_a. Use your result to rationalize Fig. 22.2.

Solution

Equations (22.15) and (22.16) are

$$\hat{c}_{\text{HA}} = \hat{c}_{\text{total}} - \hat{c}_{\text{A}^-} \quad \text{and} \quad K_a = \frac{\hat{c}_{\text{H}^+} \cdot \hat{c}_{\text{A}^-}}{\hat{c}_{\text{total}} - \hat{c}_{\text{A}^-}}$$

First we solve Equation (22.16) for $\hat{c}_{A^-}$ and find

$$\hat{c}_{A^-} = \frac{\hat{c}_{total} K_a}{\hat{c}_{H^+} + K_a}$$

By substituting into Equation (22.15), we find that

$$\hat{c}_{HA} = \hat{c}_{total} - \frac{\hat{c}_{total} K_a}{\hat{c}_{H^+} + K_a} = \frac{\hat{c}_{total} \cdot \hat{c}_{H^+}}{\hat{c}_{H^+} + K_a}$$

From these relationships, it is clear that as $\hat{c}_{H^+}$ becomes much larger than K_a ($\hat{c}_{H^+} \gg K_a$) that $\hat{c}_{A^-} \rightarrow 0$ and $\hat{c}_{HA} \rightarrow \hat{c}_{total}$. Correspondingly, as $\hat{c}_{H^+}$ becomes much smaller than K_a ($\hat{c}_{H^+} \ll K_a$) that $\hat{c}_{A^-} \rightarrow \hat{c}_{total}$ and $\hat{c}_{HA} \rightarrow \hat{c}_{total} \cdot \hat{c}_{H^+}/K_a$. These limits are in correspondence with the behavior in Fig. 22.2.

E22.18 Use the half reactions

$$I_2 + 2e^- \rightarrow 2I^- \quad E^\ominus = 0.6197\ \text{V}$$
$$I_3^- + 2e^- \rightarrow 3I^- \quad E^\ominus = 0.5355\ \text{V}$$

to determine ΔG and K for the equilibrium

$$I_2 + I^- \rightarrow I_3^-$$

Is the net reaction a redox reaction? If so, can you identify the reductant and oxidant?

Solution

It is a redox reaction. I_2 is the oxidant and I^- is the reductant. To find the Gibbs energy change, we write that

$$I_2 + 2e^- \rightarrow 2I^- \qquad E^\ominus = 0.6197\ \text{V}$$
$$3I^- \rightarrow I_3^- + 2e^- \quad E^\ominus = -0.5355\ \text{V}$$

so that

$$I_2 + I^- \rightarrow I_3^- \quad E^\ominus = 0.0842\ \text{V}$$

and

$$\begin{aligned}\Delta G^\ominus &= -\boldsymbol{nFE}^\ominus \\ &= (2\ \text{mol})(96485\ \text{C mol}^{-1})(0.0842\ \text{V}) \\ &= -16248\ \text{C V} = -16.25\ \text{kJ}\end{aligned}$$

E22.19 The formation of a peptide bond is an essential step in protein synthesis. What is the ΔG for the reaction

alanine + glycine → alylglycine

under standard aqueous conditions? What is the minimum number of ATP molecules that would need to be converted into ADP molecules to provide the necessary Gibbs energy to drive this reaction?

Solution

This reaction is the opposite of that considered in Example 18.17 of the text. Using the values for the Gibbs energy of formation from Appendix A.2, we find that

$$\begin{aligned}\Delta_r G_{aq}^\ominus &= \Delta_f G_{aq}^\ominus(\text{alylglycine}) - \Delta_f G_{aq}^\ominus(\text{alanine}) - \Delta_f G_{aq}^\ominus(\text{glycine}) \\ &= (-733.9 + 379.9 + 371.3)\ \text{kJ mol}^{-1} = 17.3\ \text{kJ mol}^{-1}\end{aligned}$$

Given that hydrolysis of ATP has $\Delta_r G_{aq}^\ominus = -30.5\ \text{kJ mol}^{-1}$ at pH = 7, we see that a single ATP molecule stores enough Gibbs energy to build this peptide bond.

E22.20 Calculate the Gibbs energy change for the reactions

$$\text{a}: \ 2H_2O \rightarrow O_2 + 2H_2$$
$$\text{b}: \ CH_3COOH + Zn + H_2 \rightarrow CH_3CHO + Zn^{2+} + H_2O$$
$$\text{c}: \ 2H_2O + 2Li \rightarrow 2OH^- + H_2 + 2Li^+$$

Determine whether these are redox reactions and identify the reductant and oxidant if they are. You may use the value $\Delta_f G^{\ominus}_{aq}(CH_3CHO) = -139.24\,kJ\,mol^{-1}$; the other values may be found in the appendix.

Solutions

We use the Gibbs energies of formation that are provided in the Appendix. For reaction a: $2H_2O \rightarrow O_2 + 2H_2$, we find

$$\Delta_r G = 2 \cdot \Delta_f G^{\ominus}_g(H_2) + \Delta_f G^{\ominus}_g(O_2) - 2 \cdot \Delta_f G^{\ominus}_l(H_2O)$$
$$= [0 + 0 - 2 \cdot (-237.19)]\ kJ\,mol^{-1} = 474.38\,kJ\,mol^{-1}$$

This disproportionation reaction is a redox reaction in which the hydrogen atoms of water are reduced to form H_2 and the oxygen atom of water is oxidized to form O_2.

For reaction b: $CH_3COOH + Zn + H_2 \rightarrow CH_3CHO + Zn^{2+} + H_2O$, we find

$$\Delta_r G = \begin{bmatrix} \Delta_f G^{\ominus}_l(H_2O) + \Delta_f G^{\ominus}_{aq}(Zn^{2+}) + \Delta_f G^{\ominus}_{aq}(CH_3CHO) - \Delta_f G^{\ominus}_g(H_2) \\ - \Delta_f G^{\ominus}_s(Zn) - \Delta_f G^{\ominus}_{aq}(CH_3COOH) \end{bmatrix}$$
$$= [-237.19 - 147.3 - 139.24 - 0 - 0 + 396.46]\ kJ\,mol^{-1}$$
$$= -127.3\,kJ\,mol^{-1}$$

This reaction is a redox reaction. The Zn is the reductant and the acetic acid is the oxidant.

For reaction c: $2H_2O + 2Li \rightarrow 2OH^- + H_2 + 2Li^+$, we find

$$\Delta_r G = \begin{bmatrix} 2 \cdot \Delta_f G^{\ominus}_{aq}(Li^+) + \Delta_f G^{\ominus}_g(H_2) + 2 \cdot \Delta_f G^{\ominus}_{aq}(OH^-) - 2 \cdot \Delta_f G^{\ominus}_s(Zn) \\ -2 \cdot \Delta_f G^{\ominus}_l(H_2O) \end{bmatrix}$$
$$= [2(-293.8) + 0 + 2(-157.3) - 0 + 2(237.19)]\ kJ\,mol^{-1}$$
$$= -427.8\,kJ\,mol^{-1}$$

This reaction is a redox reaction in which Li is the reductant and water is the oxidant.

E22.21 In Example 22.6 the oxidation of FeS_2 to ferric ion Fe^{3+} and sulfate SO_4^{2-}, important in the generation of acid mine drainage, was discussed. Streambeds suffering from this form of pollution have a red/orange color caused by ferric hydroxide

$$Fe^{3+}(aq) + 3H_2O \rightleftarrows Fe(OH)_3\ (s) + 3H^+(aq)$$

How does the concentration of the ferric ion to its hydroxide depend on a solution's pH, under standard conditions? (Take the K_{sp} of $Fe(OH)_3$ to be 4×10^{-38}.)

Solution

We can write the K_{sp} of $Fe(OH)_3$ as

$$K_{sp} = \hat{a}_{Fe^{3+}} \cdot \hat{a}^3_{OH^-} = 4 \times 10^{-38}$$

In aqueous media we have that $\hat{a}_{OH^-} = 10^{-14}/\hat{a}_{H^+}$ so that

$$K_{sp} = \hat{a}_{Fe^{3+}} \cdot \frac{10^{-42}}{\hat{a}^3_{H^+}} = 4 \times 10^{-38}$$

If we rearrange this result, we see that the ferric ion activity is

$$\hat{a}_{Fe^{3+}} = \hat{a}^3_{H^+} \cdot 4 \times 10^4 = 4 \times 10^4 \cdot 10^{-3 \cdot pH}$$

If we make the approximation that $\hat{a}_{Fe^{3+}} = \hat{c}_{Fe^{3+}}$, we find that

$$\hat{c}_{Fe^{3+}} = 4 \times 10^4 \cdot 10^{-3 \cdot pH}$$

At a pH = 2, the ferric ion concentration is $\hat{c}_{Fe^{3+}} = 4 \times 10^{-2}$, but at a pH = 5 it is already extremely low $\hat{c}_{Fe^{3+}} = 4 \times 10^{-11}$.

E22.22 Alcohol dehydrogenase removes ethanol from your bloodstream and generates acetaldehyde. In simplified form this reaction is

$$NAD^{+} + C_2H_5OH \rightarrow CH_3C(O)H + NADH$$

What is the Gibbs energy of this reaction under standard conditions? What is the Gibbs energy of this reaction at a typical body temperature 37 °C?

	NAD^{+}	C_2H_5OH	$CH_3C(O)H$	NADH
$\Delta_f G^{\ominus\prime}$ (kJ mol^{-1})	1059.1	63.0	24.1	1120.1
$\Delta_f H^{\ominus\prime}$ (kJ mol^{-1})	−10.3	−290.8	−213.6	−41.4

1

Solution

Using the data given we find that

$$\Delta_r G^{\ominus\prime} = [24.1 + 1120.1 - 63 - 1059.1]\ \text{kJ mol}^{-1}$$
$$= 22.1\ \text{kJ mol}^{-1}$$

To estimate the Gibbs energy of reaction at a higher temperature, we use the fundamental relation that $\Delta_r G = \Delta_r H - T \cdot \Delta_r S$. First we use the data given to find $\Delta_r S'$ by way of

$$\Delta_r S' = \frac{\Delta_r H' - \Delta_r G'}{T}$$

Calculating

$$\Delta_r H^{\ominus\prime} = (-14.4 - 213.6 + 290.8 + 10.3)\ \text{kJ mol}^{-1}$$
$$= 73.1\ \text{kJ mol}^{-1}$$

we find that

$$\Delta_r S^{\ominus\prime} = \frac{\Delta_r H^{\ominus\prime} - \Delta_r G^{\ominus\prime}}{T}$$
$$= \frac{(73.1 - 22.1)\ \text{kJ mol}^{-1}}{298\ \text{K}} = 171 \frac{\text{J}}{\text{mol K}}$$

Assuming that $\Delta_r S^{\ominus\prime}$ and $\Delta_r H^{\ominus\prime}$ do not change significantly over the temperature range from 298 K to 310 K, we find that

$$\Delta_r G' = 73.1\ \text{kJ mol}^{-1} - 310\ \text{K} \cdot 171\ \text{J mol}^{-1}\text{K}^{-1}$$
$$= 20.1\ \text{kJ mol}^{-1}$$

Note that the reaction itself is not exoergic at 37 C, and a determination of how much ethanol is reacted will rely on the concentrations of the ethanol as well as that of NADH and NAD^{+} in solution.

E22.23 In Section 22.4.1 we discussed the reaction of a Zn rod immersed in a solution which is 1 M in H^{+} and in Zn^{2+} and found that the reaction is strongly exoergic. Perform the same calculation for the Zn rod immersed in pure water at pH 7. Zn metal is used as a material in construction, e.g., as a coating on iron or steel. Comment on why you think it might resist corrosion.

Solution

We can directly use Equation (22.41),

$$\Delta_r G = \Delta_r G^{\ominus} + RT \ln\left(\frac{\left(\hat{c}_{Zn^{2+}}^{1/1}\right)\left(\hat{P}_{H_2}^{1/2}\right)}{\hat{c}_{H^{+}}}\right)$$

1 These data are taken from R. A. Alberty, *J. Phys. Chem. B* 2001, **105**, 7865.

For $\Delta_r G^\ominus = -73.6\text{ kJ mol}^{-1}$, $\hat{c}_{Zn^{2+}} = 1$, $\hat{P}_{H_2} = 1$ as described in Section 22.4.1, we find that

$$\Delta_r G = \left[-73{,}600\text{ J mol}^{-1} - (8.314\text{ J mol}^{-1}\text{K}^{-1})(298\text{ K})\ln\left(10^{-7}\right)\right]$$
$$= -33.7\text{ kJ mol}^{-1}$$

This result means that Zn reacts to form Zn^{2+} in water. Nevertheless Zn is used as a material for rain pipes, as well as other applications. This is only possible, because Zn forms a stable layer of a mixed oxide/carbonate on its surface which inhibits the oxidation process propagating to the bulk material. Although it is thermodynamically unstable, it is kinetically stable.

E22.24 One of the steps in the chain of coupled bio-reactions for respiration is the transfer of an electron from cytochrome b [cyt b(Fe^{2+})] to the oxidized cytochrome c_1 [cyt c_1(Fe^{3+})]. For the reaction

$$\text{cyt c}_1(\text{Fe}^{2+}) + \text{H}^+ \rightarrow \text{cyt c}_1(\text{Fe}^{3+}) + \tfrac{1}{2}\text{H}_2$$

$\Delta_r G^{\ominus\prime} = 21\text{ kJ mol}^{-1}$ and for the reaction

$$\text{cyt b}(\text{Fe}^{2+}) + \text{H}^+ \rightarrow \text{cyt b}(\text{Fe}^{3+}) + \tfrac{1}{2}\text{H}_2$$

$\Delta_r G^{\ominus\prime} = 4\text{ kJ mol}^{-1}$. Assuming that one molecule of ATP is produced by coupling to the oxidation of three cytochrome c_1 molecules in this way, calculate the $\Delta_r G^{\ominus\prime}$ for the formation of ATP.

Solution

Combining these two reaction equations for the total reaction

$$\text{cyt b}(\text{Fe}^{2+}) + \text{cyt c}_1(\text{Fe}^{3+}) \rightarrow \text{cyt b}(\text{Fe}^{3+}) + \text{cyt c}_1(\text{Fe}^{2+})$$

we obtain $\Delta_r G^{\ominus\prime} = (4 - 21)\text{ kJ mol}^{-1} = -17\text{ kJ mol}^{-1}$. In the respiratory chain of biological systems, ATP is formed by coupling this reaction with the synthesis of ATP from ADP and phosphate. Three molecules of cyt b(Fe^{2+}) and three molecules of cyt c_1(Fe^{3+}) are used to form one molecule of ATP, and we find that

$$\Delta_r G^{\ominus\prime} = (30.5 - 3 \times 17)\text{ kJ mol}^{-1} = -20.5\text{ kJ mol}^{-1}$$

E22.25 Consider the electron transfer reaction between N,N-dimethylaniline ($\Delta_r G(D/D^+) = 51.1\text{ kJ mol}^{-1}$) and 9,10-dimethylanthracene ($\Delta_r G(A/A^-) = 185.3\text{ kJ mol}^{-1}$) in acetonitrile. Is this reaction spontaneous in the ground state? Is it spontaneous upon photoexcitation of the 9,10-dimethylanthracene? See the text for additional data.

Solution

To find the Gibbs energy change for the ground state we consider the reaction

$$\text{D} + \text{A} \rightarrow \text{D}^+ + \text{A}^-$$

and its Gibbs energy change will be

$$\Delta_r G = \Delta_r G(D/D^+) + \Delta_r G(A/A^-) - \frac{\boldsymbol{N}_A\, e^2}{4\pi\varepsilon\varepsilon_0 R}$$
$$= 236.4\text{ kJ mol}^{-1} - \frac{\boldsymbol{N}_A\, e^2}{4\pi\varepsilon\varepsilon_0 R}$$

where the last term is the attraction energy of the positively and negatively charged reaction products at a distance R.

To be spontaneous would require that $\Delta_r G < 0$ or

$$\Delta_r G(D/D^+) + \Delta_r G(A/A^-) < \frac{\boldsymbol{N}_A\, e^2}{4\pi\varepsilon\varepsilon_0 R}$$

so that

$$R < \frac{\boldsymbol{N}_A\, e^2}{4\pi\varepsilon\varepsilon_0\left[\Delta_r G(D/D^+) + \Delta_r G(A/A^-)\right]}$$
$$< \frac{6.022 \times 10^{23}\text{ mol}^{-1} \cdot \left(1.602 \times 10^{-19}\text{ C}\right)^2}{4\pi \cdot 37 \cdot 8.854 \times 10^{-12}\text{ C}^2\text{ J}^{-1}\text{ m}^{-1} \cdot 236400\text{ J mol}^{-1}} = 15.9\text{ pm}$$

where we have used $\Delta_r G(D/D^+) + \Delta_r G(A/A^-) = 236.4\,\text{kJ mol}^{-1}$. This interparticle distance is significantly smaller than a characteristic bond distance in these molecules; hence, we conclude that the ground-state reaction will not be spontaneous.

If however, the dimethylanthracene is photoexcited ($\Delta_r G_{0,0} = -299.7\,\text{kJ mol}^{-1}$), we find that

$$\Delta G = \Delta_r G(D/D^+) + \Delta_r G(A/A^-) + \Delta_r G_{0,0} - \frac{N_A e^2}{4\pi\varepsilon\varepsilon_0 R}$$
$$= -63.3\,\text{kJ mol}^{-1} - \frac{6.022\times10^{23}\,\text{mol}^{-1}\cdot\left(1.602\times10^{-19}\,\text{C}\right)^2}{4\pi\cdot 37\cdot 8.854\times10^{-12}\,\text{C}^2\,\text{J}^{-1}\,\text{m}^{-1}\cdot R/\text{m}}$$

which is spontaneous for all realizable values of the inter-ion distance R.

E22.26 Show by insertion into Equation (22.14), that the degree of dissociation of a weak acid is $\alpha = 0.618$ for $\hat{c}_{\text{total}} = K_a$ and $\alpha = 0.5$ for $\hat{c}_{\text{total}} = 2K_a$.

Solution

Equation (22.14) is

$$\alpha = \frac{-K_a/2 + \sqrt{K_a\cdot\hat{c}_{\text{total}} + K_a^2/4}}{\hat{c}_{\text{total}}}$$

For the case where $\hat{c}_{\text{total}} = K_a$, we can substitute and find that

$$\alpha = \frac{-K_a/2 + \sqrt{K_a^2 + K_a^2/4}}{K_a}$$
$$= \sqrt{5/4} - 1/2 = \frac{\sqrt{5}-1}{2} = 0.618$$

For the case where $\hat{c}_{\text{total}} = 2K_a$, we can substitute and find that

$$\alpha = \frac{-K_a/2 + \sqrt{2K_a^2 + K_a^2/4}}{2K_a}$$
$$= \sqrt{9/16} - 1/4 = \frac{3}{4} - \frac{1}{4} = \frac{1}{2}$$

22.2 Problems

P22.1 The combustion of fossil fuels contributes to the buildup of CO_2 and oxides of sulfur and nitrogen in the atmosphere. As we saw in Example 22.2 CO_2 in water can modify the pH. In an analogous fashion the dissolution of SO_2 can lead to the formation of sulfonic acid H_2SO_3. If the partial pressure of SO_2 in the atmosphere is 2×10^{-8} bar, and you expose a solution of pure water to this partial pressure of SO_2, the solution becomes acidic. Use the equilibria below to calculate the pH, $[HSO_3^-]$, and $[SO_3^{2-}]$ in the solution.

$$H_2O\,(l) + SO_2\,(g) \rightleftarrows H_2SO_3\,(aq) \quad pK = -0.097$$

$$H_2SO_3\,(aq) \rightleftarrows H^+\,(aq) + HSO_3^-\,(aq) \quad pK_{a1} = 1.79$$

$$HSO_3^-\,(aq) \rightleftarrows H^+\,(aq) + SO_3^{2-}\,(aq) \quad pK_{a2} = 9.00$$

Solution

Using the reaction for sulfonic acid formation,

$$H_2O\,(l) + SO_2\,(g) \rightleftarrows H_2SO_3\,(aq) \quad pK = -0.097$$

and $K = \hat{c}_{H_2SO_3}/\left[\hat{c}_{H_2O}\cdot\hat{c}_{SO_2}\right]$ we can write that

$$\hat{c}_{H_2SO_3} = K\cdot\hat{c}_{H_2O}\hat{P}_{SO_2}$$
$$= 10^{0.097}\cdot(55.5)\left(2\times10^{-8}\right) = 1.39\times10^{-6}$$

Since pure water has a pH of 7 and H_2SO_3 is an acid, we should expect the pH of the solution to be reduced somewhat. Sulfonic acid has two possible equilibria,

$$H_2SO_3 \rightleftharpoons H^+ + HSO_3^- \quad \text{with } K_{a1} = \frac{\hat{c}_{H^+} \cdot \hat{c}_{HSO_3^-}}{\hat{c}_{H_2SO_3}} = 1.62 \times 10^{-2}$$

and

$$HSO_3^- \rightleftarrows H^+ + SO_3^{2-} \quad \text{with } K_{a2} = \frac{\hat{c}_{H^+} \cdot \hat{c}_{SO_3^{2-}}}{\hat{c}_{HSO_3^-}} = 1.00 \times 10^{-9}$$

Because the second equilibrium constant is far smaller than the first one let us neglect it for the moment. From the first equilibrium we can determine the proton concentration; using Equation (22.12) we find

$$\hat{c}_{H^+} = -\frac{1}{2}K_{a1} + \sqrt{\hat{c}_{H_2SO_3}K_{a1} + \frac{1}{4}K_{a1}^2}$$

With $\hat{c}_{H_2SO_3} = 1.39 \times 10^{-6}$ we obtain $\hat{c}_{H^+} = 1.39 \times 10^{-6}$ or a pH of 5.9.

Now we use these values in the equilibrium constant expression for the second equilibrium,

$$K_{a2} = \frac{\hat{c}_{H^+} \cdot \hat{c}_{SO_3^{2-}}}{\hat{c}_{HSO_3^-}}$$

Given that H_2SO_3 is fully ionized, consider the limiting case where $\hat{c}_{HSO_3^-} = 1.39 \times 10^{-6} - \hat{c}_{SO_3^{2-}}$ and $\hat{c}_{H^+} = 1.39 \times 10^{-6} + \hat{c}_{SO_3^{2-}}$, so that

$$K_{a2} = \frac{\left(1.39 \times 10^{-6} + \hat{c}_{SO_3^{2-}}\right) \cdot \hat{c}_{SO_3^{2-}}}{\left(1.39 \times 10^{-6} - \hat{c}_{SO_3^{2-}}\right)}$$

which simplifies to

$$\hat{c}^2_{SO_3^{2-}} + \left(1.39 \times 10^{-6} + 1.00 \times 10^{-9}\right)\hat{c}_{SO_3^{2-}} - 1.39 \times 10^{-16} = 0$$

Hence we find that

$$\hat{c}_{SO_3^{2-}} = \frac{-1.39 \times 10^{-6} \pm \sqrt{\left(1.39 \times 10^{-6}\right)^2 + 4.56 \times 10^{-16}}}{2} \approx 0$$

and it is reasonable to neglect this equilibrium.

P22.2 In Example 22.2 we computed the pH of pure water when exposed to a partial pressure of 3.89×10^{-3} bar in CO_2. If you add an excess of $CaCO_3(s)$ to the water solution to set the carbonate concentration, what will the new pH of the solution be? Use the solubility product of $CaCO_3$ in water of $\hat{c}_{Ca^{2+}} \cdot \hat{c}_{CO_3^{2-}} = 4.0 \times 10^{-9}$.

Solution

Because the $CaCO_3$ exists in excess, its dissolution will control the equilibrium so that

$$\hat{c}_{CO_3^{2-}} = \sqrt{4.0 \times 10^{-9}} = 6.3 \times 10^{-5}$$

Next we use the equilibria

$$HCO_3^- \rightleftarrows H^+ + CO_3^{2-} \quad \text{with } K_{a2} = \frac{\hat{c}_{H^+} \cdot \hat{c}_{CO_3^{2-}}}{\hat{c}_{HCO_3^-}} = 4.68 \times 10^{-11}$$

and write that

$$\hat{c}_{HCO_3^-} = \frac{\hat{c}_{H^+} \cdot \hat{c}_{CO_3^{2-}}}{K_{a2}}$$

$$= \frac{\hat{c}_{H^+} \cdot \sqrt{4.0 \times 10^{-9}}}{4.68 \times 10^{-11}} = 1.35 \times 10^6 \cdot \hat{c}_{H^+}$$

Inserting this result into the equilibrium

$$H_2CO_3 \rightleftharpoons H^+ + HCO_3^- \quad \text{with } K_{a1} = \frac{\hat{c}_{H^+} \cdot \hat{c}_{HCO_3^-}}{\hat{c}_{H_2CO_3}} = 4.47 \times 10^{-7}$$

gives

$$\hat{c}^2_{H^+} = \left(3.31 \times 10^{-13}\right) \cdot \hat{c}_{H_2CO_3}$$

From the calculation in Example 22.2 of the text, we know that the partial pressure of CO_2 in the atmosphere sets the H_2CO_3 concentration at 4.2×10^{-6}, so that

$$\hat{c}_{H^+} = \sqrt{\left(3.31 \times 10^{-13}\right) \cdot 4.2 \times 10^{-6}} = 1.18 \times 10^{-9}$$

or pH = 8.9.

Note: An alternative reaction analysis can be used to illustrate why the solution is basic. For an aqueous solution of carbonate and carbon dioxide we can consider the reaction

$$CO_3^- + CO_2 + H_2O \rightarrow 2HCO_3^-$$

followed by the equilibrium

$$HCO_3^- + H^+ \rightarrow H_2CO_3$$

which removes protons from the solution, making it alkaline. The reaction shown above represents the corrosion of limestone. The formation of limestone, a material used for building houses, proceeds by

$$CaO + CO_2 \rightarrow CaCO_3 \qquad \text{formation of limestone}$$

and its corrosion reaction is

$$CaCO_3 + CO_2 + H_2O \rightarrow Ca^2 + 2HCO_3^- \qquad \text{corrosion process}$$

P22.3 The sulfurous smell that arises from wastewater in sewers is often created by the reduction of sulfur oxides to hydrogen sulfide (H_2S) by bacteria. Using the equilibria

$$H_2S \rightarrow HS^- + H^+, \quad HS^- \rightarrow S^{2-} + H^+$$

$$K_{a1} = \frac{\hat{c}_{HS^-}\hat{c}_{H^+}}{\hat{c}_{H_2S}} = 8.9 \times 10^{-8}, \quad K_{a2} = \frac{\hat{c}_{S^{2-}}\hat{c}_{H^+}}{\hat{c}_{HS^-}} = 1 \times 10^{-19}$$

calculate the concentrations of H_2S, HS^-, and S^{2-} as a function of pH. Plot the fraction of each species present over a pH range from 5 to 10. What can you say about the pH of solution from which you can smell H_2S?

Solution

We proceed here by using the result derived in Foundation 22.2 and making the correspondence of H_2A with H_2S, of HA^- with HS^-, and A^{2-} with S^{2-}. Hence the equilibria are

$$H_2S \rightleftharpoons HS^- + H^+, \quad HS^- \rightleftharpoons S^{2-} + H^+$$

with the equilibrium constants

$$K_{a1} = \frac{\hat{c}_{H^+} \cdot \hat{c}_{HS^-}}{\hat{c}_{H_2S}} = 8.9 \times 10^{-8} \text{ and } K_{a2} = \frac{\hat{c}_{H^+} \cdot \hat{c}_{S^{2-}}}{\hat{c}_{HS^-}} = 1 \times 10^{-19}$$

and the total concentration

$$\hat{c}_{\text{total}} = \hat{c}_{H_2S} + \hat{c}_{HS^-} + \hat{c}_{S^{2-}}$$

From the solution in the Foundation 22.2, we can write expressions for each of the species concentrations; namely

$$\hat{c}_{H_2S} = \hat{c}_{\text{total}} \frac{\hat{c}^2_{H^+}}{\hat{c}^2_{H^+} + K_{a1}\hat{c}_{H^+} + K_{a1}K_{a2}};$$

$$\hat{c}_{HS^-} = \hat{c}_{\text{total}} \frac{K_{a1}\hat{c}_{H^+}}{\hat{c}^2_{H^+} + K_{a1}\hat{c}_{H^+} + K_{a1}K_{a2}}$$

and

$$\widehat{c}_{S^{2-}} = \widehat{c}_{total} \frac{K_{a1}K_{a2}}{\widehat{c}^2_{H^+} + K_{a1}\widehat{c}_{H^+} + K_{a1}K_{a2}}$$

The fractions of each species are evaluated as

$$f_{H_2S} = \frac{\widehat{c}^2_{H^+}}{\widehat{c}^2_{H^+} + K_{a1}\widehat{c}_{H^+} + K_{a1}K_{a2}}; \quad f_{HS^-} = \frac{K_{a1}\widehat{c}_{H^+}}{\widehat{c}^2_{H^+} + K_{a1}\widehat{c}_{H^+} + K_{a1}K_{a2}}$$

and

$$f_{S^{2-}} = \frac{K_{a1}K_{a2}}{\widehat{c}^2_{H^+} + K_{a1}\widehat{c}_{H^+} + K_{a1}K_{a2}}$$

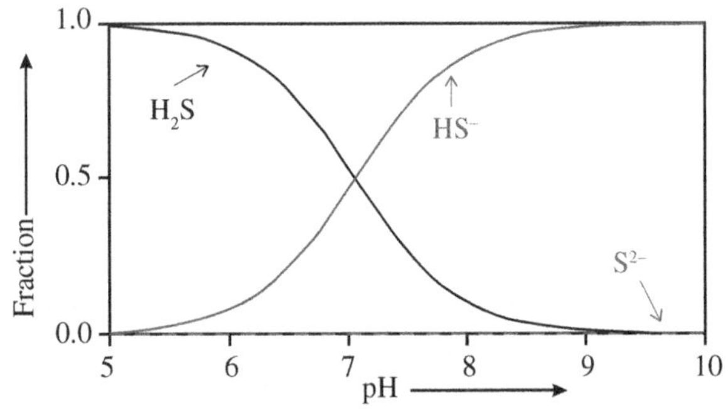

Figure P22.3 Fraction of H_2S and HS^- versus pH.

From Fig. P22.3 we see that only the species H_2S and HS^- have a significant population over the pH range of 5 to 10. At pH values less than 7 the H_2S species is dominant; hence, a solution from which we can detect H_2S being evolved must also be acidic.

Extension: Our analysis has ignored the autoionization of H_2O. How would you expect the conclusions to change if this feature of the system were included?

P22.4 Consider the acid dissociation

$$CH_3COOH \rightleftarrows CH_3COO^- + H^+$$

and take $\widehat{c}_{CH_3COO^-} = 0.10$. Calculate the $\Delta_r G^\ominus$ for this reaction under standard conditions. Compute the concentrations of acetate ions and protons assuming that the solution is ideally diluted. Account for nonideality of the solution by using the extended Debye–Hückel model for the activity coefficients of the ions and compute corrected ion concentrations. How much does the solution's pH change between your two estimates?

Solution

Using the data given in the appendix, the $\Delta_r G^\ominus$ is

$$\Delta_r G^\ominus = \Delta_r G^\ominus_{aq}(H^+) + \Delta_r G^\ominus_{aq}(CH_3COO^-) - \Delta_r G^\ominus_{aq}(CH_3COOH)$$
$$= (-376.9 + 0 + 404.1)\ \text{kJ mol}^{-1} = 27.2\ \text{kJ mol}^{-1}$$

and hence the acid dissociation constant is

$$K_a = \exp\left(-\frac{\Delta_r G^\ominus}{RT}\right)$$
$$= \exp\left(-\frac{27200\ \text{J mol}^{-1}}{8.314\ \text{J mol}^{-1}\text{K}^{-1} \cdot 298\ \text{K}}\right) = 1.71 \times 10^{-5}$$

If we neglect the autoionization of water and consider the ideally diluted case, we can use Equation (22.12)

$$\widehat{c}_{H^+} = -K_a/2 + \sqrt{0.1 \cdot K_a + K_a^2/4}$$

to write that

$$\widehat{c}_{H^+} = -8.55 \times 10^{-6} + \sqrt{1.71 \times 10^{-6}} = 1.30 \times 10^{-3}$$

or a pH of 2.89.

If we neglect the autoionization of water but account for nonidealities we must write that

$$K_a = \frac{\hat{a}_{H^+} \cdot \hat{a}_{CH_3COO^-}}{\hat{a}_{CH_3COOH}} = \frac{\hat{c}_{H^+} \cdot \hat{c}_{CH_3COO^-}}{\hat{c}_{CH_3COOH}} \frac{\hat{\gamma}_{H^+} \cdot \hat{\gamma}_{CH_3COO^-}}{\hat{\gamma}_{CH_3COOH}}$$

By realizing that $\hat{c}_{H^+} = \hat{c}_{CH_3COO^-}$, defining $\gamma^2 = \hat{\gamma}_{H^+} \cdot \hat{\gamma}_{CH_3COO^-}$, and making the approximation that $\hat{\gamma}_{CH_3COOH} = 1$, we can simplify this result

$$K_a = \frac{\hat{c}^2_{H^+}\gamma^2}{0.1 - \hat{c}_{H^+}}$$

which can be solved to yield

$$\hat{c}_{H^+} = -\frac{K_a}{2\gamma^2} + \frac{\sqrt{0.4 \cdot K_a\gamma^2 + K_a^2}}{2\gamma^2}$$

Now we can use Equation (21.93) (Debye–Hückel result) to evaluate γ as

$$\ln\gamma = -\frac{A \cdot \sqrt{I}}{1 + B \cdot R\sqrt{I}}$$

with $A = 1.175\,\text{mol}^{-1/2}\,\text{L}^{1/2}$, $B = 0.330 \times 10^{10}\text{m}^{-1}\,\text{mol}^{-1/2}\,\text{L}^{1/2}$. If we use $R = 500$ pm and calculate the ionic strength by way of our concentrations found for the ideally diluted case ($\hat{c}_{H^+} = \hat{c}_{CH_3COO^-} = 1.30 \times 10^{-3}$). We find that $I = 1.30 \times 10^{-3}\,\text{mol L}^{-1}$, and

$$\ln\gamma = -\frac{4.24 \times 10^{-2}}{1 + 0.0595} = -4.00 \times 10^{-2} \quad \text{so that} \quad \gamma = 0.96$$

Substituting this result into our equation above for $\hat{c}_{H^+}$, we find that

$$\hat{c}_{H^+} = -9.28 \times 10^{-6} + 1.36 \times 10^{-3} = 1.35 \times 10^{-3}$$

which gives a $\text{pH} = -\log\left(\hat{a}_{H^+}\right) = 2.87$.

The change in solution pH is minor, about 1%.

P22.5 Consider the dissociation of HF, HCl, and HI into protons and anions in the gas phase. What determines the trend in acidity? Contrast this result with the dissociation constants of these molecules in water. How does the acidity change and what determines the trend in acidity? Some of the facts you need can be found in the appendix; you may also use the fact that the electron affinity is $EA = 349\,\text{kJ mol}^{-1}$ for Cl, $EA = 328\,\text{kJ mol}^{-1}$ for F, and $EA = 324.7\,\text{kJ mol}^{-1}$ for Br and that the ionization energy of H is $IP = 1312.0\,\text{kJ mol}^{-1}$. (You may also need to use the *CRC Handbook of Chemistry and Physics* for some facts.)

Solution

The gas-phase reaction would be written as

$$HX(g) \rightarrow H^+(g) + X^-(g)$$

and can be understood if we build a cycle (Fig. P22.5).

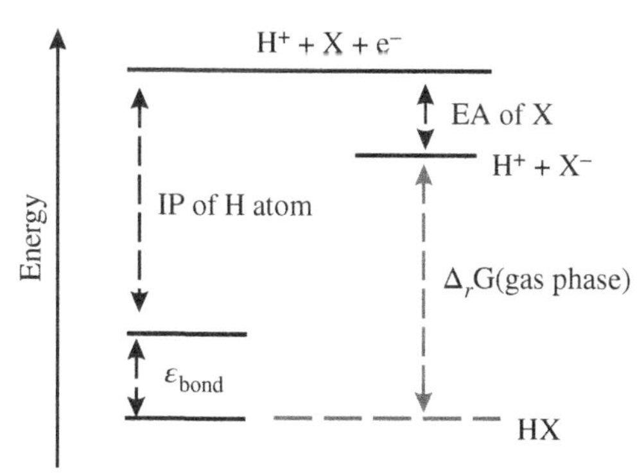

Figure P22.5 Cycle for a gas-phase reaction. *EA* = electron affinity, *IP* = ionization potential.

from which we can write that

$$\Delta_r G_{\text{gas}} = \Delta_{\text{bond}} G + \Delta_{\text{IP}} G - \Delta_{\text{EA}} G$$

For the case of HF, we find a bond energy of $\Delta_{\text{bond}} G = 569.87$ kJ mol^{-1}, from the *CRC Handbook of Chemistry and Physics*, and calculate a reaction Gibbs energy of

$$\Delta_r G_{\text{gas}} = (569.87 + 1312.0 - 328)\ \text{kJ mol}^{-1} = 1554\ \text{kJ mol}^{-1}$$

We can proceed in a likewise manner for the other molecules and find $\Delta_r G_{\text{gas}} = 1354$ kJ mol^{-1} (using $\Delta_{\text{bond}} G = 366.35$ kJ mol^{-1}) for HBr and $\Delta_r G_{\text{gas}}^{-1} = 1418.9$ kJ mol^{-1} (for $\Delta_{\text{bond}} G = 431.625$ kJ mol^{-1}) for HCl. The table reports these values for the gas phase along with those for solution, from the Appendix.

To find the reaction Gibbs energy in solution we consider the reaction

$$\text{HX(g)} \rightarrow \text{H}^+\text{(aq)} + \text{X}^-\text{(aq)}$$

and calculate

$$\Delta_r G_{\text{aq}} = \Delta G_{\text{aq}}(\text{H}^+) + \Delta G_{\text{aq}}(\text{X}^-) - \Delta G_{\text{g}}(\text{HX})$$

For HF, we find that

$$\Delta_r G_{\text{aq}} = (0 + -276.5 + 270.6)\ \text{kJ mol}^{-1} = -5.9\ \text{kJ mol}^{-1}$$

In a corresponding way we find −35.9 kJ mol^{-1} for HCl and −53.1 kJ mol^{-1} for HBr.

	HF	HCl	HBr
$\Delta_r G_{\text{gas}}$/kJ mol^{-1}	1554	1354	1418.9
$\Delta_{\text{bond}} G$/kJ mol^{-1}	569.87	431.63	366.35
$\Delta_r G_{\text{aq}}$/kJ mol^{-1}	−5.9	−35.9	−53.1

From the Gibbs energy change we see that in solution the acids show a monotonic trend of increasing exoergodicity through the series HF to HCl to HBr in water, but that in the gas phase the HCl would have the lowest $\Delta_r G$, then HBr and lastly HF. The difference between the behaviors is linked to the differences in the solvation energies of the halide ions.

P22.6 Consider the acid-base equilibria for the amino acid glycine, for which $K_1 = 3.98 \times 10^{-3}$ and $K_2 = 1.58 \times 10^{-10}$. Given a total concentration of 0.10 mol L^{-1}, plot the concentrations for each of glycine's forms as a function of the solution pH. (Hint: Refer to Foundation 22.2).

Solution

We proceed here by using the result derived in Foundation 22.2 and making the correspondence of H_2A with H^+-glycine, of HA^- with glycine, and A^{2-} with glycinate. Hence the equilibria are

$$\text{H}^+\text{-glycine} \rightleftarrows \text{glycine} \rightleftarrows \text{glycinate}$$

with the equilibrium constants

$$K_{a1} = \frac{\hat{c}_{\text{H}^+} \cdot \hat{c}_{\text{glycine}}}{\hat{c}_{\text{H}^+\text{-glycine}}} = 3.93 \times 10^{-3} \quad \text{and} \quad K_{a2} = \frac{\hat{c}_{\text{H}^+} \cdot \hat{c}_{\text{glycinate}}}{\hat{c}_{\text{glycine}}} = 1.58 \times 10^{-10}$$

and the total concentration

$$\hat{c}_{\text{total}} = \hat{c}_{\text{H}^+\text{-glycine}} + \hat{c}_{\text{glycine}} + \hat{c}_{\text{glycinate}} = 0.10$$

From the solution in Foundation 22.2 we can write expressions for each of the species concentrations; namely

$$\hat{c}_{\text{H}^+\text{-glycine}} = \hat{c}_{\text{total}} \frac{\hat{c}_{\text{H}^+}^2}{\hat{c}_{\text{H}^+}^2 + K_{a1}\hat{c}_{\text{H}^+} + K_{a1}K_{a2}}$$

$$\hat{c}_{\text{glycine}} = \hat{c}_{\text{total}} \frac{K_{a1}\hat{c}_{\text{H}^+}}{\hat{c}_{\text{H}^+}^2 + K_{a1}\hat{c}_{\text{H}^+} + K_{a1}K_{a2}}$$

and

$$\hat{c}_{\text{glycinate}} = \hat{c}_{\text{total}} \frac{K_{a1}K_{a2}}{\hat{c}^2_{\text{H}^+} + K_{a1}\hat{c}_{\text{H}^+} + K_{a1}K_{a2}}$$

From Fig. P22.6 we see that the glycine form is dominant near neutral/physiological pH.

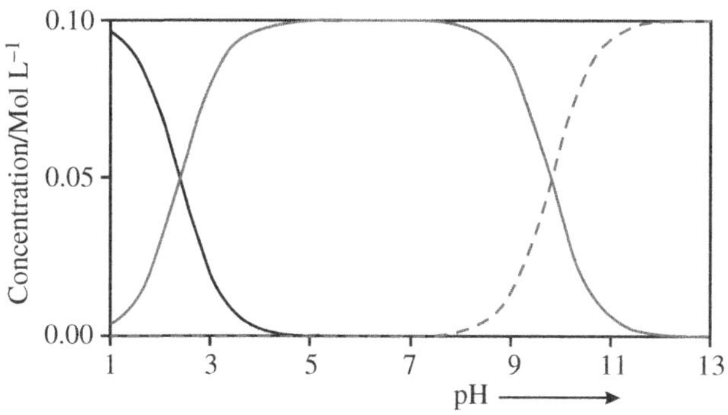

Figure P22.6 Different forms of glycine versus pH. Black solid line: glycinate form, dark gray solid line: glycine form, dark gray dashed line: H^+-glycine form.

P22.7 In Fig. 22.5 the titration curve of glycine ($\boldsymbol{n}_{\text{total}} = 0.1$ mol) with a NaOH solution is shown, calculated according to Equation (22.34). In analogy, calculate the titration curve of glycine with an HCl solution. Make the corresponding plot to Fig. 22.5.

Solution

We proceed here by using the result derived in Foundation 22.2 and making the correspondence of H_2A with H^+-glycine, of HA^- with glycine, and A^{2-} with glycinate. Hence the equilibria are

$$\text{H}^+\text{-glycine} \rightleftarrows \text{glycine} \rightleftarrows \text{glycinate}$$

with the equilibrium constants

$$K_{a1} = \frac{\hat{c}_{\text{H}^+} \cdot \hat{c}_{\text{glycine}}}{\hat{c}_{\text{H}^+\text{-glycine}}} = 3.93 \times 10^{-3} \quad \text{and} \quad K_{a2} = \frac{\hat{c}_{\text{H}^+} \cdot \hat{c}_{\text{glycinate}}}{\hat{c}_{\text{glycine}}} = 1.58 \times 10^{-10}$$

and the total concentration

$$\hat{c}_{\text{total}} = \hat{c}_{\text{H}^+\text{-glycine}} + \hat{c}_{\text{glycine}} + \hat{c}_{\text{glycinate}}$$

From the solution in Foundation 22.2 we can write expressions for each of the species concentrations; namely

$$\hat{c}_{\text{H}^+\text{-glycine}} = \hat{c}_{\text{total}} \frac{\hat{c}^2_{\text{H}^+}}{\hat{c}^2_{\text{H}^+} + K_{a1}\hat{c}_{\text{H}^+} + K_{a1}K_{a2}}$$

and

$$\hat{c}_{\text{glycine}} = \hat{c}_{\text{total}} \frac{K_{a1}\hat{c}_{\text{H}^+}}{\hat{c}^2_{\text{H}^+} + K_{a1}\hat{c}_{\text{H}^+} + K_{a1}K_{a2}}$$

and

$$\hat{c}_{\text{glycinate}} = \hat{c}_{\text{total}} \frac{K_{a1}K_{a2}}{\hat{c}^2_{\text{H}^+} + K_{a1}\hat{c}_{\text{H}^+} + K_{a1}K_{a2}}$$

In addition, we use the electroneutrality condition

$$\hat{c}_{\text{H}^+} + \hat{c}_{\text{H}^+\text{-glycine}} = \hat{c}_{\text{OH}^-} + \hat{c}_{\text{Cl}^-} + \hat{c}_{\text{glycinate}}$$

where $\hat{c}_{\text{Cl}^-}$ arises from addition of the HCl solution. With

$$\hat{c}_{\text{OH}^-} = \frac{K_w}{\hat{c}_{\text{H}^+}}$$

we can write that

$$\hat{c}_{Cl^-} = \hat{c}_{H^+} + \hat{c}_{H^+\text{-glycine}} - \hat{c}_{\text{glycinate}} - \frac{K_w}{\hat{c}_{H^+}}$$

At the beginning of the titration, $\hat{c}_{Cl^-} = 0$, so that

$$0 = \hat{c}_{H^+} + \frac{\hat{c}_{\text{total}}\left(\hat{c}^2_{H^+} - K_{a1}K_{a2}\right)}{\hat{c}^2_{H^+} + K_{a1}\hat{c}_{H^+} + K_{a1}K_{a2}} - \frac{K_w}{\hat{c}_{H^+}}$$

This is an equation for the proton concentration of the pure glycine dissolved in water. This equation can be solved, either graphically or numerically, and we find a pH value of 6.107 for the glycine in water (no HCl added). Because we are making the solution more acidic in the titration, and the pH is already mildly acidic we can neglect the second equilibrium K_{a2} (let $\hat{c}_{\text{glycinate}} = 0$) without incurring a significant error. Thus, we have that

$$\hat{c}_{Cl^-} = \hat{c}_{H^+} + \hat{c}_{H^+\text{-glycine}} - \frac{K_w}{\hat{c}_{H^+}} \quad \text{and} \quad \hat{c}_{H^+\text{-glycine}} = \hat{c}_{\text{total}}\frac{\hat{c}_{H^+}}{\hat{c}_{H^+} + K_{a1}}$$

If we write that $\boldsymbol{n}_{HCl}$ is the amount of added HCl, then it equals the amount of Cl^- ions in the solution and we can write that

$$\boldsymbol{n}_{HCl} = \boldsymbol{n}_{Cl^-} = Vc_{Cl^-} = Vc^{\ominus}\hat{c}_{Cl^-}$$

with V = volume of the solution and $c^{\ominus} = 1\,\text{mol L}^{-1}$ (standard concentration). Then with

$$\boldsymbol{n}_{\text{total}} = c_{\text{total}}V = \hat{c}_{\text{total}}Vc^{\ominus}$$

($\boldsymbol{n}_{\text{total}}$ = total amount of glycine at the start of the titration), we can write that

$$\boldsymbol{n}_{HCl} = Vc^{\ominus}\left(\hat{c}_{H^+} - \frac{K_w}{\hat{c}_{H^+}}\right) + \boldsymbol{n}_{\text{total}} \cdot \frac{\hat{c}_{H^+}}{\hat{c}_{H^+} + K_{a1}}$$

This equation relates the proton concentration in the solution to $\boldsymbol{n}_{HCl}$ and allows us to plot the pH versus the amount of acid added, if we know V.

To find V, we use the fact that

$$\boldsymbol{n}_{HCl} = \boldsymbol{n}_{Cl^-} = \left(0.1\text{ mol L}^{-1}\right) \cdot V_{\text{added}}$$

so that

$$V = 0.1\text{ L} + V_{\text{added}} = 0.1\text{ L} + \frac{\boldsymbol{n}_{HCl}}{0.1\text{ mol L}^{-1}}$$

Substituting into our previous expression we find that

$$\boldsymbol{n}_{HCl} = 0.1\text{ L}\cdot c^{\ominus}\left(\hat{c}_{H^+} - \frac{K_w}{\hat{c}_{H^+}}\right) + \frac{\boldsymbol{n}_{HCl}}{0.1\text{ mol L}^{-1}} \cdot c^{\ominus}\left(\hat{c}_{H^+} - \frac{K_w}{\hat{c}_{H^+}}\right) + \boldsymbol{n}_{\text{total}} \cdot \frac{\hat{c}_{H^+}}{\hat{c}_{H^+} + K_{a1}}$$

If we let

$$A = c^{\ominus}\left(\hat{c}_{H^+} - \frac{K_w}{\hat{c}_{H^+}}\right) \quad \text{and} \quad B = \frac{\hat{c}_{H^+}}{\hat{c}_{H^+} + K_{a1}}$$

we find that

$$\boldsymbol{n}_{HCl}\left(1 - \left(10\text{ mol}^{-1}\text{ L}\right)\cdot A/\left(\text{mol L}^{-1}\right)\right) = (0.1\text{ L})\cdot A/\left(\text{mol L}^{-1}\right) + \boldsymbol{n}_{\text{total}} \cdot B$$

or

$$\boldsymbol{n}_{HCl} = \frac{(0.1\text{ L})\cdot A/\left(\text{mol L}^{-1}\right) + \boldsymbol{n}_{\text{total}} \cdot B}{\left(1 - \left(10\text{ mol}^{-1}\text{ L}\right)\cdot A/\left(\text{mol L}^{-1}\right)\right)}$$

Figure P22.7 shows a plot of the solution pH as a function of the moles of HCl that is added to the glycine solution.

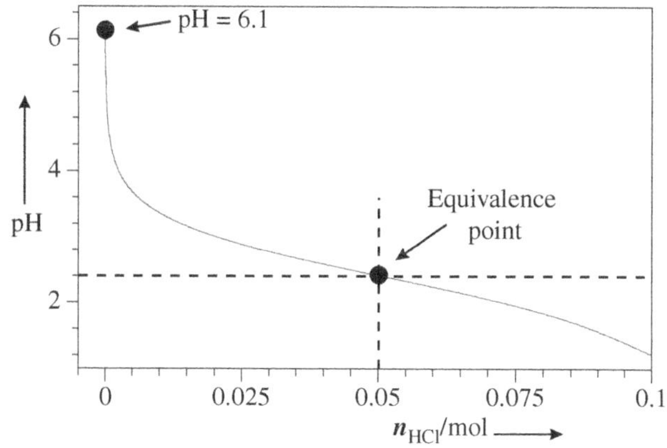

Figure P22.7 Titration of $n_{total} = 0.1$ mol of glycine with HCl. pH versus n_{HCl} of added HCl. This graph neglects the change in volume of the solution with added HCl. For sufficiently large concentrations of the HCl titrant, this approximation is acceptable.

How does this graph relate to that in Fig. 22.5? We note that the pH at the value of zero acid added, corresponds to the pH of zero base added in Fig. 22.5. The graph here would correspond to performing a titration that moves to the left in Fig. 22.5.

P22.8 Consider the equilibria of the amino acid glycine,

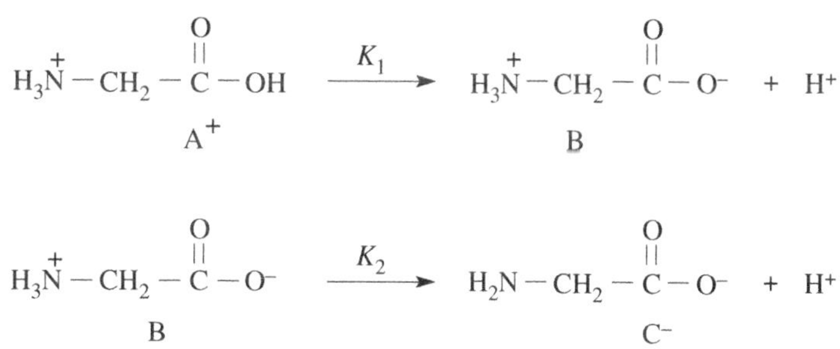

which was discussed in Section 22.3.2, and calculate the pH for two limiting cases: 1) Equilibrium $A^+ \rightleftharpoons B$ is neglected and 2) Equilibrium $B \rightleftharpoons C^-$ is neglected.

Solution

1) Equilibrium $A^+ \rightleftharpoons B$ is neglected.

In this case we have to consider the dissociation reaction $B \rightarrow C^-$, and the pH is obtained in analogy to the procedure for acetic acid in Section 22.2.2, Equation (22.12)

$$\hat{c}_{H^+} = -\frac{1}{2}K_2 + \sqrt{\hat{c}_{total}K_2 + \frac{1}{4}K_2^2}$$

With $K_2 = 10^{-9.8}$ and $\hat{c}_{total} = 0.1\ \text{mol L}^{-1}$ we obtain

$$\begin{aligned}\hat{c}_{H^+} &= -\frac{1}{2} \cdot 10^{-9.8} + \sqrt{0.1 \cdot 10^{-9.8} + \frac{1}{4}10^{-19.6}} \\ &= -7.93 \times 10^{-11} + \sqrt{1.58 \times 10^{-11} + 6.29 \times 10^{-21}} \\ &= -7.93 \times 10^{-11} + 3.97 \times 10^{-6} = 3.97 \times 10^{-6}\end{aligned}$$

This concentration corresponds to

$$pH = -\log\left(3.97 \times 10^{-6}\right) = 5.40$$

2) Equilibrium $B \rightleftharpoons C^-$ is neglected
In this case B reacts with one H^+ to form one A^+. In this process the H^+ is taken from the water, and one OH^- is being formed. Thus

$$\hat{c}_{A^+} = \hat{c}_{OH^-}$$

Then we obtain

$$K_1 = \frac{\hat{c}_{H^+} \cdot \hat{c}_B}{\hat{c}_{A^+}} = \frac{\hat{c}_{H^+} \cdot \hat{c}_B}{\hat{c}_{OH^-}} = \frac{\hat{c}_{H^+} \cdot \hat{c}_{total}}{K_w / \hat{c}_{H^+}}$$

and

$$\hat{c}_{H^+} = \sqrt{\frac{K_1 K_w}{\hat{c}_{total}}}$$

With $K_1 = 10^{-2.4}$ and $\hat{c}_{total} = 0.1\ \text{mol L}^{-1}$ we obtain

$$\hat{c}_{H^+} = \sqrt{\frac{10^{-2.4} \cdot 1 \times 10^{-14}}{0.1}} = 1.99 \times 10^{-8}$$

This corresponds to

$$pH = -\log\left(1.99 \times 10^{-8}\right) = 7.70$$

P22.9 pH change in buffer solutions: In Example 22.1 in Section 22.2.1 we derived the equation

$$pH = pK - \log \frac{\hat{c}_0 + \hat{c}_a}{\hat{c}_0 - \hat{c}_a} \tag{22.1}$$

for the pH as a function of the concentration c_a of added acid for different buffer concentrations c_0. In this problem you are asked to extend the calculations for the example of HCl and Acetic acid: Acetate buffer given in Example 22.1; for four different values of $\hat{c}_0$ (0.1, 0.2, 0.3, and 0.4) and five different values of $\hat{c}_a$ (0.00, 0.01, 0.02, 0.03, and 0.04). Plot the pH change as a function of $\hat{c}_a$ for each of the buffer concentrations $\hat{c}_0$. Comment on your findings.

Solution

We use Equation (22.1) to calculate the pH and $\Delta pH = pH - pK$; see Table P22.8 and Fig. P22.8. Note that Example 22.1 gives an example calculation.

Table P22.8 Change ΔpH of buffer solutions versus concentration c_a of added acid for different buffer concentrations c_0, according to Equation (22.1) with pK = 4.62.

	$\hat{c}_0$			
$\hat{c}_a$	**0.1**	**0.2**	**0.3**	**0.4**
0.00	0	0	0	0
0.01	−0.09	−0.04	−0.03	−0.02
0.02	−0.18	−0.09	−0.06	−0.04
0.03	−0.27	−0.13	−0.09	−0.07
0.04	−0.37	−0.18	−0.12	−0.09

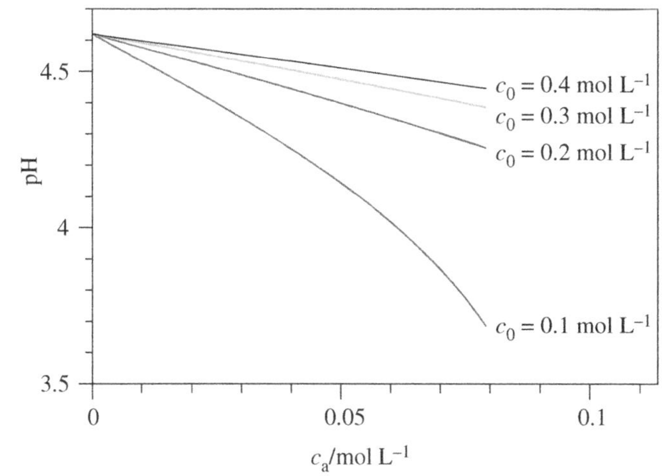

Figure P22.8 pH of a buffer solution as a function of the concentration c_a of added acid for different buffer concentrations c_0, according to Equation (22.1). pK = 4.62. Note that Equation (22.1) is only valid for $c_{H^+} \ll c_a$.

As the concentration of the buffer becomes more concentrated, it is able to inhibit better any change in the pH that results from the addition of the strong acid HCl.

P22.10 Consider the acid dissociation of oxalic acid

$$\text{COOH-COOH} \rightleftarrows \text{COOH-COO}^- + \text{H}^+$$

and take $\hat{c}_{\text{COOH-COOH}} = 0.10$. The p$K_a$ value for this reaction is p$K = 1.23$. Neglect the second dissociation step and compute the concentrations of oxalic acid ions and protons assuming that the solution is ideally diluted. Account for nonideality of the solution by using the extended Debye–Hückel model for the activity coefficients of the ions and compute corrected ion concentrations. How much does the solution's pH change between your two estimates?

Solution

From pK_a = 1.23 we calculate $K_a = 5.89 \times 10^{-2}$. If we neglect the autoionization of water and consider the ideally diluted case, we can use Equation 22.12

$$\hat{c}_{H^+} = -K_a/2 + \sqrt{0.1 \cdot K_a + K_a^2/4}$$

to write that

$$\hat{c}_{H^+} = -\frac{1}{2}5.89 \times 10^{-2} + \sqrt{0.1 \cdot 5.89 \times 10^{-2} + 8.67 \times 10^{-4}} = 5.27 \times 10^{-2}$$

or pH $= -\log \hat{c}_{H^+} = 1.28$.

If we neglect the autoionization of water but account for nonidealities we must write that

$$K_a = \frac{\hat{a}_{H^+} \cdot \hat{a}_{C_2O_4H^-}}{\hat{a}_{C_2O_4H_2}} = \frac{\hat{c}_{H^+} \cdot \hat{c}_{C_2O_4H^-}}{\hat{c}_{C_2O_4H_2}} \frac{\hat{\gamma}_{H^+} \cdot \hat{\gamma}_{C_2O_4H^-}}{\hat{\gamma}_{C_2O_4H_2}}$$

By realizing that $\hat{c}_{H^+} = \hat{c}_{C_2O_4H^-}$, defining $\gamma^2 = \hat{\gamma}_{H^+} \cdot \hat{\gamma}_{C_2O_4H^-}$, and making the approximation that $\hat{\gamma}_{C_2O_4H_2} = 1$, we can simplify this result

$$K_a = \frac{\hat{c}^2_{H^+}\gamma^2}{0.1 - \hat{c}_{H^+}}$$

which can be solved to yield

$$\hat{c}_{H^+} = -\frac{K_a}{2\gamma^2} \pm \sqrt{\frac{K_a \cdot 0.1}{\gamma^2} + \frac{K_a^2}{4\gamma^4}} = -\frac{K_a}{2\gamma^2} \pm \frac{1}{2\gamma^2}\sqrt{0.4 \cdot K_a\gamma^2 + K_a^2}$$

Now we can use Equation (21.93) (Debye–Hückel result) to evaluate γ as

$$\ln \gamma = -\frac{A \cdot \sqrt{I}}{1 + B \cdot R\sqrt{I}}$$

with $A = 1.175\,\text{mol}^{-1/2}\,\text{L}^{1/2}$ and $B = 0.330 \times 10^{10}\,\text{m}^{-1}\,\text{mol}^{-1/2}\,\text{L}^{1/2}$. If we use $R = 500$ pm and calculate the ionic strength by way of our concentrations found for the ideally diluted case ($\hat{c}_{H^+} = \hat{c}_{C_2O_4H^-} = 5.27 \times 10^{-2}$), we find that $I = 10.54 \times 10^{-2}\,\text{mol}\,\text{L}^{-1}$, and

$$\begin{aligned}\ln \gamma &= -\frac{1.175\text{ mol}^{-1/2}\text{L}^{1/2} \cdot 0.325\text{ mol}^{1/2}\text{ L}^{-1/2}}{1 + \left(0.330 \times 10^{10}\text{ m}^{-1}\text{ mol}^{-1/2}\text{L}^{1/2}\right) \cdot 500\text{ pm} \cdot 0.325\text{ mol}^{1/2}\text{ L}^{-1/2}} \\ &= -\frac{0.382}{1 + 0.536} = -0.249 \text{ so that } \gamma = 0.779\end{aligned}$$

Substituting this result into our equation above for $\hat{c}_{H^+}$, we find that

$$\begin{aligned}\hat{c}_{H^+} &= -\frac{5.89 \times 10^{-2}}{2 \cdot 0.779^2} \pm \frac{1}{2 \cdot 0.779^2}\sqrt{0.4 \cdot 5.89 \times 10^{-2} \cdot 0.779^2 + \left(5.89 \times 10^{-2}\right)^2} \\ &= -4.85 \times 10^{-2} \pm 0.824\sqrt{1.43 \times 10^{-2} + 3.46 \times 10^{-3}} \\ &= -4.85 \times 10^{-2} \pm 0.824 \times 0.133 = 6.13 \times 10^{-2}\end{aligned}$$

Lastly, we calculate the proton activity via

$$\hat{a}_{H^+} = \gamma \cdot \hat{c}_{H^+} = 0.779 \cdot 6.13 \times 10^{-2} = 4.78 \times 10^{-2}$$

which gives a pH $= -\log\left(\hat{a}_{H^+}\right) = 1.32$.

The change in solution pH compared to the ideal case is about 3%.

23

Chemical Reactions in Electrochemical Cells

23.1 Exercises

E23.1 Prove that the $\Delta G^{\ominus}$ calculated from the cell potential of the electrochemical cell in Fig. 23.1 are the same as calculated from $\Delta H^{\ominus}$, $\Delta S^{\ominus}$, and $\Delta G^{\ominus}$ in Table A.1 for 298 K.

Solution

For the reaction

$$AgI + \frac{1}{2}\ Pb \rightarrow Ag + \frac{1}{2}\ PbI_2$$

we find from Appendix A.1 that

$$\Delta_r H^{\ominus} = \frac{1}{2}\Delta_f H^{\ominus}(PbI_2) - \Delta_f H^{\ominus}(AgI)$$
$$= -\frac{1}{2}175.1\ \text{kJ mol}^{-1} - (-62.2\ \text{kJ mol}^{-1}) = -25.4\ \text{kJ mol}^{-1}$$

and

$$\Delta_r S^{\ominus} = \frac{1}{2}S_m^{\ominus}(PbI_2) + S_m^{\ominus}(Ag) - S_m^{\ominus}(AgI) - \frac{1}{2}S_m^{\ominus}(Pb)$$
$$= (\frac{1}{2}176.9 + 42.7)\ \text{J K}^{-1}\ \text{mol}^{-1} - (114.2 + \frac{1}{2}64.9)\ \text{J K}^{-1}\ \text{mol}^{-1}$$
$$= -15.5\ \text{J K}^{-1}\ \text{mol}^{-1}$$

Then

$$\Delta_r G^{\ominus} = \Delta_r H^{\ominus} - T\Delta_r S^{\ominus} = -20.8\ \text{kJ mol}^{-1} \quad \text{at 298 K}$$

which corresponds to a value of −20.8 kJ for 1 mol of AgI. This result differs by about 2% from the corresponding value calculated from the cell potential; namely

$$\Delta G^{\ominus} = -\boldsymbol{nF}E^{\ominus}$$
$$= -1\ \text{mol} \cdot 96,485\ \text{C mol}^{-1} \cdot 0.221\ \text{V}$$
$$= 21,300\ \text{J mol}^{-1} = 21.3\ \text{kJ mol}^{-1}$$

E23.2 Consider the reaction

$$2\ Ag(s) + Cl_2\ (g) \rightleftarrows 2\ Ag^+(aq) + 2\ Cl^-\ (aq)$$

which may be used to construct an electrochemical cell. The electrical potential of this cell is 1.1372 V at 298 K, and its potential has a temperature dependence of −0.595 mV K^{-1}. What is the Gibbs energy change associated with this cell when 1 mol of electrons have passed the cell by the conversion of reactants into products? What is the change in enthalpy and the change in entropy for the reaction?

Solutions Manual for Principles of Physical Chemistry, Third Edition. Edited by Hans Kuhn, David H. Waldeck, and Horst-Dieter Försterling.

Solution

To determine the Gibbs free energy change we use the fundamental relation

$$\Delta G^\ominus = -\boldsymbol{n}\boldsymbol{F}E^\ominus$$

For $\boldsymbol{n} = 1$ mole (that is 1 mol of electrons), we find

$$\begin{aligned}\Delta G^\ominus &= -1\ \text{mol} \cdot 96,485\,\text{C mol}^{-1} \cdot 1.1372\ \text{V}\\ &= 109,723\ \text{J mol}^{-1} = -109.72\ \text{kJ mol}^{-1}\end{aligned}$$

We can determine the entropy change directly from the temperature dependence because

$$\Delta S^\ominus = -\left(\frac{\partial \Delta G^\ominus}{\partial T}\right)_P = \boldsymbol{n}\boldsymbol{F}\left(\frac{\partial E^\ominus}{\partial T}\right)_P$$

so that

$$\Delta S^\ominus = 96,485\ \text{C mol}^{-1} \cdot 0.595 \times 10^{-3}\ \text{V K}^{-1} = 57.4\ \text{J K}^{-1}\ \text{mol}^{-1}$$

With $\Delta G^\ominus$ and $\Delta S^\ominus$ in hand, we find

$$\begin{aligned}\Delta H^\ominus &= \Delta G^\ominus + T\Delta S^\ominus\\ &= -109,720\ \text{kJ mol}^{-1} + (298\ \text{K})\left(57.4\ \text{J K}^{-1}\ \text{mol}^{-1}\right)\\ &= -92.6\ \text{kJ mol}^{-1}\end{aligned}$$

E23.3 Using a standard potential of $E^\ominus = -0.113$ V versus NHE for the half-cell reaction

$$NAD^+ + 2e^- + H^+ \rightarrow NADH$$

calculate $\Delta_r G^{\ominus\prime}(\text{pH} = 7)$ for the reaction

$$NAD^+ + H_2 \rightarrow NADH + H^+$$

Solution

When combined with the half-cell reaction for the NHE electrode, we find the following net reaction

$$\begin{array}{rl} NAD^+ + 2e^- + H^+ \rightarrow NADH & E^\ominus = -0.113\ \text{V}\\ H_2 \rightarrow 2e^- + 2H^+ & E^\ominus = 0.000\ \text{V}\\ \hline NAD^+ + H_2 \rightarrow NADH + H^+ & \end{array}$$

For the NADH half-cell, we find

$$\begin{aligned}\Delta G^\ominus &= -\boldsymbol{n}\boldsymbol{F}E^\ominus\\ &= 2 \cdot 96,485\ \text{C mol}^{-1} \cdot 0.113\ \text{V}\\ &= 21,805\ \text{J mol}^{-1} = 21.8\ \text{kJ mol}^{-1}\end{aligned}$$

A corresponding calculation for the NHE half-cell reaction gives a Gibbs energy change of zero. Hence at pH = 0 we find a molar Gibbs energy change of $\Delta_r G^\ominus = 21.8$ kJ mol^{-1} for this reaction.

At pH = 7 we have a $\Delta_r G^{\ominus\prime}$ that is shifted from $\Delta_r G^\ominus$. Using the law of mass action

$$\Delta_r G^{\ominus\prime} = \Delta_r G^\ominus + \boldsymbol{R}T \ln\left(\frac{\widehat{c}_{H^+} \cdot \widehat{c}_{NADH}}{\widehat{P}_{H_2} \cdot \widehat{c}_{NAD^+}}\right)$$

and substituting our standard state values of $\widehat{P}_{H_2} = \widehat{c}_{NAD^+} = \widehat{c}_{NADH} = 1$ and $\widehat{c}_{H^+} = 10^{-7}$, we find that

$$\begin{aligned}\Delta_r G^{\ominus\prime} &= \Delta_r G^\ominus + \boldsymbol{R}T \ln\left(10^{-7}\right)\\ &= [21.8 - 39.930]\ \text{kJ mol}^{-1} = -18.1\ \text{kJ mol}^{-1}\end{aligned}$$

E23.4 The redox protein Cytc$_1$ functions by oxidation and reduction of an iron ion which is bound by a heme unit of the protein. Use its reduction potential of $E^\ominus = 0.220$ V to calculate $\Delta_r G^{\ominus\prime}$ for the reaction

$$\text{Cytc}_1\ \text{Fe(II)} + \text{H}^+ \rightarrow \text{Cytc}_1\ \text{Fe(III)} + \frac{1}{2}\text{H}_2$$

Solution

This reaction is comprised of the two half-cells

$$\begin{array}{ll} \text{Cytc}_1\ \text{Fe(II)} \rightarrow \text{Cytc}_1\ \text{Fe(III)} + \text{e}^- & E^\ominus = -0.220\ \text{V} \\ \text{e}^- + \text{H}^+ \rightarrow \frac{1}{2}\text{H}_2 & E^\ominus = 0.000\ \text{V} \\ \hline \text{Cytc}_1\ \text{Fe(II)} + \text{H}^+ \rightarrow \text{Cytc}_1\ \text{Fe(III)} + \frac{1}{2}\text{H}_2 & \end{array}$$

Combining these half-cell reactions, we find that

$$\begin{aligned} \Delta G^\ominus &= -\boldsymbol{nFE}^\ominus \\ &= (1\ \text{mol})(96,485\ \text{C mol}^{-1})(0.220\ \text{V}) \\ &= 21,200\ \text{J mol}^{-1} = 21.2\ \text{kJ mol}^{-1} \end{aligned}$$

which corresponds to a molar Gibbs energy change of $\Delta_r G^\ominus = 21.2$ kJ mol^{-1}. To find the Gibbs energy under the conditions of pH = 7 we use the law of mass action

$$\Delta_r G^{\ominus\prime} = \Delta_r G^\ominus + \boldsymbol{RT} \ln \left(\frac{\hat{P}_{H_2}^{1/2} \cdot \hat{c}_{\text{Cytc}_1\ \text{Fe(III)}}}{\hat{c}_{H^+} \cdot \hat{c}_{\text{Cytc}_1\ \text{Fe(II)}}} \right)$$

At the standard conditions of $\hat{P}_{H_2} = \hat{c}_{\text{Cytc}_1\ \text{Fe(III)}} = \hat{c}_{\text{Cytc}_1\ \text{Fe(II)}} = 1$ and $\hat{c}_{H^+} = 10^{-7}$, we find

$$\begin{aligned} \Delta_r G^{\ominus\prime} &= \Delta_r G^\ominus - \boldsymbol{RT} \ln\left(\hat{c}_{H^+}\right) \\ &= [21,200 + 39,934]\ \text{J mol}^{-1} = 61.1\ \text{kJ mol}^{-1} \end{aligned}$$

E23.5 In the cell in Fig. 23.5, we used different H$^+$ concentrations in the two containers, but the hydrogen is under the same pressure P. Alternatively, we can choose equal H$^+$ concentrations, but different H$_2$ pressures P_1 and P_2. This can be achieved by bubbling H$_2$ at atmospheric pressure (P_1) on the left side and, say, a 1:9 mixture of H$_2$ and argon (as an inert gas) on the right side; then $P_1/P_2 = 10$. Derive an expression for the cell potentials in terms of the partial pressure of hydrogen on each side.

Solution

In this case the reaction is

$$\frac{1}{2}\text{H}_2(\text{Pt}_{\text{left}}) + \text{H}^+(c_{\text{right}}) \rightarrow \frac{1}{2}\text{H}_2(\text{Pt}_{\text{right}}) + \text{H}^+(c_{\text{left}})$$

As stated in the problem we have $c_{\text{left}} = c_{\text{right}}$, so that

$$\begin{aligned} E &= -\frac{\Delta_r G}{\boldsymbol{F}} = -\frac{\boldsymbol{RT}}{\boldsymbol{F}} \ln \left(\frac{P_{H_2,\text{right}}^{1/2}}{P_{H_2,\text{left}}^{1/2}} \right) \\ &= \frac{\boldsymbol{RT}}{2\boldsymbol{F}} \ln \left(\frac{P_{H_2,\text{right}}}{P_{H_2,\text{left}}} \right) \end{aligned}$$

If we consider the 10-fold difference in partial pressure of hydrogen that is stated in the problem, we find that

$$\begin{aligned} E &= -\frac{\left(8.314\ \text{J mol}^{-1}\text{K}^{-1}\right)(298\ \text{K})}{2(96,485\ \text{C mol}^{-1})} \ln\left(\frac{1}{10}\right) \\ &= -29.6 \cdot 10^{-3}\ \text{V} = -29.6\,\text{mV} \end{aligned}$$

E23.6 Write the chemical reaction for an electrochemical cell consisting of an SHE connected to a half-cell with a platinum electrode immersed in a solution containing a mixture of Fe^{2+} and Fe^{3+}. Write a mathematical expression for the standard potential of this cell and determine its value for $c_{Fe^{2+}} = 0.1 \cdot c_{Fe^{3+}}$.

Solution

This reaction comprises the two half-cells

$$Fe^{3+} + e^- \rightarrow Fe^{2+} \quad E^\ominus = 0.771\text{V} \quad \text{and} \quad \frac{1}{2}H_2 \rightarrow e^- + H^+ \quad E^\ominus = 0.000 \text{ V}$$

to give the net reaction

$$Fe^{3+} + \frac{1}{2}H_2 \rightarrow Fe^{2+} + H^+ \qquad E^\ominus = 0.771 \text{ V}$$

for which the Nernst equation has the form

$$E_{cell} = E^\ominus - \frac{RT}{F}\ln\left(\frac{\hat{c}_{Fe^{2+}}\hat{c}_{H^+}}{\hat{c}_{Fe^{3+}}\hat{P}^{1/2}_{H_2}}\right)$$

For the standard cell $\hat{c}_{Fe^{2+}} = \hat{c}_{Fe^{3+}} = \hat{c}_{H^+} = 1$, so that $E_{cell} = E^\ominus = 0.771$ V.

For the cell described in the problem statement, we have $\hat{c}_{Fe^{2+}} = 0.1 \cdot \hat{c}_{Fe^{3+}}$, and $\hat{P}^{1/2}_{H_2} = 1$ so that

$$\begin{aligned}E_{cell} &= E^\ominus - \frac{RT}{F}\ln(0.1) \\ &= \left[0.771 \text{ V} - \frac{8.314 \text{ J mol}^{-1}\text{ K}^{-1} \cdot 298 \text{ K}}{96,485 \text{ C mol}^{-1}}\ln(0.1)\right] \\ &= 0.771 \text{ V} + 59.1 \text{ V} = 0.830 \text{ V}\end{aligned}$$

That is, a 10-fold change in concentration ratio causes a 59 mV shift in the operating potential.

E23.7 Write the Nernst equation for the electrochemical cell reaction

$$\frac{1}{2}\text{ Cu}^{2+} + \frac{1}{2}\text{ H}_2 \rightarrow \frac{1}{2}\text{Cu} + \text{H}^+$$

in which $c_{HCl} = 1.184$ mol L^{-1}, $P_{H_2} = 1$ bar (i.e., conditions for an SHE). Determine the cell potential for $c_{Cu^{2+}}$ of 1.0×10^{-3} mol L^{-1}, 1.0×10^{-6} mol L^{-1}, and 1.0×10^{-9} mol L^{-1}.

Solution

In this case the Nernst equation takes the form

$$E_{cell} = E^\ominus - \frac{RT}{F}\ln\left(\frac{\hat{c}_{H^+}}{\hat{c}^{1/2}_{Cu^{2+}}}\right)$$

and for the conditions given here it transforms to

$$E_{cell} = E^\ominus_{Cu^{2+}/Cu} + \frac{RT}{2F}\ln\left(\hat{c}_{Cu^{2+}}\right) - E^\ominus_{SHE}$$

Using the value of $E^\ominus_{Cu^{2+}/Cu} = 0.340$ V, we find

$$\begin{aligned}E_{cell} &= 0.340 \text{ V} + \left(\frac{8.314 \text{ J mol}^{-1}\text{ K}^{-1} \cdot 298 \text{ K}}{2 \cdot 96,485 \text{ C mol}^{-1}}\ln\left(\hat{c}_{Cu^{2+}}\right)\right) - 0.00 \text{ V} \\ &= 0.340 \text{ V} + \left[0.0128 \cdot \ln\left(\hat{c}_{Cu^{2+}}\right)\right] \text{ V}\end{aligned}$$

Thus, for the case of $c_{Cu^{2+}} = 1.0 \times 10^{-3}$ mol L^{-1} ($\hat{c}_{Cu^{2+}} = 10^{-3}$), we find

$$E_{cell} = 0.340 \text{ V} + \left(0.0128 \cdot \ln\left(10^{-3}\right)\right) \text{ V} = 0.252 \text{ V}$$

If we proceed in like manner for the other concentrations we find the results in the table

$\hat{c}_{Cu^{2+}}$	10^{-3}	10^{-6}	10^{-9}
E_{cell}/V	0.252	0.163	0.075

E23.8 As for the quinone/hydroquinone reaction the cell potential for the reaction

$$\frac{1}{2}\text{NAD}^+ + \frac{1}{2}\text{H}^+ + \frac{1}{2}\text{H}_2(\text{SHE}) \rightarrow \frac{1}{2}\ \text{NADH} +\ \text{H}^+(\text{SHE})$$

depends on pH. Using $E^{\ominus} = -0.113$ V, evaluate the cell potential at pH values of 5.0, 7.0, and 9.0. Comment on the change in cell potential and compare it to the pH dependence of the quinone/hydroquinone cell potential.

Solution

In this case we write the half-cell reactions as

$$\tfrac{1}{2}\text{NAD}^+ + \text{e}^- + \tfrac{1}{2}\text{H}^+ \rightarrow \tfrac{1}{2}\ \text{NADH} \qquad E^{\ominus} = -0.113\ \text{V}$$

$$\tfrac{1}{2}\ \text{H}_2 \rightarrow \text{e}^- + \text{H}^+ \qquad E^{\ominus} = 0.000\,\text{V}$$

The Nernst equation for the NAD^+/NADH half-cell takes the form

$$E_{\text{cell}} = E^{\ominus} - \frac{\boldsymbol{RT}}{\boldsymbol{F}} \ln\left(\frac{\widehat{c}^{\,1/2}_{\text{NADH}}}{\widehat{c}^{\,1/2}_{\text{NAD}^+} \cdot \widehat{c}^{\,1/2}_{\text{H}^+}}\right)$$

If we use the value $E^{\ominus} = -0.113$ V and take the concentrations of the NADH and NAD^+ to be $\widehat{c}_{\text{NAD}^+} = \widehat{c}_{\text{NADH}} = 1$, we find that

$$E_{\text{cell}} = -0.113\ \text{V} + \left(0.01284 \cdot \ln\left(\widehat{c}_{\text{H}^+}\right)\right)\ \ \text{V}$$

Using the definition of the pH as $\text{pH} = -\log(\widehat{c}_{\text{H}^+}) = -0.4343 \cdot \ln(\widehat{c}_{\text{H}^+})$, we find that

$$E_{\text{cell}} = -0.113\ \text{V} - (0.0296 \cdot \text{pH})\ \text{V}$$

At a value of pH = 5, we find that

$$E_{\text{cell}} = -0.113\ \text{V} - 0.0296 \cdot (5)\ \ \ \text{V} = -0.261\ \text{V}$$

Proceeding in a like manner for the other values we find

pH	5	7	9
E_{cell}/V	−0.261	−0.320	−0.379

E23.9 Use the data point at $c = 0.00573$ mol L^{-1} in Table 23.3 to calculate the mean activity coefficient of the ions. Compare this experimental result to the prediction of the Debye–Hückel law and the Debye Hückel limiting law, Equations (23.63) and (23.64).

Solution

Using this data point and the value $E' = 0.2165$ V in Equation (23.60),

$$\ln\gamma_{\pm} = \frac{\boldsymbol{F}}{2\boldsymbol{RT}}\left(E' - E\right) - \ln\widehat{c}$$

we find that

$$\ln\gamma_{\pm} = \frac{96,485\ \text{C mol}^{-1} \cdot (0.2165\ \text{V} - 0.1940\ \text{V})}{2 \cdot 8.314\ \text{J mol}^{-1}\ \text{K}^{-1} \cdot 308\ \text{K}} - \ln(0.00573)$$
$$= -5.2280 + 5.1620 = -0.0660$$

so that $\gamma_{\pm} = 0.936$.

For comparison we calculate the activity coefficient using the Debye–Hü ckel limiting law,

$$\ln\gamma = -A \cdot \sqrt{I}$$

where $A = 1.175\ \text{mol}^{-1/2}\text{L}^{1/2}$ and I is the ionic strength, $I = \frac{1}{2}\sum c_i z_i^2$. For $c_{\text{H}^+} = c_{\text{Cl}^-} = 0.00573$ mol L^{-1}, we find that

$$I = \frac{1}{2}\left[(1)^2 \cdot 0.00573\ \text{mol L}^{-1} + (1)^2 \cdot 0.00573\ \text{mol L}^{-1}\right] = 0.00573\,\text{mol L}^{-1}$$

and

$$\ln\gamma = -\left(1.175\ \text{mol}^{-1/2}\ \text{L}^{1/2}\right) \cdot \sqrt{0.00573\ \text{mol L}^{-1}} = -0.0889$$

so that $\gamma = 0.915$; a value about 2% lower than the experimental value.

For a more precise estimate we can use the full Debye–Hückel expression, namely

$$\ln \gamma = -\frac{A \cdot z^2 \cdot \sqrt{I}}{1 + B \cdot R \cdot \sqrt{I}}$$

If we use $B = 0.330 \times 10^{10}$ mol$^{-1/2}$ L$^{1/2}$ m^{-1} and $R = 500$ pm (as given in the text), we find that

$$\ln \gamma = -\frac{1.175 \text{ mol}^{-1/2} \text{ L}^{1/2} \cdot (1)^2 \cdot \sqrt{0.00573 \text{ mol L}^{-1}}}{1 + 0.330 \times 10^{10} \text{ mol}^{-1/2} \text{ L}^{1/2} \text{ m}^{-1} \cdot 500 \times 10^{-12} \text{ m} \cdot \sqrt{0.00573 \text{ mol L}^{-1}}}$$
$$= -0.0791$$

so that $\gamma = 0.924$; a value that is within 1% of the experimental value.

E23.10 Calculate the cell potential of a calomel electrode. Compare this calculated value with the experimental value of 281 mV versus NHE, and determine the activity coefficient of the chloride ion.

Solution

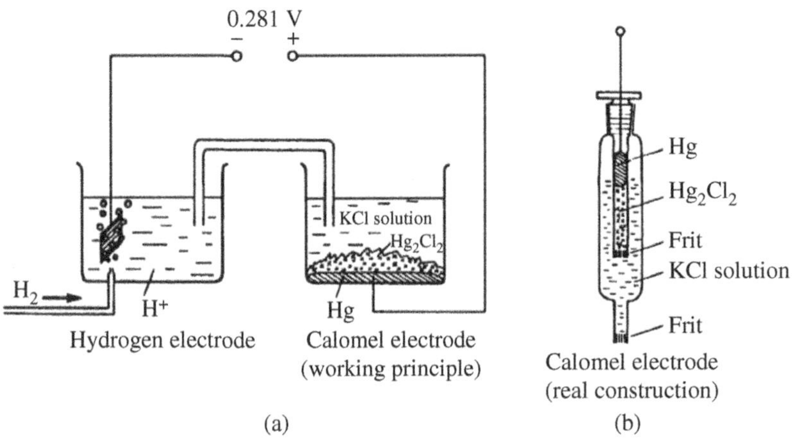

Figure E23.10 Calomel electrode. (a) Calomel electrode measured against a standard hydrogen electrode and (b) real construction of a calomel electrode.

The half-cell reaction in the calomel electrode (see Figure E23.10) is

$$\frac{1}{2} \text{Hg}_2^{2+} + e^- \rightarrow \text{Hg}$$

and the potential relative to the standard hydrogen electrode is

$$E_{\text{calomel}} = E^{\ominus} - \frac{RT}{F} \cdot \ln \frac{1}{\left(\hat{c}_{\text{Hg}_2^{2+}}\right)^{1/2}}$$

As the electrode contains an excess of chloride ions, the concentration of Hg_2^{2+} is determined by the equilibrium

$$\text{Hg}_2^{2+} + 2\,\text{Cl}^- \rightleftharpoons \text{Hg}_2\text{Cl}_2$$

$$\hat{c}_{\text{Hg}_2^{2+}} \cdot \left(\hat{c}_{\text{Cl}^-}\right)^2 = K_s = 2.038 \times 10^{-18}$$

and

$$E_{\text{calomel}} = E^{\ominus} - \frac{RT}{F} \cdot \ln \frac{\hat{c}_{\text{Cl}^-}}{\sqrt{K_s}}$$

For a calomel electrode with $\hat{c}_{\text{Cl}^-} = 1$ (1 M calomel electrode) we calculate

$$E_{\text{calomel,1M}} = 789 \text{ mV} - 25.69 \text{ mV} \cdot \ln \frac{1.0}{1.428 \times 10^{-9}} = 266 \text{ mV}$$

Actually, the voltage is 281 mV. The small difference results from the fact that the calculated value is valid only for ideally diluted solutions. To account for nonideality, we replace the concentration by the activity.

$$E_{\text{calomel}} = E^{\ominus} - \frac{RT}{F} \cdot \ln \frac{\hat{a}_{\text{Cl}^-}}{\sqrt{K_s}} = E^{\ominus} - \frac{RT}{F} \cdot \ln \frac{\hat{c}_{\text{Cl}^-} \cdot \gamma}{\sqrt{K_s}}$$

$$= E^{\ominus} - \frac{RT}{F} \cdot \ln \frac{\hat{c}_{\text{Cl}^-}}{\sqrt{K_s}} - \frac{RT}{F} \cdot \ln \gamma$$

Hence we have that

$$E_{\text{expt,calomel}} = E^{\ominus} - \frac{RT}{F} \ln \left(\frac{\hat{c}_{\text{Cl}^-}}{\sqrt{K_s}} \right) - \frac{RT}{F} \ln (\gamma)$$

$$= E_{\text{ideal,calomel}} - \frac{RT}{F} \ln (\gamma)$$

Solving for $\ln(\gamma)$, we find that

$$\ln (\gamma) = \frac{F}{RT} \left(E_{\text{ideal,calomel}} - E_{\text{expt,calomel}} \right)$$

$$= \frac{96,485 \text{ C mol}^{-1}}{8.314 \text{ J mol}^{-1} \text{ K}^{-1} \cdot 298 \text{ K}} \cdot (0.266 \text{ V} - 0.281 \text{ V}) = -0.584$$

so that $\gamma = 0.58$; a value quite different from one.

E23.11 Determine the ideal cell potential for a Mallory Cell which operates with the cell reaction

$$\text{Zn} + \text{HgO} \rightarrow \text{ZnO} + \text{Hg}$$

(Hint: Identify the two half-cell reactions and find their potentials).

Solution

This reaction can be written as the sum of two half-cell reactions; namely

$HgO + 2e^- + H_2O \rightarrow Hg + 2OH^-$	$E^{\ominus} = 0.0977$ V
$Zn + 2OH^- \rightarrow ZnO + 2e^- + H_2O$	$E^{\ominus} = 1.260$ V
$Zn + HgO \rightarrow ZnO + Hg$	$E^{\ominus} = 1.358$ V

Standard potentials of the two half-cell reactions were taken from the *CRC Handbook of Chemistry and Physics* (75th ed., 1995).

E23.12 Upon loading the Ni/Cd accumulator Ni^{3+} is reduced to Ni^{2+}, and Cd is oxidized to Cd^{2+} under basic conditions. For the Ni half-cell reaction use

$$2\text{NiO(OH)} + 2\text{H}_2\text{O} + 2\text{e}^- \rightarrow 2\text{Ni(OH)}_2 + 2\text{OH}^- \qquad E^{\ominus}(\text{Ni}) = 0.48 \text{ V}$$

Propose a reaction for the Cd half-cell and use it to compute a cell voltage. Compare your result to the 1.29 V EMF of the battery. (Note: You can find the relevant half-cell reactions in the *CRC Handbook of Chemistry and Physics*).

Solution

As given in the problem statement, the Cd is oxidized to Cd^{2+} upon loading of the cell; hence, we propose a reaction of the type

$$\text{Cd} + 2\text{OH}^- \rightarrow \text{Cd(OH)}_2 + 2\text{e}^-$$

for the Cd half-cell. Combining this with the Ni half-cell reaction that is given allows us to write the net reaction

$$2\ \text{NiO(OH)} + \text{Cd} + 2\text{H}_2\text{O} \rightarrow 2\ \text{Ni(OH)}_2 + \text{Cd(OH)}_2$$

which seems plausible.

To find the potential for the Cd half-cell, we must consider the half-cell reactions

$$Cd(Hg) + 2OH^- \rightarrow Cd(OH)_2 + 2e^- \quad E^\ominus = 0.761\ V$$
$$Cd^{2+} + 2e^- \rightarrow Cd(Hg) \quad E^\ominus = -0.352\ V$$
$$Cd \rightarrow Cd^{2+} + 2e^- \quad E^\ominus = 0.403\ V$$

so that

$$Cd + 2OH^- \rightarrow Cd(OH)_2 + 2e^- \quad E^\ominus\ (Cd) = 0.812\ V$$

Hence we calculate a total cell potential of

$$E_{cell} = 0.48\ V + 0.812\ V = 1.29\ V$$

E23.13 One can build an oxygen electrode by bubbling oxygen over a platinum foil and coupling it to a standard hydrogen electrode. Write a cell reaction for such an electrochemical cell. Using your result, write an expression for the electrode potential in terms of the pH and the oxygen pressure. What is the Gibbs energy change for this cell? What is its standard potential?

Solution

We couple an oxygen electrode (oxygen bubbling over a platinum foil) to a standard hydrogen electrode. The cell potential for the reaction can be written as

$$\frac{1}{4}\ O_2 + H^+ + \frac{1}{2}\ H_2\ (SHE) \rightarrow \frac{1}{2}\ H_2O + H^+\ (SHE)$$

So that the Nernst equation takes the form

$$E_{cell} = E^\ominus - \frac{RT}{F} \cdot \ln\left(\frac{1}{\widehat{c}_{H^+} \cdot \widehat{P}_{O_2}^{1/4}}\right)$$
$$= E^\ominus + \frac{RT}{F} \cdot \ln\left(\widehat{c}_{H^+}\right) + \frac{RT}{4F} \cdot \ln\left(\widehat{P}_{O_2}\right)$$

Using the definition of the pH as $pH = -\log\left(\widehat{c}_{H^+}\right) = -0.4343 \cdot \ln(\widehat{c}_{H^+})$, we find that

$$E_{cell} = E^\ominus - 2.303 \cdot \frac{RT}{F} \cdot pH + \frac{RT}{4F} \cdot \ln\left(\widehat{P}_{O_2}\right)$$

The reaction in this electrode generates a half mole of H_2O from a half mole of H_2 and a quarter mole of O_2. Under standard conditions ($P = P^\ominus = 1$ bar) and $T = 298$ K we can use Table A.1 in the appendix to write

$$\Delta_r G^\ominus = \frac{1}{2} \Delta_f G^\ominus (H_2O) = -118.6\ kJ\ mol^{-1}$$

Using the fundamental result

$$E^\ominus = -\frac{\Delta_r G^\ominus}{F} = \frac{118.6\ kJ\ mol^{-1}}{F} = 1.23\ V$$

Hence we find that

$$E_{cell} = \left[1.23 - 0.059 \cdot pH + 0.00642 \cdot \ln\left(\widehat{P}_{O_2}\right)\right]\ V$$

E23.14 Acceleration of Na^+ Ions. What is the time required for Na^+ ions to reach a constant speed after switching on an electrical field strength of $1\ V\ cm^{-1}$?

Solution

According to Newton's law and Equations (23.77) and (23.78), the acceleration of the ion is

$$\frac{dv}{dt} = \frac{1}{m} \cdot \left(f_{el} + f_{fric}\right) = \frac{ze}{m} \cdot F - \frac{6\pi\eta r}{m} \cdot v$$

or

$$\frac{dv}{dt} + av = b$$

with

$$a = \frac{6\pi\eta r}{m} \quad \text{and} \quad b = \frac{ze}{m} \cdot F$$

The solution of this inhomogeneous differential equation with the initial condition $v = 0$ at $t = 0$ is

$$v = \frac{b}{a} \cdot \left(1 - e^{-at}\right)$$

With $\eta = 0.89 \times 10^{-2}$ g s^{-1} cm^{-1}, $r = 200$ pm, $m = 4 \times 10^{-26}$ kg, and $F = 1$ V cm^{-1} we obtain

$$a = 8.4 \times 10^{13}\ \text{s}^{-1}\ ,\ b = 4.0 \times 10^{8}\ \text{m s}^{-2}$$

The characteristic decay time τ for this process is

$$\tau = \frac{1}{a} = \frac{m}{6\pi\eta r} \sim 10^{-14}\ \text{s}$$

After that time the Na^+ ion moves with a constant speed of

$$v = \frac{b}{a} = \frac{\frac{ze}{m} \cdot F}{\frac{6\pi\eta r}{m}} = \frac{ze \cdot F}{6\pi\eta r} = 4.8 \times 10^{-6}\ \text{ms}^{-1}.$$

E23.15 Plot the molar conductivity Λ data in the table for KCl solutions at 298 K to determine the limiting conductance (at infinite dilution) Λ_0. Show that these data agree with Kohlrausch's law, namely $\Lambda = \Lambda_0 - A \cdot \sqrt{c}$ in which A is an empirical constant and c is the concentration.[1]

$c/10^{-4}$ mol L^{-1}	0.32576	2.6570	3.5217	6.0895	11.321	20.568	32.827
Λ/cm^2 Ω^{-1} mol^{-1}	149.33	148.38	148.12	147.52	146.76	145.71	144.64

Solution

Figure E23.15a plots the experimental conductance versus the square root of the concentration, according to Kohlrausch's law. A best fit line gives a slope of $A = 90.905$ cm^2 Ω^{-1} mol$^{-3/2}$ L$^{1/2}$ and an intercept of $\Lambda_0 = 149.83$ cm^2 Ω^{-1} mol^{-1}. Figure E23.15b shows a plot of the conductance versus the concentration. Although the data range is limited the systematic nature of the deviations makes it evident that the plot is not linear.

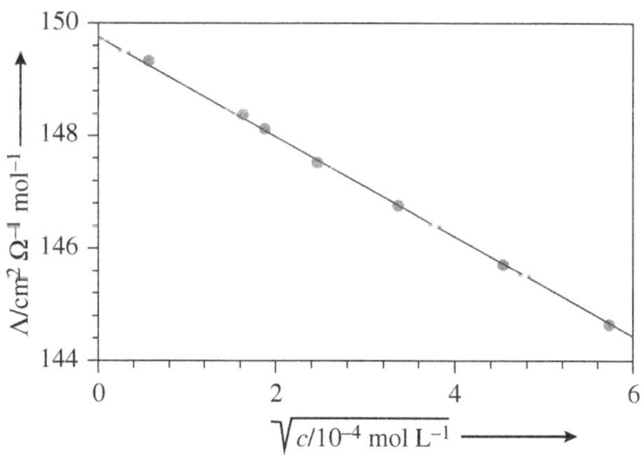

Figure E23.15a Conductance Λ versus $\sqrt{c}$ for solutions of KCl in water at 298 K.

1 Data are taken from T. Shedlovsky, *J. Am. Chem. Soc.* 1932, **54**, 1411.

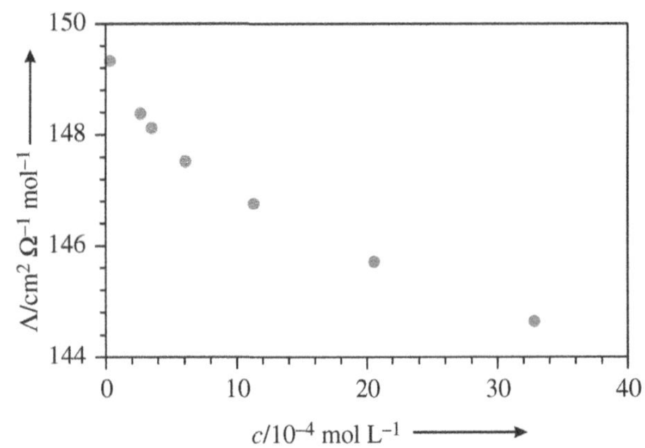

Figure E23.15b Conductance Λ versus concentration *c*.

E23.16 The limiting conductance for a salt is equal to the sum of the conductance for each ion; i.e., for a 1:1 salt we have $\Lambda_0 = \lambda_+^0 + \lambda_-^0$. If the limiting conductance of chloride is $\lambda_{Cl^-}^0 = 76.35\ \text{cm}^2\ \Omega^{-1}\ \text{mol}^{-1}$ at 298 K, determine the limiting conductance for potassium and sodium ions from the data: $\Lambda_0(\text{NaCl}) = 126.45\ \text{cm}^2\ \Omega^{-1}\ \text{mol}^{-1}$ and $\Lambda_0(\text{KCl}) = 149.85\ \text{cm}^2\ \Omega^{-1}\ \text{mol}^{-1}$.

Solution

For the sodium ion, we can write that

$$\lambda_{Na^+}^0 = \Lambda_0 - \lambda_{Cl^-}^0$$
$$= (126.45 - 76.35)\ \text{cm}^2\ \Omega^{-1}\ \text{mol}^{-1} = 50.10\ \text{cm}^2\ \Omega^{-1}\ \text{mol}^{-1}$$

In a corresponding way for the potassium ion we find $\lambda_{K^+}^0 = 73.50\ \text{cm}^2\ \Omega^{-1}\ \text{mol}^{-1}$.

Although the bare radius of the sodium ion is expected to be smaller than that of bare potassium (98 pm versus 133 pm), the sodium ion has a higher charge density and this leads to a higher solvation energy in solution and a stronger attraction to the water molecules. Hence, it will experience more drag as it moves through the water and have a higher effective (hydrodynamic) radius.

E23.17 Explain how you might use conductivity data for a weak acid (e.g., acetic acid) to determine the acid's pK_a.

Solution

The dissociation of acetic acid generates ions

$$CH_3COOH \rightleftarrows CH_3COO^- + H^+$$

and causes a concomitant change in the solution's conductivity. Hence by monitoring the conductivity we can monitor the fraction of acetic acid that is dissociated, and from this the K_a of the acid.

To be more explicit: If we consider the acid dissociation alone we can write

$$K_a = \frac{\hat{c}_{H^+}\hat{c}_{A^-}}{\hat{c}_{HA}} = \frac{\hat{c}_{H^+}^2}{\hat{c}_{total} - \hat{c}_{H^+}}$$

where we have taken $\hat{c}_{H^+} = \hat{c}_{A^-}$ from the reaction equation and $\hat{c}_{HA} = \hat{c}_{total} - \hat{c}_{H^+}$. If we write the conductivity κ as $\kappa = \hat{c}_{H^+}\Lambda$, then we find that

$$K_a = \frac{\hat{c}_{H^+}^2}{\hat{c}_{total} - \hat{c}_{H^+}} = \frac{(\kappa/\Lambda)^2}{\hat{c}_{total} - \kappa/\Lambda}$$

For example, consider a 1.00 mM solution of acetic acid (which has a molar conductivity at infinite dilution of 390 S cm^2 mol^{-1}) to have a conductivity of $\kappa = 5.10 \times 10^{-5}$ S cm^{-1}. Then we would find

$$K_a = \frac{\left[5.1 \times 10^{-5}\ \mathrm{S\ cm^{-1}} / \left(390\ \mathrm{S\ cm^2\ mol^{-1}}\right)\right]^2}{1.0 \times 10^{-3}\ \mathrm{M} - 5.1 \times 10^{-5}\ \mathrm{S\ cm^{-1}} / \left(390\ \mathrm{S\ cm^2\ mol^{-1}}\right)}$$
$$= \frac{\left(1.31 \times 10^{-4}\ \mathrm{M}\right)^2}{1.0 \times 10^{-3}\ \mathrm{M} - 1.31 \times 10^{-4}\ \mathrm{M}} = 1.97 \times 10^{-5}\ \mathrm{M}$$

which gives a pKa value of 4.7.

E23.18 In Section 23.1 we considered the reaction

$$\mathrm{AgI} + \frac{1}{2}\ \mathrm{Pb} \rightarrow \mathrm{Ag} + \frac{1}{2}\ \mathrm{PbI_2}$$

and used the electrochemical cell properties to find its $\Delta_r G^\ominus$, $\Delta_r H^\ominus$, and $\Delta_r S^\ominus$. Use the thermodynamic data in Appendix A.1 to calculate $\Delta_r G^\ominus$ for the reaction and compare them to the values found electrochemically.

Solution

From the data in Appendix A.1, we find

$$\Delta_r H^\ominus = \frac{1}{2}\Delta_f H^\ominus(\mathrm{PbI_2}) - \Delta_f H^\ominus(\mathrm{AgI})$$
$$= -\frac{1}{2}175.1\ \mathrm{kJ\ mol^{-1}} - (-62.2\ \mathrm{kJ\ mol^{-1}}) = -25.4\ \mathrm{kJ\ mol^{-1}}$$

$$\Delta_r S^\ominus = \frac{1}{2}S_m^\ominus(\mathrm{PbI_2}) + S_m^\ominus(\mathrm{Ag}) - S_m^\ominus(\mathrm{AgI}) - \frac{1}{2}S_m^\ominus(\mathrm{Pb})$$
$$= \left(\frac{1}{2}176.9 + 42.7\right)\ \mathrm{J\ K^{-1}\ mol^{-1}} - \left(114.2 + \frac{1}{2}64.9\right)\ \mathrm{J\ K^{-1}\ mol^{-1}}$$
$$= -15.5\ \mathrm{J\ K^{-1}\ mol^{-1}}$$

Then

$$\Delta_r G^\ominus = \Delta_r H^\ominus - T\Delta_r S^\ominus = -20.8\ \mathrm{kJ\ mol^{-1}} \quad \text{at 298 K}$$

These values agree with the corresponding values calculated from the cell potential.

E23.19 The chemical reaction

$$2\mathrm{PbSO_4} + 2\mathrm{H_2O} \rightarrow \mathrm{Pb} + \mathrm{PbO_2} + 2\mathrm{H_2SO_4}$$

occurs when charging a lead rechargeable battery. Calculate $\Delta_r G^\ominus$ for this reaction.

Solution

In Appendix A.1 we find the following data:

	$\Delta_f H_{298}^\ominus$/kJ mol^{-1}	$S_{298}^\ominus$/J K^{-1} mol^{-1}
Pb	0	64.9
PbO_2	−276.6	76.4
HSO_4^-/aq	−885.8	126.9
$PbSO_4$	−918.1	147.2
H_2O	−285.9	69.9

and we calculate

$$\Delta_r H_{298}^\ominus/\mathrm{kJ\ mol^{-1}} = (-276.6 - 2 \cdot 885.8) - (2 \cdot 918.1 - 2 \cdot 285.9) = +385.8$$

and

$$\Delta_r S^{\ominus}_{298}/\text{J K}^{-1}\ \text{mol}^{-1} = (64.9 + 76.4 + 2 \cdot 126.9) - (2 \cdot 147.2 + 2 \cdot 69.9) = -39.1$$

Thus

$$\begin{aligned}\Delta_r G^{\ominus}_{298} &= \Delta_r H^{\ominus}_{298} - T \cdot \Delta_r S^{\ominus}_{298} \\ &= (385.8 - 298 \cdot (-39.1 \times 10^{-3}))\ \text{kJ mol}^{-1} \\ &= (385.8 + 11.7)\ \text{kJ mol}^{-1} = 397.5\ \text{kJ mol}^{-1}\end{aligned}$$

$\Delta_r G^{\ominus}_{298}$ is positive, that is, the reaction does not occur spontaneously. Note: In the charging process it is driven by an electrical energy source.

23.2 Problems

P23.1 The cells discussed in the chapter rely on activity differences or chemical differences to generate the Gibbs energy difference that drives the reaction, hence the EMF of the cell. An electrochemical cell can also be generated by mechanical energy differences between the two half-reactions. Consider the case of two Hg electrodes, of which one of the mercury columns has a 10 cm height and the other has a 110 cm height (in the gravitational field of the earth, 9.8 m s^{-2}), in contact with the same solution of 0.1M KI. In this arrangement the liquid mercury in the 110 cm column dissolves in the solution as Hg_2^+ and diffuses to the 10 cm column where it is reduced to Hg. The process proceeds until the columns have equal height. Determine the cell potential for this mechanical system (you may assume the KI behaves ideally and the cross-sectional areas A of the columns are 1 cm^2; $\rho(\text{Hg}) = 13,546$ kg m^{-3}).

Solution

The difference in the height of the mercury columns gives rise to a difference in the Gibbs energy, namely

$$dG = m \cdot g \cdot dh = (\rho \cdot A \cdot h) \cdot g \cdot dh = (\rho \cdot A \cdot g) \cdot h \cdot dh$$

where m is the mass of the mercury column, $g = 9.8$ m s^{-2}, and h is the height of the mercury column initially at 100 cm. The Gibbs energy difference between the two column heights will be

$$\Delta G = \frac{(\rho \cdot A \cdot g)}{2} h^2 \Big|_{0.010\ \text{m}}^{0.110\ \text{m}}$$

Using this expression for ΔG and the fundamental relation between the Gibbs energy and the cell potential, namely $\Delta G = -\boldsymbol{nF}E$, we find that

$$E = -\frac{\rho \cdot A \cdot g}{2\boldsymbol{nF}} h^2 \Big|_{0.010\ \text{m}}^{0.110\ \text{m}}$$

To find the amount n of transferred electrons, we consider that the mercury transfers as Hg_2^+ and that the reaction proceeds until the Hg columns are of equal height. Thus the height of the higher column decreases by 50 cm. Taking the molar mass of Hg to be $\boldsymbol{M}_{\text{Hg}} = 200.6$ g mol^{-1}, we find that

$$\frac{\rho \cdot A \cdot \Delta h}{\boldsymbol{M}_{\text{Hg}}} = \frac{(13.546\ \text{g cm}^{-3})(1\ \text{cm}^2)(50\ \text{cm})}{200.6\ \text{g mol}^{-1}} = 3.376$$

moles is transferred. From the reaction each mole of mercury transferred corresponds to a half mole of electrons, so that

$$\boldsymbol{n} = \frac{1}{2}\frac{\rho \cdot A \cdot \Delta h}{\boldsymbol{M}_{\text{Hg}}} = \frac{1}{2}(3.376\ \text{mol}) = 1.688\ \text{mol}$$

Hence we find that

$$\begin{aligned}E &= -\frac{(13,546\ \text{kg m}^{-3})(10^{-4}\ \text{m}^2)(9.8\ \text{m s}^{-2})}{2(1.688\ \text{mol})(96,485\ \text{C mol}^{-1})}\left[(0.110\ \text{m})^2 - (0.010\ \text{m})^2\right] \\ &= 1.22 \times 10^{-7}\ \text{V} = 0.122\ \mu\text{V}\end{aligned}$$

This effect is small and can be neglected in most typical experiments.

P23.2 Consider the experiment depicted by Fig. 23.4. a) How does E change if KI is added to the $AgNO_3$ solution instead of KCl? The solubility product of AgI is $K_s = 1.6 \times 10^{-16}$. b) What will E be when a KCl solution (0.1 mol L^{-1}) is added to the left compartment and a KCN solution to the right compartment? The complex formation constant of $[Ag(CN)_2]^-$ is $K = 10^{21}$. In all cases the concentration of excess reagent is 0.1 mol L^{-1}.

Solution

a) Precipitate of AgI: $\hat{c}_{Ag^+} = 1.6 \times 10^{-16}/0.1 = 1.6 \times 10^{-15}$. Then

$$E = 59.1 \text{ mV} \cdot \log\left(\frac{0.1}{1.6 \times 10^{-15}}\right) = 815 \text{ mV}$$

which should be contrasted with the value of 461 mV in Example 23.2

b) Using the fact that

$$K = \frac{\hat{c}_{[Ag(CN)_2]^-}}{\hat{c}_{Ag^+} \cdot \hat{c}^2_{CN^-}} = 10^{21}$$

and that the Ag^+ is in equilibrium with the cyano complex, the right compartment has a concentration of

$$\hat{c}_{Ag^+} = \frac{0.1}{10^{21} \times (0.1)^2} = 10^{-20}$$

In the left compartment the concentration is $\hat{c}_{Ag^+} = 1.6 \times 10^{-9}$; this value is calculated in Example 23.2. Hence the cell potential will be

$$E = 59.1 \text{ mV} \cdot \log\left(\frac{10^{-20}}{1.6 \times 10^{-9}}\right) = -662 \text{ mV}$$

P23.3 Compare the calculated values for the saturated calomel and the silver chloride electrode with the experimental values and determine the activity coefficients of the ions. Compare this experimental activity coefficient to that you calculate from the extended Debye–Hü ckel model.

Solution

In Exercise E23.10 we showed how to calculate the activity coefficient of the chloride anion in the calomel electrode and found it to be $\gamma = 0.58$. We can proceed in a similar manner for the saturated calomel electrode, in which the chloride ion concentration is determined by the solubility product of KCl; namely $\hat{c}_{Cl^-} = 4.65$, hence we find

$$\hat{c}_{Hg_2^{2+}} = \frac{K_s}{\hat{c}^2_{Cl^-}} = \frac{2.038 \times 10^{-18}}{(4.65)^2} = 9.43 \times 10^{-20}$$

so that

$$E_{SCE} = E^{\ominus} + \frac{RT}{2F} \ln\left(\hat{c}_{Hg_2^{2+}}\right)$$
$$= [0.789 - 0.562] \quad \text{V} = 0.227 \text{ V}$$

If we write

$$E_{expt,SCE} = E_{ideal,SCE} - \frac{RT}{F} \ln(\gamma)$$

we find that

$$\ln(\gamma) = \frac{F}{RT}\left(E_{ideal,SCE} - E_{expt,SCE}\right)$$
$$= \frac{96,485 \text{ C mol}^{-1} \cdot (0.227 \text{ V} - 0.241 \text{ V})}{8.314 \text{ J mol}^{-1} \text{ K}^{-1} \cdot 298 \text{ K}} = -0.545$$

so that $\gamma = 0.580$. For comparison we calculate the activity coefficient using the Debye–Hückel formula,

$$\ln \gamma = -\frac{A \cdot z^2 \cdot \sqrt{I}}{1 + B \cdot R \cdot \sqrt{I}}$$

where $A = 1.175\ \text{mol}^{-1/2}\ \text{L}^{1/2}$, $B = 0.330 \times 10^{10}\ \text{mol}^{-1/2}\ \text{L}^{1/2}\ \text{m}^{-1}$, and we take $R = 500$ pm. We calculate I via $I = \frac{1}{2}\sum c_i z_i^2$ and find that

$$I = \frac{1}{2}\left[4.65 \cdot (1)^2 + 4.65 \cdot (1)^2 + 9.43 \times 10^{-20}\,(2)^2\right]\ \text{mol L}^{-1} = 4.65\ \text{mol L}^{-1}$$

Using this value of I gives

$$\ln \gamma = -\frac{1.175\ \text{mol}^{-1/2}\ \text{L}^{1/2} \cdot (1)^2 \cdot \sqrt{4.65\ \text{mol L}^{-1}}}{1 + 0.330 \times 10^{10}\ \text{mol}^{-1/2}\ \text{L}^{1/2}\ \text{m}^{-1} \cdot 500 \times 10^{-12}\ \ \text{m} \cdot \sqrt{4.65\ \text{mol L}^{-1}}}$$
$$= -0.556$$

leading to $\gamma = 0.574$ in excellent agreement with the experimental result.

For the Ag/AgCl electrode we have that $K_s = 1.56 \times 10^{-10}$, so that with $\hat{c}_{\text{Cl}^-} = 4.65$ Equation (23.39) becomes

$$E_{\text{AgCl}} = E^{\ominus} - \frac{RT}{F} \ln\left(\frac{\hat{c}_{\text{Cl}^-}}{K_{\text{AgCl}}}\right)$$
$$= [0.7996 - 0.6198]\ \ \text{V} = 0.1798\ \text{V}$$

Using

$$\ln(\gamma) = \frac{F}{RT}\left(E_{\text{ideal,Ag/AgCl}} - E_{\text{expt,Ag/AgCl}}\right)$$
$$= \frac{96,485\ \text{C mol}^{-1} \cdot (0.1798\ \text{V} - 0.197\ \text{V})}{8.314\ \text{J mol}^{-1}\ \text{K}^{-1} \cdot 298\ \text{K}} = -0.671$$

so that $\gamma = 0.511$. For comparison we calculate the activity coefficient using the Debye–Hückel result,

$$\ln \gamma = -\frac{A \cdot z^2 \cdot \sqrt{I}}{1 + B \cdot R \cdot \sqrt{I}}$$

where $A = 1.175\ \text{M}^{-1/2}$, $B = 0.330 \times 10^{10}\ \text{M}^{-1/2}\text{m}^{-1}$, and we take $R = 500$ pm. We calculate I via $I = \frac{1}{2}\sum c_i z_i^2$ and find that

$$I = \frac{1}{2}\left[4.65 \cdot (1)^2 + 4.65 \cdot (1)^2 + 3.35 \times 10^{-11}(1)^2\right]\ \text{mol L}^{-1} = 4.65\ \text{mol L}^{-1}$$

Using this value of I gives

$$\ln \gamma = -\frac{1.175\ \text{mol}^{-1/2}\ \text{L}^{1/2} \cdot (1)^2 \cdot \sqrt{4.65\ \text{mol L}^{-1}}}{1 + 0.330 \times 10^{10}\ \text{mol}^{-1/2}\ \text{L}^{1/2}\ \text{m}^{-1} \cdot 500 \times 10^{-12}\ \ \text{m} \cdot \sqrt{4.65\ \text{mol L}^{-1}}}$$
$$= -0.556$$

leading to $\gamma = 0.574$ in reasonable agreement with the experimental result.

Note that our analysis here and for the case of the SCE use the same value of R; however, this parameter is expected to depend on the ion sizes.

The activity coefficient of KCl solution does not change very strongly[2] for concentrations between 0.5 and 5 mol L^{-1}.

P23.4 The temperature dependence of the standard potential for a silver-silver chloride cell (versus the SHE) was determined by Harned and Ehlers (*J. Am. Chem. Soc.* 1933, **55**, 2179). Use these data to determine the enthalpy and entropy change for this cell.

T/K	273	283	293	303	313	323	333
$E^{\ominus}$/mV	236.34	231.26	225.51	219.12	212.00	204.37	196.20

2 H.S. Owen, B.B. Owen, *The Physical Chemistry of Electrolytic Solutions* (Reinhold, NY, 1950).

Solution

Figure P23.4 plots $E^{\ominus}$ versus T. We use the slope to obtain the entropy change (see Equation (23.16)).

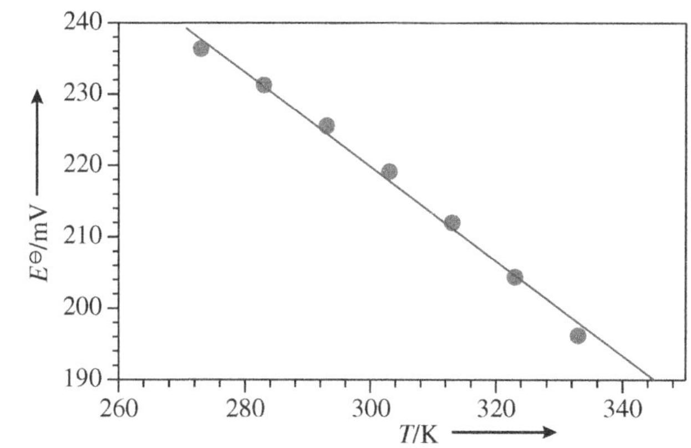

Figure P23.4 Standard potential $E^{\ominus}$ of a silver/silver chloride cell versus temperature T.

From the best fit line we obtain a slope of -6.70×10^{-4} V K^{-1} and this gives an entropy change of

$$\Delta_r S = \boldsymbol{F}\left(\frac{\partial E^{\ominus}}{\partial T}\right)_P$$
$$= -96,485\,\text{C mol}^{-1}\ 6.70 \times 10^{-4}\,\text{V K}^{-1} = -64.6\ \text{J K}^{-1}\ \text{mol}^{-1}$$

If we use the measured value of the potential at 303 K, we calculate

$$\Delta_r G = -\boldsymbol{F}E^{\ominus}$$
$$= -96,485\,\text{C mol}^{-1} \cdot (0.21912\,\text{V}) - 21.142\ \text{kJ mol}^{-1}$$

and hence the enthalpy change at 303 K is

$$\Delta_r H = \Delta_r G + T\Delta_r S$$
$$= -21,142\ \text{J mol}^{-1} - 303\ \text{K} \cdot 64.6\ \text{J K}^{-1}\ \text{mol}^{-1} = -40.7\ \text{kJ mol}^{-1}$$

P23.5 In the same study as that cited in P23.4, the workers measured the cell potential for a silver-silver chloride cell (versus the SHE) as a function of HCl concentration, at 298 K. Analyze these data to determine the activity of HCl and compare your result with the predictions of the Debye–Hückel model.

$\hat{m}$	E/V	$\hat{m}$	E/V
0.0001	0.69620	0.02	0.43022
0.0002	0.66083	0.05	0.38587
0.0005	0.61421	0.1	0.35239
0.001	0.57912	0.2	0.31871
0.002	0.54421	0.5	0.27229
0.005	0.49841	1	0.23328
0.01	0.46416	2	0.18634

Solution

To analyze these data we proceed in the manner used in Section 23.5.2. First we evaluate the quantity $E + 2\frac{RT}{F}\ln(\hat{m})$ and plot it as a function of $\sqrt{\hat{m}}$ in the region of small $\hat{m}$ so that Kohlrausch's law apply. From the intercept (limit of $\hat{m} \to 0$) we can determine the quantity E'. With E' in hand we can use Equation (23.60) to evaluate the activity coefficients at each molality. Figure P23.5a shows such a plot for the data and the best fit line has an intercept of 0.2227 V.

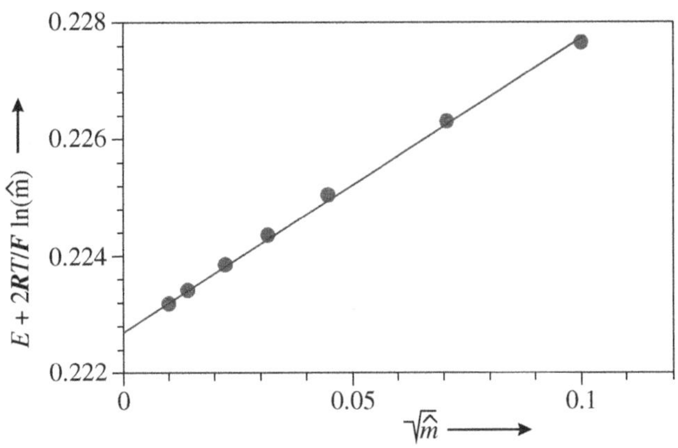

Figure P23.5a $E + 2\frac{RT}{F}\ln(\hat{m})$ versus $\sqrt{\hat{m}}$ for a Ag^+/AgCl cell for different HCl concentrations.

From Equation (23.60)

$$\ln \gamma_\pm = \frac{F}{2RT}(E' - E) - \ln \hat{m}$$

we can calculate $\gamma_\pm$. By way of example, for $\hat{m} = 0.1$ we find that

$$\ln \gamma_\pm = 19.472 \cdot (0.2227 - 0.35239) - \ln(0.1) = -0.2227$$

so that $\gamma_\pm = 0.800$. In a corresponding manner for the other molalities we find the data in the table.

$\hat{m}$	E/V	$\gamma_\pm$	$\hat{m}$	E/V	$\gamma_\pm$
0.0001	0.69620	0.991	0.02	0.43022	0.879
0.0002	0.66083	0.986	0.05	0.38587	0.834
0.0005	0.61421	0.978	0.1	0.35239	0.800
0.001	0.57912	0.968	0.2	0.31871	0.771
0.002	0.54421	0.955	0.5	0.27229	0.762
0.005	0.49841	0.932	1	0.23328	0.814
0.01	0.46416	0.908	2	0.18634	1.015

Figure P23.5b plots the experimentally determined activity coefficient and the prediction of the Debye–Hückel law (via Equation (23.63)). The circles are the experimental data and the light gray curve is the Debye–Hückel law prediction. We see that the experimental values go through a minimum at high molalities but that the Debye–Hückel values systematically decrease. The model agrees reasonably well with the data up to 0.01 molal or higher.

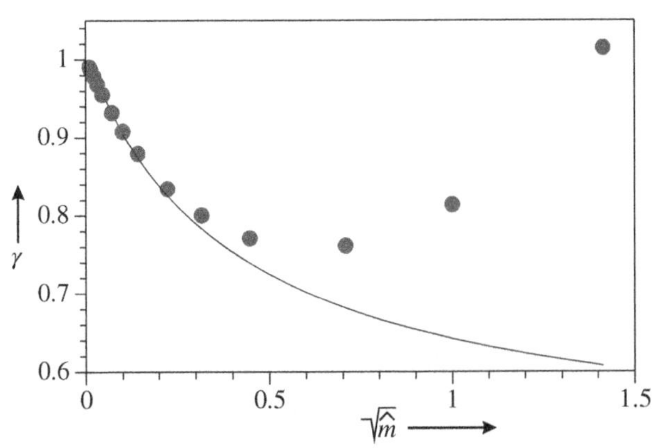

Figure P23.5b Activity coefficient γ versus $\sqrt{\hat{m}}$ for a Ag^+/AgCl cell for different HCl concentrations. Dark gray circles: experiment and light gray solid line: calculated.

P23.6 A variation of the silver-silver chloride electrode (Problems 23.4 and 23.5) can be used to determine the dissociation constant of acetic acid. In this case known amounts of CH_3COOH (molality $\hat{m}_1$, HAc), CH_3COONa (molality $\hat{m}_2$, NaAc), and NaCl (molality $\hat{m}_3$) are used for the acid solution, and the cell potential is measured as a function of concentration and temperature. For this case, show that the cell potential may be written as

$$E - E^{\ominus} + \frac{RT}{F}\ln\left(\frac{\hat{m}_{\mathrm{HAc}}\hat{m}_{\mathrm{Cl}}}{\hat{m}_{\mathrm{Ac}}}\right) = -\frac{RT}{F}\ln\left(\frac{\gamma_{\mathrm{HAc}}\gamma_{\mathrm{Cl}}}{\gamma_{\mathrm{Ac}}}\right) - \frac{RT}{F}\ln K_a$$

where K_a is the acid dissociation constant of acetic acid. At low ionic strength (infinite dilution) the ratio of the activity coefficients on the right-hand side goes to unity. Use the data in the table to determine the acid dissociation constant of acetic acid by extrapolating to the case of infinite dilution. Perform the analysis for each of the three temperatures (278, 293, and 308 K) and then analyze the enthalpy and entropy change for the ionization of acetic acid. Use the values $E^{\ominus}(278\ \mathrm{K}) = 0.23386$ V; $E^{\ominus}(293\ \mathrm{K}) = 0.2250$ V; and $E^{\ominus}(308\ \mathrm{K}) = 0.21591$ V. Compare your result to literature values.

$\hat{m}_1$	$\hat{m}_2$	$\hat{m}_3$	I	$E_{T=278}$	$E_{T=293}$	$E_{T=308}$
	mol kg^{-1}				V	
0.004779	0.004599	0.004896	0.00951	0.62392	0.63580	0.64722
0.012035	0.011582	0.012426	0.02403	0.60183	0.61241	0.62264
0.021006	0.020216	0.021516	0.04175	0.58855	0.59840	0.60792
0.04922	0.04737	0.05042	0.09781	0.56833	0.57699	0.58529
0.08101	0.07796	0.08297	0.16095	0.55667	0.56456	0.57213
0.09056	0.08716	0.09276	0.17994	0.55397	0.56171	0.56917

Solution

For this cell, we have the two half-cell reactions

$$H^+ + e^- \rightarrow \frac{1}{2}H_2 \qquad E_{H^+}$$
$$AgCl + e^- \rightarrow Ag + Cl^- \qquad E_{Ag^+/Ag}$$

and can write the cell potential as

$$E = E_{Ag^+/Ag} - E_{H^+}$$
$$= E^{\ominus}_{Ag+/Ag} - \frac{RT}{F}\ln(\hat{a}_{Cl^-}) - E^{\ominus}_{H+/H_2} - \frac{RT}{F}\ln(\hat{a}_{H^+})$$

The proton activity is determined by the acetic acid dissociation constant K_a, such that

$$\hat{a}_{H^+} = K_a\frac{\hat{a}_{\mathrm{HAc}}}{\hat{a}_{\mathrm{Ac}}}$$

Using this result with $E^{\ominus}_{H+/H_2} = 0$ and the definition of $\hat{a}_i = \hat{\gamma}_i \cdot \hat{m}_i$, we find that

$$E = E^{\ominus}_{Ag+/Ag} - \frac{RT}{F}\ln\left(\frac{\hat{a}_{\mathrm{HAc}}\hat{a}_{Cl^-}}{\hat{a}_{\mathrm{Ac}}}\right) - \frac{RT}{F}\ln(K_a)$$
$$= E^{\ominus}_{Ag+/Ag} - \frac{RT}{F}\ln\left(\frac{\hat{m}_{\mathrm{HAc}}\hat{m}_{Cl^-}}{\hat{m}_{\mathrm{Ac}}}\right) - \frac{RT}{F}\ln\left(\frac{\hat{\gamma}_{\mathrm{HAc}}\hat{\gamma}_{Cl^-}}{\hat{\gamma}_{\mathrm{Ac}}}\right) - \frac{RT}{F}\ln(K_a)$$

which rearranges to

$$E - E^{\ominus}_{Ag+/Ag} + \frac{RT}{F}\ln\left(\frac{\hat{m}_{\mathrm{HAc}}\hat{m}_{Cl^-}}{\hat{m}_{\mathrm{Ac}}}\right) = -\frac{RT}{F}\ln\left(\frac{\hat{\gamma}_{\mathrm{HAc}}\hat{\gamma}_{Cl^-}}{\hat{\gamma}_{\mathrm{Ac}}}\right) - \frac{RT}{F}\ln(K_a)$$

To find a value for K_a we consider the highly dilute limit, and plot

$$y = F\left(E - E^{\ominus}_{Ag+/Ag}\right)/(RT) + \ln\left(\frac{\hat{m}_{\mathrm{HAc}}\hat{m}_{Cl^-}}{\hat{m}_{\mathrm{Ac}}}\right) \qquad (23.1)$$

versus I in Fig. P23.6a. In the dilute limit these plots should be linear and the intercept at zero ionic strength will be equal to $-\ln(K_a)$.

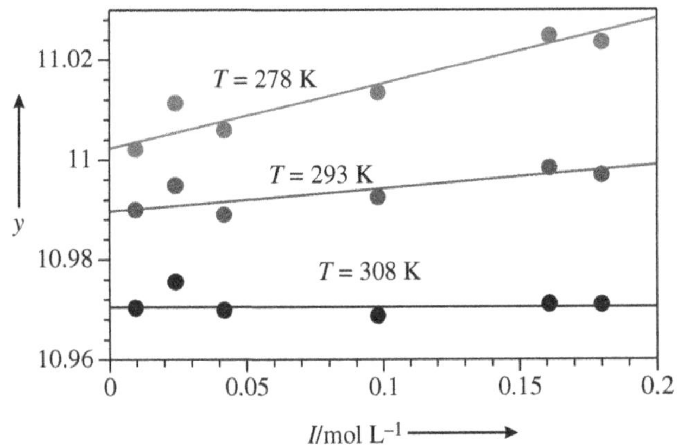

Figure P23.6a *y* in Equation (23.1) versus the ionic strength *I*.

From the intercepts we can determine the three different K_a values; namely

T/K	278	293	308
$\ln(K_a)$	−11.003	−10.990	−10.970
K_a	1.67×10^{-5}	1.69×10^{-5}	1.72×10^{-5}

A van't Hoff plot for these data in shown in Fig. P23.6b. The best fit slope is −85.6 K so that $\Delta_r H = 712$ kJ mol^{-1}. By using the relationship $\Delta_r G = -\boldsymbol{R}T \ln K_a = \Delta_r H - T\Delta_r S$, we find $\Delta_r S = 88.9$ J mol^{-1}K^{-1}.

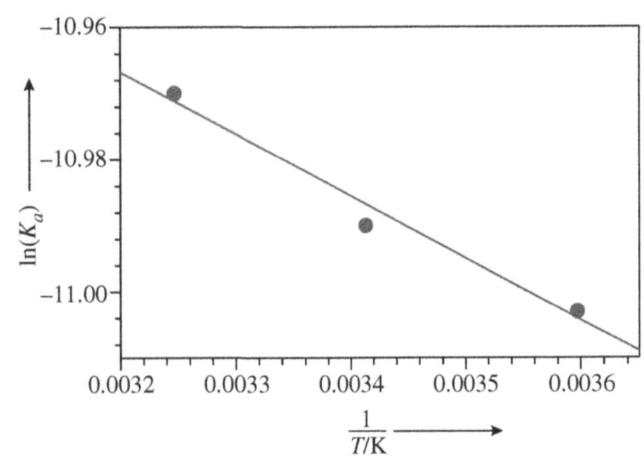

Figure P23.6b $\ln(K_a)$ versus $1/T$.

These finding are in reasonable agreement with the literature K_a value. See *Thermodynamics* by Lewis and Randall or H.S. Harned, R.W. Ehlers, *J. Am. Chem. Soc.* 1932, **54**, 1350.

P23.7 According to Equation (23.73) the work done by an idealized heat engine is

$$w_{\text{heat engine}} = \eta \cdot q_{\text{rev},2} = \left(1 - \frac{T_1}{T_2}\right) \cdot q_{\text{rev},2}$$

where T_1 and T_2 are the working temperatures of the heat engine, and $q_{\text{rev},2}$ is the heat supplied by the combustion of the fuel at temperature T_2. Compare $w_{\text{heat engine}}$ to the work done by performing the same chemical reaction in a fuel cell.

Solution

Let us consider the combustion of carbon

$$C + O_2 \rightarrow CO_2$$

in an ideal heat engine operating at the temperatures $T_1 = 300$ K and $T_2 = 800$ K. Then

$$q_{\text{rev},2} = -\Delta H^{\ominus}_{800} = 394 \text{ kJ mol}^{-1} \cdot \boldsymbol{n}_C$$

and the work done by the engine is

$$\left(1 - \frac{300}{800}\right) \times 394 \text{ kJ mol}^{-1} \cdot \boldsymbol{n}_C \text{ mol}^{-1} = 246 \text{ kJ mol}^{-1} \cdot \boldsymbol{n}_C$$

$\boldsymbol{n}_C$ is the amount of carbon being oxidized. In principle, the same process can be carried out in an electrochemical cell (fuel cell). The work done by an idealized electrochemical cell equals $-\Delta G$. Then

$$\begin{aligned} w_{\text{fuel cell}} &= -\Delta G^{\ominus} = -\left(\Delta H^{\ominus} - T\Delta S^{\ominus}\right) \\ &= \left(394 \times 10^3 + 298 \times 2.9\right) \text{ J mol}^{-1} \cdot \boldsymbol{n}_C \\ &= 395 \text{ J mol}^{-1} \cdot \boldsymbol{n}_C \qquad \text{at 298 K} \end{aligned}$$

Thus, an ideal fuel cell can generate 60% more work than can be obtained by burning coal in a heat engine.

P23.8 In Section 23.5.4, we discussed ion selective electrodes. A common issue with such devices is that they can respond to the presence of ions other than the desired one, i.e., the analyte. A common way to account for this behavior is through the empirical relationship

$$E = E^{\ominus} - \frac{\boldsymbol{RT}}{\boldsymbol{F}} \ln\left(a_1 + K \cdot a_2\right)$$

where a_1 is the activity of the analyte ion, a_2 is the activity of some other ion to which the electrode responds, and K is an empirical constant called a selectivity ratio. Provide a rationalization for this equation.

An iodide selective electrode, constructed using a silicone rubber membrane that is soaked in silver iodide solution, also responds to other anions in test solutions. The data in the table reports the change in electrode response between a 0.1 molar solution of KI and the corresponding 0.1 molar solution of some other anion. Use these data to determine the selectivity ratio for the electrode against the anions.

Anion	Br^-	Cl^-	SO_4^{2-}	PO_4^{3-}
ΔE/V	−0.140	−0.314	−0.451	−0.341

Solution

From the definition of ΔE we can write that

$$\begin{aligned} \Delta E &= E(\text{KI}) - E(\text{interferant}) \\ &= -\frac{\boldsymbol{RT}}{\boldsymbol{F}}\left[\ln\left(a_1\right) - \ln(K \cdot a_2)\right] = \frac{\boldsymbol{RT}}{\boldsymbol{F}} \ln\left(K \cdot \frac{a_2}{a_1}\right) \end{aligned}$$

Making the approximation that $a_1 = a_2$ for this test system, we find

$$\exp\left(\frac{\Delta E \cdot \boldsymbol{F}}{\boldsymbol{RT}}\right) = K$$

For the case of bromide we find that

$$K = \exp\left(\frac{-0.140 \text{ V} \cdot 96{,}485 \text{ C mol}^{-1}}{8.314 \text{ J mol}^{-1} \text{ K}^{-1} \cdot 298 \text{ K}}\right) = 4.29 \times 10^{-3}$$

Anion	Br^-	Cl^-	SO_4^{2-}	PO_4^{3-}
ΔE/V	−0.140	−0.314	−0.451	−0.341
K	$4.29 \cdot 10^{-3}$	4.90×10^{-6}	2.36×10^{-8}	1.71×10^{-6}

This electrode has an excellent selectivity for iodide ion (see G.A. Rechnitz et al., *Anal. Chem.* 1966, **38** 973).

P23.9 Consider the experimental reduction potentials and electron affinities for the aromatic hydrocarbons in the table. Determine if the observed reduction potential E (actually a half-wave potential measured in dimethylformamide) correlates with the electron affinity E_A. Use a thermodynamic cycle to explain your findings and discuss any approximations that are important for your analysis. Values are taken from the NIST Webbook and from S.L. Murov, I. Carmichael, G.L. Hug *Handbook of Photochemistry*, 2nd ed (Marcel Dekker, NY, 1993).

Molecule	E/V, versus SCE	E_A/eV
Pyrene	−2.09	0.58
Anthracene	−1.95	0.52
1,2-Benzanthracene	−2.0	0.63
Perylene	−1.67	0.97
Tetracene	−1.58	1.07

Solution

Figure P23.9a indicates a clear correlation. When all of the data are fit by a line the correlation coefficient of only 0.898, however the elimination of the anthracene value from the data set gives a correlation coefficient of 0.996.

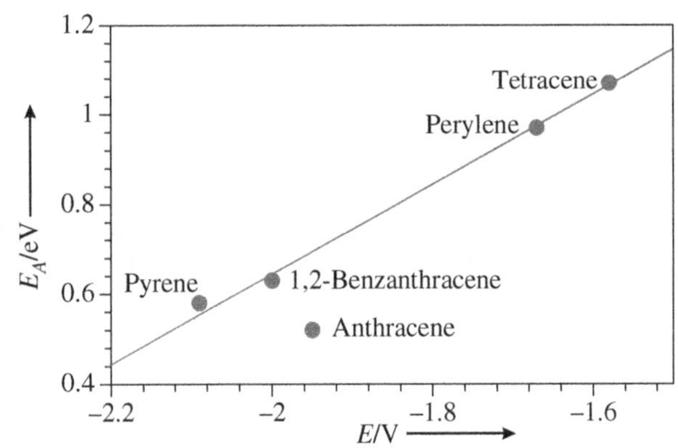

Figure P23.9a Electron affinity E_A versus reduction potential E for the molecules shown in the table.

The reduction potential is directly related to the Gibbs energy change, by way of the relation $\Delta G_{\text{reduction}} = -nFE$. Using this knowledge we can build a thermodynamic cycle (see Figure P23.9b) that shows the relationship between the Gibbs energy for reduction (hence the reduction potential) and the electron affinity. From this cycle we see the Gibbs energy for the reduction and the Electron Affinity (E.A.) are directly related to each other, but are offset by energies associated with the electrode's work function and the solvation of the molecule and molecular anion.

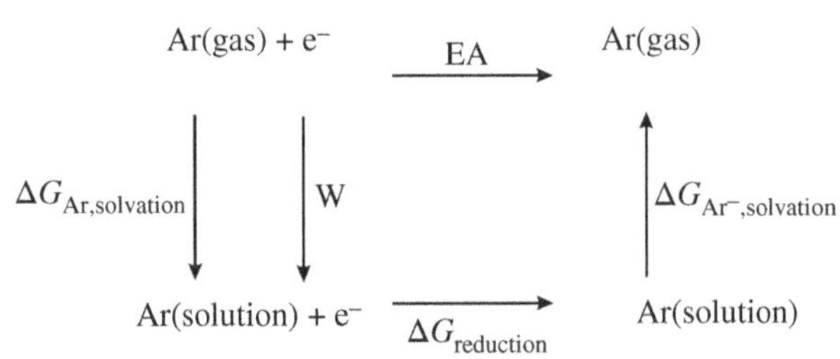

Figure P23.9b Thermodynamic cycle to show the relationship between the electron affinity *EA* and the reduction potential.

P23.10 Use the conductance data in the table to determine the limiting conductances of KCl and NaCl at T = 298 K.

$c/10^{-3}$ mol L^{-1}	0.50	1.0	2.0	5.0	10.0
Λ/cm^2 Ω^{-1} mol^{-1} for KCl	147.79	146.95	145.79	143.59	141.30
Λ/cm^2 Ω^{-1} mol^{-1} for NaCl	124.54	123.77	122.69	120.67	118.55

Solution

Figure P23.10 shows plots for the conductivity data versus the square root of the concentration for the KCl and the NaCl solution at T = 298 K. The limiting conductances can be determined from the intercepts and are Λ_0 = 149.5 cm^2 Ω^{-1} mol^{-1} for KCl and Λ_0 = 126.5 cm^2 Ω^{-1} mol^{-1} for NaCl.

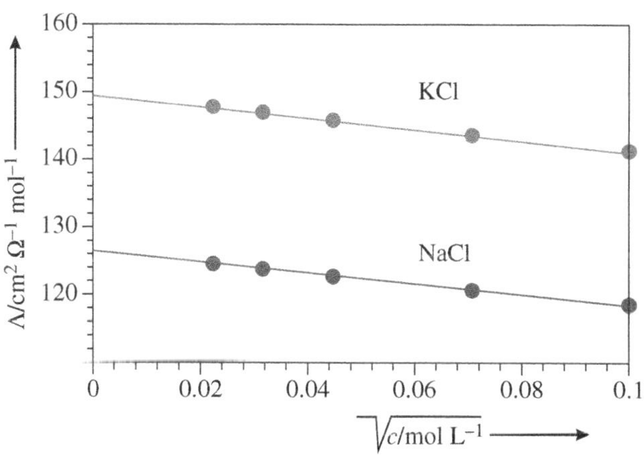

Figure P23.10 Molar conductivity Λ for KCl and NaCl versus the square root of the concentration c for T = 298 K.

We observe that the limiting conductance of NaCl is lower than that of KCl. Although the bare radius of the sodium ion is expected to be smaller than that of bare potassium (98 versus 133 pm), the sodium ion has a higher charge density and this leads to a higher solvation energy in solution and a stronger attraction to the water molecules. Hence, it will experience more drag as it moves through the water and have a higher effective (hydrodynamic) radius.

The additivity cannot be rigorously assessed with just these data, but if we assume that the limiting conductances are simple sums of the individual ion conductances, we find that

$$\Lambda_0(\text{KCl}) - \Lambda_0(\text{NaCl}) = \Lambda_0(\text{K}^+) - \Lambda_0(\text{Na}^+) = 23.4 \text{ cm}^2\ \Omega^{-1}\ \text{mol}^{-1}$$

The difference in the ion conductivities is 10–20% of the limiting molar conductances. On the basis of these calculations it is not possible to obtain conductance values for the separate ions K^+, Na^+, and Cl^-. This is possible only by using an additional experiment, namely, the measurement of the transport numbers of the ions.

P23.11 Analyze the Gibbs energy for ion solvation of the alkali metals in water by using the Born model with different measures of the ion's radius. In the first case, determine if the ion solvation energy is linear with $1/r_{\text{crystal}}$ and compare your best fit slope to the value predicted by the Born model (Section 23.4.2.5). In the second case, determine if the solvation energy is linear with $1/r_{\text{ion-H}_2\text{O}}$ and compare your best fit slope to the value predicted by the Born model. The Born model treats the solvent as dielectric continuum with a relative permittivity of $\hat{\varepsilon}_s$. Discuss this approximation in terms of the two different cavity radii used in your analysis and contrast it with what you would expect from a more physical/molecular perspective.

Ion	Cs^+	Rb^+	K^+	Na^+	Li^+
r_{crystal}/pm	184	163	152	116	88
$r_{\text{ion-H}_2\text{O}}$/pm^3	314	289	280	236	208
$\Delta_{\text{solv}}G^{\ominus}(\text{M}^+)$/kJ mol^{-1}	−278.2	−315.2	−333.8	−405.6	−505.8

Lastly, analyze the ion solvation energy by assuming that the Born model applies with an effective radius for each ion of $r_{\text{crystal}} + C$, where C is an adjustable parameter.

Solution

If we plot $\Delta_{\text{solv}}G^\ominus(\text{M}^+)$ versus $1/r_{\text{crystal}}$ and fit the data by a line, we find that the slope of the best fit line is 2.6 times shallower than the Born model prediction. If instead we choose the ion to oxygen distance (as given in the table), we find that the slope will be twice as steep as the Born model prediction. Even though the magnitude of the slopes disagrees, in each case we find that the solvation data appear to scale linearly with $1/r$, see Fig. P23.11.

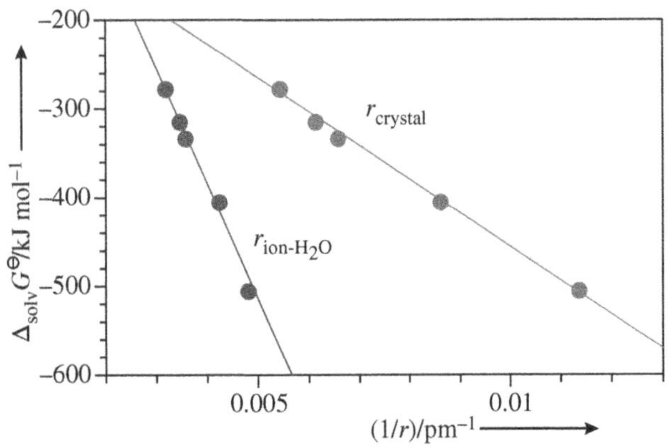

Figure P23.11 $\Delta_{\text{solve}}G^\ominus$ versus $1/r$, where r is the distance between two adjacent ions.

The continuum model assumes that the dielectric response of the solvent proceeds all the way up to the surface of the spherical charge given by the radius parameter. The crystal ionic radius is clearly too small a measure for the dielectric response of the water molecules, because their electron clouds keep their center of charge and dipole moment displaced away from the edge of the ion's cavity. The ion to water distance is a somewhat better measure because it accounts for the displacement of the water from the ion's surface; however, it is a distance which largely includes the ion and the water molecules.

A more appropriate distance for the ionic radius is likely to lie between these two limits. If we rearrange the Born model expression we find that

$$-\frac{N_A}{\Delta_{\text{solv}}G^\ominus(\text{M}^+)}\frac{e^2}{8\pi\varepsilon_0}\left(1-\frac{1}{\hat{\varepsilon}_s}\right) = r_{\text{crystal}} + C$$

Hence we can obtain a value of C for each ion; it ranges from 64 to 74 pm and has an average value of 68 pm. This value is somewhat larger than the 54 pm value used by the analysis in the text. The 54 pm value was derived from a molecular-based treatment (treats the solvent molecules as discrete) of the fluid.

P23.12 In Exercise E23.14 we found that the speed of an ion depends inversely on η, the viscosity of the solvent. Use the experimental data for Λ_0 for KCl (molar conductivity at infinite dilution) and η for water to prove this dependency.

T/K	288	298	308	318
Λ_0/cm^2 Ω^{-1} mol^{-1}	120.8	149.6	181.9	214.0
η/g s^{-1} cm^{-1}	1.138×10^{-2}	0.890×10^{-2}	0.719×10^{-2}	0.596×10^{-2}

Solution

in Equation (23.80) we found that the speed υ of an ion in an electric field of field strength ϕ/l is

$$\upsilon = \frac{ze}{6\pi\eta r}\frac{\phi}{l}$$

3 $r_{\text{ion-H}_2\text{O}}$ is an empirically determined mean distance between the metal ion and the first shell of the water molecules; see Y. Marcus, *Chem. Rev.* 1988, **88** 1475.

where ze is the charge of the ion, η is the viscosity of the solvent, and r is the radius of the charge sphere around the ion. According to Equations (23.87) and (23.91) the molar conductivity Λ is

$$\Lambda = \boldsymbol{F}\frac{ze}{6\pi\eta r}$$

or

$$\Lambda_{\text{ionpair}} = \boldsymbol{F}\frac{e}{6\pi\eta r_+}z^+ + \boldsymbol{F}\frac{e}{6\pi\eta r_-}z^-$$
$$= \boldsymbol{F}\frac{e}{6\pi\eta}\frac{z^+}{r_+} + \boldsymbol{F}\frac{e}{6\pi\eta r_-}\frac{z^-}{r_-} = \boldsymbol{F}\frac{e}{6\pi\eta}\left(\frac{z^+}{r_+} + \frac{z^-}{r_-}\right)$$

for an ion pair with charges z^+e and z^-e. In our example of KCl we have $z^+ = z^- = 1$, thus

$$\Lambda_{\text{KCl}} = \boldsymbol{F}\frac{e}{6\pi\eta}\left(\frac{1}{r_+} + \frac{1}{r_-}\right) \tag{23.2}$$

Note that this relation is valid only for the case that the ions are moving independently, that is, if the salt is sufficiently diluted ($\Lambda = \Lambda_0$). To prove this relation we plot $\Lambda_{0,\text{KCl}}$ as a function of $1/\eta$ in Fig. P23.12, where $\Lambda_{0,\text{KCl}}$ and η have been measured at different temperatures T.

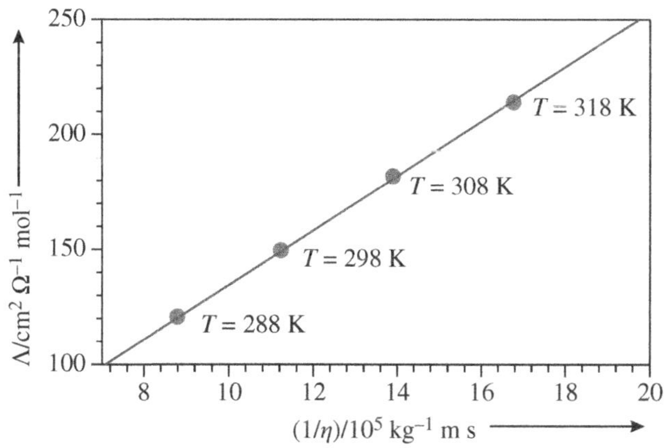

Figure P23.12 Conductivity $\Lambda_{0,\text{KCl}}$ as a function of $1/\eta$ at different temperatures T.

The plot shows an excellent straight line with a slope of

$$12.1 \times 10^{-6}\ \text{cm}^2\ \Omega^{-1}\ \text{mol}^{-1}\ \text{kg s}^{-1}\ \text{m}^{-1} = 12.1 \times 10^{-6}\ \text{C}^2\ \text{m}^{-1}\ \text{mol}^{-1}$$

According to Equation (23.2) the theoretical slope should be

$$\textit{slope} = \boldsymbol{F}\frac{e}{6\pi}\left(\frac{1}{r_+} + \frac{1}{r_-}\right)$$

and we can calculate

$$\left(\frac{1}{r_+} + \frac{1}{r_-}\right) = \textit{slope}\frac{6\pi}{\boldsymbol{F}e} = \frac{12.1 \times 10^{-6}\ \text{C}^2\ \text{m}^{-1}\ \text{mol}^{-1}\ 6\pi}{96,485 \times 1.602 \times 10^{-19}\ \text{C}^2\ \text{mol}^{-1}}$$
$$= 1.48 \times 10^{10}\ \text{m}^{-1} = 0.0148\ \text{pm}^{-1}$$

The literature values for the ionic radii in KCl are

$$r_{K^+} = 150\ \text{pm}, r_{\text{Cl}^-} = 150\ \text{pm}$$

Thus

$$\left(\frac{1}{r_+} + \frac{1}{r_-}\right) = 0.0133\ \text{pm}^{-1}$$

in reasonable agreement with our result.

24

Chemical Kinetics

24.1 Exercises

E24.1 Many organic reactions, which are used in laboratory synthesis, have activation barriers around 50 kJ mol^{-1}. Use the Arrhenius equation with a frequency factor of 10^{13} s^{-1} and a temperature of 300 K, to estimate the rate of reaction for a few different activation barriers: 30, 50, 70, and 90 kJ mol^{-1}. Which of these reactions would be most convenient to operate?

Solution
Using the Arrhenius equation we can evaluate the rate constants for these different barriers. For example, the 30 kJ mol^{-1} barrier gives

$$k_r = 10^{13}\ \text{s}^{-1} \cdot \exp\left(-\frac{30\ \text{kJ mol}^{-1}}{8.314\ \text{J mol}^{-1}\ \text{K}^{-1} \cdot 300\ \text{K}}\right) = 6.0 \times 10^{7}\ \text{s}^{-1}$$

In a corresponding way, we can calculate rate constants for the other activation barriers; these values are reported in the table.

$E_{a,m}$/kJ mol^{-1}	30	50	70	90
k_r/s^{-1}	6.0×10^{7}	2.0×10^{4}	6.5	2.1×10^{-3}

The reaction with the 90 kJ mol^{-1} barrier would be quite slow and not convenient to operate; however, the 70 and 50 kJ mol^{-1} barrier reactions would proceed at rapid but quite controllable rates at room temperature.

E24.2 The molar activation energy for the Diels–Alder reaction (ethylene reacting with 1,3-butadiene) is $E_{a,m}$ = 115 kJ mol^{-1}

$$H_2C{=}CH_2 + H_2C{=}CH{-}CH{=}CH_2 \rightarrow C_6H_{10}$$

What is the probability of the reactants exceeding this energy at T = 760 K? Calculate the rate constant using the collision theory model. Using the experimental value of k_r = 0.384 L mol^{-1} s^{-1}, calculate the steric factor.

Solution
The probability $\mathcal{P}$ of a successful collision is determined by the activation term

$$\mathcal{P} = \exp\left(-\frac{115\ \text{kJ mol}^{-1}}{8.314\ \text{J mol}^{-1}\ \text{K}^{-1} \cdot 760\ \text{K}}\right) = 1.2 \times 10^{-8}$$

To calculate the collision theory rate constant, we use Equations (24.19) and (24.20) to write

$$k_r = \sqrt{\frac{8kT}{\pi\mu}} \cdot \pi\left(r_A + r_B\right)^2 \cdot \boldsymbol{N}_A \cdot \alpha \cdot \exp\left(-\frac{E_{a,m}}{\boldsymbol{R}T}\right)$$

Solutions Manual for Principles of Physical Chemistry, Third Edition. Edited by Hans Kuhn, David H. Waldeck, and Horst-Dieter Försterling.

Using the parameters $\mu = 3.07 \times 10^{-26}$ kg, $(r_A + r_B) = 542$ pm we find

$$k_r = 933 \text{ m s}^{-1} \cdot \pi \cdot (542 \times 10^{-12} \text{ m})^2 \cdot 6.022 \times 10^{23} \text{ mol}^{-1} \cdot \alpha \cdot (1.2 \times 10^{-8})$$
$$= \alpha \cdot 6.22 \frac{\text{m}^3}{\text{mol s}} = \alpha \cdot 6.22 \times 10^3 \frac{\text{L}}{\text{mol s}}$$

Comparison of this result with the experimental value allows us to calculate the steric factor, namely

$$\alpha = \frac{0.384 \text{ L mol}^{-1} \text{ s}^{-1}}{6.22 \times 10^3 \text{ L mol}^{-1} \text{ s}^{-1}} = 6.2 \times 10^{-5}$$

Comparison of this steric factor with that found in the text for this reaction at 800 K implies that it has a significant temperature dependence.

E24.3 Perform the analysis of the Lindemann–Hinshelwood mechanism but remove the restriction that $k_1 \approx k_3$ and $k_{-1} \approx k_{-3}$.

Solution

Following the derivation in Section 24.7.2 for the L–H mechanism

$$\text{A} + \text{A} \underset{k_{-1}}{\overset{k_1}{\rightleftarrows}} \text{A}^* + \text{A}$$

$$\text{A}^* \overset{k_2}{\rightarrow} \text{B}$$

$$\text{A} + \text{B} \underset{k_{-3}}{\overset{k_3}{\rightleftarrows}} \text{A}^* + \text{B}$$

we must solve the equation

$$0 = k_1 \cdot c_A^2 - k_{-1} \cdot c_A \cdot c_{A^*} - k_2 \cdot c_{A^*} + k_3 \cdot c_A \cdot c_B - k_{-3} \cdot c_B \cdot c_{A^*}$$

Isolating c_{A^*}, we find

$$c_{A^*} = \frac{k_1 \cdot c_A^2 + k_3 \cdot c_A \cdot c_B}{k_{-1} \cdot c_A + k_2 + k_{-3} \cdot c_B}$$

Using the initial condition that $c_A + c_B = c_{A,0}$, we can eliminate c_B and find

$$c_{A^*} = \frac{k_1 \cdot c_A^2 + k_3 \cdot c_A \cdot (c_{A,0} - c_A)}{(k_{-1} \cdot c_A + k_2 + k_{-3} \cdot (c_{A,0} - c_A))} = c_A \left[\frac{(k_1 - k_3) \cdot c_A + k_3 \cdot c_{A,0}}{(k_{-1} - k_{-3}) \cdot c_A + k_2 + k_{-3} \cdot c_{A,0}} \right]$$

Hence the rate law becomes

$$\frac{dc_B}{dt} = k_2 \cdot c_{A^*} = k_2 \cdot c_A \left[\frac{(k_1 - k_3) \cdot c_A + k_3 \cdot c_{A,0}}{(k_{-1} - k_{-3}) \cdot c_A + k_2 + k_{-3} \cdot c_{A,0}} \right]$$

This expression shows that the rate law is not first order in the concentration c_A unless $k_1 = k_3$ and $k_{-1} = k_{-3}$. If we consider the limit where $k_1 = k_3$ and $k_{-1} = k_{-3}$, we find that

$$\frac{dc_B}{dt} = c_A \left[\frac{k_2 \cdot k_1 \cdot c_{A,0}}{k_2 + k_{-1} \cdot c_{A,0}} \right] = \left[\frac{k_2 \cdot k_1 \cdot c_{A,0}}{k_2 + k_{-1} \cdot c_{A,0}} \right] (c_{A,0} - c_B)$$

which is identical to Equation (24.110).

E24.4 Reaction Leading to Equilibrium.

$$\text{A} + \text{B} \underset{k_{-1}}{\overset{k_1}{\rightleftarrows}} \text{C} + \text{D}$$

For the special case of $k_1 = k_{-1}$, integrate the differential equation (24.67)

$$\frac{dc_A}{dt} = -k_1 \cdot c_A^2 + k_{-1} \cdot (c_{A,0} - c_A)^2$$

Solution

By inserting $k_1 = k_{-1} = k_r$ and separating the variables we obtain

$$\frac{dc_A}{dt} = k_r \cdot c_{A,0}^2 - 2k_r \cdot c_{A,0} \cdot c_A$$

which simplifies to

$$\frac{dc_A}{2c_A - c_{A,0}} = -k_r c_{A,0} \cdot dt$$

Then

$$\int \frac{dc_A}{2c_A - c_{A,0}} = \frac{1}{2} \cdot \ln(2c_A - c_{A,0}) = -k_r c_{A,0} t + constant$$

At $t = 0$ the concentration c_A equals $c_{A,0}$, thus $constant = \frac{1}{2} \ln c_{A,0}$ and

$$\ln\left(\frac{2c_A - c_{A,0}}{c_{A,0}}\right) = -2k_r c_{A,0} t \quad \text{or} \quad \frac{2c_A - c_{A,0}}{c_{A,0}} = e^{-2k_r c_{A,0} t}$$

Then

$$c_A = \frac{1}{2} c_{A,0} \cdot \left(1 + e^{-2k_r c_{A,0} t}\right)$$

For $t \to \infty$ (equilibrium) we obtain $c_A = \frac{1}{2} c_{A,0}$.

E24.5 Derive Equation (24.146) for the inhibited enzyme reaction.

Solution

Here we consider the reactions

$$E + S \underset{k_{-1}}{\overset{k_1}{\rightleftarrows}} (ES) \overset{k_2}{\rightarrow} P + E$$

and

$$E + I \underset{k'_{-1}}{\overset{k'_1}{\rightleftarrows}} (EI)$$

for which we wish to find

$$\frac{dc_P}{dt} = k_2 \cdot c_{ES}$$

In the steady state approximation we use

$$\frac{dc_{ES}}{dt} = 0 = k_1 \cdot c_E \cdot c_S - \left(k_2 + k_{-1}\right) \cdot c_{ES}$$

so that

$$\left(k_2 + k_{-1}\right) \cdot c_{ES} = k_1 \cdot c_E \cdot c_S$$
$$= k_1 \cdot c_S. \left[c_{E,0} - c_{ES} - c_{EI}\right]$$

where we have applied the constraint that enzyme is conserved. Solving for c_{ES} we find that

$$c_{ES} = \frac{k_1 \cdot c_S \left[c_{E,0} - c_{EI}\right]}{k_2 + k_{-1} + k_1 \cdot c_S}$$

In addition we use the condition

$$\frac{dc_{EI}}{dt} = 0 = k'_1 \cdot c_E \cdot c_I - k'_{-1} \cdot c_{EI}$$

which leads to

$$k'_{-1} \cdot c_{EI} = k'_1 \cdot c_I \cdot \left[c_{E,0} - c_{ES} - c_{EI}\right]$$

so that

$$c_{EI} = \frac{k'_1 \cdot c_I}{k'_{-1} + k'_1 \cdot c_I} \cdot \left[c_{E,0} - c_{ES}\right] = K' \cdot \left[c_{E,0} - c_{ES}\right]$$

A result that defines the parameter K'.

With this result in hand we can substitute into our expression for c_{ES}, and find that

$$c_{ES} = \frac{k_1 \cdot c_S \cdot c_{E,0}\left(1 - K'\right)}{k_2 + k_{-1} + k_1 \cdot c_S \cdot (1 - K')}$$

Now we can evaluate the rate as

$$\frac{dc_P}{dt} = k_2 \cdot c_{ES} = \frac{k_2 \cdot k_1 \cdot c_S \cdot c_{E,0}\left(1 - K'\right)}{k_2 + k_{-1} + k_1 \cdot c_S \cdot (1 - K')}$$

To compare to Equation (24.146) of the text, we rearrange this expression and write

$$\frac{dc_P}{dt} = k_2 \cdot c_{ES} = \frac{k_2 \cdot c_S\left(1 - K'\right)}{\frac{k_2+k_{-1}}{k_1} + c_S \cdot (1 - K')} \cdot c_{E,0}$$

$$= \frac{k_2 \cdot c_S\left(1 - K'\right)}{K_m + c_S \cdot (1 - K')} \cdot c_{E,0} = \frac{k_2 \cdot c_S}{\frac{K_m}{(1-K')} + c_S} \cdot c_{E,0}$$

In the limit of $c_I = 0$, $K' = 0$ and this result reproduces Equation (24.137) in the text. To compare to Equation (24.146), we convert

$$\frac{1}{(1 - K')} = \frac{1}{1 - \frac{k'_1 \cdot c_I}{k'_{-1} + k'_1 \cdot c_I}} = \frac{k'_{-1} + k'_1 \cdot c_I}{k'_{-1} + k'_1 \cdot c_I - k'_1 \cdot c_I}$$

$$= \frac{k'_{-1} + k'_1 \cdot c_I}{k'_{-1}} = 1 + \frac{k'_1 \cdot c_I}{k'_{-1}} = 1 + \frac{c_I}{K_I}$$

and we obtain Equation (24.146)

$$\frac{dc_P}{dt} = \frac{k_2 \cdot c_S}{c_S + K'_m} \cdot c_{E,0} \text{ with } K'_m = K_m \cdot \left(1 + \frac{c_I}{K_I}\right)$$

E24.6 Consider the reaction

$$2NO + 2H_2 \rightarrow N_2 + 2H_2O$$

The initial rate of this reaction was measured by following the total pressure change of the system [see C.N. Hinshelwood, T.E. Green, *J. Chem. Soc.* 1926, **730**]. Use the data below (taken at 1099 K) for this reaction to determine the rate law.

$P_{H_2}(t = 0)$/atm	0.380	0.270	0.193	0.526	0.526	0.526
$P_{NO}(t = 0)$/ atm	0.526	0.526	0.526	0.472	0.395	0.200
initial rate/10^{-4} atm s^{-1}	2.11	1.45	1.04	1.97	1.36	0.33

Solution

From the data given in the table we see that we can evaluate the initial rate for a fixed initial pressure of P_{H_2} at 0.526 atm, and we can perform a corresponding analysis for the case where the initial pressure of NO is 0.526 atm. We start with the rate equation

$$rate = k \cdot P_{H_2}^x \cdot P_{NO}^y$$

To obtain the reaction orders x and y we consider

$$\ln \frac{rate}{rate_1} = const \cdot x \cdot \ln \frac{P_{H_2}}{P_{H_2,1}} \quad \text{for } P_{NO} = 0.256 \text{ atm}$$

and

$$\ln \frac{rate}{rate_1} = const \cdot y \cdot \ln \frac{P_{NO}}{P_{NO,1}} \quad \text{for } P_{H_2} = 0.256 \text{ atm}$$

These expressions are plotted in Fig. E24.6. For the plot for P_{H_2} we find a slope of $x = 1.04$, and for the corresponding plot for P_{NO} we find a slope of $y = 2.08$. Hence the rate law is first order in hydrogen and second order in NO; namely

$$rate = k \cdot P_{H_2} \cdot P_{NO}^2$$

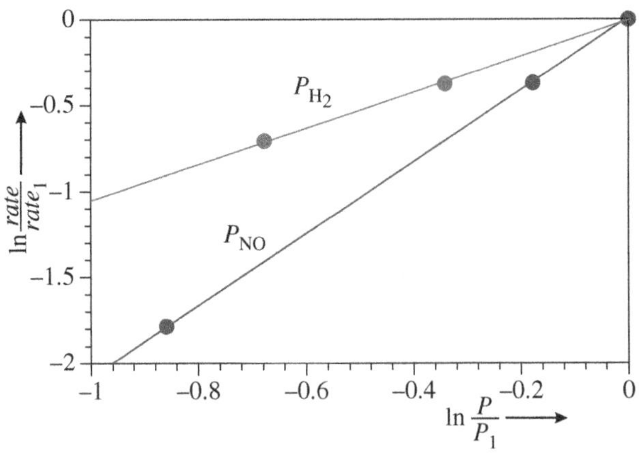

Figure E24.6 ln($rate/rate_1$) versus ln(P/P_1) for H_2 (dark gray) and NO (light gray).

E24.7 Consider the reaction

$$O + HCl \rightarrow OH + Cl$$

Measured rate constants for this reaction are $k_r = 7.83 \times 10^4$ L mol^{-1}s^{-1} at 293 K and $k_r = 1.00 \times 10^6$ L mol^{-1}s^{-1} at 403 K [see R.H. Brown, I.W.M. Smith, *Int. J. Chem. Kinetics VII*, 1975, **301**]. Perform an Arrhenius analysis and find the activation energy and the pre-exponential factor for the reaction. Use your Arrhenius parameters to compute the rate constant at 298 K.

Solution

Because we only have two data points, we directly use the Arrhenius equation in the form

$$\ln\left(\frac{k_r}{A}\right) = \frac{-E_{a,m}}{\boldsymbol{R}T}$$

We can find the activation energy by examining the ratio of the rate constants; namely

$$\ln\left(\frac{k_r(T_1)}{k_r(T_2)}\right) = \frac{-E_{a,m}}{\boldsymbol{R}}\left(\frac{1}{T_1} - \frac{1}{T_2}\right)$$

Hence from the data given we find that

$$-2.547 = \frac{-E_{a,m}}{8.314 \text{ J mol}^{-1}\text{ K}^{-1}}\left(9.32 \times 10^{-4}\text{ K}^{-1}\right)$$

so that $E_{a,m} = 22.73$ kJ mol^{-1}.

To find A we use this activation energy with one of the rate constants and find from

$$k_r = A \cdot \exp\left(\frac{-E_{a,m}}{\boldsymbol{R}T}\right)$$

that

$$\begin{aligned} A &= k_r \cdot \exp\left(\frac{E_{a,m}}{\boldsymbol{R}T}\right) \\ &= \left(7.83 \times 10^4\text{ L mol}^{-1}\text{ s}^{-1}\right) \cdot \exp\left(\frac{22,730\text{ J mol}^{-1}}{8.314\text{ J mol}^{-1}\text{ K}^{-1} \cdot 293\text{ K}}\right) \\ &= 8.83 \times 10^8\text{ L mol}^{-1}\text{ s}^{-1}\,. \end{aligned}$$

E24.8 Use the simple collision theory to calculate the frequency factor and rate constant for the reaction

$$D + H_2 \rightarrow HD + H$$

at 298 K. The measured[1] molar activation energy is $E_{a,m} = \boldsymbol{N}_A \cdot 5.28 \times 10^{-20}$ J = 31.8 kJ mol^{-1}.

1 J.P. Toennies, J. Arnold, J. Wolfrum, *Ber. Bunsenges Phys. Chem.* 1990, **94**, 1231.

Solution

The activation factor is $e^{-E_{a,m}/RT} = 2.7 \times 10^{-6}$ at 298 K, corresponding to a probability of

$$\mathcal{P} = 2.67 \times 10^{-6}$$

About 10^6 collisions have to occur before two molecules react with one another. Using $m_D = m_{H_2} = 3.3 \times 10^{-27}$ kg ($\mu = 1.6 \times 10^{-27}$ kg), $r_D = 128$ pm, and $r_{H_2} = 138$ pm, we calculate

$$A_f = 2.6 \times 10^3 \text{ m s}^{-1} \cdot \pi \cdot 7.1 \times 10^{-20} \text{ m}^2 \cdot 6.02 \times 10^{23} \text{ mol}^{-1} \cdot \alpha$$
$$= 3.5 \times 10^8 \text{ m}^3 \text{ mol}^{-1} \text{ s}^{-1} \cdot \alpha = 3.5 \times 10^{11} \text{ L mol}^{-1} \text{ s}^{-1} \cdot \alpha$$

for $T = 298$ K. Hence the rate constant is

$$k_r = 3.5 \times 10^{11} \text{ L mol}^{-1} \text{ s}^{-1} \cdot \alpha \cdot 2.7 \times 10^{-6} = 9.5 \times 10^5 \cdot \alpha \text{ L mol}^{-1} \text{ s}^{-1}$$

The experimental value[2] is 1.68×10^5 L mol^{-1} s^{-1}, which suggests that the steric factor α is 0.2. This small value suggests that details of the collision geometry are important.

E24.9 For the reaction

$$H^+ + OH^- \underset{k_{-1}}{\overset{k_1}{\rightleftharpoons}} H_2O$$

a value $k_r = 2.9 \times 10^4$ s^{-1} is obtained from a temperature jump experiment with $c_{H^+} = c_{OH^-} = 10^{-7}$ mol L^{-1} at $T = 298$ K. Given that the ionization constant of water is

$$K_w = \hat{c}_{H^+} \cdot \hat{c}_{OH^-} = 1.0 \times 10^{-14}$$

determine the forward and back reaction rate constants.

Solution

From Equation (24.178) we know that

$$k_r = k_1 \cdot (c_{H^+} + c_{OH^-}) + k_{-1}$$

On the other hand, Equation (24.181) is

$$\frac{k_1}{k_{-1}} \cdot c^\ominus = \frac{1}{\hat{c}_{H^+} \cdot \hat{c}_{OH^-}} = \frac{1}{K_w} \quad \text{or} \quad k_{-1} = k_1 \cdot c^\ominus K_w$$

Thus

$$k_r = k_1 \cdot (c_{H^+} + c_{OH^-}) + k_1 \cdot c^\ominus K_w$$
$$= k_1 \cdot \left[(c_{H^+} + c_{OH^-}) + c^\ominus K_w\right]$$

We resolve for k_1

$$k_1 = \frac{k_r}{(c_{H^+} + c_{OH^-}) + c^\ominus K_w} = \frac{2.9 \times 10^4 \text{ s}^{-1}}{2 \cdot 10^{-7} \text{ mol L}^{-1} + 1 \text{ mol L}^{-1} \cdot 1 \times 10^{-14}} = 1.45 \times 10^{11} \text{ L mol s}^{-1}$$

and for k_{-1}

$$k_{-1} = k_1 \cdot c^\ominus K_w = 1.45 \times 10^{11} \text{ L mol} \cdot 1 \text{ mol L}^{-1} \cdot 1 \times 10^{-14} = 1.45 \times 10^{-3} \text{ s}^{-1}$$

E24.10 Parallel reactions:

$A+B \rightarrow C$	main reaction
$2A \rightarrow D$	side reaction

where B is assumed to be in large excess. Solve the differential rate equation for A

$$\frac{dc_A}{dt} = -k_1' \cdot c_A - 2k_2 \cdot c_A^2$$

that is found below Equation (24.78).

2 A.A. Westenberg, N. de Haas, *J. Chem. Phys.* 1967, **47**, 1393.

Solution

We separate the variables and integrate

$$\int \frac{\mathrm{d}c_A}{k_1' \cdot c_A + 2k_2 \cdot c_A^2} = -\int \mathrm{d}t$$

According to Appendix B we can use the general integral

$$\int \frac{\mathrm{d}x}{ax^2 + bx} = \frac{1}{b} \ln\left(\frac{ax}{ax+b}\right) + const$$

so that

$$\int \frac{\mathrm{d}c_A}{k_1' \cdot c_A + 2k_2 \cdot c_A^2} = \frac{1}{k_1'} \ln\left(\frac{2k_2 c_A}{2k_2 c_A + k_1'}\right) + const$$

and our equation becomes

$$\frac{1}{k_1'} \ln\left(\frac{2k_2 c_A}{2k_2 c_A + k_1'}\right) = -t + const$$

For $t = 0$ we have $c_A = c_{A,0}$, and we can solve for the constant

$$const = \frac{1}{k_1'} \ln\left(\frac{2k_2 c_{A,0}}{2k_2 c_{A,0} + k_1'}\right)$$

and obtain

$$\frac{1}{k_1'} \ln\left(\frac{2k_2 c_A}{2k_2 c_A + k_1'}\right) = -t + \frac{1}{k_1'} \ln\left(\frac{2k_2 c_{A,0}}{2k_2 c_{A,0} + k_1'}\right)$$

By rearranging, we find that

$$\ln\left(\frac{c_A\left(2k_2 c_{A,0} + k_1'\right)}{c_{A,0}\left(2k_2 c_A + k_1'\right)}\right) = -k_1' t$$

By exponentiating both sides, we can write

$$\frac{c_A\left(2k_2 c_{A,0} + k_1'\right)}{c_{A,0}\left(2k_2 c_A + k_1'\right)} = \mathrm{e}^{-k_1' t}$$

Now we wish to solve for c_A. First we rearrange the expression

$$c_A\left(2k_2 c_{A,0} + k_1'\right) = c_{A,0}\left(2k_2 c_A + k_1'\right) \cdot \mathrm{e}^{-k_1' t} = c_{A,0} 2k_2 c_A \cdot \mathrm{e}^{-k_1' t} + c_{A,0} k_1' \cdot \mathrm{e}^{-k_1' t}$$

Then we factor c_A

$$c_A\left(2k_2 c_{A,0} + k_1' - c_{A,0} 2k_2 \cdot \mathrm{e}^{-k_1' t}\right) = c_{A,0} k_1' \cdot \mathrm{e}^{-k_1' t}$$

and isolate it to find

$$c_A = \frac{c_{A,0} k_1' \cdot \mathrm{e}^{-k_1' t}}{2k_2 c_{A,0} + k_1' - c_{A,0} 2k_2 \cdot \mathrm{e}^{-k_1' t}} = \frac{c_{A,0} k_1' \cdot \mathrm{e}^{-k_1' t}}{2k_2 c_{A,0}\left(1 - \mathrm{e}^{-k_1' t}\right) + k_1'}$$

or

$$c_A = c_{A,0} \cdot \frac{k_1' \cdot \mathrm{e}^{-k_1' t}}{k_1' + 2k_2 c_{A,0} \cdot \left(1 - \mathrm{e}^{-k_1' t}\right)}$$

E24.11 Autocatalytic Reaction:

$$A + B \rightarrow C + 2\,B$$

Solve the differential equation (24.150)

$$\frac{\mathrm{d}c_B}{\mathrm{d}t} = k_r \cdot c_B \cdot (c_{A,0} + c_{B,0} - c_B)$$

Solution

To solve this equation we separate the variables

$$\frac{dc_B}{c_B \cdot (c_{A,0} + c_{B,0} - c_B)} = k_r \cdot dt$$

or

$$\frac{dc_B}{(c_{A,0} + c_{B,0}) \cdot c_B - c_B^2} = k_r \cdot dt$$

We integrate

$$\int \frac{dc_B}{(c_{A,0} + c_{B,0}) \cdot c_B - c_B^2} = \int k_r \cdot dt$$

taking the general integral from Appendix B:

$$\int \frac{1}{ax - x^2} \cdot dx = -\frac{1}{a} \cdot \ln\left(1 - \frac{a}{x}\right)$$

Thus, we find that

$$-\frac{1}{(c_{A,0} + c_{B,0})} \cdot \left|\ln\left(1 - \frac{c_{A,0} + c_{B,0}}{c_B}\right)\right|_{c_{B_0}}^{c_B} = k_r \cdot t$$

$$\left[\ln\left(\frac{c_B - (c_{A,0} + c_{B,0})}{c_B}\right) - \ln\left(-\frac{c_{A,0}}{c_{B,0}}\right)\right] = -(c_{A,0} + c_{B,0})\, k_r \cdot t$$

$$\ln\left[-\frac{c_{B,0}}{c_{A,0}} \frac{c_B - (c_{A,0} + c_{B,0})}{c_B}\right] = -(c_{A,0} + c_{B,0})\, k_r \cdot t$$

$$\ln\left[\frac{c_{B,0}}{c_{A,0}} \frac{(c_{A,0} + c_{B,0}) - c_B}{c_B}\right] = -(c_{A,0} + c_{B,0})\, k_r \cdot t$$

or

$$\ln\left[\frac{c_{AB,0}}{c_{B,0}} \frac{c_B}{(c_{A,0} + c_{B,0}) - c_B}\right] = (c_{A,0} + c_{B,0})\, k_r \cdot t$$

By exponentiating both sides, we find that

$$\frac{c_{AB,0}}{c_{B,0}} \frac{c_B}{(c_{A,0} + c_{B,0} - c_B)} = e^{(c_{A,0}+c_{B,0})k_r t}$$

Isolating c_B, we obtain

$$c_B = \frac{c_{B,0}(c_{A,0} + c_{B,0})}{c_{A,0}\, e^{-(c_{A,0}+c_{B,0})k_r t} + c_{B,0}} = \frac{c_{A,0} + c_{B,0}}{1 + \frac{c_{A,0}}{c_{B,0}}\, e^{-(c_{A,0}+c_{B,0})k_r t}}$$

E24.12 Consider the reaction

$$N_2O_5 \rightarrow 2NO_2 + \frac{1}{2} O_2$$

for the thermal decomposition of N_2O_5 [F. Daniels, E.H. Johnston, *J. Am. Chem. Soc.* 1921, **43**, 53] which is first order in the N_2O_5 concentration. The temperature dependence of the rate constant is given by the following data:

$k_r/10^{-5}\,s^{-1}$	0.0787	3.46	49.8	487
T/K	273	298	318	338

Determine the Arrhenius parameters for this reaction.

Solution

With four data points we proceed here by performing an Arrhenius plot. Using

$$k_r = A \exp\left[-\frac{E_{a,m}}{RT}\right]$$

we can write that

$$\frac{k_r}{k_{r,1}} = \exp\left[-\frac{E_{a,m}}{R}\left(\frac{1}{T} - \frac{1}{T_1}\right)\right] \quad \text{or} \quad \ln\frac{k_r}{k_{r,1}} = -\frac{E_{a,m}}{R}\left(\frac{1}{T} - \frac{1}{T_1}\right)$$

We let $k_{r,1} = 7.87 \times 10^{-7} \text{s}^{-1}$ and evaluate $\ln(k_r/k_{r,1})$ and $1/T$ to find

$\ln(k_r / k_{r,1})$	0.00	3.78	6.45	8.73
$1/T/10^{-3}\text{K}^{-1}$	3.66	3.36	3.14	2.96

A plot is shown in Fig. E24.12. This plot has a slope of -1.25×10^4 K. Thus

$$\frac{E_{a,m}}{R} = 12.5 \times 10^3 \text{ K} \quad \text{so that} \quad E_{a,m} = 103.9 \text{ kJ mol}^{-1}$$

and

$$A = k_{r,1} \exp\left[\frac{E_{A,m}}{RT_1}\right] = 0.0787 \times 10^{-5} \text{ s}^{-1} \exp\left[\frac{103.9 \text{ kJ mol}^{-1}}{8.314 \text{ J K}^{-1} \text{ mol}^{-1} \cdot 273 \text{ K}}\right] = 5.98 \times 10^{13} \text{ s}^{-1}$$

resulting in a best fit activation barrier of $E_{a,m} = 103.9 \text{ kJ mol}^{-1}$ and a pre-exponential factor of $6.0 \times 10^{13} \text{ s}^{-1}$.

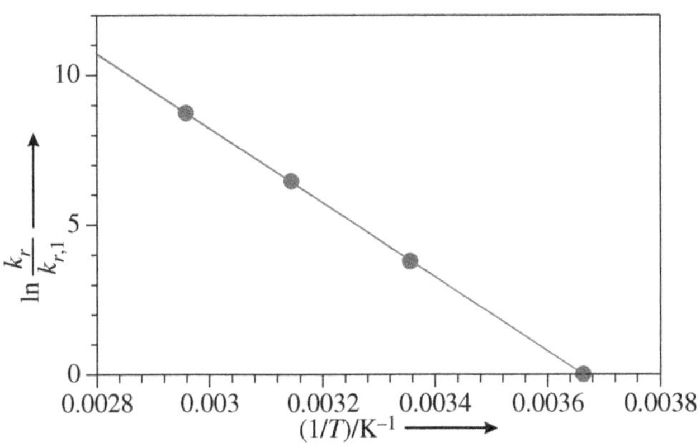

Figure E24.12 $\ln(k_r/k_{r_1})$ versus $1/T$ for the decomposition of N_2O_5.

E24.13 Consider the gas-phase thermal reaction

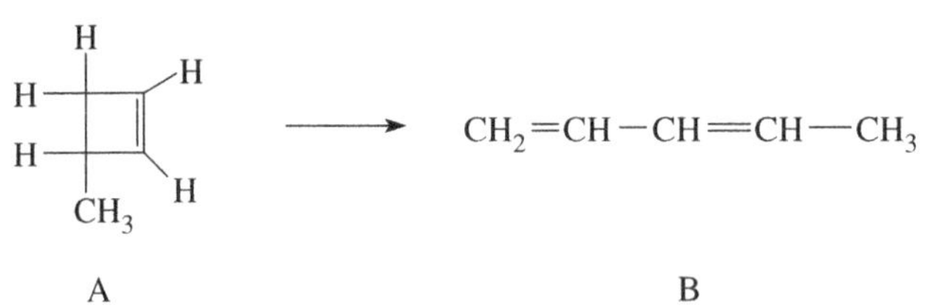

for which the rate constant at different initial concentrations of A and at 396.6 K is

$c_{A,0}/10^{-3}$ mol L^{-1}	4.03	8.06	21.8	41.1	78.6	105	203
$k_r/10^{-4}$ s^{-1}	0.48	0.57	0.79	0.92	1.03	1.10	1.17

a) Plot these data and comment on how the rate constant changes with the initial concentration of A. b) Use the Lindemann–Hinshelwood mechanism to describe this reaction and to explain the dependence of the measured rate constant on the initial concentration of species A.

Solution

A plot of these data is shown in Fig. E24.13 (dark gray solid circles).

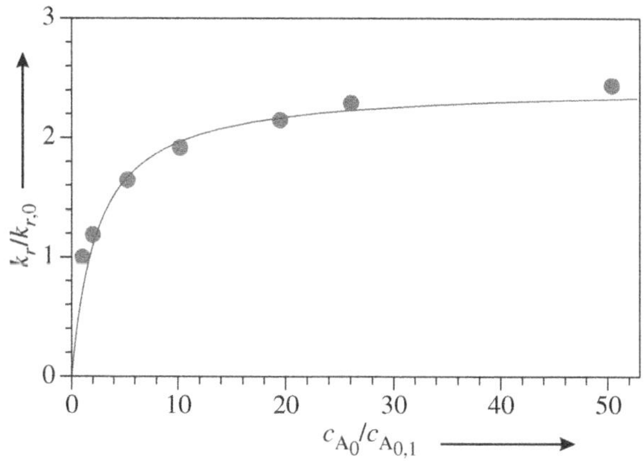

Figure E24.13 $k_r/k_{r,1}$ versus $c_{A_0}/c_{A_{0,1}}$ for the gas-phase decomposition of the cyclobutene. Solid dark gray circles: experiment, light gray solid line: Lindemann–Hinshelwood theory, Equation (24.111), with the parameters $k_1 = 1.19 \times 10^{-2}$ L mol^{-1}s^{-1}, $k_{-1} = 101$ s^{-1}, and $k_2 = 1.0$ s^{-1}.

At low initial concentration the reaction increases almost linearly with concentration, but at high initial concentration it becomes independent of the concentration.

We can understand this behavior by using the following Lindemann–Hinshelwood type of mechanism

$$\mathrm{A} + \mathrm{M} \underset{k_{-1}}{\overset{k_1}{\rightleftarrows}} \mathrm{A}^* + \mathrm{M}$$

$$\mathrm{A}^* \xrightarrow{k_2} \mathrm{B}$$

where M is either an A molecule, a B molecule, or perhaps a buffer gas molecule. For the situation described here, we know that $c_\mathrm{M} = c_\mathrm{A,0}$ because it is a unimolecular transformation and we begin with only A molecules. This mechanism gives the rate law

$$\frac{\mathrm{d}c_\mathrm{B}}{\mathrm{d}t} = \frac{k_2 \cdot k_1 \cdot c_\mathrm{A,0}}{k_2 + k_{-1} \cdot c_\mathrm{A,0}} \cdot c_\mathrm{A} = k_r \cdot c_\mathrm{A} \quad \text{with} \quad k_r = \frac{k_2 \cdot k_1 \cdot c_\mathrm{A,0}}{k_2 + k_{-1} \cdot c_\mathrm{A,0}}$$

At low concentration the rate of reaction is limited by the rate of activation collisions. Hence k_r increases with concentration of the gas (pressure of the gas). At high concentrations the reaction rate is limited by the ring opening elementary step, not the activation of the molecules. Hence the k_r does not change with $c_\mathrm{A,0}$.

To compare the experimental data with the mechanism, we consider the limiting case

$$k_r = k_1 \cdot c_\mathrm{A,0} \qquad \text{for small } c_\mathrm{A,0}$$

From the values $c_{\mathrm{A}_0} = 4.03 \times 10^{-3}$ mol L^{-1}, $k_r = 0.48 \times 10^{-4}$ s^{-1} it follows that $k_1 = 1.19 \times 10^{-2}$ L mol^{-1} s^{-1}. Next we consider the limiting case

$$k_r = \frac{k_2 \cdot k_1}{k_{-1}} \qquad \text{for large } c_\mathrm{A,0}$$

From the value $k_r = 1.17 \times 10^{-4}$ s^{-1} it follows that

$$\frac{k_2}{k_{-1}} = \frac{k_r}{k_1} = \frac{1.17 \times 10^{-4}\ \mathrm{s}^{-1}}{1.19 \times 10^{-2}\ \mathrm{L\ mol^{-1}\ s^{-1}}} = 9.83 \times 10^{-3}\ \mathrm{mol\ L^{-1}}$$

With the assumption $k_2 = 1.0$ s^{-1} we obtain

$$k_{-1} = \frac{1.0\ \mathrm{s}^{-1}}{9.83 \times 10^{-3}\ \mathrm{mol\ L^{-1}}} = 1.2 \times 10^2\ \mathrm{L\ mol^{-1}\ s^{-1}}$$

Figure E24.13 plots Equation (24.111) with $k_1 = 1.19 \times 10^{-2}$ L mol^{-1}s^{-1}, $k_2 = 1.0$ s^{-1}, and $k_{-1} = 1.2 \times 10^2$ L mol^{-1}s^{-1} (light gray solid line).

E24.14 The reaction between hypochlorite ion and iodide ion in alkaline solution,

$$ClO^-(aq) + I^-(aq) \rightarrow Cl^-(aq) + IO^-(aq)$$

has been studied by Y.T. Chia, R.E. Connick, *J. Phys. Chem.* 1959, **63**, 1518. They found the following initial rate data for different initial concentrations of the reactants.

	$c_{OH^-,0}$/mol L^{-1}	$c_{ClO^-,0}$/mol L^{-1}	$c_{I^-,0}$/mol L^{-1}	*Rate*/mol L^{-1}s^{-1}
A	1.0×10^{-5}	4.7×10^{-3}	2.2×10^{-3}	62.0
B	1.0×10^{-5}	2.5×10^{-3}	4.7×10^{-3}	70.5
C	1.0×10^{-5}	1.9×10^{-3}	2.1×10^{-3}	23.9
D	4.5×10^{-6}	1.9×10^{-3}	2.1×10^{-3}	53.2

Use these data to determine the rate law and the rate constants. Describe an experimental method that you would use to measure the initial rate in this reaction.

Solution

First we consider cases A and C for which the initial concentrations, $c_{OH^-,0}$ and $c_{I^-,0}$, are nearly constant. For these cases we expect that the rate law will take the form

$$rate \sim c^{\alpha}_{ClO^-,0} \text{ where } \alpha \text{ is a power to be determined.}$$

Using the data we find that

$$\frac{62.0}{23.9} = \left[\frac{4.7 \times 10^{-3}}{1.9 \times 10^{-3}}\right]^{\alpha} \quad \text{so that } \alpha = \frac{\ln[2.59]}{\ln[2.47]} = 1.05 \approx 1$$

Next we consider the cases C and D, for which the initial concentrations $c_{I^-,0}$ and $c_{ClO^-,0}$, are constant. For these cases we expect that the rate law will take the form,

$$rate \sim c^{\beta}_{OH^-,0} \text{ where } \beta \text{ is a power to be determined.}$$

Using the data we find that

$$\frac{23.9}{53.2} = \left[\frac{1.0 \times 10^{-5}}{0.45 \times 10^{-5}}\right]^{\beta} \quad \text{so that } \beta = \frac{\ln[0.449]}{\ln[2.22]} = -1.00 \approx -1$$

Lastly, we use the cases A and B, for which $c_{OH^-,0}$ is constant, to find that

$$rate \sim c_{ClO^-,0} \cdot c^{\gamma}_{I^-,0} \text{ where } \gamma \text{ is a power to be determined.}$$

Using the data we find that

$$\frac{62.0}{70.5} = \frac{4.7 \times 10^{-3}}{2.5 \times 10^{-3}}\left[\frac{2.2 \times 10^{-3}}{4.7 \times 10^{-3}}\right]^{\gamma} \quad \text{so that } \gamma = \frac{\ln[0.467]}{\ln[0.468]} = 1.00 \approx 1$$

With these findings, we can write the rate law as

$$rate = k_r \cdot \frac{c_{ClO^-} \cdot c_{I^-}}{c_{OH^-}}$$

To find the rate constant, we can substitute the data in each case, and we find that $k_r = 60\ s^{-1}$.

Chia and Connick monitored the reaction solution's absorbance at 400 nm as a function of time for the IO^- to follow this reaction. Other methods might also be used.

E24.15 Consider the electron transfer reaction

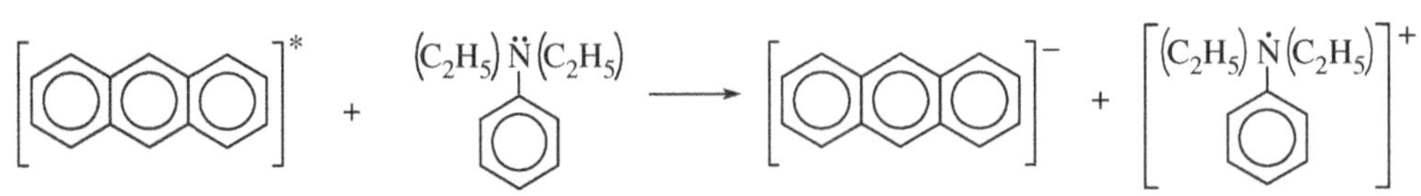

Anthracene molecules (molecules A) dissolved in acetonitrile are excited by light (the asterisk symbolizes the excited state). They react with diethylaniline molecules (molecules B). Estimate the sizes of the two molecules,

r_A and r_B. Combine your estimate with the measured value $(D_A + D_B) = 3.8 \times 10^{-5}$ cm^{2}s^{-1}, and calculate the rate constant from Equation (24.41). Compare your result to the experimental value $k_r = 2.0 \times 10^7$ m^3 mol^{-1} s^{-1} (D. Rehm, A. Weller, *Ber. Bunsenges. Phys. Chem.* 1969, **73**, 834).

Solution

To estimate the sizes we use bond increments and account for the van der Waals radius of the atoms along the molecular periphery. While this method is somewhat rough it provides a reasonable order of magnitude. We find that $r_A + r_B = 700$ pm. Using this value with Equation (24.41)

$$k_r = 4\pi \cdot (r_A + r_B) \cdot (D_A + D_B) \cdot \boldsymbol{N}_A$$

we find that $k_r = 4\pi$ (700 pm) $(3.8 \times 10^{-5}$ cm^2 s^{-1}) 6.02×10^{23} mol^{-1} $= 2.0 \times 10^7$ m^3 mol^{-1} s^{-1}, which is in excellent agreement with the experimental value reported by Rehm and Weller.

E24.16 Consider the equilibrium of the indicator dye alizarin yellow (A) with OH^- to make the deprotonated form (C) which was studied using a temperature jump method.

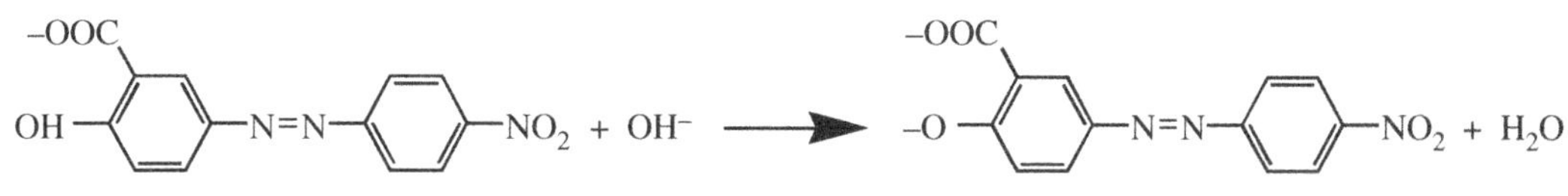

$$A(aq) + OH^-\ (aq) \rightarrow\ C(aq).$$

At a pH of 10.30 and $c_{A,0} = 1.26 \times 10^{-4}$ mol L^{-1}, the relaxation rate constant is $k_r = 4.43$ s^{-1}, and at a pH of 10.70 and $c_{A,0} = 2.01 \times 10^{-4}$ mol L^{-1}, the relaxation rate constant is $k_r = 5.92$ s^{-1}. Use these data to determine k_1 and k_{-1}.

Solution

For a pH of 10.30, the pOH is 3.70, and hence the concentration of hydroxide is

$$c_{OH^-} = 10^{-pOH} \text{ mol L}^{-1} = 10^{-3.70} \text{ mol L}^{-1} = 2.00 \times 10^{-4} \text{ mol L}^{-1}$$

In a corresponding manner for a pH of 10.70, we find that $c_{OH^-} = 5.01 \times 10^{-4}$ mol L^{-1}.

First, we find the value of k_1 from the relation

$$k_r = k_{-1} + k_1 \left(c_{A,0} + c_{OH^-}\right)$$

Because we have relaxation rates at two different c_{OH^-} concentrations, we can determine k_1 by subtracting the two results, which eliminates k_{-1}; namely we find that

$$\begin{aligned} k_r(10.7) - k_r(10.3) &= k_1 \cdot \left[\left(c_{A,0} + c_{OH^-}\right)_{10.7} - \left(c_{A,0} + c_{OH^-}\right)_{10.3}\right] \\ &= k_1 \cdot \left[(1.26 \times 10^{-4} + 5.01 \times 10^{-4}) - (1.26 \times 10^{-4} + 2.00 \times 10^{-4})\right] \text{ mol L}^{-1} \\ &= k_1 \cdot 3.01 \times 10^{-4} \text{ mol L}^{-1} \end{aligned}$$

Thus

$$k_1 = \frac{\left[k_r(10.7) - k_r(10.3)\right] \text{ s}^{-1}}{3.01 \times 10^{-4} \text{ mol L}^{-1}} = \frac{5.92 \text{ s}^{-1} - 4.43 \text{ s}^{-1}}{3.01 \times 10^{-4} \text{ mol L}^{-1}} = 4.95 \times 10^3 \text{ mol L}^{-1} \text{ s}^{-1}$$

With k_1 in hand, it is straightforward to evaluate k_{-1}, as

$$\begin{aligned} k_{-1} &= k_r - k_1 \left(c_{A,0} + c_{OH^-}\right) \\ &= \left[5.92 \text{ s}^{-1} - \left(4.95 \times 10^3 \text{ mol}^{-1} \text{ L s}^{-1}\right) \left(5.01 + 2.0\right) \times 10^{-4} \text{ mol L}^{-1}\right)\right] \\ &= 3.5 \text{ s}^{-1} \end{aligned}$$

E24.17 In Fig. 24.10a, $k_r' = k_r \cdot c_{OH^-}$ is plotted versus T and Fig. 24.10b plots $\ln(k_r'/k_{r,1}')$ versus $1/T$ for reaction (24.62)

$$R^+ + OH^- \rightarrow ROH$$

(experimental data points see Table 24.2). Compare the experimental frequency factor to that you estimate from Equation (24.34). Using Fig. 24.10b the slope −7640 K of the straight line gives $E_{a,m} = \boldsymbol{R} \cdot 7640$ K $= 63.5$ kJ mol^{-1} and the intercept gives a frequency factor $A_f = 1.8 \times 10^{10}$ L mol^{-1}s^{-1}.

Solution

Equation (24.34)

$$A_f = 4\pi \cdot r_A^2 \cdot \boldsymbol{N}_A \cdot \sqrt{\frac{8kT}{\pi m_A}} \cdot \alpha$$

where α is the steric factor. As given in Example 24.7, we use $r_A = r_{OH-} = 250 \times 10^{-12}$ m and $m_A = 3 \times 10^{-26}$ kg, so that

$$\begin{aligned} A_f &= 4\pi \cdot \left(250 \times 10^{-12}\ \text{m}\right)^2 \cdot 6.022 \times 10^{23} \text{mol}^{-1} \cdot 609\ \text{m s}^{-1} \cdot \alpha \\ &= 2.88 \times 10^8 \cdot \alpha\ \text{m}^3\ \text{mol}^{-1}\ \text{s}^{-1} \\ &= 2.88 \times 10^{11} \cdot \alpha\ \text{L mol}^{-1}\ \text{s}^{-1} \end{aligned}$$

Comparison of this estimated value with that determined from the graph can be used to determine α, namely we find that

$$\alpha = \frac{1.8 \times 10^{10}}{2.88 \times 10^{11}} = 0.063$$

E24.18 Consider the recombination reaction of two separated iodine atoms to make diatomic iodine

$$\text{I} + \text{I} \rightleftarrows \text{I}_2$$

Find an expression for the rate law of iodine recombination, by using the following mechanism

$$\text{I} + \text{I} \underset{k_{-1}}{\overset{k_1}{\rightleftarrows}} \text{I}_2^*$$

$$\text{I}_2^* + \text{Ar} \xrightarrow{k_2} \text{I}_2 + \text{Ar}$$

Using your expression for the rate law, describe, verbally, how the formation rate of I_2 will depend on the pressure of Ar.

Solution

If we use the last step in the mechanism, we see that the rate law for the I_2 formation can be written as

$$\frac{dc_{I_2}}{dt} = k_2 \cdot c_{Ar} \cdot c_{I_2^*}$$

Because c_{Ar} is constant, we need only determine the time dependence of $c_{I_2^*}$ and integrate. Using the mechanism, we can write that

$$\frac{dc_{I_2^*}}{dt} = k_1 \cdot c_I^2 - k_{-1} \cdot c_{I_2^*} - k_2 \cdot c_{Ar} \cdot c_{I_2^*}$$

If we use the steady-state approximation then we find that

$$0 = k_1 \cdot c_I^2 - k_{-1} \cdot c_{I_2^*} - k_2 \cdot c_{Ar} \cdot c_{I_2^*}$$

and we can solve for $c_{I_2^*}$ to find

$$c_{I_2^*} = \frac{k_1 \cdot c_I^2}{k_{-1} + k_2 \cdot c_{Ar}}$$

Hence we find that

$$\frac{dc_{I_2}}{dt} = k_2 \cdot c_{Ar} \frac{k_1 \cdot c_I^2}{k_{-1} + k_2 \cdot c_{Ar}}$$

within the limit that the steady-state approximation applies.

At low pressures of argon (that is $k_2 \cdot c_{Ar} \ll k_{-1}$), the rate is proportional to c_{Ar}. Under these circumstances the collisional deactivation of the nascent I_2 is rate limiting.

At high pressures of argon (that is $k_2 \cdot c_{Ar} \gg k_{-1}$), the rate is independent of c_{Ar}. Under these circumstances the rate is limited by the collision of the two iodine atoms, not the deactivating collisions.

E24.19 In Equation (24.32), we estimate the mean free path for a molecule in a liquid to be one-third the radius of the molecule. Given that the density of water is $\rho = 1.00$ g cm^{-3} and a water molecule's van der Waals radius is approximately $r_W = 0.15$ nm radius, estimate the distance between two water molecules in solution. Comment on the validity of our estimate.

Solution

Using the given density we find a molar volume of

$$V_m = \frac{\boldsymbol{M}}{\rho} = \frac{18.0\ \text{g mol}^{-1}}{1.00\ \text{g cm}^{-3}} = 18.0\ \frac{\text{cm}^3}{\text{mol}} = 0.0180\ \text{L mol}^{-1}$$

and a molecular volume of

$$\frac{V_m}{\boldsymbol{N}_\text{A}} = \frac{0.0180\ \text{L mol}^{-1}}{6.02 \cdot 10^{23}\ \text{mol}^{-1}} = 2.99 \cdot 10^{-26}\ \text{L} = 0.0299\ \text{nm}^3$$

Assuming a spherical volume of this size around a water molecule, we find that it has an effective radius r of

$$r = \left(\frac{3}{4\pi}\frac{V_m}{\boldsymbol{N}_\text{A}}\right)^{1/3} = \left(\frac{0.090}{4\pi}\right)^{1/3}\ \text{nm} = 0.19\ \text{nm}$$

Hence a water molecule travels about $r - r_W = (0.19 - 0.15)$ nm $= 0.04$ nm, or one-third to one-fourth of its van der Waals radius, before it enters the volume of its neighbor.

Although a more realistic analysis would account for a distribution of mean free paths, we see that the simple estimate of one-third is reasonable. The essential features of the collision model do not depend on the exact choice of the numerical factor, and one might just as readily use one-half or one-fourth as an estimate.

E24.20 Stern–Volmer Kinetics. An excited electronic state of a molecule can be relaxed (quenched) by collisions with appropriate reaction partners. Often this process leads to de-excitation of the excited state by transfer of the energy to the acceptor molecule. For example, electronically exciting a tryptophan molecule in solution leads to a long lived excited state with an average lifetime of 2.5 ns. By increasing the concentration of oxygen in solution one can reduce the molecule's excited state lifetime and quench the fluorescence. Assume that the reaction mechanism is

$$\text{tryptophan}^* + O_2 \rightarrow \text{tryptophan} + O_2^*$$

and show that the fluorescence intensity decreases linearly with the concentration of O_2.

Solution

Under constant illumination of the sample (hence excitation of the tryptophan), a steady state condition is reached and we observe a steady state level of fluorescence F. The rate law for the decay of excited tryptophan molecules at steady state is

$$\frac{dc_{\text{tryptophan}^*}}{dt} = G(t) - k_\text{rad} \cdot c_{\text{tryptophan}^*} - k_r \cdot c_{O_2} \cdot c_{\text{tryptophan}^*} = 0$$

where $G(t)$ describes the generation of excited states by the illuminating light, $k_\text{rad} c_{\text{tryptophan}^*}$ describes the depopulation of excited states by fluorescence, and the last term describes the nonradiative quenching of the excited states by oxygen. When no oxygen is present, this equation simplifies to

$$\frac{dc_{\text{tryptophan}^*}}{dt} = G(t) - k_\text{rad} \cdot c_{\text{tryptophan}^*,0} = 0$$

which has some fluorescence intensity F_0. Combining these two equations allows us to eliminate $G(t)$, and we find that

$$\frac{c_{\text{tryptophan}^*,0}}{c_{\text{tryptophan}^*}} = 1 + \frac{k_r}{k_\text{rad}} \cdot c_{O_2} = 1 + k_r \cdot \tau_0 \cdot c_{O_2}$$

where $\tau_0 = 1/k_\text{rad}$. The ratio of the concentrations of the excited states can be measured by a ratio of their fluorescence intensity or the fluorescence relaxation times. Hence we can write that

$$\frac{I_{\text{fluorescence},0}}{I_\text{fluorescence}} = 1 + k_r \cdot \tau_0 \cdot c_{O_2}$$

or

$$\frac{I_{\text{fluorescence}}}{I_{\text{fluorescence},0}} = \frac{1}{1 + k_r \cdot \tau_0 \cdot c_{O_2}}$$

E24.21 Temperature jump experiments on the reaction

$$Br^- + HOBr + H^+ \underset{k_{-1}}{\overset{k_1}{\rightleftharpoons}} Br_2 + H_2O$$

gives the overall rate constant values $k_1 = 1.61 \times 10^{10}\ L^2\ mol^{-2} s^{-1}$ and $k_{-1} = 110\ s^{-1}$ at 20 °C (M. Eigen, K. Kustin, *J. Am. Chem. Soc.* 1962, **84**, 1355). Assume that bromine is injected into pure water so that the pH and the relative concentrations of Br^-, HOBr are determined by the chemical equilibrium, that is

$$(c_{HOBr})_\infty = (c_{H^+})_\infty = (c_{Br^-})_\infty$$

and take the steady-state concentration of bromine to be $(c_{Br_2})_\infty = 2 \times 10^{-3}\ mol\ L^{-1}$.What is the observed relaxation rate constant in the measurement?

Solution

At steady state we can write that

$$\frac{dc_{Br_{2,\infty}}}{dt} = 0 = -(c_{Br_2})_\infty \cdot k_{-1} + (c_{HOBr})_\infty \cdot (c_{H^+})_\infty \cdot (c_{Br^-})_\infty \cdot k_1$$
$$\approx -(c_{Br_2})_\infty \cdot k_{-1} + y_\infty^3 \cdot k_1$$

where we write that $(c_{HOBr})_\infty = (c_{H^+})_\infty = (c_{Br^-})_\infty = y_\infty^3$.
Then for the equilibrium we have

$$y_\infty^3 = (c_{Br_2})_\infty \cdot \frac{k_{-1}}{k_1}$$

If we perturb the system away from equilibrium by an amount x, then

$$\frac{dc_{Br_2}}{dt} = -\frac{dx}{dt} = -\left(c_{Br_{2,\infty}} - x\right) \cdot k_{-1} + \left(y_\infty + x\right)^3 \cdot k_1$$
$$= y_\infty^3 \cdot k_1 - \hat{c}_{Br_{2,\infty}} \cdot k_{-1} + x \cdot k_{-1} + 3k_1 \cdot y_\infty^2 \cdot x$$

Here we neglect higher terms of x (x^2 or x^3). Because of the equilibrium condition we have

$$y_\infty^3 \cdot k_1 - \hat{c}_{Br_{2,\infty}} \cdot k_{-1} = 0$$

and the result for dx/dt simplifies to

$$-\frac{dx}{dt} = \left(k_{-1} + 3k_1 \cdot y_\infty^2\right) \cdot x = k_r \cdot x$$

which defines

$$k_r = k_{-1} + 3k_1 \cdot y_\infty^2$$

Because of the equilibrium condition

$$y_\infty^3 = (c_{Br_2})_\infty \cdot \frac{k_{-1}}{k_1}$$

we conclude that k_r depends on the equilibrium concentration of bromine. Let us assume that $(c_{Br_2})_\infty = 2 \times 10^{-3}$ mol L $^{-1}$. Then

$$y_\infty^2 = \left[(c_{Br_2})_\infty \cdot \frac{k_{-1}}{k_1}\right]^{2/3} = \left[\left(2 \times 10^{-3}\right)\ mol\ L^{-1} \cdot \frac{110\ s^{-1}}{1.61 \times 10^{10}\ L^2\ mol^{-2}\ s^{-1}}{}_1\right]^{2/3}$$
$$= 5.73 \times \times 10^{-8}\ mol^2\ L^{-2}$$

and

$$k_r = k_{-1} + 3k_1 \cdot 5.73 \times 10^{-8}\ mol^2\ L^{-2}$$

Using the values for k_1 and k_{-1}, we find that

$$k_r = 110\ \text{s}^{-1} + 3 \cdot 1.61 \times 10^{10}\ \text{L}^2\text{mol}^{-2}\ \text{s}^{-1} \cdot 5.73 \times \times 10^{-8}\ \text{mol}^2\ \text{L}^{-2}$$
$$= 110\ \text{s}^{-1} + 2.77 \times 10^3\ \text{s}^{-1} = 2.88 \times 10^3\ \text{s}^{-1}$$

Note: In this consideration we assumed that

$$(c_{\text{HOBr}})_\infty = (c_{\text{H}^+})_\infty = (c_{\text{Br}^-})_\infty$$

that is, we injected bromine into pure water, and the pH of the solution was determined by the amount of injected bromine. For experimental reasons, a better approach is to add bromine to a buffer solution with a fixed pH. In this case we can treat c_{H^+} as a constant and replace k_1 by $k_1' = k_1 \cdot c_{\text{H}^+}$. Eigen and Kustin choose this latter way, using buffer solutions with different pH.

E24.22 Use the simple collision theory to calculate the frequency factor and rate constant for the reaction

$$\text{D} + \text{H}_2 \ \rightarrow\ \text{HD} + \text{H}$$

at 298 K. The measured[3] molar activation energy is $E_{a,m} = \boldsymbol{N}_\text{A} \cdot 5.28 \times 10^{-20}\ \text{J} = 31.8\ \text{kJ mol}^{-1}$. You can take the radii of D and H_2 to be $r_\text{D} = 128$ pm, and $r_{\text{H}_2} = 138$ pm. Compare your result with the experimental value of $1.68 \times 10^5\ \text{L mol}^{-1}\ \text{s}^{-1}$ and calculate a steric factor for the reaction.

Solution

The activation factor is $e^{-E_{a,m}/RT} = 2.7 \times 10^{-6}$ at 298 K, corresponding to a collision probability of

$$\mathcal{P} = 2.7 \times 10^{-6}$$

About 10^6 collisions have to occur before two molecules react with one another. Using $m_\text{D} = m_{\text{H}_2} = 3.3 \times 10^{-27}$ kg ($\mu = 1.6 \times 10^{-27}$ kg), $r_\text{D} = 128$ pm, and $r_{\text{H}_2} = 138$ pm, we calculate

$$A_f = 2.6 \times 10^3\ \text{m s}^{-1} \cdot \pi \cdot 7.1 \times 10^{-20}\ \text{m}^2 \cdot 6.02 \times 10^{23}\ \text{mol}^{-1} \cdot \alpha$$
$$= 3.5 \times 10^8\ \text{m}^3\ \text{mol}^{-1}\ \text{s}^{-1} \cdot \alpha = 3.5 \times 10^{11}\ \text{L mol}^{-1}\ \text{s}^{-1} \cdot \alpha$$

for $T = 298$ K. Hence the rate constant is

$$k_r = 3.5 \times 10^{11}\ \text{L mol}^{-1}\ \text{s}^{-1} \cdot \alpha \cdot 2.7 \times 10^{-6} = 9.5 \times 10^5 \cdot \alpha\ \text{L mol}^{-1}\ \text{s}^{-1}$$

The experimental value[4] is $1.68 \times 10^5\ \text{L mol}^{-1}\ \text{s}^{-1}$, which suggests that the steric factor is $\alpha = 0.2$. This small value suggests that details of the experimental geometry are important.

24.2 Problems

P24.1 Jeong and Grady (*Biochemistry* 1995, **34**, 3734) investigated the enzyme kinetics of dihydrofolate reductase (DHFR) with the substrate 6,8-dimethylpterin (S). Both the enzyme DHFR and the substrate S have a carboxy group that is involved in the reaction. They used the Michaelis–Menten mechanism to describe the reaction

$$\text{DHFR} + \text{S} \underset{k_{-a}}{\overset{k_a}{\rightleftharpoons}} (\text{DHFR-S}) \overset{k_b}{\rightarrow} \text{DHFR} + \text{P}$$

where

$$\frac{dc_\text{P}}{dt} = \frac{k_b \cdot c_{\text{DHFR},0} \cdot c_\text{S}}{K_\text{M} + c_\text{S}}$$

By controlling the reaction conditions and measuring the initial rate of the reaction, they were able to determine the rate constant k_b. Using the rate law, show mathematically and explain verbally how this is possible. In their conclusions they state: "The active complex in the reduction of the mechanism-based substrate 6,8-diMe-Pterin by DHFRs with NADPH is characterized by a deprotonated enzyme carboxy group which binds protonated

3 J.P. Toennies, J. Arnold, J. Wolfrum, *Ber. Bunsenges Phys. Chem.* 1990, **94**, 1231.
4 A.A. Westenberg, N. de Haas, *J. Chem. Phys.* 1967, **47**, 1393.

6,8-diMe-Pterin. The rate-limiting step for the reaction is hydride-ion transfer, with no perturbation on the catalytic rate by product release."

Based on this statement, describe how you think the rate constant might change with pH.

Solution

For the case of an initial rate $c_S \sim c_{S,0}$. Hence, for $c_S \gg K_m$ we find that

$$\frac{dc_P}{dt} \approx k_b \cdot c_{DHFR,0}$$

By performing initial rate measurements at different concentrations of $c_{DHFR,0}$ and high enough values of $c_{S,0}$, the value of k_b can be determined from this relation.

The rate constant should be sensitive to pH over a limited range of values. The pH needs to be high enough to keep the carboxy group of the enzyme deprotonated but low enough to have a protonated substrate.

P24.2 Consider the unimolecular decomposition of di-*t* -butylperoxide in the gas phase,

$$(CH_3)_3COOC(CH_3)_3 \rightarrow 2(CH_3)_2CO + C_2H_6$$

The following table gives the data for the total pressure of the system versus time:

t/min	0	6	12	18
P_{total}/Torr	173.5	211.3	244.4	273.9

Find a mathematical expression for the time dependence of the total pressure in terms of the initial pressure, the time, and the rate constant. Make a plot which shows that the data follow your mathematical expression. In particular, rearrange the expression and perform a plot, which should be linear. Obtain a value for the rate constant by analyzing your plot. Use the value of your rate constant to determine the half-life for the reaction.

Solution

First we find a mathematical expression for how the total pressure changes as the reaction proceeds by assuming the gas behaves ideally. We write the initial pressure P_0 as the pressure of the di-*t*-butylperoxide before decomposition. Because each molecule of peroxide that decomposes generates three molecules we can write the total pressure P_{total} at time t as

$$P_{total} = (P_0 - \Delta P) + 3\Delta P = P_0 + 2\Delta P$$

where ΔP is the pressure change caused by decomposition of the reactant. Namely, the partial pressure of reactant at any time t is given by

$$P_{reactant} = (P_0 - \Delta P) = P_0 \exp(-k_r t)$$

where k_r is the unimolecular decay rate of peroxide. Rearranging this expression we find that

$$\Delta P = P_0 \left[1 - \exp(-k_r t)\right]$$

so that

$$P_{total} = 3P_0 - 2P_0 \exp(-k_r t)$$

Next we rearrange this expression so that it is linearized. By isolating the exponential term we find that

$$\frac{3}{2} - \frac{P_{total}}{2P_0} = \exp(-k_r t)$$

and then we can take its logarithm to find that

$$y = \ln\left(\frac{3}{2} - \frac{P_{total}}{2P_0}\right) = -k_r t \qquad (24P2.1)$$

Hence we can linearize the data and find

t/min	0	6	12	18
$\ln\left(\frac{3}{2} - \frac{P_{total}}{2P_0}\right)$	0	−0.115	−0.229	−0.342

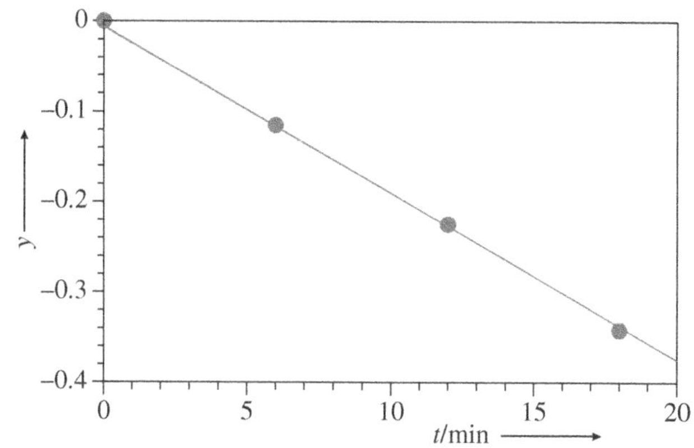

Figure P24.2 Plot of the quantity y in Equation (24P2.1) versus time t.

If we restrict the intercept to be zero, then the best fit line in Fig. P24.2 gives a slope of $-0.0190\ \text{min}^{-1}$, hence a rate constant of $k_r = 0.0190\ \text{min}^{-1} = 3.2 \times 10^{-4}\text{s}^{-1}$.

P24.3 Campbell and Plane (M.L. Campbell, J.M.C. Plane, *J. Phys. Chem. A* 2003, **107** 3747) studied the gas-phase recombination reaction kinetics between Pd and O_2. The Pd atoms were generated in the gas phase by photodissociation of a palladium(II) trifluoroacetate compound in the gas phase. The concentrations of O_2 and gaseous Pd were followed spectroscopically. The bimolecular recombination reaction may be written as

$$\text{Pd(gas)} + O_2 \rightarrow PdO_2$$

where the observed second-order rate constant k_{obs} is defined through the relation

$$\frac{dP_{O_2}}{dt} = -k_{\text{obs}}\ P_{O_2} \times P_{\text{Pd}}$$

The table summarizes their findings for the observed rate constant as a function of the pressure and temperature of an argon buffer gas. Explain why the observed rate constant k_{obs} depends on the presence of a buffer gas. Propose a set of elementary steps for the reaction mechanism and explain how your mechanism accounts for the observed pressure dependence at 294 K.
Second-Order Rate Constants for Pd Reacting with O_2.

T/K	P/Torr	k_{obs} $(10^8 \frac{\text{L}}{\text{mol s}})$	T/K	P/Torr	k_{obs} $(10^8 \frac{\text{L}}{\text{mol s}})$
294	5.0	0.8	294	20.0	3.7
294	10.0	1.5	348	20.0	2.5
294	15.0	3.3	373	20.0	2.2
294	40.0	8.4	398	20.0	1.8
294	60.0	13	423	20.0	1.3
294	80.0	16	448	20.0	1.1
294	100.0	20	473	20.0	0.72
294	150.0	30	498	20.0	0.54
294	200.0	36	523	20.0	0.45
294	250.0	45			
294	300.0	49			

The temperature dependence for this reaction is unusual. Using the temperature dependence of the k_{obs} data at 20 Torr, make an Arrhenius plot for the rate constant and comment on the result. By analyzing their data with a detailed molecular model, they found that the rate constant at low Ar pressure $k_{rec,0}$ was given by $2.71 \times 10^{18}/(T/K)^{2.76}$ $L^2\ mol^{-2}\ s^{-1}$ and that the rate constant at high Ar pressures was given by

$$k_{rec,\infty} = 1.52 \times 10^{11} \exp\left[-69.3/(T/K)\right]\ L\ mol^{-1}\ s^{-1}.$$

Discuss these qualitatively different temperature dependencies in terms of your proposed mechanism.

Solution

We propose a Lindemann–Hinshelwood (LH) mechanism of the sort

$$a: Pd + O_2 \overset{k_1}{\rightarrow} PdO_2^*$$

$$b: PdO_2^* \overset{k_{-1}}{\rightarrow} Pd + O_2$$

and

$$c: PdO_2^* + Ar \overset{k_2}{\rightarrow} PdO_2 + Ar$$

When the Pd and O_2 collide they create a PdO_2 species in a highly excited state that can either dissociate or proceed to product. Collision with Ar buffer gas deactivates this excited molecule and thereby traps the PdO_2 product. This Lindemann–Hinshelwood mechanism leads to a reaction rate that increases with Ar pressure at low pressure because step c) becomes the rate limiting step at low pressure. At high argon pressures ($k_{rec,\infty}$) the rate limiting step is the creation of PdO_2^* by way of the equilibrium represented by steps a) and b).

An Arrhenius plot of the data at 20 Torr is shown in Fig. P24.3. The graph is not linear and it has a positive slope which means that the rate constant decreases as the temperature rises. The LH mechanism is consistent with the unusual temperature dependence reported by Campbell and Plane. At low pressure ($k_{obs,0}$) the rate limiting step is c), in which internal energy of the PdO_2^* is transferred to the Ar buffer gas and the PdO_2 buffer gas is stabilized. As T increases at a fixed pressure the Ar atoms carry more kinetic energy and the de-excitation of PdO_2^* is less effective; that is, the higher energy collisions are less effective at "cooling" the PdO_2^*. At high Ar pressures ($k_{obs,\infty}$), however, the rate limiting step is the PdO_2^* formation (step a), and the overall rate is not limited by the Ar collisions and their efficiency for stabilizing the PdO_2^*. In fact, they find that step a) is activated, i.e., has an activation energy, albeit small ~0.58 kJ mol^{-1}.

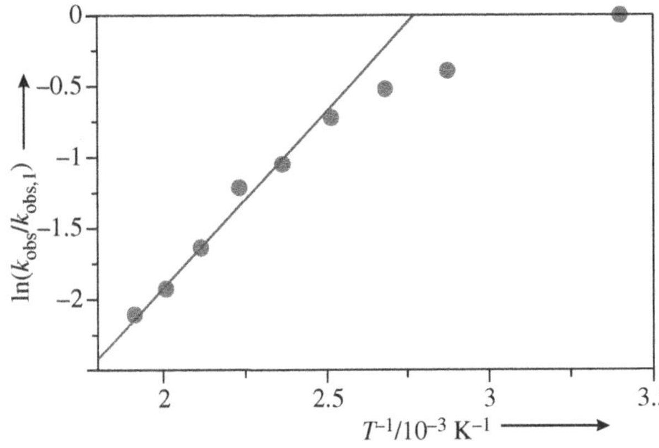

Figure P24.3 Temperature dependence of reaction 24.2. $\ln(k_{obs}/k_{obs,1})$ versus $1/T$ at $P = 20$ Torr.

P24.4 Consider the self-exchange reaction

$$^{55}Fe^{2+} + {}^{56}Fe^{3+} \rightarrow {}^{55}Fe^{3+} + {}^{56}Fe^{2+}$$

which occurs spontaneously and can be followed by chemically identifying a particular oxidation state and monitoring the radioactive decay of the ^{55}Fe nuclide (see J. Silverman, R.W. Dodson, *J. Am. Chem. Soc.* 1952, **56**, 846). For example, one would begin an experimental run with $^{55}Fe^{2+}$ and no $^{55}Fe^{3+}$ present. The formation of $^{55}Fe^{3+}$ would reflect the reaction rate. From the following rate data determine the rate law for the reaction, and find a value for the rate constant, k_r.

Run	A	B	C	D	E	F	G	H
$c_{^{55}Fe^{3+},0}/10^{-4}$ mol L^{-1}	1.03	1.05	1.06	1.08	2.10	3.11	4.18	4.24
$c_{^{55}Fe^{2+},0}/10^{-4}$ mol L^{-1}	0.998	1.97	2.94	3.92	3.94	3.95	3.92	3.85
Rate/10^{-8} mol L^{-1} s^{-1}	1.38	2.85	4.08	5.63	11.17	16.46	23.10	21.87

Given the results of your determination for the rate law, state an alternative experimental test you could perform to determine the rate law. Be specific about how you would proceed with an experimental procedure.

Solution

First we consider the variation of the rate with the concentration of ferrous ion. For example, we find that

$$\frac{rate(\mathrm{A})}{rate(\mathrm{B})} = \frac{c^{\alpha}_{^{55}Fe^{2+},0}(\mathrm{A})}{c^{\alpha}_{^{55}Fe^{2+},0}(\mathrm{B})} \quad \text{or} \quad \frac{1.38}{2.85} = \left(\frac{0.998}{1.97}\right)^{\alpha}$$

which implies that $\alpha = 1.06$. A similar analysis for runs A and C gives $\alpha = 1.0$.
Next we consider the variation of the rate with the concentration of ferric ion. For example, we find that

$$\frac{rate(\mathrm{D})}{rate(\mathrm{G})} = \frac{c^{\beta}_{Fe^{3+},0}(\mathrm{D})}{c^{\beta}_{Fe^{3+},0}(\mathrm{G})} \quad \text{or} \quad \frac{1.08}{4.18} = \left(\frac{5.63}{23.10}\right)^{\beta}$$

which implies that $\beta = 0.960$. A similar analysis for runs E and F gives $\beta = 1.01$.
Hence we conclude that the rate law has the form

$$rate = k_r \cdot c_{Fe^{2+}} \cdot c_{Fe^{3+}}$$

If we use the run A data, we find a rate constant of

$$k_r = \frac{c_{Fe^{2+}} \cdot c_{Fe^{3+}}}{rate}$$
$$= \frac{(1.03 \times 10^{-4})\,(0.998 \times 10^{-4})}{1.38 \times 10^{-8}} \text{ mol L}^{-1}\text{ s}^{-1} = 1.34 \text{ mol L}^{-1}\text{ s}^{-1}$$

A couple of possibilities exist. One possibility is to begin the experiment with a large excess of $c_{^{55}Fe^{2+}}$ so that the rate of reaction for this isotope with an Fe^{3+} is quasi-first order. In this the appearance of $c_{^{55}Fe^{3+}}$ should grow exponentially with time. A corresponding reaction could be done in which $c_{^{55}Fe^{3+}}$ is in excess so that the reaction is quasi-first order in Fe^{2+}.

P24.5 The rate for the thermal decomposition of methyl ethyl ether has been determined using the method of initial rates by E.W.R. Steacie, *J. Chem. Phys.* 1933, **1**, 618. Figure P24.5 plots some of their data at three temperatures as a function of pressure. Postulate a reaction mechanism for the decomposition of methyl ethyl ether. Why does the rate increase as the pressure of the gas increases?

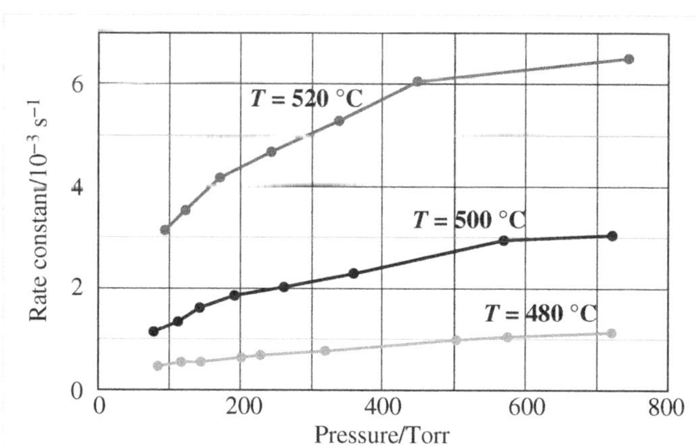

Figure P24.5 Dependence of the first-order rate constant for the thermal decomposition of methylethylether on the pressure.

Solution

We postulate the following reaction mechanism

$$CH_3\text{–}CH_2\text{–}O\text{–}CH_3 \xrightarrow{q} CH_2\text{=}CH_2 + HOCH_3$$

where q represents the addition of heat (q positive); or to simplify the notation

$$A \xrightarrow{q} B + C$$

The rate increases with pressure because the collision rate increases with pressure and hence the number of activated molecules per unit time increases; a Lindemann–Hinshelwood mechanism. In the limit of high pressure the rate constant is independent of the pressure, because the creation of activated species no longer limits the rate.

P24.6 Consecutive reaction:

$$A + B \xrightarrow{k_1} C \quad C + B \xrightarrow{k_2} D$$

where B is assumed to be in large excess. Solve the corresponding differential equation (24.84)

$$\frac{dc_C}{dt} + k_2' c_C = k_1' c_{A,0} \cdot e^{-k_1' t}$$

Solution

To begin, we set

$$c_C = y \cdot f$$

where y is the solution of the corresponding homogeneous differential equation

$$\frac{dy}{dt} + k_2' \cdot y = 0$$

Then

$$y = \alpha \cdot e^{-k_2' t} \quad \text{and} \quad c_C = \alpha \cdot e^{-k_2' t} \cdot f$$

(α = integration constant). Using this result, the derivative of c_C with respect to t becomes

$$\frac{dc_C}{dt} = -k_2' \alpha \cdot e^{-k_2' t} \cdot f + \alpha \cdot e^{-k_2' t} \cdot \frac{df}{dt}$$

Insertion into the original differential equation results in a differential equation for f:

$$-k_2' \alpha \cdot e^{-k_2' t} \cdot f + \alpha \cdot e^{-k_2' t} \cdot \frac{df}{dt} + k_2' \alpha \cdot e^{-k_2' t} \cdot f = k_1' c_{A,0} \cdot e^{-k_1' t}$$

or

$$\frac{df}{dt} = \frac{k_1'}{\alpha} c_{A,0} \cdot e^{-(k_1' - k_2')t}$$

By integration we obtain

$$f = -\frac{k_1'}{\alpha} \cdot \frac{c_{A,0}}{k_1' - k_2'} \cdot e^{-(k_1' - k_2')t} + \beta$$

where β is an integration constant. Then

$$\begin{aligned} c_C &= y \cdot f \\ &= -\alpha e^{-k_2' t} \cdot \frac{k_1' c_{A,0}}{\alpha\left(k_1' - k_2'\right)} \cdot e^{-(k_1' - k_2')t} + \beta \cdot \alpha e^{-k_2' t} \\ &= -\frac{k_1' c_{A,0}}{\left(k_1' - k_2'\right)} \cdot e^{-k_1' t} + \beta \cdot \alpha\, e^{-k_2' t} \end{aligned}$$

To eliminate the integration constant we apply the initial conditions. If at time $t = 0$ the concentration c_C equals zero, then

$$\beta \cdot \alpha = \frac{k_1' c_{A,0}}{k_1' - k_2'}$$

Substituting this result gives

$$c_C = -\frac{k_1' c_{A,0}}{(k_1' - k_2')} \cdot e^{-k_1' t} + \frac{k_1' c_{A,0}}{k_1' - k_2'} e^{-k_2' t}$$

and

$$c_C = -\frac{k_1' c_{A,0}}{k_1' - k_2'} \cdot \left(e^{-k_1' t} - e^{-k_2' t}\right)$$

P24.7 Given the expression for the concentration of C in the reaction sequence

$$A + B \xrightarrow{k_1} C, C + B \xrightarrow{k_2} D$$

$$c_C = -\frac{k_1' c_{A,0}}{k_1' - k_2'} \cdot \left(e^{-k_1' t} - e^{-k_2' t}\right)$$

which you show in P24.6, determine the time t_{max} required to reach the maximum concentration of c_C. What is the result for c_C and for t_{max} in the special case $k_1' = k_2' = k'$?

Solution

i) To find the maximum concentration, we differentiate c_C with respect to t and set it equal to zero:

$$-k_1' \cdot e^{-k_1' t_{max}} + k_2' \cdot e^{-k_2' t_{max}} = 0$$

This equation is satisfied when

$$e^{-(k_1' - k_2') t_{max}} = \frac{k_2'}{k_1'}$$

Taking the logarithm of both sides, we find

$$-\left(k_1' - k_2'\right) t_{max} = \ln\left(\frac{k_2'}{k_1'}\right) \quad \text{and} \quad t_{max} = \frac{\ln(k_2'/k_1')}{k_2' - k_1'}$$

ii) What happens in the special case $k_2' = k_1'$? In this case numerator and denominator become zero, and we have to evaluate the expression for t_{max} by using l'Hopital's rule. If we make the substitution

$$k_2' = k_1' + \Delta x$$

then we can take the limit as Δx goes to zero in order to find our result. Upon substitution we find that

$$c_C = \left(k_1' c_{A,0}\right) \frac{e^{-k_1' t} - e^{-(k_1' + \Delta x)t}}{\Delta x} = \left(k_1' c_{A,0}\right) e^{-k_1' t} \frac{1 - e^{-\Delta x \cdot t}}{\Delta x}$$

and the limit is

$$\lim_{\Delta x \to 0} c_C = \lim_{\Delta x \to 0} \left(k_1' c_{A,0} e^{-k_1' t} \left[\frac{1 - e^{-\Delta x t}}{\Delta x}\right]\right)$$

$$= k_1' c_{A,0} e^{-k_1' t} \lim_{\Delta x \to 0} \left[\frac{1 - e^{\Delta x t}}{\Delta x}\right]$$

Now we apply l'Hopital's rule and find

$$\lim_{\Delta x \to 0} c_C = k_1' c_{A,0} e^{-k_1' t} \lim_{\Delta x \to 0} \left[\frac{1 - e^{-\Delta x t}}{\Delta x}\right]$$

$$= k_1' c_{A,0} e^{-k_1' t} \lim_{\Delta x \to 0} \left[\frac{t \cdot e^{-\Delta x t}}{1}\right] = k_1' c_{A,0} t \cdot e^{-k_1' t}$$

Accordingly, we have to evaluate the expression for t_{max}

$$t_{max} = \frac{\ln(k_2'/k_1')}{k_2' - k_1'}$$

to obtain

$$\lim_{\Delta x \to 0} t_{max} = \lim_{\Delta x \to 0} \frac{\ln\left(\frac{k_1' + \Delta x}{k_1'}\right)}{\Delta x} = \lim_{\Delta x \to 0} \frac{\frac{k_1'}{k_1' + \Delta x} \cdot \frac{1}{k_1'}}{1} = \frac{1}{k_1'}$$

P24.8 General Form of Second-Order Reaction. Consider the general case of the bimolecular reaction

$$A + B \rightarrow C \quad \text{with the rate} \quad \frac{dc_A}{dt} = -k_r \cdot c_A \cdot c_B$$

in which $c_{A,0}$ and $c_{B,0}$ can be different.

Step 1: Using the definitions $c_A = c_{A,0} - x$, and $c_B = c_{B,0} - x$ where x is the amount of reacted species, show that the integrated rate law can be written as

$$\int \frac{dx}{(c_{A,0} - x) \cdot (c_{B,0} - x)} = k_r \int dt$$

Step 2: Integrate and apply the boundary condition of $x = 0$ at $t = 0$ and show that

$$\frac{c_{B,0} - x}{c_{A,0} - x} = \frac{c_{B,0}}{c_{A,0}} \cdot e^{(c_{B,0} - c_{A,0})k_r t}$$

Step 3: Use the definition $x = c_{A,0} - c_A$ to write the rate law in terms of c_B

Step 4: Show that the half-life is given by the expression

$$\tau_{1/2} = \frac{\ln\left(2 - \frac{c_{A,0}}{c_{B,0}}\right)}{\left(c_{B,0} - c_{A,0}\right) k_r} \quad \text{for } c_{A,0} \leq c_{B,0}$$

Solution

Step 1: If we substitute $c_A = c_{A,0} - x$, and $c_B = c_{B,0} - x$ into the rate expression we find that

$$\frac{dc_A}{dt} = -k_r \cdot c_A \cdot c_B = -k_r \cdot \left(c_{A,0} - x\right) \cdot \left(c_{B,0} - x\right)$$

or

$$-\frac{dx}{dt} = -k_r \cdot \left(c_{A,0} - x\right) \cdot \left(c_{B,0} - x\right)$$

To solve this equation, we must separate the variables and integrate:

$$\int \frac{1}{\left(c_{A,0} - x\right) \cdot \left(c_{B,0} - x\right)} dx = k_r \cdot \int dt'$$

The solution of the left integral is found in Appendix B and we obtain

$$\int \frac{1}{\left(c_{A,0} - x\right) \cdot \left(c_{B,0} - x\right)} dx = \frac{1}{c_{B,0} - c_{A,0}} \cdot \ln\left(\frac{c_{B,0} - x}{c_{A,0} - x}\right)$$

Step 2: Applying the boundary conditions gives

$$\left| \frac{1}{c_{B,0} - c_{A,0}} \cdot \ln \frac{c_{B,0} - x}{c_{A,0} - x} \right|_{x=0}^{x} = k_r \cdot \left|t'\right|_0^t$$

$$\frac{1}{c_{B,0} - c_{A,0}} \cdot \left[\ln \frac{c_{B,0} - x}{c_{A,0} - x} - \ln \frac{c_{B,0}}{c_{A,0}}\right] = k_r \cdot t$$

$$\ln \left[\frac{c_{B,0} - x}{c_{A,0} - x} \cdot \frac{c_{A,0}}{c_{B,0}}\right] = (c_{B,0} - c_{A,0}) \cdot k_r \cdot t$$

and

$$\frac{c_{B,0} - x}{c_{A,0} - x} \cdot \frac{c_{A,0}}{c_{B,0}} = \exp\left[(c_{B,0} - c_{A,0}) \cdot k_r \cdot t\right]$$

or

$$\frac{c_{B,0} - x}{c_{A,0} - x} = \frac{c_{B,0}}{c_{A,0}} \cdot \exp\left[(c_{B,0} - c_{A,0}) \cdot k_r \cdot t\right]$$

Step 3: We resubstitute x in the expression for $c_{B,0} - x$

$$c_B = (c_{A,0} - x)\frac{c_{B,0}}{c_{A,0}} \cdot \exp\left[(c_{B,0} - c_{A,0}) \cdot k_r \cdot t\right]$$

Because of $c_B = c_{B,0} - x$ we find that

$$c_{A,0} - x = c_{A,0} - c_{B,0} + c_B$$

and

$$c_B = (c_{A,0} - c_{B,0} + c_B)\frac{c_{B,0}}{c_{A,0}} \cdot \exp\left[(c_{B,0} - c_{A,0}) \cdot k_r \cdot t\right]$$
$$c_B = (c_{A,0} - c_{B,0})\frac{c_{B,0}}{c_{A,0}} + c_B\frac{c_{B,0}}{c_{A,0}} \cdot \exp\left[(c_{B,0} - c_{A,0}) \cdot k_r \cdot t\right]$$

Finally we resolve for c_B

$$c_B\left[1 - \frac{c_{B,0}}{c_{A,0}} \cdot \exp\left[(c_{B,0} - c_{A,0}) \cdot k_r \cdot t\right]\right] = (c_{A,0} - c_{B,0})\frac{c_{B,0}}{c_{A,0}} \cdot \exp\left[(c_{B,0} - c_{A,0}) \cdot k_r \cdot t\right]$$

and

$$c_B = \frac{(c_{A,0} - c_{B,0})\frac{c_{B,0}}{c_{A,0}} \cdot \exp\left[(c_{B,0} - c_{A,0}) \cdot k_r \cdot t\right]}{1 - \frac{c_{B,0}}{c_{A,0}} \cdot \exp\left[(c_{B,0} - c_{A,0}) \cdot k_r \cdot t\right]}$$
$$= \frac{(c_{A,0} - c_{B,0})}{\frac{c_{A,0}}{c_{B,0}} \cdot e^{(c_{B,0} - c_{A,0}) \cdot k_r \cdot t} - 1}$$

Accordingly, we obtain

$$c_A = \frac{c_{B,0} - c_{A,0}}{\frac{c_{B,0}}{c_{A,0}} \cdot e^{(c_{B,0} - c_{A,0})k_r t} - 1}$$

We can check the result for $t = 0$ and $t \to \infty$:

$$c_B(t = 0) = \frac{(c_{A,0} - c_{B,0})}{\frac{c_{AB,0}}{c_{B,0}} - 1} = \frac{(c_{A,0} - c_{B,0})}{\frac{c_{A,0} - c_{B,0}}{c_{B,0}}} = c_{B,0}$$
$$c_B(t \to \infty) = \frac{(c_{A,0} - c_{B,0})}{-1} = c_{B,0} - c_{A,0}$$

Step 4: To obtain the half-time $\tau_{1/2}$ we set $c_B = \frac{1}{2}c_{B,0}$ and write

$$\frac{1}{2}c_{B,0} = \frac{(c_{A,0} - c_{B,0})}{\frac{c_{A,0}}{c_{B,0}} \cdot \exp\left[-(c_{B,0} - c_{A,0}) \cdot k_r \cdot \tau_{1/2}\right] - 1}$$

so that

$$\frac{1}{2}c_{B,0}\left[\frac{c_{A,0}}{c_{B,0}} \cdot \exp\left[-(c_{B,0} - c_{A,0}) \cdot k_r \cdot \tau_{1/2}\right] - 1\right] = c_{A,0} - c_{B,0}$$

which rearranges in the following manner

$$\frac{1}{2}c_{B,0}\frac{c_{A},0}{c_{B,0}} \cdot \exp\left[-(c_{B,0} - c_{A,0}) \cdot k_r \cdot \tau_{1/2}\right] - \frac{1}{2}c_{B,0} = c_{A,0} - c_{B,0}$$
$$\frac{1}{2}c_{A,0} \cdot \exp\left[-(c_{B,0} - c_{A,0}) \cdot k_r \cdot \tau_{1/2}\right] = \frac{1}{2}c_{B,0} + c_{A,0} - c_{B,0}$$
$$\exp\left[-(c_{B,0} - c_{A,0}) \cdot k_r \cdot \tau_{1/2}\right] = 2 - \frac{c_{B,0}}{c_{A,0}}$$
$$\left[-(c_{B,0} - c_{A,0}) \cdot k_r \cdot \tau_{1/2}\right] = \ln\left(2 - \frac{c_{B,0}}{c_{A,0}}\right)$$
$$\tau_{1/2} = -\frac{\ln\left(2 - \frac{c_{B,0}}{c_{A,0}}\right)}{(c_{B,0} - c_{A,0}) \cdot k_r}$$

and finally

$$\tau_{1/2} = -\frac{\ln\left(2 - \frac{c_{B,0}}{c_{A,0}}\right)}{(c_{B,0} - c_{A,0}) \cdot k_r} \quad \text{for } c_{B,0} < c_{A,0}$$

P24.9 In this problem you show that the general expression for the second-order rate law where $c_{B,0} \neq c_{A,0}$,

$$c_A = \frac{c_{B,0} - c_{A,0}}{\frac{c_{B,0}}{c_{A,0}} \cdot e^{(c_{B,0}-c_{A,0})k_r t} - 1}$$

reduces to the expression

$$c_A = \frac{1}{\frac{1}{c_{A,0}} + k_r t} \quad \text{for } c_{B,0} = c_{A,0}$$

and to the quasi-first order expression

$$c_A = c_{A,0} \cdot e^{-c_{B,0} k_r t} \quad \text{for } c_{B,0} \gg c_{A,0}$$

Make plots of c_A versus time t plots of the three expressions using $k_r = 2\ \text{L mol}^{-1}\ \text{s}^{-1}$ and $c_{A,0} = 0.1\ \text{mol L}^{-1}$ for Curve 1: $c_{B,0} = 0.15\ \text{mol L}^{-1}$. Curve 2: $c_{B,0} = 0.10\ \text{mol L}^{-1}$. Curve 3: $c_{B,0} = 1.0\ \text{mol L}^{-1}$.

(Hint: To show the simplification of the general expression to the special case of $c_{B,0} = c_{A,0}$, you must be careful to avoid divergences. Use l'Hopital's rule, and proceed by defining $c_{B,0} = c_{A,0} + z$ and taking the limit as z approaches zero.)

Solution

First we show that the general result reduces to the special case when $c_{B,0} = c_{A,0}$. We set $c_{B,0} = c_{A,0} + z$ in the rate equation, and consider the limit

$$\lim_{z\to 0} c_A = \lim_{z\to 0} \frac{z}{\left(1 + \frac{z}{c_{A,0}}\right) \cdot e^{z k_r t} - 1}$$

Because the numerator and denominator each go to zero as z goes to zero, we use l'Hospital's rule; namely

$$\lim_{z\to 0} c_A = \lim_{z\to 0} \frac{1}{e^{z k_r t} \frac{1}{c_{A,0}} + k_r t \left(1 + z c_{A,0}\right) e^{z k_r t}}$$

$$= \lim_{z\to 0} \frac{1}{e^{z k_r t}\left[\frac{1}{c_{A,0}} + k_r t + \frac{z k_r t}{c_{A,0}}\right]} = \frac{1}{\frac{1}{c_{A,0}} + k_r t}$$

Next we consider the limit of this result, where $c_{B,0} \gg c_{A,0}$. In this case $c_{B,0} - c_{A,0} \approx c_{B,0}$ and we can neglect the term −1 in the denominator of the rate equation. Thus

$$c_A = \frac{c_{B,0}}{\frac{c_{B,0}}{c_{A,0}} \cdot e^{c_{B,0} k_r t}} = c_{A,0} \cdot e^{-c_{B,0} k_r t}$$

Lastly, Fig. P24.9 plots c_A for the general case and for the two limiting cases considered above.

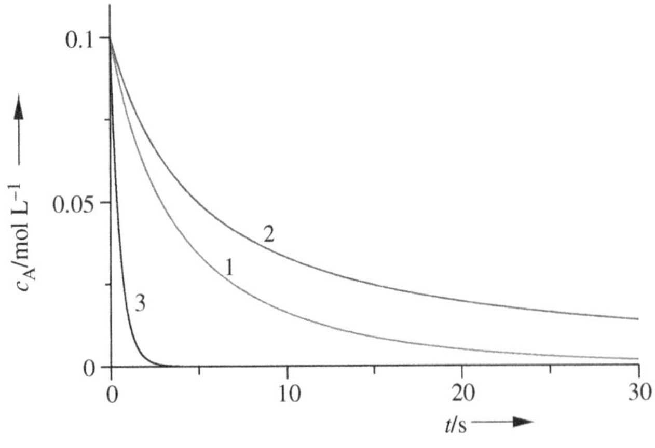

Figure P24.9 c_A versus time t plot according to Equation (24.58) with $k_r = 2\ \text{L mol}^{-1}\ \text{s}^{-1}$ and $c_{A,0} = 0.1\ \text{mol L}^{-1}$. Curve 1: $c_{B,0} = 0.15\ \text{mol L}^{-1}$. Curve 2: $c_{B,0} = 0.10\ \text{mol L}^{-1}$. Curve 3: $c_{B,0} = 1.0\ \text{mol L}^{-1}$.

P24.10 In Equation (24.150) in Section 24.7.5 for an autocatalytic reaction

$$\mathrm{A} + \mathrm{B} \overset{k_1}{\to} \mathrm{C} + 2\mathrm{B}$$

we found for the differential equation

$$\frac{\mathrm{d}c_\mathrm{B}}{\mathrm{d}t} = k_r \cdot c_\mathrm{B} \cdot (c_{\mathrm{A},0} + c_{\mathrm{B},0} - c_\mathrm{B})$$

Solve this equation and show that

$$c_\mathrm{B} = \frac{c_{\mathrm{A},0} + c_{\mathrm{B},0}}{1 + \frac{c_{\mathrm{A},0}}{c_{\mathrm{B},0}}\,\mathrm{e}^{-(c_{\mathrm{A},0}+c_{\mathrm{B},0})k_r t}}$$

Solution

One way to prove this result is to substitute the solution given into the differential equation and to show that the equality holds. This is left for the reader to confirm.

Here we show how to solve the equation directly. To solve this equation we separate the variables, by writing

$$\frac{\mathrm{d}c_\mathrm{B}}{c_\mathrm{B} \cdot \left(c_{\mathrm{A},0} + c_{\mathrm{B},0} - c_\mathrm{B}\right)} = k_r \cdot \mathrm{d}t$$

and integrate

$$\int \frac{\mathrm{d}c_\mathrm{B}}{c_\mathrm{B} \cdot \left(c_{\mathrm{A},0} + c_{\mathrm{B},0} - c_\mathrm{B}\right)} = k_r t \quad \text{or} \quad \int \frac{\mathrm{d}c_\mathrm{B}}{\left(c_{\mathrm{A},0} + c_{\mathrm{B},0}\right) \cdot c_\mathrm{B} - c_\mathrm{B}^2} = k_r t$$

In the appendix we find the general solution for the integral

$$\int \frac{1}{ax - x^2} \cdot \mathrm{d}x = -\frac{1}{a} \cdot \ln\left(1 - \frac{a}{x}\right)$$

Thus we obtain

$$\int \frac{\mathrm{d}c_\mathrm{B}}{\left(c_{\mathrm{A},0} + c_{\mathrm{B},0}\right) \cdot c_\mathrm{B} - c_\mathrm{B}^2} = -\frac{1}{c_{\mathrm{A},0} + c_{\mathrm{B},0}} \ln\left(1 - \frac{c_{\mathrm{A},0} + c_{\mathrm{B},0}}{c_\mathrm{B}}\right)$$

and

$$-\frac{1}{c_{\mathrm{A},0} + c_{\mathrm{B},0}} \left|\ln\left(1 - \frac{c_{\mathrm{A},0} + c_{\mathrm{B},0}}{c_\mathrm{B}}\right)\right|_{c_{\mathrm{B}0}}^{c_\mathrm{B}} = k_r t$$

$$\left|\ln\left(\frac{c_\mathrm{B} - (c_{\mathrm{A},0} + c_{\mathrm{B},0})}{c_\mathrm{B}}\right)\right|_{c_{\mathrm{B}_0}}^{c_\mathrm{B}} = -\frac{1}{c_{\mathrm{A},0} + c_{\mathrm{B},0}} k_r t$$

$$\left[\ln\left(\frac{c_\mathrm{B} - (c_{\mathrm{A},0} + c_{\mathrm{B},0})}{c_\mathrm{B}}\right) - \ln\left(\frac{c_{\mathrm{B}_0} - (c_{\mathrm{A},0} + c_{\mathrm{B},0})}{c_{\mathrm{B}_0}}\right)\right] = -(c_{\mathrm{A},0} + c_{\mathrm{B},0})k_r t$$

$$\left[\ln\left(\frac{c_\mathrm{B} - (c_{\mathrm{A},0} + c_{\mathrm{B},0})}{c_\mathrm{B}}\right) - \ln\left(-\frac{c_{\mathrm{A},0}}{c_{\mathrm{B}_0}}\right)\right] = -(c_{\mathrm{A},0} + c_{\mathrm{B},0})k_r t$$

$$\ln\left[-\frac{c_\mathrm{B} - (c_{\mathrm{A},0} + c_{\mathrm{B},0})}{c_\mathrm{B}} \frac{c_{\mathrm{B}_0}}{c_{\mathrm{A},0}}\right] = -(c_{\mathrm{A},0} + c_{\mathrm{B},0})k_r t$$

$$\ln\left[-\frac{c_\mathrm{B}}{c_\mathrm{B} c_\mathrm{B} - (c_{\mathrm{A},0} + c_{\mathrm{B},0})} \frac{c_{\mathrm{A}_0}}{c_{\mathrm{B},0}}\right] = (c_{\mathrm{A},0} + c_{\mathrm{B},0})k_r t$$

$$\ln\left[\frac{c_{\mathrm{A},0}}{c_{\mathrm{B},0}} \frac{c_\mathrm{B}}{(c_{\mathrm{A},0} + c_{\mathrm{B},0}) - c_\mathrm{B}}\right] = (c_{\mathrm{A},0} + c_{\mathrm{B},0})k_r t$$

By exponentiating both sides, we find that

$$\frac{c_{\mathrm{A},0}}{c_{\mathrm{B},0}} \frac{c_\mathrm{B}}{(c_{\mathrm{A},0} + c_{\mathrm{B},0}) - c_\mathrm{B}} = \mathrm{e}^{(c_{\mathrm{A},0}+c_{\mathrm{B},0})k_r t}$$

Isolating c_B, we find

$$c_\mathrm{B} = \frac{c_{\mathrm{B},0}\left(c_{\mathrm{A},0} + c_{\mathrm{B},0}\right)}{c_{\mathrm{A},0}\,\mathrm{e}^{-(c_{\mathrm{A},0}+c_{\mathrm{B},0})k_r t} + c_{\mathrm{B},0}} = \frac{c_{\mathrm{A},0} + c_{\mathrm{B},0}}{1 + \frac{c_{\mathrm{A},0}}{c_{\mathrm{B},0}}\,\mathrm{e}^{-(c_{\mathrm{A},0}+c_{\mathrm{B},0})k_r t}}$$

P24.11 Critical Bromide Concentration. Estimate c_{crit} in the inhibition of the autocatalytic oxidation of Ru^{2+} by considering the rate equations for the autocatalytic process and include the inhibition process. Assume that the pH is low, so that $k_3' = k_3 \cdot c_{H^+}$ and use the steady-state approximation to find an expression for c_{BrO_2}. Combine this result with the results in Example 24.15 to find an expression for $c_{Ru^{2+}}$. If we define the critical bromide concentration to be the transition point between an exponentially decreasing $c_{Ru^{2+}}$ and an exponentially increasing $c_{Ru^{2+}}$, show that $c_{crit} = k_2'/k_3'$.

Solution

We begin with the differential rate expression

$$\frac{dc_{HBrO_2}}{dt} = k_1' \cdot c_{Ru^{2+}} \cdot c_{BrO_2} - k_2' \cdot c_{HBrO_2} - k_3' \cdot c_{HBrO_2} \cdot c_{Br^-}$$

Using the steady state approximation for c_{BrO_2} (Equation (24.157))

$$\frac{dc_{BrO_2}}{dt} = -k_1' c_{Ru^{2+}} \cdot c_{BrO_2} + 2k_2' c_{HBrO_2} = 0$$

$$c_{BrO_2} = \frac{2k_2' c_{HBrO_2}}{k_1' c_{Ru^{2+}}}$$

we find

$$\frac{dc_{HBrO_2}}{dt} = k_1' \cdot c_{Ru^{2+}} \cdot \frac{2k_2' c_{HBrO_2}}{k_1' c_{Ru^{2+}}} - k_2' \cdot c_{HBrO_2} - k_3' \cdot c_{HBrO_2} \cdot c_{Br^-}$$
$$= c_{HBrO_2} \cdot \left(k_2' - k_3' \cdot c_{Br^-}\right)$$

$$\frac{dc_{HBrO_2}}{c_{HBrO_2}} = \left(k_2' - k_3' \cdot c_{Br^-}\right) \cdot dt$$

$$\ln \frac{c_{HBrO_2}}{c_{HBrO_2,0}} = \left(k_2' - k_3' \cdot c_{Br^-}\right) \cdot t$$

and

$$c_{HBrO_2} = c_{HBrO_2,0} \cdot e^{(k_2' - k_3' \cdot c_{Br^-})t}$$

For the concentration of Ru^{2+}, in analogy to the consideration in Example 24.15, we obtain

$$\frac{dc_{Ru^{2+}}}{dt} = -2k_2' c_{HBrO_2} = -2k_2' \cdot c_{HBrO_2,0} \cdot e^{(k_2' - k_3' \cdot c_{Br^-})t}$$

$$c_{Ru^{2+}} = c_{Ru^{2+},0} - 2k_2' \cdot c_{HBrO_2,0} \cdot \left[e^{(k_2' - k_3' c_{Br^-})t} - 1\right]$$

For $k_2' > k_3' \cdot c_{Br^-}$ the exponent is positive: exponential growth of $HBrO_2$. For $k_2' < k_3' \cdot c_{Br^-}$ we expect an exponential decay. The critical bromide concentration is defined by the concentration at the transition point

$$k_2' - k_3' c_{crit} = 0, \; c_{crit} = \frac{k_2'}{k_3'}$$

P24.12 The $H_2 + I_2 \rightarrow 2\,HI$ reaction proceeds by the formation of H_2I adducts, rather than through a chain mechanism.[5] Using the reaction mechanism

$$I_2 \underset{k_{-1}}{\overset{k_1}{\rightleftarrows}} 2\,I \qquad (24P12.1)$$

$$H_2 + I \underset{k_{-2}}{\overset{k_2}{\rightleftarrows}} H_2I \qquad (24P12.2)$$

$$H_2I + I \overset{k_3}{\rightarrow} 2\,HI \qquad (24P12.3)$$

5 J.H. Sullivan, Mechanism of the "bimolecular" hydrogen-iodine reaction, *J. Chem. Phys.* 1967, **46**, 73.

show that the rate law is bimolecular can be written as

$$\frac{dc_{HI}}{dt} = k_{eff} \cdot c_{I_2} \cdot c_{H_2}$$

and find an expression for k_{eff} in terms of the rate constants given in the fundamental steps of the reaction mechanism.

Solution

From Equation (24P12.3), we can write that

$$\frac{dc_{HI}}{dt} = 2k_3 \cdot c_{H_2I} \cdot c_I$$

The equilibria represented by (24P12.1) and (24P12.2) allow us to write that

$$\frac{k_1}{k_{-1}} = \frac{c_I^2}{c_{I_2}} \text{ and } \frac{k_2}{k_{-2}} = \frac{c_{H_2I}}{c_{H_2} \cdot c_I}$$

If we solve the equilibrium (24P12.2) for the concentration of the adduct H_2I, we can substitute it into the differential rate law for HI and formation and find

$$\frac{dc_{HI}}{dt} = 2k_3 \cdot \frac{k_2}{k_{-2}} \cdot c_{H_2} \cdot c_I \cdot c_I = 2k_3 \cdot \frac{k_2}{k_{-2}} \cdot c_{H_2} \cdot c_I^2$$

Next we solve the equilibrium (24P12.1) for the concentration of I squared

$$\frac{c_I^2}{c_{I_2}} = \frac{k_1}{k_{\ 1}} \qquad c_I^2 = \frac{k_1}{k_{-1}} c_{I_2}$$

and substitute it into the differential rate law and find

$$\frac{dc_{HI}}{dt} = 2k_3 \cdot \frac{k_2}{k_{-2}} \cdot c_{H_2} \cdot \frac{k_1}{k_{-1}} \cdot c_{I_2}$$

$$= 2k_3 \cdot \frac{k_1}{k_{-1}} \cdot \frac{k_2}{k_{-2}} \cdot c_{I_2} \cdot c_{H_2}$$

This result confirms that the reaction is formally second order, although the mechanism is much more complicated than a second-order elementary step.

Comparison of this result with that shown in the problem statement,

$$\frac{dc_{HI}}{dt} = k_{eff} \cdot c_{I_2} \cdot c_{H_2}$$

allows us to deduce that the effective rate constant k_{eff} is

$$k_{eff} = 2k_3 \cdot \frac{k_1}{k_{-1}} \cdot \frac{k_2}{k_{-2}}$$

25

Transition States and Chemical Reactions

25.1 Exercises

E25.1 The gas-phase reaction between F_2 and SF_5 is first order in each of the reactants. At 298 K the rate constant is 4.39×10^5 L mol^{-1} s^{-1} and the empirically determined activation energy is 20.29 kJ mol^{-1}. Use the definition of the empirical activation energy $E_{a,m}^{\text{empirical}}$ given in the text and the transition state theory expression for a bimolecular gas-phase rate constant, namely

$$k_{\text{TST}} = \left(\frac{kT}{h}\right)\left(\frac{1}{c^{\ominus}}\right) \cdot \exp\left(\frac{\Delta_r S^{\ddagger\ominus}}{\boldsymbol{R}}\right) \cdot \exp\left(-\frac{\Delta_r H^{\ddagger\ominus}}{\boldsymbol{R}T}\right)$$

to calculate the entropy of activation and the enthalpy of activation at 298 K.

Solution

We start with the transition state equation for a bimolecular reaction

$$k_{\text{TST}} = \frac{1}{c^{\ominus}} \cdot \frac{kT}{h} \cdot K^{\ddagger} = \frac{1}{c^{\ominus}} \cdot \frac{kT}{h} \cdot e^{\Delta_r S^{\ddagger\ominus}/\boldsymbol{R}} \cdot e^{\Delta_r H^{\ominus}/(\boldsymbol{R}T)} \tag{25.1}$$

In the text (see Section 25.2.3) we derive a relation between $E_{a,m}^{\text{empirical}}$ and $\Delta_r H^{\ddagger\ominus}$, namely,

$$E_{a,m}^{\text{empirical}} = \Delta_r H^{\ddagger\ominus} + \boldsymbol{R}T$$

where we have assumed that $\Delta_r H^{\ddagger\ominus}$ and $\Delta_r S^{\ddagger\ominus}$ do not change with temperature. Solving for $\Delta_r H^{\ddagger\ominus}$, we find that

$$\begin{aligned}\Delta_r H^{\ddagger\ominus} &= E_{a,m}^{\text{empirical}} - \boldsymbol{R}T \\ &= \left[20.29 \text{ kJ mol}^{-1} - 8.314 \text{ J mol}^{-1}\text{K}^{-1} \cdot 298 \text{ K}\right] = 17.81 \text{kJ mol}^{-1}\end{aligned}$$

Using this value and our value of the rate constant at 298 K, we can find the entropy of activation. Using Equations (25.24–25.36) of the textbook, we solve for $\exp\left(\Delta_r S^{\ddagger\ominus}/\boldsymbol{R}\right)$:

$$\exp\left(\frac{\Delta_r S^{\ddagger\ominus}}{\boldsymbol{R}}\right) = k_r \exp\left(\frac{\Delta_r H^{\ddagger\ominus}}{\boldsymbol{R}T}\right)\left(\frac{h}{kT}\right)c^{\ominus}$$

or

$$\Delta_r S^{\ddagger\ominus} = \frac{\Delta_r H^{\ddagger\ominus}}{T} + \boldsymbol{R}\ln\left[k_r\left(\frac{h}{kT}\right)c^{\ominus}\right]$$

Substituting the given values we find that

$$\frac{\Delta_r H^{\ddagger\ominus}}{T} = \frac{17.81 \text{ kJ mol}^{-1}}{298 \text{ K}} = 59.76 \text{ J mol}^{-1} \text{ K}^{-1}$$

$$\begin{aligned}&\boldsymbol{R}\ln\left[k_r\left(\frac{h}{kT}\right)c^{\ominus}\right] \\ &= \boldsymbol{R}\ln\left[4.39 \times 10^5 \text{ L mol}^{-1} \text{ s}^{-1} \frac{6.626 \times 10^{-34} \text{ J s}}{1.38 \times 10^{-23} \text{ J K}^{-1} \text{ 298 K}} \cdot (1 \text{ mol L}^{-1})\right]\end{aligned}$$

Solutions Manual for Principles of Physical Chemistry, Third Edition. Edited by Hans Kuhn, David H. Waldeck, and Horst-Dieter Försterling.

$$= \boldsymbol{R} \ln \left[4.39 \times 10^{5} \cdot 1.61 \times 10^{-13}\right]$$
$$= \boldsymbol{R} \ln \left[7.068 \times 10^{-8}\right] = -16.5\boldsymbol{R} = -137 \text{ J mol}^{-1} \text{ K}^{-1}$$

Thus

$$\Delta_r S^{\ddagger\ominus} = (59.76 - 137) \text{ J mol}^{-1} \text{ K}^{-1} = -77.2 \text{ J mol}^{-1} \text{ K}^{-1}$$

The large, negative activation entropy implies that the transition state has a constrained geometry, e.g., a colinear F_5S–F–F.[1]

E25.2 Compare the unimolecular transition state theory expression for the rate constant to the Arrhenius expression for the rate constant

$$k_r = A \cdot \exp(-E_{a,m}^{\text{empirical}}/\boldsymbol{R}T)$$

and show that

$$E_{a,m}^{\text{empirical}} = \Delta_r H^{\ddagger\ominus} + \boldsymbol{R}T$$

Solution

From the Arrhenius expression we find, if the preexponential factor A is assumed to be independent of temperature, that

$$\ln\left(\frac{k_r}{k_{r,1}}\right) = -\frac{E_{a,m}^{\text{empirical}}}{\boldsymbol{R}}\left(\frac{1}{T} - \frac{1}{T_1}\right) \tag{25.2}$$

where k_r is the rate constant at temperature T and $k_{r,1}$ is the rate constant at temperature T_1.

The unimolecular transition state theory expression for the rate constant is

$$k_r = A_f \cdot e^{-\Delta_r H^{\ddagger\ominus}/(\boldsymbol{R}T)} = \frac{kT}{h} \cdot e^{\Delta_r S^{\ddagger\ominus}/\boldsymbol{R}} \cdot e^{-\Delta_r H^{\ddagger\ominus}/(\boldsymbol{R}T)}$$

We see that in this case the preexponential factor is proportional to the temperature T and we have

$$A_f = A_{f,0} \cdot T$$

Then we can write that

$$k_r = A_f \cdot \exp\left(-\frac{\Delta_r H^{\ddagger\ominus}}{\boldsymbol{R}T}\right) = A_{f,0} \cdot T \cdot \exp\left(-\frac{\Delta_r H^{\ddagger\ominus}}{\boldsymbol{R}T}\right)$$

or

$$k_{r,1} = A_{f,0} \cdot T_1 \cdot \exp\left(-\frac{\Delta_r H^{\ddagger\ominus}}{\boldsymbol{R}T_1}\right)$$

at a temperature T_1. Thus, we can write that

$$\frac{k_r}{k_{r,1}} = \frac{T}{T_1} \cdot \exp\left(-\frac{\Delta_r H^{\ddagger\ominus}}{\boldsymbol{R}T}\right) / \exp\left(-\frac{\Delta_r H^{\ddagger\ominus}}{\boldsymbol{R}T_1}\right)$$

Finally we take the logarithm of both sides and find

$$\ln\left(\frac{k_r}{k_{r,1}}\right) = \ln\left(\frac{T}{T_1}\right) - \frac{\Delta_r H^{\ddagger\ominus}}{\boldsymbol{R}T} + \frac{\Delta_r H^{\ddagger\ominus}}{\boldsymbol{R}T_1}$$
$$= \ln\left(\frac{T}{T_1}\right) - \frac{\Delta_r H^{\ddagger\ominus}}{\boldsymbol{R}}\left(\frac{1}{T} - \frac{1}{T_1}\right)$$

Reorganization of the terms yields

$$\ln\left(\frac{k_r}{k_{r,1}}\right) - \ln\left(\frac{T}{T_1}\right) = -\frac{\Delta_r H^{\ddagger\ominus}}{\boldsymbol{R}}\left(\frac{1}{T} - \frac{1}{T_1}\right) \tag{25.3}$$

1 C. Buendía-Atencio, G.P. Pieffet, A.E. Croce, C.J. Cobos, *Comput. Theor. Chem.* 2016, doi: http://dx.doi.org/10.1016/j.comptc.2016.05.015.

How is the slope $-\Delta_r H^{\ddagger\ominus}/R$ related to the empirical activation energy $E_{a,m}^{\text{empirical}}$? Using Equations (25.39) and (25.40), we find that

$$\frac{\mathrm{d}\ln(k_r/k_{r,1})}{\mathrm{d}(1/T)} = -\frac{E_{a,m}^{\text{empirical}}}{R} \qquad \text{(Arrhenius)}$$

and

$$\frac{\mathrm{d}\left[\ln(k_r/k_{r,1}) - \ln(T/T_1)\right]}{\mathrm{d}(1/T)} = -\frac{\Delta_r H^{\ddagger\ominus}}{R} \qquad \text{(Unimolecular transition state)}$$

Thus we obtain

$$\begin{aligned}-\frac{\Delta_r H^{\ddagger\ominus}}{R} &= -\frac{E_{a,m}^{\text{empirical}}}{R} - \frac{\mathrm{d}\left[\ln(T/T_1)\right]}{\mathrm{d}(1/T)} \\ &= -\frac{E_{a,m}^{\text{empirical}}}{R} - \frac{\mathrm{d}\left[\ln(T/T_1)\right]}{\mathrm{d}T}\cdot\frac{\mathrm{d}T}{\mathrm{d}(1/T)} \\ &= -\frac{E_{a,m}^{\text{empirical}}}{R} - \frac{1}{T}\cdot\frac{\mathrm{d}T}{\mathrm{d}(1/T)}\end{aligned}$$

With the substitution $1/T = x$ it follows that

$$\frac{\mathrm{d}T}{\mathrm{d}(1/T)} = \frac{\mathrm{d}\frac{1}{x}}{\mathrm{d}x} = -\frac{1}{x^2} = -T^2$$

and we find

$$-\frac{\Delta_r H^{\ddagger\ominus}}{R} = -\frac{E_{a,m}^{\text{empirical}}}{R} + T \quad \text{or} \quad \Delta_r H^{\ddagger\ominus} = E_{a,m}^{\text{empirical}} - RT$$

Note that this is the same result as we find in the text for the bimolecular transition state theory expression.

E25.3 For the reaction in Exercise E24.12, perform a transition state theory analysis, using the bimolecular expression for the rate constant, to determine $\Delta_r H^{\ddagger\ominus}$ and $\Delta_r S^{\ddagger\ominus}$. Comment on the sign of your value for $\Delta_r S^{\ddagger\ominus}$ and on its magnitude.

Solution

We consider the thermal decomposition of N_2O_5,[2] which is first order in the N_2O_5 concentration

$$N_2O_5 \rightarrow 2NO_2 + \frac{1}{2}O_2$$

and analyze its rate constant data in Table E25.3 by way of the transition state theory rate constant expression,

$$k_{\text{TST}} = \frac{kT}{h}\exp\left(\frac{\Delta_r S^{\ddagger\ominus}}{R}\right)\exp\left(-\frac{\Delta_r H^{\ddagger\ominus}}{RT}\right)$$

Using the definitions $k_{r,1}/T_1 = 2.88\cdot10^{-9}\ \mathrm{s^{-1}\ K^{-1}}$ and $T_1 = 273$ K, we find the data in Table E25.3, columns 4 and 5.

Table E25.3 Rate constants for Reaction (25.4).

$k_r/(10^{-5}\,\mathrm{s^{-1}})$	T/K	$k_r/T\,/(\mathrm{s^{-1}\,K^{-1}})$	$\ln(k_r/T/(k_{r,1}/T_1))$	$1/T\,/(10^{-3}\,\mathrm{K^{-1}})$
0.0787	273	$2.88\cdot10^{-9}$	0	3.66
3.46	298	$1.16\cdot10^{-7}$	3.696	3.36
49.8	318	$1.57\cdot10^{-6}$	6.298	3.15
487	338	$1.44\cdot10^{-5}$	8.517	2.96

2 F. Daniels and E. H. Johnston, *J. Am. Chem. Soc.* 1921, **43**, 53.

To evaluate the data, we identify the experimental rate constant k_r with the theoretical rate constant k_{TST}. Then we obtain

$$\ln\left(\frac{k_r/T}{k_{r,1}/T_1}\right) = \ln\left(\frac{k_{\mathrm{TST}}/T}{k_{\mathrm{TST},1}/T_1}\right) = -\frac{\Delta_r H^{\ddagger\ominus}}{\boldsymbol{R}}\left(\frac{1}{T}-\frac{1}{T_1}\right)$$

and plot $\ln(k_r/T/(k_{r,1}/T_1))$ versus $1/T$; see Fig. E25.3.

Figure E25.3 $\ln(k_r/T/(k_{r,1}/T_1))$ versus $1/T$.

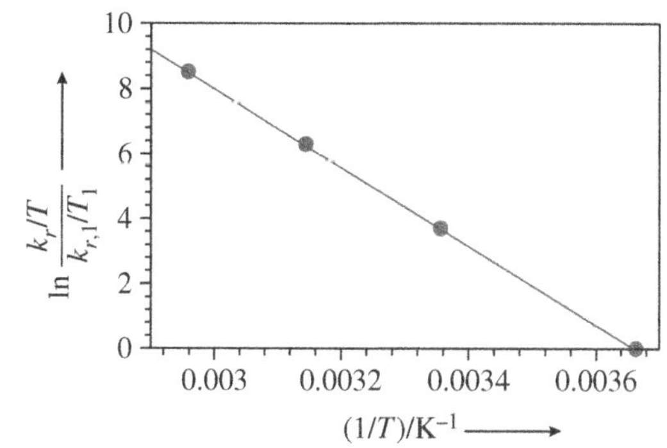

A fit of these data to a line gives a slope of −12, 199 K. Hence we have that

$$\Delta_r H^{\ddagger\ominus} = 12,199\ \mathrm{K}\cdot 8.314\ \mathrm{J\,mol^{-1}\,K^{-1}} = 101.4\ \mathrm{kJ\,mol^{-1}}$$

and find activation entropy values from the relation

$$k_{\mathrm{TST}} = \frac{kT}{h}\exp\left(\frac{\Delta_r S^{\ddagger\ominus}}{\boldsymbol{R}}\right)\exp\left(-\frac{\Delta_r H^{\ddagger\ominus}}{\boldsymbol{R}T}\right)$$

Rearranging this expression, we can write

$$\exp\left(\frac{\Delta_r S^{\ddagger\ominus}}{\boldsymbol{R}}\right) = \frac{k_{\mathrm{TST}}h}{kT}\exp\left(\frac{\Delta_r H^{\ddagger\ominus}}{\boldsymbol{R}T}\right)$$

so that

$$\frac{\Delta_r S^{\ddagger\ominus}}{\boldsymbol{R}} = \ln\left(\frac{k_{\mathrm{TST}}h}{kT}\right) + \frac{\Delta_r H^{\ddagger\ominus}}{\boldsymbol{R}T}$$

If we identify k_{TST} with k_r then we obtain

$$\Delta_r S^{\ddagger\ominus} = \boldsymbol{R}\ln\left(\frac{hk_r}{kT}\right) + \frac{\Delta_r H^{\ddagger\ominus}}{T}$$

For example, at 273 K we find

$$\boldsymbol{R}\ln\left(\frac{hk_r}{kT}\right) = \boldsymbol{R}\ln\left(\frac{6.626\cdot 10^{-34}\ \mathrm{J\,s}\ 7.87\cdot 10^{-7}\mathrm{s^{-1}}}{1.3802\cdot 10^{-23}\ \mathrm{J\,K^{-1}}\ 273\ \mathrm{K}}\right) = -361\ \mathrm{J\,K^{-1}\,mol^{-1}}$$

so that

$$\frac{\Delta_r H^{\ddagger\ominus}}{T} = \frac{101.4\ \mathrm{kJ\,mol^{-1}}}{273\ \mathrm{K}} = 371\ \mathrm{J\,K^{-1}\,mol^{-1}}$$

and

$$\Delta_r S^{\ddagger\ominus} = (-361+371)\ \mathrm{J\,K^{-1}\,mol^{-1}} = 10.0\ \mathrm{J\,K^{-1}\,mol^{-1}}$$

Proceeding similarly for the other temperatures, we find that

$k_r/(10^{-5}\mathrm{s^{-1}})$	T/K	$\Delta_r S^{\ddagger\ominus}/(\mathrm{J\,mol^{-1}\,K^{-1}})$
0.0787	273	10.0
3.46	298	9.97
49.8	318	10.2
487	338	9.78

The entropic term is positive and corresponds to an increase in the total number of available states in the activated complex, but does not depend on temperature within the experimental error.

E25.4 Consider the transition state theory expression for the rate constant in a unimolecular reaction. Using a potential energy diagram, describe the primary assumptions used in deriving this expression and provide the important highlights of the derivation.

Solution

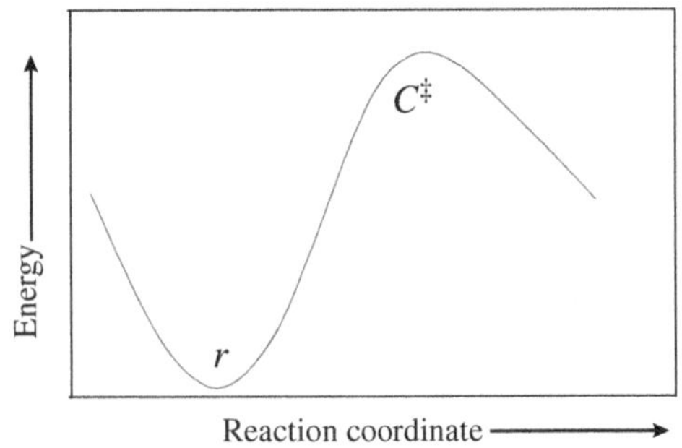

Figure E25.4 Energy versus reaction coordinate diagram showing the transition state.

Here we list six primary assumptions for the transition state theory, see Fig. E25.4. First, we assume that the transition state, $C^\ddagger$, is in "equilibrium" with the reactant r, so that we can write an equilibrium constant $K^\ddagger$ between the activated complex and the reactant as a ratio of their concentrations; namely

$$K^\ddagger = \frac{c(C^\ddagger)}{c(r)}$$

Second, we write the rate law as

$$\frac{dc_{\text{product}}}{dt} = k^\ddagger\ c(C^\ddagger) = k^\ddagger\ K^\ddagger\ c(r)$$

where $k^\ddagger$ is the rate constant for leaving the transition state. Third, we associate the overall rate constant k_r with

$$k_r = k^\ddagger\ K^\ddagger$$

Fourth, we take $k^\ddagger = \nu$ where ν is the characteristic frequency for the reactant to move across the barrier top. Fifth we write $K^\ddagger$ as

$$K^\ddagger = \frac{Z^\ominus_{C^\ddagger}/\sigma_{C^\ddagger}\ Z^\ominus_{C^\ddagger,\text{reaction coordinate}}}{Z^\ominus_r/\sigma_r\ Z^\ominus_{r,\text{reaction coordinate}}}$$

Lastly, if we consider the limit where $Z^\ominus_{C^\ddagger,\text{reaction coordinate}} \approx kT/(h\nu)$ and $Z^\ominus_{r,\text{reaction coordinate}} \approx 1$, we find that

$$k_r = k^\ddagger\ K^\ddagger\ = \nu K^\ddagger = \frac{kT}{h}\frac{\frac{Z^\ominus_{C^\ddagger}}{\sigma_{C^\ddagger}}}{\frac{Z^\ominus_r}{\sigma_r}} = \frac{kT}{h}\exp\left(-\Delta_r G^{\ddagger\ominus}/\boldsymbol{RT}\right)$$

E25.5 Consider the thermodynamic form of the transition state theory expression for a unimolecular reaction. Describe how each of the different degrees of freedom of the reactant contributes to the $\Delta_r S^{\ddagger\ominus}$ term and the $\Delta_r H^{\ddagger\ominus}$ term. Be explicit and mathematically exact. Provide an example of a unimolecular reaction in which you expect $\Delta_r S^{\ddagger\ominus}$ to be positive. Provide an example of a unimolecular reaction in which you expect $\Delta_r S^{\ddagger\ominus}$ to be negative.

Solution

Use

$$k_r = \nu K^\ddagger\ = \frac{kT}{h}\frac{\frac{Z^\ominus_{C^\ddagger}}{\sigma_{C^\ddagger}}}{\frac{Z^\ominus_r}{\sigma_r}}$$

$$= \frac{kT}{h}\left(\frac{Z^\ominus_{C^\ddagger,\text{trans}}}{Z^\ominus_{r,\text{trans}}}\right)\left(\frac{Z^\ominus_{C^\ddagger,\text{rotn}}/\sigma_{C^\ddagger}}{Z^\ominus_{r,\text{rotn}}/\sigma_r}\right)\left(\frac{Z^\ominus_{C^\ddagger,\text{vibn}}}{Z^\ominus_{r,\text{vibn}}}\right)\left(\frac{Z^\ominus_{C^\ddagger,\text{elec}}}{Z^\ominus_{r,\text{elec}}}\right)$$

The first term in parentheses gives the translational contribution to the rate constant expression. For a unimolecular reaction the transition state and the reactant state have the same total mass so that the partition function ratio is 1.

The second term in parentheses is the rotational contribution. If the molecule is nonlinear in both the reactant and transition state it does not contribute to $\Delta_r H^{\ddagger\ominus}$ but does contribute to $\Delta_r S^{\ddagger\ominus}$. The same logic works if it is linear in both cases. If the molecule's geometry changes from linear to nonlinear it will contribute to $\Delta_r H^{\ddagger\ominus}$.

The third term is the contribution from the vibrations, excluding that associated with the reaction coordinate, and they contribute to both $\Delta_r H^{\ddagger\ominus}$ and $\Delta_r S^{\ddagger\ominus}$.

The last term is the contribution from the electronic degrees of freedom. As long as only one electronic state is involved, or if the electronic energy level's degeneracy does not change, then it contributes only to $\Delta_r H^{\ddagger\ominus}$ and not $\Delta_r S^{\ddagger\ominus}$.

Most dissociation reactions would provide an example where we would expect a positive entropy of activation; e.g., $H_3C\text{–}CN \rightarrow CH_3 + CN$. For a negative entropy of activation, we must have a transition state that is more constrained than the reactant state, for example

E25.6 Consider a potential energy surface for a C–H bond cleavage and a C–D bond cleavage – of some large organic molecule. If we take a typical C–H vibrational frequency to be 3500 cm^{-1}, what is the zero point energy shift between the two bonds? Assume that the zero point energy shift at the transition state can be neglected. How would you justify this assumption? Ignoring any other differences in the two molecules' structures and molecular parameters, estimate the isotope effect on the bond cleavage; i.e., determine $k_r(\mathrm{CH})/k_r(\mathrm{CD})$ using the zero point energy shifts.

Solution

First we estimate the zero point energy difference that arises from the isotope effect. Within the harmonic oscillator approximation, we know that $2\pi\nu = \sqrt{k_f/\mu}$, so that we can write

$$\frac{\nu_{\text{C-D}}}{\nu_{\text{C-H}}} = \frac{\sqrt{k_f/\mu}_{\text{C-D}}}{\sqrt{k_f/\mu}_{\text{C-H}}} = \sqrt{\frac{\mu_{\text{C-H}}}{\mu_{\text{C-D}}}} = \sqrt{\frac{1}{2}} \approx 0.71$$

where we have assumed that the force constants for the two bonds are the same (i.e., the electronic energy is independent of isotope).

In the approximation that only this zero point energy shift affects the reaction rate constant, we can write that

$$\frac{k_{r,\text{C-H}}}{k_{r,\text{C-D}}} = \exp\left[-\left(h\nu_{0,\text{C-H}} - h\nu_{0,\text{C-D}}\right)/(2kT)\right]$$
$$= \exp\left[\frac{h\nu_{0,\text{C-H}}}{2kT}\left(1 - \sqrt{\frac{\mu_{\text{C-H}}}{\mu_{\text{C-D}}}}\right)\right]$$

For a $\tilde{\nu}_{0,\text{C-H}} = 3500\ \text{cm}^{-1}$ mode, we find that $h\nu_{0,\text{C-H}} = 6.96 \times 10^{-20}$ J, so that at a temperature of $T = 300$ K we find that

$$\frac{k_{r,\text{C-H}}}{k_{r,\text{C-D}}} \approx \exp\left(\frac{6.96 \times 10^{-20}\ \text{J}}{2 \cdot 1.38 \times 10^{-23}\ \text{J K}^{-1} \cdot 300\ \text{K}}(1 - 0.71)\right) = 11.4$$

E25.7 Consider the reaction

$$H_2 + {}^{79}Br \rightarrow H^{79}Br + H$$

This reaction is quite sensitive to the isotope of hydrogen used. The data in the graph provide measured reaction rate constants for this reaction with different isotopes of hydrogen; XY indicates HD, HT, and D_2. Note that the data are plotted as ln $[k_r(H_2)/k_r(XY)]$. Use transition state theory to explain the trend in the rate constant ratios. The H_2 vibrational wavenumber is 4401 cm^{-1}.

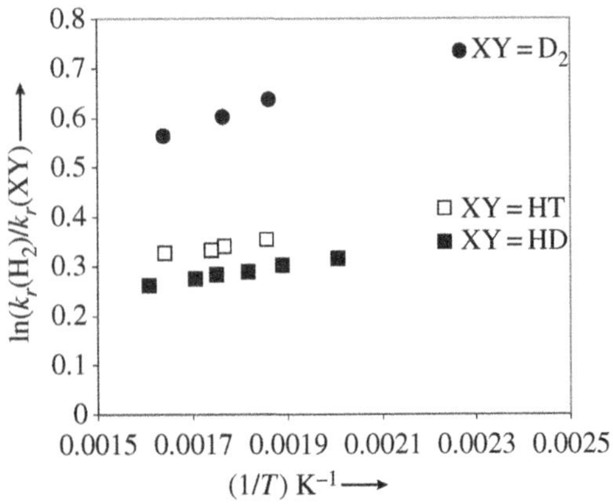

Figure E25.7 Plot of ln $[k_r(H_2)/k_r(XY)]$ versus $1/T$ for the reaction (25.5) with different isotopes.

Solution

We expect that the primary change in the rate constants with isotope for these reactions can be explained by considering the zero point energy for the different cases. In a first approximation, we make the approximation that only the zero point energy shift affects the rate constant. Hence we can write that

$$\frac{k_{r,\text{H-H}}}{k_{r,\text{X-Y}}} = \frac{\exp\left[-\left(h\nu_{0,\text{H-H-Br}} - h\nu_{0,\text{H-H}}\right)/(2kT)\right]}{\exp\left[-\left(h\nu_{0,\text{X-Y-Br}} - h\nu_{0,\text{X-Y}}\right)/(2kT)\right]}$$

Because the Br atom is so much heavier than the hydrogen atom and its isotopes (D and T), we make the further approximation that the differences in the activated complexes vibrational spectrum can be neglected. Hence we find that

$$\frac{k_{r,\text{H-H}}}{k_{r,\text{X-Y}}} \approx \exp\left[\left(h\nu_{0,\text{H-H}} - h\nu_{0,\text{X-Y}}\right)/(2kT)\right]$$

or

$$\ln\left(\frac{k_{r,\text{H-H}}}{k_{r,\text{X-Y}}}\right) \approx \frac{h\nu_{0,\text{H-H}} - h\nu_{0,\text{X-Y}}}{2k}\frac{1}{T}$$

Using the expression $2\pi\nu_0 = \sqrt{k_f/\mu}$, and assuming that the force constant k_f does not change with the isotope, we find that

$$\ln\left(\frac{k_{r,\text{H-H}}}{k_{r,\text{X-Y}}}\right) \approx \frac{h\nu_{0,\text{H-H}}}{2k}\left(1 - \sqrt{\frac{\mu_{\text{H-H}}}{\mu_{\text{X-Y}}}}\right)\frac{1}{T}$$

Thus we see that the rate constant ratio (and the slopes of the plots) differs by the term

$$\left(1 - \sqrt{\frac{\mu_{\text{H-H}}}{\mu_{\text{X-Y}}}}\right)$$

which takes the value 0.134 for HD, the value 0.184 for HT, and 0.293 for D_2. Thus for $T = 500$ K we obtain

$$\ln\left(\frac{k_{r,\text{H-H}}}{k_{r,\text{X-Y}}}\right)_{T=500\text{ K}} = 6.34\cdot\left(1 - \sqrt{\frac{\mu_{\text{H-H}}}{\mu_{\text{X-Y}}}}\right)$$

Table E25.7 compares the calculated data to the experiment (see Fig. E25.7).

Table E25.7 Calculation of Rate Constant Ratios and Comparison to Experiment at $T = 500$ K.

XY	$1-\sqrt{\mu_{H-H}/\mu_{X-Y}}$	$\ln\left(k_{r,H-H}/k_{r,X-Y}\right)$	$k_{r,H-H}/k_{r,X-Y}$	
			Calculated	Experiment
HD	0.134	0.849	2.33	1.42
HT	0.184	1.167	3.21	1.49
D_2	0.293	1.858	6.41	2.01

This trend in the values is exactly that shown in the graph. A fit to the rate constant data corroborates this conclusion (R. E. Weston Jr., *Science* 1967, **158**, 332).

Hence we see that although the trend in the values calculated here are consistent with the data, the absolute values are a factor of two or three higher than the data. This finding suggests that zero point energy considerations of the hydrogen reactant are not the only contribution to the isotope effect. A next step in developing a more accurate treatment would be to account for the differences in the vibrational frequencies of the activated complex with the isotope.

E25.8 Consider the bimolecular reaction

$$OH\,(g) + H_2\,(g) \rightarrow H_2O\,(g) + H\,(g)$$

and take the Arrhenius parameters for this reaction to be $A = 3.0 \times 10^9\ \text{L mol}^{-1}\ \text{s}^{-1}$ and $E_{a,m}^{empirical} = 16.5\ \text{kJ mol}^{-1}$[3]. Find the values of $\Delta_r H^{\ddagger}$ and $\Delta_r S^{\ddagger}$ at 298 K. Use a standard state of $P^{\ominus} = 1$ bar in the expression for the transition state theory rate constant of this gas-phase reaction, rather than the $c^{\ominus} = 1\ \text{mol L}^{-1}$ used for the solution rate constant expression that is given in the text.

Solution

Using the transition state theory expression for the bimolecular rate law, we can write that

$$k_{TST} = \left(\frac{kT}{h}\right) \cdot \frac{1}{c^{\ominus}} \cdot \exp\left(\frac{\Delta_r S^{\ddagger\ominus}}{\boldsymbol{R}}\right) \cdot \exp\left(-\frac{\Delta_r H^{\ddagger\ominus}}{\boldsymbol{R}T}\right)$$

The Arrhenius activation energy is equivalent to the empirical activation energy, so that we can write $E_{a,m}^{empirical} = 16.5\ \text{kJ mol}^{-1}$ from the problem statement. In the text (see Section 25.2.3) we find that

$$E_{a,m}^{empirical} = \boldsymbol{R}T + \Delta_r H^{\ddagger\ominus}$$

Solving for $\Delta_r H^{\ddagger\ominus}$, we find that

$$\begin{aligned}\Delta_r H^{\ddagger\ominus} &= E_{a,m}^{empirical} - \boldsymbol{R}T \\ &= \left[16.5\ \text{kJ mol}^{-1} - 8.314\ \text{J mol}^{-1}\ \text{K}^{-1} \cdot 298\ \text{K}\right] \\ &= 14.0\ \text{kJ mol}^{-1}\end{aligned}$$

Using this result for $\Delta_r H^{\ddagger\ominus}$, we can determine the activation entropy by way of the Arrhenius rate law determination. If we write

$$\begin{aligned}k_{Arr} &= A\exp\left(-\frac{E_{a,m}^{empirical}}{\boldsymbol{R}T}\right) = A = 3.0 \times 10^9\ \text{L mol}^{-1}\ \text{s}^{-1} \cdot \exp\left(-\frac{16.5\ \text{kJ mol}^{-1}}{\boldsymbol{R}T}\right) \\ &= 3.0 \times 10^9\ \text{L mol}^{-1}\ \text{s}^{-1} \cdot \exp(-6.66) = 3.0 \times 10^9\ \text{L mol}^{-1}\ \text{s}^{-1} \cdot 1.28 \times 10^{-3} \\ &= 3.84 \times 10^6\ \text{L mol}^{-1}\ \text{s}^{-1}\end{aligned}$$

3 A. R. Ravishankara et al., *J. Phys. Chem.* 1981, **85**, 2498–2503

then we can equate the two rate constant expressions (i.e., $k_{Arr} = k_{TST}$) and find that

$$k_{\mathrm{Arr}} = \frac{kT}{h}\frac{1}{c^{\ominus}}\exp\left(\frac{\Delta_r S^{\ddagger\ominus}}{R}\right)\exp\left(-\frac{\Delta_r H^{\ddagger\ominus}}{RT}\right)$$

which rearranges to

$$\exp\left(\frac{\Delta_r S^{\ddagger\ominus}}{R}\right) = k_{\mathrm{Arr}} \cdot c^{\ominus} \cdot \frac{h}{kT}\exp\left(\frac{\Delta_r H^{\ddagger\ominus}}{RT}\right)$$

Solving for $\Delta_r S^{\ddagger\ominus}$ we find

$$\Delta_r S^{\ddagger\ominus} = R\ln\left[k_{\mathrm{Arr}} \cdot c^{\ominus} \cdot \frac{h}{kT}\right] + \frac{\Delta_r H^{\ddagger\ominus}}{T}$$

$$R\ln\left[3.84\times10^{6}\ \mathrm{L\ mol^{-1}\ s^{-1}} \cdot 1\ \mathrm{mol\ L^{-1}} \cdot \frac{6.626\times10^{-34}\ \mathrm{J\ s}}{1.38\times10^{-23}\ \mathrm{J\ K^{-1}} \cdot 298\ \mathrm{K}}\right] + \frac{14.0\ \mathrm{kJ\ mol^{-1}}}{298\ \mathrm{K}}$$

$$= R \cdot \ln(6.19\times10^{-7}) + 46.9\ \mathrm{J\ mol^{-1}K^{-1}} = (-118.9 + 46.9)\ \mathrm{J\ mol^{-1}K^{-1}} = -71.9\ \mathrm{J\ mol^{-1}K^{-1}}$$

The large, negative activation entropy implies that the transition state has a constrained geometry.

E25.9 Consider the recombination of ethyl radicals in the gas phase. The reaction

$$^{\bullet}CH_2 - CH_3 + {}^{\bullet}CH_2 - CH_3 \rightarrow CH_3CH_2CH_2CH_3$$

is diffusion-controlled ($E_a = 0$). Estimate the rate constant at $T = 298$ K. Given that the experimental rate constant is 1.6×10^{11} L mol $^{-1}$s^{-1} calculate the steric factor α.

Solution

We use Equation (24.21) to calculate k_r

$$k_r = \sqrt{\frac{8kT}{\pi\mu}} \cdot \pi \cdot (r_A + r_B)^2 \cdot N_A \cdot \alpha$$

With $\mu = 24\times10^{-27}$ kg, $(r_A + r_B) = 500$ pm we obtain

$$k_r = \sqrt{\frac{8.318\times10^{-23}\ \mathrm{J\ K^{-1}} \cdot 298\ \mathrm{K}}{\pi \cdot 24\times10^{-27}\ \mathrm{kg}}} \cdot \pi \cdot (500\times10^{-12}\ \mathrm{m})^2 \cdot 6.02\times10^{23}\ \mathrm{mol^{-1}} \cdot \alpha$$

$$= \sqrt{3.29\times10^{5}}\ \mathrm{m\ s^{-1}} \cdot 4.72\times10^{5}\ \mathrm{m^2\ mol^{-1}} \cdot \alpha = 2.71\times10^{11}\ \mathrm{L\ mol^{-1}\ s^{-1}} \cdot \alpha$$

The experimental value is $k_{r,\mathrm{expt}} = 1.6\times10^{11}$ L mol^{-1} s^{-1}, thus

$$\alpha = \frac{1.6\times10^{11}\ \mathrm{L\ mol^{-1}s^{-1}}}{2.71\times10^{11}\ \mathrm{L\ mol^{-1}s^{-1}}} = 0.6$$

E25.10 Consider the artificial data given in the table, which is calculated from the relation

$$k_r = A_{f,0} \cdot T \cdot \exp\left(-\frac{\Delta_r H^{\ddagger\ominus}}{RT}\right)$$

for the transition state theory using $\Delta_r H^{\ddagger\ominus} = 50$ kJ mol^{-1}, $A_{f,0} = 1\times10^{7}\,\mathrm{s^{-1}}$, and different ranges of the temperature.

T/K	300	305	310	320	340	350	400	450	500
k_r/s^{-1}	5.90	8.34	11.6	22.0	70.7	1.21×10^{2}	1.18×10^{3}	7.07×10^{3}	2.99×10^{4}

Perform an Arrhenius analysis for these data and compare the result to Equation

$$\Delta_r H^{\ddagger\ominus} = E_{a,m}^{\mathrm{empirical}} - RT$$

in which you use the mean value,

$$T_{\mathrm{mean}} = \frac{1}{2}(T_{\mathrm{end}} + T_{\mathrm{start}})$$

To observe a significant trend in how the difference changes, consider temperature ranges of 10, 50, 100, and 200 K.

Solution

According to the Arrhenius equation

$$\ln\frac{k_r}{k_{r,1}} = \frac{E_{a,m}^{\text{empirical}}}{\boldsymbol{R}T}\left(\frac{1}{T}-\frac{1}{T_1}\right)$$

we plot $\ln(k_r/k_{r,1})$ versus $1/T$. From the slopes we obtain the values of $E_{a,m}^{\text{empirical}}$ given in Table E25.10a.

Table E5.10a Evaluation of $E_{a,m}^{\text{empirical}}$ Obtained from the Arrhenius Plots for Different Ranges of the Temperature.

Range /K	300 – 310	300 – 350	300 – 400	300 – 500
$E_{a,m}^{\text{empirical}}$ kJ mol^{-1}	52.54	52.69	52.86	53.14
T_{mean}	305	325	350	400

The bottom row of Table E25.10a reports the mean value of the temperature for the different temperature ranges. In the problem statement we are given $\Delta_r H^{\ddagger\ominus} = 50$ kJ mol^{-1}, and we compare this value to that computed from Equation

$$\Delta_r H^{\ddagger\ominus} = E_{a,m}^{\text{empirical}} - \boldsymbol{R}T$$

in Table E25.10b.

Table E25.10b Comparison of $E_{a,m}^{\text{empirical}}$ and $\Delta_r H^{\ddagger\ominus}$.

Range /K	300 – 310	300 – 350	300 – 400	300 – 500
$E_{a,m}^{\text{empirical}}$ kJ mol^{-1}	52.54	52.69	52.86	53.14
$\Delta_r H^{\ddagger\ominus} = E_{a,m}^{\text{empirical}} - \boldsymbol{R}T_{mean}$	50.00	49.99	49.95	49.81

We see that the correction term $\boldsymbol{R}T$ works well up to a temperature range of 50 K; however, with an increasing temperature range noticeable deviations occur.

E25.11 Consider the reaction: $O(^3P) + Cl_2 \rightarrow OCl + Cl$, where $O(^3P)$ means that the oxygen atom is in its triplet state rather than its singlet state. Consider the potential energy surface shown in Fig. E25.11. The energies assigned to the contours are labeled in kJ mol^{-1}, and the surface contours assume a linear collision geometry. Provide a verbal description of what is happening for the trajectory drawn in each diagram – as the system moves from points A to B to C to D. (Figure E25.11 is sketched from that given by A.M. Kosmas, *J. Chem. Soc. Faraday Trans.* 1993, **89**, 2999)

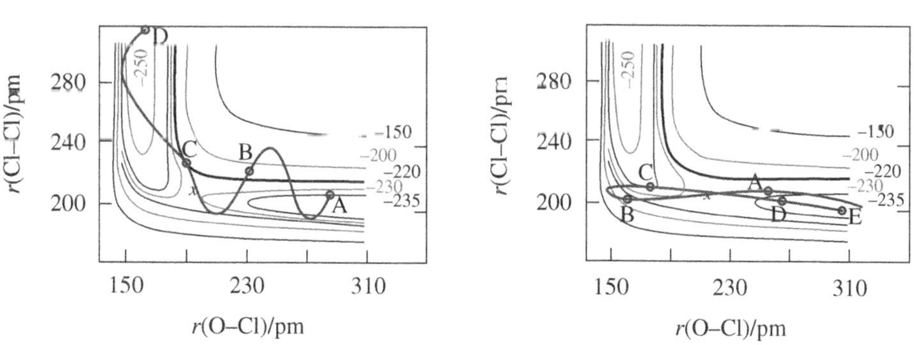

Figure E25.11 Potential energy surface for the reaction of O with Cl_2.

Solution

The panel on the left shows a reactive trajectory. In the entrance channel the system appears to be vibrating considerably (i.e., it is vibrationally excited). Between the positions B and C the system moves through the transition state and proceeds to product. In the exit channel the system has gained considerably in translational energy, because it appears to move farther during the product channel's vibrational oscillation.

The panel on the right shows a nonreactive trajectory. The molecule does not have much vibrational excitation (compare to the panel on the left). After passing through the transition state, the system "bounces" off of the potential wall and "recrosses" the transition state to form the reactant.

E25.12 Consider the reaction: $O(^3P) + Cl_2 \rightarrow OCl + Cl$, where $O(^3P)$ means that the oxygen atom is in its triplet state rather than its singlet state. Two different potential energy contours are shown here. Describe how you expect the distribution of energy in the final products to change as the input energy changes. In particular, the distribution of energy in the translational, rotational, and vibrational degrees of freedom. (Note the shift in the transition state's location, labeled by a cross.) (Figure E25.12 is sketched from those given by A.M. Kosmas, *J. Chem. Soc. Faraday Trans.* **89**, 2999 (1993).)

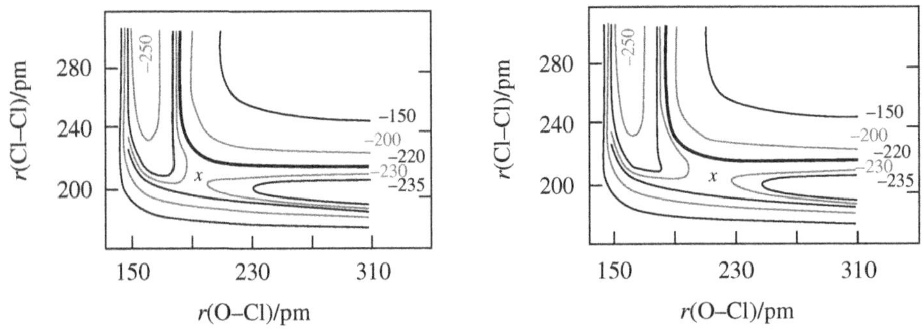

Figure E25.12 Potential energy surface for the reaction of O with Cl_2.

Solution

The primary difference in these two sketches of the potential energy surface is the shift in the transition state energy position. The shift in the transition state position from the surface in the left panel to that in the right panel causes the incident translational energy to be more important in the entrance channel (r(O–Cl)), and vibrational energy to be more important in the exit channel (r(Cl–Cl)). Hence, we expect an increase in vibrational excitation of the products with increasing collision energy, and we expect that the vibrational excitation will be more important for the right contour.

E25.13 Given that $A = 7.0 \times 10^{10}$ L mol^{-1} s^{-1} for the reaction

$$D + H_2 \rightarrow HD + H$$

under standard conditions. Determine $\Delta_r S^{\ddagger\ominus}$.

Solution

The rate equation for this reaction in terms of the concentration is

$$\frac{dc_H}{dt} = k_r \cdot c_D \cdot c_{H_2}$$

We use the fact that

$$\Delta_r H^{\ddagger\ominus} = E_{a,m}^{\text{empirical}} - RT$$

Using this result for $\Delta H^{\ddagger}$, we can equate the two rate constant expressions (i.e., $k_{\text{Arr}} = k_{\text{TST}}$). Following the derivation in Section 25.2.1 we find that

$$A \exp\left(-\frac{E_{a,m}^{\text{empirical}}}{RT}\right) = \left(\frac{kT}{h}\right)\left(\frac{1}{c^{\ominus}}\right)\exp\left(\frac{\Delta_r S^{\ddagger\ominus}}{R}\right)\exp\left(-\frac{\Delta_r H^{\ddagger\ominus}}{RT}\right)$$

where $c^{\ominus} = 1 \text{ mol L}^{-1}$ is the standard concentration. This expression rearranges to

$$\exp\left(\frac{\Delta_r S^{\ddagger\ominus}}{\boldsymbol{R}}\right) = \left[A \exp\left(-\frac{E_{a,m}^{\text{empirical}}}{\boldsymbol{R}T}\right)\right] \Big/ \left[\frac{kT}{h}\frac{1}{c^{\ominus}} \exp\left(-\frac{\Delta_r H^{\ddagger\ominus}}{\boldsymbol{R}T}\right)\right]$$

$$= \frac{A \cdot h}{kT} \cdot c^{\ominus} \cdot \exp\left(-\frac{E_{a,m}^{\text{empirical}} - \Delta_r H^{\ddagger\ominus}}{\boldsymbol{R}T}\right)$$

$$= \frac{A \cdot h}{kT} \cdot c^{\ominus} \cdot \exp\left(-\frac{\boldsymbol{R}T}{\boldsymbol{R}T}\right) = \frac{A \cdot h}{kT} \cdot c^{\ominus} \cdot \exp(-1)$$

Solving for $\Delta_r S^{\ddagger\ominus}$ we find

$$\Delta_r S^{\ddagger\ominus} = \boldsymbol{R} \ln\left[\frac{A \cdot h}{kT} \cdot c^{\ominus}\right] - \boldsymbol{R}$$

so that

$$\frac{A \cdot h}{kT} \cdot c^{\ominus} = \frac{7.0 \times 10^{10} \text{ L mol}^{-1} \text{ s}^{-1} \cdot 6.626 \cdot 10^{-34} \text{ J s} \cdot 1 \text{ mol L}^{-1}}{1.38 \cdot 10^{-23} \text{ J K}^{-1} \cdot 298 \text{ K}}$$
$$= 1.12 \times 10^{-2}$$

$$\Delta_r S^{\ddagger\ominus} = \boldsymbol{R} \cdot \ln\left(1.12 \times 10^{-2}\right) - \boldsymbol{R} = (-37.3 - 8.314) \text{ J mol}^{-1} \text{ K}^{-1}$$
$$= -45.6 \text{ J mol}^{-1} \text{ K}^{-1}$$

E25.14 The excited molecule ICN* is created with an internuclear distance of about 275 pm and the molecule is fully dissociated at about 400 pm. Use these distances and the 200 fs risetime for the CN formation (Fig. 25.9) to estimate the relative velocity of the dissociating fragments.

Solution

Using a simple classical treatment to estimate the speed, we find that

$$speed = \frac{(400 - 275) \times 10^{-12}}{200 \times 10^{-15}} \frac{\text{m}}{\text{s}} = 625 \frac{\text{m}}{\text{s}}$$

E25.15 The formation of the CN radical from the bond dissociation in Fig. 25.9 occurs in 200 fs. Compare this time to the vibrational period of the CN radical which has a vibrational frequency of $\nu = 6.2 \times 10^{13} \text{ s}^{-1}$. Compare this time to the thermally averaged rotational period of the ICN molecule ($T = 298$ K and $I = 2.6 \times 10^{-45}$ kg m^{-2}).

Solution

We estimate the vibrational period as

$$T_{\text{vib}} = \frac{1}{\nu} = \frac{1}{6.2 \times 10^{13} \text{ s}^{-1}} = 16 \text{ fs}$$

so that we could have about 13 to 14 vibrational periods of CN during the bond dissociation of ICN.
We estimate the rotational period by applying the equipartition theorem

$$\frac{1}{2} I\omega^2 = \frac{kT}{2}$$

and solving for the angular frequency

$$\omega = \sqrt{\frac{kT}{I}} = 1.3 \times 10^{12} \text{ s}^{-1}$$

Hence we find an the angular period of

$$T_{\text{rot}} = \frac{1}{\nu} = \frac{2\pi}{\omega} = 4.8 \times 10^{-12} \text{ s} = 4.8 \text{ ps}$$

This value is considerably slower (about 20 times) than the characteristic time for dissociation.

E25.16 In the gas-phase reaction

$$K + Br_2 \rightarrow KBr + Br$$

$E_{a,m} = 0$ and the steric factor is $\alpha = 4.8$. The large α is explained by a harpooning mechanism, which can be represented by

$$K + Br_2 \rightarrow K^+ + Br_2^- \rightarrow KBr + Br$$

When a K atom approaches a Br_2 molecule and a critical distance $\boldsymbol{R}$ is reached, an electron can be transferred from K to Br_2. Then the Coulomb attraction initiates a fast reaction of both ions, and the effective collision cross section is $\pi \boldsymbol{R}^2$ instead of $\pi(r_K + r_{Br_2})^2$, where $r_K + r_{Br_2} = 400$ pm. Use the steric factor to estimate the distance of approach at which the electron "jumps" from the potassium atom to the bromine molecule. What is the Coulomb force between the two ions at this distance?

Solution

We can identify the steric factor with the ratio of the two collision cross sections.

$$\alpha = \frac{\pi \boldsymbol{R}^2}{\pi\left(r_K + r_{Br_2}\right)^2}$$

Then we obtain $\boldsymbol{R} = \left(r_K + r_{Br_2}\right)\sqrt{\alpha} = 400\ \text{pm}\cdot\sqrt{4.8} = 876$ pm. The Coulomb force at this distance is

$$\begin{aligned} f &= \frac{e^2}{4\pi\varepsilon_0 \boldsymbol{R}^2} \\ &= \frac{\left(1.602\times10^{-19}\ \text{C}\right)^2}{4\pi\cdot 8.85\times10^{-12}\ \text{C}^2\ \text{J}^{-1}\ \text{m}^{-1}\cdot\left(876\times10^{-12}\ \text{m}\right)^2} \\ &= 3.0\times10^{-10}\ \text{N} = 0.30\ \text{nN} \end{aligned}$$

From Chapter 7, recall that a 1 nN force is strong enough to displace a C–C–C bending mode by 30 pm. For the reaction here, the Coulomb force acts to pull the anion and cation toward each other and promotes reaction.

E25.17 In contrast to the slow vibrational relaxation for the ground electronic state of I_2 in Xe (> 3 ns), the vibrational relaxation is nearly complete within 200 ps in CCl_4. Provide an explanation for this difference.

Solution

For the case of Xe the excess vibrational energy in the iodine molecule must be transferred to the Xe fluid's translational degrees of freedom. This process is relatively inefficient because of the relatively large difference between the magnitude of the vibrational energy level spacings and the amount of translational energy imparted in a collision (collision energies are of order kT). For the molecular liquid CCl_4 a larger variety of energy states are available for the transfer of energy in a collision. In particular, the collision can transfer vibrational energy from the iodine molecule to rotational energy of molecules in the fluid and/or vibrational energy of molecules in the fluid. Hence the large number of motions that can accept the energy lost by the iodine and their broader distribution in energy leads to a more facile collisional energy exchange. See the paper A. L. Harris, C. B. Brown, and C. B. Harris, *Ann. Rev. Phys. Chem.* 1988, **39,** 341 (1988) for a more refined discussion.

E25.18 In Section 25.2.1 we discussed the transition state theory for the

$$D + H_2 \rightarrow H + DH$$

reaction and identified the asymmetric stretch motion with the reaction coordinate. Why is this an appropriate choice? Why is the symmetric stretch an inappropriate choice? Cast your discussion in terms of the energy profile for the reaction in Fig. 25.1.

Solution

The assignment of the asymmetric stretch motion of the triatomic transition with the reaction coordinate makes physical sense because it can be viewed as a motion that leads to a compression of two of the atomic nuclei toward a bond distance that approaches the equilibrium bond distance of a diatomic and the increase in the distance between the remaining atom and the two that form the nascent diatomic molecule. In contrast, the symmetric stretch motion has both bond distances either increasing (or decreasing) in phase; hence, it does not drive one pair of the atoms toward a geometry that might have a more stable energy. This behavior is evident in Fig. 25.1b in which the reaction proceeds to product by an increase in the B–C distance at a nearly fixed value of the A–B distance.

E25.19 In Example 25.5 we found that $\sigma_{r,\max} = 1.57 \times 10^4$ pm^2 and $E_{\text{thresh}} = 0.38$ eV. Use these values in Equation (25.57) to compute the thermal rate constant.

Solution

Using $E_{\text{thresh}} = 0.38 \text{ eV} = 6.08 \times 10^{-20}$ J, we obtain $E_{\text{thresh}}/(kT) = 14.7$. Taking the reduced mass of the collision to be $\mu = ((1.67 \times 6.69)/(8.36)) \times 10^{-27}$ kg $= 1.34 \times 10^{-27}$ kg, we can calculate

$$k_r(T) = \boldsymbol{N}_A \sqrt{\frac{8 \cdot 1.38 \times 10^{-23}\ \text{J K}^{-1} \cdot 298\ \text{K}}{\pi \cdot 1.34 \times 10^{-27}\ \text{kg}}} \cdot 1.57 \times 10^{-20}\ \text{m}^2 \cdot \exp(-14.7)$$
$$= 1.1 \times 10^4\ \text{L mol}^{-1}\ \text{s}^{-1}$$

The experimentally observed[4] thermal rate constant, 1.6×10^4 L mol^{-1} s^{-1}, is only slightly larger than this value.

E25.20 Calculate the moment of inertia I_{DHH} of the transition state complex D–H–H that is discussed in the text and extract an expression for the reduced mass $\mu^{\ddagger}$.

Solution

We consider the following body diagram.

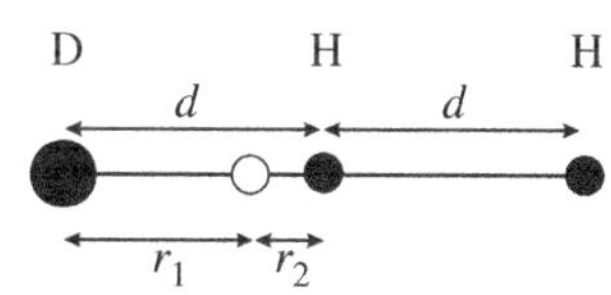

Figure E25.20 Activated complex D–H–H, calculation of center of mass. The open circle indicates the center of mass.

We calculate the reduced mass $\mu^{\ddagger}$ for the rotational motion of the activated complex

D–H–H

First we calculate the distances r_1 and r_2 in Fig. E25.20, where the open circle indicates the center of mass. From the relation

$$r_1 m_{\text{D}} = r_2 m_{\text{H}} + (d + r_2) m_{\text{H}}$$

we obtain

$$r_1 m_{\text{D}} = m_{\text{H}} \left(d + 2r_2\right)$$

Using the additional condition $r_1 + r_2 = d$, we find

$$r_1 = \frac{3m_{\text{H}}d}{m_{\text{D}} + 2m_{\text{H}}} = \frac{3m_{\text{H}}d}{2m_{\text{H}} + 2m_{\text{H}}} = \frac{3}{4}d,\ r_2 = \frac{1}{4}d$$

Next we calculate the moment of inertia.

$$\begin{aligned} I &= m_{\text{D}} r_1^2 + m_{\text{H}} r_2^2 + m_{\text{H}}(d + r_2)^2 \\ &= 2m_{\text{H}} r_1^2 + m_{\text{H}} r_2^2 + m_{\text{H}}(d + r_2)^2 \\ &= m_{\text{H}}\left[2r_1^2 + r_2^2 + (d + r_2)^2\right] \\ &= m_{\text{H}}\left[2r_1^2 + 2r_2^2 + d^2 + 2d \cdot r_2\right] \\ &= m_{\text{H}} d^2 \left[2\frac{9}{16} + 2\frac{1}{16} + \frac{16}{16} + \frac{8}{16}\right] = m_{\text{H}} d^2 \cdot \frac{11}{4} \end{aligned}$$

4 A.A. Westenberg, N. de Haas, *J. Chem. Phys.* 1967, **47**, 1393.

This result corresponds to $I = \mu^{\ddagger} d^2$; thus, for the reduced mass we find

$$\mu^{\ddagger} = m_{\mathrm{H}} \cdot \frac{11}{4}$$

E25.21 Assume that a gas-phase I_2 molecule has an excess vibrational energy of $E = 3.80 \times 10^{-20}$ J and is in its ground electronic state with a characteristic vibrational frequency of $\nu_0 = 6.45 \times 10^{12}\ \mathrm{s}^{-1}$. Given that the equilibrium bond distance is $x_{\mathrm{eq}} = 266$ pm, estimate the maximum distance of the I atoms from each other during the molecule's vibrational motion. Compare this value to the maximum displacement that was estimated in Example 25.6.

Solution

Assume a harmonic potential so that

$$\Delta E = \frac{1}{2} k \left(x - x_{\mathrm{eq}}\right)^2 = \frac{1}{2} \mu \nu_0^2 \left(x - x_{\mathrm{eq}}\right)^2$$

Hence we wish to find the energy ΔE where

$$\Delta E = E + h\nu_0/2 = 3.80 \times 10^{-20}\,\mathrm{J} + \frac{\left(6.626 \times 10^{-34}\ \mathrm{J\,s}\right)\left(6.45 \times 10^{12}\ \mathrm{s}^{-1}\right)}{2}$$
$$= 4.01 \times 10^{-20}\ \mathrm{J}$$

Solving for the displacement from equilibrium, we find that

$$\left(x - x_{\mathrm{eq}}\right) = \pm \sqrt{\frac{2 \cdot 4.01 \times 10^{-20}\ \mathrm{J}}{\left(6.45 \times 10^{12}\,\mathrm{s}^{-1}\right)^2 \left(1.05 \times 10^{-25}\ \mathrm{kg}\right)}}$$
$$= \pm 1.35 \times 10^{-10}\mathrm{m} = \pm 135\ \mathrm{pm}$$

Using $x_{\mathrm{eq}} = 266$ pm, we find that $x = 401$ pm, which is somewhat smaller than the value calculated in Example 25.6.

E25.22 In the collision theory model, the rate constant is described by

$$\boxed{k_r = A_f \cdot \mathrm{e}^{-E_{a,m}/RT}}$$

and

$$\boxed{A_f = \sqrt{\frac{8kT}{\pi \mu}} \cdot \pi \cdot \left(r_A + r_B\right)^2 \cdot \boldsymbol{N}_A \cdot \alpha}$$

Describe how to rigorously extract the activation energy $E_{a,m}$ from experimental rate constant data.

Solution

The analysis given here is similar to that given in Section 25.3.3 for the activation enthalpy in the transition state theory. To begin we write the rate equation as

$$k_r = \sqrt{\frac{8kT}{\pi \mu}} \cdot \pi \cdot \left(r_A + r_B\right)^2 \cdot \boldsymbol{N}_A \cdot \alpha \cdot \mathrm{e}^{-E_{a,m}/RT}$$
$$= A_{f,0} \cdot \sqrt{T} \cdot \mathrm{e}^{-E_{a,m}/RT}$$

which defines $A_{f,0}$. Then we have, according to Equations (25.25) and (25.26)

$$k_r = A_f \cdot \exp\left(-\frac{E_{a,m}}{RT}\right) = A_{f,0} \cdot \sqrt{T} \cdot \exp\left(-\frac{E_{a,m}}{RT}\right)$$
$$k_{r,1} = A_{f,0} \cdot \sqrt{T_1} \cdot \exp\left(-\frac{E_{a,m}}{RT_1}\right)$$

and

$$\frac{k_r}{k_{r,1}} = \sqrt{\frac{T}{T_1}} \cdot \exp\left(-\frac{E_{a,m}}{\boldsymbol{R}T}\right) / \exp\left(-\frac{E_{a,m}}{\boldsymbol{R}T_1}\right)$$

Finally we take the logarithm of both sides and find

$$\begin{aligned}\ln\left(\frac{k_r}{k_{r,1}}\right) &= \frac{1}{2}\ln\left(\frac{T}{T_1}\right) - \frac{E_{a,m}}{\boldsymbol{R}T} + \frac{E_{a,m}}{\boldsymbol{R}T_1} \\ &= \frac{1}{2}\ln\left(\frac{T}{T_1}\right) - \frac{E_{a,m}}{\boldsymbol{R}}\left(\frac{1}{T} - \frac{1}{T_1}\right)\end{aligned}$$

Reorganization of the terms yields

$$\ln\left(\frac{k_r}{k_{r,1}}\right) - \frac{1}{2}\ln\left(\frac{T}{T_1}\right) = -\frac{E_{a,m}}{\boldsymbol{R}}\left(\frac{1}{T} - \frac{1}{T_1}\right)$$

Thus if we plot the quantity

$$y = \ln\left(\frac{k_r}{k_{r,1}}\right) - \frac{1}{2}\ln\left(\frac{T}{T_1}\right)$$

versus $1/T$, then we expect a straight line with slope $-E_{a,m}/\boldsymbol{R}$.

How is this slope related to the empirical activation energy? We start with

$$\frac{\mathrm{d}\ln(k_r/k_{r,1})}{\mathrm{d}(1/T)} = -\frac{E_{a,m}^{\text{empirical}}}{\boldsymbol{R}} \qquad (A_f \text{ assumed to be constant})$$

and

$$\frac{\mathrm{d}\left[\ln(k_r/k_{r,1}) - \frac{1}{2}\ln(T/T_1)\right]}{\mathrm{d}(1/T)} = -\frac{E_{a,m}}{\boldsymbol{R}} \qquad (\text{Temperature dependence of } A_f \text{ taken into account})$$

Thus we obtain

$$\begin{aligned}-\frac{E_{a,m}}{\boldsymbol{R}} &= -\frac{E_{a,m}^{\text{empirical}}}{\boldsymbol{R}} - \frac{1}{2}\frac{\mathrm{d}\left[\ln(T/T_1)\right]}{\mathrm{d}(1/T)} \\ &= -\frac{E_{a,m}^{\text{empirical}}}{\boldsymbol{R}} - \frac{1}{2}\frac{\mathrm{d}\left[\ln(T/T_1)\right]}{\mathrm{d}T} \cdot \frac{\mathrm{d}T}{\mathrm{d}(1/T)} \\ &= -\frac{E_{a,m}^{\text{empirical}}}{\boldsymbol{R}} - \frac{1}{2}\frac{1}{T} \cdot \frac{\mathrm{d}T}{\mathrm{d}(1/T)}\end{aligned}$$

We calculate the last term by setting $x = 1/T$

$$\frac{\mathrm{d}T}{\mathrm{d}(1/T)} = \frac{\mathrm{d}(1/x)}{\mathrm{d}x} = -\frac{1}{x^2} = -T^2$$

to obtain

$$-\frac{E_{a,m}}{\boldsymbol{R}} = -\frac{E_{a,m}^{\text{empirical}}}{\boldsymbol{R}} + \frac{1}{2}T \quad \text{or} \quad E_{a,m} = E_{a,m}^{\text{empirical}} - \frac{1}{2}\boldsymbol{R}T$$

25.2 Problems

P25.1 Consider the photodissociation of the CF_3I molecule into a CF_3 radical and an I* radical (the * indicates that the I atom is electronically excited) with a 275 nm photon. Given that the binding energy (ε_B) of the C–I bond corresponds to a wavenumber of 18,600 cm^{-1}, and the energy difference between I and I* is 7603 cm^{-1}, compute the excess energy left in the CF_3 and I* after the photodissociation. *In part 1*: Assume that all the excess energy goes into the relative translational motion of the CF_3 and I* fragments and find the relative velocity of the two fragments. Then determine how long it takes for them to reach a separation distance of 840 pm – assuming they began at the equilibrium bond distance of 210 pm. *In part 2*: Assume that five quanta of the CF_3 umbrella motion (580 cm^{-1}) are excited and the rest of the excess energy is statistically distributed among the translational and rotational degrees of freedom of the fragments, estimate the relative translational velocity of the CF_3 and I* fragments and compute the time it takes for the I* atom and the CF_3 fragment to reach a distance of 840 pm. Provide a verbal explanation for why you might expect the umbrella motion of CF_3 to be highly excited.

Solution

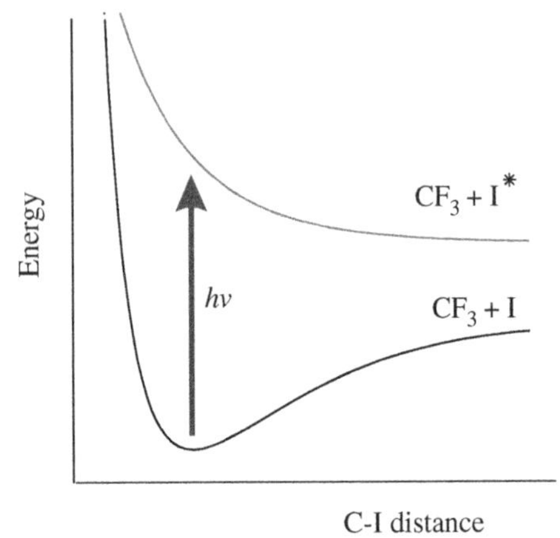

Figure P25.1 Potential energy curves for the ground state and the first excited state.

First we compute the excess energy in the CF_3 and the I* after excitation by the photon, see Fig. P25.1. To do this we need to find the energy difference between the asymptote of the dissociative curve shown in the figure and the energy for the C-I distance upon excitation by the light. The 275 nm photon carries an energy

$$h\nu = \frac{hc}{\lambda} = 7.23 \times 10^{-19} \text{ J}$$

From the wavenumbers given in the problem we find 3.70×10^{-19} J for the bond energy ε_B and 1.51×10^{-19} J for the energy difference between I and I*. From these facts, we find the excess energy to be

$$E_{\text{excess}} = \left(7.23 \times 10^{-19} - 1.51 \times 10^{-19} - 3.70 \times 10^{-19}\right) \text{ J} = 2.02 \times 10^{-19} \text{ J}$$

If all of this excess energy is imparted into the relative translation of the two fragments, we can calculate their relative speed. This value will provide an upper limit on the speed. To find the relative speed, we first calculate the reduced mass μ between the two fragments and find

$$\mu = \frac{(69)(127)}{(69+127)} \cdot 1.66 \times 10^{-27} \text{ kg} = 7.42 \times 10^{-26} \text{ kg}$$

Then we equate the relative kinetic energy for the particles with the excess energy and find the relative speed c_{rel} to be

$$c_{\text{rel}} = \sqrt{\frac{2E_{\text{excess}}}{\mu}}$$

$$= \sqrt{\frac{2\left(2.02 \times 10^{-19}\right) \text{ J}}{7.42 \times 10^{-26} \text{ kg}}} = 2.33 \times 10^{3} \text{ m s}^{-1}$$

With the relative speed in hand, it is straightforward to estimate the time to reach 840 pm as

$$t = \frac{(840 - 210) \times 10^{-12}}{2.33 \times 10^{3}} \text{s} = 270 \text{ fs}$$

Now we consider a scenario in which part of the excess energy is imparted to five quanta of the umbrella vibration (580 cm^{-1} or 1.15×10^{-20} J each). Hence we lose 5.75×10^{-20} J from the excess energy and only have $(20.2 - 5.75) \times 10^{-20}$ J $= 14.5 \times 10^{-20}$ J to be statistically distributed between the six translational degrees of freedom and three rotational degrees of freedom. Dividing this energy equally among the different translational and rotational modes, we find that 1.6×10^{-20} J is available for each of these motions. Using 1.6×10^{-20} J for the translational energy in the particles' relative motion, we find that

$$c_{rel} = \sqrt{\frac{2E}{\mu}} = \sqrt{\frac{2\left(1.6 \times 10^{-20}\right)\ \text{J}}{7.42 \times 10^{-26}\ \text{kg}}} = 657\ \text{m s}^{-1}$$

With the relative speed in hand, we estimate the time to reach 840 pm as

$$t = \frac{(840 - 210) \times 10^{-12}}{657}\text{s} = 959\ \text{fs}$$

nearly a factor of four longer.

We expect the initial geometry of the CF_3I molecule to have I-C-F bond angles of more than 109 degrees (based on simple VSEPR arguments); however, upon dissociation we expect that the trifluoromethyl radical would be more planar in geometry. Not unlike that known for the methyl radical. The displacement of the atoms in the umbrella vibration of CF_3 would move the molecular fragment between these geometries.

P25.2 You wish to follow the thermal isomerization reaction

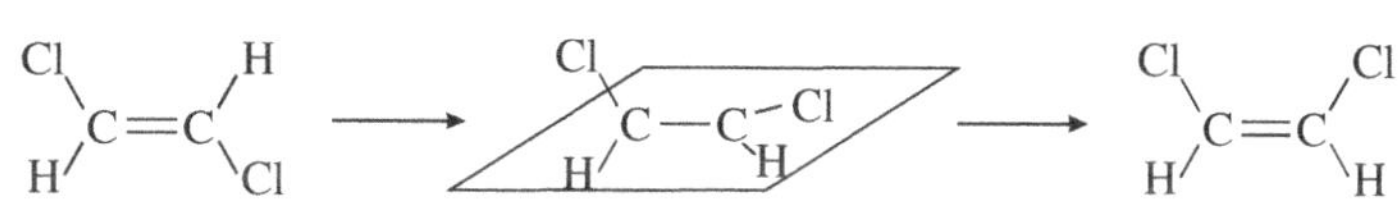

trans-Form   Transition state   *cis*-Form

This unimolecular reaction proceeds with a rate constant of about 2 s^{-1} at 500 K and 43 s^{-1} at 600 K. Describe in detail how you would characterize the kinetics of this reaction over the temperature range of 500 to 600 K. In particular, describe the experimental method that you would use to determine the rate law, including a) what quantity, or quantities, you would measure and b) how you would use these quantities to determine the rate law. Provide a detailed sketch of your experimental apparatus(es). Indicate clearly how you would perform the measurement and control the system's properties.

Solution

A variety of different answers are possible for ways to monitor this reaction. For example, one could use infrared spectroscopy to monitor the *trans* and *cis* populations in real time (e.g., the *cis* form has a strong absorbance near 550 cm^{-1} that is not present for the *trans* form). Alternatively, one could use a sample and quench approach, as long as the analysis method analyzed the quenched sample under conditions such as a lower temperature (e.g., 300 K) for which the reaction rate is slow enough.

Using the quench method you could place your initial reactant *trans* form into a nonreactive vessel that can be heated to the appropriate temperatures. The sample could be placed in a furnace at the temperature of reaction to be studied for some time *t* and then withdrawn and cooled rapidly to room temperature, where the isomerization is quite slow. Once at room temperature, the spectrum could be measured or the sample could be transferred to an appropriate optical cell or device for characterization. If a series of measurements are performed that have the same conditions but different residence times (different values of *t*) in the furnace then it would be possible to determine the time evolution of the *trans* isomer concentration, from which the rate law can be determined. See J. L. Jones and R. L. Taylor *JACS* 1940, **62**, 3480.

P25.3 (Adapted from S. Benson, Foundations of Chemical Kinetics) Consider the reaction

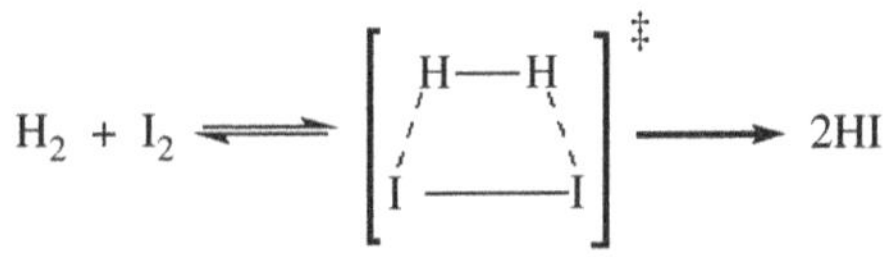

Assume that the transition state is planar and has an axis of symmetry. Show using transition state theory that the rate constant may be written as

$$k_r = \frac{\boldsymbol{N}_A h^3}{4\pi^2 kT}\left(\frac{m^\ddagger}{m_{H_2}m_{I_2}}\right)^{3/2}\left(\frac{\sqrt{I_A I_B I_C}}{I_{H_2}I_{I_2}}\right)\left[\frac{\left(1-\exp\left(-\frac{h\nu_{H_2}}{kT}\right)\right)\left(1-\exp\left(-\frac{h\nu_{I_2}}{kT}\right)\right)}{\prod_{i=1}^{5}\left(1-\exp\left(-\frac{h\nu_i^\ddagger}{kT}\right)\right)}\right]\exp\left(-\frac{E_{0,m}^\ddagger}{\boldsymbol{RT}}\right)$$

Note that you should assume that $\sigma_{H_2} = \sigma_{I_2} = \sigma^\ddagger = 2$. Now use the following parameters to estimate the reaction rate constant. a) $I_{I_2} = 7.485 \times 10^{-45}$ kg^{-1} m^2; $\nu_{I_2} = 214.57$ cm^{-1}. b) $I_{H_2} = 4.56 \times 10^{-48}$ kg^{-1}m^2; $\nu_{H_2} = 4395.2$ cm^{-1}. c) The bond distances in the activated complex are estimated to be H–H: 97 pm, I–I: 295 pm, and H–I: 175 pm. d) The normal mode wavenumbers of the activated complex are estimated to be 994cm^{-1}, 86cm^{-1}, 1280cm^{-1}, 965‡ cm^{-1}, 1400cm^{-1}, and 1730cm^{-1}. The reaction coordinate is associated with the 965 cm^{-1} mode. e) The activation barrier is $E_{0,m}^\ddagger = 170.5$ kJ mol^{-1} at a temperature of 600K. Compare your calculated rate constant value to the experimental value of 9.0×10^{-6} M^{-1} s^{-1}.

Solution

We write the reaction rate constant k_r as

$$k_r = \frac{kT}{h}\left(\frac{RT}{p^\ominus}\right)\boldsymbol{N}_A \frac{Z_{AC}^{\ddagger\ominus}/\sigma^\ddagger}{\left(Z_{H_2}^\ominus/\sigma_{H_2}\right)\left(Z_{I_2}^\ominus/\sigma_{I_2}\right)}\exp\left(-\frac{E_{0,m}^\ddagger}{\boldsymbol{RT}}\right)$$

for which the partition functions are

$$\frac{Z_{H_2}^\ominus}{\sigma_{H_2}} = \frac{1}{2}\left(\frac{2\pi m_{H_2}kT}{h^2}\right)^{3/2} V\left(\frac{T}{\theta_{rot,H_2}}\right)\frac{\exp\left(\varepsilon_{B,H_2}/(kT)\right)}{1-\exp\left(-h\nu_{H_2}/(kT)\right)}$$

$$\frac{Z_{I_2}^\ominus}{\sigma_{I_2}} = \frac{1}{2}\left(\frac{2\pi m_{I_2}kT}{h^2}\right)^{3/2} V\left(\frac{T}{\theta_{rot,I_2}}\right)\frac{\exp\left(\varepsilon_{B,I_2}/(kT)\right)}{1-\exp\left(-h\nu_{I_2}/(kT)\right)}$$

and

$$\frac{Z_{AC}^\ominus}{\sigma^\ddagger} = \frac{1}{2}\left(\frac{2\pi m_{AC}^\ddagger kT}{h^2}\right)^{3/2} V\left(\frac{T^{3/2}\sqrt{\pi}}{\sqrt{\theta_{rot,A}\theta_{rot,B}\theta_{rot,C}}}\right)\frac{\exp\left(\varepsilon_{B,AC}/(kT)\right)}{\prod_{j=1}^{5}\left(1-\exp\left(-h\nu_j/(kT)\right)\right)}$$

Substituting into the rate constant expression we find that

$$k_r = \frac{h^2}{\pi T\sqrt{2k}}\boldsymbol{N}_A\left(\frac{\theta_{rot,H_2}\theta_{rot,I_2}}{\sqrt{\theta_{rot,A}\theta_{rot,B}\theta_{rot,C}}}\right)\left(\frac{m_{AC}^\ddagger}{m_{H_2}m_{I_2}}\right)^{3/2} f_{vib}\exp\left(-\frac{E_{0,m}^\ddagger}{\boldsymbol{RT}}\right)$$

where

$$f_{vib} = \frac{\left(1-\exp\left(-h\nu_{I_2}/(kT)\right)\right)\left(1-\exp\left(-h\nu_{H_2}/(kT)\right)\right)}{\prod_{j=1}^{5}\left(1-\exp\left(-h\nu_j/(kT)\right)\right)}$$

and

$$E_{0,m}^\ddagger = \varepsilon_{B,H_2} + \varepsilon_{B,I_2} - \varepsilon_{B,AC}$$

If we use the definition of the characteristic rotational temperature, $\theta_{rot} = h^2/\left(8\pi^2 Ik\right)$ this expression can be simplified further, and we find

$$k_r = \frac{h^3}{4\pi^2 kT}\boldsymbol{N}_A\left(\frac{\sqrt{I_A I_C I_B}}{I_{H_2} I_{I_2}}\right)\left(\frac{m^{\ddagger}_{AC}}{m_{H_2} m_{I_2}}\right)^{3/2} f_{vib}\exp\left(-\frac{E^{\ddagger}_{0,m}}{\boldsymbol{R}T}\right)$$

To evaluate this result we must find the molecular parameters. All of them are directly given in the problem statement, except for the inertial moments of the transition state. Using the geometry shown below we find that

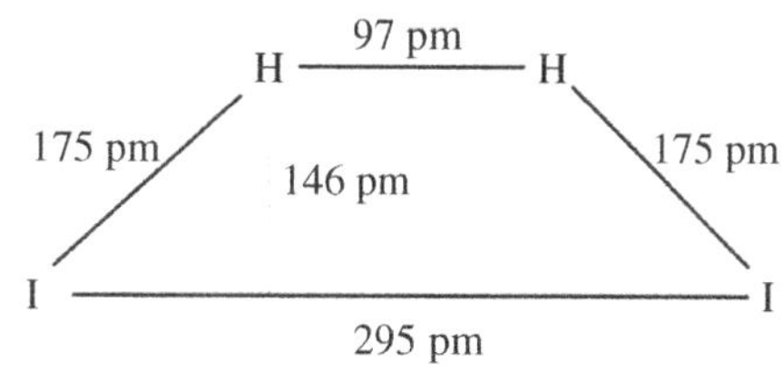

$I_C = 1.35 \times 10^{-46}$ kg m^2, $I_B = 9.18 \times 10^{-45}$ kg m^2, and $I_A = 9.31 \times 10^{-45}$ kg m^2. Using these parameters, we provide the following quantities

$$\frac{\sqrt{I_A I_C I_B}}{I_{H_2} I_{I_2}} = 3.22 \times 10^{24}\ \text{kg}^{\ 1/2}\ \text{m}^{-1} \quad \text{and} \quad \left(\frac{m^{\ddagger}_{AC}}{m_{H_2} m_{I_2}}\right)^{3/2} = 5.27 \times 10^{39}\ \text{kg}^{-3/2}$$

and

$$E^{\ddagger}_{0,m} = 170.5\ \text{kJ mol}^{-1}$$

At 600 K we find that

$$f_{vib} = \frac{(0.402)(1)}{(0.186)(0.909)(0.954)(1)(1)} = 2.49$$

so that

$$k_r = \left[\begin{matrix}\left(5.36 \times 10^{-58}\text{kg}^2\ \text{m}^4\ \text{s}^{-1}\text{mol}^{-1}\right)\left(3.22 \times 10^{24}\ \text{kg}^{-1/2}\ \text{m}^{-1}\right) \\ \cdot\left(5.27 \times 10^{39}\ \text{kg}^{-3/2}\right)(2.49)\end{matrix}\right]\exp(-34.18)$$

$$= 3.24 \times 10^{-8}\frac{\text{m}^3}{\text{s mol}} = 3.24 \times 10^{-5}\text{L mol}^{-1}\text{s}^{-1}$$

For such a "simple" model, the correspondence to the experimental value is quite good, about a factor of three. If we assign this difference to the transmission coefficient κ, we find that

$$\kappa = \frac{9.0 \times 10^{-6}}{3.24 \times 10^{-5}} = 0.27$$

P25.4 Consider the reaction

$$\text{HO(g)} + \text{HO(g)} \rightarrow [\text{HO–HO}]^{\ddagger} \rightarrow \text{H}_2\text{O(g)} + \text{O(g)}$$

The measured thermal rate constant for this reaction is 8.7×10^8 L mol^{-1}s^{-1} at 298 K.[5] Take the energy barrier $\Delta_r H^{\ddagger\ominus}$ for this reaction to be 2.0 kJ mol^{-1} and consider the transition state to be a linear geometry, as shown above. Using the data given below and assuming that the frequency for the reaction coordinate mode is $\nu_0 = 9.37 \times 10^{13}$ s^{-1} and $\kappa = 1$, compute the transition state theory rate constant for this reaction at 298 K.

Species	σ	θ_{rot} (K)	θ_{vib} (K)
OH	1	27	5384
HO—HO	1	5	5400(ν_s), 4500(ν_a), 1500, 1100, 800, 100, 100

Compare your calculated value of the rate constant to the experimental value.

5 Y. Bedjanian, G. LeBras, and G. Poulet, *J. Phys. Chem. A.* 1999, **103**, 7017.

Solution

In this case the transition state theory rate constant expression becomes

$$k_{\mathrm{TST}} = \boldsymbol{N}_A \frac{kT}{h}\left(\frac{Z^{\ddagger\ominus}/V}{\left(Z^{\ominus}_{\mathrm{OH}}/V\right)^2}\right)\exp\left(-\frac{\Delta_r H^{\ddagger\ominus}}{\boldsymbol{R}T}\right)$$

At 298 K, we find that

$$\frac{Z^{\ominus}_{\mathrm{OH}}}{V} = \left(\frac{2\pi m_{\mathrm{OH}}kT}{h^2}\right)^{3/2}\left(\frac{T}{\theta_{\mathrm{rot,OH}}}\right)\left(\frac{1}{1-\exp\left(-\theta_{\mathrm{vib,OH}}/T\right)}\right)$$
$$= (6.77\times 10^{31}\,\mathrm{m}^{-3})(11.04)(1.0) = 7.47\times 10^{32}\ \mathrm{m}^{-3}$$

and

$$\frac{Z^{\ddagger\ominus}}{V} = \left(\frac{2\pi m^{\ddagger}kT}{h^2}\right)^{3/2}\left(\frac{T}{\theta^{\ddagger}_{\mathrm{rot}}}\right)\left[\prod_{j=1}^{7}\left(\frac{1}{1-\exp\left(-\theta_{\mathrm{vib,j}}/T\right)}\right)\right]$$
$$= (1.91\times 10^{32}\ \mathrm{m}^{-3})(59.6)\,[13.6] = 1.55\times 10^{35}\ \mathrm{m}^{-3}$$

so that the rate constant becomes

$$k_{\mathrm{TST}} = 3.74\times 10^{36}\ \mathrm{mol}^{-1}\ \mathrm{s}^{-1}\cdot\frac{1.55\times 10^{35}\ \mathrm{m}^{-3}}{\left(7.47\times 10^{32}\ \mathrm{m}^{-3}\right)^2}\exp(-0.807)$$
$$= 4.63\times 10^{5}\ \frac{\mathrm{m}^3}{\mathrm{mol\ s}} = 4.63\times 10^{8}\ \frac{\mathrm{L}}{\mathrm{mol\ s}}$$

This calculated value differs by about a factor of two from the experimental value.

P25.5 Consider the reaction

$$\mathrm{F} + \mathrm{H_2} \rightarrow \mathrm{HF} + \mathrm{H}$$

which has been studied by Polanyi and coworkers (D. H. Maylotte, J. C. Polanyi and K. B. Woodall, *J. Chem. Phys.* 1972, **57**, 1547). Two observations are noteworthy. First, it is possible to determine the relative populations of the vibrational states formed for HF in this reaction; see the table. The vibrational wavenumber of HF is 4138 cm^{-1} and of H_2 is 4401 cm^{-1}.

Table: Vibrational Energy Distribution of Product HF at 298 K.

Vibrational quantum number	1	2	3
Fraction of molecules	0.174	0.562	0.264

Second, this reaction is exothermic and the energy is partitioned into the products as 66% vibration, 8% rotation, and 26% translation. Use the potential energy surface in the figure to explain these two observations.

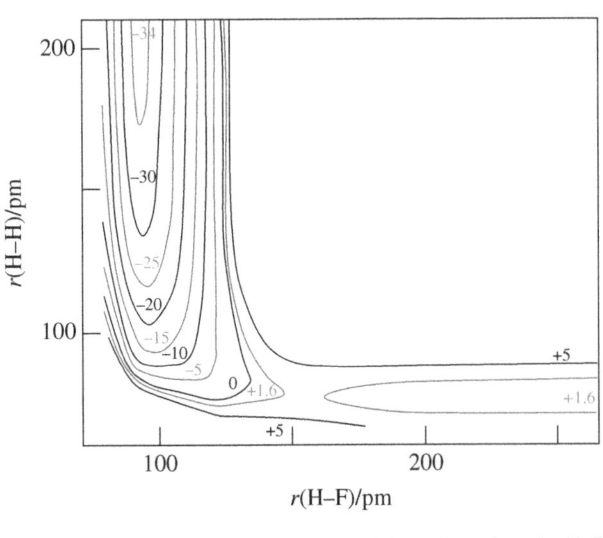

Figure P25.5a This surface is sketched from that given by H. F. Schaeffer III, *J. Phys. Chem.* 1985, **89**, 5336. The contour labels have units of kcal mol^{-1}.

Describe an experimental arrangement that you might use to investigate the vibrational distribution of the HF product. Provide a sketch of an apparatus. Describe in words your experimental method.

Solution

For the reaction $F + H_2$, the transition state is "early"; i.e., before the turn. Hence the reaction is enhanced by relative translational energy. The products, however, are expected to be vibrationally excited. The low amount of rotational energy implies a collinear or nearly collinear transition state.

The schematic diagram in Fig. P25.5a illustrates the geometry of a reaction. Each of the reactants expands from a source into a vacuum chamber to form beams of reactants. The reactant beams collide near the center of the vacuum chamber and form products. The infrared chemiluminescence of the product species is monitored from the top (Fig. P25.5b).

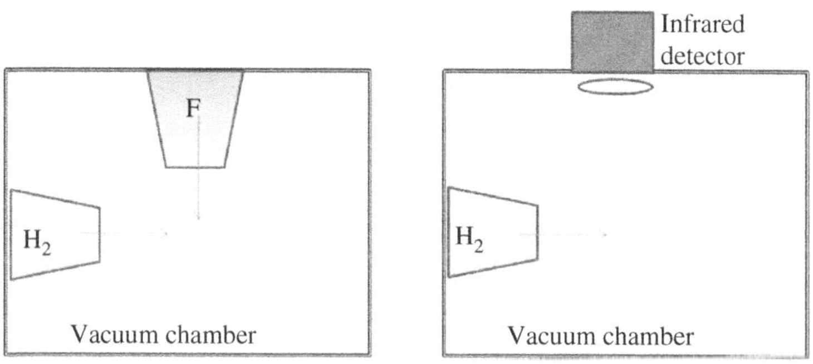

Figure P25.5b Experimental setup for a molecular beam apparatus.

A number of other approaches to monitoring the reaction product distribution are possible, such as using laser-induced fluorescence to study the HF product.

P25.6 Consider the reaction

$$H_2 + I_2 \underset{k_{-1}}{\overset{k_1}{\rightleftarrows}} 2HI \qquad (25.6)$$

The forward rate constant is $k_1 = 10^{11} \exp\left(-E_{1,m}/RT\right)$ L mol^{-1} s^{-1} where $E_{1,m} = 165$ kJ mol^{-1}, and the backward rate constant is $k_{-1} = 6 \times 10^{10} \exp\left(-E_{2,m}/RT\right)$ L mol^{-1} s^{-1} where $E_{2,m} = 185$ kJ mol^{-1}. What is the equilibrium constant for a mixture of H_2, I_2 and HI at 1000 K? What is the standard Gibbs energy change at this temperature? Determine the activation entropy and the activation enthalpy for the forward reaction at 1000 K. Perform the same calculation for the backward reaction rate constant. Sketch the Gibbs energy versus the reaction coordinate and label any of the energies that you know.

Solution

The equilibrium constant for the mixture is given by the ratio of the rate constants, so that at 1000 K we find

$$K = \frac{k_1}{k_{-1}} = \frac{2.40 \times 10^5}{1.30 \times 10^4} = 18.4$$

and hence the reaction Gibbs energy is

$$\begin{aligned}\Delta_r G &= -RT \ln K \\ &= -(8.314 \text{ J mol}^{-1}\text{K}^{-1})(1000 \text{ K}) \ln(18.4) \text{ J mol}^{-1} \\ &= -\left[(8314)\ln(18.4)\right] \text{ J mol}^{-1} = -24.2 \text{ kJ mol}^{-1}\end{aligned}$$

To determine the activation parameters we first write that

$$k_1 = B \exp\left(\frac{\Delta_r S^{\ddagger}_{k_1,m}}{R}\right) \exp\left(-\frac{\Delta_r H^{\ddagger}_{k_1,m}}{RT}\right)$$

where

$$B = \frac{kT}{h}\frac{\boldsymbol{R}T}{P^{\ominus}} \text{ and } \Delta_r H^{\ddagger}_{k_1,m} = E_{1,m} - 2\boldsymbol{R}T$$

so that the pre-exponential factor in the Arrhenius expression A, is given by

$$A = B\exp\left(2 + \frac{\Delta_r S^{\ddagger}_{k_1,m}}{\boldsymbol{R}}\right) \text{ and } \Delta_r S^{\ddagger}_{k_1,m} = \boldsymbol{R}\left[\ln(A/B) - 2\right]$$

At 1000 K, we find that

$$\Delta_r H^{\ddagger}_{k_1,m} = E_{1,m} - 2\boldsymbol{R}T$$
$$= (165,000 - 16,628) \text{ J mol}^{-1} = 148.4 \text{ kJ mol}^{-1}$$

and

$$\Delta S^{\ddagger}_{k_1,m} = \boldsymbol{R}\left[\ln\left(\frac{10^{11}}{1.73 \times 10^{15}}\right) - 2\right] = -97.8 \text{ J mol}^{-1} \text{ K}^{-1}$$

This combination gives a value of $\Delta_r G^{\ddagger}_{k_1,m} = \Delta_r H^{\ddagger}_{k_1,m} - T\Delta_r S^{\ddagger}_{k_1,m} = 246.1 \text{ kJ mol}^{-1}$.

A similar calculation for the reverse reaction gives $\Delta_r H^{\ddagger}_{k_{-1},m} = 168.4$ kJ mol^{-1} and $\Delta_r S^{\ddagger}_{k_{-1},m} = -102.0$ J mol^{-1}K^{-1}, so that $\Delta_r G^{\ddagger}_{k_{-1},m} = \Delta_r H^{\ddagger}_{k_{-1},m} - T\Delta_r S^{\ddagger}_{k_{-1},m} = 270.4$ kJ mol^{-1}. Using these values along with those calculated above we can sketch a reaction coordinate diagram like that in Fig. P25.6.

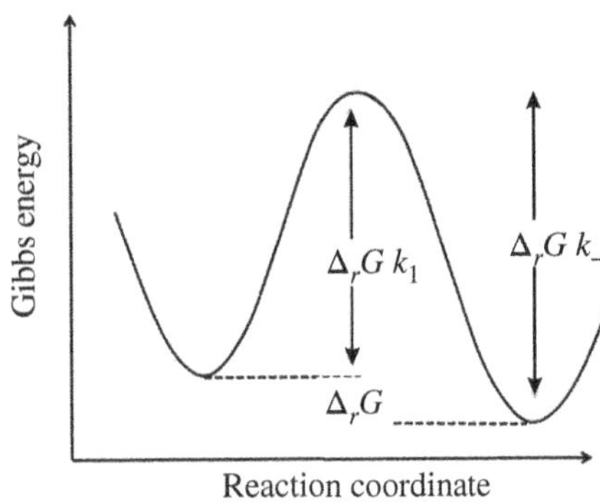

Figure P25.6 Gibbs energy versus reaction coordinate for reaction (25.6).

P25.7 The enol compound E was investigated in a hydrocarbon glass at low temperature (B. Nickel, K.H. Grellmann, J.S. Stephan, P.J. Walla, Ber. Bunsenges. 199, **102**, 436). When exciting by light, E goes over into the excited singlet state of the keto form K, then into the triplet state of K, into the triplet state of E and finally back into the ground state of E:

$$\text{E} \xrightarrow{\text{light}} {}^1\text{K}^* \rightarrow {}^3\text{K}^* \rightarrow {}^3\text{E}^* \rightarrow \text{E}$$

Below 25 K, the rate of reaction ${}^3\text{K}^* \rightarrow {}^3\text{E}^*$ is independent of temperature. Explain why replacing the proton by a deuteron reduces the rate by 2000 times. Use a simple calculation to estimate the change in rate constant.

Solution

When replacing the proton by a deuteron the rate of this reaction is 2000 times slower. We estimate the ratio k_H/k_D by using the values given in Section 13.5.2 for the hydrogen bond in water: $E_a = 0.5 \times 10^{-19}$ J $\times \boldsymbol{N}_A =$ 30 kJ mol^{-1} and $r = (180 - 97)$ pm ≈ 80 pm.

Following the procedure in the Example 25.7,we write the ratio of the rate constants $k_{\text{H}}/k_{\text{D}}$ as

$$\frac{k_{\text{H}}}{k_{\text{D}}} = \frac{\exp\left(-C\sqrt{m_{\text{H}}}\right)}{\exp\left(-C\sqrt{m_{\text{D}}}\right)} = \exp\left(-C \cdot \left[\sqrt{m_{\text{H}}} - \sqrt{m_{\text{D}}}\right]\right)$$
$$= \exp\left(-C\sqrt{m_{\text{H}}} \cdot \left[1 - \sqrt{2}\right]\right)$$

and

$$C\sqrt{m_{\mathrm{H}}} = 4.8 \times 10^{14} \times 4.09 \times 10^{-14} = 19.6$$

so that

$$\frac{k_{\mathrm{H}}}{k_{\mathrm{D}}} = \exp\left(-19.6 \times \left[1 - \sqrt{2}\right]\right) \approx 3000$$

in good agreement with the experiment.

P25.8 Consider the H atom transfer reaction

$$\mathrm{F + HI \rightarrow \ FHI \ \rightarrow HF + I \ or \ F + DI \ \rightarrow \ FDI \ \rightarrow DF + I}$$

which shows an isotope effect on the rate constant that is $k_r(\mathrm{HI})/k_r(\mathrm{DI}) = 1.29$ at $T = 298$ K. In their paper (J. Chem. Phys. 1979, **70**, 1759), Mei and Moore use a bent transition state to calculate this isotope ratio. Perform a related calculation but assume that the transition state is linear. Compare your calculated isotope ratio to that of Moore and Mei's for the bent transition state. Comment on which geometry you think is more reasonable.

Solution

Using Equation (25.20) and

$$N(\mathrm{F}) = N(\mathrm{HI}) = N(\mathrm{FHI}) = N(\mathrm{DI}) = N(\mathrm{FDI}),$$

we find that

$$\frac{k_r(\mathrm{HI})}{k_r(\mathrm{DI})} = \frac{Z^{\ominus}_{\mathrm{FHI}}}{Z^{\ominus}_{\mathrm{F}} Z^{\ominus}_{\mathrm{HI}}} \frac{Z^{\ominus}_{\mathrm{F}} Z^{\ominus}_{\mathrm{DI}}}{Z^{\ominus}_{\mathrm{FDI}}} = \frac{Z^{\ominus}_{\mathrm{FHI}}}{Z^{\ominus}_{\mathrm{HI}}} \frac{Z^{\ominus}_{\mathrm{DI}}}{Z^{\ominus}_{\mathrm{FDI}}}$$

Rather than evaluate this expression explicitly, we make an estimate by neglecting differences in molecular partition functions of the FHI and FDI activated complexes. This approximation should work well for the translational and rotational contributions since the H and D atoms are in the center of the molecule and of small mass compared to the F and I atoms. Since the asymmetric stretch vibrational mode is the reaction coordinate, it does not contribute to $Z^{\ominus}$. This approximation may not work so well for the bending vibrational modes and the symmetric stretch vibrational modes, because their resonant frequencies will be different (higher for FHI than FDI). Hence

$$\frac{Z^{\ominus}_{\mathrm{FHI}}}{Z^{\ominus}_{\mathrm{FDI}}} < 1$$

and our approximation will provide an upper bound on the rate constant ratio. Then

$$\frac{k_r(\mathrm{HI})}{k_r(\mathrm{DI})} \leq \frac{Z^{\ominus}_{\mathrm{DI}}}{Z^{\ominus}_{\mathrm{HI}}}$$

In the evaluation of the remaining molecular partition functions of HI and DI we neglect the difference in the translational contribution (the masses m_{HI} and m_{DI} differ only slightly). Also we assume that the electronic energies of HI and DI are the same. With these approximations, the ratio of both rate constants is determined by the rotational and vibrational contributions to the molecular partition functions. We consider the high temperature limit for the rotational partition function and a temperature range where the molecules are in their vibrational ground states, thus we find the rotational partition functions to be

$$Z_{\mathrm{rot}}(\mathrm{DI}) = \frac{kT8\pi^2\mu_{\mathrm{DI}}d^2}{h^2}, \; Z_{\mathrm{rot}}(\mathrm{HI}) = \frac{kT8\pi^2\mu_{\mathrm{HI}}d^2}{h^2}$$

and the vibrational partition functions to be

$$Z_{\mathrm{vib}}(\mathrm{DI}) = \exp\left[-\frac{1}{2}h\nu_{\mathrm{DI}}\right], \; Z_{\mathrm{vib}}(\mathrm{HI}) = \exp\left[-\frac{1}{2}h\nu_{\mathrm{HI}}\right]$$

For the ratio of the two rate constants we obtain

$$\frac{k_r(\mathrm{HI})}{k_r(\mathrm{DI})} \leq \frac{Z_{\mathrm{rot}}(\mathrm{DI})Z_{\mathrm{vib}}(\mathrm{DI})}{Z_{\mathrm{rot}}(\mathrm{HI})Z_{\mathrm{vib}}(\mathrm{HI})} = \frac{\mu_{\mathrm{DI}}}{\mu_{\mathrm{HI}}} \cdot \exp\left(-\frac{h\nu_{\mathrm{DI}} - h\nu_{\mathrm{HI}}}{2kT}\right)$$

where v is the vibration frequency. Using the reduced masses $\mu_{DI} = m_D m_I/(m_D + m_I)$, $\mu_{HI} = m_H m_I/(m_H + m_I)$, and the vibrational frequencies of $v_{HI} = 6.92 \times 10^{13}\ s^{-1}$ and $v_{DI} = 4.90 \times 10^{13}\ s^{-1}$ (both frequencies differ by a factor of $\sqrt{2}$), we find for $T = 298$ K

$$\frac{k_r(\mathrm{HI})}{k_r(\mathrm{DI})} \leq \left(\frac{3.39 \times 10^{-27}\ \mathrm{kg}}{1.66 \times 10^{-27}\ \mathrm{kg}}\right) \cdot 5.09 = 10$$

This ratio is much larger than that calculated for the bent geometry by Mei and Moore ($k_r(\mathrm{HI})/k_r(\mathrm{DI}) = 1.29$), so that it seems the bent geometry is a more reasonable transition state geometry.

P25.9 Recently Baldwin and Chapman (*J. Org. Chem.* 2005, **70**, 377) studied 1,5 sigmatropic shift of H atoms for 1,3-cyclohexadienes by synthesizing the monodeuterated molecule and studying its interconversion between the three forms:

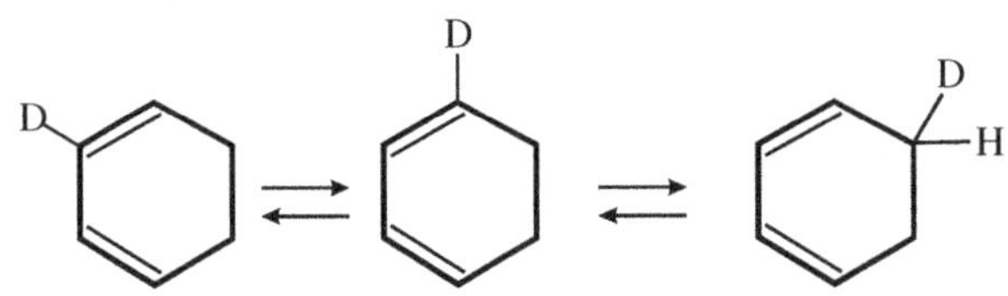

In their study they measured the time for the system to equilibrate between the three forms, and they found the following temperature dependence for this reaction

T/K	527	557
k/s^{-1}	0.394×10^{-4}	3.00×10^{-4}

a) From their data, determine the activation enthalpy for this reaction. Comment on the size of the activation enthalpy you find; i.e., is it what you would expect and why (or why not)? b) From their data, determine the activation entropy for the reaction. Comment on the sign of the activation entropy; i.e., is it what you expect and why (or why not)? c) In their analysis they assume that the H atoms move but that the D atom stays in place. Is this a good assumption? Justify your answer with an estimate of the relative rates in the two cases.

Solution

a) We assume that the bond breaking C–H is the rate limiting step. We can estimate the activation enthalpy from the TST rate constant expression, via

$$\frac{k_{\mathrm{TST}}}{T} = \frac{k}{h} \exp\left(\frac{\Delta_r S^{\ddagger\ominus}}{R}\right) \exp\left(-\frac{\Delta_r H^{\ddagger\ominus}}{RT}\right)$$

and using our rate constant data at the two temperatures to write

$$\frac{0.394 \times 10^{-4}}{3.00 \times 10^{-4}} \cdot \frac{557}{527} = \exp\left(-\frac{\Delta_r H^{\ddagger\ominus}/(\mathrm{J\ mol^{-1}})}{R/(\mathrm{J\ mol^{-1}\ K^{-1}})}\left[\frac{1}{527\ \mathrm{K}} - \frac{1}{557\ \mathrm{K}}\right]\right) = 0.139$$

so that

$$\Delta_r H^{\ddagger\ominus} = -\left(8.314\ \mathrm{J\ mol^{-1}\ K^{-1}}\right) \frac{\ln(0.139)}{1.022 \times 10^{-4}\ \mathrm{K^{-1}}}$$
$$= 160,527\ \mathrm{J\ mol^{-1}} = 161\ \mathrm{kJ\ mol^{-1}}$$

This enthalpy of activation is about half that required to break a typical C–H bond. It is reasonable that the activation enthalpy is this large because the H atom migrates pretty far, yet is below that for full C–H bond cleavage.

b) To estimate the activation entropy, we assume that the transmission coefficient is $\kappa = 1$, so that

$$\exp\left(\frac{\Delta_r S^{\ddagger\ominus}}{R}\right) = k_{\mathrm{TST}} \frac{h}{kT} \exp\left(\frac{\Delta_r H^{\ddagger\ominus}}{RT}\right)$$

and

$$\frac{\Delta_r S^{\ddagger\ominus}}{\boldsymbol{R}} = \ln\left(\frac{6.626\times 10^{-34}\ \text{J s}\cdot 0.394\times 10^{-4}\ \text{s}^{-1}}{1.38\times 10^{-23}\ \text{J K}^{-1}\cdot 527\ \text{K}}\right) + \frac{161.0\ \text{kJ mol}^{-1}}{8.314\ \text{J mol}^{-1}\ \text{K}^{-1}\cdot 527\ \text{K}}$$

Hence we find that

$$\Delta_r S^{\ddagger\ominus} = \boldsymbol{R}(-3.65) = -30.4\ \text{J mol}^{-1}\ \text{K}^{-1}$$

The negative activation entropy implies that the transition state requires a preferred set of geometries and/or protons. This finding is consistent with the idea that the H-atom moves across the ring rather than around the ring and seems plausible.

c) The assumption that the H-atom migrates, and not D, originates from the difference in zero point energies for these two bond types (C–H) versus (C–D). To estimate the difference in rate constants we use the k_{TST} expression to write

$$\frac{k_{\text{TST}}(\text{C–H})}{k_{\text{TST}}(\text{C–D})} \approx \exp\left(-\frac{\Delta E_0(\text{C–H}) - \Delta E_0(\text{C–D})}{kT}\right)$$

We can simplify the energy difference if we assume that the electronic contributions are the same, namely

$$\Delta E_0(\text{C–H}) - \Delta E_0(\text{C–D}) = \left[\Delta E_{\text{elec}}(\text{C–H}) - \frac{h\nu(\text{C–H})}{2}\right] - \left[\Delta E_{elec}(\text{C–D}) - \frac{h\nu(\text{C–D})}{2}\right]$$
$$= \frac{1}{2}h\nu(\text{C–D}) - \frac{1}{2}h\nu(\text{C–H})$$

so that

$$\frac{k_{\text{TST}}(\text{C–H})}{k_{\text{TST}}(\text{C–D})} \approx \exp\left(-\frac{h\nu(\text{C–D}) - h\nu(\text{C–H})}{2kT}\right)$$

For $\nu(\text{C–H}) = 1.08\times 10^{14}\text{s}^{-1}$ and $\nu(\text{C–D}) = 7.64\times 10^{13}\text{s}^{-1}$, we find a ratio at 557 K to be

$$\frac{k_{\text{TST}}(\text{C–H})}{k_{\text{TST}}(\text{C–D})} \approx 3.9$$

that is a factor of four different.

P25.10 Use transition state theory to rationalize the observed isotope ratios in Table 25.1.

Solution

We compare the rate constants for the reactions

$$\text{Cl} + \text{H}_2 \rightarrow \text{ClHH} \rightarrow \text{HCl} + \text{H and Cl} + \text{D}_2 \rightarrow \text{ClDD} \rightarrow \text{DCl} + \text{D}$$

discussed in Section 25.2.4. Using Equation (25.20) in Section 25.2.1, we find that

$$\frac{k_r(\text{H}_2)}{k_r(\text{D}_2)} = \frac{Z^\ominus_{\text{ClHH}}}{Z^\ominus_{\text{Cl}} Z^\ominus_{\text{H}_2}} \frac{Z^\ominus_{\text{Cl}} Z^\ominus_{\text{D}_2}}{Z^\ominus_{\text{ClDD}}} = \frac{Z^\ominus_{\text{ClHH}}}{Z^\ominus_{\text{ClDD}}}\cdot\frac{Z^\ominus_{\text{D}_2}}{Z^\ominus_{\text{H}_2}}$$

The ratio of the partition functions for D_2 and H_2 follows in a manner very similar to that we used for the exchange equilibrium in Section 18.5.7.1. Briefly, we can write

$$\frac{Z^\ominus_{\text{D}_2}}{Z^\ominus_{\text{H}_2}} = \frac{Z^\ominus_{\text{D}_2,\text{trans}}}{Z^\ominus_{\text{H}_2,\text{trans}}}\frac{Z^\ominus_{\text{D}_2,\text{rot}}}{Z^\ominus_{\text{H}_2,\text{rot}}}\frac{Z^\ominus_{\text{D}_2,\text{vib}}}{Z^\ominus_{\text{H}_2,\text{vib}}}\frac{Z^\ominus_{\text{D}_2,\text{el}}}{Z^\ominus_{\text{H}_2,\text{el}}}$$
$$= \left(\frac{m_{\text{D}_2}}{m_{\text{H}_2}}\right)^{3/2}\left(\frac{m_{\text{D}}/2}{m_{\text{H}}/2}\right)\cdot\exp\left(-\frac{h}{2kT}\left(\nu_{0,\text{D}_2} - \nu_{0,\text{H}_2}\right)\right)$$
$$= 2^{3/2}\cdot 2\cdot\exp\left(-\frac{h}{2kT}\left(\nu_{0,\text{D}_2} - \nu_{0,\text{H}_2}\right)\right)$$

We calculate the vibration frequencies according to

$$\nu_{0,\text{H}_2} = \frac{1}{2\pi}\sqrt{\frac{k_f}{\mu_{\text{H}_2}}} = \frac{1}{2\pi}\sqrt{\frac{574\ \text{N m}^{-1}}{0.837\times 10^{-27}\ \text{kg}}} = 1.38\times 10^{14}s^{-1}$$

and

$$\nu_{0,D_2} = \frac{\nu_{0,H_2}}{\sqrt{2}} = 0.932 \times 10^{14}\ s^{-1}$$

Thus

$$-\frac{h}{2k}\left(\nu_{0,D_2} - \nu_{0,H_2}\right) = \frac{6.626 \times 10^{-34}\ J\ s}{2 \cdot 1.38 \times 10^{-23}\ J\ K^{-1}} 0.386 \times 10^{14}\ s^{-1} = 926.7\ K$$

and for $\frac{Z^\ominus_{D_2}}{Z^\ominus_{H_2}}$ at $T = 295.2$ K we obtain

$$\frac{Z^\ominus_{D_2}}{Z^\ominus_{H_2}} = 4\sqrt{2} \cdot \exp\left(\frac{926.7\ K}{295.2\ K}\right) = 130$$

The ratio of the partition functions at the transition state is more difficult to treat rigorously, because of our lack of knowledge about the structures. Hence we use recent theoretical calculations of the transition state structural parameters, assuming a linear geometry. The parameters for the transition state are given in the table

ABC	R_{AB}, pm	R_{BC}, pm	I, 10^{-46} kg m 2
ClHH	140.1	99.0	1.74
ClDD	140.1	99.0	2.33

ABC	$\left(\frac{1}{\lambda}\right)_{stretch}$, cm^{-1}	$\left(\frac{1}{\lambda}\right)_{bend}$, cm^{-1}	$E_{0,m}$, kJ mol^{-1}
ClHH	1358	581	49.4
ClDD	984	412	44.5

where E_0 is the energy at the barrier top which includes the zero point vibrational energies of the symmetric stretch mode and of the doubly degenerate bending mode. Using these parameters we can write that

$$\frac{Z^\ominus_{ClHH}}{Z^\ominus_{ClDD}} = \frac{Z^\ominus_{ClHH,trans}}{Z^\ominus_{ClDD,trans}} \frac{Z^\ominus_{ClHH,rot}}{Z^\ominus_{ClDD,rot}} \frac{Z^\ominus_{ClHH,vib,el}}{Z^\ominus_{ClDD,vib,el}}$$
$$= \left(\frac{m_{ClHH}}{m_{ClDD}}\right)^{3/2} \left(\frac{\mu_{ClHH}}{\mu_{ClDD}}\right) \cdot \frac{1 - \exp(-h\nu_{0,stretch},_{ClDD}/kT)}{1 - \exp(-h\nu_{0,stretch},_{ClHH}/kT)}$$
$$\cdot \left(\frac{1 - \exp(-h\nu_{bend},_{ClDD}/kT)}{1 - \exp(-h\nu_{bend},_{ClHH}/kT)}\right)^2 \cdot \exp\left[-\frac{(E_0(\mathrm{ClHH}) - E_0(\mathrm{ClDD}))}{\boldsymbol{RT}}\right]$$

From the table for $T = 295.2$ K we calculate

	ClHH	ClDD
$h\nu_{0,stretch}/10^{-20}$ J	2.69	1.96
$h\nu_{0,bend}/10^{-20}$ J	1.15	0.82
$\frac{h\nu_{0,stretch}}{kT}$	6.60	4.81
$\frac{h\nu_{0,bend}}{kT}$	2.82	2.01

$$\frac{Z^\ominus_{ClHH}}{Z^\ominus_{ClDD}} = \left(\frac{37.4658}{39.4816}\right)^{3/2} \left(\frac{1.74}{2.33}\right) \cdot \frac{1 - \exp(-4.81)}{1 - \exp(-6.60)} \cdot \left(\frac{1 - \exp(-2.01)}{1 - \exp(-2.82)}\right)^2 \cdot \exp[-1.99]$$
$$= 0.924 \cdot 0.747 \cdot 0.993 \cdot (0.921)^2 \cdot 0.136 = 0.080$$

Hence for the ratio of the rate constants at $T = 295.2$ K we find

$$\frac{k_r(H_2)}{k_r(D_2)} = \frac{Z^\ominus_{ClHH}}{Z^\ominus_{ClDD}} \frac{Z^\ominus_{D_2}}{Z^\ominus_{H_2}} = 0.080 \cdot 130 = 10.4$$

Accordingly, for $T = 384.7$ K and 498.6 K we find the ratios 7.3 and 5.5, respectively, in reasonable agreement with the data in Table 25.1.

P25.11 Consider the reaction of $H + D_2$ and assume that the reaction barrier has a functional form of

$$E_a(\gamma) = E_a + E'_a(1 - \cos\gamma)$$

where γ is the angle that the H-atom trajectory makes with the D_2 molecule's bond axis.

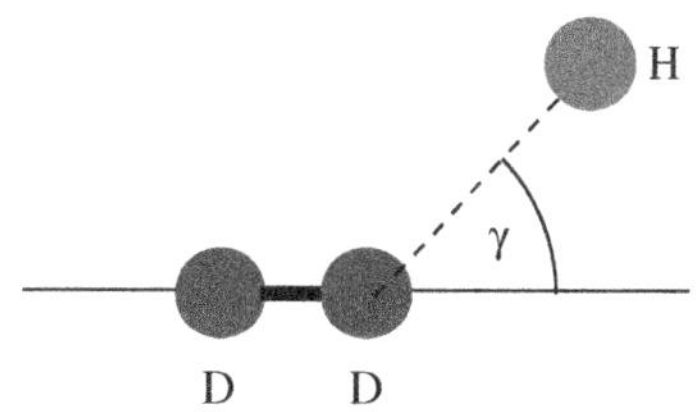

Combine this expression with the angle-dependent cross-section

$$\sigma_r(E,\gamma) = 0 \quad \text{for} \quad E < E_a(\gamma)$$
$$\sigma_r(E,\gamma) = \pi d^2 \cdot \left(1 - \frac{E_a(\gamma)}{E}\right) \quad \text{for} \quad E > E_a(\gamma)$$

to obtain an explicit expression for the dependence of $\sigma_r(E,\gamma)$ on γ. Compare your result to the angle dependence of the cross-section found from quasiclassical trajectory calculations at $E = 0.55$ eV (see Table below) by plotting the computational data[6] and comparing it to the model.

$\sigma_r(E = 0.55\text{ eV})/\text{Å}^2$	1.1	0.80	0.56	0.22	0.02	< 0.01	< 0.01
$\cos\gamma$	0.96	0.84	0.74	0.64	0.54	0.45	0.35

Lastly, perform an average over the angle $\cos(\gamma)$, to show that

$$\langle\sigma_r(E)\rangle_\gamma = \pi d^2 \cdot \frac{(E - E_a)^2}{4E \cdot E_a}$$

Solution

By substitution it is straightforward to show that

$$\sigma_r(E,\gamma) = \pi d^2 \cdot \left(1 - \frac{E_a + E'_a}{E}\right) + \pi d^2 \cdot \frac{E'_a}{E}\cos\gamma \text{ for } E > E_a + E'_a(1 - \cos\gamma)$$

If we use $\pi d^2 = 21.6\,\text{A}^2$ (see Example 25.5) and fit to the data we find that the best fit parameters are $E_a = 0.51$eV and $E'_a = 0.07$ eV (see Fig. P25.11).

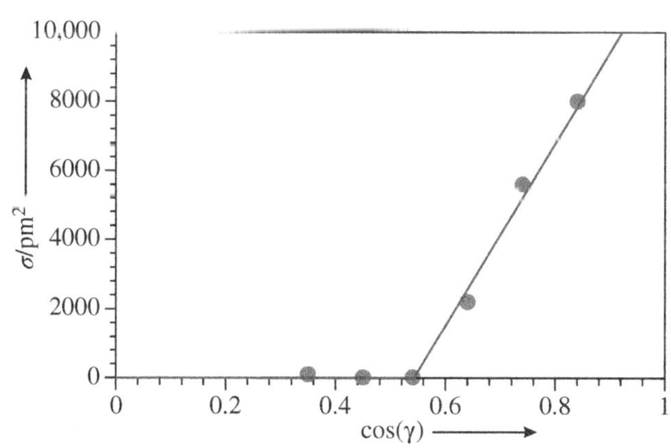

Figure P25.11 Cross section σ versus $\cos(\gamma)$.

6 N. C. Blais, R. B. Bernstein, and R. D. Levine *J. Phys. Chem.* 1985, **89**, 10.

To derive the equation for $\langle\sigma_r(E)\rangle_\gamma$ we perform the integral

$$\langle\sigma_r(E)\rangle_\gamma = \frac{1}{2}\cdot\int_{-1}^{1}\sigma_r(E,\gamma)H\left(E-E_a(\gamma)\right)\mathrm{d}(\cos\gamma)$$

where $H\left(E-E_a(\gamma)\right)$ is the Heaviside function; i.e., $H\left(E-E_a(\gamma)\right)=0$ for $E\le E_a(\gamma)$ and $H\left(E-E_a(\gamma)\right)=1$ for $E\ge E_a(\gamma)$.

The condition from the Heaviside function requires that the integrand be zero for

$$\cos\gamma \le \frac{E_a+E'_a-E}{E'_a}$$

and have the value

$$\pi d^2\cdot\left(1-\frac{E_a+E'_a}{E}\right)+\pi d^2\cdot\frac{E'_a}{E}\cos\gamma$$

for

$$\cos\gamma \ge \frac{E_a+E'_a-E}{E'_a}$$

Hence, we can write the angle-averaged cross-section as

$$\langle\sigma_r(E)\rangle_\gamma = \frac{1}{2}\cdot\int_{\frac{E_a+E'_a-E}{E'_a}}^{1}\left[\pi d^2\cdot\left(1-\frac{E_a+E'_a}{E}\right)+\pi d^2\cdot\frac{E'_a}{E}\cos\gamma\right]\mathrm{d}(\cos\gamma)$$

Performing the integration we find that

$$\begin{aligned}\langle\sigma_r(E)\rangle_\gamma &= \frac{\pi d^2}{2}\cdot\left[\begin{array}{c}\left(1-\frac{E_a+E'_a}{E}\right)\left(1-\frac{E_a+E'_a-E}{E'_a}\right)\\ +\frac{E'_a}{2E}\left(1-\left(\frac{E_a+E'_a-E}{E'_a}\right)^2\right)\end{array}\right]\\
&= \frac{\pi d^2}{2}\cdot\left(1-\frac{E_a+E'_a-E}{E'_a}\right)\left[\begin{array}{c}\left(1-\frac{E_a+E'_a}{E}\right)\\ +\frac{E'_a}{2E}\left(1+\frac{E_a+E'_a-E}{E'_a}\right)\end{array}\right]\\
&= \frac{\pi d^2}{2}\cdot\left(\frac{E-E_a}{E'_a}\right)\left[\left(\frac{2E-2E_a-2E'_a}{2E}\right)+\left(\frac{E_a+2E'_a-E}{2E}\right)\right]\\
&= \frac{\pi d^2}{2}\cdot\left(\frac{E-E_a}{E'_a}\right)\left[\frac{E-E_a}{2E}\right]=\frac{\pi d^2}{4}\cdot\left(\frac{\left(E-E_a\right)^2}{E'_a\cdot E}\right)\end{aligned}$$

This result shows that the cross-section for the reaction decreases as the component of the collision energy that is perpendicular to the bond axis, E'_a, increases.

26

Macromolecules

26.1 Exercises

E26.1 (a) Show that the ratio I/I_{forward} in Equation (26.11) is one for $\varphi = 0$ (forward scattering). (b) Use Equation (26.11) for backward scattering ($\varphi = 180°$) at $\alpha < 1$ to find the ratio of backward to forward scattering. Compare to the result obtained in Equation (26.10).

Solution

Equation (26.11) is

$$\frac{I_\varphi}{I_{\text{forward}}} = \frac{2}{C^2}\left(e^{-C} - 1 + C\right) \quad \text{with} \quad C = \frac{8\pi^2}{3}\frac{\overline{h^2}}{\lambda^2}\sin^2\left(\frac{\varphi}{2}\right)$$

For $\varphi = 0°$ (the case of forward scattering) we find that $C = 0$; however, we must be careful in evaluating this limit for the intensity ratio. If we consider the limit of $C < 1$, then we can expand the exponential so that the term in parentheses takes the form

$$e^{-C} - 1 + C \approx 1 - C + \frac{1}{2}C^2 - \frac{1}{6}C^3 - 1 + C = C^2\left(\frac{1}{2} - \frac{1}{6}C\right)$$

and we find that

$$\frac{I_\varphi}{I_{\text{forward}}} \approx 1 - \frac{1}{3}C$$

For $\varphi = 0$ (forward scattering) we obtain $C(0) = 0$ and $I/I_{\text{forward}} = 1$.
For $\varphi = 180°$ (backward scattering) we obtain

$$C = \frac{8\pi^2}{3}\frac{\overline{h^2}}{\lambda^2}$$

and

$$\frac{I_{\text{backward}}}{I_{\text{forward}}} \approx 1 - \frac{1}{3}\frac{8\pi^2}{3}\frac{\overline{h^2}}{\lambda^2} = 1 - \frac{8\pi^2}{9}\frac{\overline{h^2}}{\lambda^2}$$

This is only slightly less than expected from Equation (26.10).

E26.2 From Equation (26.11) derive an expression for the ratio I_{135}/I_{45} plotted in Fig. 26.6. What do you expect for this ratio in the cases $\sqrt{\overline{h^2}}/\lambda = 1$ and $\sqrt{\overline{h^2}}/\lambda \gg 1$?

Solution

We begin with Equation (26.11),

$$\frac{I_\varphi}{I_{\text{forward}}} = \frac{2}{C^2}\left(e^{-C} - 1 + C\right) \quad \text{with} \quad C = \frac{8\pi^2}{3}\frac{\overline{h^2}}{\lambda^2}\sin^2\frac{\varphi}{2},$$

Solutions Manual for Principles of Physical Chemistry, Third Edition. Edited by Hans Kuhn, David H. Waldeck, and Horst-Dieter Försterling.

and write the ratio $I_{135°}/I_{45°}$ as

$$\frac{I_{135°}}{I_{45°}} = \frac{\frac{I_{135°}}{I_{\text{forward}}}}{\frac{I_{45°}}{I_{\text{forward}}}} = \frac{\left[\frac{2}{C^2}\left(e^{-C} - 1 + C\right)\right]_{\varphi=135°}}{\left[\frac{2}{C^2}\left(e^{-C} - 1 + C\right)\right]_{\varphi=45°}}$$

$$= \frac{\frac{2}{C^2_{135°}}\left(e^{-C_{135°}} - 1 + C_{135°}\right)}{\frac{2}{C^2_{45°}}\left(e^{-C_{45°}} - 1 + C_{45°}\right)} = \frac{C^2_{45°}\left(e^{-C_{135°}} - 1 + C_{135°}\right)}{C^2_{135°}\left(e^{-C_{45°}} - 1 + C_{45°}\right)}$$

with

$$C_{135°} = \frac{8\pi^2}{3}\frac{\overline{h^2}}{\lambda^2}\sin^2\frac{135°}{2} = \frac{8\pi^2}{3}\frac{\overline{h^2}}{\lambda^2}\cdot 0.853$$

and

$$C_{45°} = \frac{8\pi^2}{3}\frac{\overline{h^2}}{\lambda^2}\sin^2\frac{45°}{2} = \frac{8\pi^2}{3}\frac{\overline{h^2}}{\lambda^2}\cdot 0.146$$

In the case where $\sqrt{\overline{h^2}}/\lambda = 1$, $C_{135°} = \frac{8\pi^2}{3}\cdot 0.853$, $C_{45°} = \frac{8\pi^2}{3}\cdot 0.146$ we obtain

$$\frac{I_{135°}}{I_{45°}} = \left(\frac{0.146}{0.835}\right)^2 \frac{\left(e^{-C_{135°}} - 1 + C_{135°}\right)}{\left(e^{-C_{45°}} - 1 + C_{45°}\right)}$$

$$= \left(\frac{0.146}{0.835}\right)^2 \frac{21.450}{2.864} = 0.2290$$

In the case where $\sqrt{\overline{h^2}}/\lambda \gg 1$, $C_{135°} \gg 1$, $C_{45°} \gg 1$ we obtain

$$\frac{I_{135°}}{I_{45°}} = \frac{C^2_{45°}\left(e^{-C_{135°}} - 1 + C_{135°}\right)}{C^2_{135°}\left(e^{-C_{45°}} - 1 + C_{45°}\right)}$$

$$\simeq \frac{C^2_{45°}\left(C_{135°}\right)}{C^2_{135°}\left(C_{45°}\right)} = \frac{C_{45°}}{C_{135°}} = \frac{0.146}{0.835} = 0.175$$

E26.3 The following data have been obtained for ribosomes: sedimentation constant $s = 82.6 \times 10^{13}$ s, diffusion coefficient $D = 1.52 \times 10^{-7}$ cm^2 s^{-1}, partial molar volume $\upsilon_{\text{partial}} = 0.61$ cm^3 g^{-1}, and a density $\rho = 1$ g cm^{-3} at 20°C. Calculate the molecular mass m of the ribosomes.

Solution

According to Equation (26.24) we have

$$\boldsymbol{M} = \frac{\boldsymbol{RT}\cdot s}{D\cdot(1 - \upsilon_{\text{partial}}\cdot\rho)} = 3.4 \times 10^6 \text{ g mol}^{-1}$$

so that the mass of a ribosome is

$$m = \frac{\boldsymbol{M}}{\boldsymbol{N_A}} = \frac{3.4 \times 10^6 \text{ g mol}^{-1}}{6.02 \times 10^{23} \text{ mol}^{-1}} = 5.6 \times 10^{-18} \text{ g}$$

E26.4 Normalize the probability distribution for the end-to-end distance of a random coil, which is given by Equation (26.6).

Solution

We are asked to normalize the probability distribution

$$\mathcal{P}(h)\,\mathrm{d}h = const\cdot 4\pi h^2\cdot\exp\left(-\frac{3h^2}{2N_{\text{stat}}\cdot l^2}\right)\mathrm{d}h$$

Hence we find the value for *const*, such that

$$1 = \int_0^\infty \mathcal{P}(h)\,\mathrm{d}h = const\cdot 4\pi\int_0^\infty h^2\cdot\exp\left(-\frac{3h^2}{2N_{\text{stat}}\cdot l^2}\right)\mathrm{d}h$$

If we let $x^2 = 3h^2/\left(2N_{\text{stat}} \cdot l^2\right)$ then $x\,dx = \left[3/\left(2N_{\text{stat}} \cdot l^2\right)\right] h\,dh$ so that

$$\begin{aligned} 1 &= const \cdot 4\pi \frac{2N_{\text{stat}} \cdot l^2}{3} \sqrt{\frac{2N_{\text{stat}} \cdot l^2}{3}} \int_0^\infty x^2 \cdot \exp\left(-x^2\right)\,dx \\ &= const \cdot 4\pi \frac{2N_{\text{stat}} \cdot l^2}{3} \sqrt{\frac{2N_{\text{stat}} \cdot l^2}{3}} \frac{\sqrt{\pi}}{4} \\ &= const \cdot \left(\frac{2N_{\text{stat}} \cdot l^2}{3}\pi\right)^{3/2} \end{aligned}$$

where we have used the integral as given in Appendix B. The normalization constant is

$$const = \left(\frac{3}{2\pi N_{\text{stat}} \cdot l^2}\right)^{3/2}$$

E26.5 Evaluate the integral for the average of h^2 in Equation (26.8), by using the distribution function given in Equation (26.6).

Solution

Here we use our normalized result for the distribution function from E26.4 and evaluate $\overline{h^2}$, so that

$$\overline{h^2} = \left(\frac{3}{2\pi N_{\text{stat}} \cdot l^2}\right)^{3/2} \cdot 4\pi \int_0^\infty h^4 \cdot \exp\left(-\frac{3h^2}{2N_{\text{stat}} \cdot l^2}\right)\,dh$$

If we let $x^2 = 3h^2/\left(2N_{\text{stat}} \cdot l^2\right)$ then $x\,dx = \left[3/\left(2N_{\text{stat}} \cdot l^2\right)\right] h\,dh$ so that

$$\begin{aligned} \overline{h^2} &= \left(\frac{3}{2\pi N_{\text{stat}} \cdot l^2}\right)^{3/2} \cdot 4\pi \left[\frac{\left(2N_{\text{stat}} \cdot l^2\right)}{3}\right]^2 \sqrt{\frac{\left(2N_{\text{stat}} \cdot l^2\right)}{3}} \int_0^\infty x^4 \cdot \exp\left(-x^2\right)\,dx \\ &= \frac{4}{\sqrt{\pi}} \left[\frac{\left(2N_{\text{stat}} \cdot l^2\right)}{3}\right] \frac{3\sqrt{\pi}}{8} \\ &= N_{\text{stat}} \cdot l^2 \end{aligned}$$

where we have used the table in Appendix B to evaluate the integral. This result matches that given in Box 26.1.

E26.6 Using the probability distribution for a random coil, Equation (26.6), find the most probable end-to-end distance of a polymer chain. Compare this result to that we found for the mean square end-to-end distance. Provide a verbal explanation for why these two quantities are different.

Solution

In order to find the maximum of the distribution function, we take its derivative and solve for the value of h where it is zero. Hence we write that

$$\begin{aligned} \frac{d\mathcal{P}(h)}{dh} = 0 &= \frac{d}{dh}\left[h^2 \cdot \exp\left(-\frac{3h^2}{2N_{\text{stat}} \cdot l^2}\right)\right]_{h=h_{\max}} \\ &= \left|2h\exp\left(-\frac{3h^2}{2N_{\text{stat}} \cdot l^2}\right) - \frac{3h}{N_{\text{stat}} \cdot l^2}h^2 \cdot \exp\left(-\frac{3h^2}{2N_{\text{stat}} \cdot l^2}\right)\right|_{h=h_{\max}} \\ &= h\exp\left(-\frac{3h^2}{2N_{\text{stat}} \cdot l^2}\right) \cdot \left|2 - \frac{3}{N_{\text{stat}} \cdot l^2}h^2\right|_{h=h_{\max}} \end{aligned}$$

so that

$$0 = 2 - \frac{3}{N_{\text{stat}} \cdot l^2}h_{\max}^2 \text{ or } h_{\max}^2 = \frac{2}{3}N_{\text{stat}} \cdot l^2 = \frac{2}{3}\overline{h^2}$$

which is two-thirds of the value $\overline{h^2}$. The most probable distance will be

$$h_{\max} = \sqrt{\frac{2N_{\text{stat}} \cdot l^2}{3}}$$

The most probable distance and the root-mean-square distance are different because of the asymmetric shape of the distribution function which results from the fact that there are more ways for the ends to be far apart than close together.

E26.7 Consider the scattering data for a 0.20 g L^{-1} solution of polystyrene in toluene at 293 K, for which the excitation wavelength is 546.0 nm. Analyze these data[1] to determine the mean square end-to-end distance of the polymer chain.

$I(\varphi)/I_{forward}$	0.606	0.502	0.378	0.302	0.212	0.165	0.132
$\varphi/°$	25.8	36.9	53.0	66.4	90.0	113.6	143.1

Solution

We plot these data in Fig. E26.7 (dark gray solid circles) and fit them by Equation (26.11); namely

$$\frac{I(\varphi)}{I_{forward}} = \frac{2}{C^2}\left[\exp(-C) - 1 + C\right] \qquad \text{where} \quad C = \frac{8\pi^2}{3}\frac{\overline{h^2}}{\lambda^2}\sin^2(\varphi/2)$$

The best fit curve (light gray solid line) is plotted with the data points in Fig. E26.7 using the fit parameters

$$\frac{8\pi^2}{3}\frac{\overline{h^2}}{\lambda^2} = 22.4 \text{ or } h_{rms} = \sqrt{\overline{h^2}} = 500 \text{ nm}$$

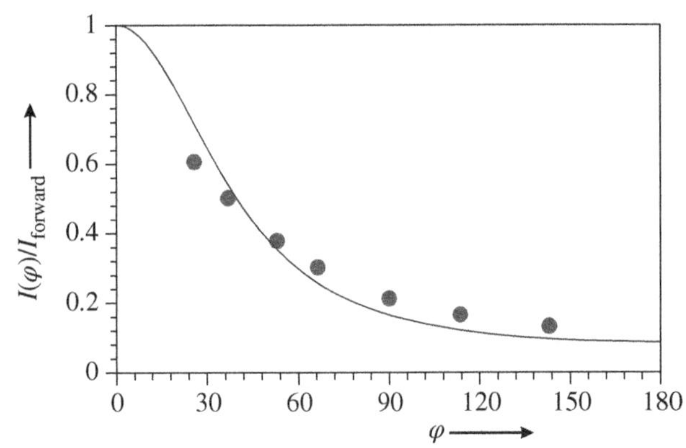

Figure E26.7 $I(\varphi)/I_{forward}$ versus angle φ. Dark gray solid circles: experiment; light gray solid line: calculation.

E26.8 Measurements on dilute solutions of polyisobutylene (monomer molecular mass of 54 g mol^{-1}) in chloroform (shear viscosity η of 0.537 mPa s at the temperature of these measurements T = 298 K) give the diffusion coefficient data in the table. Analyze these data[2] to determine the statistical length for the polymer.

$D/10^{-11} m^2 s^{-1}$	8.56	6.79	3.19	1.97	1.37	1.24	0.67
$\mathbf{M}$/kg mol^{-1}	57.1	95.0	247	610	1100	1900	3800

Solution

We use Equation (26.17)

$$D = \frac{kT}{3\pi\eta}\frac{1}{\sqrt{bl}}\frac{1}{\sqrt{N}}$$

to analyze these data. First we convert the data to monomer number by using a monomer mass of 0.0541 kg mol^{-1} and find that

1 Data are adapted from B.H. Zimm, *J. Chem. Phys.* 1948, **16**, 1099.
2 Data are from W. Brown, P. Zhou, *Macromolecules* 1991, **24**, 5151.

$D/10^{-11}\,m^2 s^{-1}$	8.56	6.79	3.19	1.97	1.37	1.24	0.67
N	1056	1756	4567	11, 278	20, 337	35, 128	70, 256

Next we plot D versus $1/\sqrt{N}$ (see Fig. E26.8) and fit it to a line with a zero intercept. The figure shows the best fit line which has a slope of 2.66×10^{-9} m^2 s^{-1}.

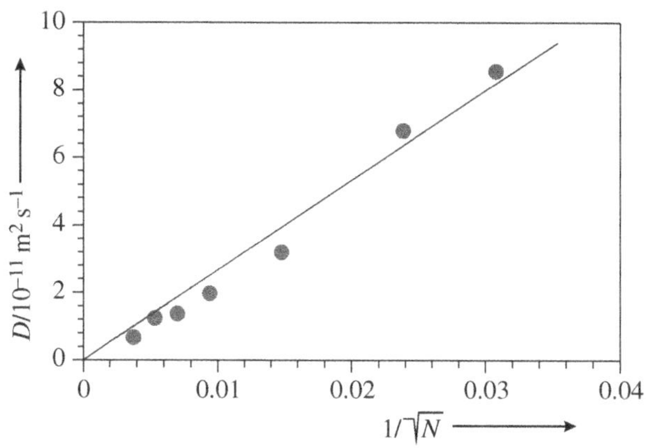

Figure E26.8 Diffusion coefficient D versus $1/\sqrt{N}$, where N is the number of monomer units in polyisobutylene. Dark gray solid circles: experiment and light gray solid line:calculated.

From Equation (26.17), we expect that

$$2.66 \times 10^{-9}\ \text{m}^2\ \text{s}^{-1} = slope = \frac{kT}{3\pi\eta}\frac{1}{\sqrt{bl}}$$

Using the parameters given above we find that

$$\sqrt{bl} = \frac{1.3806 \times 10^{-23}\ \text{J K}^{-1} \cdot 298\ \text{K}}{3\pi \cdot 5.37 \times 10^{-4}\ \text{Pa s}\ \ 2.66 \times 10^{-9}\ \text{m}^2\ \text{s}^{-1}} = 3.05 \times 10^{-10}\ \text{m}$$

If we take $b = 0.25$ nm, we find that

$$l = \frac{(3.05 \times 10^{-10}\ \text{m})^2}{0.25 \times 10^{-9}\ \text{m}} = 3.7 \times 10^{-10}\ \text{m}$$

This value is about the size of an individual monomer unit, indicating that the chain is highly flexible.

E26.9 Measurements on dilute solutions of polyisobutylene (monomer molecular mass of $\boldsymbol{M}_{\text{monomer}} = 54$ g mol^{-1} and $b = 250$ pm, where b is the contribution of each monomer to the contour length) in chloroform (at $T = 298$ K) give the Staudinger index $[\eta]$ data in the table. [3] Analyze these data to determine the mean square end-to-end distance $\overline{h^2}$ of the polymer chains and then extract the statistical length for the polymer by plotting $\overline{h^2}$ versus N_{monomer}.

$[\eta]/\text{cm}^3\ \text{g}^{-1}$	29.3	38.8	65.7	107.9
$\boldsymbol{M}/\text{kg mol}^{-1}$	57.1	95.0	247	610

Solution

Here we use Equation (26.33) to determine $\overline{h^2}$ for each molecular weight; namely

$$\overline{h^2} = \left(\frac{m \cdot [\eta]}{1.31}\right)^{2/3} = \left(\frac{\boldsymbol{M} \cdot [\eta]}{1.31 \cdot \boldsymbol{N}_A}\right)^{2/3}$$

3 Data are from W. Brown, P. Zhou, *Macromolecules* 1991, **24**, 5151.

By way of example, for $[\eta] = 29.3\ \text{cm}^3\ \text{g}^{-1}$, we find

$$\overline{h^2} = \left(\frac{57.1 \times 10^3\ \text{g mol}^{-1} \cdot 29.3 \times 10^{-6}\ \text{m}^3\ \text{g}^{-1}}{1.31 \cdot 6.022 \times 10^{23}\ \text{mol}^{-1}}\right)^{2/3} = 1.65 \times 10^{-16}\ \text{m}^2$$

We can also calculate the monomer number via $N_{\text{monomer}} = \boldsymbol{M}/\boldsymbol{M}_{\text{monomer}}$; by way of example for 57.1 kg mol^{-1} we find

$$N_{\text{monomer}} = \frac{57.1\ \text{kg mol}^{-1}}{0.054088\ \text{kg mol}^{-1}} = 1057$$

In a corresponding manner we find the values in the table.

$[\eta]/\text{cm}^3\ \text{g}^{-1}$	29.3	38.8	65.7	107.9
$\overline{h^2}/10^{-16}\ \text{m}^2$	1.65	2.79	7.51	19.1
N_{monomer}	1057	1756	4567	11, 278

Next we use the result that

$$\overline{h^2} = l \cdot b \cdot N_{\text{monomer}}$$

where l is the statistical length and b is the contribution of each monomer to the contour length (e.g., see Example 26.1 of the text). A plot of the mean square end-to-end distance versus the monomer number is shown in Fig. E26.9.

The best fit line has a slope of $0.168 \times 10^{-18}\ \text{m}^2$ and should be equal to the product $b \cdot l$. Hence, if we use $b = 250$ pm, we find that $l = 0.67$ nm. This value is two to three monomer units in length.

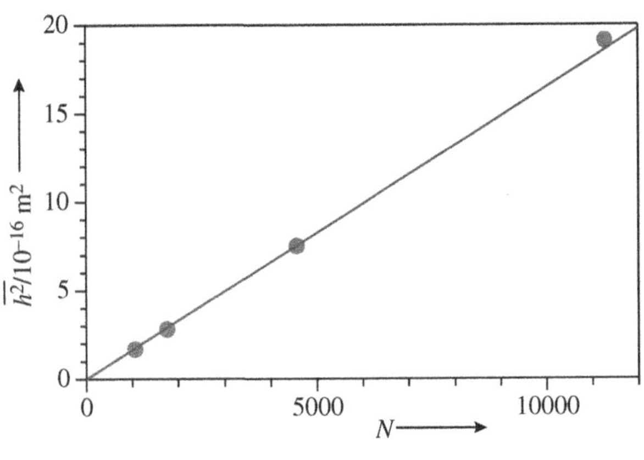

Figure E26.9 $\overline{h^2}$ versus the number N of monomer units for polybutylene in chloroform at 298 K. Dark gray solid circles: experiment and light gray solid line: regression line.

E26.10 Consider the Staudinger index data for different molar masses of polyisobutylene (monomer molecular mass of $\boldsymbol{M}_{\text{monomer}} = 54.088\ \text{g mol}^{-1}$ and $b = 250$ pm) in cyclohexane at 303 K. Use these data to determine the mean square end-to-end distance of the polymer and to determine the diameter of the random coil, as a function of the polymer's average molecular mass (see *J. Am. Chem. Soc.* 1953, **75,** 1775)

$\boldsymbol{M}/\text{kg mol}^{-1}$	37.8	167	333	710
$[\eta]/\ \text{cm}^3\ \text{g}^{-1}$	38.8	112	181	287

Solution

Here we use Equation (26.33) to determine $\overline{h^2}$ for each molecular weight; namely

$$\overline{h^2} = \left(\frac{m \cdot [\eta]}{1.31}\right)^{2/3} = \left(\frac{\boldsymbol{M} \cdot [\eta]}{1.31 \cdot \boldsymbol{N}_A}\right)^{2/3}$$

By way of example, for $[\eta] = 38.8\ \mathrm{cm^3\ g^{-1}}$, we find

$$\overline{h^2} = \left(\frac{37.8 \times 10^3\ \mathrm{g\ mol^{-1}} \cdot 38.8 \times 10^{-6}\ \mathrm{m^3\ g^{-1}}}{1.31 \cdot 6.022 \times 10^{23}\ \mathrm{mol^{-1}}} \right)^{2/3} = 1.51 \times 10^{-16}\ \mathrm{m^2}$$

We can calculate the monomer number by way of $N_{\mathrm{monomer}} = \boldsymbol{M}/\boldsymbol{M}_{\mathrm{monomer}}$; by way of example for 37.8 kg mol^{-1} we find

$$N_{\mathrm{monomer}} = \frac{37.8\ \mathrm{kg\ mol^{-1}}}{0.054088\ \mathrm{kg\ mol^{-1}}} = 699$$

In a corresponding manner we find the values in the table.

$[\eta]/\mathrm{cm^3\ g^{-1}}$	38.8	112	181	287
$\overline{h^2}/10^{-16}\ \mathrm{m^2}$	1.51	8.25	18.0	40.6
N_{monomer}	699	3088	6157	13, 127

A plot of the mean square end-to-end distance versus the monomer number N_{number} is shown in Fig. E26.10. The best fit line has a slope of $3.04 \times 10^{-19}\mathrm{m^2}$. As already shown in Exercise 26.9 this slope is equal to the product $b \cdot l$. Hence, if we use $b = 250$ pm, we find that $l = 1.2$ nm, or four to five monomer units.

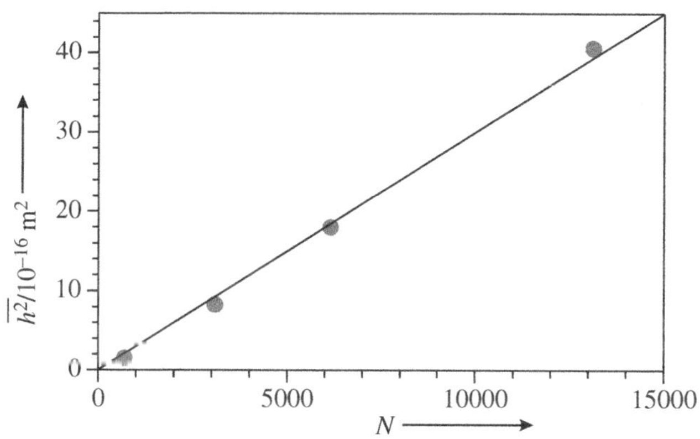

Figure E26.10 $\overline{h^2}$ versus the number N of monomer units for polybutylene in cyclohexane at 303 K. Dark gray solid circles: experiment and light gray solid line: regression line.

Using Equation (26.12), we estimate the diameter of the random coil as

$$h_{\mathrm{rms}} = \sqrt{\overline{h^2}}$$

which ranges from 12 to 64 nm for this range of molecular weight.

E26.11 Example 26.3 analyzes diffusion data for cellulose acetate in acetone using the refined model with the translational form factor C_{transl} (Equation (26.37)). Analyze these same data with the simpler formula given by Equation (26.17) and compare your analysis to that of Example 26.3.

Solution

Equation (26.17) is

$$D = \frac{kT}{3\pi\eta} \frac{1}{\sqrt{bl}} \frac{1}{\sqrt{N}}$$

In this case we use the values given in the table, namely,

N_{monomer}	40	196	550	746
$\sqrt{N_{\text{monomer}}}$	6.33	14.0	23.5	27.3
$D\ 10^{-11}$ m^2 s^{-1}	20.7	7.7	4.0	3.4

and plot D versus $1/\sqrt{N}$ and fit it to a line with a zero intercept. Figure E26.11 shows the best fit line which has a slope of 1.24×10^{-9}m^{2}s^{-1}.

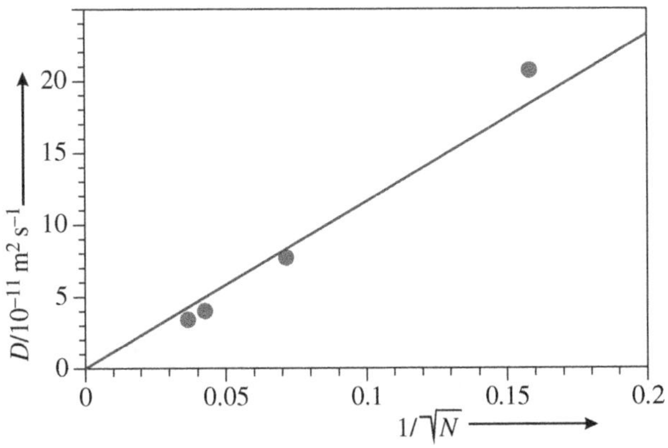

Figure E26.11 Diffusion coefficient D versus $1/\sqrt{N}$, where N is the number of monomer units in cellulose acetate in acetone. Dark gray solid circles: experiment and light gray solid line: calculated.

From Equation (26.17), we expect that

$$1.24 \times 10^{-9}\ \text{m}^2\ \text{s}^{-1} = slope = \frac{kT}{3\pi\eta}\frac{1}{\sqrt{bl}}$$

Using the parameters given above we find that

$$\sqrt{bl} = \frac{1.3806 \times 10^{-23}\ \text{J K}^{-1} \cdot 298\ \text{K}}{3\pi \cdot 3.0 \times 10^{-4}\ \text{Pa s} \cdot 1.24 \times 10^{-9}\ \text{m}^2\ \text{s}^{-1}} = 1.17 \times 10^{-9}\ \text{m}$$

If we take $b = 0.515$ nm, we find that

$$l = \frac{(1.17 \times 10^{-9}\ \text{m})^2}{0.515 \times 10^{-9}\ \text{m}}\text{m} = 2.66 \times 10^{-9}\ \text{m}$$

This value is about one-fourth of the value found by using the refined model.

E26.12 Example 26.4 analyzes sedimentation data for cellulose acetate in acetone using the refined model with the translational form factor C_{transl} (26.37). Analyze these same data with the simpler formula given by Equation (26.23) and compare your analysis to that of Example 26.4.

Solution

Equation (26.23) is

$$s = \frac{m_{\text{monomer}}(1 - v_{\text{part}}\rho)}{3\pi\eta\sqrt{bl}}\sqrt{N_{\text{monomer}}}$$

In this case we use the values given in the table, namely

N_{monomer}	40	196	550	746
$\sqrt{N_{\text{monomer}}}$	6.33	14.0	23.5	27.3
$s\ 10^{-13}$ s	4.07	7.5	10.9	12.5

and make a plot of s versus $\sqrt{N_{\text{monomer}}}$, see Fig. E26.12.

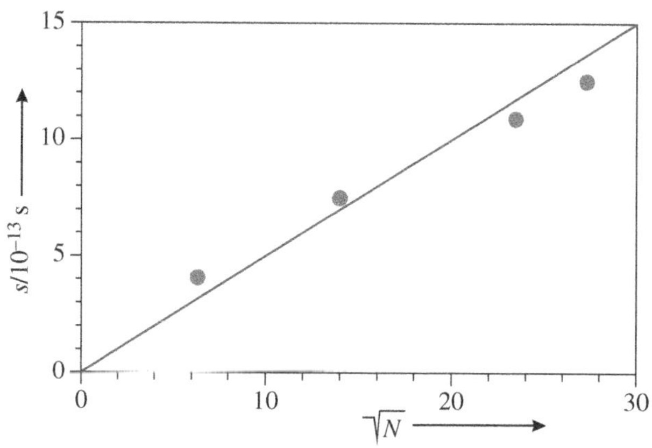

Figure E26.12 Sedimentation constant s versus $\sqrt{N}$ where N is the number of monomer units for cellulose acetate in acetone.

If we restrict the intercept to be zero, then the best fit line has a slope of 4.75×10^{-14} s and comparison with Equation (26.23) shows that

$$\sqrt{b \cdot l} = \frac{m \cdot \left(1 - V_{\text{part}}\rho\right)}{3\pi\eta \cdot (slope)}$$

From Example 26.4, we take $m = 4.32 \times 10^{-25}$ kg, $b = 0.515$ nm, $\rho = 0.79\,\text{g cm}^{-3}$, $\eta = 0.3 \times 10^{-2}$ Poise $= 0.3 \times 10^{-3}$ N m^{-2} s (viscosity of acetone) and $V_{\text{part}} = 0.68\ \text{cm}^3\ \text{g}^{-1}$, so that

$$\sqrt{b \cdot l} = \frac{4.32 \times 10^{-25}\ \text{kg} \cdot \left(1 - 0.68\ \text{cm}^3\ \text{g}^{-1} \cdot 0.79\ \text{g cm}^{-3}\right)}{3\pi \cdot 0.3 \times 10^{-3}\ \text{N m}^{-2}\ \text{s} \cdot 4.75 \times 10^{-14}\ \text{s}}$$
$$= 1.49 \times 10^{-9}\ \text{m}$$

Then $l = \left(1.49 \times 10^{-9}\ \text{m}\right)^2 / 0.515 \times 10^{-9}\ \text{m} = 4.3$ nm. A value that is one-third of that obtained from the more refined model.

E26.13 The sedimentation constant s was measured for different molar masses of polyisobutylene in cyclohexane at 293 K (see table[4]). Use these data and the facts $\upsilon_{\text{part}} = 1.091\ \text{cm}^3/\text{g}$ and $\rho = 0.779\ \text{g/cm}^3$ to determine D for each of these polymers. Does the diffusion coefficient D change with $\boldsymbol{M}$ in a way you would expect?

$\boldsymbol{M}$/kg mol^{-1}	30.9	172.	672	1420
$s/10^{13}$ s	0.925	1.94	3.33	4.45

Solution

Equation (26.24) relates the diffusion coefficient D directly to the sedimentation constant s, and we can use it with the values given in the problem statement to find D; namely we write that

$$D = \frac{kT}{m}\frac{s}{1 - \upsilon_{\text{part}}\rho} = \frac{RT}{\boldsymbol{M}}\frac{s}{1 - \upsilon_{\text{part}}\rho}$$

Hence we find that

$\boldsymbol{M}$/kg mol^{-1}	30.9	172.	672	1420
$D/10^{-11}$ m^2 s^{-1}	4.87	1.84	0.807	0.510

Clearly, the diffusion coefficient decreases as the polymer size increases; however, the decrease is somewhat stronger than the $1/\sqrt{N}$ dependence that is predicted by Equation (26.17).

4 Data are from L. Mandelkern, W. Krigbaum, H.A. Scheraga, P.J. Flory, *J. Chem. Phys.* 1952, **20**, 1392.

E26.14 Using your knowledge of osmotic pressure from Chapter 18 and the data in the table for polyisobutylene in cyclohexane, determine its molar mass. The temperature is 303 K.

${}^{\text{osm}}P/\text{N m}^{-2}$	64.8	151.0	255.0
$c'/\text{g L}^{-1}$	1.9	4.0	5.9

Solution

For an ideal solution, we can use the osmotic pressure formula (see Chapter 18)

$$^{\text{osm}}P = RT \cdot c$$

To write that

$$c = \frac{n}{V} = \frac{m}{MV} = \frac{1}{M}c'$$

and

$$M = \frac{c'}{c} = \frac{c'RT}{^{\text{osm}}P}$$

For the case of ${}^{\text{osm}}P = 64.8$ Pa, we find

$$M = \frac{1.9\,\text{kg m}^{-3} \cdot 8.314\ \text{J mol}^{-1}\ \text{K}^{-1} \cdot 303\ \text{K}}{64.8\ \text{N m}^{-2}} = 74\frac{\text{kg}}{\text{mol}}$$

Correspondingly, we find the values in Table E26.14.

Table E26.14 Molar Mass M Calculated from the Osmotic Pressure.

${}^{\text{osm}}P/\text{N}$			
${}^{\text{osm}}P/\text{N m}^{-2}$	64.8	151.0	255.0
$c'/\text{g L}^{-1}$	1.9	4.0	5.9
$M/\text{kg mol}^{-1}$	74	67	59

E26.15 Consider Equation (26.11), for the case of backward scattering and $C < 1$. Show that

$$\frac{I_{\text{backward}}}{I_{\text{forward}}} \simeq 1 - \frac{8\pi^2}{9}\frac{\overline{h^2}}{\lambda^2}$$

Note: This exercise is a shorter version of E26.1.

Solution

For backward scattering we have that $\varphi = 180°$, so that Equation (26.11) becomes

$$\frac{I_{\text{backward}}}{I_{\text{forward}}} = \frac{2}{(C)^2}(\exp(-C) - 1 + C) \quad \text{with} \quad C = \frac{8\pi^2}{3}\frac{\overline{h^2}}{\lambda^2}$$

If we expand the exponential in a series, we find that

$$\begin{aligned}\exp(-C) - 1 + C &\approx 1 - C + \frac{1}{2}C^2 - \frac{1}{6}C^3 + \cdots - 1 + C \\ &= C^2\left(\frac{1}{2} - \frac{1}{6}C\right) - \cdots\end{aligned}$$

If we only retain the terms up to third order, then we find that

$$\frac{I_{\text{backward}}}{I_{\text{forward}}} \approx 1 - \frac{1}{3}C + \cdots = 1 - \frac{8\pi^2}{9}\frac{\overline{h^2}}{\lambda^2}$$

E26.16 The table provides data on small 4.3 kDa DNA double helices in a 1.5% agarose gel (taken from *Biopolymers* 1994, **34** 249–259). Show that the velocity in an electric field changes linearly with the applied field strength.

υ/cm h^{-1}	0.0090	0.014	0.017	0.023	0.035	0.052	0.080	0.137	0.180
F/V cm^{-1}	0.12	0.20	0.24	0.34	0.50	0.72	1.0	1.5	2.0

Solution

Figure E26.16 shows a plot of the speed versus the applied field strength and a best fit of these data by a line.

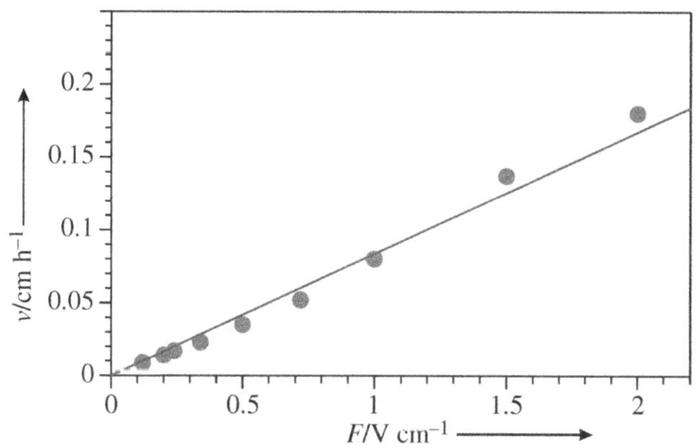

Figure E26.16 Speed υ versus applied field strength F for the movement of DNA double helices (4.3 kDa) in a 1.5% agarose gel.

E26.17 In Box 26.2 we consider the bending of a DNA filament. Use that analysis to estimate the length of a statistical chain element for actin filaments of radius $r = 3.5$ nm.

Solution

As in Box 26.2 we assume a value of

$$A_{\text{Young}} = 5.5 \times 10^7 \text{ N m}^{-2}$$

If we rearrange the expression for the energy of bending, we find that

$$l = A_{\text{Young}} \frac{3\pi^3}{128} \frac{r^4}{E_{\alpha=\pi/2}} = A_{\text{Young}} \frac{3\pi^3}{128} \frac{r^4}{kT}$$

Using $r = 3.5$ nm we find a statistical chain length element at $T = 300$ K of

$$l = 5.5 \times 10^7 \text{ N m}^{-2} \cdot \frac{3\pi^3}{128} \cdot \frac{(3.5 \times 10^{-9} \text{ m})^4}{1.38 \times 10^{-23} \text{ J K}^{-1} \cdot 300 \text{ K}}$$
$$= 1.4 \times 10^{-6} \text{ m} = 1.4\,\mu\text{m}$$

E26.18 Simplify the empirical equation for the force f in Box 26.3 in the case that $h \ll L$.

Solution

The empirical equation for the force f in Box 26.3 may be written as

$$f = -\frac{kT}{2l}\left[\frac{1}{(1-x)^2} - 1 + 4x\right]$$

where we define $x = h/L \ll 1$. If we expand $1/(1-x)^2$ in a Taylor series, we find that

$$\frac{1}{(1-x)^2} = 1 + 2x + \cdots$$

so that

$$f = -\frac{kT}{2l}[1 + 2x + \cdots - 1 + 4x]$$
$$= -\frac{3kT}{l}x = -\frac{3kT}{l}\frac{h}{L}$$

E26.19 For incident light polarized in the z-direction, the intensity I_φ of the light scattered in the xy-plane in a direction forming an angle φ with the direction of the incident beam ($I_{\varphi=0} = I_{\text{forward}}$), is given by Equation (26.11)

$$\frac{I_\varphi}{I_{\text{forward}}} = \frac{2}{C^2}\left(\mathrm{e}^{-C} - 1 + C\right) \quad \text{with} \quad C = \frac{8\pi^2}{3}\frac{\overline{h^2}}{\lambda^2}\sin^2\left(\frac{\varphi}{2}\right)$$

Show that this equation reduces to

$$\frac{I_{\text{backward}}}{I_{\text{forward}}} = 1 - \frac{1}{3}C = 1 - \frac{8\pi^2}{9}\frac{\overline{h^2}}{\lambda^2}$$

when the parameter C is smaller than unity.

Solution

For $C < 1$ we expand the exponential function in a power series

$$\mathrm{e}^{-C} = 1 - C + \frac{1}{2!}C^2 - \frac{1}{3!}C^3$$

to find

$$\frac{I_\varphi}{I_{\text{forward}}} = \frac{2}{C^2}\left(1 - C + \frac{1}{2!}C^2 - \frac{1}{3!}C^3 - 1 + C\right)$$

$$= \frac{2}{C^2}\left(\frac{1}{2!}C^2 - \frac{1}{3!}C^3\right) = 2\left(\frac{1}{2} - \frac{1}{6}C\right) = 1 - \frac{1}{3}C$$

In the range $0.8 < C < 1$ this approximation differs by only 10% from the exact value. We apply this expression to the case of backward scattering: $\varphi = 180°$, $\sin^2(\varphi/2) = 1$. Then we obtain

$$C = \frac{8\pi^2}{3}\frac{\overline{h^2}}{\lambda^2}$$

and

$$\frac{I_{\text{backward}}}{I_{\text{forward}}} = 1 - \frac{1}{3}C = 1 - \frac{8\pi^2}{9}\frac{\overline{h^2}}{\lambda^2}$$

This expression is practically the same as that in Equation (26.10) in Section 26.3.1. We find that the intensity of backward scattering is smaller than that for forward scattering, and that the actual ratio depends on the ratio $\overline{h^2}/\lambda^2$. For $\overline{h^2}/\lambda^2 = 0.05$ we obtain $I_{\text{backward}}/I_{\text{forward}} = 0.44$, that is, the intensity of backward scattering is about one-half of the intensity of forward scattering.

E26.20 Using

$$\frac{I_\varphi}{I_{\text{forward}}} = \frac{2}{C^2}\left(\mathrm{e}^{-C} - 1 + C\right) \quad \text{with} \quad C = \frac{8\pi^2}{3}\frac{\overline{h^2}}{\lambda^2}\sin^2\frac{\varphi}{2}$$

find the ratio $I_{135°}/I_{45°}$.

Solution

The ratio $I_{135°}/I_{45°}$ is

$$\frac{I_{135°}}{I_{45°}} = \frac{\frac{I_{135°}}{I_{\text{forward}}}}{\frac{I_{45°}}{I_{\text{forward}}}} = \frac{\left[\frac{2}{C^2}\left(\mathrm{e}^{-C} - 1 + C\right)\right]_{\varphi=135°}}{\left[\frac{2}{C^2}\left(\mathrm{e}^{-C} - 1 + C\right)\right]_{\varphi=45°}}$$

$$= \frac{\frac{2}{C^2_{135°}}\left(\mathrm{e}^{-C_{135°}} - 1 + C_{135°}\right)}{\frac{2}{C^2_{45°}}\left(\mathrm{e}^{-C_{45°}} - 1 + C_{45°}\right)}$$

with

$$C_{135°} = \frac{8\pi^2}{3}\frac{\overline{h^2}}{\lambda^2}\sin^2\frac{135°}{2} = \frac{8\pi^2}{3}\frac{\overline{h^2}}{\lambda^2}\cdot 0.853$$

and

$$C_{45°} = \frac{8\pi^2}{3}\frac{\overline{h^2}}{\lambda^2}\sin^2\frac{45°}{2} = \frac{8\pi^2}{3}\frac{\overline{h^2}}{\lambda^2}\cdot 0.146$$

26.2 Problems

P26.1 The polymer description in this chapter was cast in terms of the mean square end-to-end distance, $\overline{h^2}$, of a polymer strand. Often, it is more convenient to use a different quantity called the radius of gyration, R_g, because it is related more directly to experimental measurables. The radius of gyration is defined as

$$R_g^2 = \frac{1}{2N^2}\sum_{j=1}^{N}\sum_{i=1}^{N}\overline{\left(\vec{r}_j - \vec{r}_i\right)^2}$$

where $\vec{r}_i$ is the vector specifying the position of statistical segment i in the chain and N is the number of statistical segments. Using this definition show that the radius of gyration for the Gaussian chain is given by

$$R_g^2 = \frac{1}{6}Nl^2$$

(Hint: Use the fact that each segment is statistically related to each other one.).

Solution

Because the positions of the segments are randomly placed with respect to each other, we see that the mean square distance between segment i and segment j is given by the same relationship as the end-to-end distance, but for $|j - i|$ segments; i.e.,

$$\overline{\left(\vec{r}_j - \vec{r}_i\right)^2} = |j - i| \cdot l^2$$

Substituting this result into the expression for the radius of gyration, we find that

$$R_g^2 = \frac{1}{2N^2}\sum_{j=1}^{N}\sum_{i=1}^{N}|j-i|\cdot l^2 = \frac{l^2}{2N^2}\sum_{j=1}^{N}\sum_{i=1}^{N}|j-i|$$

For large N we can replace the summation by an integral, so that

$$\begin{aligned}
R_g^2 &= \frac{l^2}{2N^2}\int_0^N\left(\int_0^N |y-x|\cdot dy\right)\cdot dx \\
&= \frac{l^2}{2N^2}\int_0^N\left(\int_0^x (x-y)\cdot dy\right)\cdot dx + \frac{l^2}{2N^2}\int_0^N\left(\int_x^N (y-x)\cdot dy\right)\cdot dx \\
&= \frac{l^2}{2N^2}\int_0^N \left[xy - \frac{y^2}{2}\right]_{y=0}^{y=x}\cdot dx + \frac{l^2}{2N^2}\int_0^N\left[\frac{y^2}{2} - xy\right]_{y=x}^{y=N}\cdot dx \\
&= \frac{l^2}{2N^2}\int_0^N\left[\left(x^2 - \frac{x^2}{2}\right) + \left(\frac{N^2}{2} - xN - \frac{x^2}{2} + x^2\right)\right]\cdot dx \\
&= \frac{l^2}{2N^2}\int_0^N\left[-xN + \frac{N^2}{2} + x^2\right]\cdot dx = \frac{l^2}{2N^2}\left[-\frac{x^2}{2}N + x\frac{N^2}{2} + \frac{x^3}{3}\right]_{x=0}^{x=N} \\
&= \frac{l^2}{2N^2}\left[-\frac{N^3}{2} + \frac{N^3}{2} + \frac{N^3}{3}\right] = \frac{l^2}{2N^2}\left[\frac{N^3}{3}\right] = \frac{N\,l^2}{6}
\end{aligned}$$

where we have replaced the discrete variables i and j with x and y, which are continuous.

If we use the fact that

$$\overline{h^2} = Nl^2$$

we find that

$$R_g^2 = \frac{1}{6}\overline{h^2}$$

P26.2 Using the definition of the radius of gyration show that it may also be written as

$$R_g^2 = \frac{1}{N}\sum_{j=1}^{N}\overline{\left(\vec{r}_j - \vec{r}_{\text{center of mass}}\right)^2}$$

where $\vec{r}_{\text{cm}}$ is the vector identifying the center of mass in the laboratory frame of reference.

Solution

Using the definition for the center of mass vector,

$$M \cdot \vec{r}_{\text{center of mass}} = \sum_{i=1}^{N} m_i \cdot \vec{r}_i$$

we find that

$$\vec{r}_{\text{center of mass}} = \frac{1}{N}\sum_{i=1}^{N} \vec{r}_i \text{ since } N = \frac{M}{m_i}$$

If we rewrite this expression for R_g, we find that

$$R_g^2 = \frac{1}{N}\sum_{j=1}^{N} \overline{\left(\vec{r}_j - \vec{r}_{\text{center of mass}}\right)^2} = \frac{1}{N}\sum_{j=1}^{N} \overline{\left(\vec{r}_j^{\,2} - 2\vec{r}_j\vec{r}_{\text{center of mass}} + \vec{r}^{\,2}_{\text{center of mass}}\right)}$$

$$= \frac{1}{N}\sum_{j=1}^{N} \overline{\vec{r}_j^{\,2}} - 2\vec{r}_{\text{center of mass}} \frac{1}{N}\sum_{i=1}^{N} \vec{r}_i + \vec{r}^{\,2}_{\text{center of mass}} = \frac{1}{N}\sum_{j=1}^{N} \overline{\vec{r}_j^{\,2}} - \vec{r}^{\,2}_{\text{center of mass}}$$

This result is identical to our original definition of R_g, since

$$R_g^2 = \frac{1}{2N^2}\sum_{j=1}^{N}\sum_{i=1}^{N} \overline{\left(\vec{r}_j - \vec{r}_i\right)^2}$$

$$= \frac{1}{2N}\sum_{j=1}^{N} \overline{\vec{r}_j^{\,2}} - \frac{1}{N^2}\sum_{j=1}^{N}\sum_{i=1}^{N} \overline{\vec{r}_j\vec{r}_i} + \frac{1}{2N}\sum_{i=1}^{N} \overline{\vec{r}_i^{\,2}}$$

$$= \frac{1}{N}\sum_{j=1}^{N} \overline{\vec{r}_j^{\,2}} - \left(\frac{1}{N}\sum_{j=1}^{N} \overline{\vec{r}_j}\right)\left(\frac{1}{N}\sum_{i=1}^{N} \overline{\vec{r}_i}\right) = \frac{1}{N}\sum_{j=1}^{N} \overline{\vec{r}_j^{\,2}} - \vec{r}^{\,2}_{\text{center of mass}}$$

P26.3 Use the Einstein relation $D = kT/\zeta$ for the diffusion coefficient D to prove that Equation (26.24) holds generally.

Solution

Equation (26.24) is

$$m = \frac{kT}{D} \frac{s}{1 - V_{\text{part}}\rho}$$

where m is the mass of the polymer, s is the sedimentation constant, V_{part} is the partial volume, and ρ is the density of the solvent. As in the original derivation of the sedimentation rate we require force balance between the frictional damping and the motion in the gravitational field. Accordingly we write

$$f_{\text{friction}} = \zeta \cdot v$$

where ζ is the friction coefficient and v is the speed. In addition, we have that

$$f_{\text{gravitation}} = m \cdot g_{\text{centrifuge}} \cdot \left(1 - V_{\text{part}}\rho\right)$$

where $g_{\text{centrifuge}}$ is the acceleration in the ultracentrifuge. Applying the force balance condition

$$f_{\text{friction}} = f_{\text{gravitation}}$$

we obtain

$$\zeta \cdot v = m \cdot g_{\text{centrifuge}} \cdot \left(1 - V_{\text{part}}\rho\right)$$

Substituting for the friction coefficient with the Einstein relation

$$D = \frac{kT}{\zeta}$$

we find

$$\frac{kT}{D} \cdot v = m \cdot g_{\text{centrifuge}} \cdot \left(1 - V_{\text{part}}\rho\right)$$

and

$$v = \frac{D}{kT} m \cdot g_{\text{centrifuge}} \cdot \left(1 - V_{\text{part}}\rho\right)$$

Rearranging this expression to find s with Equation (26.23) we obtain

$$s = \frac{v}{g_{\text{centrifuge}}} = \frac{D}{kT} \frac{m \cdot g_{\text{centrifuge}} \cdot \left(1 - V_{\text{part}}\rho\right)}{g_{\text{centrifuge}}} = \frac{D}{kT} m \cdot \left(1 - V_{\text{part}}\rho\right)$$

or

$$m = \frac{kT}{D} \frac{s}{1 - V_{\text{part}}\rho}$$

which is our result.

P26.4 Combine some of our developments in Section 26.3.2 to show that $s \cdot [\eta]/N_{\text{monomer}}$ should be constant for a given polymer solution. Use the data in the table[5] for different molecular weights of polyisobutylene in cyclohexane at 293 K and the facts $v_{\text{part}} = 1.091\ \text{cm}^3/\text{g}$, $\eta = 0.97$ mPa s, $\rho = 0.779\ \text{g/cm}^3$, and $b = 250$ pm to determine this constant and evaluate the statistical length l.

M (kg/mol)	30.9	172.	672	1420
$[\eta]$ (cm^3/g)	34.2	112	287	489
s (10^{13} s)	0.925	1.94	3.33	4.45

Solution

We start with Equations (26.24) and (26.33)

$$s = m_{\text{coil}} \frac{D}{kT} \left(1 - v_{\text{part}}\rho\right)$$

$$[\eta] = 1.31 \cdot \left(\overline{h^2}\right)^{3/2} \frac{1}{m_{\text{coil}}}$$

Combining these equations we obtain

$$\begin{aligned} s \cdot [\eta]/N_{\text{monomer}} &= \left[\frac{m_{\text{coil}} \cdot D}{kT}(1 - V_{\text{part}}\rho)\right] \cdot \left[1.31 \cdot \left(\overline{h^2}\right)^{3/2} \frac{1}{m_{\text{coil}}}\right] \cdot \frac{1}{N_{\text{monomer}}} \\ &= \frac{D}{kT}(1 - V_{\text{part}}\rho) \cdot 1.31 \cdot \left(\overline{h^2}\right)^{3/2} \cdot \frac{1}{N_{\text{monomer}}} \end{aligned}$$

Next we use Equation (26.14)

$$D = \frac{kT}{6\pi\eta r_{\text{coil}}}$$

to obtain

$$s \cdot [\eta]/N_{\text{monomer}} = \frac{1}{6\pi\eta r_{\text{coil}}}(1 - V_{\text{part}}\rho) \cdot 1.31 \cdot \left(\overline{h^2}\right)^{3/2} \cdot \frac{1}{N_{\text{monomer}}}$$

With Equations (26.16) and (26.8)

$$r_{\text{coil}} = \frac{1}{2}\sqrt{L \cdot l} \qquad \overline{h^2} = L \cdot l$$

5 Data are from L. Mandelkern, W. Krigbaum, H. A. Scheraga, and P. J. Flory J. Chem. Phys. **20**, 1392 (1952).

we find that

$$s \cdot [\eta]/N_{\text{monomer}} = \frac{1}{6\pi\eta}\frac{2}{\sqrt{L \cdot l}}(1 - V_{\text{part}}\rho) \cdot 1.31 \cdot (L \cdot l)^{3/2} \cdot \frac{1}{N_{\text{monomer}}}$$
$$= \frac{1}{3\pi\eta}(1 - V_{\text{part}}\rho) \cdot 1.31 \cdot L \cdot l \cdot \frac{1}{N_{\text{monomer}}}$$

Because of

$$L = b \cdot N_{\text{monomer}}$$

where b is the contribution of the monomer to the contour length, this equation becomes

$$s \cdot [\eta]/N_{\text{monomer}} = \frac{1}{3\pi\eta}(1 - V_{\text{part}}\rho) \cdot 1.31 \cdot b \cdot N_{\text{monomer}} \cdot l \cdot \frac{1}{N_{\text{monomer}}}$$
$$= \frac{1}{3\pi\eta}(1 - V_{\text{part}}\rho) \cdot 1.31 \cdot b \cdot l$$

For a given solvent and polymer these parameters are fixed, and hence the value of $s \cdot [\eta]/N_{\text{monomer}}$ is constant. Alternatively we can rearrange this expression as

$$b \cdot l = \frac{3\pi\eta}{(1 - V_{\text{part}}\rho) \cdot 1.31} s \cdot [\eta]/N_{\text{monomer}}$$

Using the values in the table in the problem statement and $\boldsymbol{M}_{\text{monomer}} = 0.05409$ kg/mol, we find that

$\boldsymbol{M}$/kg mol^{-1}	30.9	172.	672	1420
N_{monomer}	632	3180	12, 424	26, 252
$(s \cdot [\eta]/N_{\text{monomer}})/10^{13}$s cm^3 g^{-1}	0.0501	0.0683	0.0769	0.0829
$b \cdot l$/nm^2	0.233	0.318	0.357	0.385

If we use a value of $b = 0.250$ nm, we find statistical lengths that range from 0.93 nm to 1.5 nm:

$b \cdot l$/nm^2	0.233	0.318	0.357	0.385
l/nm	0.93	1.27	1.42	1.54

P26.5 Perform Exercises E26.9 and E26.10 with data for polyisobutylene in different solvents. How much does the characteristic size of the polymer change between these solvents? Provide a verbal explanation for any difference you observe.

Solution

See the solutions to E26.9 and E26.10 for details. In chloroform we found a statistical length of 0.67 nm and

$[\eta]$/ cm^3 g^{-1}	29.3	38.8	65.7	107.9
$\overline{h^2}/10^{-16}$ m^2	1.65	2.79	7.51	19.1
N_{monomer}	1056	1756	4567	11, 278

whereas in cyclohexane we found a statistical length of 1.2 nm and

$[\eta]$/ cm^3 g^{-1}	38.8	112	181	287
$\overline{h^2}/10^{-16}$ m^2	1.51	8.25	18.0	40.6
N_{monomer}	699	3088	6157	13, 127

It is evident that some difference exists for the statistical lengths and the mean square end-to-end distances in these solvents and that for a given contour length that the polymer in cyclohexane has a larger mean square end-to-end distance, see Fig. P26.5.

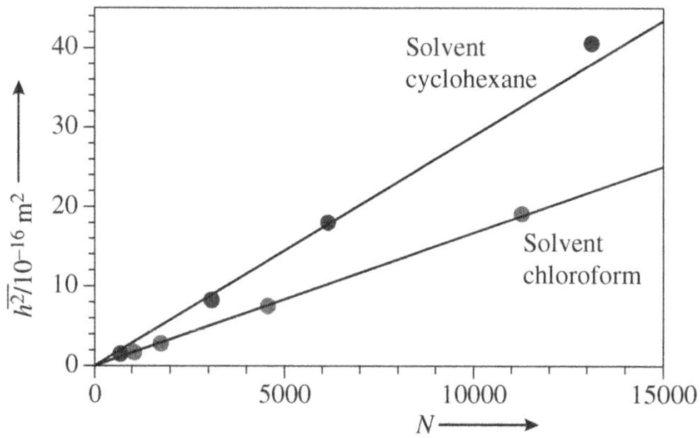

Figure P26.5 $\overline{h^2}$ versus the number N of monomer units for polybutylene in chloroform at 298 K (dark gray solid circles) and in cyclohexane at 303 K (light gray solid circles): experiment. Black solid lines: regression lines.

The larger characteristic size $\overline{h^2}$ of the polymer in cyclohexane, compared to that in chloroform, suggests that the polymer is better solvated in cyclohexane, hence less compact.

P26.6 The data in the table[6] provide characteristic sizes of proteins when they are denatured in solution. Plot the radius of gyration, R_g, versus the amino acid number and compare it to the prediction of a random coil model, for which $R_g^2 = \overline{h^2}/6$ (see Problem P26.1). From a best fit to the random coil model extract a statistical chain length.

Protein	GroEL	yPGK	Creatine Kinase	α-TS	Carbonic Anhydrase	Apomyo-globin
$N_{monomer}$	549	416	380	268	260	154
R_g/nm	8.2	7.1	4.6	4.9	5.9	4.0

Protein	ctACP	Protein L	Fyn SH3	Ubiquitin	GCN4-p2	drK SH3
$N_{monomer}$	98	79	78	76	66	59
R_g/nm	3.05	2.6	2.6	2.5	2.4	2.2

Protein	Snase	Lysozyme	RNase A	pI3K SH2	pI3K SH3	mAcp
$N_{monomer}$	149	129	124	112	103	98
R_g/nm	3.7	3.6	3.3	3.0	3.1	3.0

Protein	Protein G	Cyt c	AK-37	AK-32	AK-27	AK-16
$N_{monomer}$	52	39	37	32	27	16
R_g/nm	2.3	1.8	1.7	1.5	1.3	1.0

Solution

Figure P26.6 shows a plot of the radius of gyration versus the amino acid (monomer) number N. The solid curve shows a square root dependence of the form

$$R_g = C \cdot \sqrt{N} = \sqrt{\frac{b \cdot l}{6}}\sqrt{N}$$

with a best fit value of $C = 0.311$ nm; so that $b \cdot l = 0.582$ nm^2. If we use a b value of 0.5 nm, then we find $l = 1.2$ nm, which suggests a very flexible chain.

6 The data are taken from J.E. Kohn, *PNAS* 2004, **101**, 12491.

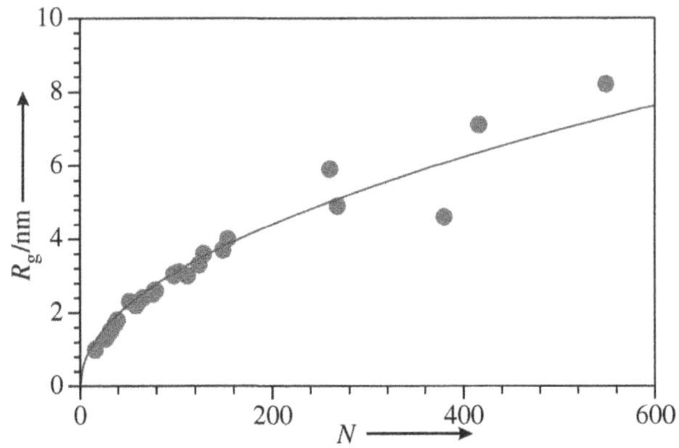

Figure P26.6 Radius of gyration R_g versus the number N of monomers for denatured proteins. Dark gray solid circles: experiment and light gray solid line: calculation.

P26.7 Obtain a rubber band and perform an experiment that will determine the heat flow, q, and the entropy change, ΔS, for the following process. In step A, stretch the rubber band six inches at constant temperature. In step B, release the rubber band so that it "snaps back" instantaneously and then allow it to warm up to the initial temperature. By performing a measurement, determine the amount of heat absorbed by the rubberband, q, in returning to its starting temperature (step B). If step A is assumed to be reversible, estimate the entropy change, which occurs in step B. From the entropy change estimate the change in the number of configurations.

Solution

The mass of the rubber band used in this experiment was found to be $m = 1$ g and the *CRC Handbook of Chemistry and Physics* reported a heat capacity of 1.74 J g^{-1}K^{-1} for rubber. The force constant k of the rubber band was measured to be ~ 60 N/m (performed by adding different masses to the rubber band and measuring the change in its length x; *force* $= -k \cdot x$).

In step A we stretch the rubber band isothermally to 0.15 m. For this step the heat q_A is not known, but the work is found from

$$w_A = \int_{\text{initial}}^{\text{final}} -F\,\mathrm{d}x = k\int_0^{0.15\text{ m}} x\,\mathrm{d}x = \frac{1}{2}kx^2$$
$$= \frac{1}{2}\cdot 60\text{ N m}^{-1}\cdot(0.15\text{ m})^2\text{ J} = 0.675\text{ J}$$

A measurement of the temperature of the rubber band at this point shows that it is 295 K.

In step B the rubber band is released and it "snaps back." Because this process proceeds so quickly there is no time for the flow of heat; hence, it is adiabatic and $q_B = 0$. When the rubber band is released it is no longer acting under an external tension and its contraction does not perform any work, hence $w_B = 0$. A measurement of the temperature of the rubber band at this point shows that it is 294 K; i.e., the rubber band has cooled by 1°.

In the third step (step C) the rubber band warms back to room temperature, with $\Delta T = 1$ K. No change in length occurs so the work is zero; $w_C = 0$. Heat flow occurs and we can calculate the heat by

$$q_C = m\cdot C\cdot\Delta T$$
$$= 1\text{ g}\cdot 1.74\text{ J K}^{-1}\text{ g}^{-1}\cdot 1\text{ K} = 1.74\text{ J}$$

For the entire cycle $\Delta U = 0$, so that

$$\Delta U = 0 = q_A + w_A + q_B + w_B + q_C + w_C$$
$$= q_A + 0.675\text{ J} + 0 + 0 + 1.74\text{ J} + 0$$

and $q_A = -2.415$ J; that is heat flows out of the rubber band in step A.

Because the entropy is a state function, we can use the cycle to determine the entropy change for the irreversible step, step B. If step A is reversible then its entropy change is

$$\Delta S_A = \int_{\text{initial}}^{\text{final}} \frac{\mathrm{d}q_{\text{rev}}}{T} = \frac{q_A}{T}$$

because the step is isothermal. Hence we find that

$$\Delta S_A = (-2.415\ \text{J}) / (295\ \text{K}) = -8.19 \times 10^{-3}\ \text{J K}^{-1}$$

For step C, the entropy change is

$$\Delta S_C = \int_{\text{initial}}^{\text{final}} \frac{dq_{\text{rev}}}{T} = mC \int_{294}^{295} \frac{dT}{T}$$
$$= (1\ \text{g})\ (1.74\ \text{J K}^{-1}\ \text{g}^{-1}) \ln\left(\frac{295}{294}\right) = 5.91 \times 10^{-3}\ \text{J K}^{-1}$$

Because entropy is a state function, we can write that

$$\Delta S_B = -\Delta S_A - \Delta S_C$$
$$= 8.19 \times 10^{-3}\ \text{J K}^{-1} - 5.91 \times 10^{-3}\ \text{J K}^{-1} = 2.28 \times 10^{-3}\ \text{J K}^{-1}$$

Hence the entropy change in the transformation from state A to state B is positive; i.e., the rubber band contracts but its entropy increases. In terms of configurations we find that

$$\frac{\Omega_{\text{contracted}}}{\Omega_{\text{stretched}}} = \exp\left(\frac{\Delta S_B}{k}\right) = \exp\left(\frac{2.28 \times 10^{-3}\ \text{J K}^{-1}}{1.38 \times 10^{-23}\ \text{J K}^{-1}}\right) = \exp(1.65 \times 10^{20})$$

Each of the different polymer chains of the rubber has many more configurations available when it is contracted than when it is stretched and aligned.

P26.8 Consider the light scattering experiment whose geometry is shown in Fig. 26.6. Determine the ratio of the intensities of the scattered light arriving at points P_1 and P_2, respectively.

Solution

The wave scattered from centers A and B arrive at P_2 with the same phase:

$$F(P_2) = 2F_0 \cos\left[\omega(t - \frac{s}{c_0})\right]$$

At point P_1 it arrives with a phase shift $2a/c_0$ (c_0 = velocity of light):

$$F(P_1) = F_0 \cos\left[\omega(t - \frac{s-a}{c_0})\right] + F_0 \cos\left[\omega(t - \frac{s+a}{c_0})\right]$$
$$= 2F_0 \cos\left[\omega(t - \frac{s}{c_0})\right] \cos\left(\frac{\omega a}{c_0}\right)$$

where $\omega/c_0 = 2\pi/\lambda$. This gives an intensity ratio for the light arriving in P_1 and P_2

$$\frac{I(P_1)}{I(P_2)} = \frac{F^2(P_1)}{F^2(P_2)} = \cos^2\left(\frac{2\pi a}{\lambda}\right)$$

Because of

$$\cos x = 1 - \frac{1}{2}x^2 + \cdots$$

for sufficiently small x we have

$$\cos^2 x = \left(1 - \frac{1}{2}x^2\right)^2 = 1 - x^2$$

Then we obtain

$$\frac{I(P_1)}{I(P_2)} \approx 1 - \left(\frac{2\pi a}{\lambda}\right)^2$$

Note that in our model $2a = \sqrt{h^2}$ where a is the distance between points P_1 and P_2. Also see Foundation 26.1.

P26.9 In Section 26.5 (see Fig. 26.16) we showed that a protein can be considered as a compact sphere. For comparison, assume that the proteins are statistical coils and calculate the coil radius r_{coil}, using Equation (26.16) as a function of the molar mass $\boldsymbol{M}$ (use $T = 295$ K and $\eta = 10^{-3}$ Pa s). Assume the values $bl = 0.08$ nm^2 and $\boldsymbol{M}_{monomer} = 120$ g mol^{-1}.

Protein	M g mol^{-1}	D_{trans} 10^{-11} m^2 s^{-1}
Lactate dehydrogenase	145,169	5.05
Ribonuclease A	13,700	10.7
Nitrogenase MoFe	220,000	4
Citrate Synthase	97,938	5.8
Chymotrypsinogen	25,660	9.3
β-Lactoglobulin	36,730	7.82
Ovalbumin	43,500	7.96
Myoglobin	17,190	10.8
GPD	142,868	5
BPTI(q)	6158	12.9
Aldolase	156,000	4.45
Lysozyme	14,320	10.9
Catalase	230,340	4.1

Solution

First we use the measured diffusion coefficient to calculate the hydrodynamic radius of the protein by way of Equation (26.14). For example, in the case of the protein BPTI we find

$$r = \frac{kT}{D_{trans} 6\pi\eta}$$
$$= \frac{1.38 \times 10^{-23} \text{ J K}^{-1} \cdot 295 \text{ K})}{12.9 \times 10^{-11} \text{ m}^2 \text{ s}^{-1} \cdot 6\pi \cdot 10^{-3} \text{ Pa s}} \text{ m} = 1.67 \times 10^{-9} \text{ m}$$

In a similar manner, we find that

Protein	M/ g mol^{-1}	r/nm
Lactate dehydrogenase	145,169	4.24
Ribonuclease A	13,700	2.00
Nitrogenase MoFe	220,000	5.35
Citrate Synthase	97,938	3.69
Chymotrypsin	25,660	2.30
β-Lactoglobulin	36,730	2.74
Ovalbumin	43,500	2.69
Myoglobin	17,190	1.98
GPD	142,868	4.28
BPTI(q)	6158	1.67
Aldolase	156,000	4.81
Lysozyme	14,320	1.96
Catalase	230,340	5.22

Figure P26.9 shows a plot of the hydrodynamic radius of the protein as a function of the average amino acid number N where we define

$$N \equiv \frac{\boldsymbol{M}/\ \mathrm{g\ mol^{-1}}}{120\ \mathrm{g\ mol^{-1}}}$$

The black curve shows a fit by Equation (26.16)

$$r_{\mathrm{coil}} = \frac{1}{2}\sqrt{N_{\mathrm{mono}}}bl \tag{26P9.1}$$

with the parameters given in the problem statement ($bl = 0.32\ \mathrm{nm}^2$).
For example, for $N_{\mathrm{mono}} = 2000$ we obtain

$$r_{\mathrm{coil}} = 0.5 \times \sqrt{2000 \times 0.08}\ \mathrm{nm} = 6.3\ \mathrm{nm}$$

It is evident that the model does not curve strongly enough with increasing N to fit the data set. This observation should be contrasted with that for the compacted sphere model shown in Fig. 26.16, which curves too strongly to fit the data set.

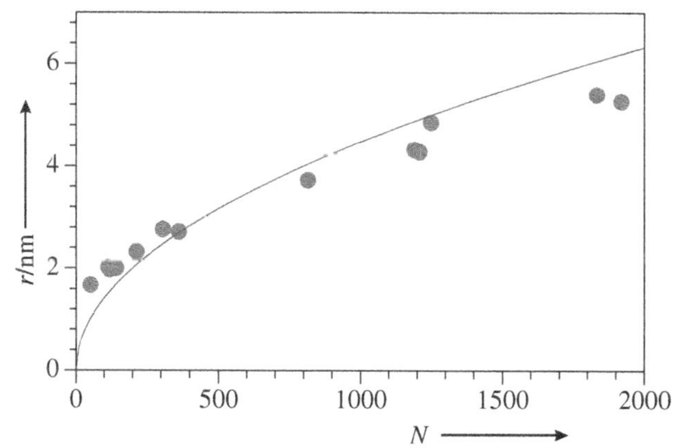

Figure P26.9 Coil radius r versus the number N of monomer units for proteins. Dark gray solid circles: experiment and light gray solid line: calculated from Equation (26P9.1) with $bl = 0.08\mathrm{nm}^2$.

P26.10 Extension of an Unraveled Coil: Consider Equation (26.51)

$$\frac{kT}{2l}\left[\frac{1}{\left(1-\overline{\Delta x_n}/l\right)^2} - 1 + \frac{4\overline{\Delta x_n}}{l}\right] = 6\pi\eta \cdot l \cdot \frac{1}{2}\sqrt{N_{\mathrm{stat}} - n} \cdot v$$

where l is the statistical chain element, $\overline{\Delta x_n}$ is the average extension of the chain element n in the x-direction (see Fig. 26.14b), η is the viscosity, N_{stat} is the number of statistical chain elements, and v is the flow speed of the fluid. We can rewrite this equation as

$$\frac{1}{\left(1-\overline{\Delta x_n}/l\right)^2} - 1 + \frac{4\overline{\Delta x_n}}{l} = \frac{6\pi\eta \cdot l^2 \cdot v}{kT}\sqrt{N_{\mathrm{stat}} - n}$$

a) Using the values of $L = 50\ \mu\mathrm{m}$, $l = 100\ \mathrm{nm}$ (thus $N_{\mathrm{stat}} = L/l = 500$), $v = 14\ \mu\mathrm{m\ s^{-1}}$, $\eta = 1 \times 10^{-3}\ \mathrm{N\ m^{-2}\ s}$, and $T = 300\ \mathrm{K}$, show that

$$\frac{1}{\left(1-\overline{\Delta x_n}/l\right)^2} - 1 + \frac{4\overline{\Delta x_n}}{l} = 0.637 \cdot \sqrt{N_{\mathrm{stat}} - n}$$

b) Solve this equation, either graphically or numerically, for n values of 1, 100, 200, 300, 400, and 500. Using the value of l given in part a) calculate $\overline{\Delta x_n}$ for each n value.
c) Using the $\overline{\Delta x_n}$ and the fact that the total extension x is obtained by summing up over all elements n, find the fractional extension x/L as a function of flow sped v.

Solution

Using the parameter values given in part a), we find that

$$\frac{6\pi\eta \cdot l^2 \cdot v}{kT} = \frac{6\pi \cdot (1\times 10^{-3}\ \mathrm{N\ m^{-2}\ s}) \cdot (100\times 10^{-9}\ \mathrm{m})^2 \cdot (14\times 10^{-6}\ \mathrm{m\ s^{-1}}\ )}{(1.38\times 10^{-23}\ \mathrm{N\ m\ K^{-1}}) \cdot 300\ \mathrm{K}}$$

$$= 0.637 \frac{\mathrm{N\ m^{-2}\ s\ m^2\ m\ s^{-1}}}{\mathrm{N\ m\ K^{-1} \cdot K}} = 0.637$$

and the equation reduces to

$$\frac{1}{\left(1 - \overline{\Delta x}_n/l\right)^2} - 1 + \frac{4\overline{\Delta x}_n}{l} = 0.637 \cdot \sqrt{N_{\mathrm{stat}} - n}$$

for a flow speed of 14 μm s^{-1}.

Part b) This equation is solved numerically. We proceed in the following manner: for each value of n in the range between 1 and N_{stat} we calculate the right-hand side of this equation.

$$u = 0.637 \cdot \sqrt{N_{\mathrm{stat}} - n}$$

Next we calculate values for the left-hand side of the equation in the range between 0 and 1.0 for $\overline{\Delta x}_n/l$.

$$w = \frac{1}{\left(1 - \overline{\Delta x}_n/l\right)^2} - 1 + \frac{4\overline{\Delta x}_n}{l}$$

Finally we search in the data file, for what value of $\overline{\Delta x}_n/l$ the condition

$$w - u = 0$$

is fulfilled. For example, if $N_{\mathrm{stat}} = 500$ and $n = 100$ we have $u = 12.74$ and we obtain the data points in the vicinity of $w - u = 0$ (see Table P26.10a). We conclude that $\overline{\Delta x}_n/l = 0.70$ for $n = 100$.

Table P26.10a w, $u - w$, and $\overline{\Delta x}_n/l$ Calculated for $N_{\mathrm{stat}} = 500$, $n = 100$, and $l = 100$ nm.

w	$u - w$	$\overline{\Delta x}_n/l$
12.309	0.431	0.6930
12.455	0.285	0.6950
12.604	0.136	0.6960
12.680	**0.060**	**0.6970**
12.756	**−0.016**	**0.6980**
12.911	−0.171	0.7000

In a like manner we proceed for other values of n, and we find the results displayed in Table P26.10b. These data are illustrated in Fig. P26.10a. Note that the dark gray curve in the figure corresponds to the same calculation performed for a much higher number of n values ($n = 500$) so that the smooth behavior is apparent.

Table P26.10b n, $\overline{\Delta x}_n/l$, and $\overline{\Delta x}_n$ Calculated for $N_{\mathrm{stat}} = 500$, $v = 14$ μm s^{-1}, and $l = 100$ nm.

n	$\overline{\Delta x}_n/l$	$\overline{\Delta x}_n$/nm
1	0.71	71
100	0.70	70
200	0.67	67
300	0.64	64
400	0.56	56
500	0.00	0

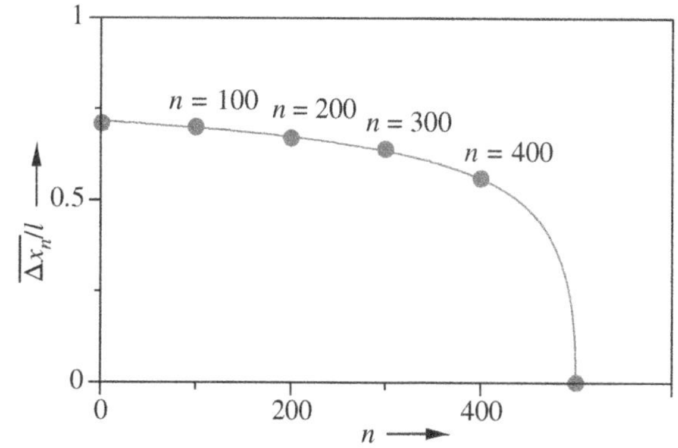

Figure P26.10a $\overline{\Delta x_n}/l$ versus n for $L = 50$ μm, $v = 14$ μm s^{-1}, $\eta = 1 \times 10^{-3}$ N m^{-2}, and $T = 300$ K. Dark gray filled circles: data points from Table P26.10b and solid dark gray line: same, but 500 data points calculated.

Part c) The total extension x is obtained by summing up over all elements n

$$x = \sum_{n=1}^{N_{stat}} \overline{\Delta x_n}$$

This is done numerically for n in the range from 1 to 500. For the conditions in part b) ($v = 14\,\mu m\ s^{-1}$, $l = 100$ nm, $L = 500 \quad l = 50\,\mu m$) we obtain

$$\sum \frac{\overline{\Delta x_n}}{l} = 310 \qquad x = \sum \overline{\Delta x_n} = 310 \cdot 100\ \text{nm} = 3.10 \times 10^{-5}\ \text{m} \, x/L = \frac{3.10 \times 10^{-5}\ \text{m}}{50 \times 10^{-6}\ \text{m}} = 0.62$$

Accordingly, values for different flow speeds v can be obtained, and they are displayed in Table P26.10c. The fractional extension x/L is plotted in Fig. P26.10b as a function of the flow speed v (see also Fig. F26.14d of the textbook, where this calculation is compared to the experiment). Note that the dark gray curve in the figure corresponds to the same calculation performed for a much higher number of v values so that the smooth behavior is apparent.

Table P26.10c Flow Speed v and Fractional Extension x/L for $L = 50$ μm.

v/μm s^{-1}	x/L
0	0
14	0.62
28	0.74
56	0.82
85	0.85

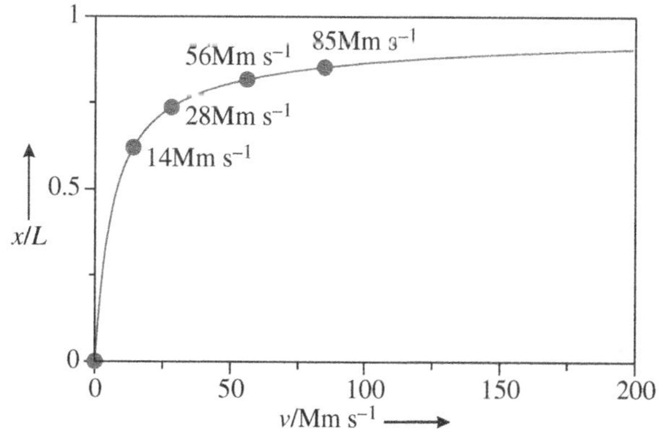

Figure P26.10b The fractional extension x/L as a function of flow speed v for $L = 50$ μm, $N_{stat} = 500$, $\eta = 1 \times 10^{-3}$ N m^{-2} s, and $T = 300$ K. Filled circles: data from Table P26.10c; solid dark gray line: same, but 200 data points calculated. Note that the dark gray curve corresponds to the same calculation as for the data in Table P26.10c performed for a much higher number of v values so that the smooth behavior is apparent.

P26.11 Time to Restore Random Coil from Unraveled Chain (see Fig. 26.15a and its discussion): After stopping the current, coiling begins at the open end where the tension along the chain is smallest. After some time a coil, including a part of the chain of contour length $L - x$, has formed. This coil grows and the displacement x of its edge from the fixed end of the chain (see Fig. 26.15a) moves with speed υ.

a) Approximate the coil as a sphere of radius

$$r_{\text{coil}} = \frac{1}{2}\sqrt{(L-x)l}$$

and use Stokes law to estimate the force acting on it.

b) Thermal collisions provide a force f' between the molecular chain and the molecules of the solvent, which acts to randomize the orientation of the statistical chain element. For keeping the chain element well oriented (probability for $\varphi = 0°$ considerably larger than for $\varphi = 180°$, see Fig. 26.14b) the energy $f' \cdot l$ to turn the element must be about kT. By assuming that f' and f balance, show that the speed v is given by

$$\upsilon = \frac{kT}{3\pi\eta\sqrt{L-x} \cdot l^{3/2}}$$

c) Integrate the velocity expression to show that

$$x = L - \left(\frac{kT}{2\pi\eta}\right)^{2/3} \frac{1}{l} \cdot t^{2/3}$$

Solution

a) The hydrodynamic resistance is $f = 6\pi\eta r_{\text{coil}} \cdot \upsilon$, so that the force acting on the statistical chain element connecting the coiled with the uncoiled portion is given by

$$f = 6\pi\eta\frac{1}{2}\sqrt{(L-x)l} \cdot \upsilon$$

b) From the problem statement, we deduce that

$$f' = \frac{kT}{l}$$

If the forces balance, we find

$$6\pi\eta\frac{1}{2}\sqrt{(L-x)l} \cdot \upsilon = \frac{kT}{l}$$

Hence the speed υ is

$$\upsilon = \frac{kT}{3\pi\eta\sqrt{L-x} \cdot l^{3/2}}$$

c) We can rewrite the velocity expression as

$$\upsilon = \frac{kT}{3\pi\eta\sqrt{L-x} \cdot l^{3/2}} = -\frac{\mathrm{d}x}{\mathrm{d}t}$$

which can be rearranged to give

$$-\sqrt{L-x} \cdot \mathrm{d}x = \frac{kT}{3\pi\eta l^{3/2}} \cdot \mathrm{d}t$$

By integration we obtain

$$\frac{2}{3}(L-x)^{3/2} = \frac{kT}{3\pi\eta l^{3/2}} \cdot t$$

Thus

$$x = L - \left(\frac{kT}{2\pi\eta}\right)^{2/3} \frac{1}{l} \cdot t^{2/3}$$

P26.12 Stretching a Chain: Here you estimate the force to stretch a piece of rubber along the z-direction. For simplicity, assume that the lines connecting the end points of the chains (net points) have the direction of the x, y, and z axes, $N_{\text{thread}}/3$ in each axis (Fig. P26.12).

Figure P26.12 Rubber idealized by molecular chains, $N_{thread}/3$ oriented in the x direction, and $N_{thread}/3$ each in the y and z directions. Dots: net points.

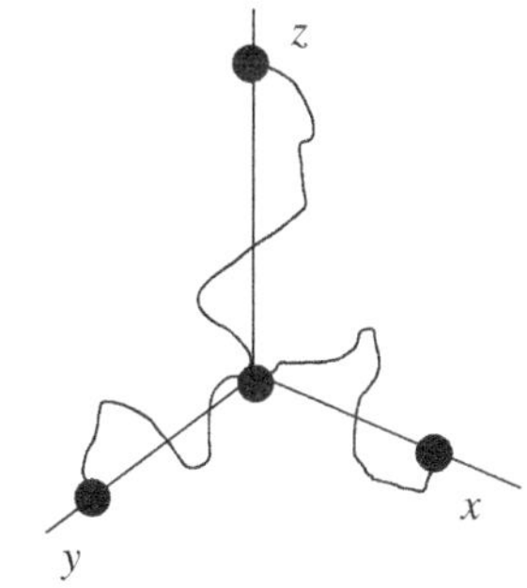

Furthermore, assume that h has the average value $h_1 = \sqrt{Ll}$ before stretching; and after stretching in the z-direction, it has

$$h_2 = \sqrt{Ll} \cdot \left(1 + \frac{\Delta z}{z}\right)$$

a) Assume that the force constant for stretching is $k_f = (3kT)/(L \cdot l)$ (Equation (26B3.5) in the text); and calculate the work that is performed on the z directed chains ($\frac{1}{3}N_{thread}$ chains).
b) Assume that the volume of the rubber is unchanged by stretching along z and use it to find the change in size along x and y. Use this change in the cross sectional area to find the work performed on the x and y directed chains.
c) Combine your results from parts a) and b) to show that the total work is given by

$$w = kT\frac{3}{2}N_{thread} \cdot \left(\frac{\Delta z}{z}\right)^2$$

for $\Delta z/z \ll 1$.
d) Using the result in c), find an expression for the force.

Solution

a) The work on the z-directed chains is

$$w_z = \frac{N_{thread}}{3} \cdot \int_{h_1}^{h_2} k_f\ h \cdot dh = \frac{3kT}{2Ll}\left(h_2^2 - h_1^2\right) \cdot \frac{N_{thread}}{3}$$

$$= \frac{3}{2}kT \cdot \left[\left(1 + \frac{\Delta z}{z}\right)^2 - 1\right] \cdot \frac{N_{thread}}{3}$$

b) The volume is unchanged when stretching the sample. This means that the cross sectional diameter is diminished by a factor α where

$$\left(1 + \frac{\Delta z}{z}\right)\alpha^2 = 1 \quad \text{and} \quad \alpha = \frac{1}{\sqrt{1 + \Delta z/z}}$$

Then, the work will be

$$w_x = w_y = \frac{3}{2}kT \cdot \left(\frac{1}{1 + \frac{\Delta z}{z}} - 1\right) \cdot \frac{N_{thread}}{3}$$

c) We sum the work terms to find

$$w = w_x + w_y + w_z = \frac{3}{2}kT \cdot \left[\left(1 + \frac{\Delta z}{z}\right)^2 + \frac{2}{1 + \frac{\Delta z}{z}} - 3\right] \cdot \frac{N_{thread}}{3}$$

For $\Delta z/z \ll 1$ we obtain

$$w = kT\frac{N_{thread}}{2}\left[\left(\frac{\Delta z}{z}\right)^2 + 2\left(\frac{\Delta z}{z}\right)^2\right] = kT\frac{3}{2}N_{thread} \cdot \left(\frac{\Delta z}{z}\right)^2$$

d) The force is

$$f_{total} = \frac{dw}{d\Delta z} = 3kT \cdot N_{thread} \cdot \frac{\Delta z}{z^2}$$

27

Organized Molecular Assemblies

27.1 Exercises

E27.1 In Fig. 27.2 a lamella is withdrawn from a soap solution. Given the surface tension of water $\gamma = 72.8$ mN m^{-1} calculate the work to pull the movable frame of width $L = 3$ cm to a 1 cm height, $h = 1$ cm. Compare this result to the work needed to pull the moveable frame against the gravitational force (assume the frame is made from platinum wire of thickness 0.2 mm). In addition, compare your results to the work needed to pull the liquid within the lamella against the gravitational force. Comment on the relative magnitudes.

Solution

For the work of withdrawing the lamella we find

$$\Delta G = \gamma A$$
$$= 72.8 \times 10^{-3}\text{N m}^{-1} \cdot 2 \cdot 3 \times 10^{-4}\ \text{m}^2 = 4.36 \times 10^{-5}\ \text{J}$$

To calculate the work of lifting the frame in the gravitational field, we begin with the density of Pt ($\rho_{\text{Pt}} = 21.5\ \text{g cm}^{-3}$) and multiply by its volume ($V = \pi(0.01\ \text{cm})^2 \cdot 3\ \text{cm} = \pi \cdot 3 \times 10^{-4}\ \text{cm}^3$) to get the frame's mass

$$m = \rho_{\text{Pt}} V$$
$$= 21.5\ \text{g cm}^{-3} \cdot \pi \cdot 3 \times 10^{-4}\ \text{cm}^3$$
$$= 2.03 \times 10^{-2}\ \text{g} = 2.03 \times 10^{-5}\ \text{kg}$$

Then we calculate the work as

$$\Delta G = mgh$$
$$= 2.03 \times 10^{-5}\ \text{kg} \cdot 9.81 \frac{\text{m}}{\text{s}^2} \cdot 10^{-2}\ \text{m} = 1.99 \times 10^{-6}\ \text{J}$$

which is less than one-tenth of the work required to create the lamella.

To calculate the work of lifting the water within the film, we begin with the density for water, $\rho_{\text{H}_2\text{O}} = 1.0\ \text{g cm}^{-3}$, and calculate the mass of the water by assuming that the lamella has a thickness of 0.01 mm, so that

$$m = \rho_{\text{H}_2\text{O}} V$$
$$= 1\ \text{g cm}^{-3} \cdot 3\ \text{cm}^2 \cdot 0.001\ \text{cm}$$
$$= 3 \times 10^{-3}\ \text{g} = 3 \times 10^{-6}\ \text{kg}$$

With the mass in hand, we can calculate the work as

$$\Delta G = mgh$$
$$= 3 \times 10^{-6}\ \text{kg} \cdot 9.81 \frac{\text{m}}{\text{s}^2} \cdot 10^{-2}\ \text{m} = 3 \times 10^{-7}\ \text{J}$$

This value is only a fraction of a percent of the work required to make the lamella.

Solutions Manual for Principles of Physical Chemistry, Third Edition. Edited by Hans Kuhn, David H. Waldeck, and Horst-Dieter Försterling.

E27.2 Calculate the contact angles α and β in Fig. 27.4 from Equations (27.5) and (27.6), using the data given in the text.

Solution

With the notations $a = \gamma_{n-decane/air} = 23.9\ \mathrm{mN\ m^{-1}}$, $b = \gamma_{n-decane/water} = 51.2\ \mathrm{mN\ m^{-1}}$, and $c = \gamma_{\mathrm{water/air}} = 72.8\ \mathrm{mN\ m^{-1}}$ and Equations (27.5) and (27.6) we obtain

$$a \cdot \cos\alpha + b \cdot \cos\beta = c \quad \text{and} \quad a \cdot \sin\alpha = b \cdot \sin\beta$$

We replace $\sin\alpha$ by $\sqrt{1-\cos^2\alpha}$ in the second equation, eliminate $\cos\alpha$ from both equations and solve for $\cos\beta$:

$$\cos\beta = \frac{1-(b/a)^2-(c/a)^2}{-2\left(bc/a^2\right)} = 0.986, \quad \text{thus we find} \quad \beta = 9.6\,^\circ$$

Inserting this result into the second equation, we obtain

$$\sin\alpha = \frac{b}{a} \cdot \sin\beta = 0.357, \quad \text{so that} \quad \alpha = 21^{\mathrm{o}}$$

E27.3 Calculate the work to spread a water film of 1 cm² area on the surface of glass (take its contact angle to be 10°). Compare your result to the case of spreading water on a 1 cm² area of graphite (take its contact angle to be 86°) and a 1 cm² area of paraffin (take its contact angle to be 110°).[1] Using these results, calculate the work per molecule of water and comment on the trend through the series of surfaces.

Solution

The differential work to spread the film would be given by

$$\delta w = \left(\gamma_{\mathrm{solid,water}} - \gamma_{\mathrm{solid,gas}}\right)\ \mathrm{d}A$$

which can be rearranged using Equation (27.18) to give

$$\delta w = -\gamma_{\mathrm{water,gas}}\ \cos\Theta\ \mathrm{d}A$$

Taking $\gamma_{\mathrm{water,gas}}\ \cos\Theta$ to be independent of area we find that

$$w = -\gamma_{\mathrm{water,gas}}\ \cos\Theta\ \Delta A$$

Hence for water such that $\gamma_{\mathrm{water,air}} = 72.8\ \mathrm{mN\ m^{-1}}$, we find that

$$w_{\mathrm{glass}} = -72.8 \times 10^{-3}\ \mathrm{N\ m^{-1}}\ \cos(10^{\circ}) \cdot 10^{-4}\ \mathrm{m^2} = -7.2 \times 10^{-6}\ \mathrm{J}$$

A corresponding calculation for graphite gives $w_{\mathrm{graphite}} = -0.51 \times 10^{-6}$ J and for paraffin gives $w_{\mathrm{graphite}} = 2.5 \times 10^{-6}$ J.

If we approximate the diameter of a water molecule as 400 pm, then we estimate that 6.25×10^{14} water molecules would form a monolayer coverage on a 1 cm² area. For the case of water on glass this would give an average work of -1.2×10^{-20} J per water molecule, whereas it would give -0.085×10^{-20} J per water molecule on graphite and 0.42×10^{-20} J per water molecule on paraffin.

The trend in the energy values implies that glass is more hydrophilic than graphite, which in turn is more hydrophilic than paraffin. The glass with a large number of oxygen and hydroxyl binding sites is expected to be very hydrophilic, whereas paraffin comprising aliphatic chains does not have sites to bind and stabilize water molecules through hydrogen bonding. While graphite is commonly thought of as pure carbon, exposed sites on graphite can display oxygen containing functional groups (carboxylic acids, etc.) from reaction with air and the π−electron systems can act as weak hydrogen-bond acceptor sites. These attributes likely contribute to it being slightly hydrophilic.

E27.4 For low enough concentrations of surfactant c the surface pressure Π in a Langmuir trough varies linearly with the surfactant concentration, i.e.,

$$\Pi = \gamma_0 - \gamma = b\,c$$

1 The value measured for the contact angle is very sensitive to surface structure, surface impurities, and surface hydration. The values used in this exercise are taken from Table X-2 of A.W. Adamson, *Physical Chemistry of Surfaces* (Wiley, 1990).

where b is a constant. Show that $b = \boldsymbol{RTn}_{\text{surfactant}}/(cA)$ and the surface pressure equation has the form of a two-dimensional ideal gas law.

Solution

Rearranging this result, we find that

$$\gamma = \gamma_0 - bc$$

so that

$$\frac{d\gamma}{dc} = -b$$

Using Equation (27.15) we know that

$$\frac{d\gamma}{dc} = -\frac{\boldsymbol{RT}}{c}\frac{\boldsymbol{n}_{\text{surfactant}}}{A}$$

where A is the surface area and $\boldsymbol{n}_{\text{surfactant}}$ is the amount of surfactant. Combining these results we see that

$$b = \frac{\boldsymbol{RT}}{c}\frac{\boldsymbol{n}_{\text{surfactant}}}{A}$$

so that

$$\Pi = \boldsymbol{RT}\frac{\boldsymbol{n}_{\text{surfactant}}}{A}$$

and

$$\Pi A = \boldsymbol{RT} \cdot \boldsymbol{n}_{\text{surfactant}}$$

This has the form of the ideal gas law where the surface pressure Π replaces the pressure P and the surface area A replaces the volume V.

E27.5 In the limit that the radius of a glass capillary tube r is smaller than the height h of the liquid water inside, J.W. Rayleigh[2] found an approximate, but more exact, solution to the capillary rise, namely

$$h = \frac{2\gamma}{\rho g r} - \frac{r}{3} + \frac{0.1288 r^2}{h} - \frac{0.1312 r^3}{h^2} + \cdots$$

This equation is valid in the range

$$r < 0.46\sqrt{rh}$$

In this treatment, the first correction term $(r/3)$ accounts for the mass of the fluid above the meniscus and the other two terms account for the deviation of the meniscus from sphericity. Evaluate h for the case used in Example 27.1, and determine the percentage error in our approximate result.

Solution

Example 27.1 used the approximate equation

$$h_{\text{approx}} = \frac{2\gamma}{\rho g r} = \frac{2 \cdot 72.8 \text{ mN m}^{-1}}{1.00\text{g cm}^{-3} \cdot 9.807 \text{ m s}^{-2} \cdot 0.5 \text{ mm}} = 2.97 \text{ cm}.$$

with a tube radius of $r = 0.5$ mm, a surface tension of $\gamma = 72.8$ mN m^{-1}, a gravitational constant of $g = 9.807$ m s^{-2}, and a density of $\rho = 1.00$ g cm^{-3}, to calculate a maximum height of 2.97 cm. It is evident that the more exact Rayleigh equation is cubic in form, and it can be solved by a number of different methods. We choose to plot the function

$$y = \frac{2\gamma}{\rho g r} - \frac{r}{3} + \frac{0.1288 r^2}{h} - \frac{0.1312 r^3}{h^2} - h \quad (27.1)$$

versus h and find the value where it crosses $y = 0$; this occurs at $h = 29.53$ mm

2 L. Rayleigh, On the theory of the capillary tube, *Proc. R. Soc. A* 1916, **92**, 184–195.

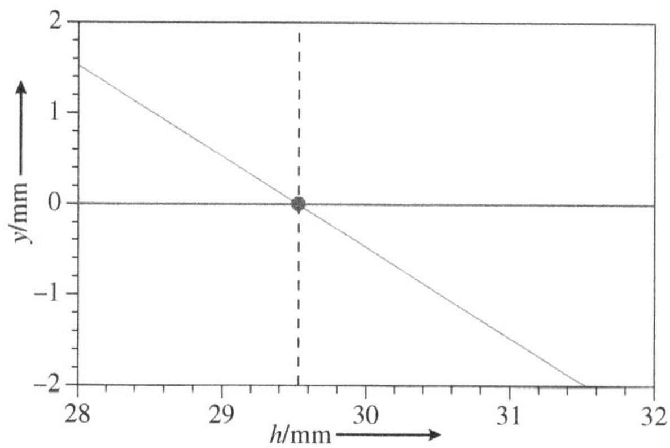

Figure E27.5 Quantity y in Equation (27.1) versus the height h.

With this value we can compute a percentage error for our approximate solution as

$$\text{percent error} = \frac{29.69 - 29.53}{29.69} 100\% = 0.5\%$$

Alternatively, one could proceed in an iterative manner to find the error. If we begin with the value of $h_{\text{approx}} = 2.97$ cm, we find the correction as

$$-\frac{r}{3} + \frac{0.1288r^2}{h} - \frac{0.1312r^3}{h^2} = \left(-\frac{0.05}{3} + \frac{0.1288 \cdot (0.05)^2}{2.97} - \frac{0.1312(0.05)^3}{2.97^2}\right) \text{ cm}$$
$$= (-0.0167 + 0.00010 - 0.0000018) \text{ cm} = -0.0166 \text{ cm}$$

resulting in an error of $-0.017/2.97 = -0.5\%$. We see that only the first correction term is essential, we can neglect the following terms completely.

In E27.6 you consider the case of larger diameter capillaries in which the error in the approximate solution is more important.

E27.6 Calculate the rise of water in a glass capillary of different radius values and compare your findings with experiment. You should use the approximate formula $h_{\text{approx}} = 2\gamma/\rho g r$ and the more accurate formula derived by Rayleigh (see Exercise E27.5). Comment on how the error increases with the capillary radius. For water at 20° use a surface tension of $\gamma = 72.8$ mN m^{-1}, a gravitational constant of $g = 9.807$ m s^{-2}, and a density of $\rho = 1.00$ g cm^{-3}.

Solution

Using the approximate Equation (27.10)

$$h_{\text{approx}} = \frac{2\gamma}{\rho g r}$$

with a surface tension of $\gamma = 72.8$ mN m^{-1}, a gravitational constant of $g = 9.807$ m s^{-2}, and a density of $\rho = 1.00$ g cm^{-3}, we find

$$h_{\text{approx}} = \frac{2 \cdot 72.8 \times 10^{-3}\text{N m}^{-1}}{1.00 \times 10^3 \text{ kg m}^{-3} \cdot 9.807 \text{ m s}^{-2} \cdot 2 \times 10^{-3} \text{ m}}$$
$$= 7.42 \times 10^{-3} \text{ m} = 7.42 \text{ mm}$$

for a 2 mm radius capillary. Next we use the method of successive approximations

$$h = \frac{2\gamma}{\rho g r} - \frac{r}{3} + \frac{0.1288r^2}{h_{\text{approx}}} - \frac{0.1312r^3}{h^2_{\text{approx}}} + \cdots$$

For the 2 mm case and using $h_{\text{approx}} = 7.43$ mm in the correction terms, we find that $h = 6.81$ mm; if we iterate a second time by using $h = 6.81$ mm in the correction term we find that $h = 6.82$ mm, indicating good convergence. Thus we find an error of about 9%.

Performing this procedure for the capillary radii given in the problem statement, we find the values given in Table E27.6.

Table E27.6 Comparison of Calculated and Measured Rise Height h for Different Radii r, Using the Surface Tension of Water.

r	h_{approx}	$corr1$	$corr2$	$corr3$	h	h_{exp}	
mm	mm	mm	mm	mm	mm	mm	
0.4	37.12	−0.133	0.001	−0.000	36.98	35.9	S. Baek Appl. Sci. 11 (2021) 3533
0.4313	34.42	−0.144	0.001	−0.000	34.28	34.33	Richards, JACS 43 (1921) 827
0.6	24.74	−0.200	0.002	−0.000	24.55	24.1	S. Baek Appl. Sci. 11 (2021) 3533
0.8	18.56	−0.267	0.004	−0.000	18.30	16.1	S. Baek Appl. Sci. 11 (2021) 3533
1.0099	14.70	−0.337	0.009	−0.001	14.37	14.34	Richards, JACS 43 (1921) 827
4.66	3.19	−1.553	0.878	−1.308	1.20	0.076	Richards, JACS 43 (1921) 827

Note that the first correction term $r/3$ dominates, similar to what was found in E27.5. The application of the correction factor shows a much better agreement of the calculated value with the experimental values reported by Richards for the two cases where $r/h < 1$. For the case where r/h exceeds one the calculation still shows a strong deviation from the prediction, in accordance with the limit $r < 0.46\sqrt{rh}$.

E27.7 Derive Equation (27.18) by starting from Equation (27.5).

Solution
Because the solid is rigid, we know that $\beta = 0°$ in Fig. 27.4 and the water phase of Fig. 27.4 corresponds to the solid phase of Fig. 27.12; hence, Equation (27.5) becomes

$$\gamma_{decane/air} \cdot \cos\alpha dl + \gamma_{decane/solid}\, dl = \gamma_{solid/air}\, dl$$

which rearranges to

$$\cos\alpha = \frac{\gamma_{solid/air} - \gamma_{decane/solid}}{\gamma_{decane/air}}$$

A comparison of Figs. 27.4 and 27.12 shows that α corresponds to the angle Θ in Equation (27.18), hence, we obtain the result.

E27.8 In Example 27.2 we discussed micelles composed of sodium dodecylsulfate in water with an optimal aggregation number of $n^* = 58$. Rather than assuming a spherical shape for the micelle, assume it has an ellipsoidal shape in which the short axis diameter is twice the length l_c and the long axis has a radius b (for a prolate ellipsoid $n^*v = 4\pi b\, l_c^2/3$ and for an oblate ellipsoid $n^*v = 4\pi b^2\, l_c/3$). Calculate the area per head group for this case.

Solution
As shown in the example $l_c \approx 1.546$ nm and $V \approx 0.345$ nm^3. For a prolate ellipsoid, we find

$$b = \frac{3n^*v}{4\pi l_c^2} = \frac{3 \cdot 58 \cdot 0.345\,\text{nm}^3}{4\pi \cdot (1.546\,\text{nm})^2} = 2.00\,\text{nm}$$

The surface area per head group is

$$A_0 = \frac{2\pi}{n^*}\left(\frac{(b \cdot l_c \cdot \cos^{-1}(l_c/b))}{\sin\left(\cos^{-1}(l_c/b)\right)} + l_c^2\right) = 0.622\,\text{nm}^2$$

so that the shape factor is very similar to that found in the spherical approximation, namely

$$\frac{V}{A_0 \cdot l_c} = \frac{0.345\,\text{nm}^3}{1.546\,\text{nm} \cdot 0.622\,\text{nm}^2} = 0.36$$

For an oblate ellipsoid, we find

$$b = \sqrt{\frac{3n^*v}{4\pi l_c}} = \sqrt{\frac{3 \cdot 58 \cdot 0.345}{4\pi \cdot 1.546}}\,\text{nm} = 1.76\,\text{nm}$$

The surface area per head group is

$$A_0 = \frac{2\pi}{n^*}\left[b^2 + \frac{l_c^2}{\sin\left(\cos^{-1}(l_c/b)\right)}\ln\left(\frac{l_c/b}{1-\sin\left(\cos^{-1}(l_c/b)\right)}\right)\right]$$
$$= \frac{2\pi}{58}\left[1.99^2 + \frac{1.546^2}{\sin(\cos^{-1}(1.546/1.76))}\ln\left(\frac{1.546/1.76}{1-\sin(\cos^{-1}(1.546/1.76))}\right)\right]\ \text{nm}^2$$
$$= 0.711\ \text{nm}^2$$

which gives a shape factor of 0.31 and is once again very close to that obtained for the sphere model. We conclude that the micelle shape for sodium dodecylsulfate in water is very nearly spherical.

E27.9 Consider a system like Fig. 27.6b, which shows the rise of liquid water in a glass capillary for a situation in which the water pool and the capillary are in contact with the air, but connect the capillary tube to an external pressure source P. What pressure must be applied to push the water back down to the level of that in the water pool?

Solution

In this exercise we use the same considerations as in deriving Equation (27.10) but we allow for the pressure inside the capillary P_{in} to be different from that outside P_{out}, acting on the water pool. For the condition that the pressures are equal $P_{in} = P_{out}$, Equation (27.10) gives the height h_{max} as

$$h_{max} = \frac{2\gamma}{g\rho r}$$

For a height of h_{max} the mass of the fluid in the tube is $m = \pi r^2 \cdot \rho \cdot h_{max}$ and it exerts a force on the liquid pool of $m \cdot g = \pi r^2 \cdot \rho \cdot h_{max} \cdot g$; hence, the pressure that the liquid column exerts on the pool of water at its base is

$$P = \frac{force}{area} = \frac{\pi r^2 \cdot \rho \cdot h_{max} \cdot g}{\pi r^2} = \rho \cdot h_{max} \cdot g$$

Hence if we apply a pressure $P = \rho \cdot h_{max} \cdot g$, we can push the liquid downward to the level of the pool. We find that the required pressure is given by

$$P = \rho \cdot h_{max} \cdot g = \rho \cdot \frac{2\gamma}{g\rho r} \cdot g = \frac{2\gamma}{r}$$

Hence the pressure required is proportional to the surface tension and inversely proportional to the capillary radius.

E27.10 Equation (27.20) gives the distribution of aggregate sizes for an amphiphile. Use the fact that this equation also holds for the optimum aggregate size n^* and eliminate x_1 from this equation to show that

$$x_n = n \cdot \left(\frac{x_{n^*}}{n^*}\right)^{n/n^*} \cdot \exp\left(-\frac{n \cdot \left(\Delta\mu_n^{\ominus} - \Delta\mu_{n^*}^{\ominus}\right)}{RT}\right)$$

Solution

For the optimum aggregate size Equation (27.20) becomes

$$x_{n^*} = n^* \cdot x_1^{n^*} \cdot \exp\left(-\frac{n^* \cdot \Delta\mu_{n^*}^{\ominus}}{RT}\right)$$

Dividing Equation (27.20) by this equation raised to the power n/n^*, we find that

$$\frac{x_n}{\left(x_{n^*}\right)^{n/n^*}} = \frac{n \cdot x_1^n \cdot \exp\left(-\frac{n \cdot \Delta\mu_n^{\ominus}}{RT}\right)}{(n^*)^{n/n^*} \cdot x_1^n \cdot \exp\left(-\frac{n \cdot \Delta\mu_{n^*}^{\ominus}}{RT}\right)}$$

which rearranges to the final result

$$x_n = n \cdot \left(\frac{x_{n^*}}{n^*}\right)^{n/n^*} \cdot \exp\left(-\frac{n \cdot \left(\Delta\mu_n^{\ominus} - \Delta\mu_{n^*}^{\ominus}\right)}{RT}\right)$$

E27.11 Explain why the critical micelle concentration of alkyl sulfates decreases with increasing alkyl chain length (an octyl chain is $130 \cdot 10^{-3}$ mol L^{-1}, decyl chain is $33.2 \cdot 10^{-3}$ mol L^{-1}, dodecyl chain is $8.1 \cdot 10^{-3}$ mol L^{-1}, and a tetradecyl chain is $2.0 \cdot 10^{-3}$ mol L^{-1}).

Solution

As we showed in Section 27.4, a number of terms contribute to the change in chemical potential for forming a micelle with n amphiphile molecules; namely the head-to-head repulsion terms which act to destabilize the micelle and the hydrophobic terms which act to stabilize the micelles. As the alkane chain of the amphiphile molecules that comprise the micelle increases in length, the magnitude of the hydrophobic term increases leading to a greater stability. In contrast, the head group's contribution does not change. Hence the aggregation of amphiphile molecules is enhanced and occurs at a lower average concentration of amphiphile molecules; i.e., tetradecyl chains aggregate at a somewhat lower concentration than dodecyl chains, and so forth.

The additional chain length increases the hydrophobic stabilization. As shown by Equation (27.23), the chemical potential of a micelle with n-aggregated molecules can be written as

$$\Delta\mu_n^\ominus = \Delta\mu_{\text{HP,bulk}}^\ominus - \gamma A' \cdot \boldsymbol{N}_A + C_1 \cdot n^{-1/3} + C_2 \cdot n^{1/3}$$

where C_1 and C_2 are constants, the $\gamma A'$ is the area of the amphiphile per head group, and the $\Delta\mu_{\text{HP,bulk}}^\ominus$ term is the hydrophobic stabilizing terms which becomes more strongly stabilizing as the alkane chain length increases (see Section 27.4). Thus the increasing stability, represented by $\Delta\mu_n^\ominus$, with increasing alkane chain length gives rise to micelle formation at lower concentrations.

E27.12 Explain why the change in the critical micelle concentration per methylene for alkyl sulfates (see Exercise E27.11) is smaller than the corresponding change in the case of dialkylsulfates (e.g., dioctylmethylsulfate has a $x_{\text{critical}} = 2.35 \cdot 10^{-3}$ mol L^{-1} and ditetradecylmethylsulfate has a $x_{\text{critical}} = 0.08 \cdot 10^{-3}$ mol L^{-1}).

Solution

To better compare let us calculate the average change per methylene unit for the single chains and the double chains. For the double chains, we see that x_{critical} decreases by 30 times for a change of 6 methylenes; i.e., by 5 times per methylene unit. In contrast, the single chains change by 65 times for 6 methylenes (see data in E27.11) or nearly 11 times per methylene unit. Thus the two cases differ by about a factor of two. This difference can be understood to arise from the average surface area of alkane chain that is exposed in the two cases. For the double chain surfactants, part of the chains are already stabilized by interaction with their partner chain; hence, the stabilization per methylene unit is less than in the corresponding case of the single chain amphiphile molecules.

E27.13 Discuss the interference colors that appear when forming a black lipid membrane (see Section 27.5.3).

Solution

In the initial stages of formation, the films may be thin but thicker than the wavelength of the light. In this regime they give rise to an etalon effect and cause particular colors of the spectrum to be observed because of constructive and destructive interference. These interference fringes should also depend on the angle of illumination (observation). As the film becomes thinner than the wavelength of the visible light the colors disappear and the film becomes gray, getting darker and darker with decreasing thickness until it appears black.

E27.14 A black lipid membrane has a capacitance of $C = 0.4$ μF for an area of 1 cm^2. Calculate the thickness of the membrane (see Section 27.5.3).

Solution

If we model the film as a parallel plate capacitor then we can use Equation (27.25) and rearrange it to solve for the thickness d; namely

$$\begin{aligned} d &= \hat{\varepsilon}_r \varepsilon_0 \frac{A}{C} \\ &= 2.5 \cdot 8.854 \times 10^{-12} \text{F m}^{-1} \frac{10^{-4}\ \text{m}^2}{0.4 \times 10^{-6}\ \text{F}} \\ &= 5.5 \times 10^{-9}\ \text{m} = 5.5\ \text{nm} \end{aligned}$$

E27.15 Cho et al. (*Biophys. J.* 1999, **76,** 1136) measured the diffusion constant for an aquaporin protein (assume 30 nm radius) in a red blood cell membrane to be 3.1×10^{-11} cm^2 s^{-1}. Estimate the time it takes for a mobile aquaporin to diffuse 3μm, about the radius of the cell.

Solution

The problem statement provides the diffusion coefficient and the root mean square displacement, from which we can estimate the mean time τ as

$$\tau = \frac{x_{RMS}^2}{2D}$$
$$= \frac{\left(3 \times 10^{-6}\ \text{m}\right)^2}{2 \cdot 3.1 \times 10^{-15}\ \text{m}^2\ \text{s}^{-1}}\text{s} = 1450\ \text{s} = 24\ \text{min}$$

E27.16 Consider the placement of a drop of liquid A on a surface of liquid B; consider A and B to be immiscible. It is common to define a spreading coefficient Υ by the relation

$$\Upsilon_{A/B} = \gamma_B - \gamma_A - \gamma_{A/B}$$

Show that a positive spreading coefficient corresponds to a negative Gibbs energy for spreading. What is the spreading coefficient for water on benzene, given that $\gamma_{H_2O} = 72.8 \times 10^{-3}$ J m^{-2}, $\gamma_{C_6H_6} = 28.8 \times 10^{-3}$ J m^{-2}, and $\gamma_{H_2O/C_6H_6} = 35 \times 10^{-3}$ J m^{-2}?

Solution

The Gibbs energy depends on the surface areas so that

$$dG = \left(\frac{\partial G}{\partial A_A}\right) dA_A + \left(\frac{\partial G}{\partial A_B}\right) dA_B + \left(\frac{\partial G}{\partial A}\right) dA$$

where A is the A/B interfacial area, see Fig. E27.16.

Figure E27.16 Drop of liquid A on a surface of liquid B.

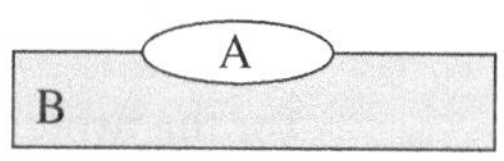

From the geometry we see that

$$dA_A = -dA_B = dA$$

that is the surface area of B decreases as the surface area of A increases and the interfacial area AB increases. From these relations we can write that

$$dG = \left(\gamma_A - \gamma_B + \gamma_{A/B}\right) dA$$

which implies that

$$-\frac{dG}{dA} = \Upsilon_{A/B} = \left(\gamma_B - \gamma_A - \gamma_{A/B}\right)$$

When $\Upsilon_{A/B} > 0$, then $(dG/dA) < 0$, so that A spreads on B.
For the special case of water on benzene, we find that

$$\Upsilon_{A/B} = (72.8 - 28.8 - 35) \times 10^{-3}\ \text{J m}^{-2} = 9.0 \times 10^{-3}\ \text{J m}^{-2}$$

hence it spreads.

E27.17 Use the diffusion coefficient D in Example 27.4 and the Einstein relation, $D = kT/\zeta$, to calculate the effective viscosity of a membrane $\eta_{membrane}$ (use $T = 298$ K and take the radius of the dye molecule to be $r = 0.5$ nm). Saffman and Delbrueck[3] derived an expression for the translational diffusion of a particle in a viscous planar

3 P.G. Saffman, M. Delbrueck, Brownian motion in biological membranes, *PNAS* 1975, **72**, 3111.

membrane of thickness h that is surrounded by a lower viscosity medium (e.g., water which has a shear viscosity of about 0.01 N m^{-2} s). They found that the friction coefficient ζ is given by

$$\zeta = \frac{4\pi\, h\, \eta_{\text{membrane}}}{\log\left(\frac{h\,\eta_{\text{membrane}}}{r\,\eta_{\text{solvent}}} - 0.5772\right)}$$

Determine the value of η_{membrane} for the case where h is the thickness of the membrane (~5 nm) and η_{solvent} is the viscosity of the surrounding fluid (~0.001 N m^{-2} s). Also calculate the value of η_{membrane} by using the formula for the translational friction of a sphere in three dimensions, i.e.,

$$\zeta = 6\pi\, r\, \eta_{\text{membrane}}$$

Compare the value you find from the three-dimensional model to that for the more appropriate two-dimensional model.

Solution
Using the diffusion coefficient in the Stokes–Einstein relation

$$D = \frac{kT}{6\pi\eta r} = \frac{kT}{\zeta}$$

we can find the friction coefficient as

$$\zeta = \frac{kT}{D}$$

For the three-dimensional model, we find a viscosity of

$$\eta_{\text{membrane}} = \frac{\zeta}{6\pi r} = \frac{1}{6\pi r}\frac{kT}{D}$$
$$= \frac{1.3806 \times 10^{-23}\text{J K}^{-1} \cdot 298\text{ K}}{8 \times 10^{-13}\text{ m}^2\text{ s}^{-1}\; 6\pi \cdot 0.5 \times 10^{-9}\text{ m}} = 0.5\text{ N m}^{-2}\text{ s}$$

For the two-dimensional model we find

$$\zeta = \frac{kT}{D} = \frac{4\pi\, h\, \eta_{\text{membrane}}}{\log\left(\frac{h\cdot\eta_{\text{membrane}}}{r\,\eta_{\text{solvent}}} - 0.5772\right)}$$

or

$$\frac{kT}{4\pi\, h\, D} - \frac{\eta_{\text{membrane}}}{\log\left(\frac{h\,\eta_{\text{membrane}}}{r\,\eta_{\text{solvent}}} - 0.5772\right)} = 0$$

Note that

$$\frac{kT}{4\pi\, h\, D} = \frac{1.3806 \times 10^{-23}\text{ J K}^{-1} \cdot 298\text{ K}}{8 \times 10^{-13}\text{ m}^2\text{ s}^{-1}\; 4\pi \cdot 5.0 \times 10^{-9}\text{ m}} = 0.0818\text{ N m}^{-2}\text{ s}$$

We solve this equation numerically by plotting the equation for a range of η_{membrane} values and determining where it crosses zero. Using the values given for the other parameters we can write that

$$y = 0.0818\text{ N m}^{-2}\text{ s} - \frac{\eta_{\text{membrane}}}{\log\left(10{,}000\text{ N}^{-1}\text{ m}^2\text{ s}^{-1} \cdot \eta_{\text{membrane}} - 0.5772\right)} \quad \text{(27E17.1)}$$

Figure E27.17 shows a plot of y versus η_{membrane}. The plot gives a value of

$$\eta_{\text{membrane}} = 0.728\text{ N m}^{-2}\text{ s}$$

This value, which accounts for the essential structural features of the physical process (diffusion in the disc-like viscous medium surrounded by a less viscous medium), is similar to the value 0.5 N m^{-2} s deduced from a three-dimensional model.

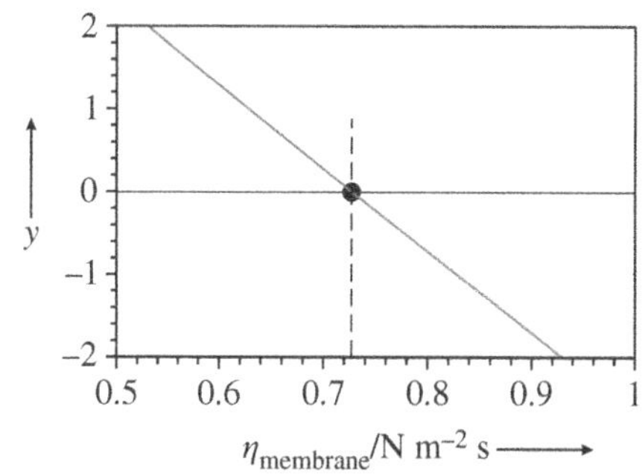

Figure E27.17 Quantity y in Equation (E27E17.1) versus $\eta_{membrane}$.

E27.18 In 1937 Blodgett and Langmuir (*Phys. Rev.* 1937, **51,** 964) measured optical interference fringes through Langmuir–Blodgett films of calcium stearate on glass. For light at normal incidence they showed that minima of the intensity of the reflected light occur at

$$\hat{n} N \delta = m\frac{\lambda}{4}\ , \ m = 1, 3, 5, \ldots$$

where $\lambda = 589.3$ nm is the light wavelength, $\hat{n} = 1.462$ is the refractive index, N is the number of monolayers, and δ is the thickness of a monolayer.

m	1	3	5	7	9	11	13
N	39	120	202	284	365	447	528

Solution

Their measurement of the light minima is explained in Fig. E27.18a. The monolayer assembly is illuminated with the light of a sodium vapor lamp (wavelength of 589.3 nm). Part of the light is reflected from the surface of the assembly (ray a), whereas another part is reflected from the bottom of the assembly (ray b). The phase shift between both rays is $2N\delta$ (the incident light and the reflected light need an additional path of $N\delta$) for vacuum conditions (refractive index $\hat{n} = 1$).

The interference from the phase shift of the two rays gives an intensity minimum at $m\lambda/2$ ($m = 1, 3, 5, \ldots$). The assembly has a refractive index $\hat{n} = 1.462$, and the condition for an intensity minimum is

$$2N\delta\,\hat{n} = m\frac{\lambda}{2} \quad \text{or} \quad N\delta\,\hat{n} = m\frac{\lambda}{4}\ , \ m = 1, 3, 5, \ldots$$

According to the table a first intensity minimum ($m = 1$) is observed for $N = 39$, that is, for an assembly consisting of 39 monolayers. In the experiment more and more monolayers are added and for an assembly consisting of 120 monolayers the next intensity minimum is observed ($m = 3$).

Figure E27.18a Monolayer assembly with N monolayers deposited on a glass plate.

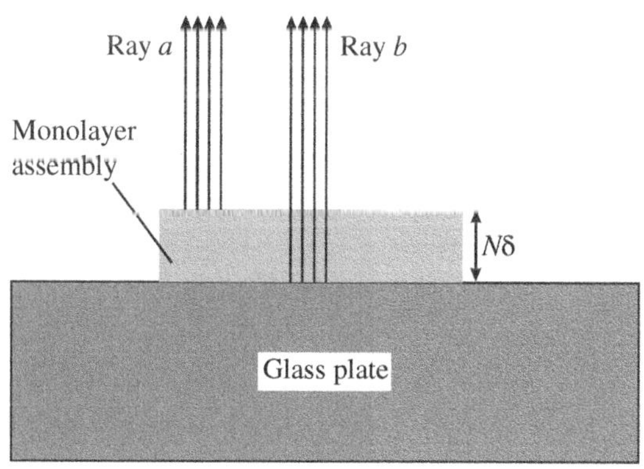

In the actual experiment many assemblies with different numbers of monolayers are deposited on one glass plate and the intensity of the reflected light along this arrangement is observed.

The relation given in the problem statement predicts a linear dependence for a plot of m versus N with a slope of $4\hat{n}\delta/\lambda$; hence, we plot the data and fit it to a line (see Fig. E27.18b).

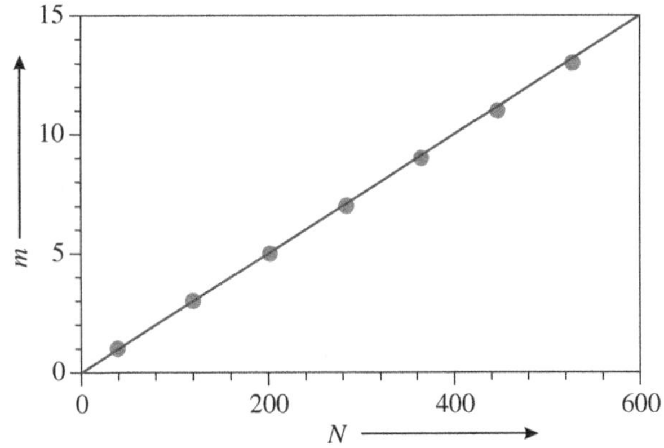

Figure E27.18b Integer m versus the number N of monolayers.

The fit of these data is excellent and yields a slope of 0.0246; hence, we find that

$$\delta = \frac{slope \cdot \lambda}{4\hat{n}}$$
$$= \frac{0.0246 \cdot 5.893 \times 10^{-7}\text{ m}}{4 \cdot 1.462} = 2.48 \times 10^{-9}\text{ m} = 2.48\text{ nm}$$

That is, the average thickness of one monolayer is 2.48 nm.

E27.19 Given that the average thickness of an octadecanethiol film is $d = 2.1$ nm, estimate the average tilt angle of the octadecane chains with respect to the surface normal of the gold surface. Assume that the thickness of a methylene group is 0.14 nm and that of the S atom is 0.2 nm.

Solution

Using the parameters given above, the chain length should be $l = (0.2\text{ nm} + (0.13\text{ nm})\ 18)\text{ nm} = 2.4$ nm. If the chains were oriented normal to the surface of the Au film then the measured thickness should be 2.4 nm, which is larger than the measured thickness. By assuming that the decrease in film thickness arises solely from tilting of the chains at an angle θ, we can write

$$\cos\theta = \frac{d}{l} = \frac{2.1\text{ nm}}{2.4\text{ nm}} = 0.875$$

and $\theta = 29^{\text{o}}$.

E27.20 Calculate the work required to transfer a charged group (such as the COO^- side group of the amino acid aspartic acid) from water into a membrane.

Solution

We approximate the charged COO^- side group of aspartate, including a solvation shell, by a sphere of radius $r = 0.5$ nm. From the Born model (see Section 23.4.2.5) the work required to transfer a charge from vacuum to a dielectric is

$$w = \frac{-1}{4\pi\varepsilon_0}\frac{e^2}{2r}\left(1 - \frac{1}{\varepsilon_r}\right) = -3.6\text{ eV}\left(1 - \frac{1}{\varepsilon_r}\right)$$

Hence the work required to bring a sphere of radius r from a medium with relative dielectric constant ε_{r_1} (in this case water, $\varepsilon_{r_1} = 80$) to a medium with relative dielectric constant ε_{r_2} (hydrocarbon portion, $\varepsilon_{r_2} = 2.5$) is

$$w = \frac{e^2}{4\pi\varepsilon_0 2r}\left(\frac{1}{\varepsilon_{r_2}} - \frac{1}{\varepsilon_{r_1}}\right) = 0.6\text{ eV} = 9.8 \times 10^{-20}\text{ J}$$

E27.21 Consider a Na^+ ion moving through the gramicidin pore of a membrane. Compute the difference in solvation energy for a Na^+ ion in water solution (relative permittivity of 80) and that inside of a membrane (relative permittivity of 2.5). Comment on your result.

Solution

This calculation is very similar to that in E27.20, where we consider the work required to bring a sphere of radius r from a medium with relative dielectric constant ε_{r_1} (in this case water, $\varepsilon_{r_1} = 80$) to a medium with relative dielectric constant ε_{r_2} (hydrocarbon portion, $\varepsilon_{r_2} = 2.5$) is

$$w = \frac{e^2}{4\pi\varepsilon_0 2r}\left(\frac{1}{\varepsilon_{r_2}} - \frac{1}{\varepsilon_{r_1}}\right)$$

We take the radius of a hydrated sodium ion to be 300 pm, so that the work required to transfer the ion is

$$w = \frac{e^2}{4\pi\varepsilon_0 2r}\left(\frac{1}{\varepsilon_{r_2}} - \frac{1}{\varepsilon_{r_1}}\right) = 0.9 \text{ eV} = 1.5 \times 10^{-19} \text{ J}$$

or on a per molar basis 96.8 kJ mol^{-1} = 20.7 kcal mol^{-1}.

This energy barrier is quite high; hence, an ion channel such as gramicidin contains residues and water molecules that can act to stabilize the ion below the value estimated by this simple dielectric model.

E27.22 Calculate the contact angles α and β in Fig. 27.4 (Section 27.2.1) from Equation (27.5)

$$\gamma_{\text{decane/air}} \cdot \cos\alpha \cdot dl + \gamma_{\text{decane/water}} \cdot \cos\beta \cdot dl = \gamma_{\text{water/air}} \cdot dl$$

for the horizontal components of the force f and from Equation (27.6)

$$\gamma_{\text{decane/air}} \cdot \sin\alpha = \gamma_{\text{decane/water}} \cdot \sin\beta$$

for the vertical components, using the data given in the text. Use notations $\gamma_{\text{decane/air}} = 23.9$ mN m^{-1}, $\gamma_{\text{decane/water}} =$ 51.2 mN m^{-1}, and $\gamma_{\text{water/air}} = 72.8$ mN m^{-1}.

Solution

Figure 27.4 is

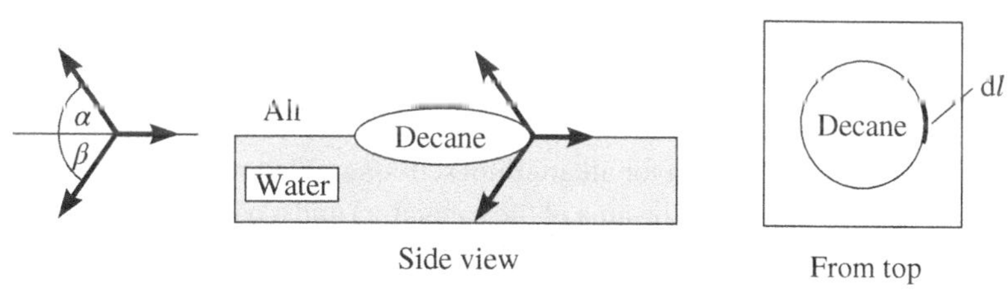

With the notations $a = \gamma_{\text{decane/air}} = 23.9$ mN m^{-1}, $b = \gamma_{\text{decane/water}} = 51.2$ mN m^{-1}, and $c = \gamma_{\text{water/air}} = 72.8$ mN m^{-1} we obtain

$$a \cdot \cos\alpha + b \cdot \cos\beta = c \quad \text{and} \quad a \cdot \sin\alpha = b \cdot \sin\beta \tag{27.1}$$

In the second equation we replace $\sin\alpha$ by $\sqrt{1-\cos^2\alpha}$, and we isolate the $\cos\alpha$ term in each equation

$$\cos\alpha = \frac{c - b\cdot\cos\beta}{a} \quad \text{and} \quad \cos^2\alpha = 1 - \frac{b^2\sin^2\beta}{a^2}$$

Now we eliminate $\cos\alpha$ from these equations and find

$$\left(\frac{c - b\cdot\cos\beta}{a}\right)^2 = 1 - \frac{b^2\sin^2\beta}{a^2}$$

Rearranging to solve for $\cos\beta$, we find

$$c^2 - 2bc\cdot\cos\beta + b^2\cdot\cos^2\beta = a^2 - b^2\sin^2\beta = a^2 - b^2\left(1-\cos^2\beta\right)$$

and

$$\cos\beta = -\frac{a^2 - b^2 - c^2}{2bc}$$

Substituting the values for a, b, and c, we find that

$$\cos\beta = 0.986 \quad \text{and} \quad \beta = 10°$$

Inserting β in the Equation (27.4) we obtain

$$\sin\alpha = \frac{b\cdot\sin\beta}{a} = \frac{51.2\cdot 0.174}{23.9} = 0.372,\ \alpha = 22\ °$$

27.2 Problems

P27.1 Consider a spherical soap bubble with surface tension γ. Assuming static equilibrium, show that the pressure difference ΔP between the inside and the outside of the bubble is given by

$$\Delta P = 2\gamma/r$$

where r is the radius of the bubble.

Solution

At static equilibrium the forces acting on the bubble's surface must balance, i.e.,

$$P_{\text{inside}}\cdot A = P_{\text{outside}}\cdot A + 8\pi\gamma r$$

where $P_{\text{inside}}\cdot dA$ is the force on the differential area dA from the gas molecules inside striking the bubble wall, $P_{\text{outside}}\cdot dA$ is the force on the differential area dA from the gas molecules outside striking the bubble wall, and $8\pi\gamma r$ is the force associated with the surface tension. When the bubble changes its radius by dr, the differential work required is *force*$\cdot dr = \gamma\cdot dA$ where

$$dA = 4\pi(r+dr)^2 - 4\pi r^2 = 8\pi r\cdot dr + 4\pi(dr)^2 \simeq 8\pi r\cdot dr$$

for small enough dr. Hence, we see that the differential work is $8\pi r\gamma\cdot dr$ and the force is $8\pi r\gamma$. Using $A = 4\pi r^2$ and rearranging, we find that

$$\Delta P = P_{\text{inside}} - P_{\text{outside}} = \frac{2\gamma}{r}$$

Hence the gas pressure inside the bubble is larger than the gas pressure outside the bubble.

P27.2 The table provides capacitance data for alkanethiol self-assembled monolayer films on a gold electrode. Use these data (C is the capacitance and A is the area of the capacitor) and a parallel plate capacitor model to find the monolayer film thickness as a function of n; n is the number of methylene groups in the alkanethiol $HS(CH_2)_nCH_3$. Use the relative permittivity $\hat{\varepsilon} = 2.3$ (that of polyethylene). From the bond parameters of an alkane chain, each methylene should contribute 130 pm to the length. Use this fact and the findings of your analysis to estimate the average tilt angle of the alkane chains on the gold electrode.

n	1	3	5	7	9	11	15	17
C/A (μF cm^{-2})	4.6	3.9	2.6	2.1	1.7	1.3	1.1	0.98

Solution

We assume that the thickness of the film d increases linearly with n, so that $d = a\cdot n + b$ where a and b are constants. The parallel plate model tells us that

$$\frac{A}{C} = \frac{d}{\hat{\varepsilon}\varepsilon_0} = \frac{b}{\hat{\varepsilon}\varepsilon_0} + n\cdot\frac{a}{\hat{\varepsilon}\varepsilon_0}$$

The fit to the data gives

$$a = 0.0555\,\mu\text{F}^{-1}\,\text{cm}^2\cdot\hat{\varepsilon}\varepsilon_0 = 0.0555\times 10^6\frac{\text{V}}{\text{C}}\cdot 1\times 10^{-4}\,\text{m}^2\ 8.854\times 10^{-12}\,\text{C}^2\,\text{J}\cdot 2.3$$

$$= 1.13\cdot 10^{-10}\frac{\text{V m}^2\text{C}^2}{\text{C J m}} = 1.13\cdot 10^{-10}\frac{\text{V m}^2\,\text{C}^2}{\text{C A V s m}} = 113\,\text{pm}$$

Hence, each methylene group contributes 113 pm to the average thickness and the thiol and methyl groups contribute 270 pm.

Figure P27.2 The quantity A/C for a capacitor made of alkanethiol self-assembled monolayer films with n methyl groups is plotted versus n.

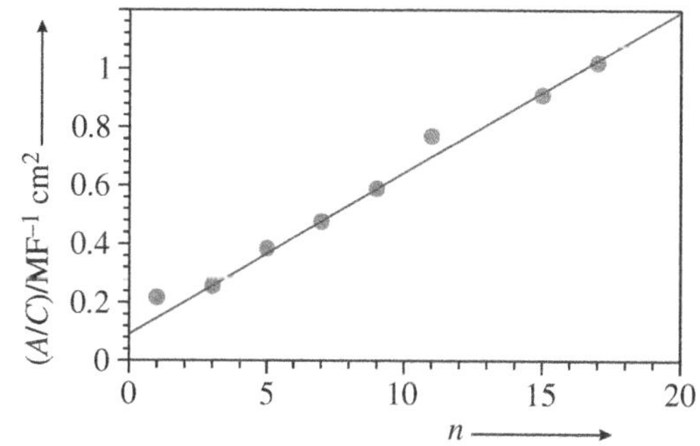

The methylenes contribute less to the thickness than that expected from their overall length. If we assume that this reduction results entirely from the tilt angle of the chain we find that the tilt angle is (see E27.19)

$$\theta = \cos^{-1}\left(\frac{1.1}{1.3}\right) = 32\,^\circ$$

P27.3 In this model, you are asked to evaluate a simple model for the diffusion of ions through a membrane. Assume that the concentration of sodium ions inside of a liposome with an inner diameter of 100 nm is 0.1 mM and that the thickness of the bilayer is 3 nm. Estimate the number of collisions of sodium ions with the inner wall of the liposome if its diffusion coefficient is 1.3×10^{-5} cm^2 s^{-1} (see Chapter 14). Use this "collision rate" to estimate the rate of ions through the bilayer if the activation energy for the transport of an ion through a membrane is the energy that you calculated in E27.20.

Solution

To estimate the collision frequency, we will calculate the root-mean-square displacement of an ion and compare it to the internal diameter of the liposome. For the diffusion coefficient given, we evaluate the root-mean-square displacement as

$$x_{\text{rms}} = \sqrt{2D \cdot \Delta t}$$

For a characteristic time of one second, we find that a sodium ion would have a root-mean-square average displacement of 51 μm, if it were not confined by the membrane. Comparing to the diameter of the liposome of 100 nm, we estimate that about 510 collisions would occur with the walls in this time. Because the liposome is taken to be spherical, all of the directions (x, y, and z) are equal and we need not consider other displacements explicitly. Hence the collision frequency is 510 s^{-1}.

We are told that the internal concentration is 0.1×10^{-3} mol L^{-1} and we can calculate its volume to be 5.2×10^{-19} L (about one-half of an attoliter), so that there are about 31 sodium ions in the liposome. Assuming that these ions act independently gives an average collision rate of 15, 000 s^{-1} = 1.5×10^4 s^{-1} for ions with the membrane's inner wall.

Each of these collisions has a probability for the ion to successfully permeate the membrane. We estimate that probability by a Boltzmann factor with an activation energy of 9.8×10^{-20} J, as calculated in E27.20. Hence the probability for permeating the layer in any given collision is

$$\exp\left(-\frac{9.8 \times 10^{-20}\ \text{J}}{1.3806 \times 10^{-23}\ \text{J K}^{-1} \cdot 298\ \text{K}}\right) = 4.5 \times 10^{-11}$$

and we estimate the permeation rate to be

$$\begin{aligned} rate &= 1.5 \times 10^4\ \text{s}^{-1} \cdot 4.5 \times 10^{-11} \\ &= 6.8 \times 10^{-7}\ \text{s}^{-1} = 2.4 \times 10^{-3}\ \text{hr}^{-1} \end{aligned}$$

Hence it is quite unlikely that the ion will permeate the membrane in this manner.
If water permeated the membrane or a pore formed, then the activation energy could be dramatically lower and the permeation rate would increase accordingly.

P27.4 Calculate the electrostatic repulsion energy between head groups for an idealized micelle structure, like that shown in Fig. P27.4. Model the micelle interface (left) as two charged concentric spheres (right), and make the approximation that $\delta \ll R$ so that it can be treated as a parallel plate capacitor.

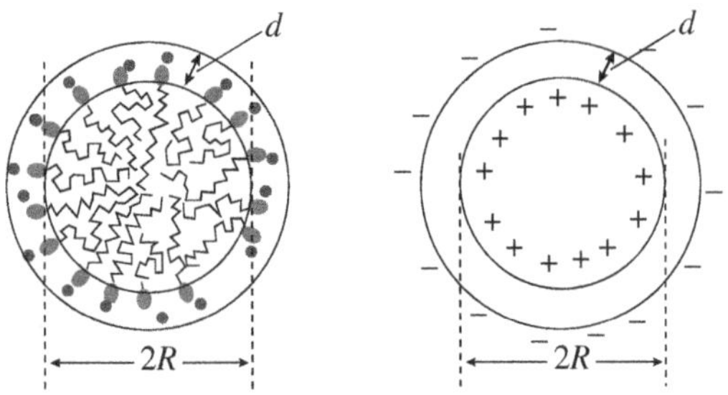

Figure P27.4 Diagram of a spherical micelle whose hydrophobic core has the radius R. We model the interfacial space charge by two concentric charge spheres with a gap of distance δ and a relative permittivity $\hat{\varepsilon}$. Dark gray filled circles: negatively charged SO_4^- headgroups in sodium dodecyl sulfate, light gray filled circles: positive charges of the Na^+ counterions.

Part 1: Perform derivation for the electrostatic energy E in the capacitor and show that it is

$$E = \frac{Q^2}{2\hat{\varepsilon}\varepsilon_0}\frac{\delta}{4\pi R^2}$$

where Q is the charge.
Part 2: Make the correspondence with the micelle model and show that

$$E_{HR} = \frac{E}{n} = \frac{e^2}{2\hat{\varepsilon}\varepsilon_0}\frac{\delta}{A}$$

where n is the aggregation number and A is the area of the head group per molecule.

Solution
Part 1: The voltage drop V across a parallel plate capacitor is the product of the electric field F and the distance δ; i.e.,

$$V = F\,\delta$$

and the electric field between the plates is

$$F = \frac{Q}{\hat{\varepsilon}\varepsilon_0 \cdot area}$$

where Q is the charge, and $\hat{\varepsilon}$ is the relative permittivity. With this field, the voltage drop across the plates is

$$V = \frac{Q}{\hat{\varepsilon}\varepsilon_0 \cdot area}\,\delta$$

The energy in the capacitor is

$$E = \int_0^Q \frac{Q'}{\hat{\varepsilon}\varepsilon_0 \cdot area}\,\delta \mathrm{d}Q' = \frac{\delta}{\hat{\varepsilon}\varepsilon_0 \cdot area}\,\frac{Q^2}{2}$$

In our case, the area is that for the surface of a sphere with radius R, so that $area = 4\pi R^2$, and we find that

$$E = \frac{\delta}{\hat{\varepsilon}\varepsilon_0 \cdot 4\pi R^2}\,\frac{Q^2}{2}$$

Part 2: In our model, the total charge in the double layer is assumed to be $Q = ne$ and the total area of the surface is $area = nA$, where A is the area per amphiphilic head group. Making these substitutions, we find that

$$E = \frac{(ne)^2}{2\hat{\varepsilon}\varepsilon_0}\frac{\delta}{nA} \quad \text{so that} \quad \frac{E}{n} = \frac{e^2}{2\hat{\varepsilon}\varepsilon_0}\frac{\delta}{A}$$

P27.5 Consider the case of a very thin (i.e., ignore thickness of plate) rectangular plate of width L that is suspended above the surface of a liquid. As it is just placed in contact with the surface of the liquid, capillary forces cause the liquid to rise. By balancing the capillary force with the gravitational force of the liquid pulled above the bulk liquid surface, show that

$$\gamma \cos\theta = weight/(2L)$$

where γ is the surface tension and θ is the contact angle. Next, assume that the surface of the liquid, which is pulled above the surface, has the shape of a cylinder (length L) and derive an expression for the contact angle θ. Given that the density ρ of hexadecane is 0.773×10^{-3} kg m^{-3} and its surface tension γ is 27.2 mN m^{-1}, calculate the contact angle for a plate with length $L = 3$ cm at a height of $h = 0.1$ cm.

Solution

Figure P27.5a provides a side-view and a front-view of the plate (black) and the liquid layer (gray).

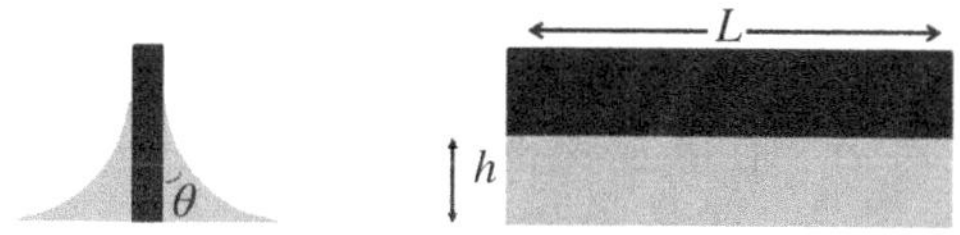

Figure P27.5a Side-view and front-view of a plate (black) immersed in a liquid layer (gray).

We analyze the situation by balancing the capillary force $L \cdot \gamma \cos\theta$ with the force due to gravity for the liquid pulled above the surface. In the limit where the plate's thickness is very small, as compared to its length L, we find that

$$2L \cdot \gamma \cos\theta = weight$$

which is the result asked for in the first part of the problem.

For the liquid reaching a height h above the surface and having a cylindrical surface profile, we find that its volume is given by

$$V = 2\left[L \cdot h^2\left(1 - \frac{\pi}{4}\right)\right] = 2\left[\frac{3\pi}{4} L \cdot h^2\right]$$

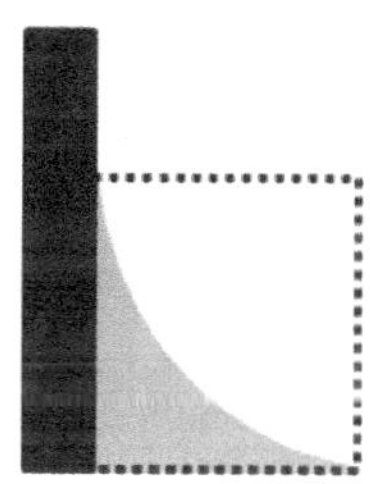

Figure P27.5b The sketch illustrates the shape for the liquid volume which arises from capillarity on the Wilhelmy plate.

so that its total mass will be

$$mass = \rho\left[\frac{3\pi}{2} L \cdot h^2\right]$$

where ρ is the density of the liquid. Now we balance the forces and write

$$2L \cdot \gamma \cos\theta = \rho\left[\frac{3\pi}{2} L \cdot h^2\right] g$$

where g is the acceleration due to gravity. Isolating $\cos\theta$, we find that

$$\cos\theta = \left[\frac{3\pi}{4}\frac{h^2 \rho\, g}{\gamma}\right]$$

so that

$$\theta = \arccos\left[\frac{3\pi}{4}\frac{h^2 \rho\, g}{\gamma}\right]$$

Using the formula found for θ and the parameters given for hexadecane (with g = 9.8 m s^{-2}), we find that

$$\theta = \arccos\left[\frac{3\pi}{4}\frac{(0.1\text{cm})^2 \cdot 0.773\text{ g cm}^{-3} \cdot 980\text{ cm s}^{-2}}{27.2\text{ g s}^{-2}}\right]$$

$$= 49^\circ$$

P27.6 Consider a soap bubble. In this problem you will show that the pressure of air inside of a small soap bubble is larger than that inside of a larger soap bubble; i.e., the difference in pressure ΔP across the soap film decreases with its curvature. Consider a curved surface that has a surface tension γ and show that the pressure difference ΔP across the surface is

$$\Delta P = \gamma \left(\frac{1}{R_x} + \frac{1}{R_y} \right)$$

where R_x is the radius of curvature along the x-direction and R_y is the radius of curvature along the y-direction. Hint: Solve by finding an expression for the work of expanding the surface from the plane (x, y) outward to $(x + \delta x, y + \delta y)$, and equate this work to that for the displacement outward given by $work = force \times \delta z$, where δz is the displacement outward.

For a spherical soap bubble ($R_x = R_y$), which has two surfaces (an inner and outer surface of the soap film), and a surface tension of $0.025\ \text{N m}^{-1}$, calculate the pressure differential across the soap film for a radius of 5 cm and a radius of 1 cm.

Solution

The diagram in Fig. P27.6 shows two surfaces that are displaced by an incremental amount outward, dz.

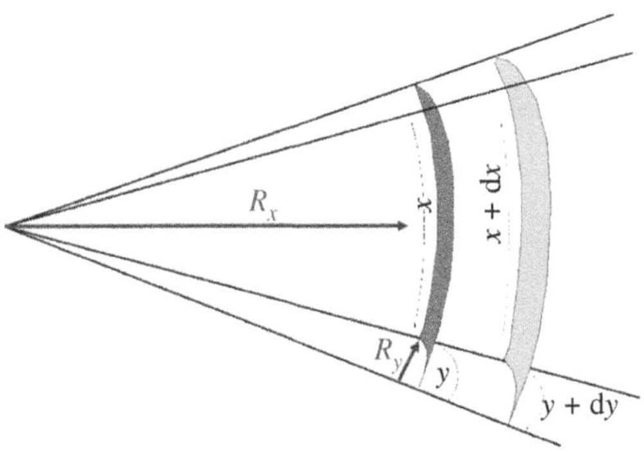

Figure P27.6 The dark light gray area represents a surface element for a smaller bubble, with a radius R_x along the x-direction and radius R_y along the y-direction. The lighter light gray area shows the element after the bubble expands by dz.

For a surface with a curvature R_x along the x-direction, we can write

$$\frac{x}{R_x} = \frac{x + \delta x}{R_x + \delta z} \quad \text{or} \quad \delta x = \frac{x}{R_x} \cdot \delta z$$

Similarly for a curvature R_y along the y-direction, we can write

$$\frac{y}{R_y} = \frac{y + \delta y}{R_y + \delta z} \quad \text{or} \quad \delta y = \frac{y}{R_y} \cdot \delta z$$

When the surface area expands from $A = x \cdot y$ to $A + \delta A$, we find that

$$\begin{aligned} A + \delta A &= (x + \delta x)(y + \delta y) \\ &= x \cdot y + y \cdot \delta x + x \cdot \delta y + \delta x \cdot \delta y \end{aligned}$$

If we neglect the second-order term $\delta x \cdot \delta y$, we find that

$$\begin{aligned} \delta A &= y \cdot \delta x + x \cdot \delta y = y \cdot \frac{x}{R_1} \cdot \delta z + x \cdot \frac{y}{R_2} \cdot \delta z \\ &= A \cdot \delta z \left(\frac{1}{R_x} + \frac{1}{R_y} \right) \end{aligned}$$

The work to expand the surface by δA is

$$work = \gamma \cdot \delta A$$

At equilibrium a pressure difference ΔP exists across the surface. The work for the incremental expansion is provided by $\Delta P \cdot A \cdot \delta z$, so that

$$\Delta P \cdot A \cdot \delta z = \gamma \cdot \delta A = \gamma \cdot A \cdot \delta z \left(\frac{1}{R_x} + \frac{1}{R_y} \right)$$

and

$$\Delta P = \gamma \cdot \left(\frac{1}{R_x} + \frac{1}{R_y}\right)$$

For the case of a spherical bubble of radius R, we find that

$$\Delta P = \Delta P_{\text{inner}} + \Delta P_{\text{outer}} = \gamma \cdot \frac{2}{R_{\text{inner}}} + \gamma \cdot \frac{2}{R_{\text{outer}}}$$

Assuming that $R_{\text{inner}} \simeq R_{\text{outer}}$ we find that

$$\Delta P = 4 \cdot \gamma \cdot \frac{1}{R} = \frac{0.100\text{N m}^{-1}}{R}$$

where we have used $\gamma = 0.025$ N m^{-1}. For a 1 cm radius we find $\Delta P = 10$ Pa and for a 5 cm radius we find $\Delta P = 2$ Pa. These pressures differences are quite small; for $R = 1$ cm it is 10^{-4} of atmospheric pressure.

P27.7 Using the result of P27.6 and elementary considerations show that the vapor pressure over a liquid with a spherically curved surface of radius a is given by

$$P = P_0 \cdot \exp\left(\frac{2V_m\gamma}{a\boldsymbol{R}T}\right)$$

where P_0 is the vapor pressure above a flat surface, V_m is the molar volume of the liquid, $\boldsymbol{R}$ is the gas constant, T is the temperature, and γ is the surface tension.

Use the formula you derive to calculate the percent increase in the vapor pressure above a spherical drop of cyclohexane at 293 K which has a $\gamma = 24.95$ mN m^{-1} and a molar volume of 0.108 L mol^{-1}. Consider the case of a 0.1 mm diameter drop, a 0.2 μm diameter drop, and a 20 nm diameter drop.

Solution

At a temperature T, the change in Gibbs energy to curve the surface is

$$\Delta_m G = \int V_m \, dP$$

For a small enough change, we can write that $\int V_m \, dP \simeq V_m \cdot \Delta P$, and find that

$$\Delta_m G = V_m \cdot \Delta P = V_m \cdot \gamma \cdot \left(\frac{1}{R_x} + \frac{1}{R_y}\right)$$

where the pressure difference ΔP is considered to be that between the pressure for a flat surface $P^{\ominus}$ and that for a curved surface P. In the case of an ideal gas, we know that

$$G_m = G_{0,m} + \boldsymbol{R}T \ln\left(\frac{P}{P_0}\right)$$

so that

$$\Delta_r G = \boldsymbol{R}T \ln\left(\frac{P}{P_0}\right) - V_m \cdot \gamma \cdot \left(\frac{1}{R_x} + \frac{1}{R_y}\right)$$

Hence we find that

$$P = P_0 \exp\left[\frac{V_m\gamma}{\boldsymbol{R}T}\left(\frac{1}{R_x} + \frac{1}{R_y}\right)\right]$$

For a spherical surface $R_x = R_y = r$, so that

$$P = P_0 \exp\left(\frac{2V_m\gamma}{\boldsymbol{R}T} \cdot \frac{1}{r}\right)$$

For cyclohexane

$$\frac{P}{P_0} = \exp\left(\frac{2 \cdot 1.08 \times 10^{-4}\text{m}^3 \text{ mol}^{-1} \cdot 2.495 \times 10^{-2} \text{ N m}^{-1}}{8.314 \text{ J mol}^{-1}\text{K}^{-1} \cdot 293 \text{ K}} \cdot \frac{1}{r}\right)$$
$$= \exp\left(2.21 \times 10^{-9} \text{ m} \cdot \frac{1}{r}\right)$$

or

$$\frac{\Delta P}{P_0} = \exp\left(2.21\times10^{-9}\ \mathrm{m}\cdot\frac{1}{r}\right) - 1$$

For the three drop diameters specified we find

a	0.10 mm	1.0 μm	10 nm
$\frac{\Delta P}{P_0}$	0.002%	0.22%	24.7%

Thus, we see that the curvature has a very weak effect on the vapor pressure unless the droplets are very small. This prediction of the Kelvin equation has been largely confirmed by experiments; see L.R. Fisher, J.N. Israelachvilli, Experimental studies on the applicability of the Kelvin equation to highly curved concave menisci, *J. Colloid Interface Sci.*, 1981, **80**, 528–541.

P27.8 Figure 27.16 shows conductivity data for a solution of sodium dodecylsulfate in water as a function of the amount of dodecylsulfate. Some of these data are reproduced in the table. Read the paper by N. Jalsenjak, D. Tezak (*Chem. Eur. J.* 2004, **10**, 5000) and use the model they develop to analyze the data and obtain the aggregation number.

c_{total}/mmol L^{-1}	0	0.150	0.280	0.420	0.550	0.700	0.830	0.927
κ/mS cm^{-1}	0	0.10	0.20	0.29	0.37	0.46	0.54	0.58

c_{total}/mmol L^{-1}	1.04	1.16	1.33	1.50	1.65	1.81	1.96	2.15
κ/mS cm^{-1}	0.61	0.64	0.68	0.72	0.75	0.80	0.84	0.88

Solution

In this example solution the data are analyzed using the model in the paper with two free parameters. As described in the paper the data points were fit locally to a quadratic function and then its derivative was used to determine the slope. For sake of illustration and comparison some, but not all, of the determined slopes are reported here.

c_{total}/mmol L^{-1}	0.150	0.420	0.830	1.04	1.33	1.65	1.96
$\frac{d\kappa}{dc_{total}}$/S cm^2 mol^{-1}	71.3	64.9	50.2	27.3	23.2	25.5	23.4

The analysis in the paper assumes that the solution's conductivity can be written as a sum of contributions from the surfactant, the counterion, and the micelle, namely that

$$\kappa = \lambda_s\cdot c_{total} - \xi\left[\lambda_s - (1-\alpha)\,\lambda_{Na^+} - \lambda_{micelle}/N\right]$$

where N is the aggregation number, α is the counter-ion binding parameter, λ_s is the molar conductivity of dodecylsulfate, and ξ is the extent-of-reaction (i.e., micelle formation), which is given by $\xi = N\cdot c_{micelle}$. By modeling the micelle conductivity as $\lambda_{micelle} = \lambda_{Na^+}\cdot N^{5/3}(1-\alpha)^2$, they find that

$$\kappa = \lambda_s\cdot c_{total} - \xi\left(\lambda_s\left(1-N^{2/3}(1-\alpha)^2\right)\right) - \lambda_{Na^+}\left((1-\alpha)-N^{2/3}(1-\alpha)^2\right)$$

They then proceed to find an expression for the derivative of κ with respect to the concentration c_{total}. This procedure resulted in Equation (14) of the paper by Jalsenjak, namely

$$\frac{d\kappa}{dc} = \lambda_s - \frac{A\cdot c\left(\lambda_s\cdot c-\kappa\right) - B\left(\lambda_s\cdot c-\kappa\right)^2/Q}{c^2 - D\cdot\left(\lambda_s\cdot c-\kappa\right)^2/Q^2 + E\cdot\left(\lambda_s\cdot c-\kappa\right)/Q}$$

and the slope values that are plotted in Fig. P27.8. In this equation,

$$A = N(1+\alpha);\ B = 2N\alpha;\ D = N\alpha(1+\alpha-1/N);\ E = N(1+\alpha^2-(1-\alpha)/N)$$
$$Q = \lambda\left(1-N^{2/3}(1-\alpha)^2\right) - \lambda_{Na^+}\left((1-\alpha)-N^{2/3}(1-\alpha)^2\right)$$

where the molar conductivity of sodium λ_{Na^+} is taken to be 50.1 S cm^2 mol^{-1}. The curve in Fig. P27.8 shows a best fit by four concentrations greater than the critical micelle concentration. This fit gives an aggregation number $N = 42$, and a counter ion empirical binding parameter α of 0.72, when a molar conductivity of $\lambda_S = 63.5$ S cm^2 mol^{-1} is used for the sodium dodecylsulfate surfactant ion.

Figure P27.8 dκ/dc versus the concentration c.

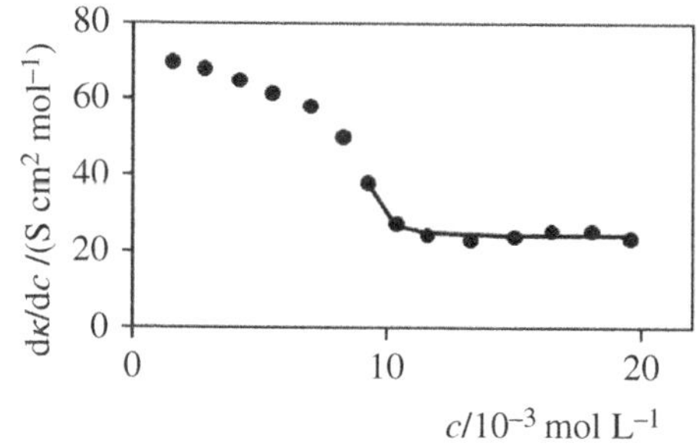

These parameters are comparable to those reported in Table 2 of the paper. They differ somewhat from those fits because not all of the data shown in the paper are used in the analysis here. In particular we note that the molar conductivity of the surfactant shows excellent agreement with that reported in the paper and by earlier workers, suggesting that the analysis is robust in this regard. By contrast the aggregation number depends more sensitively on the data density and on the fitting range chosen (see the fitting values reported in Table 2 of the paper). The value of $N = 42$ found in the analysis here is somewhat lower than the value for model A in the paper by Jalsenjak, but is in good agreement with the aggregation numbers found for fits by models B and C.

28

Supramolecular Machines

28.1 Exercises

E28.1 Using Equation (28.3) for the distance d_0 in a donor–acceptor system, calculate the quantum yield ϕ from the measured $d_0 = 7$ nm. The wavelength of the light emitted by the donor is $\lambda = 480$ nm, the absorbance of the layer is $A_A = 0.012$, and the index of refraction is $\hat{n} = 1.5$.

Solution

We solve Equation (28.3) for the quantum yield ϕ and find

$$\phi = \left(\frac{d_0 \hat{n} 4\pi}{\lambda_D}\right)^4 \frac{4}{9A_A} = \left(\frac{7\ \text{nm} \cdot 1.5 \cdot 4\pi}{480\ \text{nm}}\right)^4 \frac{4}{9 \cdot 0.012} = 0.2$$

E28.2 Consider the two molecules fluorescein (emitter) and tetramethylrhodamine (absorber) and calculate the characteristic Förster distance, r_0, using Equation (28.9). Assume that the quantum yield of fluorescein is 0.95, that the extinction coefficient of the tetramethylrhodamine is $72,000\ \text{mol}^{-1}\ \text{L cm}^{-1}$, and the wavelength at the absorbance maximum is $\lambda = 550$ nm.

Solution

Using Equation (28.9), namely

$$r_0 = \left(\frac{9\varepsilon_A \cdot 2.303 \cdot c_0^4}{128\pi^5 \mathbf{N}_A \nu_0^4} \cdot \frac{\kappa^2 \phi}{\hat{n}^4}\right)^{1/6}$$

where ε_A is the extinction coefficient of the absorber, $\kappa = 2\cos\Theta_1 \cos\Theta_2 - \sin\Theta_1 \sin\Theta_2 \cos\xi$, is a measure of the orientation of the transition moments of absorber and emitter, ϕ is the quantum yield of the emitter, ν_0 is the frequency of the emitted light wave, and $\hat{n} = 1.5$ is the refractive index of the medium. For two parallel transition moments we find $\Theta_1 = \Theta_2 = \pi/2$ and $\xi = 0$; thus, $\kappa^2 = 1$.

With $\nu_0 = c_0/\lambda$ we find that

$$r_0 = \left(\frac{9\varepsilon_A \cdot 2.303 \cdot \lambda^4}{128\pi^5 \mathbf{N}_A} \cdot \frac{\kappa^2 \phi}{\hat{n}^4}\right)^{1/6}$$

$$= \left(\frac{9 \cdot 72,000\ \text{mol}^{-1}\ \text{L cm}^{-1} \cdot 2.303 \cdot (550 \times 10^{-9}\ \text{m})^4}{128\pi^5 \cdot 6.022 \times 10^{23}\ \text{mol}^{-1}} \cdot \frac{1 \cdot 0.95}{(1.5)^4}\right)^{1/6}$$

$$= \left(1.086 \times 10^{-48}\ \frac{\text{L m}^4}{\text{cm}}\right)^{1/6} = (1.086 \times 10^{-49}\ \text{m}^6)^{1/6} = 6.9 \times 10^{-9}\ \text{m} = 6.9\ \text{nm}$$

E28.3 Consider the electric field that acts on a chromophore from nearby charges, e.g., charged amino acids. Assume that the negatively charged amino acids D212 and D85 (denoted as 1 and 2 in Fig. E28.3) are at distance $d = 0.4$ nm from the N atom and from C atom 14, respectively. What is their field strength F at the chromophore?

Solutions Manual for Principles of Physical Chemistry, Third Edition. Edited by Hans Kuhn, David H. Waldeck, and Horst-Dieter Försterling.

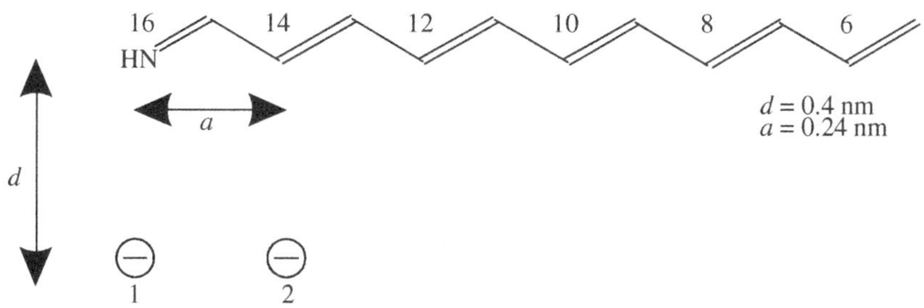

Figure E28.3 Chromophore in the electric field of two negative charges.

Solution

For $a = 0.24$ nm and $d = 0.4$ nm, the potential energy of a π electron at the N atom in the field of charges 1 (distance d) and 2 (distance $\sqrt{d^2 + a^2}$) in vacuum is

$$E_N = \frac{e^2}{4\pi\varepsilon_0}\left(\frac{1}{d} + \frac{1}{\sqrt{d^2 + a^2}}\right) = 2.30 \times 10^{-28}\ \text{J m} \cdot \left[2.50 \times 10^9\ + 2.14 \times 10^9\right]\ \text{m}^{-1}\ = 1.07 \times 10^{-18}\ \text{J} = 6.66\ \text{eV}$$

If we account for the local environment by a relative dielectric constant of $\hat{\varepsilon}_r = 2$, then we find that

$$E_N = \frac{e^2}{4\pi\hat{\varepsilon}_r\varepsilon_0}\left(\frac{1}{d} + \frac{1}{\sqrt{d^2 + a^2}}\right) = 1.15 \times 10^{-28}\ \text{J m} \cdot \left[2.50 \times 10^9\ + 2.14 \times 10^9\right]\ \text{m}^{-1} = 0.35 \times 10^{-18}\ \text{J} = 3.34\ \text{eV}$$

We find the energy at the C atom 12 in the field of charges 1 (distance $\sqrt{d^2 + (2a)^2}$) and 2 (distance $\sqrt{d^2 + a^2}$) as

$$E_{C12} = \frac{e^2}{4\pi\hat{\varepsilon}_r\varepsilon_0}\left(\frac{1}{\sqrt{d^2 + (2a)^2}} + \frac{1}{\sqrt{d^2 + a^2}}\right) = 1.15 \times 10^{-28}\ \text{J m} \cdot \left[1.60 \times 10^9\ + 2.14 \times 10^9\right]\ \text{m}^{-1}$$
$$= 0.430 \times 10^{-18}\ \text{J} = 2.68\ \text{eV}$$

Correspondingly, we can find the values at each of the sites on the chromophore. The table shows the values for E at N, C_{14}, C_{12}, C_{10}, C_8, and C_6.

Site	16	14	12	10	8	6
$E/10^{-19}$ J	5.35	5.35	4.29	3.25	2.51	2.01
E/eV	3.34	3.34	2.68	2.03	1.57	1.26

The field acting between C_{14} and C_8 with $a = 0.24$ nm is then

$$F = \frac{E_N - E_{C8}}{3 \cdot a \cdot e} = \frac{(5.35 - 2.51) \times 10^{-19}\ \text{J}}{3 \cdot 0.24\ \text{nm} \cdot 1.602 \times 10^{-19}\ \text{C}} = 2.46 \times 10^9 \frac{\text{J}}{\text{C m}} = 2.46\ \text{V nm}^{-1} = 2.46 \times 10^9\ \text{V m}^{-1}$$

E28.4 The circular dichroism of chlorosomes changes sign within the absorption band at 740 nm (panel c of Fig. E28.4), similar to the example in Section 9.3. However, the linear dichroism of stretched samples is constant within the absorption band[1] —in contrast to the situation in Section 9.3 where the coupled oscillations are polarized perpendicular to each other. Try to rationalize this surprising effect and the values

$$\left(\frac{\Delta\varepsilon}{\varepsilon}\right)_{750\ \text{nm}} = -10^{-2}, \quad \left(\frac{\Delta\varepsilon}{\varepsilon}\right)_{720\ \text{nm}} = +10^{-2}$$

taken[2] from the figure. Assume that the exciton has the same size as the exciton in the brickstone work aggregate in Section 28.3 (distributed over about 10 molecules at 300 K). For simplicity consider the exciton as confined to molecules 1 and 2 (panel a of Fig. E28.4; that is, replace the excited portion by two coupled oscillators (panel b).

1 F. van Mouvik, K. Griebenow, B. van Haeringen, A.R. Holzwarth, R. van Grondelle, in: *Current Research in Photosynthesis*, ed. M. Baltscheffsky, Vol. **11**, 141 (Kluver Academic Publishers, 1990.

2 H. Tamiaki, M. Amakawa, Y. Shimono, R. Tanikaga, A.R. Holzwarth, K. Schaffner, *Photochem. Photobiol.* 1996, **63**, 92.

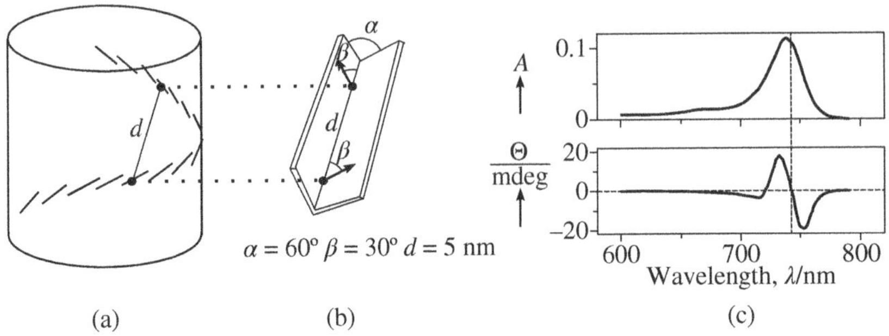

Figure E28.4 Model for the circular dichroism of chlorosomes.

Solution

Using the values given in panel b of Fig. E28.4 and relation (9.68, case a, λ_1), we find

$$g = \frac{-2\pi d}{\lambda_{\min}} = \frac{-2\pi \cdot 5\ \text{nm}}{720\ \text{nm}} = -4 \times 10^{-2}$$

for the right-handed helix (panel a); correspondingly, from relation ((9.68), case b, λ_1) we find

$$g = \frac{+2\pi d}{\lambda_{\min}} = \frac{+2\pi \cdot 5\ \text{nm}}{720\ \text{nm}} = +4 \times 10^{-2}$$

$g = +4 \times 10^{-2}$ for the corresponding left-handed helix. These g -values are in agreement with the corresponding measured ratios $\Delta\varepsilon/\varepsilon$. The result indicates that cylinders of left- and right-handed helices are present in a 1:1 ratio.

However, what is the reason for the splitting of the circular dichroism band? Such a splitting should not occur for mirror images. Indeed, the two forms are not exactly mirror images: The chlorophyll molecule is chiral, and for that reason the excitation energy for the left-handed helix should be somewhat different from that for the right-handed helix.

E28.5 The data in the table[3] show the temperature dependence for the exciton of a dye molecule (pseudocyanine bromide) in an ethylene glycol/water glass as a function of temperature, for which the lifetime is $\tau_0 = 3.7 \times 10^{-9}$ s. Analyze this temperature dependence using Equation (28.14) and determine the binding energy of the exciton.

$k_r/10^9\text{s}^{-1}$	0.90	1.5	2.2	2.9	4.1	3.8	5.4
T/K	222	200	174	143	133	114	114

$k_r/10^9\text{s}^{-1}$	6.4	9.0	11.6	14.8	14.7	14.7	15.0
T/K	105	80	62	45	32	19	10

Solution

We transform Equation (28.14) into a form that allows us to plot the observed k_r versus $1/T$; we proceed by

$$\tau_{\text{agg}} = \frac{\tau_0}{N}$$

where τ_{agg} is the fluorescence lifetime of the aggregate, τ_0 is the fluorescence lifetime of a monomer, and N is the number of monomers in the aggregate.

We assume that the binding energy between the N coherent dipole oscillators is

$$-\Delta E = NkT \qquad \text{or} \qquad N = \frac{-\Delta E}{kT}$$

Thus

$$\tau_{\text{agg}} = \frac{\tau_0}{N} = -\frac{kT}{\Delta E}\tau_0$$

3 These data are read from the graph in Figure E28.5 of S. DeBoer, D.A. Wiersma, *Chem. Phys. Lett.* 1990, **165**, 45.

From the relation $k_r = 1/\tau_{agg}$, we find that

$$k_r = -\frac{\Delta E}{k\tau_0} \cdot \frac{1}{T}$$

Figure E28.5 shows a plot of k_r versus $1/T$. The fit to the linear region has a slope of 9.45×10^{11} K s^{-1}. Hence we find that

$$\begin{aligned} -\Delta E &= (slope) \cdot k\tau_0 \\ &= 9.45 \times 10^{11}\ \text{K s}^{-1} \cdot 1.3806 \times 10^{-23}\ \text{J K}^{-1} \cdot 3.7 \times 10^{-9}\ \text{s} \\ &= 1.3 \times 10^{-19}\ \text{J} = 0.81\ \text{eV} \end{aligned}$$

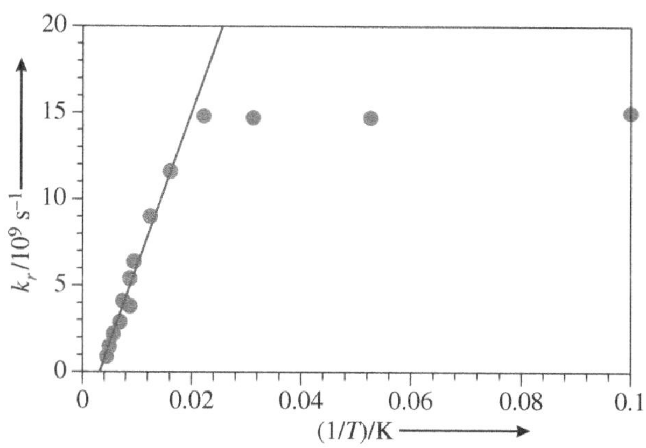

Figure E28.5 Exciton; k_r for a dye molecule (pseudocyanine bromide) in an ethylene glycol/water glass as a function of $1/T$.

The emission rate of the exciton increases as the temperature decreases and reaches a maximum value at low temperatures. The rate stops changing at low enough temperatures because all of the oscillators in the aggregate are coherent at these temperatures. Then we can calculate the number N of monomers by

$$N = \frac{\tau_0}{\tau_{agg}} = \tau_0 \cdot k_r$$

If we take the low temperature limit of the radiative rate to be 15.0×10^9 s^{-1} we estimate that

$$N = 15.0 \times 10^9\ \text{s}^{-1} \cdot 3.7 \times 10^{-9}\ \text{s} = 55$$

Alternatively, we can calculate N by

$$N = \frac{-\Delta E}{k} \frac{1}{T}$$

Using $1/T = 0.02$ K^{-1} taken from Fig. E28.5 we obtain

$$N = \frac{1.3 \times 10^{-19}\ \text{J}}{1.38 \times 10^{-23}\ \text{J K}^{-1}} 0.02\ \text{K}^{-1} = 43$$

E28.6 In Section 28.3.2 we calculate the speed of an exciton as $\upsilon = 2.4$ km s^{-1} and its length as $l = 5$ nm. How long does it take for the exciton to move a distance of 10 times its own length? Compare this time to the period of a 1000 cm^{-1} vibration; a 100 cm^{-1} vibration; and a 10 cm^{-1} intermolecular motion.

Solution

Given a 5 nm length for the exciton we are asked to find the time t it takes for it to travel a distance of 50 nm. Assuming a constant speed we find

$$t = \frac{10 \cdot l}{\upsilon} = \frac{50 \times 10^{-9}\ \text{m}}{2.4 \times 10^3\ \text{m s}^{-1}} = 2.1 \times 10^{-11}\ \text{s} = 21\ \text{ps}$$

Taking the vibration period $T_{vib} = 1/\nu$ where ν is the frequency of the vibration, we find that

$$T_{vib} = \frac{1}{\nu} = \frac{1}{c_0\,\tilde{\nu}}$$

and $\tilde{\nu}$ is the wavenumber of the vibration. For the 1000 cm^{-1} vibration we find that

$$T_{vib} = \frac{1}{3\times 10^{10}\ \mathrm{cm\,s^{-1}} \cdot 1000\ \mathrm{cm^{-1}}} = 0.33\times 10^{-13}\ \mathrm{s^{-1}}$$

Correspondingly, we find $3.3\times 10^{-13}\mathrm{s}^{-1}$ for a 100 cm^{-1} vibration and $3.3\times 10^{-12}\mathrm{s}^{-1}$ for a 10 cm^{-1} vibration.

Each of the vibrational periods is significantly shorter than the characteristic time for the exciton to move 10 times its own length; however, we note that the characteristic time for the exciton to move its own length would be 2.1 ps which is comparable to the period of a 10 cm^{-1} intermolecular vibration.

E28.7 Consider a monolayer thick tunneling barrier that contains a certain fraction f of defects. Consider the electrical current through the defects to have the same magnitude as the tunneling current through a 0.5 nm thick tunneling barrier. Consider the true reciprocal decay length of the current to be $b = 14\ \mathrm{nm^{-1}}$. Calculate the current as a function of film thickness (from 0 to 10 nm) for the cases of no defects, 0.1% defects, and 1% defects.

Solution

We use the general tunneling formula that

$$i(d) = i_0 \cdot \exp(-bd)$$

where i_0 is the current at zero distance. First we find the tunneling current through the defect, i_{defect}, as

$$i_{defect} = i_0 \cdot \exp\left[-14\ \mathrm{nm^{-1}} \cdot 0.5\ \mathrm{nm}\right] = i_0 \cdot 9.1\times 10^{-4}$$

For a fraction f of defects, we can write the total (composite) current i_{total} as

$$\frac{i}{i_0} = f \cdot \frac{i_{defect}}{i_0} + (1-f)\cdot \exp(-bd)$$

or

$$\ln\frac{i}{i_0} = \ln\left[f \cdot \frac{i_{defect}}{i_0} + (1-f)\cdot \exp(-bd)\right]$$
$$= \ln\left[f \cdot 9.1\times 10^{-4} + (1-f)\cdot \exp(-bd)\right]$$

For $d \to \infty$ we obtain $\ln(i/i_0) = \ln\left[f \cdot 9.1\times 10^{-4}\right]$, for $d = 0$ we obtain $\ln(i/i_0) = 0$.

Figure E28.7 plots this function versus d for each of the three defect fractions given in the problem statement. For the case of zero defects the current falls exponentially over the whole range. For the case of defects present we see that the current decays exponentially to a baseline value that corresponds to the total defect current and is then independent of distance. For the conditions chosen this occurs between 1 and 1.5 nm.

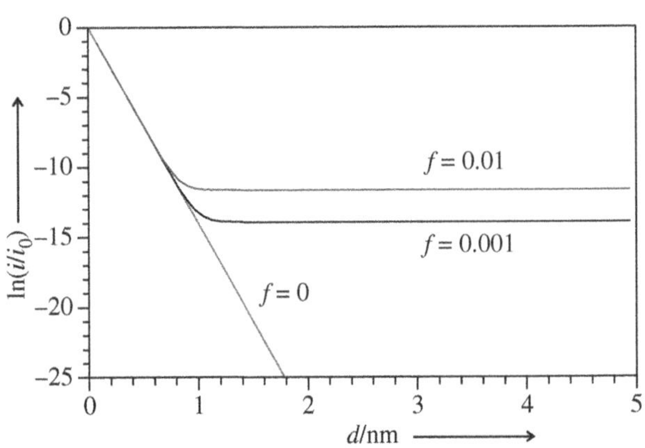

Figure E28.7 $\ln(i/i_0)$ versus the distance d. Dark gray solid line: $f = 0$, black solid line: $f = 0.001$, light gray solid line: $f = 0.01$.

E28.8 An important technical issue in electron tunneling experiments is the possibility of tunneling through defect sites. Consider a tunneling experiment for a SAM (decanethiol, C10) coated Au electrode (1 mm^2 area) in contact with the $Fe(CN)_6^{3-}/Fe(CN)_6^{4-}$ redox couple and an observed tunneling current of 100 nA at an applied bias potential of 0.25 V versus Ag/AgCl. Assume that some number of defects exist in the SAM film and that a defect has a current that is 10 times higher than that which occurs by tunneling through a chain. What percentage of defects can exist in the film before they contribute 1% to the current? Use this result to calculate the current per defect site. If you assume that the defects have the same tunneling efficiency independent of the carbon chain length of the SAM, what fraction of defects are needed to contribute 1% to the current through an octadecanethiol (C18) film?

Solution

Part 1: The total measured current i_{meas} is the sum of the defect current i_{defect} and the tunneling current i_{tunnel}, so that

$$i_{meas} = (1-f) \cdot i_{tunnel} + f \cdot i_{defect}$$

where f is the fraction of the electrode area that are defect sites. To find an upper bound on the number of defects, we wish to consider the case where $f \cdot i_{defect} = 0.01 \cdot i_{meas}$, and we proceed by eliminating i_{meas} from these equations so that

$$100 \cdot f \cdot i_{defect} = (1-f) \cdot i_{tunnel} + f \cdot i_{defect}$$

It is given in the problem statement that the current through a defect site is 10 times that through tunneling by a C10 chain; i.e., $i_{defect} = 10 \cdot i_{tunnel}$. For a C10 film, we can write that

$$100 \cdot f \cdot i_{defect} = 0.1\,(1-f) \cdot i_{defect} + f \cdot i_{defect}$$

and solving for f we find that

$$99.1 \cdot f = 0.1 \quad \text{or} \quad f = 0.0010$$

Hence 0.1% of the sites must be defects before they would contribute 1% of the current.

From the problem statement, it is evident that if the defects contribute 1% to the current that their total current contribution will be 1 nA, and from the assumption that the defect current is 10 times the C10 tunneling current we find that 0.1% of the sites on the surface contribute to this 1 nA. If we take the cross-sectional area of a hydrocarbon chain to be 0.2 nm ^{2} then the number of sites in a 1 mm^2 surface is given by

$$N_{sites} = \frac{10^{-6}\ m^2}{0.2 \times 10^{-18}\ m^2} = 5 \times 10^{12}$$

Hence we calculate 5×10^9 defect sites and 2×10^{-18} A per defect site. Using the charge on an electron this current corresponds to 12–13 electrons per second through a defect site. As a word of caution we note that this analysis assumes that the area of a defect site is the same as the area of a hydrocarbon chain, and this is likely to not be true. For example a defect site might correspond to a chain that is lying down on the surface, so that it occupies significantly more surface area. It is likely that a distribution of different defect structures exist on the surface.

Part 2: Here we assume a decay constant b of 1 per methylene so that we drop from 100 nA to

$$\exp(-b \cdot 8) \cdot 100\ nA = 0.034\ nA$$

which is much below the 1 nA current that arose from 0.1% defects through a C10 film that was calculated in part 1. To calculate the percentage of defects in the C18 film that would give rise to 1% of 0.034 nA we find that the total defect current will be

$$0.34\ pA = 3.4 \times 10^{-13}\ A$$

Using our value of 2×10^{-18} A per defect site from part 1, we find that the number of defect sites must be

$$N = \frac{3.4 \times 10^{-13}\ A}{2 \times 10^{-18}\ A} = 1.7 \times 10^5$$

or fewer. Clearly the constraints on the longer carbon chain experiments are much more severe than those on the C10 experiment.

E28.9 Estimate the height of a tunneling barrier formed by a dielectric medium, by determining how far it lies below the vacuum level. In particular, find the polarization energy for transferring an electron from vacuum into a dielectric medium of relative permittivity $\varepsilon_r = 2.5$. Estimate this polarization energy by modeling the electron as a charged sphere of radius a) $r = 0.2$ nm and b) $r = 0.5$ nm.

Solution

We calculate the energy w to transfer the sphere with charge e from vacuum into the dielectric medium with relative permittivity $\hat{\varepsilon}_r$. The energy dw to transfer the charge dQ is

$$\delta w = -\frac{1}{4\pi\varepsilon_0}\frac{e-Q}{r}\cdot dQ + \frac{1}{4\pi\varepsilon_0}\frac{1}{\varepsilon_r}\frac{Q}{r}\cdot dQ$$

Thus by integration we obtain

$$w = \int_0^e \delta w = -\frac{1}{4\pi\varepsilon_0 r}\left(-e^2 + \frac{1}{2}e^2 + \frac{1}{\varepsilon_r}\cdot\frac{1}{2}e^2\right) = \frac{-1}{4\pi\varepsilon_0}\frac{e^2}{2r}\left(1 - \frac{1}{\hat{\varepsilon}_r}\right)$$

For the case of $r = 0.2$ nm, we find that

$$w = \frac{-1}{4\pi\varepsilon_0}\frac{e^2}{2r}\left(1 - \frac{1}{\hat{\varepsilon}_r}\right) = -3.6\ \text{eV}\left(1 - \frac{1}{\hat{\varepsilon}_r}\right)$$

In the case $\hat{\varepsilon}_r = 2.5$ (hydrocarbon portion, $w = -2.2$ eV) the energy barrier is about 2.2 eV below vacuum.

For the case of $r = 0.5$ nm, we find that

$$w = \frac{-1}{4\pi\varepsilon_0}\frac{e^2}{2r}\left(1 - \frac{1}{\hat{\varepsilon}_r}\right) = -1.4\ \text{eV}\left(1 - \frac{1}{\hat{\varepsilon}_r}\right)$$

In the case $\hat{\varepsilon}_r = 2.5$ (hydrocarbon portion, $w = -0.84$ eV), the energy barrier is about 0.84 eV below vacuum.

E28.10 In Exercise E28.9, you estimate the polarization energy for an electron that is localized to a certain size in a dielectric medium. By making the charged sphere smaller you can increase the polarization stabilization; however, this decrease in size increases the electron's kinetic energy. Use the de Broglie relation to estimate the kinetic energy of the electron for the two cases in Exercise E28.9. Is it possible to find an optimal size for the electron?

Solution

Using the de Broglie relation, we estimate the kinetic energy E_{kin} as

$$E_{\text{kin}} = \frac{h^2}{2m_e\Lambda^2}$$

If we replace the sphere by a cube of side length $2r$, then the kinetic energy for the ground state will be

$$E_{\text{kin}} = \frac{h^2}{2m_e\Lambda^2}\cdot 3 \quad \text{with} \quad \Lambda = 4r$$

so that

$$E_{\text{kin}} = \frac{h^2}{2m_e 16r^2}\cdot 3 = \frac{h^2}{m_e r^2}\frac{3}{32}$$

For $r = 0.2$ nm, we find $E_{\text{kin}} = 7.1$ eV; and for $r = 0.5$ nm, we find $E_{\text{kin}} = 1.1$ eV.

If we write the total energy as $E_{\text{kin}} + w$, we find that

$$E_{\text{total}} = \frac{3h^2}{32m_e r^2} - \frac{1}{4\pi\varepsilon_0}\frac{e^2}{2r}\left(1 - \frac{1}{\hat{\varepsilon}_r}\right)$$

where we have used the result of E28.9 for w. This function clearly has a minimum at a given value of r, and we can find that value by taking the derivative of E_{total} and setting it equal to zero, so that

$$\left.\frac{dE_{\text{total}}}{dr}\right|_{r=r_{\text{opt}}} = 0 = \left.\left(-\frac{3h^2}{16m_e r^3} + \frac{1}{4\pi\varepsilon_0}\frac{e^2}{2r^2}\left(1 - \frac{1}{\hat{\varepsilon}_r}\right)\right)\right|_{r=r_{\text{opt}}}$$

Solving for r_{opt}, we find that

$$\frac{1}{r_{\text{opt}}} = \frac{2}{3\pi^2}\frac{\pi m_e e^2}{\varepsilon_0 h^2}\left(1 - \frac{1}{\hat{\varepsilon}_r}\right) = \frac{2}{3\pi^2}\frac{1}{a_0}\left(1 - \frac{1}{\hat{\varepsilon}_r}\right)$$

which becomes

$$r_{opt} = \frac{3\pi^2 a_0}{2\left(1 - 1/\hat{\varepsilon}_r\right)}$$

where we have used the definition of the Bohr radius. We see that as the dielectric constant becomes small (approaches 1) the optimal radius becomes very large, and as the dielectric constant becomes large the optimal radius approaches $3\pi^2 a_0/2$.

E28.11 Consider the case of a monolayer film in which the hydrocarbon chains of length l are perpendicular to the layer plane ($d = l$) and $2r = 0.475$ nm is the distance between the axes of adjacent chains (Fig. E28.11a). What is expected to happen to the current through the film when the hydrocarbon chains are inclined and the distance becomes $d' = l\cos\varphi$, where φ is the tilt angle (panel b)?

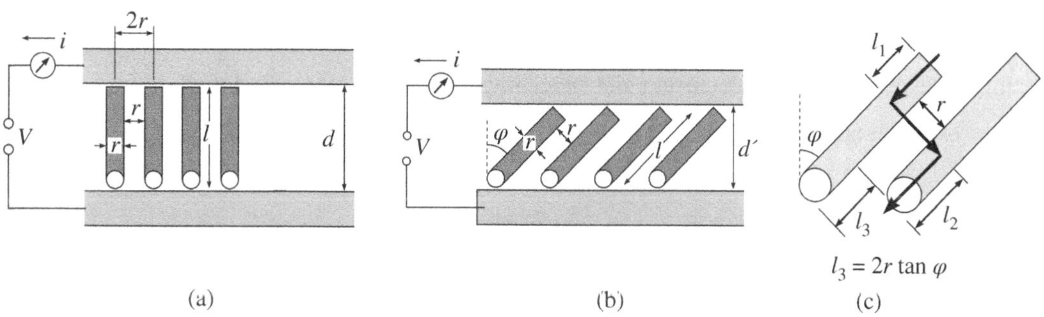

Figure E28.11 Monolayer film with hydrocarbon chains perpendicular (a) and inclined (b) to the layer plane. A close up view of the geometry for two chains is shown in (c).

Solution

In Section 28.4.2 we found that for the situation $\varphi = 0$ the tunneling current i along the C–C bonds is

$$i = \frac{1}{\kappa}\cdot\frac{A}{d}\cdot V = \frac{1}{\kappa}\cdot\frac{A}{l}\cdot V \qquad \text{with} \qquad \kappa = \alpha \cdot e^{-bd}$$

where κ is the conductivity. A is the area between the metal electrodes, and d is their distance (Fig. E28.11a), according to the length l of the carbon chains.

If the tilt angle φ is different from zero and tunneling along the highly polarizable region of the C–C bonds dominates (through-bond tunneling, Fig. E28.11b), the current per chain is unchanged despite the smaller distance d' between the electrodes. Consider that the field strength F in Fig. E28.11b is

$$F = \frac{V}{d'} \quad \text{and} \quad d' = d\cdot\cos\varphi$$

Thus

$$F = \frac{V}{d\cdot\cos\varphi}$$

On the other hand, the component F_{chain} of the field strength in direction of the chain is

$$F_{chain} = F\cdot\cos\varphi = \frac{V}{d\cdot\cos\varphi}\cdot\cos\varphi = \frac{V}{d} = F$$

Thus the current $i_{through\ bond}$, which is proportional to the field strength, does not depend on the angle φ, and we find

$$i_{through\ bond} = V\frac{A}{d}\cdot\alpha\cdot\exp\left[-bl\right]$$

If the electron can tunnel between chains, however, then the tunneling distance along the chain is decreased by $l_3 = 2r\tan\varphi$ (see Fig. E28.11c) and the interchain tunneling might dominate at large tilt angles. If one hop between chains, of distance r, is allowed, then the tunneling distance along the chain is reduced to

$$l_1 + l_2 = l - l_3 = l - 2r\tan\varphi$$

On the other hand, tunneling through nonbonded contacts (through-space tunneling) is less probable because of the higher effective tunneling barrier. In this case the estimated tunneling distance is r and its characteristic decay length is b_{vacuum}, so that the tunneling current with a through space hop is given by

$$i_{through\ space} = V\frac{A}{d}\alpha \cdot \exp\left[-b(l-2r\tan\varphi)\right] \cdot \exp\left[-b_{vacuum} \cdot r\right]$$

where we have assumed that the other factors (in the preexponential term) are the same as in the expression for $i_{through\ bond}$. Assuming a rectangular barrier and a tunneling barrier height of about 4 eV, we obtain $b_{vacuum} = 22\ \text{nm}^{-1}$. Assuming that the tunneling pathways contribute independently, the sum of both contributions to the current is

$$i = i_{through\ bond} + i_{through\ space}$$

The second term increases with increasing tilt angle.

At what angle does the tunneling current for the pathways become equal, that is

$$i_{through\ bond} = i_{through\ space}$$

Inserting our expressions for each current and cancelling identical terms we find that

$$1 = \exp\left[b(2r\tan\varphi)\right] \cdot \exp\left[-b_{vacuum} \cdot r\right]$$

or

$$(2b\tan\varphi)\, r = b_{vacuum} \cdot r \quad \text{so that} \quad \tan\varphi = \frac{b_{vacuum}}{2b}$$

By way of example consider the data in Fig. 28.23 for the length-dependent tunneling through methylene chains, for which we found $b = 10.5\ \text{nm}^{-1}$. Using this value with $b_{vacuum} = 22\ \text{nm}^{-1}$, we find that the tunneling currents are equal at an angle of $\varphi = 46°$.

The effect of through space tunneling has been observed for an arrangement in which one metal electrode is replaced by an electrolyte with a redox couple and the fatty acid monolayer is replaced by an alkanethiol monolayer. The metal electrode is a mercury drop. It is covered by the monolayer. When extending the drop a continuous increase of the tilt angle φ is achieved and a corresponding increase in current is measured.[4] The through bond tunneling makes the largest contribution to the current for angles less than 15°. Yamamoto[5] extends this model to the case of more than one tunneling hop between alkane chains.

E28.12 Tivanski et al.[6] measured the conductivity through conjugated molecules with two different types of linkers. Use their results to estimate the difference in the resistance of the two types of thiol contacts.

Solution

Tivanski studied the conductivity through the two molecules shown in Fig. E28.12.

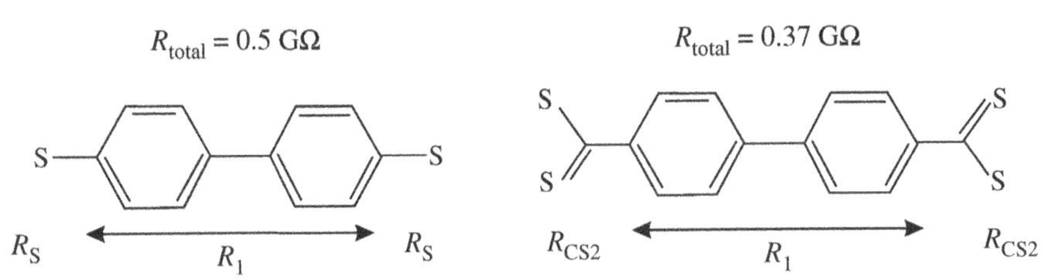

Figure E28.12 Two types of thiol contacts.

If we assume that the components of the molecule, biphenyl unit and the contact functionalities on each end, constitute a serial resistor path, then for the dithiol compound we can write the resistance as

$$0.5\ \text{G}\Omega = R_1 + 2R_S$$

4 K. Slowinski, R.V. Chamberlain, C.J. Miller, M. Majda, *J. Am. Chem. Soc.* 1997, **119**, 11910.
5 H. Yamamoto, D.H. Waldeck, *J. Phys. Chem. B* 2002, **106**, 7469.
6 A.V. Tivanski, Y. He, E. Borguet, H. Liu, G.C. Walker, D.H. Waldeck, *J. Phys. Chem. B* 2005, **109**, 5398.

and for the second molecule we can write the resistance as

$$0.37\ \text{G}\Omega = R_1 + 2R_{CS_2}$$

Taking the difference, we find that the difference in their resistances is

$$0.065\ \text{G}\Omega = R_S - R_{CS_2}$$

Even though the CS_2 unit is significantly larger than the thiol, it has a smaller contact resistance, presumably because of its electronic character (π-system).

E28.13 The pseudorotaxane system shown in Fig. E28.13a forms a bound complex of the TTF-polyether dumbbell with the cationic cyclophane when the TTF is neutral. Upon oxidation of the TTF unit, the system dissociates with a rate constant of $0.2\,\text{s}^{-1}$ at 298 K in CH_3CN ($\eta = 0.37$ mPa s). Use the Smoluchowski limit of the Kramers model (see Foundation 25.2) and this rate constant to estimate the barrier to dissociation.

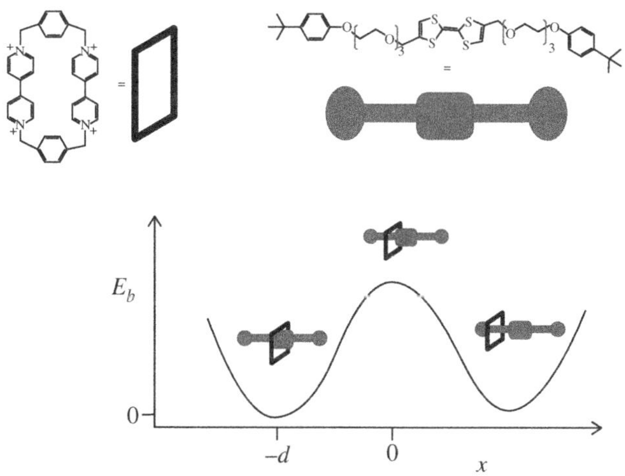

Figure E28.13a Pseudorotaxane system.

Solution

We begin by using the high friction (Smoluchowski) limit of the Kramers equation which is reported in Chapter 25, Equations (25.61) and (25.63); namely

$$k_{KR} = \frac{\sqrt{k_0 k_b}}{2\pi\zeta} \exp\left(-\frac{E_b}{kT}\right)$$

where k_0 and k_b are the curvatures of the potential curve, and E_b is the barrier height. We use Stoke's law $\zeta = 6\pi\eta r$ and find that

$$k_{KR} = \frac{\sqrt{k_0 k_b}}{12\pi^2 \eta r} \exp\left(-\frac{E_b}{kT}\right)$$

Because only one rate constant value at one temperature is available, we will need to make some approximations and estimate some molecular parameters.

In order to simplify, we make the approximation that the well force constant k_0 and the barrier force constant k_b are equal; i.e., $k_0 = k_b$. In this case the potential surface shown in Fig. 25F.1 gives the relationship

$$k_0 = k_b = \frac{4E_b}{d^2}$$

To see this, consider that the potential energy near the reactant minimum is parabolic, $\frac{1}{2}k_0 x^2$, and that near the barrier top (at $x = d$) is given by $E_b - \frac{1}{2}k_b(x-d)^2 = E_b - \frac{1}{2}k_0(x-d)^2$, with the latter equality holding because of our assumption that $k_0 = k_b$. The potential energy surface should be single-valued, so we require that these two energy expressions match at the halfway point of $x = d/2$; that is

$$\frac{1}{2}k_0 x^2\bigg|_{x=d/2} = E_b - \frac{1}{2}k_0(x-d)^2\bigg|_{x=d/2}$$

so that

$$\frac{1}{8}k_0 d^2 = E_b - \frac{1}{8}k_0 d^2 \quad \text{or} \quad k_0 = \frac{4E_b}{d^2}$$

Then our rate constant expression simplifies to

$$k_{KR} = \frac{E_b}{3\pi^2 d^2 \eta r} \exp\left(-\frac{E_b}{kT}\right)$$

Here we estimate d to be 40 pm and we model the radius r of the moving unit to be 100 pm. Figure E28.13b plots this function and shows where it crosses the measured rate constant value, at 0.2 s^{-1}. The crossing occurs at $E_b = 1.38 \times 10^{-19}$ J, a value that corresponds to 83 kJ mol^{-1}.

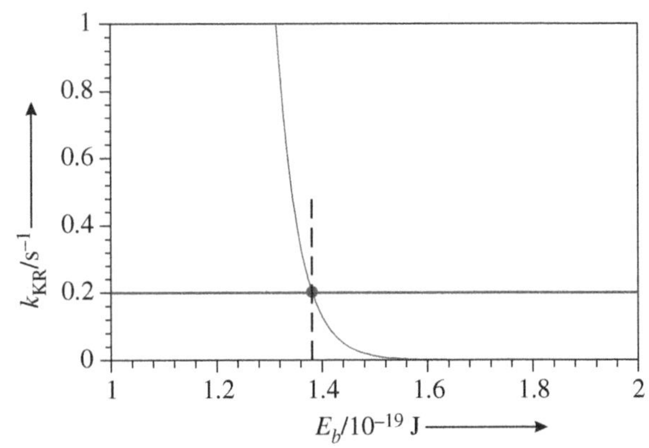

Figure E28.13b The rate constant k_{KR} is plotted versus the barrier height E_b.

E28.14 For the catenane ring rotation in Fig. 28.39 the switching time for (b) to (c) is 14 hours at 293 K in CH_3CN and from (d) to (a) is a few seconds. Rationalize why these times are so different. If the catenane changes its orientation by diffusion, with a rotational diffusion coefficient given approximately by $D_{rotn} = kT/6V\eta$ (where V is the hydrodynamic volume and η is the viscosity of the solvent), estimate the time for it to switch. This expression for the diffusion coefficient is found by using the Einstein relation $D_{rotn} = kT/\zeta_{rotn}$ and using a hydrodynamic (Stokes model) for the rotational friction coefficient.[7] Compare your result to the observed rates in CH_3CN at 293 K ($\eta = 0.38$ mPa s). What is likely to limit the rate?

Solution

To calculate the rotational diffusion coefficient we first estimate the hydrodynamic volume of the catenane by adding up the van der Waals volume increments for the different groups (see A. Bondi, J. Phys. Chem. 1964, **68**, 441) so that $V_m \simeq 300\ cm^3\ mol^{-1}$ or $V \simeq 500\ Å^3 = 5.0 \times 10^8\ pm^3$. Hence we find that

$$D_{rotn} = \frac{kT}{6\eta V} = \frac{1.3806 \times 10^{-23}\ J\ K^{-1} \cdot 293\ K}{6 \cdot 5.0 \times 10^{-28}\ m^3 \cdot 0.38 \times 10^{-3}\ N\ s\ m^{-2}} = 3.5 \times 10^9\ s^{-1}$$

which suggests a relaxation time of 100 ps—much faster than either of the measured switching rates. Hence we expect that each of the different forms is stable and a significant barrier exists to their interchange. One could proceed along the lines of analysis in E28.13 and use Kramers model to estimate a barrier.

One could also model this system using the relation

$$\overline{x^2} = 2D_{catenane} \cdot \tau$$

for translational diffusion, by assuming that the end groups in the catenane ring change position by a diffusion path. If we estimate the mean square diffusion length as the square of the distance between 18 C–C bonds, we find

$$\overline{x^2} = (18 \cdot 140\ pm)^2 = 6.4 \times 10^6 \times 10^{-24}\ m^2$$

7 P. Debye, *Polar Molecules* (Dover, NY, 1929).

Using a typical value for the translational diffusion constant, $D_{catenane} = 10^{-5}\ cm^2\ s^{-1}$, we find

$$\tau = \frac{\overline{x^2}}{2D_{catenane}} = \frac{6.4 \times 10^{-18}\ m^2}{2 \times 10^{-9}\ m^2\ s^{-1}} = 3.2 \times 10^{-9}\ s$$

This result should be a lower limit for τ, because both end groups must diffuse into the same direction and is of the same order as that found using the rotational diffusion approximation.

E28.15 Consider a kinesin protein that is transporting a 200 nm diameter vesicle through a cell at 1 μm s^{-1}. Estimate a viscosity for the medium and compare the frictional force to that provided by the kinesin, ca. 5 pN. At what viscosity would you expect these values to be equal?

Solution

We estimate a lower bound for the viscosity of a cell to be that of water, $\eta = 1$ cP. We take this as a lower bound because the cell consists of organized molecular assemblies that will provide structural elements that act to increase the viscosity above that of neat water. If we use Stokes law to estimate the friction coefficient ζ, namely $\zeta = 6\pi\eta R$ then we find that $\zeta = 1.9 \times 10^{-6}\ g\ s^{-1}$. Note that if Stokes law holds then the friction coefficient will increase in direct proportion to the viscosity. For a mean speed of 1 μm s^{-1} we calculate a force by $force = -\zeta v$, so that the force is equal to 1.9×10^{-15} N; this value is 2000–3000 times smaller than the 5 pN value that is given for kinesin. Note that a larger choice for the shear viscosity would result in a proportionally larger force.

Here we consider the limit that the kinesin provides a steady 5 pN force and the vesicle moves at a steady speed of 1 μm s^{-1}. In this case we can estimate the effective viscosity as

$$\eta = \frac{force}{6\pi R v}$$

$$= \frac{5 \times 10^{-12}\ N}{6\pi \times 10^{-7}\ m \times 10^{-6}\ m\ s^{-1}} = 2.6\ kg\ m^{-1}\ s^{-1} = 26\ poise = 2600\ cP$$

This viscosity is about three times larger than that of glycerol at 298 K.

28.2 Problems

P28.1 S.D. Straight et al. (*J. Am. Chem. Soc.* 2005, **127**, 9403) synthesized and studied a molecule that performs the AND logic operation. Describe how their system works. Propose an alternative molecular logic element that uses chemical binding events for inputs.

Solution

They synthesized a supermolecule, consisting of three different chromophores; their system covalently attaches a porphyrin fluorophore (P) to two chromophores (Q1 and Q2) that undergo isomerization, in which one of the isomer forms quenches the porphyrin and the other does not. When Q1 and Q2 are both in their on state (isomer forms that do not quench the porphyrin), then the porphyrin P can be excited and it fluoresces. If either, or both, Q1 or Q2 are in their off state (the isomer form that does quench the porphyrin emission), then the fluorescence of P is turned off. In this study the chromophores Q1 and Q2 are modulated by light and/or heat.

Many answers are possible here. We can use a similar design but have one of the quenchers modulated by a chemical binding event. For example, 4-cyanobenzoic acid has a similar reduction potential to that of 1,4-dicyanobenzene (discussed in Chapter 22). Hence one could attach a receptor functionality for the cyanobenzoic acid on the porphyrin-based supermolecule, so that when the cyanobenzoic acid is bound the photoexcited porphyrin will be quenched by electron transfer and when the cyanobenzoic acid is not bound the porphyrin could radiate (if the other unit is in the on state).

P28.2 The review article V. Balzani, A. Credi, F.M. Raymo, J. F. Stoddart, Artificial molecular machines, *Angew. Chem.* 2000, **39,** 3348 contains descriptions of a number of different supramolecular devices. Investigate one of these in detail, explain its operation principles, and delineate its specifications; i.e., performance characteristics. Lastly, propose one way to improve the machine.

Solution

A number of answers are possible for this problem.

P28.3 In Section 28.3, we discussed the effect of a mirror on an Eu luminophore. Consider the situation in which the Eu is placed at the center of two opposing perfect mirrors, i.e., in an optical cavity, and describe how the radiative rate of the Eu* is expected to change as the distance between the mirrors changes. Consider the mirror distances to vary over a range of 10 nm to 10 μm and only consider cavity modes along the axis which is normal to the mirror surfaces.

Solution

This situation is somewhat different than that discussed in the text because the two mirrors create an optical cavity (or etalon) in which only particular electric field modes (standing waves) are supported. These modes will have a wavelength λ that is given by the condition

$$\lambda = \frac{2d}{q}$$

where $q = 1, 2, \ldots$ is the mode number and we take the refractive index to be unity (assume in vacuum). From this condition we see that the longest wavelength mode that can be supported in the cavity is $\lambda = 2d$. Given that the wavelength for the Eu* emission is 620 nm, we see that distances between the mirrors that are smaller than 310 nm will not be able to support electric fields near the resonance of the Eu*. This fact means that the excited state will not be able to relax by radiating; i.e., the fluorescence emission will be shut-off. Hence we should see a lengthening of the molecules excited state lifetime and a spatial anisotropy in its emission profile. As a caveat we note that we are only considering modes along the axis between the two mirrors; appropriate modes may exist in the other directions.

For cavity spacings that are larger than 310 nm, modes can be supported that affect the excited state emission of the Eu*. We would expect to see an oscillation of the excited state lifetime (shorter to longer) as the cavity modes move in and out of resonance with the Eu*; however, we expect that the amplitude of these modulations will decrease as d becomes larger and hence the density of modes near 620 nm increases.

Lastly, we note that at the very short distance of 10 nm the nonradiative energy transfer from the Eu* to the mirror can occur and lead to a shortening of the lifetime, in the same manner as discussed in the text.

P28.4 Catalytic Nanomotors: Y. Wang, R.M. Hernandez, D.J. Bartlett, Jr., J.M. Bingham, T.R. Kline, A. Sen, and T.E. Mallouk reported the autonomous movement of bimetallic nanorods (e.g., Pt: Au dyad) in 5 wt. % aqueous peroxide solutions, see *Langmuir* 2006, **22**, 10451. Read this paper and explain the mechanism by which these rods move preferentially toward their one end (e.g., Pt end of Pt:Au).

Solution

From the discussion in the introduction to this paper it is evident that a number of different mechanisms may contribute to the autonomous propulsion of bimetallic nanorods. Wang and co-authors provide a detailed study in which they report the direction of motion observed for a number of different metal combinations and compare it to that predicted by an electrokinetic mechanism. Their mechanism requires electron flow from the anode end to the cathode end of the bimetallic rod and a flow of protons and water in the same direction. The displacement of water from the anode to the cathode corresponds to motion in the direction of the anode. For each of the different metal combinations studied they find that the rod moves in the direction of the anode. Replacement of one of the metals by a polymer film containing an enzyme for oxygen evolution did not result in any directional motion.

P28.5 S. Bhosale et al., *Science* 2006 **313,** 84 reported an artificial light driven proton pump, which is based on a π-stacked assembly of chromophores that are inserted into the lipid bilayer of a vesicle. Read and analyze this paper. Provide a brief (2 paragraph or less) description of its operation.

Solution

This paper describes the preparation of a π-stacked multichromophore scaffold (composed of naphthalene diimides) that intercalates into the lipid bilayer of a vesicle. These workers placed a quinone as an electron acceptor inside the vesicle and EDTA as an electron donor outside of the vesicle. Upon photo-illumination the transmembrane chromophore drives the reduction of the quinone on the interior of the vesicle, which they detect by the increase in pH of the solution inside the vesicle (the quinone reaction is a proton-coupled electron transfer; see Section 22.5). The hole on the chromophore is compensated by the sacrificial oxidation of EDTA. This process continues and creates a proton gradient across the bilayer of the vesicle.

An interesting feature of their system is that the addition of a ligand can cause the transmembrane chromophore stacks to assemble into a different structure that opens a channel between the inside and outside of the vesicle. This transformation discharges the proton gradient.

P28.6 Artificial Bacterial Photosynthesis: G. Steinberg-Yfrach, J.-L. Rigaud, E.N. Durantini, A.L. Moore, D. Gust, T.A. Moore, Light-driven production of ATP catalyzed by F0F1-ATP synthase in an artificial photosynthetic membrane, *Nature* 1998, **392,** 478 report an artificial supramolecular assembly that mimics bacterial photosynthesis. Read and describe the findings of this paper. Be sure to report which aspects of the system are artificial and which are natural.

Solution

This work describes the creation of an artificial supramolecular assembly that converts light energy into ATP from ADP; hence, it mimics the photosynthetic process. These workers created a liposome that had a number of different components imbedded in its membrane. These included a synthetic supermolecule composed of three units that upon photoexcitation creates a charge separated state with a radical anion at the external side of the membrane and a radical cation at the internal side of the membrane. Also imbedded in the bilayer are numerous lipophilic 2,5-diphenylbenzoquinone (Qs) molecules, which can undergo a proton coupled electron reduction at the outer interface and then migrate through the membrane (or transfer via other Qs molecules in the membrane) to the internal interface where it undergoes a proton coupled oxidation, thereby injecting a proton to the interior of the liposome. This aspect of the device is artificial and acts to use the light energy to create a proton gradient between the inside and the outside of the liposome.

In addition to the above elements, the workers incorporate the protein ATP synthase into the membrane. This natural *trans*-membrane protein exploits the proton gradient to synthesize ATP from ADP in the solution that is external to the liposome. See section 28.5.5.3 for a discussion of the operation of ATP synthase.

P28.7 Switching Device. J. He, F. Chen, P.A. Liddell, J. Andréasson, S.D. Straight, D. Gust, T.A. Moore, A.L. Moore, J. Li, O.F. Sankey, S.M. Lindsay, *Nanotechnology* 2005, **16**, 695 reported a light driven electrical switch. They investigated a molecule that exists in an open form (a) and a closed form (b):

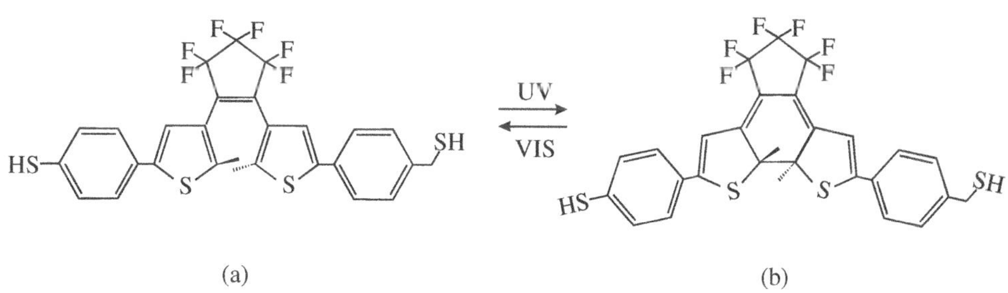

By binding the thiol groups at both ends to small gold particles they were able to measure the change in the molecule's resistance. Read and analyze this paper. Provide a short (two paragraphs or less) explanation of their claim and the observations which support it.

Solution

The molecule is photoisomerized from (a) to (b) by ultraviolet light (UV) and from (b) to (a) by visible light (VIS). Thus the system acts as a switch. A single molecule is spanned between a gold STM tip and a gold surface in a break junction arrangement. The tip is repeatedly pushed into the gold surface in the presence of a solution of the molecules and then molecules adsorb across the STM/gold surface junction as the tip is withdrawn. An ohmic behavior is found when applying a voltage up to 0.2 V and the molecules are in the form (a) with a resistance of 500 MΩ; for the case of molecules in the form (b) an ohmic behavior is observed for a voltage up to 0.12 V with a resistance of 4 MΩ.

P28.8 Sequencing Individual DNA Molecules: W.J. Greenleaf and S. M. Block report a method for measuring the sequence of oligonucleotides in "Single-Molecule, Motion-Based DNA Sequencing Using RNA Polymerase",

Science 2006, **313,** 801. Provide a short (two paragraphs or less) explanation of their claim and the observations which support it.

Solution

These workers use optical tweezers to hold two polystyrene beads in space. One of the beads holds an RNA polymerase and the other contains the DNA template, which will be copied. As the copy is made the beads are pulled closer together, and this displacement is monitored. By manipulating the concentration of nucleobase reagent in the solution the authors are able to make one of the nucleobases have a significantly lower availability; hence, the reaction pauses when this nucleobase is added to the chain in the polymerization reaction and it is possible to read out the displacement at which each of these bases is added. By performing the experiment four times with the reduction in concentration of each of the possible naturally occurring four nucleobases it is possible to sequence the DNA. The repetition of the experiment requires that the registry between the experiments be exact. The authors achieved this by creating an artificial primer unit on the end of the DNA to provide a frame for the registry.

P28.9 R.A. van Delden, M.K.J. ter Wiel, M.M. Pollard, J. Vicario, N. Koumura, B.L. Feringa (Unidirectional molecular motor on a gold surface, *Nature* 2005, **437,** 1337) have reported a molecular motor which is driven by a combination of photochemical and thermal isomerization processes. Read and analyze this paper. Provide a short (less than or equal to two paragraphs) explanation of the motor's operation principle and their observations which support it.

Solution

These authors have created a chiral molecule that contains an ethylenic linkage about which isomerization can be driven by light (either at $\lambda > 280$ nm or $\lambda = 365$ nm). The 280 nm excitation drives an isomerization from the M form with an axial methyl group to a P form with an equilateral methyl group. This form transforms by heating to an M form with an axial methyl group which is exactly 180° from that of the original form. If the molecule were not tethered to a gold surface by two thiol linkages it would not be possible to distinguish these two forms. A second excitation causes isomerization to the P form with an equilateral methyl group which upon heating transforms to the M form with an axial methyl group.

The authors demonstrate the rotational nature of the molecules motion in two different ways. In one case they monitor the CD spectrum of the molecule at each of the stable states, and this clearly shows the different conformations but not necessarily the unidirectional nature of the motion. In a second experiment, they perform a C13 labeled NMR study that does provide evidence for the unidirectional nature.

P28.10 In the paper H. Oevering et al., *J. Am. Chem. Soc.* 1987, **109,** 3258, Paddon-Row and coworkers demonstrated the tunneling nature of an electron transfer reaction by studying how the reaction rate changes as a function of the distance between a naphthalene (electron donor) and an acceptor (electron acceptor). However, the reaction Gibbs energy and the activation energy for the reaction changes with the donor-to-acceptor distance. Evaluate this effect for the system studied by Paddon-Row and assess whether it can be neglected.

System	k_{CS}/s^{-1}	R/pm
6σ	3.3×10^{11}	680
8σ	3.0×10^{10}	940
10σ	2.4×10^{9}	1150
12σ	1.6×10^{8}	1350

Solution

As we showed in Chapter 17, the reaction Gibbs energy for charge separation is given by

$$\Delta_r G = \Delta_r G^{A/A^-} + \Delta_r G^{D/D^+} - \Delta_r G_{0,0} - \frac{N_A e^2}{4\pi\hat{\varepsilon}\varepsilon_0 R}$$

Using the values reported in the JACS paper, namely $\Delta_r G^{A/A^-} = 164.0$ kJ mol^{-1}, $\Delta_r G^{D/D^+} = 106.1$ kJ mol^{-1}, $\Delta_r G_{0,0} = 366.6$ kJ mol^{-1} and $\hat{\varepsilon} = 37$ in acetonitrile, we find that

$$\Delta_r G = -96.5 \text{ kJ mol}^{-1} - \frac{3768}{R/\text{pm}} \text{kJ mol}^{-1}$$

Similarly, the reorganization energy is calculated from

$$\lambda = \lambda_{\text{in}} + \frac{N_A e^2}{4\pi\varepsilon_0}\left(\frac{1}{r} - \frac{1}{R}\right)\left(1 - \frac{1}{\hat{\varepsilon}}\right)$$

with $r = 450$ pm and $\lambda_{\text{in}} = 57.9$ kJ mol^{-1}, so that we can calculate the activation Gibbs energy by

$$\Delta_r G^{\ddagger\ominus} = \left(\Delta_r G + \lambda\right)^2/4\lambda$$

and the corresponding activation term, $\exp(-\Delta_r G^{\ddagger\ominus}/RT)$. The values we find are reported in the table.

System	R pm	$\Delta_r G$	λ kJ mol^{-1}	$\Delta_r G^{\ddagger\ominus}$	$\exp(-\Delta_r G^{\ddagger\ominus}/RT)$	k_{ET} s^{-1}
6σ	680	−102.0	60.7	7.0	0.059	3.3×10^{11}
8σ	940	−100.5	62.1	5.9	0.092	3.0×10^{10}
10σ	1150	−99.8	62.9	5.4	0.11	2.4×10^{9}
12σ	1350	−99.3	63.3	5.1	0.13	1.6×10^{8}

Hence we find that the activation term changes by a factor of 2 through this series of compounds, whereas the rate constant is changing by a factor of 20,000. Hence their conclusion is robust and demonstrates the dramatic effect of tunneling distance on the rate constant for electron transfer.

P28.11 Here you consider the Brownian motion problem in Box 28.2, in which the Brownian "particle" is driven toward smaller values of x by a force f. In the spirit of Section 14.3.3, the computer program "P28.11" divides the range from $x = 0$ to $x = x_1$ into 10 equal intervals of length a. The "particle" starts between $x = 0$ and $x = a$; and it moves stochastically from interval to interval with a probability of $\mathcal{P}_+$ in the positive direction and a probability of $\mathcal{P}_-$ in the negative direction. In the course of the process the "particle" reaches the first interval (between $x = 0$ and $x = a$) many times before reaching the last interval between $x = x_1 - a$ to $x = x_1$. We assume that each time it is reflected at this position according to Equation (28.48); hence, it leaves this range with a very small probability. In the computer program, we ask for the number n_1 of steps until the interval between $x_1 - a$ and x_1 is reached, on average. We also ask for the average number n_0 of reflections at $x = 0$ during that process.

Solution

In the program you enter the value of the force f (in units of pN). From this value, the program calculates the probability $\mathcal{P}_+$ in the positive direction and the probability $\mathcal{P}_-$ in the negative direction, according to the derivation given in Box 28.3. With these values in hand, the program calculates the number of diffusion steps necessary until the position $x = x_1$ is reached. This process is repeated 1000 times and the mean values n_1 and n_0 are obtained.

For example, in Box 28.3 the calculation is performed for a value of $f = -6$ pN, resulting in a value of $\mathcal{P}_+ = 0.3188$. In this case the program calculates the average number of steps as $n_1 = 10,372$ and the average number of reflections as $n_0 = 3751$, see also Section 28.5.5 (Mechanism of F_0 motor). Figure P28.11a plots the position of the particle versus the number of diffusion steps. This figure and Table P28.11 show that in most cases the position of the particle is at $x = 0$ or $x = a$. Figure P28.11b shows the number of steps necessary to reach the position $x = x_1$ in 1000 different runs. This number lies in a wide range between a minimum number of 18 and a maximum number of 77,127 steps, compared to the average of $n_1 = 7259$.

These results depend strongly on the value of the force f. For example, with $f = -5.5$ pN the program calculates the averages $n_1 = 3810$ and $n_0 = 1176$.

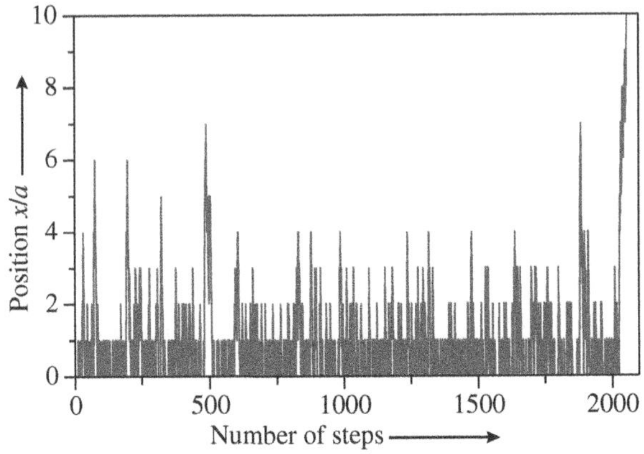

Figure P28.11a The position of the particle is plotted versus the number of steps in run 1 for $f = -6$ pN.

Table 28.11 Number of Cases When the Particle Is Found at Position x/a.

Position x/a	0	1	2	3	4	5	6	7	8	9	10
Number of cases	1154	514	214	88	35	18	12	10	7	3	1

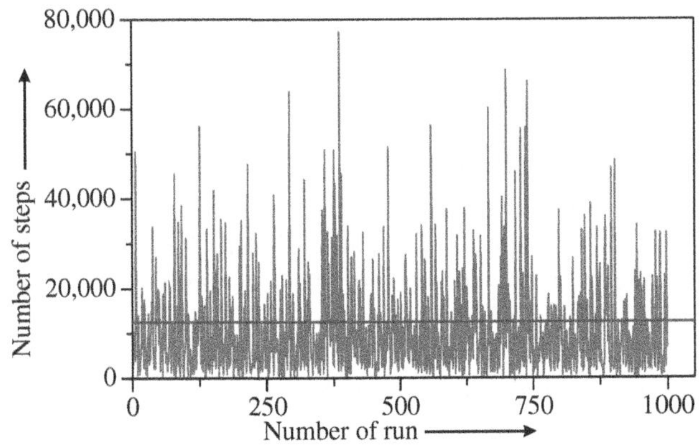

Figure P28.11b The number of steps necessary to reach the position $x = x_1$ in the different runs (1000 runs total), dark gray solid lines. The horizontal line indicates the average number of steps, $n_1 = 10,327$.

1) Run the program with $f = -6$ pN to reproduce the values in Box 28.3.
2) Run the program with different values of f in the range from $f = 0$ (no external force acting on the particle) to $f = -10$ pN (strong external force acting on the particle).
3) Determine the external force for which the difference in probabilities is 3% compared to the case $f = -6$ pN.

29

Origin of Life. Matter Carrying Information

29.1 Exercises

E29.1 In Foundation 29.4 we consider the entropy change

$$\Delta S = k \cdot \left(\mathcal{P} \cdot \ln \frac{1}{\mathcal{P}} + \mathcal{Q} \cdot \ln \frac{1}{\mathcal{Q}} \right)$$

where $\mathcal{Q} = 1 - \mathcal{P}$ and $\mathcal{P}$ is a probability with a value between 0 and 1. Show that ΔS has a maximum for $\mathcal{P} = 1/2$.

Solution

First we discuss the function

$$y = \mathcal{P} \cdot \ln \frac{1}{\mathcal{P}} + \mathcal{Q} \cdot \ln \frac{1}{\mathcal{Q}} = \mathcal{P} \cdot \ln \frac{1}{\mathcal{P}} + (1 - \mathcal{P}) \cdot \ln \frac{1}{1 - \mathcal{P}}$$
$$= -\mathcal{P} \cdot \ln \mathcal{P} - (1 - \mathcal{P}) \cdot \ln(1 - \mathcal{P})$$

For $\mathcal{P} \to 0$ we obtain, using L'Hôpital's rule,

$$\lim_{\mathcal{P}\to 0} y = \lim_{\mathcal{P}\to 0} [-\mathcal{P} \cdot \ln \mathcal{P} - (1) \cdot \ln(1)] = -\lim_{\mathcal{P}\to 0} [\mathcal{P} \cdot \ln \mathcal{P}]$$
$$= -\lim_{\mathcal{P}\to 0} \left[\frac{\ln \mathcal{P}}{1/\mathcal{P}} \right] = -\lim_{\mathcal{P}\to 0} \left[\frac{1/\mathcal{P}}{1/\mathcal{P}^2} \right] = -\lim_{\mathcal{P}\to 0} [\mathcal{P}] = 0$$

For $\mathcal{P} \to 1$ we obtain

$$y = \lim_{\mathcal{P}\to 1} [-1 \cdot \ln 1 - (1 - \mathcal{P}) \ln(1 - \mathcal{P})] = -\lim_{\mathcal{P}\to 1} (1 - \mathcal{P}) \ln(1 - \mathcal{P})$$
$$= -\lim_{\mathcal{P}\to 1} \frac{\ln(1 - \mathcal{P})}{1/(1 - \mathcal{P})} = \lim_{\mathcal{P}\to 1} \frac{1/(1 - \mathcal{P})}{1/(1 - \mathcal{P})^2} = \lim_{\mathcal{P}\to 1} (1 - \mathcal{P}) = 0$$

For $\mathcal{P} \to \frac{1}{2}$ we obtain

$$y = \frac{1}{2} \ln \frac{1}{1/2} + \left(1 - \frac{1}{2}\right) \cdot \ln \frac{1}{1 - \frac{1}{2}} = \frac{1}{2} \ln 2 + \frac{1}{2} \ln 2 = \ln 2$$

Figure E29.1 plots y versus $\mathcal{P}$. According to the plot,we have a maximum at $\mathcal{P} = 0.5$.

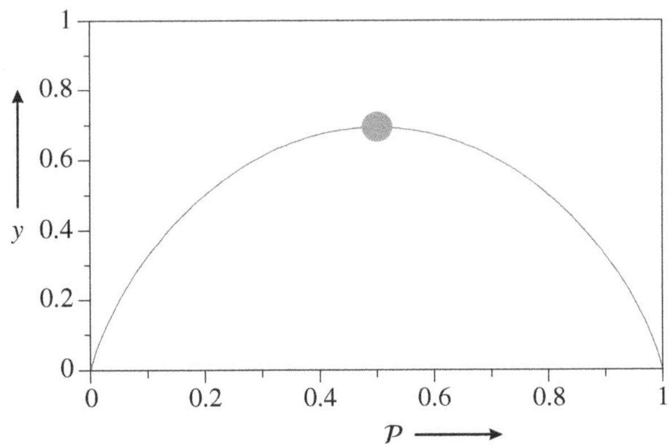

Figure E29.1 Plot of y versus $\mathcal{P}$.

We can determine the maximum of y analytically, taking the first derivative with respect to $\mathcal{P}$.

$$y = -\mathcal{P} \cdot \ln \mathcal{P} - (1 - \mathcal{P}) \cdot \ln(1 - \mathcal{P})$$

$$\frac{dy}{d\mathcal{P}} = -\ln \mathcal{P} - \mathcal{P}\frac{1}{\mathcal{P}} + \ln(1 - \mathcal{P}) + (1 - \mathcal{P})\frac{1}{(1 - \mathcal{P})}$$
$$= -\ln \mathcal{P} + \ln(1 - \mathcal{P}) = \ln\left(\frac{1 - \mathcal{P}}{\mathcal{P}}\right)$$

This expression becomes zero for

$$\frac{1 - \mathcal{P}_{\text{extremum}}}{\mathcal{P}_{\text{extremum}}} = 1 \quad \text{or} \quad \mathcal{P}_{\text{extremum}} = \frac{1}{2}$$

in accordance with the maximum in Fig. E29.1. Thus we have proved that $\Delta S = k \cdot y$ has a maximum at $\mathcal{P} = \frac{1}{2}$.

29.2 Problems

P29.1 Aggregation of Folded Strands. A requirement for selection is that after the growth phase the folded strands must reach each other and precisely interlock in a time span short compared to the decay phase. If we estimate a time $t = 10$ s required for aggregating two molecules, use Fig. P29.1 and the Einstein–Smoluchowski equation to estimate the volume V of a compartment that would include both molecules to allow the process in the given time. Suppose that the aggregation of the folded strands A and B requires that the points a and b approach each other more closely than a critical distance d of about 0.1 nm. Assume that strand A remains stationary and that b is at the center of strand B; then point b participates only in the translational diffusion of B. The strands must also have the correct mutual orientation; perform your analysis for the case of $\alpha = 1$ and $\alpha = 0.1$, where α is the steric factor, i.e., the probability (when the points a and b have reached the critical distance d) that the strands A and B are oriented correctly to allow bonding.

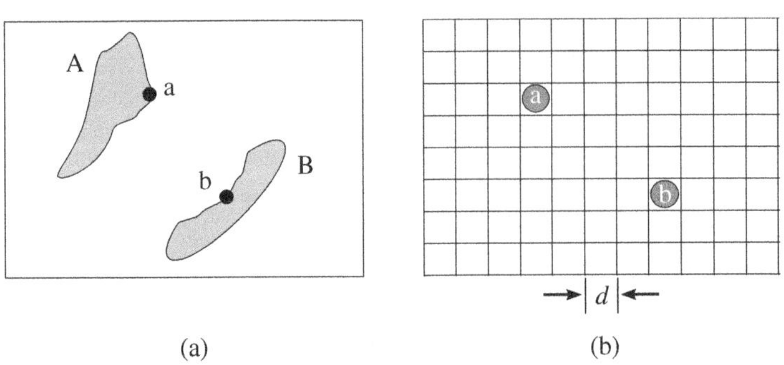

Figure P29.1 Diffusion of two folded strands to form an aggregate. Panel (a) illustrates the importance of a steric factor, whereas panel (b) neglects the sterics and only considers the diffusion.

Solution

To estimate the average time it takes for point b to approach point a within a distance d, volume V (assumed to be cubic) is divided into cubic cells of edge length d. The average time τ it takes for point b to move from one cell to the next is then, according to the Einstein–Smoluchowski Equation,

$$\tau = \frac{d^2}{2D}$$

where D is the diffusion constant of strand B. The average time t required for point b to reach the cell containing point a is

$$t = \tau \cdot \frac{V}{d^3} = \frac{d^2}{2D} \cdot \frac{V}{d^3} = \frac{V}{2D \cdot d}$$

Aggregation of strands A and B still requires that their mutual orientation be appropriate (steric factor α). With $D = 5 \times 10^{-11}\,\text{m}^2\,\text{s}^{-1}$, $d = 0.1$ nm, $t = 10$ s and $\alpha = 0.01$ we obtain

$$V = t \cdot 2D \cdot d \cdot \alpha = 1 \times 10^{-21}\ \text{m}^3 = 10^6\ \text{nm}^3$$

This is the estimated pore volume (cube of an edge measuring roughly 100 nm); for comparison, the diameter of a bacterium is roughly 1000 nm = 1 μm.

P29.2 Replication Error Rate of a Bacterium and a Human. Estimate the replication error probability per nucleotide base in a bacterium consisting of 10^3 proteins (three nucleotides code for a single amino acid). What is different in a human with about 40×10^3 proteins? Assume the protein has about 300 amino acids and estimate the replication probability per base to be 10^{-6}.

Solution

A protein consisting of about 300 amino acids, each requiring three bases for its establishment. Therefore, the DNA strand of a bacterium should consist of about

$$N_{\text{total}} = 3 \times 300 \times 10^3 = 9 \times 10^5 \approx 10^6$$

bases; this is indeed the case. The information given by the sequence of bases in the DNA should be transferred to the next generation. Thus the replication error probability $\mathcal{P}$ per base should not be larger than 10^{-6} to allow for a sufficient amount of error-free copies. Furthermore, $\mathcal{P}$ should not be much smaller than 10^{-6} to allow for as many mutations as possible to allow adaptability. Thus we estimate $\mathcal{P} \approx 10^{-6}$.

In the case of a human it must be taken into account that the information

$$N_{\text{total}} = 3 \times 300 \times 40 \times 10^3 = 36 \times 10^6 \approx 4 \times 10^7$$

must be transferred by the germ cell lineage (about 100 cell divisions in a row) to the next generation. Thus, $\mathcal{P}$ should be about

$$\mathcal{P} \approx \frac{1}{100} \cdot \frac{1}{N_{\text{total}}} \approx 2 \times 10^{-10}$$

This is actually the case. This accuracy is achieved by sophisticated repair mechanisms.

P29.3 Time Needed to Evolve a Bacterium. Evolution up to the period in which genetic machinery appeared must have occurred rapidly, because of the large error frequency that dominated these early stages. This time is small compared to the time required for the instruction of the about 1000 proteins of a bacterium. Estimate the time needed to evolve a bacterium.

Assume that one protein after another is inserted into the form existing at the time. This is achieved by lengthening of the DNA strand by $N = 10^3$ nucleotides for each new protein (about 300 amino acids). To simplify, assume that each protein is instructed in some hundred steps of optimization. Between each of these steps we must wait until a new random sequence in the non-instructed section of the DNA strand is reached, i.e., $1000 \cdot \frac{1}{\mathcal{P}}$ generations. Then each base in that section had a chance to mutate. This implies $1/\mathcal{P} = 10^6$ generations per step, if $\mathcal{P} = 10^{-6}$ is the replication error probability per base estimated for an early bacterium (see Problem 29.2).

Solution

Thus $10^6 \times 10^2 = 10^8$ generations are needed per protein and therefore a total of $10^8 \times 10^3 = 10^{11}$ generations to instruct the 1000 proteins. This takes 3×10^8 years assuming one day per generation. This time is smaller than the about 10^9 years available for the process in geological history. This is a pessimistic estimate since proteins develop largely by gene duplication which strongly decreases the time for instructing each additional protein.

P29.4 Maximum Genetic Information Carried by DNA. Assume that the replication error probability per base in a human is about $\mathcal{P} = 10^{-10}$ (see Problem 29.2). Also assume that the loss of information, caused by thermal processes $\mathcal{P}'$, during time t between two subsequent cell divisions is not larger than the loss by replication $\mathcal{P}$. Assume that the $\mathcal{P}'$ error probability provides an upper bound on the information storage and estimate the maximum genetic information that can be stored in a human by DNA.

Solution

According to Problem 29.2, the replication error probability per base in a human to copy the genetic information is required to be about $\mathcal{P} = 10^{-10}$. Another requirement is that the loss of information during time t between two subsequent cell divisions caused by thermal collisions is not larger than the loss by replication.

We consider the probability $\mathcal{P}'$ that a given base is exchanged by another base during time t due to an occasional strong collision. $\mathcal{P}'$ cannot be larger than about 10^{-10} in order to store the genetic information of a human or a higher organism. $\mathcal{P}'$ must be smaller than $\mathcal{P}$:

$$\mathcal{P}' < \mathcal{P}$$

We can estimate $\mathcal{P}'$ from the relation

$$\mathcal{P}' = t \cdot \nu \cdot e^{-\Delta E/(kT)}$$

where ν is the collision frequency (thermal process) and ΔE is the activation energy. We take the collision frequency to be on the order of 10^{13}s^{-1}, and we consider a time span $t = 0.1$ year $= 3 \times 10^6$ s for the time between cell divisions in the germ cell lineage. The activation energy is certainly smaller than the bond energy of a C–C-bond (6×10^{-19} J); assuming that DNA is well protected the activation energy might be as high as half of the bond energy: $\Delta E = 3 \times 10^{-19}$ J. Then for $T = 300$ K

$$\mathcal{P}' = 3 \times 10^6 \ \text{s} \times 10^{13} \ \text{s}^{-1} \times 3 \times 10^{-32} \approx 10^{-12}$$

Thus the condition $\mathcal{P}' < \mathcal{P}$ is fulfilled.

This estimate indicates that a barrier is reached in the genetic information to be stored in a human or in any higher organism. More information cannot be stored by DNA at room temperature.

How to overcome this barrier? Another mechanism is needed to transfer a larger amount of information from one generation to the next. Artificial information storage (writing, computer) provides a way to overcome this basic barrier in the evolution of life. Through the creation of civilizations, mankind has passed the critical stage to transfer larger amounts of information from one generation to the next than can be transferred by DNA.

Greek Alphabet

Alpha	A	α	Eta	H	η	Nu	N	ν	Tau	T	τ
Beta	B	β	Theta	Θ	θ	Xi	Ξ	ξ	Upsilon	Υ	υ
Gamma	Γ	γ	Iota	I	ι	Omnicron	O	o	Phi	Φ	ϕ
Delta	Δ	δ	Kappa	K	κ	Pi	Π	π	Chi	X	χ
Epsilon	E	ε	Lambda	Λ	λ	Rho	P	ρ	Psi	Ψ	ψ
Zeta	Z	ς	Mu	M	μ	Sigma	Σ	σ	Omega	Ω	ω

Mathematical Relations

$\pi = 3.14159265359$ $\quad$ $\mathrm{e} = \lim_{n\to\infty}\left(1+\frac{1}{n}\right)^n = 2.71828182846$

$\ln x = \ln(10)\cdot \log x = 2.302585 \cdot \log x$ $\quad$ $\ln(\mathrm{e}) = 1$

$N! = 1\cdot 2\cdot 3\ldots\cdot N$ $\quad$ Stirling's formula: $\ln N! = N\ln N - N$

$\ln(xy) = \ln x + \ln y$ $\quad$ $\ln\left(\frac{x}{y}\right) = \ln x - \ln y$ $\quad$ $\ln x^y = y\ln x$

$\mathrm{e}^x\,\mathrm{e}^y = \mathrm{e}^{(x+y)}$ $\quad$ $\frac{\mathrm{e}^x}{\mathrm{e}^y} = \mathrm{e}^{(x-y)}$ $\quad$ $(\mathrm{e}^x)^y = \mathrm{e}^{xy}$ $\quad$ $\mathrm{e}^{\pm ix} = \cos x \pm i\sin x$

$\tan\alpha = \frac{\sin\alpha}{\cos\alpha}$ $\quad$ $\cot\alpha = \frac{\cos\alpha}{\sin\alpha}$ $\quad$ $\sin^2\alpha + \cos^2\alpha = 1$

$\sin(\alpha\pm\beta) = \sin\alpha\cos\beta \pm \cos\alpha\sin\beta$ $\quad$ $\cos(\alpha\pm\beta) = \cos\alpha\cos\beta \mp \sin\alpha\sin\beta$

$\mathrm{d}(f+g) = \mathrm{d}f + \mathrm{d}g$ $\quad$ $\mathrm{d}(fg) = g\mathrm{d}f + f\mathrm{d}g$ $\quad$ $\mathrm{d}\left(\frac{f}{g}\right) = \frac{g\mathrm{d}f - f\mathrm{d}g}{g^2}$

$\frac{\mathrm{d}}{\mathrm{d}x}(x^n) = nx^{n-1}$ $\quad$ $\frac{\mathrm{d}}{\mathrm{d}x}(\mathrm{e}^{ax}) = a\mathrm{e}^{ax}$ $\quad$ $\frac{\mathrm{d}}{\mathrm{d}x}(\ln x) = \frac{1}{x}$

$\frac{\mathrm{d}}{\mathrm{d}x}\sin(ax) = a\cos(ax)$ $\quad$ $\frac{\mathrm{d}}{\mathrm{d}x}\cos(ax) = -a\sin(ax)$

$\int x^n\mathrm{d}x = \frac{x^{n+1}}{n+1} + const$ $\quad$ $\int \frac{1}{x}\mathrm{d}x = \ln x + const$

$\int \sin(ax)\,\mathrm{d}x = -\frac{1}{a}\cos(ax) + const$ $\quad$ $\int \cos(ax)\,\mathrm{d}x = \frac{1}{a}\sin(ax) + const$

$\int \sin^2(ax)\,\mathrm{d}x = \frac{1}{2}x - \frac{a\sin(2ax)}{4} + const$ $\quad$ $\int_0^\infty x^n\,\mathrm{e}^{-ax}\mathrm{d}x = \frac{n!}{a^{n+1}}$

Taylor series: $\quad f(x) = \sum_{n=0}^{\infty}\frac{1}{n!}\left(\frac{\mathrm{d}^n f}{\mathrm{d}x^n}\right)_{x=x_0}(x-x_0)^n$

$\mathrm{e}^x = 1 + x + \frac{1}{2}x^2 + \ldots$ $\quad$ $\ln(1+x) = x - \frac{1}{2}x^2 + \frac{1}{3}x^3 - \ldots$

$\frac{1}{1+x} = 1 - x + x^2 - \ldots$ $\quad$ $\sqrt{1+x} = 1 + \frac{1}{2}x - \frac{1}{8}x^2 + \ldots$